Anatomy, Age and Ecology of High Mountain Plants in Ladakh, the Western Himalaya

Jiří Doležal • Miroslav Dvorský • Annett Börner
Jan Wild • Fritz Hans Schweingruber

Anatomy, Age and Ecology of High Mountain Plants in Ladakh, the Western Himalaya

Springer

Jiří Doležal
Faculty of Science, Department of Botany
University of South Bohemia
České Budějovice, Czech Republic

Miroslav Dvorský
Institute of Botany
The Czech Academy of Sciences
Průhonice, Czech Republic

Annett Börner
Calyptra Pty Ltd
Adelaide, Australia

Jan Wild
Institute of Botany
The Czech Academy of Sciences
Průhonice, Czech Republic

Fritz Hans Schweingruber
Swiss Federal Research Institute WSL
Birmensdorf, Switzerland

ISBN 978-3-030-08766-1 ISBN 978-3-319-78699-5 (eBook)
https://doi.org/10.1007/978-3-319-78699-5

Cover illustration: *Haloxylon thomsonii*, family of Amaranthaceae. Stem cross-section of a small shrub on rocks in the desert of northern Ladakh. Single vascular bundles are embedded in aerenchymatic and sclerenchymatic tissues.

Printed on acid-free paper

This Springer imprint is published by the registered company Springer International Publishing AG part of Springer Nature.
The registered company address is: Gewerbestrasse 11, 6330 Cham, Switzerland

Contents

Abbreviations

ae	aerenchym
bpit	bordered pit
ca	cambium
cal	callus, parenchymatic cells
clu	cell lumen, cell lumina
co	cortex
cork	
cry	crystal
csi	collapsed sieve tubes
ct	conjunctive tissue
cu	cuticula
di	(ray) dilatation
ds	dark-stained substances
duct	
en	endodermis
ep	epidermis
ew	earlywood
ewt	earlywood tracheid
ewv	earlywood vessel
f	fiber
ft	fiber tracheid
ge	gelatinous fibers
gr	growth ring
grb	growth ring boundary
he	helical thickenings
ivp	intervessel pit
la	laticifers
lcw	lignified cell wall
lf	libriform fiber
lw	latewood
lwt	latewood tracheid
lwv	latewood vessel
mu	mucilage
nu	nucleus
p	perforation
pa	parenchyma
ph	phloem
phe	phellem
phg	phellogen
pit	
pith	
r	ray
rd	resin duct
sc	sclereid
sf	septate fibers
shc	sheet cell
si	sieve tube, sieve element
spit	simple pit
ta	tannins
te	tension wood
tr	tracheid
ty	tylosis
ulcw	unlignified cell wall
v	vessel
vab	vascular bundle
vat	vascular tracheid
xy	xylem

JIŘÍ DOLEŽAL has been working in the Himalayas for more than ten years and tackled the challenge of the field plant collection together with many students from the University of South Bohemia in České Budějovice. Determining and sorting of the collected plant material from dry and cold Himalayan deserts were necessary steps before laboratory sectioning and analysis of the rich anatomical structures. Jiří lives in České Budějovice, Czech Republic, with three kids and wife Lucka, whom he thanks for her constant support.

FRITZ SCHWEINGRUBER sectioning a tiny stem of a herb with a „fossil" of a microtome (Reichert constructed approx. 1940) in his laboratory at the Research Institute for Forest, Snow and Landscape at Birmensdorf, Switzerland. Anatomical work requires primarily a steady hand, very sharp microtome blades and a lot of patience. To prepare the micro-sections for comparative analysis they are stained, dehydrated and preserved in a resin.

MIROSLAV DVORSKÝ took part in five Himalayan expeditions and helped collecting plant samples and vegetation data as a PhD student. He soon became increasingly keen on taking photographs of all plant species encountered, and this was not long before the idea of a field guide to the local flora was conceived. Plant species identification in the absence of any standard determination key was often a tricky part of the fieldwork, and required gathering of herbarium specimens and cooperation with taxonomy experts.

JAN WILD's career is already more than twenty years related to geoinformatics or broadly speaking to spatial data. Production of pretty maps is thus his hobby as well as a way of communicating a scientific work to the public. Jan lives with his wife Hanka and two children on the southern border of Prague, Czech Republic, close to Průhonice, where is the headquarter of the Institute of Botany, his long-term employer.

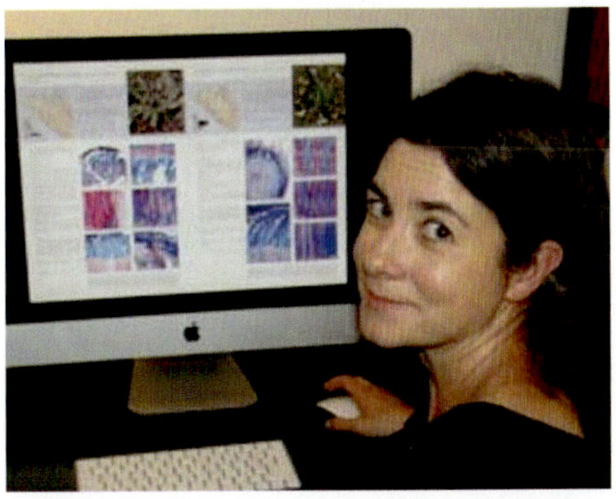

ANNETT BÖRNER has been working in the field of scientific publishing for more than fifteen years and took up the challenge of the book's design and editorial process. Presenting the material in a clear and structured way demands a high standard of technical and botanical knowledge. Annett lives in Adelaide, Australia, and when she gets tired of looking at plants on a computer screen she loves going on field trips and looking at them in real life.

Tso Moriri Lake is at an altitude of 4522 m a.s.l. It is the largest of the high altitude lakes entirely within India.

INTRODUCTION

Despite widespread interest in alpine ecology, we have little information on anatomy of high-mountain plants. Most studies are based on data from European mountains (Schweingruber et al. 2011; Schweingruber and Büntgen 2013), while the largest and highest mountainous areas are located in Asia. On the Tibetan Plateau and the neighboring Himalayan crests, which cover an area almost 13 times larger than the area of the European Alps, extensive areas of alpine climate and vegetation host a unique and species-rich flora (Miehe et al. 2011) and serve as grazing land for yaks, wild sheep and other mammals, while many plant species are utilised by local people for medicinal purposes. No comprehensive monography concerning both ecology and anatomy of alpine plants is available from these areas. However, about 70% of all glaciers, representing the only reliable water reservoir for local people and their livelihoods, are presently retreating (Xu et al. 2009; Schmidt and Nüsser 2017), indicating that a considerable environmental change in this high altitude region is underway.

The general lack of studies on anatomy of high-mountain plants may reflect working habits of biologists. While ecologists tend to use outdoors as a laboratory, for the plant anatomist, laboratory work is done indoors, and the habit of working both in the field and in the laboratory apparently does not come easily to most plant anatomists, who commonly do comparative work on material they seldom collect. High-mountain areas in the Himalayas and Tibet are often difficult to access, and so plant anatomists are less likely to visit these areas. Perhaps more significantly, plant anatomists tend not to be interested in plant ecology and physiology (and *vice versa* for plant ecologists), such that those kinds of data needed for structure/function correlations are often unavailable in tandem. Collaboration among researchers from different fields seems a desirable way of proceeding with studies on anatomy, ecology and evolutionary adaptations of high-mountain plants. The present book is a result of such collaboration among plant ecologists, anatomist and graphic and GIS experts in the high and sacred realm of Ladakh, located in a mountain landscape in the Transhimalaya of Northwest India.

"Leave the forest, leave the temperate climate and explore the anatomy of high-mountain plants in relation to their morphological and ecophysiological adaptations to frost and to drought in cold desert ecosystems" are the basic ideas of the present book. The book introduces plants of very broad taxonomic spectrum and distribution, ecological habitat preferences, morphology, longevity and evolutionary history, all growing naturally in Ladakh. Xylem and bark structures of selected representatives are described for Equisetophyta (1 species), Pteridophyta (10), Gnetophyta (3), Pinophyta (1), monocotyledons (155) and dicotyledons (345 species) belonging to 63 plant families and 236 genera. A total of about 1250 species of vascular plants are known to occur in the wild in Ladakh. The number seems to be surprisingly low as compared to, for example, the Czech Republic, which has a slightly smaller surface area, but about four times more plant species. While the low species richness obviously relates to a generally adverse, cold, arid, high altitude environment, the flora encountered in Ladakh is still speciose, phylogenetically diverse and ecologically impressive.

This book illustrates and describes more than 500 species of Ladakh's vascular flora. We have tried to select a set of common, widespread species as well as some rarer ones, which are considered endemic, phytogeographically interesting or potentially endangered. All described plants grow in high, largely treeless altitudes in a cold arid region with short vegetation periods. Most species are shown through photos in their natural habitats in Ladakh. Where appropriate photos were not available (mostly of grasses and sedges) we provided images of scanned herbarium specimens. Annotated for each species are the accepted Latin name and its synonyms, key identification features, geographical distribution and habitat. The range of each species in Ladakh is shown as a distribution map, and its frequency along the altitudinal gradient is depicted as a diagram.

Anatomical descriptions are based on root collars (transition between roots and shoots) or rhizomes of several individuals

J. Doležal et al., *Anatomy, Age and Ecology of High Mountain Plants in Ladakh, the Western Himalaya*, https://doi.org/10.1007/978-3-319-78699-5_1

collected for each species in the field. These parts of plants provide the maximum number of annual rings (Schweingruber and Poschlod 2005). Most of the material was preserved in 40% ethanol and sectioned without embedding in transverse, tangential and radial directions with a sliding microtome, stained with Astrablue/Safranin and preserved in Canada Balsam. Herbarium material was used for all Monocots and a few annual Amaranthaceae. Preparation methods are described in detail by Gärtner and Schweingruber (2013).

Anatomical features of xylem of dicotyledonous trees and shrubs are principally defined by the IAWA List (Wheeler et al. 1989). After sectioning and inspecting herbaceous species from non-forest regions the number of features in the xylem had to be expanded and bark features newly defined (Crivellaro and Schweingruber 2016). The given anatomical feature codes in this book are based on Schweingruber et al. (2011). Anatomical characteristics of Monocots are mostly based on Metcalfe (1960) and Schweingruber and Berger (2017).

The detailed knowledge of plant anatomical characters and their variation among different species and habitats is key to understanding plant evolution and function (Niklas 1985). Variation in anatomical structure is a result of several forces such as: the adaptation of species to the prevailing conditions in their habitats (Schweingruber et al. 2014, Schweingruber and Börner 2018), phenotypic plasticity as an ability of individuals with an identical genotype to develop differently — based on specific conditions during their ontogeny (Sultan 2000), and evolutionary constraints (phylogenetic inertia) in which taxa that share part of their evolutionary history possess similar 'blue-prints' (Ackerly 2000). The present book compares variation in more than 120 anatomical features of Dicots and 42 features of Monocots among (1) dominant habitat types in Ladakh such as cold deserts, alpine steppes, saline wetlands, alpine grasslands, ruderal vegetation, alpine screes and subnival zones, (2) with respect to species optima on the broad elevation gradient in Ladakh from 2650 to 6150 m, (3) among different growth and life forms such as short-lived plants, perennial plants with main tap roots, various clonal rhizomatous plants, compact alpine cushion plants and rare arborescent species, (4) and among plant height classes. Patterns of species altitudinal and geographical distributions were derived from more than 122,000 records of occurrence of vascular plant species in Ladakh collected by Dr. Leoš Klimeš in the period 1997–2007.

The present book has four major goals: a) Describing the stem-anatomical structures of vascular plants from Ladakh, b) Providing basic information about species morphology, geographical and elevational distribution and habitat preferences, c) Constructing a phylogenetic tree for target species from nucleotide sequences, d) Relating stem anatomical features to phylogeny, ecology, morphology and ecophysiology.

Ladakh is a mountainous region located in NW India (Jammu & Kashmir State), ranging between 2600 and 7672 m a.s.l. Its nearly 87,000 km^2 include a considerable part of the transitional zone between outer Himalayan ranges and Western Tibet, forming a naturally delimited area of Transhimalaya. The interface between the humid outer Himalaya and dry cold plains of central and western Tibet is generally quite sharp. For example, in Nepal, this transitional zone is usually not more than a few kilometers broad. In such conditions flora of the transitional zone is not particularly interesting, as it is formed by depauperate floras of outer Himalayas and Tibet. Only in the Western Himalayas (and to some extent in the easternmost Himalayas) the transitional zone is broad enough to host a specific flora, relatively rich in species restricted to this zone. In spite of the high elevation and severity of the environment, local flora shows a remarkable phylogenetical and ecological diversity. Nevertheless, in terms of species richness per area the whole region belongs among biodiversity coldspots. Plants in Ladakh are generally subjected to multiple stresses, such as low precipitation, extreme diurnal temperature fluctuation, strong winds, solifluction at the higher and salinity at lower elevations (Stewart 1916–1917).

Geomorphologically, western and southern parts of Ladakh differ considerably from Eastern Ladakh. They are dominated by deep, narrow valleys, and glaciated peaks. Precipitation is much higher, especially at higher elevations, and the snow line is situated by 1000 m lower than in the east. Consequently, the flora of Western Ladakh differs considerably, and true Himalayan species are more prominent. Eastern Ladakh represents an area which is well delimited climatically, biogeographically and culturally, and its flora is mostly related to that of the Tibetan Plateau. Northern Ladakh, encompassing a part of Eastern Karakorum Range, has dry climate and rugged terrain, and the highest summits of Ladakh are located there.

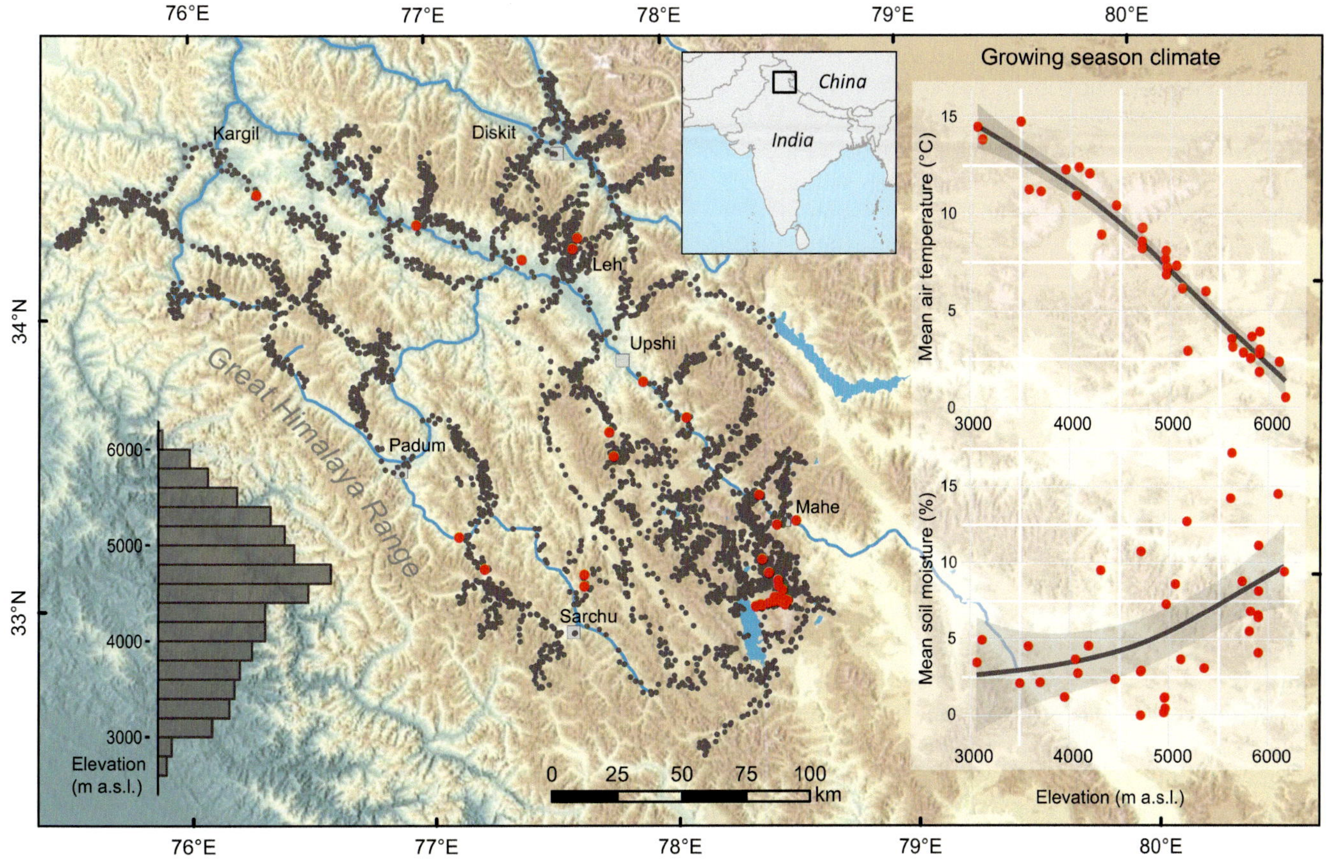

Distribution of vegetation plots and in situ measured climate in the study region in Ladakh, NW Himalayas. Floristic data were collected in 4,150 plots (100× 100 m, black dots) and climate measured at 36 sites (red dots) between 2013 and 2014. Histogram on the left side shows the frequency of vegetation plots within respective elevational bands. Mean air temperature of the growing season decreases with elevation (Pearson's r = -0.96), while mean soil moisture increases (Pearson's r = 0.49; graphs on the right side). The trend lines were fitted with GAM and shaded regions represent 95% confidence intervals. Background map is based on SRTM (*http://www2.jpl.nasa.gov/srtm/*) and OpenStreetMap data.

Biological soil crusts, Nubra Valley. These crusts develop in arid regions, resulting from an intimate association between soil particles and cyanobacteria, algae, microfungi, lichens and bryophytes. They play a key role in retaining carbon, nitrogen and liquid water, and protect against erosion, which is important for the subsequent plant and animal succession.

Climate and soil

Ladakh is situated in the rain-shadow of the High Himalayas and is consequently rarely affected by monsoonal precipitation, otherwise a prominent feature in much of India and southeast Asia. As a result, the climate of Ladakh is predominantly arid, with an annual precipitation of 50–300 mm, and cold due to high elevation, with an annual mean temperature well below 10 °C. At low elevations, intense drought prevails due to lower precipitation and relatively high summer temperatures. On the other hand, high-elevation zones are more humid because of a decrease in evapotranspiration and increase in precipitation. The length of the growing season decreases linearly with increasing elevation, from 270 to 30 frost-free days, while the mean growing season temperature decreases from 13 to 3 °C between 3400 and 6150 m (Karger et al. 2016; Dvorský et al. 2017). Temperatures in the study area vary from 0 to 30 °C in summer and from −40 to −10 °C in winter. Annual precipitation is about 50 mm in deserts and semi-deserts (falling mostly during summer season), 100 mm in the alpine steppes, and 150–250 mm in the alpine and subnival zones. Precipitation above 5000 m is falling mostly as snow and is more frequent in summer than in winter.

Soils have a coarse-grained structure, with a high percentage of large gravel, low water and organic matter content, high pH (7–8) and relatively high concentrations of total N and P (Řeháková et al. 2011; Janatková et al. 2013). The organic matter content and concentrations of total nitrogen, ammonia, nitrates and phosphorus increase with increasing elevation, with the highest values in the alpine zone and the lowest in deserts and semi-deserts (Doležal et al. 2016). The concentration of calcium and magnesium and the pH decreases with increasing elevation. The coarsest soils are found in subnival zone, where the fraction >0.5 mm represents up to 60% of the soil particles. Soils with the highest proportion of fine particles are found in alpine meadows.

Phytogeography

Phytogeographically, Ladakh is situated in the great high mountain knot of western Central Asia, near the southern edge of the Holarctic or Boreal Realm towards the Paleotropic Realm (Good 1964, Takhtajan 1986). Due to the present high elevation and cold climate, and probably also reflecting its paleogeography, the flora of Ladakh is of an almost entirely Holarctic affinity. Ladakh is also located close to the meeting point of three major floristic regions, which are roughly characterised by contrasting precipitation regimes and having widely diverging floristic compositions: the almost permantly arid Central Asian Region, the Irano-Turanian Region with winter-rain, and the Sino-Himalayan Region with summer rain (Nüsser and Dickoré 2002). Central Asian and Sino-Himalayan elements make up about equal shares in the flora of Ladakh. There is a primary, generally sharp line of divergence between these two floristic regions, which runs along the crest of the High Himalayas and excludes almost the whole subtropical and temperate forest flora of the Himalayas from Ladakh.

The area of transition in Ladakh is apparently big and diverse enough, to be tentatively divided into three or four neighboring phytogeographical provinces. The northern section, north of the Ladakh Range, but extending in the east into the upper Indus valley, belongs to the *Karakorum Province* with a very sparse Central Asian flora including few high-elevation endemics. The second, comprising the south and western part of Ladakh and including the lower Indus Valley, Suru/Dras and Zanskar, can be referred to as the Inner *Northwest Himalayan* or Western Transhimalayan Province. This province has a generally depauperate, but still relatively large, share of Himalayan herbs and shrubs. The third *Southwest Tibetan Province* includes the southeastern part of Ladakh (Ruphu, Tso Moriri) with its high-elevation rolling hills and basins, contiguous to the Tibetan Plateau. The entire species composition of southeast Ladakh is essentially Tibetan and widely distinct from the high elevation flora of the Karakorum. A fourth *North Tibetan* or Changtang Province might accommodate the relatively small or disputed areas on the eastern border of Ladakh north of the Indus, including Pangong, the Upper Shyok Valley, Depsang and Aksai Chin, with a typical, very depauperate North Tibetan flora.

Bright blue flowers of *Delphinium brunonianum* decorate the rugged slopes in Zanskar, central Ladakh.

Species richness and diversity

Despite the efforts of many naturalists, no precise or up to-date checklist of Ladakh's flora exists, not to mention the exhaustive Standard Flora. As a result, basic information, such as species number, remains unknown or is based on widely diverging estimates. Estimated totals of vascular plant species in Ladakh range from about 611 (Kachroo et al. 1977) or 880 (Kachroo 1993) to approximately 1250 species (Klimeš and Dickoré 2006). We refer here to the latter number, which seems to fit into large scale gradients. Further research might result in higher numbers for Ladakh, as a result of geographic border effects or neophytes, or on account of apomictic or spurious taxa. Taxonomic fuzz causes serious difficulties not only to scientists, but also to natural habitat protection activities, and research focused on medicinal plants, which is a traditional topic of Indian botanists.

Overall species richness in Ladakh is clearly governed by elevation. It culminates between 4500 and 5000 m, and decreases towards both ends of the elevation gradient (Klimeš 2003).

Vegetation and elevational belts

Due to mainly climatic constraints, the vegetation cover of Ladakh is generally sparse and over wide tracts spurious or almost absent. Especially at lower elevations, it may be only more lush where some water is available. Natural forests are absent in Ladakh, except for some small juniper groves (*Juniperus semiglobosa*) in few favorable mid-elevation areas, and scarce birch stands (*Betula jacquemontii*) at the potential upper tree line near the Himalayan southwest border. Still, the vegetation of Ladakh displays a striking vertical zoning, which may be further diversified by aspect, exposure and substrate. According to the gradient of increasing overall surface elevation and decreasing humidity in Ladakh, the elevational limits of individual species and vegetation belts generally ascend towards the north and east.

Desert occupies only a small part of the arid gorge section in the lower Indus valley close to the Pakistani border and up to approximately 2900 m; the relatively high summer temperatures provide occasional habitats for species typical of adjacent lower Baltistan and other taxa of subtropical affinities, usually not ascending to much higher elevations. Much of the main Indus valley to beyond the capital Leh, and reaching far into the tributary valleys, including Dras, Suru, Zanskar, and the Shyok valley system up to Nubra hosts a *semi-desert* vegetation, extending up to 4200 m in the west and 4400 m in the east. *Alpine steppe* represents the zonal vegetation and prevails widely, extending up to 4700 m in the west, and to 5400 m in the east. Higher up, a narrow belt of *alpine grasslands* occurs, including the characteristic moist alpine turf of *Kobresia pygmaea*. Above between 5200 m in the west and south, and 5700 m in the east, a very sparse *subnival vegetation* is developed; the upper vegetation limit lies at 6150 m. Vegetation of snow beds and springs can be found near the snow line and on slopes with surface water. Vegetation of screes is frequent throughout the region. Azonal vegetation includes wetlands, occupying relatively large areas of flat and broad valley bottoms in the whole area; these habitats are often saline. Vegetation of river banks and river beds often supports large woody species, and can be found along major rivers and streams. Synanthropic vegetation includes plant assemblages developed on eutrophicated ground by stables of domestic animals and near villages up to 5400 m; weedy assemblages grow in arable fields up to 4700 m.

A re-inspection of a historical transplant site on the slopes of Chalung Peak (6546 m a.s.l.), eastern Ladakh.

The history of botanical exploration

Until quite recently, Ladakh was difficult to access and relatively unknown. Travels in Ladakh, the Himalayas and Tibet have always been physically demanding, full of privation and often dangerous. Several explorers have died in High Asia. However, many plant species (but also animals, fossils or rock formations) described from this part of the world bear their finders' names until today. Access, transport logistics and obtaining permits from local authorities were further common limitations for exploration and collecting in these remote and wild areas. The first foreign traveler to enter Ladakh with a profound knowledge of various branches of Natural History, including botany, was *William Moorcroft* in 1820 (Moorcroft and Trebeck 1841). Among the most famous explorers of Ladakh until the end of the 19th century were *Henry Strachey*, *Thomas Thomson*, the *Schlagintweit* brothers (*Adolph*, *Hermann* & *Robert*), *John Lindsay Stewart*, *John Firminger Duthie* and *Ferdinand Stolička* (*Stoliczka*).

Few botanists visited Ladakh during the first half of the 20th century. *Ralph Randles Stewart* published the first, brief Flora of Ladakh (Stewart 1916–1917) while *Frank Ludlow* and *Walter Koelz* were probably the most notable plant collectors. Due to border concerns and armed conflicts Ladakh was effectively closed to foreigners and most Indian citizens from the late 1940s to the mid-1970s.

Botanical research only resumed when parts of Ladakh were successively opened. *Hans Hartmann*, a Swiss geobotanist, visited Ladakh six times between 1974 and 1997 and systematically expanded the knowledge of its vegetation (Hartmann 1983–2009). At around the same time Indian botanists started to explore Ladakh along the main roads (Kachroo et al. 1977; Murti 2001; Singh et al. 2002). The comprehensive, but still incomplete, Flora of Pakistan (Nasir and Ali 1970–) covers Ladakh to a variable extent. While the northern part of Ladakh had already been covered by Pampanini's "La Flora del Caracorum" (1930), the Monocots were revised, covering the total area of Ladakh in the "Flora Karakorumensis" (Dickoré 1995). A full account, including specimen and observation data of all vascular plants, has also been published as a web-database under the Global Biodiversity Information Facility (Dickoré 2011).

In the line of modern explorers, *Leoš Klimeš* (1960–2007), a Czech botanist of the Třeboň Institute of Botany, was exceptional in his zeal and determination. Over the course of 12 extensive field campaigns between 1997 and 2007, he hiked to the remotest and highest locations all over Ladakh and systematically worked on a floristic inventory for the purpose of a complete standard flora. He cooperated extensively with *Bernhard Dickoré* and other specialists worldwide in the identification and taxonomy of his specimens. *Leoš Klimeš* found many species that were not previously known from the area and increased the knowledge of plant distribution in Ladakh significantly. He also discovered, mainly at extreme elevations, several species new to science which were subsequently described in cooperation with taxonomic specialists (Al-Shehbaz 2002). These include *Aphragmus ladakianus* Al-Shehbaz, *Draba alshehbazii* Klimeš & D.A.German, *Ladakiella klimesii* (Al-Shehbaz) D.A.German & Al-Shehbaz, *Taraxacum candidatum* Kirschner, Štěpánek & Klimeš, *T. virgineum* Kirschner, Štěpánek & Klimeš, and several more. On his last expedition in 2007 Leoš disappeared somewhere in Ladakh. All investigations about his whereabouts remained unsuccessful. His plant collection, with about 7000 specimens is probably the largest individual collection from Ladakh worldwide and is stored at the Herbarium of the Institute of Botany in Průhonice (PRA).

In the field of ecology, the scientific legacy of *Leoš Klimeš* (Klimeš 2003, 2008; Klimeš and Doležal 2010) has been continued by a team led by *Jiří Doležal* of the Institute of Botany, Czech Academy of Sciences. Owing to the cooperation with experts from various fields of botany, the scientific scope of the team has covered a wide range of plant life aspects. Starting with a description of main vegetation types (Dvorský et al. 2011) and clonality (Klimešová et al. 2011), the research later focused on plant-plant interactions (de Bello et al. 2011; Dvorský et al. 2013; Liancourt et al. 2017), nutrient status of high-elevation species (Macek et al. 2012; Chlumská et al. 2014) and herbochronology (Schweingruber et al. 2014). Besides, a series of studies has been performed on practically unknown soil microbial life (Řeháková et al. 2011; Aschenbach et al. 2013; Janatková et al. 2013; Čapková et al. 2016; Kotilínek et al. 2017; Řeháková et al. 2017) along with its interaction and association with vascular plants (Řeháková et al. 2015; Angel et al. 2016). Particular attention has been paid to the upper elevational limit of plant life (Dvorský et al. 2015) and its dynamics (Doležal et al. 2016; Dvorský et al. 2016). In addition, general ecological hypotheses concerning plant coexistence (de Bello et al. 2011) or niche shape (Dvorský et al. 2017) have been studied on distributional data from Ladakh. Eventually, the Czech research team shed light on the growth dynamics of woody species and its relationship with climate (Doležal et al. 2016; Sohar et al. 2017).

Vegetation – basic types and zonation

The prevailing arid climate of Ladakh with annual precipitation around 100 mm is the key factor determining the vegetation aspect, and the edaphic qualities together with specific microclimate govern its local patterns. Semideserts occupy the relatively lowest elevations of the Indus Valley, up to 4200 m a.s.l. in the west and 4400 m a.s.l. in the east. Alpine steppe represents the zonal vegetation and prevails widely, extending up to 4700 m a.s.l. in the west, and to 5400 m a.s.l. in the east. Higher up, a narrow belt of alpine grasslands occurs, including the characteristic moist alpine turf of *Kobresia pygmaea*. Above between 5200 m a.s.l. in the west and south, and 5700 m a.s.l. in the east, a very sparse subnival vegetation is developed; vascular plants occur regularly up to 5900 m a.s.l. Vegetation of snow beds and springs can be found near the snow line and on slopes with surface water. Vegetation of screes is frequent throughout the region. Azonal vegetation includes wetlands, occupying relatively large areas of flat and broad valley bottoms in the whole area; these habitats are often saline. Vegetation of river banks and river beds often supports large woody species, and can be found along major rivers and streams. Synanthropic vegetation includes plant assemblages developed on eutrophicated ground by stables of domestic animals and near villages up to 5400 m a.s.l. Weedy assemblages grow in arable fields up to 4700 m a.s.l.

Here, we present a selection of the main vegetation types in Ladakh. They are arranged by their elevation, from the lowest and driest parts up to the highest and coldest locations. Within the respective vegetation types, we introduce some of the characteristic species with notes on their distribution and habitat properties, and with further information on co-occurring species.

Semideserts

with *Haloxylon thomsonii*, *Halogeton* spp., *Kochia* spp., *Bassia dasyphylla*, *Ephedra* spp. occur at the lowest elevations of Nubra Valley.

Semideserts.

The relatively lowest elevations of the Indus Valley support very sparse subtropical semi-desert vegetation with scattered *Arnebia guttata*, *Echinops cornigerus*, *Erodium tibetanum* or *Peganum harmala*. Near Spituk.

Semideserts.

Echinops cornigerus always grows in very dry habitats, and is common in sandy semideserts at the lowest elevation in the Indus River valley. Near Leh.

Semideserts.

Lycium ruthenicum colonizes sand dunes in Nubra Valley. Hundar.

Semideserts.

Ajania fruticulosa is common in semi-deserts and steppes of the lowest elevations in the Indus valley and locally at middle elevations of Rupshu. Occurring together with *Arnebia euchroma*, *Lonicera spinosa*, *Stipa orientalis*, *S. caucasica*, *Alyssum canescens*, *Askellia flexuosa*, *Salsola jacquemontii*, *Oxytropis microphylla*. Upper Indus Valley.

Semideserts.

Nepeta floccosa and *Corydalis flabellata* are abundant in semideserts on unstable loamy and sandy slopes. Indus Valley.

Steppes.

Dry steppe vegetation with *Acantholimon lycopodioides*, *Askellia flexuosa*, *Campanula pallida*, *Chesneya cuneata*, *Dictyolimon macrorrhabdos*, *Ephedra intermedia*, *Melica persica* and *Menitskia tibetica*. Near Miru.

Steppes.

Menitskia tibetica is frequent on dry unstable slopes and on unstable stony screes and other stony habitats, in the valley of the Indus River only. Here with *Ajania fruticulosa*.

Steppes.

Acantholimon lycopodioides is common on dry non-mobile slopes in SE Zanskar. Here with *Eritrichium* sp., *Ephedra* sp., *Krascheninnikovia ceratoides*. Purfi La.

Steppes.

Krascheninnikovia pungens is one of the dominant species in dry stony steppe, on sandy plains and on stabilized slopes at lower and middle elevations. Occurring together with *Stipa caucasica*, *Oxytropis microphylla*, *Elymus jacquemontii*, *Oxytropis tatarica*, *Stipa subsessiliflora*, *Alyssum canescens*, *Salsola jacquemontii*, *Caragana versicolor*, *Artemisia minor*, *Euphorbia tibetica*, *Artemisia salsoloides*, *Dracocephalum heterophyllum*. Here also with *Askellia flexuosa*. Indus Valley.

Steppes.

Rosa webbiana grows on stony river terraces, on dry stony slopes and in rock crevices at the lowest altitudes of the Indus River valley and in SE Zanskar. Here with *Ajania fruticulosa* and *Morina coulteriana*. Near Kargil.

Steppes.

Stipa purpurea grows in semi-deserts on sandy substrates, on stony slopes with steppe vegetation, at higher elevations infrequent but locally dominant on dry stabilized stony slopes. Occurring together with *Allium przewalskianum*, *Stipa subsessiliflora*. Here with *Schistophyllidium bifurcum* and *Artemisia minor*. Rupshu.

Steppes.

Askellia flexuosa is very frequent and abundant on river beds with low-density vegetation, on stabilized gravel slopes, along roads. Always in dry habitats, common on stony sediments and river gravel deposits. Occurring together with *Scrophularia dentata*, *Lonicera spinosa*, *Stipa orientalis*, *Senecio dubitabilis*, *Hippophae tibetana*, *Arnebia euchroma*, *Ajania fruticulosa* and *Stipa caucasica*. Tso Moriri.

Steppes.

Arnebia euchroma is common on screes, stabilized slopes, and along roads in disturbed habitats. It can be a dominant species especially in SE Zanskar and locally in the area of Kiagar Tso and Tso Moriri in Rupshu. Occurring together with *Ajania fruticulosa*, *Crepis flexuosa*, *Lonicera spinosa*, *Stipa orientalis*, *Salsola jacquemontii*, *Scrophularia dentata*, *Stipa caucasica*, *Artemisia salsoloides*, *Senecio dubitabilis* and *Physochlaina praealta*. Here also with *Aconogonon tortuosum*, *Chamaerhodos sabulosa* and *Leymus secalinus*. Tso Moriri.

Steppes.

Stipa caucasica is one of the main indicators of dry alpine steppes. Frequent on sandy plains and stony slopes, semi-deserts, stony steppes, stony and rocky slopes, rare on pure sand. Occurring together with *Alyssum canescens, Oxytropis microphylla, Krascheninnikovia pungens, Stipa subsessiliflora, Euphorbia tibetica, Salsola jacquemontii, Artemisia santolinifolia, Ajania fruticulosa, Artemisia salsoloides, Oxytropis tatarica, Dracocephalum heterophyllum, Arnebia euchroma, Askellia flexuosa, Artemisia minor.* Here also with *Aconogonon tortuosum, Dracocephalum heterophyllum* and *Saussurea glanduligera.* Tso Moriri.

Steppes.

Artemisia salsoloides is a typical species of almost desertic fluvio-glacial high plains, dominant on stony steppes, dried river beds. Occurring together with *Euphorbia tibetica, Knorringia pamirica, Nepeta discolor, Alyssum canescens, Oxytropis microphylla, Stipa caucasica, Artemisia santolinifolia, Scrophularia dentata, Stipa subsessiliflora, Dracocephalum heterophyllum, Arnebia euchroma* and *Krascheninnikovia pungens.* Tso Moriri.

Steppes.

Alyssum canescens grows on sandy plains and stony steppes, rare up to high altitudes on stabilized gentle slopes. In Rupshu and neighbouring areas of Leh and E Zanskar very common. Occurring together with *Stipa caucasica, Oxytropis microphylla, Euphorbia tibetica, Stipa subsessiliflora, Artemisia salsoloides, Artemisia santolinifolia, Ajania fruticulosa, Krascheninnikovia pungens, Aconogonon tortuosum, Dracocephalum heterophyllum, Artemisia sieversiana, Salsola jacquemontii, Oxytropis tatarica.* Rupshu.

Steppes.

Oxytropis tatarica is common on dry stabilized slopes, in semi-deserts of sandy plains, in steppes at lower to higher elevations. Occurring together with *Krascheninnikovia pungens*, *Stipa subsessiliflora*, *S. caucasica*, *Arenaria bryophylla*, *Elymus jacquemontii*, *Artemisia minor*, *Oxytropis microphylla*, *Alyssum canescens*, *Nepeta discolor*. Rupshu.

Steppes.

Ephedra gerardiana occurs on stabilized slopes, often gravel, stony and rocky habitats at lower to middle elevations in SE Zanskar, along the Indus River and in Rupshu. Occurring together with *Artemisia stracheyi*, *A. santolinifolia*, *Euphorbia tibetica*, *Caragana versicolor*. From Shenge La to Photaksar.

Steppes.

Caragana versicolor is one of the dominant shrubs growing on dry slopes, in dry watersheds, on older gravels replaces *Kraschenninikovia pungens* in succession. Common in the whole area, less frequently in the narrow valleys of SE Zanskar. Occurring together with *Krascheninnikovia pungens*. Markha Valley.

Steppes.

Juniperus semiglobosa naturally occurs in SE Zanskar only, on dry slopes of the narrow valleys of the Tsarap River, but it is also planted on sacred places. Here with *Acantholimon lycopodioides* and *Ephedra intermedia*. Purfi La.

Steppes.

Betula utilis subsp. *jacquemontii* forms the treeline in W Himalayas. In Ladakh, it can be found on stabilized stony screes well supplied by fresh water, along the Tsarap River only. Purfi La.

Steppes.

Semenovia lasiocarpa is common on steep steppe slopes in Zanskar. Here with *Artemisia* sp., *Arnebia euchroma*, *Rubia tibetica*, *Stipa* sp. Near Purne.

Steppes.

Euphorbia tibetica grows on unstable loamy, gravel and sandy slopes, sandy plains, stony steppes, more abundant on disturbed places. At lower and middle elevations, in northern parts of Rupshu common. Occurring together with *Artemisia salsoloides*, *Artemisia santolinifolia*, *Alyssum canescens*, *Oxytropis microphylla*, *Stipa subsessiliflora*, *Stipa caucasica*, *Artemisia stracheyi*, *Nepeta discolor*, *Allium przewalskianum*, *Ephedra gerardiana*, *Knorringia pamirica*, *Krascheninnikovia pungens*, *Salsola jacquemontii*, *Dracocephalum heterophyllum*. Here also with *Leymus secalinus* and *Chamaerhodos sabulosa*. Tso Moriri.

Steppes.

Artemisia santolinifolia is a frequent, but rarely dominating species, with a wide ecological amplitude, at lower and medium elevations. Most abundant on stabilized slopes and stony steppes, stony slopes, dry, rarely slightly mesic habitats. Occurring together with *Euphorbia tibetica*, *Ephedra gerardiana*, *Stipa caucasica*, *Artemisia salsoloides*, *Silene nepalensis*, *Alyssum canescens*, *Elsholtzia eriostachya*, *Artemisia stracheyi*, *Urtica hyperborea*, *Knorringia pamirica*, *Artemisia minor* and *Oxytropis microphylla*. Tso Moriri.

Steppes

with a Tibetan species *Artemisia stracheyi* can be found on flat river terraces and lakesides on eolitic sediments, in Rupshu only. The long-term stability of this ecosystem supports longevity in plants; here the oldest individuals were found (up to 70 years). Tso Moriri.

Steppes.

Carex moorcroftii is a typical plant of sandy plains, abundant and dominant on sandy deposits with unstable soil surface, less abundant on stabilized slopes. Occurring together with *Potentilla bifurca* and *Elsholtzia eriostachya*. Here also with *Elymus schrenkianus* and *Arenaria bryophylla*. Changthang.

Steppes.

Cold alpine steppe on a gravel slope with *Saussurea glanduligera*, *Stipa subsessiliflora*, *Alyssum canescens* and *Chamaerhodos sabulosa*. Changthang.

Steppes.

Stipa subsessiliflora is one of the dominants in dry alpine steppes and on sandy plains, frequent in stony steppes and stabilized slopes with scattered vegetation. Reaching the highest elevations among *Stipa* species. Occurring together with *Oxytropis microphylla*, *Stipa caucasica*, *Alyssum canescens*, *Euphorbia tibetica*, *Dracocephalum heterophyllum*, *Oxytropis tatarica*, *Artemisia minor*, *A. salsoloides*, *Krascheninnikovia pungens*, *Elymus jacquemontii*, *Artemisia stracheyi*, *Nepeta discolor*, *Stipa purpurea*. Changthang.

Alpine grasslands.

Allium przewalskianum is forming large and dense patches on dry to mesic slopes, in rock crevices, among stones, often near streams. Frequent in SE Zanskar, rare in NE Zanskar at lower elevations, in Rupshu at middle altitudes. Occurring together with *Artemisia sieversiana*, *Euphorbia tibetica*, *Artemisia stracheyi*, *Elymus schrenkianus*, *Artemisia demissa* and *Stipa purpurea*. Descent from Shingo La.

Alpine grasslands.

Festuca olgae is a stout hemicryptophyte, forming dense tussocks. Common on relatively moist stony and gravel slopes at middle elevations in Zanskar, missing in Rupshu and Leh. Here with *Bistorta affinis*, *Erigeron* sp., *Leontopodium ochroleucum*, *Potentilla* sp. Zanskar.

Alpine grasslands.

High-alpine pastures with *Astragalus confertus*, *Carex borii*, *Kobresia royleana*, *Leontopodium ochroleucum*, *Poa attenuata*, *Potentilla pamirica* and *Sibbaldia tetrandra*. Nubra Valley.

Alpine grasslands

with dominant forbs can be found in S and SW Zanskar on the relatively moist northern slopes of the Himalaya Range. Here with *Corydalis meifolia*, *C. thyrsiflora*, *Erigeron uniflorus* agg., *Eritrichium villosum*, *Kobresia nitens*, *Pedicularis bicornuta*, *P. cheilanthifolia*, *P. rhinanthoides*, *Potentilla sojakii*, *Saussurea schultzii* etc. Descent from Shingo La.

Alpine grasslands.

Bistorta affinis is forming dense stands on mesic slopes and in river beds, dominating large areas in Zanskar. Descent from Shingo La.

Alpine grasslands

with the striking plants of *Pleurospermum candollei*, together with *Corydalis moorcroftiana*, *Draba* sp., *Eritrichium villosum*, *Erigeron* sp., *Kobresia* sp. Descent from Shingo La.

Alpine grasslands.

Carex pseudofoetida grows on river banks and on lake margins, forming mesic turf and large mat where it may strongly dominate, in waterlogged and flooded soils. From the lowest to high elevations, where it is an important component of summer pastures. Nubra Valley.

Alpine grasslands.

Carex sagaensis grows on river banks and in alpine grasslands where it may strongly dominate, in Rupshu and neighbouring areas of Leh and Zanskar only. Occurring together with *Lomatogonium carinthiacum*, *Kobresia pygmaea*, *Stellaria depressa*, *Poa calliopsis*, *Kobresia royleana*, *K. schoenoides*, *Ranunculus sarmentosus*, *Thalictrum alpinum* and *Potentilla pamirica*. Tso Moriri.

Alpine grasslands.

Juncus thomsonii grows in mesic to wet habitats situated closely to watersheds, sometimes with permanent water level up to 5 cm above soil surface. Dominant in salty wetlands. Occurring together with *Carex orbicularis*, *Carex microglochin*, *Carex atrofusca*, *Deschampsia cespitosa*, *Stellaria irrigua*, *Comastoma pedunculatum*, *Pedicularis longiflora* subsp. *tubiformis*, *Saxifraga hirculus* s.l., *Arenaria stracheyi*, *Ranunculus sarmentosus*, *Lomatogonium carinthiacum*, *Oxytropis pusilla*, *Gentiana leucomelaena*, *Poa calliopsis*, *Ranunculus pulchellus*, *Thalictrum alpinum*. Nubra Valley.

Alpine grasslands.

Kobresia schoenoides grows in mesic to wet habitats of river banks, on stabilized stony mesic slopes. Often in saline pastures. Occurring together with *Kobresia royleana, Poa calliopsis, Thalictrum alpinum, Lomatogonium carinthiacum, Comastoma pedunculatum, Kobreasia pygmea, Saussurea leontodontoides, Potentilla anserina, Potentilla multifida, Leontopodium leontopodinum, Carex sagaensis, Festuca nitidula, Parnassia laxmannii, Rhodiola tibetica, Aster flaccidus, Stellaria depressa, Pleurospermum hookeri, Stracheya tibetica, Calamagrostis holciformis, Gentiana leucomelaena.* Tso Moriri.

Alpine grasslands.

Kobresia pygmaea is often a strong dominant on moist alpine turf on mesic river banks, usually close to rivers or brooks. Occurring together with *Thalictrum alpinum, Saxifraga hirculus* s.l., *Carex atrofusca, Carex sagaensis, Oxytropis platysema, Potentilla pamirica, Kobresia schoenoides, Poa calliopsis, Koenigia islandica, Kobresia royleana, Saussurea leontodontoides, Stellaria depressa, Rhodiola tibetica, Lomatogonium carinthiacum, Astragalus confertus, Stellaria irrigua.* Tso Moriri.

Alpine grasslands.

Moist alpine turf of *Kobresia pygmaea* is strongly resistant to trampling. Possibly the most characteristic species of the Tibetan Plateau, where it dominates the high alpine pastures, and its total area is estimated 450,000 km^2. Tso Moriri.

Alpine grasslands

dominated by *Carex sagaensis*, *Kobresia pygmaea* and *Poa attenuata*. Near Tso Moriri.

Alpine grasslands.

Astragalus confertus grows in open grasslands, and on stabilized gentle slopes. Occurring together with *Saussurea gnaphalodes*, *Potentilla pamirica*, *Saxifraga hirculus* s.l., *Draba altaica*, *Thylacospermum caespitosum*, *Ranunculus lobatus*, *Saxifraga cernua*, *Kobresia pygmaea* and *Trisetum spicatum*. Tso Moriri.

Alpine grasslands.

Potentilla pamirica is frequent on mesic stony slopes and on mesic to dry gravel sediments. Reaching the highest elevations among *Potentilla* species. Occurring together with *Poa attenuata*, *Stellaria depressa*, *Draba altaica*, *Arenaria bryophylla*, *Thylacospermum caespitosum*, *Kobresia pygmaea*, *Saussurea gnaphalodes*, *S. leontodontoides*, *S. bracteata*, *Astragalus strictus*, *Poa pratensis* subsp. *staintonii*, *Trisetum spicatum*, *Astragalus confertus*, *Thalictrum alpinum*, *Delphinium brunonianum*, *Carex sagaensis*. Near Tso Moriri.

Alpine grasslands

may replace in succession the *Thylacospermum caespitosum* stage on old glacial moraines, especially under grazing pressure of domestic animals. Here, the graminoids include *Carex borii, C. pseudofoetida, Elymus schrenkianus, Kobresia royleana, K. schoenoides, Poa attenuata* and *Trisetum spicatum*. Nubra Valley.

Subnival vegetation.

Poa attenuata is forming dense tussocks in stony steppes, dry river beds, screes, mesic turf at higher elevations, ubiquitous at about 5000–5600 m a.s.l.. At all elevations, but rare below 4500 m a.s.l., in Rupshu markedly frequent. Reaching the highest elevations among monocotyledonous plants in the world. Occurring together with *Potentilla pamirica, Arenaria bryophylla, Thylacospermum caespitosum, Saussurea gnaphalodes, Draba altaica, Urtica hyperborea, Astragalus strictus, Elymus jacquemontii, Artemisia minor*. Here also with *Desideria pumila, Ladakiella klimesii, Saussurea glacialis, Stellaria decumbens* and *Waldheimia tridactylites*. Changthang.

Subnival vegetation.

Thylacospermum caespitosum is one of the most conspicuous cushion plants of the Tibetan Plateau. It is frequent on mesic to wet stabilized slopes in the subnival zone, on old stabilized stony river sediments, rarely on plains. Occurring together with *Saussurea gnaphalodes, S. bracteata, Arenaria bryophylla, Potentilla pamirica, Pleurospermum stellatum, Androsace baltistanica, Eritrichium hemisphaericum, Gentiana prostrata, Leontopodium pusillum, Primula macrophylla, Saxifraga cernua, Waldheimia tridactylites, Draba altaica, Poa attenuata, Trisetum spicatum, Aphragmus oxycarpus, Aster flaccidus, Astragalus confertus, Saxifraga nanella, Stellaria depressa*. Near Tso Moriri.

Subnival vegetation.

Species reaching the highest elevations are usually 1–3 cm in height, e.g. *Draba altaica, Saussurea gnaphalodes, Saxifraga stella-aurea, Stellaria decumbens*. Near Tso Moriri.

Subnival vegetation.

The highest elevations with vascular plants life in Ladakh are located east of Tso Moriri lake in Changthang, on the high-elevation plateau between Chalung and Chamser Kangri Peaks (growing as high as 6150 m a.s.l.). Here at 5900 m a.s.l., 20 species can still be found, including *Aphragmus oxycarpus, Draba altaica, D. oreades, Ladakiella klimesii, Poa attenuata, Saussurea glacialis, S. inversa, Stellaria decumbens* and *Waldheimia tridactylites*.

Springs and snow beds.

Primula macrophylla grows in snow beds, on solufluction soils and flushes, and in wet spring areas at the highest elevations only. Occurring together with *Aphragmus oxycarpus, Waldheimia tridactylites, Saxifraga nanella, Saussurea gnaphalodes, Draba altaica, Thylacospermum caespitosum, Festuca hartmanii, Koenigia islandica*. Tso Moriri.

Springs and snow beds.
Cremanthodium ellisii is frequent in snow beds and moss-rich patches close to streams and springs at higher elevations. Zanskar.

Springs and snow beds.
Saxifraga nanella grows in snow beds and in wet rock crevices at high elevations. Occurring together with *Waldheimia tridactylites*, *Saussurea gnaphalodes*, *Eritrichium hemisphaericum*, *Primula macrophylla*, *Draba altaica*, *Thylacospermum caespitosum*, *Pegaeophyton scapiflorum*. Here also with *Aphragmus oxycarpus*, *Poa attenuata*, *Saussurea inversa* and *Stellaria decumbens*. Near Tso Moriri.

Screes.
Aconogonon tortuosum is dominating on stable sandy slopes with stones, on unstable loamy slopes, often on screes. Occurring together with *Artemisia salsoloides*, *Dracocephalum heterophyllum*, *Corispermum tibeticum*, *Euphorbia tibetica*, *Artemisia santolinifolia*, *Alyssum canescens*, *Artemisia minor*, *Scrophularia dentata*, *Oxytropis microphylla*, *Salsola jacquemontii*. Kiupa La.

Screes.

Fine screes and unstable loamy slopes with *Rheum tibeticum*, *Aconogonon tortuosum* and *Thermopsis inflata*. A habitat typical also for *Astragalus munroi*, *Biebersteinia odora* or *Dasiphora dryadanthoides*. Stok.

Screes.

Delphinium brunonianum grows on stony and gravel slopes, both stabilized and unstable, common in the whole area, localities below 5000 m are usually on steep slopes of rocky gorges. Occurring together with *Arabis tibetica*, *Ajania tibetica*, *Elsholtzia eriostachya*, *Trisetum spicatum*, *Carex stenophylla*, *Polygonum cognatum*, *Silene nepalensis*, *Rhodiola tibetica*, *Pleurospermum stellatum*, *Urtica hyperborea*, *Stellaria depressa*, *Potentilla pamirica* and *Elymus schrenkianus*. Here with *Urtica hyperborea* and *Artemisia stracheyi*. Tso Moriri.

Screes.

Elymus schrenkianus inhabits dry to mesic slopes, gravel deposits, unstable screes, stony habitats. In Rupshu and neighbouring areas of Leh and Zanskar very frequent. Occurring together with *Allium przewalskianum*, *Delphinium brunonianum*, *Hippophae tibetana*, *Polygonum cognatum*, *Silene gonosperma*, *Silene nepalensis*, *Leontopodium leontopodinum*, *Scrophularia dentata* and *Potentilla multifida*. Tso Moriri.

Screes.

Stabilized scree with a spring of water with exceptionally ornamental and species-rich vegetation (*Cremanthodium ellisii, Draba altaica, Festuca tibetica, Gentiana nubigena, Poa attenuata, Potentilla pamirica, Primula macrophylla, Saussurea inversa, Saxifraga cernua, S. nanella, Stellaria decumbens, Waldheimia tridactylites*). Mentok, Rupshu.

Screes.

Carex nivalis colonizes glacier forefields and moraines, also grows on mesic turf on slopes and river banks. Here with *Waldheimia tridactylites* and *Trisetum spicatum*. Nubra Valley.

Wetlands.

Pedicularis longiflora var. *tubiformis* is common in spring areas and on mesic to wet river banks. Typical also for salt marshes and salt meadows. Occurring together with *Carex microglochin, C. orbicularis, Triglochin palustre, Aster flaccidus, Oxytropis pusilla, Ranunculus pulchellus, Comastoma pedunculatum, Argentina anserina, Hippuris vulgaris, Triglochin maritima, Zannichellia palustris*. From Shingo La to Kargyak.

Wetlands.

Freshwater wetlands with *Carex pamirensis* are rare in Ladakh, confined to the flat valley bottoms in western Zanskar. Zanskar.

Wetlands.

Rumex tibeticus together with *Ranunculus natans* and *Potamogeton filiformis* in an eutrophicated freshwater pool. Indus Valley.

Wetlands.

Saline wetlands may be relatively species-rich (*Blysmus compressus*, *Carex orbicularis*, *Gentiana leucomelaena*, *Halerpestes sarmentosa*, *Oxytropis hypoglottoides*, *Pedicularis longiflora*, *Primula pamirica*, *Puccinellia himalaica*, *Triglochin maritima*). Tso Kar Basin.

Wetlands.

Triglochin maritima and *T. palustre* may dominate in marshes, irrigation canals, in open saline vegetation, on river banks in turf. Occurring together with *Primula pamirica*, *Silene nepalensis*, *Hordeum brevisubulatum*, *Pedicularis cheilanthifolia*, *Aster flaccidus*, *Pedicularis longiflora* subsp. *tubiformis*, *Puccinellia thomsonii*, *Carex orbicularis*, *Blysmus compressus*. Tso Kar Basin.

Saline vegetation.

Knorringia pamirica is common in open-canopy vegetation, often disturbed, on loamy and sandy sediments, dry to wet, in salty wetlands and grasslands, tolerates trampling. Occurring together with *Puccinellia himalaica*, *Suaeda olufsenii*. Puga Valley.

Saline vegetation.

Suaeda olufsenii is a common indicator of halophilous habitats in the whole area at lower and middle elevations; always in species poor open-canopy vegetation. Occurring together with *Puccinellia himalaica*, *Knorringia pamirica*, *Atriplex tatarica*. Here also with *Glaux maritima*. Rupshu.

Saline vegetation.

Puccinellia himalaica grows in marshes and open saline vegetation, on emerged sandy and loamy bottoms often as a pioneer, tolerates high levels of salt in the soil and persists in habitats covered by a crust of salt or borax. From the lowest to high elevations, but between 4500 and 5000 m a.s.l. nearly ubiquitous. Occurring together with *Ranunculus sarmentosus*, *Suaeda olufsenii*, *Calamagrostis holciformis*, *Oxytropis hypoglottoides*, *Knorringia pamirica*, *Glaux maritima*, *Taraxacum* sect. *Leucantha*, *Hymenolobus procumbens*. Puga Valley.

Saline vegetation.

Puccinellia thomsonii can be found around the hot spring in the Puga Valley (the type locality), along the brook with masses of *Zannichellia palustris*, with species-poor large stands at the E margin of the locality where borax crusts are deposited at soil surface, and at wetlands situated at the foot of the road in N margin of the plain. Occurring together with *Blysmus compressus*, *Triglochin palustre*, *T. maritima*. Endemic, known from the type locality only.

River banks and river beds.

Hippophae tibetana is colonizing river gravel deposits at lower and middle elevations, often on salty gravels. Occurring together with *Askellia flexuosa*, *Myricaria germanica*, *Silene gonosperma*, *Elymus schrenkianus*. Zanskar.

River banks and river beds.

Myricaria germanica subsp. *alopecuroides* can be found in river beds at lower and middle elevations. Occurring together with *Hippophae tibetana*, *Silene gonosperma*, *Pleurospermum hookeri*. Here also with *Aconogonon tortuosum*, *Ajania fruticulosa*, *Artemisia salsoloides* and *Kobresia schoenoides*. Rupshu.

River banks and river beds.

Epilobium latifolium colonizes stony habitats and river gravel deposits in SE Zanskar. Here it occurs with *Hippophae tibetana*, *Askellia flexuosa*, *Myricaria germanica*, *Silene gonosperma*, *Elymus schrenkianus*. Descent from Shingo La.

River banks and river beds.

Calamagrostis pseudophragmites grows on river beds, especially on sandy sediments and on slopes situated closely to streams. At lower elevations in the Indus River valley and in SE Zanskar. Upper Indus near Mahe.

River banks and river beds.

Myricaria elegans grows on river banks and on stream sides. Along glacier streams ascents to the highest elevations among large woody species in Ladakh, up to 4400 m a.s.l. Nubra Valley.

River banks and river beds.

Hippophae rhamnoides subsp. *turkestanica* often dominates on river gravel deposits at lower elevations. Nubra Valley.

River banks and river beds.

Salix schugnanica is a tree (up to about 6 m in height) or shrub, common in Zanskar. Growing in river beds, along streams, frequently cultivated with other *Salix* and *Populus* species. Mulbekh.

River banks and river beds.

Salix sericocarpa is a tree (up to about 8 m in height) or shrub, growing on river banks, along irrigation canals, planted in gardens and along roads, cultivated up to nearly 4700 m a.s.l. The most common willow in the Indus River valley and also in side valleys of Indus. Near Leh.

Animal resting places.

Plant assemblages developed on eutrophicated ground include *Atriplex tatarica*, *Axyris hybrida*, *Artemisia macrocephala*, *Chenopodium karoi*, *Hymenolobus procumbens*, *Microula tibetica*, *Microgynoecium tibeticum*, *Physochlaina praealta*, *Smelowskia tibetica*, many of which are characteristic annual species of animal resting places throughout the Tibetan Plateau. Tso Moriri.

Weedy field margins.

Thanks to water supply from irrigation canals, field margins may host relatively rich plant assemblages (*Conioselinum vaginatum*, *Gentiana leucomelaena*, *Gentianella moorcroftiana*, *Lepyrodiclis holosteoides*, *Pedicularis cheilanthifolia*, *Oxytropis hypoglottoides*). Zanskar.

Field work

requires trained muscles, high-quality tools, and a portion of patience. Jiří Doležal takes advantage of a natural landslide to excavate the massive root system of *Aconogonon tortuosum*. Ascent to Shingo La.

Anatomical descriptions of horsetails and ferns

J. Doležal et al., *Anatomy, Age and Ecology of High Mountain Plants in Ladakh, the Western Himalaya*, https://doi.org/10.1007/978-3-319-78699-5_2

Equisetum ramosissimum Desf.

Equisetaceae

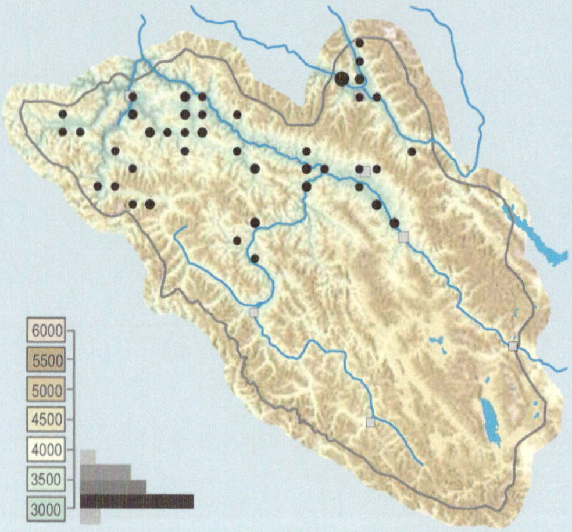

Description: Perennial herbs, 20–60 cm tall, rhizomatous, stems green, monomorphic, ridged, lower part often branched, strobilus shortly clavate, 0.5–2.5 cm long.

Distribution: Africa, Eurasia.

Habitat: Forest margins, sandy plains.

Rhizome round, with a diameter of 1.5–4 mm. A large aerenchymatic cortex surrounds a central part with vascular bundles.

Central part

Center hollow, surrounded by 5–8 vascular bundles. Each bundle surrounded by an unlignified, thin-walled, indistinct endodermis. Large Carinal canals (air ducts), distinct phloem and indistinct or absent xylem characteristic for vascular bundles. Xylem, if present, consists of a very few tracheids, diameter 25–30 μm. Phloem well-developed, contains some large polygonal, unlignified sieve cells and a few companion cells.

Cortex

Outer cortex consists of a lignified external and an unlignified internal compact cellular band. Inner part contains very large intercellulars (aerenchyma).

Analyzed: Rhizomes of 2 individuals.

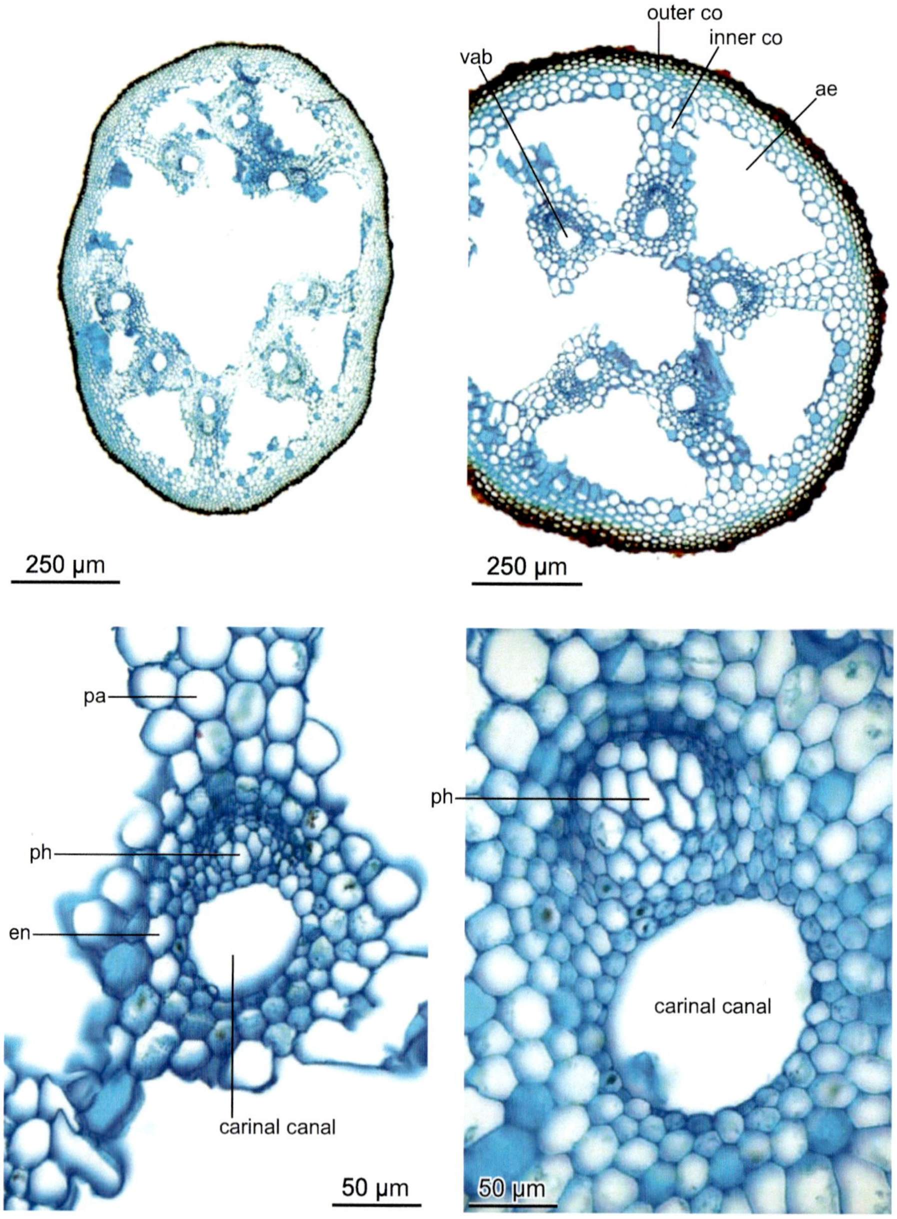

Athyrium distans (D.Don) T.Moore

Athyriaceae

Syn.: *Asplenium distans* D.Don

Description: Perennial ferns, 40–60 cm tall, rhizomes creeping, with brown scales, fronds 3–pinnatisect.

Distribution: N India, Nepal, China, Japan.

Habitat: Forest margins, alpine shrub meadows.

Petiole ribbed, with a diameter of 1–1.3 mm. A large cortex surrounds two reniform amphiversal vascular bundles.

Central part

Vascular bundles composed of a xylem with polygonal, lignified tracheids, diameters up to 40 μm. Surrounding phloem consists of unlignified sieve tubes and companion cells. Endodermis absent.

Cortex

Outer cortex consists of an unlignified epidermis and a 2–3-seriate, lignified cellular belt. Large inner part is composed of thin-walled, unlignified cells and small, triangular intercellulars.

Analyzed: Petiole of 1 herbarium specimen.

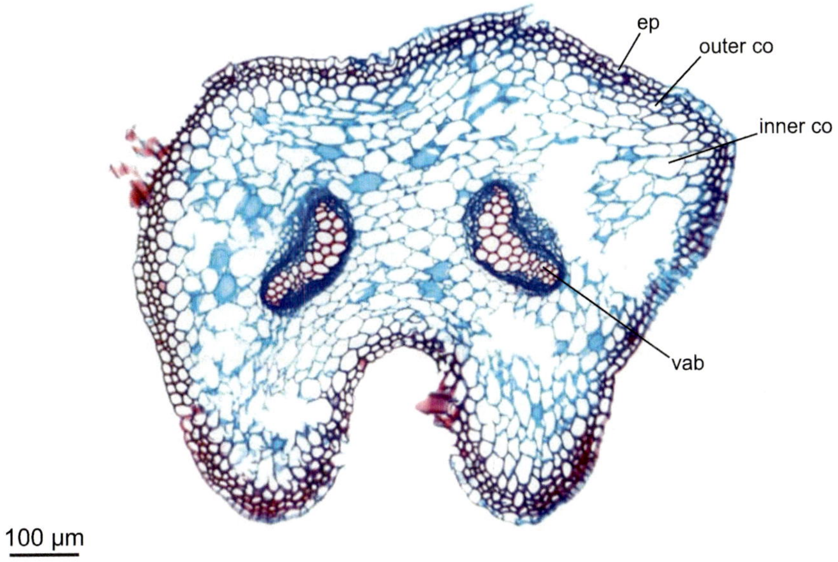

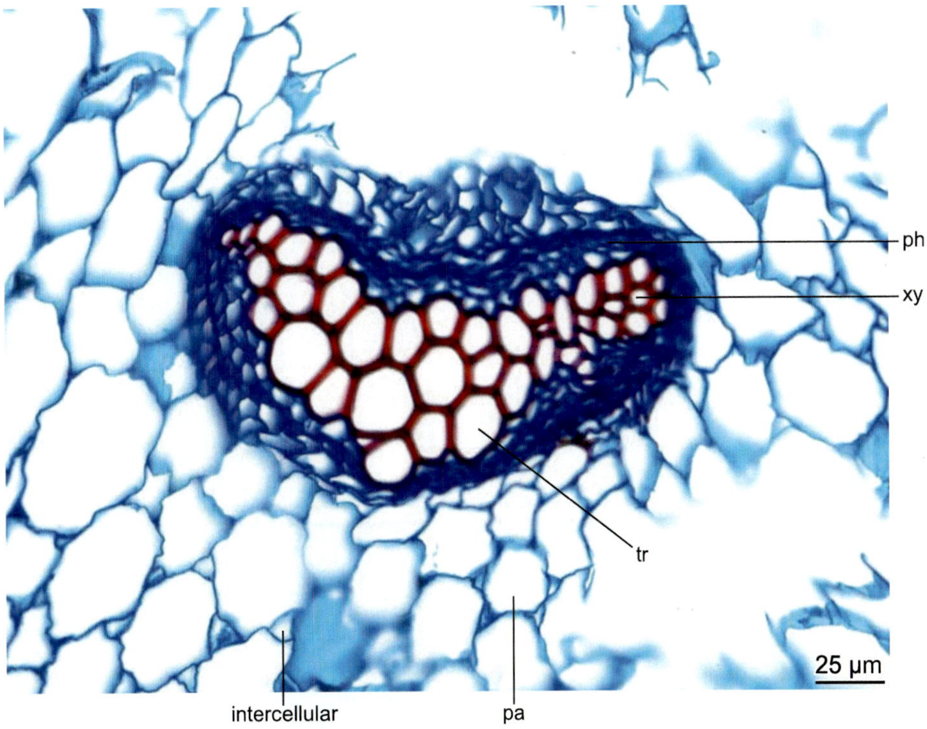

Deparia acuta (Ching) Fraser-Jenk.

Athyriaceae

Syn.: *Lunathyrium acutum* Ching

Description: Perennial ferns, 30–80 cm tall, rhizomes erect, with red-brown scales, fronds pinnate-pinnatipartite, narrowly elliptic to lanceolate.

Distribution: Pakistan, India (Jammu and Kashmir, Himachal Pradesh, Uttarakhand), China (Sichuan, Xizang).

Habitat: Rocky meadows, mountain slopes.

Petiole ribbed, with a diameter of 1.8–2.2 mm. A large cortex surrounds two wing-shaped amphivasal vascular bundles.

Central part

Vascular bundles composed of a xylem with polygonal, lignified tracheids, diameters up to 60 µm. Surrounding phloem consists of unlignified sieve tubes and companion cells. Endodermis consists of small, thin-walled, rectangular, unlignified cells.

Cortex

Outer cortex consists of an unlignified epidermis and a 4–6-seriate lignified cellular belt. Large inner part is composed of thin-walled, unlignified cells.

Analyzed: Petiole of 1 herbarium specimen.

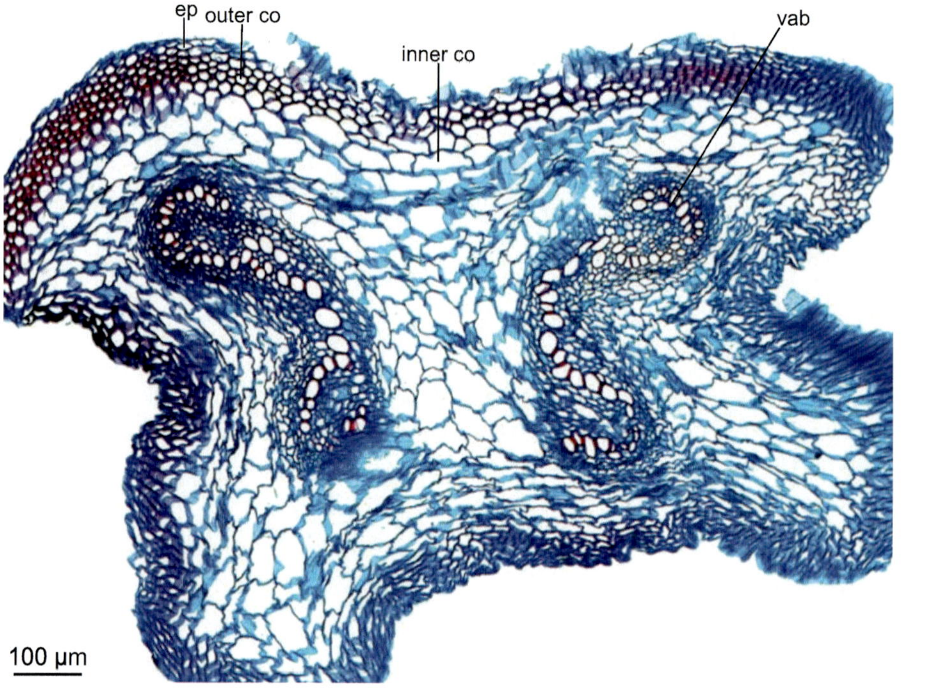

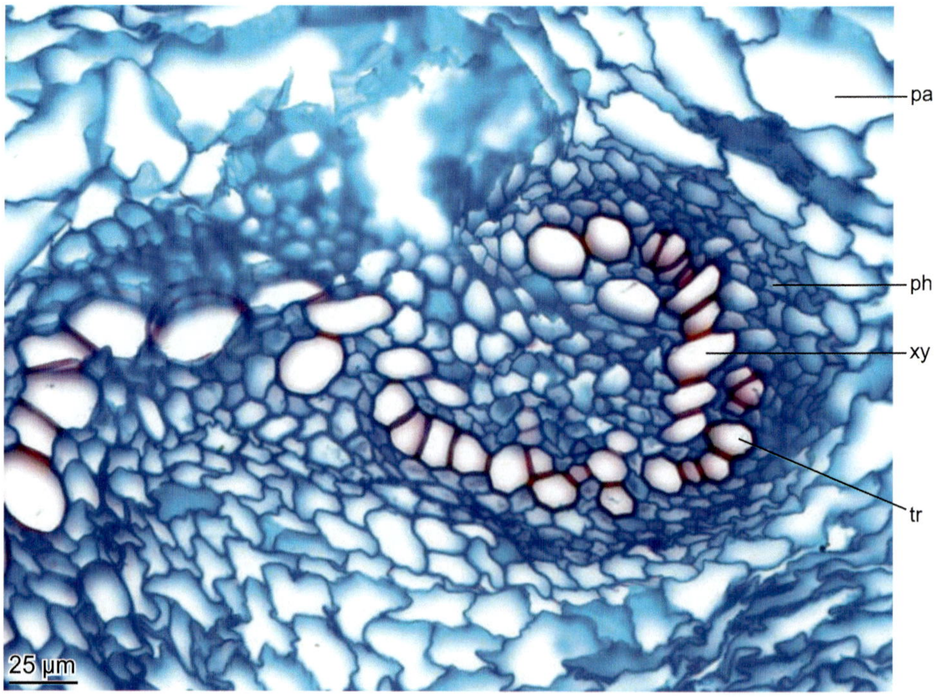

Cystopteris dickieana R.Sim

Cystopteridaceae

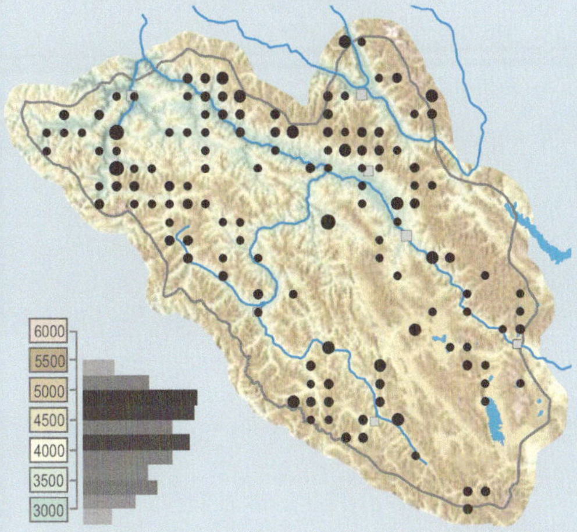

Syn.: *C. fragilis* subsp. *dickieana* (R.Sim) Hyl.

Description: Perennial ferns, 10–30 cm tall, rhizome creeping, sparsely scaly, fronds 2-pinnate, aggregated, petioles 5–20 cm long.

Distribution: North America, Africa, Europe, Afghanistan, Pakistan, India, Nepal, China (Gansu, Hebei, Qinghai, Shaanxi, Sichuan, Xinjiang, Xizang, Yunnan).

Habitat: Rocky crevices, stone walls.

Petiole ribbed, with a diameter of 0.9–1.0 mm. A large cortex surrounds two round to reniform amphiversal vascular bundles.

Central part

Vascular bundles composed of a xylem with polygonal, lignified tracheids, diameters up to 40 µm. Surrounding phloem consists of unlignified sieve tubes and companion cells. Endodermis, if present, unlignified and thin-walled.

Cortex

Outer cortex consists of an unlignified epidermis and a 2–3-seriate, lignified cellular belt. Large inner part is composed of thin-walled, unlignified cells and small, triangular intercellulars.

Analyzed: Petiole of 1 herbarium specimen.

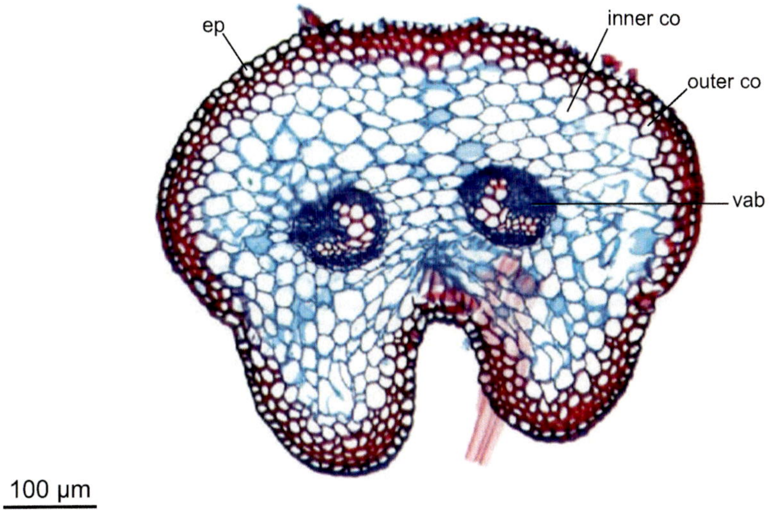

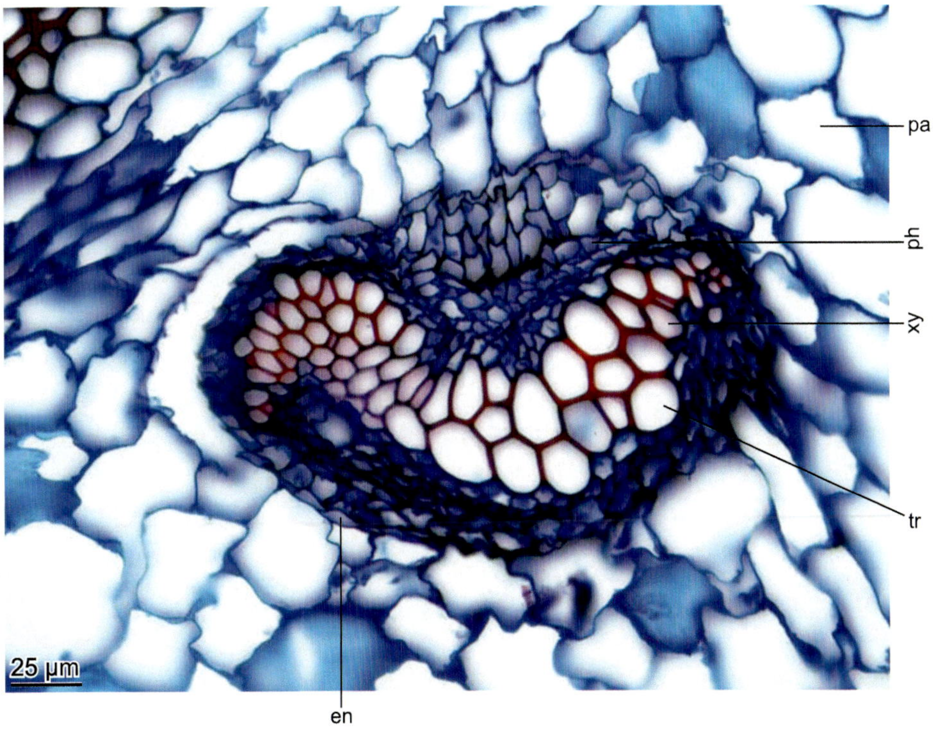

Cystopteris fragilis (L.) Bernh.

Cystopteridaceae

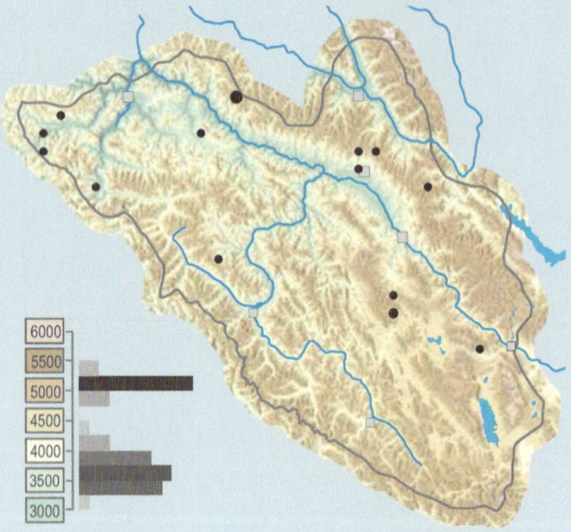

Description: Perennial ferns, 10–30 cm tall, rhizome creeping, sparsely scaly, fronds 2-pinnate, aggregated, petioles 5–20 cm long.

Distribution: Temperate regions and tropical mountains worldwide.

Habitat: Rock crevices, shaded slopes.

Petiole ribbed, with a diameter of 1.0–1.1 mm. A large cortex surrounds two elongated reniform amphiversal vascular bundles.

Central part

Vascular bundles composed of a xylem with polygonal, lignified tracheids, diameters up to 50 µm. Surrounding phloem consists of unlignified sieve tubes and companion cells. Endodermis consists of small, thin-walled, rectangular, unlignified cells.

Cortex

Outer cortex consists of an unlignified epidermis and a 2–3-seriate, lignified cellular belt. Large inner part is composed of thin-walled, unlignified cells and small, triangular intercellulars.

Analyzed: Petioles of 2 herbarium specimens.

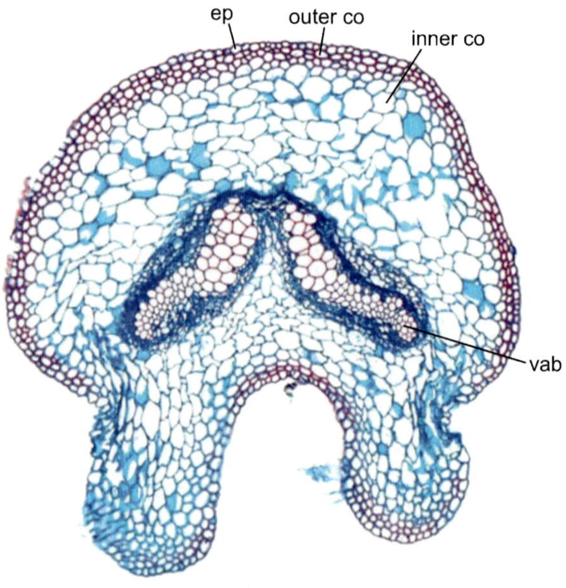

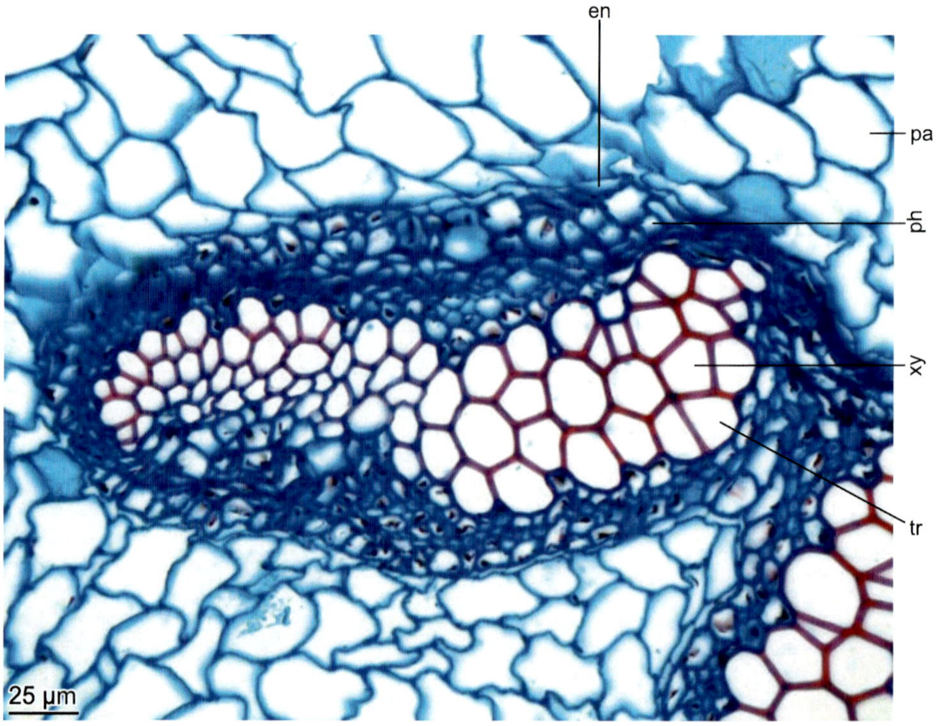

Gymnocarpium fedtschenkoanum Pojark.

Cystopteridaceae

Description: Perennial ferns, 20–50 cm tall, rhizome creeping, fronds 2-pinnate-pinnatipartite, triangular.

Distribution: Tajikistan, Afghanistan, N Pakistan, India (Jammu and Kashmir, Himachal Pradesh, Uttar Pradesh, Sikkim), Nepal.

Habitat: Rocky meadows, rock crevices.

Petiole ribbed, with a diameter of 1.3–1.5 mm. A cortex surrounds 2 round, large, amphivasal vascular bundles.

Central part

Vascular bundles composed of a xylem with polygonal, lignified tracheids, diameters up to 70 µm. Surrounding phloem consists of unlignified sieve tubes and companion cells. Endodermis consists of cells with an extremely thick, dark-stained centripetal wall, and a very thin, unlignified centrifugal wall.

Cortex

Outer cortex consists of a lignified epidermis and a 6–8-seriate, lignified, thick-walled cellular belt. Inner part is composed of thin-walled, unlignified cells.

Analyzed: Petioles of 2 herbarium specimens.

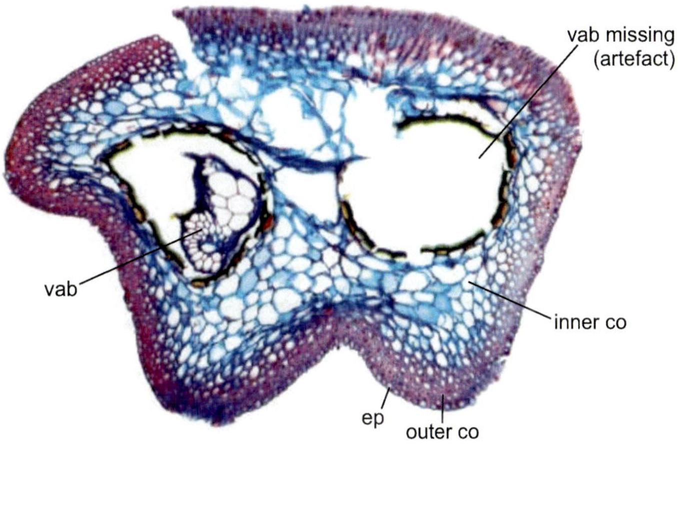

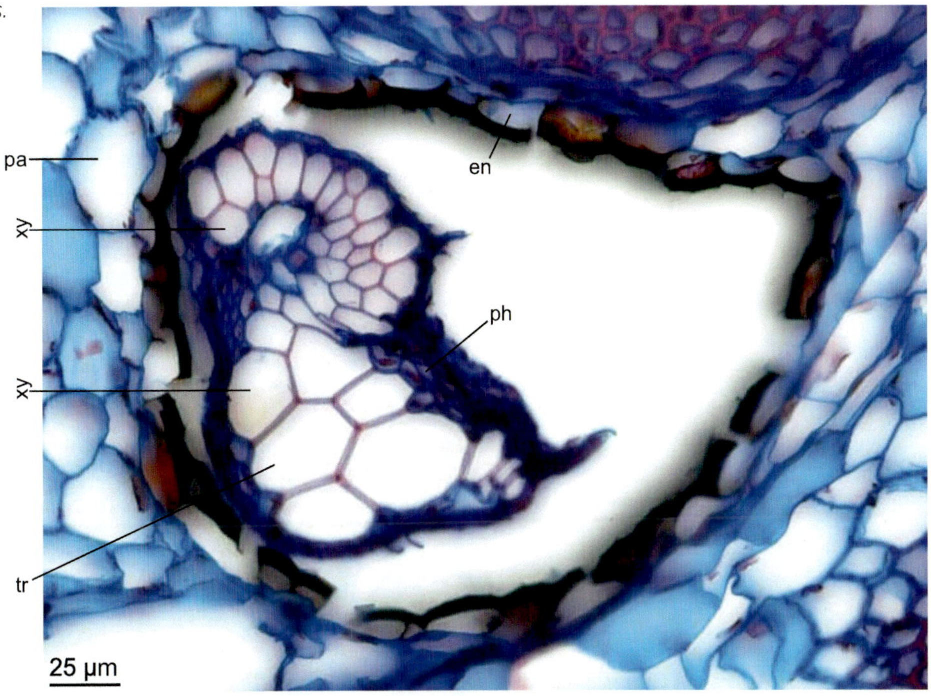

Dryopteris blanfordii (C.Hope) C.Chr.

Dryopteridaceae

Syn.: *D. gongboensis* Ching, *Nephrodium blanfordii* C.Hope

Description: Perennial ferns, 40–90 cm tall, rhizomes creeping, with brown scales, fronds tripinnatipartite.

Distribution: Afghanistan, Pakistan, India (Jammu and Kashmir, Himachal Pradesh, Uttarakhand), Nepal, China (Gansu, Sichuan, Xizang, Yunnan).

Habitat: Forest floors.

Petiole with a wavy circumference, with a diameter of 1.5–2 mm. A large cortex surrounds 5 round, amphivasal vascular bundles.

Central part

Vascular bundles composed of a xylem with polygonal, lignified tracheids, diameters up to 35 µm, and small, unlignified parenchyma cells. Surrounding phloem consists of unlignified sieve tubes and companion cells. Endodermis consists of cells with an extremely thick, dark-stained centripetal wall, and a very thin, unlignified centrifugal wall.

Cortex

Outer cortex consists of an unlignified epidermis and a 6–8-seriate, lignified cellular belt. Large inner part is composed of thin-walled, unlignified cells and small, triangular intercellulars.

Analyzed: Petiole of 1 herbarium specimen.

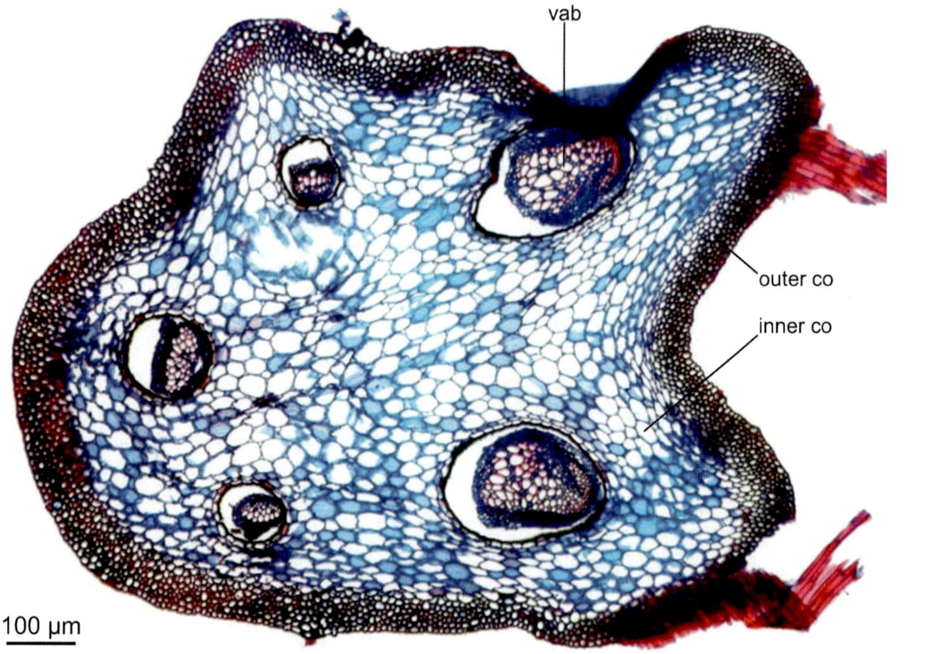

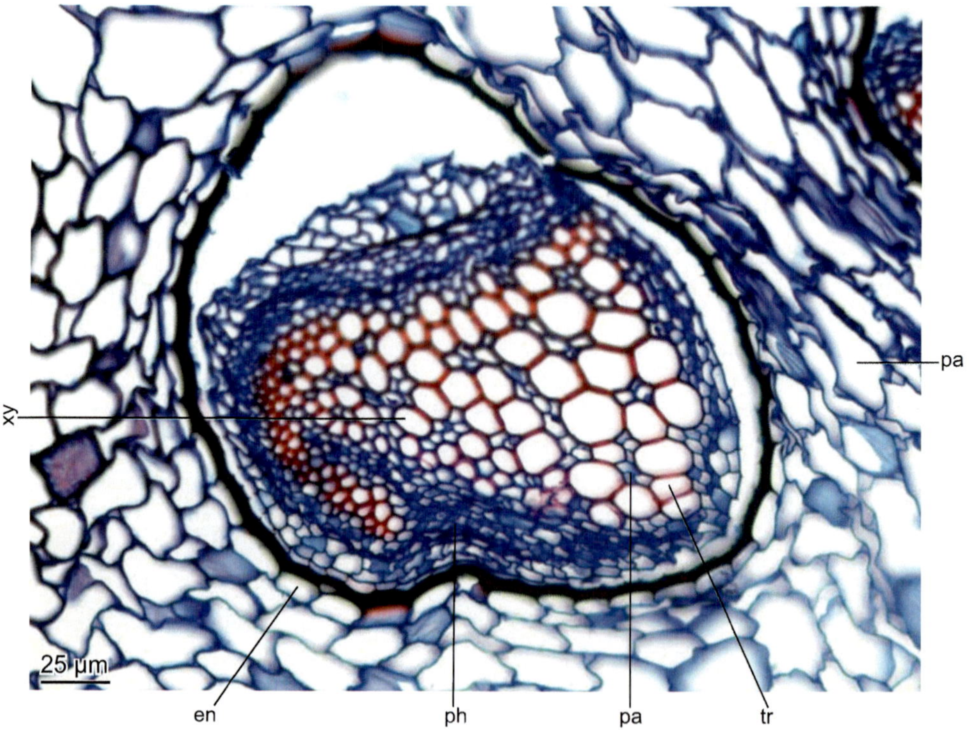

Dryopteris komarovii Kossinsky

Dryopteridaceae

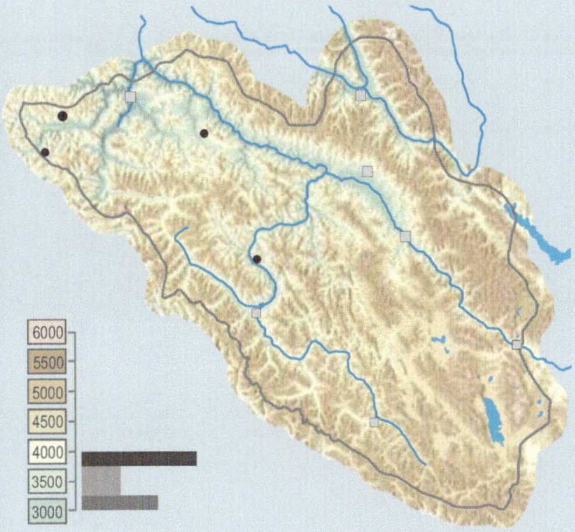

Syn.: *D. barbigera* subsp. *komarovii* (Kossinsky) Fraser-Jenk., *D. subbarbigera* Ching

Description: Perennial ferns, 30–50 cm tall, rhizome erect, densely scaly, fronds 2-pinnate, cespitose.

Distribution: Russia, Pakistan, India, Nepal, China (Gansu, Qinghai, Shaanxi, Sichuan, Taiwan, Xizang, Yunnan), Bhutan, Myanmar.

Habitat: Rock crevices.

Petiole with a wavy circumference, with a diameter of 3–4 mm. A large cortex surrounds 5 round, amphivasal vascular bundles.

Central part

Vascular bundles composed of a xylem with polygonal, lignified tracheids, diameters up to 40 µm, and small, unlignified parenchyma cells. Surrounding phloem consists of unlignified sieve tubes and companion cells. Endodermis consists of cells with an extremely thick, dark-stained centripetal wall, and a very thin, unlignified centrifugal wall.

Cortex

Outer cortex consists of an unlignified epidermis and a 6–8-seriate, lignified cellular belt. Large inner part is composed of thin-walled, unlignified cells and small, triangular intercellulars.

Analyzed: Petioles of 2 herbarium specimens.

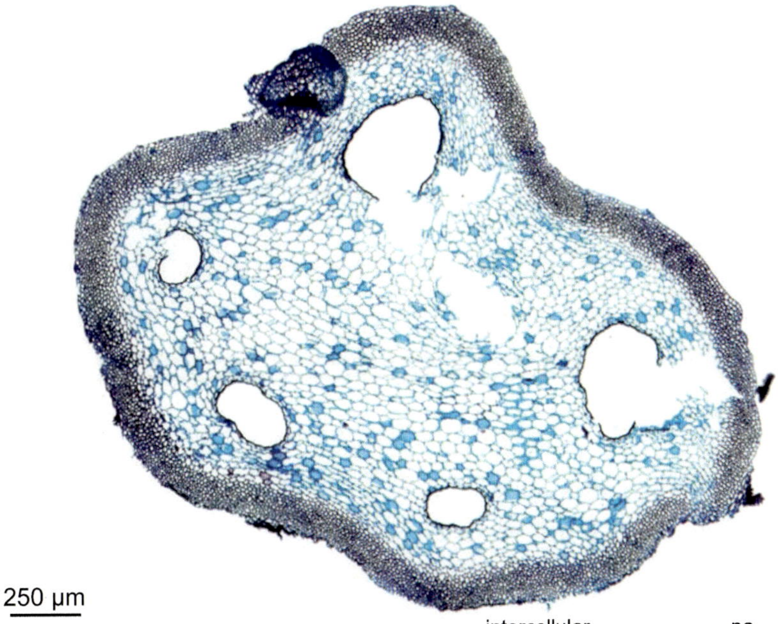

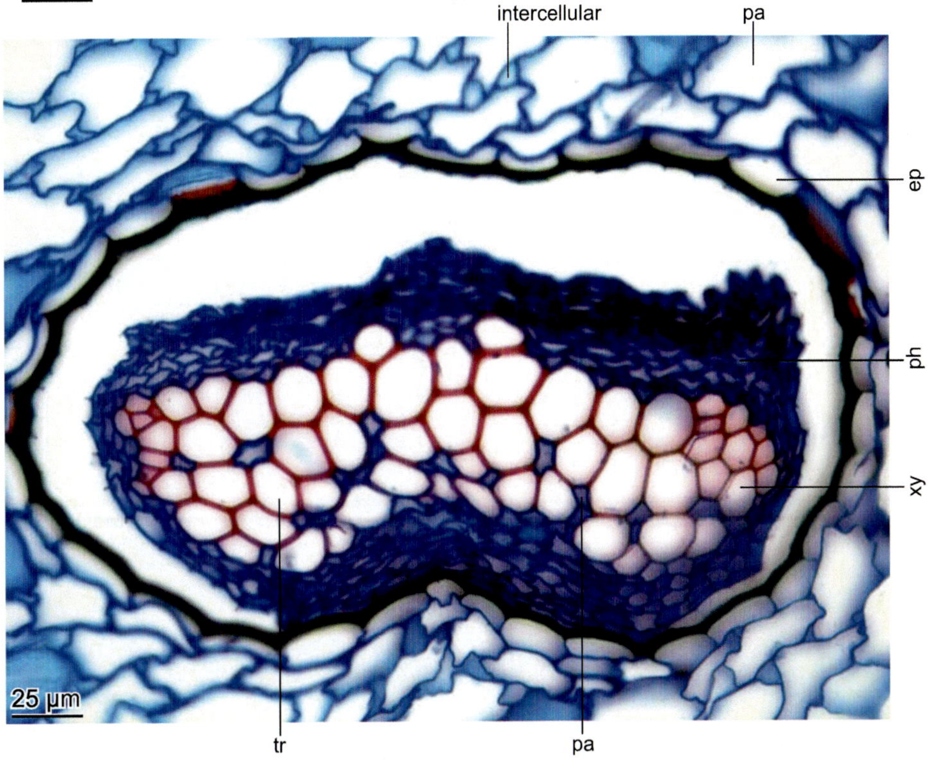

Dryopteris ramosa (C.Hope) C.Chr.

Dryopteridaceae

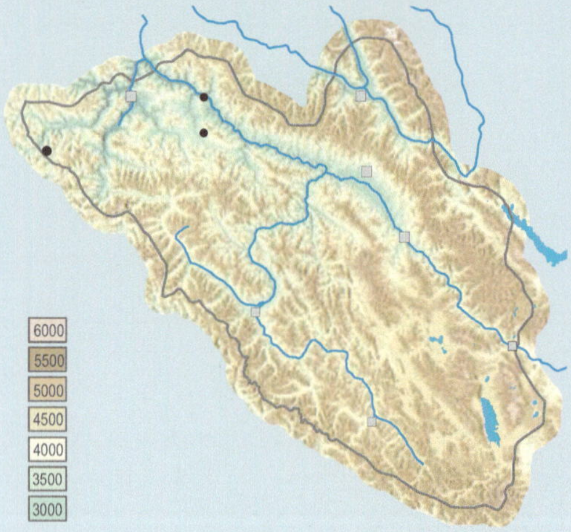

Syn.: *Nephrodium ramosum* C.Hope

Description: Perennial ferns, 50–120 cm tall, rhizome densely scaly, fronds tripinnate, deltate.

Distribution: Afghanistan, Pakistan, Bhutan, India (Jammu and Kashmir, Himachal Pradesh, Uttarakhand), Nepal.

Habitat: Forest floors.

Petiole ribbed, with a diameter of 2–3 mm. A large cortex surrounds 5 round, amphivasal vascular bundles.

Central part

Vascular bundles composed of a xylem with polygonal, lignified tracheids, diameters up to 50 µm, and small, unlignified parenchyma cells. Surrounding phloem consists of unlignified sieve tubes and companion cells. Endodermis consists of cells with an extremely thick, dark-stained centripetal wall, and a very thin, unlignified centrifugal wall.

Cortex

Outer cortex consists of an unlignified epidermis and a 6–8-seriate, lignified cellular belt. Large inner part is composed of thin-walled, unlignified cells and small, triangular intercellulars.

Analyzed: Petiole of 1 herbarium specimen.

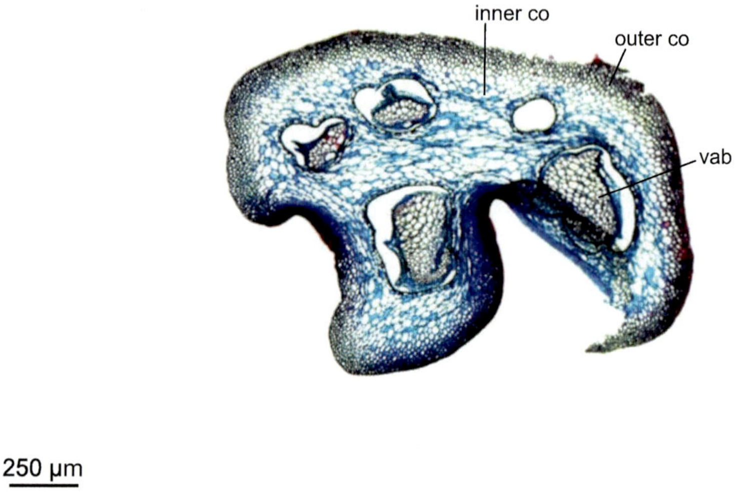

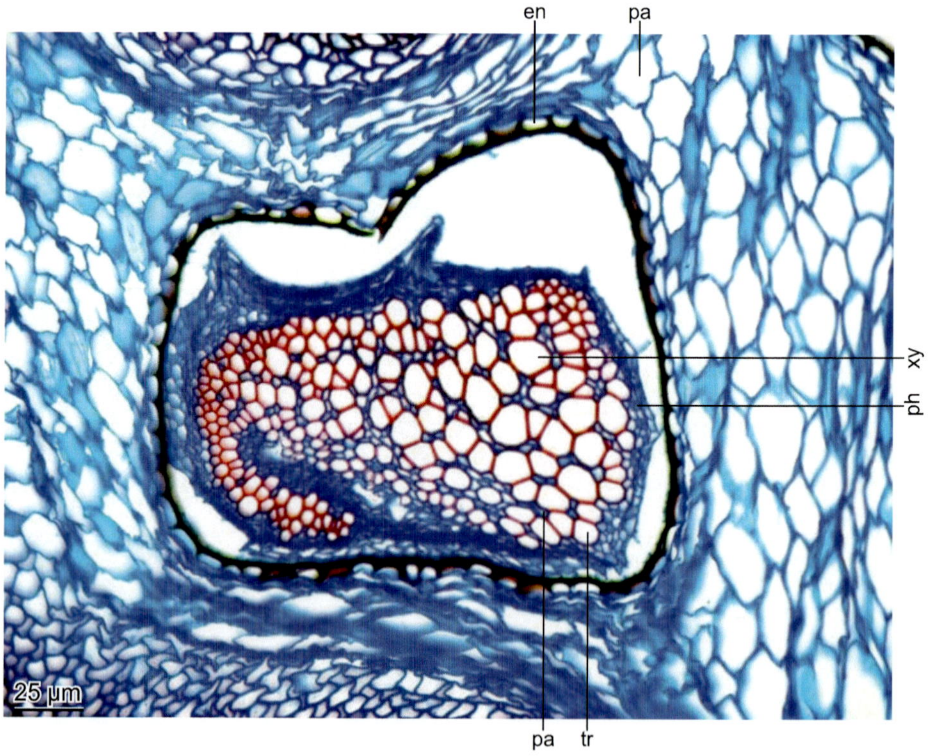

Polystichum prescottianum (Wall. ex Mett.) T.Moore

Dryopteridaceae

Syn.: *Aspidium prescottianum* Wall. ex Mett., *P. erinaceum* Ching & S.K.Wu

Description: Perennial ferns, 30–40 cm tall, rhizome erect, with broadly lanceolate dark brown scales, fronds bipinnatifid, oblanceolate.

Distribution: Afghanistan, Pakistan, N India, Nepal, China (Xizang, Taiwan), Bhutan, Myanmar.

Habitat: Alpine meadows.

Petiole ribbed, with a diameter of 1.5–2.5 mm. A large cortex surrounds 4 round, amphivasal vascular bundles.

Central part

Vascular bundles composed of a xylem with polygonal, lignified tracheids, diameters up to 50 µm, and small, unlignified parenchyma cells. Surrounding phloem consists of unlignified sieve tubes. Endodermis consists of cells with an extremely thick, dark-stained centripetal wall, and a very thin, unlignified centrifugal wall.

Cortex

Outer cortex consists of a lignified epidermis and a 6–8-seriate, lignified, thick-walled cellular belt. Inner part is composed of thin-walled, unlignified cells and small, triangular intercellulars.

Analyzed: Petiole of 1 herbarium specimen.

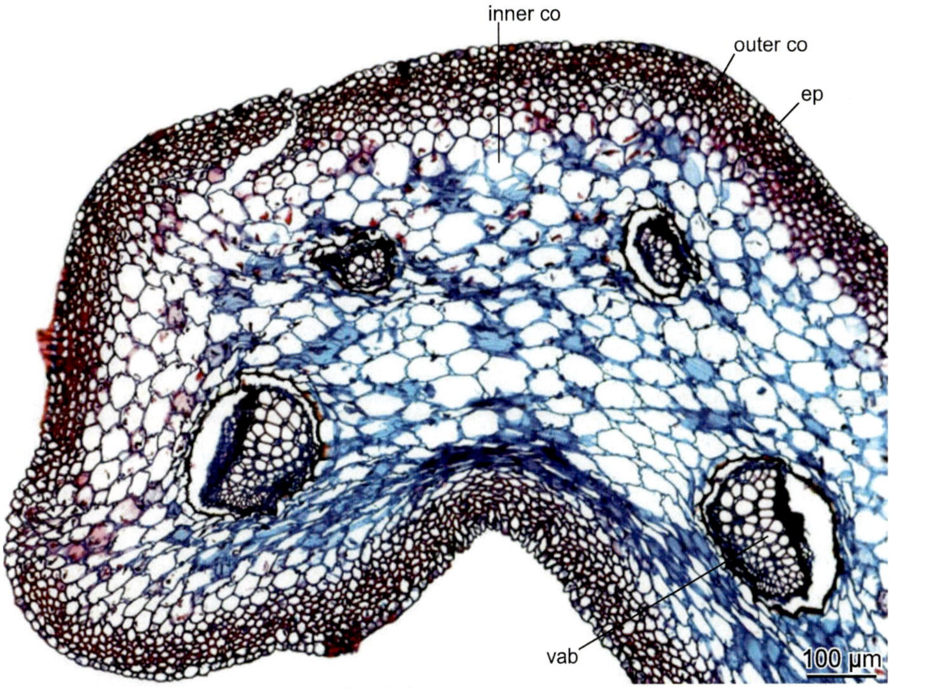

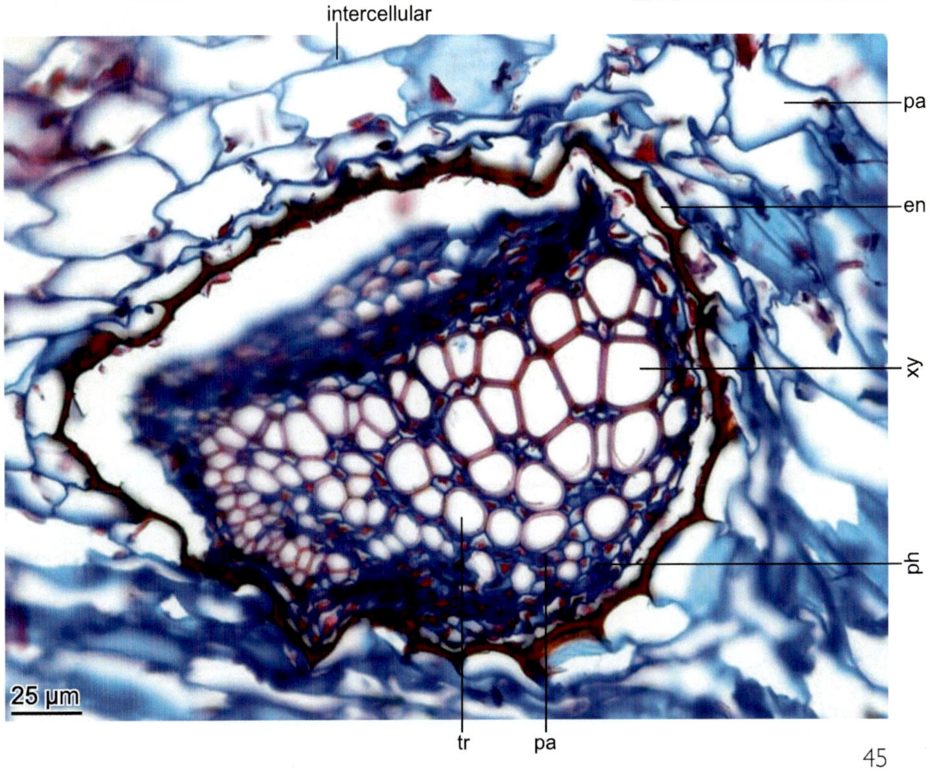

Woodsia glabella R.Br. ex Richardson
Woodsiaceae

Syn.: *W. alpina* var. *glabella* (R.Br. ex Richardson) D.C.Eaton, *W. hyperborea* var. *glabella* (R.Br. ex Richardson) Watt

Description: Perennial ferns, 2–10 cm tall, rhizome erect, with brown lanceolate scales, fronds pinnate-pinnatifid, linear to lanceolate.

Distribution: Circumpolar.

Habitat: Calcareous rock crevices, screes.

Petiole ribbed, with a diameter of 0.4–0.6 mm. A thick-walled cortex surrounds one wing-shaped amphivasal vascular bundle.

Central part
Vascular bundle composed of a xylem with polygonal, lignified tracheids, diameters up to 30 µm. Surrounding phloem consists of unlignified sieve tubes. Endodermis consists of thin-walled cells.

Cortex
Outer cortex consists of a thin- to thick-walled epidermis and a 2–3-seriate, lignified, thick-walled cellular belt. Inner part is composed of thin-walled, unlignified cells and small, triangular intercellulars.

Analyzed: Petiole of 1 herbarium specimen.

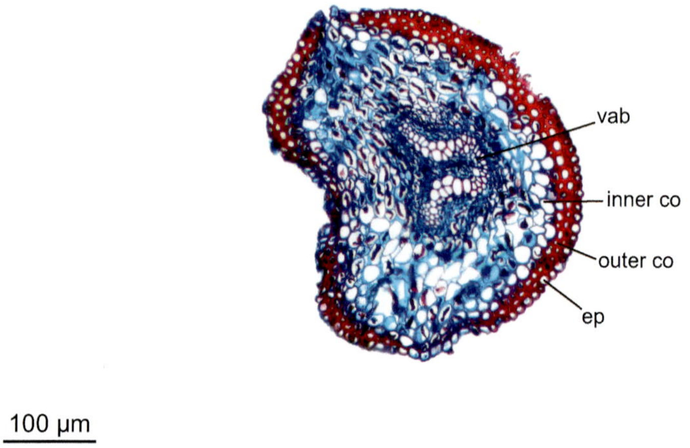

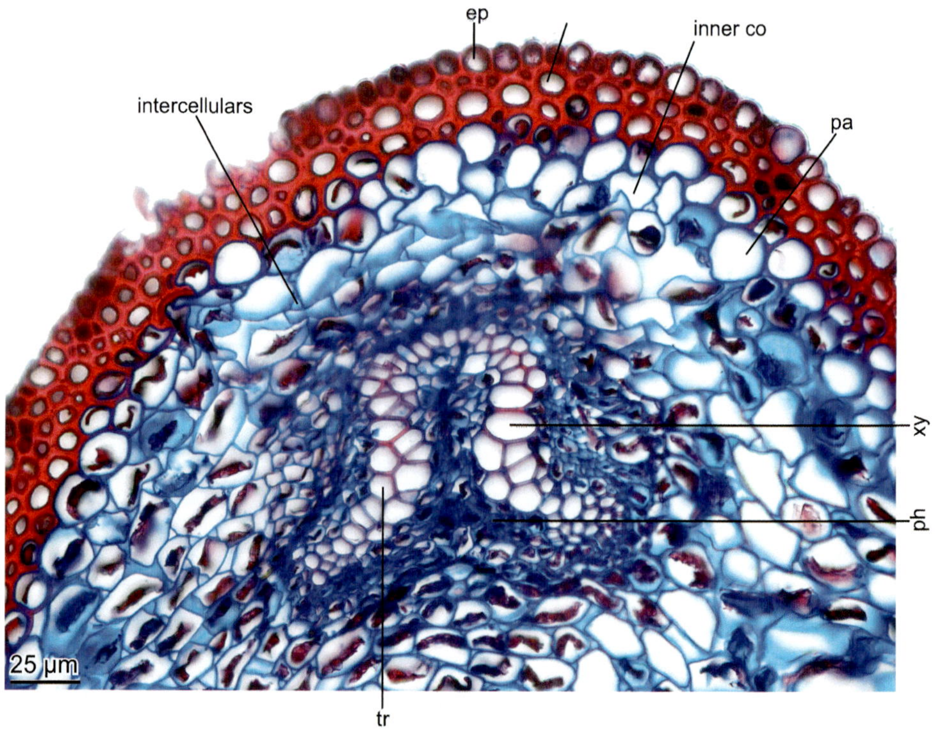

Anatomical descriptions of dicotyledons

J. Doležal et al., *Anatomy, Age and Ecology of High Mountain Plants in Ladakh, the Western Himalaya*, https://doi.org/10.1007/978-3-319-78699-5_3

Amaranthus powellii S.Watson

Amaranthaceae

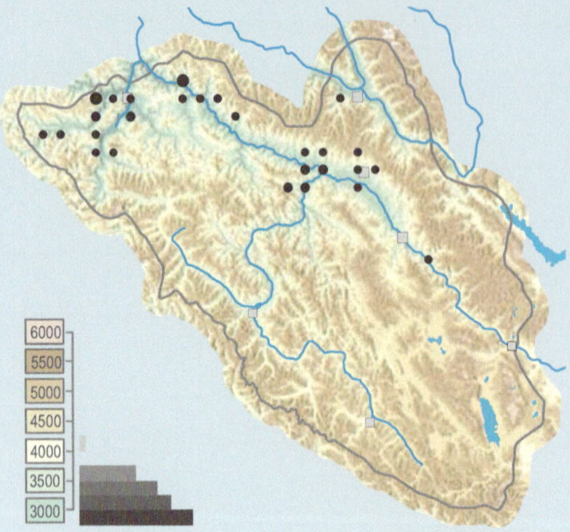

Syn.: *A. bracteosus* Uline & W.L.Bray, *A. retroflexus* var. *powellii* (S.Watson) B.Boivin

Description: Annuals, 30–150 cm tall, stems stiff, branched, flowers minute, arranged in spikes, composing erect and rigid inflorescences.

Distribution: Originally native to SW United States and adjacent regions of Mexico,, now widely naturalized in temperate regions.

Habitat: A weed, colonizing field margins.

Stem diameter 8 mm. Bark/xylem proportion 1:10. Annual plant. Rings absent.

Xylem

Transverse section

Root collar: Vessels solitary and in short radial rows. Vessel diameter 40–50 µm. Fibers thin- to thick-walled. Parenchyma paratracheal. Successive cambia diffuse foraminate. Conjunctive tissue lignified, thin-walled. Isolated vascular bundles arranged in tangential rows. With some crystal sand grains in conjunctive tissue.

Tangential section

Rays 2–3 seriate; >10-seriate between vascular bundles in the center of the stem.

Radial section

Perforation plates simple. Intervessel pits round and horizontally enlarged, in alternating position. Vessel length 80–120 µm.

Bark

Root collar: Phloem homogeneous with distinct companion cells. Cortex with very large parenchyma cells. Crystal sand in the cortex. Phellem absent.

Pith

No slide available.

Analyzed: Root collar of polar root and shoot of 1 individual.

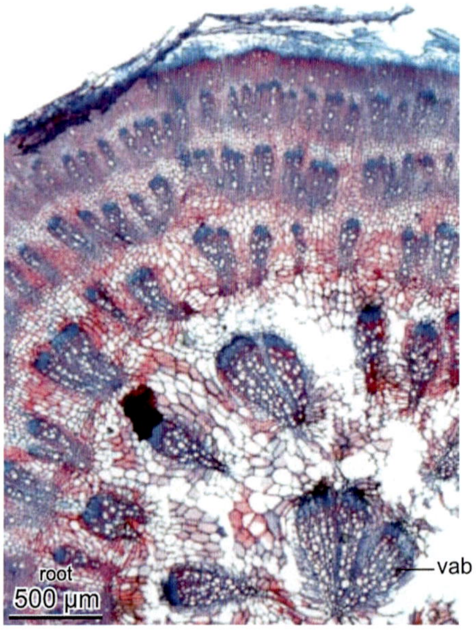

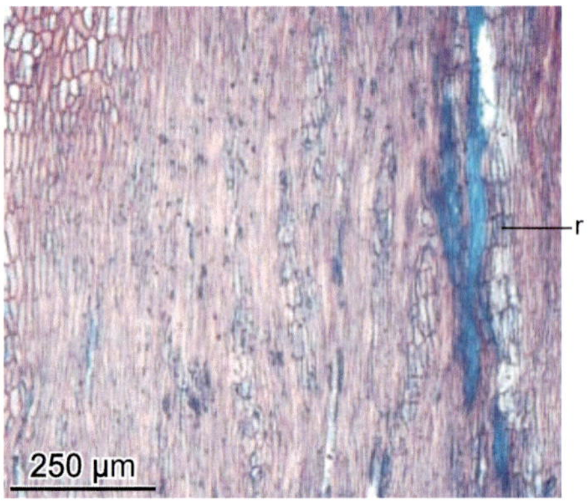

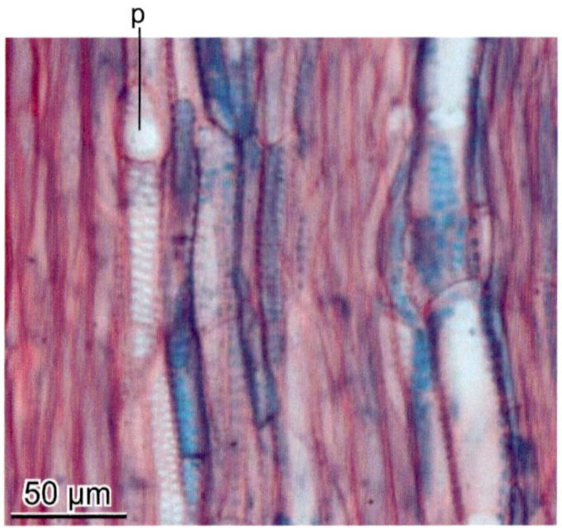

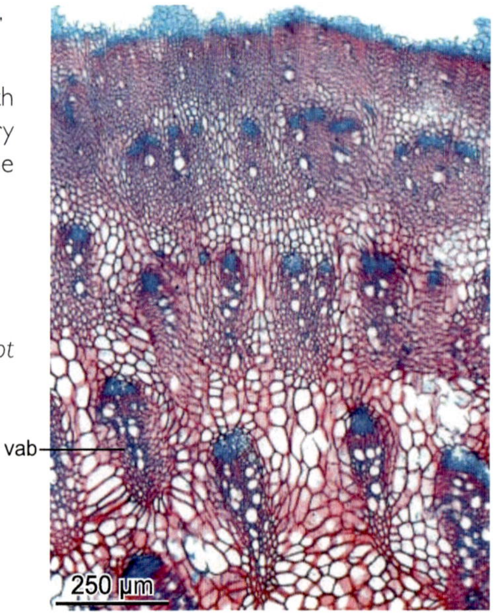

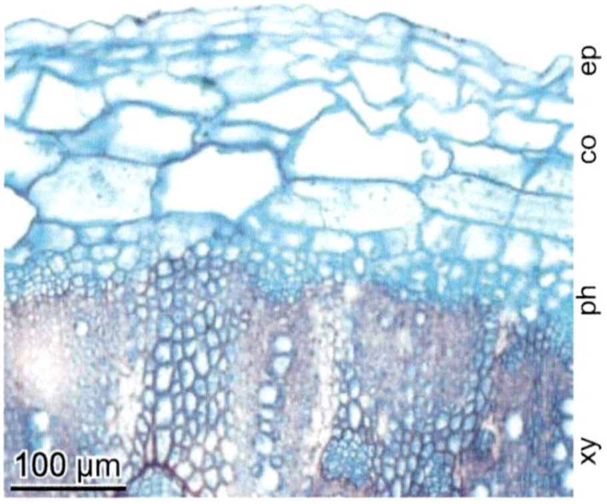

Features according to IAWA

Xylem 2.1 - 9 - 9.1 - 13 - 22 - 40.2 - 50.1 - 52.3 - 61 - 70 - 79 - 97 - 99 - 99.1 - 105 - 134 - 134.1 - 153

Atriplex schugnanica Iljin

Amaranthaceae

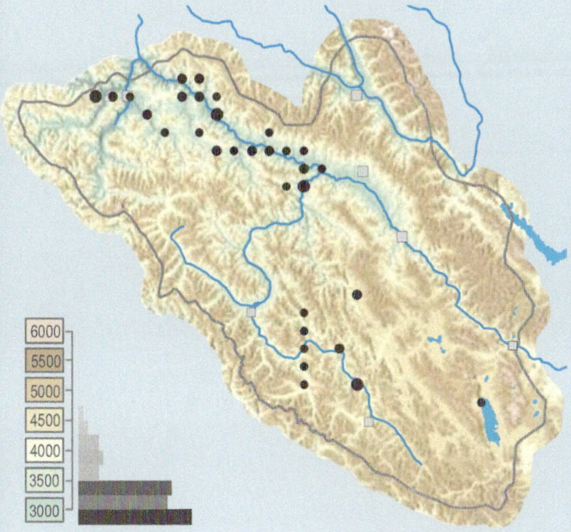

Description: Annuals, 15–50 cm tall, branched, leaves alternate, ovate, inflorescence interrupted, spike-like, leafless.

Distribution: C Asia, Tajikistan, Pakistan, India.

Habitat: Sandy and stony places.

Stem diameter 3 mm. Bark/xylem proportion 1:10. Mean ring width 1.5 mm. Maximal 1 annual ring.

Xylem

Transverse section

Root collar: Vessels in short radial rows. Vessel diameter 40–50 µm. Fibers thin- to thick-walled. Parenchyma paratracheal. Successive cambia continuous. Conjunctive tissue not lignified, thin-walled. With some crystal sand in conjunctive tissue.

Shoot: Vessels in groups. Successive cambia foraminate.

Tangential section

Rays 2–3-seriate.

Radial section

Perforation plates simple. Intervessel pits round in alternating position. Vessel length 80–120 µm.

Bark

Root collar: Phloem homogeneous. Phellem consists of a few layers of thin-walled, square cells.

Shoot: Cortex consists of large, thin-walled parenchyma cells containing crystal druses. Phellem absent.

Pith

Round, filled with very large, thin-walled, polygonal parenchyma cells and a few vascular bundles.

Analyzed: Root collar of polar root and shoot of 1 individual.

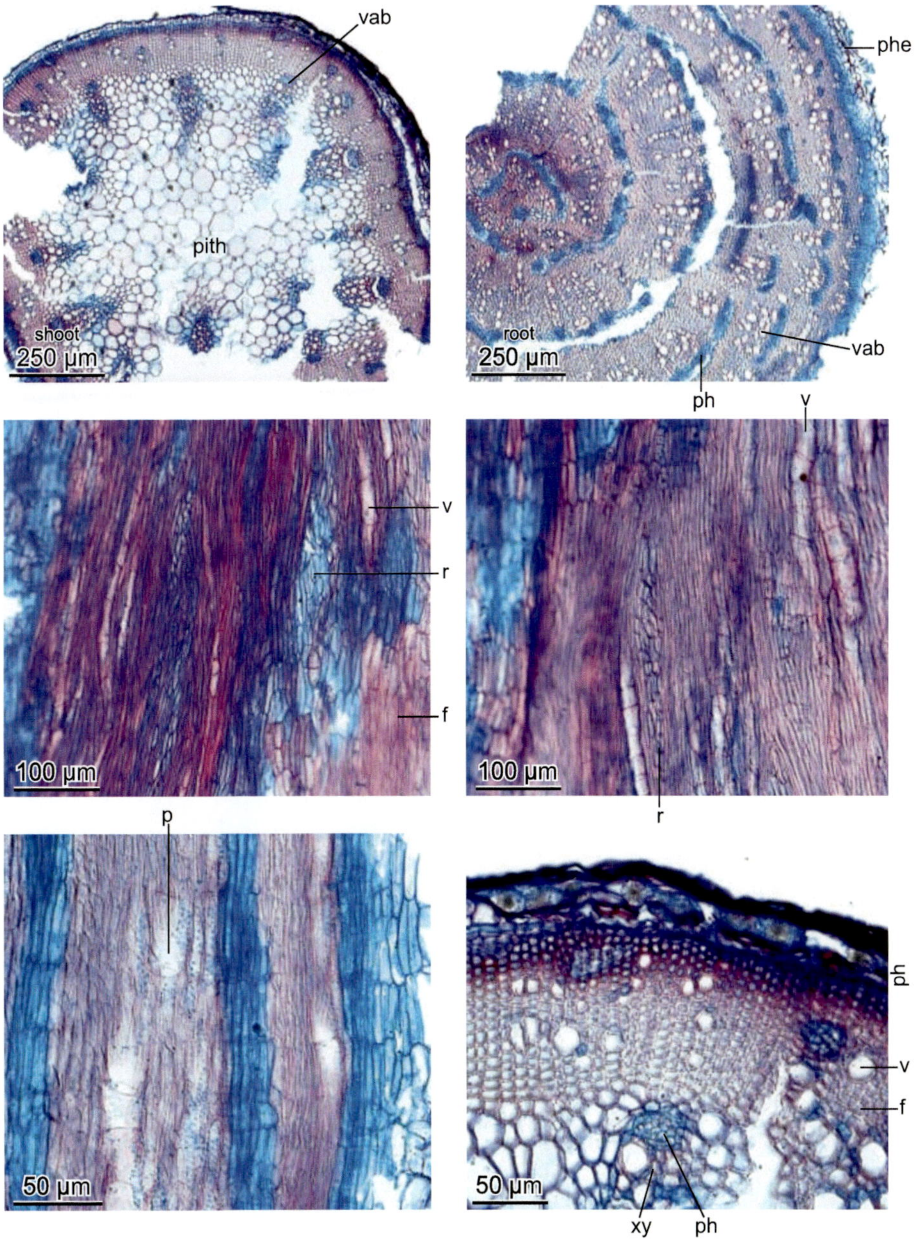

Features according to IAWA

Xylem *Root* *2.1 - 9.1 - 13 - 22 - 40.2 - 50.2 - 52.3 - 61 - 70 - 79 - 97 - 105 - 133.2 - 134.1 - 153*
Shoot *2.1 - 9 - 11 - 13 - 22 - 40.2 - 50.2 - 52.3 - 61 - 70 - 79 - 97 - 105 - 134 - 134.1 - 153*

Axyris hybrida L.

Amaranthaceae

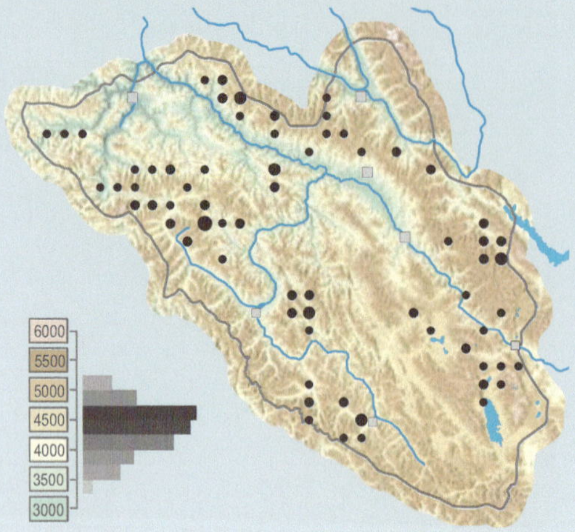

Description: Annuals, 5–30 cm tall, hairy, stems erect, branched from the base, leaves ovate to elliptic, flowers unisexual, minute, in short spikes.

Distribution: SE Russia, C and SW Asia, Pakistan, N India, Nepal, China (Gansu, Hebei, Heilongjiang, Henan, Nei Mongol, Qinghai, Shanxi, Xinjiang, Xizang, Yunnan), Mongolia.

Habitat: Sandy and gravel slopes, field margins.

Stem diameter 0.9 mm. Bark/xylem proportion 1:4. Maximal 1 annual ring.

Xylem

Transverse section

Vessels slightly angular, solitary and in radial groups. Vessel diameter 10–15 µm. Fibers thin- to thick-walled. Parenchyma paratracheal. Successive cambia, xylem continuous, phloem foraminate. Conjunctive tissue absent. Crystals absent.

Tangential section

Rays absent.

Radial section

Perforation plates simple. Intervessel pits round in alternating position. Vessel length 60–80 µm.

Bark

Phloem mostly uniform or with a few small sieve tubes. Phellem consists of a few layers of thin-walled, square cells. Crystal sand in the phellem.

Pith

No slide available.

Analyzed: Root collar of 1 individual (herbarium specimen).

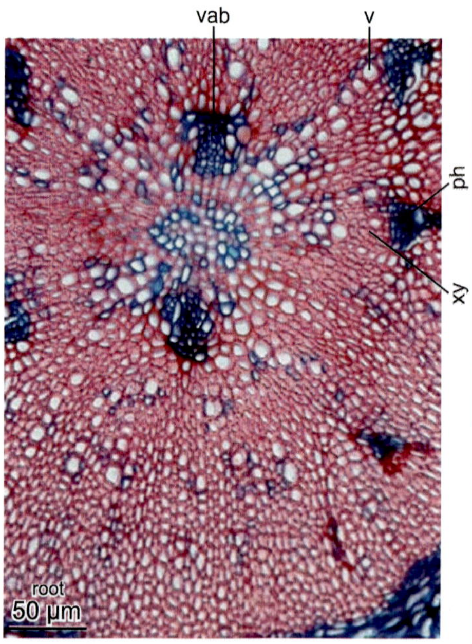

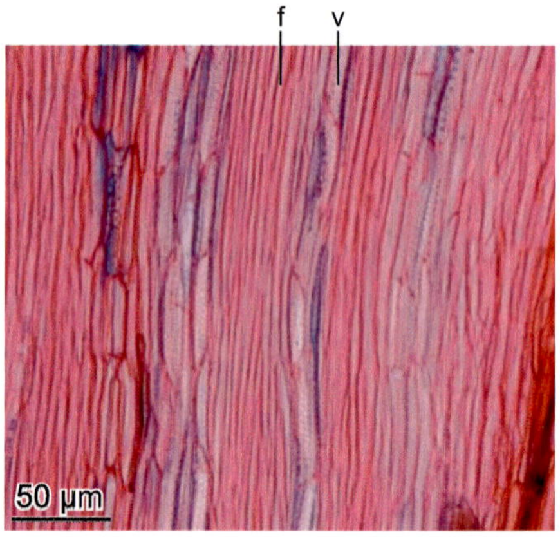

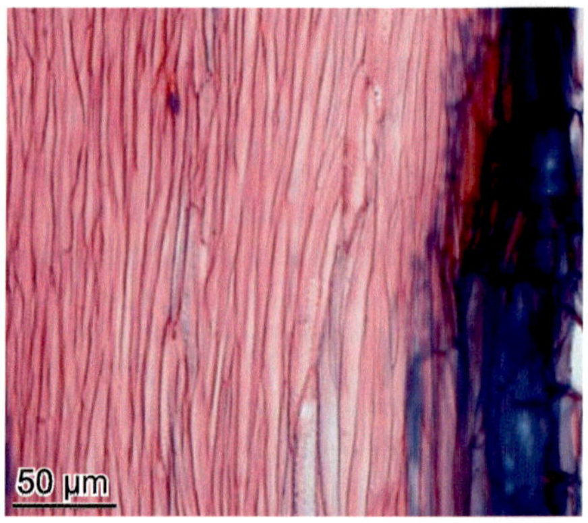

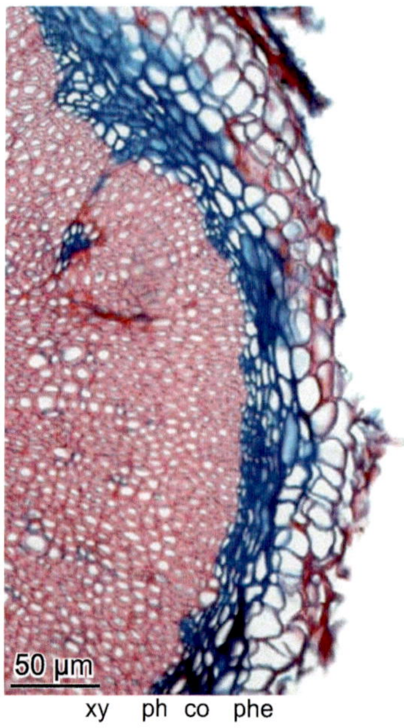

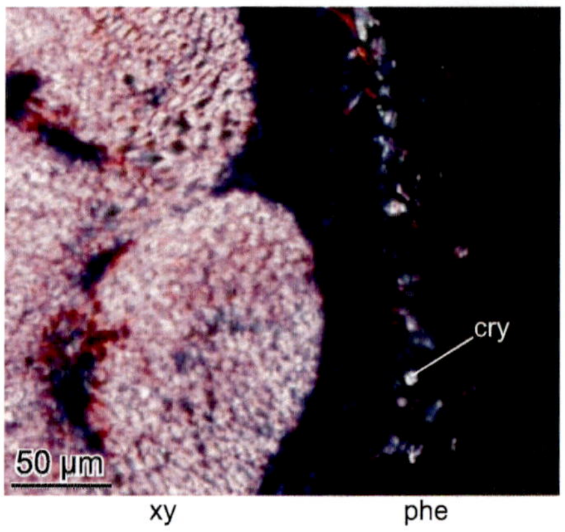

Features according to IAWA

Xylem 2.1 - 9 - 9.1 - 12 - 13 - 22 - 40.1 - 50.2 - 52.2 - 61 - 70 - 79 - 117 - 133.1 - 134.2

Axyris prostrata L.

Amaranthaceae

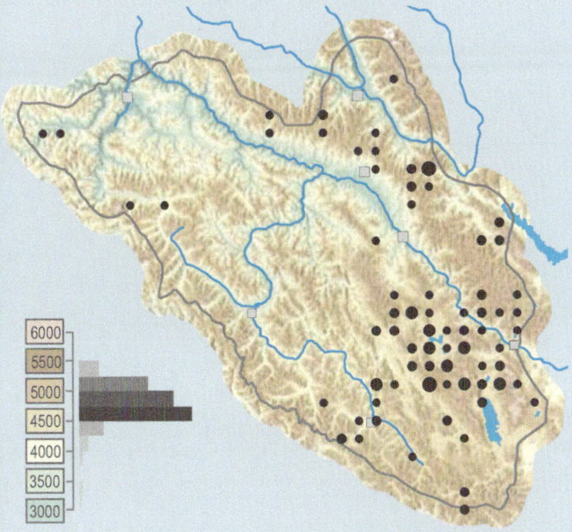

Syn.: *A. pamirica* B.Fedtsch.

Description: Annuals, 2–15 cm tall, stems and branches reddish, prostrate or ascending, densely stellate pubescent, flowers minute, in short spikes.

Distribution: Russia (S Siberia), Tajikistan, India (Jammu and Kashmir, Sikkim), Nepal, China, (Qinghai, Xinjiang, Xizang), Mongolia.

Habitat: Eutrophicated places, sandy plains.

Stem diameter 1.5–2.1 mm. Bark/xylem proportion 1:5. Maximal 1 annual ring.

Xylem

Transverse section

Vessels angular, solitary and in small groups. Vessel diameter 30–40 µm. Fibers thin- to thick-walled. Parenchyma paratracheal. Successive cambia continuous. Conjunctive tissue small, cells thin-walled, unlignified. Crystals absent.

Tangential section

Rays absent.

Radial section

Perforation plates simple. Intervessel pits round in alternating position. Vessel length 100–110 µm.

Bark

Phloem mostly uniform or with a few small sieve tubes. Phellem consists of a few layers of thin-walled, square cells. Crystals absent.

Pith

No slide available.

Analyzed: Root collar of 2 individuals (herbarium specimens).

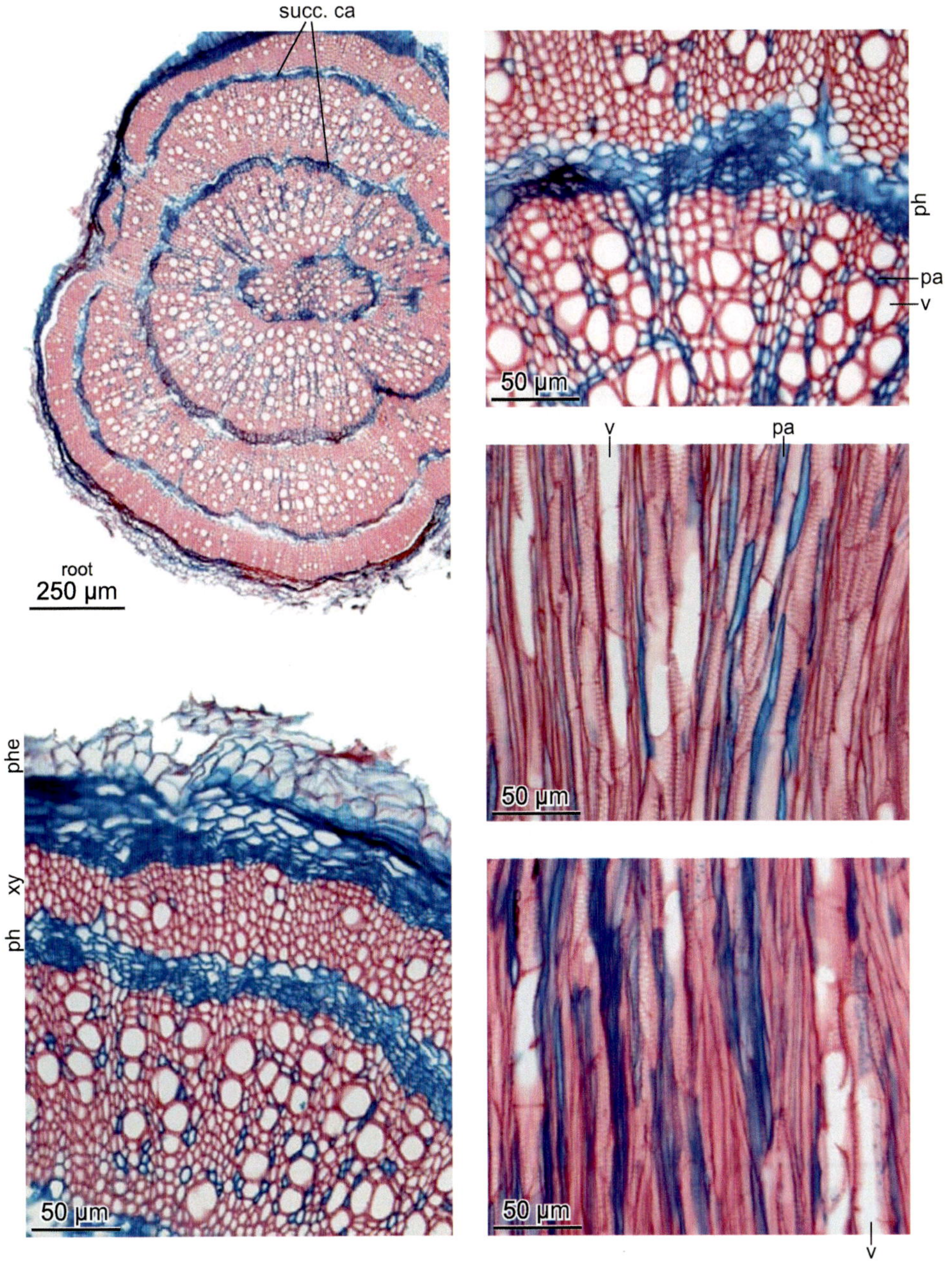

Features according to IAWA

Xylem 2.1 - 9 - 11 - 12 - 13 - 22 - 40.2 - 50.2 - 52.3 - 61 - 70 - 79 - 117 - 133.2 - 134.1

Bassia dasyphylla (Fisch. & C.A.Mey.) Kuntze

Amaranthaceae

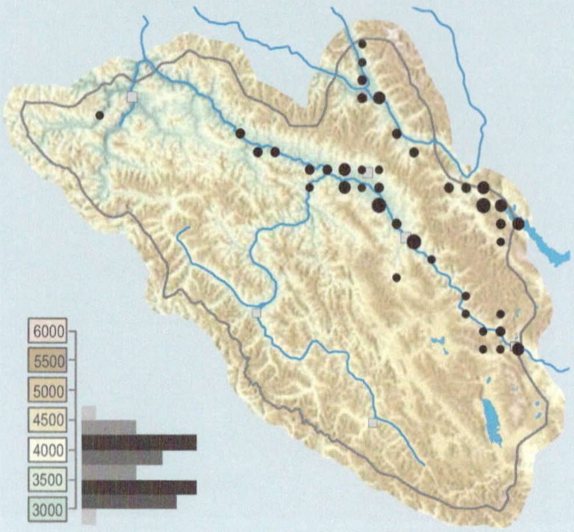

Syn.: *Kochia dasyphylla* Fisch. & C.A.Mey., *B. divaricata* (Kar. & Kir.) Kuntze

Description: Annuals, 10–50 cm tall, divaricately branched from the base, densely villous, leaves alternate, terete, obtuse, flowers solitary in axils, with star-shaped outgrowths in fruit.

Distribution: Russia (S Siberia), C Asia, India, China (Gansu, Hebei, Heilongjiang, Jilin, Liaoning, Nei Mongol, Qinghai, Shandong, Shanxi, Xinjiang, Xizang), Mongolia.

Habitat: Deserts, along roads, resting places of domestic animals, saline-alkaline habitats.

Stem diameter 3–5 mm. Bark/xylem proportion 1:12. Mean ring width 1.5 mm. Maximal 1 annual ring.

Xylem

Transverse section

Root collar: Vessels in short radial rows. Vessel diameter 30–50 µm. Fibers thin- to thick-walled. Parenchyma paratracheal. Successive cambia concentric continuous. Conjunctive tissue not lignified, thin-walled. With few crystal druses.

Tangential section

Rays 1–2-seriate.

Radial section

Perforation plates simple. Intervessel pits round in alternating position. Vessel length 60–90 µm.

Bark

Phloem homogeneous. Phellem consists of a few layers of thin-walled, rectangular cells.

Pith

Round, filled with very large, thin-walled, polygonal parenchyma cells.

Analyzed: Root collar of polar root and shoots of 3 individuals.

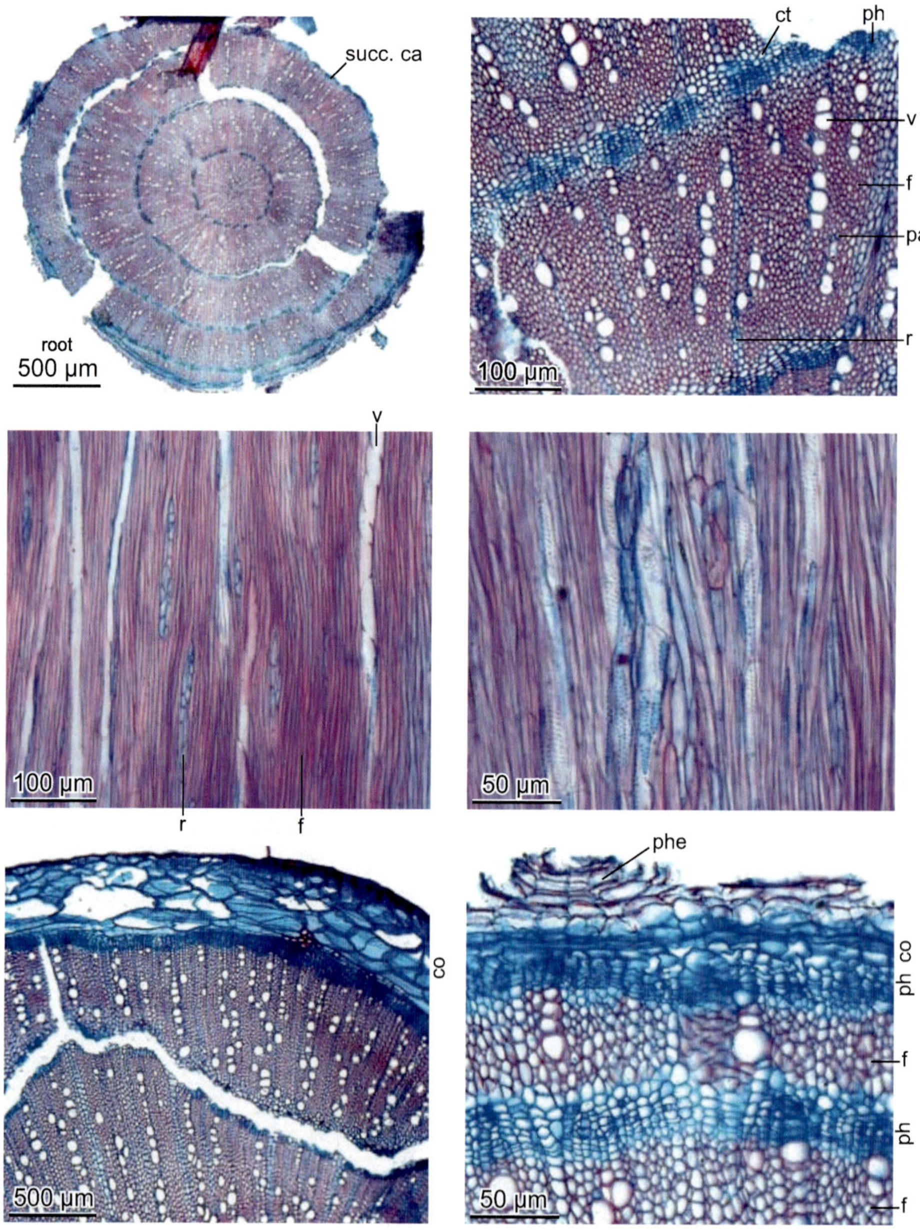

Features according to IAWA

Xylem 2.1 - 9.1 - 13 - 22 - 40.2 - 50.2 - 52.2 - 61 - 70 - 79 - 97 - 105 - 133.2 - 134.1 - 144

Chenopodium foliosum (Moench) Aschers.

Amaranthaceae

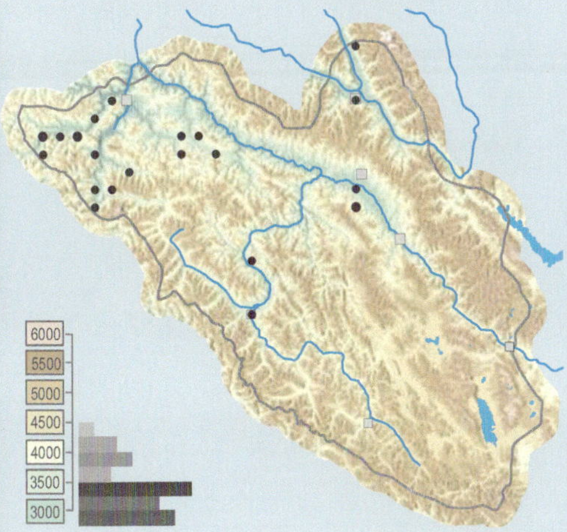

Syn.: *Blitum virgatum* L., *C. korshinskyi* Litv., *Monocarpus foliosus* Moench

Description: Annuals, 20–70 cm tall, glabrous, stems erect, branched from the base, leaves narrowly triangular, margin irregularly dentate, flowers light green, in axillary glomerules, becoming red and succulent in fruit, ca. 4 mm across.

Distribution: N Africa, Europe, C and SW Asia; introduced in North America.

Habitat: A weed in disturbed habitats and cultivated land.

Stem diameter 8–15 mm. Bark/xylem proportion 1:4. Maximal 1 annual ring.

Xylem

Transverse section
Vessels round, solitary and in short radial multiples. Vessel diameter 25–35 µm. Fibers thin- to thick-walled. Parenchyma paratracheal. Successive cambia continuous. Conjunctive tissue large, cells thin-walled, unlignified. Crystal druses in conjunctive tissue.

Tangential section
A few large radial unlignified parenchyma zones and 2–4-seriate rays, lignified or unlignified.

Radial section
Perforation plates simple. Intervessel pits round in alternating position. Vessel length 100–120 µm.

Bark

Phloem mostly uniform or with a few small sieve tubes. Crystal druses in the cortex. Phellem absent. Crystals absent.

Pith

No slide available.

Analyzed: Root collar of 4 individuals.

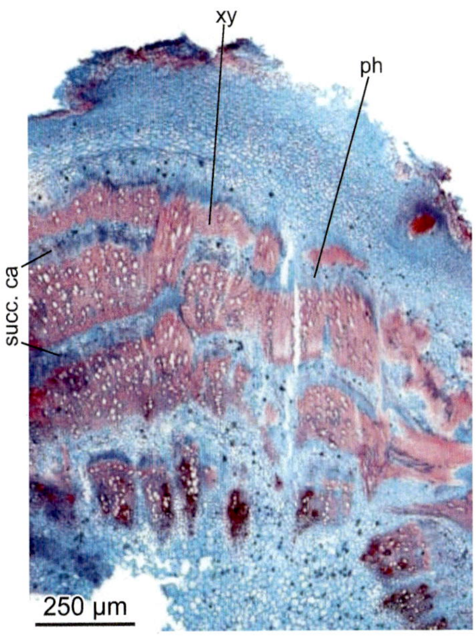

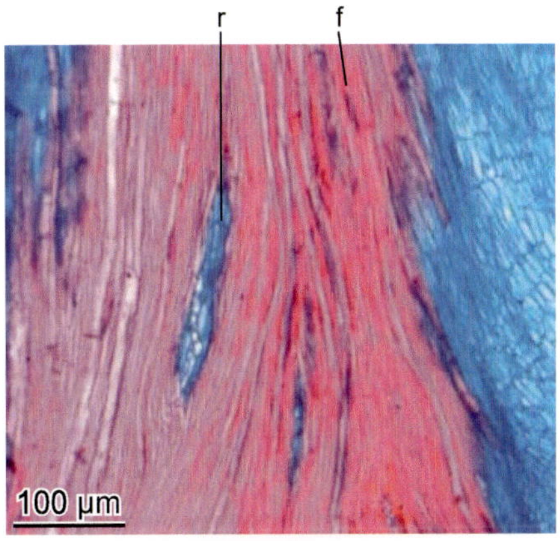

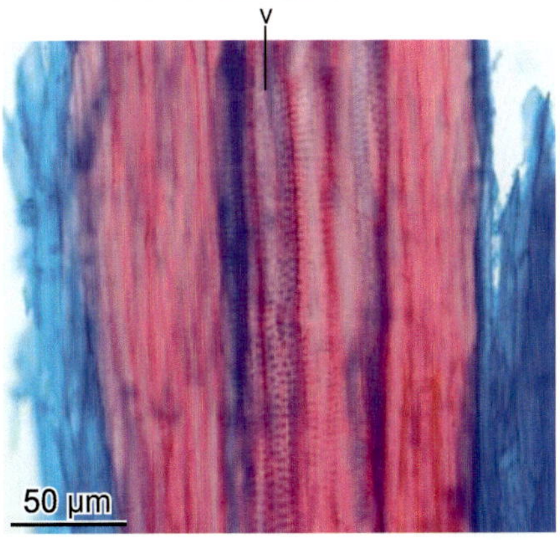

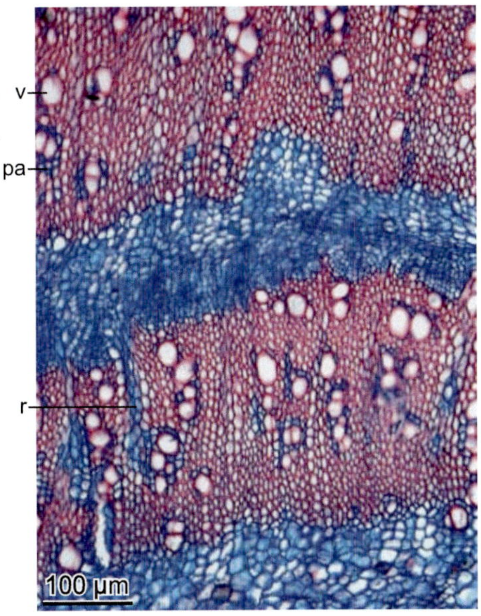

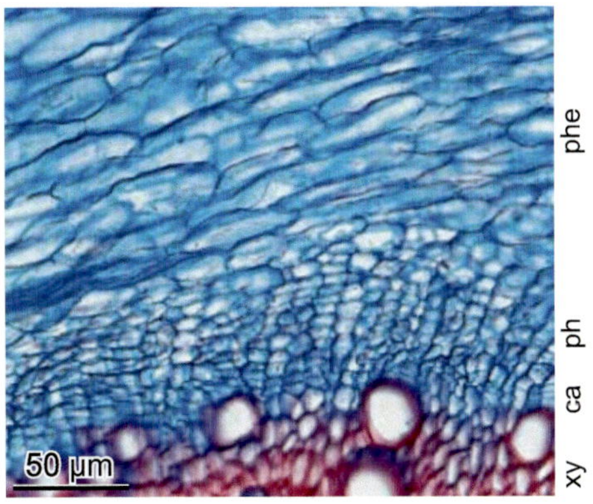

Features according to IAWA

Xylem 2.1 - 9 - 9.1 - 11 - 13 - 22 - 40.2 - 50.2 - 52.3 - 61 - 70 - 79 - 97 - 99 - 100.2 - 105 - 133.2 - 134.1 - 144

Chenopodium glaucum L.

Amaranthaceae

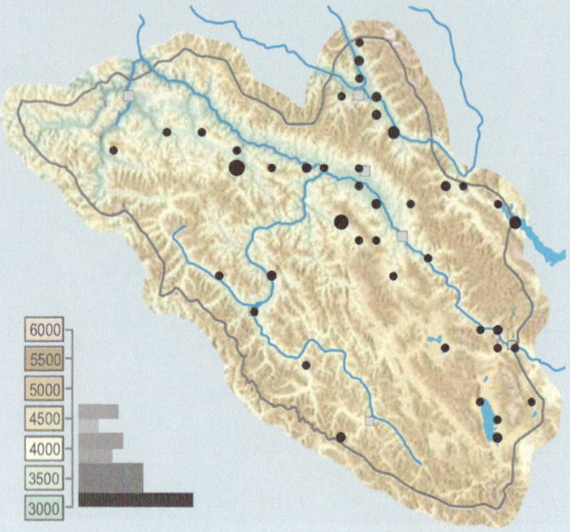

Syn.: *Oxybasis glauca* (L.) S.Fuentes, Uotila & Borsch

Description: Annuals, 5–50 cm tall, stems often much branched, leaves elliptic to ovate, mostly coarsely sinuate, sometimes subentire or slightly 3-lobed, green and glabrescent above, glaucous below, flowers minute, composing small dense glomerules arranged spicately or paniculately.

Distribution: Temperate zone of Europe and Asia.

Habitat: A weed on arable land, disturbed marshes.

Stem diameter 1.5–1.8 mm. Bark/xylem proportion 1:2. Maximal 1 annual ring.

Xylem

Transverse section

Vessels slightly angular, in radial multiples. Vessel diameter 25–30 μm. Fibers thin- to thick-walled. Parenchyma paratracheal. Successive cambia continuous. Conjunctive tissue small, cells thin-walled, unlignified cells. Crystal sand in the conjunctive tissue.

Tangential section

Rays absent.

Radial section

Perforation plates simple. Intervessel pits round in alternating position. Vessel length 100–120 μm.

Bark

Phloem with groups of sieve tubes. Cortex cells large, lignified, thin-walled. Crystal sand in the phloem and cortex. Phellem absent.

Pith

No slide available.

Analyzed: Root collar of 2 individuals (herbarium specimens).

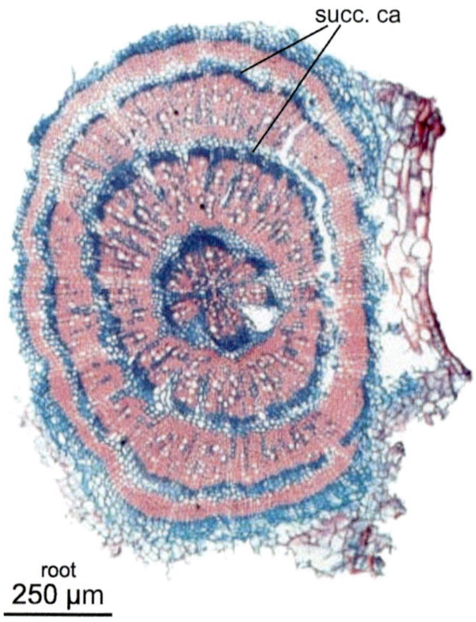

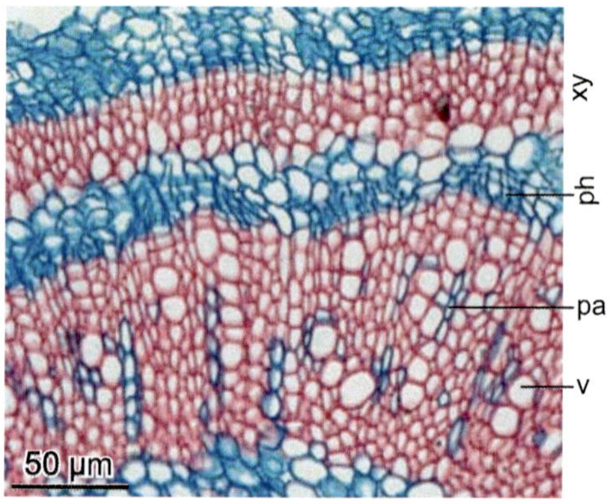

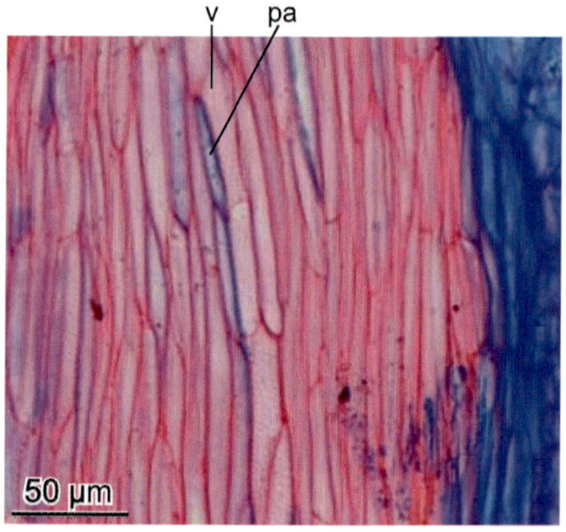

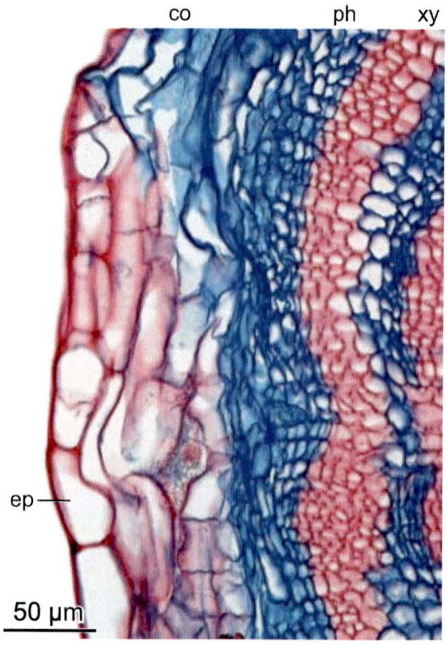

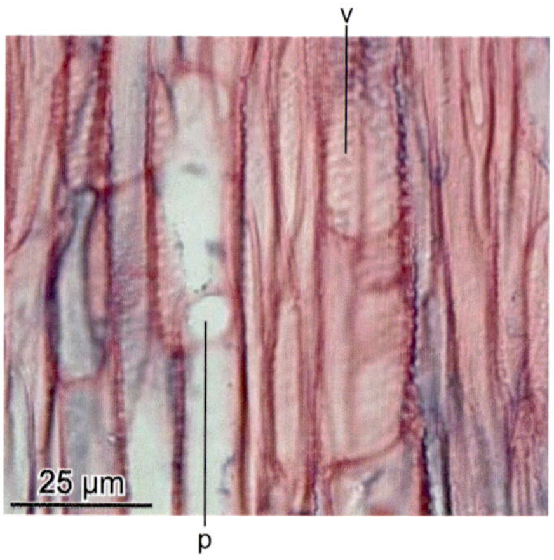

Features according to IAWA

Xylem 2.1 - 9.1 - 11 - 12 - 13 - 22 - 40.2 - 50.2 - 52.3 - 61 - 70 - 79 - 117 - 133.2 - 134.1 - 153

Chenopodium karoi (Murr) Aellen

Amaranthaceae

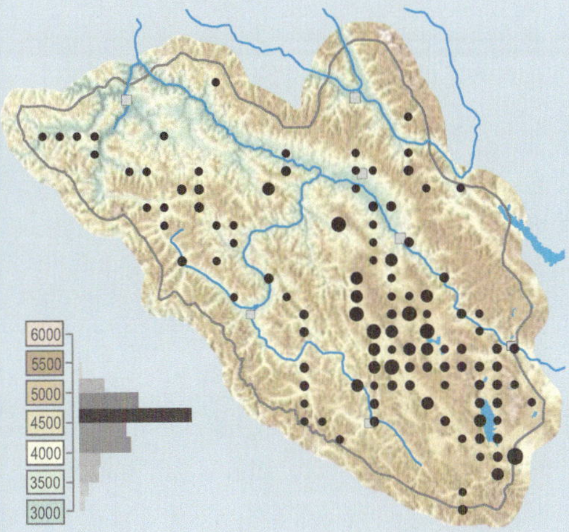

Syn.: *C. album* subsp. *karoi* Murr

Description: Annuals, 10–30 cm tall, stems erect to prostrate, yellowish to green-striped, leaves ovate, farinose below, margin usually 3-lobed, flowers in glomerules, composing a panicle.

Distribution: Russia (Far East, Siberia), C Asia, Pakistan, India, China (Gansu, Hebei, Qinghai, Sichuan, Xinjiang, Xizang), Mongolia.

Habitat: Disturbed places, a weed of arable land, eutrophicated sites in villages, disturbed sites along roads.

Stem diameter 1.1–1.2 mm. Bark/xylem proportion 1:4. Maximal 1 annual ring.

Xylem

Transverse section

Vessels slightly angular in radial multiples. Vessel diameter 20–30 µm. Fibers thin- to thick-walled. Parenchyma paratracheal. Successive cambia, xylem continuous, phloem foraminate. Conjunctive tissue absent. Crystals absent.

Tangential section

Rays absent.

Radial section

Perforation plates simple. Intervessel pits round in alternating position. Vessel length 120–140 µm.

Bark

Phloem mostly uniform or with a few small sieve tubes. Cortex cells large, thin-walled. Crystal druses in the phloem and the cortex.

Pith

No slide available.

Analyzed: Root collar of 2 individuals.

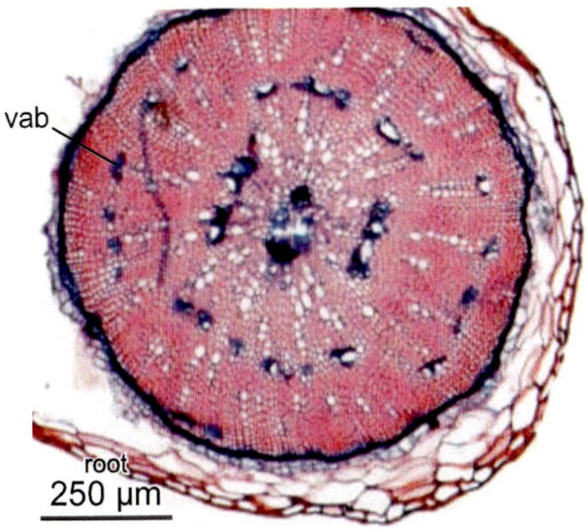

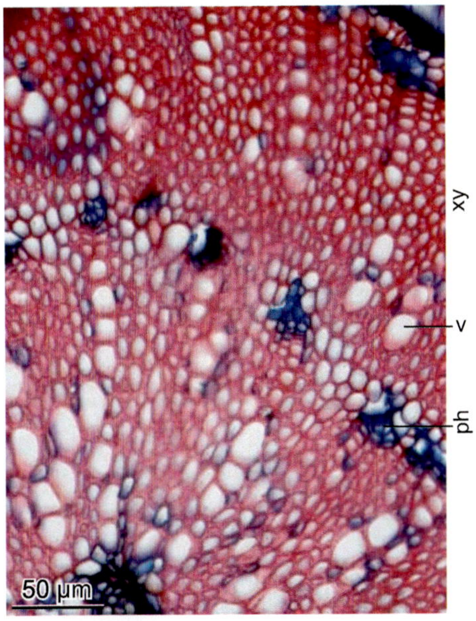

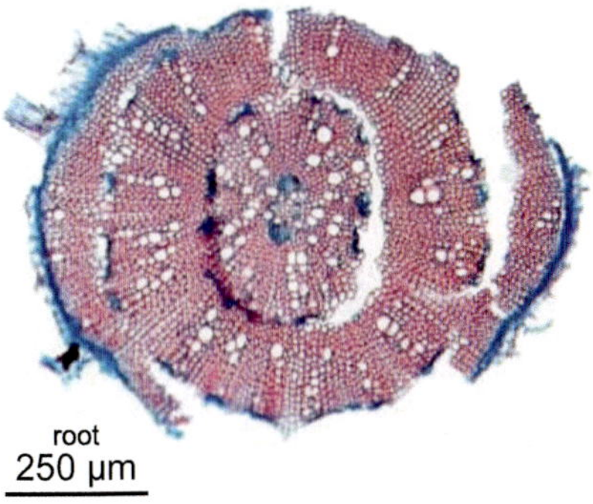

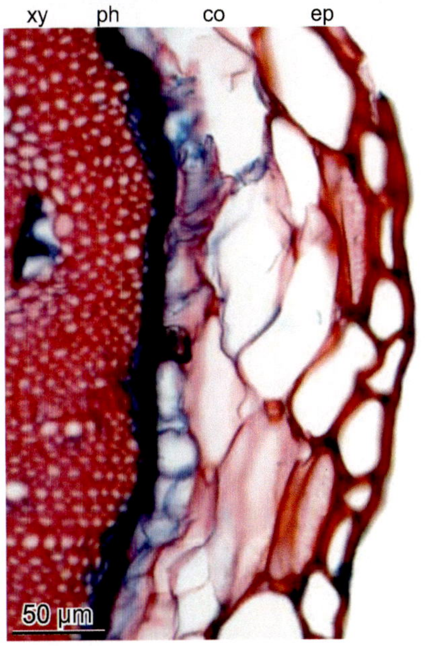

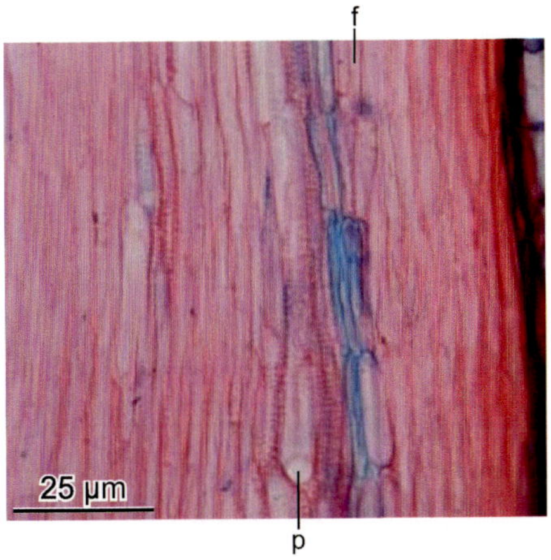

Features according to IAWA

Xylem 2.1 - 9 - 12 - 13 - 22 - 40.2 - 50.1 - 52.3 - 61 - 70 - 79 - 117 - 133.1 - 134.2

Chenopodium pamiricum Iljin

Amaranthaceae

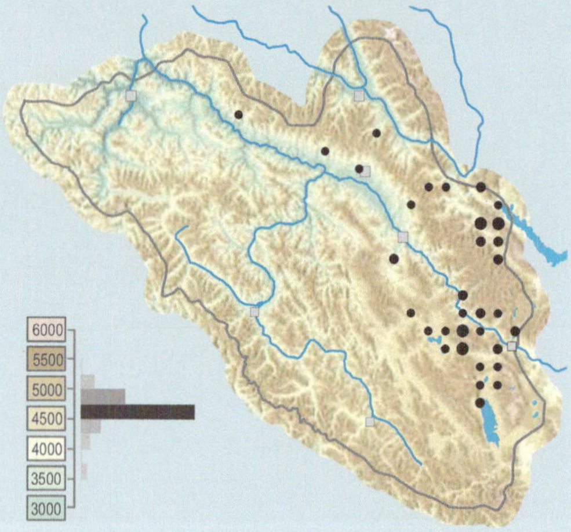

Description: Annuals, 5–20 cm tall, farinose, stems yellowish to reddish, branched, leaves ovate, entire or slightly 3-lobed, flowers in glomerules, composing a panicle.

Distribution: Afghanistan, Pakistan, N India, China, Mongolia.

Habitat: Disturbed places, a weed in fields.

Stem diameter 1–2 mm. Bark/xylem proportion 1:4. Mean ring width 1 mm. Maximal 1 annual ring.

Xylem

Transverse section

Root collar: Annual. Vessels in groups. Vessel diameter 40–60 µm. Fibers thin-walled. Parenchyma paratracheal. Successive cambia foraminate and tangentially enlarged. Conjunctive tissue absent. Crystals absent.
Shoot: Vessels in groups. Successive cambia foraminate. Crystals absent

Tangential section

Rays 3–5-seriate, confluent with axial tissue.

Radial section

Perforation plates simple. Intervessel pits round in alternating position. Vessel length 80–120 µm.

Bark

Root collar: Phloem homogeneous. Phellem consists of a few layers of thin-walled, square cells.
Shoot: Cortex consists of large, thin-walled parenchyma cells containing large crystal druses.

Pith

Round, filled with very large, thin-walled, polygonal parenchyma cells.

Analyzed: Root collar of polar root and shoot of 1 individual.

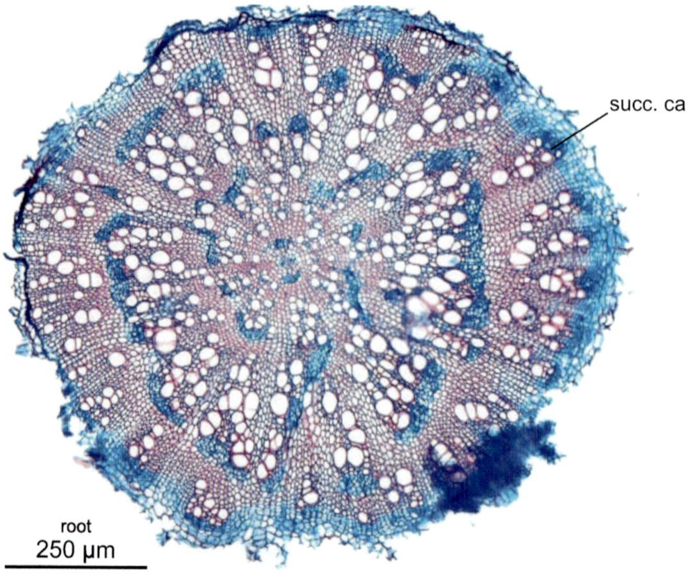

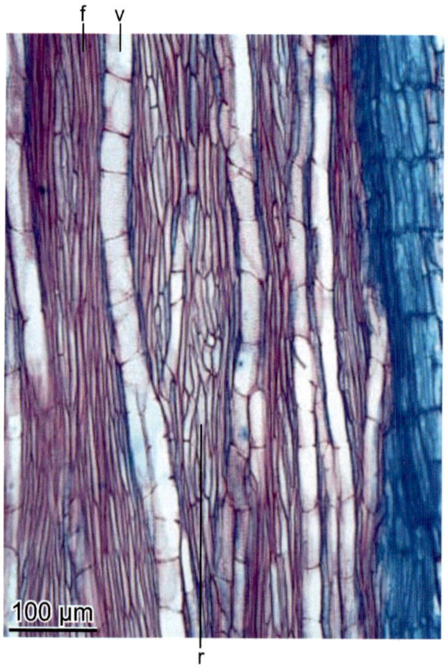

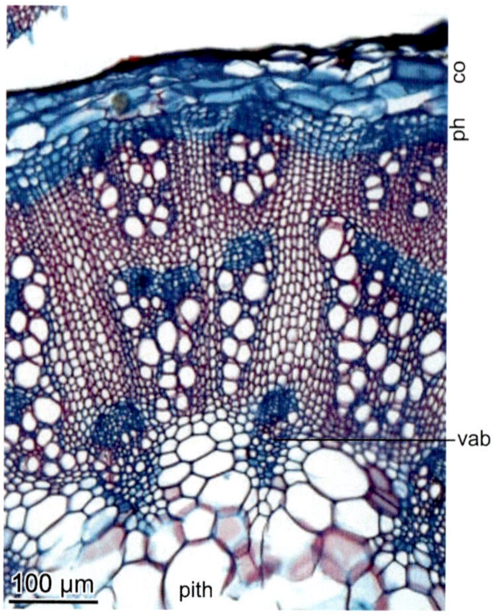

Features according to IAWA

Xylem	*Root*	*2.1 - 11 - 13 - 22 - 40.2 - 41 - 50.2 - 52.2 - 61 - 68 - 79 - 98 - 100.1 - 133.2 - 134.1*
	Shoot	*similar to root*

Corispermum tibeticum Iljin

Amaranthaceae

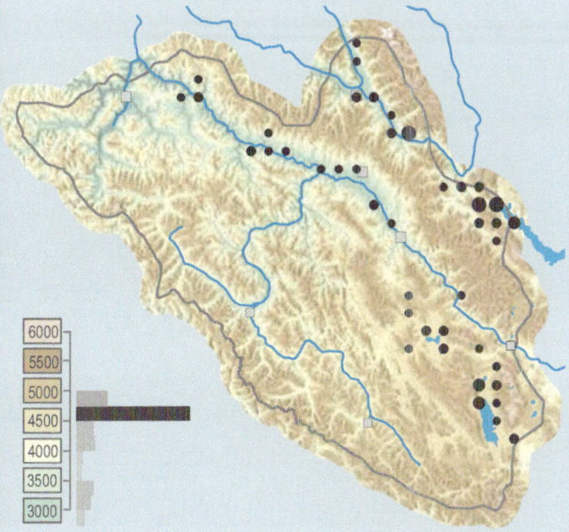

Description: Annuals, 5–20 cm tall, hairy, branched from the base, leaves oblong, fleshy, spikes loose to compact.

Distribution: C Asia, Pakistan, N India, China (Xizang).

Habitat: Synanthropic habitats, unstable and eroded sand slopes, disturbed habitats along roads.

Stem diameter 3–5 mm. Bark/xylem proportion 1:10. Mean ring width 1 mm. Maximal 1 ring.

Xylem

Transverse section

Root collar: Annual. Vessels in groups. Vessel diameter 60–90 µm. Fibers thin- to thick-walled, not stained. Parenchyma paratracheal. Successive cambia continuous. Conjunctive tissue not lignified, thin-walled. With crystal sand in conjunctive tissue.

Shoot: Successive cambia foraminate.

Tangential section

Rays absent.

Radial section

Perforation plates simple. Intervessel pits round in alternating position. Vessel length 60–100 µm.

Bark

Root collar: Phellem cells thin-walled, square.

Shoot: Cortex consists of large, thin-walled parenchyma cells containing large crystal druses.

Pith

Round, filled with very large, thin-walled, polygonal parenchyma cells.

Analyzed: Root collar of polar root and shoots of 3 individuals.

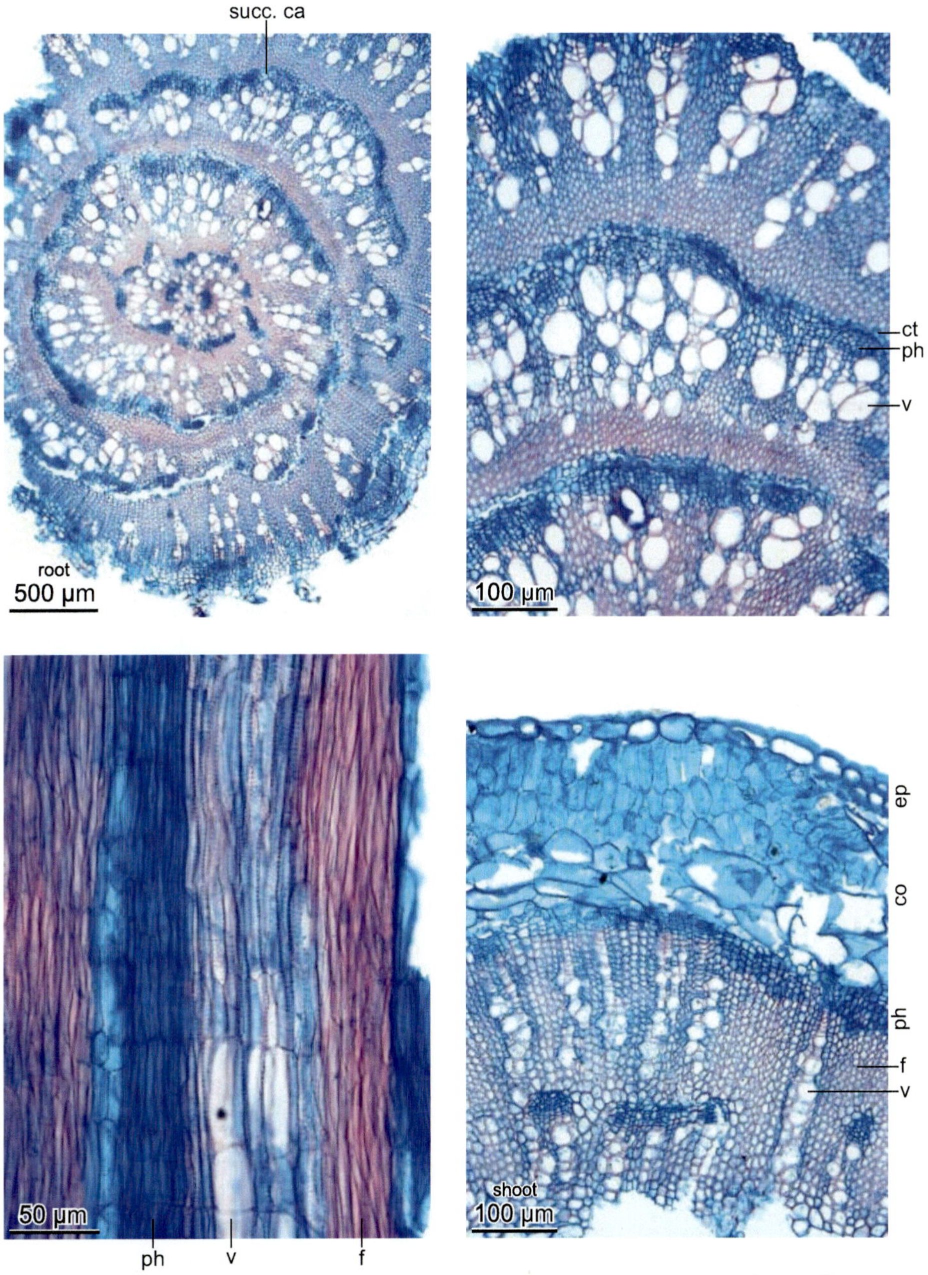

Features according to IAWA

Xylem 2.1 - 9 - 11 - 13 - 22 - 39.2 - 41 - 50.2 - 52.2 - 61 - 70 - 79 - 79.1 - 117 - 133.2 - 153

Dysphania botrys (L.) Mosyakin & Clemants

Amaranthaceae

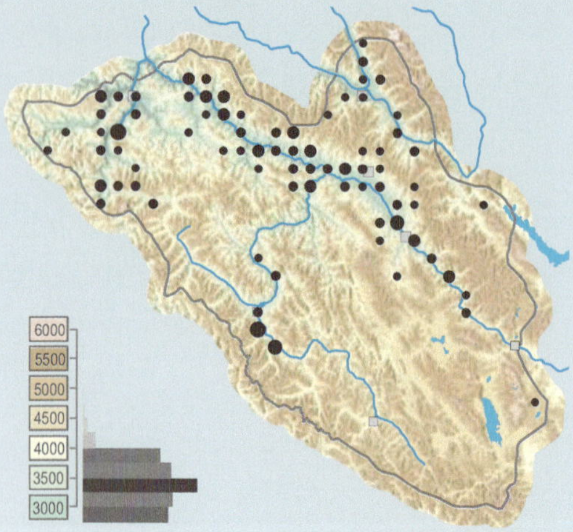

Syn.: *Chenopodium botrys* L., *Ambrina botrys* (L.) Moq.

Description: Annuals, 10–60 cm tall, aromatic, glandular hairy, stem branched, leaves oblong, margin pinnatifid, flowers yellowish-green, ca. 1 mm across, in dense axillary and terminal cymes.

Distribution: S Europe, N Africa, C and SW Asia, India, China (Xinjiang); introduced to N America, locally naturalized in other subtropical to warm-temperate regions.

Habitat: Sandy river sediments, disturbed habitats along roads, stream gravel, a weed of sandy barley fields, often synanthropic.

Stem diameter 3–4 mm. Bark/xylem proportion 1:6. Mean ring width 1.5 mm. Maximal 1 annual ring.

Xylem

Transverse section

Root collar: Annual. Isolated vascular bundles in the center. Vessels in short radial rows. Vessel diameter 30–50 µm. Fibers thin- to thick-walled. Parenchyma paratracheal. Successive cambia continuous. Conjunctive tissue not or poorly lignified, thin-walled. With crystal sand in conjunctive tissue.

Shoot: Vessels solitary. Successive cambia foraminate.

Tangential section

Rays 2–4-seriate.

Radial section

Perforation plates simple. Intervessel pits round in alternating position. Vessel length 60–100 µm.

Bark

Root collar: Phloem homogeneous. Phellem consists of a few layers of thin-walled, square cells.

Shoot: Cortex consists of large, thin-walled parenchyma cells. Phellem absent. Epidermis papilous.

Pith

Round, filled with very large, thin-walled, polygonal parenchyma cells.

Analyzed: Root collar of polar root and shoots of 3 individuals.

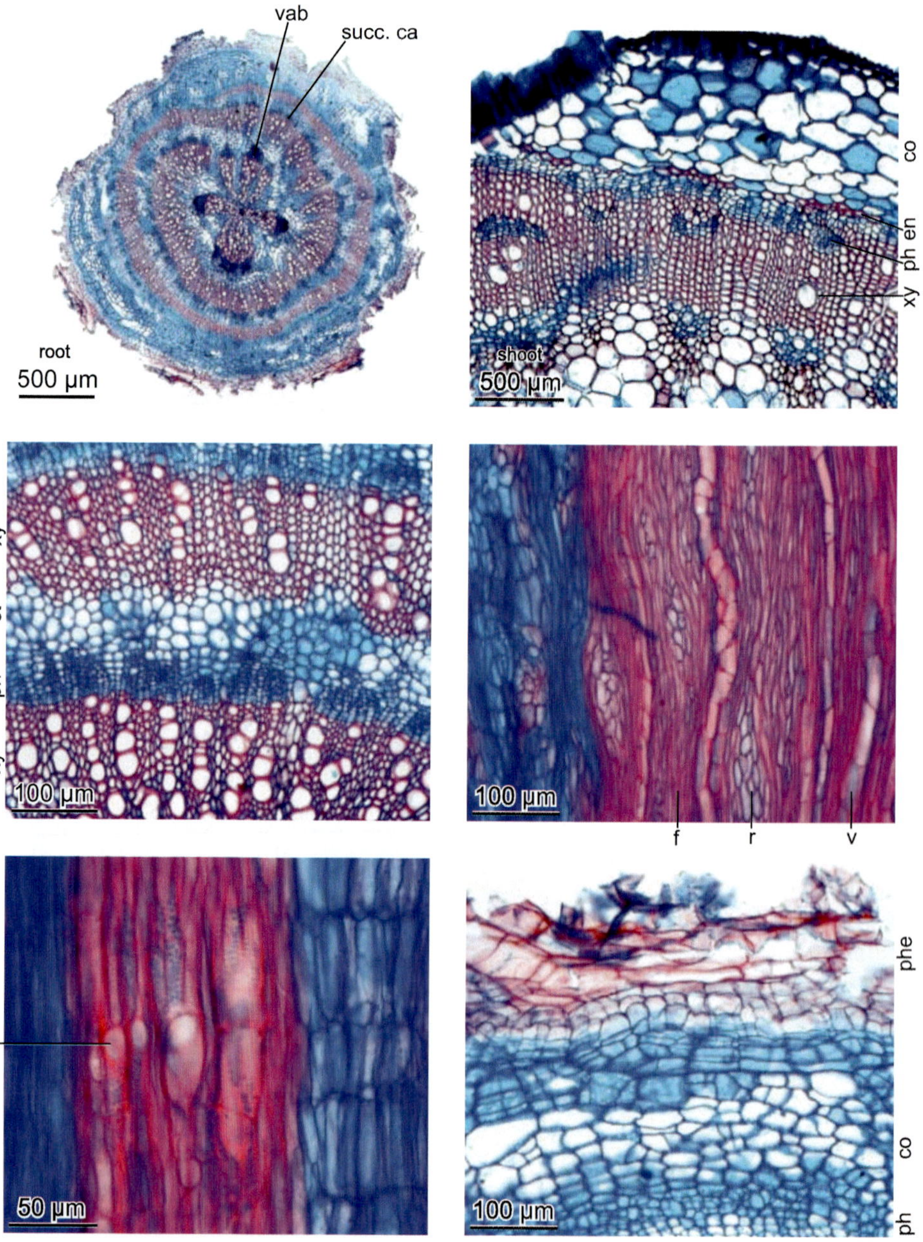

Features according to IAWA

Xylem Root *2.1 - 9.1 - 13 - 22 - 40.2 - 50.2 - 52.2 - 61 - 70 - 79 - 97 - 98 - 105 - 133.2 - 134.1 - 153*

Shoot *2.1 - 9 - 13 - 22 - 40.2 - 50.2 - 52.2 - 61 - 70 - 79 - 97 - 98 - 105 - 134 - 134.1 - 153*

Dysphania himalaica Uotila
Amaranthaceae

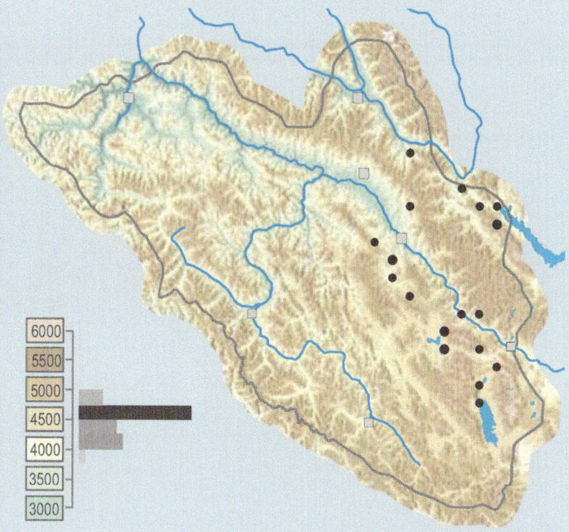

Syn.: *Chenopodium himalaicum* Klimeš, in schedis

Description: Annuals, 5–10 cm tall, aromatic, glandular hairy, stems erects, much branched basally, leaves narrowly elliptic to lanceolate, margin sinuate, flowers ca. 1 mm across, solitary in branch axils, composing a diffuse inflorescence.

Distribution: India (Ladakh in Jammu and Kashmir), Nepal, China (Xizang).

Habitat: Grassy slopes, pastures.

Stem diameter 0.7 mm. Bark/xylem proportion 1:2. Maximal 1 annual ring.

Xylem

Transverse section

Vessels slightly angular, solitary. Vessel diameter 20–30 µm. Fibers thin- to thick-walled. Parenchyma paratracheal. Successive cambia, xylem continuous, phloem foraminate. Conjunctive tissue absent. Crystals absent.

Tangential section

Rays absent.

Radial section

Perforation plates simple. Intervessel pits round in alternating position. Vessel length 80–100 µm.

Bark

Phloem mostly uniform or with a few small sieve tubes. Cortex cells large, thin-walled. Crystal sand in the cortex.

Pith

No slide available.

Analyzed: Root collar of 2 individuals (herbarium specimens).

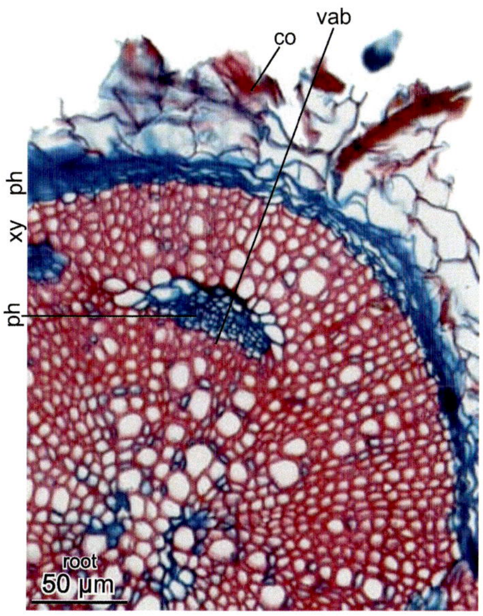

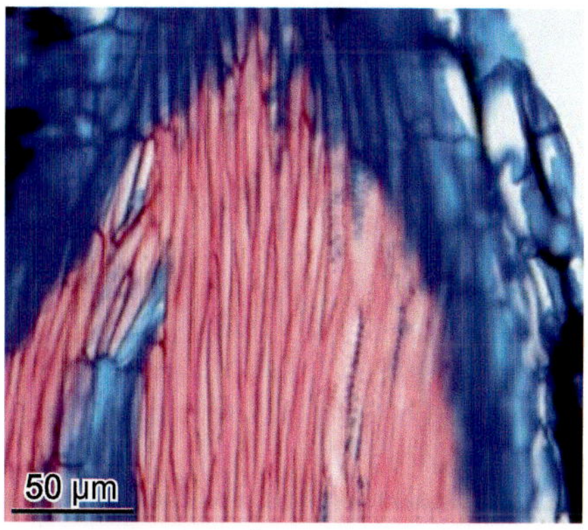

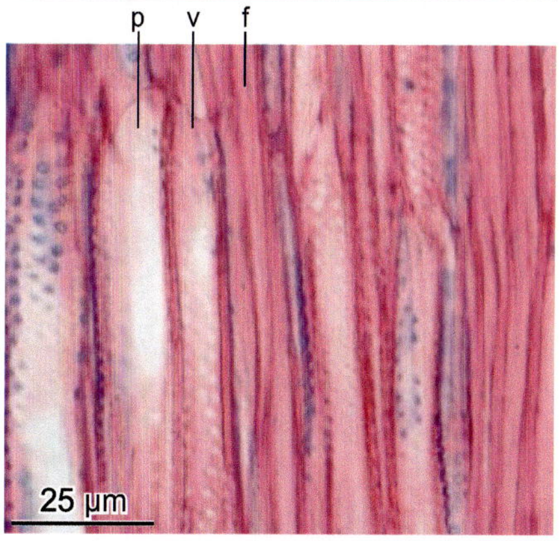

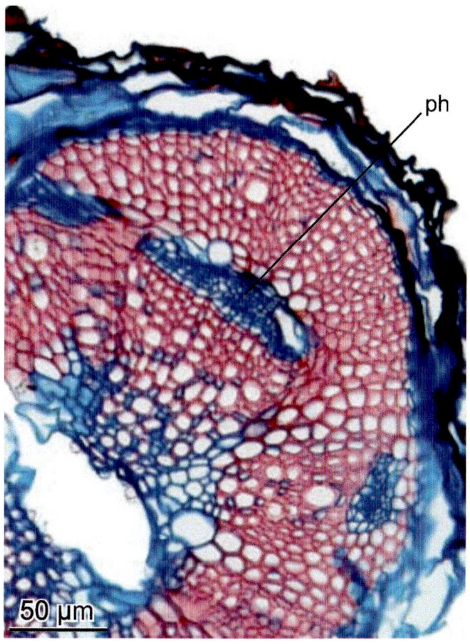

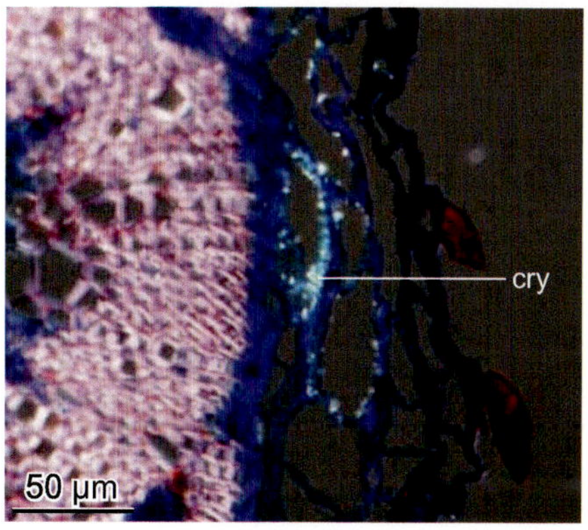

Features according to IAWA

Xylem 2.1 - 9 - 12 - 13 - 22 - 40.2 - 50.1 - 52.3 - 61 - 70 - 79 - 117 - 133.1 - 134.2

Dysphania nepalensis (Colla) Mosyakin & Clemants
Amaranthaceae

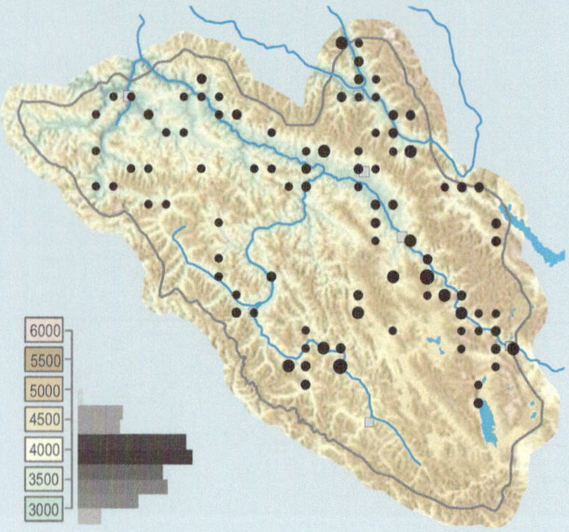

Syn.: *Chenopodium nepalense* Colla

Description: Annuals, 5–40 cm tall, aromatic, glandular hairy, stems erect, leaves elliptic-ovate, pinnatifid with 2–5 broad lobes, flowers yellowish-green, ca. 1 mm across, in lax axillary and terminal cymes.

Distribution: Afghanistan, Pakistan, India, Nepal, China (Xizang, Yunnan, Sichuan), Bhutan.

Habitat: Disturbed places, a weed of arable land, eutrophicated habitats in villages, disturbed sites along roads.

Stem diameter 3–5 mm. Bark/xylem proportion 1:12. Ring width 0.1–0.15 mm. Maximal 1 annual ring.

Xylem

Transverse section

Root collar: Annual. Vessels solitary. Vessel diameter 20–40 µm. Fibers thin- to thick-walled. Parenchyma paratracheal. Successive cambia foraminate. Conjunctive tissue absent. Crystals absent.

Shoot: Vessels in groups. Successive cambia foraminate. With crystal sand.

Tangential section

Rays absent.

Radial section

Perforation plates simple. Intervessel pits round in alternating position. Vessel length 60–100 µm.

Bark

Root: Phloem homogeneous. Phellem consists of a few layers of thin-walled, square cells.

Shoot: Cortex consists of large, thin-walled parenchyma cells containing large crystal druses.

Pith

Round, filled with very large, thin-walled, polygonal parenchyma cells.

Analyzed: Root collar of polar root and shoots of 2 individuals.

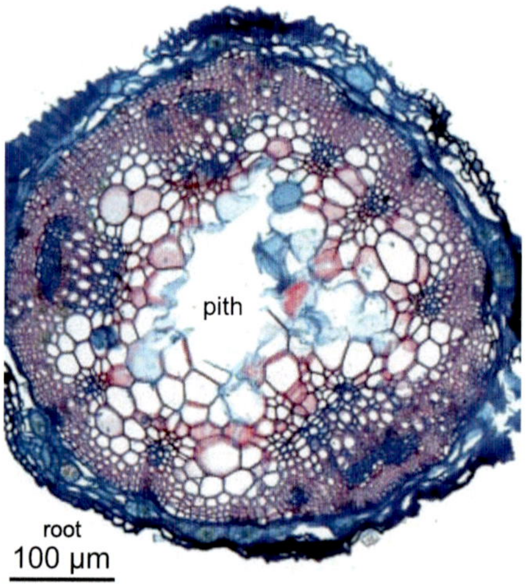

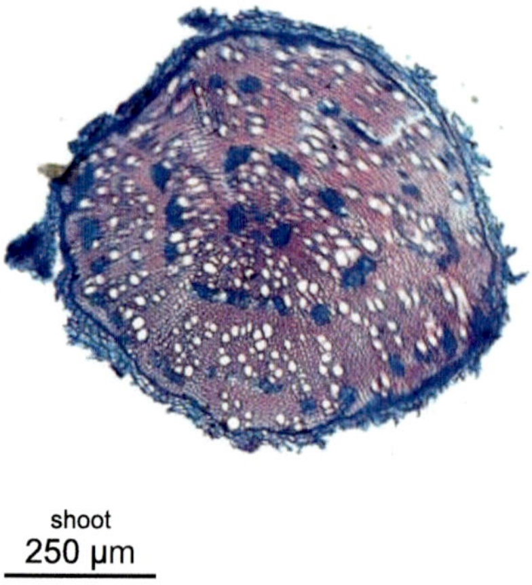

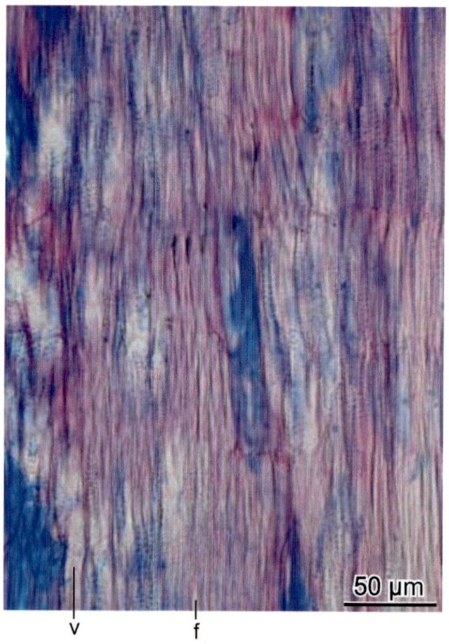

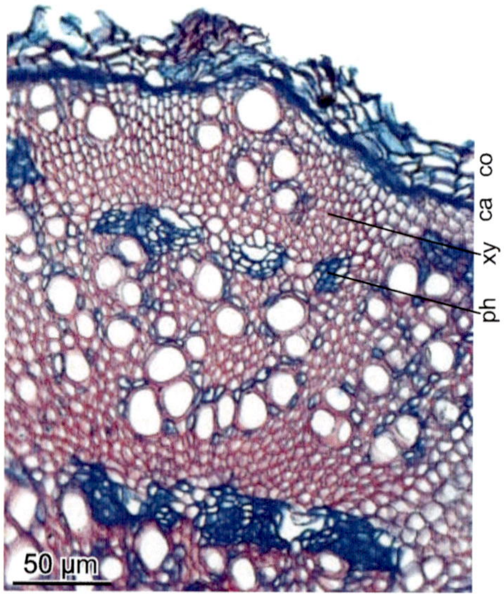

Features according to IAWA

Xylem Root 2.1 - 9 - 13 - 22 - 40.2 - 50.2 - 52.2 - 61 - 70 - 79 - 117 - 133.2 - 134.1 - 153

Shoot 2.1 - 9 - 13 - 22 - 40.2 - 50.2 - 52.2 - 61 - 70 - 79 - 117 - 134 - 134.1 - 153

Dysphania tibetica (A.J.Li) Uotila

Amaranthaceae

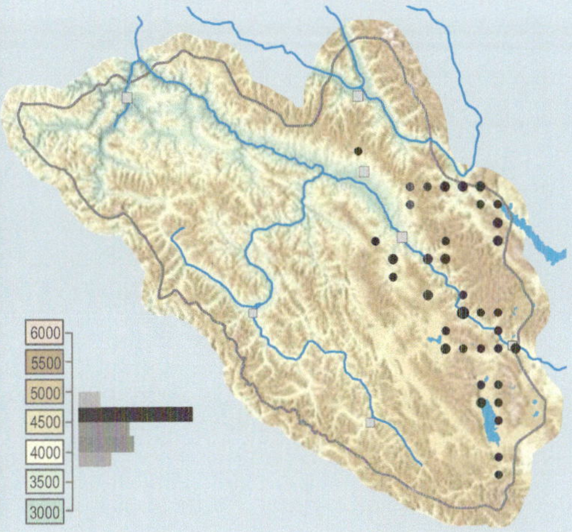

Syn.: *Chenopodium tibeticum* A.J.Li

Description: Annuals, 5–40 cm tall, aromatic, glandular hairy, stems prostrate to ascending, leaves linear to narrowly lanceolate, margin erose-dentate, flowers yellowish-green, ca. 1 mm across, solitary in branch axils or in dense axillary glomerules.

Distribution: NW India, W China.

Habitat: Disturbed habitats, a weed in fields.

Stem diameter 2.2–2.5 mm. Bark/xylem proportion 1:5. Maximal 1 annual ring.

Xylem

Transverse section

Vessels slightly angular, solitary and in small groups. Vessel diameter 25–30 µm. Fibers thin- to thick-walled. Parenchyma paratracheal. Successive cambia, xylem continuous, phloem foraminate. Conjunctive tissue small, cells thin-walled, lignified. Crystal sand in the cortex.

Tangential section

Rays absent.

Radial section

Simple perforation plates. Intervessel pits round in alternating position. Vessel length 120–150 µm.

Bark

Phloem mostly uniform or with a few small sieve tubes. Phellem consists of a few layers of thin-walled, square cells. Crystals absent.

Pith

No slide available.

Analyzed: Root collar of 2 individuals (herbarium specimens).

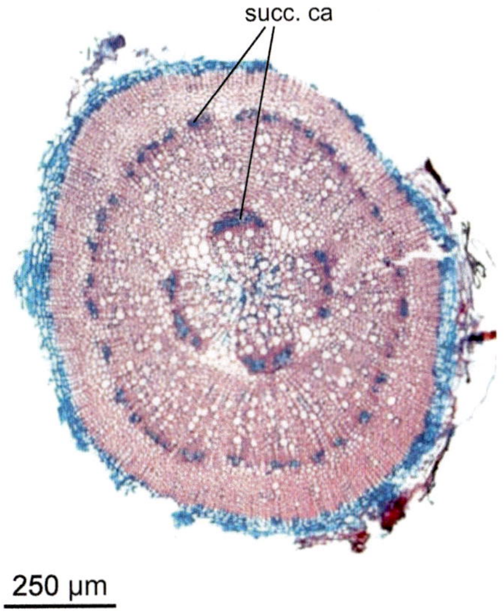

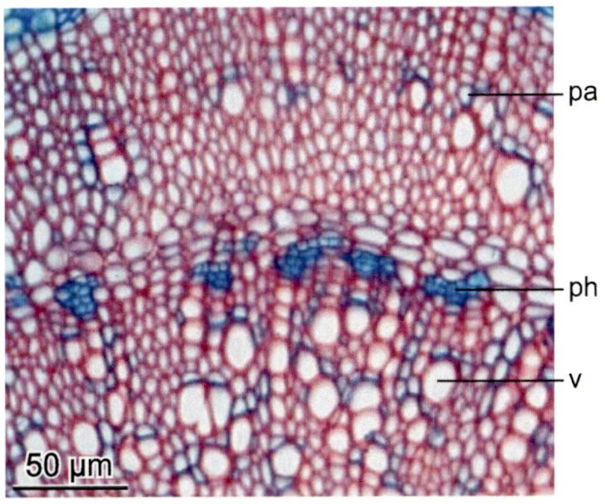

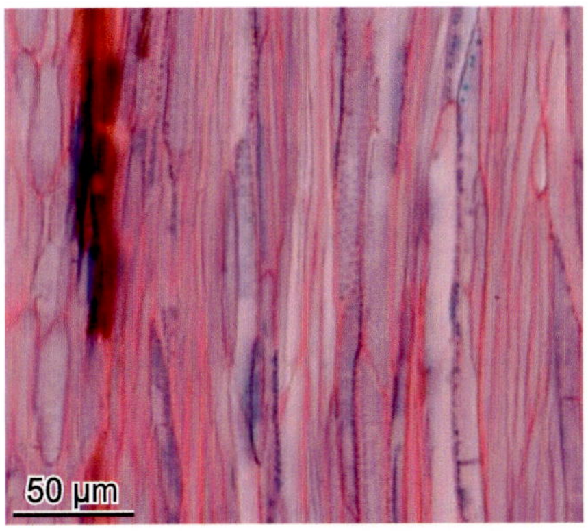

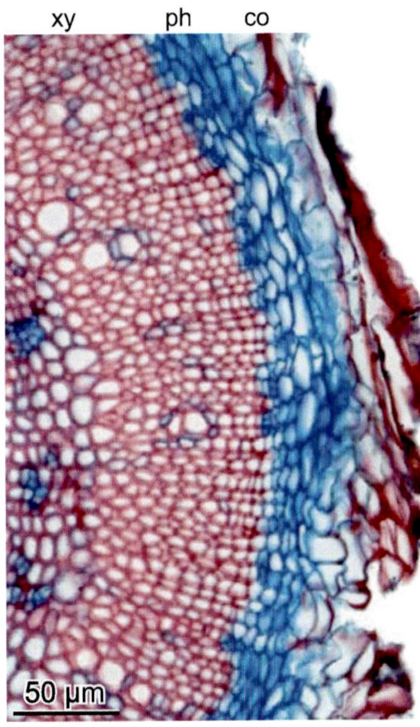

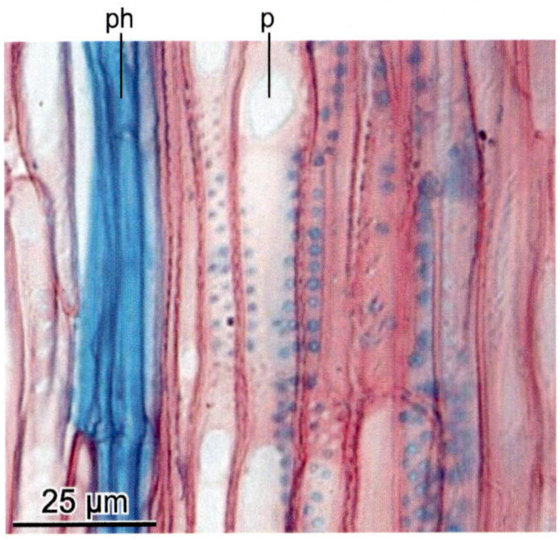

Features according to IAWA

Xylem 2.1 - 9 - 11 - 13 - 22 - 40.2 - 50.2 - 52.3 - 61 - 70 - 79 - 117 - 133.2 - 134.1

Halogeton arachnoideus Moq.

Amaranthaceae

Syn.: *Micropeplis arachnoidea* (Moq.) Bunge, *Salsola aptera* Hand.-Mazz.

Description: Annuals, 10–40 cm tall, arachnoid hairy when young, later glabrescent, leaves linear-cylindrical, usually without a terminal seta, flowers in 2–3-flowered axillary glomerules.

Distribution: C Asia, India, China (Gansu, Nei Mongol, Ningxia, Qinghai, Shaanxi, Shanxi, Xinjiang).

Habitat: Deserts, rocks and mesic habitats.

Stem diameter 3–4 mm. Bark/xylem proportion 1:10. Mean ring width 1 mm. Maximal 1 annual ring.

Xylem

Transverse section

Vessels in short radial rows. Vessel diameter 40–60 µm. Fibers thin- to thick-walled, not stained. Parenchyma paratracheal. Successive cambia continuous and foraminate at the periphery. Conjunctive tissue not lignified, thin-walled. With very few prismatic crystals in conjunctive tissue.

Tangential section

Rays 1–3-seriate, aggregate.

Radial section

Perforation plates simple. Intervessel pits round in alternating position. Vessel length 100–160 µm.

Bark

Phloem homogeneous. Phellem consists of a few layers of thin-walled, square cells.

Pith

No slide available.

Analyzed: Root collars and shoots of 1 individual.

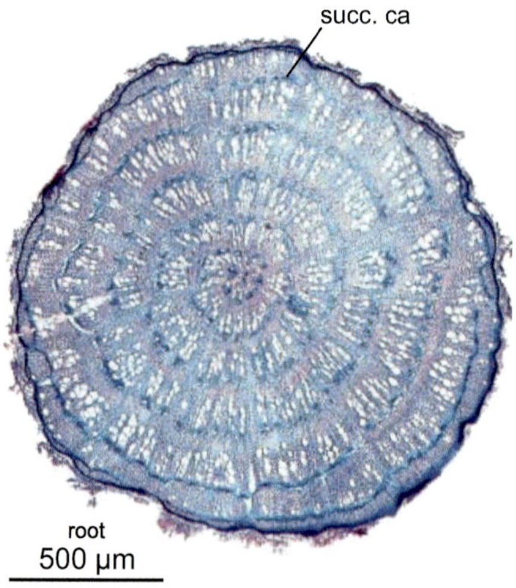

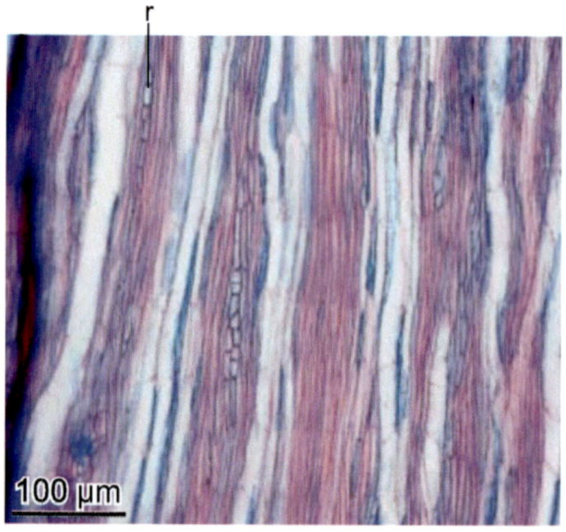

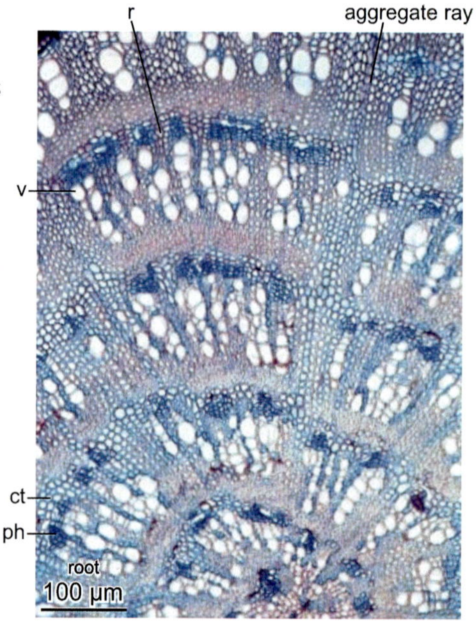

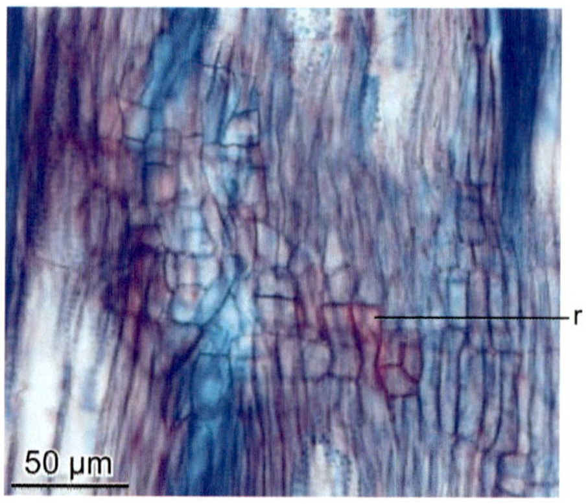

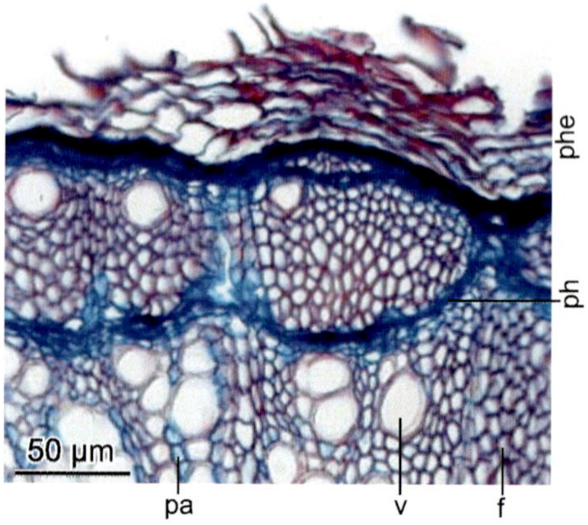

Features according to IAWA

Xylem 2.1 - 9.1 - 11 - 13 - 22 - 40.2 - 41 - 50.2 - 52.3 - 61 - 70 - 79 - 97 - 105 - 133.2 - 134 - 134.1 - 136

Halogeton glomeratus (M.Bieb.) Ledeb.

Amaranthaceae

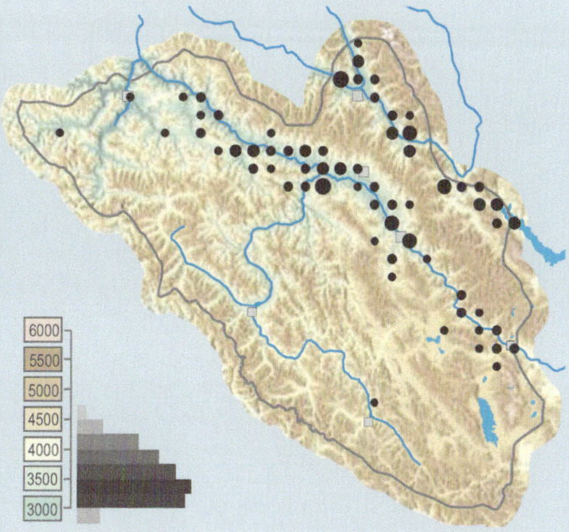

Syn.: *Anabasis glomerata* M.Bieb.

Description: Annuals, 5–40 cm tall, main stem erect, lateral stems spreading, densely hairy, leaves linear, terminated with awn, flowers in axillary glomerules, perianth segments membranous and transparent.

Distribution: Russia (Siberia), Kazakhstan, Kyrgyzstan, Tajikistan, Afghanistan, Uzbekistan, N Pakistan, N India, China (Gansu, Qinghai, Xinjiang, Xizang), Mongolia; naturalized in North America.

Habitat: Disturbed, barren, alkaline soils, withstands high concentrations of salinity.

Stem diameter 3–4 mm. Bark/xylem proportion 1:10. Mean ring width 1 mm. Maximal 1 annual ring.

Xylem

Transverse section

Root collar: Annual. Vessels solitary and in short radial rows. Vessel diameter 60–90 µm. Vessel walls not stained. Fibers thin- to thick-walled, not stained. Parenchyma paratracheal. Successive cambia continuous. Conjunctive tissue not lignified, thin-walled. With crystal sand in conjunctive tissue.

Shoot: Successive cambia foraminate.

Tangential section

Rays 1–5-seriate, confluent with axial tissue.

Radial section

Perforation plates simple. Intervessel pits round in alternating position. Vessel length 100–160 µm.

Bark

No slide available.

Pith

Round, filled with very large, thin-walled, polygonal parenchyma cells.

Analyzed: Root collars and shoots of 3 individuals.

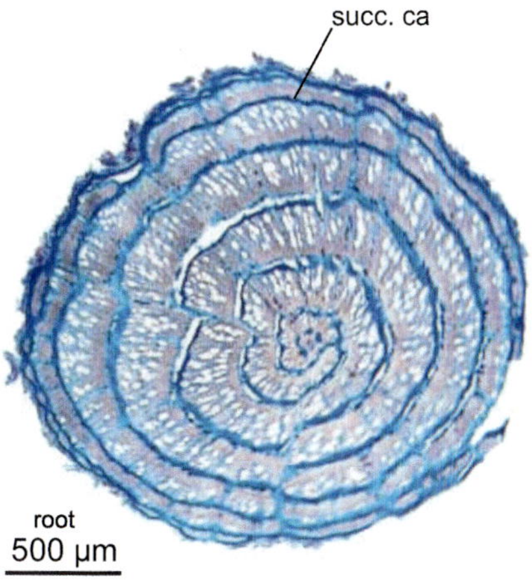

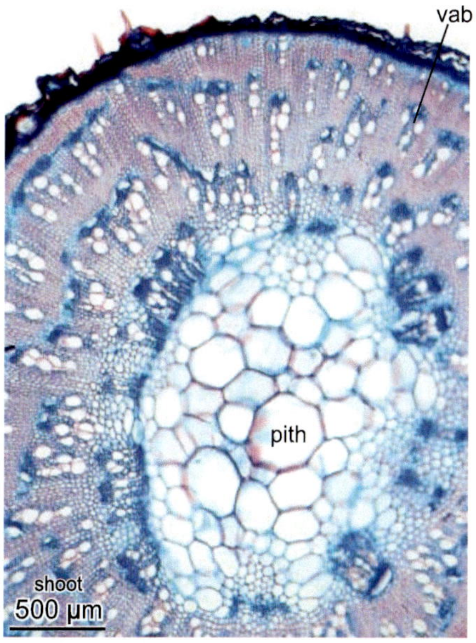

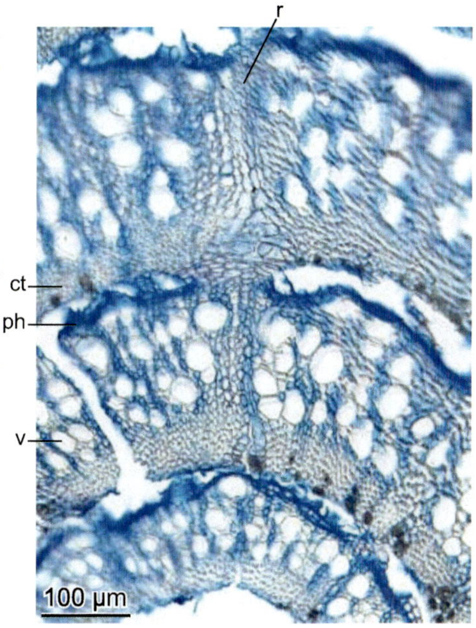

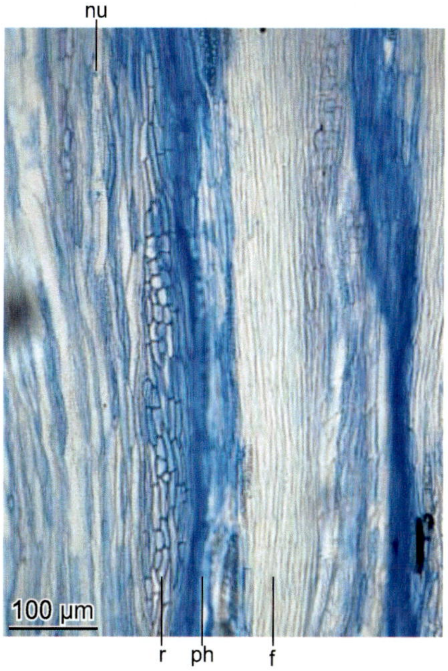

Features according to IAWA

Xylem 2.1 - 9 - 11 - 13 - 22 - 39.2 - 41 - 50.2 - 52.3 - 61 - 70 - 70.4 - 79 - 97 - 105 - 133.2 - 134.1 - 144

Haloxylon thomsonii Bunge ex Bois

Amaranthaceae

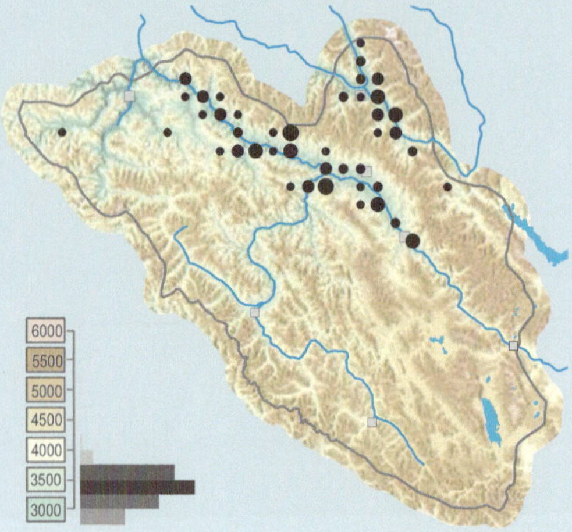

Description: Subshrubs, 10–40 cm tall, glaucous, branched from the base, leaves linear, 4–5 mm long, flowers in terminal spikes, perianth segments ovate, fruiting perianth wings white to pink, ca. 8 mm across.

Distribution: Pakistan (Gilgit, Baltistan), N India (Jammu & Kashmir), ?China (?Xizang)

Habitat: Deserts, rocks, always very dry habitats.

Stem diameter 5–15 mm. Bark/xylem proportion 1:20. Rings absent.

Xylem

Transverse section

Stem center: With growth ring boundaries in the xylem. Vessels thick-walled, solitary or in small groups. Vessel diameter 30–40 µm. Fibers thick-walled. Parenchyma paratracheal. Successive cambia concentric foraminate. Conjunctive tissue barely lignified, thin- to thick-walled. Crystals absent.

Stem periphery: Growth ring boundaries absent. Vessels thick-walled, solitary or in small groups. Vessel diameter 40–60 µm. Vessels towards the center with dark-staining substances. Fibers thick-walled. Parenchyma paratracheal. Successive cambia concentric foraminate. Conjunctive tissue thin-walled. Crystals absent.

Tangential section

Rays absent or locally present, 5–7-seriate, confluent with the axial tissue.

Radial section

Perforation plates simple. Intervessel pits round in alternating position. Vessel length 100–150 µm.

Bark

Stem: Phellem consists of many layers of thin-walled, square and rectangular cells.

Pith

No slide available.

Analyzed: Stems of 10 individuals.

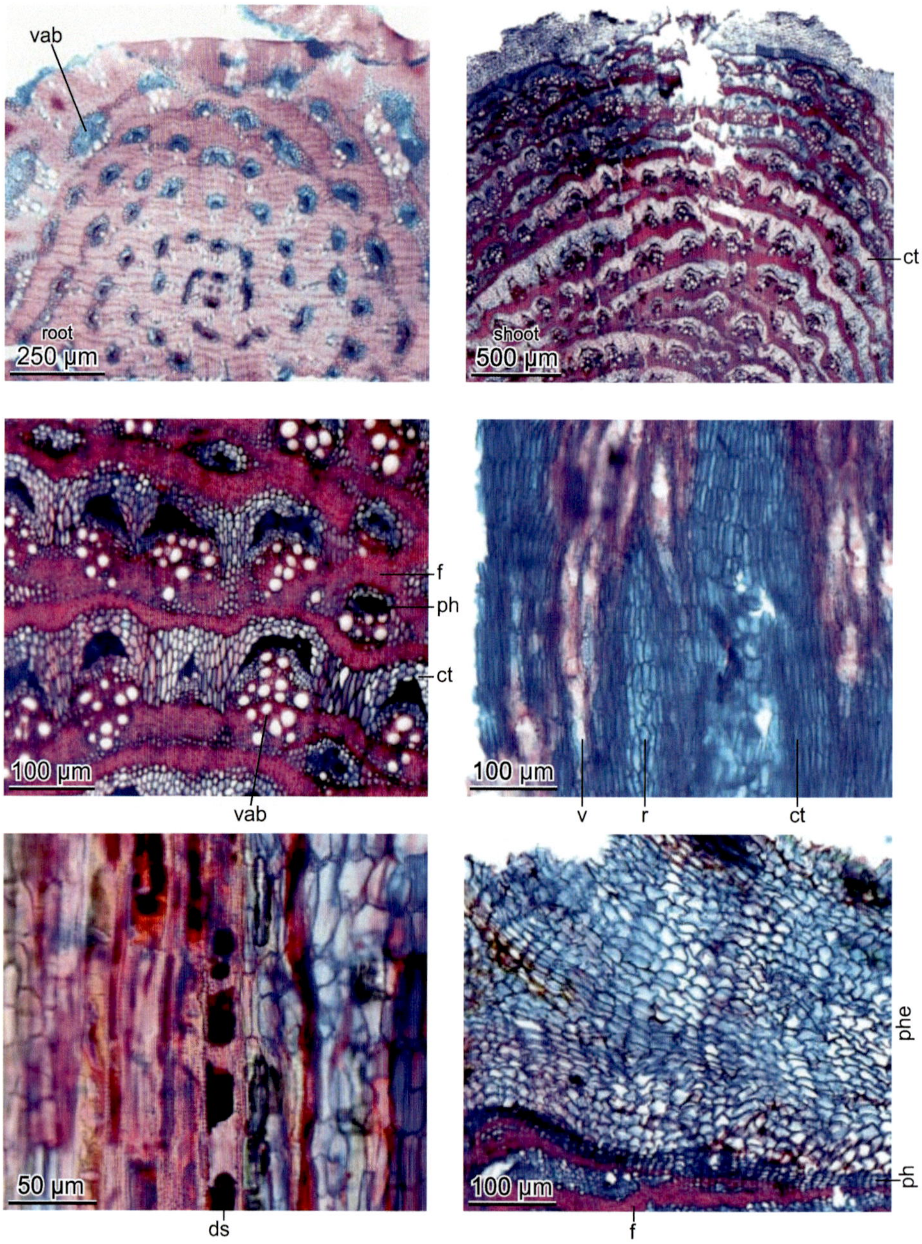

Features according to IAWA

Xylem	*Adult stem*	*2.1 - 9.1 - 10 - 13 - 22 - 39.1 - 40.2 - 50.2 - 52.3 - 61 - 69 - 79 - (98 - 100.1 - 100.2) - 117 - 133.2 - 134.1*
	Juvenile stem	*1 - 2.1 - 9.1 - 10 - 13 - 22 - 39.1 - 40.2 - 50.2 - 52.3 - 61 - 69 - 79 - (98 - 100.1 - 100.2) - 117 - 133.1 - rare 134.1*

Kochia prostrata (L.) Schrad.

Amaranthaceae

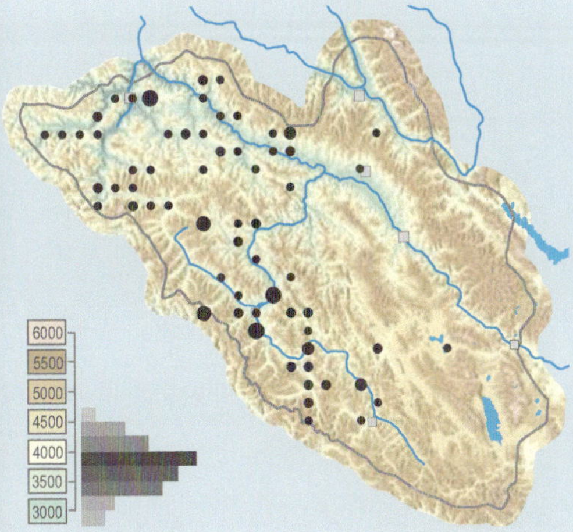

Syn.: *Bassia prostrata* (L.) A.J.Scott

Description: Subshrubs, 20–80 cm tall, pilose, leaves alternate, usually clustered on short axillary branchlets, sessile, linear, flowers in glomerules arranged in spikes on upper part of annual branches.

Distribution: N Africa, SE Europe, Caucasus, Russia (Siberia), Turkey, Iran, Kazakhstan, Turkmenistan, Kyrgyzstan, Tajikistan, Afghanistan, Pakistan, NW India, China (Gansu, Hebei, Heilongjiang, Liaoning, Nei Mongol, Ningxia, Shaanxi, Shanxi, Xinjiang, Xizang).

Habitat: Dry steppes.

Stem diameter 3–6 mm. Bark/xylem proportion 1:4. Mean ring width 0.1 mm. Probably >5 annual rings. The number is uncertain in material with successive cambia.

Xylem

Transverse section

Root collar: Isolated vascular bundles arranged in tangential rows. Distinct rings within the xylem of the vascular bundles. Semi-ring-porous. Vessels solitary. Vessel diameter 15–20 µm. Fibers thin- to thick-walled. Parenchyma absent, partially pervasive. Successive cambia foraminate. Conjunctive tissue not lignified, thin-walled. A lot of crystal sand in large rays and conjunctive tissue.

Tangential section

Rays 1–10-seriate.

Radial section

Perforation plates simple. Intervessel pits round in alternating position. Vessel length 100–120 µm.

Bark

Phloem with tangential layers of sieve tubes. With groups of sclereids. Phellem consists of layers of thin-walled, rectangular cells.

Pith

Round, filled with small round cells.

Analyzed: Root collar of polar roots and shoots of 3 individuals.

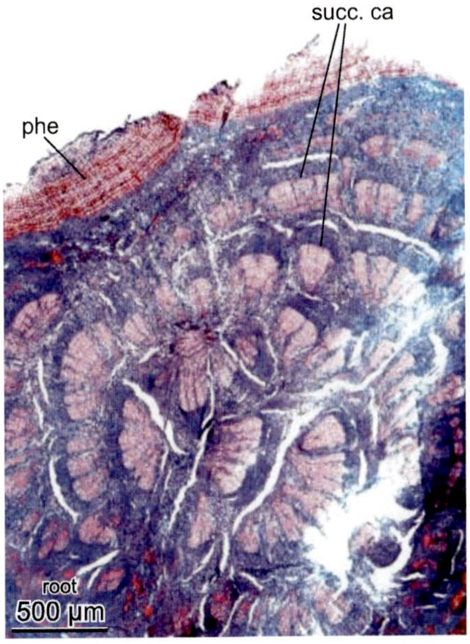

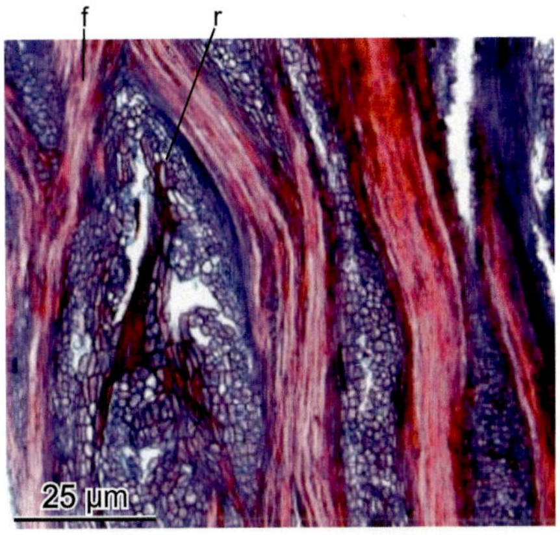

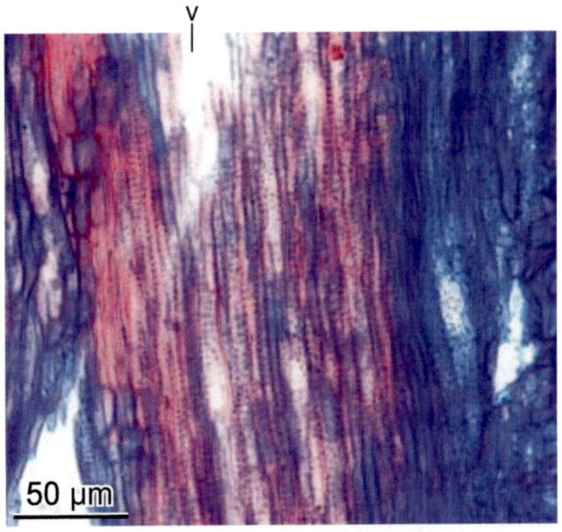

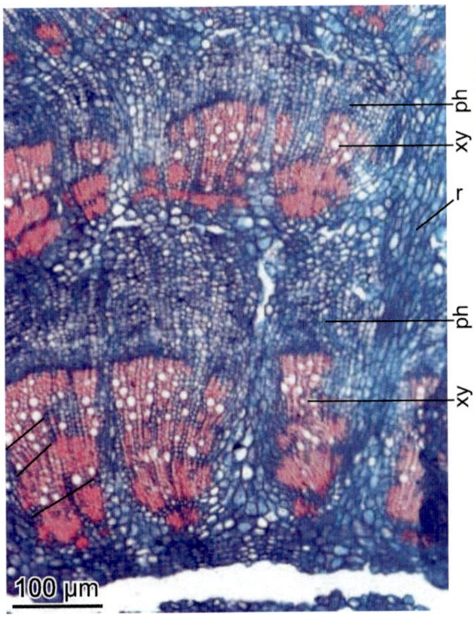

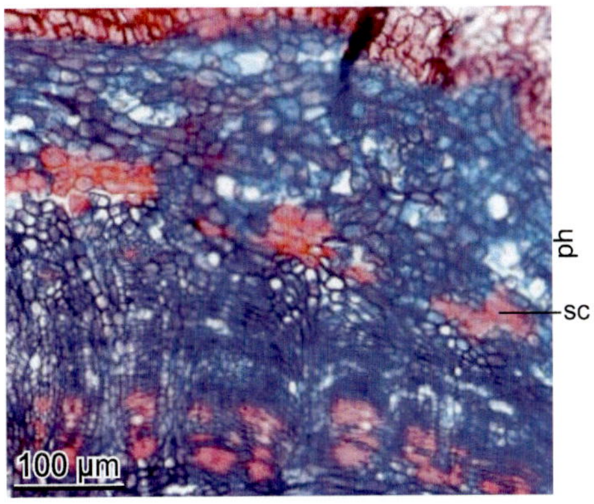

Features according to IAWA

Xylem 1 - 9 - 13 - 22 - 40.1 - 50.2 - 52.3 - 61 - 70 - 75 - 79.1 - 97 - 98 - 105 - 133.2 - 134.1 - 153

Krascheninnikovia ceratoides (L.) Gueldenst.

Amaranthaceae

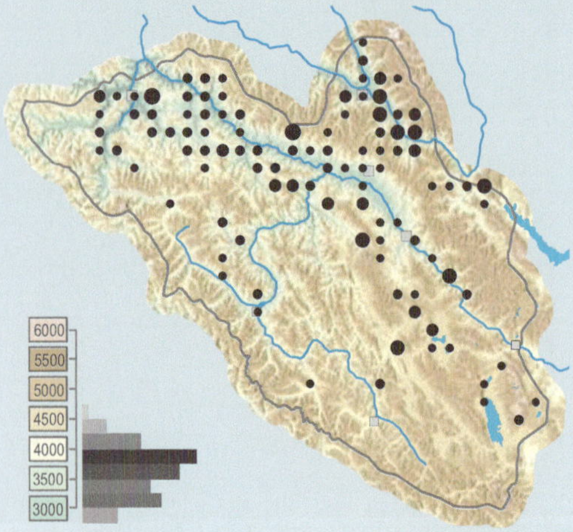

Syn.: *Axyris ceratoides* L., *Eurotia ceratoides* (L.) C.A.Mey

Description: Shrubs, 100–150 cm tall, hoary with stellate hairs, much branched, branches flexuous and spreading, leaves linear to lanceolate, flowers minute, in axillary clusters forming terminal spikes.

Distribution: N Africa, SE Europe, Russia (Siberia), Iran, Afghanistan, China (Gansu, Nei Mongol, Qinghai, Xinjiang, Xizang), Mongolia.

Habitat: Semideserts, steppes.

Stem diameter 10–20 mm. Bark/xylem proportion 1:20. Rings absent.

Xylem

Transverse section

Growth ring boundaries absent. Vessels thick-walled, solitary or in small groups. Vessel diameter 20–40 µm. Vessels in the center with dark-staining substances. Fibers thick-walled. Parenchyma paratracheal. Successive cambia foraminate and concentric. Conjunctive tissue thin-walled. With crystal druses.

Tangential section

Rays absent.

Radial section

Perforation plates simple. Intervessel pits round in alternating position. Vessel length 80–120 µm.

Bark

Stem: Phellem consists of many layers of thin-walled, square and rectangular cells.

Pith

No slide available.

Analyzed: Stems of 3 individuals.

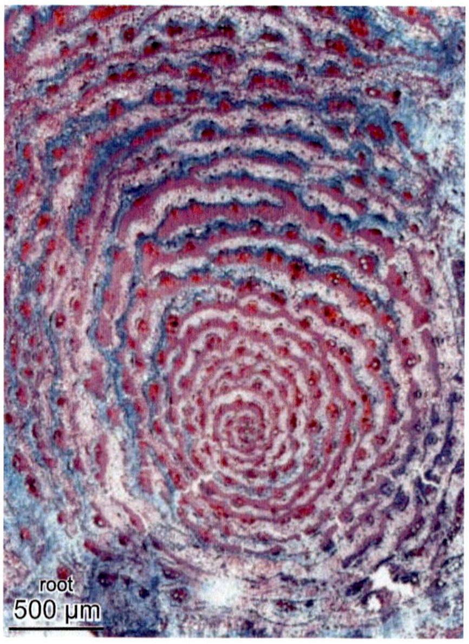

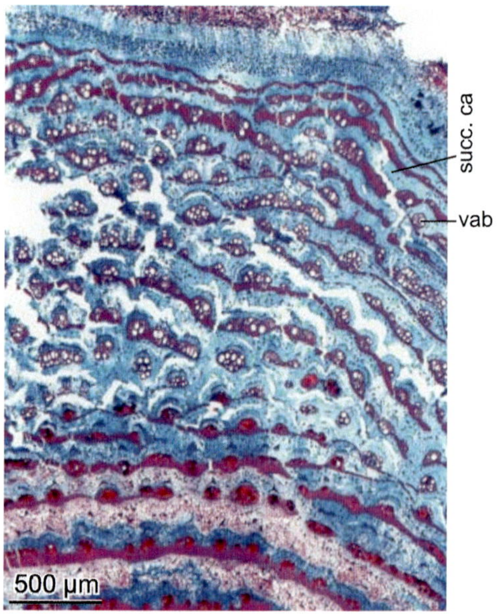

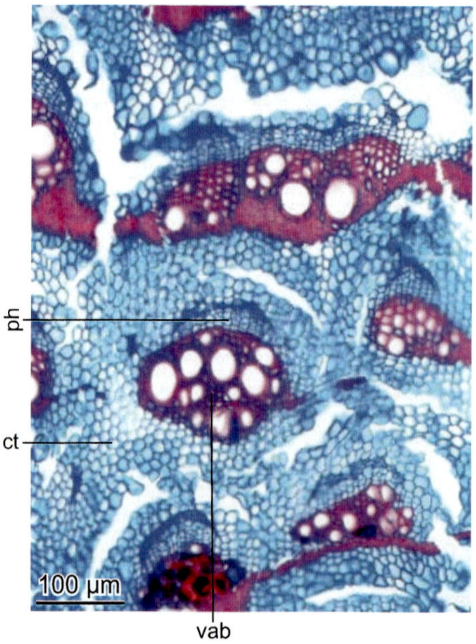

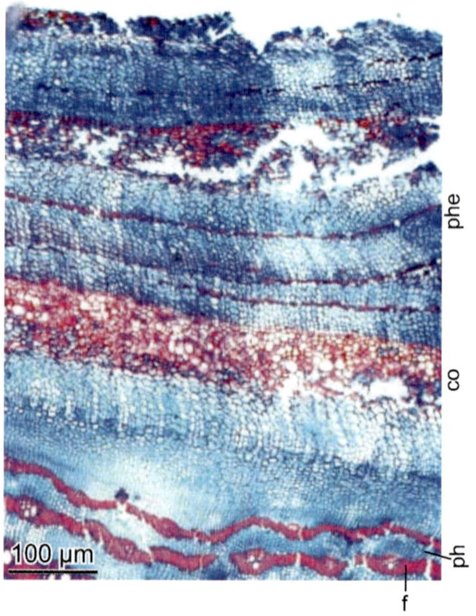

Features according to IAWA

Xylem	*Adult stem*	*2.1 - 9.1 - 10 - 13 - 22 - 39.1 - 40.2 - 50.2 - 52.3 - 61 - 69 - 79 - 117 - 133.2 - 134 - 134.1 - 144*
	Juvenile stem	*1 - 2.1 - 9.1 - 10 - 13 - 22 - 39.1 - 40.2 - 50.2 - 52.3 - 61 - 69 - 79 - 117 - 133.1 - 134 - rare 134.1*

Krascheninnikovia pungens (Pazij) Podlech

Amaranthaceae

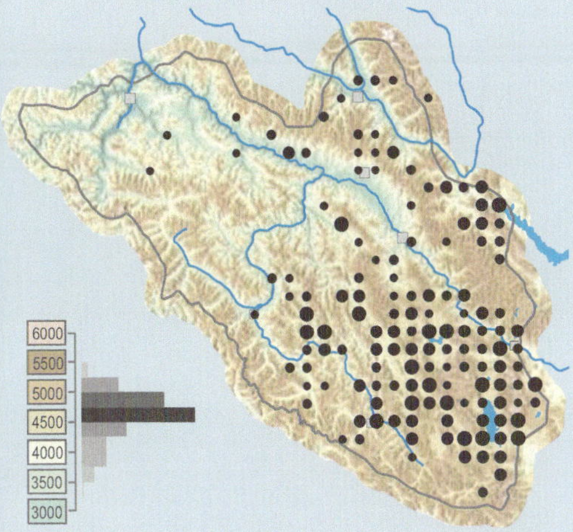

Description: Shrubs, 15–80 cm tall, hoary with stellate hairs, base woody and often splitting, branches rigid, pungent, leaves linear to lanceolate, flowers minute, in axillary clusters forming terminal spikes.

Distribution: Tajikistan, Afghanistan, N Pakistan, NW India, Nepal, China.

Habitat: Dry stony steppes, sandy plains, stabilized slopes and semideserts.

Stem diameter 10–40 mm. Bark/xylem proportion 1:20. Rings absent.

Xylem

Transverse section

Growth ring boundaries absent. Vessels thin- to thick-walled, generally in small groups. Vessel diameter 30–40 µm. Vessels in the center with dark-staining substances. Fibers thick-walled. Parenchyma paratracheal. Successive cambia foraminate and concentric. Conjunctive tissue unlignified and thin-walled. With crystal druses.

Tangential section

Rays absent. Parenchyma in the conjunctive tissue storied.

Radial section

Simple perforation plates. Intervessel pits round in alternating position. Vessel length 40–60 µm.

Bark

Phloem with irregularly distributed sieve elements. Phellem consists of many layers of thin-walled, square and rectangular cells.

Pith

No slide available.

Analyzed: Stems of 5 individuals.

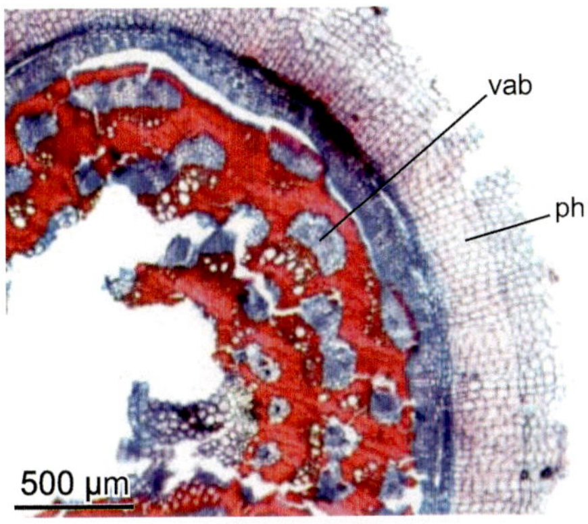

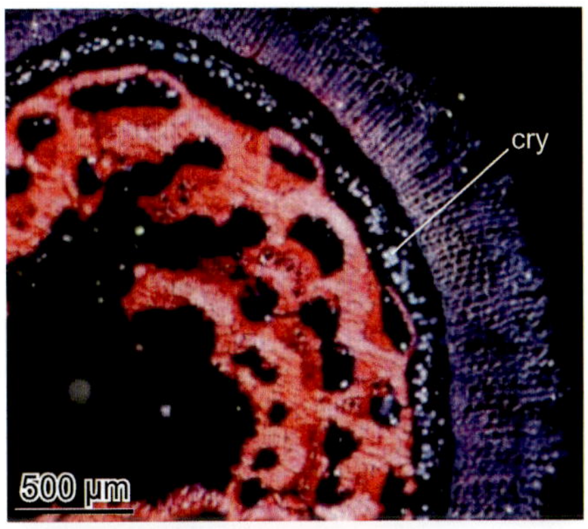

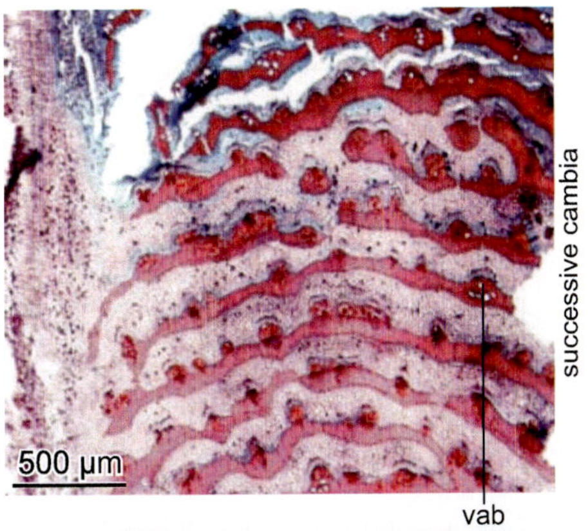

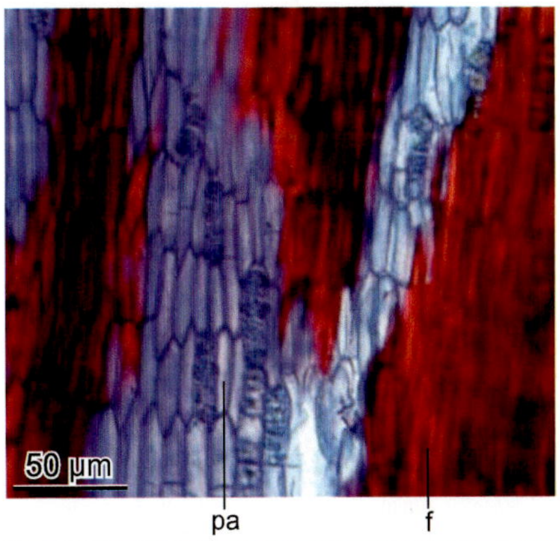

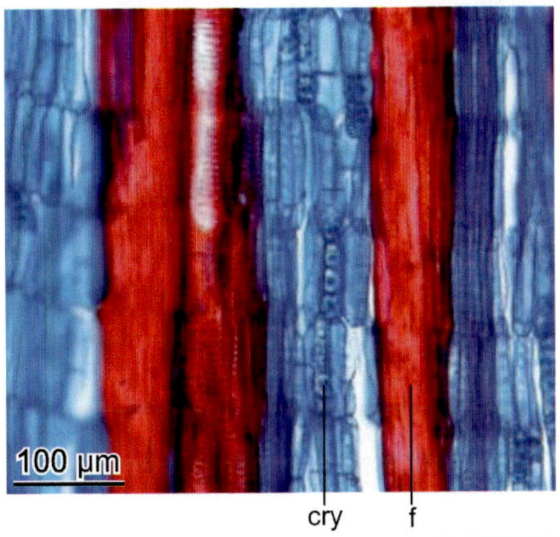

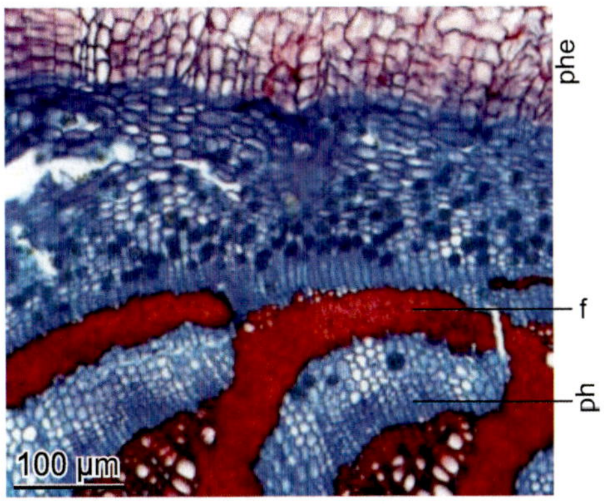

Features according to IAWA

Xylem	*Adult stem*	*1 - 2.1 - 9.1 - 11 - 13 - 22 - 40.2 - 50.2 - 52.2 - 61 - 69 - 79 - 117 - 120 - 133.2 - 134 - 134.1 - 144*

Salsola jacquemontii Moq.

Amaranthaceae

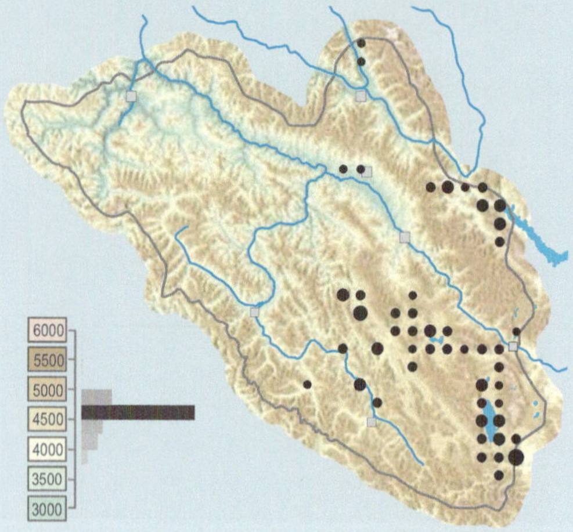

Syn.: *S. nepalensis* Grubov, *S. sinkiangensis* A.J.Li

Description: Annuals, 5–30 cm tall, hispid, stems erect or prostrate, leaves linear with a terminal spine, fleshy, flowers in dense spikes, fruits winged, 4–7 mm across.

Distribution: E Tajikistan, NE Afghanistan, N Pakistan, N India, Nepal, China (Xinjiang), W Mongolia.

Habitat: Saline and disturbed habitats, deserts and stony steppes.

Stem diameter 3–5 mm. Bark/xylem proportion 1:10. Mean ring width 1.5 mm. Maximal 1 ring.

Xylem

Transverse section

Root collar: Annual. Vessels in radial rows. Vessel diameter 30–50 µm. Fibers thin- to thick-walled. Parenchyma paratracheal. Successive cambia continuous. Conjunctive tissue lignified, thin-walled. Crystals absent.

Tangential section

Rays 2–3-seriate.

Radial section

Perforation plates simple. Intervessel pits round in alternating position. Vessel length 80–120 µm.

Bark

Root collar: Phloem homogeneous. Phellem consists of a few layers of thin-walled, rectangular cells.

Pith

No slide available.

Analyzed: Root collar of polar root of 2 individuals.

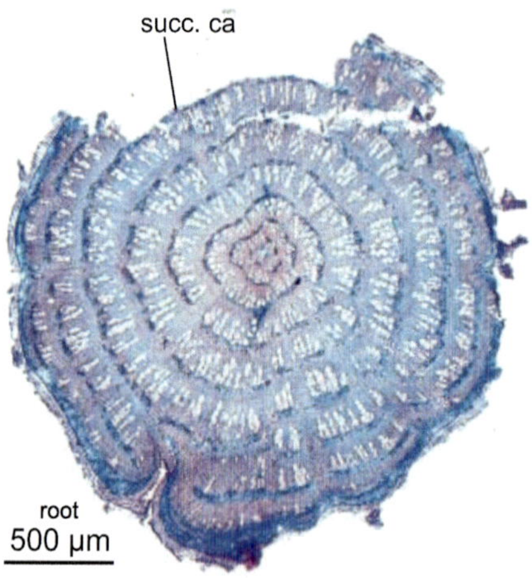

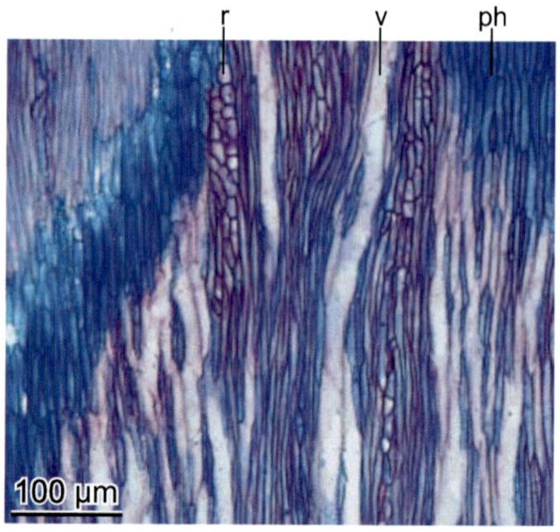

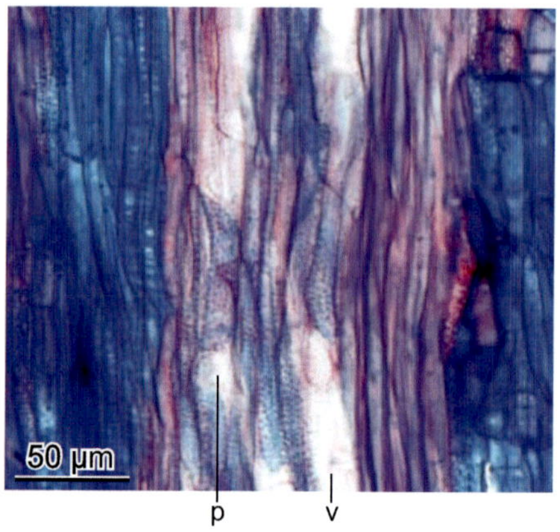

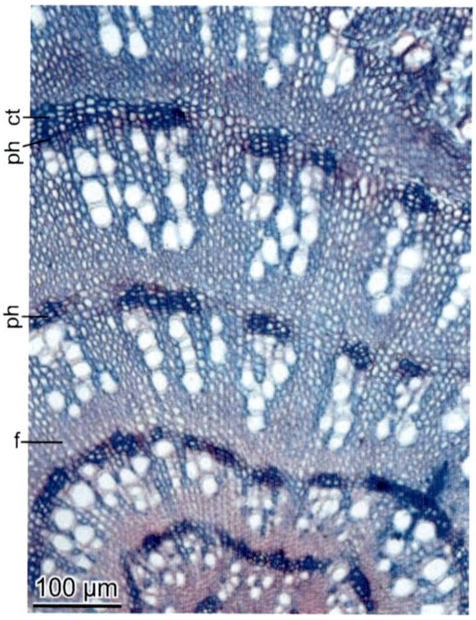

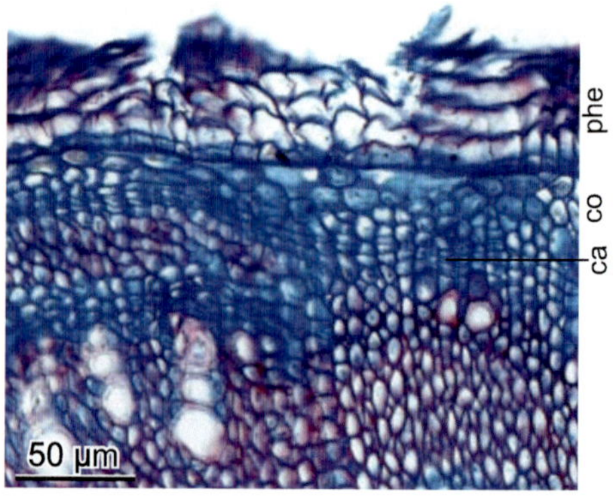

Features according to IAWA

Xylem Root 2.1 - 9.1 - 10 - 13 - 22 - 40.2 - 50.2 - 52.3 - 61 - 70 - 79 - 97 - 105 - 133.2 - 134.1

Bupleurum gracillimum Klotzsch

Apiaceae

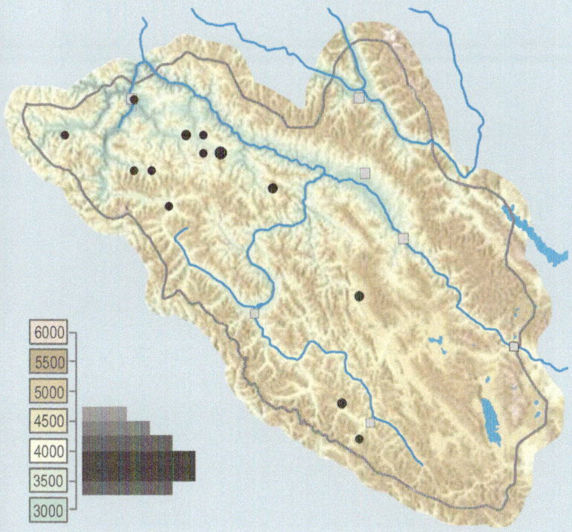

Syn.: *B. falcatum* var. *gracillimum* (Klotzsch) H.Wolff

Description: Perennial herbs, 5–35 cm tall, cespitose, stems decumbent, leaves linear, flowers yellow, arranged in umbells 1–3 cm across.

Distribution: Pakistan, India, Nepal, China (Sichuan), Bhutan, N Myanmar.

Habitat: Stabilized slopes, alpine meadows, along streams.

Stem diameter 5 mm. Bark/xylem proportion 1:0.5. Mean ring width 0.2 mm. Maximal 5 annual rings.

Xylem

Transverse section

Semi-ring-porous. Earlywood vessel diameter 30–50 µm. Vessels mainly solitary or in short radial multiples. Vessels thin- to thick-walled, not stained. Ring variability often indicated by larger (50 µm) or smaller (20 µm) vessel diameters. Thin- to thick-walled, unstained fibers occur in a peripheral ring. Parenchyma paratracheal and pervasive.

Tangential section

Rays absent.

Radial section

Perforation plates simple. Intervessel pits scalariform and round. Vessel length 100–150 µm. External part of phloem compressed (wavy, collapsed sieve tubes) due to the mechanical strength of the phellem. Phellem cells irregularly formed, thin-walled.

Bark

Sieve tubes in indistinct groups. Sclerenchyma ducts and crystals absent. Rays slightly dilated. Phellem consists of one or many rows of thin-walled cells.

Pith

No slide available.

Analyzed: Root collar of polar root of 1 individual.

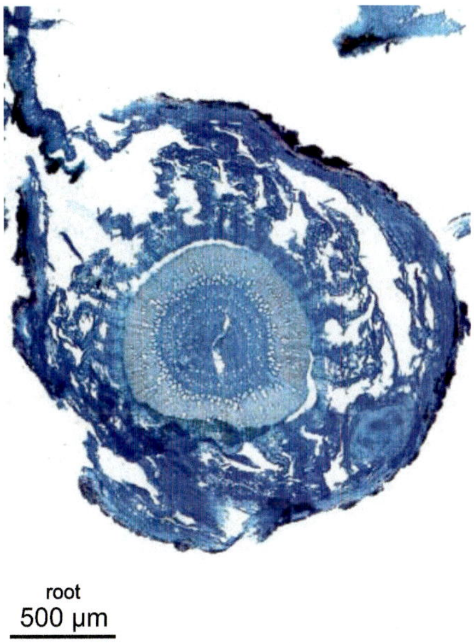

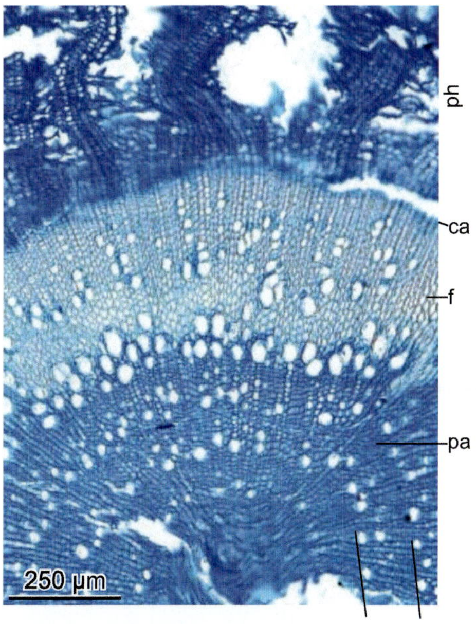

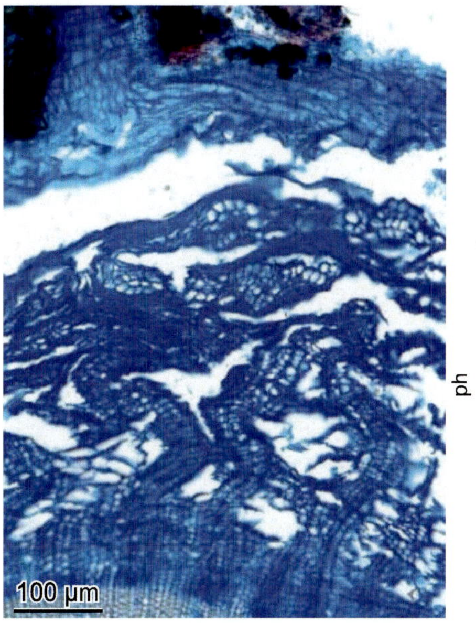

Features according to IAWA

Xylem 1 - 4 - 9 - 9.1 - 13 - 20 - 22 - 39.2 - 40.2 - 50.2 - 52.3 - 61 - 70 - 70.3 - 70.4 - 79 - 79.1 - 117

Bupleurum marginatum Wall. ex DC.

Apiaceae

Syn.: *B. falcatum* var. *hoffmeisteri* (Klotzsch) C.B.Clarke

Description: Perennial herbs, 25–100 cm tall, stems rigid, leaves linear, margin conspicuously white-cartilaginous, flowers pale yellow, arranged in umbells 1.5–4 cm across.

Distribution: Pakistan, N India, Nepal, China (Gansu, Guizhou, Hubei, Qinghai, Sichuan, Xizang, Yunnan), Bhutan, Myanmar.

Habitat: Forests, alpine meadows, river banks.

Stem diameter 10 mm. Bark/xylem proportion 1:3. Mean ring width 0.3 mm. Maximal 19 annual rings.

Xylem

Transverse section

Semi-ring-porous. Earlywood vessel diameter 30–50 µm, latewood vessel diameter 10–15 µm. Vessels mainly solitary. Vessels thin- to thick-walled, lignified. Thin- to thick-walled fibers mainly in the earlywood, or absent. Fibers with small pits. Parenchyma absent in the fiber zone and pervasive in fiber-less zones.

Tangential section

Rays 3–10-seriate.

Radial section

Perforations mainly simple, rarely irregularly scalariform. Intervessel pits horizontally enlarged and scalariform. Vessel length 80–120 µm. Ray cells mainly square, rarely prostrate. A few crystal sand grains in large rays.

Bark

Sieve tubes in young phloem in indistinct groups. Sclerenchyma absent. External part of the phloem compressed due to the mechanical strength of the phellem. Compressed sieve tubes appear as tangential dark zones. Ducts present and surrounded by 2 rows of cells. Rays intensively bent. Crystals absent. Phellem consists of one to many rows of thin-walled, rectangular cells. Secondary central cork layers delimitate old, dead xylem from the younger, living part.

Pith

No slide available.

Analyzed: Root collar of polar root of 4 individuals.

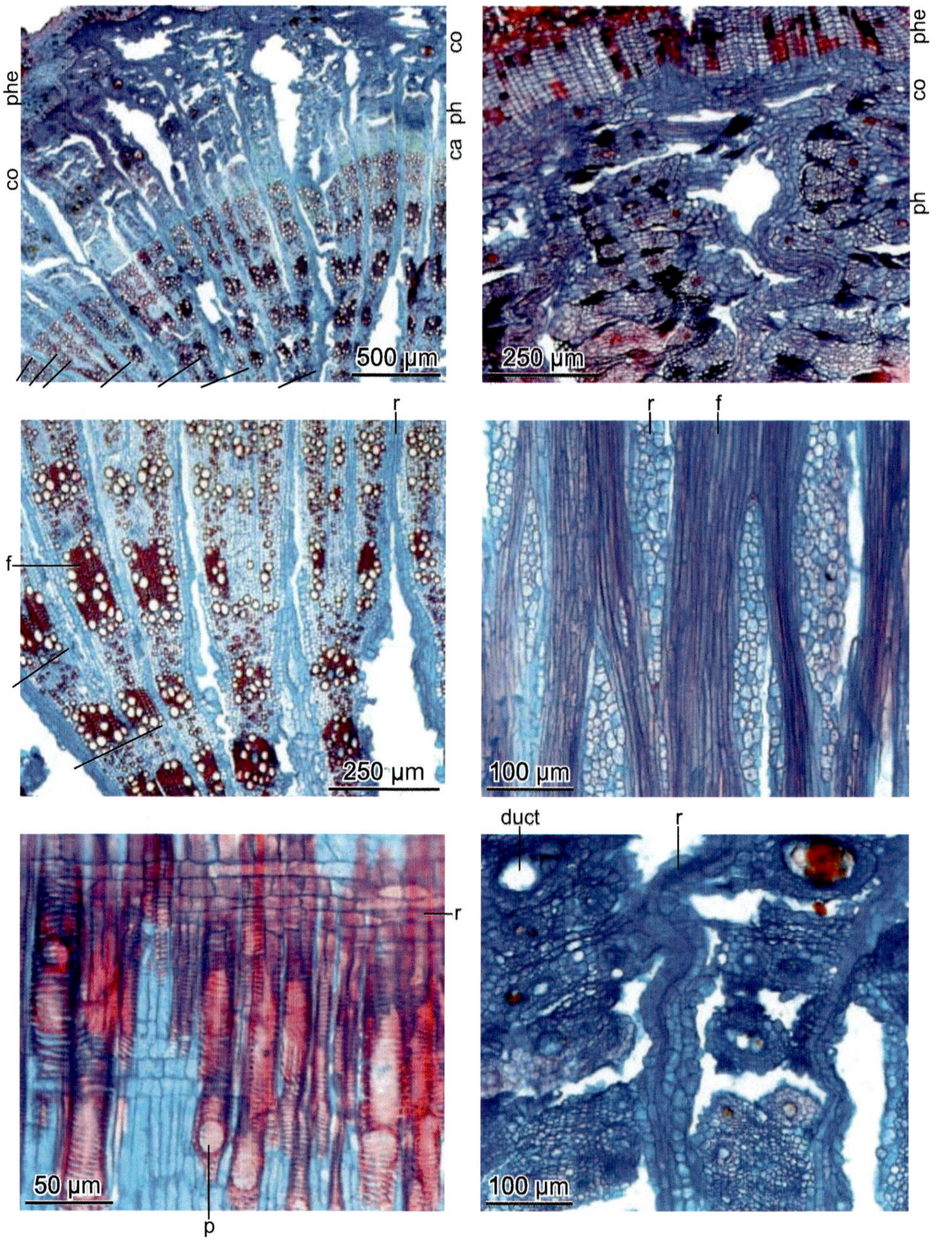

Features according to IAWA

Xylem 1 - 4 - 9 - 13 - 14 - 20 - 22 - 40.2 - 50.2 - 52.2 - 52.3 - 61 - 70 - 70.3 - 70.4 - 75 - 79.1 - 98 - 100.2 - 104 - 108 - 153

Conioselinum vaginatum (Spreng.) Thell.

Apiaceae

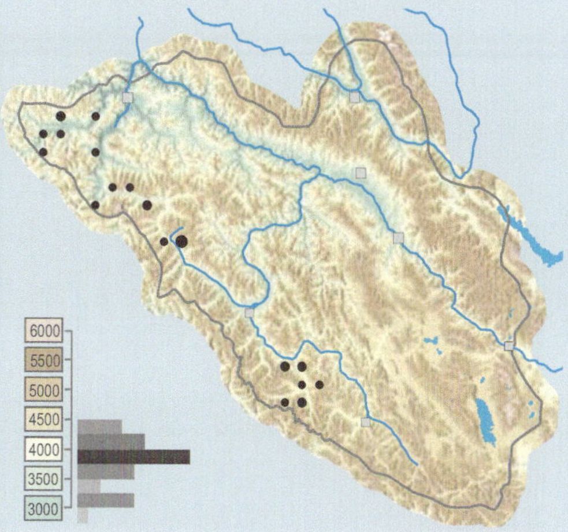

Syn.: *Ligusticum vaginatum* Spreng., *Selinum papyraceum* C.B.Clarke, *C. papyraceum* (C.B.Clarke) Pimenov & Kljuykov

Description: Perennial herbs, 30–90 cm tall, stem branched, leaves 2–3-pinnate, ultimate segments lanceolate, pinnatifid, flowers white, in umbels 5–10 cm across.

Distribution: C Europe, Russia (Siberia), SW Asia, Kazakhstan, Turkmenistan, Uzbekistan, Kyrgyzstan, India (Jammu and Kashmir, Himachal Pradesh, Uttarakhand, Sikkim), China (Xinjiang).

Habitat: Grasslands, alpine meadows.

Stem diameter 9 mm. Bark/xylem proportion 1:1. Mean ring width 0.2 mm. Maximal 4 annual rings.

Xylem

Transverse section

Semi-ring-porous. Ring boundaries marked by a fiber-free zone. Vessels mainly solitary or in small groups. Earlywood vessel diameter 40–50 µm. Vessels thin- to thick-walled, lignified. Thick-walled fibers in large groups. Parenchyma paratracheal and pervasive.

Tangential section

Rays with 5–8 cells, with sheath cells. Ray height >1 mm. Rays disappear in polarized light.

Radial section

Perforation plates simple. Intervessel pits scalariform and round. Vessel length 120–180 µm.

Bark

Sieve tubes indistinct. With many large ducts. Sclerenchyma and crystals absent. Rays slightly dilated. Phellem consists of one to many rows of thin-walled cells.

Pith

No slide available.

Analyzed: Root collar of polar root of 1 individual.

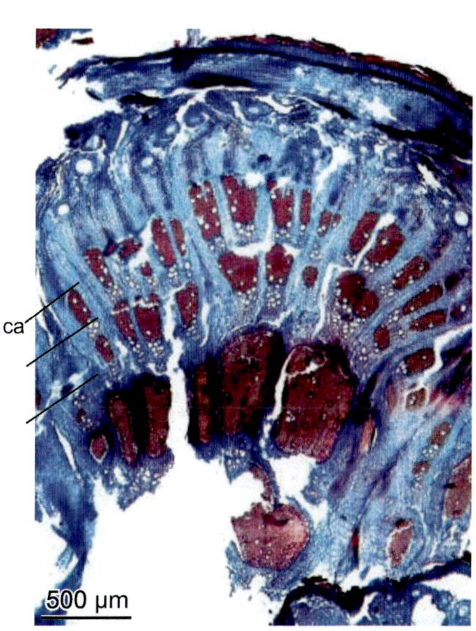

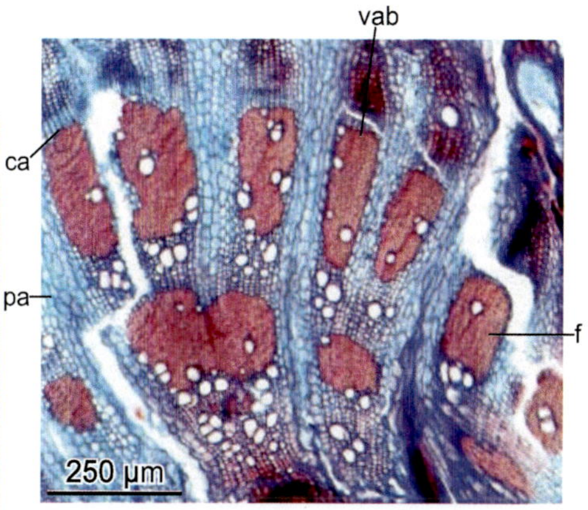

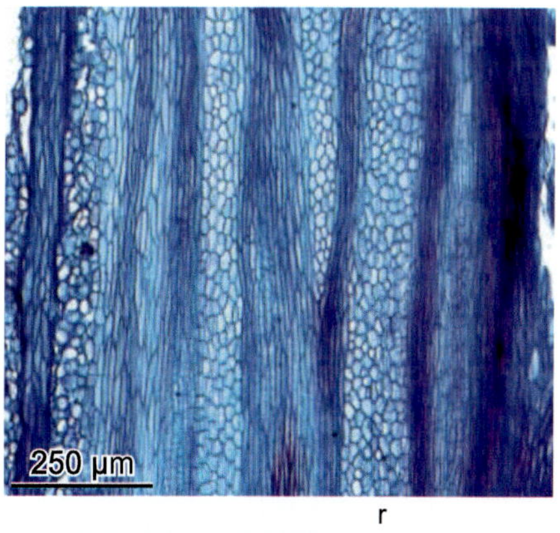

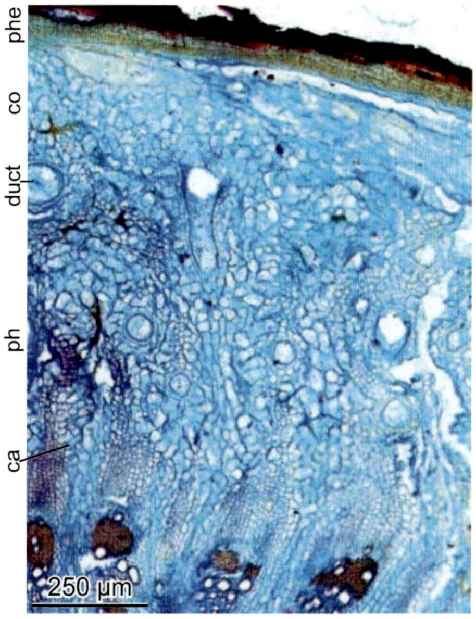

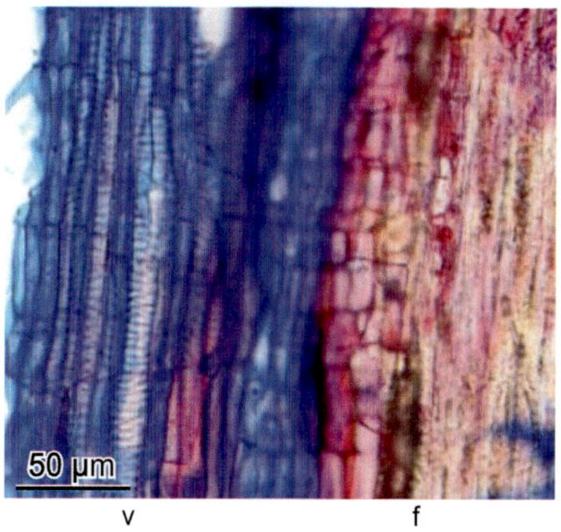

Features according to IAWA

Xylem 1 - 4 - 9 - 13 - 20 - 22 - 40.2 - 41 - 50.2 - 52.3 - 61 - 79 - 79.1 - 98 - 100.2 - 105 - 135.2

Heracleum pinnatum C.B.Clarke
Apiaceae

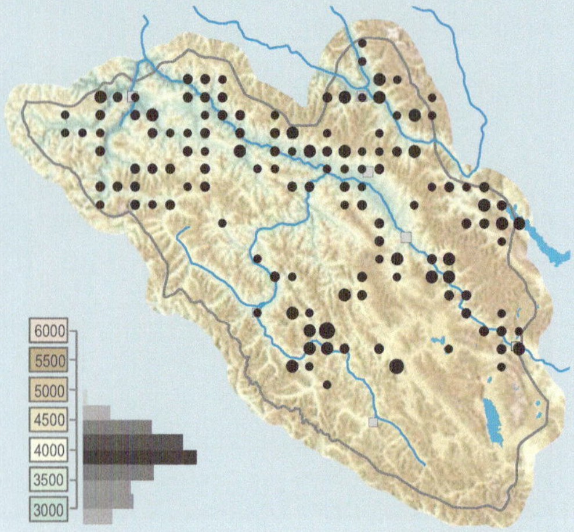

Description: Perennial herbs, 30–60 cm tall, leaves mostly basal, pinnate, margin of segments serrate, flowers whitish, arranged in composed umbels up to 10 cm across.

Distribution: Pakistan, India, China (Xizang).

Habitat: Stony habitats at river banks, stony slopes.

Stem diameter 12–16 mm. Bark/xylem proportion 1:1. Mean ring width 0.7 mm. Maximal 24 annual rings.

Xylem

Transverse section

Semi-ring-porous. Rings hardly recognizable, marked by a row of large earlywood vessels (diameter 80 µm) and smaller latewood vessels (diameter 30 µm). Ring variability often indicated by larger (90 µm) or smaller (60 µm) earlywood vessel diameters. Vessels mainly solitary. Vessels thick-walled, mostly unstained. Fibers absent. Parenchyma pervasive.

Tangential section

Ray width 2–3 cells. Rays disappear in polarized light.

Radial section

Perforation plates simple. Intervessel pits scalariform. Vessel length 90–110 µm. Ray cells square. Tissue of the central part of the root collar zone wavy due to root contraction.

Bark

Phloem large, consists of small, irregularly distributed groups of sieve tubes. Large oil ducts arranged in tangential lines. External part of phloem compressed (wavy rays and collapsed sieve tubes) due to the mechanical strength of the phellem. Sclerenchyma and crystals absent. Phellem consists of one to many rows of thin-walled cells.

Pith

No slide available.

Analyzed: Root collar of polar root of 4 individuals.

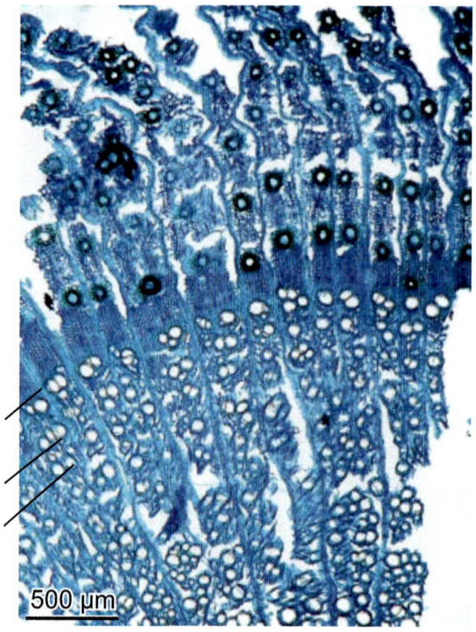

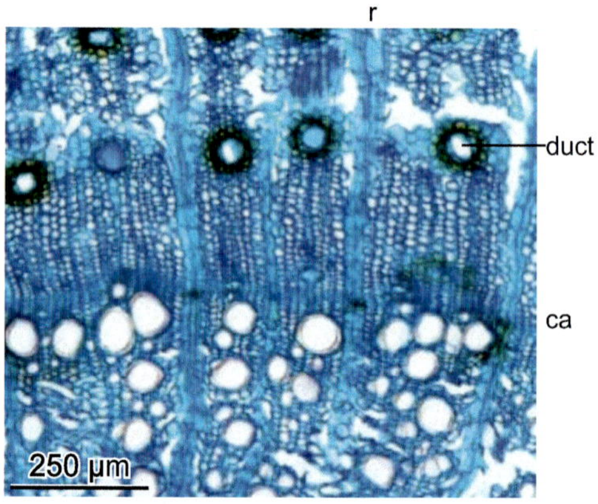

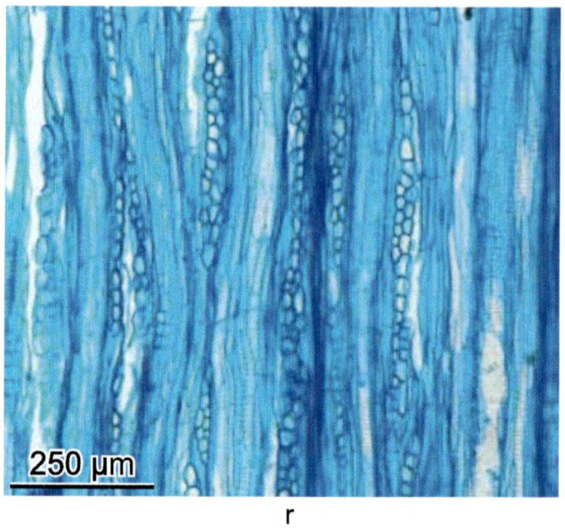

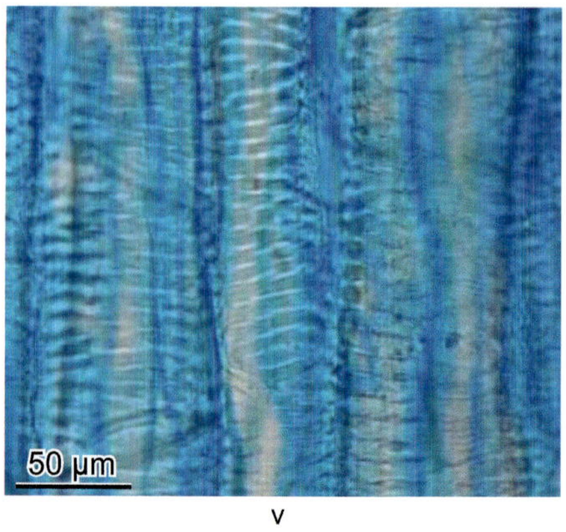

Features according to IAWA

Xylem 1 - 4 - 9 - 13 - 20 - 39.2 - 41 - 50.1 - 52.2 - 52.3 - 60.1 - 79.1 - 97 - 100.2 - 105

Pleurospermum candollei (DC.) C.B.Clarke

Apiaceae

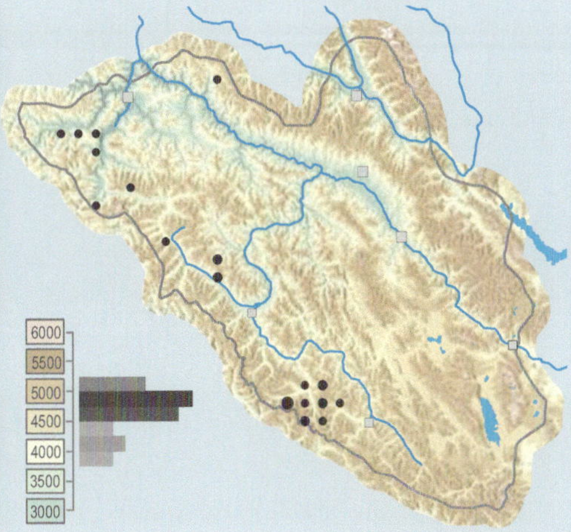

Syn.: *Hymenolaena candollei* DC.

Description: Biennial or perennial herbs, 10–30 cm tall, glabrous, stems fistular, up to 2 cm wide in older plants, leaves 1–2-pinnate, segments deeply incised, flowers white, arranged in umbels 2–4 cm across, forming a composed umbel.

Distribution: Pakistan, NW India, C and W Nepal, China (Xizang).

Habitat: Stabilized stony screes, alpine meadows.

Stem diameter 9 mm. Bark/xylem proportion 1:1. Mean ring width 0.25 mm. Maximal 13 annual rings.

Xylem

Transverse section

Semi-ring-porous. Rings not very distinct. Earlywood vessel diameter 40–50 µm. Vessels mainly solitary. Vessels thin- to thick-walled, lignified. Fibers present in a small peripheral band but absent in the largest part of the section. Parenchyma mostly pervasive, rarely paratracheal.

Tangential section

Ray width 5–8 cells, confluent with axial tissue. Rays disappear in polarized light.

Radial section

Perforation plates simple. Intervessel pits scalariform. Vessel length 100–200 µm. Tissue of the central part of the root collar zone wavy due to root contraction.

Bark

Phloem large, consists of indistinct, small, radially arranged groups of sieve tubes. Very small ducts. Sclerenchyma and crystals absent. Rays slightly dilated. Phellem consists of one to many rows of thin-walled cells.

Pith

No slide available.

Analyzed: Root collar of polar root of 4 individuals.

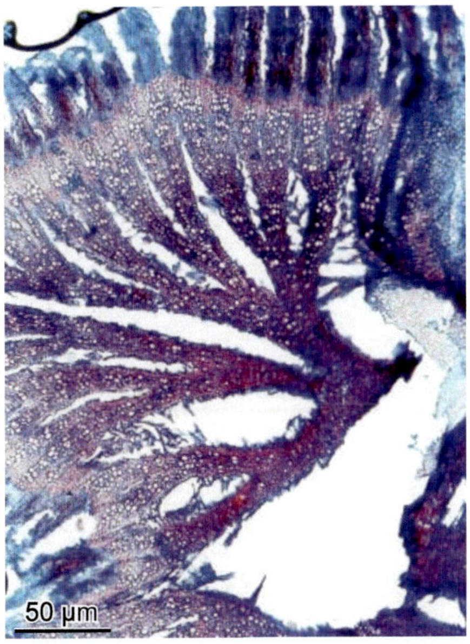

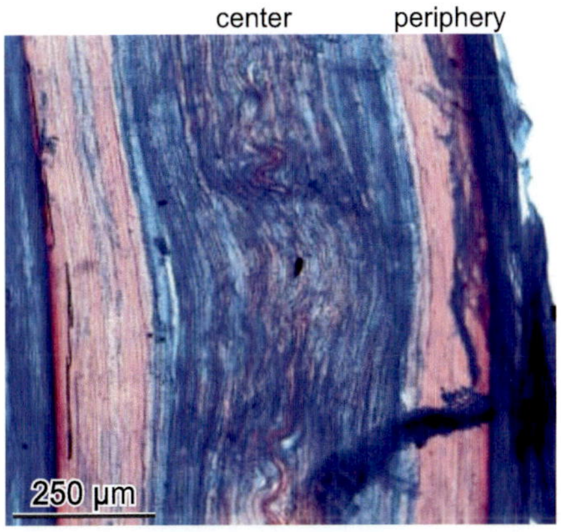

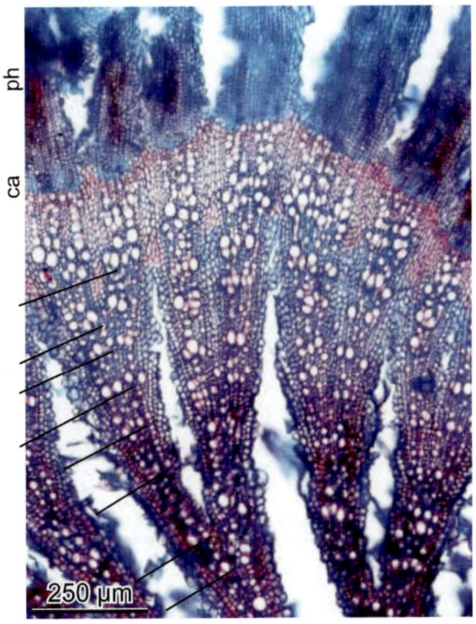

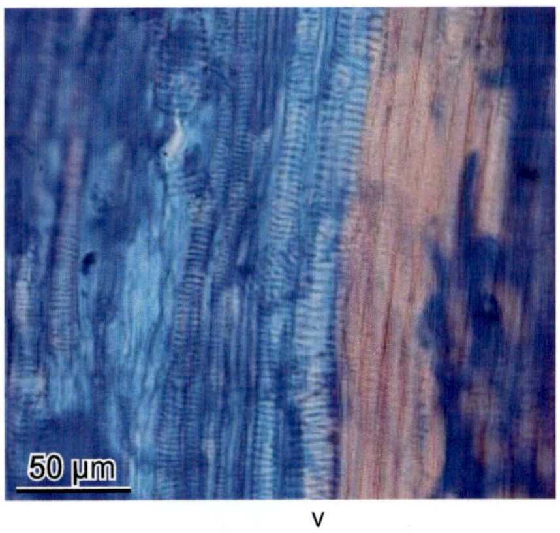

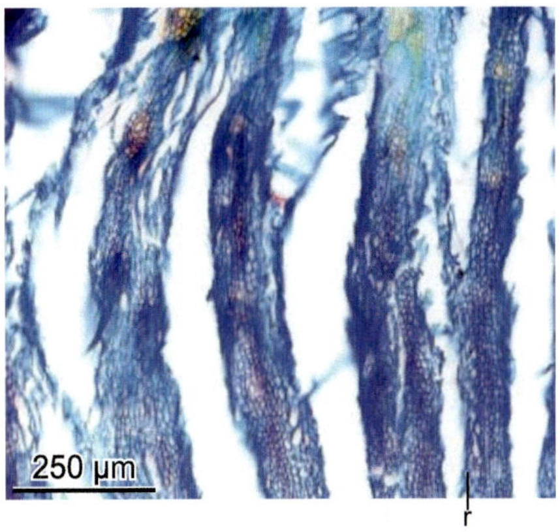

Features according to IAWA

Xylem 1 - 4 - 9 - 13 - 20 - 40.2 - 50.1 - 52.3 - 60.1 - 61 - 70 - 79 - 79.1 - 98 - 100.1 - 100.2 - 105

Pleurospermum hookeri C.B.Clarke

Apiaceae

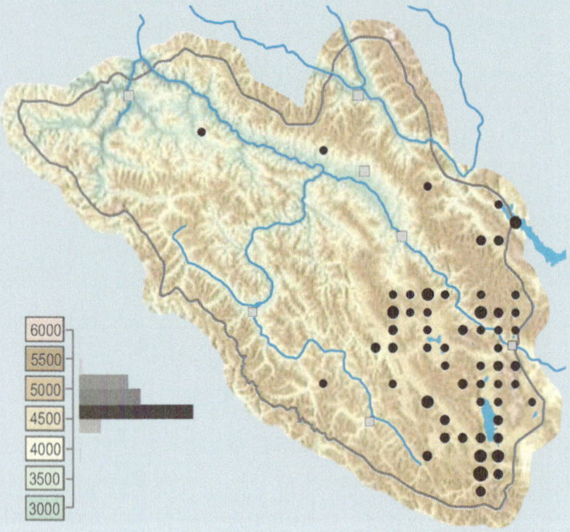

Description: Biennial or perennial herbs, 10–40 cm tall, glabrous, stems ribbed, leaves 2–4-pinnate, gradually reduced upwards, flowers white, in umbels 6–7 cm across, bracts with membranous margin.

Distribution: N India, Nepal, China (Gansu, Qinghai, Sichuan, Xizang, Yunnan), Bhutan.

Habitat: River banks, alpine meadows.

Stem diameter 6 mm. Bark/xylem proportion 1:0.8. Mean ring width 0.3 mm. Maximal 10 annual rings.

Xylem

Transverse section

Semi-ring-porous. Rings marked by a row of large earlywood vessels (diameter 40 µm) and slightly smaller latewood vessels (diameter 20 µm). Vessels mainly solitary. Ring variability often indicated by larger (50 µm) or smaller (30 µm) earlywood vessel diameters. Vessels thin- to thick-walled, red-stained. Fibers absent. Parenchyma pervasive.

Tangential section

Ray width 5–8 cells, confluent with the axial tissue. Rays disappear in polarized light.

Radial section

Perforation plates simple. Intervessel pits scalariform. Vessel length 50–100 µm. Tissue of the central part of the root collar zone wavy due to root contraction.

Bark

Phloem large, consists of small, radially arranged groups of sieve tubes. External part of phloem compressed (wavy rays and collapsed sieve tubes) due to the mechanical strength of the phellem. Sclerenchyma and crystals absent. Small ducts. Rays slightly dilated. Phellem consists of one to many rows of thin-walled cells.

Pith

No slide available.

Analyzed: Root collar of polar root of 1 individual.

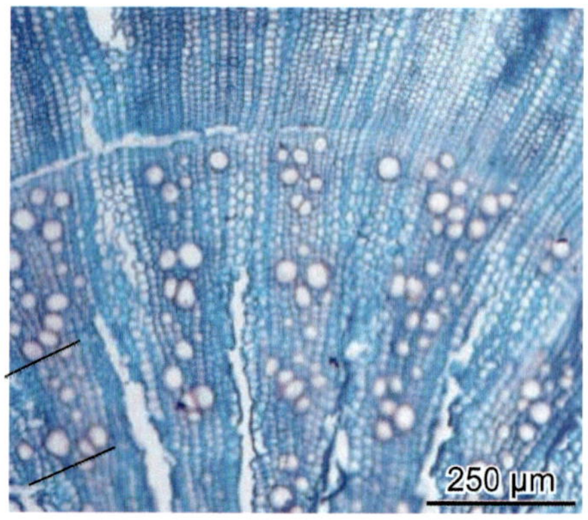

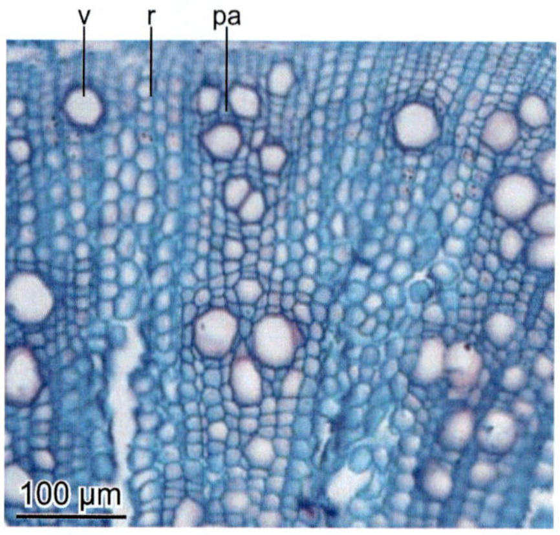

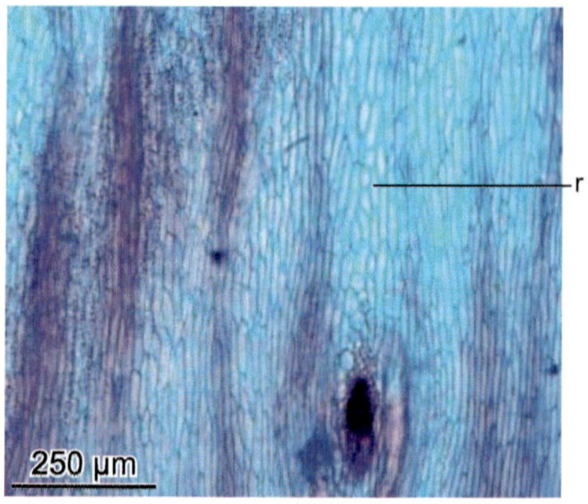

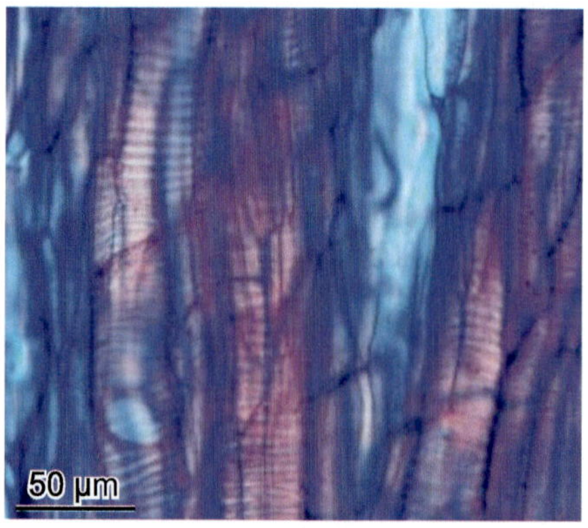

root contraction

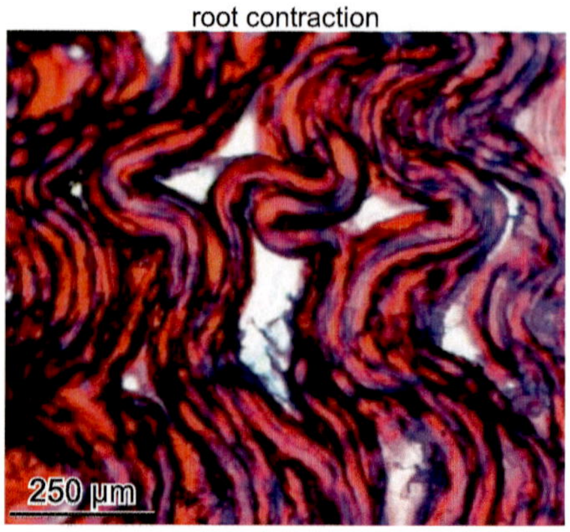

Features according to IAWA

Xylem 1 - 4 - 9 - 13 - 20 - 40.2 - 50.1 - 52.2 - 60.1 - 79.1 - 98 - 100.2 - 105

Pleurospermum lindleyanum (Klotzsch) B.Fedtsch.

Apiaceae

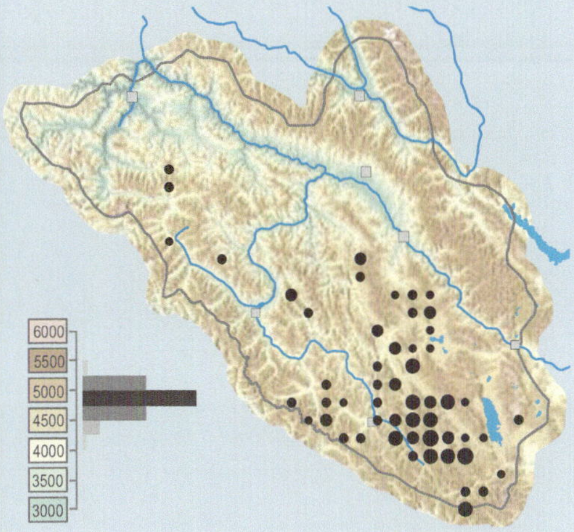

Syn.: *Hymenolaena lindleyana* Klotzsch, *P. stellatum* var. *lindleyanum* (Klotzsch) C.B.Clarke

Description: Biennial or perennial herbs, 5–30 cm tall, stem inconspicuous, elongating in fruit, tinged purple-red, leaves mostly basal, 2–3-ternate-pinnate, ultimate segments oblong to linear, flowers whitish, arranged in umbels 2–4 cm across.

Distribution: Pakistan, NW India, China (Xinjiang, Xizang).

Habitat: Alpine meadows, grassy and gravel slopes.

Stem diameter 9 mm. Bark/xylem proportion 1:1. Ring width 0.1–0.2 mm. Maximal 15 annual rings.

Xylem

Transverse section

Semi-ring-porous, rings distinct. Earlywood vessel diameter 40–50 µm. Vessels in tangential groups, thin- to thick-walled, lignified. Fibers present in a small peripheral band but absent in the largest part of the section. Parenchyma pervasive in the stem center and paratracheal in the peripheral fiber belt.

Tangential section

Ray width 5–8 cells, confluent with the axial tissue.

Radial section

Perforation plates simple. Intervessel pits scalariform in the stem center and round at the periphery. Vessel length 100–200 µm.

Bark

Phloem large, consists of indistinct, small, radially-arranged groups of sieve tubes. With small ducts. Sclerenchyma and crystals absent. Rays slightly dilated. Phellem consists of few rows of thin-walled cells.

Pith

Round.

Analyzed: Root collar of polar root of 5 individuals.

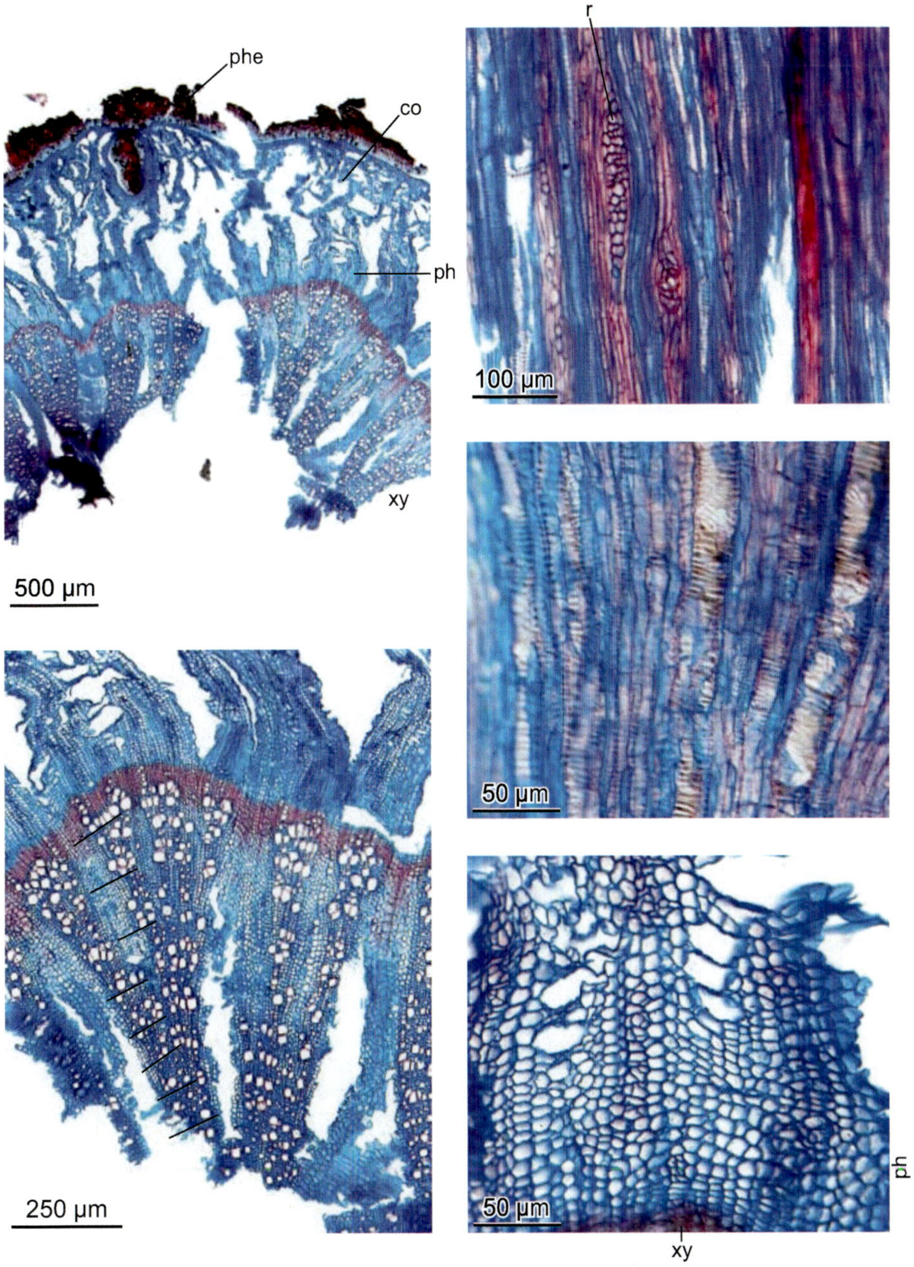

Features according to IAWA

Xylem 1 - 4 - 6 - 11 - 13 - 20 - 40.2 - 50.1 - 52.3 - 60.1 - 61 - 70 - 79 - 79.1 - 98 - 100.1 - 105

Semenovia lasiocarpa (Boiss.) Manden.

Apiaceae

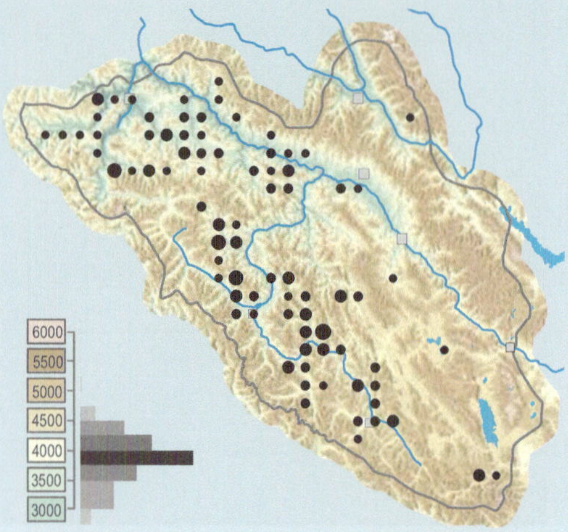

Syn.: *Peucedanum lasiocarpum* Boiss., *Heracleum thomsonii* C.B.Clarke, *S. lasiocarpa* subsp. *thomsonii* (C.B.Clarke) Rech.f. & Riedl, *S. thomsonii* (C.B.Clarke) Manden.

Description: Perennial herbs, 20–50 cm tall, leaves 2–3-pinnatisect, flowers white, arranged in umbels 3–5 cm across, outer flowers of umbels usually with enlarged petals.

Distribution: Iran, Afghanistan, Pakistan, NW India.

Habitat: Stabilized steppe slopes.

Stem diameter 16 mm. Bark/xylem proportion 1:1. Mean ring width 0.15 mm. Maximal 20 annual rings.

Xylem

Transverse section

Ring-porous to semi-ring-porous. Rings marked by a row of large earlywood vessels (diameter 50–60 µm) and small latewood vessels (diameter 20 µm). Vessels in groups, thin- to thick-walled, lignified. Fibers absent. Parenchyma pervasive.

Tangential section

Ray width 3–5 cells, confluent with the axial tissue. Height >1 mm. Rays disappear in polarized light.

Radial section

Perforation plates simple. Intervessel pits scalariform. Vessel length 200–300 µm.

Bark

Sieve tubes indistinct. External part of phloem compressed (wavy rays and collapsed sieve tubes) due to the mechanical strength of the phellem. Many large ducts. Sclerenchyma and crystals absent. Rays not dilated. Phellem consists of one to many rows of thin-walled, rectangular cells.

Pith

No slide available.

Analyzed: Root collar of polar root of 1 individual.

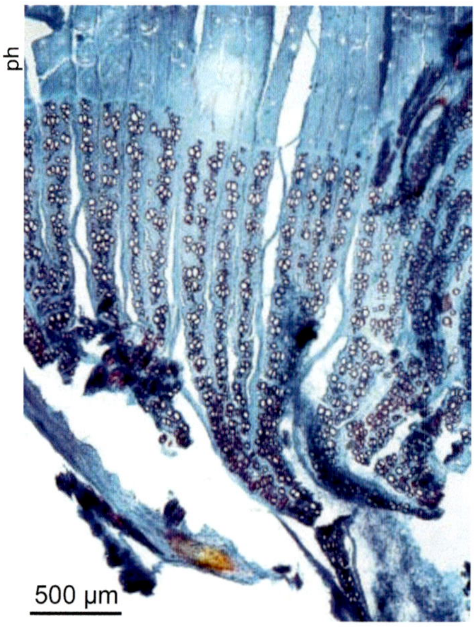

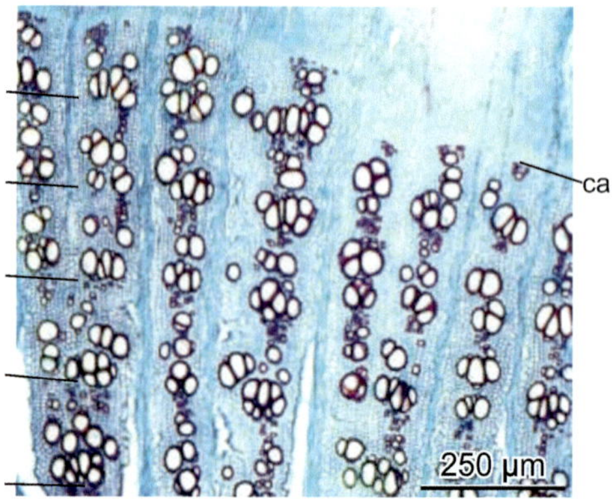

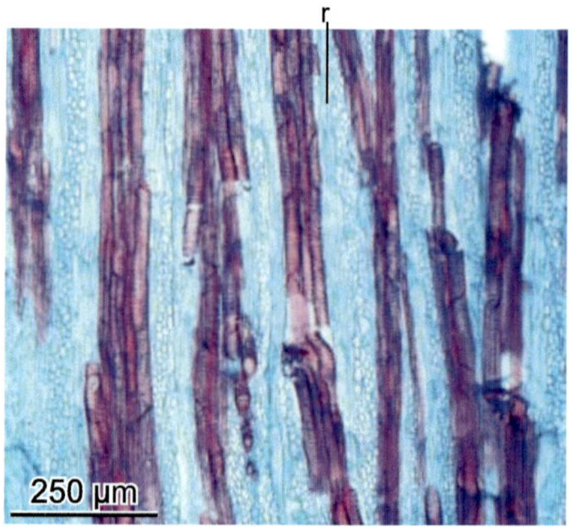

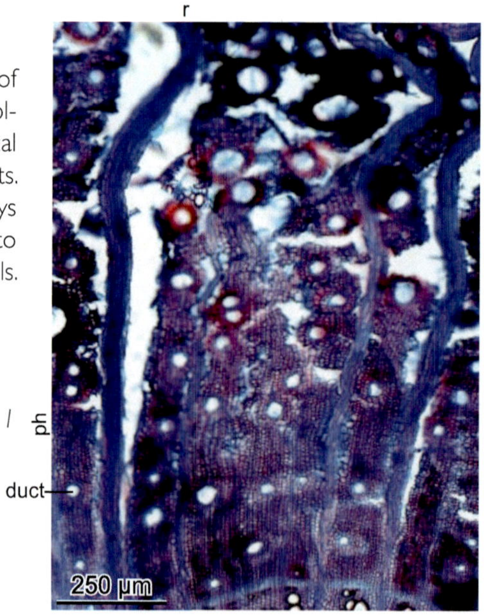

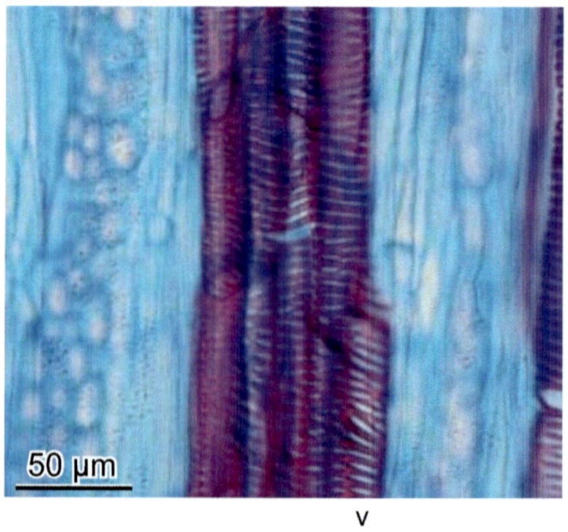

Features according to IAWA

Xylem 1 - 3 - 4 - 11 - 13 - 20 - 41 - 50.2 - 52.3 - 60.1 - 79.1 - 98 - 100.1 - 100.2 - 105

Seseli mucronatum (Schrenk) Pimenov & Sdobnina

Apiaceae

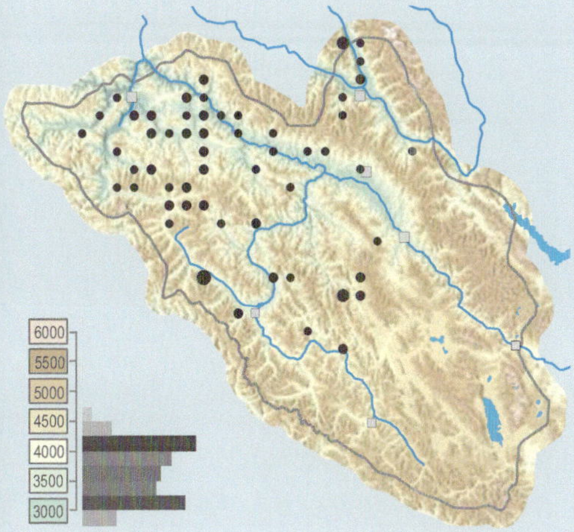

Syn.: *Neogaya mucronata* Schrenk, *Ligusticum mucronatum* (Schrenk) Leute, *L. thomsonii* C.B.Clarke, *Pachypleurum mucronatum* (Schrenk) Schischk.

Description: Perennial herbs, 15–80 cm tall, basal leaves oblong, pinnate, pinnae 5–7 pairs, oblong-ovate, pinnatifid to pinnatipartite, stem leaves much reduced, flowers white, arranged in terminal and lateral umbels, 2–7 cm across.

Distribution: Russia, Kazakhstan, Kyrgyzstan, Afghanistan, N Pakistan, NW India, China (N Xinjiang).

Habitat: Alpine meadows, along streams.

Stem diameter 15–20 mm. Bark/xylem proportion 1:1. Mean ring width 0.3 mm. Maximal 3 annual rings.

Xylem

Transverse section

Ring-porous to semi-ring-porous. Rings marked by a row of large earlywood vessels (diameter 60–70 µm) and small latewood vessels (diameter 20 µm). Vessels solitary, thin- to thick-walled, lignified. Fibers generally absent. A few groups of thick-walled fibers near the center. Parenchyma pervasive.

Tangential section

Ray width 3–5 cells, confluent with the axial tissue. Rays disappear in polarized light.

Radial section

Perforation plates simple. Intervessel pits scalariform. Vessel length 100–150 µm. Tissue of the central part of the root collar zone is wavy due to root contraction.

Bark

Sieve tubes indistinct in radial rows. External part of phloem compressed (wavy rays and collapsed sieve tubes) due to the mechanical strength of the phellem. With large ducts. Crystals absent. Rays not dilated. Phellem consists of one to many rows of thin-walled, rectangular cells.

Pith

No slide available.

Analyzed: Root collar of polar root of 3 individuals.

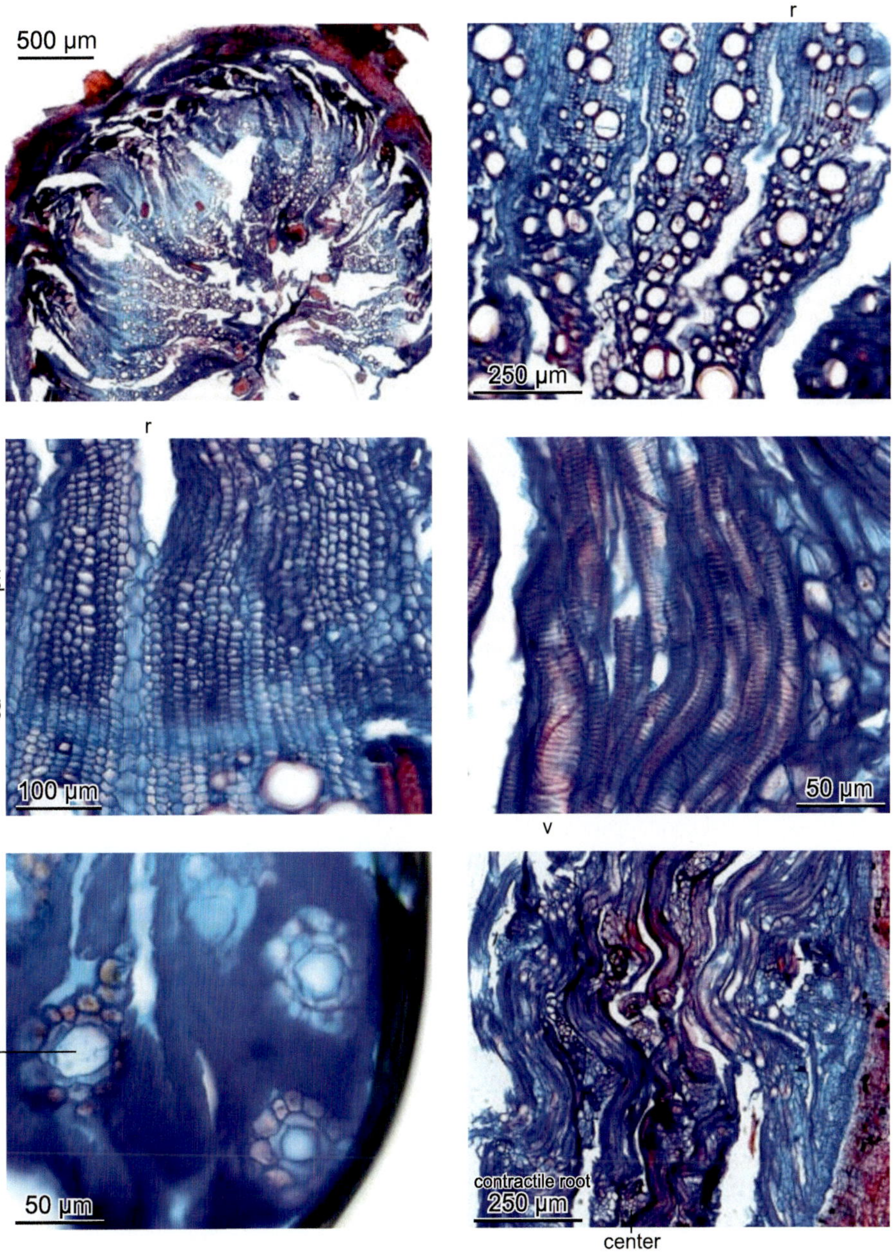

Features according to IAWA

Xylem 1 - 4 - 9 - 13 - 20 - 41 - 50.1 - 52.2 - 60.1 - 61 - 69 - 79.1 - 98 - 100.1 - 100.2 - 105

Cynanchum acutum subsp. *sibiricum* (Willd.) Rech.f.

Apocynaceae

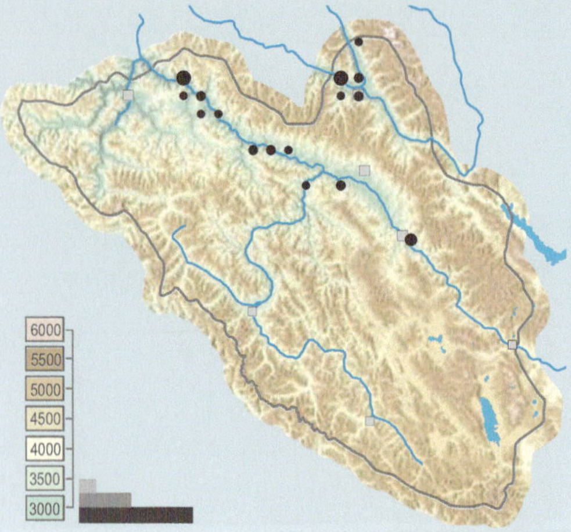

Syn.: *C. sibiricum* Willd.

Description: Perennial climbers, up to 3 m long, stems woody, branched, villous, leaves cordate to hastate, flowers white outside, white to purple inside, 6–12 mm across, in condensed racemes.

Distribution: Russia, SW Asia, Kazakhstan, Turkmenistan, Afghanistan, Pakistan, India, China (Gansu, Hebei, Nei Mongol, Ningxia, Xinjiang, Xizang), Mongolia.

Habitat: Sandy slopes, steppes, along rivers.

Stem diameter 10 mm. Bark/xylem proportion of adult idividuals 1:2. Mean ring width 0.4 mm. Maximal 9 rings.

Xylem

Transverse section

Ring-porous. Vessels solitary. Ring boundaries marked by large earlywood vessels (diameter 100–150 μm) and smaller latewood vessels (diameter 40–60 μm) and marginal parenchyma. Fibers thick-walled. Parenchyma paratracheal and marginal. Rays unlignified. Many crystal druses in large rays.

Tangential section

Ray width 5–8 cells, often with two distinct sizes, unlignified.

Radial section

Perforation plates simple. Intervessel pits round in alternating position. Vessel length 70–100 μm. Fibers with small round pits. Ray cells mainly square, occasionally prostrate. Crystal druses very frequent.

Bark

Phloem with radial and tangential layers of compressed sieve tubes. Large rays dilated. With laticifers and crystal druses. Cortex with irregular groups of sclereids. Phellem cells thin-walled, rectangular.

Pith

Square, with round edges. With many crystal druses and laticifers.

Analyzed: Stem of 4 individuals.

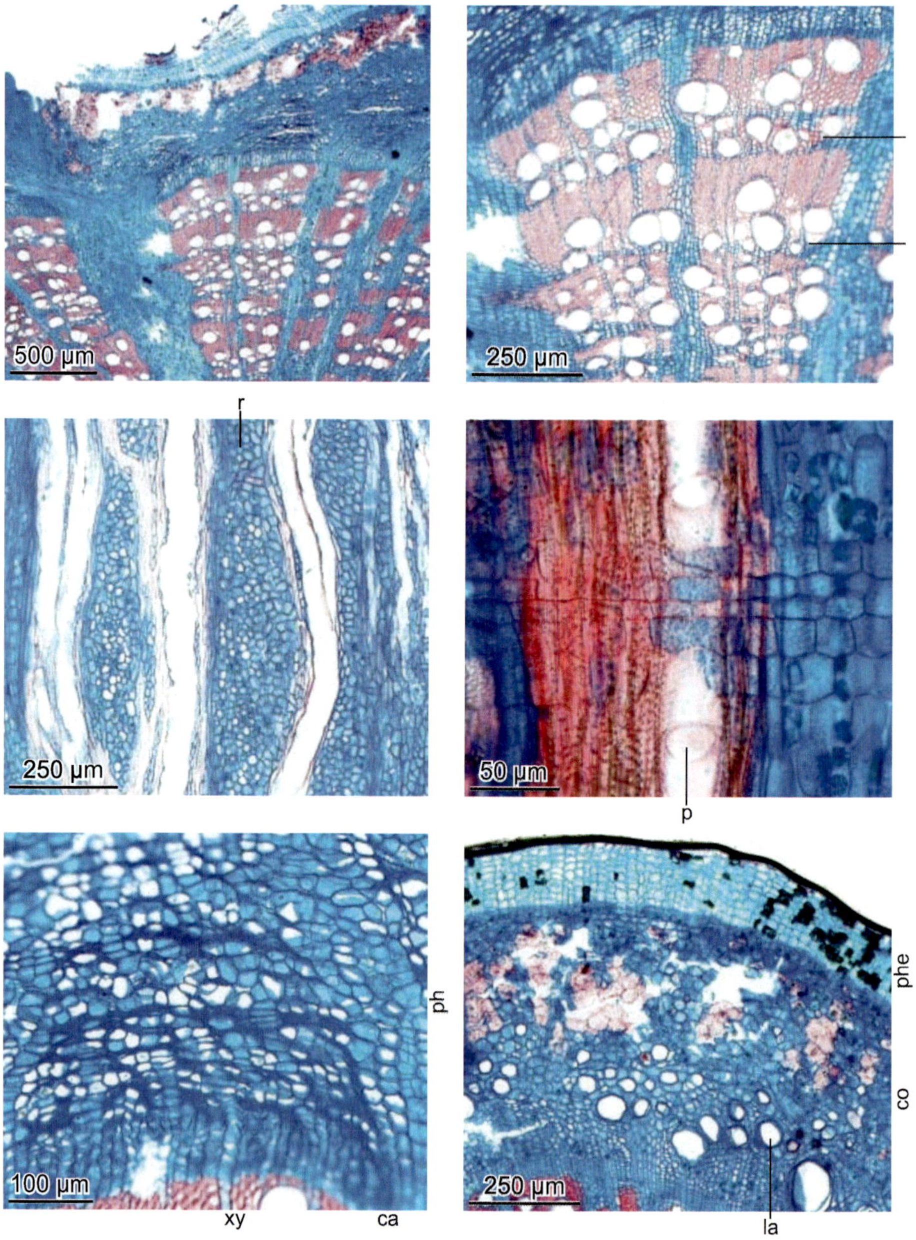

Features according to IAWA

Xylem 1 - 3 - 9 - 13 - 22 - 42 - 50.1 - 52.2 - 61 - 69 - 79 - 89.1 - 98 - 100.2 - 103 - 105 - 108

Ajania fruticulosa (Ledeb.) Poljak.

Asteraceae

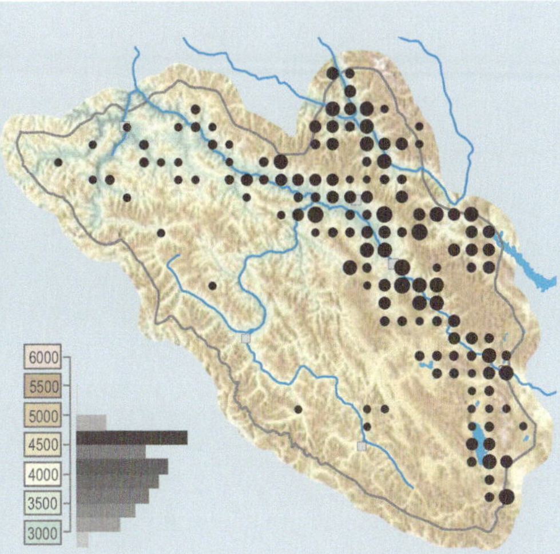

Syn.: *Tanacetum fruticulosum* Ledeb., *T. gracile* Hook.f. & Thomson

Description: Subshrubs, 10–45 cm tall, forming dense groups of shoots, grayish pubescent, leaves pinnatisect, ultimate segments linear, capitula yellow, 4–5 mm across, in terminal flat-top panicle.

Distribution: Russia (Siberia), Iran, Kazakhstan, Turkmenistan, Afghanistan, Pakistan, India, China (Gansu, Jiangsu, Nei Mongol, Qinghai, Shaanxi, Xinjiang, Xizang), Mongolia.

Habitat: Steppes and semideserts, open stony slopes.

Stem diameter 4–5 mm. Bark/xylem proportion 1:0.3. Ring width 0.2–0.7 mm. Maximal 4 annual rings.

Xylem

Transverse section

Diffuse-porous. Vessels in short radial multiples or in groups. Ring boundary poorly marked by few layers of flat fibers. Diameter of vessels 30–40 µm. Fibers thick-walled. Parenchyma paratracheal. Many layers of intra-xylary phellem.

Tangential section

Ray width 2–4 cells. Rays often indistinct.

Radial section

Perforation plates simple. Intervessel pits round in alternating position. Vessel length 120–150 µm.

Bark

Phloem with groups of thick-walled sclereids. Groups of sieve tubes absent. With oil ducts. Phellem consists of many layers of square cells.

Pith

No slide available.

Analyzed: Root collar of polar root of 4 individuals.

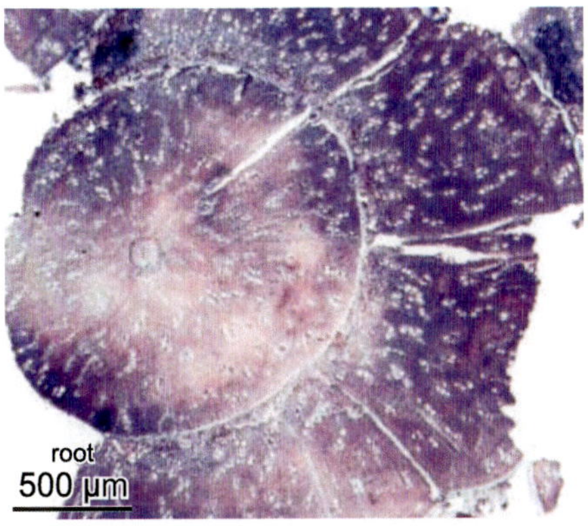

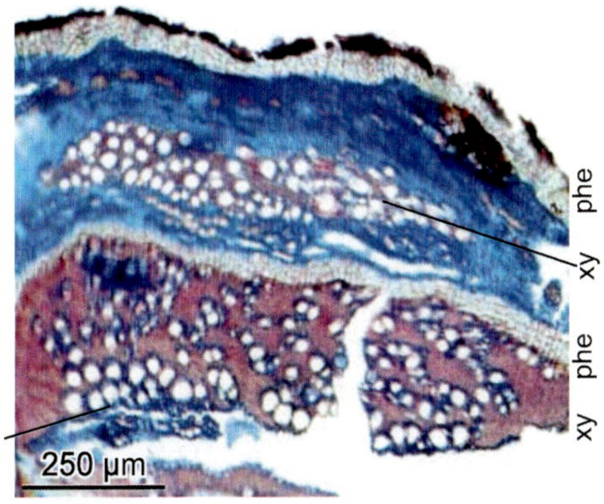

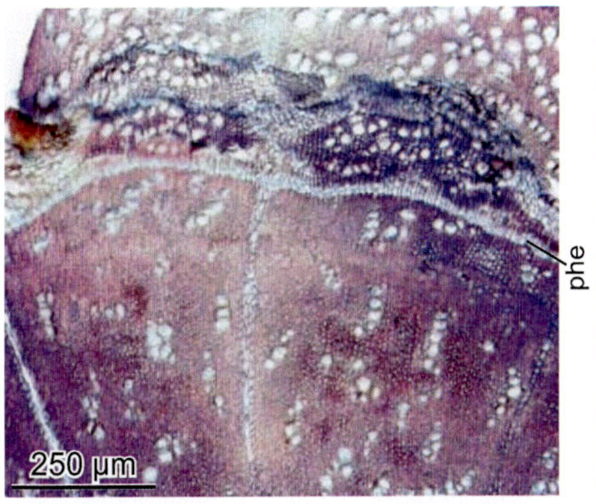

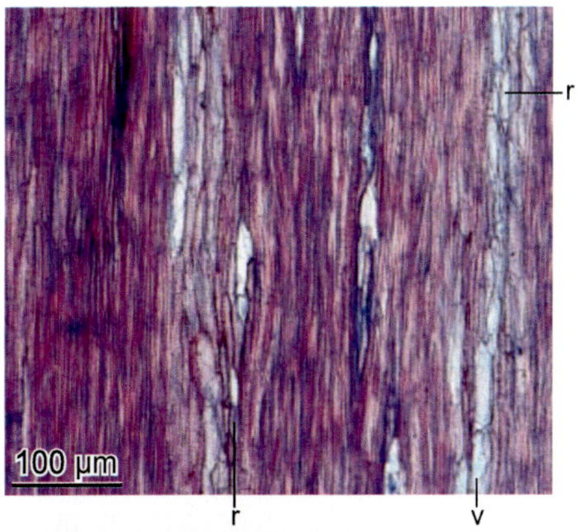

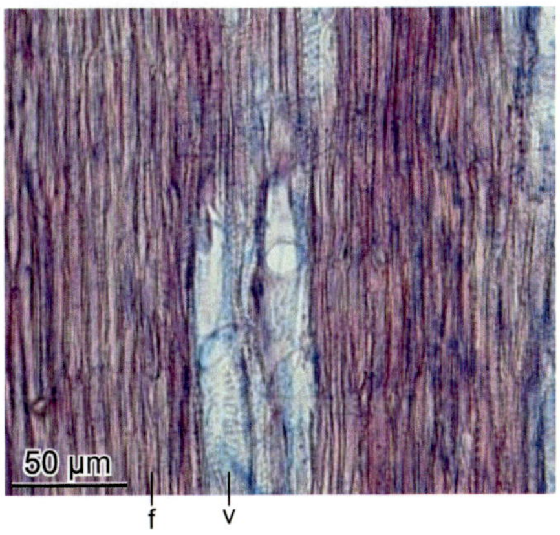

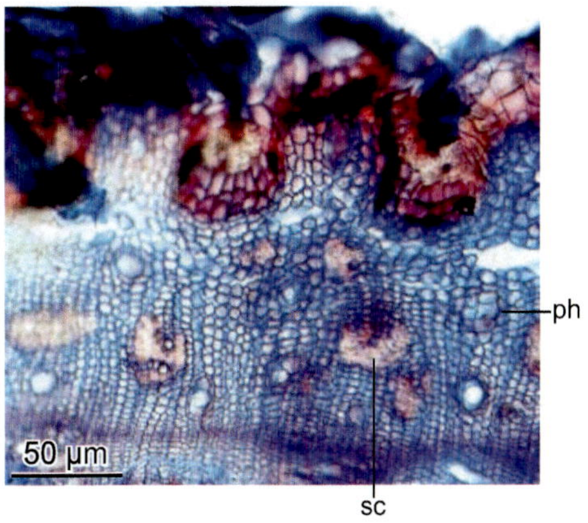

Features according to IAWA

Xylem 1 - 5 - 9.1 - 11 - 13 - 20 - 40.2 - 50.2 - 52.3 - 61 - 69 - 79 - 97 - 98 - 105 - 135.1 - 135.2

Ajania tibetica (Hook.f. & Thomson) Tzvelev

Asteraceae

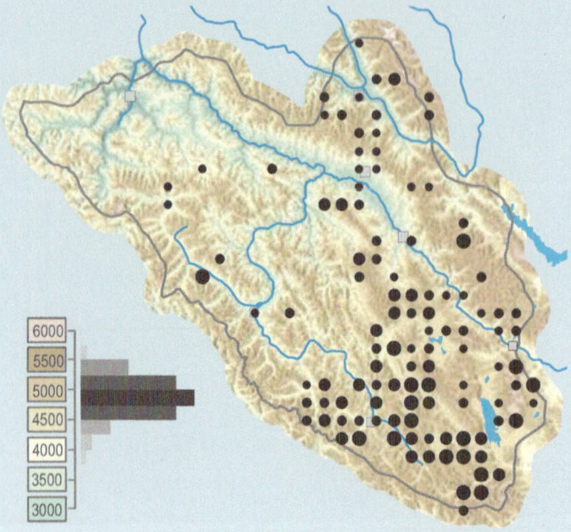

Syn.: *Tanacetum tibeticum* Hook.f. & Thomson ex C.B.Clarke

Description: Subshrubs, 4–20 cm tall, cespitose, grayish tomentose, leaves 2-pinnatisect, capitula yellow, ca. 1 cm across, solitary or few in compound corymbs.

Distribution: Kazakhstan, Pakistan, NW India, China (Sichuan, Xizang).

Habitat: Stable loam and gravel slopes, moraines, open turf, dry river beds.

Stem diameter 6–14 mm. Bark/xylem proportion 1:1. Ring width 0.25–0.6 mm. Maximal 5 annual rings.

Xylem

Transverse section

Semi-ring-porous. Vessels solitary and in small groups. Ring boundary marked by large earlywood (50–60 µm) and small latewood vessel diameters (20 µm). Thick-walled fibers occur in small groups. Parenchyma pervasive. With intra-xylary phellem.

Tangential section

Rays absent.

Radial section

Perforation plates simple. Intervessel pits round in alternating position. Vessel length 80–100 µm.

Bark

Sclereids in irregular groups. Groups of sieve tubes absent. Cortex consists of round parenchyma cells (100–130 µm) and small intercellulars. With oil ducts. Phellem consists of many layers of square cells.

Pith

Round. With irregular intercellulars.

Analyzed: Root collar of polar root of 3 individuals.

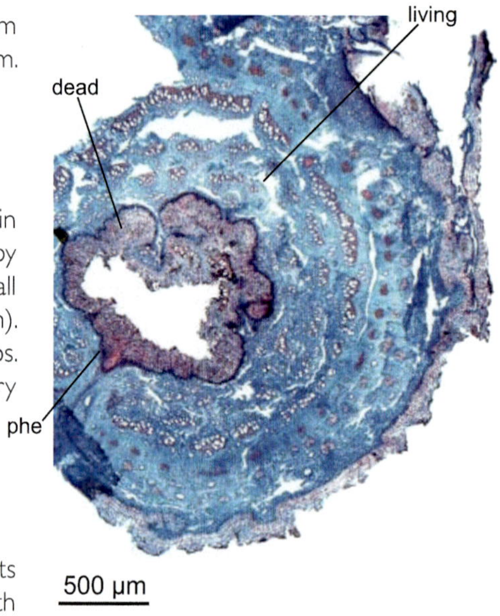

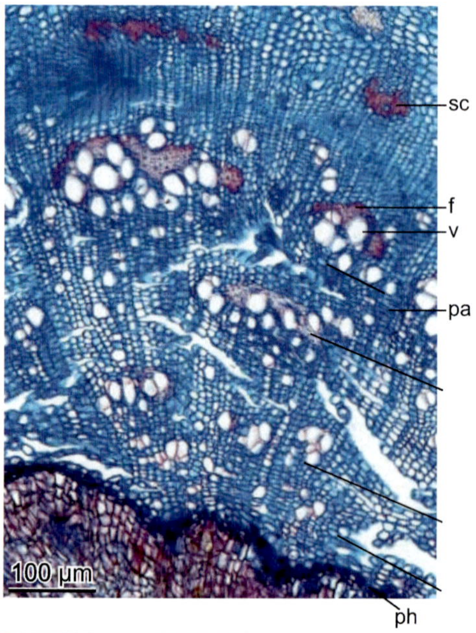

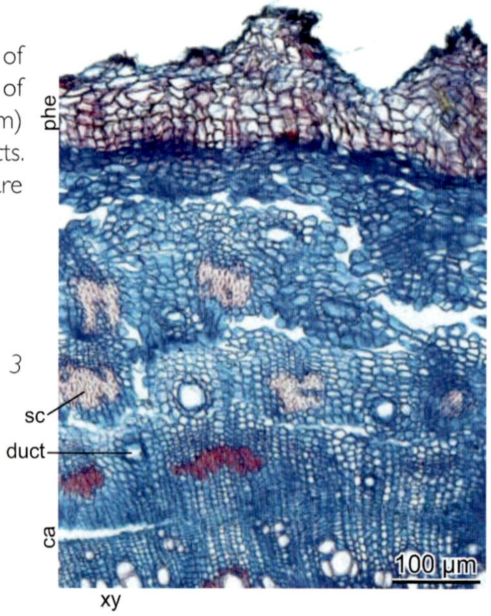

Features according to IAWA

Xylem 1 - 4 - 9 - 11 - 13 - 20 - 40.2 - 50.2 - 52.2 - 61 - 69 - 79.1 - 117 - 135.2

Anaphalis nubigena DC.

Asteraceae

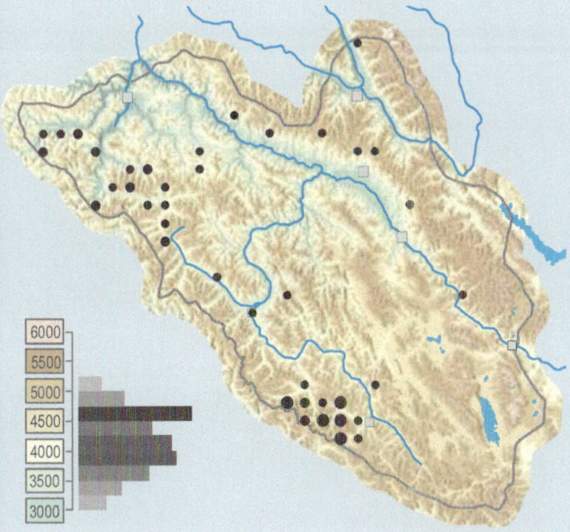

Syn.: *A. nubigena* var. *monocephala* (DC.) C.B.Clarke

Description: Perennial herbs, 5–20 cm tall, white tomentose, leaves elliptic to oblanceolate, capitula 1–1.5 cm across, ray florets white, disc florets yellow, in compact corymbs.

Distribution: Pakistan, India, Nepal, Bhutan, China.

Habitat: Stabilized mesic and dry slopes.

Stem diameter up to 4 mm. Bark/xylem proportion 1:4. Ring width 0.07–0.12 mm. Maximal 26 annual rings.

Xylem

Transverse section

Diffuse-porous. Rings marked by a band of marginal thick-walled fibers. Vessels solitary. Earlywood vessel diameter 15–25 µm, latewood with very small vessels (diameter 10–15 µm). Central vessels with dark-stained substances. Fibers thin- to thick-walled. Parenchyma and crystals absent.

Tangential section

Rays rare, 1–3-seriate, often confluent.

Radial section

Perforation plates simple. Intervessel pits round in alternating position. Vessel length 60–80 µm.

Bark

Phloem with small groups of sieve elements. Cortex with tangential groups of fibers. Epidermis cells round, lignified. Phellem layered.

Pith

Round to polygonal, with angular, lignified parenchyma cells.

Analyzed: Stem of 5 individuals.

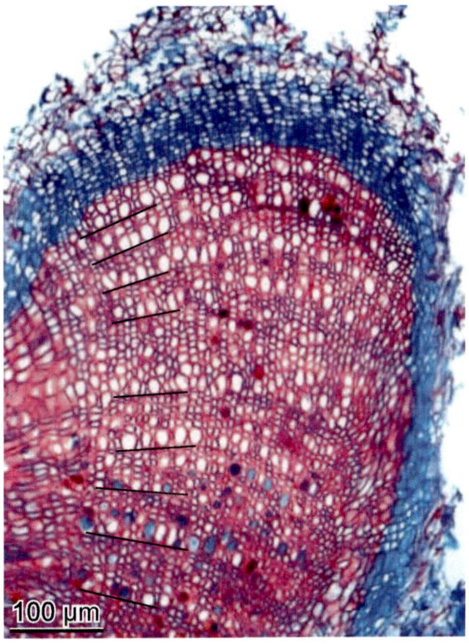

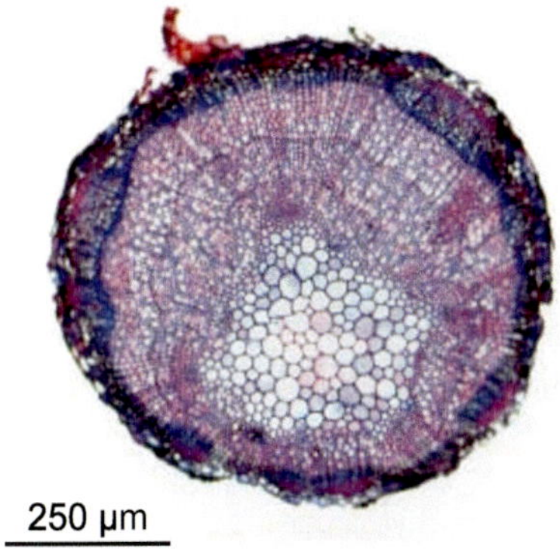

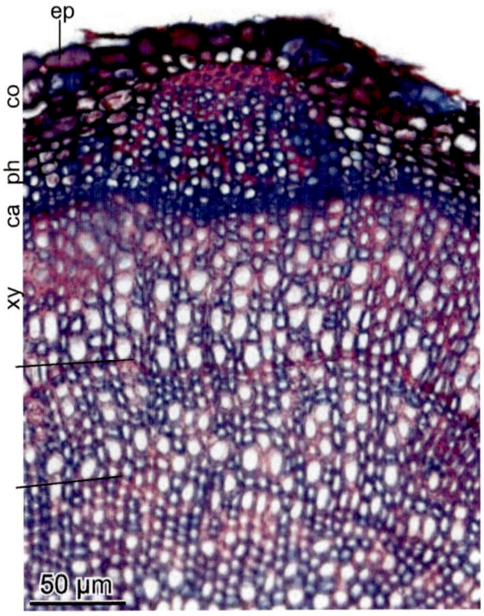

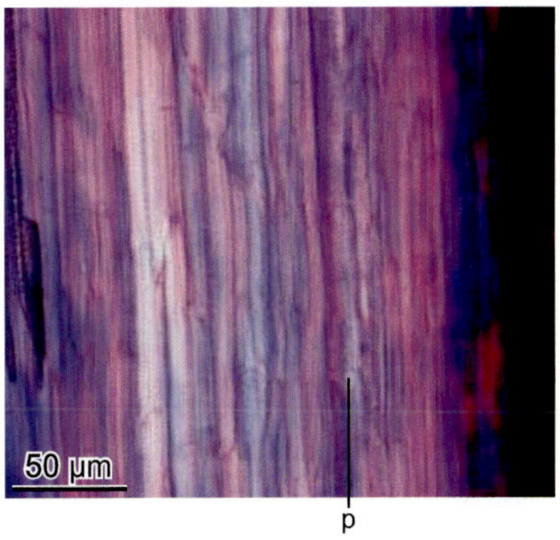

Features according to IAWA

Xylem 1 - 5 - 9 - 12 - 13 - 22 - 40.1 - 50.2 - 52.2 - 58 - 61 - 70 - 75 - 100.1 - 105

Anaphalis virgata Thomson

Asteraceae

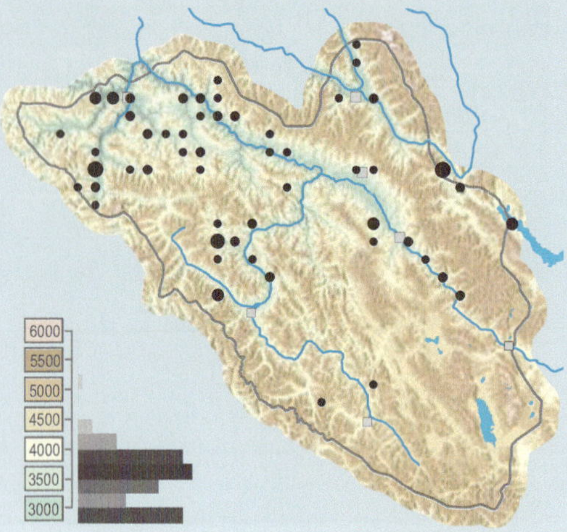

Syn.: *A. stoliczkae* C.B.Clarke, *A. aurora* Rech.f. & Edelb., *A. candollei* Georgiadou

Description: Perennial herbs, 20–45 cm tall, white tomentose, leaves linear-lanceolate, capitula 2–6 mm across, ray florets white, disc florets yellow, in compact corymbs.

Distribution: C Asia, SW Asia (Iran), Tajikistan, Afghanistan, Pakistan, India, Nepal, China (Xinjiang, Xizang), Bhutan.

Habitat: Stabilized slopes.

Stem diameter 12 mm. Bark/xylem proportion 1:6. Ring width 0.15–0.4 mm. Maximal 26 annual rings.

Xylem

Transverse section

Semi-ring-porous. Rings indicated by larger, thick-walled earlywood vessels (diameter 40–60 µm) and smaller latewood vessels (diameter 20–30 µm) and one layer of terminal parenchyma. Fibers thin- to thick-walled. Parenchyma paratracheal and marginal. Crystals absent. Center with dark-staining substances.

Tangential section

Rays absent, some individuals with aggregate rays.

Radial section

Perforation plates simple. Intervessel pits round in alternating position. Vessel length 100–150 µm.

Bark

Phloem layered with alternating continuous bands of parenchyma and fibers. Sieve tubes not differentiated from parenchyma cells. Phellem cells irregularly formed, thin-walled.

Pith

No slide available.

Analyzed: Root collars of 5 individuals.

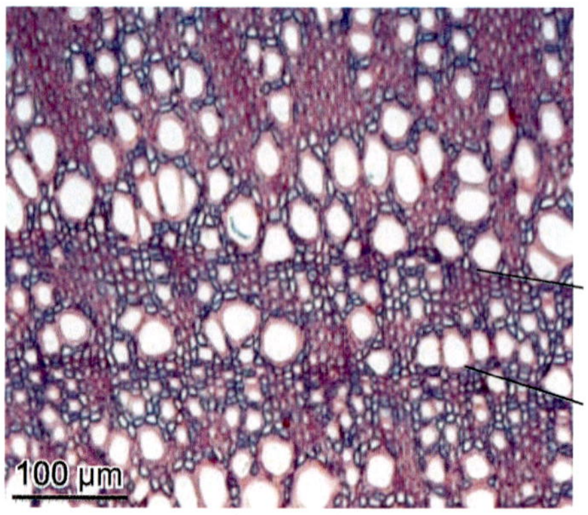

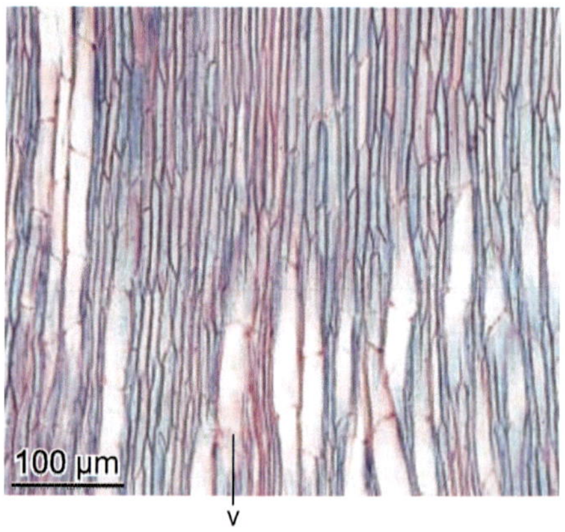

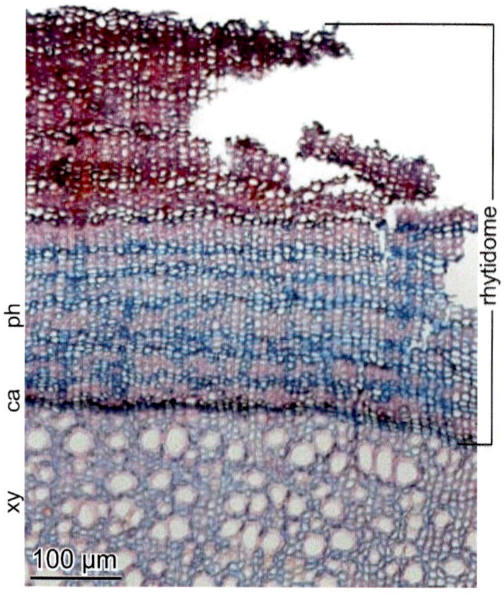

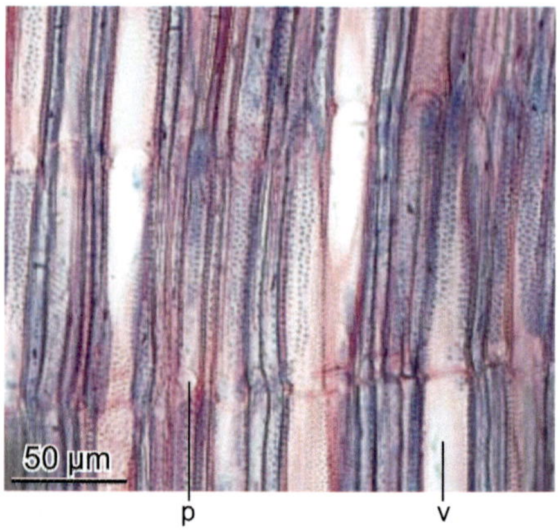

Features according to IAWA

Xylem 1 - 4 - 9 - 13 - 22 - 39.1 - 40.2 - 41 - 50.2 - 52.2 - 52.3 - 58 - 61 - 70 - 79 - 89 - 101 - 117

Artemisia brevifolia Wall. ex DC.

Asteraceae

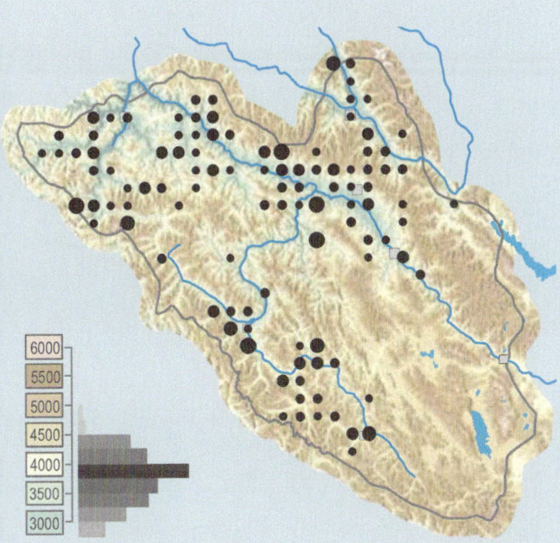

Syn.: *Seriphidium thomsonianum* (C.B.Clarke) Y.Ling & Y.R.Ling

Description: Subshrubs, 20–40 cm tall, cespitose, leaves pinnatisect, capitula ca. 2 mm across, in terminal panicles.

Distribution: NE Afghanistan, N Pakistan, NW India (Jammu and Kashmir, Himachal Pradesh), China (Xizang).

Habitat: Steppes on stabilized slopes.

Stem diameter up to 10 mm. Bark/xylem proportion 1:5. Ring width 0.3–0.6 mm. Maximal 14 annual rings.

Xylem

Transverse section

Semi-ring-porous. Rings marked by bands of marginal parenchyma containing small vessels. Ring variability often indicated by larger (70 μm) or smaller (40 μm) earlywood vessel diameters and a latewood zone with very small vessel diameters (10–20 μm). Fibers thick-walled. Parenchyma paratracheal. With intra-xylary periderm.

Tangential section

Rays width 4–6 cells, with sheath cells. Rays in the latewood disappear in polarized light.

Radial section

Perforation plates simple. Intervessel pits round in alternating position. Vessel length 80–120 μm.

Bark

Phloem with tangentially arranged groups of thick-walled fiber. Oil ducts not observed. Phelloderm layered. Phellem consists of few rows of large, thin-walled cells.

Pith

No slide available.

Analyzed: Stem of 2 individuals.

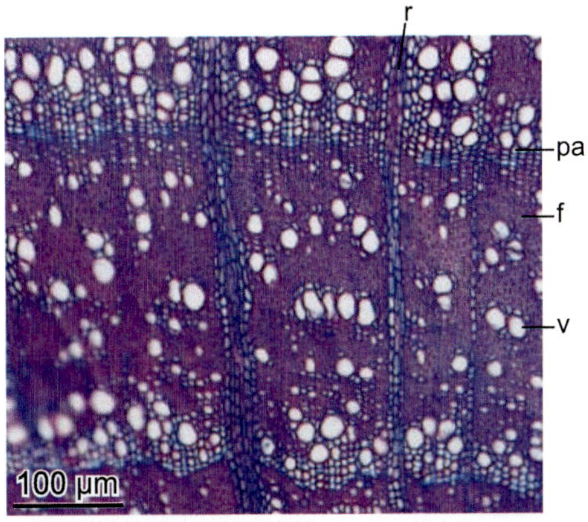

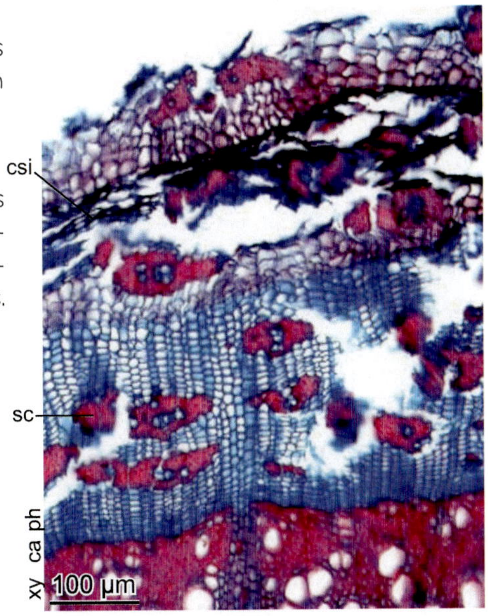

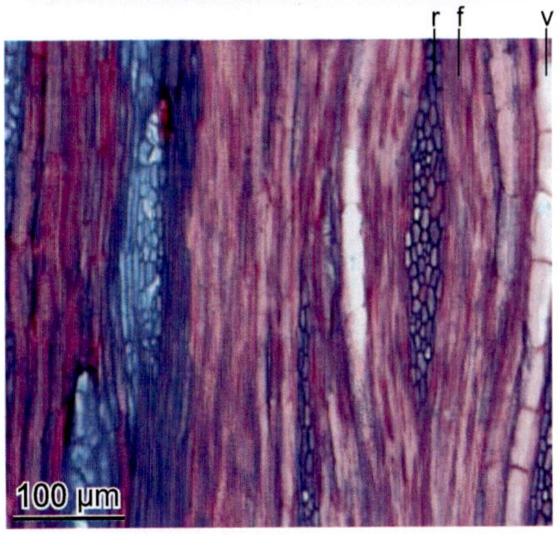

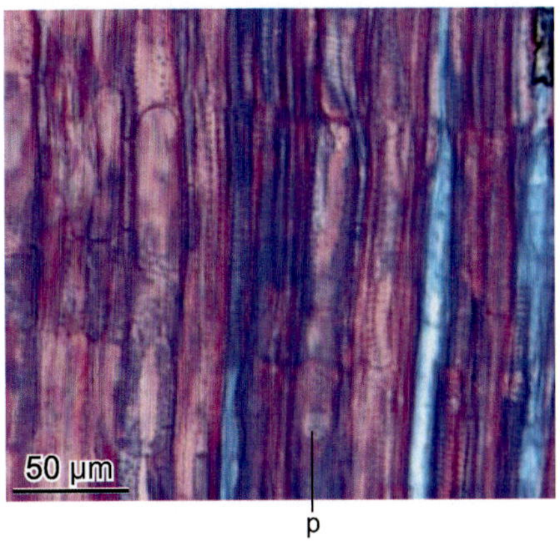

Features according to IAWA

Xylem 1 - 4 - 9 - 13 - 22 - 40.2 - 41 - 50.1 - 52.2 - 52.3 - 61 - 69 - 79 - 98 - 100.2 - 105 - 110 - 135.2

Artemisia demissa Krasch.

Asteraceae

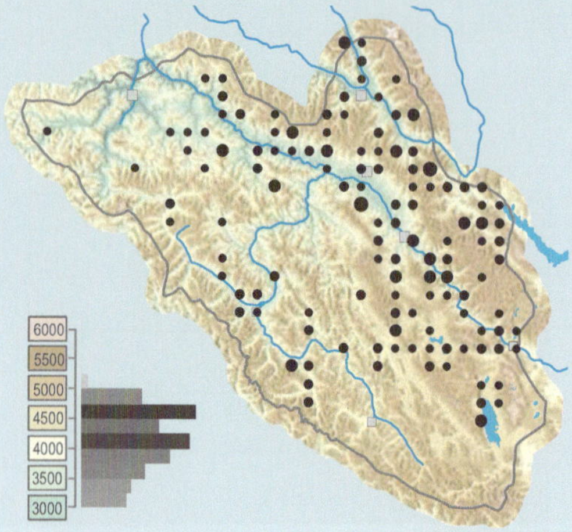

Syn.: *A. stricta* Edgew. 1846, non B.Heyne ex DC. 1836

Description: Annuals or biennials, 5–20 cm tall, usually branched from the base, lower branches ascending, leaves oblong to ovate, pinnatisect, capitula 1.5–2 mm across, in narrow spike-like panicles.

Distribution: Tajikistan, Afghanistan, India (Jammu and Kashmir), China (Gansu, Nei Mongol, Qinghai, Sichuan, Xinjiang, Xizang).

Habitat: Sandy steppes and disturbed habitats, gravel river banks, resting places of domestic animals.

Stem diameter up to 3 mm. Bark/xylem proportion 1:3. Ring width 0.3–0.6 mm. Maximal 2 annual rings.

Xylem

Transverse section

With only one ring, or semi-ring-porous with 2 rings. Earlywood vessel diameter 15–25 µm. Vessels in long radial multiples. Fibers thin- to thick-walled. Parenchyma paratracheal.

Tangential section

Ray width 1–2 cells.

Radial section

Perforation plates simple. Intervessel pits round in alternating position. Vessel length 100–180 µm.

Bark

Phloem with tangentially arranged groups of thick-walled fibers. Phellem layered, consisting of few rows of large, thin-walled cells. Cortex with a few oil ducts.

Pith

Star-shaped.

Analyzed: Stem of 4 individuals.

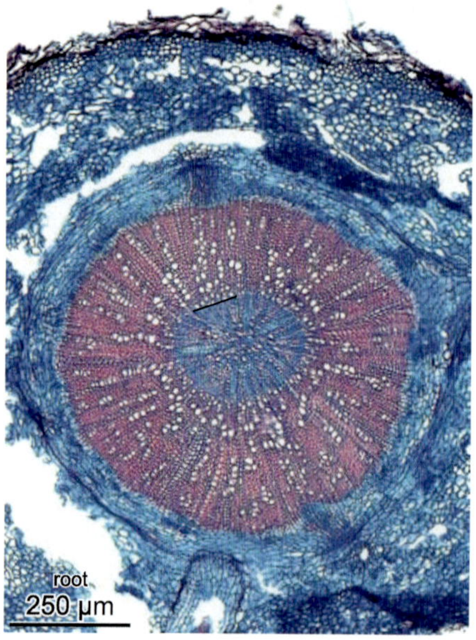

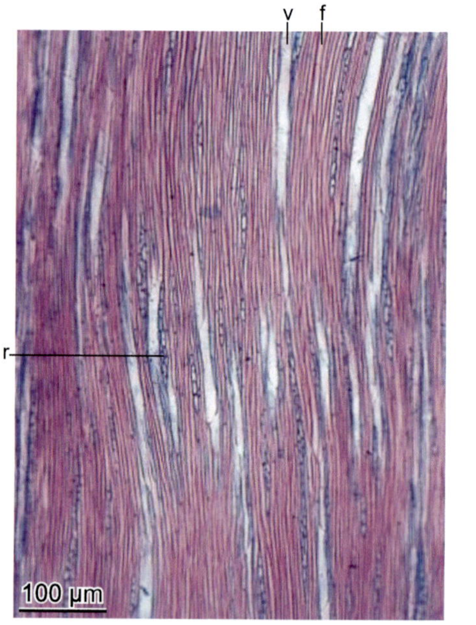

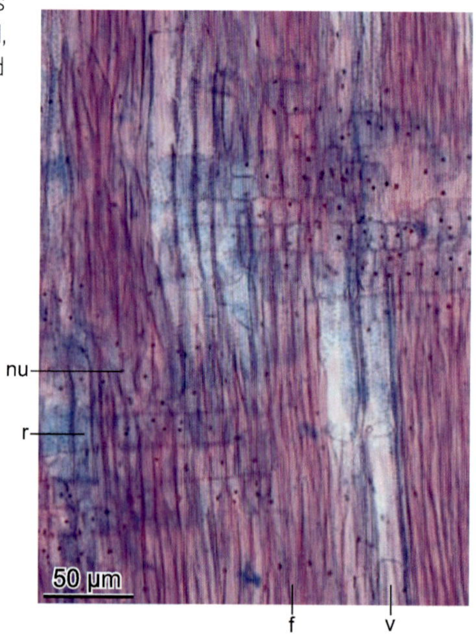

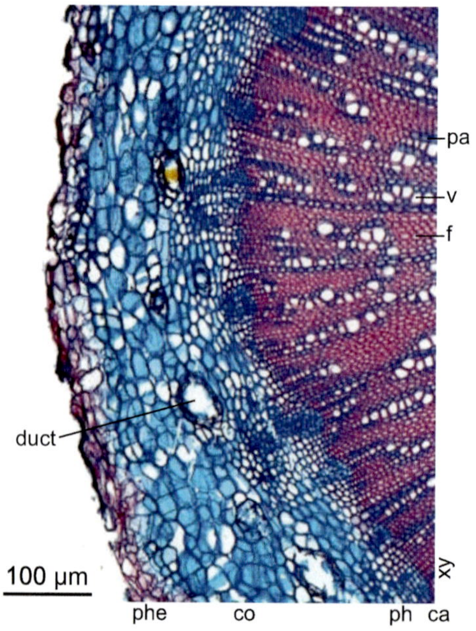

Features according to IAWA

Xylem 1 - 2.1 - 4 - 10 - 13 - 22 - 40.1 - 40.2 - 50.2 - 52.3 - 61 - 70 - 79 - 97 - 100.2 - 105

Artemisia desertorum Spreng.

Asteraceae

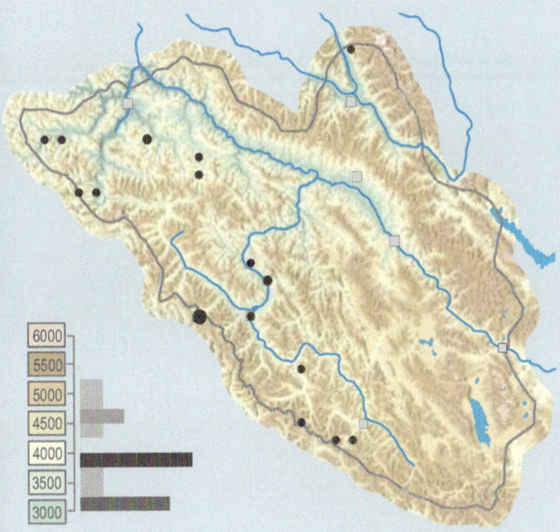

Syn.: *A. foetida* Jacquem. ex DC., *A. jacquemontiana* Bess.

Description: Subshrubs, 10–70 cm tall, glabrous, stems several, leaves 2–3-pinnatisect, capitula 1.5–3 mm across, subsessile, nodding, in branched panicles.

Distribution: E Russia, N Pakistan, India (Jammu and Kashmir), Nepal, China (Gansu, Guizhou, Hebei, Heilongjiang, Jilin, Liaoning, Nei Mongol, Ningxia, Qinghai, Shaanxi, Shanxi, Sichuan, Xinjiang, Xizang, Yunnan), Mongolia, Korea, Japan.

Habitat: Stony steppes, dry stony river beds.

Stem diameter up to 15 mm. Bark/xylem proportion 1:2. Ring width 0.4–0.7 mm. More than 6 rings. Center rotten.

Xylem

Transverse section

Rings present. Vessel diameter 50–80 µm. Vessels in groups. Fibers thin- to thick-walled, arranged in small groups within vessel groups. Parenchyma pervasive. With intra-xylary periderm bands. Tissue dead inside the bands. Rays indistinct. Vascular bundles remaining.

Tangential section

Rays >10-seriate, confluent

Radial section

Perforation plates simple. Intervessel pits round and horizontally enlarged in alternating position. Vessel length 100–200 µm. Crystals absent.

Bark

Sieve tubes and parenchyma indistinguishable. With groups of thick-walled fibers. With oil ducts. Phellem layered, consisting of a few rows of rectangular, thin-walled cells.

Pith

Polygonal, filled with unlignified parenchyma cells.

Analyzed: Stem of 4 individuals.

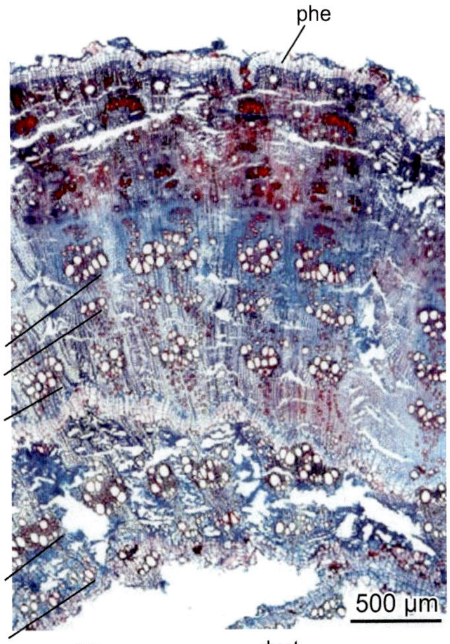

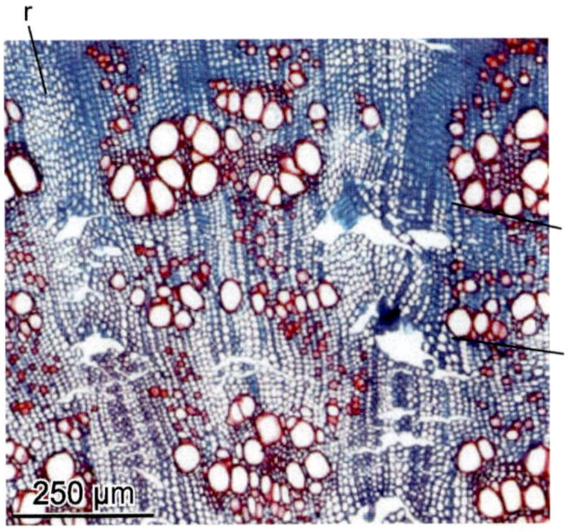

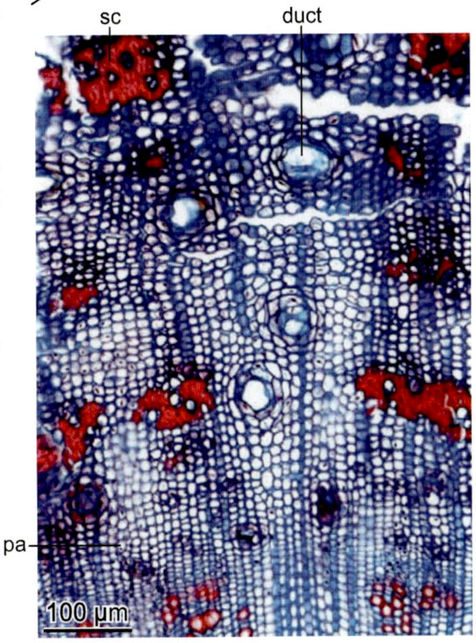

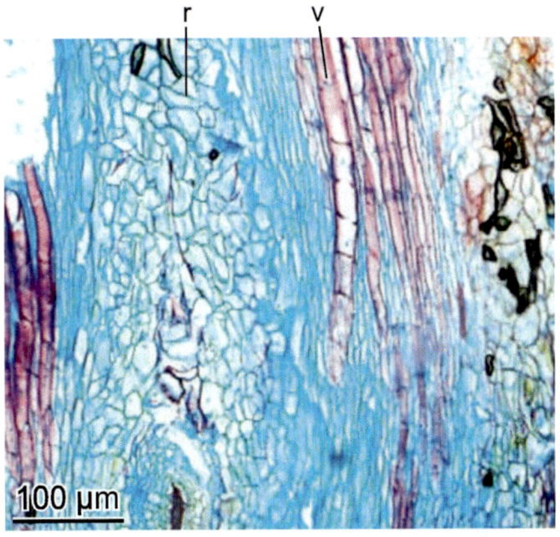

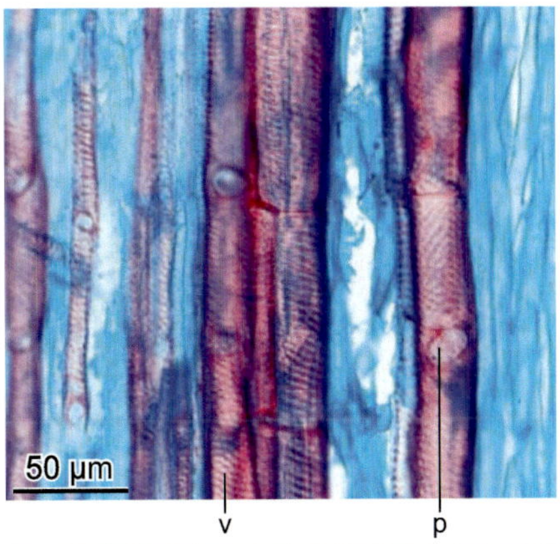

Features according to IAWA

Xylem 1 - 4 - 11 - 13 - 22 - 40.2 - 42 - 50.1 - 52.3 - 61 - 70 - 79 - 79.1 - 99 - 99.1 - 100.1 - 100.2 - 135.2

Artemisia hedinii Ostenf.

Asteraceae

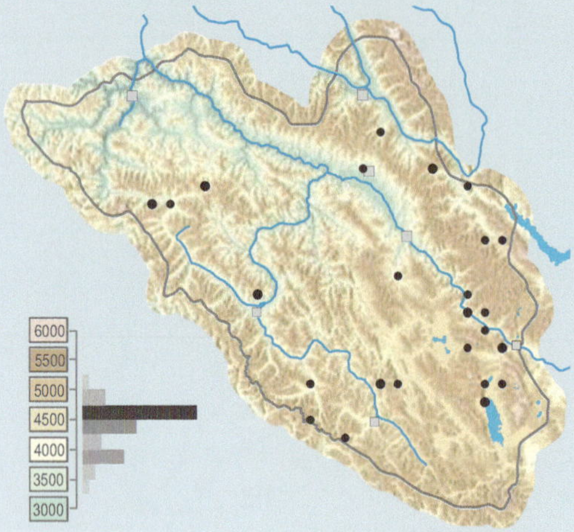

Description: Annuals, 15–60 cm tall, stems purplish, leaves 2-pinnatisect, capitula 3–4 mm across, in dense narrow panicles.

Distribution: Tajikistan, N Pakistan, N India, Nepal, China (Gansu, Nei Mongol, Qinghai, Sichuan, Xinjiang, Xizang, Yunnan).

Habitat: Rocky slopes, along streams, marshlands.

Stem diameter up to 9 mm. Bark/xylem proportion 1:3. Ring width 2–3 mm. Maximal 2 annual rings.

Xylem

Transverse section

With only one ring or with growth zones. Vessel diameter 30–50 µm. Vessels in long radial multiples. Vessels thin-walled. Fibers thin-walled. Parenchyma paratracheal.

Tangential section

Ray width 1-3 cells.

Radial section

Perforation plates simple. Intervessel pits round in alternating position. Vessel length 80–120 µm.

Bark

Phloem with radially arranged groups of sieve tubes. Cortex with a few oil ducts. Phelloderm layered. Phellem consists of few rows of large, thin-walled cells.

Pith

Star-shaped with round, unlignified parenchyma cells.

Analyzed: Stem of 2 individuals.

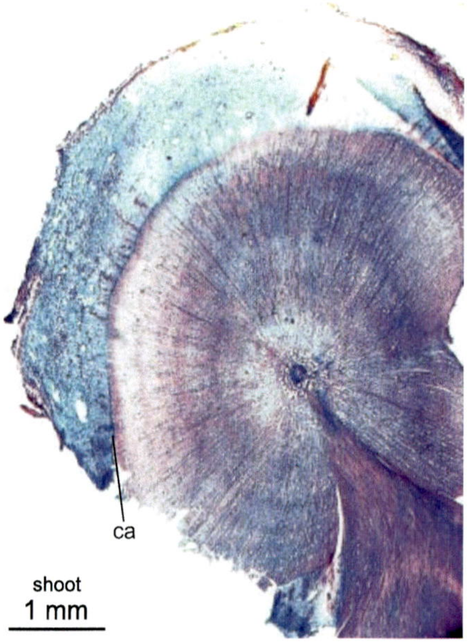

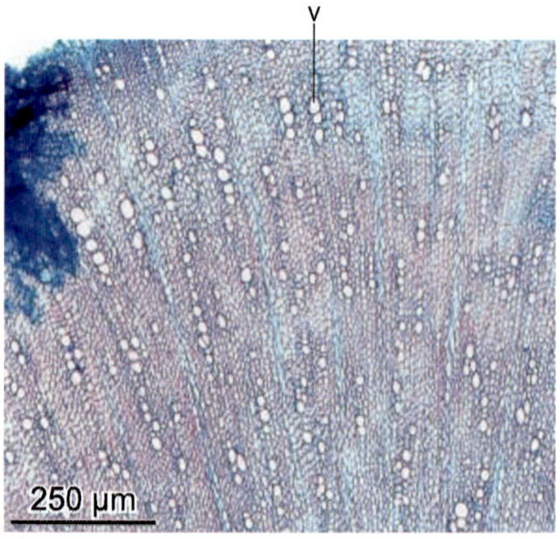

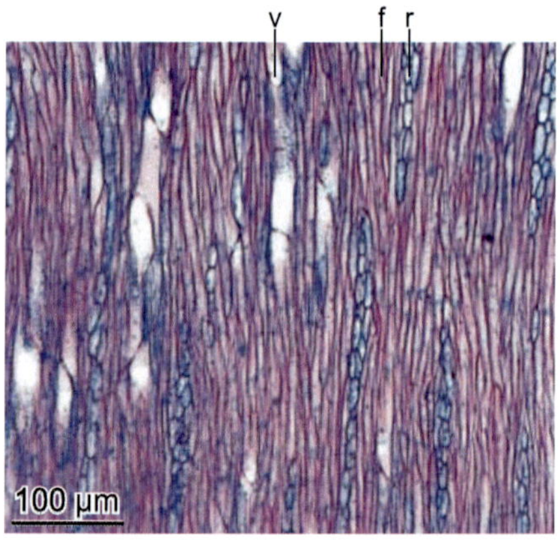

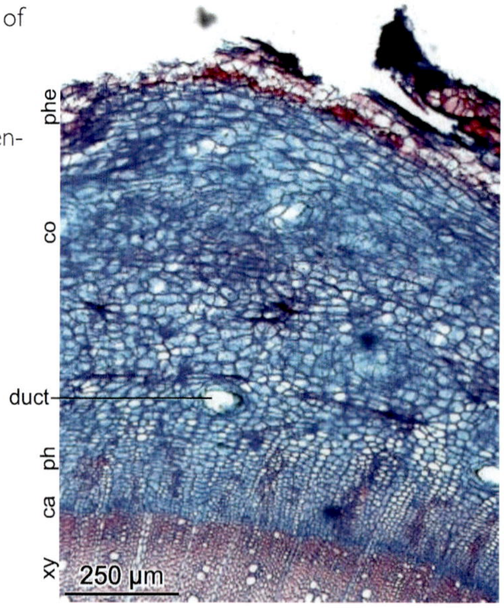

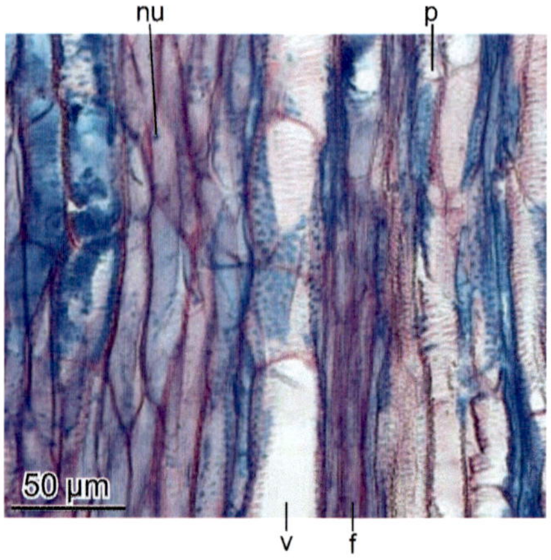

Features according to IAWA

Xylem 2.1 - 10 - 13 - 22 - 40.2 - 50.2 - 52.2 - 52.3 - 61 - 68 - 79 - 97 - 105

Artemisia japonica Thunb.

Asteraceae

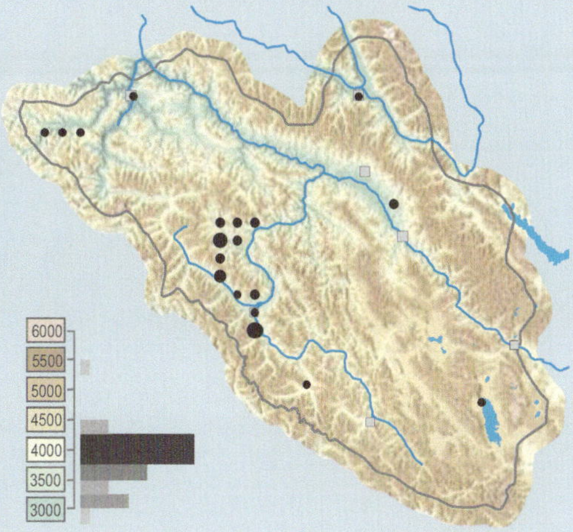

Syn.: *A. parviflora* Buch.-Ham. ex Roxb.

Description: Subshrubs, 50–90 cm tall, aromatic, leaves spatulate, pinnately lobed, hairy, capitula 1.5–2.5 mm across, shortly pedunculate, nodding, in narrow panicles.

Distribution: Afghanistan, Pakistan, N India, Nepal, China, Bhutan, Myanmar, Laos, Thailand, Vietnam, E Russia, Korea, Japan, Philippines.

Habitat: Shrublands, dry mountain slopes.

Stem diameter up to 10 mm. Bark/xylem proportion 1:5. Ring width 0.5–1 mm. With more than 10 rings. Center rotten.

Xylem

Transverse section

Rings distinct. Vessel diameter 80–150 µm. Vessels in groups. Fibers thin- to thick-walled, absent in the latewood. Parenchyma paratracheal, pervasive and marginal (terminal). With intra-xylary periderm-bands. Tissue dead inside the bands. Crystals absent.

Tangential section

Ray width 3–10 cells, often confluent.

Radial section

Perforation plates simple. Intervessel pits round in alternating position. Vessel length 120–150 µm.

Bark

Sieve elements and parenchyma cells cannot be distinguished. Phloem with tangentially arranged, irregular, small groups of thick-walled fibers. With oil ducts. Phellem layered, consisting of irregularly formed, large, thin-walled cells.

Pith

No slide available.

Analyzed: Stem of 2 individuals.

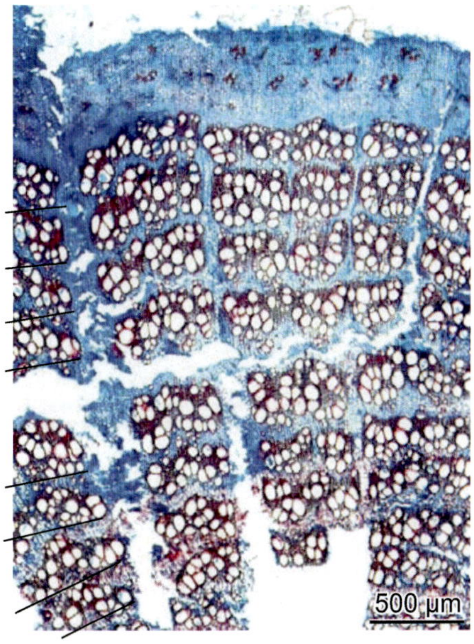

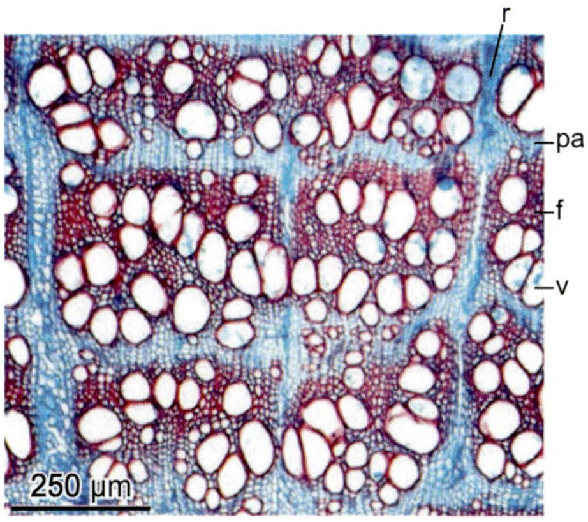

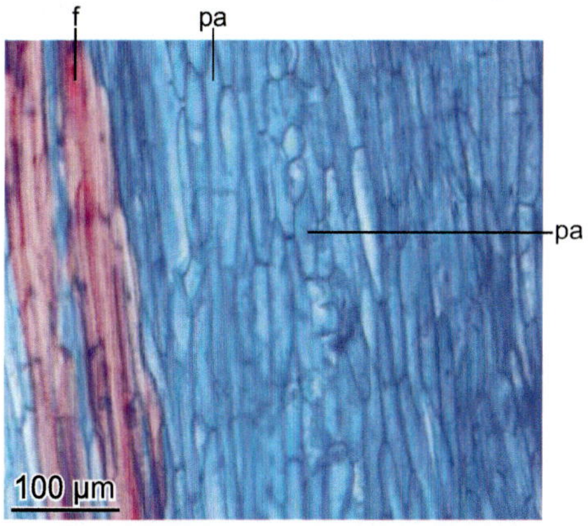

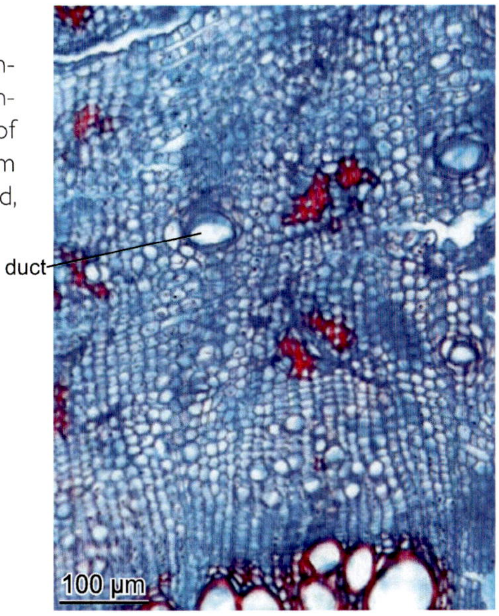

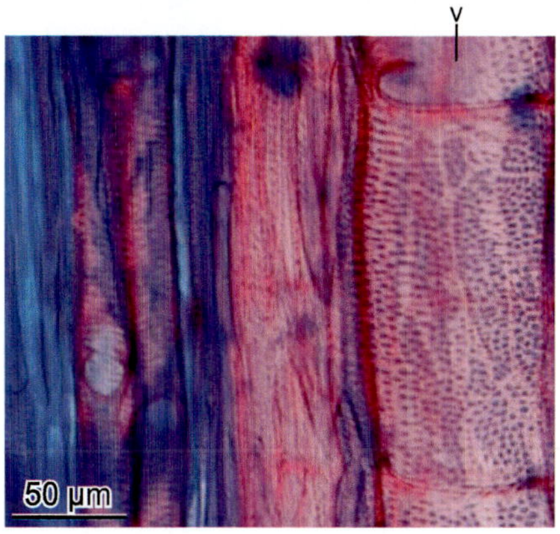

Features according to IAWA

Xylem 1 - 4 - 11 - 13 - 22 - 40.2 - 41 - 42 - 50.2 - 52.3 - 53.1 - 61 - 70 - 79 - 79.1 - 89 - 100.1 - 100.2 - 104 - 135.1 - 135.2

Artemisia macrocephala Jacquem. ex Besser

Asteraceae

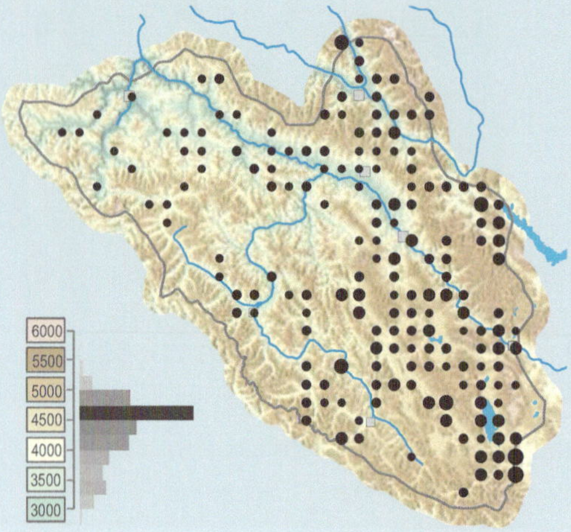

Syn.: *A. griffithiana* Boiss., *A. sieversiana* var. *pygmaea* Krylov

Description: Annuals, stems 10–30 cm tall, usually not branched, gray pubescent, leaves 2-pinnatisect, capitula 6–10 mm across, shortly pedunculate, nodding, in erect lax simple racemes.

Distribution: Russia, Kazakhstan, Kyrgyzstan, Tajikistan, Afghanistan, Pakistan, India, China (Gansu, Ningxia, Qinghai, Xinjiang, Xizang), Mongolia.

Habitat: Disturbed habitats on roads margins, drier gravel river deposits, synanthropic habitats, abandoned fields.

Stem diameter up to 2.5 mm. Bark/xylem proportion 1:3. Mean ring width 0.5 mm. Maximal 2 annual rings.

Xylem

Transverse section

With only one ring or semi-ring-porous with 2 rings. Earlywood vessel diameter 15–25 µm. Vessels solitary or in short radial multiples. Fibers thin- to thick-walled. Parenchyma paratracheal.

Tangential section

Rays uniseriate.

Radial section

Perforation plates simple. Intervessel pits round in alternating position. Vessel length 80–120 µm.

Bark

Phloem with radially arranged groups of sieve tubes. Oil ducts not observed. Phellem consists of one row of thin-walled phellem cells (phelloids).

Pith

No slide available.

Analyzed: Root of 1 individual.

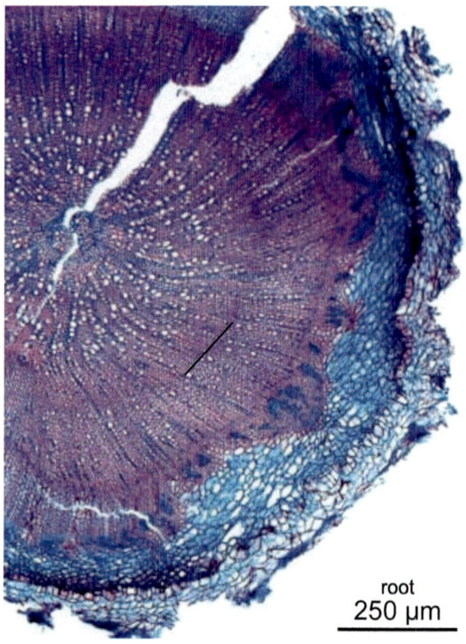

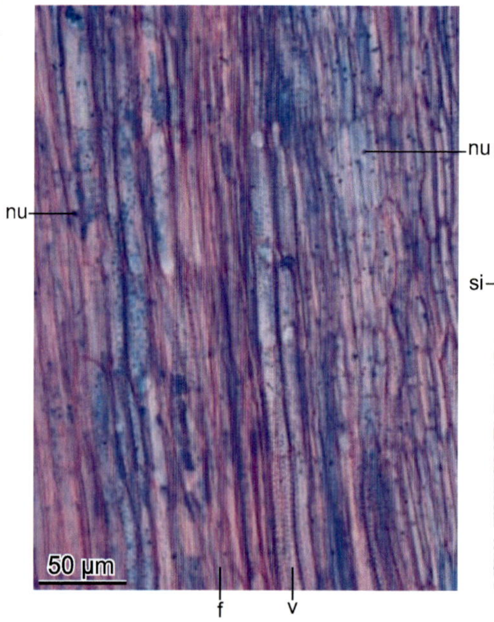

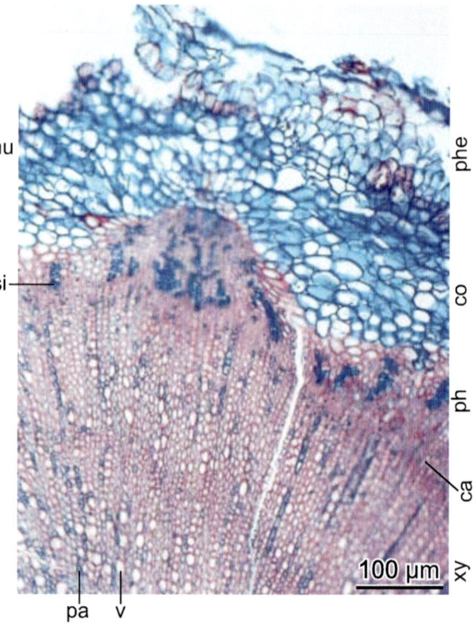

Features according to IAWA

Xylem 1 - 2.1 - 4,9 - 9.1 - 13 - 22 - 40.1 - 40.2 - 50.2 - 52.3 - 61 - 70 - 79 - 96 - 100.2 - 105

Artemisia minor Jacquem. ex Besser

Asteraceae

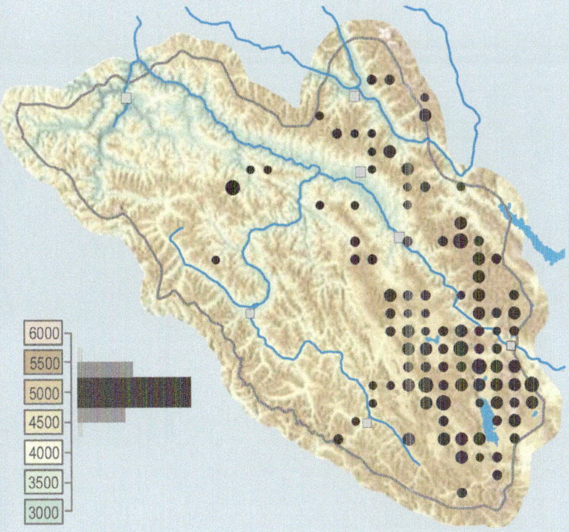

Syn.: *A. sieversiana* var. *tibetica* C.B.Clarke, *A. tibetica* (C.B.Clarke) Hook.f. & Thomson

Description: Subshrubs, 8–12 cm tall, silky villose, densely cespitose, stems numerous, ascending, leaves 2-pinnatisect, capitula 7–8 mm across, subsessile, in simple lax racemes.

Distribution: N Pakistan, NW India (Jammu and Kashmir), W Nepal, China (Xinjiang, Xizang).

Habitat: Stabilized slopes, stony steppes.

Stem diameter up to 4 mm. Bark/xylem proportion 1:2.5. Mean ring width 0.2 mm. Maximal 8 annual rings.

Xylem

Transverse section

Semi-ring-porous. Rings, if present, marked by an interrupted row of earlywood vessels. Ring variability often indicated by larger (30 μm) or smaller (20 μm) vessel diameters and a latewood zone with very small vessel diameters (<10 μm). Vessels thick-walled (2–3 μm), not stained. Fibers thick-walled, arranged in partial tangential bands. Parenchyma pervasive. With intraxylary periderm-bands. Vessels and parenchyma cells contain dark-staining substances in the stem center (heartwood).

Tangential section

Ray width 4–6 cells, confluent with the axial tissue. Rays disappear in polarized light.

Radial section

Perforation plates simple. Intervessel pits round in alternating position. Vessel length 60–120 μm.

Bark

Phloem with indistinctly tangentially arranged groups of thick-walled fiber groups. With oil ducts. Phelloderm layered. Phellem consists of many rows of flat cells.

Pith

Round, filled with large, thin-walled irregularly formed cells.

Analyzed: Stem of 1 individual.

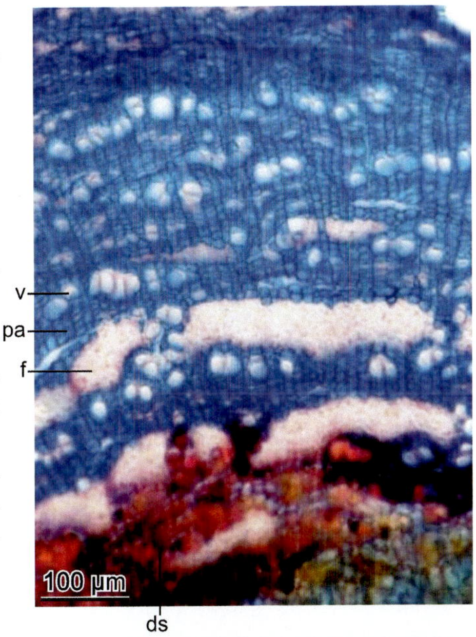

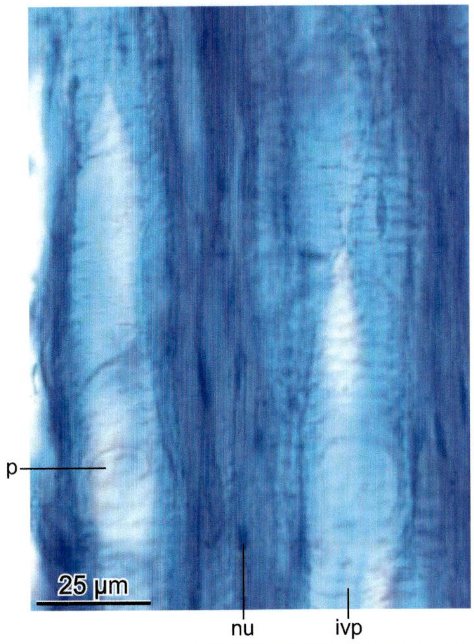

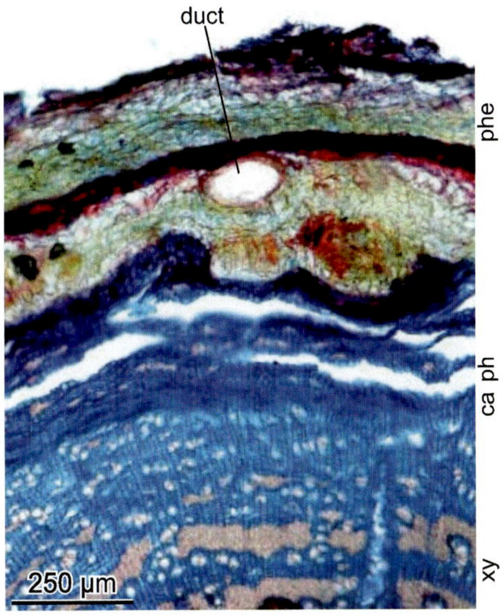

Features according to IAWA

Xylem 1 - 5 - 9 - 13 - 22 - 40.1 - 50.1 - 52.2 - 61 - 69 - 70.1 - 79.1 - 98 - 100.2 - 105 - 135.2

Artemisia moorcroftiana Wall. ex DC.

Asteraceae

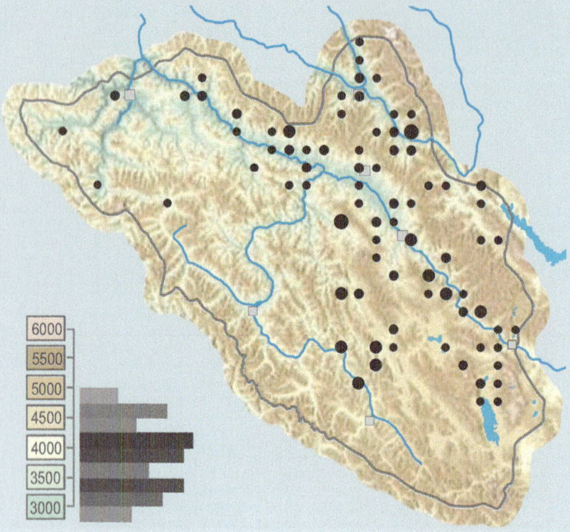

Description: Subshrubs, 50–70 cm tall, sparsely arachnoid hairy, stems several, leaves deeply 2-pinnatipartite, capitula 4–6 mm across, hemispherical, sessile, in interrupted, contracted raceme-like panicles.

Distribution: N Pakistan, NW India, Bhutan, China (Gansu, Ningxia, Qinghai, Sichuan, Xizang, Yunnan).

Habitat: Rocky hills, subalpine steppes and meadows.

Stem diameter up to 7 mm. Bark/xylem proportion 1:5. Ring width 0.2–0.6 mm. Maximal 6 annual rings.

Xylem

Transverse section

Semi-ring-porous. Rings marked by a band of marginal parenchyma. Earlywood vessel diameter 40–50 μm. Latewood zone with very small vessel diameters (<10 μm). Vessels arranged solitary or in small groups. Parenchyma paratracheal and pervasive in the latewood. Vascular bundles remaining.

Tangential section

Ray width 6–>10 cells.

Radial section

Perforation plates simple. Intervessel pits round in alternating position in the fibrous zone and scalariform in the parenchymateous latewood zone. Vessel length 80–120 μm.

Bark

Phloem with irregular groups of thick-walled fibers. Oil ducts in the cortex. Phellem consists of many rows of large, thin-walled cells.

Pith

Round, consisting of unlignified, thin-walled, round cells.

Analyzed: Stem of 4 individuals.

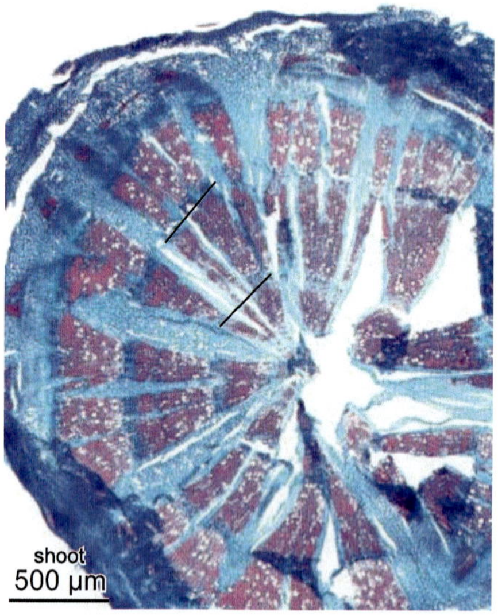

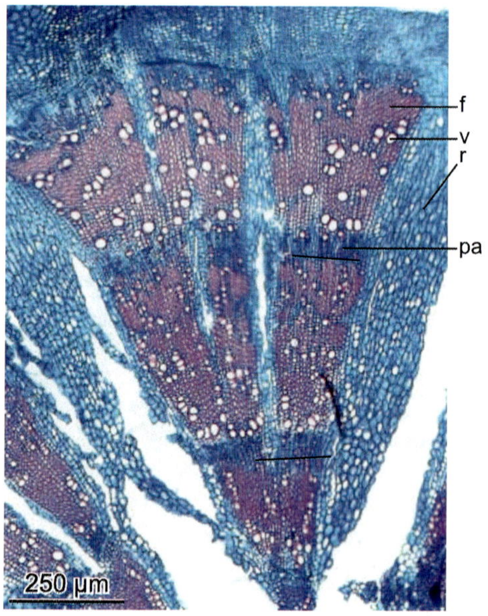

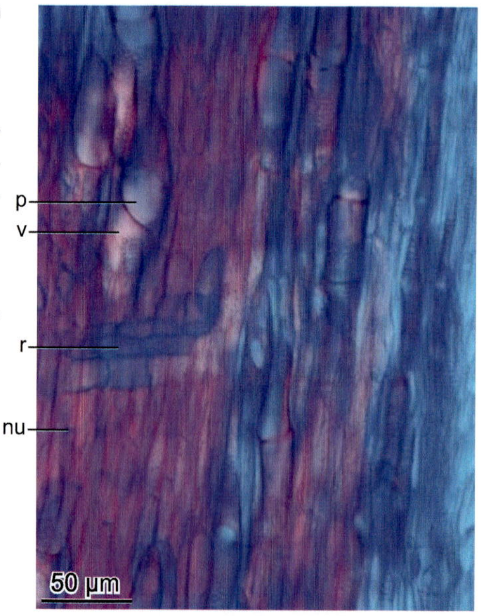

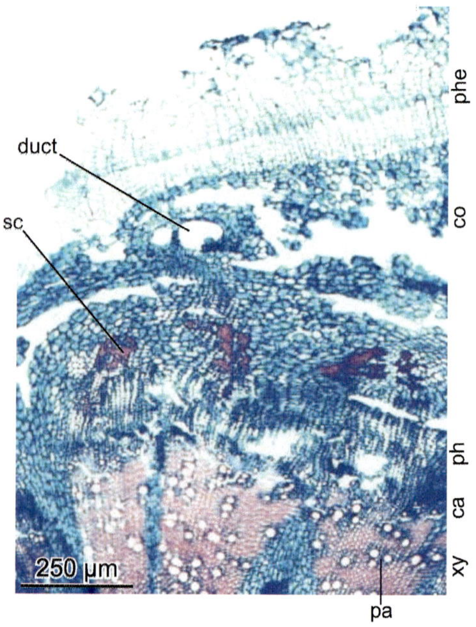

Features according to IAWA

Xylem 1 - 4 - 9 - 11 - 13 - 20 - 22 - 40.2 - 50.2 - 52.2 - 52.3 - 61 - 70 - 79 - 79.1 - 99 - 100.1 - 100.2 - 105

Artemisia rhodantha Rupr.

Asteraceae

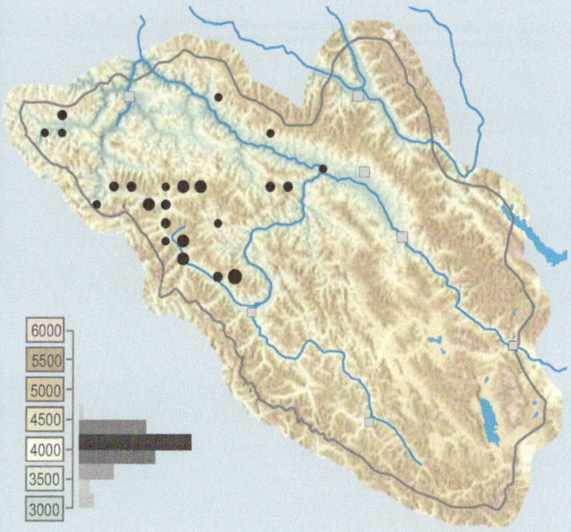

Syn.: *A. maritima* var. *thomsoniana* C.B.Clarke, *Seriphidium rhodanthum* (Rupr.) Poljak., *S. thomsonianum* (C.B.Clarke) Ling & Y.R.Ling

Description: Subshrubs, 5–30 cm tall, densely white tomentose, leaves 2–3-pinnatisect, segments linear, capitula ca. 3 mm across, sessile, in narrow, spike-like panicles.

Distribution: Kazakhstan, Kyrgyzstan, Tajikistan, NW India, China (Xinjiang).

Habitat: Rocky slopes, steppes.

Stem diameter up to 16 mm. Bark/xylem proportion 1:6. Ring width 0.2–0.6 mm. Maximal 28 annual rings.

Xylem

Transverse section

Semi-ring-porous. Rings marked by bands of thick-walled fibers. Ring variability often indicated by larger (50 µm) or smaller (30 µm) earlywood vessel diameters and a latewood zone with very small vessel diameters (<10 µm). Vessels solitary or in small groups. Fibers thick-walled. Parenchyma paratracheal. With intra-xylary periderm-bands.

Tangential section

Ray width 2–3 cells.

Radial section

Perforation plates simple. Intervessel pits round in alternating position. Vessel length 80–120 µm.

Bark

Phloem with tangentially arranged groups of thick-walled fiber groups. Oil ducts not observed. Phellem layered, consisting of many rows of large, thin-walled cells.

Pith

No slide available.

Analyzed: Stem of 4 individuals.

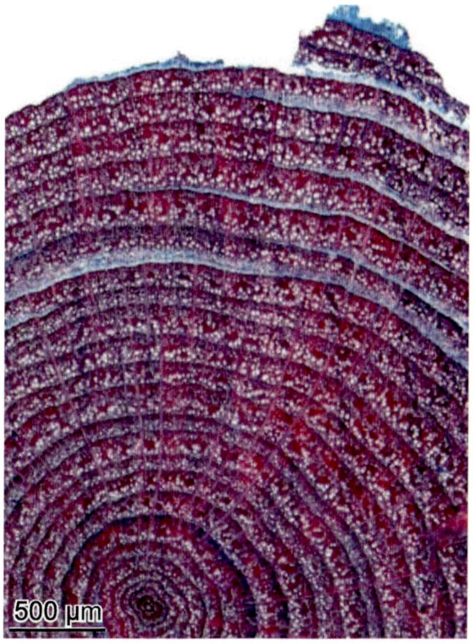

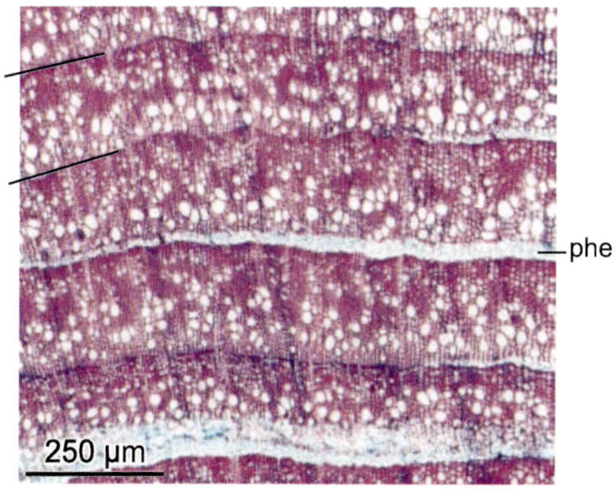

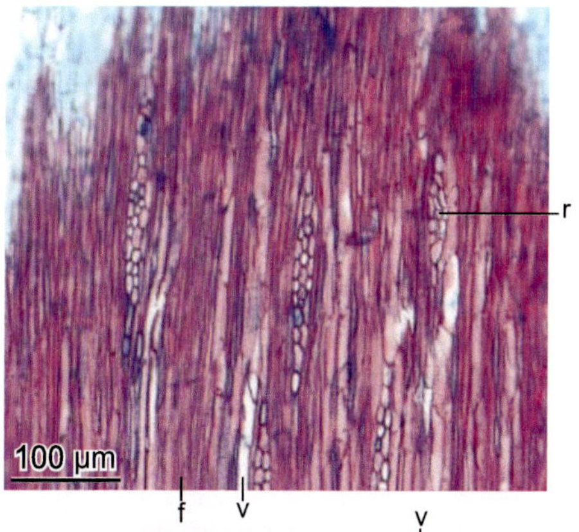

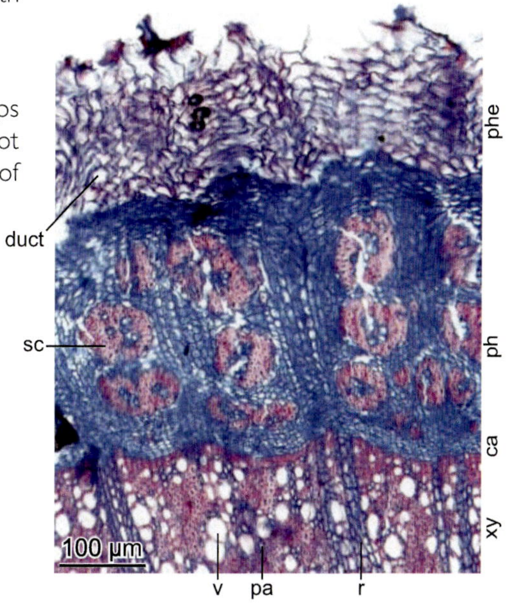

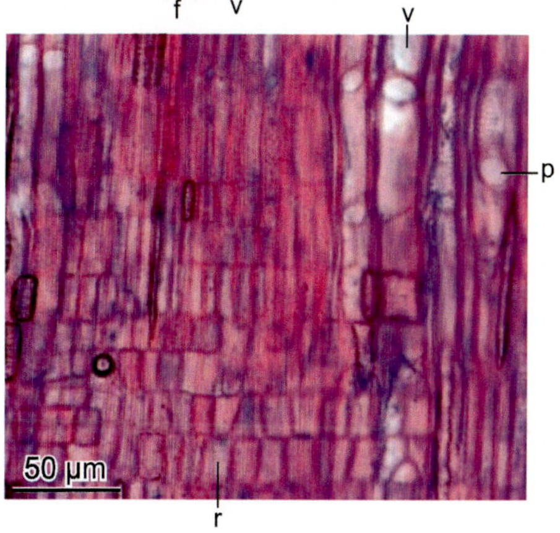

Features according to IAWA

Xylem 1 - 4 - 9 - 11 - 13 - 22 - 40.2 - 50.1 - 50.2 - 52.2 - 52.3 - 61 - 69 - 79 - 97 - 105 - 135.1

Artemisia salsoloides Willd.

Asteraceae

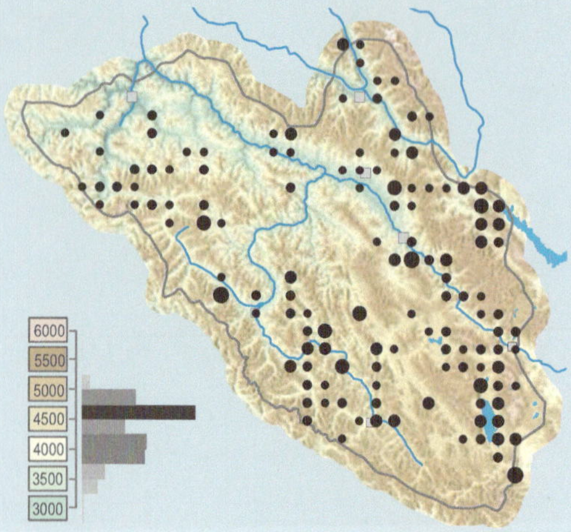

Syn.: *A. wellbyi* Hemsl. & H.Pearson

Description: Subshrubs, 20–50 cm tall, glabrous, stems several, leaves 2–3-pinnatisect, glaucous, almost sessile, capitula ca. 3 mm across, subsessile, in branched panicles.

Distribution: S Ukraine, Caucasus, Russia (Siberia), Pakistan (Baltistan), India (Jammu and Kashmir), China (Xizang).

Habitat: Stony steppes on stabilized slopes, dry stony river beds.

Stem diameter up to 10 mm. Bark/xylem proportion 1:4. Rings absent.

Xylem

Transverse section

Rings absent. Vessel diameter 40–60 µm. Vessels in groups, not stained. Fibers thick-walled, arranged in small groups around vessels. Parenchyma pervasive. With intraxylary periderm-bands. Tissue dead inside the bands. Rays absent.

Tangential section

No slide available.

Radial section

No slide available.

Bark

Phloem with irregularly arranged small groups of thick-walled fibers. With oil ducts. Phellem layered, consisting of a few rows of large, thin-walled cells.

Pith

No slide available.

Analyzed: Stem of 5 individuals.

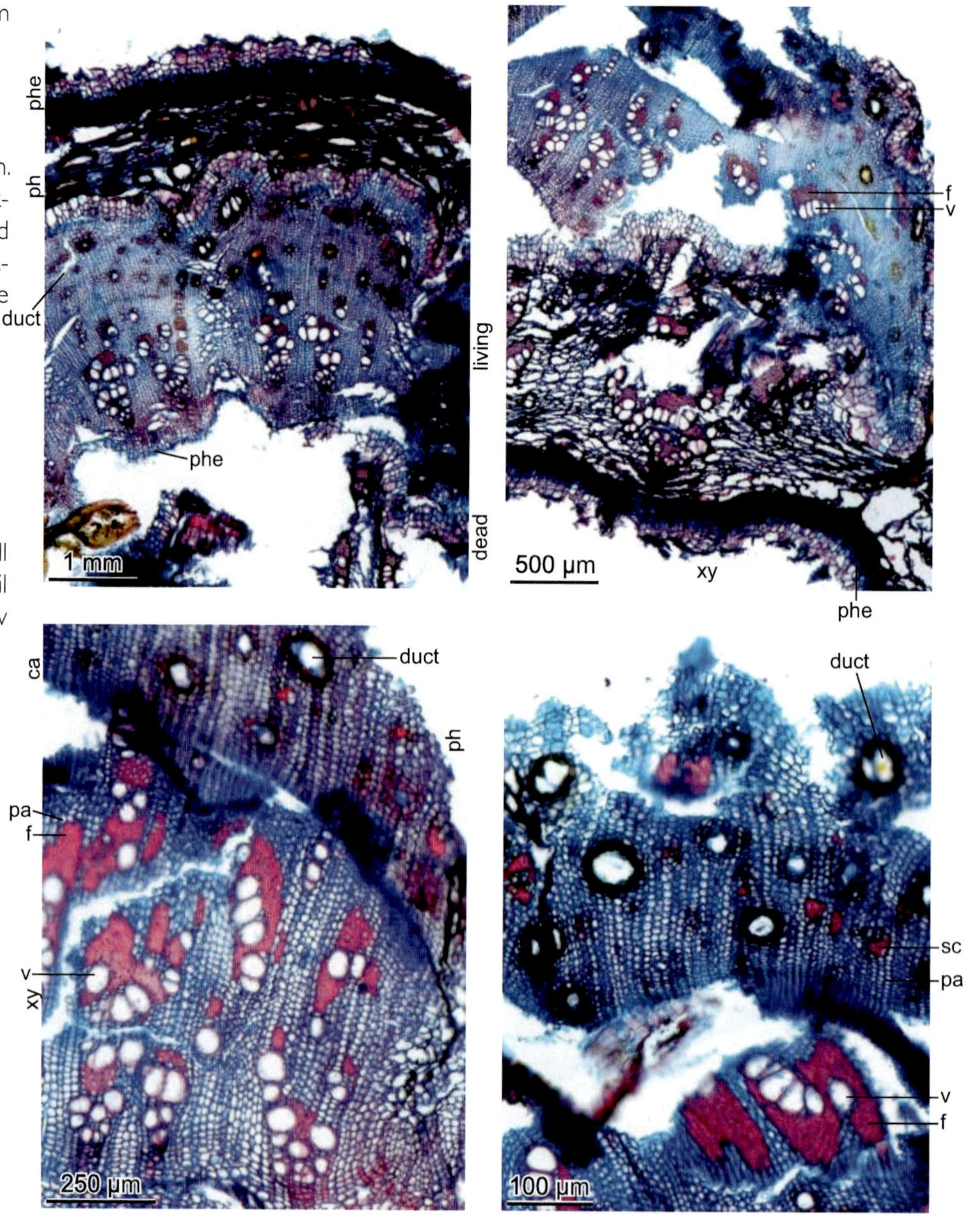

Features according to IAWA

Xylem 2 - 11 - 13 - 22 - 40.2 - 41 - 50.1 - 52.2 - 52.3 - 61 - 69 - 79 - 79.1 - 117 - 135.1

Artemisia santolinifolia Turcz. ex Krasch.

Asteraceae

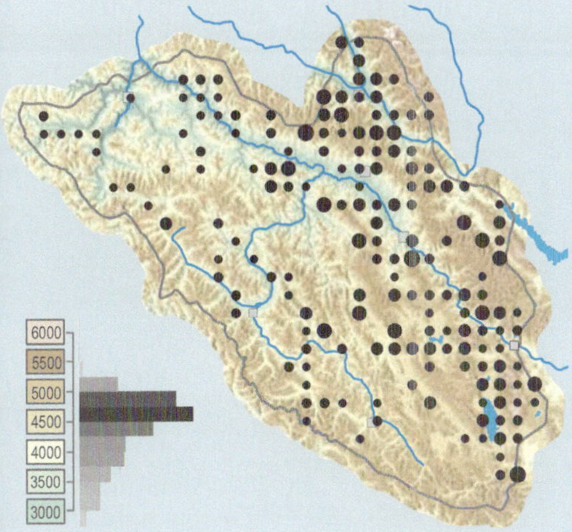

Syn.: *A. gmelinii* auct. non Weber ex Stechm.

Description: Subshrubs, 15–40 cm tall, aromatic, cespitose, leaves 3-pinnatisect, capitula 3–5 mm across, almost globose, shortly pedunculate, nodding, in lax panicles.

Distribution: Russia, C Asia, Afghanistan, Pakistan (Gilgit, Swat, Baltistan, Kashmir), India (Jammu and Kashmir), China, Mongolia.

Habitat: Stabilized slopes and stony steppes.

Stem diameter up to 7 mm. Bark/xylem proportion 1:5. Mean ring width 0.2 mm. Maximal 17 annual rings.

Xylem

Transverse section

Semi-ring-porous. Rings marked by a band of marginal parenchyma with small vessels. Ring variability often indicated by larger (50 µm) or smaller (30 µm) earlywood vessel diameters and a latewood zone with very small vessel diameters (<10 µm). Vessels mostly solitary or in small groups. Fibers thick-walled, arranged in continuous tangential bands. Parenchyma paratracheal in fiber zones and pervasive in parenchyma bands. Intra-xylary periderm-bands absent. Vessels contain dark-staining substances.

Tangential section

Ray width 4–6 cells. Rays disappear in polarized light.

Radial section

Perforation plates simple. Intervessel pits round in alternating position. Vessel length 80–110 µm.

Bark

Phloem with tangentially arranged groups of thick-walled fibers. Oil ducts not observed. Phellem layered, consisting of few rows of large, thin-walled cells (phelloids).

Pith

No slide available.

Analyzed: Stem of 2 individuals.

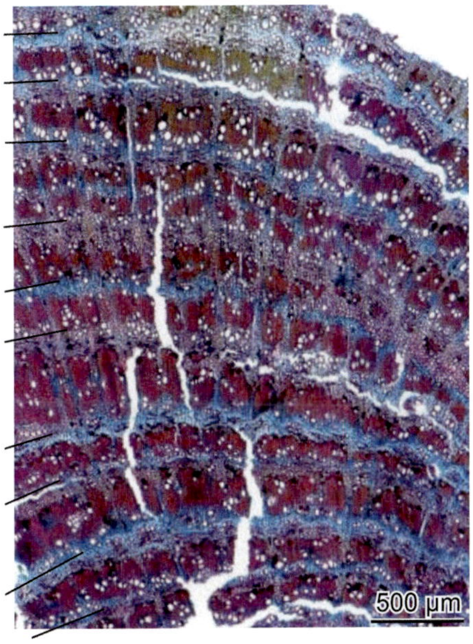

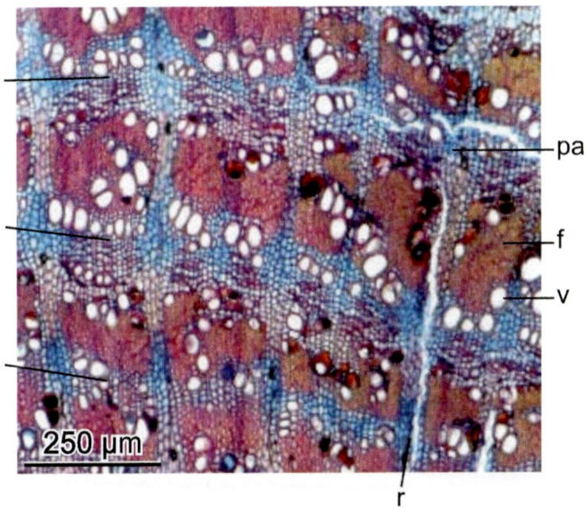

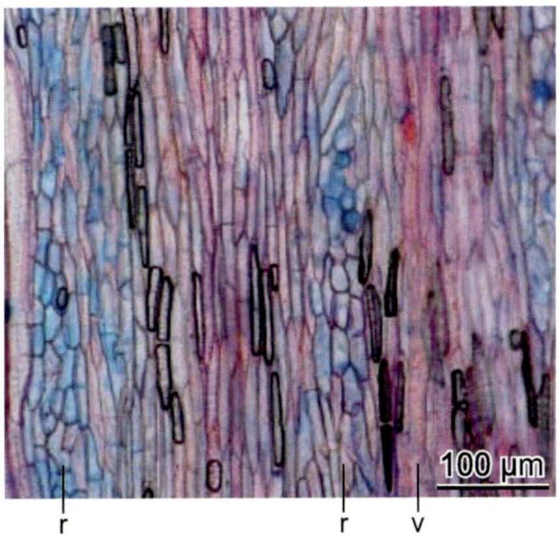

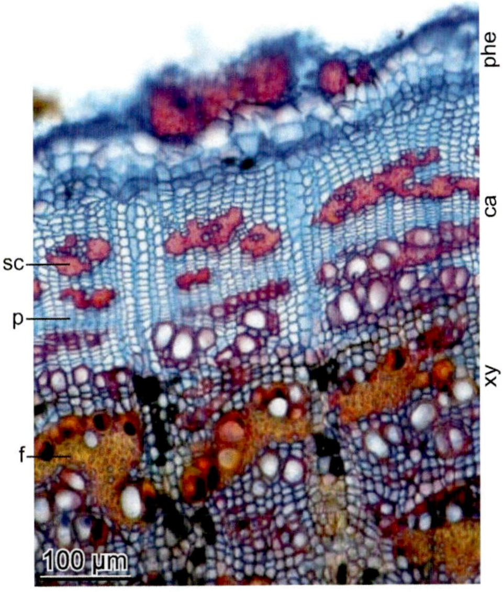

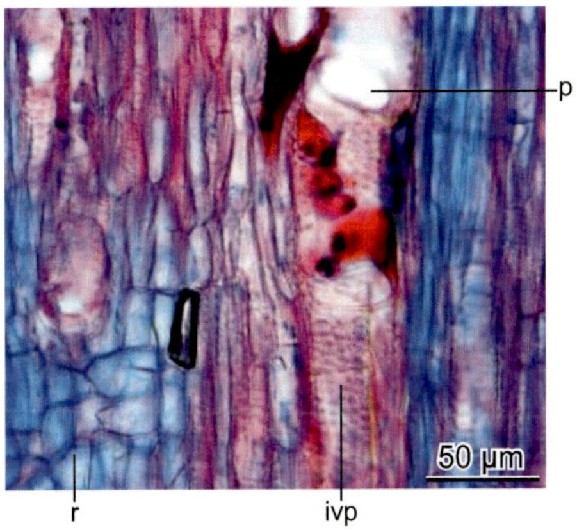

Features according to IAWA

Xylem 1 - 5 - 9 - 11 - 13 - 22 - 40.2 - 50.1 - 52.2 - 61 - 69 - 79 - 79.1 - 98 - 100.2 - 105

Artemisia scoparia Waldst. & Kit.

Asteraceae

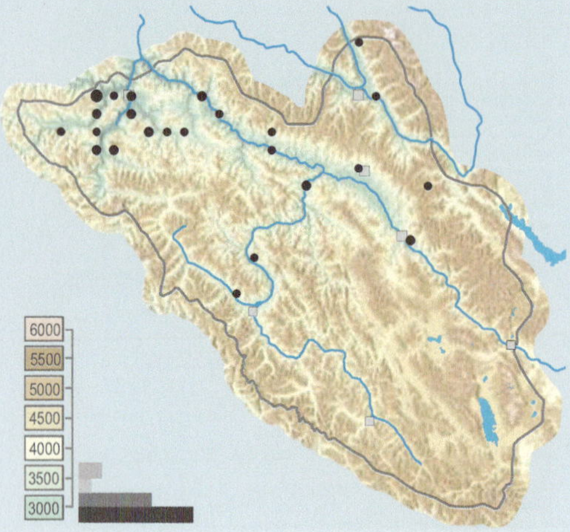

Syn.: *A. capillaris* var. *scoparia* (Waldst. & Kit.) Pamp.

Description: Annuals or biennials, 30–70 cm tall, aromatic, gray or yellowish pubescent, stems much branched, leaves 2–3-pinnatisect, capitula 1–1.5 mm across, subsessile, in broad panicles.

Distribution: N Africa, Europe, C and SW Asia, Russia, Afghanistan, Pakistan, India, China, Thailand, Korea, Japan.

Habitat: Stony steppes, dry river beds.

Stem diameter up to 5 mm. Bark/xylem proportion 1:2. Ring width 0.5–2 mm. Maximal 2 annual rings.

Xylem

Transverse section

With only one ring or semi-ring-porous with 2 rings. Earlywood vessel diameter 20–30 μm. Vessels solitary or in short radial multiples. Fibers thin- to thick-walled, parenchyma paratracheal and marginal in the external stem part. Stem center without fibers and pervasive parenchyma.

Tangential section

Ray width 1–3 cells.

Radial section

Perforation plates simple. Intervessel pits round in alternating position. Vessel length 150–200 μm.

Bark

Phloem with small groups of sieve elements. Phloem and cortex with oil ducts and groups of fibers. Phellem cells thin-walled and square.

Pith

No slide available.

Analyzed: Stem of 4 individuals.

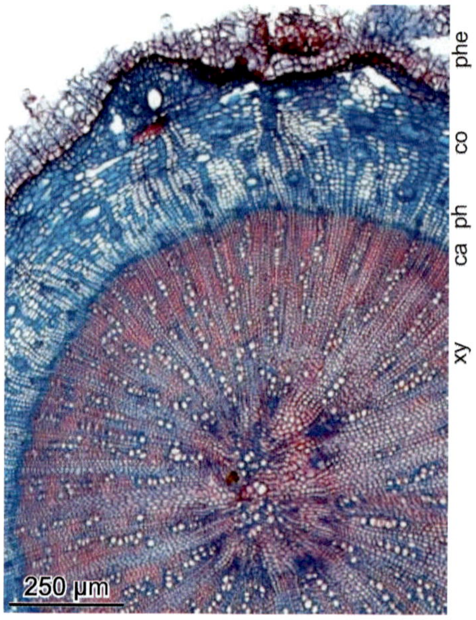

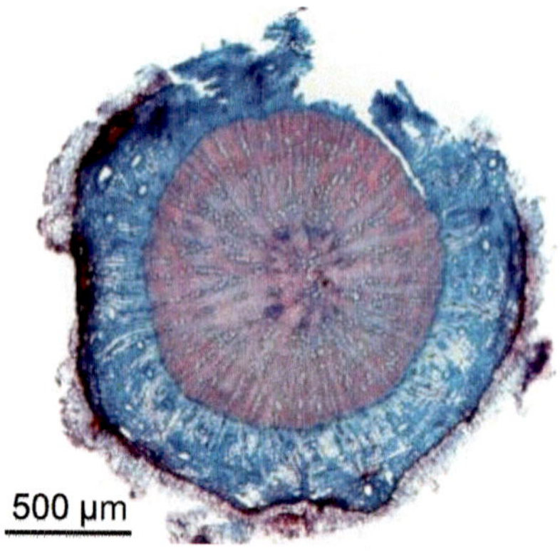

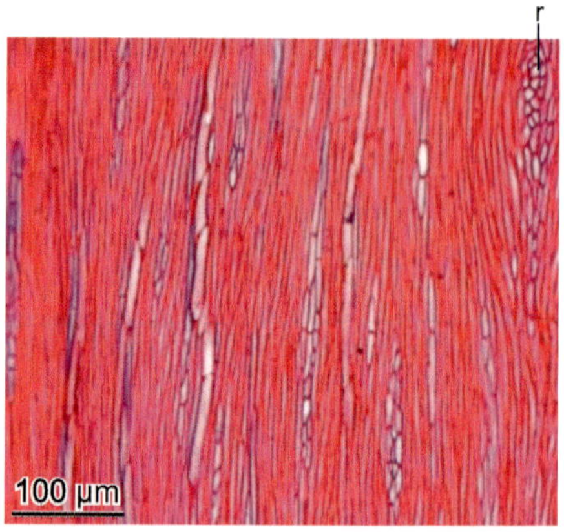

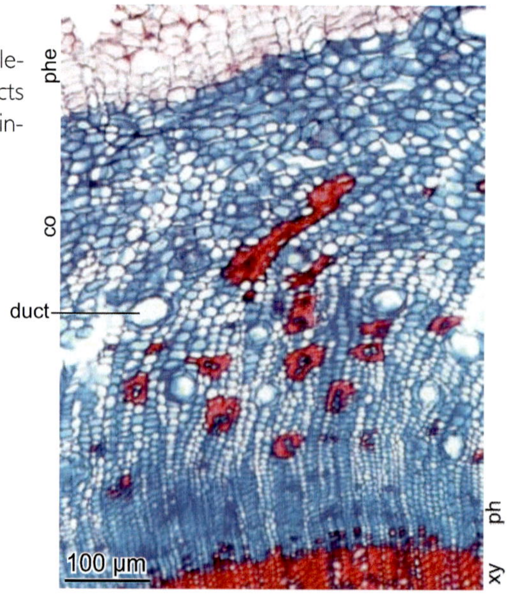

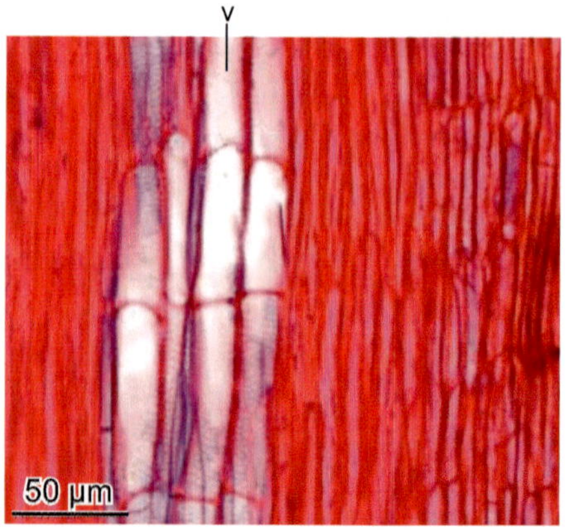

Features according to IAWA

Xylem 1 - 2.1 - 4 - 9.1 - 13 - 22 - 40.2 - 50.2 - 52.3 - 61 - 70 - 70.3 - 79 - 79.1 - 89 - 97 - 105

Artemisia stracheyi Hook.f. & Thomson ex C.B.Clarke

Asteraceae

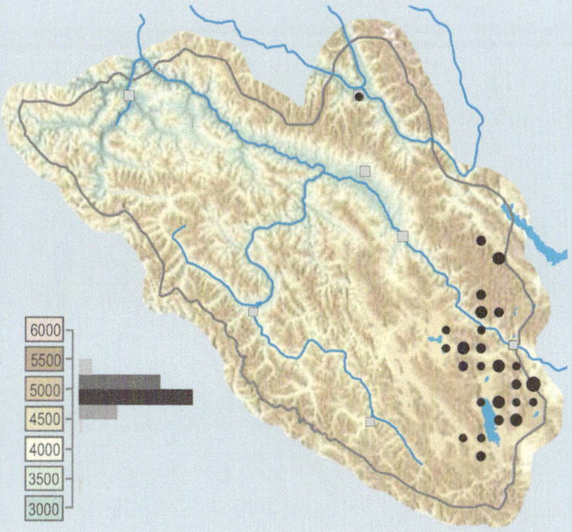

Syn.: *Artemisiella stracheyi* (Hook.f. & Thomson ex C.B.Clarke) Ghafoor

Description: Subshrubs, 15–45 cm tall, densely gray yellow tomentose, leaves 3-pinnatisect, capitula 6–10 mm across, hemispheric, nodding, in lax racemes.

Distribution: NW India (Jammu and Kashmir), China (S Xizang).

Habitat: Stony and sandy steppes.

Stem diameter up to 6 cm. Bark/xylem proportion 1:20. Mean ring width 0.5 mm. Maximal 70 annual rings.

Xylem

Transverse section

Ring-porous. Rings marked by large differences of earlywood and latewood vessel diameters. Ring variability often indicated by larger (80 µm) or smaller (40 µm) earlywood vessel diameters and a latewood zone with very small vessel diameters (<10 µm). Vessels in small groups, arranged in tangential lines. Vessels thick-walled (2–3 µm). Fibers thick-walled, arranged in groups in the latewood. Parenchyma paratracheal in fiber zones and pervasive in parenchyma bands. With very distinct intra-xylary periderm-bands.

Tangential section

Ray width 5–8 cells, with sheet cells. Rays regularly ship-like formed and regularly distributed. Rays disappear in polarized light.

Radial section

Perforation plates simple. Intervessel pits scalariform. Vessel length 90–110 µm.

Bark

Phloem with tangentially arranged groups of thick-walled fibers. With oil ducts. Phelloderm layered. Phellem consists of few rows of large, thin-walled cells (phelloids) and some rows of collapsed cells.

Pith

No slide available.

Analyzed: Stem of 2 individuals.

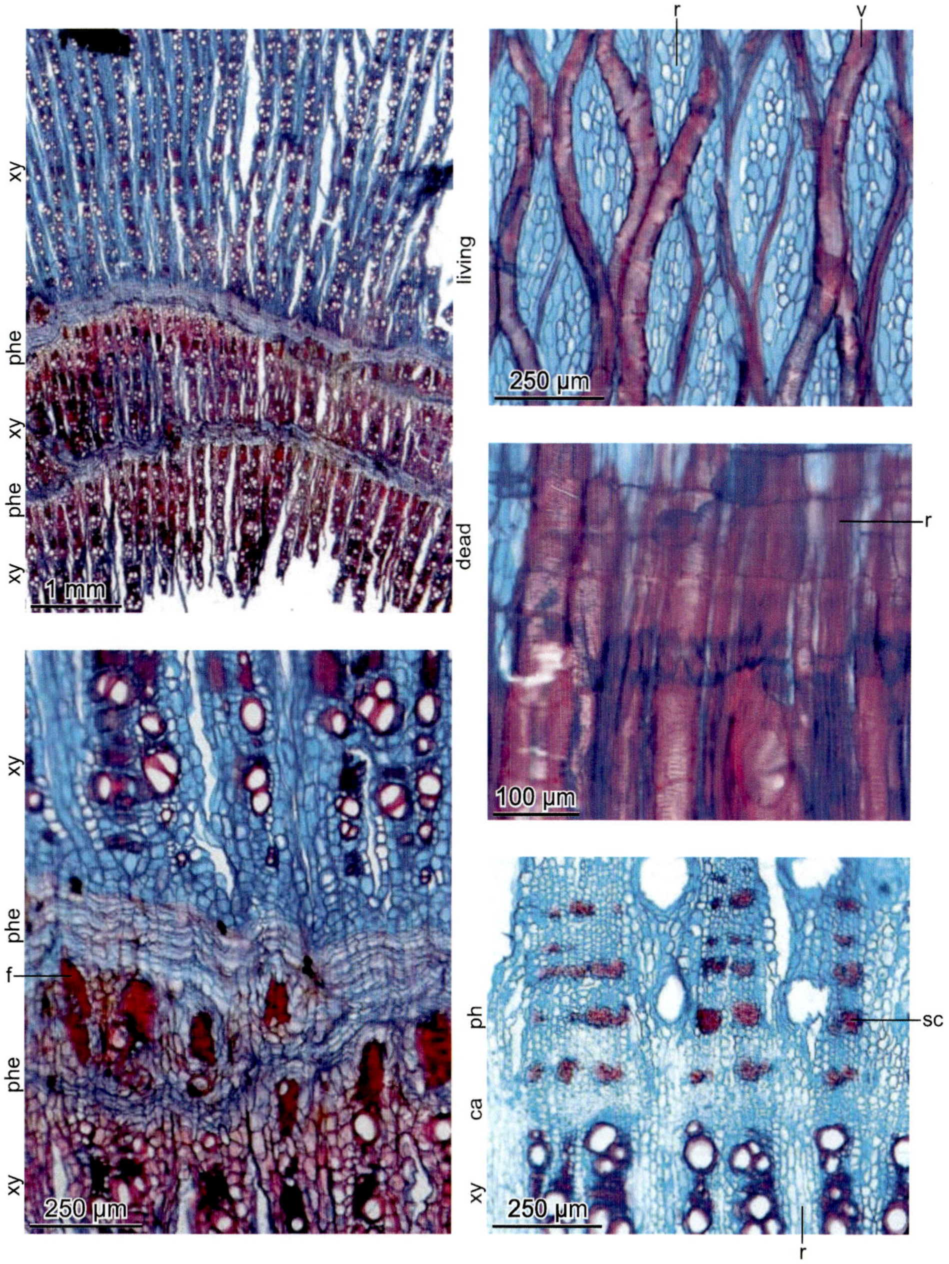

Features according to IAWA

Xylem 1 - 3 - 9 - 13 - 20 - 41 - 50.1 - 52.2 - 61 - 69 - 79 - 79.1 - 98 - 100.2 - 105 - 110 - 135.1 - 135.2

Artemisia tournefortiana Rchb.

Asteraceae

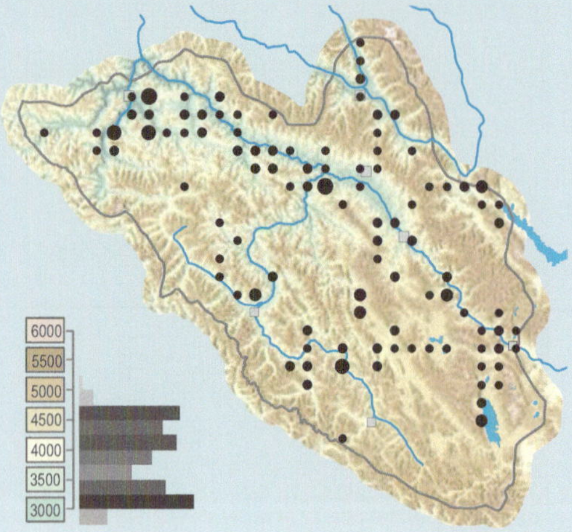

Description: Annuals, 100–150 cm tall, leaves 1–3-pinnatisect, capitula ca. 2 mm across, densely congested in cylindric panicles.

Distribution: E Turkey, Armenia, Iran, Kazakhstan, Kyrgyzstan, Tajikistan, Turkmenistan, Uzbekistan, N Pakistan, India (Jammu and Kashmir, Himachal Pradesh, Uttarakhand, Sikkim), Nepal, China (Xinjiang, Xizang), Mongolia; introduced in Europe.

Habitat: A weed in barley fields, wet to mesic loamy river sediments, muddy stream sediments, gravel in river beds, dry watersheds.

Stem diameter up to 3 mm. Bark/xylem proportion 1:4. Mean ring width 0.6 mm. Maximal 2 annual rings.

Xylem

Transverse section

Semi-ring-porous, 2 rings. Earlywood vessel diameter 30–50 µm. Vessels solitary or in short radial multiples. Fibers thin- to thick-walled, parenchyma paratracheal in the peripheral ring, pervasive in the center. Fibers absent in the center and pervasive parenchyma.

Tangential section

Ray width 1–2 cells.

Radial section

Perforation plates simple. Intervessel pits round in alternating position. Vessel length 100–150 µm.

Bark

Phloem with fibers and small groups of sieve elements. Cortex with oil ducts and few fibers. Phellem cells thin-walled and square.

Pith

No slide available.

Analyzed: Stem of 2 individuals.

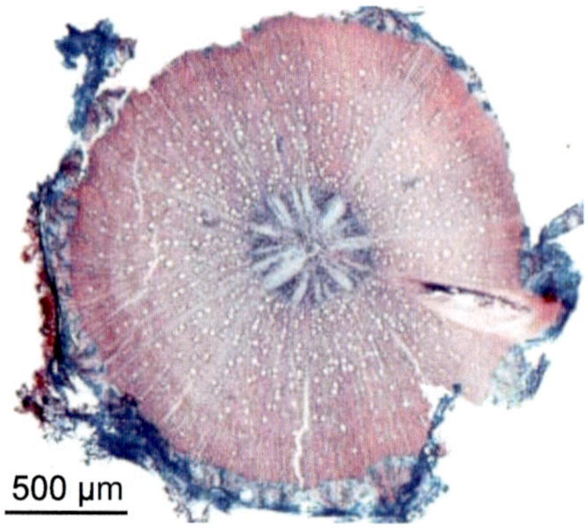

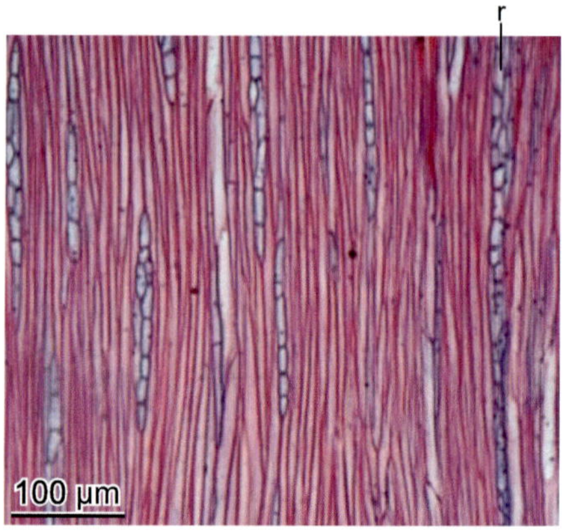

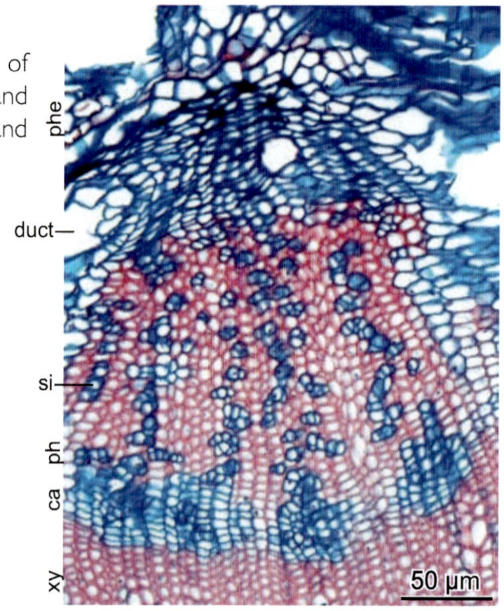

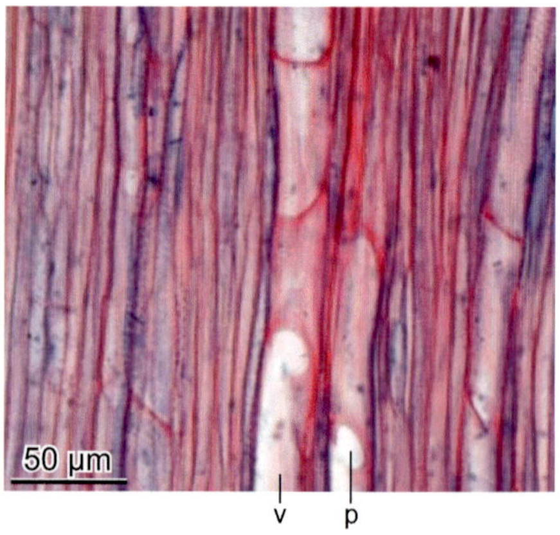

Features according to IAWA

Xylem 1 - 4 - 9.1 - 13 - 22 - 40.2 - 50.2 - 52.2 - 61 - 70 - 70.3 - 79 - 79.1 - 97 - 105

Askellia benthamii (C.B.Clarke) Sennikov

Asteraceae

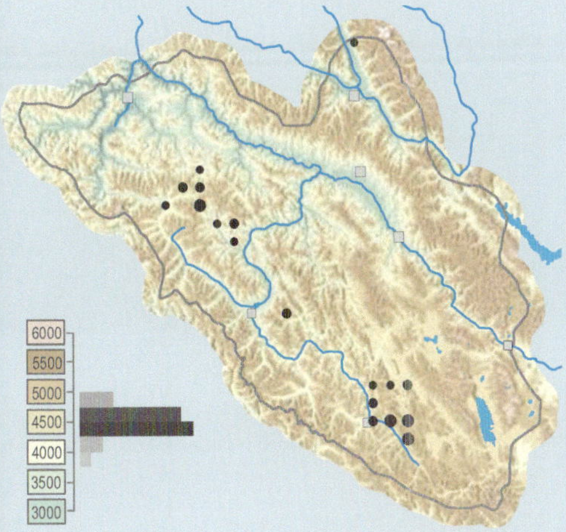

Syn.: *Crepis naniformis* Babc.

Description: Perennial herbs, 5–20 cm tall, glabrous, stems branched from the base, leaves broadly ovate to obovate, margin entire or sparsely crenate, capitula yellow, 8–10 mm across, in corymbiform synflorescences.

Distribution: NW India.

Habitat: Screes, gravel slopes.

Stem diameter up to 20 mm. Bark/xylem proportion 1:5. Rings absent.

Xylem

Transverse section

Solitary vessels with a diameter of 40–50 μm. Vessels mostly solitary or in small groups. Parenchyma pervasive. Fibers absent. With small, distinct intra-xylary periderm-bands. Extremely large, thin-walled, unlignified rays between radial vessel/parenchyma strips. All ray cells contain crystal druses.

Tangential section

Ray width >10 cells. Rays disappear in polarized light.

Radial section

Perforation plates simple. Intervessel pits scalariform. Vessel length 120–200 μm.

Bark

Phloem missing. Intra-xylary phellem bands consist of many layers of rectangular cells, mostly filled with brown substances.

Pith

No slide available.

Analyzed: Stem of 2 individuals.

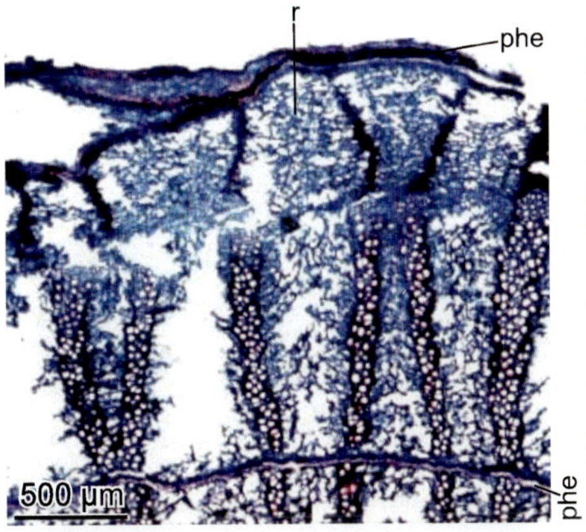

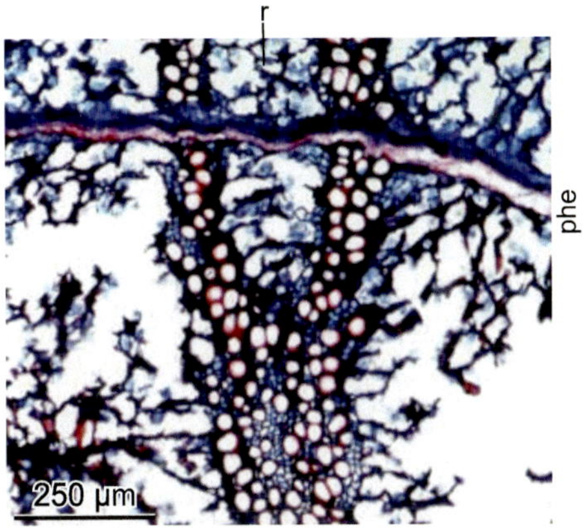

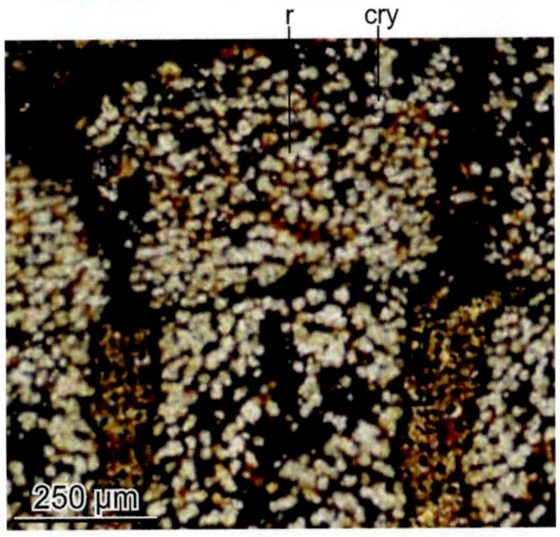

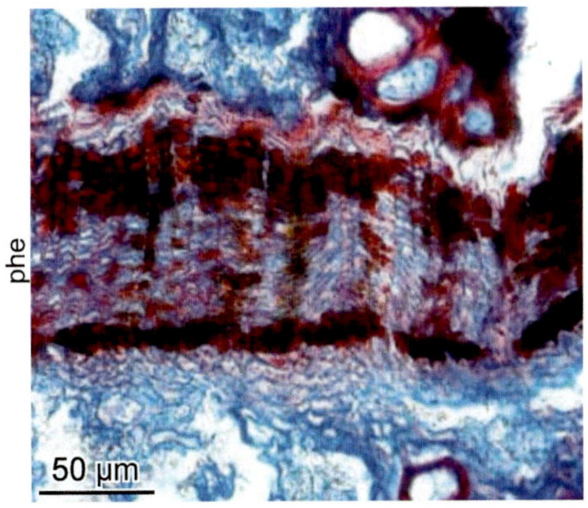

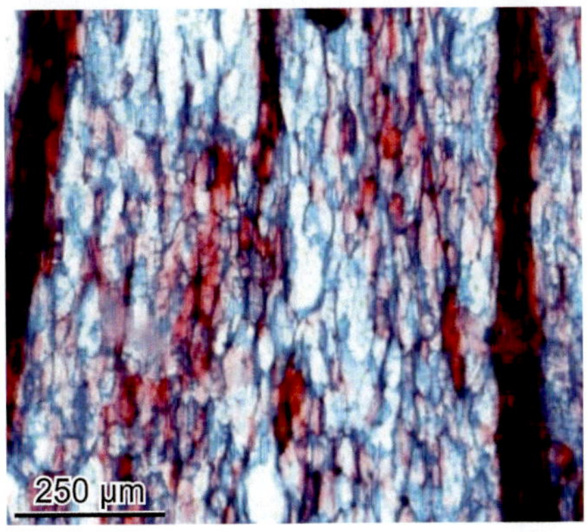

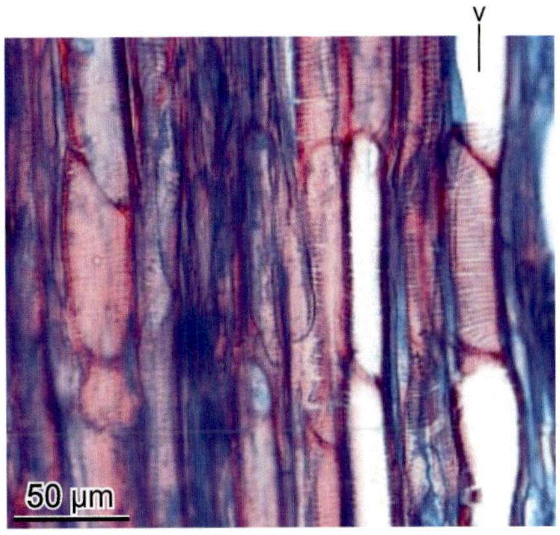

Features according to IAWA

Xylem 2 - 5 - 9 - 13 - 22 - 40.2 - 50.1 - 52.2 - 52.3 - 60.1 - 79.1 - 99 - 100.2 - 102 - 105 - 135.1 - 144

Askellia flexuosa (Ledeb.) W.A.Weber

Asteraceae

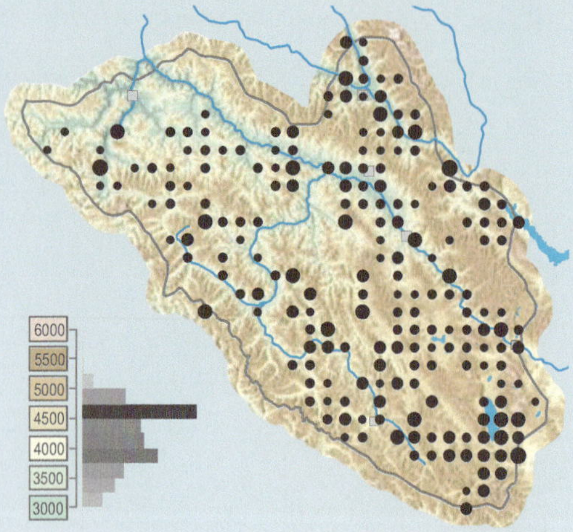

Syn.: *Prenanthes polymorpha* var. *flexuosa* Ledeb., *Crepis flexuosa* (Ledeb.) C.B.Clarke

Description: Perennial herbs, 3–30 cm tall, divaricately branched from the base, leaves ovate to linear, pinnatifid to pinnatisect, capitula yellow, ca. 1 cm across, numerous.

Distribution: SW Asia, S and C Russia, Kazakhstan, Kyrgyzstan, Tajikistan, Afghanistan, Pakistan, India, Nepal, China, Mongolia.

Habitat: Stony or gravel stream beds, stabilized slopes with low density vegetation, along roads.

Stem diameter 3–5 mm. Bark/xylem proportion 1:3. Ring width 0.8–1.1 mm. Maximal 2 annual rings.

Xylem

Transverse section

Diffuse-porous. Rings distinctly marked by unlignified marginal parenchyma. Vessels mostly solitary. Earlywood vessel diameter 50–80 μm. Fibers thin- to thick-walled. Parenchyma paratracheal and marginal.

Tangential section

Rays 3–5-seriate, disappearing in the stem center in polarized light.

Radial section

Perforation plates simple. Intervessel pits round in alternating position. Vessel length 90–130 μm. Ray cells upright.

Bark

Phloem consists of tangentially interrupted radial groups of sieve tubes between ray dilatations. Cortex cells large (50–80 μm), round to oval. Phellem cells thin-walled, irregularly formed.

Pith

Polygonal, star-shaped.

Analyzed: Rhizome of 2 individuals.

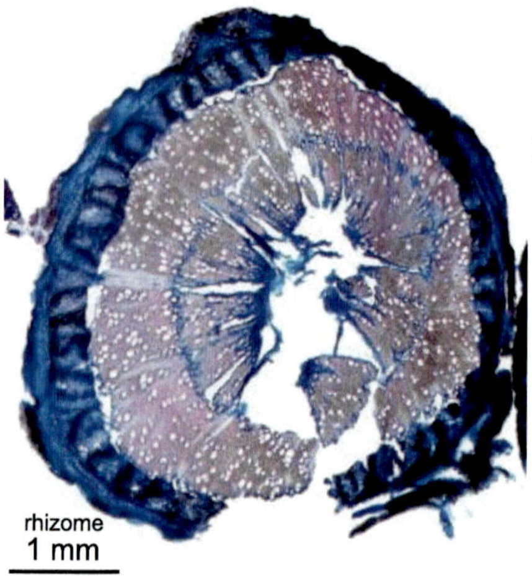

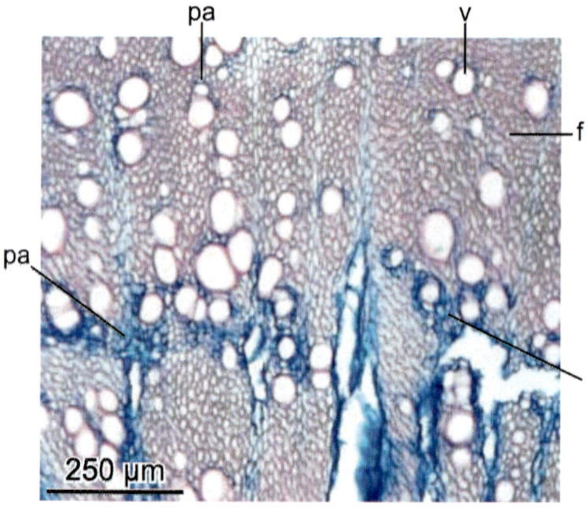

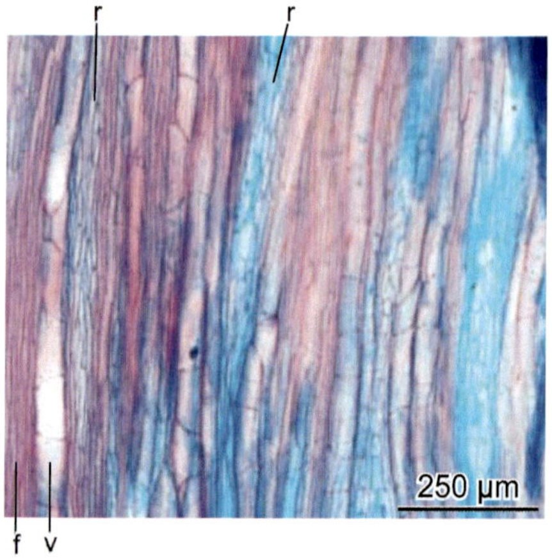

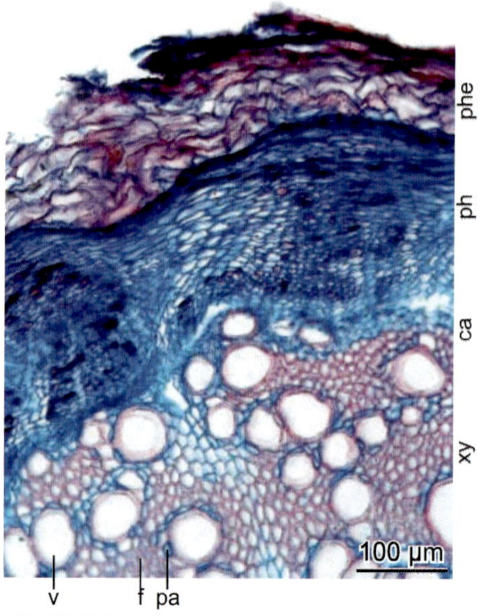

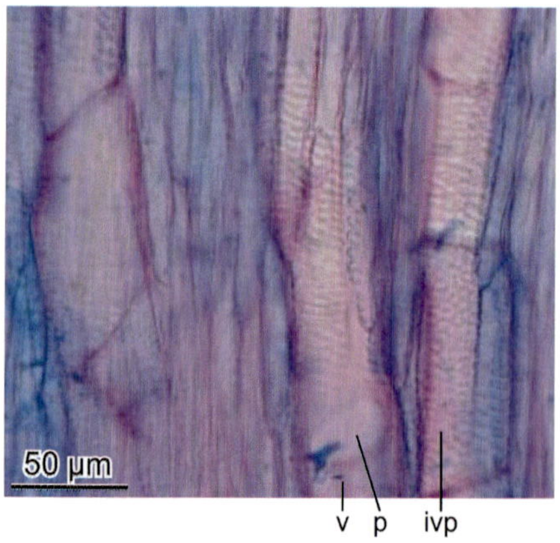

Features according to IAWA

Xylem 1 - 5 - 9 - 13 - 22 - 41 - 50.1 - 52.2 - 52.3 - 61 - 70 - 79 - 89 - 89.1 - 98 - 105

Aster flaccidus Bunge

Asteraceae

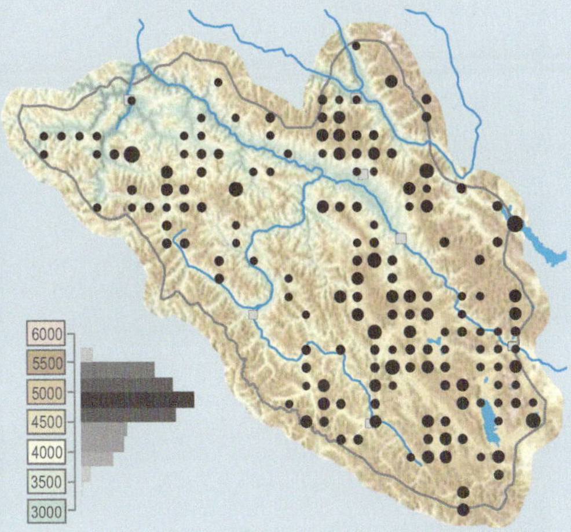

Description: Perennial herbs, 3–10 cm tall, leaves lanceolate, mostly basal, capitula 4–5 cm across, terminal, solitary, ray florets blue, disk florets orange to yellow.

Distribution: Iran, Russia (Siberia), Kazakhstan, Uzbekistan, Afghanistan, Pakistan, India, Nepal, Bhutan, China (Gansu, Hebei, Qinghai, Shaanxi, Shanxi, Sichuan, Xinjiang, Xizang, Yunnan), Mongolia.

Habitat: Screes and stabilized slopes, wet alpine grasslands and pastures.

Stem diameter 1.5–1.8 mm. Bark/xylem proportion 1:1. Mean ring width 0.08 mm. Maximal 8 annual rings.

Xylem

Transverse section

Semi-ring-porous. Rings distinctly marked by large differences between earlywood (15–20 µm) and latewood vessel diameters (5 µm). Vessels mostly solitary. Fibers absent. Parenchyma pervasive. Rays dilate within the xylem. Vascular bundles remain.

Tangential section

Rays 5–>10-seriate, disappearing in polarized light.

Radial section

Perforation plates simple. Intervessel pits round in alternating position. Vessel length 100–150 µm. Ray cells upright.

Bark

Phloem uniform. Cortex cells large (80–100 µm), round to oval. Phellem consisting of distinctly layered, thin-walled, regularly formed cells.

Pith

Polygonal, surrounded by vascular bundles.

Analyzed: Root collar of polar root of 1 individual.

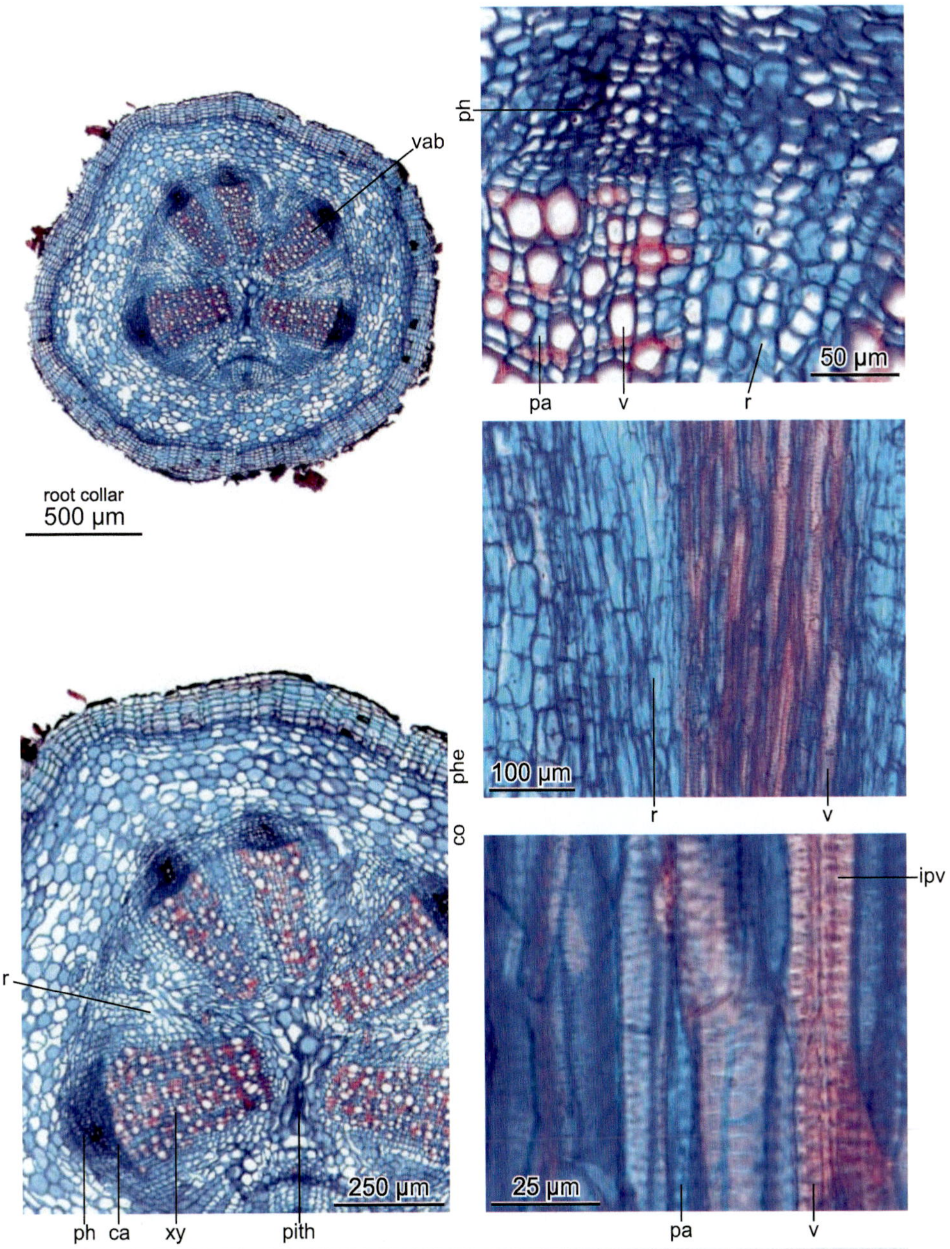

Features according to IAWA

Xylem 1 - 4 - 9 - 13 - 22 - 40.1 - 50.2 - 52.3 - 60.1 - 79.1 - 99 - 99.1 - 100.2 - 105

Brachyactis pubescens (DC.) Aitch. & C.B.Clarke
Asteraceae

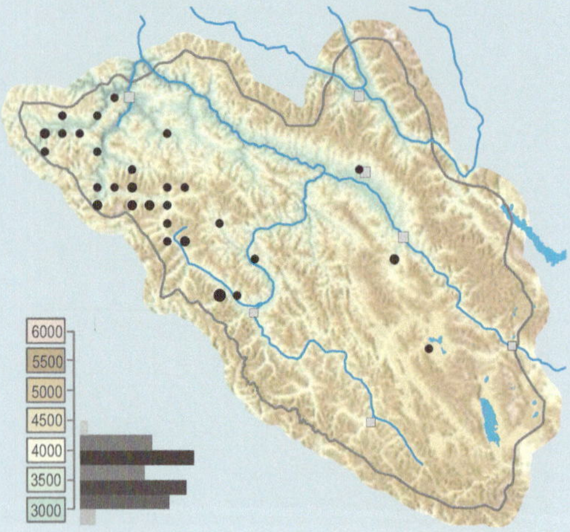

Syn.: *Conyza pubescens* DC., *Neobrachyactis pubescens* (DC.) Brouillet, *B. robusta* Benth.

Description: Annuals, 5–20 cm tall, densely glandular hairy, leaves obovate or oblong, margin sparsely serrate, capitula pale yellow, 1.5–2 cm across, in racemiform synflorescences.

Distribution: Afghanistan, Pakistan, India, Nepal, China (W Xizang).

Habitat: Mountain slopes, disturbed places.

Stem diameter 3.5–5 mm. Bark/xylem proportion 1:2. Mean ring width 0.25 mm. Maximal 4 annual rings.

Xylem

Transverse section
Semi-ring-porous. Vessels solitary and in small radial groups. Earlywood vessel diameter 20–25 µm, latewood vessel diameter 18–15 µm. Fibers thin- to thick-walled. Parenchyma paratracheal.

Tangential section
Rays of two distinct sizes: uni-seriate and 3–6-seriate, with sheath cells.

Radial section
Perforation plates simple. Intervessel pits round. Vessel length 150–180 µm.

Bark

Phloem with groups of sieve tubes and fibers. Ray dilatations. Cortex cells large, round and rectangular. Sclerenchyma and crystals absent. Phellem thick, consisting of square and rectangular cells.

Pith

Pith polygonal.

Analyzed: Rhizome of 2 individuals (herbarium specimen).

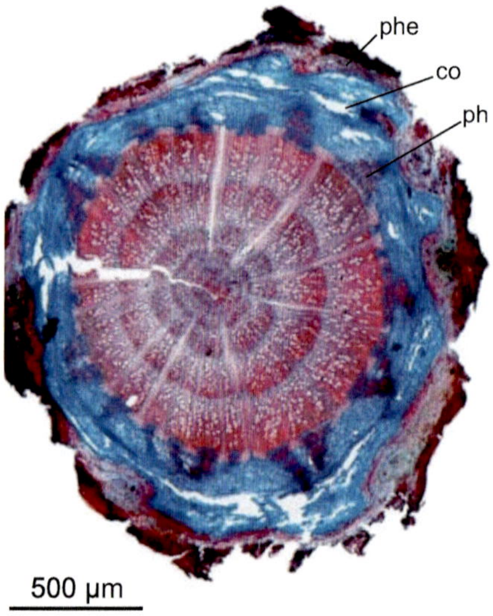

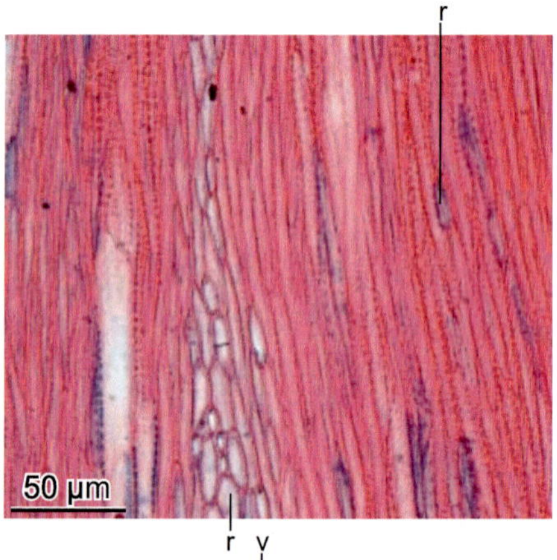

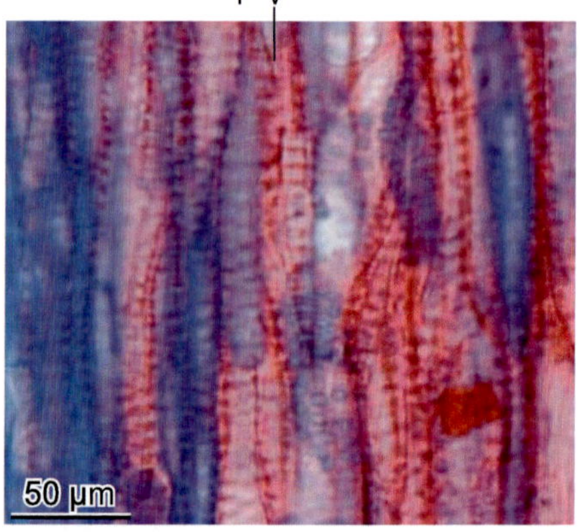

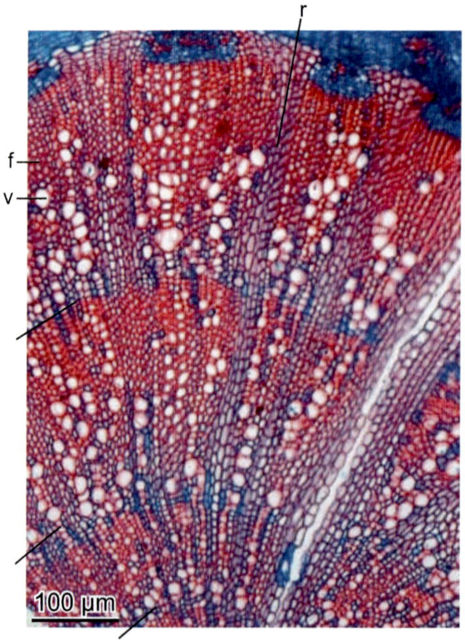

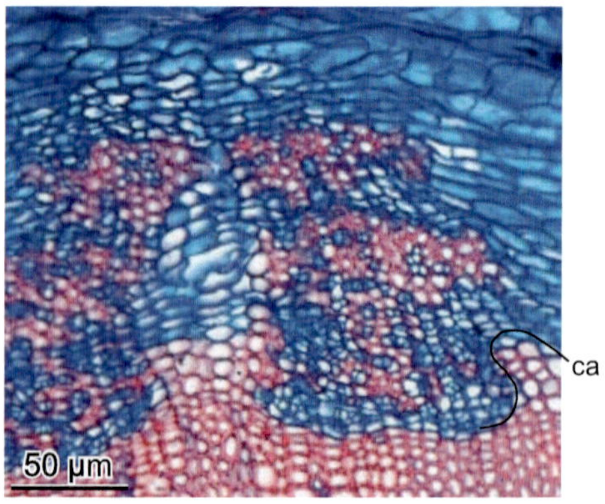

Features according to IAWA
Xylem 1 - 4 - 9 - 9.1 - 13 - 22 - 40.2 - 50.1 - 53.3 - 61 - 70 - 79 - 96 - 98 - 103 - 110

Cirsium arvense (L.) Scop.

Asteraceae

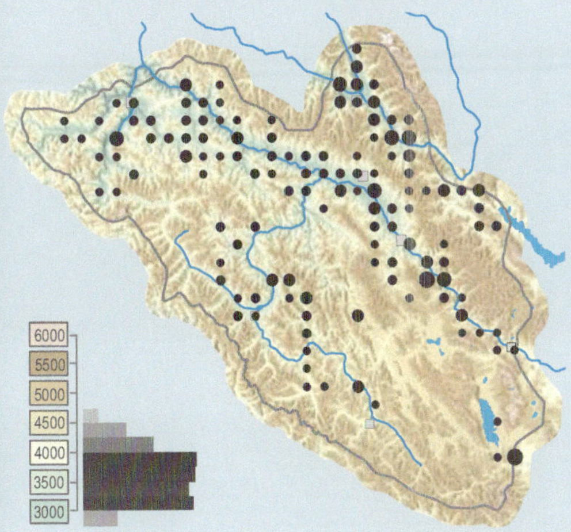

Syn.: *Breea arvensis* (L.) Less.

Description: Perennial herbs, 30–160 cm tall, dioecious, pungent, leaves lanceolate, pinnately divided or simple, capitula reddish purple or rarely white, 2–3 cm across, numerous.

Distribution: Europe, SW Asia, Russia, Kazakhstan, Afghanistan, India, Nepal, China, Mongolia, Korea, Japan; introduced worldwide.

Habitat: Sandy river sediments, eroded river terraces, a weed in barley fields.

Stem diameter 5–8 mm. Bark/xylem proportion 1:3. Ring width 0.6–3 mm. Maximal 3 annual rings.

Xylem

Transverse section

Annual or perennial. Diffuse-porous if rings are present. Rings distinctly marked by a few rows flat fibers in the latewood. Ring variability low. Vessels form short radial rows or small groups, often in intra-annual tangential rows. Earlywood vessel diameter 50–70 µm. Fibers thin- to thick-walled. Parenchyma paratracheal.

Tangential section

Rays 4–7-seriate.

Radial section

Perforation plates simple. Intervessel pits round in alternating position. Vessel length 90–130 µm. Ray cells square and upright.

Bark

Phloem with small groups of sieve tubes. Arcade-like fiber groups at the outside of the vascular bundle. Distinct ray dilatations. Cortex cells large (50–80 µm), round to oval, containing few ducts. Phellem not well developed.

Pith

Polygonal, star-shaped.

Analyzed: Root collar of polar root of 5 individuals.

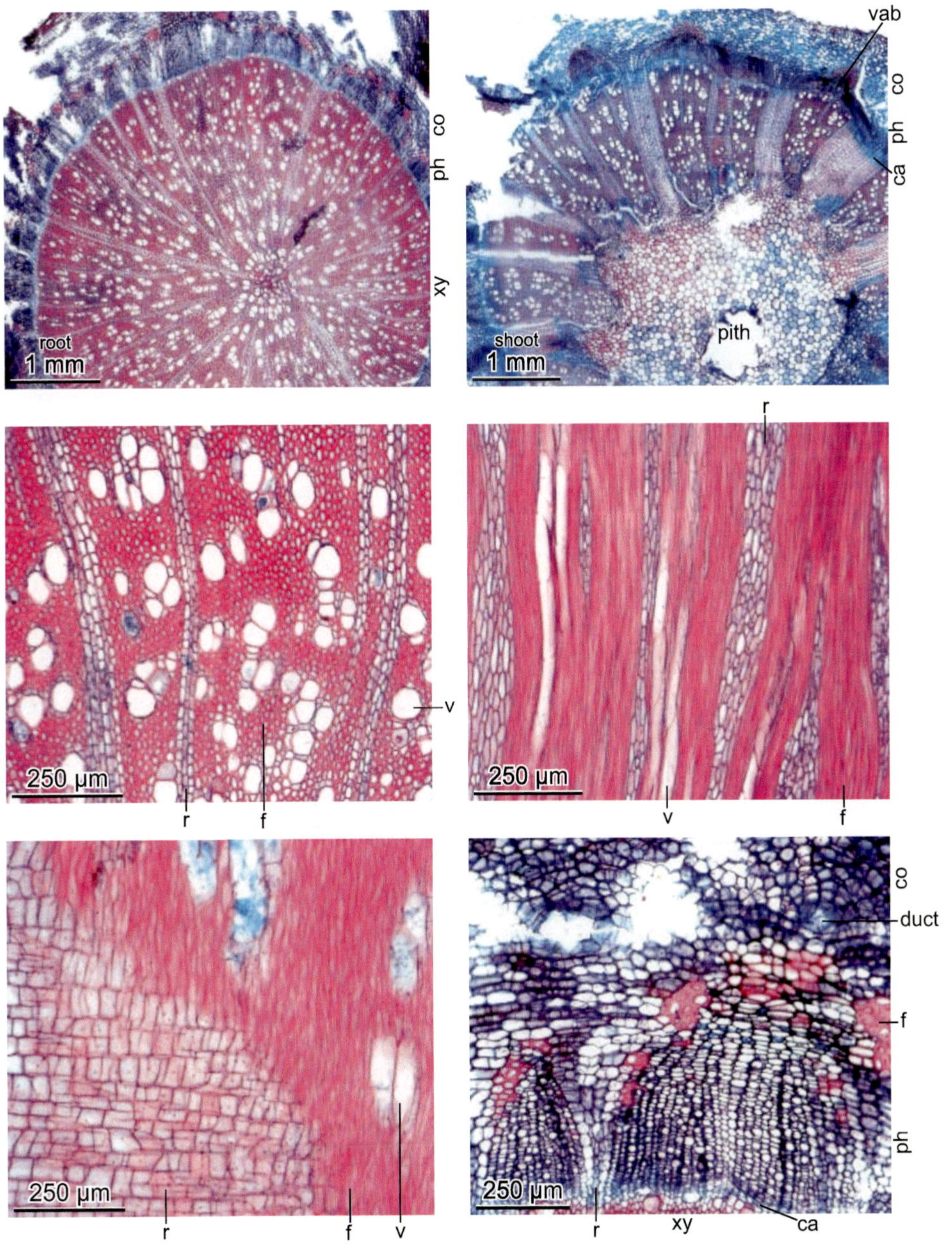

Features according to IAWA

Xylem 1 - 5 - 9.1 - 10 - 13 - 22 - 41 - 50.1 - 52.2 - 52.3 - 61 - 70 - 79 - 98 - 105

Cousinia thomsonii C.B.Clarke

Asteraceae

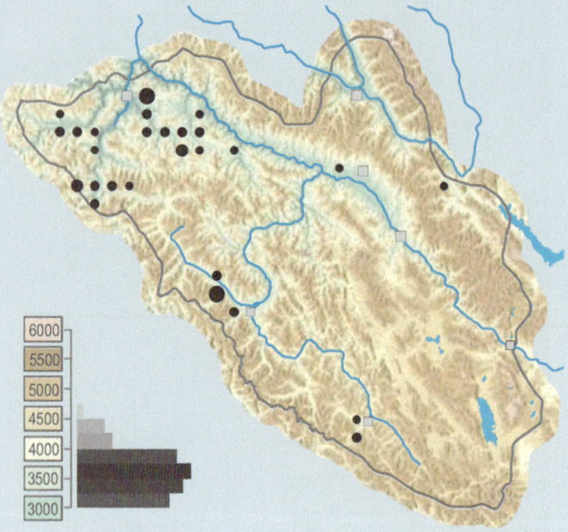

Description: Biennial herbs, 30–80 cm tall, leaves spiny, capitula purple to pink, involucre of capitula 3–4 cm across, densely cobwebby.

Distribution: N Pakistan, NW India (Jammu & Kashmir, Himachal Pradesh, Uttarakhand), W Nepal, China (Xizang).

Habitat: Meadows, alluvial gravel.

Stem diameter 7–8 mm. Bark/xylem proportion 1:1. Mean ring width 0.4 mm. Maximal 9 annual rings.

Xylem

Transverse section

Ring-porous to semi-ring-porous. Rings only marked by the difference of vessel size. Vessels in groups. Earlywood vessel diameter 80–100 µm. Fibers thin- to thick-walled, in groups. Parenchyma pervasive.

Tangential section

Rays 3–5-seriate and rare. Ray cells large and small. Rays confluent, disappearing in polarized light.

Radial section

Perforation plates simple. Intervessel pits round to slightly scalariform in alternating position. Vessel length 100–150 µm. Ray cells upright.

Bark

Phloem consists mainly of parenchyma. Sieve tubes stay in radial lines. Irregular groups of fibers. With large, irregular intercellular spaces. Phellem cells thin-walled, irregularly formed.

Pith

No slide available.

Analyzed: Rhizome of 2 individuals.

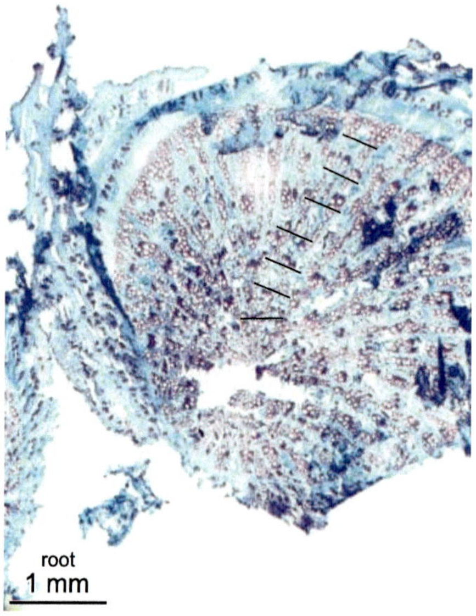

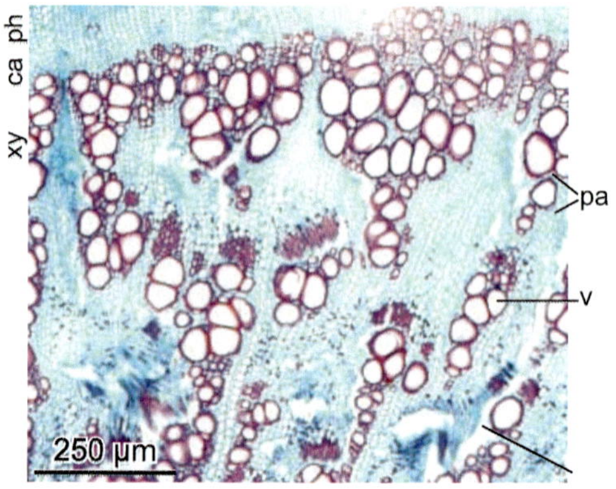

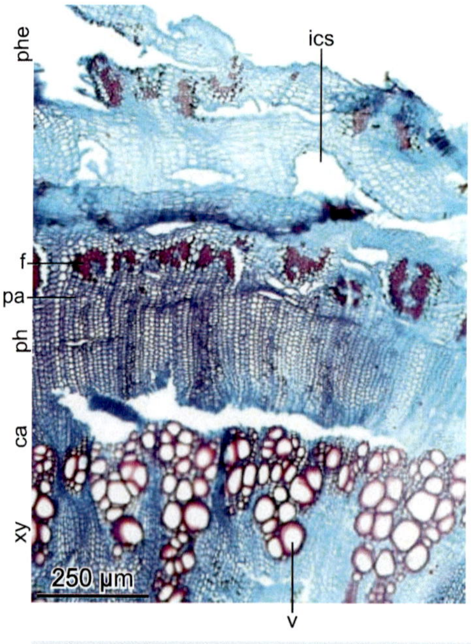

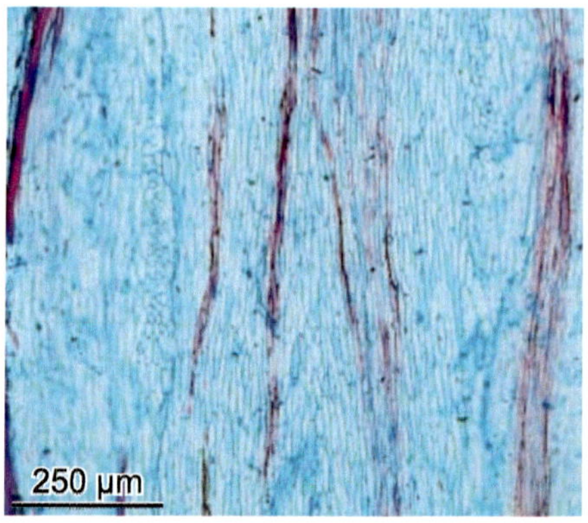

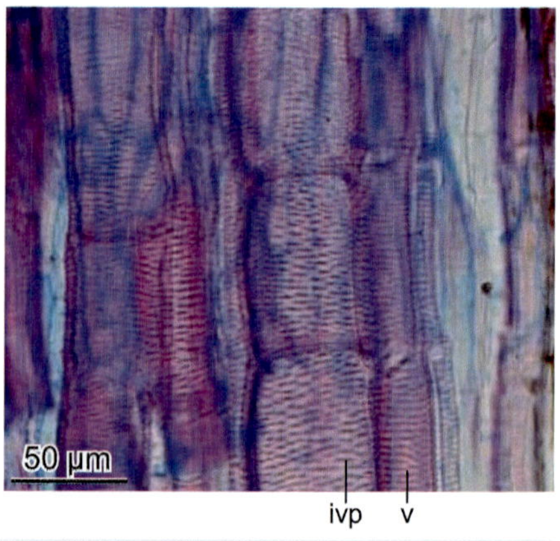

Features according to IAWA

Xylem 1 - 3 - 4 - 11 - 13 - 20 - 22 - 41 - 50.1 - 52.2 - 52.3 - 61 - 70 - 79.1 - 98 - 100.1 - 100.2 - 105

Cremanthodium decaisnei C.B.Clarke

Asteraceae

Syn.: *C. clarkei* R.Good

Description: Perennial herbs, 5–25 cm tall, leaves reniform, basal ones long-petioled, margin irregularly crenate, capitula yellow, 3–6 cm across, solitary, nodding.

Distribution: India (Jammu and Kashmir, Sikkim), Nepal, China (Gansu, Qinghai, Sichuan, Xizang, Yunnan), Bhutan.

Habitat: Along alpine streams, alpine meadows.

Stem diameter 5 mm. Bark/xylem proportion 5:1. Maximal 1 annual ring. Ring width 0.08 mm.

Xylem

Transverse section

Only one ring. With a few vascular bundles between large parenchymateous zones. Earlywood vessel diameter 8–10 µm. Fibers absent. Parenchyma pervasive.

Tangential section

Large parenchymatic zones between vascular bundles. Rays absent within the bundles.

Radial section

Perforation plates simple. Intervessel pits scalariform. Vessel length 50–60 µm.

Bark

Sieve tubes in small groups. Cortex consists of round to oval, unlignified parenchyma cells. Phellem absent.

Pith

Pith round, hollow or filled with unlignified parenchyma cells.

Analyzed: Root collar of 1 individual (herbarium specimen).

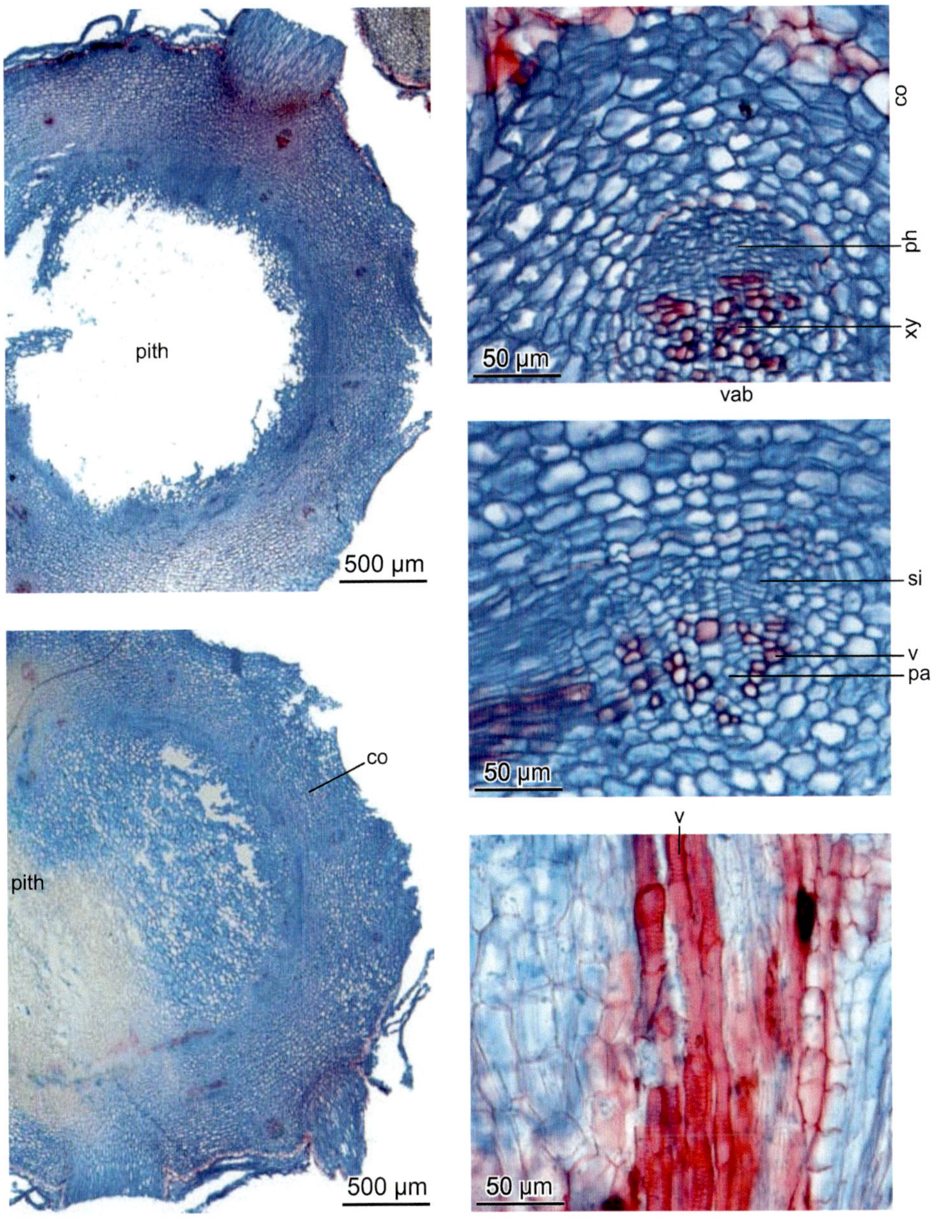

Features according to IAWA

Xylem 2 - 9 - 12 - 13 - 20 - 40.1 - 50.1 - 52.2 - 60.1 - 79.1 - 99 - 99.1 - 100.1 - 100.2 - 105

Cremanthodium ellisii (Hook.f.) Kitam.

Asteraceae

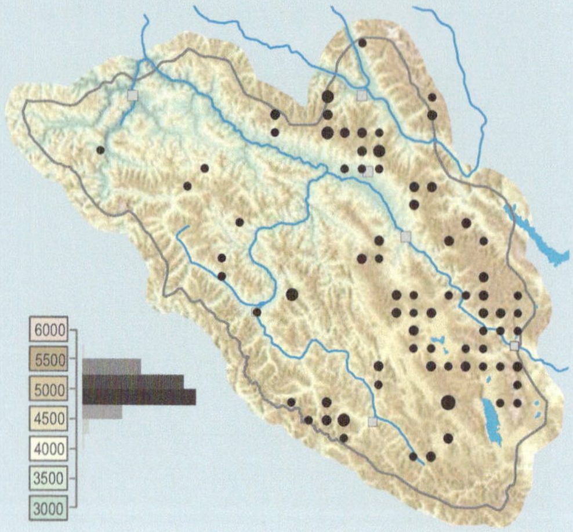

Syn.: *C. plantagineum* Maxim.

Description: Perennial herbs, 10–40 cm tall, leaves broadly elliptic to oblong, capitula ca. 5 cm across, ray florets yellow, disc florets dark yellow.

Distribution: Pakistan, India, Nepal, Bhutan, China.

Habitat: Snow beds and moss-patches close to streams and springs, river banks, alpine meadows, gravel slopes.

Stem diameter 3–5 mm. Bark/xylem proportion 1:1. Indistinct annual rings.

Xylem

Transverse section

Only one ring. With a few vascular bundles. In between large parenchymateous zones. Earlywood vessel diameter 20–30 µm. Fibers absent. Parenchyma pervasive.

Tangential section

Rays indistinct.

Radial section

Perforation plates simple. Intervessel pits scalariform. Vessel length 50–100 µm.

Bark

Sieve tubes stay in small groups lines. Cortex consists of round to oval, unlignified parenchyma cells. Phellem absent.

Pith

Round, hollow.

Analyzed: Rhizome of 3 individuals.

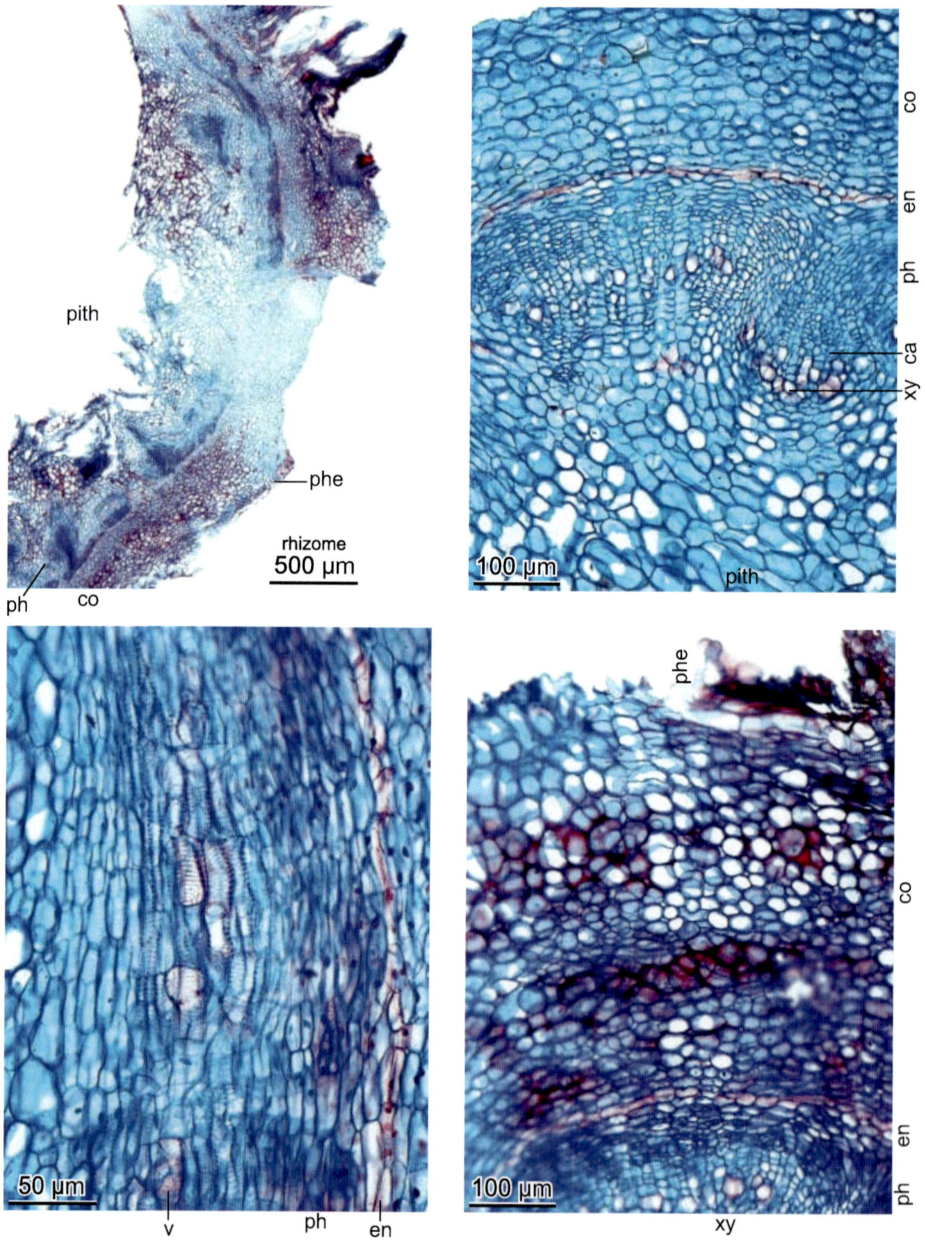

Features according to IAWA

Xylem 2.1 - 11 - 13 - 20 - 401 - 40.2 - 50.1 - 52.2 - 60.1 - 79.1 - 99 - 100.1 - 100.2 - 105

Echinops cornigerus DC.

Asteraceae

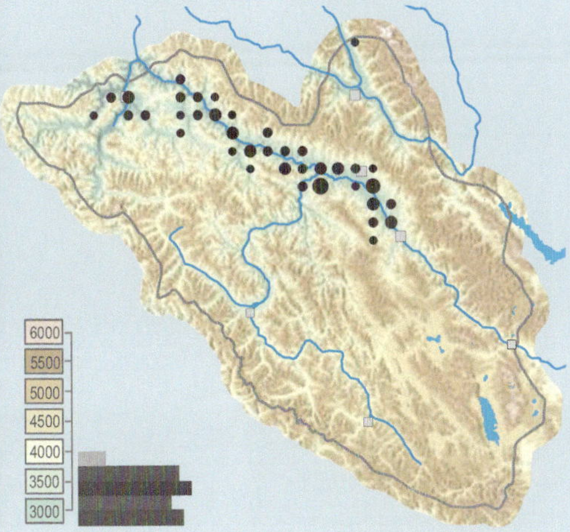

Description: Perennial herbs, 20–80 cm tall, leaves pinnatifid, spiny, silvery tomentose below, flowers blue, in a spheric synflorescence, up to 10 cm across.

Distribution: Afghanistan, NW Pakistan, N India (Jammu and Kashmir, Himachal Pradesh, Himachal Pradesh, Uttarakhand), Nepal.

Habitat: Sandy demideserts and deserts, unstable sandy slopes.

Stem diameter 10 mm. Bark/xylem proportion 1:1. Mean ring width 1.5 mm. Maximal 2 annual rings.

Xylem

Transverse section

Rings distinct. Ring boundaries indicated by marginal parenchyma. Vessel diameter 80–10 µm. Vessels in groups. Fibers thin- to thick-walled, arranged in small groups. Parenchyma paratracheal. Vascular bundles remain.

Tangential section

Rays 6–8-seriate, with sheath cells. Rays disappear in polarized light.

Radial section

Perforation plates simple. Intervessel pits round to slightly scalariform. Vessel length 120–150 µm.

Bark

Sieve tubes arranged in irregular radial groups. Phloem contains many small irregular groups of sclereids. Phellem consists of a few rows of large, square, thin-walled cells.

Pith

No slide available.

Analyzed: Root collar of polar root of 1 individual.

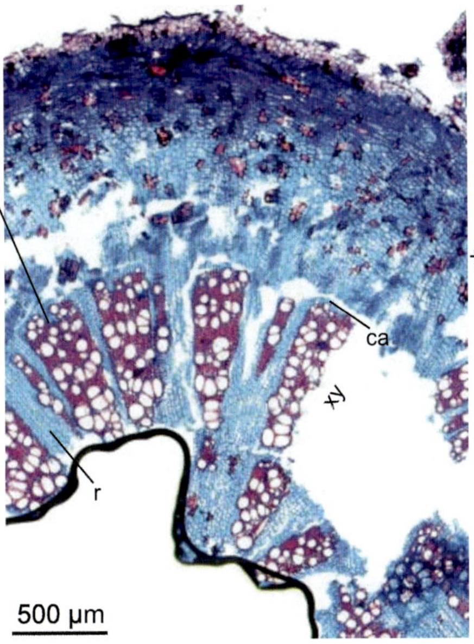

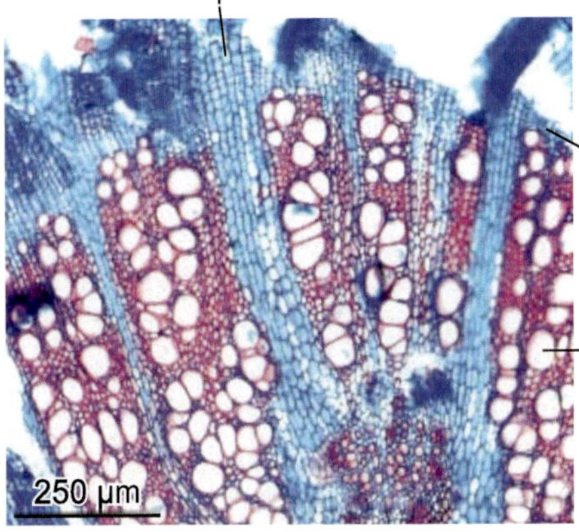

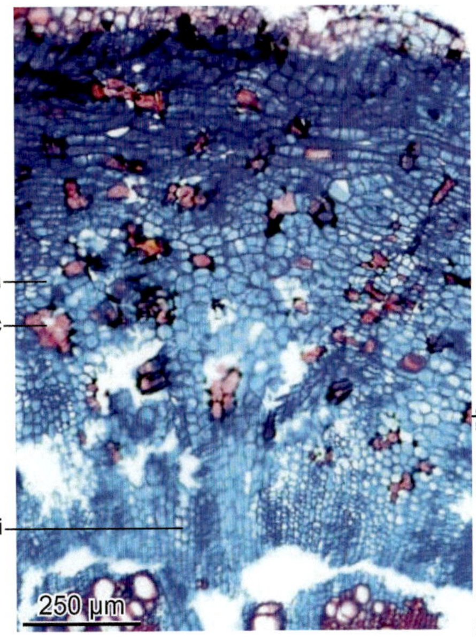

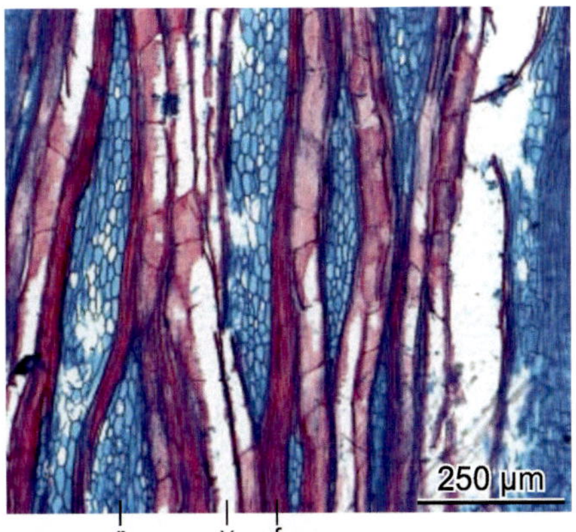

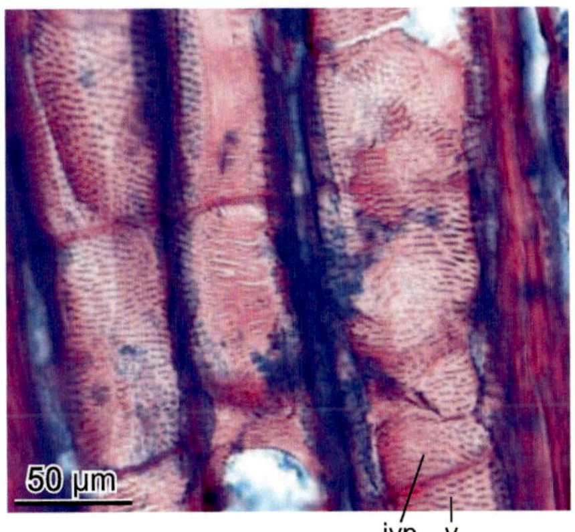

Features according to IAWA

Xylem 1 - 5 - 11 - 13 - 20 - 22 - 41 - 50.1 - 52.3 - 61 - 70 - 79 - 98 - 99.1 - 100.2 - 105

Erigeron uniflorus L.

Asteraceae

Description: Perennial herbs, 2–15 cm tall, wooly hairy, leaves both basal and cauline, oblong to oblanceolate, capitula 1–1.5 cm across, solitary, disc florets yellow, ray florets white or pink.

Distribution: N America, Europe, Asia (Caucasus, W Himalayas).

Habitat: Rock crevices, alpine meadows.

Stem diameter 3–4 mm. Bark/xylem proportion 1:1. Ring width 0.05–0.2 mm. Maximal 9 annual rings.

Xylem

Transverse section

Semi-ring-porous. Vessels solitary and in groups with partially angular outline. Ring boundary marked by large earlywood (15–20 μm) and small latewood vessel diameters (<10 μm). Fibers thin- to thick-walled, in groups. Parenchyma absent in fibrous zones and pervasive in fiber-less zones. Vascular bundles remain. Crystals absent.

Tangential section

Multiseriate rays with >10 cells width.

Radial section

Perforation plates simple. Intervessel pits round in alternating position. Vessel length 50–100 μm.

Bark

Phloem with small groups sieve tubes. Cortex consists of round parenchyma cells (100–130 μm) and small intercellulars. With large ducts. Phellem consists of many layers of rectangular cells.

Pith

Polygonal, cells large and round.

Analyzed: Root collar of polar root of 4 individuals.

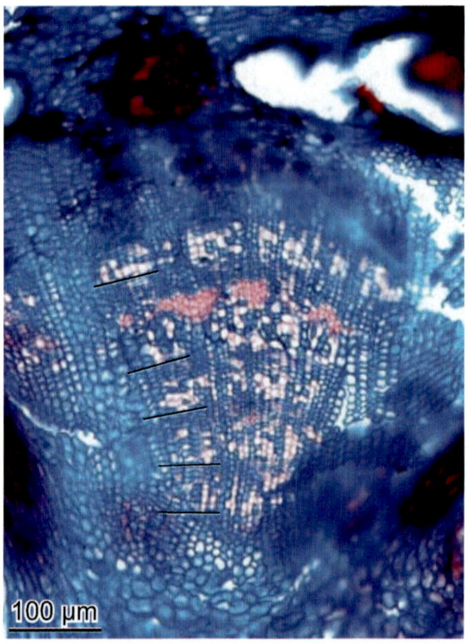

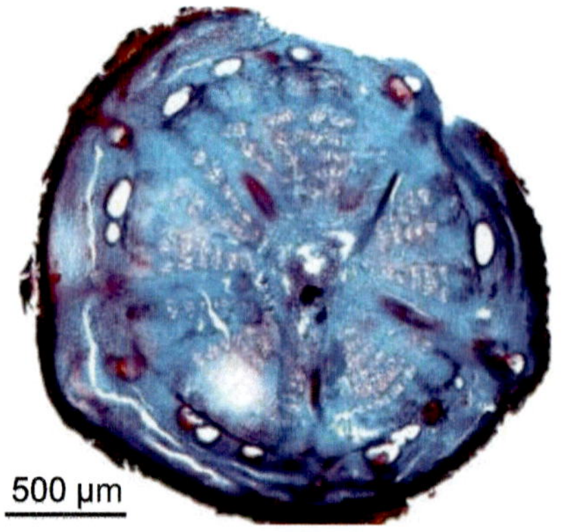

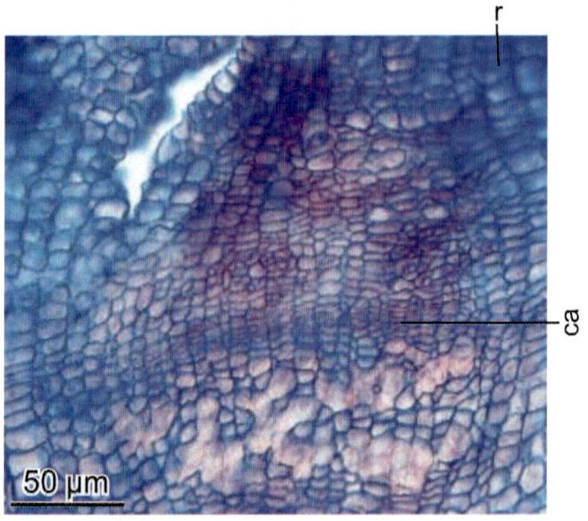

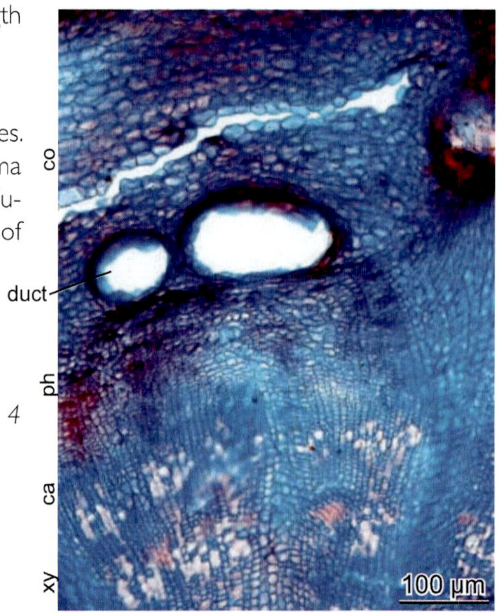

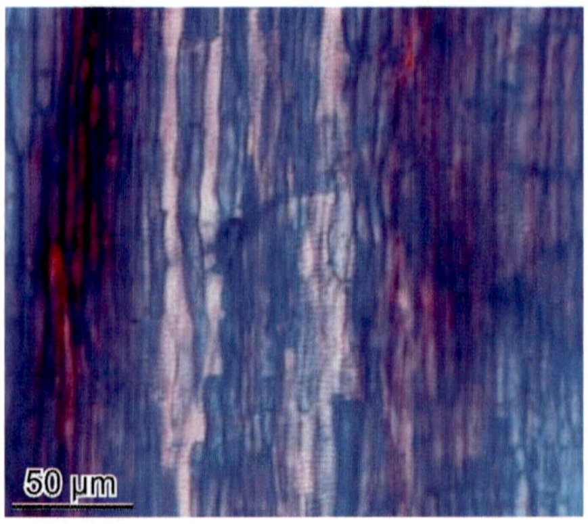

Features according to IAWA

Xylem 1 - 4 - 9 - 11 - 12 - 13 - 22 - 40.1 - 50.2 - 52.2 - 61 - 70 - 76 - 79.1 - 99 - 99.1 - 100.2 - 105

Filago paradoxa Wagenitz

Asteraceae

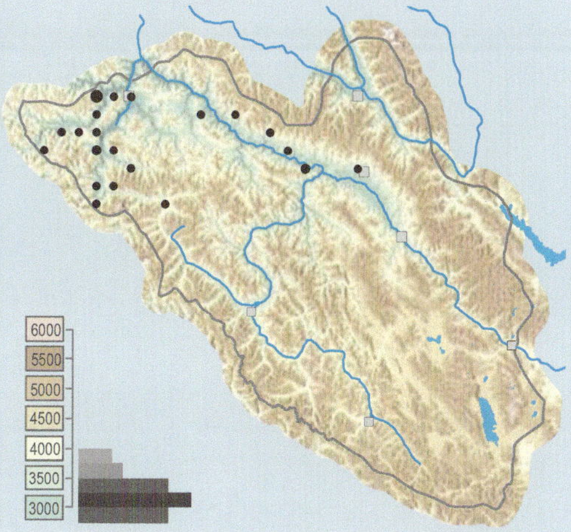

Description: Annuals, 5–20 cm tall, gray-tomentose, stems erect, branched from the base, leaves lanceolate to narrowly spathulate, heads yellowish, arranged by 6–10 in glomerules, 8–12 mm across.

Distribution: E Iran, Afghanistan, Pakistan, India (Jammu and Kashmir).

Habitat: Open sandy places.

Stem diameter 1–1.2 mm. Bark/xylem proportion 1:4. Mean ring width 0.3 mm. Maximal 1 annual ring.

Xylem

Transverse section

Vessels solitary and in radial groups. Vessel diameter 10–15 µm. Fibers thin- to thick-walled. Parenchyma paratracheal.

Tangential section

Rays absent.

Radial section

Simple perforation plates. Intervessel pits round, with slit-like apertures. Vessel length 140–160 µm.

Bark

Phloem uniform, very small. Ray dilatations. Cortex cells large (100 µm), round and rectangular. Sclerenchyma and crystals absent. Phellem absent.

Pith

Pith polygonal.

Analyzed: Rhizome of 2 individuals (herbarium specimens).

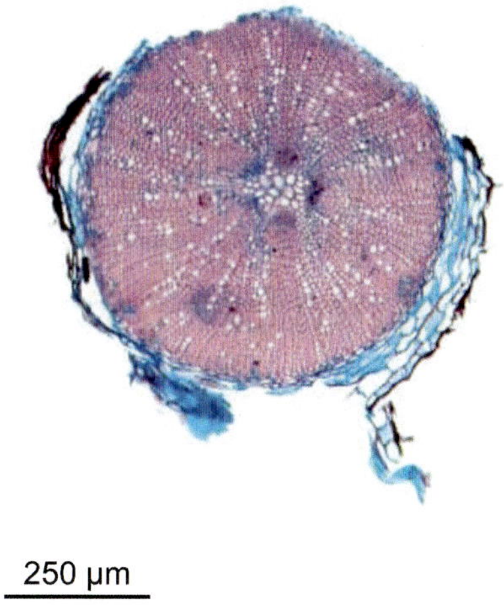

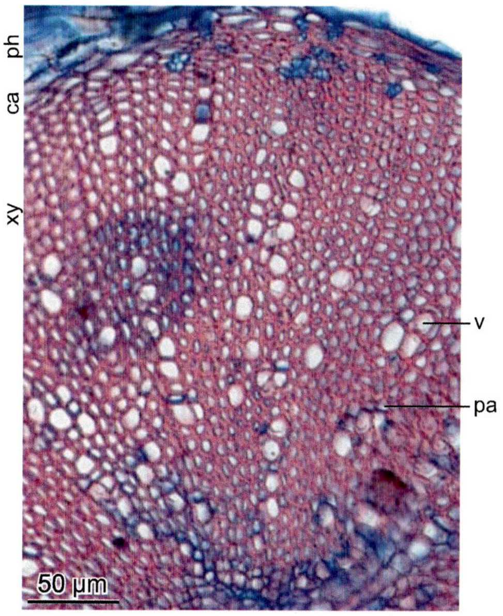

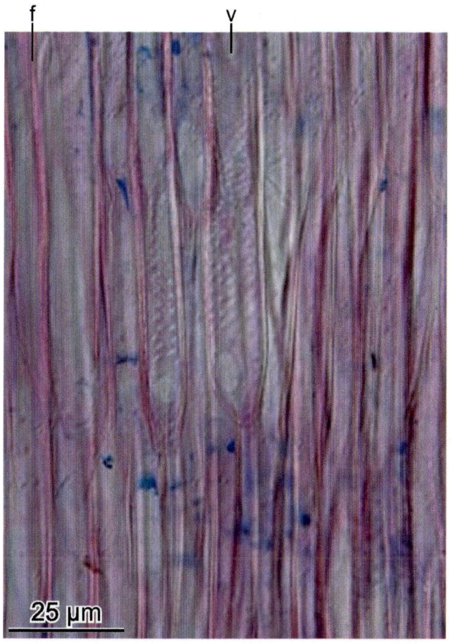

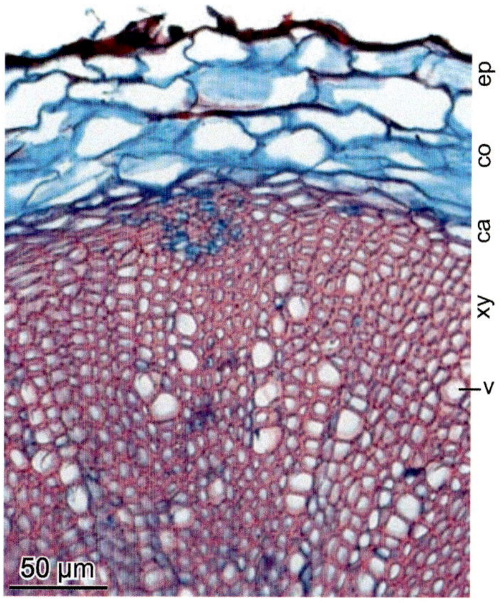

Features according to IAWA

Xylem 2.1 - 9.1 - 10 - 13 - 22 - 40.1 - 50.1 - 52.3 - 61 - 70 - 79 - 117

Galinsoga parviflora Cav.

Asteraceae

Syn.: *G. quinqueradiata* Ruiz et Pav., *Wiborgia acmella* Roth, *W. parviflora* (Cav.) Kunth

Description: Annuals, 5–60 cm tall, glabrous or sparsely hairy, leaves ovate, margin denticulate, capitula 5 mm across, arranged in terminal synflorescences, disc florets yellow, ray florets white.

Distribution: Native to Central and South America; widely naturalized.

Habitat: Fields, gardens, road sides.

Stem diameter 5–12 mm. Bark/xylem proportion 1:3. Ring width 2–6 mm. Maximal 1 annual ring.

Xylem

Transverse section

Vessels solitary and in short radial rows, often in intra-annual tangential rows. Vessel diameter 40–60 µm. Fibers thin- to thick-walled. Parenchyma paratracheal. Crystals absent.

Tangential section

Rays 4–8-seriate, with sheath cells.

Radial section

Perforation plates simple. Intervessel pits round in alternating position. Vessel length 100–120 µm. Ray cells square and upright.

Bark

Phloem with a few small, groups of sieve tubes. Slight ray dilatations. Cortex cells large (50–80 µm), round to oval. Ducts absent. Phellem cells large, thin-walled.

Pith

Round, partially with unlignified large cells.

Analyzed: Root collar of polar root of 2 individuals.

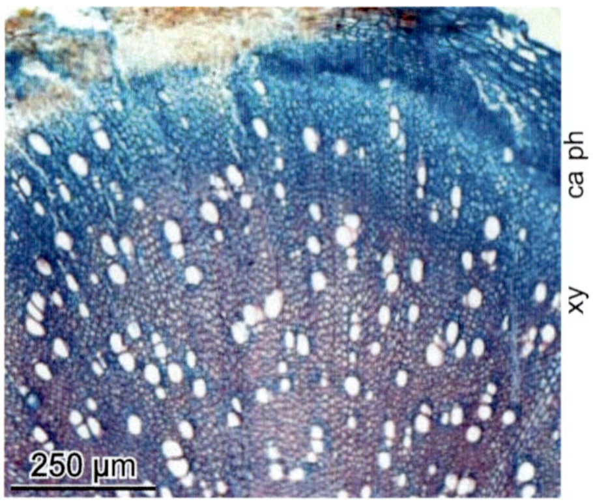

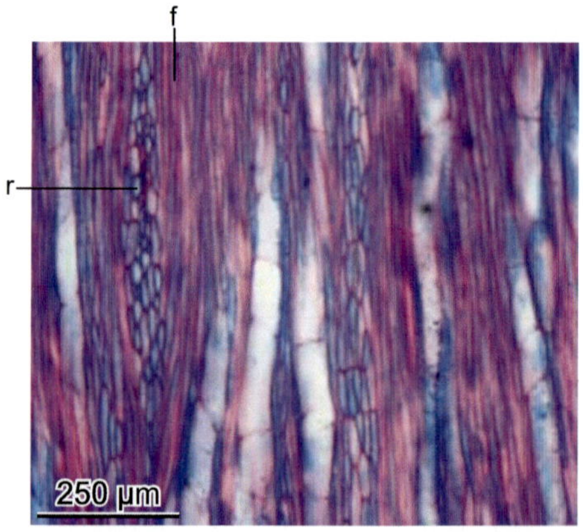

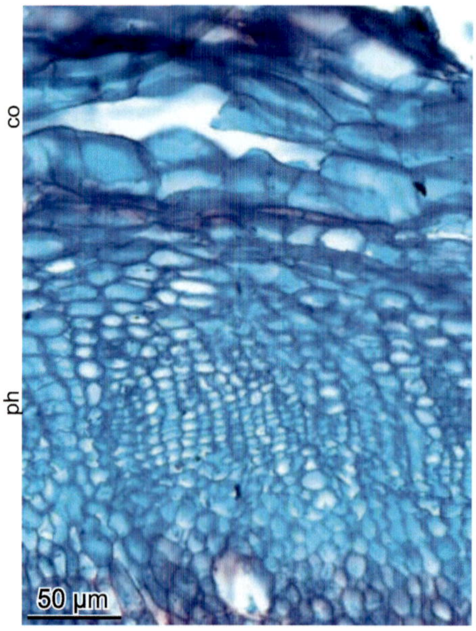

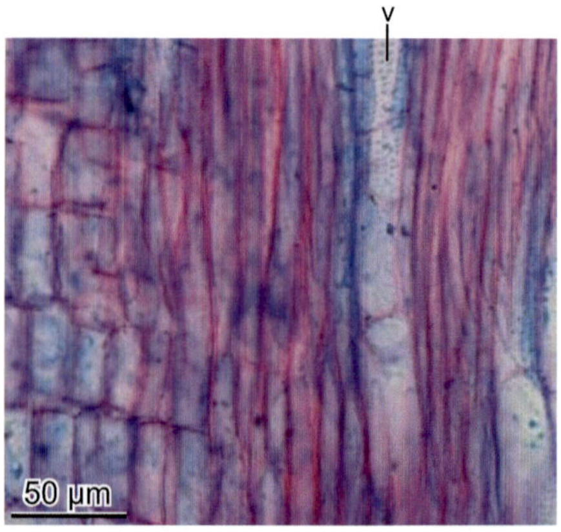

Features according to IAWA

Xylem 2.1 - 6 - 9 - 9.1 - 13 - 22 - 41 - 50.1 - 52.3 - 61 - 70 - 79 - 98 - 105 - 110

Heteropappus semiprostratus Griers.

Asteraceae

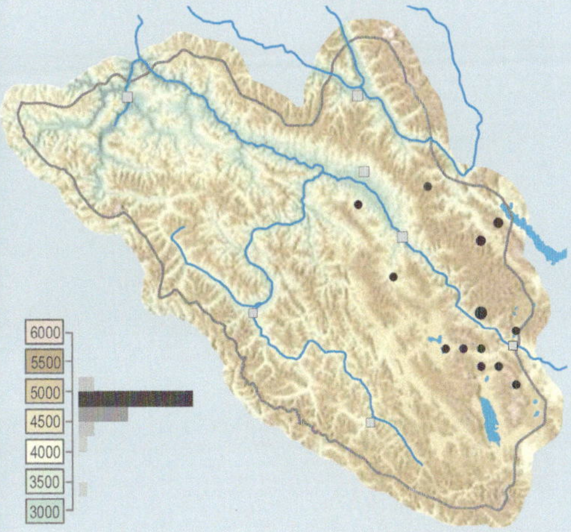

Syn.: *Aster semiprostratus* (Griers.) H.Ikeda

Description: Perennial herbs, 5–15 cm tall, cespitose, strigose and glandular, stems decumbent or ascending, leaves linear or oblanceolate to spatulate, capitula 1.5–3 cm across, solitary at the ends of stems or branches, ray florets blue or purplish, disc florets yellow.

Distribution: N India, Nepal, China (Qinghai, Xizang).

Habitat: Stony steppes, gravel and sandy slopes.

Stem diameter 2-5 mm. Bark/xylem proportion 1:2. Ring width 0.1–0.6 mm. Maximal 9 annual rings.

Xylem

Transverse section

Semi-ring-porous. Vessels solitary and in groups with angular outline. Ring boundary marked by large earlywood (25–30 µm) and small latewood vessel diameters (<10 µm). Thin- to thick-walled fibers in the earlywood. Parenchyma absent or paratracheal in the earlywood, pervasive in the latewood. Vascular bundles remain. Crystals absent.

Tangential section

Multiseriate rays with >10 cells width.

Radial section

Perforation plates simple. Intervessel pits round in alternating position. Vessel length 100–180 µm.

Bark

Phloem with small groups sieve tubes. Distinct ray dilatations. With a few isolated fibers. Cortex consists of round parenchyma cells (100–130 µm) and small intercellulars. With large ducts. Phellem consists of many layers of rectangular cells. With phellem towards the pith.

Pith

No slide available.

Analyzed: Root collar of polar root of 3 individuals.

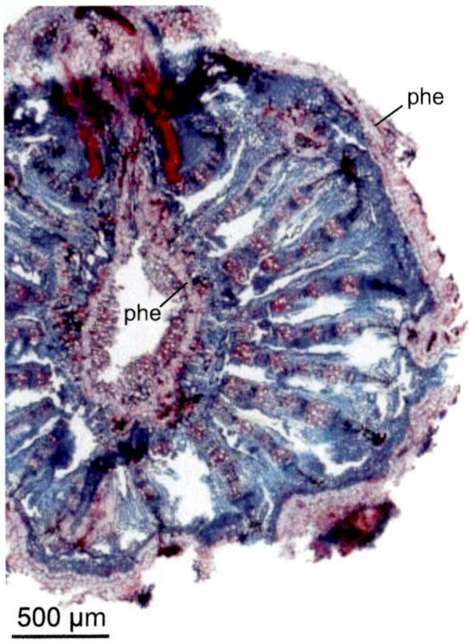

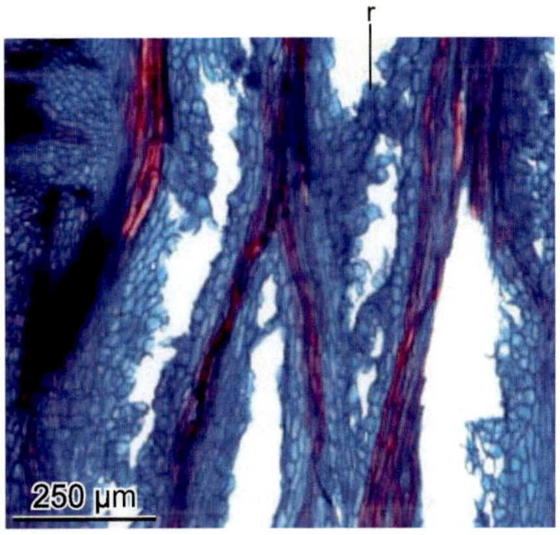

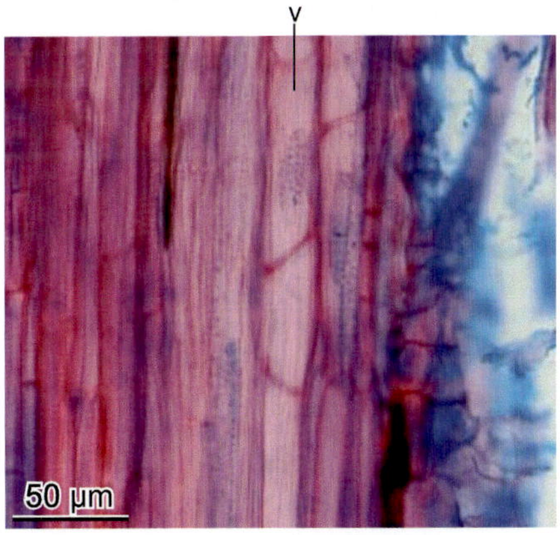

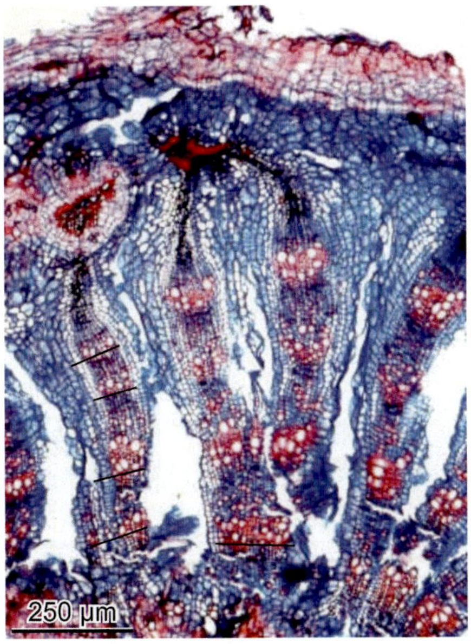

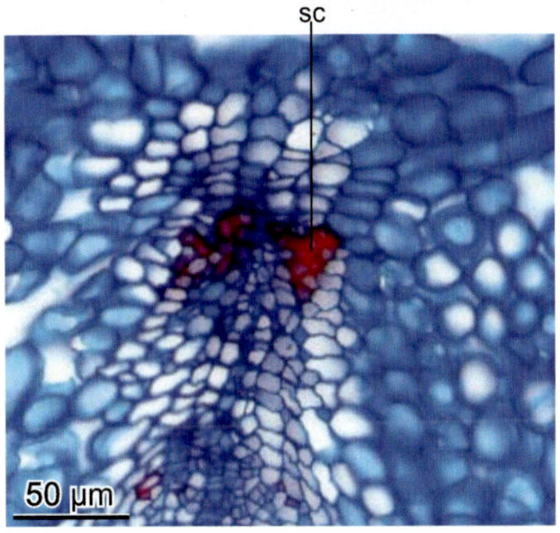

Features according to IAWA

Xylem 1 - 4 - 9 - 12 - 13 - 22 - 40.2 - 50.1 - 52.2 - 61 - 70 - 76 - 79 - 79.1 - 99 - 99.1 - 100.2 - 105

Hippolytia senecionis (Jacquem. ex Besser) Tzvelev

Asteraceae

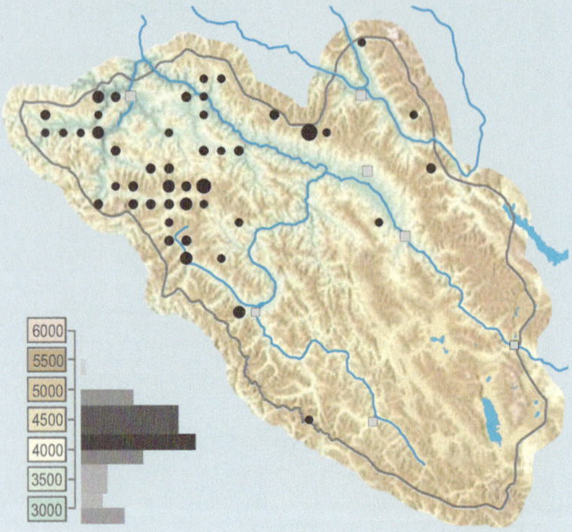

Syn.: *Artemisia senecionis* Jacquem. ex Besser, *Tanacetum senecionis* (Jacquem. ex Besser) DC.

Description: Perennial herbs, 10–40 cm tall, leaves 2-pinnatisect, ultimate segments linear-oblong, gray hairy, capitula yellow, 5–6 mm across, in congested terminal cymes.

Distribution: NE Afghanistan, N Pakistan, N India, China (Xizang).

Habitat: Desert steppes.

Stem diameter 3.5 mm. Bark/xylem proportion 1:2. Mean ring width 0.1 mm. Maximal 5 annual rings.

Xylem

Transverse section

Semi-ring-porous. Solitary vessels with angular outline. Ring boundary marked by large earlywood (40–50 µm) and small latewood vessel diameters (<10 µm). Thick-walled fibers occur in tangential groups. Parenchyma pervasive. Vascular bundles remain.

Tangential section

Multiseriate rays with >10 cells.

Radial section

Perforation plates simple. Intervessel pits round in alternating position. Vessel length 60–100 µm.

Bark

Phloem with small groups of sieve tubes. Sclereids in irregular groups. Cortex consists of round parenchyma cells (100–130 µm) and small intercellulars. With large oil ducts. Phellem consists of many layers of rectangular cells.

Pith

Round. With irregular intercellulars.

Analyzed: Root collar of polar root of 1 individual.

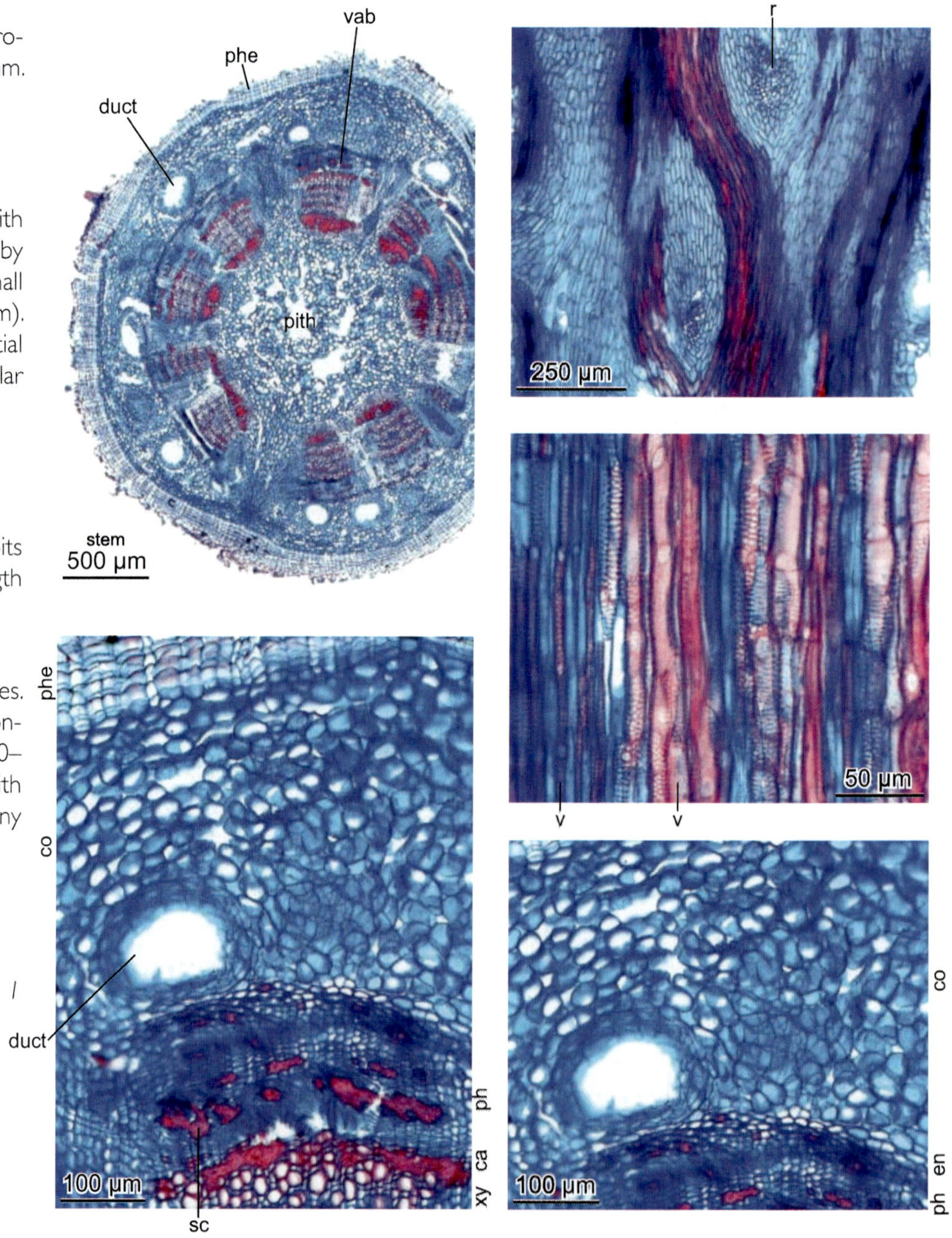

Features according to IAWA

Xylem 1 - 4 - 9 - 12 - 13 - 20 - 40.2 - 50.2 - 52.2 - 61 - 69 - 79.1 - 99 - 99.1 - 105

Inula koelzii Dawar & Qaiser.

Asteraceae

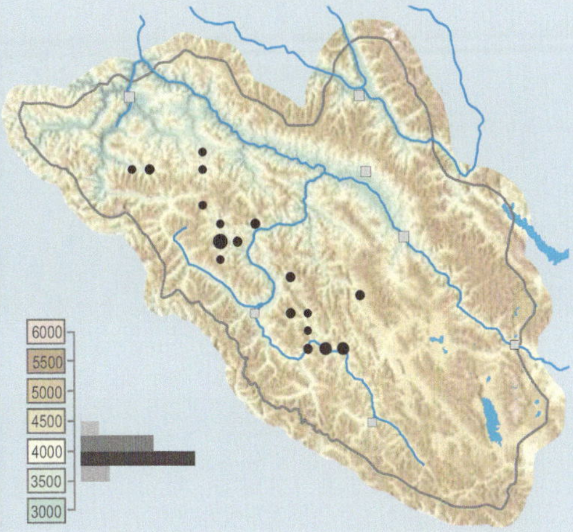

Description: Perennial herbs, 15–40 cm tall, pubescent, leaves broadly lanceolate to ovate, margin serrate, capitula yellow, 3–5 cm across, in few-flowered corymbs.

Distribution: India (Jammu and Kashmir).

Habitat: Mesic stony screes, along water streams.

Stem diameter 8–12 mm. Bark/xylem proportion 1:3. Ring width 0.1–0.4 mm. Maximal >4 annual rings. Center rotten.

Xylem

Transverse section

Semi-ring-porous. Rings distinctly marked by differences between earlywood (50–80 μm) and latewood vessel diameters (10–20 μm). Vessels are mostly solitary or in small groups. Fibers absent. Parenchyma pervasive. Vascular bundles remain.

Tangential section

Rays indistinct, confluent >10-seriate, disappearing in polarized light.

Radial section

Perforation plates simple. Intervessel pits round in alternating position to scalariform. Vessel length 80–120 μm. Ray cells upright.

Bark

Phloem contains small groups of sieve elements. With ducts. Cortex absent. Phellem towards the periphery and the center consists of a few rows or large and many rows of compressed, thin-walled cells.

Pith

No slide available.

Analyzed: Root collar of polar root of 4 individuals.

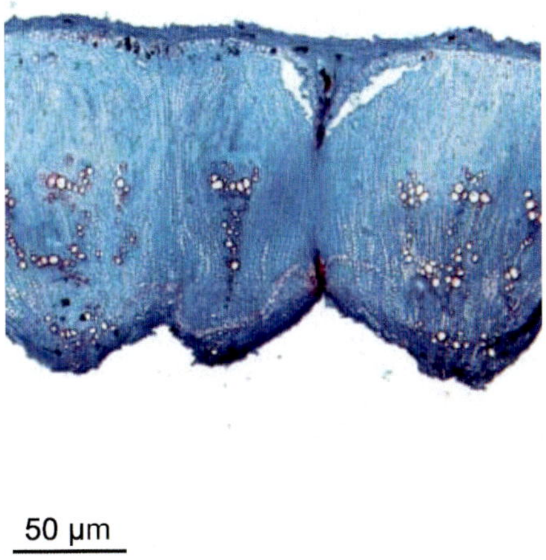

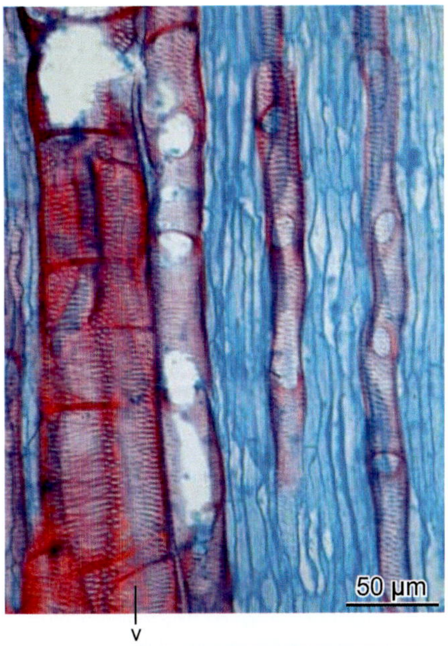

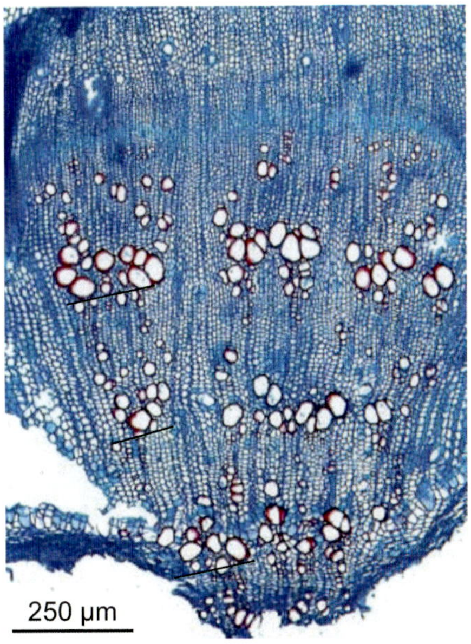

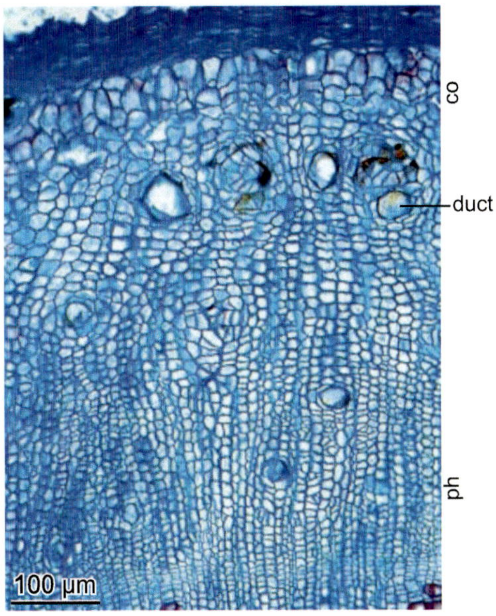

Features according to IAWA

Xylem 1 - 4 - 9 - 11 - 13 - 20 - 22 - 41 - 50.1 - 52.2 - 52.3 - 60.1 - 79.1 - 99 - 99.1 - 101.1 - 100.2 - 105

Inula obtusifolia A.Kern.
Asteraceae

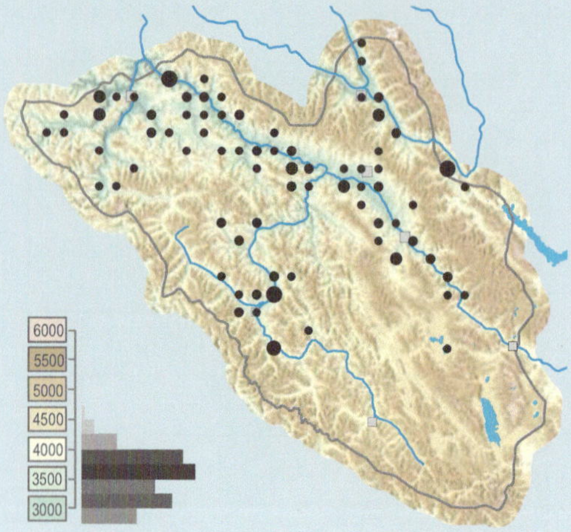

Description: Perennial herbs, 10–30 cm tall, pilose, leaves sessile, elliptic, 3–7 × 2–3 cm, capitula yellow, up to 3 cm across, terminal, usually solitary.

Distribution: E Afghanistan, Pakistan, India, China (Xizang).

Habitat: Dry cliffs, mesic stony screes, along water streams.

Stem diameter 8 mm. Bark/xylem proportion 1:5. Mean ring width 0.15 mm. Maximal 20 annual rings.

Xylem

Transverse section
Semi-ring-porous to ring-porous. Rings distinctly marked by large vessels in the earlywood (in average 70 µm diameter) and small vessels in the latewood (20 µm diameter). Ring variability mainly indicated by larger (80 µm) or smaller (40 µm) earlywood vessel diameters. Vessels solitary or in short tangential groups. Fibers thin- to thick-walled. Dark-staining substances in fibers of the center (heartwood). Parenchyma paratracheal.

Tangential section
Rays indistinct and sometimes confluent with the axial tissue.

Radial section
Perforation plates simple. Intervessel pits round in alternating position. Vessel length 110–180 µm.

Bark

Phloem consists of rectangular cells. With many small ducts. Cortex cells large, round to oval (50–80 µm). Phellem cells thin-walled, irregularly formed.

Pith

Polygonal, star-shaped.

Analyzed: Root collar of polar root of 1 individual.

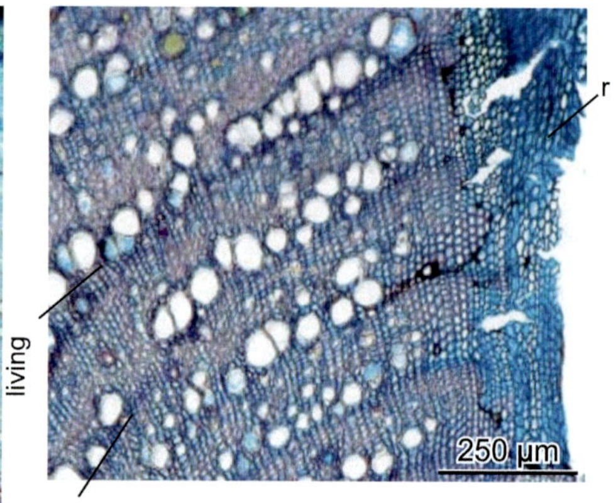

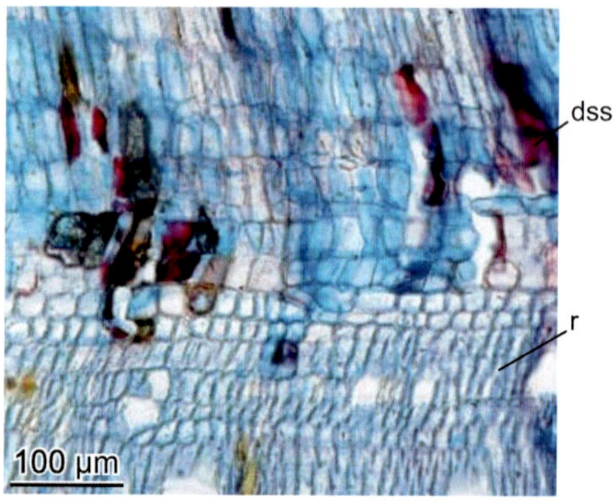

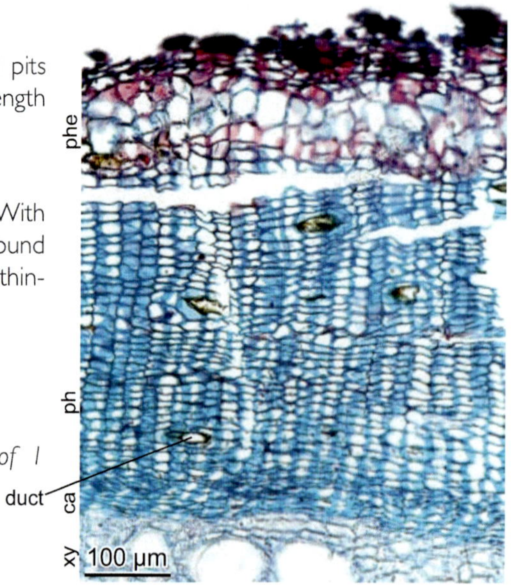

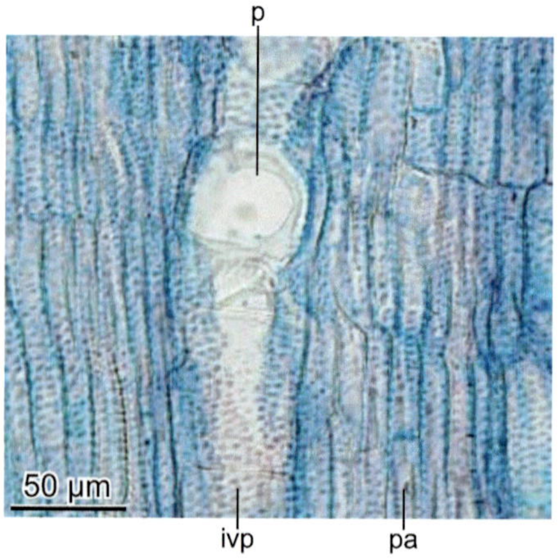

Features according to IAWA

Xylem 1 - 5 - 9 - 13 - 22 - 41 - 50.1 - 52.2 - 52.3 - 61 - 70 - 79 - 89 - 89.1 - 98 - 105

Lactuca lessertiana Wall. ex C.B.Clarke

Asteraceae

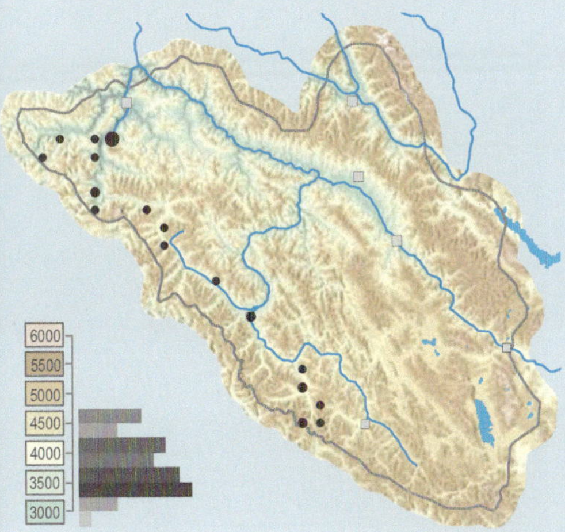

Syn.: *Cicerbita lessertiana* (DC.) Mamgain & R.R.Rao, *Melanoseris lessertiana* (DC.) Decne., *Mulgedium lessertianum* DC.

Description: Perennial herbs, 5–30 cm tall, glabrous, basal leaves elliptic to oblanceolate, undivided to lyrately pinnatisect, margin entire to shallowly sinuate-dentate, upper stem leaves narrower and smaller, capitula blue, 1.5–3 cm across, single or several in a lax racemes.

Distribution: Pakistan, N India, Nepal, China (Xizang, Yunnan), Bhutan.

Habitat: Grassy slopes.

Stem diameter 7 mm. Bark/xylem proportion 1:1. Mean ring width 0.3 mm. Maximal 1 annual ring.

Xylem

Transverse section

Semi-ring-porous. Rings indistinct, slightly marked by differences between earlywood (40–50 µm) and latewood vessel diameters (10–20 µm). Vessels mostly in small radial groups. Fibers absent. Parenchyma pervasive. Vascular bundles remain.

Tangential section

Rays indistinct, >10-seriate and disappearing in polarized light.

Radial section

Perforation plates simple. Intervessel pits round in alternating position to scalariform. Vessel length 120–180 µm. Ray cells upright.

Bark

Phloem contains small, isolated and radial groups of sieve tubes. Cortex consist of an irregularly formed association of cells and intercellular spaces. Phellem cells thin-walled, irregularly formed.

Pith

Round, surrounded by vascular bundles. Small groups of phloem at the periphery of the pith.

Analyzed: Root collar of polar root of 1 individual.

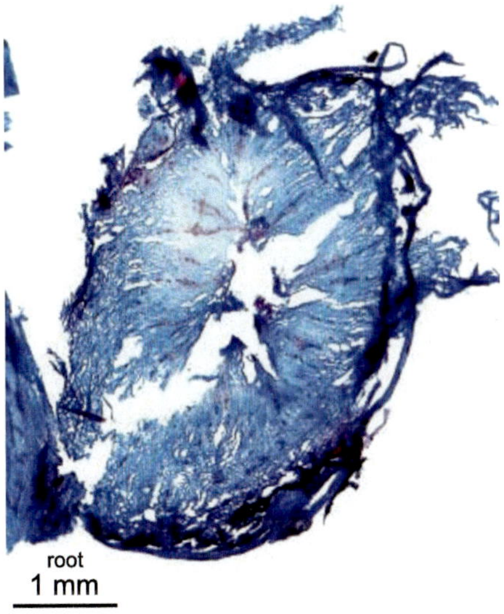

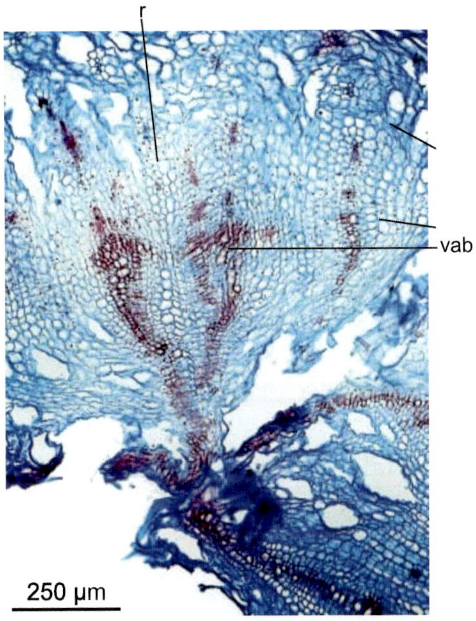

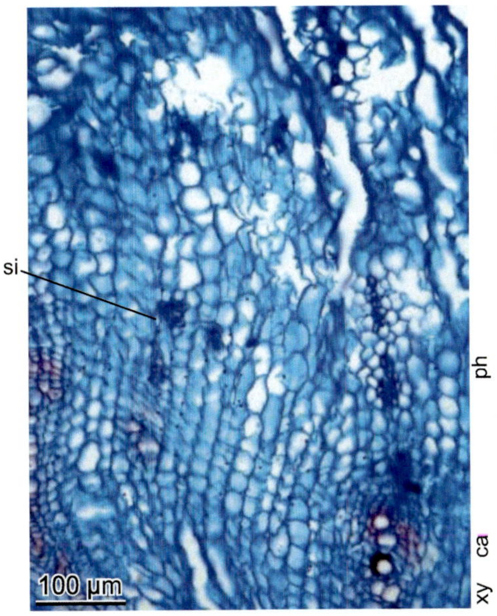

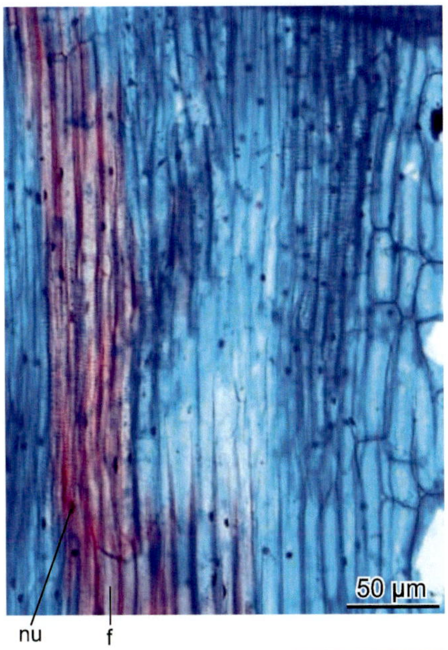

Features according to IAWA

Xylem 1 - 4 - 10 - 13 - 20 - 22 - 40.2 - 50.1 - 52.3 - 60.1 - 79.1 - 99 - 99.1 - 105

Lactuca orientalis (Boiss.) Boiss.

Asteraceae

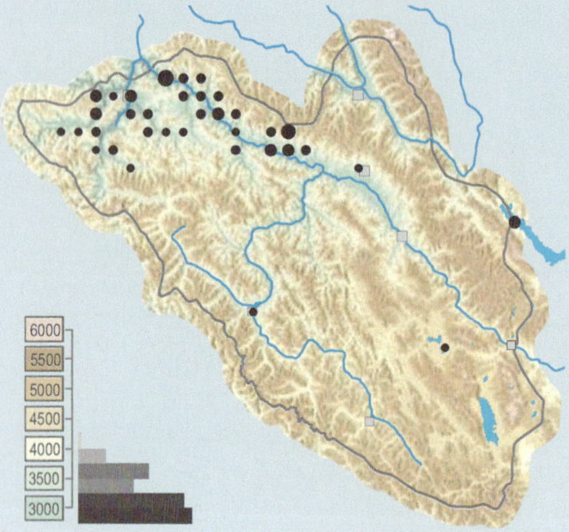

Syn.: *L. viminea* var. *erostris* Regel, *Scariola orientalis* (Boiss.) Soják

Description: Perennial herbs, 20–60 cm tall, glabrous, stems rigid, basal leaves rosulate, glaucous, sinuate-dentate, stem leaves basally with linear auricles adnately long decurrent on stem, capitula pale yellow, ca. 1 cm across, solitary.

Distribution: N Africa, SW Asia, Caucasus, E Turkey, Iran, Turkmenistan, Uzbekistan, Kyrgyzstan, Tajikistan, Afghanistan, Pakistan, N India, NW China.

Habitat: Mountain slopes and scree.

Stem diameter 3–10 mm. Bark/xylem proportion 1:10. Ring width 0.25–0.4 mm. Maximal 15 annual rings.

Xylem

Transverse section

Distinct rings. Diffuse- to semi-ring-porous, vessels solitary. Ring boundary marked by cork layers. Vessel diameter 70–100 µm. Thick-walled fibers partially in the latewood. Parenchyma pervasive.

Tangential section

Rays >10-seriate, irregularly present. Rays confluent, unlignified, disappearing in polarized light.

Radial section

Perforation plates simple. Intervessel pits round in alternating position. Vessel length 80–100 µm.

Bark

Phloem with tangentially arranged groups of sieve tubes. Ducts absent. Phellem consists of radial rows of thin-walled square and rectangular cells.

Pith

Round with medullary phloem.

Analyzed: Root collar of polar root of 3 individuals.

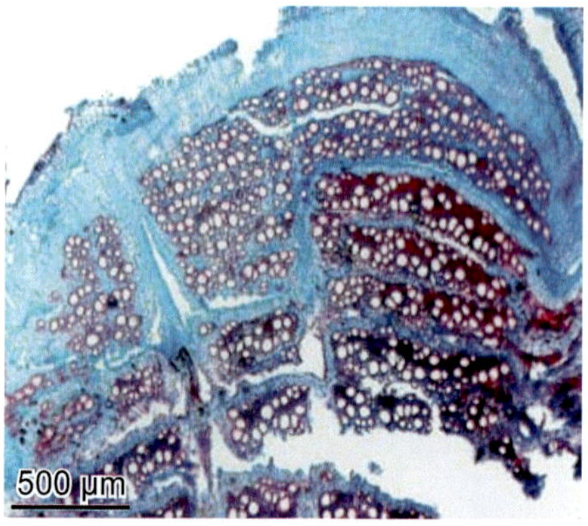

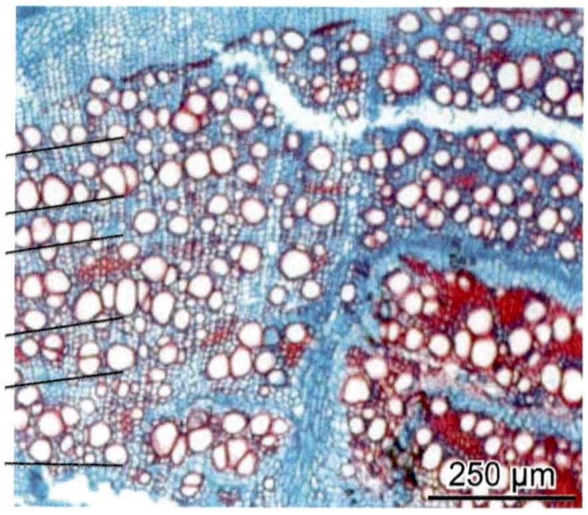

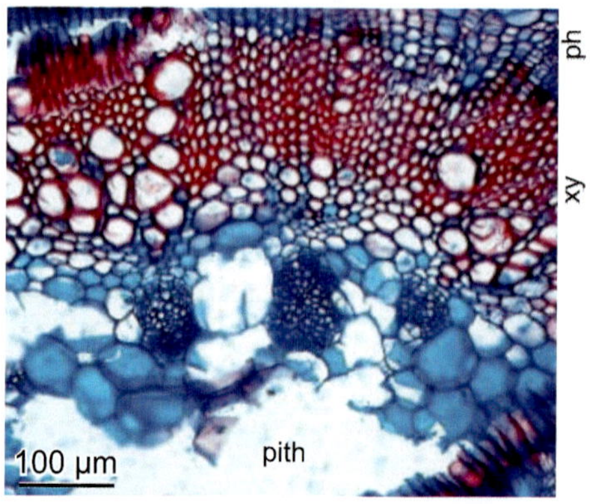

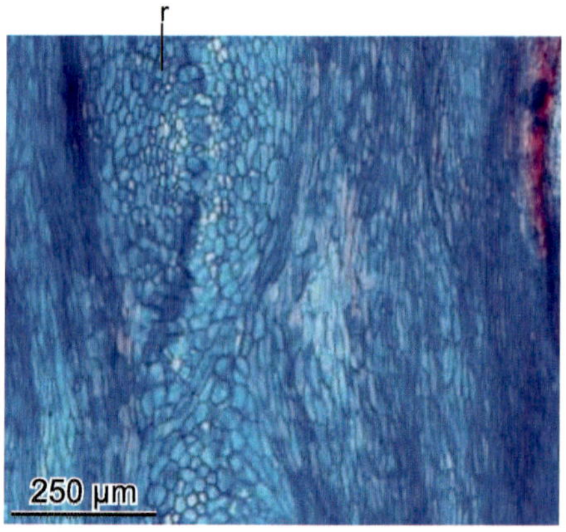

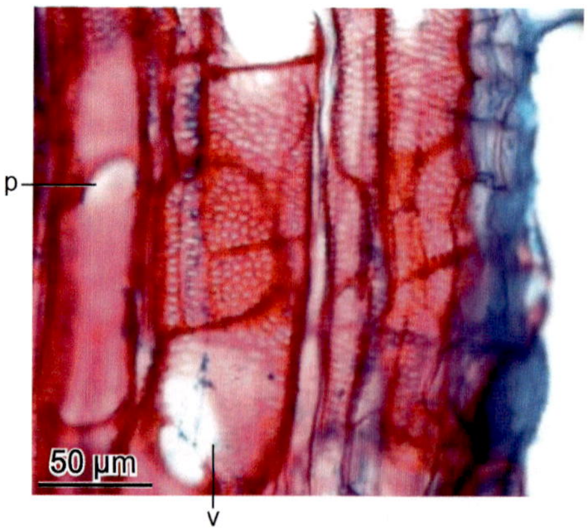

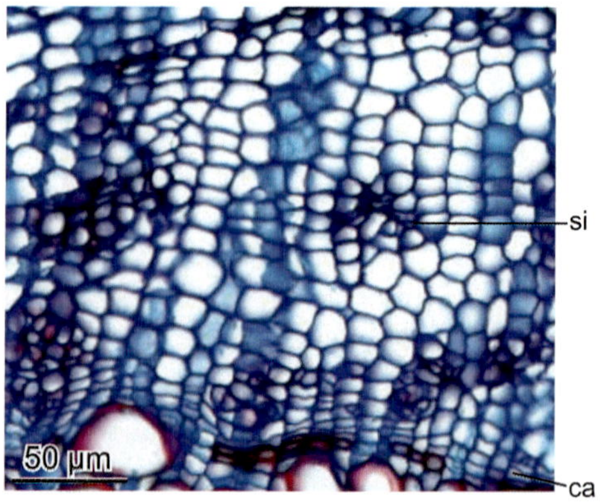

Features according to IAWA

Xylem 1 - 4 - 9 - 13 - 22 - 41 - 50.2 - 52.2 - 61 - 69 - 79.1 - 99 - 100.1 - 105 - 135.1

Lactuca tatarica (L.) C.A.Mey.

Asteraceae

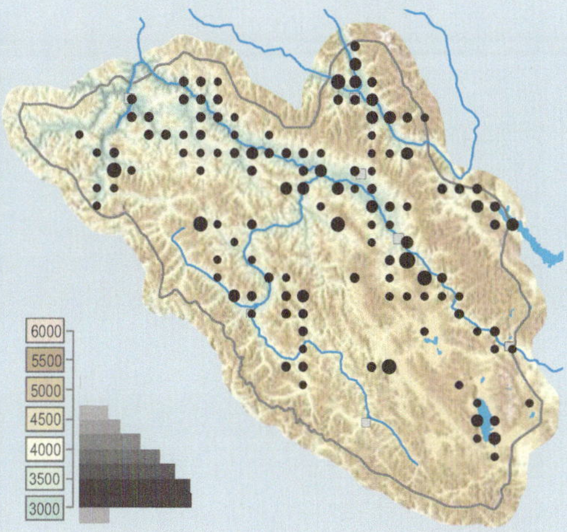

Syn.: *Sonchus tataricus* L., *Lagedium tataricum* (L.) Soják, *Mulgedium tataricum* (L.) DC.

Description: Perennial herbs, 15–60 cm tall, stems erect, leaves pinnatisect to pinnatifid,, glaucous, capitula purple blue, 1.5–2 cm across, in terminal corymbs.

Distribution: E Europe, W Asia, Russia (Siberia), Kazakhstan, Turkmenistan, Uzbekistan, Kyrgyzstan, Tajikistan, Afghanistan, N Pakistan, N India, W and N China, Mongolia; naturalized in W Europe, adventive to N America.

Habitat: Exposed loamy and saline soils, sandy soils along roads.

Stem diameter 7–15 mm. Bark/xylem proportion 1:2. Ring width 0.2–0.5 mm. Maximal 7 annual rings.

Xylem

Transverse section

Semi-ring-porous. Rings distinctly marked by large differences between earlywood (50–70 µm) and latewood vessel diameters (10–20 µm). Vessels mostly solitary. A belt of thin- to thick-walled fibers occurs only in the peripheral ring. Parenchyma paratracheal and pervasive. Vascular bundles remain.

Tangential section

Rays 5–>10-seriate, disappearing in polarized light

Radial section

Perforation plates simple. Intervessel pits round in alternating position. Vessel length 100–150 µm. Ray cells upright.

Bark

Phloem contains groups of sieve tubes. Cortex cells large (80–100 µm), round to oval. Small ducts occur only in annual shoots. Phellem cells thin-walled, irregularly formed.

Pith

Round, surrounded by vascular bundles. Small groups of phloem at the periphery of the pith (medullary phloem in annual shoots).

Analyzed: Rhizome of 2 individuals.

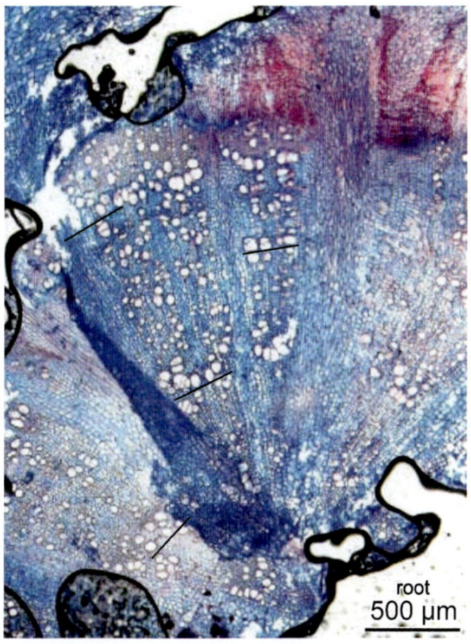

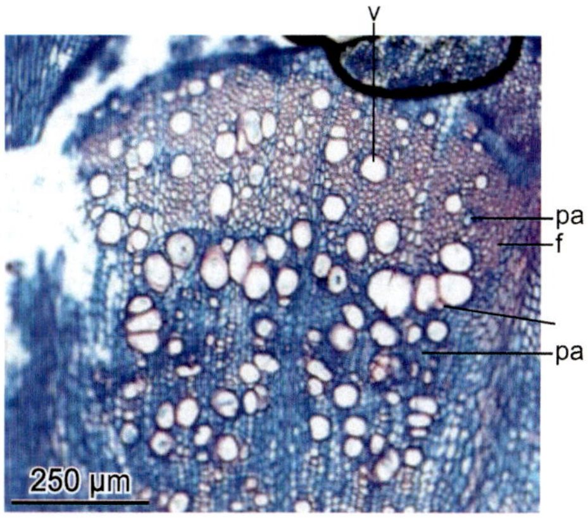

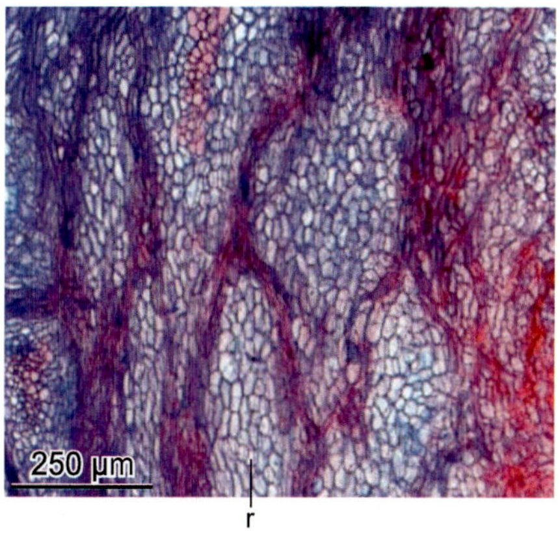

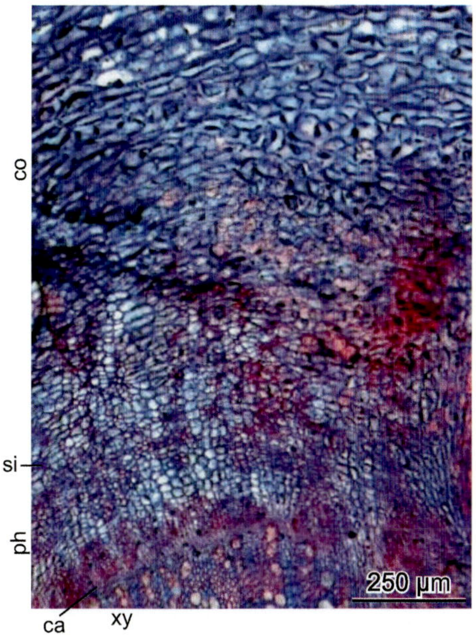

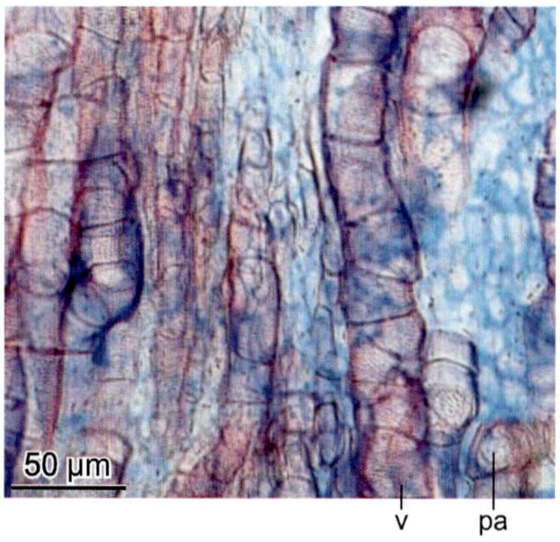

Features according to IAWA

Xylem 1 - 4 - 9 - 13 - 22 - 41 - 50.1 - 52.3 - 61 - 70.3 - 79 - 79.1 - 98 - 99 - 99.1 - 100.2 - 105

Leontopodium leontopodinum (DC.) Hand.-Mazz.

Asteraceae

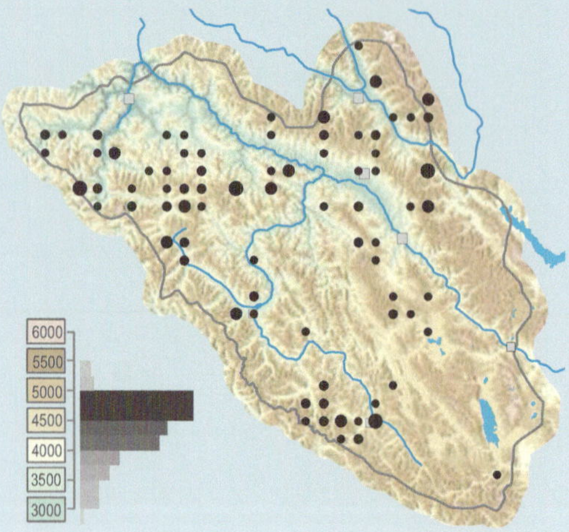

Syn.: *L. ochroleucum* Beauv.

Description: Perennial herbs, 5–15 cm tall, densely white tomentose, leaves oblong, capitula 5–7 mm across, densely aggregated by 5–7, together with bracteal leaves forming a star-shaped synflorescence, 1.5–2.5 cm across.

Distribution: Russia, Kazakhstan, Afghanistan, Pakistan, India, China (Qinghai, Xinjiang, Xizang), Butan, Mongolia.

Habitat: Stabilized mesic to dry slopes, stony deposits, dry meadows.

Stem diameter 3–5 mm. Bark/xylem proportion 1:4. Mean ring width 1.2 mm. Maximal 7 annual rings.

Xylem

Transverse section

Diffuse-porous. Vessels solitary and in small groups. Fibers and vessels are difficult to distinguish on cross-sections, because they have the same diameter. Thick-walled vessels with diameters of 5–10 µm. Ring boundary marked by large earlywood and small latewood fibers. Fibers thin- to thick-walled. Parenchyma absent or unrecognizeable.

Tangential section

Rays absent or indistinct.

Radial section

Perforation plates simple. Intervessel pits round. Vessel length 80–100 µm.

Bark

Large, uniform phloem. Cortex cells large (15–20 µm), round to oval. Sclerenchyma and crystals absent. Phellem layered.

Pith

Roundish, filled with fairly lignified, thick-walled parenchyma cells.

Analyzed: Rhizome of 1 individual.

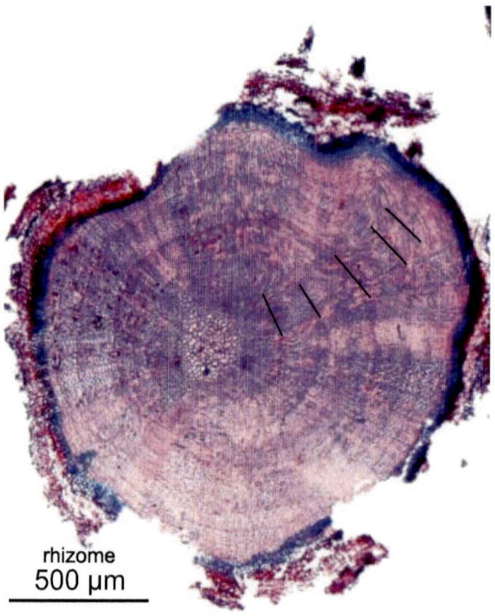

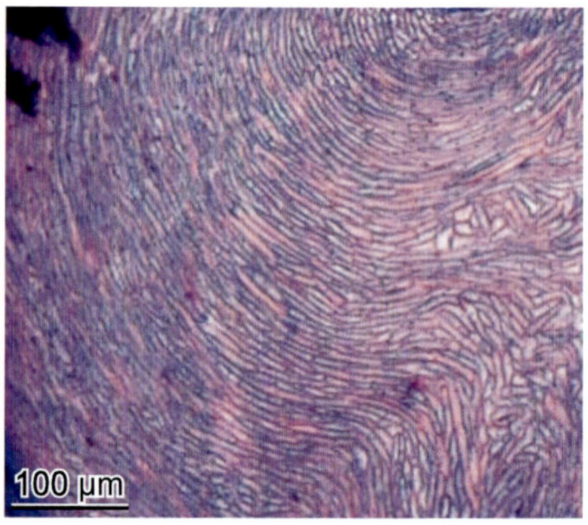

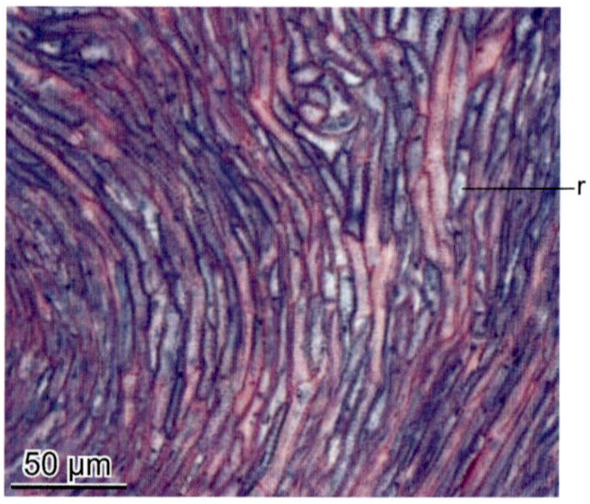

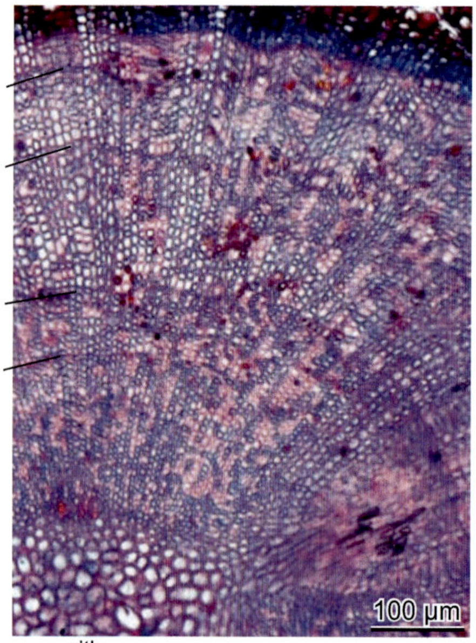

pith

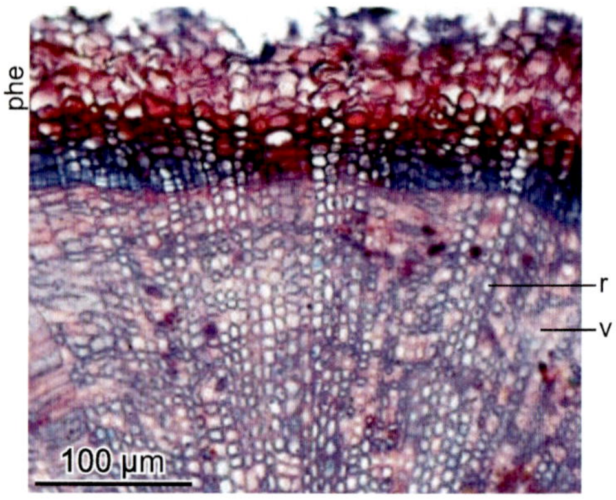

Features according to IAWA

Xylem 1 - 5 - 9 - 11 - 13 - 20 - 39.1 - 40.1 - 50.1 - 53.1 - 61 - 75 - 117

Lipschitziella stemmaphora (Klatt) Dickoré, comb. nov.

Asteraceae

Syn.: *Jurinea ceratocarpa* var. *depressa* C.B.Clarke, *Saussurea ceratocarpa* var. *depressa* (C.B.Clarke ex Hook.f.) Lipsch., *S. stemmaphora* Klatt

Description: Perennial herbs, 5–10 cm tall, stemless, leaves basal, pinnatipartite or lyrate, wooly below, capitula purplish blue, ca. 2 cm across, solitary or few.

Distribution: Pakistan, NW India.

Habitat: Wet to mesic meadows, gravel banks of streams.

Stem diameter 2 mm. Bark/xylem proportion 1:1. Ring width 0.2–1 mm. Maximal 3 annual rings.

Xylem

Transverse section

Semi-ring-porous. Earlywood vessels in radial rows or small groups. Ring boundary marked by large earlywood (30–40 µm) and small latewood vessel diameters (<10 µm). Fibers forming a peripheral continuous belt. Fibers absent in the center. Parenchyma paratracheal in the fiber belt and pervasive in the center. With remaining vascular bundles. Large ray-like parenchyma zones between vascular bundles.

Tangential section

Rays between vascular bundles 10-seriate, mostly confluent. Ray cells upright. Rays disappear in polarized light.

Radial section

Perforation plates simple. Intervessel pits round at the stem periphery and scalariformin the center. Vessel length 80–150 µm.

Bark

Phloem with scattered, small groups of sieve tubes. Cortex cells large (15–20 µm), round to oval. Sclerenchyma and crystals absent. Ducts present. Phellem consists of one layer of square cells (phelloids).

Pith

Large, round, filled with unlignified round cells.

Analyzed: Root collar of polar root of 3 individuals.

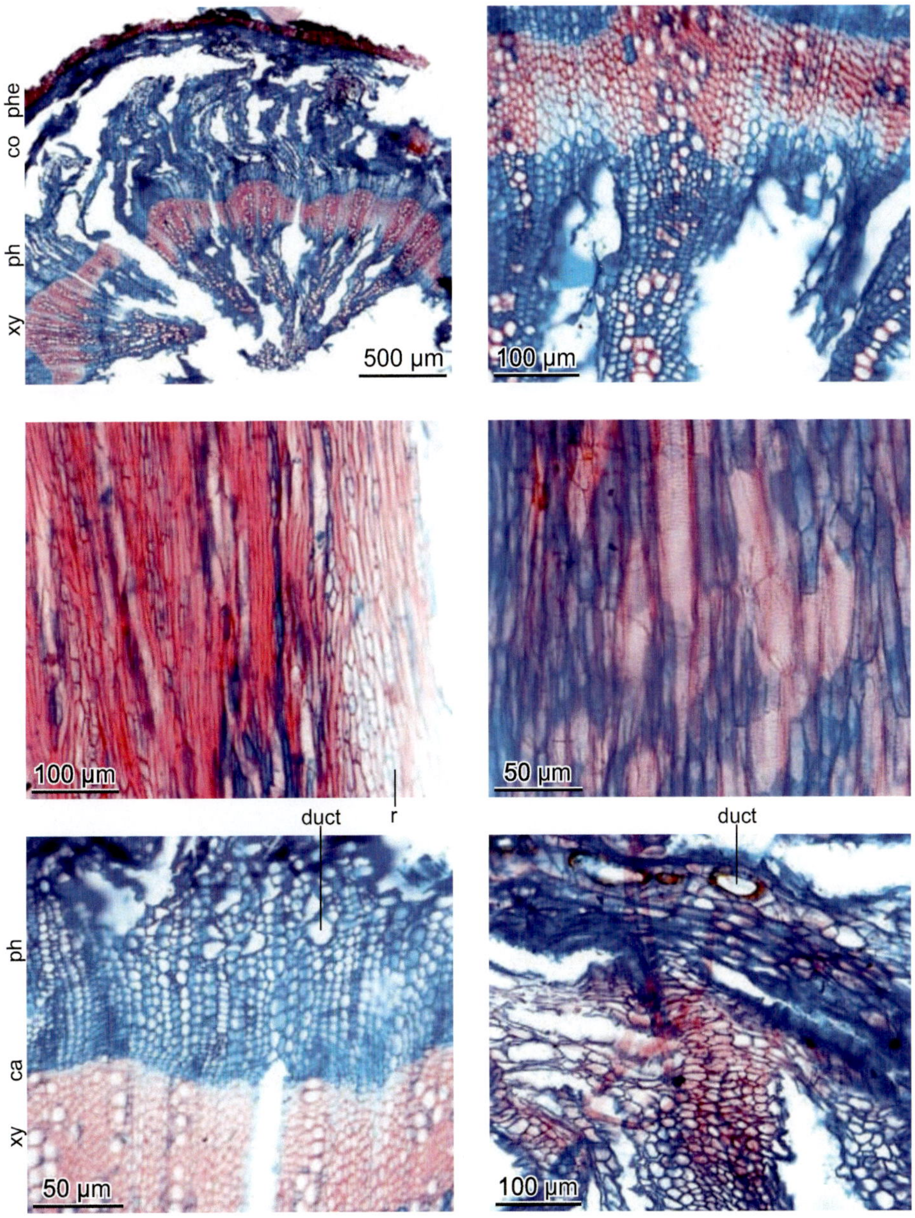

Features according to IAWA

Xylem 1 - 4 - 9 - 11 - 13 - 20 - 22 - 40.2 - 50.1 - 52.2 - 52.3 - 60.1 - 79.1 - 99 - 99.1 - 100.1 - 100.2 - 105

Psychrogeton andryaloides (DC.) Novopokr. ex Krasch.

Asteraceae

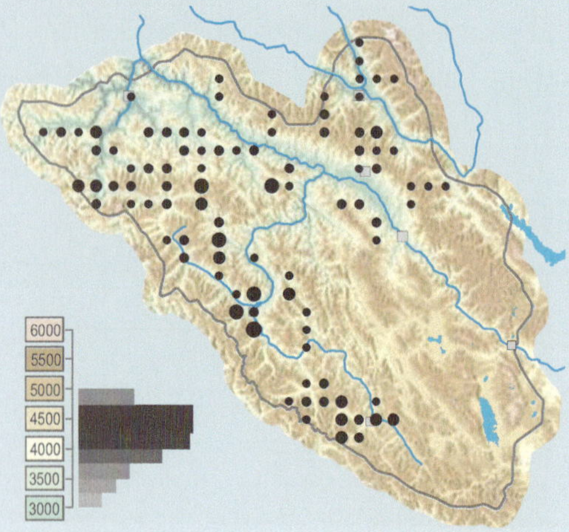

Description: Perennial herbs, 3–20 cm tall, lanate, leaves oblanceolate to obovate or spatulate, capitula white, yellow, or reddish, 1.5–2.5 cm across, solitary.

Distribution: Iran, Russia, Tajikistan, Afghanistan, Pakistan, India, China (Xinjiang, Xizang).

Habitat: Stabilized stony slopes.

Stem diameter 5–12 mm. Bark/xylem proportion 1:1. Ring width 0.8–1.1 mm. Maximal 2 annual rings.

Xylem

Transverse section

Semi-ring-porous. Rings marked by bands of thick-walled latewood fibers. Vessels angular, solitary or in small groups. Fibers thick-walled. Parenchyma paratracheal. With heartwood.

Tangential section

Rays 4–6-seriate, with sheath cells. Rays disappear in polarized light.

Radial section

Perforation plates simple. Intervessel pits round in alternating position. Vessel length 100–120 µm.

Bark

Phloem with radially arranged groups of thick-walled fiber groups. With ray dilatations. Oil ducts absent. Phellem layered, consisting of many sheets of large, thin-walled cells.

Pith

Round. Round parenchyma cells in circular, uniseriate layers.

Analyzed: Root collar of polar root of 2 individuals.

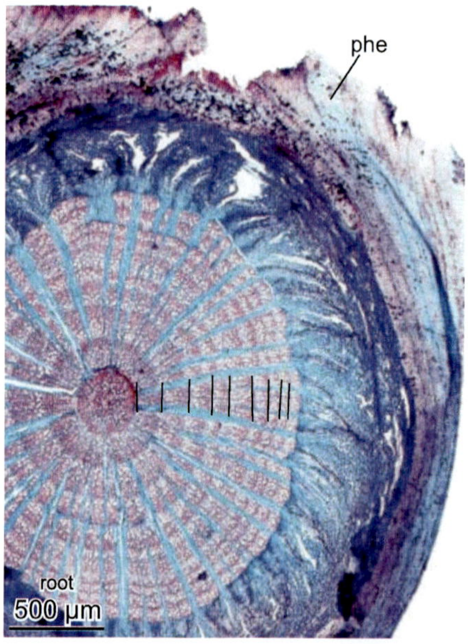

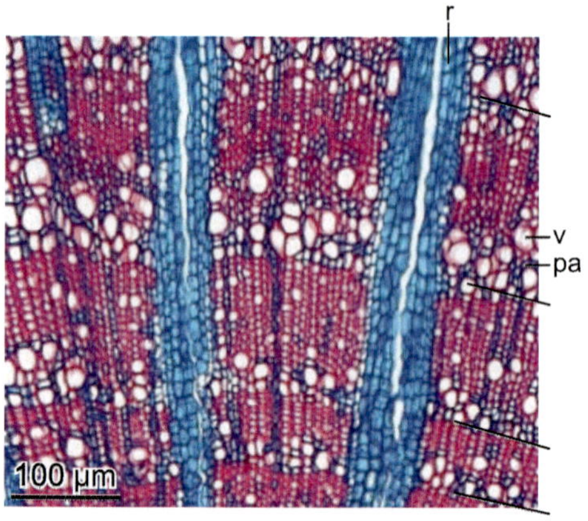

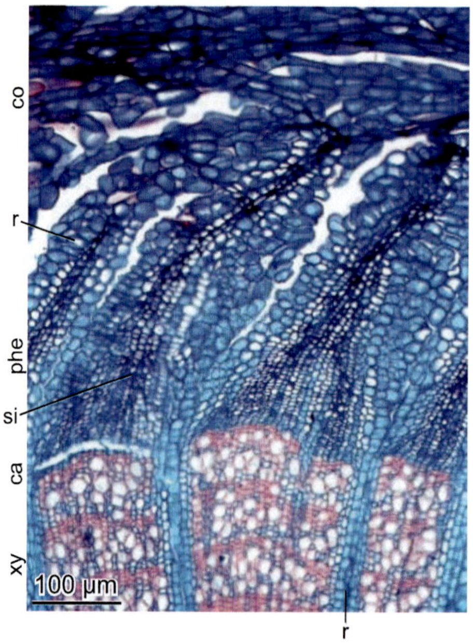

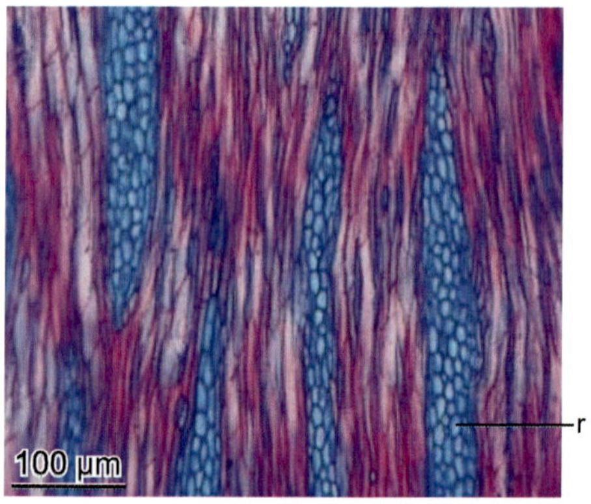

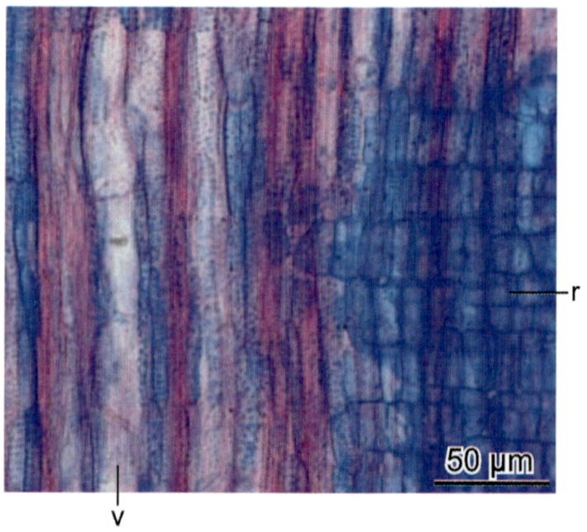

Features according to IAWA

Xylem 1 - 4 - 9 - 13 - 22 - 40.2 - 50.1 - 50.2 - 52.3 - 61 - 69 - 79 - 98 - 100.2 - 105 - 110

Richteria pyrethroides Kar. & Kir.

Asteraceae

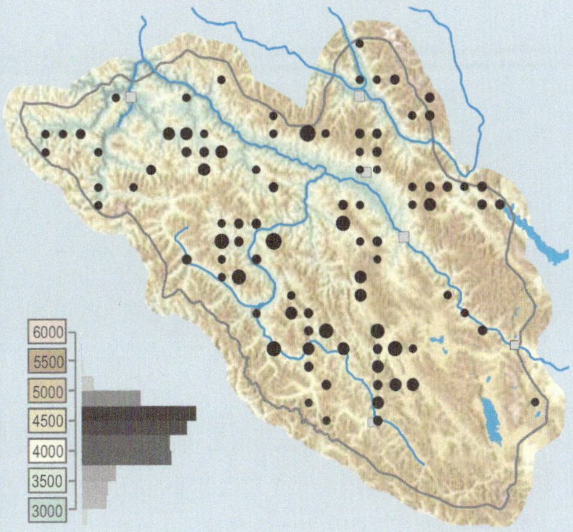

Syn.: *Chrysanthemum arassanicum* C.Winkler, *Pyrethrum pyrethroides* (Kar. & Kir.) B.Fedtsch. ex Krasch., *Tanacetum pyrethroides* (Kar. & Kir.) Sch.Bip., *T. transiliense* Herder

Description: Perennial herbs, 10–40 cm tall, cespitose, leaves 2-pinnatisect, gray-white villous, ultimate segments lanceolate, capitula 2–4 cm across, usually solitary, disc florets yellow, ray florets white or pinkish.

Distribution: Russia, Kazakhstan, Uzbekistan, Pakistan, N India, China (Xinjiang).

Habitat: Dry stony stabilized slopes.

Stem diameter 2–4 mm. Bark/xylem proportion 1:0.8. Ring width 0.1–0.2 mm. Maximal 20 annual rings.

Xylem

Transverse section

Semi-ring-porous. Solitary vessels. Ring boundary marked by large earlywood (20–25 µm) and small latewood vessel diameters (<10 µm) and layers of thick-walled, flat fibers. Vessels in the central part of the stem with dark-stained substances (heartwood). Parenchyma paratracheal and marginal. Vascular bundles remain.

Tangential section

Multiseriate, unlignified rays with >10 cells which disappear in polarized light.

Radial section

Perforation plates simple. Intervessel pits round in alternating position. Vessel length 120–250 µm.

Bark

Phloem with tangential layers of thick-walled fibers. Oil ducts absent. Phellem consists of many layers of rectangular cells and groups of sclereids.

Pith

Round.

Analyzed: Root collar of polar root of 3 individuals.

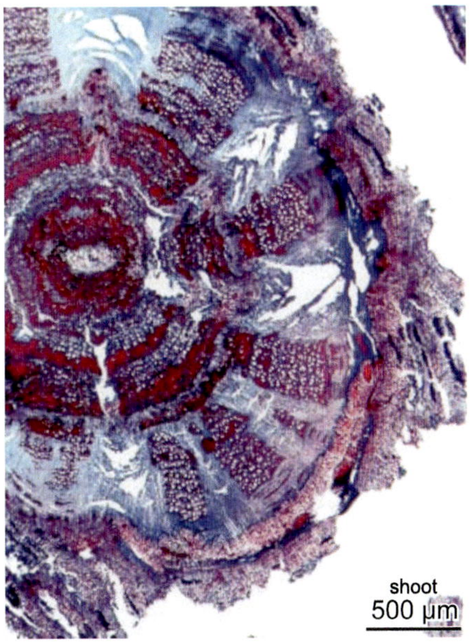

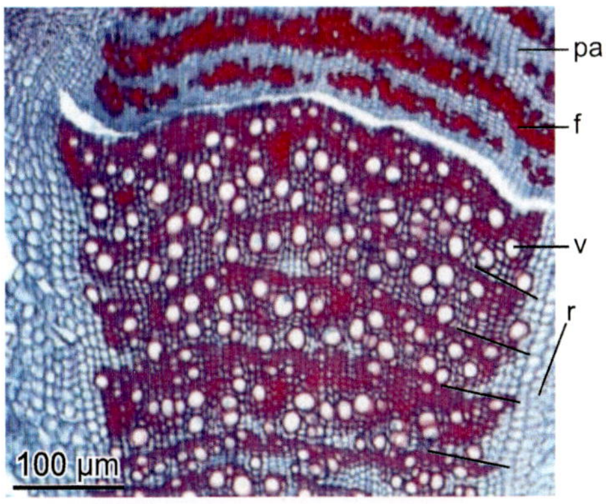

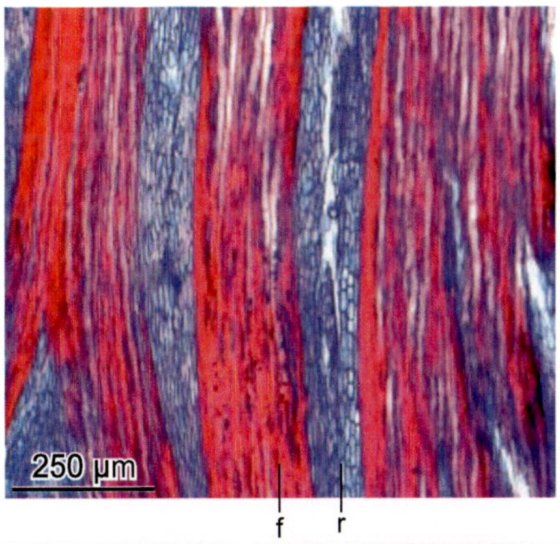

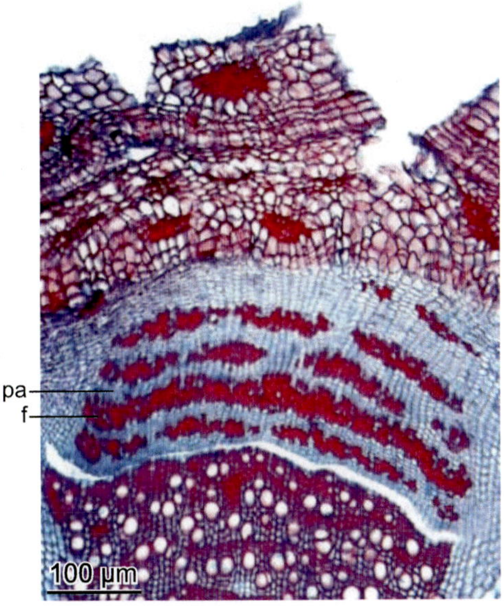

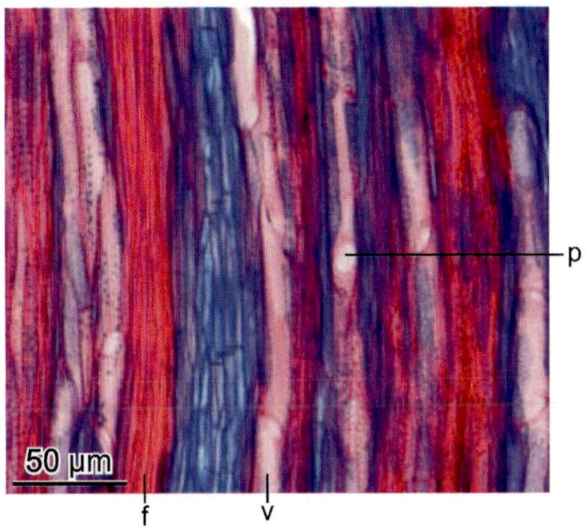

Features according to IAWA

Xylem 1 - 4 - 9 - 13 - 20 - 40.1 - 40.2 - 50.2 - 52.3 - 58 - 61 - 69 - 79 - 89 - 99 - 99.1 - 105

Saussurea andryaloides (DC.) Sch.Bip.

Asteraceae

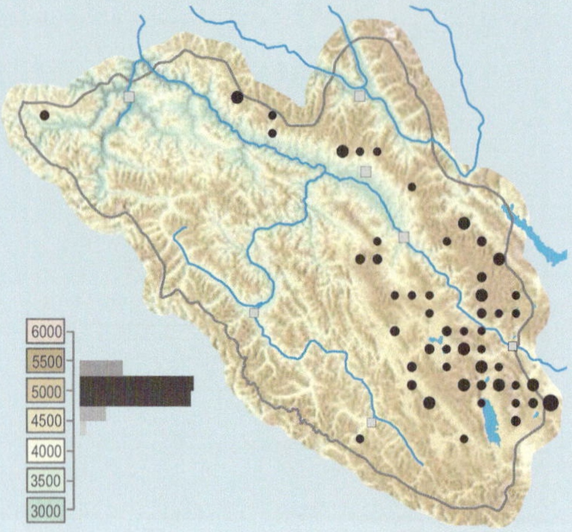

Syn.: *Aplotaxis andryaloides* DC., *S. falconeri* Hook.f., *S. stoliczkae* C.B.Clarke

Description: Perennial herbs, 2–6 cm tall, stemless or shortly stemmed, leaves broadly linear to narrowly ovate-elliptic, lyrate to pinnately lobed, or undivided and margin sinuate-dentate, densely tomentose below, arachnoid pubescent above, capitula purplish-red, ca. 2 cm across, solitary.

Distribution: NW India, China (Qinghai, Xinjiang, Xizang).

Habitat: Scree slopes, steppes.

Stem diameter 2 mm. Bark/xylem proportion 1:3. Ring width 0.1–1.5 mm. More than 3 annual rings. Center rotten. Segmented stems.

Xylem

Transverse section

Semi-ring-porous. Earlywood vessels angular and solitary or in small groups. Ring boundary marked by large earlywood (30–50 μm) and small latewood vessel diameters (<10 μm). Fibers thin- to thick-walled. Parenchyma paratracheal in the earlywood and pervasive in the latewood. With remaining vascular bundles. Large ray-like parenchyma zones between vascular bundles. With cork-layers around segements.

Tangential section

Rays between vascular bundles 10-seriate, confluent. Ray cells upright. Rays disappear in polarized light.

Radial section

Perforation plates simple. Intervessel pits scalariform. Vessel length 80–150 μm.

Bark

Phloem with few scattered, small groups of sieve tubes. Cortex cells round to oval. Sclerenchyma and crystals absent. Ducts absent. Phellem consists of one layer of square cells (phelloids). Secondary intraxylary phloem towards the pith separates living and dead tissues.

Pith

No slide available.

Analyzed: Root collar of polar root of 2 individuals.

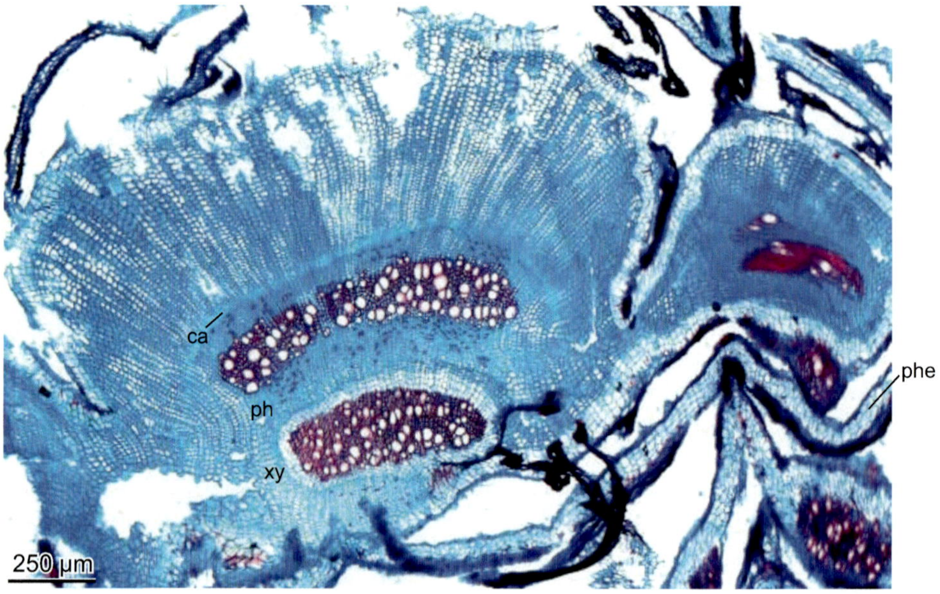

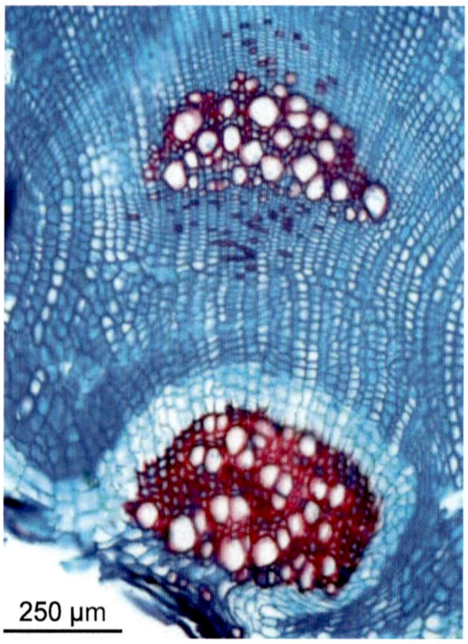

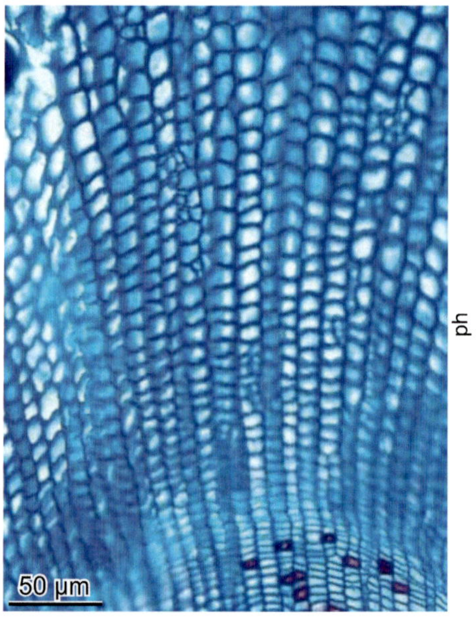

Features according to IAWA

Xylem 1 - 4 - 9 - 11 - 13 - 20 - 40.2 - 50.2 - 52.3 - 60.1 - 79.1 - 99 - 99.1 - 100.1 - 100.2 - 105 - 135.1 - 135.2

Saussurea bracteata Decne.

Asteraceae

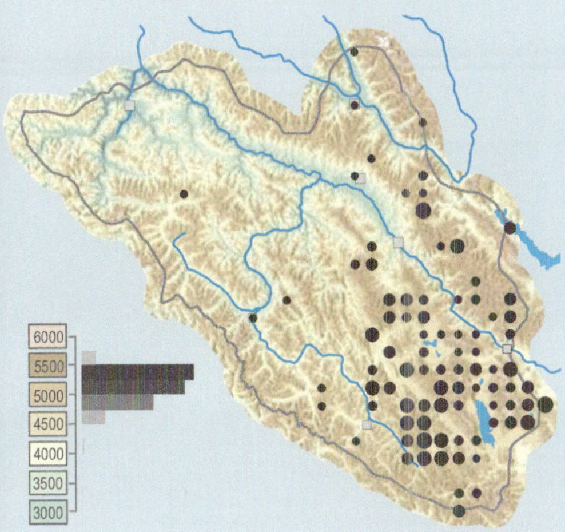

Syn.: *S. schlagintweitii* Klatt, *S. schultzii* var. *nana* Pamp., *S. nana* (Pamp.) Pamp.

Description: Perennial herbs, 3–8 cm tall, sometimes stemless, leaves narrowly elliptic, glandular hairy above, glabrous below, margin sinuate-dentate, upper cauline leaves membranous, yellowish tinged with purple, half-enclosing capitulum, flowers purplish red.

Distribution: Pakistan, NW India, China (Xinjiang, Xizang).

Habitat: Stony slopes, gravel areas, alpine meadows.

Stem diameter 2 mm. Bark/xylem proportion 1:1. Ring width 0.1–1.5 mm. Maximal 4 annual rings. Segmented stems.

Xylem

Transverse section

Semi-ring-porous. Earlywood vessels solitary or in small groups. Ring boundary marked by large earlywood (30–40 µm) and small latewood vessel diameters (<10 µm). Vessels stained and unstained. Intervessel pits scalariform. Fibers absent. Parenchyma pervasive. With remaining vascular bundles. Large ray-like parenchyma zones between vascular bundles.

Tangential section

Large, multiseriate rays between vascular bundles. Ray cells upright. Rays disappear in polarized light.

Radial section

Perforation plates simple. Intervessel pits scalariform. Vessel length 80–120 µm.

Bark

Phloem with few scattered, small groups of sieve tubes. Cortex cells large (15–20 µm), round to oval. Sclerenchyma and crystals absent. Ducts observed in all species. With oil ducts. Phellem consists of one layer of square cells (phelloids). Secondary intra-xylary phloem towards the pith separates living and dead tissues.

Pith

No slide available.

Analyzed: Root collar of polar root of 1 individual.

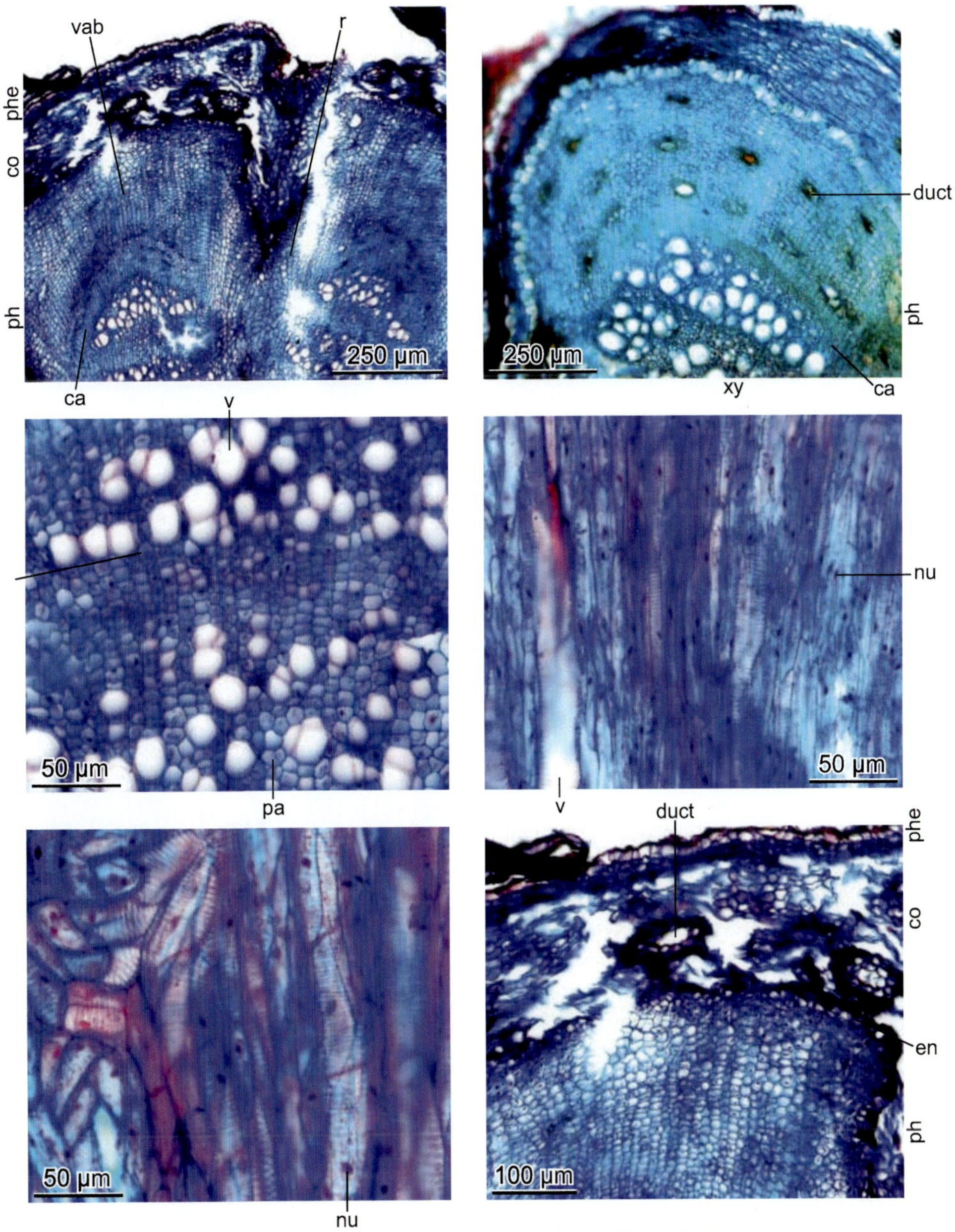

Features according to IAWA

Xylem 1 - 4 - 9 - 11 - 13 - 20 - 40.2 - 50.2 - 52.3 - 60.1 - 79.1 - 99 - 99.1 - 100.1 - 100.2 - 105 - 135.2

Saussurea elliptica C.B.Clarke ex Hook.f.

Asteraceae

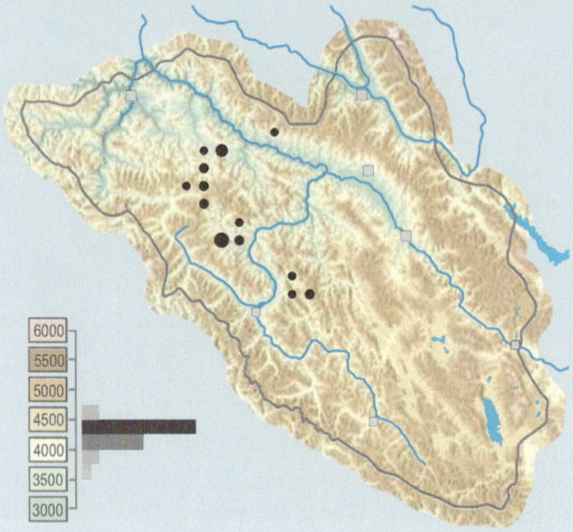

Description: Perennial herbs, 2–10 cm tall, almost stemless, arachnoid hairy, leaves ovate to elliptic, margin denticulate, capitula pale purple, 1–1.5 cm across, clustered by 2–10 in a corymbiform synflorescence.

Distribution: Kazakhstan, Kyrgyzstan, Tajikistan, Pakistan, NW India, China (W Xinjiang).

Habitat: Alpine meadows, rocky slopes, moraines.

Stem diameter 2 mm. Bark/xylem proportion 1:3. Ring width 0.2–4 mm. More than 9 annual rings. Segmented stems. Center rotten.

Xylem

Transverse section

Semi-ring-porous. Earlywood vessels solitary or in small groups. Ring boundary marked by large earlywood (30–50 µm) and small latewood vessel diameters (<10 µm). Thick-walled fibers in clusters. Parenchyma pervasive. With remaining vascular bundles. Large ray-like parenchyma zones between vascular bundles. With cork layers inside and outside of the vascular bundles.

Tangential section

Rays between vascular bundles 10-seriate, confluent. Ray cells upright. Rays disappear in polarized light.

Radial section

Perforation plates simple. Intervessel pits scalariform. Vessel length 80–150 µm.

Bark

Phloem with few scattered, small groups of sieve tubes. Cortex missing. Sclerenchyma and crystals absent. With ducts. Phellem consists of one layer of square cells (phelloids). Secondary intra-xylary phloem towards the pith separates living and dead tissues.

Pith

No slide available.

Analyzed: Root collar of polar root of 3 individuals.

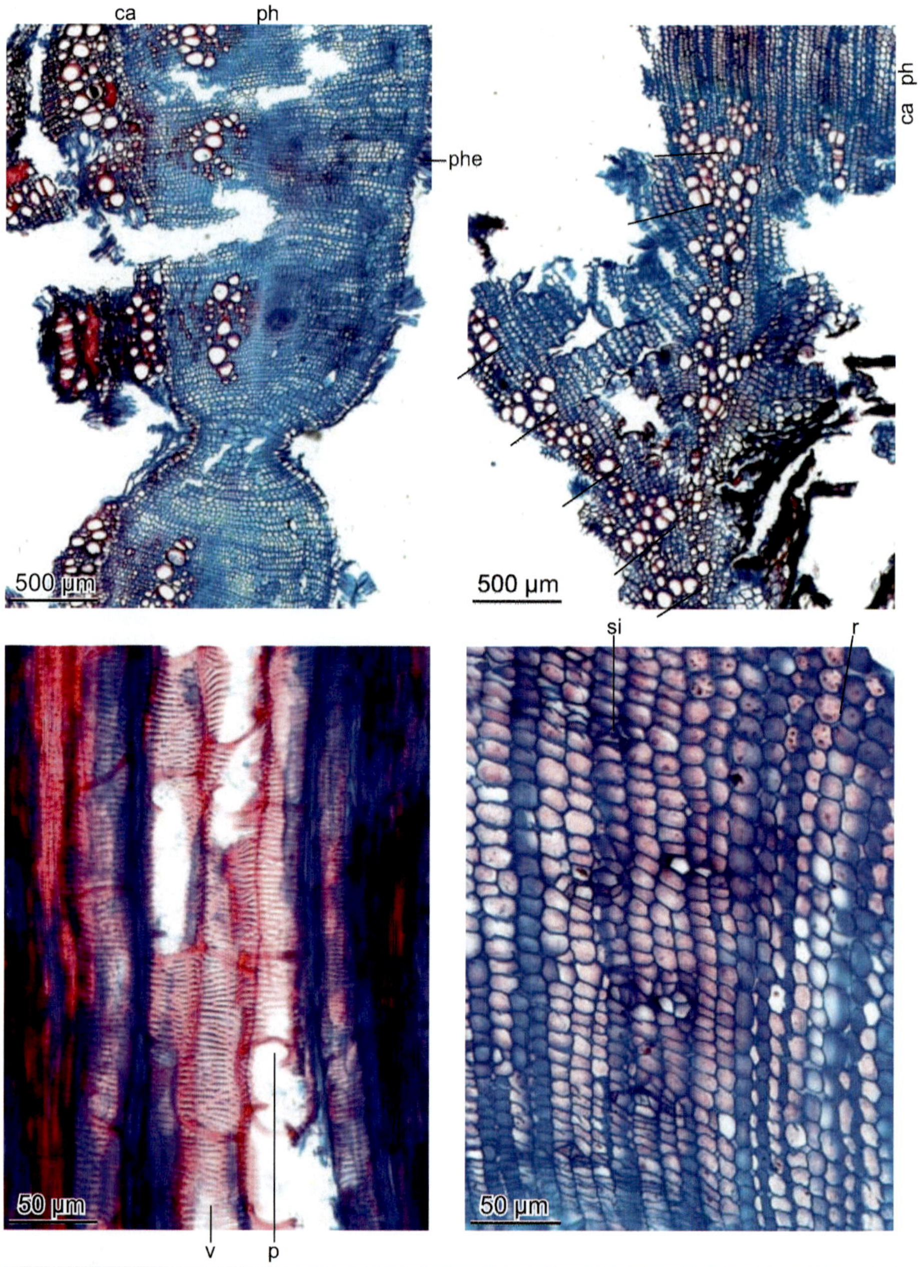

Features according to IAWA

Xylem *1 - 4 - 9 - 11 - 13 - 20 - 40.2 - 50.2 - 52.2 - 52.3 - 60.1 - 61 - 79.1 - 99 - 99.1 - 100.1 - 100.2 - 105 - 135.1 - 135.2*

Saussurea glacialis Herder

Asteraceae

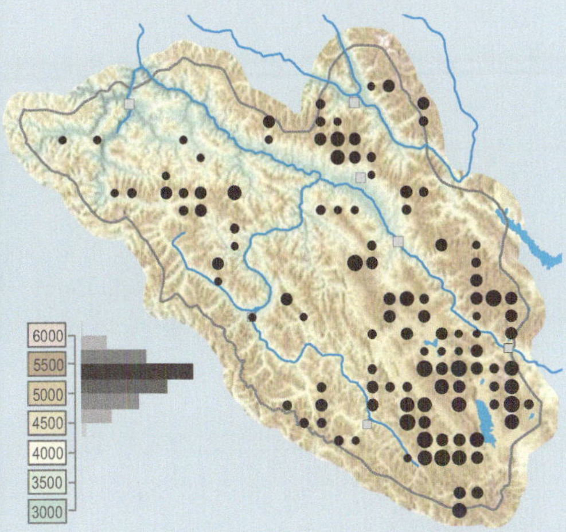

Syn.: *S. chtonocephala* Bornm., *S. pamirica* C.Winkler, *S. violacea* Pamp.

Description: Perennial herbs, 2–6 cm tall, leaves narrowly elliptic to obovate, yellowish or white lanate above, white lanate below, margin crenate or entire, flowers purple, arranged in 3–15 capitula forming a hemispheric synflorescence 2–4 cm across, pappus of achenes dirty white.

Distribution: Russia (Siberia), E Kazakhstan, Kyrgyzstan, Tajikistan, E Afghanistan, N Pakistan, NW India, China (Qinghai, Xinjiang, Xizang), Mongolia.

Habitat: Scree slopes.

Stem diameter 2–4 mm. Bark/xylem proportion 1:1. Ring width 0.05–0.15 mm. Maximal 10 annual rings. Segmented stems.

Xylem

Transverse section

Semi-ring-porous. Earlywood vessels solitary or in small groups. Ring boundary marked by large earlywood (30–40 μm) and small latewood vessel diameters (<10 μm). Vessels stained and unstained. Intervessel pits scalariform. Fibers absent. Parenchyma pervasive. With remaining vascular bundles in the upper part of the rhizome. Large ray-like parenchyma zones between vascular bundles. Secondary intra-xylary phloem towards the pith separates living and dead tissues.

Tangential section

Large, multiseriate rays between vascular bundles. Ray cells upright. Rays disappear in polarized light.

Radial section

Perforation plates simple. Intervessel pits scalariform. Vessel length 80–120 μm.

Bark

Phloem with radially and tangentially arranged groups of sieve tubes. Sclerenchyma and crystals absent. With small oil ducts. Phellem consists of one layer of square cells (phelloids). Secondary intraxylary phloem towards the pith separates living and dead tissues.

Pith

No slide available.

Analyzed: Root collar of polar root of 3 individuals.

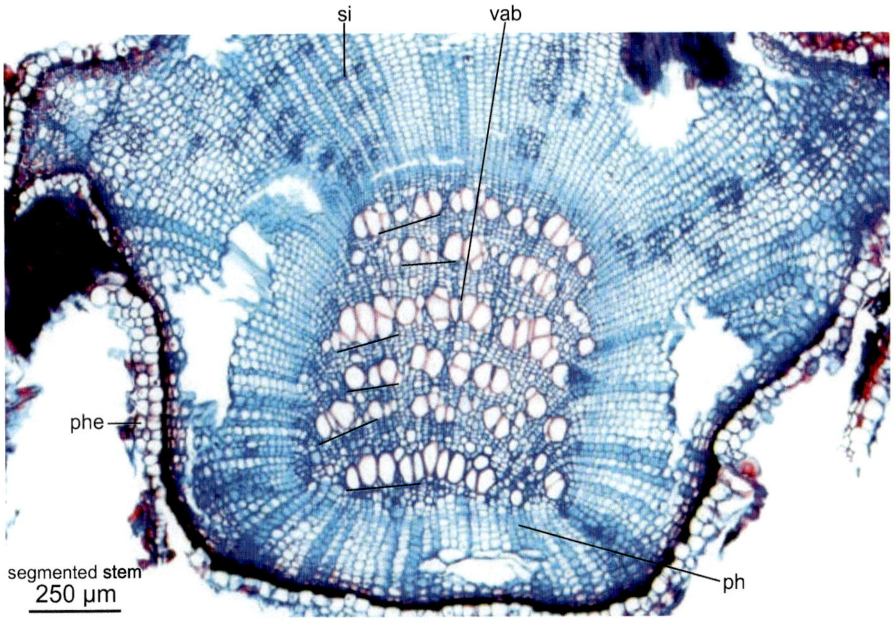

Features according to IAWA

Xylem 1 - 4 - 9 - 11 - 13 - 20 - 40.2 - 50.2 - 52.2 - 60.1 - 79.1 - 99 - 99.1 - 100.1 - 100.2 - 105 - 135.2

Saussurea glanduligera Sch.Bip.

Asteraceae

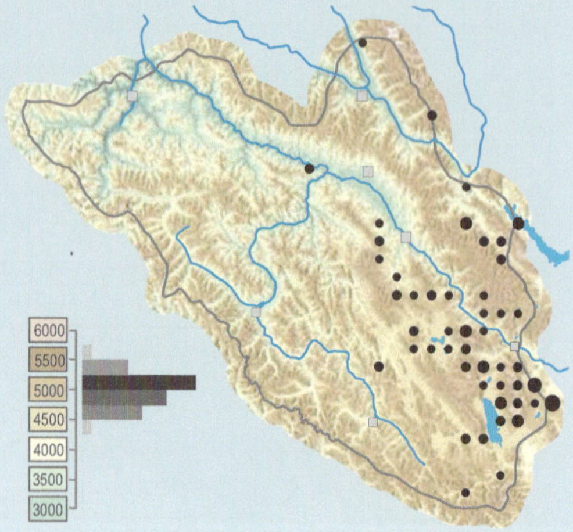

Description: Perennial herbs, 10–25 cm tall, cespitose, leaves narrowly elliptic to linear, flowers purplish red, in a terminal solitary capitulum.

Distribution: NW India, China (Xinjiang, Xizang).

Habitat: Rocky stabilized slopes, rock crevices.

Stem diameter 2–4 mm. Bark/xylem proportion 1:1. Ring width 0.04–0.05 mm. Maximal 4 annual rings. Segmented stems.

Xylem

Transverse section

Semi-ring-porous. Earlywood vessels solitary or in small groups. Ring boundary marked by large earlywood (40–50 µm) and small latewood vessel diameters (10 µm). Vessels unstained. Intervessel pits scalariform. Fibers absent. Parenchyma pervasive. With remaining vascular bundles. Large ray-like parenchyma zones between vascular bundles.

Tangential section

Large, multiseriate rays between vascular bundles. Ray cells upright. Rays disappear in polarized light.

Radial section

Perforation plates simple. Intervessel pits scalariform. Vessel length 80–120 µm.

Bark

Phloem with indistinct, small groups of sieve tubes. Sclerenchyma and crystals absent. With oil ducts. Phellem consists of one layer of square cells (phelloids). Secondary intra-xylary phloem towards the pith separates living and dead tissues.

Pith

No slide available.

Analyzed: Root collar of polar root of 2 individuals.

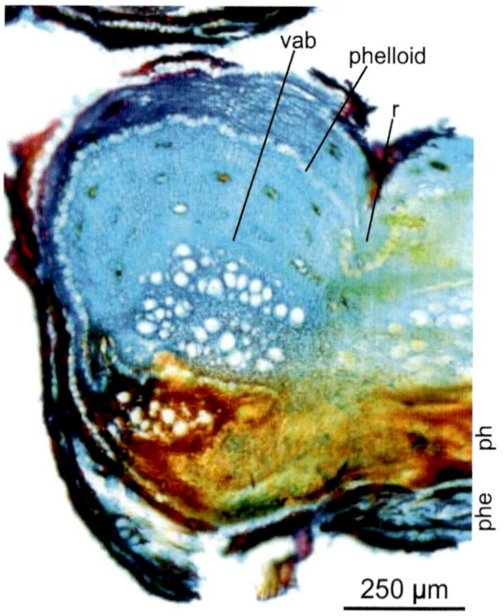

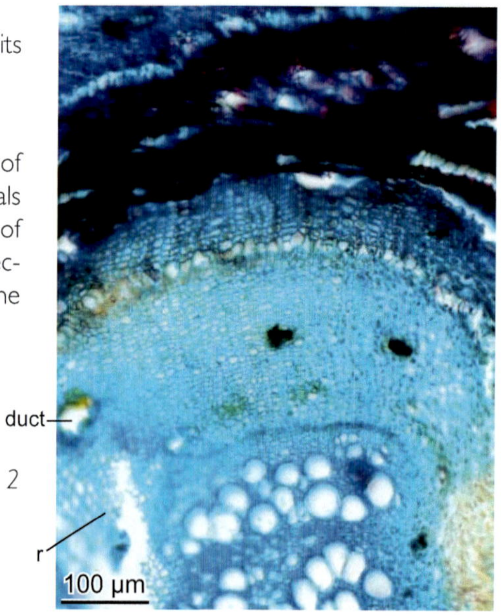

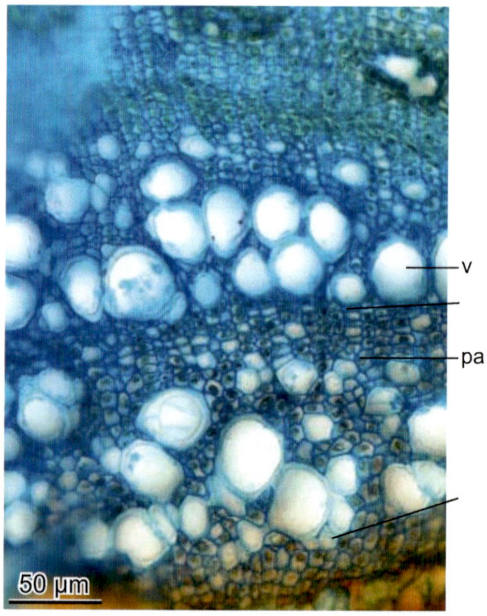

Features according to IAWA

Xylem 1 - 4 - 9 - 11 - 13 - 20 - 40.2 - 50.2 - 52.2 - 60.1 - 79.1 - 99 - 99.1 - 100.1 - 100.2 - 105 - 135.2

Saussurea gnaphalodes (Royle ex DC.) Sch.Bip.

Asteraceae

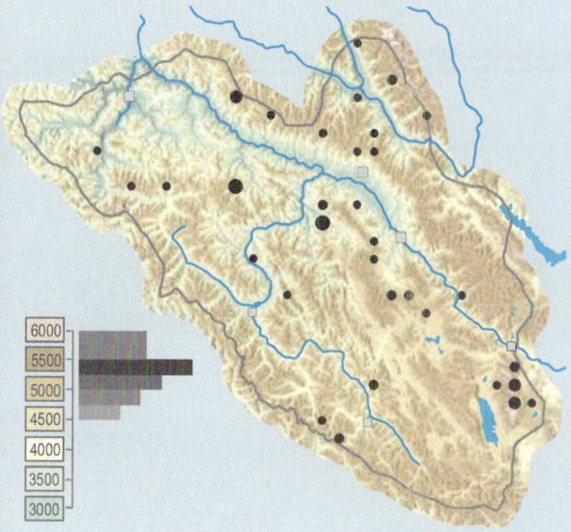

Syn.: *Aplotaxis gnaphalodes* Royle ex DC., *S. sorocephala* (Schrenk ex Fisch. & C.A.Mey.) Schrenk

Description: Perennial herbs, 2–6 cm tall, leaves obovate, spatulate, or linear-oblong, arachnoid lanate, margin entire or crenate, flowers rose-purple, in 6–20 capitula forming a hemispheric synflorescence 2–4 cm across, pappus of achenes gray to blackish.

Distribution: E Kazakhstan, Kyrgyzstan, Tajikistan, NE Afghanistan, N Pakistan, NW India, W Nepal, W China.

Habitat: Screes and unstable loamy and gravel slopes, unstable solifluction soils.

Stem diameter 2–4 mm. Bark/xylem proportion 1:1. Ring width 0.05–0.15 mm. Maximal 6 annual rings. Segmented stems.

Xylem

Transverse section

Semi-ring-porous. Earlywood vessels solitary or in small groups. Ring boundary marked by large earlywood (40–50 µm) and small latewood vessel diameters (10 µm). Vessels stained and unstained. Intervessel pits scalariform. Fibers absent. Parenchyma pervasive. With remaining vascular bundles. Large ray-like parenchyma zones between vascular bundles.

Tangential section

Large, multiseriate rays between vascular bundles. Ray cells upright. Rays disappear in polarized light.

Radial section

Perforation plates simple. Intervessel pits scalariform. Vessel length 80–100 µm.

Bark

Phloem with indistinct, small groups sieve tubes. Sclerenchyma and crystals absent. With large oil ducts, small oil ducts absent. Phellem consists of one layer of square cells (phelloids). Secondary intra-xylary phloem towards the pith separates living and dead tissues.

Pith

No slide available.

Analyzed: Root collar of polar root of 5 individuals.

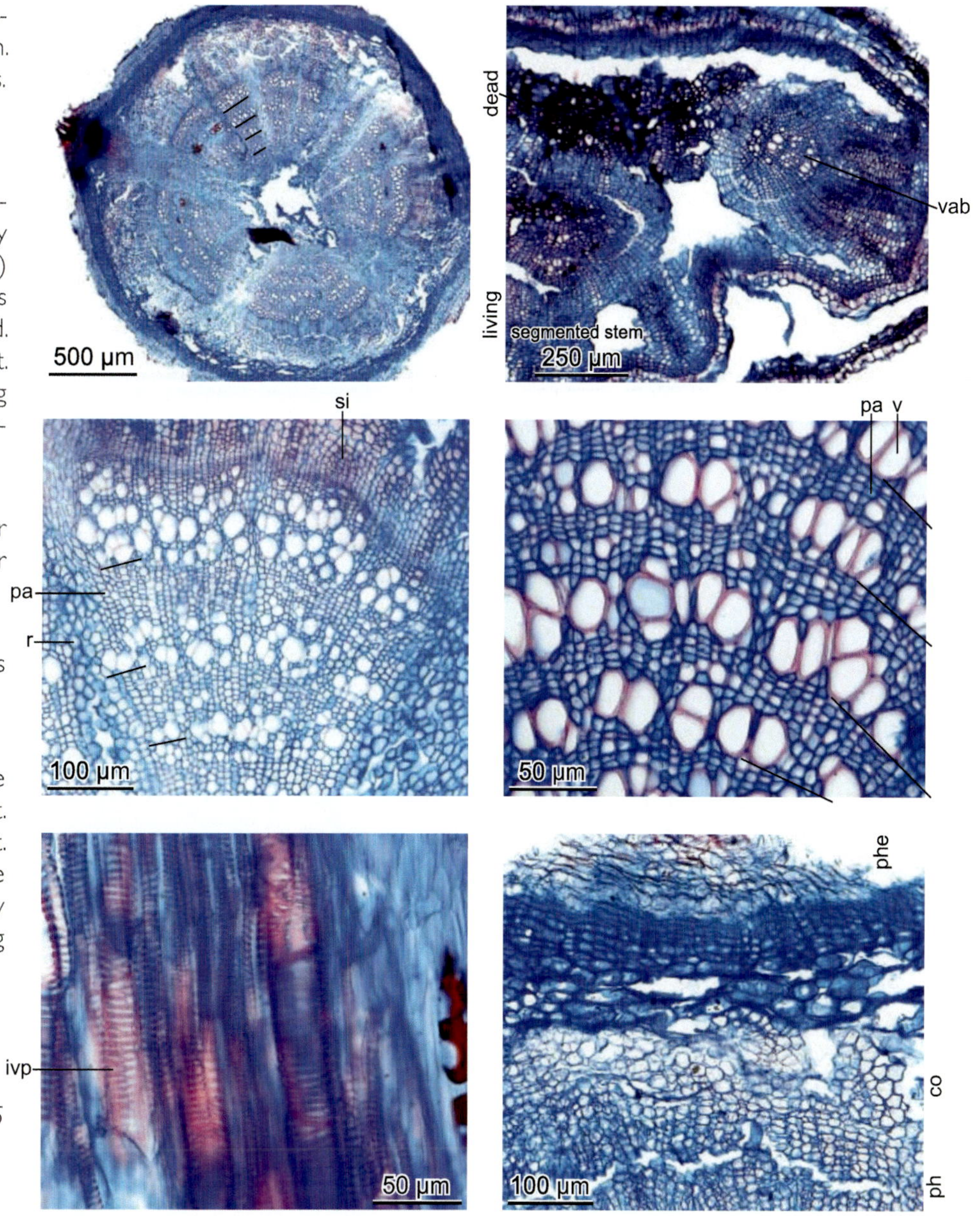

Features according to IAWA

Xylem *1 - 4 - 9 - 11 - 13 - 20 - 40.2 - 50.2 - 52.2 - 60.1 - 79.1 - 99 - 99.1 - 100.1 - 100.2 - 105 - 135.2*

Saussurea inversa Raab-Straube
Asteraceae

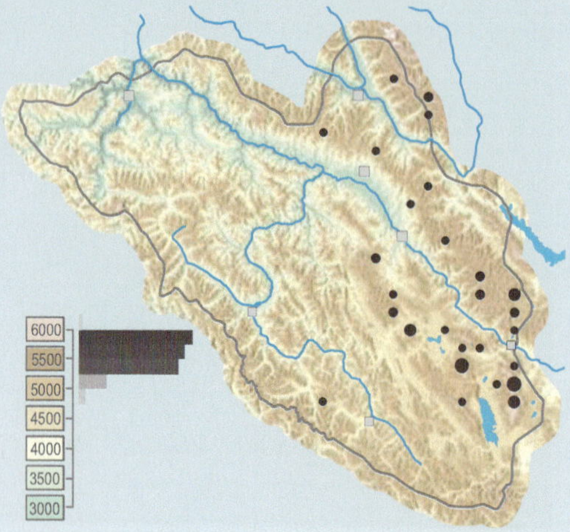

Syn.: *S. sorocephala* var. *glabrata* Hook.f., *S. hypsipeta* auct. non Diels ex H.Limpr.

Description: Perennial herbs, 3–15 cm tall, leaves narrowly obovate-spatulate, elliptic, or oblong, both surfaces glabrous or sparsely white arachnoid, green, sometimes tinged with purple, margin dentate or entire, upper stem leaves narrowly ovate, reflexed, white and blackish lanate, flowers rose-purple, arranged in 6–15 capitula forming a hemispheric synflorescence 2–4 cm across, pappus of achenes gray to blackish.

Distribution: NW India, China (Qinghai, Xinjiang, Xizang).

Habitat: Alpine scree slopes.

Stem diameter 1–2.5 mm. Bark/xylem proportion 1:1. Ring width 0.1–1.5 mm. Maximal 5 annual rings. Segmented stems.

Xylem

Transverse section

Semi-ring-porous, earlywood vessels solitary or in small groups. Ring boundary marked by large earlywood (20–30 µm) and small latewood vessel diameters (<10 µm). Vessels stained and unstained. Intervessel pits scalariform. Fibers absent. Parenchyma pervasive. With remaining vascular bundles. Large ray-like parenchyma zones between vascular bundles.

Tangential section

Large, multiseriate rays between vascular bundles. Ray cells upright. Rays disappear in polarized light.

Radial section

Perforation plates simple. Intervessel pits scalariform. Vessel length 80–100 µm.

Bark

Phloem with indistinct, small groups of sieve tubes. Sclerenchyma and crystals absent. With large oil ducts, small oil ducts absent. Phellem consists of one layer of square cells (phelloids). Secondary intraxylary phloem towards the pith separates living and dead tissues.

Pith

No slide available.

Analyzed: Root collar of polar root of 7 individuals.

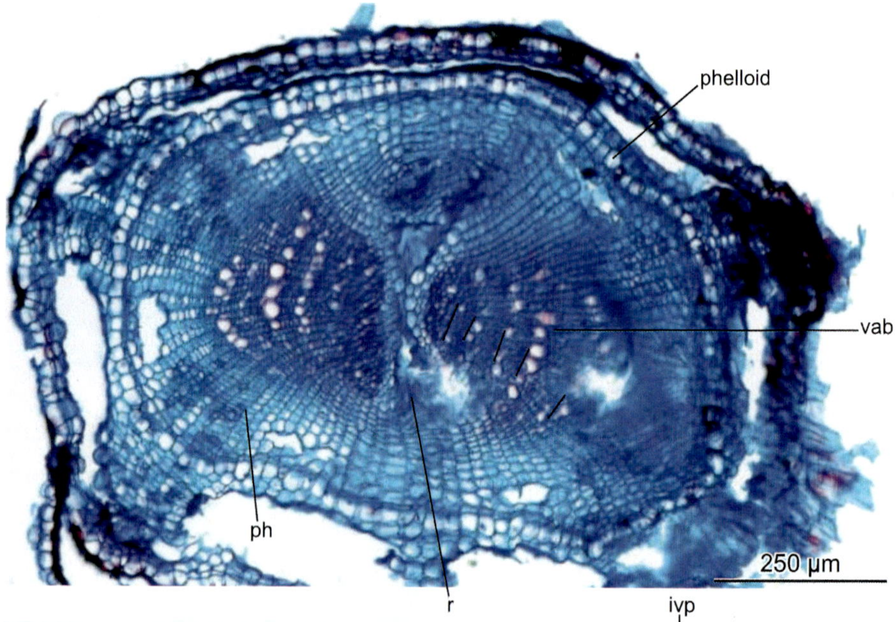

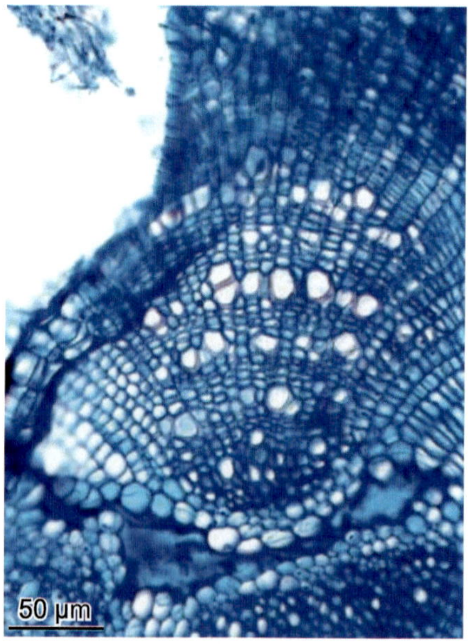

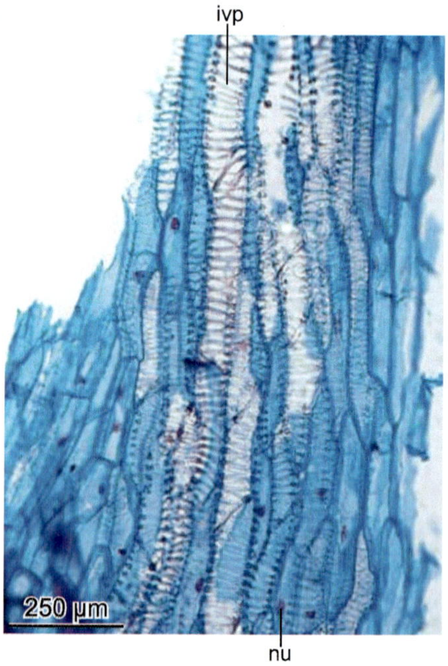

Features according to IAWA

Xylem 1 - 4 - 9 - 11 - 13 - 20 - 40.2 - 50.2 - 52.2 - 60.1 - 79.1 - 99 - 99.1 - 100.1 - 100.2 - 105 - 135.2

Saussurea jacea (Klotzsch) C.B.Clarke

Asteraceae

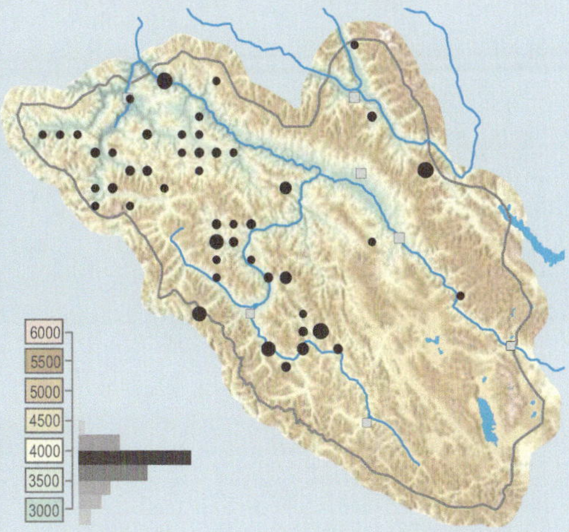

Syn.: *Aplotaxis jacea* Klotzsch

Description: Perennial herbs, 30–90 cm tall, stems solitary, densely leafy, leaves ovate to broadly elliptic, grayish white and densely tomentose below, margin serrulate, capitula purple, 1.5–2.5 cm across, solitary at the end of stems or branches.

Distribution: E Afghanistan, N Pakistan, NW India.

Habitat: Unstable slopes, steep screes.

Stem diameter 16 mm. Bark/xylem proportion 1:0.5. Mean ring width 0.6 mm. Maximal 20 annual rings.

Xylem

Transverse section

Semi-ring-porous. Earlywood vessels in small groups. Latewood vessels in radial multiples. Ring boundary marked by large earlywood (80–100 µm) and small latewood vessel diameters (20 µm). Vessels stained. Intervessel pits round. Fibers thin- to thick-walled. Parenchyma paratracheal. With remaining vascular bundles. Large ray-like parenchyma zones between vascular bundles.

Tangential section

Large, multiseriate rays between vascular bundles. Ray cells upright. Rays disappear in polarized light.

Radial section

Perforation plates simple. Intervessel pits scalariform. Vessel length 80–100 µm.

Bark

Phloem with radial groups of sieve tubes. Sclerenchyma in irregular groups. Oil ducts absent. Phellem not observed.

Pith

No slide available.

Analyzed: Root collar of polar root of 1 individual.

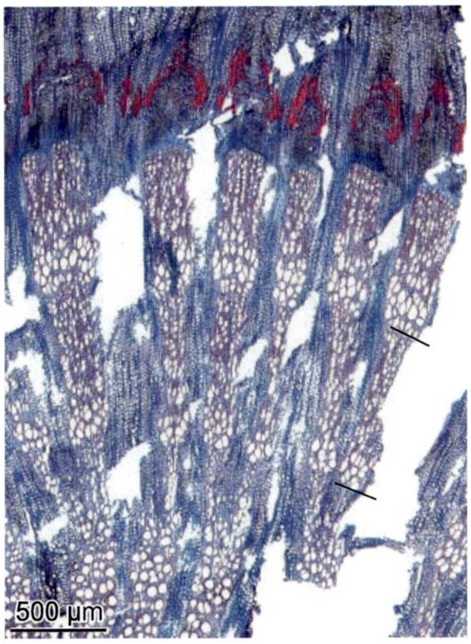

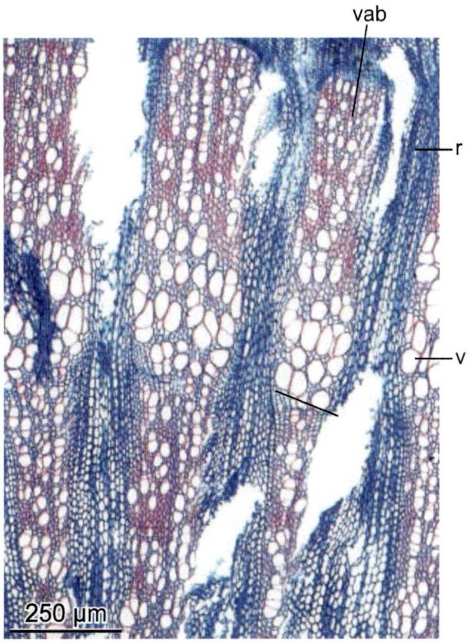

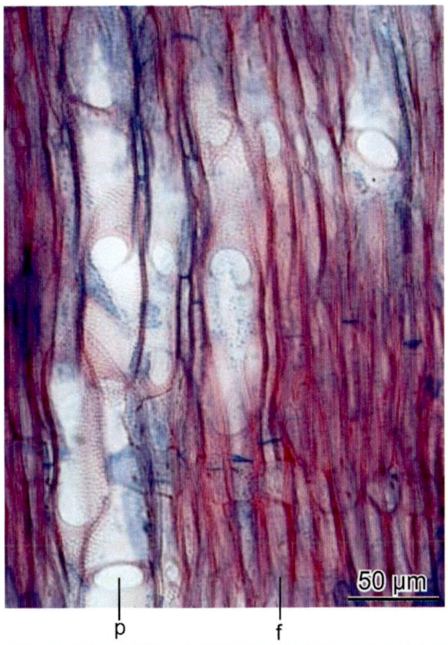

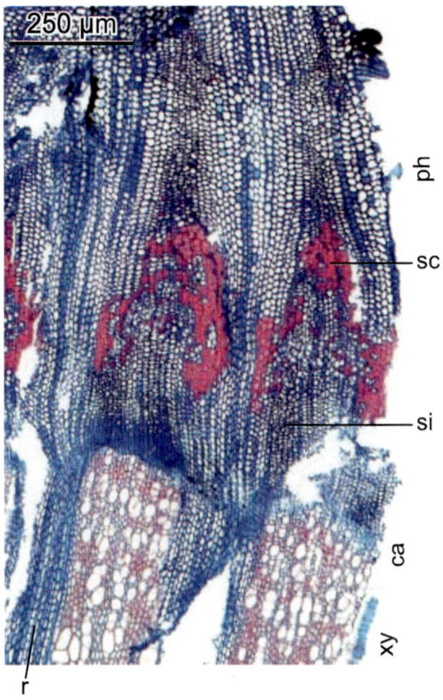

Features according to IAWA

Xylem 1 - 4 - 9.1 - 11 - 13 - 22 - 41 - 50.2 - 52.3 - 61 - 70 - 79 - 99 - 99.1 - 100.2 - 105

Saussurea leontodontoides (DC.) Sch.Bip.

Asteraceae

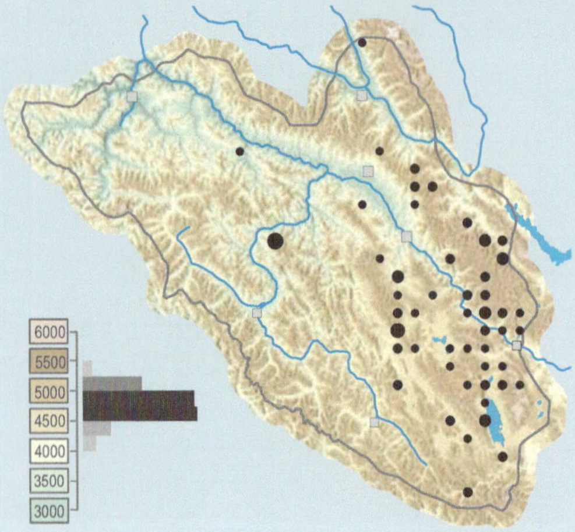

Syn.: *Aplotaxis leontodontoides* DC., *S. kunthiana* (Wall.) C.B.Clarke

Description: Perennial herbs, 3–15 cm tall, stemless, leaves in a basal rosette, linear-oblong to narrowly elliptic, pinnatisect, grayish white and densely tomentose below, capitula purplish red, 1–3 cm across, solitary.

Distribution: NW India, Nepal, China (Qinghai, Sichuan, Xizang, Yunnan).

Habitat: Wet to mesic turf of stream banks.

Diameter of stem fragments 1 mm. Bark/xylem proportion 1:1. Mean ring width 0.3 mm. Maximal 2 annual rings. Segmented stems.

Xylem

Transverse section

Semi-ring-porous. Earlywood vessels solitary. Ring boundary marked by large earlywood (20–30 µm) and small latewood vessel diameters (<10 µm). Vessels stained and unstained. Intervessel pits scalariform. Thick-walled fibers in groups. Parenchyma pervasive. With remaining vascular bundles. Large ray-like parenchyma zones between vascular bundles.

Tangential section

Large, multiseriate rays between vascular bundles. Ray cells upright. Rays disappear in polarized light.

Radial section

Perforation plates simple. Intervessel pits scalariform. Vessel length 80–100 µm.

Bark

Phloem with indistinct, small groups of sieve tubes. Sclerenchyma and crystals absent. With large oil ducts. Phellem consists of one layer of square cells (phelloids). Secondary intra-xylary phloem towards the pith separates living and dead tissues.

Pith

No slide available.

Analyzed: Root collar of polar root of 1 individual.

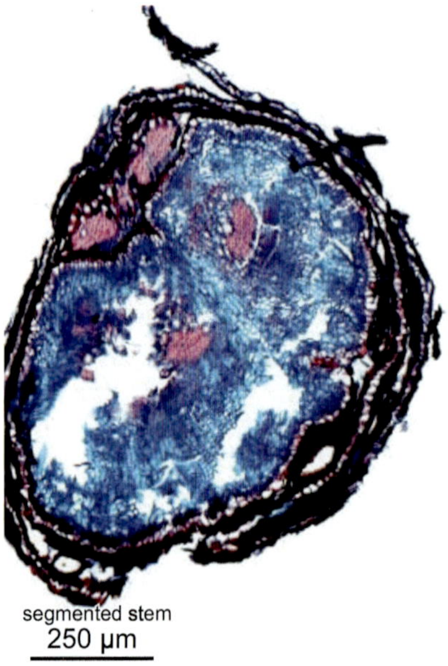

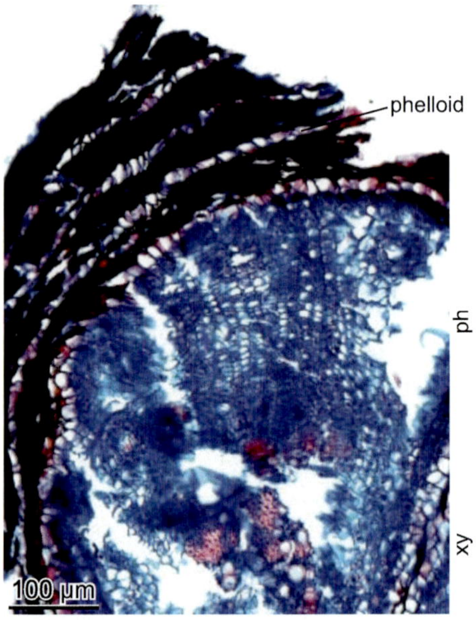

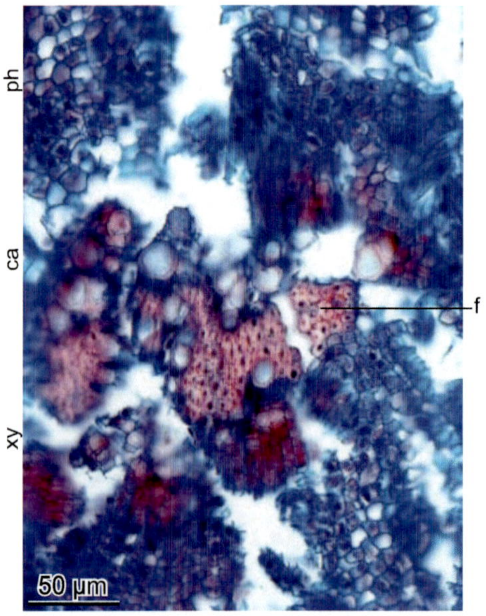

Features according to IAWA

Xylem 1 - 4 - 9 - 11 - 13 - 20 - 40.1 - 50.2 - 52.2 - 61 - 69 - 79.1 - 99 - 99.1 - 100.1 - 100.2 - 105 - 135.2

Saussurea medusa Maxim.

Asteraceae

Syn.: *S. dainellii* Pamp.

Description: Short-lived perennials, 5–10 cm tall, white or yellowish arachnoid lanate, basal leaves obovate, rhombic or orbicular, both surfaces margin dentate to pinnately lobed, uppermost leaves narrowly ovate to linear, capitula bluish purple, numerous, in a hemispheric synflorescence 3–7 cm across.

Distribution: N India (Jammu and Kashmir), Nepal, China (Gansu, Qinghai, Sichuan, Xinjiang, Xizang, Yunnan).

Habitat: Alpine scree slopes.

Stem diameter 6 mm. Bark/xylem proportion 1:0.3. Mean ring width 0.13 mm. Maximal 8 annual rings.

Xylem

Transverse section

Semi-ring-porous. Earlywood vessels solitary or in small groups. Ring boundary marked by large earlywood (40–60 μm) and small latewood vessel diameters (20 μm). Vessels unstained. Intervessel pits scalariform. Fibers absent. Parenchyma pervasive. With remaining vascular bundles. Large ray-like parenchyma zones between vascular bundles.

Tangential section

Large, multiseriate rays between vascular bundles. Ray cells upright. Rays disappear in polarized light.

Radial section

Perforation plates simple. Intervessel pits scalariform. Vessel length 80–100 μm.

Bark

Phloem with indistinct, small groups of sieve tubes. Sclerenchyma and crystals absent. With large oil ducts, small oil ducts absent. Phellem consists of one layer of square cells (phelloids). Secondary intraxylary phloem towards the pith separates living and dead tissues.

Pith

Triangular.

Analyzed: Root collar of polar root of 1 individual.

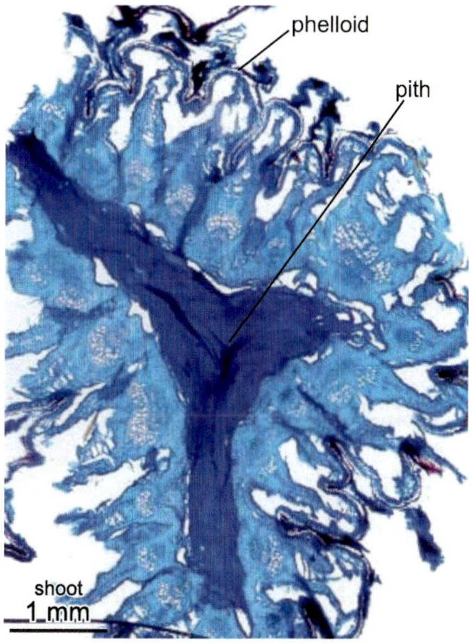

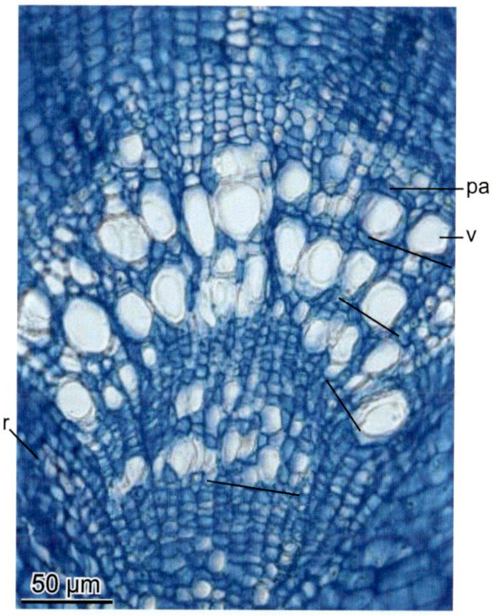

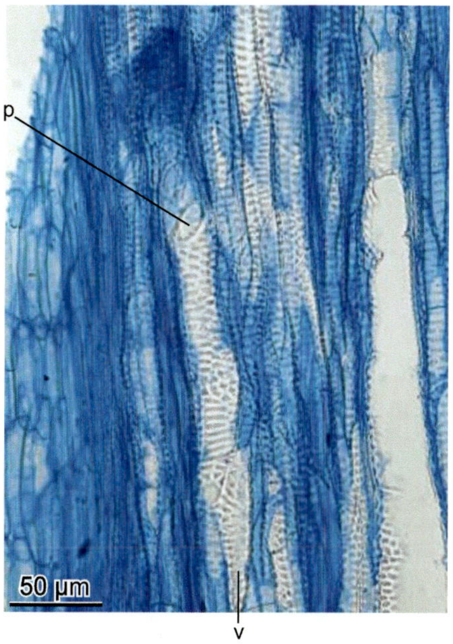

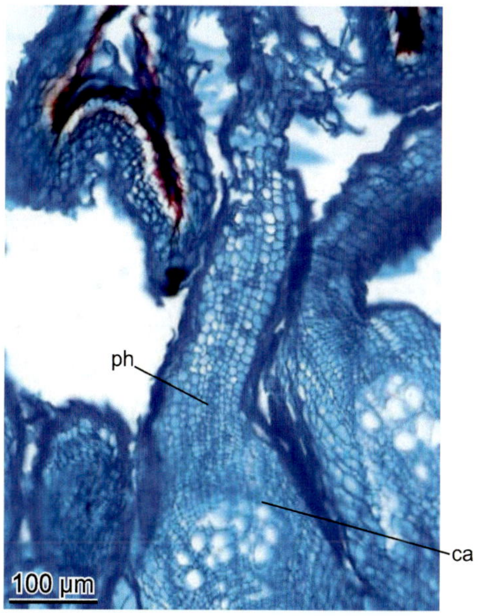

Features according to IAWA

Xylem 1 - 4 - 9 - 11 - 13 - 20 - 40.2 - 50.2 - 52.2 - 60.1 - 79.1 - 99 - 99.1 - 100.1 - 100.2 - 105 - 135.2

Saussurea schultzii Hook.f.

Asteraceae

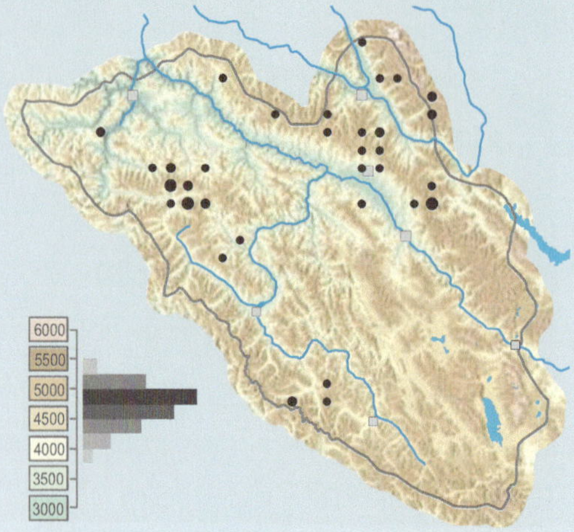

Description: Perennial herbs, 15–40 cm tall, glandular hairy, stems erect, leaves narrowly elliptic, margin sinuate-dentate, uppermost leaves membranous, yellowish to pinkish, half-enclosing synflorescence, capitula purple, in condensed 3–8-flowered synflorescences, 3–10 cm across.

Distribution: Pakistan, NW India, China (Xinjiang).

Habitat: Glacier moraines, rocky slopes, scree slopes.

Stem diameter 2 mm. Bark/xylem proportion 1:3. Ring width 0.25–3 mm. More than 3 annual rings. Segmented stems. Center rotten

Xylem

Transverse section

Semi-ring-porous. Earlywood vessels solitary or in small groups. Ring boundary marked by large earlywood (40–50 µm) and small latewood vessel diameters (<10 µm). Fibers absent. Parenchyma pervasive. With remaining vascular bundles. Large ray-like parenchyma zones between vascular bundles. With cork layers inside and outside of the vascular bundles.

Tangential section

Rays between vascular bundles 10-seriate, confluent. Ray cells upright. Rays disappear in polarized light.

Radial section

Perforation plates simple. Intervessel pits round and scalariform. Vessel length 100–150 µm.

Bark

Phloem with scattered, small groups of sieve tubes. Cortex missing. Sclerenchyma and crystals absent. Ducts absent. Phellem consists of several layers of irregularly formed cells. Secondary intra-xylary phloem towards the pith separates living and dead tissues.

Pith

No slide available.

Analyzed: Root collar of polar root of 3 individuals.

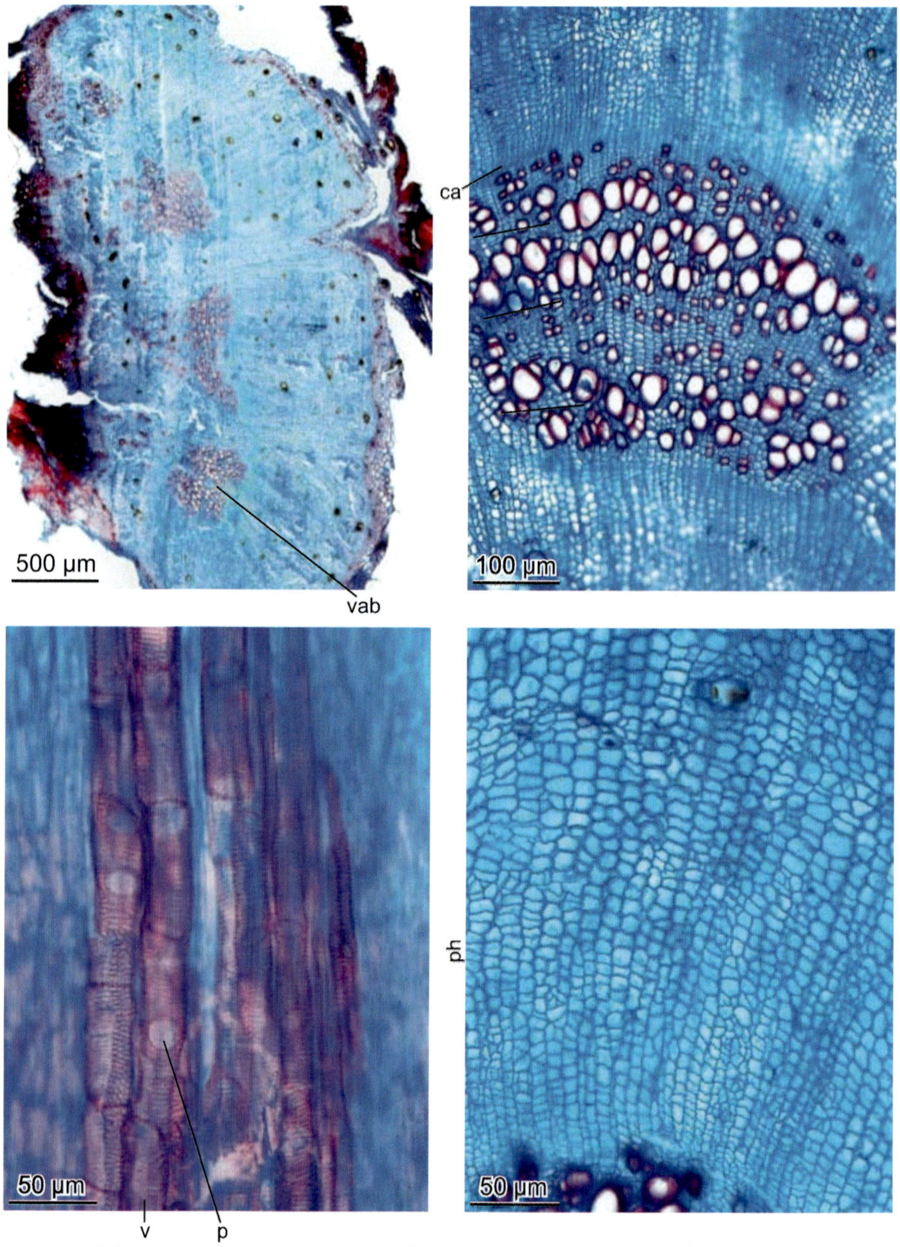

Features according to IAWA

Xylem 1 - 4 - 9 - 11 - 13 - 20 - 40.2 - 50.2 - 52.2 - 52.3 - 60.1 - 79.1 - 99 - 99.1 - 100.1 - 100.2 - 105 - 135.1 - 135.2

Scorzonera virgata DC.

Asteraceae

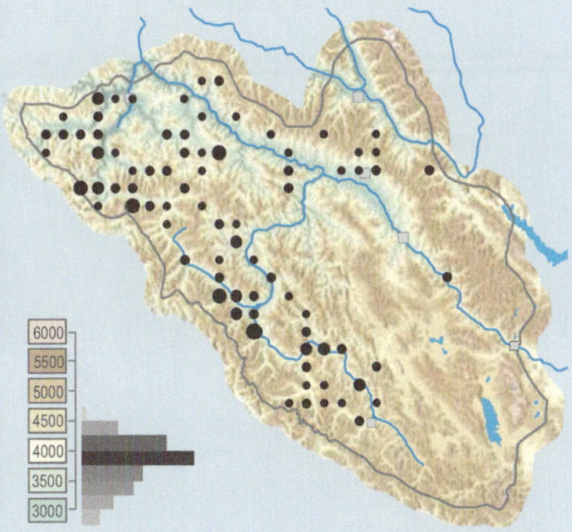

Syn.: *S. divaricata* Hook.f.

Description: Perennial herbs, 10–50 cm tall, stems rigid, leaves linear to filiform, crowded basally, capitula yellow, 2–3 cm across, terminal, numerous.

Distribution: Pakistan, NW India.

Habitat: Steppes, stabilized stony slopes.

Stem diameter 10–12 mm. Bark/xylem proportion 1:1. Ring width 0.4–0.5 mm. Maximal 14 annual rings.

Xylem

Transverse section

Semi-ring-porous. Vessels in short radial multiples. Ring boundary marked by a row of tangential flat fibers. Diameter of earlywood vessels 50–60 µm. Vessels stained. Intervessel pits round. Fibers thin- to thick-walled. Parenchyma paratracheal. With intra-xylary phellem.

Tangential section

Large, multiseriate rays between vascular bundles. Ray cells upright. Rays disappear in polarized light.

Radial section

Perforation plates simple. Intervessel pits scalariform. Vessel length 100–150 µm.

Bark

Phloem with radial groups of sieve tubes. Sclerenchyma absent. Oil ducts absent. Phellem consists of many rows of rectangular cells.

Pith

No slide available.

Analyzed: Root collar of polar root of 1 individual.

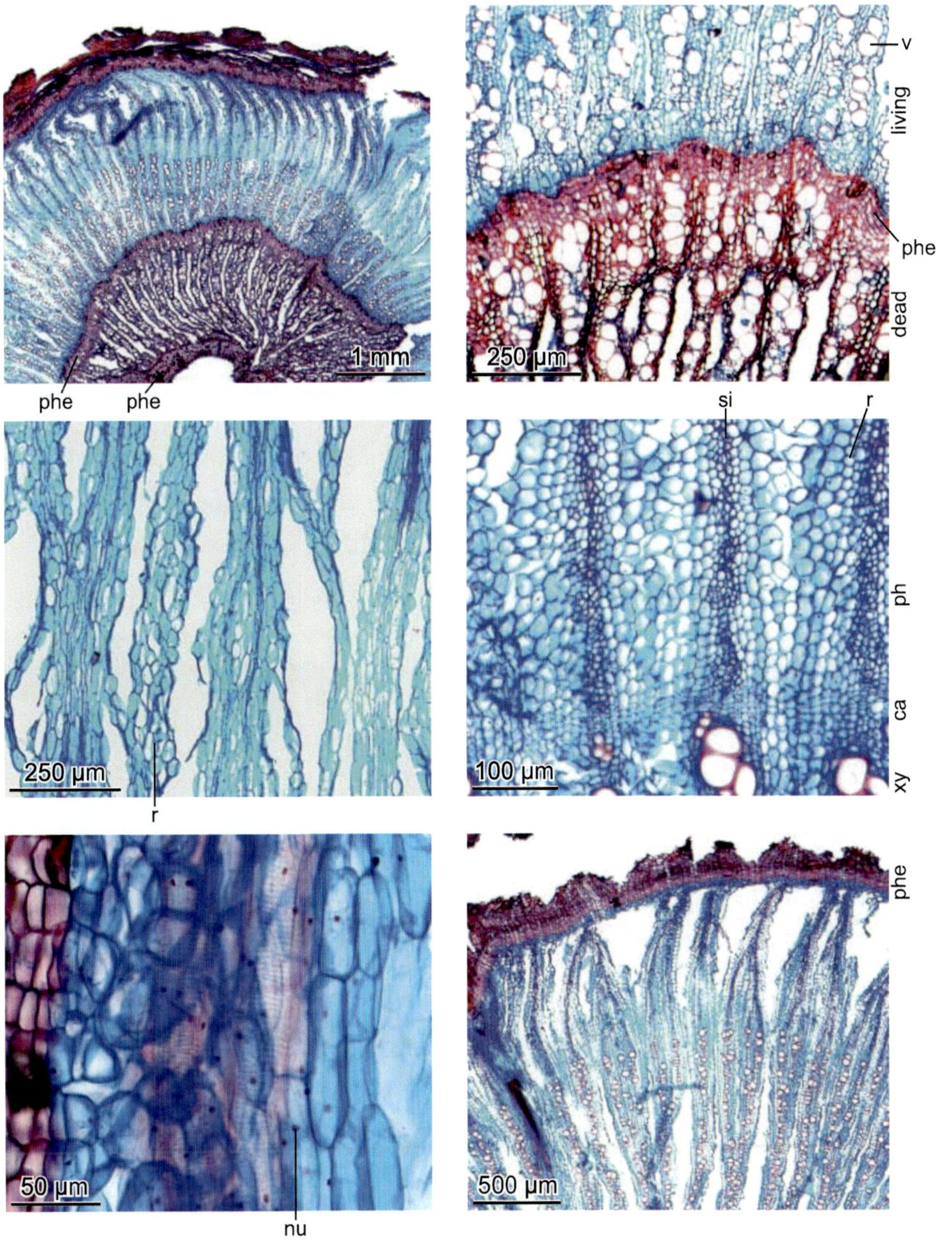

Features according to IAWA

Xylem 1 - 4 - 9.1 - 13 - 20 - 41 - 50.2 - 52.3 - 61 - 70 - 79 - 98 - 100.2 - 105 - 135.2

Senecio dubitabilis C.Jeffrey & Y.L.Chen

Asteraceae

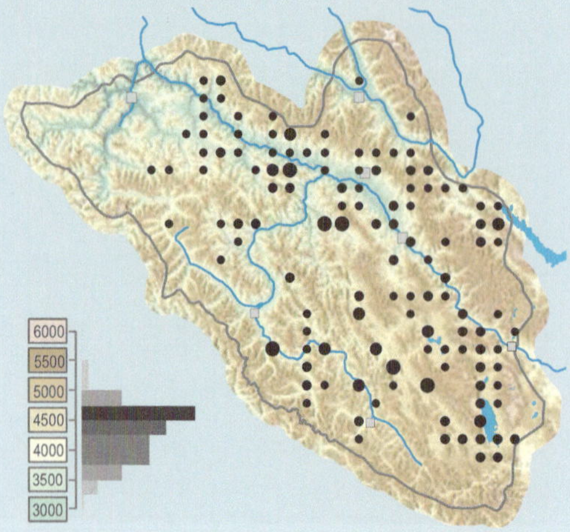

Syn.: *S. dubius* Ledeb.

Description: Annuals, 5–30 cm tall, leaves spatulate, oblong or linear, margin remotely dentate or entire, capitula yellow, ca. 6 mm across, terminal, numerous, ray florets absent.

Distribution: Russia (Siberia), Uzbekistan, Kyrgyzstan, Tajikistan, Pakistan, NW India, China (Gansu, Hebei, Nei Mongol, Qinghai, Shaanxi, Xinjiang, Xizang), Mongolia.

Habitat: Sandy plains, disturbed sites, old dry stony river beds.

Stem diameter 2.5 mm. Bark/xylem proportion 1:0.2. Mean ring width 1.5 mm. Maximal 1 annual ring.

Xylem

Transverse section

Only one ring. Vessels in short radial rows. Vessel diameter 30–50 µm. Fibers thin- to thick-walled. Parenchyma paratracheal.

Tangential section

Rays absent.

Radial section

Perforation plates simple. Intervessel pits round in alternating position. Vessel length 100–140 µm.

Bark

Phloem homogeneous. Cortex cells large (50–80 µm), oval. Phellem absent.

Pith

No slide available.

Analyzed: Root collar of polar root of 1 individual.

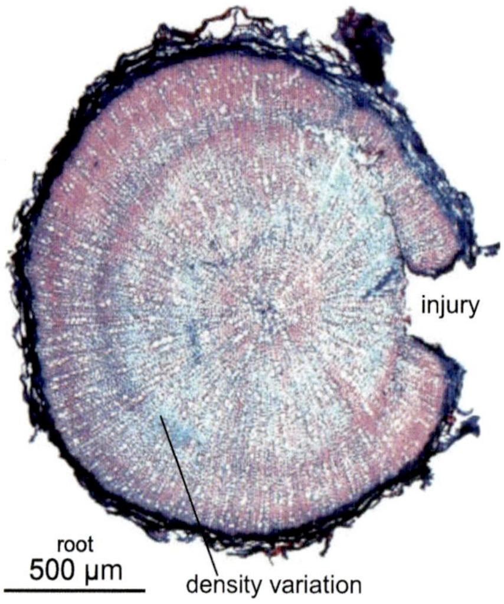

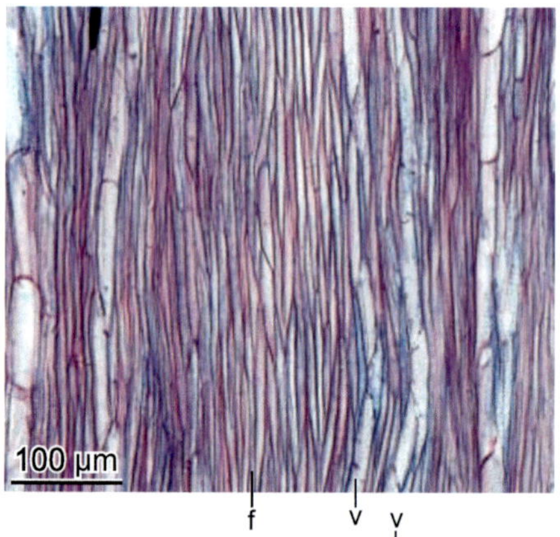

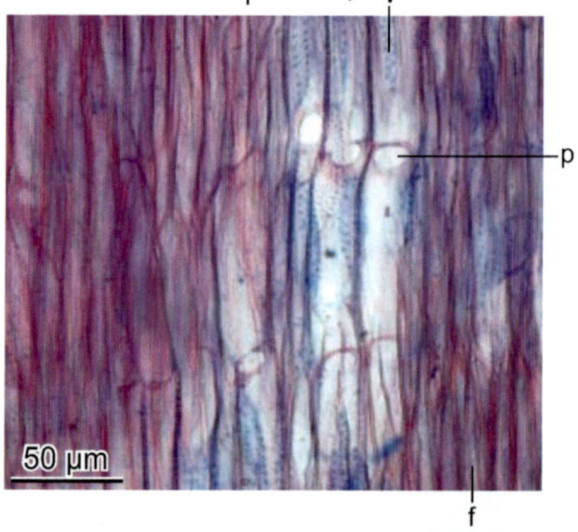

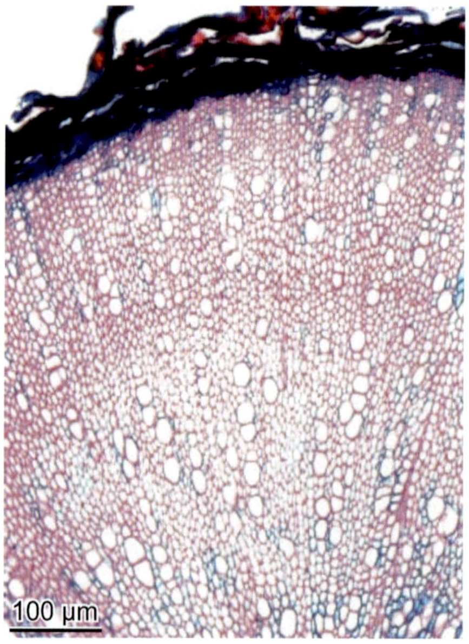

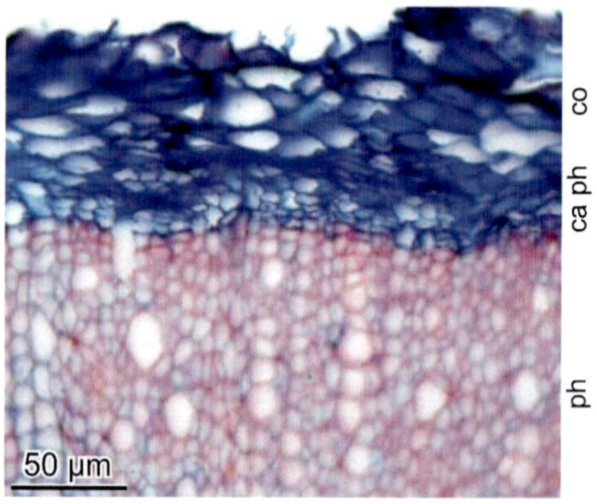

Features according to IAWA

Xylem 2.1 - 9.1 - 13 - 22 - 40.2 - 50.2 - 52.3 - 61 - 70 - 79 - 117

Senecio krascheninnikovii Schischk.

Asteraceae

Syn.: *S. pedunculatus* Edgew.

Description: Annuals, 5–30 cm tall, leaves ovate-oblong, pinnatifid to pinnatisect with 4–8 lobes, capitula yellow, ca. 6 mm across, terminal, numerous, ray florets present.

Distribution: Russia, Iran, Kazakhstan, Kyrgyzstan, Tajikistan, NE Afghanistan, N Pakistan, NW India, China (Qinghai, Xinjiang, Xizang).

Habitat: Disturbed sites, gravel slopes.

Stem diameter 1–2 mm. Bark/xylem proportion 1:0.7. Mean ring width 0.3 mm. Maximal 2 annual rings.

Xylem

Transverse section

Diffuse-porous. Vessels in short and long radial multiples. Ring boundary marked by a row of unstained fibers. Diameter of earlywood vessels 30–50 µm. Fibers thin- to thick-walled. Parenchyma paratracheal.

Tangential section

Multiseriate rays with sheath cells.

Radial section

Perforation plates simple. Intervessel pits round in alternating position. Vessel length 100–120 µm.

Bark

Phloem with groups of sieve tubes. Sclerenchyma and oil ducts absent. Cortex consists of very large, oval parenchyma cells (100–130 µm). Cortex with a few oil ducts. Phellem absent.

Pith

Round.

Analyzed: Root collar of polar root of 2 individuals.

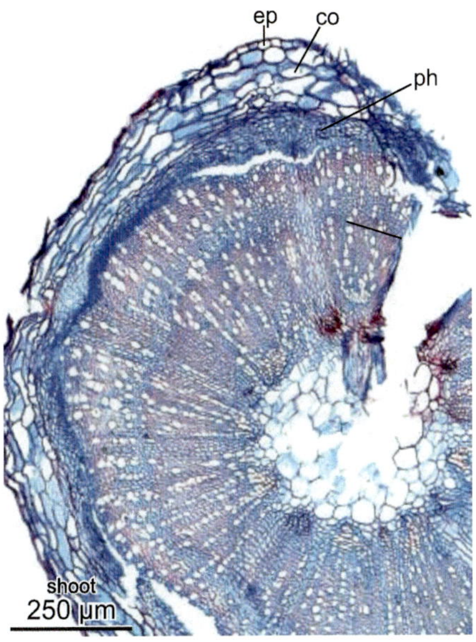

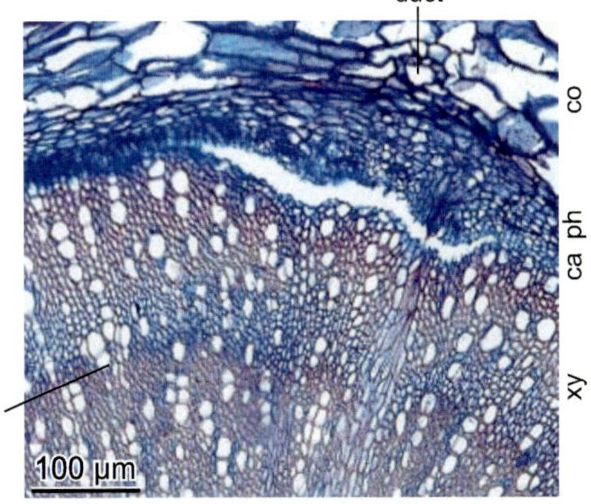

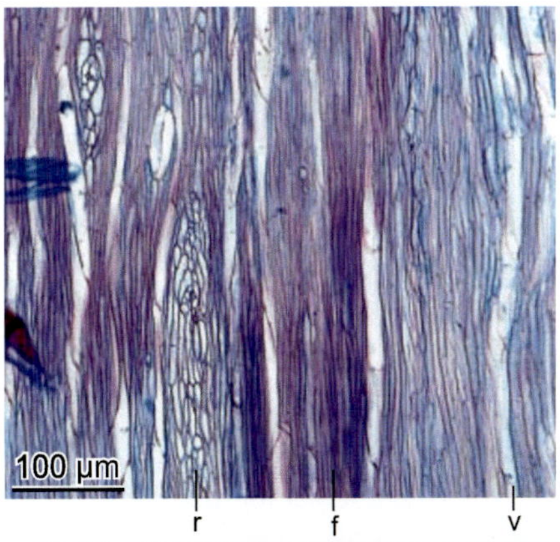

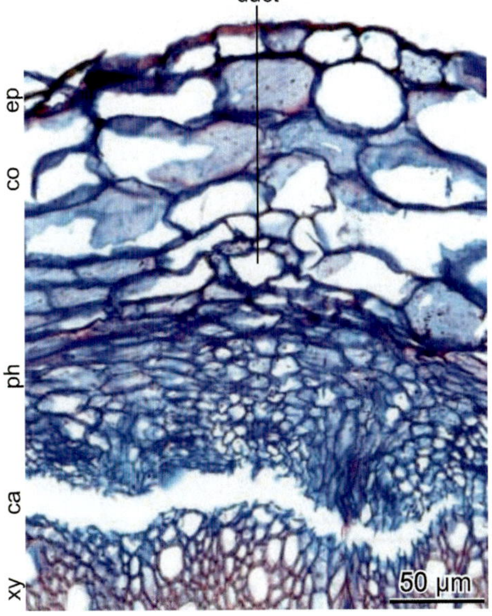

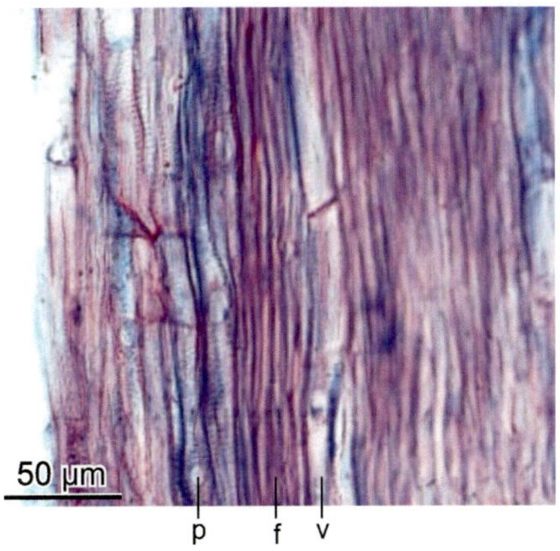

Features according to IAWA

Xylem 1 - 5 - 9.1 - 10 - 13 - 20 - 40.2 - 50.2 - 52.3 - 61 - 70 - 79 - 98 - 105 - 110

Tanacetum artemisioides Sch. Bip. ex Hk. f.

Asteraceae

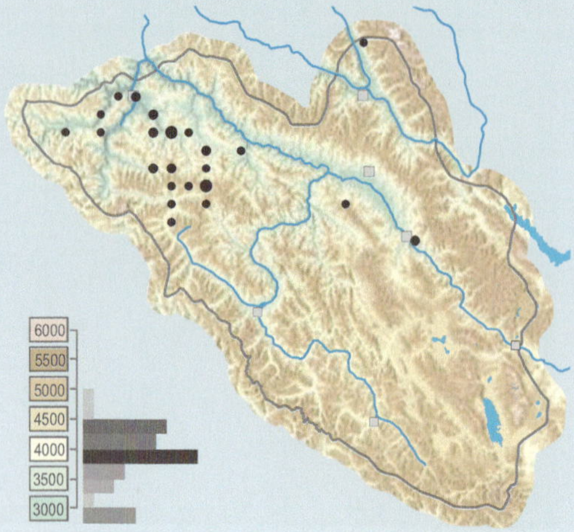

Syn.: *Artemisia fasciculata* non M. Bieb.: C. B. Clarke

Description: Subshrubs, 20–50 cm tall, appressed hairy, leaves palmately 1–2-pinnatisect, ultimate segments oblong, obtuse, capitula yellow, 4–5 mm across, in terminal flat-top lax corymbs.

Distribution: Pakistan, N India, W China, Bhutan.

Habitat: Steppes, stony slopes.

Stem diameter 3 mm. Bark/xylem proportion 1:1. Rings indistinct.

Xylem

Transverse section
Vessels in short radial multiples or in groups. Diameter of vessels 30–50 µm. Fibers thick-walled. Parenchyma paratracheal. Many layers of intra-xylary phellem.

Tangential section
Rays absent.

Radial section
Perforation plates simple. Intervessel pits round in alternating position. Vessel length 180–200 µm.

Bark

Phloem with groups of thick-walled sclereids with sieve tubes in their center. With oil ducts. Phellem consists of many layers of square cells. Oil ducts absent.

Pith

No slide available.

Analyzed: Root collar of polar root of 1 individual.

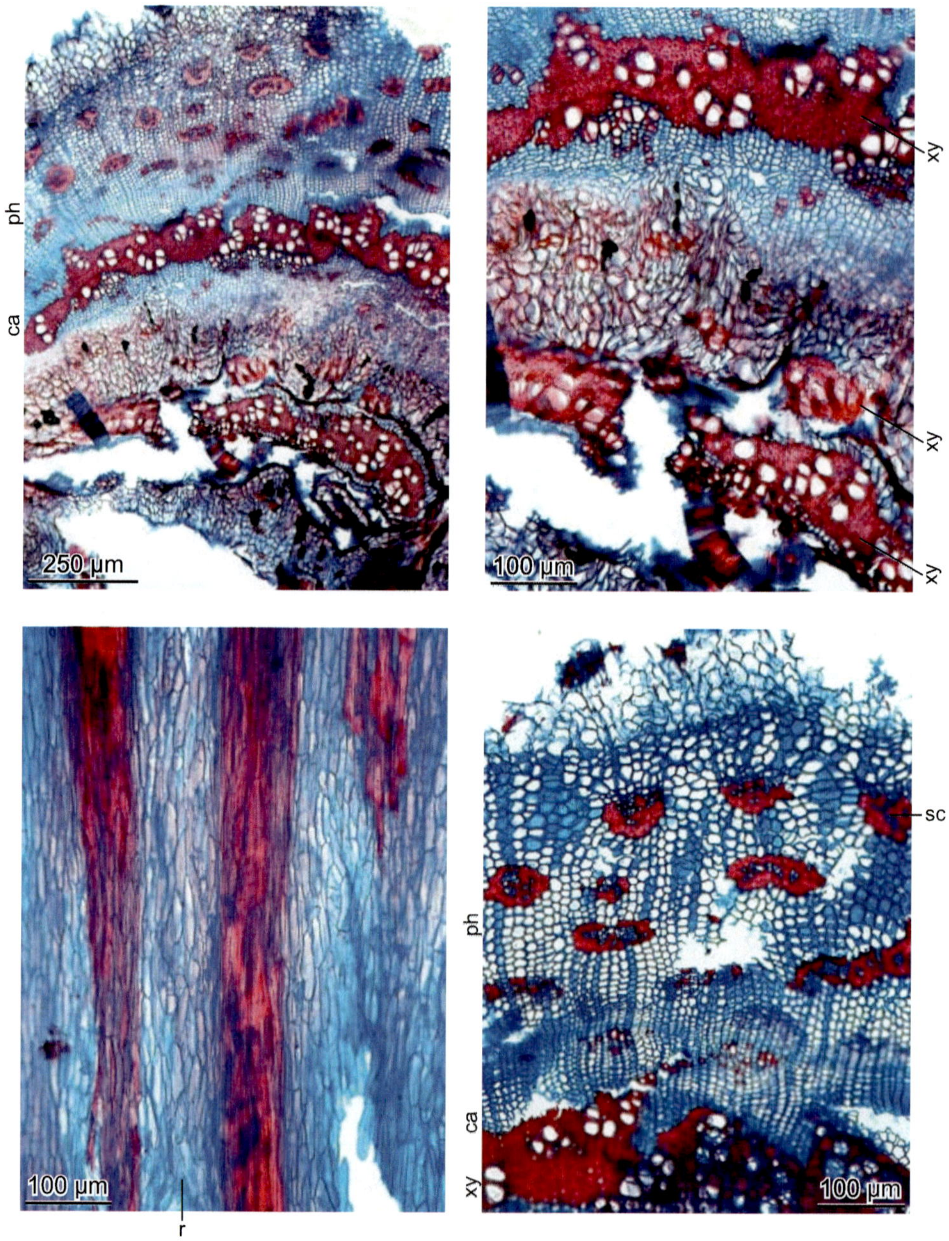

Features according to IAWA

Xylem 2 - 9.1 - 11 - 13 - 20 - 40.2 - 50.2 - 52.3 - 61 - 69 - 79 - 117 - 135.1 - 135.2

Tanacetum stoliczkae (C.B.Clarke) R.Khan

Asteraceae

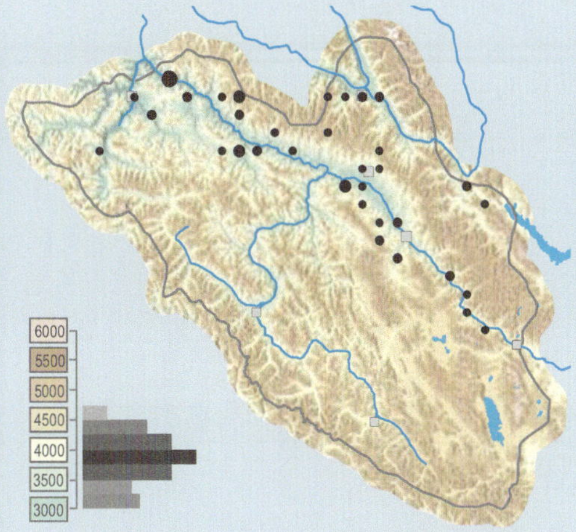

Syn.: *Chrysanthemum stoliczkae* C.B.Clarke

Description: Subshrubs, 15–35 cm tall, almost glabrous, stems woody, leaves pinnatisect, capitula 2–4 cm across, solitary, disc florets yellow, ray florets white or pinkish.

Distribution: Pakistan, NW India (Jammu and Kashmir).

Habitat: Rock crevices.

Stem diameter 2–4 mm. Bark/xylem proportion 1:0.5. Ring width 0.15–1.8 mm. Maximal 33 annual rings.

Xylem

Transverse section

Semi-ring-porous. Vessels solitary and in tangential groups. Ring boundary marked by large earlywood (50–70 µm) and small latewood vessel diameters (<20 µm) and a fiber zone with few vessels. Vessels in the central part of the stem with dark-stained substances (heartwood). Parenchyma paratracheal.

Tangential section

Rays absent.

Radial section

Perforation plates simple. Intervessel pits round in alternating position. Vessel length 100–150 µm.

Bark

Phloem with tangentially arranged groups of thick-walled fibers. Oil ducts absent. Phellem cells thin-walled, square.

Pith

No slide available.

Analyzed: Root collar of polar root of 3 individuals.

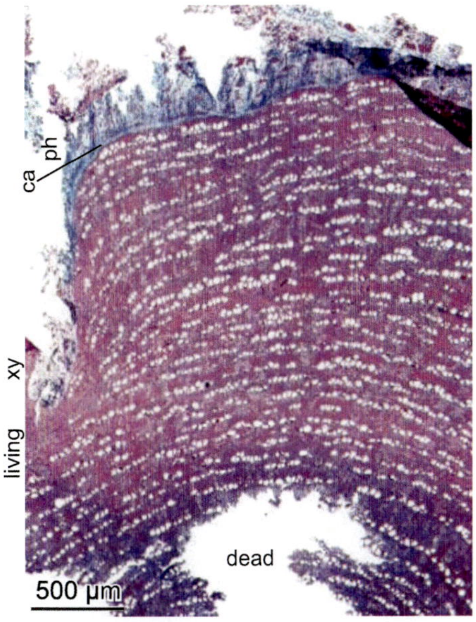

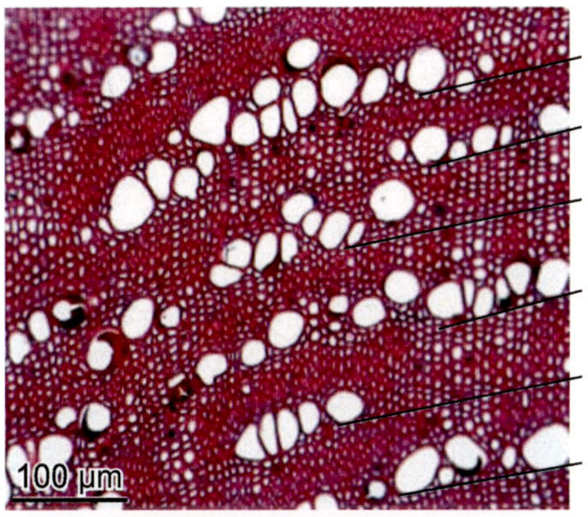

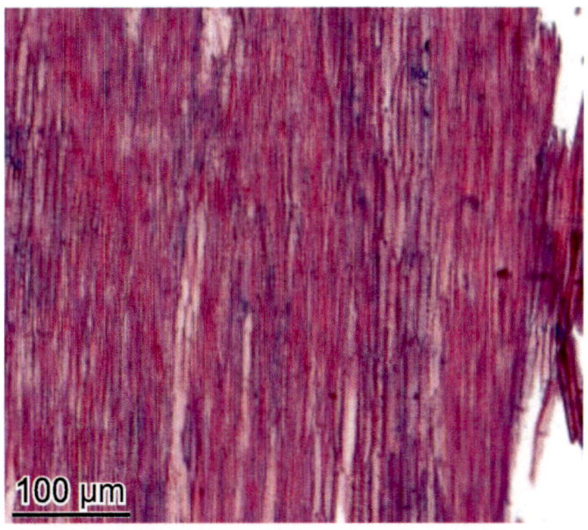

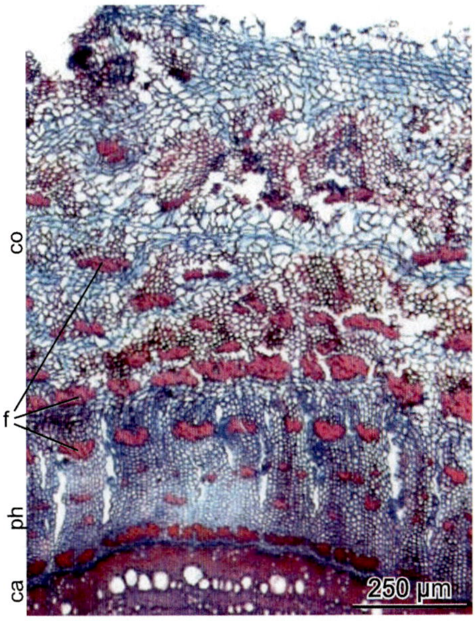

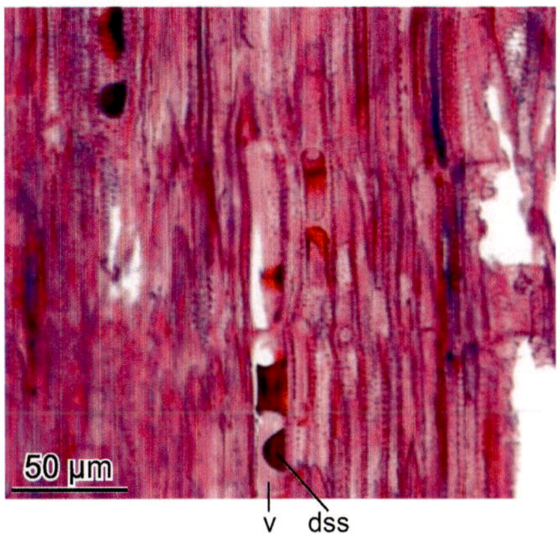

Features according to IAWA

Xylem 1 - 4 - 9 - 11 - 13 - 20 - 41 - 50.2 - 52.3 - 58 - 61 - 70 - 79 - 117

Tragopogon gracilis D.Don

Asteraceae

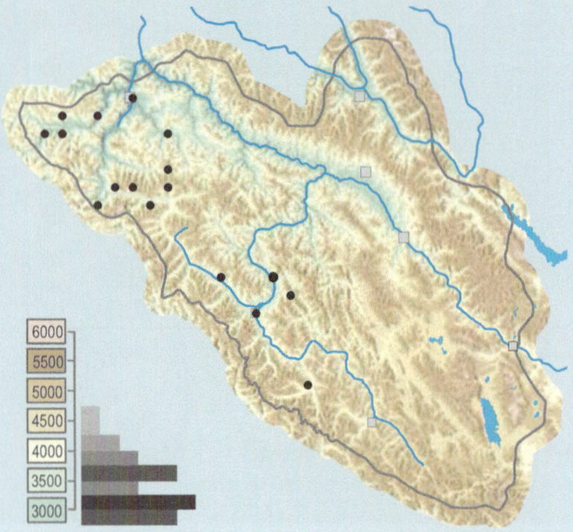

Syn.: *T. pamiricus* S.A.Nikitin

Description: Perennial herbs, 10–20 cm tall, glabrous, leaves lanceolate to subulate, capitula yellow, 2–3 cm across, solitary.

Distribution: Kazakhstan, Uzbekistan, Kyrgyzstan, Tajikistan, Afghanistan, N India, Nepal, China (Xinjiang, Xizang).

Habitat: Alpine meadows, river banks.

Stem diameter 11 mm. Bark/xylem proportion 1:2. Ring width 0.4–0.5 mm. Maximal 4 annual rings.

Xylem

Transverse section

Rings recognizable. Semi-ring-porous. Vessels in short radial multiples. Ring boundary marked by large earlywood (40–50 µm) and small latewood vessel diameters (10–20 µm). Fibers thin- to thick-walled, in a peripheral ring. Fibers absent in the large center. Parenchyma absent in the fiber belt and pervasive in the center. Vascular bundles remaining.

Tangential section

Large, multiseriate rays between vascular bundles. Ray cells upright. Rays disappear in polarized light.

Radial section

Perforation plates simple. Intervessel pits round in vessels within the fiber belt and scalariformin in those in the center. Vessel length 180–200 µm.

Bark

Phloem with radial groups of sieve tubes. Sclerenchyma absent. Ducts absent. Phellem consists of many rows of irregular formed, thin-walled cells.

Pith

Round, with round medullary phloem groups.

Analyzed: Root collar of polar root of 1 individual.

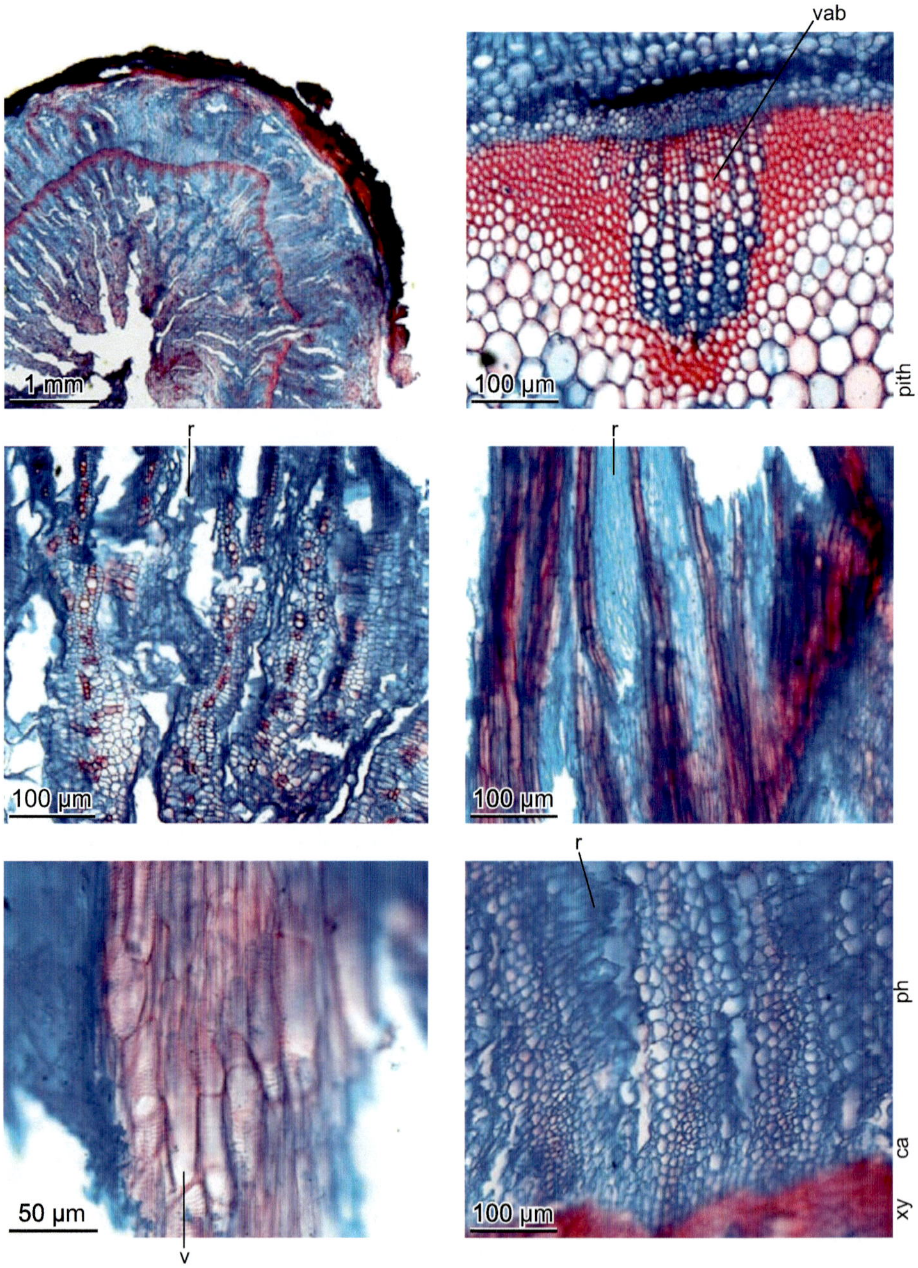

Features according to IAWA

Xylem 1 - 4 - 9.1 - 13 - 20 - 22 - 40.2 - 50.2 - 52.3 - 61 - 70 - 76 - 79 - 99 - 99.1 - 100.1 - 100.2 - 105

Waldheimia nivea (Hook.f. & Thomson ex C.B.Clarke) Regel

Asteraceae

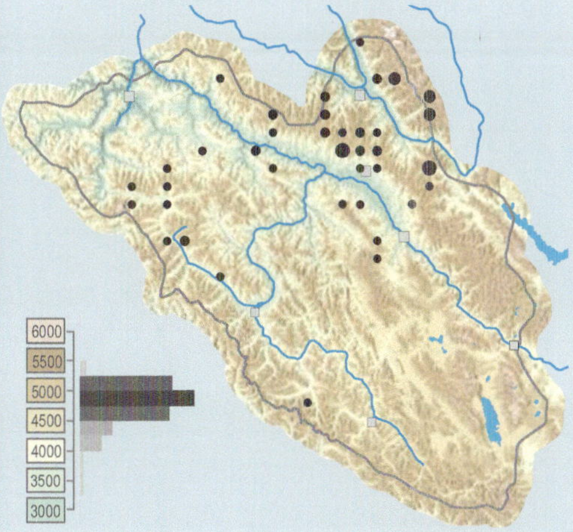

Syn.: *Allardia nivea* Hook.f. & Thomson ex C.B.Clarke

Description: Perennial herbs, 4–10 cm tall, densely white wooly hairy, stems many, loosely cespitose, leaves cuneate, palmately lobed, capitula terminal, solitary, 1.5–2 cm across, ray florets dark pink, disc florets yellow.

Distribution: Afghanistan, Pakistan, India, Nepal, China (S Xizang).

Habitat: Moraines of glaciers, gravel scree slopes.

Stem diameter 3–5 mm. Bark/xylem proportion 1:0.8. Ring width 0.05–0.06 mm. Maximal 22 annual rings.

Xylem

Transverse section

Semi-ring-porous. Vessels solitary. Ring boundary marked by large earlywood (40–50 µm) and small latewood vessel diameters (20 µm) and flat fibers in the latewood. Parenchyma paratracheal. Crystals absent.

Tangential section

Rays 1–3-seriate , not lignified, disappear in polarized light.

Radial section

Perforation plates simple. Intervessel pits round in alternating position, scalariform. Vessel length 60–80 µm.

Bark

Phloem with irregular groups of sieve tubes. Phellem consists of many layers thin-walled, rectangular cells.

Pith

Round.

Analyzed: Root collar of polar root of 3 individuals.

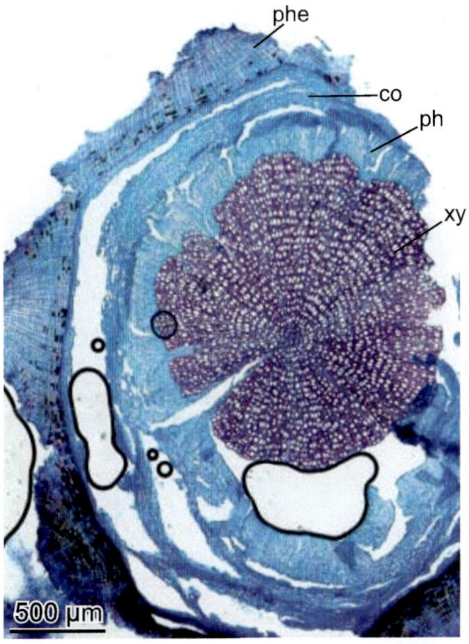

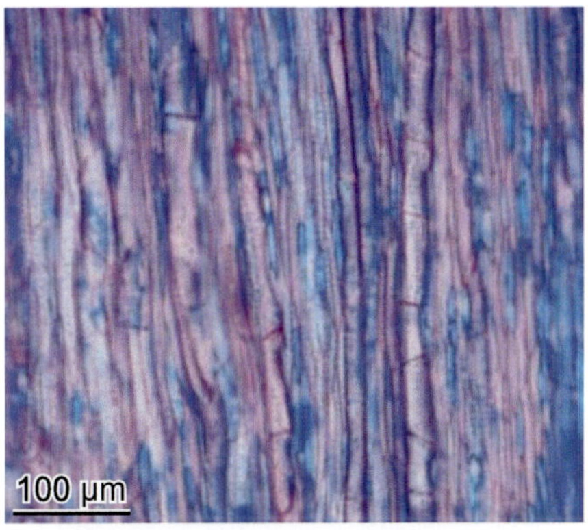

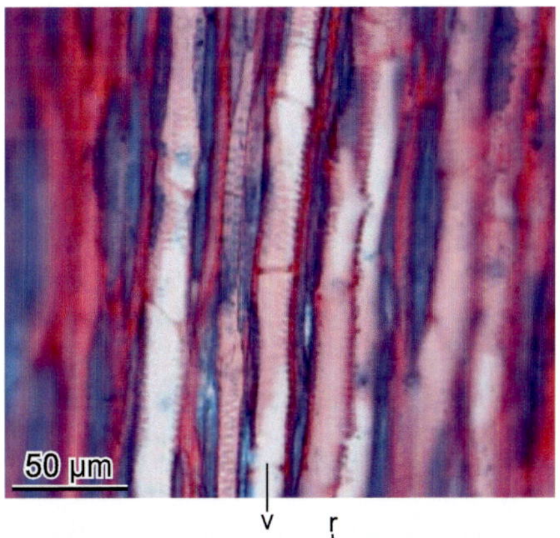

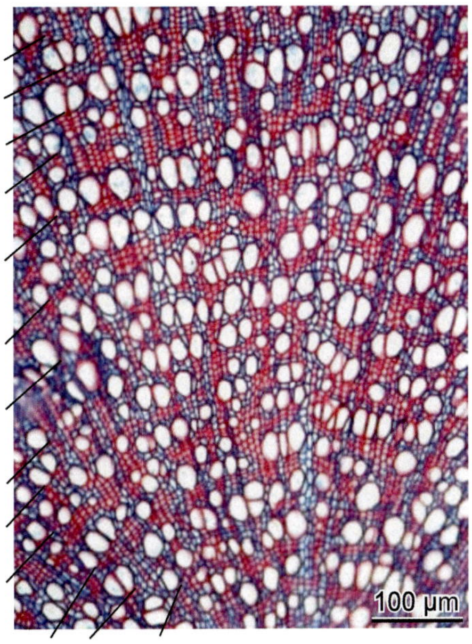

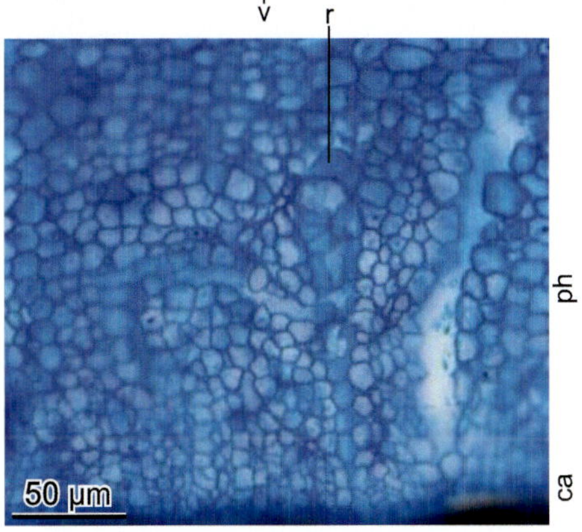

Features according to IAWA

Xylem 1 - 4 - 9 - 13 - 20 - 22 - 40.2 - 50.2 - 52.3 - 61 - 70 - 79 - 97 - 101 - 100.2 - 105

Waldheimia tomentosa (Decne.) Regel

Asteraceae

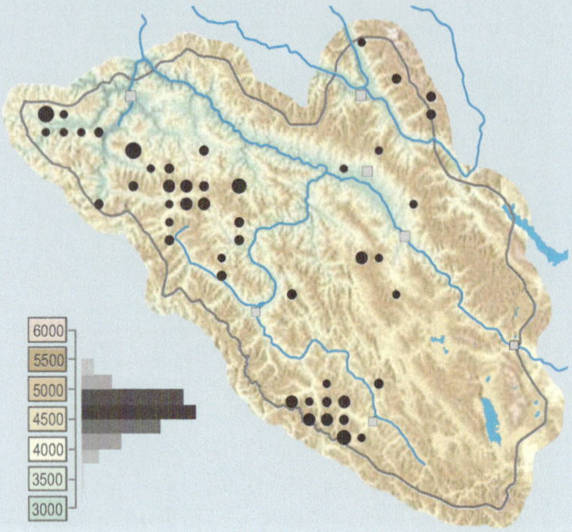

Syn.: *Allardia tomentosa* Decne.

Description: Perennial herbs, 10–15 cm tall, densely whitish tomentose, loosely cespitose, leaves 2-pinnatisect, ultimate segments linear, capitula terminal, solitary, 3–5 cm across, ray florets pink, disc florets yellow.

Distribution: Kazakhstan, Uzbekistan, Afghanistan, Pakistan, India (Jammu and Kashmir, Himachal Pradesh, Uttar Pradesh), Nepal, China (Xizang).

Habitat: Stabilized slopes, rock crevices.

Stem diameter 5–6 mm. Bark/xylem proportion 1:0.8. Ring width 0.07–0.09 mm. Maximal 20 annual rings.

Xylem

Transverse section

Semi-ring-porous. Vessels solitary, with angular outline. Ring boundary marked by large earlywood (40–50 µm) and small latewood vessel diameters (20 µm) and few flat fibers in the latewood. Parenchyma paratracheal. Fibers in the center with dark-staining substances (heartwood)

Tangential section

Rays 8–>10-seriate. Rays unlignified, disappearing in polarized light.

Radial section

Perforation plates simple. Intervessel pits round in alternating position. Vessel length 100–120 µm.

Bark

Phloem with tangentially arranged groups of sieve tubes. With ray dilatations. Oil ducts absent. Phellem cells thin-walled, square.

Pith

Round.

Analyzed: Root collar of polar root of 3 individuals.

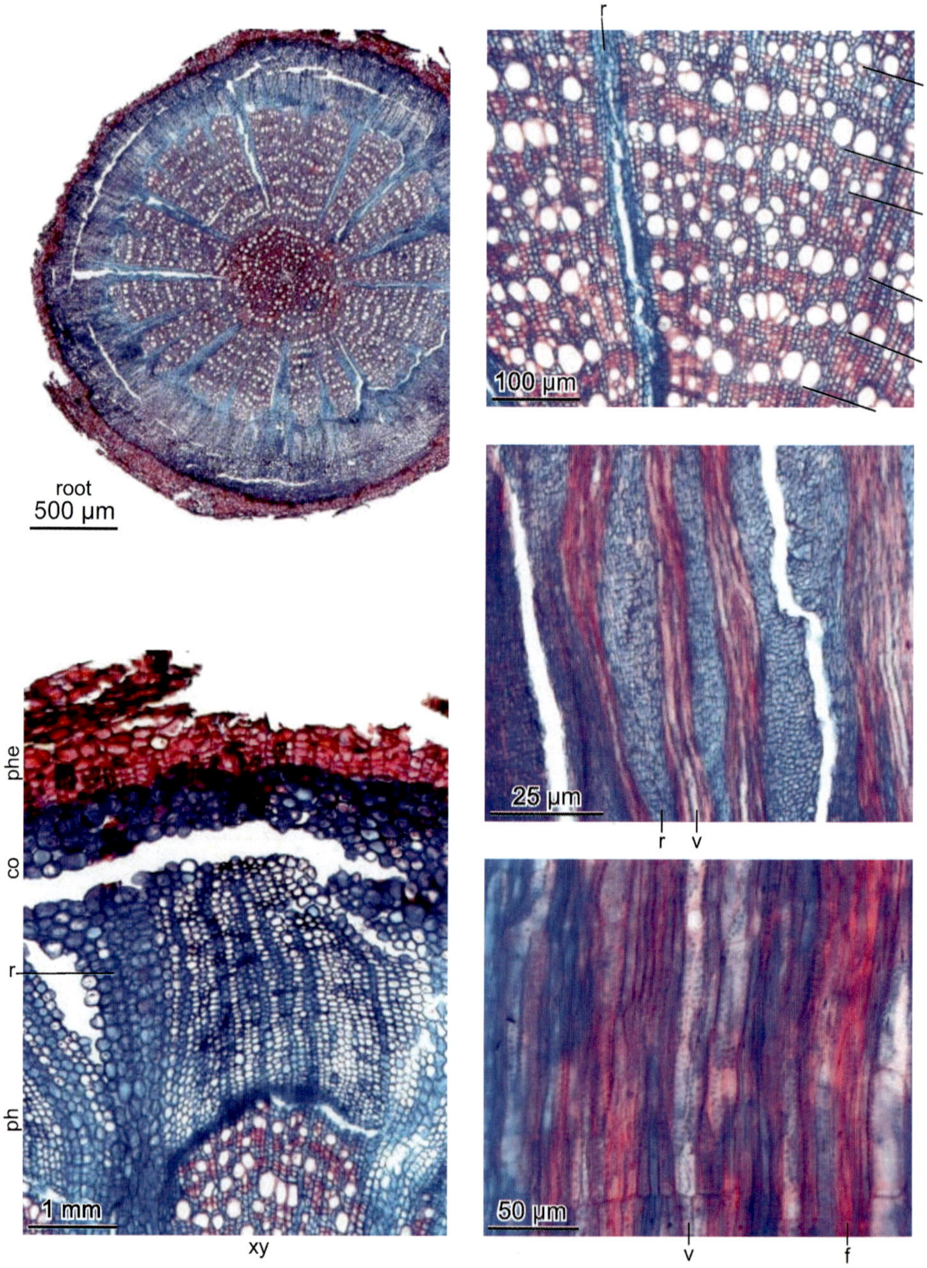

Features according to IAWA

Xylem 1 - 4 - 9 - 12 - 13 - 20 - 40.2 - 50.2 - 52.3 - 58 - 61 - 70 - 79 - 98 - 99 - 100.2 - 105

Waldheimia tridactylites Kar. & Kir.

Asteraceae

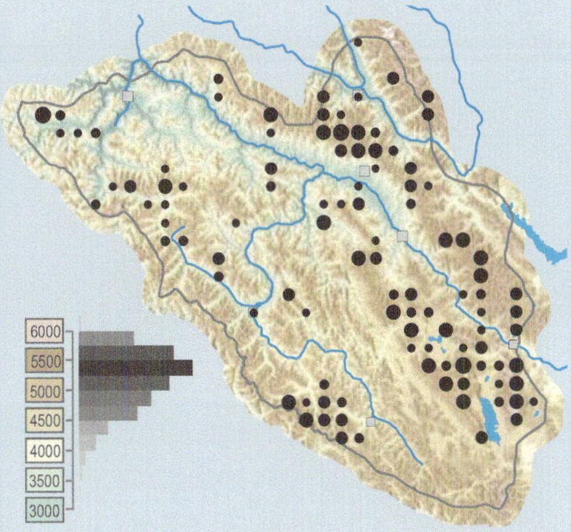

Syn.: *Allardia tridactylites* (Kar. & Kir.) Sch.Bip., *A. glabra* Decne.

Description: Perennial herbs, 1–5 cm tall, glabrous, cespitose, with many prostrate shoots from procumbent rhizome, leaves obovate to spatulate, palmately 3(–5)-lobed, 1–3 cm long, capitula solitary, terminal, 2.5–3.5 cm across, disc florets yellow, ray florets pink to whitish-pink or purplish.

Distribution: Russia, Kazakhstan, Uzbekistan, Afghanistan, Pakistan, India, China (Xinjiang, Xizang), Mongolia.

Habitat: Gravel screes, mesic slopes and rock crevices.

Stem diameter 1.5–3 mm. Bark/xylem proportion 1:2. Ring width 0.04–0.1 mm. Maximal 20 annual rings.

Xylem

Transverse section

Semi-ring-porous. Vessels solitary, with angular outline. Ring boundary not very distinct; if present, marked by a few flat fibers in the latewood. Diameter of vessels 15–20 µm. Shoots with remaining vascular bundles. Parenchyma paratracheal.

Tangential section

Rays 8–>10-seriate. Rays unlignified, disappearing in polarized light.

Radial section

Perforation plates simple. Intervessel pits mainly round in alternating position. Vessel length 80–120 µm.

Bark

Phloem with tangentially arranged groups of sieve tubes. With ray dilatations. With large oil ducts. Phellem cells thin-walled, square.

Pith

Round.

Analyzed: Root collar of polar root of 3 individuals.

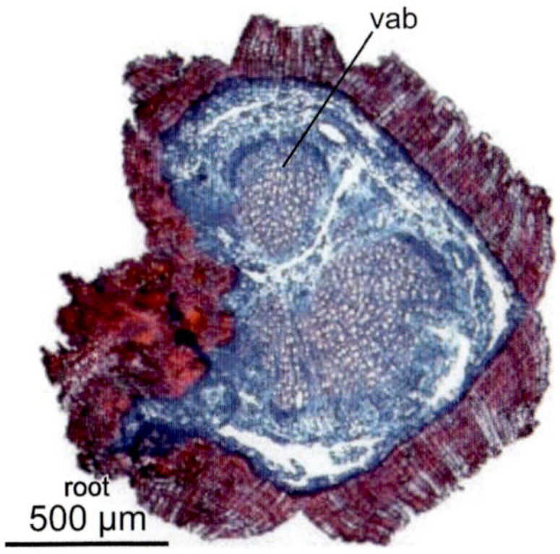

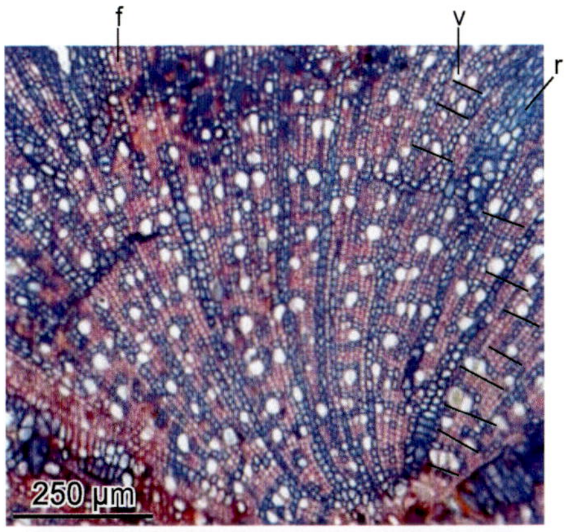

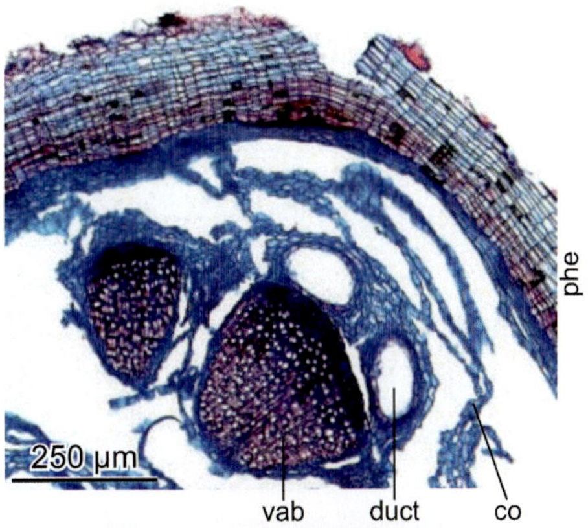

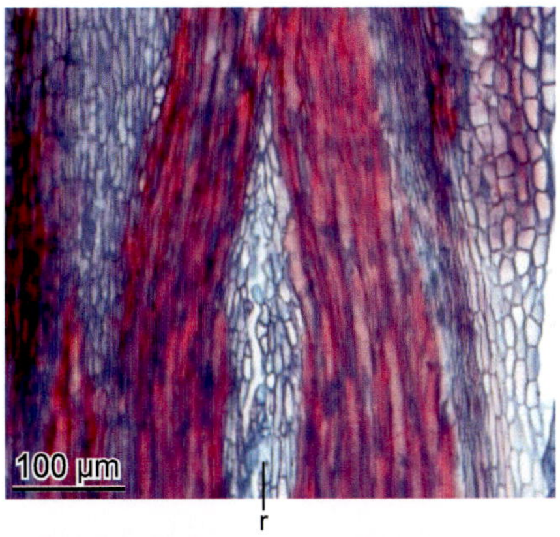

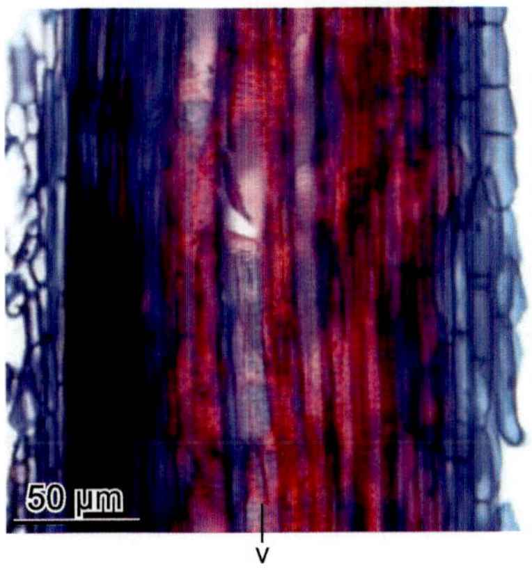

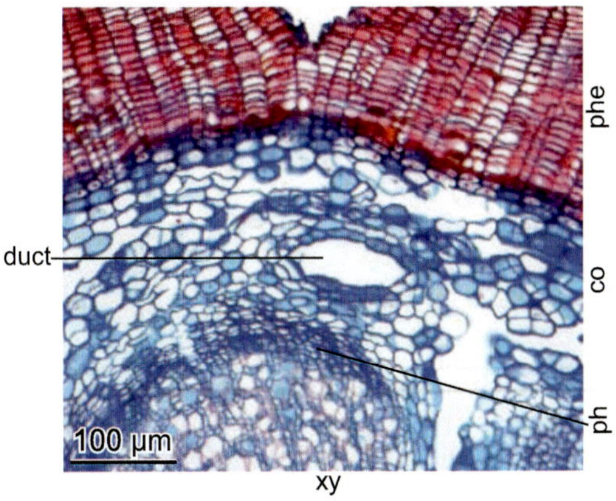

Features according to IAWA

Xylem 1 - 4 - 9 - 12 - 13 - 20 - 40.2 - 50.2 - 52.3 - 58 - 61 - 70 - 79 - 98 - 99 - 100.2 - 105

Waldheimia vestita (Hook.f. & Thomson ex C.B.Clarke) Pamp.

Asteraceae

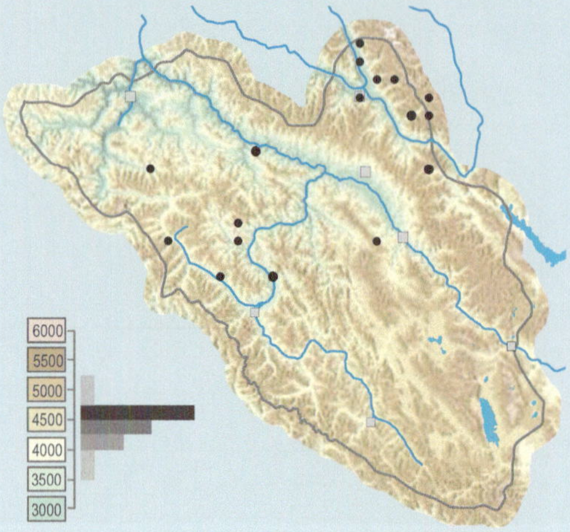

Syn.: *Allardia vestita* Hook.f. & Thomson ex C.B.Clarke

Description: Perennial herbs, 4–6 cm tall, densely yellowish or gray-white lanate, with creeping woody rhizomes, stems many, loosely cespitose, leaves cuneate, deeply palmately lobed, capitula terminal, solitary, 3.5–4 cm across, ray florets pink, disc florets yellow.

Distribution: N Pakistan, NW India, China (S Xizang).

Habitat: Mountain slopes, moraines, scree.

Stem diameter 5–6 mm. Bark/xylem proportion 1:0.8. Ring width 0.08–0.2 mm. Maximal 8 annual rings.

Xylem

Transverse section

Semi-ring-porous,. Vessels solitary, with angular outline. Ring boundary marked by large earlywood (20–30 µm) and small latewood vessel diameters (<10 µm) and flat fibers in the latewood. Parenchyma paratracheal and apotracheal.

Tangential section

Rays 8–>10-seriate. With distinct sheath cells. Rays unlignified, disappear in polarized light.

Radial section

Perforation plates simple. Intervessel pits round in alternating position. Vessel length 100–120 µm.

Bark

Phloem with tangentially arranged groups of sieve tubes. With ray dilatations. With ducts. Phellem cells thin-walled, square.

Pith

Round.

Analyzed: Polar root of 3 individuals.

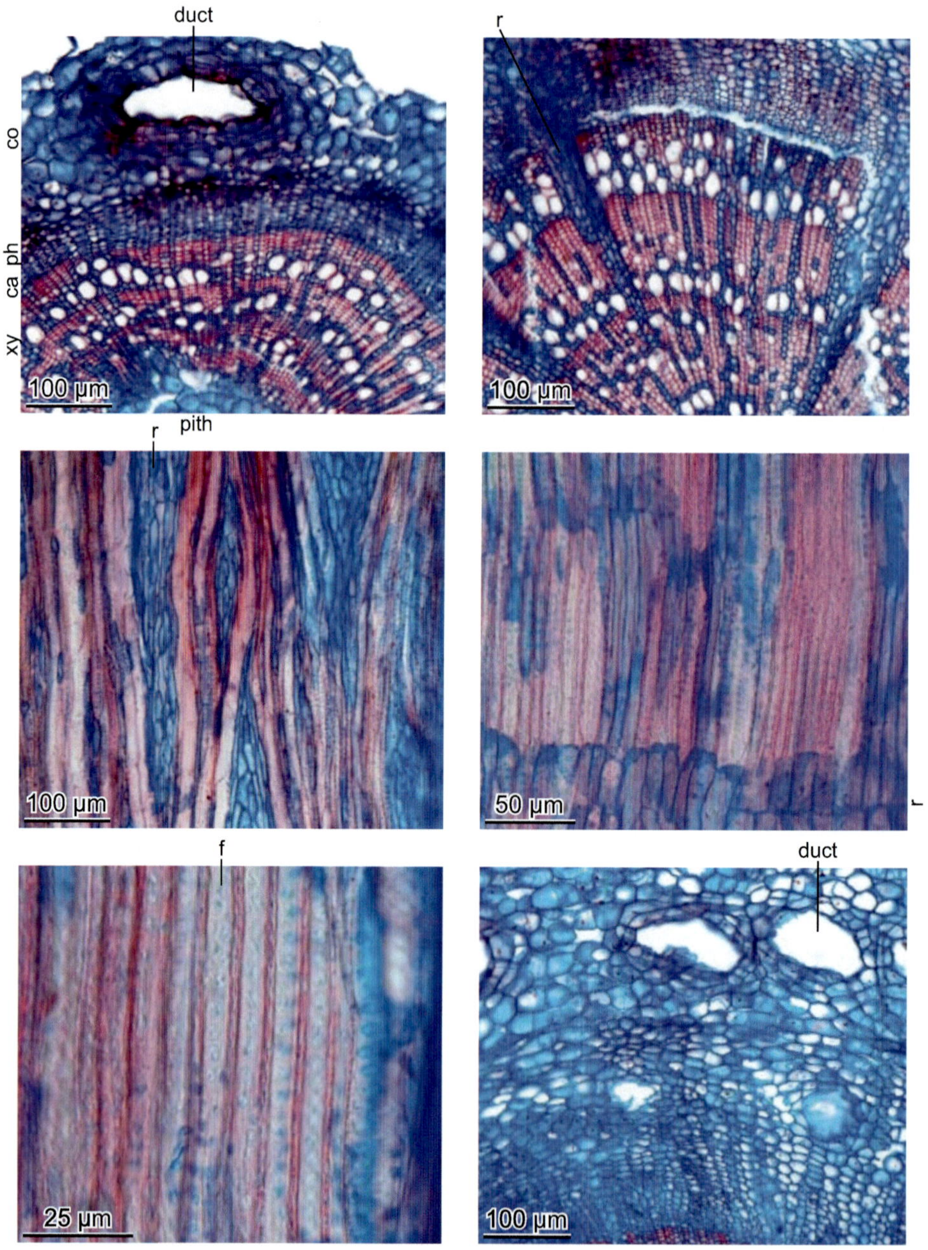

Features according to IAWA

Xylem 1 - 4 - 9 - 12 - 13 - 20 - 40.2 - 50.2 - 52.3 - 61 - 70 - 76 - 79 - 98 - 99 - 100.2 - 105 - 110

Betula jacquemontii (Spach) H.J.P.Winkl.

Betulaceae

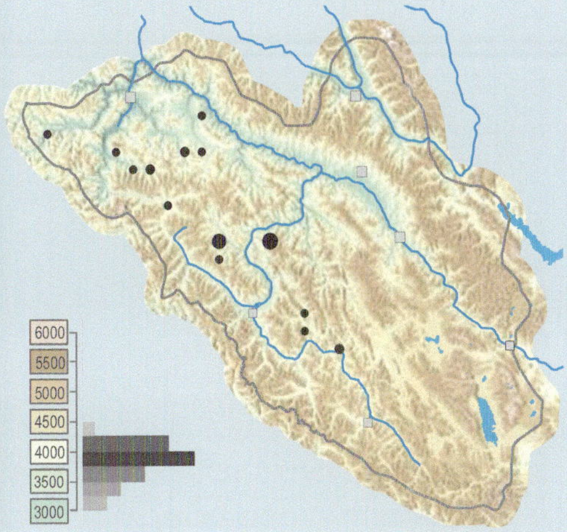

Syn:. *B. utilis* var. *jacquemontii* (Spach) H.J.P.Winkl., *B. utilis* subsp. *jacquemontii* (Spach) Kitam.

Description: Trees or small shrubs, 2–15 m tall, bark white or brownish, young shoots pubescent, leaves ovate or rhomboid, base rounded to cuneate, margin double-serrate.

Distribution: India (Jammu and Kashmir, Himachal Pradesh, Uttar Pradesh), Nepal.

Habitat: Upper limit of trees, wet slopes.

Stem diameter 30–40 cm. Bark/xylem proportion 1:20. Mean ring width 0.5–2.5 mm. Maximal 100 rings.

Xylem

Transverse section

Diffuse-porous. Rings distinct. Ring boundaries marked by a 1–3-seriate layer of terminal parenchyma. Vessels round, solitary or in short radial rows. Earlywood vessel diameter 30–60 µm. Fibers thin- to thick-walled. Parenchyma apotracheal and marginal.

Tangential section

Rays 1–3-seriate.

Radial section

Perforation scalariform with 12–20 bars. Intervessel pits small, round. Ray-vessel pits numerous with a diameter of 1.5–2 µm. Vessel length 400–600 µm.

Bark

No slide available.

Pith

No slide available.

Analyzed: Stems of 2 individuals.

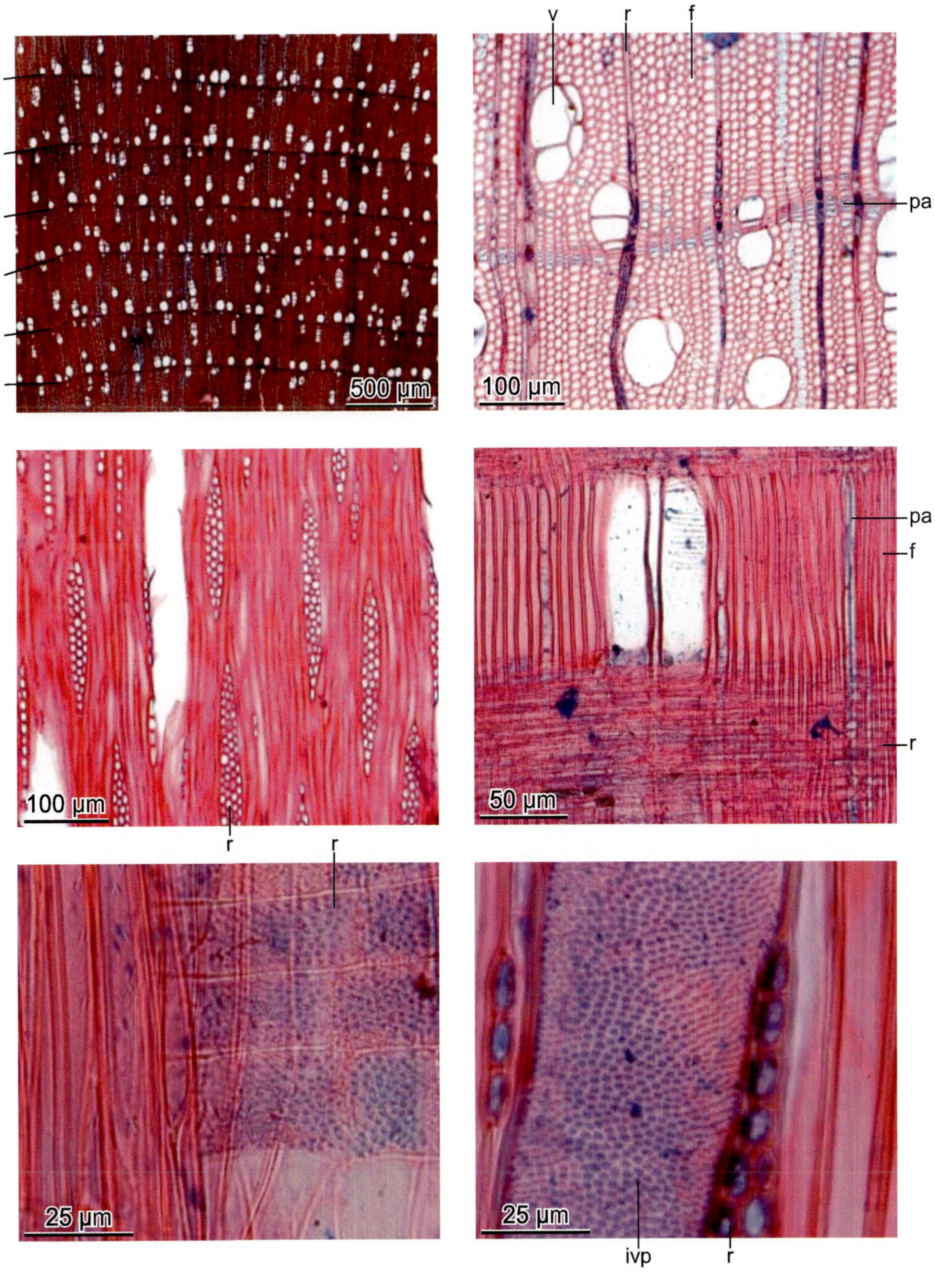

Features according to IAWA

Xylem 1 - 5 - 9 - 9.1 - 14 - 22 - 42 - 50.1 - 51.3 - 53.2 - 61 - 69 - 76 - 89 - 97 - 99 - 104

Biebersteinia odora Stephan ex Fisch.

Biebersteiniaceae

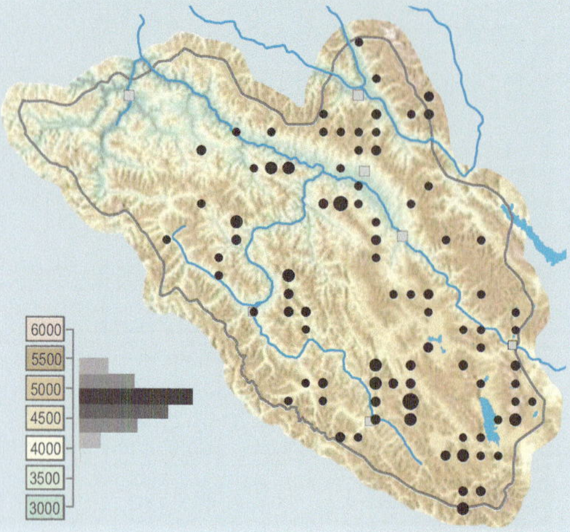

Syn.: *B. emodi* Jaub. & Spach

Description: Perennial herbs, 10–30 cm tall, aromatic, glandular-pubescent, leaves pinnatisect, flowers yellow with orange centre, 1–1.5 cm across, in racemes.

Distribution: Kazakhstan, Kyrgyzstan, Tajikistan, N Pakistan, NW India, China (Xinjiang, Xizang), Mongolia.

Habitat: Dry screes, rock crevices.

Stem diameter 9 mm. Bark/xylem proportion 1:2. Ring width 0.1–0.25 mm. Maximal 20 annual rings.

Xylem

Transverse section

Semi-ring-porous. Rings distinctly marked by large differences between earlywood (80–100 µm) and latewood vessel diameters (10–20 µm). Earlywood vessels arranged in a multiple row. Fibers absent. Parenchyma pervasive. Rays dilate within the xylem. Vascular bundles remain. With crystal druses in the rays.

Characteristic is the stem dilatation and disintegration. Soon after stem and root formation radial growth occurs lobe-like, and the upper part of the root or rhizome disintegrates by forming secondary intraxylary phellem.

Tangential section

Rays 5–>10-seriate.

Radial section

Perforation plates simple. Intervessel pits scalariform. Vessel length 120–200 µm. Ray cells upright.

Bark

Phloem uniform. Crystal druses arranged in tangential rows. Cortex cells large (80–100 µm), round to oval, and few contain crystal druses. Phellem layered, consisting of living cells with blue-stained contents (phelloids) and dead, brown-stained cells.

Pith

Pith consists of very thin-walled, large parenchyma cells.

Analyzed: Root collar of polar root of 4 individuals.

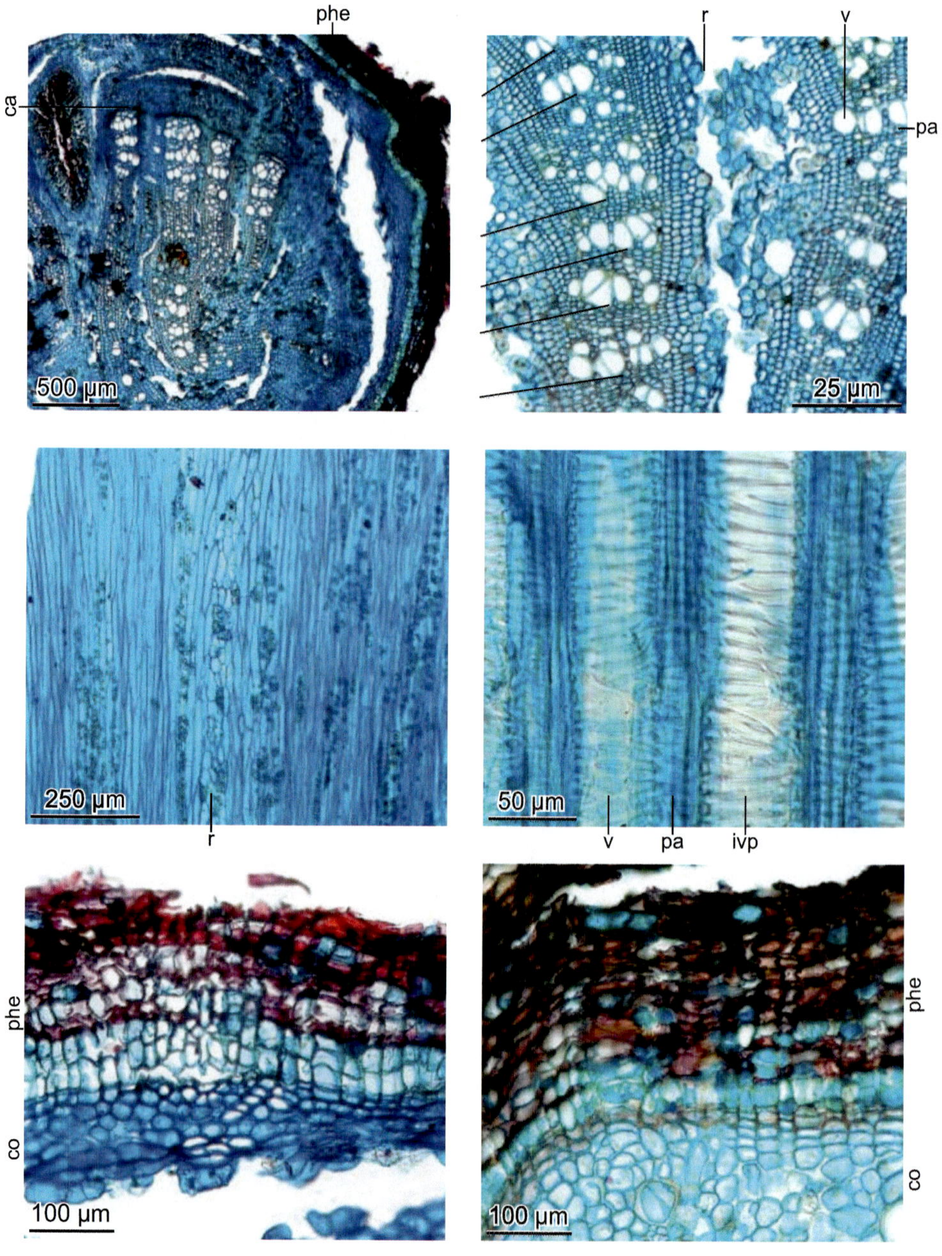

Features according to IAWA

Xylem 1 - 4 - 10 - 13 - 20 - 41 - 50.2 - 52.3 - 60.1 - 79.1 - 98 - 99 - 99.1 - 100.2 - 105

Actinocarya acaulis (W.W.Sm.) I.M.Johnst.

Boraginaceae

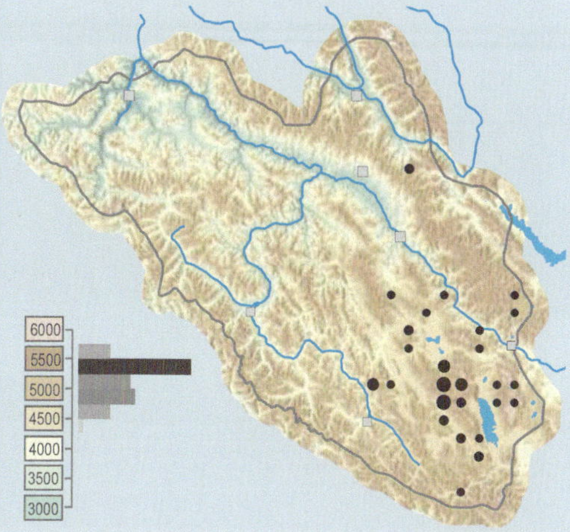

Syn.: *Metaeritrichium microuloides* W.T.Wang

Description: Annuals, 3–5 cm tall, stems branched from the base, leaves spatulate, 1–2 cm long, flowers purple-blue, 2–3 mm across, solitary on axillar pedicels ca. 2 cm long.

Distribution: India (Jammu and Kashmir, Sikkim), China (Qinghai, Xinjiang, Xizang)

Habitat: Alpine to subnival moist river beds, open turf, loamy soil deposits around marmot burrows, animal resting places.

Stem diameter 1 mm. Bark/xylem proportion (living part) 2:1. Mean ring width 0.4 mm. Maximal 1 living annual ring.

Xylem

Transverse section

One ring present. Vessels solitary with a diameter of 20 µm. Fibers absent. Parenchyma pervasive.

Tangential section

Rays absent.

Radial section

Perforation plates simple. Intervessel pits scalariform. Vessel length 120–180 µm.

Bark

Phloem consists of small groups of sieve tubes. Cortex cells large (80–100 µm), round to oval. Phellem absent.

Pith

No slide available.

Analyzed: Root collar of polar root of 1 individual.

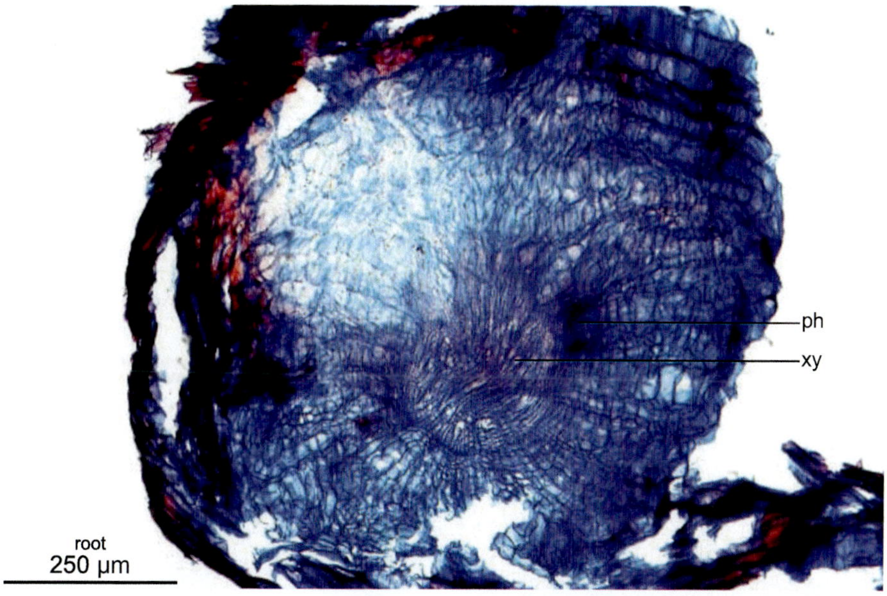

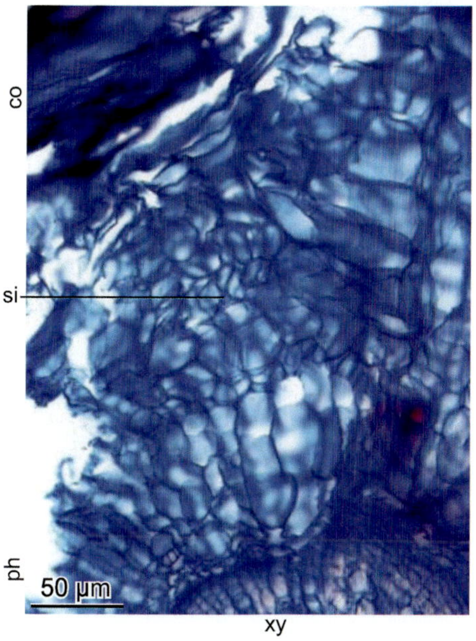

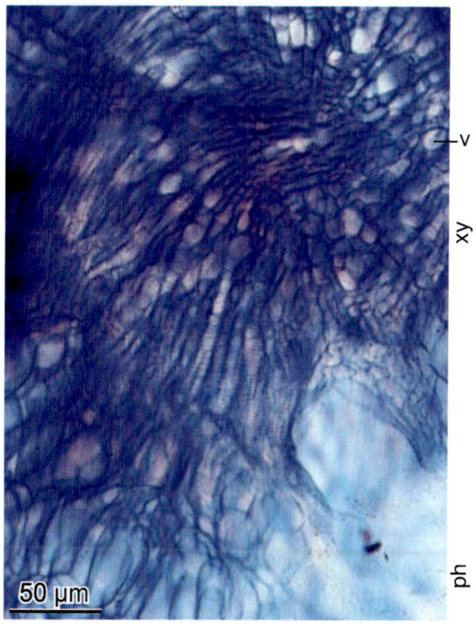

Features according to IAWA

Xylem 2.1 - 9 - 13 - 20 - 40.1 - 50.2 - 52.3 - 60.1 - 79.1 - 117

Arnebia euchroma (Royle) I.M.Johnst.

Boraginaceae

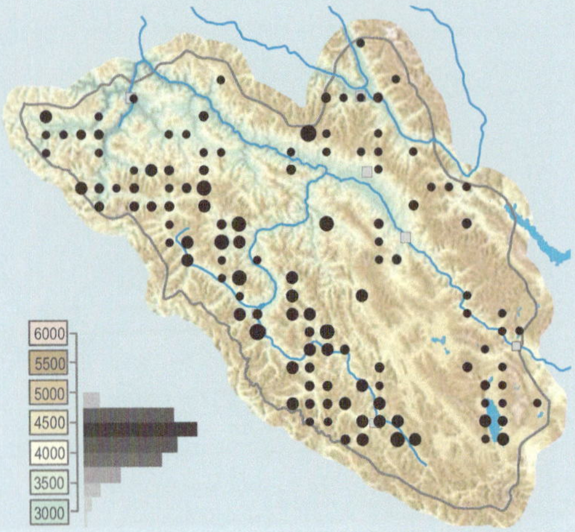

Syn.: *Lithospermum euchromon* Royle

Description: Perennial herbs, 15–40 cm tall, bristly hairy, leaves ovate to oblong, flowers dark purple, sometimes pale yellow and purple-red tinged, ca. 1 cm across, arranged in cymes.

Distribution: Iran, S Russia, Kazakhstan, Turkmenistan, Uzbekistan, Kyrgyzstan, Tajikistan, Afghanistan, Pakistan, India (Jammu and Kashmir), Nepal, China (Xinjiang, Xizang).

Habitat: Screes, stabilized slopes, along roads in disturbed habitats.

Stem diameter 10 mm. Bark/xylem proportion (living part) 1:1. Mean ring width 0.4 mm. Maximal 3 living annual rings. Total age unknown. Estimation 10 years.

Xylem

Transverse section

Semi-ring-porous. Rings distinctly marked by large differences between earlywood (60–100 μm) and latewood vessel diameters (10–20 μm). Vessels solitary. Fibers absent. Parenchyma pervasive. Vascular bundles remain. Parenchymatic zones (rays) separate vessel/parenchyma groups. Characteristic of the red-stained stem is the dilatation and the central disintegration. Few years after stem and root formation an intra-xylary phellem layer separates the living and dead part. The inner part of the root or rhizome disintegrates.

Tangential section

Rays >10-seriate, confluent with the axial tissue. Rays disappear in polarized light.

Radial section

Perforation plates simple. Intervessel pits scalariform in earlywood, roundish and in alternating position. Vessel length 80–120 μm. Ray cells upright. Rays confluent with axial tissue.

Bark

Phloem with small groups of sieve tubes. Phellem layered, consisting of rectangular cells and peals.

Pith

No slide available.

Analyzed: Root collar of polar root of 2 individuals.

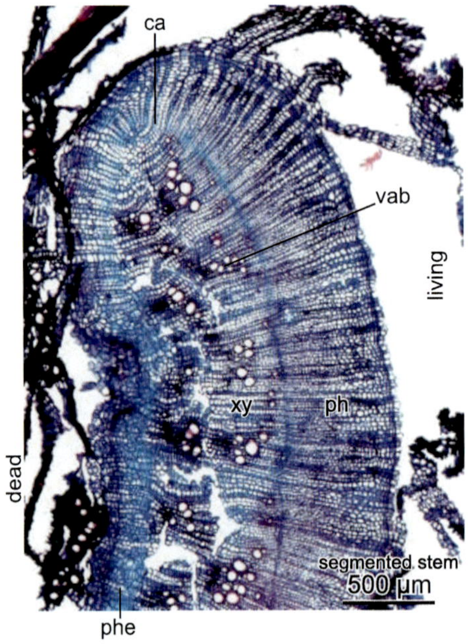

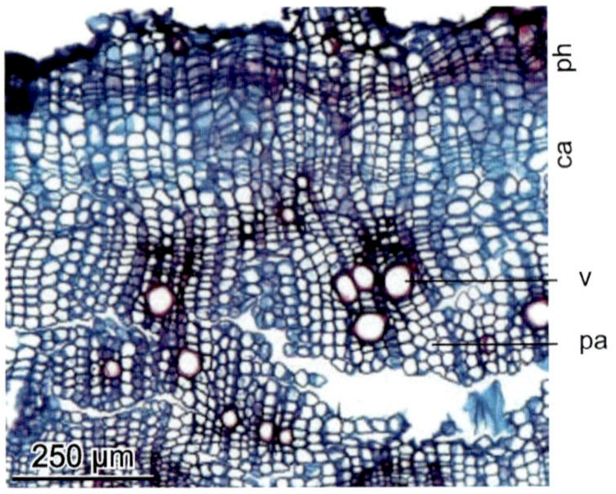

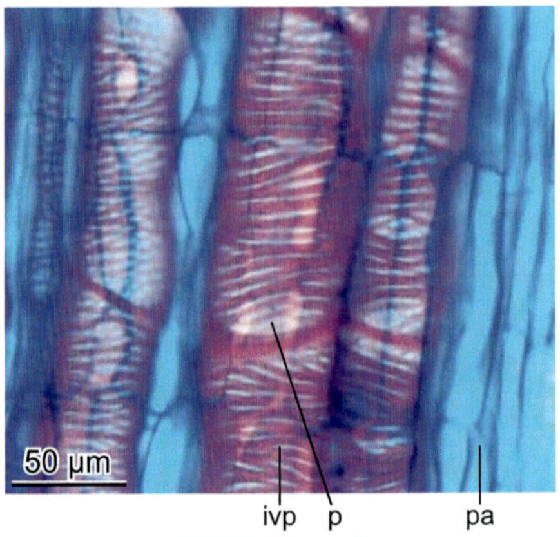

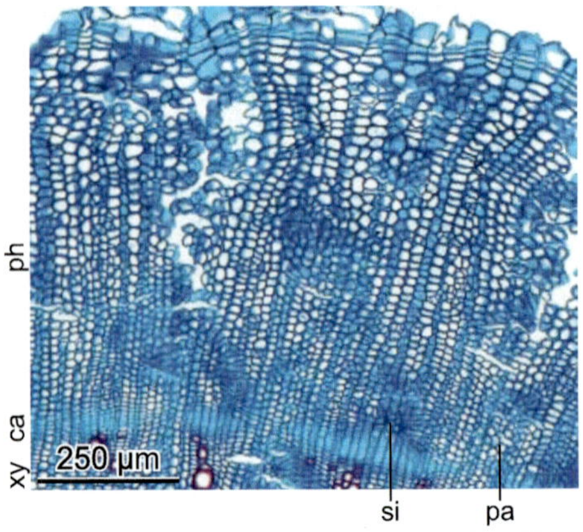

Features according to IAWA

Xylem 1 - 4 - 9 - 13 - 20 - (22) - 41 - 50.1 - 52.3 - 60.1 - 79.1 - 99 - 99.1 - 100.2 - 105

Arnebia guttata Bunge

Boraginaceae

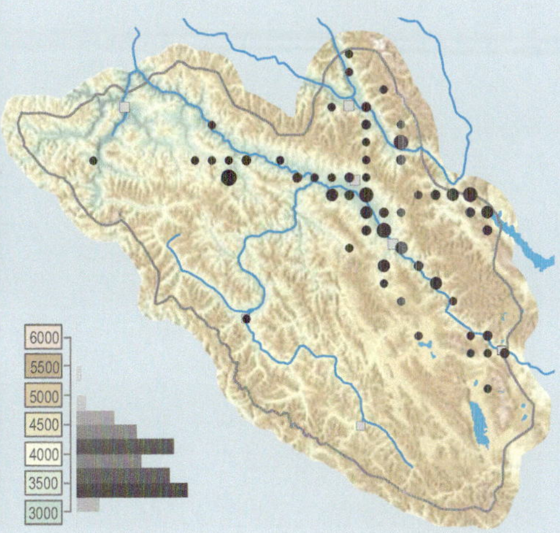

Syn.: *A. thomsonii* C.B.Clarke, *A. tibetana* Kurz, *Lithospermum guttatum* (Bunge) I.M. Johnst., *Macrotomia guttata* (Bunge) Farrer

Description: Biennial or perennial herbs, 10–25 cm tall, bristly hairy, leaves ovate to oblong, flowers yellow, sometimes with violet markings, ca. 1 cm across, arranged in cymes.

Distribution: Russia, Kazakhstan, Turkmenistan, Uzbekistan, Kyrgyzstan, Tajikistan, Afghanistan, Pakistan, India (Jammu and Kashmir), China (Gansu, Hebei, Nei Mongol, Ningxia, Xinjiang, Xizang), Mongolia.

Habitat: Sand deserts.

Stem diameter 8 mm. Bark/xylem proportion (living part) 1:1. Mean ring width 0.3 mm. Maximal 3 living annual rings. Total age unknown. Estimation 10 years.

Arnebia guttata is differentiated from *A. euchroma* mainly by the presence of fibers in the xylem.

Xylem

Transverse section

Characteristic is the stem dilatation and disintegration. Few years after stem and root formation a intra-xylary phellem layer separates the living and dead part. The inner part of the root or rhizome disintegrates.

Semi-ring-porous. Rings indistinct. Diameter of large vessels 60–70 µm and of small vessels 20 µm. Vessels solitary. Fibers thin- to thick-walled, in groups. Parenchyma pervasive. Vascular bundles remain. Parenchymatic zones (rays) separate vessel/parenchyma groups.

Tangential section

Rays >10-seriate, confluent with axial tissue. Rays disappear in polarized light.

Radial section

Perforation plates simple. Intervessel pits scalariform. Earlywood pits roundish and in alternating position. Vessel length 90–120 µm. Ray cells upright. Rays confluent with axial tissue.

Bark

Phloem with small groups of sieve tubes, partially in tangential lines. Phellem layered, consisting of rectangular cells and peals.

Pith

No slide available.

Analyzed: Root collar of polar root of 1 individual.

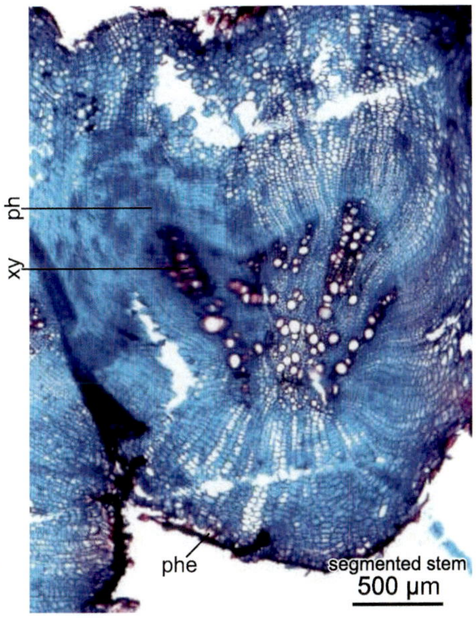

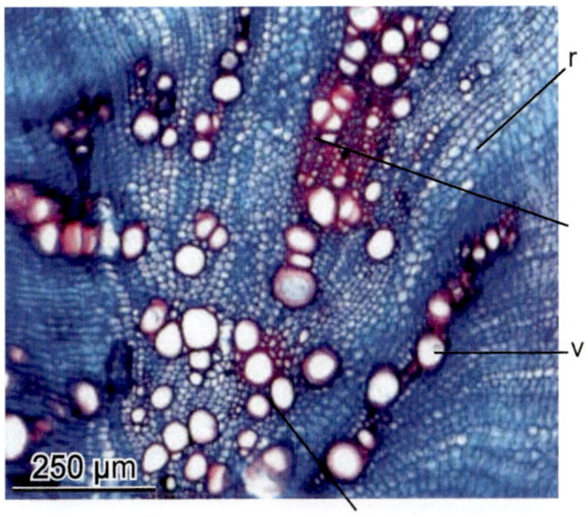

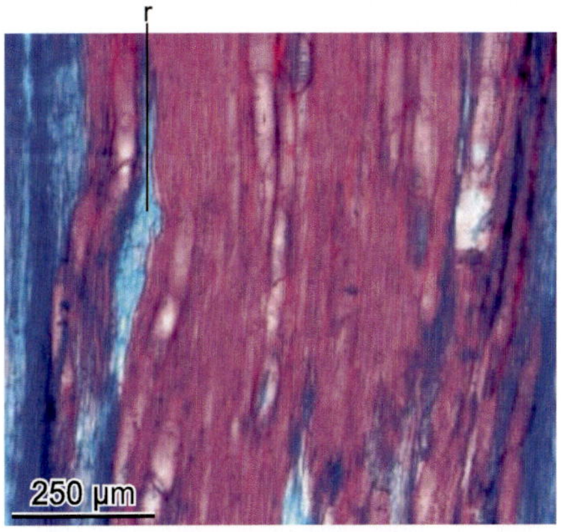

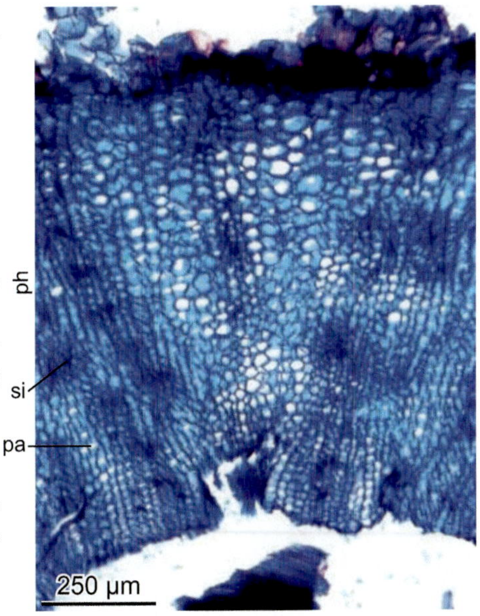

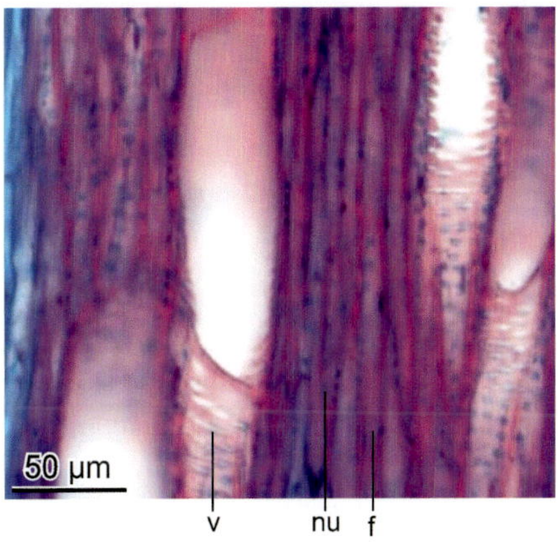

Features according to IAWA

Xylem 2 - 9 - 13 - 20 - (22) - 41 - 50.1 - 52.3 - 61 - 79.1 - 99 - 99.1 - 100.2 - 105

Eritrichium hemisphaericum W.T.Wang

Boraginaceae

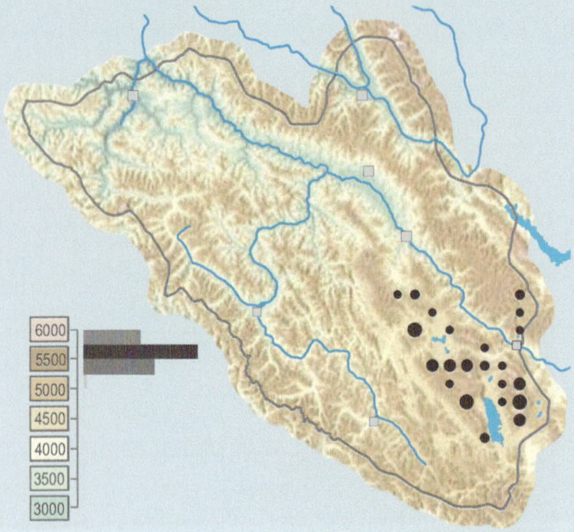

Description: Perennial herbs, 1–3 cm tall, pubescent, densely cespitose, leaves spatulate to obovate-oblong, flowers blue, ca. 2 mm across, cymes terminal, 1–2-flowered, nutlets with appendages.

Distribution: India (Jammu and Kashmir), China (Qinghai, Xizang).

Habitat: Stabilized wind-exposed slopes, among stones.

Stem diameter 1 mm. Bark/xylem proportion 1:2. Ring width 0.04–0.05 mm. Rings difficult to recognize. Maximal 9 living annual rings.

Xylem

Transverse section

Diffuse-porous. Rings indistinct. Vessels thick-walled and solitary. Vessel diameter 10–20 µm. Fibers absent. Parenchyma pervasive.

Tangential section

Rays absent.

Radial section

Perforation plates simple. Intervessel pits scalariform. Vessel length 50–100 µm.

Bark

Phloem with small groups of sieve elements. Phellem cells irregular-formed, thick-walled.

Pith

Round, filled with round, unlignified cells.

Analyzed: Root collar of polar root of 2 individuals.

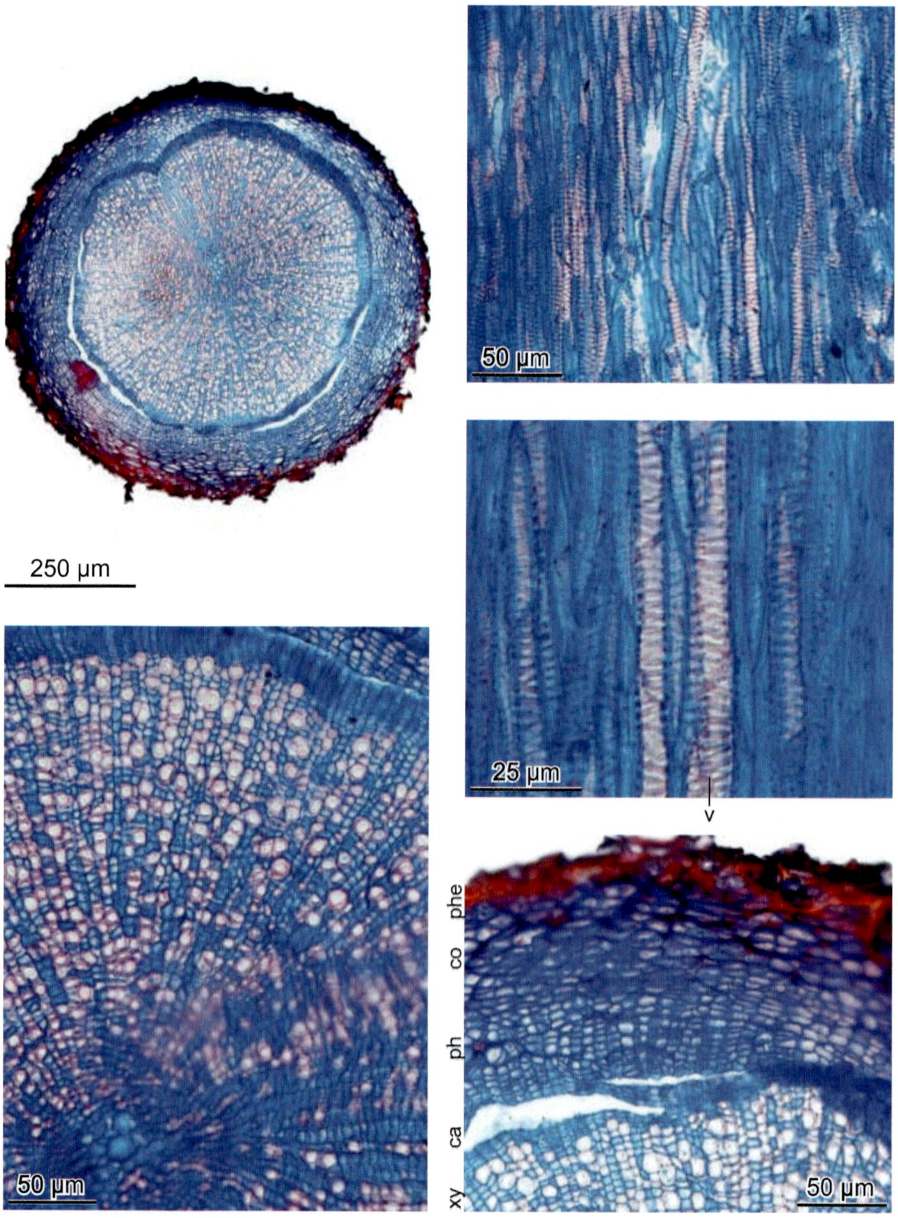

Features according to IAWA

Xylem 2 - 9 - 13 - 22 - 40.1 - 50.2 - 52.2 - 60.1 - 79.1 - 117

Eritrichium spathulatum (Benth.) C.B.Clarke

Boraginaceae

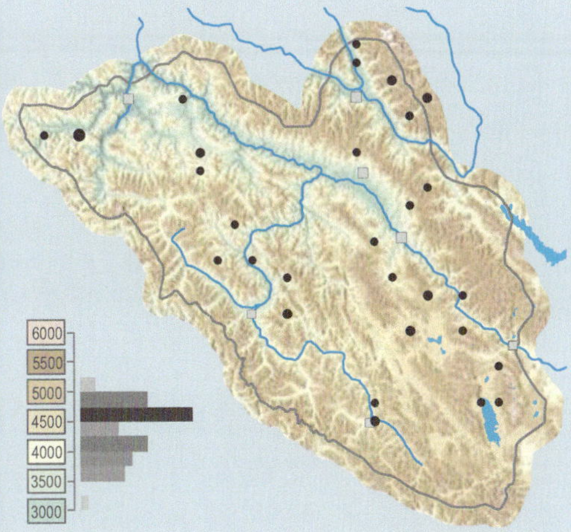

Syn.: *Echinospermum spathulatum* Benth., *Eritrichium canum* var. *spathulatum* (Benth.) Y.Nasir

Description: Perennial herbs, 10–35 cm tall, cespitose, pubescent, stems often decumbent, leaves spatulate to oblanceolate, sessile, flowers light blue with yellow throat, ca. 3 mm across, cymes many-flowered, nutlets with glochids.

Distribution: Pakistan, N India, China (Xizang).

Habitat: Hillside meadows, stabilized slopes, rock crevices, along roads.

Stem diameter 1–1.2 mm. Bark/xylem proportion 1:2. Maximal rings 7. Mean ring width 0.05 mm.

Xylem

Transverse section

Individual with distinct rings. Semi-ring-porous. Vessels angular, thick-walled and solitary. Earlywood vessel diameter 25–30 µm, latewood vessel diameter 10 µm. Fibers absent. Parenchyma pervasive.

Tangential section

Rays absent.

Radial section

Perforation plates simple. Intervessel pits round to tangentially enlarged. Vessel length 60–70 µm.

Bark

Phloem with small groups of sieve elements. Phellem cells rectangular or irregularly formed.

Pith

Pith oval.

Analyzed: Polar root of 2 individuals (herbarium specimen).

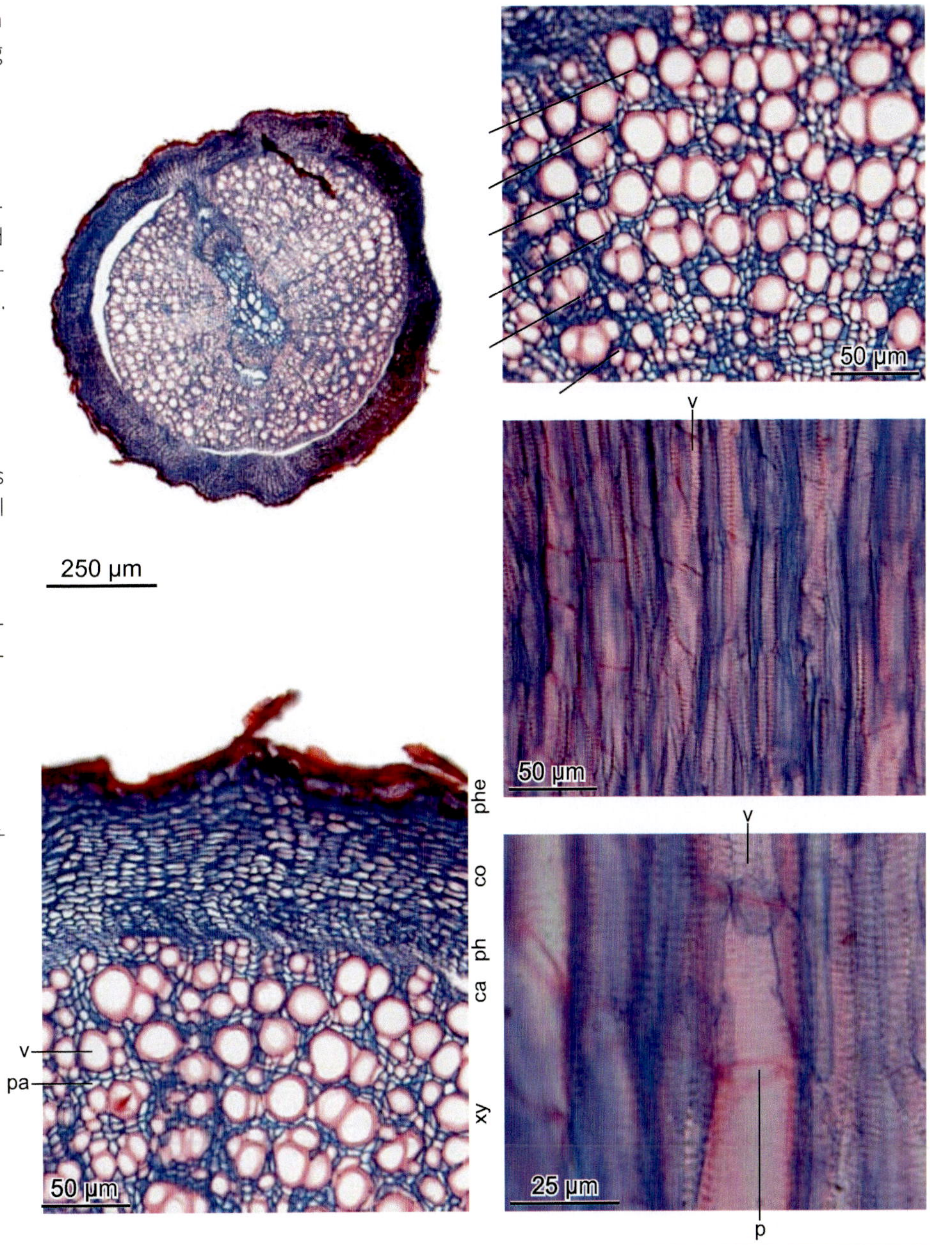

Features according to IAWA

Xylem 1 - 9 - 13 - 22 - 39.1 - 40.2 - 50.2 - 52.2 - 60.1 - 79.1 - 117

Eritrichium villosum (Ledeb.) Bunge

Boraginaceae

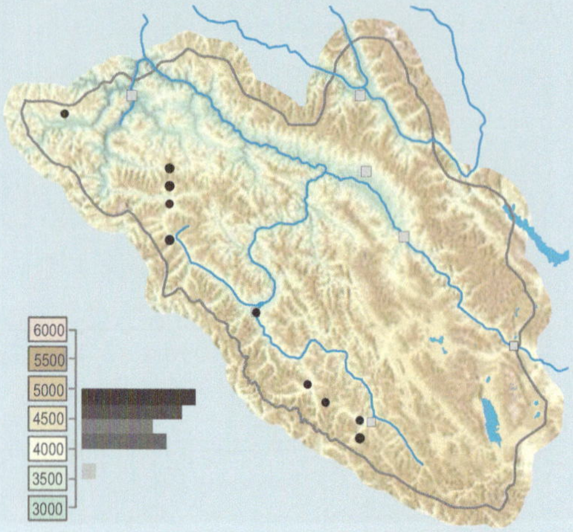

Syn.: *Myosotis villosa* Ledeb., *E. nanum* subsp. *villosum* (Ledeb.) Brand

Description: Perennial herbs, 5–20 cm tall, cespitose, pilose, basal leaves broadly lanceolate, forming a rosette, flowers blue, ca. 3 mm across, cymes many-flowered, terminal.

Distribution: Russia (Siberia), Kazakhstan, Afghanistan, Pakistan, N India, China (Heilongjiang, Xinjiang, Xizang), Mongolia.

Habitat: Alpine meadows.

Stem diameter 1 mm. Bark/xylem proportion 1:3. Ring width 0.05–0.1 mm. Rings distinct and indistinct. Maximal 14 living annual rings.

Xylem

Transverse section

Semi-ring-porous. Vessels thick-walled and solitary. Individual with distinct rings. Earlywood vessel diameter 10–20 µm, latewood vessel diameter 5 µm. Fibers absent. Parenchyma pervasive.

Tangential section

Rays absent.

Radial section

Perforation plates simple. Intervessel pits scalariform. Vessel length 50–100 µm.

Bark

Phloem with small groups of sieve elements. Phellem cells irregular-formed.

Pith

No slide available.

Analyzed: Root collar of polar root of 2 individuals.

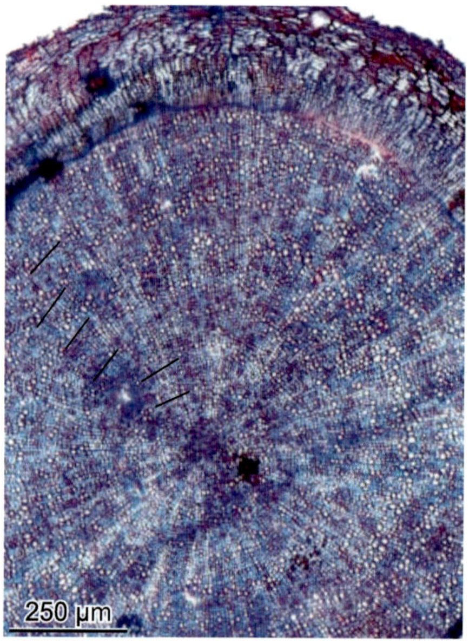

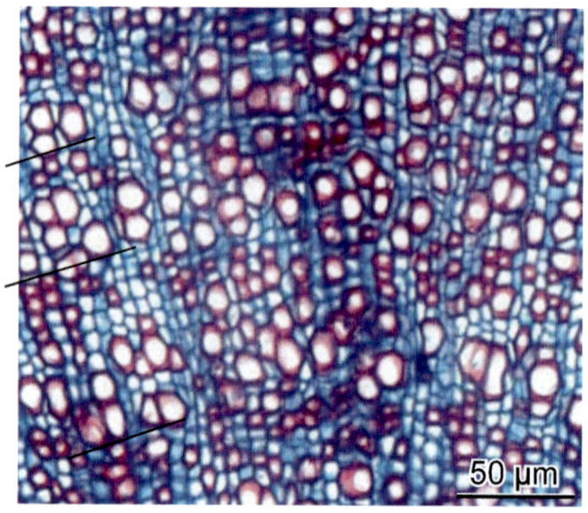

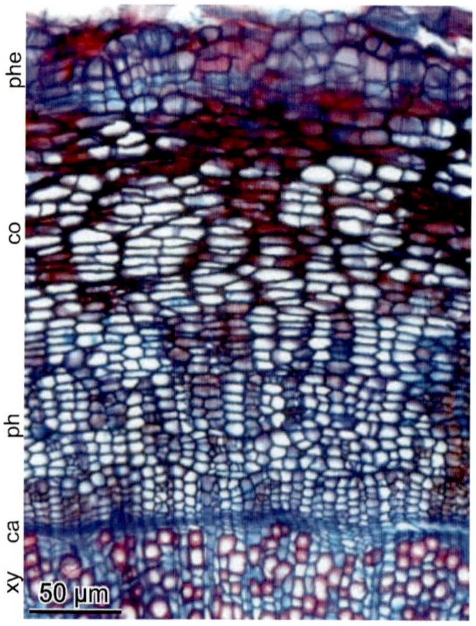

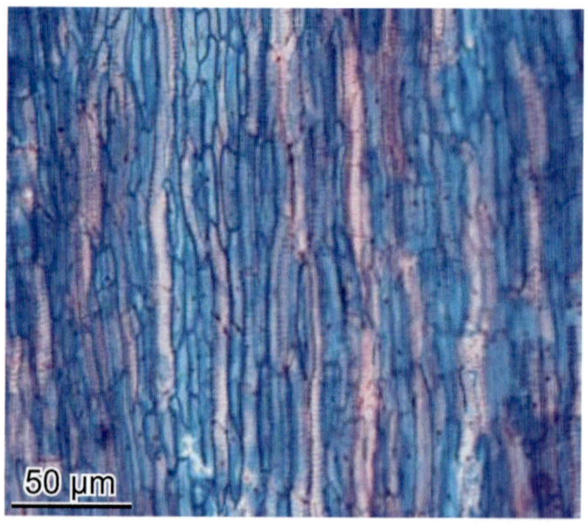

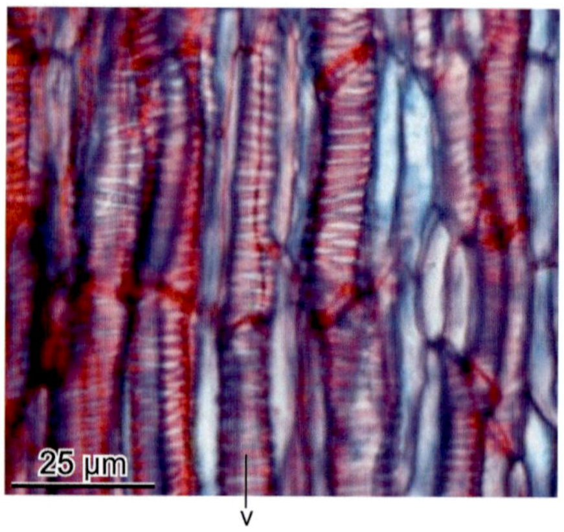

Features according to IAWA

Xylem 2 - 9 - 13 - 22 - 40.1 - 50.2 - 52.2 - 60.1 - 79.1 - 117

Lappula patula (Lehm.) Asch. ex Gürke

Boraginaceae

Syn.: *Echinospermum patulum* Lehm., *Myosotis redowskii* Hornem.

Description: Annuals, 20–35 cm tall, gray hispid, basal leaves in a rosette, stem leaves sessile, linear to spatulate, flowers light blue, ca. 3 mm across, sessile, in numerous secund inflorescences, fruits with glochids.

Distribution: N Africa, SW Asia, Russia, Kazakhstan, Turkmenistan, Uzbekistan, Kyrgyzstan, Tajikistan, Afghanistan, Pakistan, NW India, China (N Xinjiang).

Habitat: Waste places, field margins, sandy river deposits

Stem diameter 2.5–3 mm. Bark/xylem proportion 1:4. Mean ring width 0.25 mm. Maximal 2 annual rings.

Xylem

Transverse section

Semi-ring-porous. Vessels angular, solitary and in short radial multiples. Vessel diameter 25–35 µm. Fibers thin- to thick-walled. Fibers absent in the center. Parenchyma paratracheal, pervasive in the center.

Tangential section

Rays absent, or confluent with axial tissue.

Radial section

Perforation plates simple. Intervessel pits round. Vessel length 110–140 µm.

Bark

Phloem homogeneous. Sieve elements and parenchyma cells indistinguishable. Phellem absent.

Pith

Pith irregularly formed.

Analyzed: Polar root of 2 individuals (herbarium specimens).

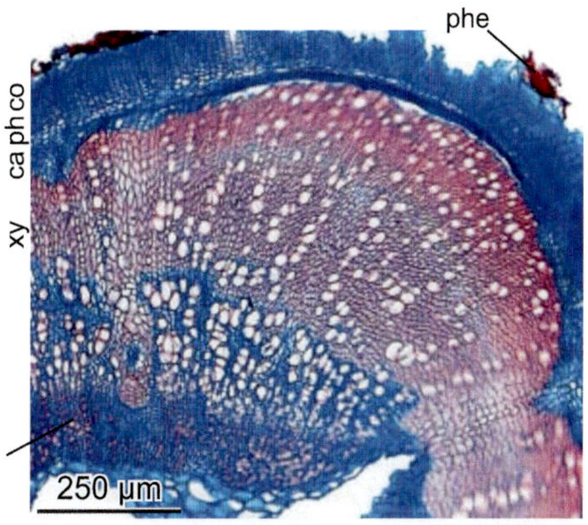

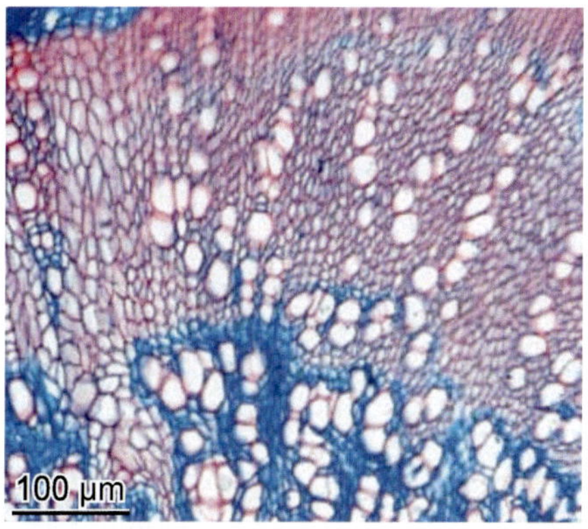

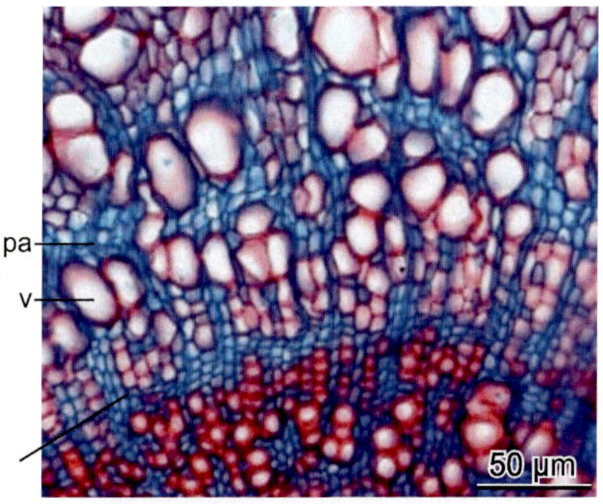

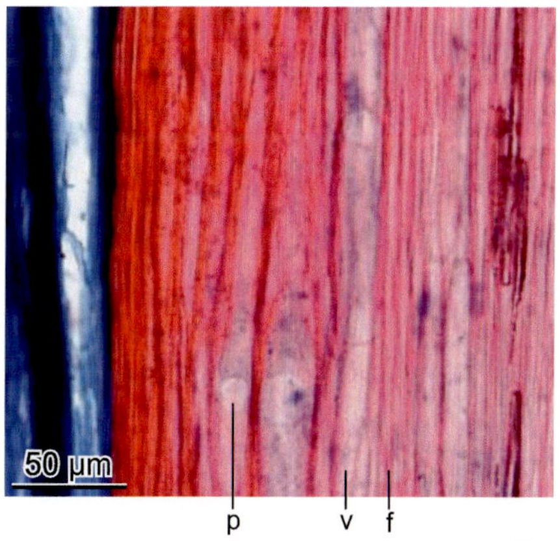

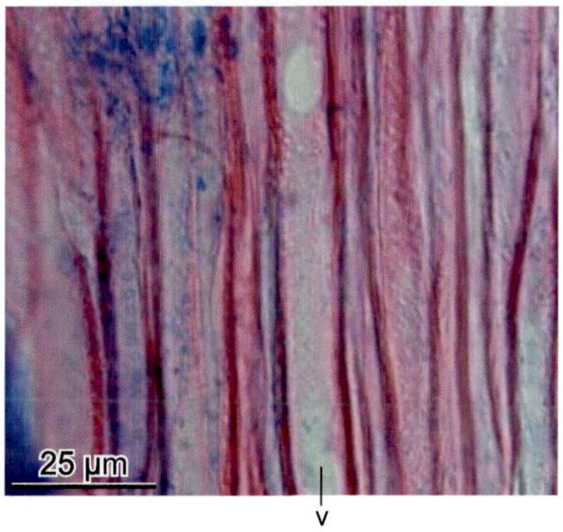

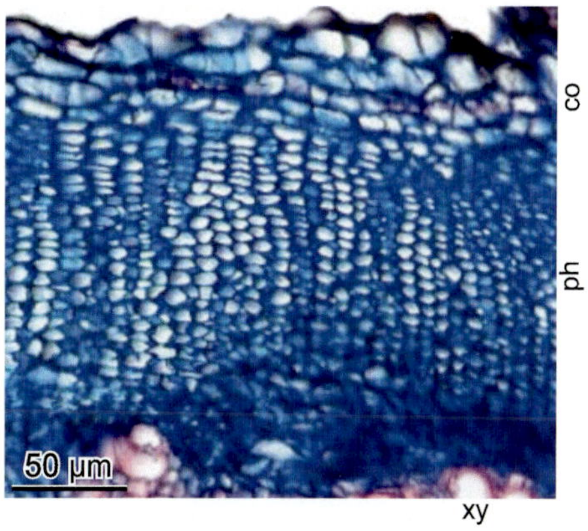

Features according to IAWA

Xylem 2 - 9 - 9.1 - 13 - 22 - 41 - 50.1 - 52.3 - 61 - 70 - 79 - 79.1 - 117

Lappula tadshikorum Popov

Boraginaceae

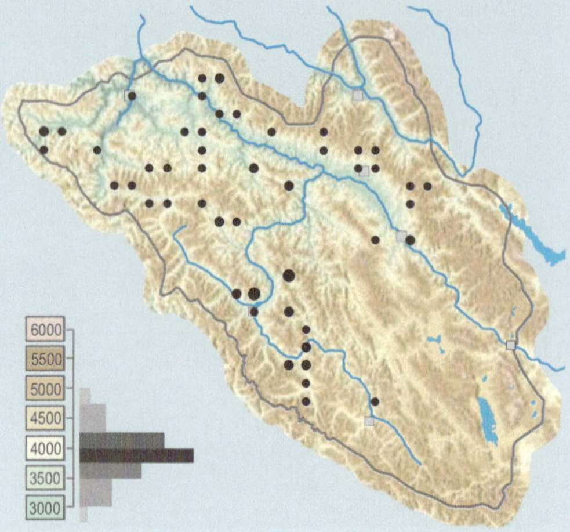

Description: Biennials, 15–40 cm tall, densely gray-white hirsute, stems erect, branched from the base, leaves linear to lanceolate, flowers pale blue, yellowish inside, ca. 2.5 mm across, inflorescence 7–15 cm in fruit, nutlets with glochids.

Distribution: Russia, Kazakhstan, Turkmenistan, Uzbekistan, Kyrgyzstan, Tajikistan, India, China (Xinjiang).

Habitat: Stony slopes.

Xylem diameter 5 mm. Bark missing. Mean ring width 0.25 mm. Maximal 2 living annual rings.

Xylem

Transverse section

One ring present. Vessels in short radial multiples. Vessel diameter 50–80 µm. Fibers thin- to thick-walled. Parenchyma paratracheal.

Tangential section

Rays 5–6-seriate, often confluent with axial tissue.

Radial section

Perforation plates simple. Intervessel pits round. Vessel length 100–150 µm. Ray cells tall and upright.

Bark

No slide available.

Pith

Round and hollow.

Analyzed: Root collar of polar root of 1 individual.

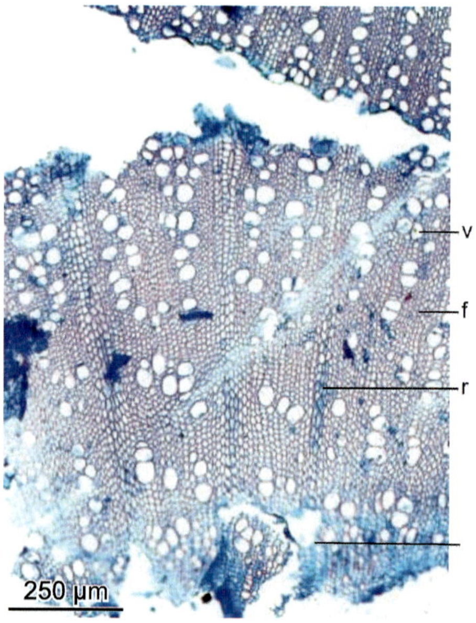

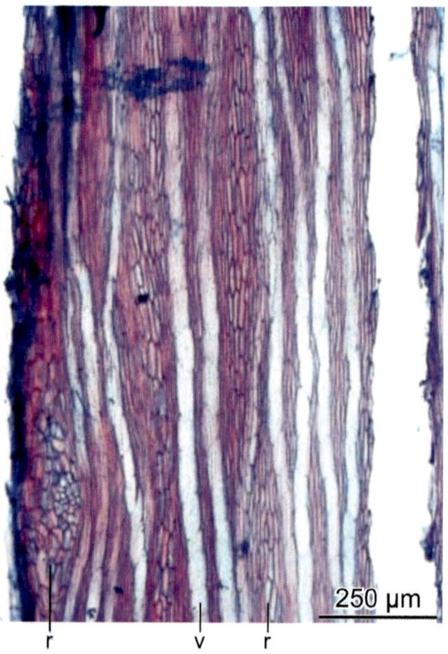

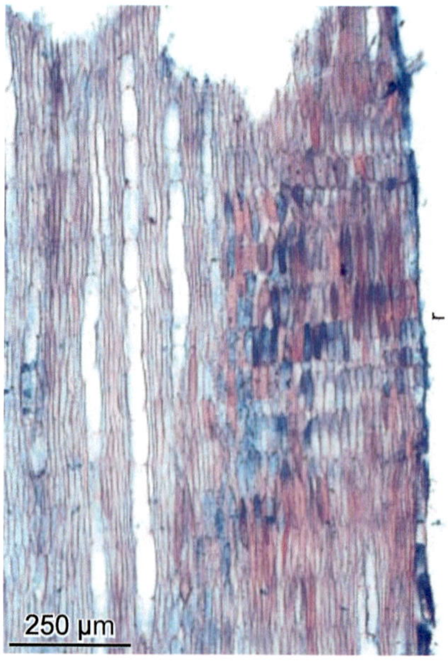

Features according to IAWA

Xylem 2.1 - 9 - 9.1 - 13 - 22 - 41 - 50.2 - 52.3 - 61 - 70 - 79 - 98 - 100.1 - 105

Lindelofia anchusoides (Lindl.) Lehm.

Boraginaceae

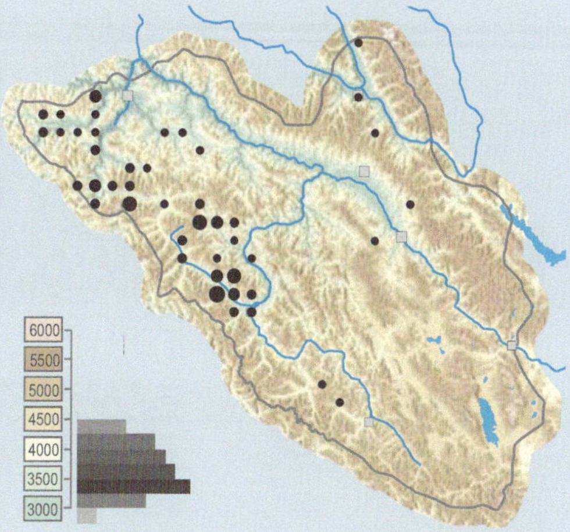

Syn.: *Cynoglossum anchusoides* Lindl.

Description: Perennial herbs, 30–75 cm tall, softly hairy, leaves lanceolate, flowers blue to purple, ca. 1 cm across, inflorescence bifurcate, nutlets with appendages.

Distribution: Afghanistan, Pakistan, NW India.

Habitat: Dry loamy, gravel and stony slopes.

Rings recognizable. Stem diameter 8–10 mm. Bark/xylem proportion 1:3. Ring width 0.3–0.4 mm. Maximal 20 living annual rings. Often with segmented stems.

Xylem

Transverse section

Semi-ring-porous. Vessels solitary and in radial groups. Earlywood vessel diameter 60–80 µm, latewood vessel diameter 10–30 µm. Fibers absent. Parenchyma pervasive. A lot of slime in parenchyma cells. Vascular bundles remaining. Secondary periderm towards the pith.

Tangential section

Rays 10-seriate and often indistinct, confluent with axial tissue.

Radial section

Perforation plates simple. Intervessel pits round to scalariform. Vessel length 50–100 µm. Ray cells upright.

Bark

Phloem with small groups of sieve elements. Contains a lot of slime.

Pith

No slide available.

Analyzed: Root collar of 4 individuals.

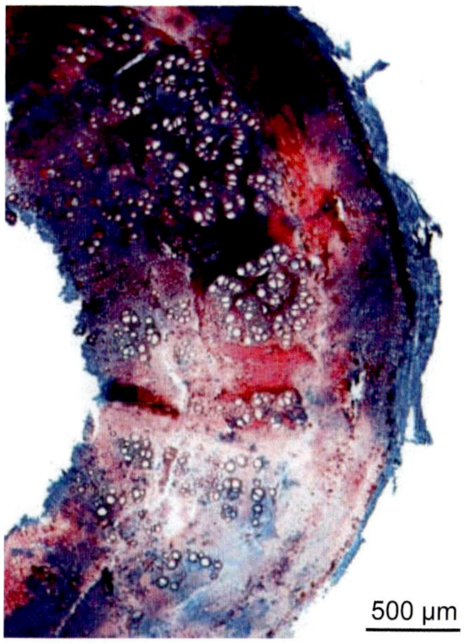

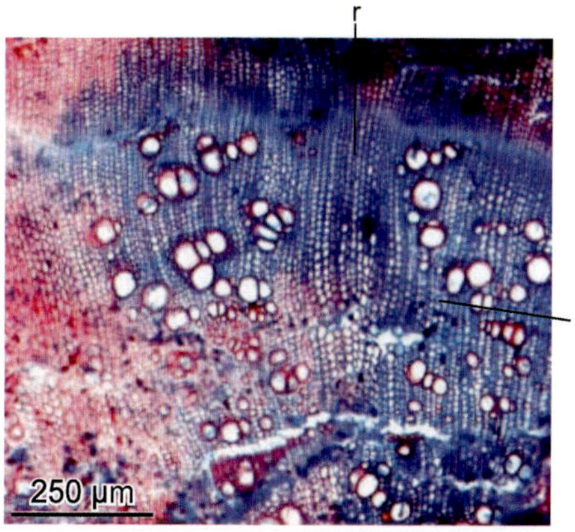

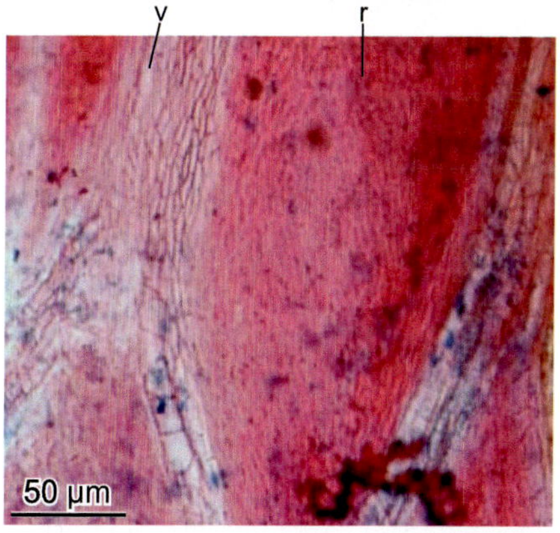

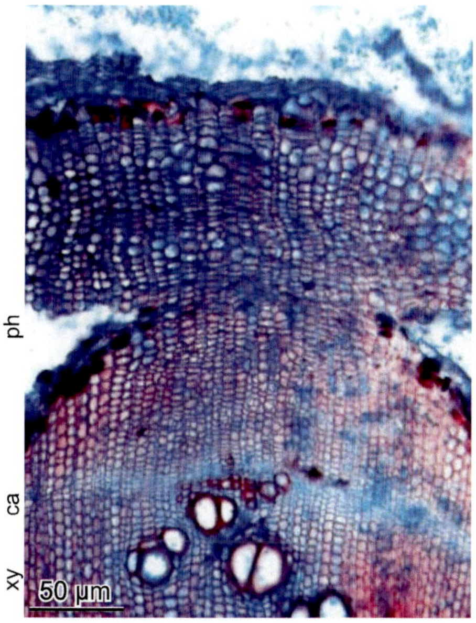

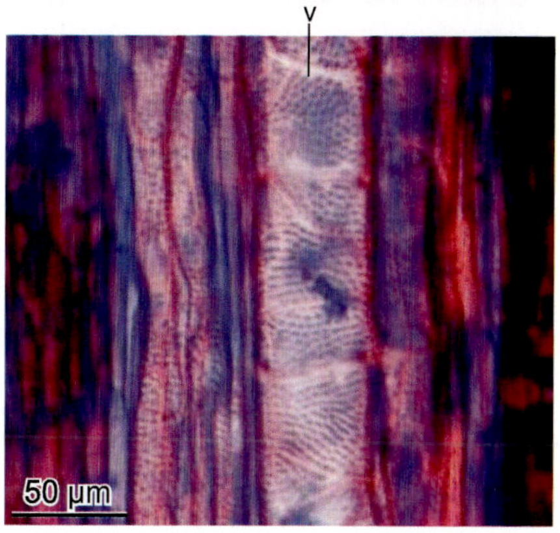

Features according to IAWA

Xylem *1 - 4 - 10 - 13 - 20 - 22 - 41 - 50.2 - 52.3 - 60.1 - 79.1 - 99 - 99.1 - 100.1 - 100.2 - 105 - 135.2*

Lindelofia stylosa (Kar. & Kir.) Brand

Boraginaceae

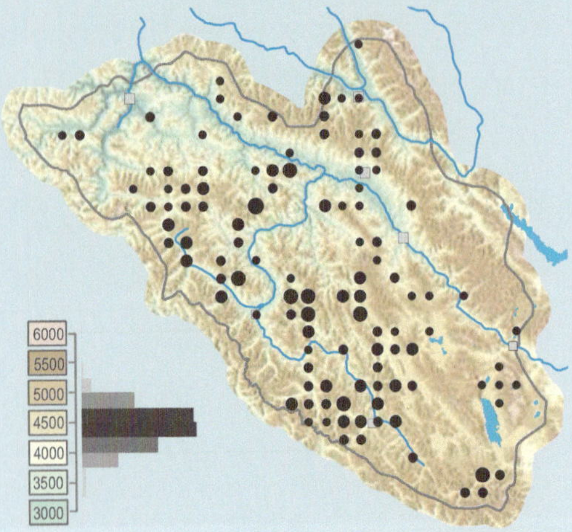

Syn.: *Cynoglossum stylosum* Kar. & Kir.

Description: Perennial herbs, 20–50 cm tall, appressed pubescent, basal leaves oblong-elliptic, 8–25 cm long, stem leaves lanceolate, 2–8 cm long, sessile, flowers purple, ca. 1 cm across, inflorescence terminal, nutlets with appendages.

Distribution: Kazakhstan, Turkmenistan, Uzbekistan, Kyrgyzstan, Tajikistan, Afghanistan, Pakistan, N India, China (Gansu, Xinjiang, Xizang), Mongolia.

Habitat: Dry loamy, gravel and stony slopes.

Stem diameter 8–10 mm. Bark/xylem proportion (living part) 1:0.5. Ring width 0.3–0.4 mm. Maximal 2 living annual rings.

Xylem

Transverse section

Diffuse-porous. Vessels in radial groups. Earlywood vessel diameter 60–80 µm. Fibers absent. Parenchyma paratracheal and pervasive. With much slime in parenchyma cells.

Tangential section

Rays 5–6-seriate, confluent with axial tissue.

Radial section

Perforation plates simple. Intervessel pits roundto scalariform. Vessel length 100–150 µm. Ray cells tall and upright.

Bark

Phloem homogeneous with a lot of slime.

Pith

No slide available.

Analyzed: Root collar of polar root of 4 individuals.

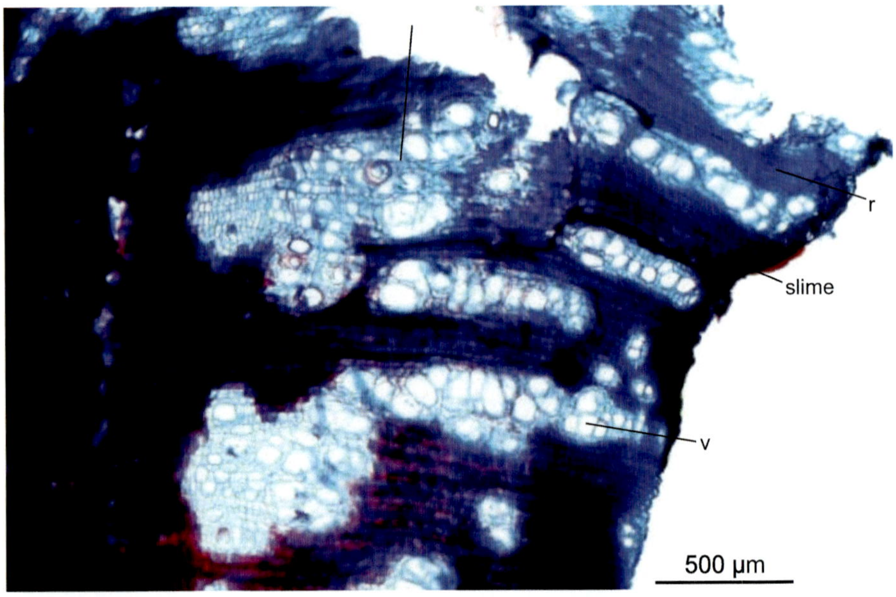

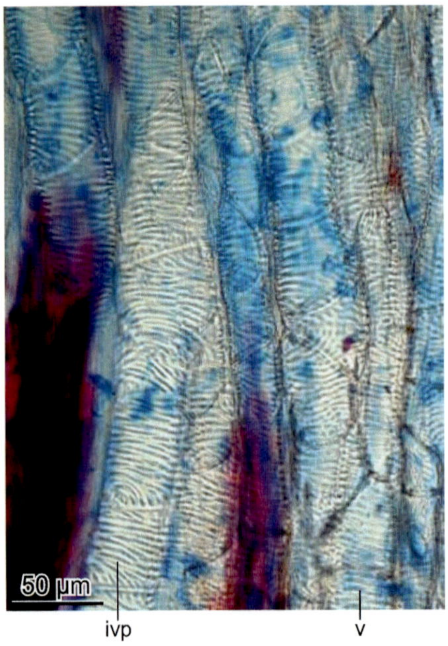

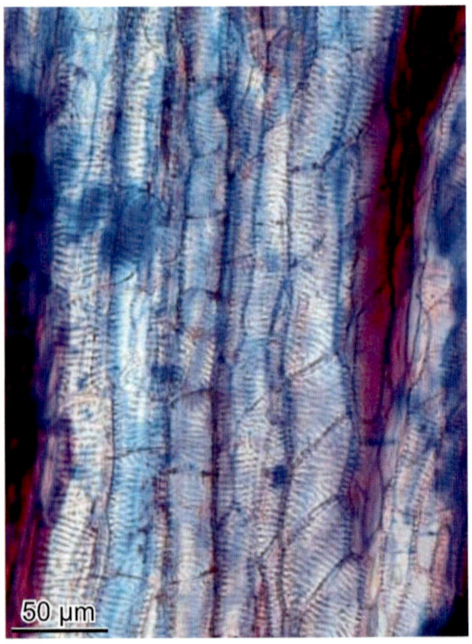

Features according to IAWA

Xylem 5 - 10 - 13 - 20 - 22 - 41 - 50.2 - 52.3 - 60.1 - 79 - 79.1 - 98 - 100.1 - 100.2 - 105

Mattiastrum himalayense (Klotzsch) Brand

Boraginaceae

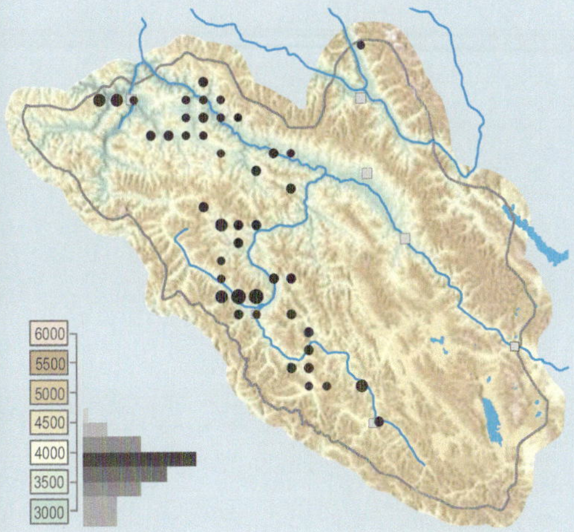

Syn.: *Mattia himalayensis* Klotzsch, *Paracaryum himalayense* (Klotzsch) C.B.Clarke

Description: Annuals or biennials, 20–30 cm tall, stems branched, erect, leaves oblong to spatulate, pubescent, flowers whitish, 3–4 mm across, subsessile, inflorescences lax, axillary and terminal, nutlets with glochids.

Distribution: Afghanistan, Pakistan, NW India, China (Xizang).

Habitat: Stabilized loamy and gravel slopes.

Stem diameter 10 mm. Bark/xylem proportion (living part) 1:1. Mean ring width 0.4 mm. Maximal 3 living annual rings. Total age unknown. Estimation 10 years.

Xylem

Transverse section

Semi-ring-porous. Rings distinctly marked by large differences between earlywood (60–80 µm) and latewood vessel diameters (20–30 µm). Vessels mostly solitary and in groups. Fibers absent. Parenchyma pervasive. Vascular bundles remain. With intra-xylary groups of sieve tubes.

Tangential section

Rays >10-seriate and confluent with axial tissue.

Radial section

Perforation plates simple. Intervessel pits round to scalariform. Vessel length 70–100 µm. Ray cells upright.

Bark

Phloem consists mainly of radially arranged parenchyma cells with small groups of sieve tubes.

Pith

No slide available.

Analyzed: Root collar of polar root of 2 individuals.

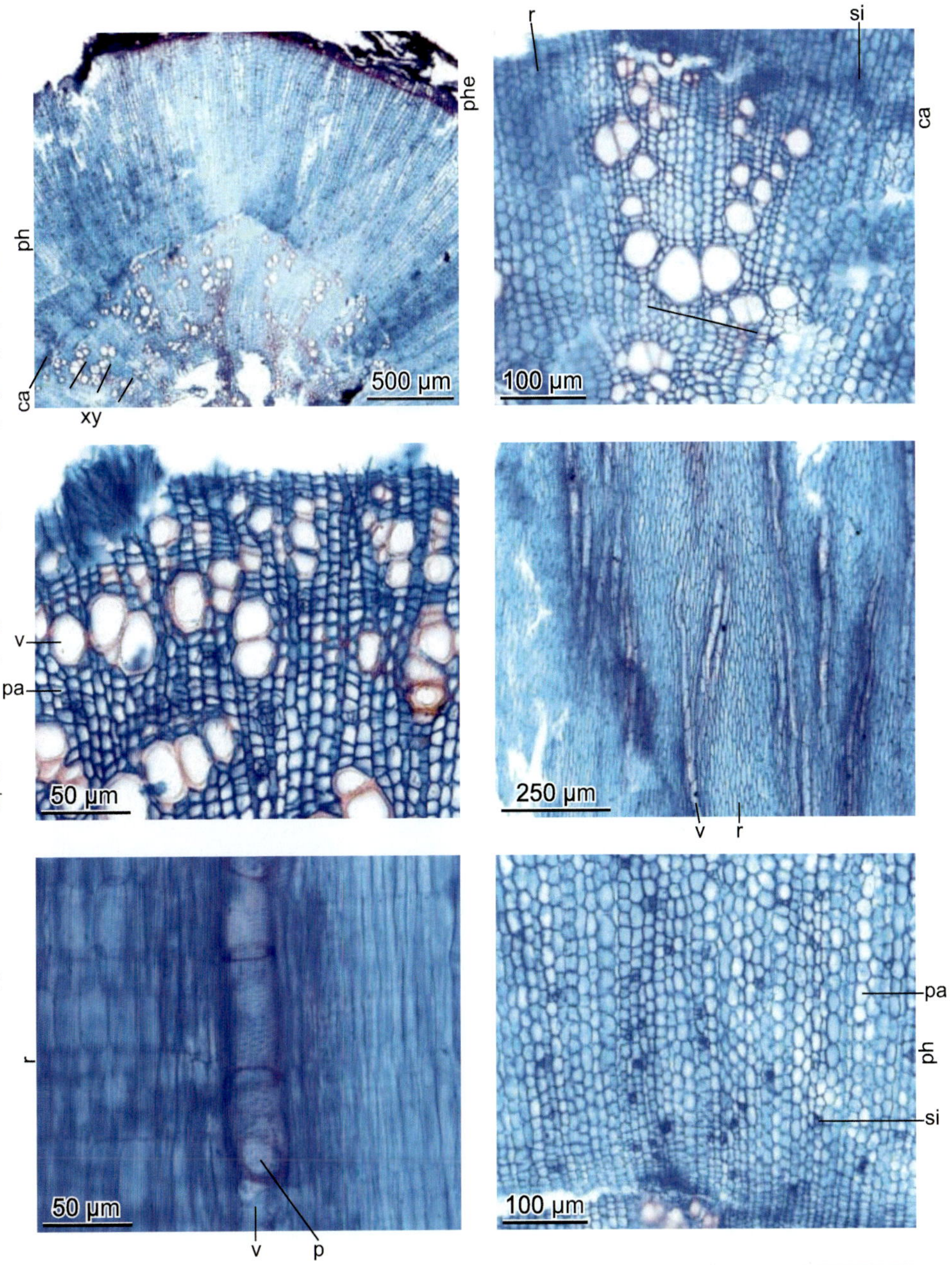

Features according to IAWA

Xylem 1 - 4 - 9 - 10 - 13 - 20 - 22 - 41 - 50.2 - 52.3 - 60.1 - 79.1 - 99 - 99.1 - 100.1 - 100.2 - 105 - 135

Rochelia disperma (L.f.) C.Koch

Boraginaceae

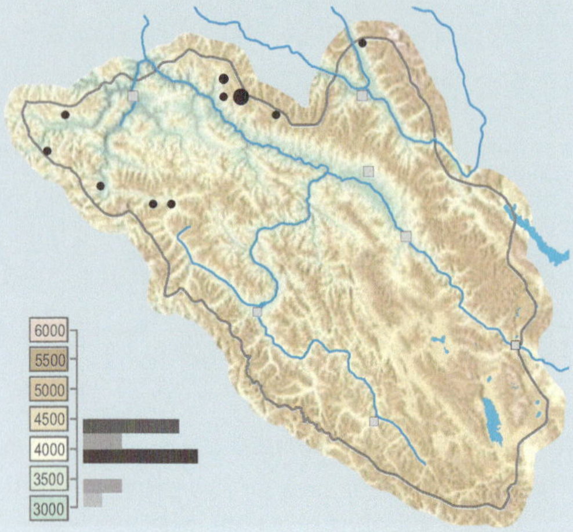

Syn.: *R. bungei* Trautv., *R. stylaris* Boiss.

Description: Perennial herbs, 10–15 cm tall, gray-hispid, stems erect, usually branched basally, basal leaves shortly petiolate, obovate, stem leaves sessile, lanceolate, flowers light blue to purple, ca. 3 mm across, in lax inflorescences, 5–10 cm long in fruit.

Distribution: Iran, Russia, Kazakhstan, Turkmenistan, Uzbekistan, Kyrgyzstan, Tajikistan, Afghanistan, Pakistan, India (Ladakh in Jammu and Kashmir), China (N Xinjiang), Mongolia.

Habitat: Saline and alkaline areas.

Stem diameter 2.5 mm. Bark/xylem proportion 1:3. Only one ring. Ring width 0.7 mm.

Xylem

Transverse section

Vessels thin- to thick-walled, solitary, with a diameter of 20–25 µm. Fibers thin- to thick-walled. Parenchyma absent.

Tangential section

Rays absent.

Radial section

Perforation plates simple. Intervessel pits small, round. Vessel length 150–180 µm.

Bark

Phloem homogeneous. Sieve elements and parenchyma cells cannot be distinguished. Phellem absent.

Pith

No slide available.

Analyzed: Polar root of 1 individual (herbarium specimen).

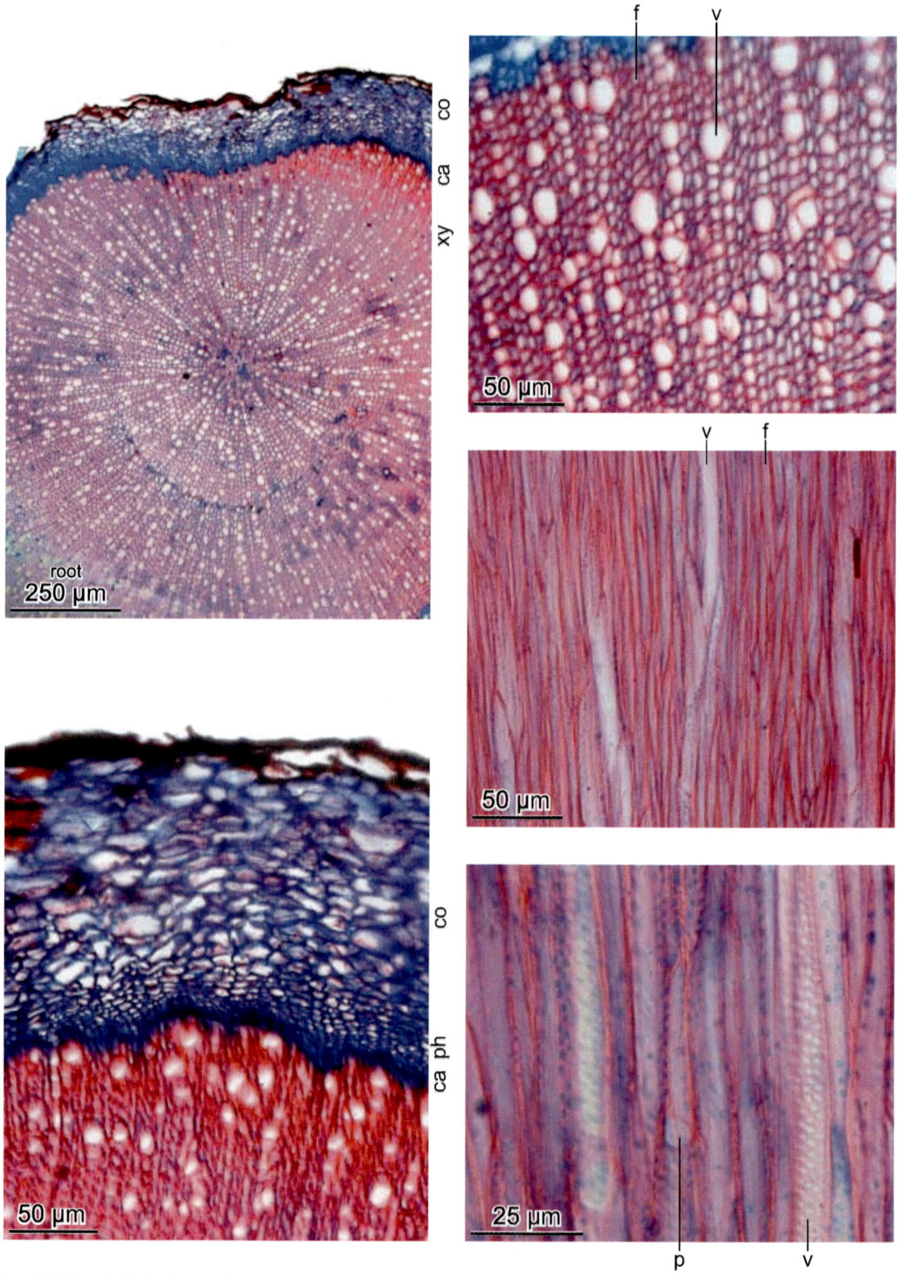

Features according to IAWA

Xylem 2.1 - 9 - 13 - 22 - 40.2 - 50.1 - 52.3 - 61 - 70 - 75 - 117

Alyssum canescens DC.

Brassicaceae

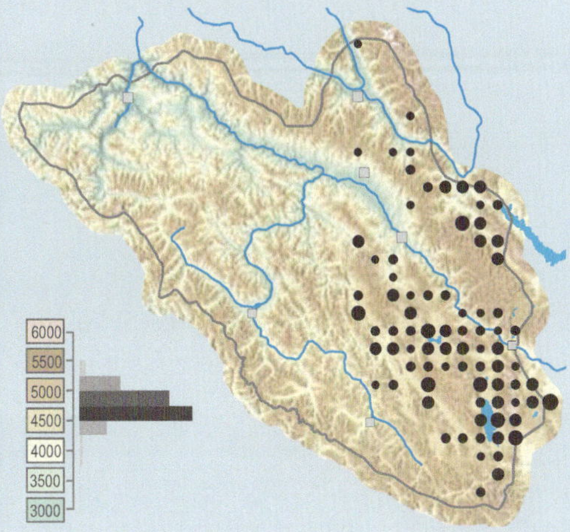

Syn.: *Ptilotrichum canescens* (DC.) C.A.Mey., *Stevenia canescens* (DC.) D.A.German

Description: Perennial herbs, 2–10 cm tall, silvery canescent, leaves oblong to linear, fleshy, sessile, flowers white or pinkish, ca. 5 mm across, in racemes.

Distribution: Russia (Siberia), Kazakhstan, Pakistan, India, China (Gansu, Hebei, Heilongjiang, Jilin, Nei Mongol, Ningxia, Qinghai, Shaanxi, Shanxi, Xinjiang, Xizang), Mongolia.

Habitat: Sandy plains and stony steppes.

Stem diameter 3–5 mm. Bark/xylem proportion 1:5. Ring width 0.1–0.15 mm. Maximal 30 annual rings.

Xylem

Transverse section

Semi-ring-porous. Ring boundaries marked by solitary, large earlywood vessels. Earlywood vessel diameter 30–50 µm. Vessels thick-walled (2–3 µm), mostly lignified. Some earlywood vessels with helical thickenings. Fibers absent. Parenchyma pervasive. With remaining vascular bundles. Large ray-like parenchyma zones between vascular bundles disappear in polarized light. Rays absent within vascular bundles.

Tangential section

Multiseriate rays consist of very thin-walled parenchyma cells. Ray cells upright.

Radial section

Perforation plates simple. Intervessel pits scalariform to round in alternating position. Vessel length 70–100 µm.

Bark

Phloem small, irregularly formed. Cortex small, cells large (15–20 µm), round to oval. Sclerenchyma and crystals absent. Phellem consists of many (up to 40) layers of collapsed cells.

Pith

Irregularly formed, cells large (30–40 µm), thin-walled, round.

Analyzed: Root collar of polar root of 3 individuals.

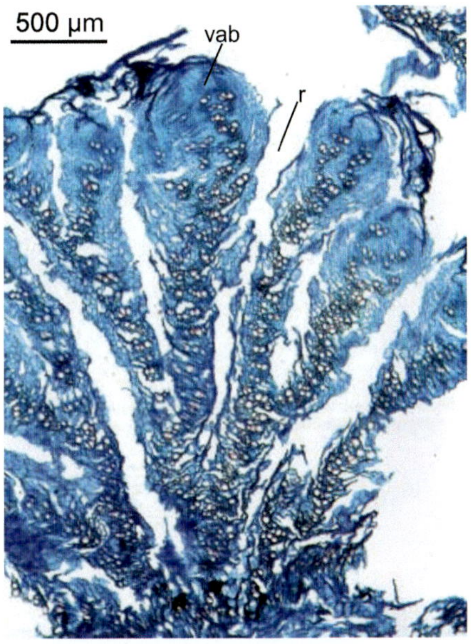

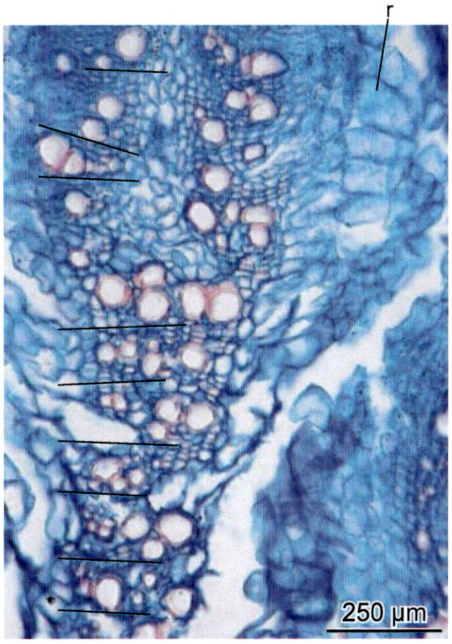

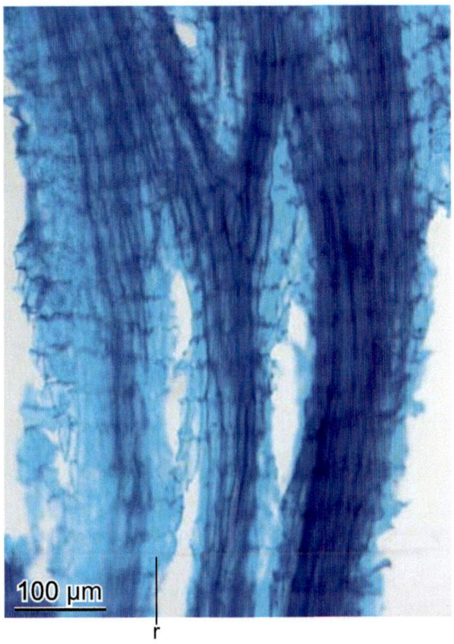

Features according to IAWA

Xylem 1 - 4 - 9 - 13 - 20 - 22 - 36 - 39.1 - 40.1 - 50.2 - 52.1 - 60.1 - 79.1 - 99 - 99.1 - 100.2 - 105

Aphragmus ladakianus Al-Shehbaz

Brassicaceae

Description: Perennial herbs, 2–5 cm tall, stems branched from the base, leaves spatulate to oblong, flowers pale lilac or white to pinkish, ca. 3 mm across, in lax racemes.

Distribution: NW India.

Habitat: River bed sediments, rock crevices, moraines, alpine pastures, open stony slopes.

Stem diameter 2–5 mm. Bark/xylem proportion 1:1. Annual rings absent

Xylem

Transverse section

Vessels angular, solitary with a diameter of 20–30 µm. Fibers absent. Parenchyma pervasive. With remaining vascular bundles. Large ray-like parenchyma zones between vascular bundles disappear in polarized light. Rays absent within vascular bundles.

Tangential section

Multiseriate rays, confluent with axial parenchyma cells. Ray cells upright.

Radial section

Perforation plates simple. Intervessel pits scalariform. Vessel length 50–80 µm.

Bark

Phloem with small groups of sieve element. Cortex absent. Phellem consists of several radially arranged cell layers.

Pith

No slide available.

Analyzed: Root collar of polar roots of 3 individuals.

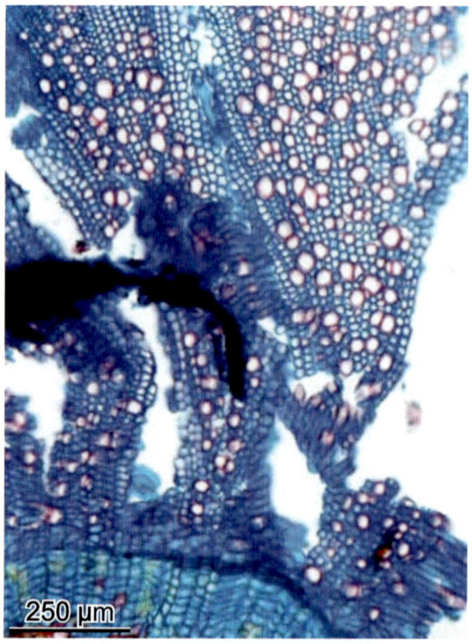

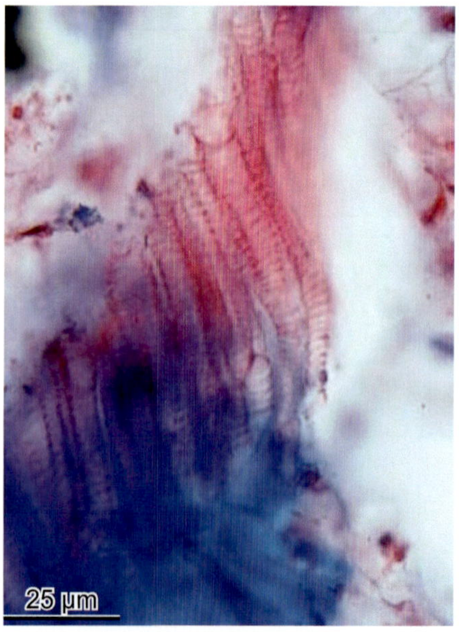

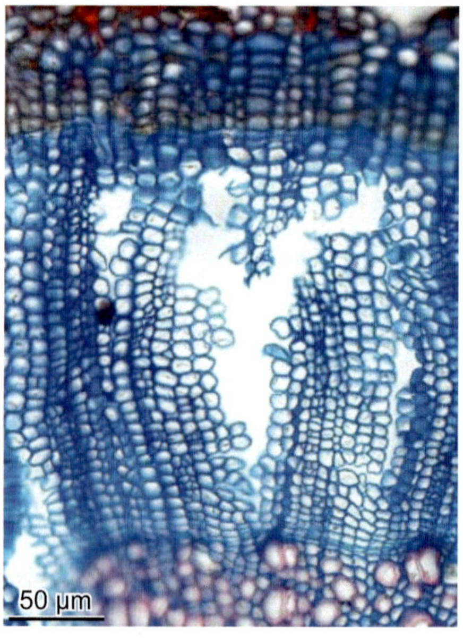

Features according to IAWA

Xylem 2 - 9 - 13 - 20 - 40.2 - 50.2 - 52.2 - 60.1 - 79.1 - 99 - 99.1 - 100.1 - 100.2 - 105

Aphragmus oxycarpus (Hook.f. & Thomson) Jafri

Brassicaceae

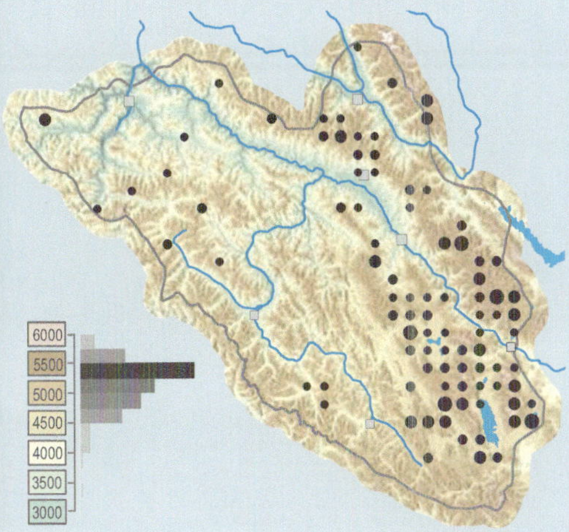

Syn.: *Braya oxycarpa* Hook.f. & Thomson

Description: Perennial herbs, 2–8 cm tall, stems branched from the base, suberect or ascending to spreading, leaves spatulate to oblong, flowers pale lilac or white to pinkish, ca. 3 mm across, in lax racemes.

Distribution: Tajikistan, NE Afghanistan, Pakistan, India (Jammu and Kashmir, Sikkim), Nepal, Bhutan, China (Qinghai, Sichuan, Xinjiang, Xizang, Yunnan).

Habitat: Mesic gravel river bed sediments, rock crevices, moraines, alpine pastures, open stony slopes.

Stem diameter 2–3 mm. Bark/xylem proportion 1:0.5. Mean ring width 0.02 mm. Maximal 16 annual rings.

Xylem

Transverse section

Semi-ring-porous. Ring boundaries marked by a row of small earlywood vessels (diameter 5–10 µm). Ring variability low. Vessels thin- to thick-walled (2–3 µm), unlignified. Fibers absent. Parenchyma pervasive. With remaining vascular bundles. Large ray-like parenchyma zones between vascular bundles disappear in polarized light. Rays absent within vascular bundles.

Tangential section

Multiseriate rays confluent with axial parenchyma cells. Ray cells upright.

Radial section

Perforation plates simple. Intervessel pits pseudoscalariform. Vessel length 50–80 µm.

Bark

Phloem large, containing irregularly distributed groups of sieve tubes. Cortex small, cells large (15–30 µm), round to oval. Sclerenchyma and crystals absent. Phellem consists of many (up to 15) collapsed layers of cells.

Pith

Pentagonal, filled with large (15–30 µm), thin-walled, round cells.

Analyzed: Root collar of polar root of 3 individuals.

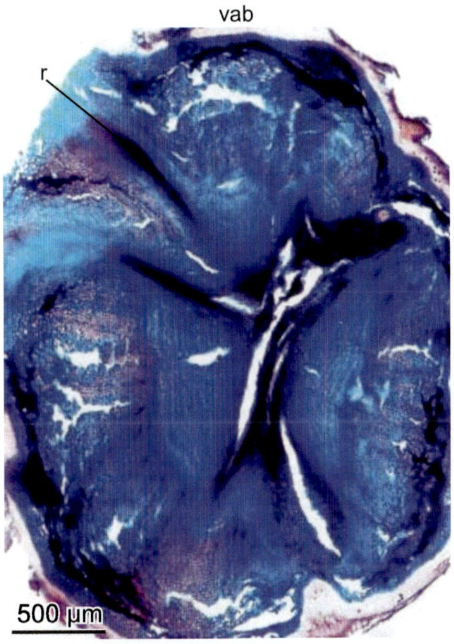

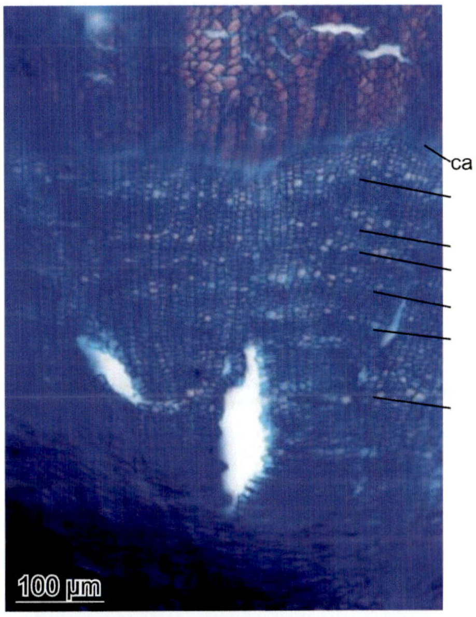

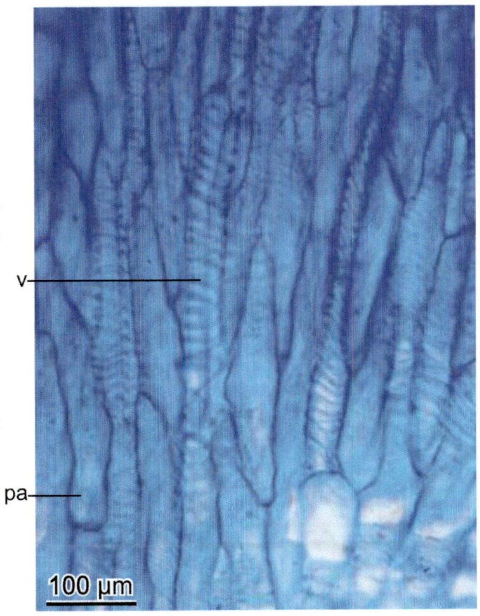

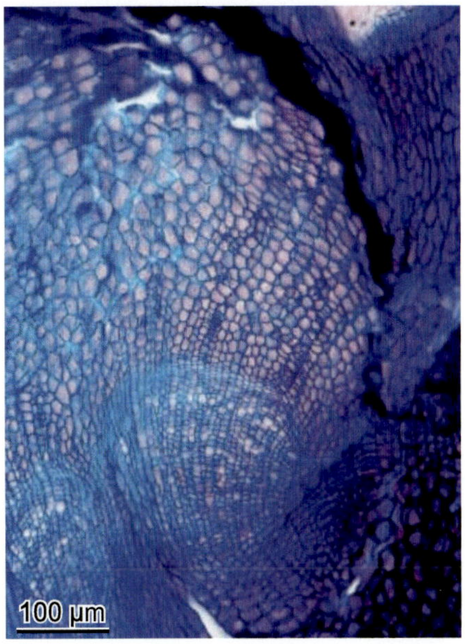

Features according to IAWA

Xylem 1 - 4 - 9 - 13 - 20.1 - 40.1 - 50.2 - 52.1 - 60.1 - 79.1 - 99 - 99.1 - 100.2 - 105

Arabis tibetica Hook.f. & Thomson

Brassicaceae

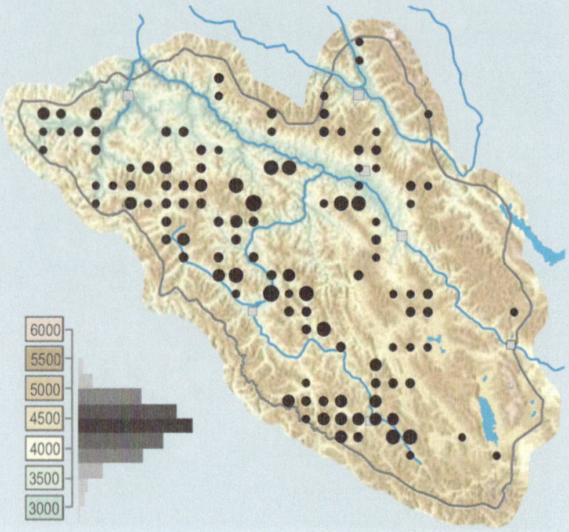

Syn: *A. multicaulis* Pamp., *Arabidopsis tibetica* (Hook.f. & Thomson) Y.C.Lan & Z.X.An

Description: Biennial herbs, 10–35 cm tall, usually branched from the base, leaves spatulate, margin dentate, flowers white or rarely pinkish, ca. 3 mm across, in racemes.

Distribution: Tajikistan, Kyrgyzstan, Afghanistan, Pakistan, India (Jammu and Kashmir), China (Xizang).

Habitat: Stabilized slopes, rock crevices, below exposed rocks.

Stem diameter 1–2 mm. Bark/xylem proportion 1:1. Ring width 0.15–0.25 mm. Maximal 3 annual rings.

Xylem

Transverse section

Semi-ring-porous. Rings distinctly marked by a row of earlywood vessels. Ring variability low. Vessels form radial rows with more than 6 cells. Earlywood vessel diameter 20–30 µm. Thin- to thick-walled fibers form a peripheral belt around a fiber-less center. Parenchyma in the fiber belt paratracheal and pervasive in the center.

Tangential section

Rays absent.

Radial section

Perforation plates simple. Intervessel pits round in the fiber belt, in alternating position, and scalariform in the fiber-less center. Vessel length 50–70 µm.

Bark

Phloem large, with groups of sieve tubes. With irregular, large intercellulars. Cortex small, cells large (50 µm), round to oval. Sclerenchyma and crystals absent. Phellem consists of few layers of square cells.

Pith

Small, round, filled with small, thin-walled, round cells.

Analyzed: Root collar of polar root of 2 individuals

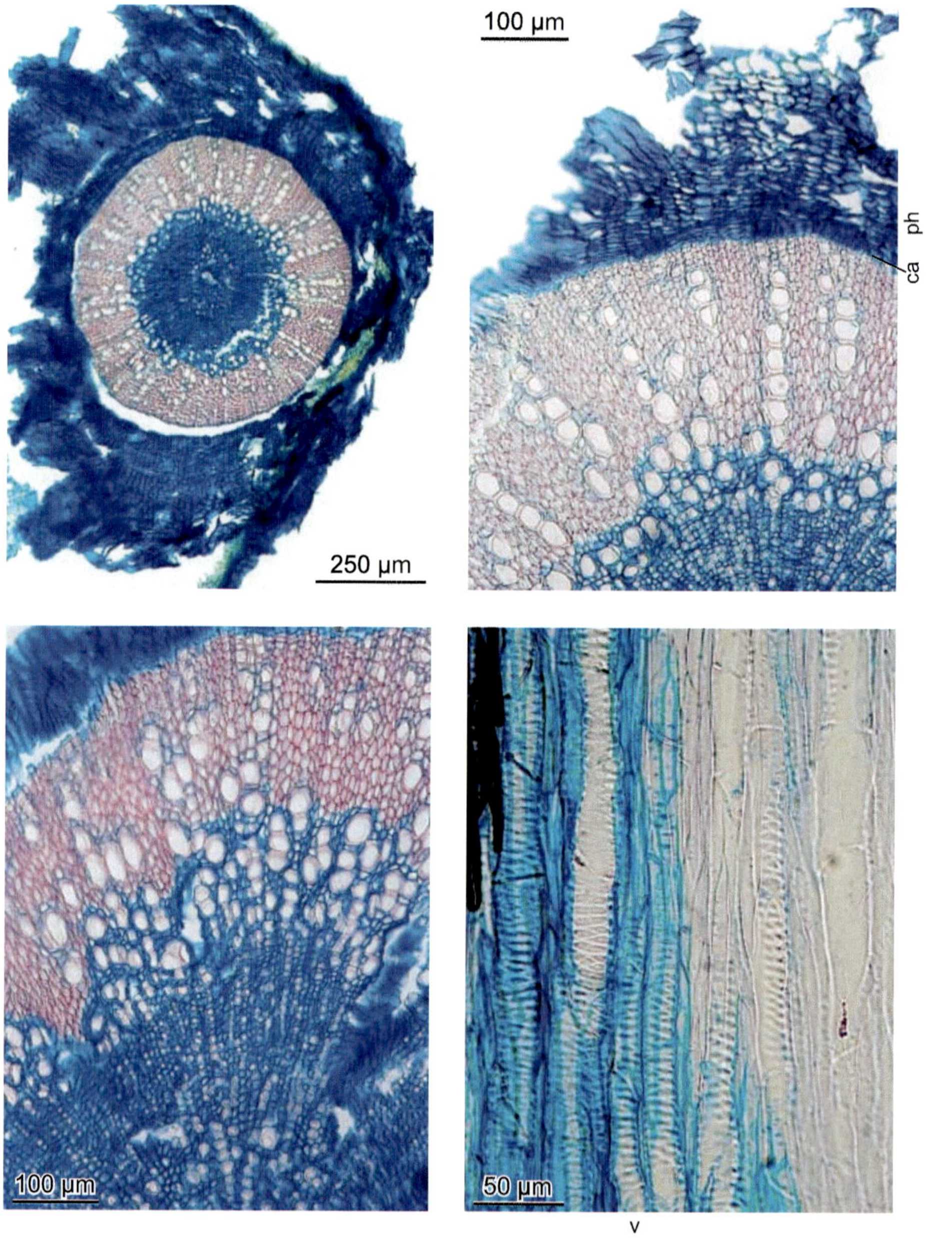

Features according to IAWA

Xylem 1 - 4 - 9.1 - 10 - 13 - 20 - 22 - 40.2 - 50.2 - 52.2 - 61 - 70 - 70.3 - 79 - 79.1 - 117

Christolea crassifolia Cambess.

Brassicaceae

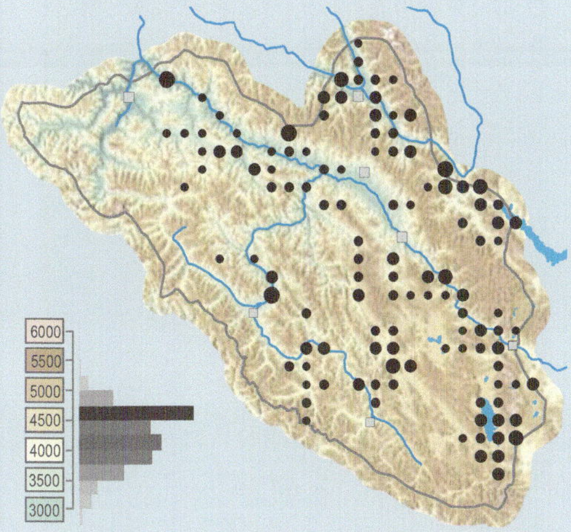

Syn.: *C. afghanica* (Rech.f.) Rech.f., *C. incisa* O.E.Schulz, *C. pamirica* Korsh., *Koelzia afghanica* Rech.f., *Parrya ramosissima* Franch.

Description: Perennial herbs, 10–40 cm tall, stems branched, suberect or decumbent, leaves variable in shape and size, obovate, usually with 3 teeth, villous to glabrous, flowers white with purplish base, 6–7 mm across, in many-flowered racemes.

Distribution: Tajikistan, Afghanistan, Pakistan, India, Nepal, China (Qinghai, Xinjiang, Xizang).

Habitat: Stabilized slopes, unstable gravel screes.

Stem diameter 5–14 mm. Bark/xylem proportion 1:3. Ring width 0.3–0.8 mm. Maximal 16 annual rings. Center rotten.

Xylem

Transverse section

Semi-ring-porous. Ring boundaries marked by a zone of small vessels in the latewood (diameter 20–40 µm) and large earlywood vessels (diameter 80–100 µm). Earlywood vessels arranged in radial rows and groups (up to 8 cells). Vessels thick-walled (2–3 µm), lignified. Small groups of fibers surround some vessels. Parenchyma pervasive. Rays disappear in polarized light. Secondary intra-xylary periderms separate periodically older, centripetal dead parts from external living tissues.

Tangential section

Three to 6-seriate rays consisting of thin-walled, non-lignified cells, confluent with the axial tissue.

Radial section

Perforation plates simple. Intervessel pits scalariform to round in alternating position. Vessel length 70–100 µm.

Bark

Adult phloem radially structured, consisting of unlignified parenchyma and sieve tubes and strips of thick-walled fibers. Phellem consists of many (up to 30) layers of regularly formed cells. Peripheral cells filled with dark substances.

Pith

Pith in young stems irregular polygonal, containing large, round, unlignified, thin-walled cells.

Analyzed: Root collar of polar root of 10 individuals.

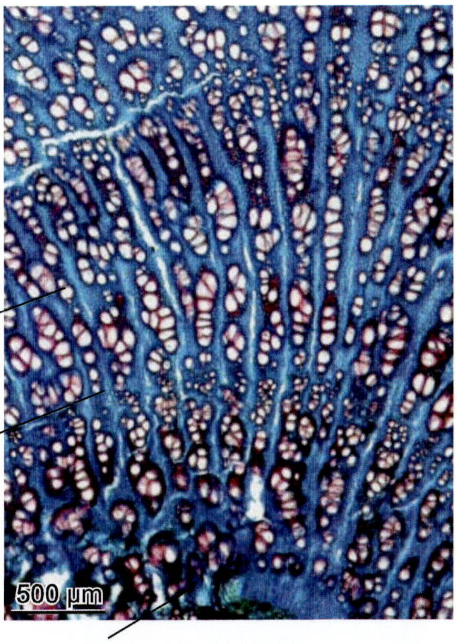

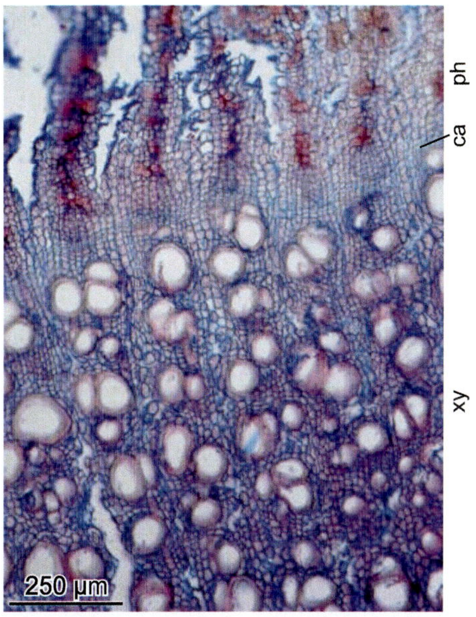

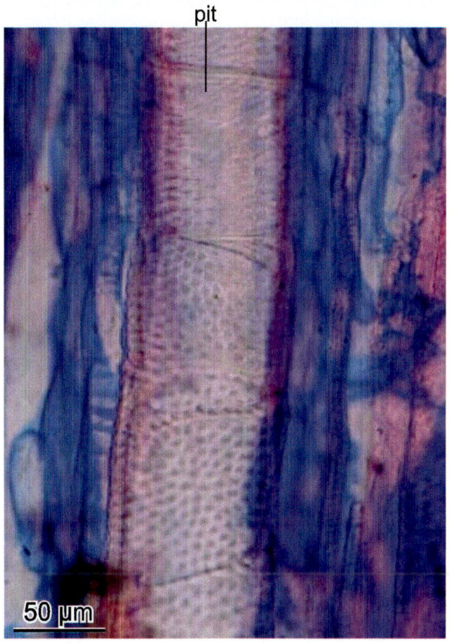

Features according to IAWA

Xylem *1 - 4 - 9 - 9.1 - 10 - 13 - 22 - 36 - 39.1 - 41 - 50.1 - 52.2 - 61 - 70 - 79 - 79.1 - 98 - 100.1 - 100.2 - 105 - 135.2*

Conringia planisiliqua Fisch. & C.A.Mey.
Brassicaceae

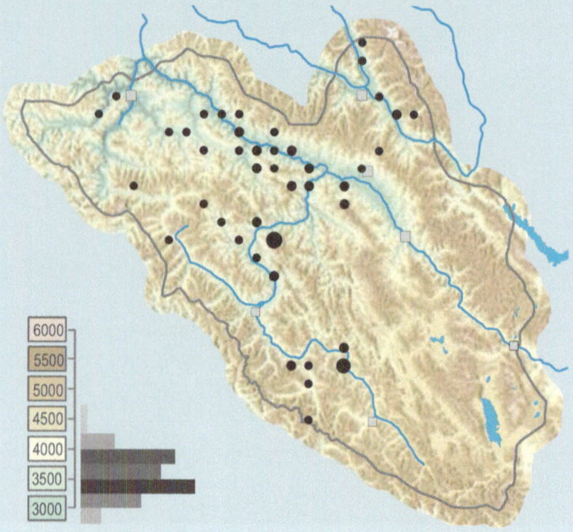

Syn.: *Erysimum planisiliquum* (Fisch. & C.A.Mey.) Steud.

Description: Annuals, 10–40 cm tall, glabrous, glaucous, leaves amplexicaul, flowers pale yellow or whitish, ca. 5 mm across, in racemes.

Distribution: Iran, Russia, Kazakhstan, Uzbekistan, Turkmenistan, Kyrgyzstan, Tajikistan, Afghanistan, Pakistan, India China, Mongolia.

Habitat: A weed in fields.

Stem diameter 10 mm. Bark/xylem proportion 1:18. Ring width 4 mm. Maximal 1 annual ring.

Xylem

Transverse section

Vessels radially in short and long multiples, in tangential rows. Vessel diameter 20–25 µm. Thin- to thick-walled fibers. Parenchyma rare, if present paratracheal

Tangential section

Rays 2–3-seriate.

Radial section

Perforation plates simple. Intervessel pits round in alternating position. Vessel length 75–100 µm.

Bark

Phloem with small groups of sieve tubes. Cortex cells round to oval. Sclerenchyma and crystals absent. Cortex consists of few layers of square cells. Phellem absent.

Pith

No slide available.

Analyzed: Root collar of polar roots of 1 individual.

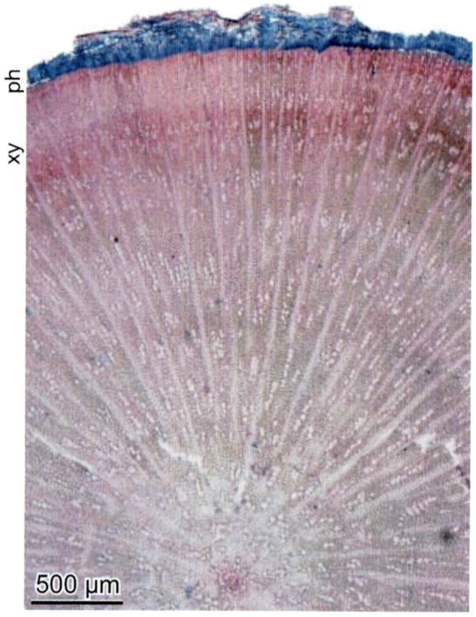

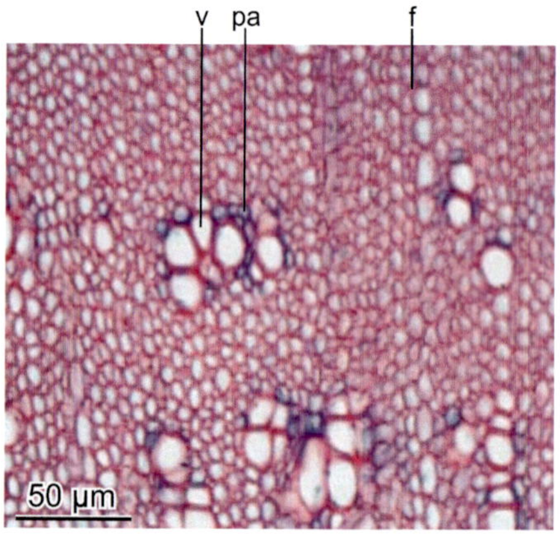

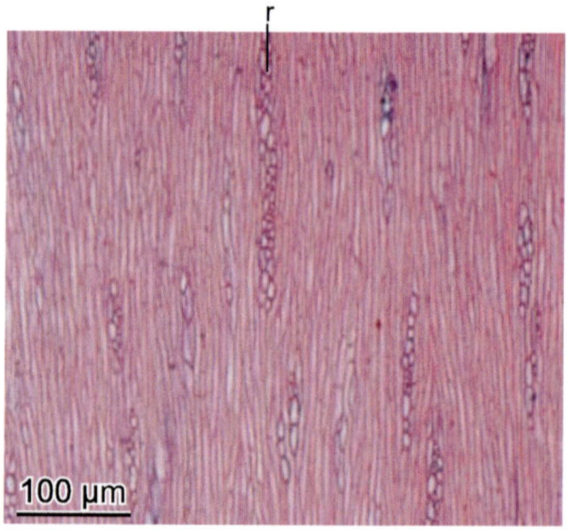

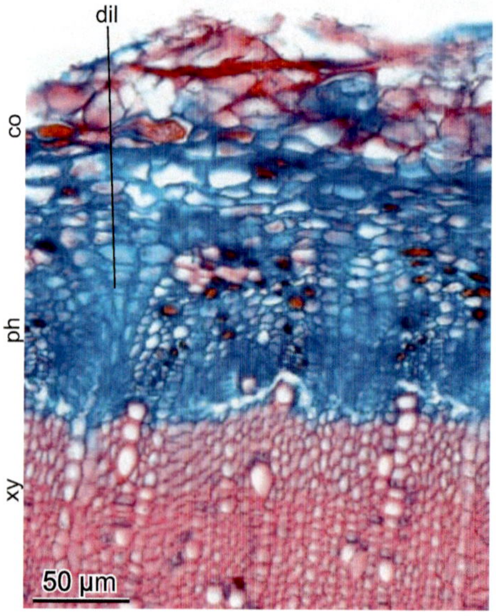

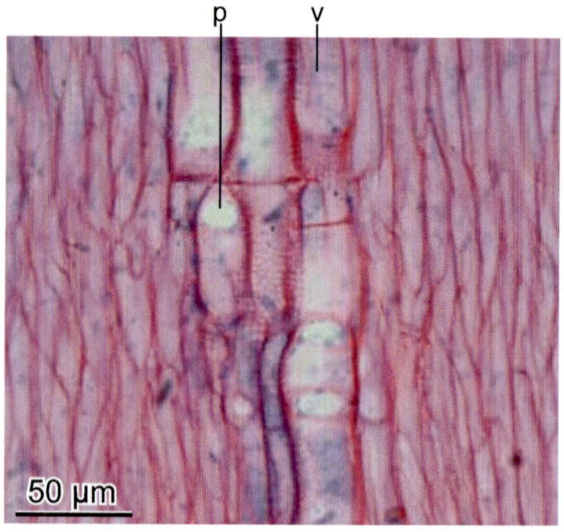

Features according to IAWA

Xylem 2.1 - 6 - 9.1 - 10 - 13 - 20 - 22 - 40.2 - 50.1 - 52.2 - 61 - 70 - 79 - 97 - 105

Crucihimalaya mollissima (C.A.Mey.) Al-Shehbaz, O'Kane & R.A.Price

Brassicaceae

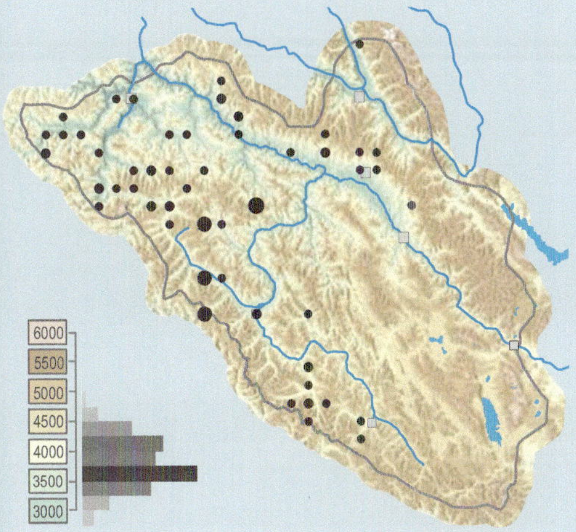

Syn.: *Sisymbrium mollissimum* C.A.Mey., *Arabidopsis mollissima* (C.A.Mey.) N.Busch

Description: Perennial herbs, 5–40 cm tall, densely pubescent, leaves spatulate to oblong, flowers white or pink, 3–4 mm across, in terminal racemes.

Distribution: Russia (Siberia), Kazakhstan, Kyrgyzstan, Tajikistan, Afghanistan, Pakistan, India, China (Gansu, Sichuan, Xinjiang, Xizang), Mongolia.

Habitat: Synanthropic on stony walls and field margins, slopes.

Stem diameter 3–6 mm. Bark/xylem proportion 1:3. Ring width 0.3–0.4 mm. Maximal 3 annual rings.

Xylem

Transverse section

Diffuse-porous. Rings distinctly marked by a large marginal paremchyma. Ring variability low. Vessels solitary and in short radial rows. Earlywood vessel diameter 15–25 µm. Thin- to thick-walled fibers form a peripheral belt around a fiber-less center. Parenchyma in the fiber belt paratracheal and pervasive in the center.

Tangential section

Rays indistinct, unlignified.

Radial section

Perforation plates simple. Intervessel pits round in alternating position and in the fiber belt and scalariform in the fiber-less center. Vessel length 60–80 µm.

Bark

Phloem large, with radial groups of sieve tubes. With irregular spaces. Cortex small, cells round to square. Sclerenchyma and crystals absent. Phellem consists of few layers of thin-walled, irregularly formed cells.

Pith

Small, round to star-shaped, mostly hollow.

Analyzed: Root collar of polar root of 3 individuals.

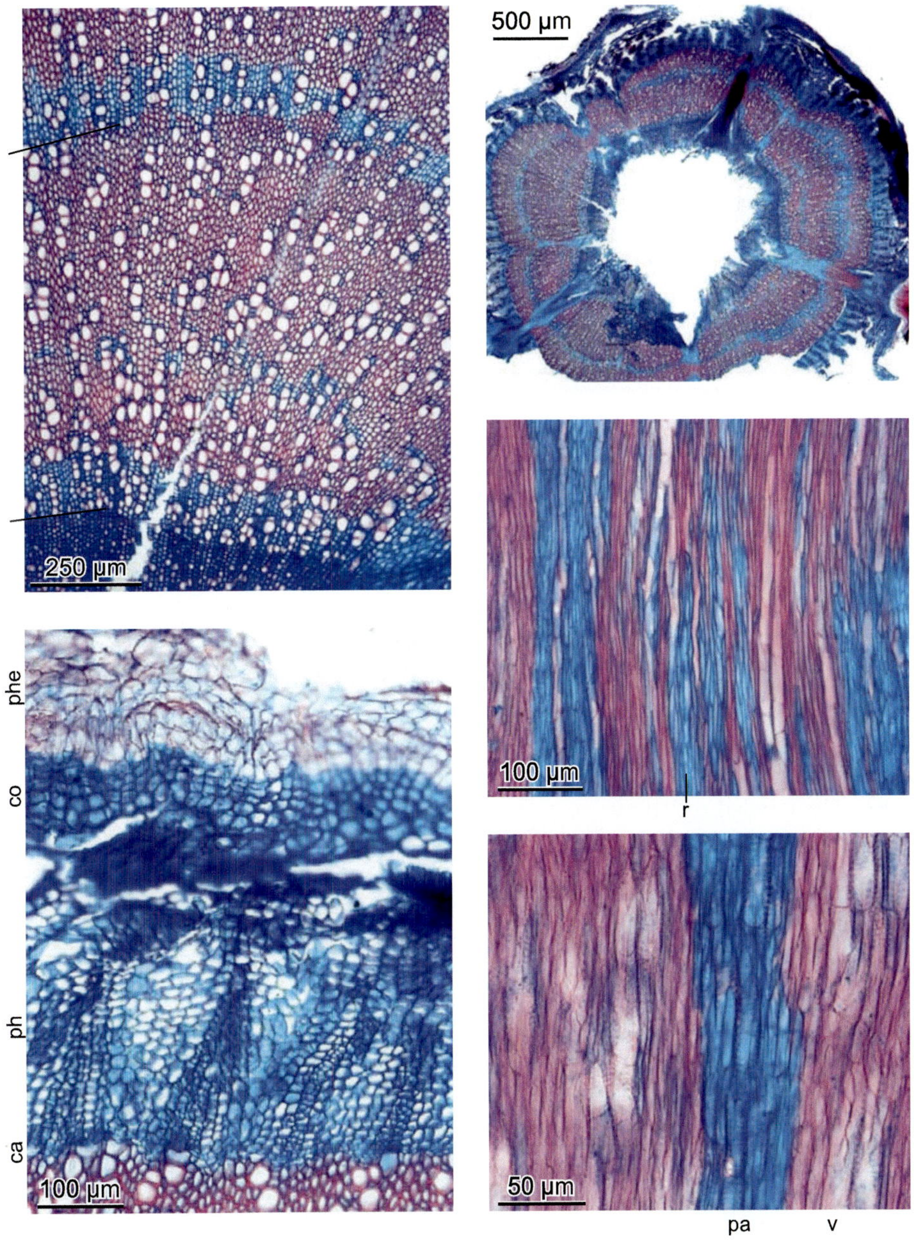

Features according to IAWA

Xylem 1 - 5 - 9 - 9.1 - 13 - 20 - 22 - 40.1 - 50.2 - 52.2 - 61 - 70 - 70.3 - 79 - 79.1 - 97 - 99 - 117

Desideria himalayensis (Cambess.) Al-Shehbaz

Brassicaceae

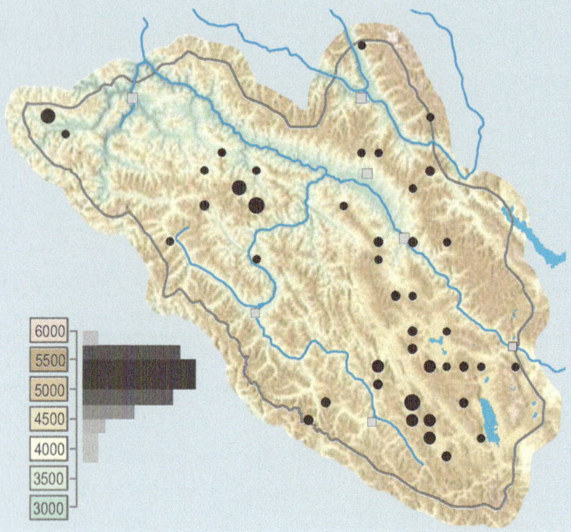

Syn.: *Cheiranthus himalayensis* Cambess., *Christolea himalayensis* (Cambess.) Jafri, *Ermania himalayensis* (Cambess.) O.E.Schulz, *Solms-laubachia himalayensis* (Cambess.) J.P.Yue, Al-Shehbaz & H.Sun

Description: Perennial herbs, 4–15 cm tall, stems spreading, pilose, leaves spatulate, usually with 3(-5) shallow teeth on the tip, flowers lilac, 5–6 mm across, in many-flowered racemes, fruits linear, 2–3 cm long, pilose.

Distribution: Pakistan, N India, Nepal, China (Qinghai, Xizang).

Habitat: Sandstone scree, alpine slopes.

Stem diameter 2–4 mm. Bark/xylem proportion 1:1. Ring width 0.2–0.6 mm. Maximal 8 annual rings.

Xylem

Transverse section

Semi-ring-porous. Rings marked by a row of earlywood vessels. Ring boundaries often indicated by larger (50 μm) or smaller (30 μm) vessel diameters. Fibers absent. Parenchyma pervasive. Very large rays. Vascular bundles remaining.

Tangential section

Ray width >10 cells, with a layer of sheath cells. Rays disappear in polarized light.

Radial section

Perforation plates simple. Intervessel pits scalariform. Vessel length 50–90 μm.

Bark

Phloem with small tangentially and radially arranged groups of sieve tubes. Cortex cells large (30–40 μm), round to oval. With irregular spaces. Sclerenchyma and crystals absent. Phellem consists of a few rows of thin-walled, rectangular cells.

Pith

Round, hollow. Few groups of fibers often at the periphery of the pith.

Analyzed: Root collar of polar root of 5 individuals.

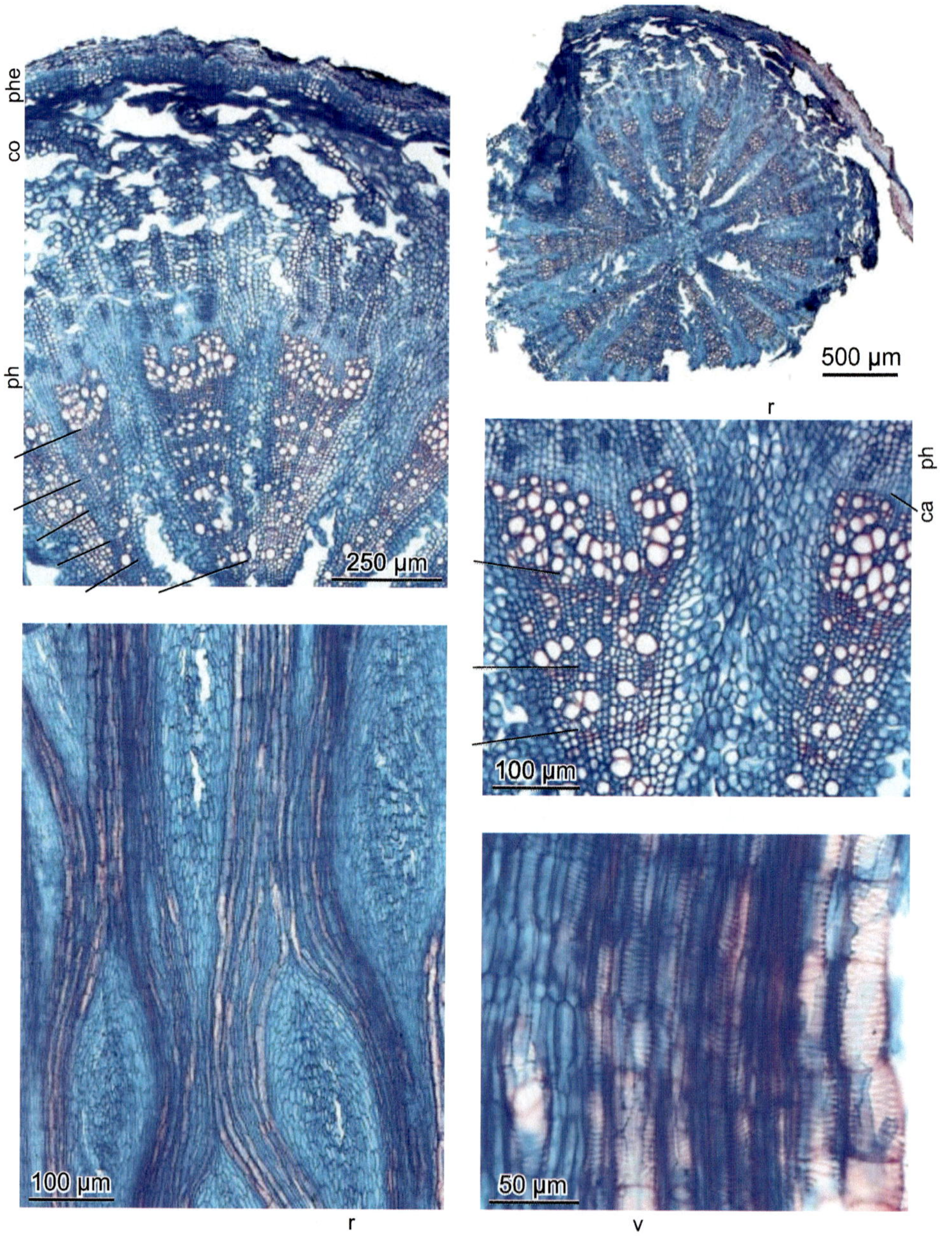

Features according to IAWA

Xylem 1 - 4 - 9 - 13 - 20.1 - 40.1 - 50.2 - 52.2 - 60.1 - 79.1 - 98 - 99.1 - 100.2 - 105

Desideria linearis (N.Busch) Al-Shehbaz
Brassicaceae

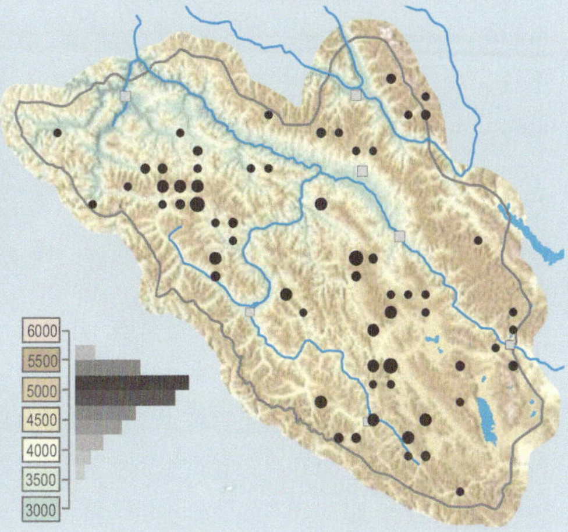

Syn.: *Christolea linearis* N.Busch, *Solms-laubachia linearis* (N. Busch) J.P.Yue, Al-Shehbaz & H.Sun

Description: Perennial herbs, 4–15 cm tall, densely pilose, leaves obovate or spatulate, usually with 3 deep teeth, flowers purple, in 8–20-flowered racemes, fruits linear, 20–35 × 1–1.7 mm, pilose.

Distribution: Tajikistan, N Pakistan, N India, Nepal, China (Xinjiang).

Habitat: Screes, moraines, sandy or gravel slopes.

Stem diameter 2–4 mm. Bark/xylem proportion 1:1. Ring width 0.08–0.13 mm. Maximal 8 annual rings.

Xylem

Transverse section
Semi-ring-porous. Rings marked by a row of earlywood vessels. Vessel walls not stained. Ring boundaries often indicated by larger (50 µm) or smaller (30 µm) vessel diameters. Fibers absent. Parenchyma pervasive. Large rays.

Tangential section
Ray width 4–8 cells. Rays confluent, disappear in polarized light.

Radial section
Perforation plates simple. Intervessel pits scalariform. Vessel length 60–90 µm.

Bark

Phloem with small tangentially and radially arranged groups of sieve tubes. Cortex cells large (30–40 µm), round to oval. With irregular spaces. Sclerenchyma and crystals absent. Phellem consists of many rows of thin-walled, rectangular cells.

Pith

In the shoot, round and hollow (figure shows a root).

Analyzed: Root collar of polar root of 9 individuals.

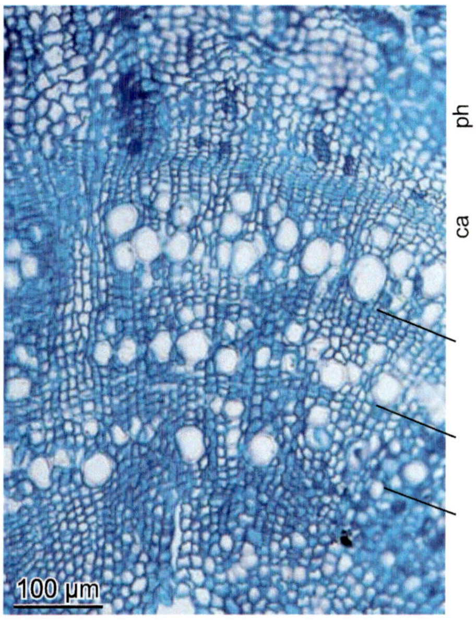

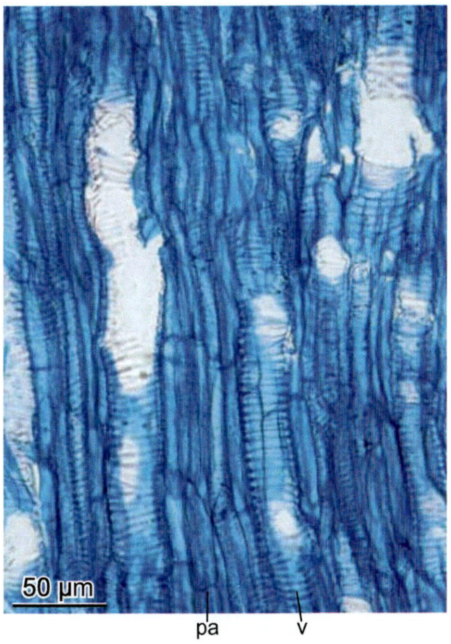

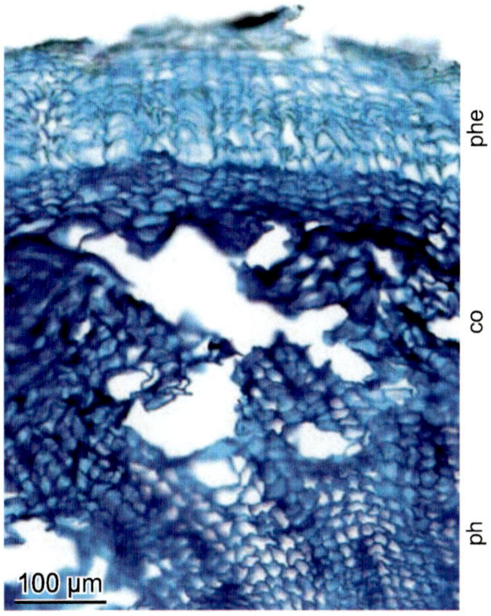

Features according to IAWA
Xylem 1 - 4 - 9 - 13 - 20.1 - 39.2 - 40.1 - 50.2 - 52.2 - 60.1 - 70.4 - 79.1 - 98 - 99.1 - 100.2 - 105

Desideria pumila (Kurz) Al-Shehbaz

Brassicaceae

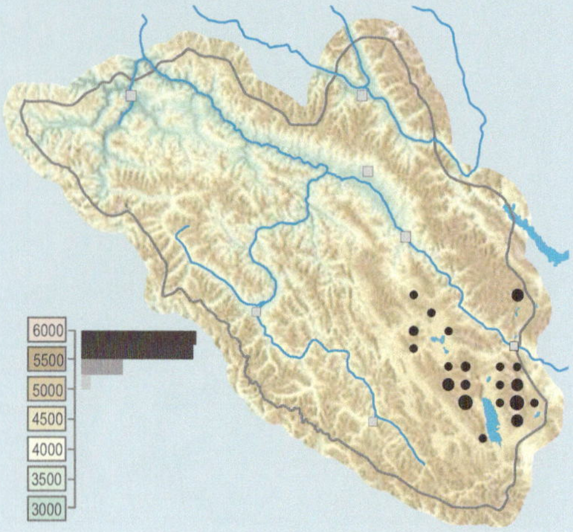

Syn.: *Christolea pumila* (Kurz) Jafri, *Ermania koelzii* O.E.Schulz, *Parrya pumila* Kurz

Description: Perennial herbs, 1–5 cm tall, leaves spatulate, entire, flowers purple, solitary from a basal rosette, fruits lanceolate, glabrous on valves and pilose on margins.

Distribution: India (E Ladakh), China (Xizang).

Habitat: Unstable permafrost stony soils.

Stem diameter 2–4 mm. Bark/xylem proportion 1:6. Ring width 0.08–0.13 mm. Maximal 18 annual rings.

Xylem

Transverse section

Semi-ring-porous. Rings marked by 1–2 rows of earlywood vessels. Vessels lignified or unlignified. Ring boundaries indicated by larger (40 µm) or smaller (25 µm) vessel diameters. Fibers absent. Parenchyma pervasive. Large rays.

Tangential section

Ray width 4–8 cells. Rays confluent with axial tissue. Rays disappear in polarized light.

Radial section

Perforation plates simple. Intervessel pits scalariform. Vessel length 60–90 µm.

Bark

Phloem with small, radially arranged groups of sieve tubes. Cortex cells large (40–60 µm), round to oval. With irregular spaces. Sclerenchyma and crystals absent. Phellem consists of many rows of thin-walled, rectangular, collapsed cells.

Pith

No slide available.

Analyzed: Root collar of polar root of 3 individuals.

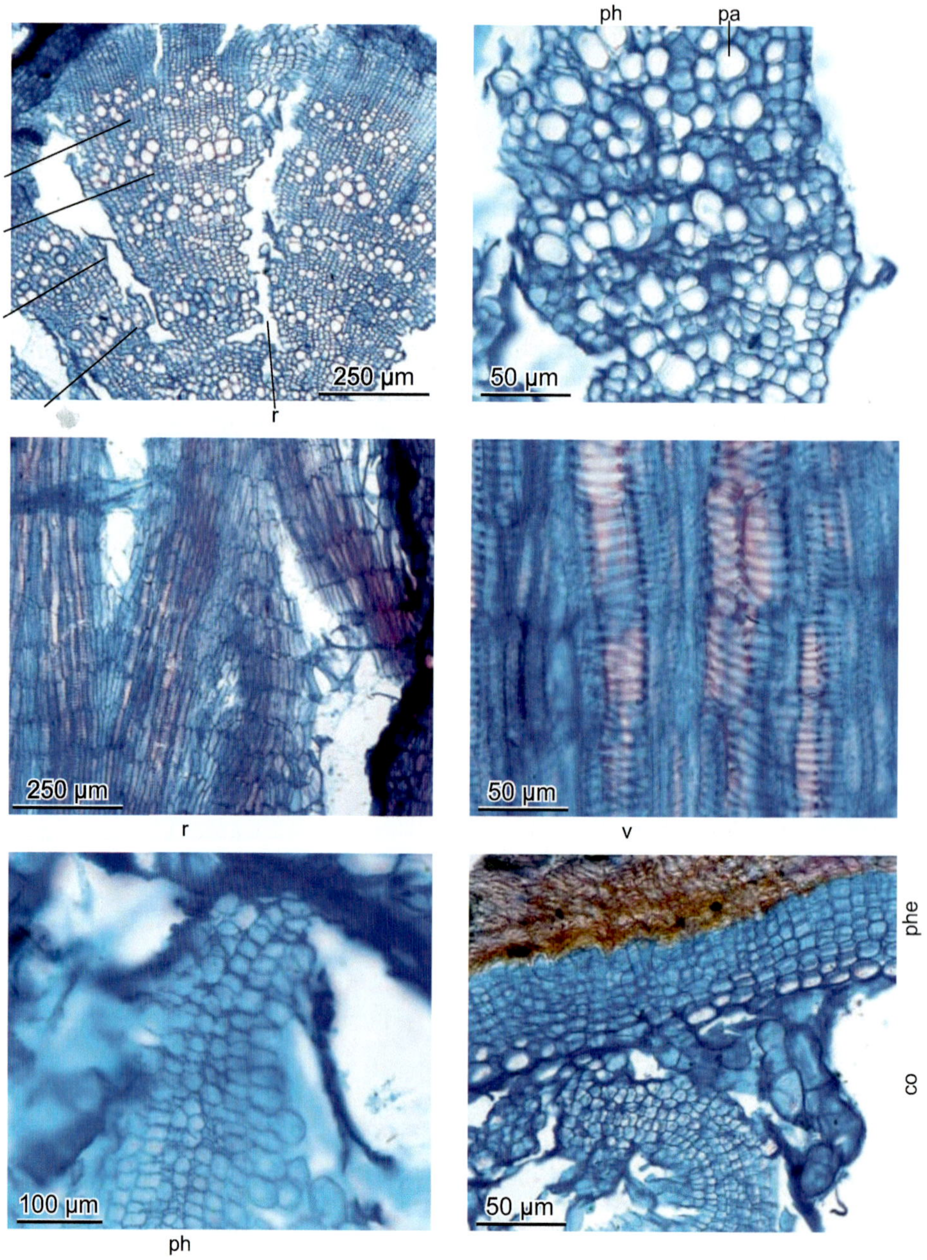

Features according to IAWA

Xylem 1 - 4 - 9 - 13 - 20.1 - 40.1 - 50.2 - 52.2 - 60.1 - 70.4 - 79.1 - 98 - 99.1 - 100.1 - 100.2 - 105

Dilophia salsa Thomson

Brassicaceae

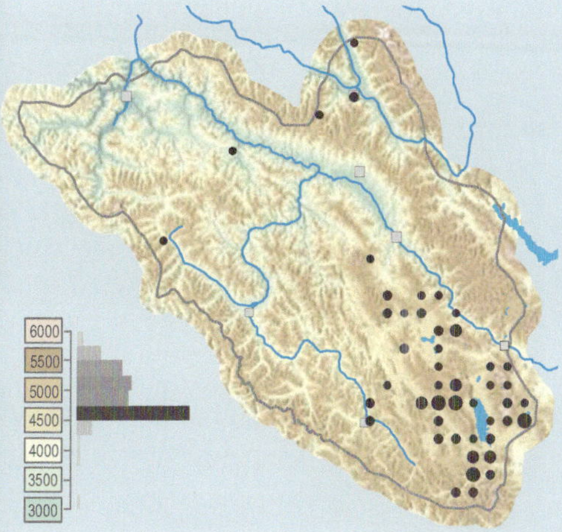

Syn.: *D. dutreuilii* Franch., *D. kaschgarica* Rupr.

Description: Annuals, 1–6 cm tall, stems branched from the base, forming small dense cushions 1–10 cm across, leaves spatulate, oblong or linear, flowers white or pink, ca. 2 mm across, in dense corymbs.

Distribution: Kyrgyzstan, Tajikistan, Pakistan, India, Nepal, China (Gansu, Qinghai, Xinjiang, Xizang), Bhutan.

Habitat: Saline wetlands, along watersheds.

Stem diameter 0.5–1 mm. Bark/xylem proportion 1:0.5. Ring width 0.06–0.15 mm. Maximal 2 annual rings.

Xylem

Transverse section

Semi-ring-porous. Rings mostly distinctly marked by an earlywood zone with vessel groups. Central part of the stem consists mainly of pervasive parenchyma with few small vessels (diameter <10 µm). Second ring contains vessels with a diameter of 10–15 µm and pervasive parenchyma cells. Vessels thin-walled (<1 µm), poorly lignified. Vascular bundles remain. Fibers absent.

Tangential section

Rays indistinct.

Radial section

Perforation plates simple. Intervessel pits pseudoscalariform. Vessel length 120–140 µm.

Bark

Phloem small, consisting of few groups of sieve tubes between large parenchyma zones. Cortex large, containing irregular, large intercellulars between large parenchyma cells. Rays dilated. Sclerenchyma and crystals absent. Phellem consists of 2–3 rows of irregularly formed cells.

Pith

Not clearly delimitated from the xylem.

Analyzed: Root collar of polar root of 8 individuals.

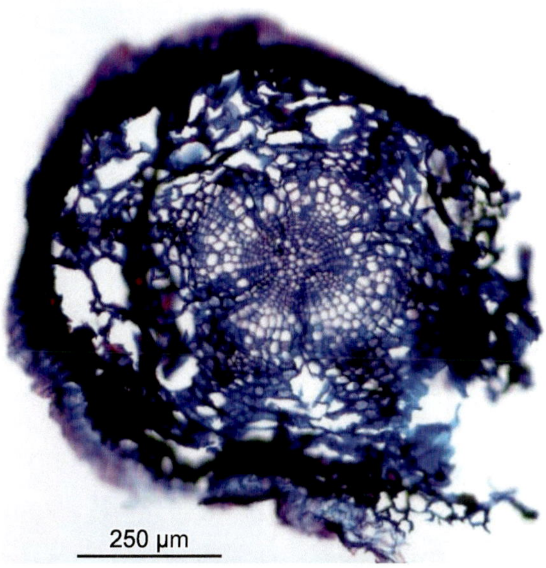

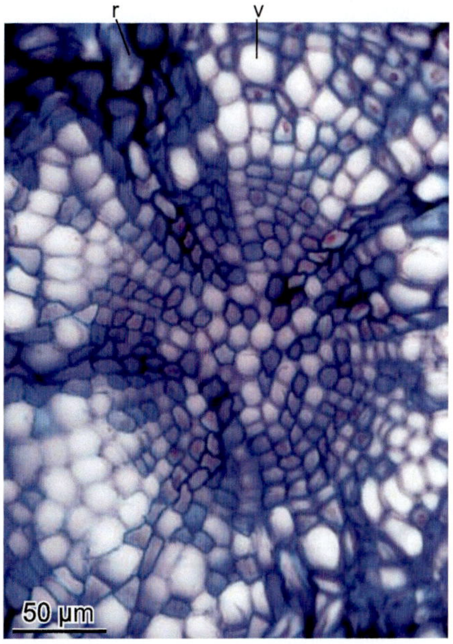

v

Features according to IAWA

Xylem 1 - 4 - 9 - 11 - 20.1 - 39.1 - 40.1 - 50.2 - 52.2 - 60.1 - 79 - 79.1 - 99 - 99.1 - 100.1 - 100.2

Dontostemon glandulosus (Kar. & Kir.) O.E.Schulz

Brassicaceae

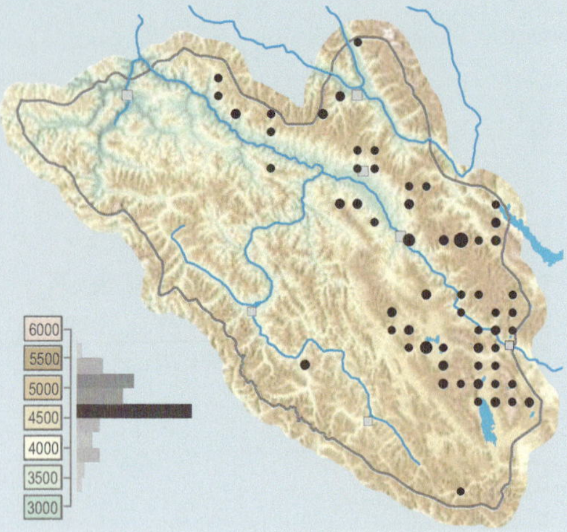

Syn.: *Arabis glandulosa* Kar. & Kir.

Description: Annuals or biennials, 4–10 cm tall, glandular, leaves mostly basal, lanceolate, flowers lilac, 3–4 mm across, in 10–25-flowered racemes.

Distribution: Russia, Kazakhstan, Tajikistan, Pakistan, India, Nepal, China.

Habitat: Stabilized slopes, gravel plains, river banks, alpine meadows.

Stem diameter 1–2 mm. Bark/xylem proportion 1:0.5. Ring width 0.2–0.3 mm. Maximal 3 annual rings.

Xylem

Transverse section

Semi-ring-porous. Vessels in short radial rows up to 4 cells. Earlywood vessel diameter 20–30 µm. Fibers thin- to thick-walled, absent in the center. Parenchyma paratracheal in the fiber belt and pervasive in the center.

Tangential section

Rays absent or indistinct, not lignified.

Radial section

Perforation plates simple. Intervessel pits round in alternating position in the fiber belt and scalariform in the fiber-less center. Vessel length 50–70 µm.

Bark

Phloem large, with small groups of sieve tubes. With irregular shizogeneous radial spaces. Sclerenchyma and crystals absent. Phellem consists of few layers of square cells.

Pith

No slide available.

Analyzed: Root collar of polar roots of 3 individuals.

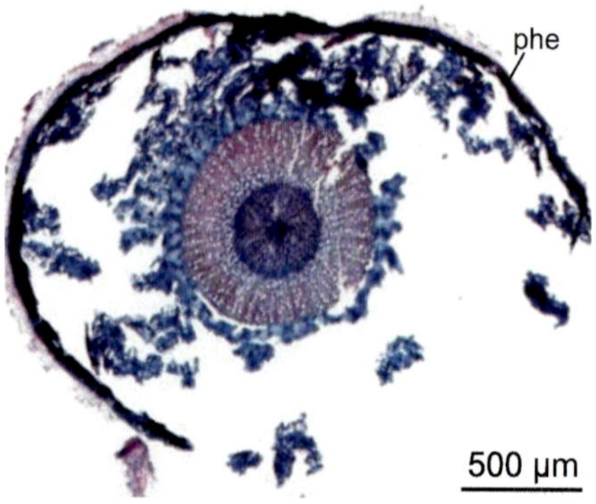

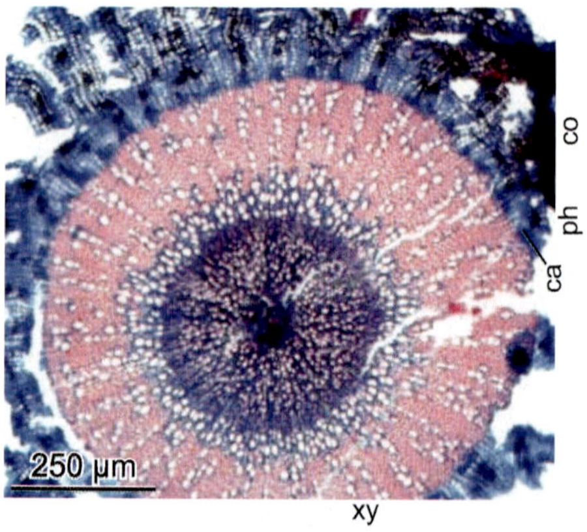

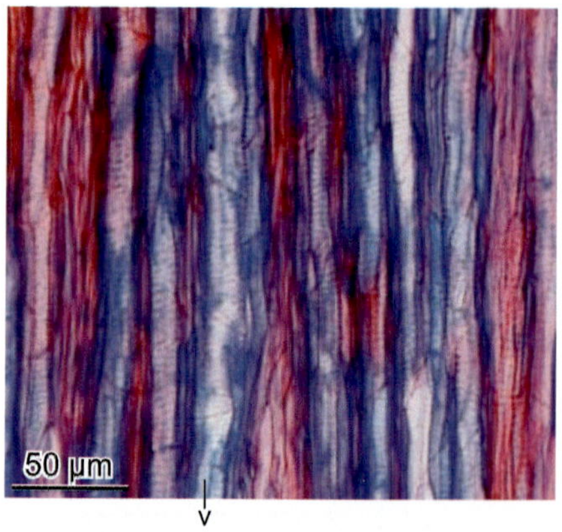

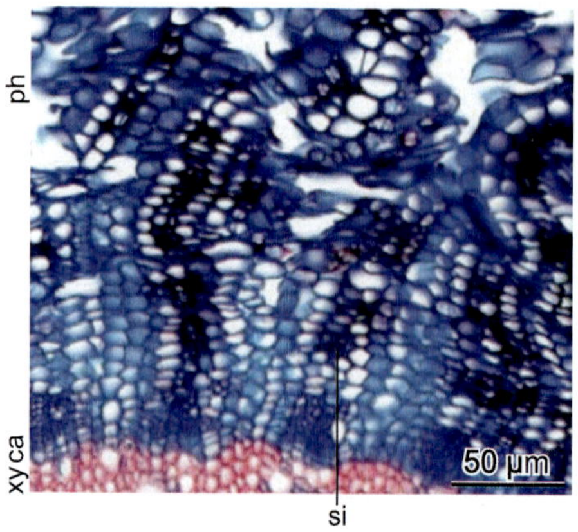

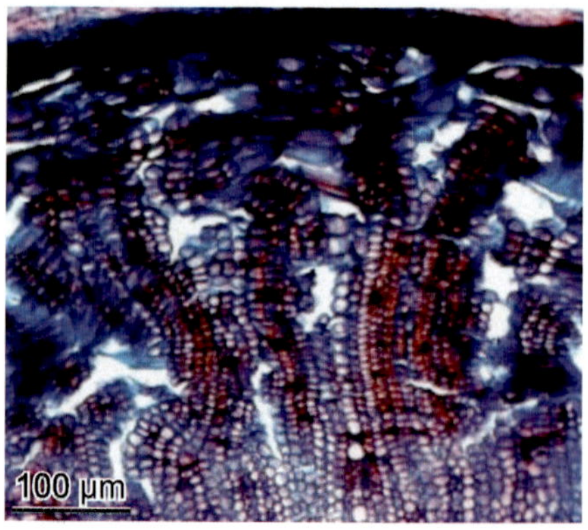

Features according to IAWA

Xylem 1 - 4 - 9.1 - 13 - 20 - 22 - 40.2 - 50.2 - 52.2 - 61 - 70 - 70.3 - 79 - 79.1 - 117

Draba altaica (C.A.Mey.) Bunge

Brassicaceae

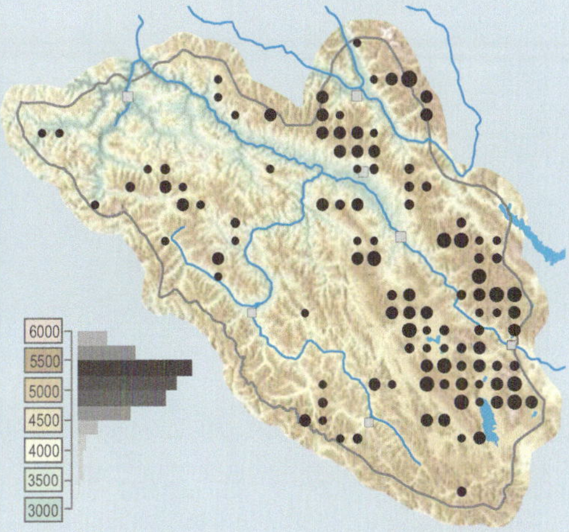

Syn.: *D. rupestris* var. *altaica* C.A.Mey., *D. modesta* W.W.Sm., *Braya glacialis* Korsh. ex Tolm.

Description: Perennial herbs, 2–8 cm tall, loosely cespitose, basal leaves lanceolate, persisting, cauline leaves 1–6, flowers white, ca. 2.5 mm across, in compact 6–15-flowered racemes, fruits ovoid, 4–7 mm long, glabrous.

Distribution: Russia, Kazakhstan, Kyrgyzstan, Afghanistan, Tajikistan, Pakistan, India, Nepal, China (Gansu, Qinghai, Sichuan, Xinjiang, Xizang, Yunnan), Mongolia.

Habitat: Rock crevices, stabilized slopes, gravel areas, moraines, streamsides.

Stem diameter 2–3 mm. Bark/xylem proportion 1:1. Mean ring width 0.03 mm. Maximal 8 annual rings.

Xylem

Transverse section

Diffuse-porous. Rings, if present, marked by a row of earlywood vessels. Ring boundaries often indicated by larger (15 µm) or smaller (10 µm) vessel diameters. Vessels thick-walled (2–3 µm), lignified. Fibers absent. Parenchyma pervasive. With remaining vascular bundles. Large ray-like parenchyma zones between vascular bundles disappear in polarized light. Rays absent within vascular bundles.

Tangential section

Multiseriate rays, confluent with axial parenchyma cells. Ray cells upright.

Radial section

Perforation plates simple. Intervessel pits pseudoscalariform. Vessel length 40–70 µm.

Bark

Phloem large, with tangentially or irregularly distributed groups of sieve tubes. Cortex small, cells large (15–20 µm), round to oval. Sclerenchyma and crystals absent. Phellem consists of many (up to 15) layers of collapsed cells.

Pith

Star-shaped, pentagonal, cells large (30–40 µm), thin-walled, round. Cells widely spaced in the center.

Analyzed: Root collar of polar root of 2 individuals.

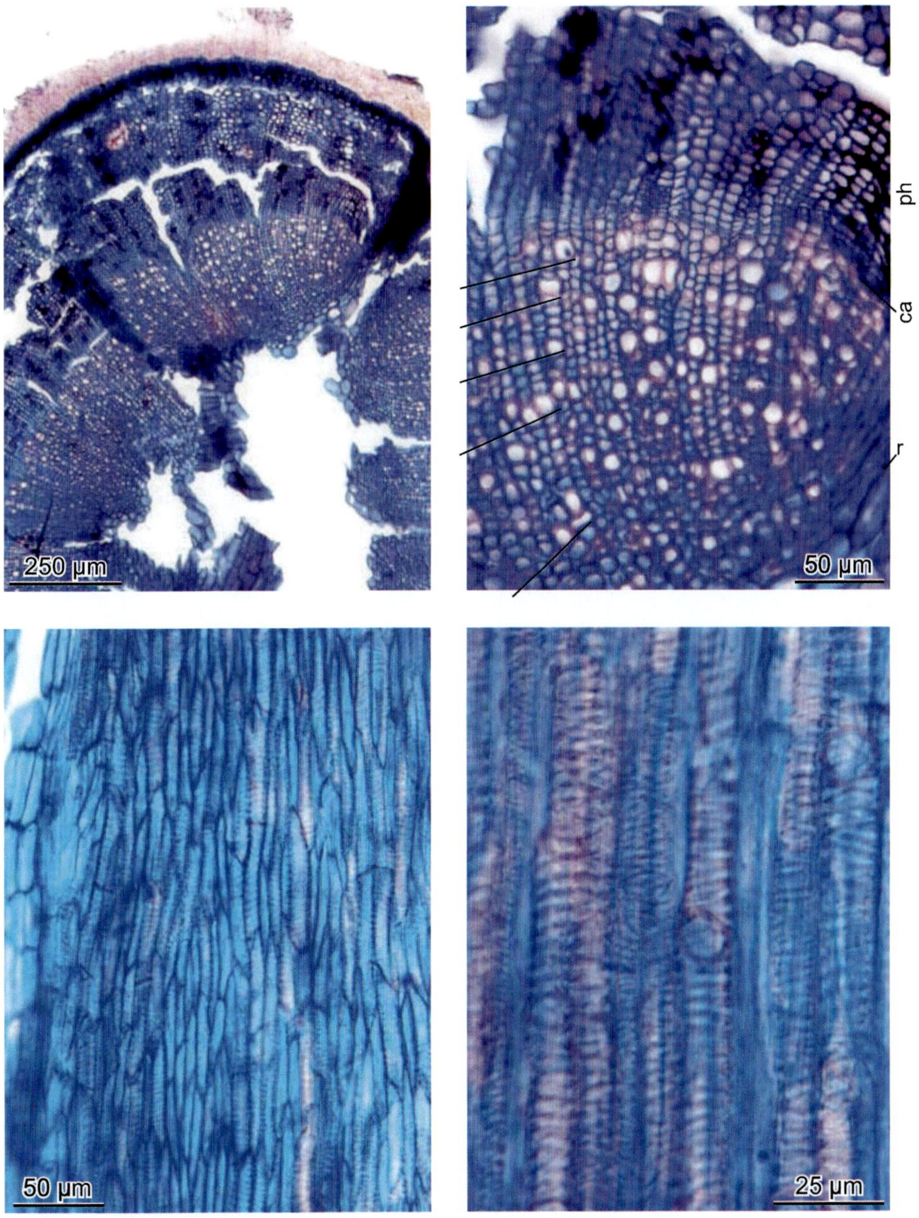

r

v

Features according to IAWA

Xylem 1 - 5 - 9 - 13 - 20.1 - 40.1 - 50.2 - 52.1 - 60.1 - 79.1 - 99.1 - 100.1 - 100.2 - 105

Draba cachemirica Gand.

Brassicaceae

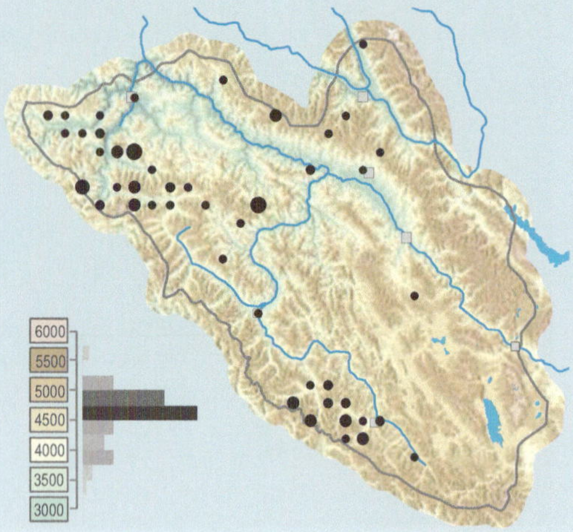

Description: Perennial herbs, 1–10 cm tall, densely cespitose, leaves in dense basal rosettes only, persistent, flowers yellow, in 3–12-flowered racemes, fruit ovate, 2.5–4.5 × 2–3.5 mm.

Distribution: N Pakistan, N India.

Habitat: Stabilized slopes, gravel areas.

Stem diameter 1.5–2 mm. Bark/xylem proportion 1:1. Ring width 0.06–0.09 mm. Maximal 19 annual rings.

Xylem

Transverse section

Semi-ring-porous. Rings, if present, marked by a row of earlywood vessels. Ring boundaries often indicated by larger (15 µm) or smaller (10 µm) vessel diameters. Vessels thick-walled (2–3 µm), lignified. Fibers absent. Parenchyma pervasive. Rays absent.

Tangential section

Rays absent.

Radial section

Perforation plates simple. Intervessel pits pseudoscalariform. Vessel length 40–60 µm.

Bark

Phloem large, with radial groups of sieve tubes. Cortex small, cells large (15–20 µm), round to oval. Sclerenchyma and crystals absent. Phellem consists of many layers of collapsed, thin-walled cells.

Pith

Star-shaped, pentagonal, cells large (30–40 µm), thin-walled, round. Cells widely spaced in the center.

Analyzed: Root collar of polar root of 11 individuals.

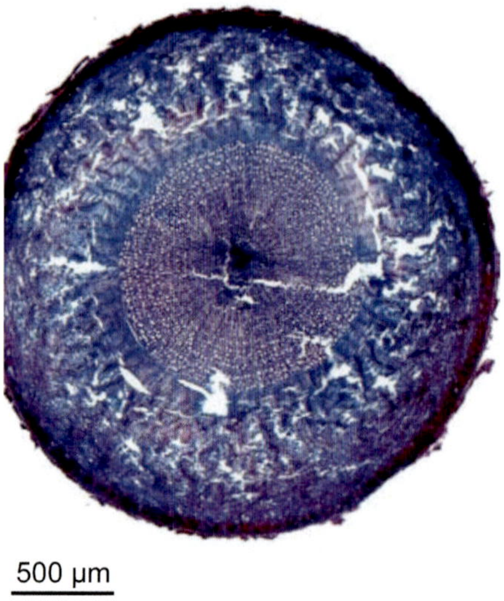

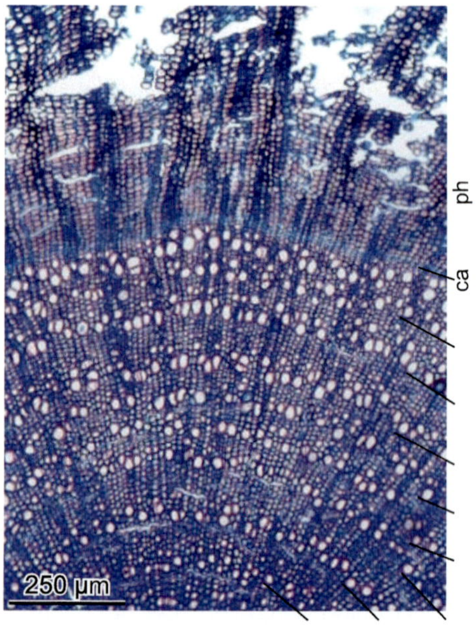

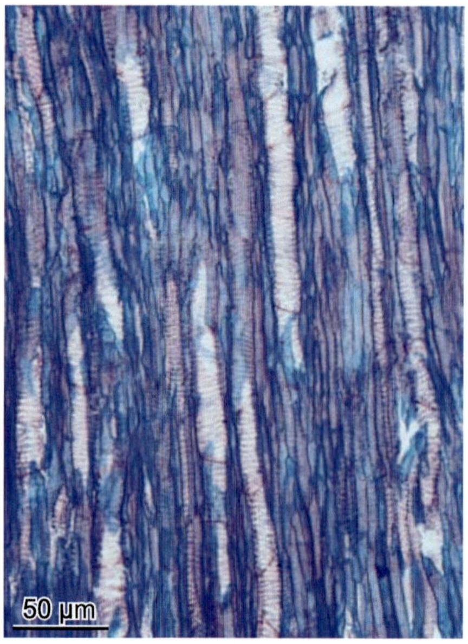

v

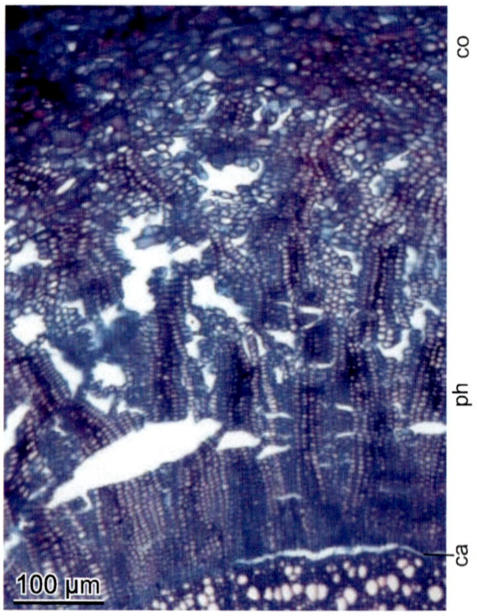

Features according to IAWA

Xylem 1 - 4 - 9 - 13 - 20.1 - 40.1 - 50.2 - 52.1 - 60.1 - 79.1 - 117

Draba glomerata Royle

Brassicaceae

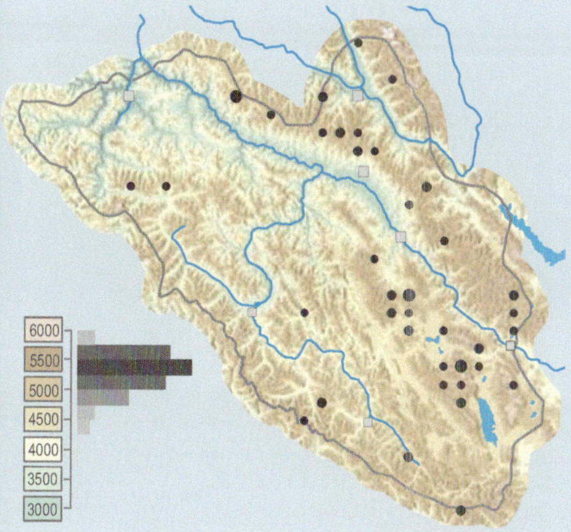

Description: Perennial herbs, 2–8 cm tall, densely cespitose, densely tomentose, leaves obovate, old leaves persistent on branches, flowers white, 2–3 mm across, in 3–10-flowered racemes.

Distribution: Pakistan, India, Nepal, China (Gansu, Sichuan, Qinghai, Xinjiang, Xizang).

Habitat: Gravelly slopes, stabilized screes.

Stem diameter 3–6 mm. Bark/xylem proportion 1:1. Ring width 0.05–0.09 mm. Maximal 8 annual rings.

Xylem

Transverse section

Distinct rings. Semi-ring-porous. Ring variability indicated by larger earlywood (15–20 μm) and smaller latewood vessel diameters (<10 μm). Vessels solitary. Vessel walls fairly thick and poorly stained. Fibers absent. Parenchyma pervasive.

Tangential section

Rays absent.

Radial section

Perforation plates simple. Intervessel pits scalariform. Vessel length 60–90 μm.

Bark

Phloem with small groups of sieve tubes. Phellem consists of many layers of square cells. Few crystal druses in the cortex.

Pith

Polygonal, filled with unlignified round cells.

Analyzed: Root collar of polar roots of 4 individuals.

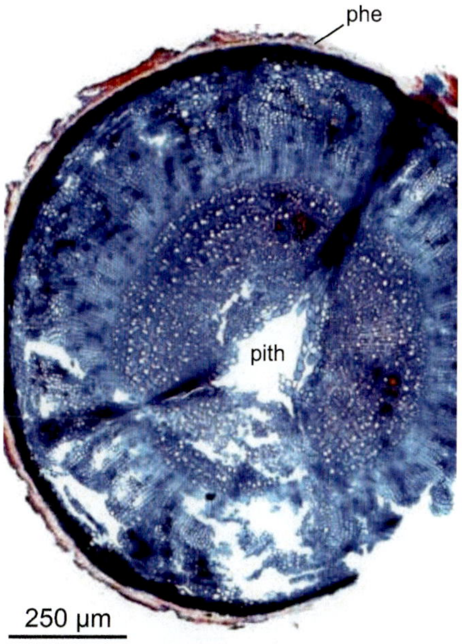

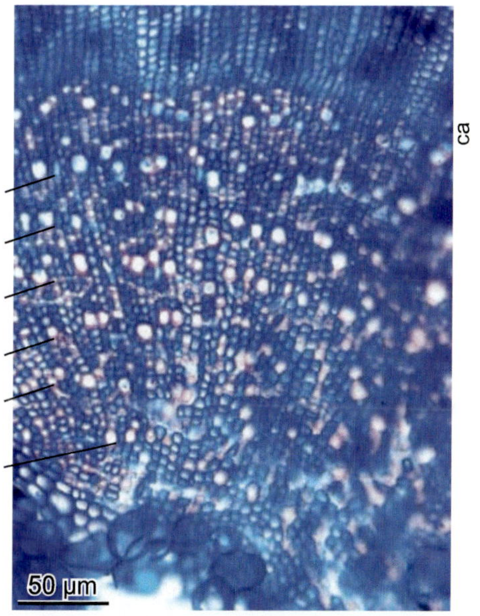

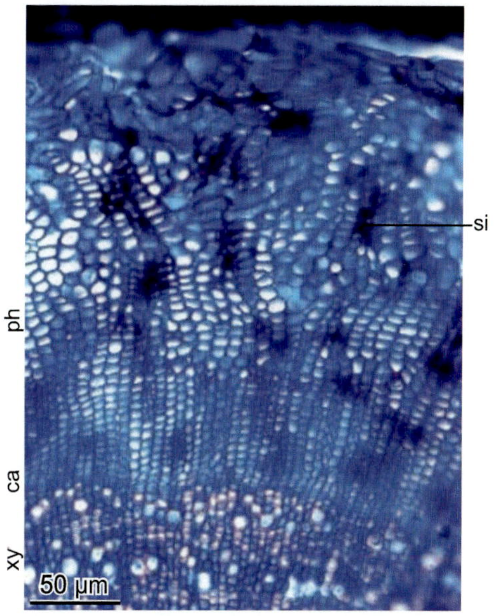

Features according to IAWA

Xylem 1 - 4 - 9 - 13 - 20 - 39.1 - 40.1 - 50.2 - 52.2 - 60.1 - 79.1 - 117

Draba himachalensis Al-Shehbaz
Brassicaceae

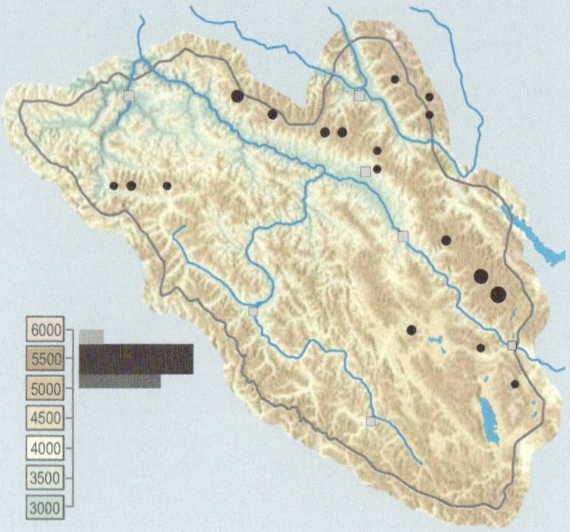

Description: Perennial herbs, 0.5–1 cm tall, cespitose, canescent tomentose, leaves in basal rosettes, spatulate to oblong, flowers white, in 2–4-flowered racemes which hardly emerge from the basal rosette, fruit orbicular, 2–3 mm across.

Distribution: India (Jammu and Kashmir, Himachal Pradesh).

Habitat: Scree slopes.

Stem diameter 1–1.5 mm. Bark/xylem proportion 1:1. Ring width 0.04–0.05 mm. Maximal 12 annual rings.

Xylem

Transverse section

Distinct rings. Semi-ring-porous. Ring variability indicated by larger earlywood (15–20 μm) and smaller latewood vessel diameters (<10 μm). Vessels solitary. Fibers absent. Parenchyma pervasive. Crystals absent. Some individuals with remaining vascular bundles.

Tangential section

Rays absent.

Radial section

Perforation plates simple. Intervessel pits scalariform. Vessel length 60–90 μm.

Bark

Phloem with small, radially arranged groups of sieve tubes. Phellem consists of many layers of square cells.

Pith

Polygonal, filled with unlignified round cells.

Analyzed: Root collar of polar roots of 4 individuals.

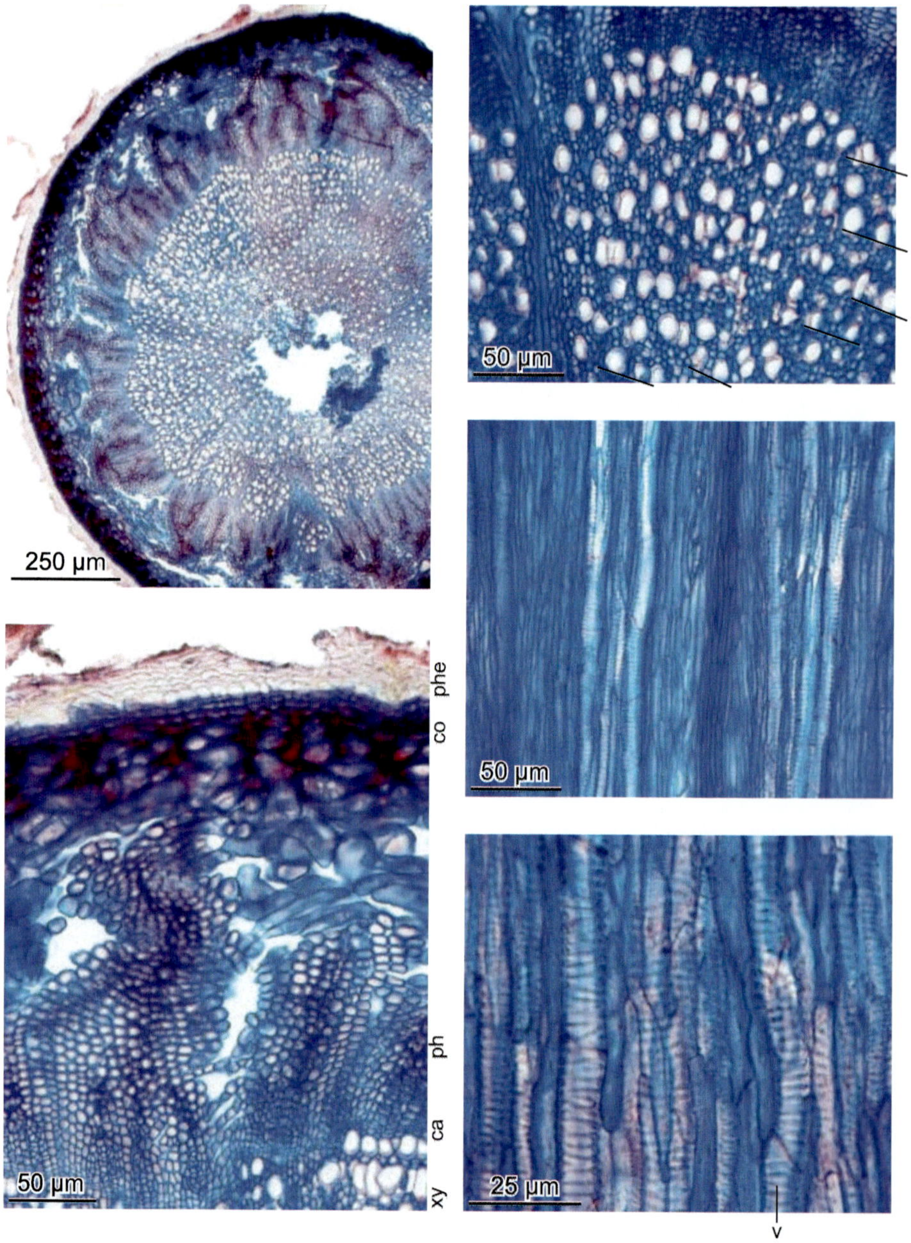

Features according to IAWA

Xylem 1 - 4 - 9 - 13 - 20 - 39.1 - 40.1 - 50.2 - 52.2 - 60.1 - 79.1 - 98 - 99.1 - 117

Draba lasiophylla Royle
Brassicaceae

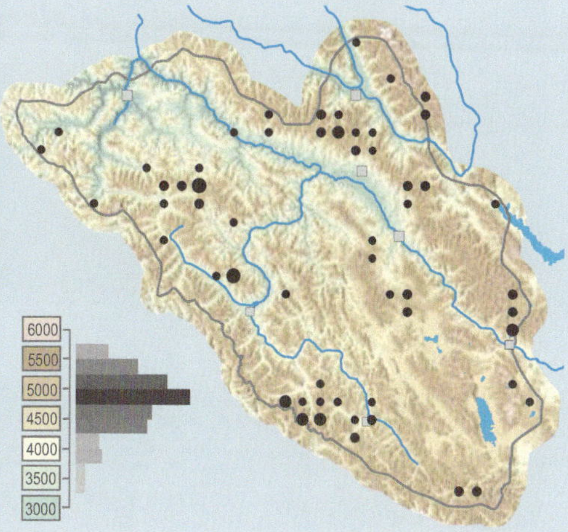

Syn.: *D. nubigena* O.E.Schulz

Description: Perennial herbs, 5–20 cm tall, cespitose, tomentose, leafy, leaves lanceolate to ovate, flowers white, 2–3 mm across, in 7–20-flowered racemes, elongating in fruit.

Distribution: Kazakhstan, Uzbekistan, Kyrgyzstan, Tajikistan, Pakistan, India, Nepal, China (Gansu, Hubei, Qinghai, Shaanxi, Sichuan, Xinjiang, Xizang), Bhutan.

Habitat: Mountain slopes, rock crevices, alpine grasslands.

Stem diameter 1–2 mm. Bark/xylem proportion 1:1. Ring width 0.05–0.1 mm. Maximal 10 annual rings.

Xylem

Transverse section

Distinct rings. Semi-ring-porous. Ring variability indicated by larger earlywood (15–25 µm) and smaller latewood vessel diameters (<10 µm). Vessels solitary. Fibers in bands in the earlywood. Parenchyma paratracheal, pervasive and marginal. With remaining vascular bundles.

Tangential section

Rays between vascular bundles 4–8-seriate.

Radial section

Perforation plates simple. Intervessel pits round and scalariform. Vessel length 50–70 µm.

Bark

Phloem with small, tangentially arranged groups of sieve tubes. Phellem consists of many layers of square cells.

Pith

Polygonal, filled with unlignified round cells.

Analyzed: Root collar of polar roots of 5 individuals.

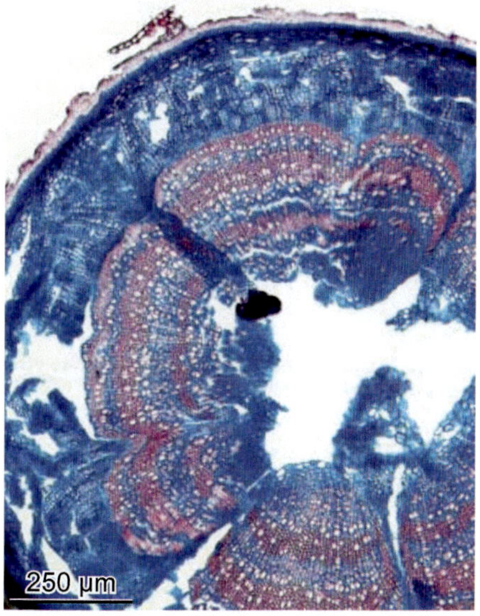

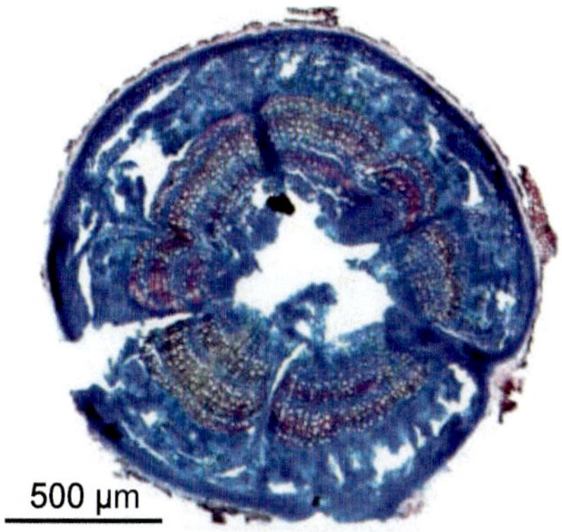

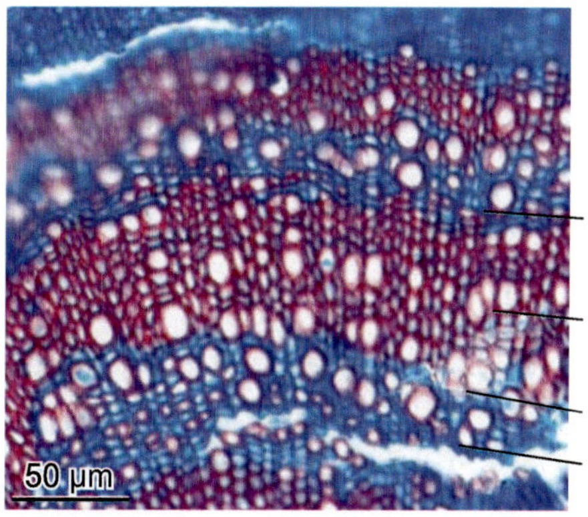

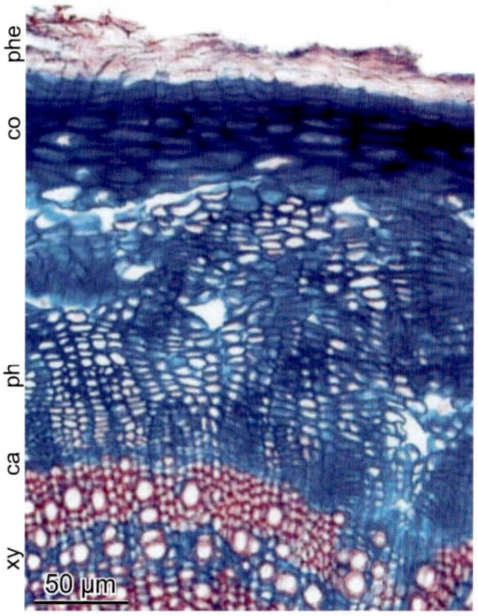

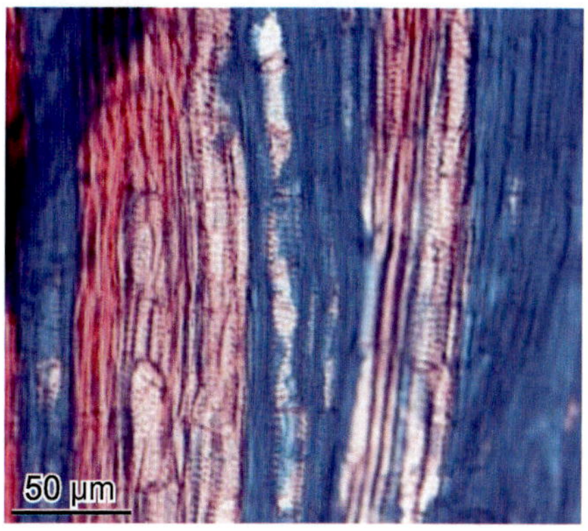

Features according to IAWA

Xylem 1 - 4 - 9 - 13 - 20 - 22 - 40.1 - 40.2 - 50.2 - 52.2 - 61 - 79 - 79.1 - 89 - 98 - 99.1 - 100.1 - 100.2 - 105

Draba oreades Schrenk

Brassicaceae

Description: Perennials, 1–15 cm tall, cespitose, leaves in rosettes, lanceolate, cauline leaves in most cases absent, flowers yellow, ca. 3 mm across, in 4–15-flowered racemes, fruits ovate, 4–9 × 1.6–4.5 mm.

Distribution: Russia, Kyrgyzstan, Kazakhstan, Tajikistan, Pakistan, India, China (Gansu, Nei Mongol, Qinghai, Shaanxi, Sichuan, Xinjiang, Xizang, Yunnan), Bhutan, Mongolia.

Habitat: Mesic to wet habitats around springs and snow beds.

Note: Probably the most variable of all Asian species in the genus.

Stem diameter 2–3 mm. Bark/xylem proportion 1:1. Mean ring width 0.08 mm. Maximal 5 annual rings.

Xylem

Transverse section

Semi-ring-porous. Rings marked by a dense row of earlywood vessels. Ring boundaries often indicated by larger (40 µm) or smaller (20 µm) vessel diameters. Vessels thick-walled (2–3 µm), lignified. Fibers absent. Parenchyma pervasive. Rays absent.

Tangential section

Rays absent.

Radial section

Perforation plates simple. Intervessel pits pseudoscalariform. Vessel length 80–150 µm.

Bark

Phloem large, with tangentially or irregularly distributed groups of sieve tubes. Cortex small, cells large (20–30 µm), round to oval. Sclerenchyma and crystals absent. Phellem consists of many (up to 10) layers of cells.

Pith

Triangular, cells large (20–30 µm), thin-walled, round. Cells widely spaced in the center. Pith and surrounding xylem cells can be filled with dark-staining substances.

Analyzed: Root collar of polar root of 8 individuals.

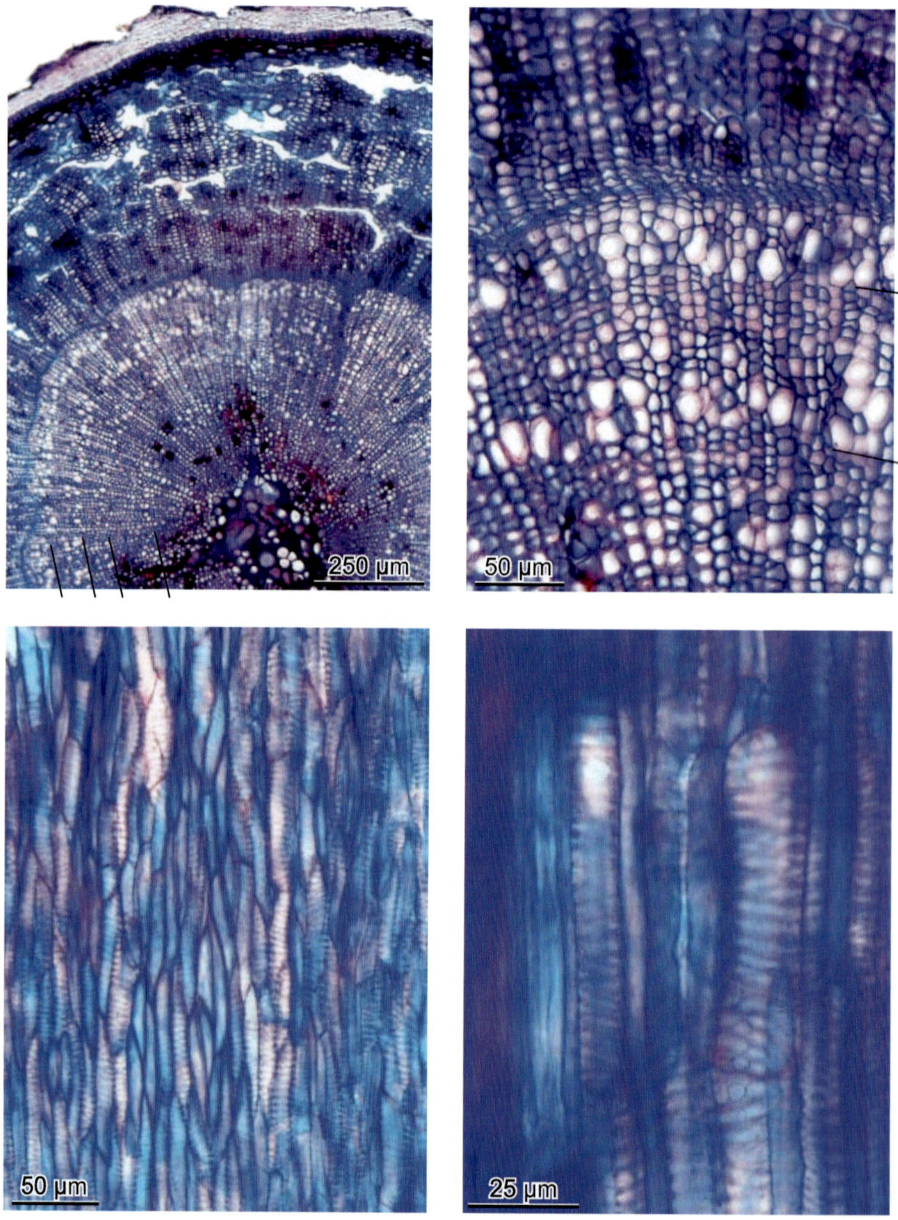

Features according to IAWA

Xylem 1 - 4 - 9 - 13 - 20.1 - 40.1 - 50.2 - 52.2 - 58 - 60.1 - 79.1 - 117

Ladakiella klimesii (Al-Shehbaz) D.A.German & Al-Shehbaz

Brassicaceae

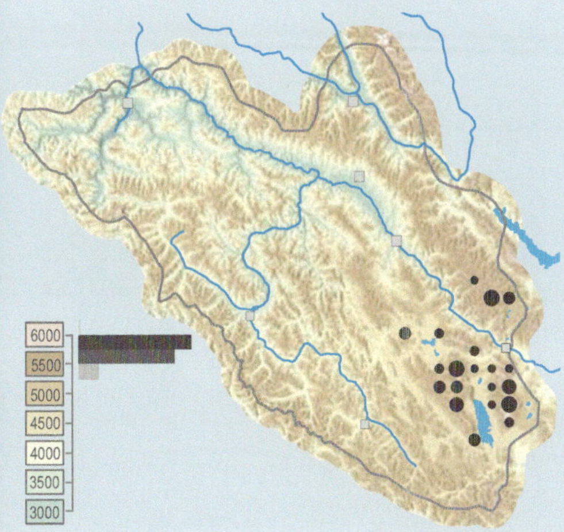

Syn.: *Alyssum klimesii* Al-Shehbaz

Description: Perennial herbs, 1–3 cm tall, densely cespitose, silvery hairy, leaves ovate, flowers pink, in 1–3-flowered racemes on very short pedicels.

Distribution: NW India (described in 2001 from Tso Moriri region), China (Xizang).

Habitat: Subnival vegetation near snowline.

Stem diameter 1–2 mm. Bark/xylem proportion 1:3. Mean ring width 0.03 mm. Maximal 22 annual rings.

Xylem

Transverse section

Semi-ring-porous. Rings mostly distinctly marked by a row of earlywood vessels. Ring boundaries indicated by larger (15 µm) or smaller (10 µm) vessel diameters. Vessels thick-walled (2–3 µm), vessel walls unstained. Fibers absent. Parenchyma pervasive. With remaining vascular bundles. Large ray-like parenchyma zones between vascular bundles disappear in polarized light. Rays absent within vascular bundles.

Tangential section

Multiseriate rays confluent with axial parenchyma cells. Ray cells upright.
See also transverse section.

Radial section

Perforation plates simple. Intervessel pits pseudoscalariform. Vessel length 40–60 µm.

Bark

Phloem large, with bent radial strips of sieve tubes between large, partially dilated parenchyma zones. Large rays and ray-less zones dilatate. Irregular, large intercellulars. Cortex small, cells large (50 µm), round to oval. Sclerenchyma and crystals absent. Phellem consists of several (up to 6) layers of square, collapsed cells.

Pith

Consists of large (40–60 µm), thin-walled, round cells. Stem center often isolated by cork-layers.

Analyzed: Root collar of polar root of 12 individuals.

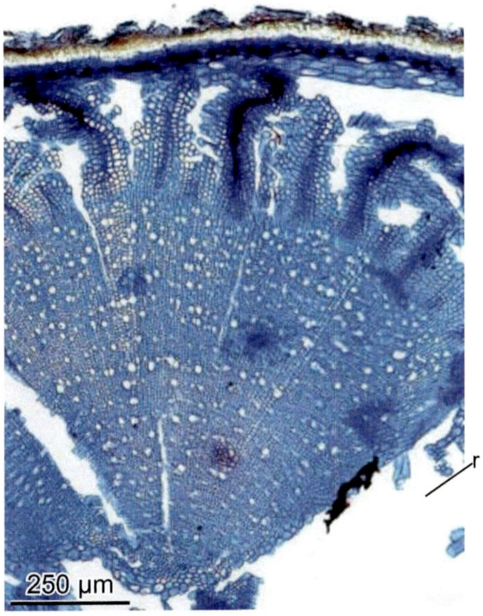

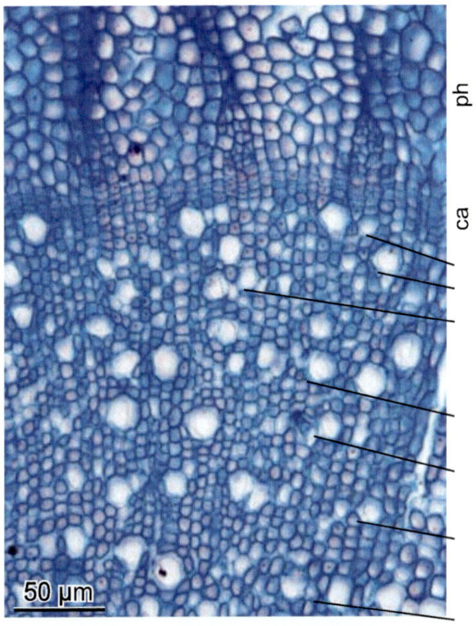

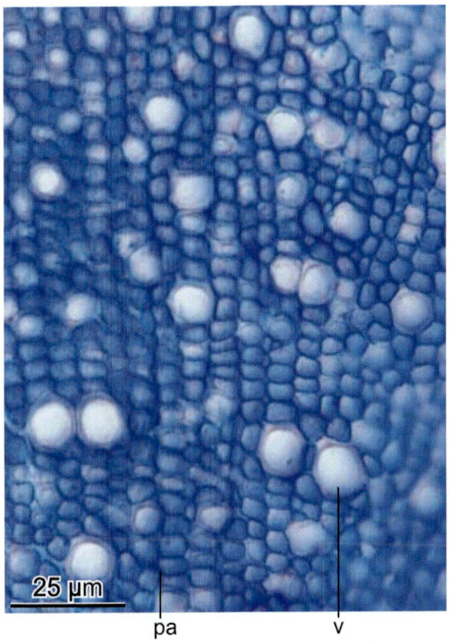

Features according to IAWA

Xylem 1 - 4 - 9 - 13 - 20.1 - 39.1 - 40.1 - 50.2 - 52.1 - 60.1 - 70.4 - 79.1 - 99 - 99.1 - 100.1 - 100.2 - 105 - 135.2

Lepidium apetalum Willd.

Brassicaeae

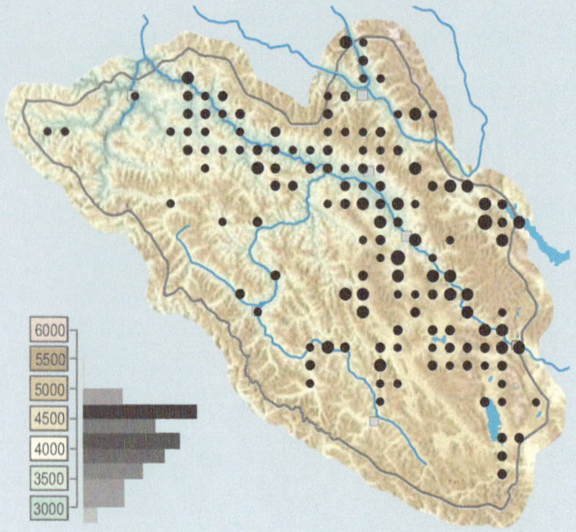

Description: Annuals or biennials, 10–30 cm tall, branched, basal and lower leaves pinnatisect, margin sinuate or dentate, upper leaves linear, entire, flowers greenish, 1 mm across, in 20–30-flowered racemes, 5–20 cm long in fruit.

Distribution: SE Kazakhstan, Tajikistan, N Pakistan, N India, Nepal, China, Mongolia, Korea, Japan.

Habitat: Arable land, disturbed habitats along roads, eutrophicated stables of domestic animals.

Stem diameter 1.5–3 mm. Bark/xylem proportion 1:3. Ring width 0.6–1.5 mm. Maximal 1 annual ring.

Xylem

Transverse section

With one ring. Vessels in long radial multiples (2–6) with a diameter of 20–30 µm. Vessels solitary or in short radial rows and groups. Fibers thin- and thin- to thick-walled, in tangential zones. Parenchyma paratracheal. With intra-xylary groups of sieve elements.

Tangential section

Rays absent.

Radial section

Perforation plates simple. Intervessel pits small and round, in alternating position. Vessel length 130–180 µm.

Bark

Adult phloem radially and tangentially structured. Sieve tubes in radial and round groups. Fibers in small groups within the cortex. Parenchyma dilated. Phellem consists of few layers of thin-walled, irregularly formed cells.

Pith

Round to polygonal, filled with unlignified, thin-walled cells.

Analyzed: Root collar of polar roots of 2 individuals.

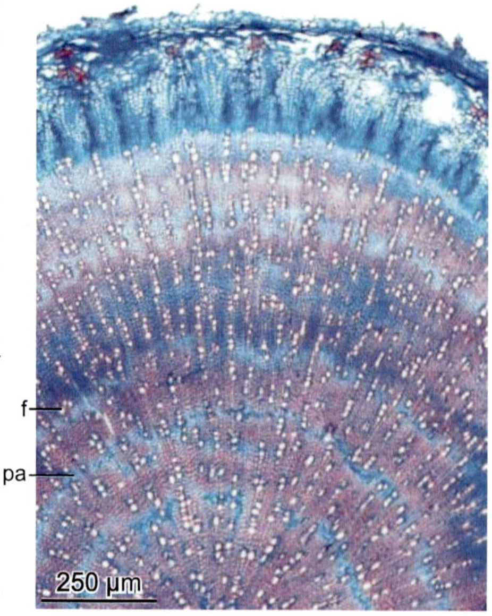

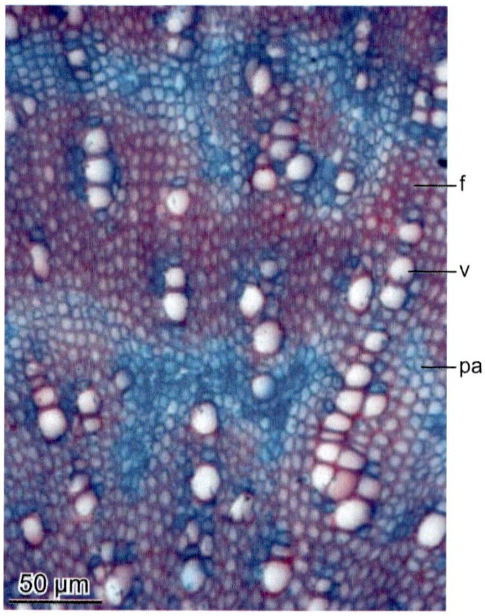

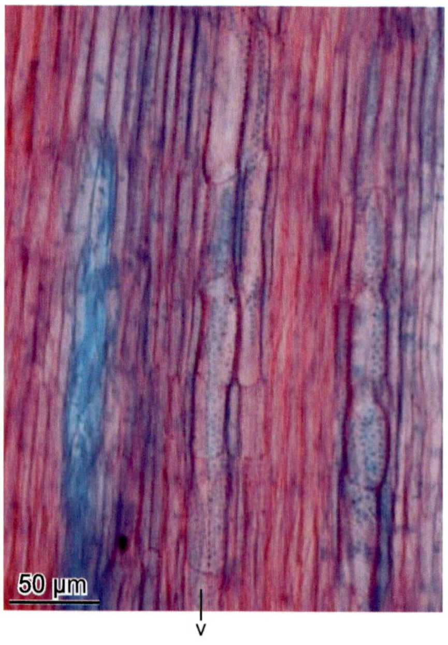

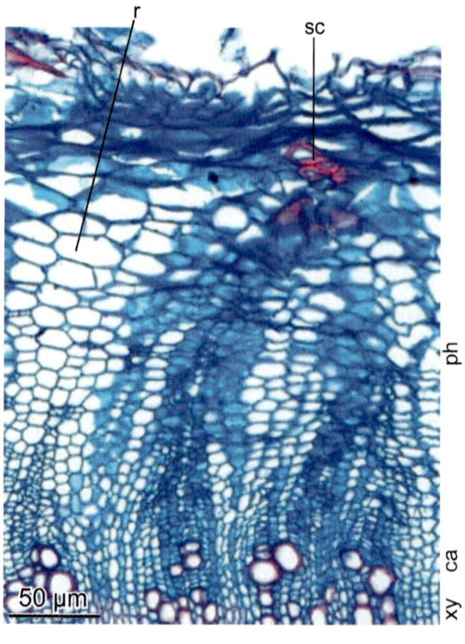

Features according to IAWA

Xylem 2.1 - 9.1 - 10 - 13 - 22 - 40.2 - 50.2 - 52.3 - 61 - 67 - 70 - 79 - 117 - 135

Lepidium capitatum Hook.f. & Thomson

Brassicaeae

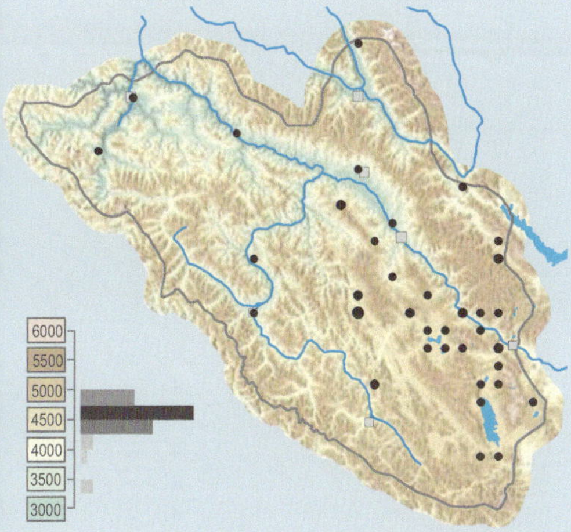

Syn.: *L. kunlunshanicum* G.L.Zhou & Z.X.An

Description: Annuals or biennials, 5–30 cm tall, hairy, stems prostrate, leaves oblong, 1–2-pinnatifid, flowers white, ca. 2 mm across, in capitate racemes.

Distribution: Pakistan, NW India, Nepal, China (Gansu, Qinghai, Sichuan, Xinjiang, Xizang, Yunnan), Bhutan.

Habitat: Arable land, disturbed habitats along roads, eutrophicated stables of domestic animals, mountain slopes.

Stem diameter up to 16 mm. Bark/xylem proportion 1:1.5. Ring width 0.4–0.6 mm. Maximal 14 annual rings.

Xylem

Transverse section

Semi-ring-porous. Ring boundaries marked by a zone of marginal parenchyma, containing small vessels with a diameter of 20–40 µm and large zone of earlywood vessels of a diameter of 80–100 µm. Earlywood vessels arranged solitary or in short radial rows and groups. Vessels thick-walled (2–3 µm), lignified. Fibers thick-walled. Parenchyma paratracheal in the earlywood and pervasive in the latewood. Rays disappear in polarized light. Dark-staining substances occur in the latewood of some individuals.

Tangential section

Four to 6-seriate rays consisting of thin-walled, non-lignified cells, surrounded by distinctly larger cells (sheath cells).

Radial section

Perforation plates simple. Intervessel pits small and round in alternating position. Vessel length 130–200 µm.

Bark

Adult phloem radially and tangentially structured. Sieve tubes in radial and round groups of thick-walled sclereids in tangential lines. Rays dilated. Phellem consists of many (up to 30) layers of thin-walled, regularly formed rectangular cells.

Pith

Pith in young stems round, containing unlignified, thin-walled cells.

Analyzed: Root collar of polar root of 3 individuals.

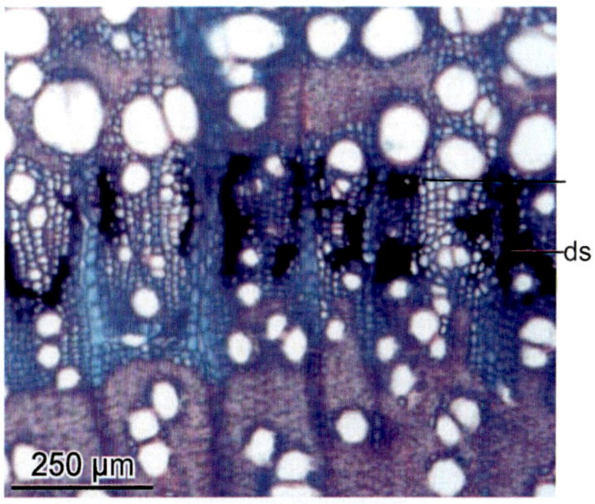

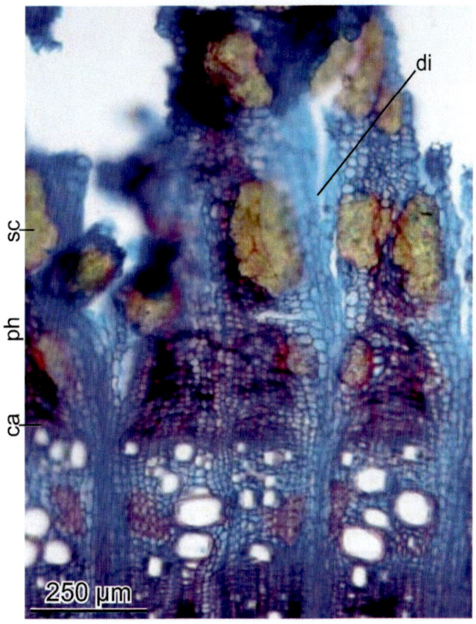

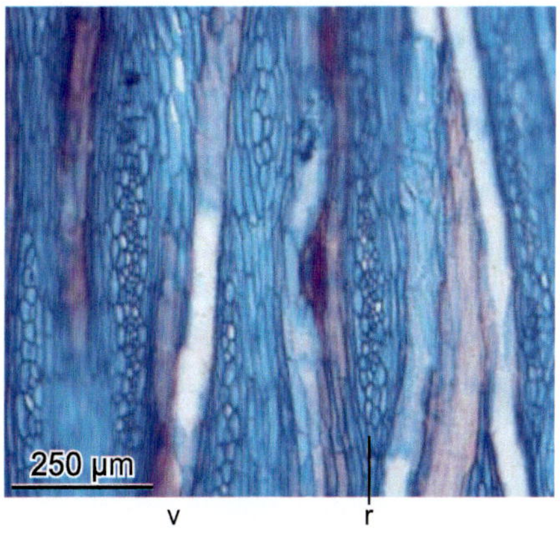

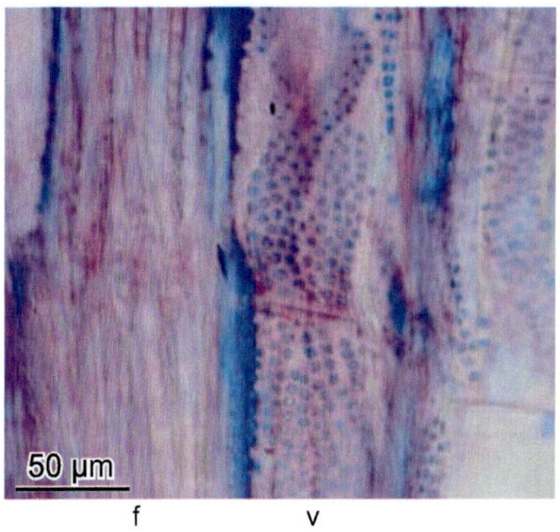

Features according to IAWA

Xylem 1 - 4 - 9 - 9.1 - 13 - 22 - 39.1 - 41 - 50.1 - 52.3 - 61 - 69 - 79 - 79.1 - 98 - 100.2 - 105 - 110

Lepidium latifolium L.

Brassicaeae

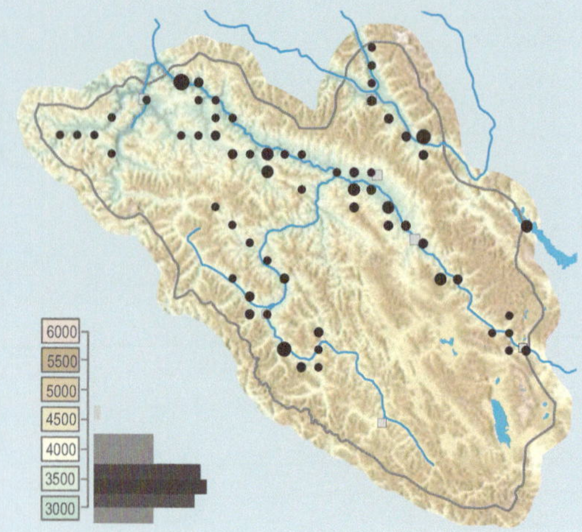

Syn.: *Cardaria latifolia* (L.) Spach, *L. obtusum* Basiner, *L. sibiricum* Schweigg.

Description: Perennial herbs, 20–100 cm tall, often branched, leaves ovate-lanceolate, margin entire to serrate, flowers white, 2.5 mm across, in 25–50-flowered corymbose racemes.

Distribution: N Africa, Europe, W Asia, Caucasus, Russia (Siberia), Iran, Turkey, Kazakhstan, Turkmenistan, Uzbekistan, Kyrgyzstan, Tajikistan, Afghanistan, N Pakistan, India (Jammu and Kashmir, Himachal Pradesh), China, Mongolia; naturalized elsewhere.

Habitat: Field margins, dumps and abandoned places in settlements.

Stem diameter up to 16 mm. Bark/xylem proportion 1:2. Ring width 0.2–0.9 mm. Maximal 16 annual rings.

Xylem

Transverse section

Semi-ring-porous to ring-porous. Ring boundaries marked by a zone of marginal parenchyma, containing small vessels of 20–40 μm diameter, and a large zone of earlywood with vessels of 60–80 μm diameter. Earlywood vessels arranged solitary or in short radial rows and groups. Vessels thick-walled (2–3 μm), lignified. Fibers thick-walled, in bands in large, and in groups in small rings. Parenchyma paratracheal in earlywood and pervasive in latewood. Rays disappear in polarized light. Dark-staining substances in the latewood of some individuals.

Tangential section

Four to 6-seriate rays, consisting of thin-walled, unlignified cells, surrounded by sheath cells.

Radial section

Perforation plates simple. Intervessel pits scalariform in marginal parenchyma zones, and small and round in alternating position in fiber zones. Vessel length 150–200 μm.

Bark

Adult phloem radially and tangentially structured. Sieve tubes and thick-walled sclereids in radial and tangential round groups. Rays dilated. Phellem consists of many (up to 20) layers of thin-walled, regularly formed rectangular cells.

Pith

Pith in young stems round, containing unlignified, thin-walled cells.

Analyzed: Root collar of polar root of 4 individuals.

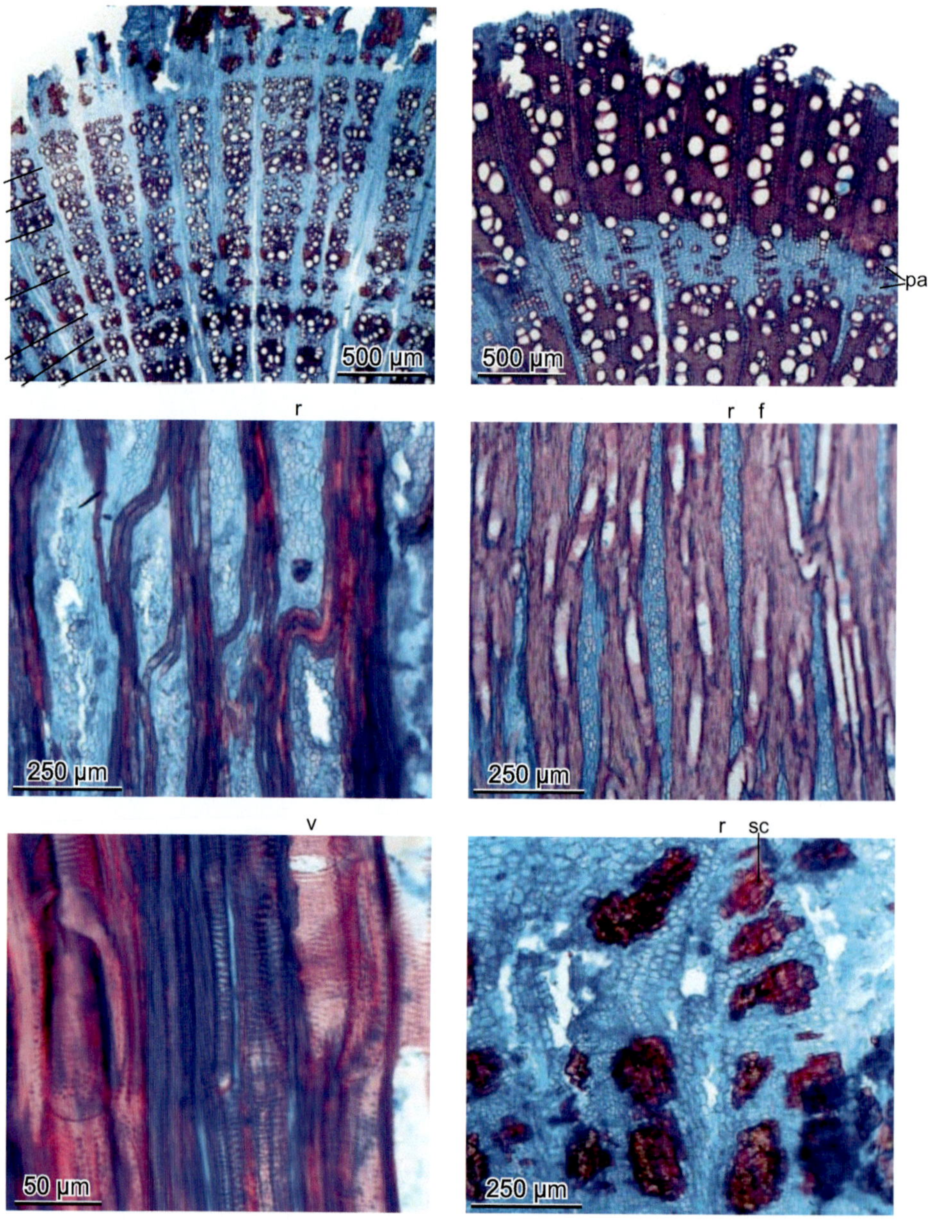

Features according to IAWA

Xylem 1 - 3 - 4 - 9 - 9.1 - 13 - 22 - 39.1 - 41 - 50.1 - 52.3 - 61 - 69 - 79 - 79.1 - 98 - 100.2 - 105 - 110

Matthiola chorassanica Bunge ex Boiss.

Brassicaceae

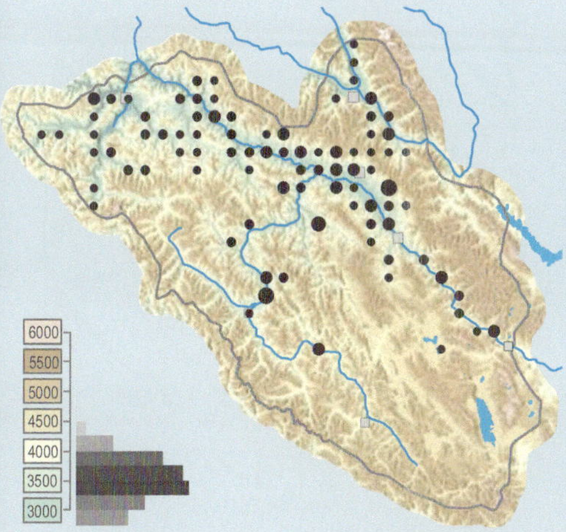

Syn.: *M. flavida* var. *integrifolia* (Kom.) O.E.Schulz, *M. integrifolia* Kom., *M. tenera* Rech.f.

Description: Perennial herbs, 16–50 cm tall, tomentose, stems erect, branched, leaves ovate to elliptic, margin entire or dentate, flowers greenish, purplish or brownish, ca. 1.5 cm across, petals circinately involute, fruits narrowly linear, up to 10 cm long.

Distribution: SW Asia, Uzbekistan, Tajikistan, Afghanistan, Pakistan, N India, China (Xinjiang, Xizang).

Habitat: Dry rocky slopes.

Stem diameter 5–7 mm. Bark/xylem proportion 1:4. Ring width 0.3–0.8 mm. Maximal 10 annual rings.

Xylem

Transverse section

Semi-ring-porous. Rings, if present, marked by an interrupted row of earlywood vessels. Ring boundaries often indicated by larger (70 µm) or smaller (50 µm) vessel diameters and a latewood zone with small vessels (20–30 µm). Vessels thick-walled (2–3 µm), vessel walls not stained. Fibers thin- to thick-walled, arranged in partial tangential bands. Parenchyma pervasive and paratracheal within fiber zones.

Tangential section

Ray width 5–8 cells, confluent with the axial tissue. Rays disappear in polarized light.

Radial section

Perforation plates simple. Intervessel pits round, in alternating position. Vessel length 60–120 µm.

Bark

Phloem large, with indistinct, tangentially arranged groups of sieve tubes. Small groups of thick-walled fibers. Indistinct ray dilatations and irregular, large intercellulars. Cortex small, cells small (10–20 µm), round to oval. Crystals absent. Phellem consists of many rows of flat cells.

Pith

Round, filled with thin-walled, round cells with 40–60 µm diameter.

Analyzed: Root collar of polar root of 10 individuals.

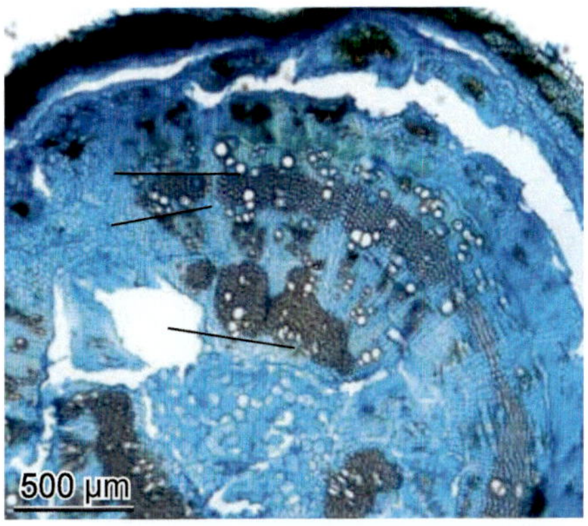

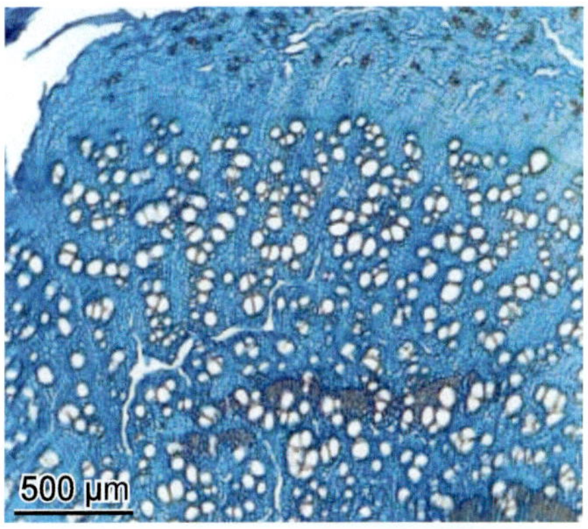

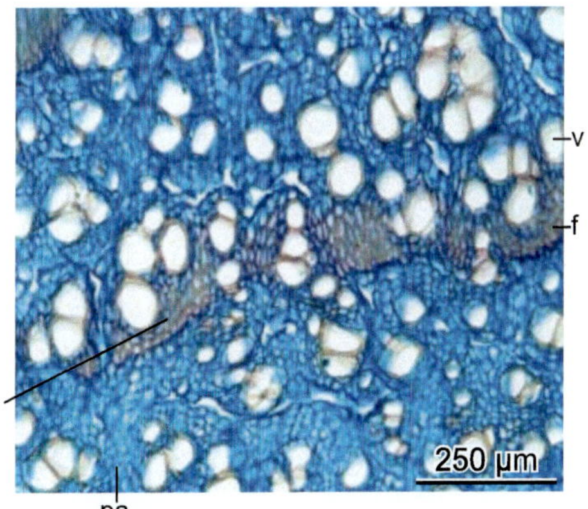

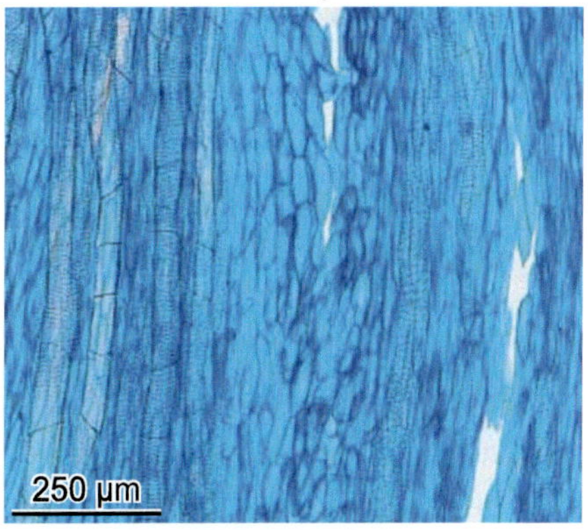

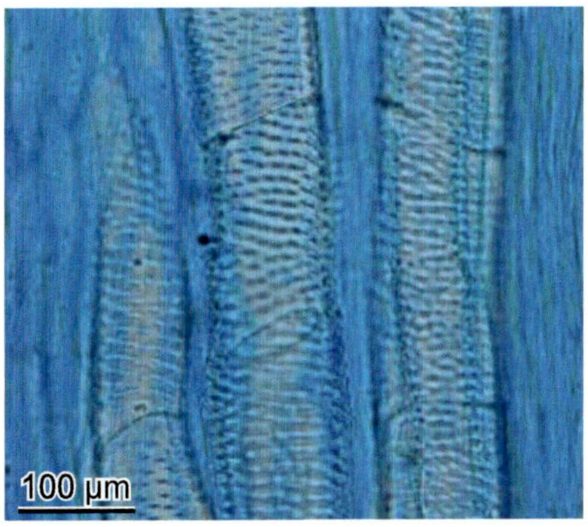

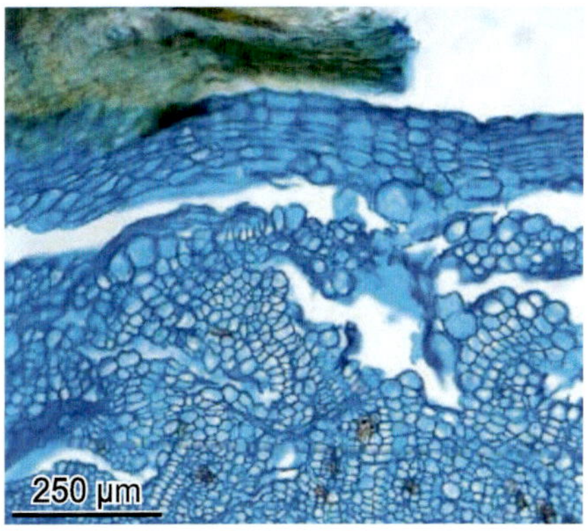

Features according to IAWA

Xylem *1 - 4 - 9 - 9.1 - 13 - 22 - 40.1 - 50.2 - 52.2 - 61 - 70 - 79.1 - 98 - 100.1 - 100.2 - 105*

Neotorularia humilis (C.A.Mey.) Hedge & J.Léonard
Brassicaceae

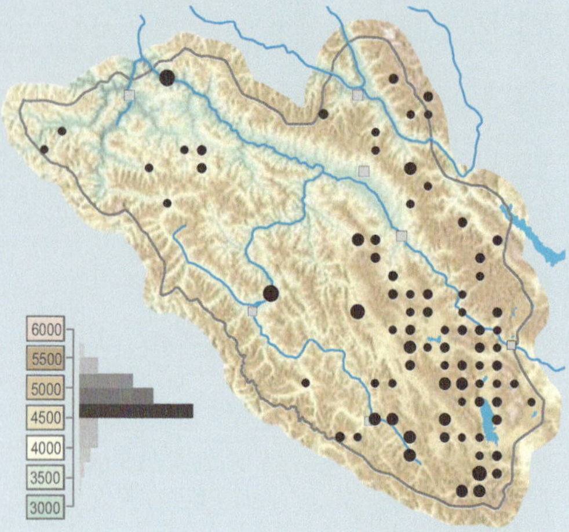

Syn.: *Sisymbrium humile* C.A.Mey., *Torularia humilis* (C.A.Mey.) O.E.Schulz, *Braya humilis* (Ledeb.) B.L.Rob.

Description: Perennial herbs, 5–25 cm tall, pubescent, leaves obovate to oblanceolate, margins sinuate-dentate, apex acute, flowers white, pink or purple, 0.5–1 cm across, arranged in terminal racemes.

Distribution: N America (Canada, USA, Greenland), Russia (Siberia), India (Jammu and Kashmir), China (Gansu, Hebei, Henan, Nei Mongol, Qinghai, Shaanxi, Xinjiang, Xizang).

Habitat: Alpine grasslands, river beds.

Stem diameter 2–5 mm. Bark/xylem proportion 1:2. Ring width 0.4–0.6 mm. Maximal 3 annual rings.

Xylem

Transverse section

Semi-ring-porous. Rings distinct. Vessels solitary or in short radial rows. Earlywood vessel diameter 30–45 µm. Fibers thin- to thick-walled in a peripheral lobed belt around a fiber-less center. Parenchyma paratracheal in the fiber belt and pervasive in the center. Rays absent. Vascular bundles remaining.

Tangential section

Rays >10-seriate, confluent, not lignified.

Radial section

Perforation plates simple. Intervessel pits round in alternating position in the fiber belt, scalariform in the fiber-less center. Vessel length 80–100 µm.

Bark

Phloem large, with tangentially arranged groups of sieve tubes. Cortex cells irregularly formed. Sclerenchyma and crystals absent. Phellem consists of layers of round cells.

Pith

No slide available.

Analyzed: Root collar of polar root of 2 individuals.

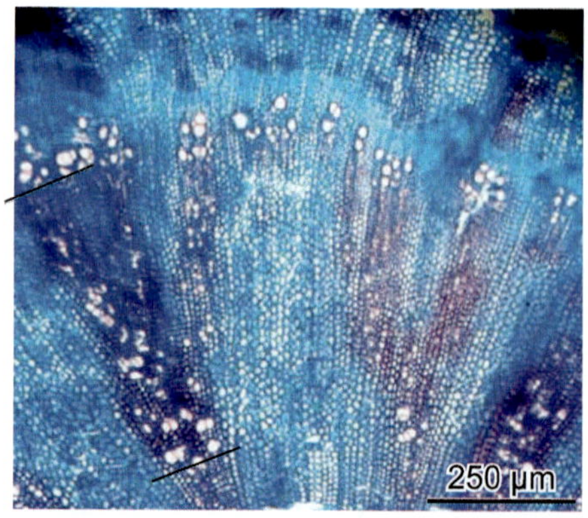

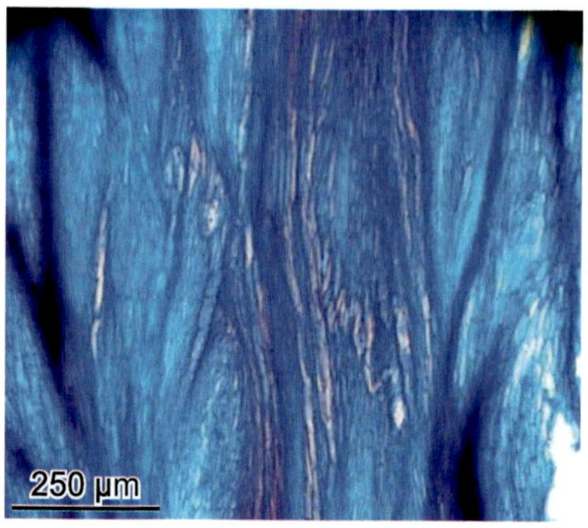

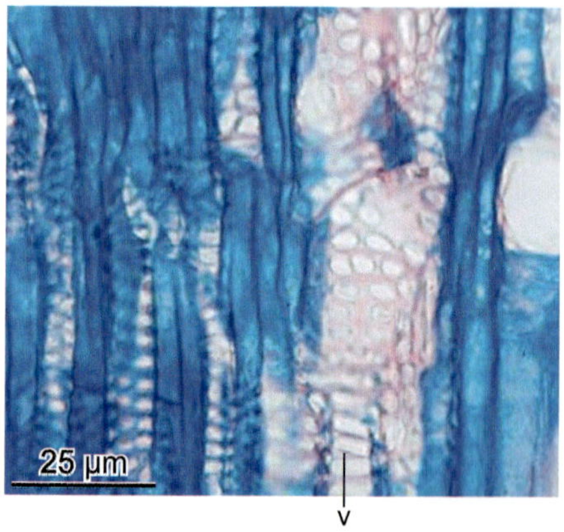

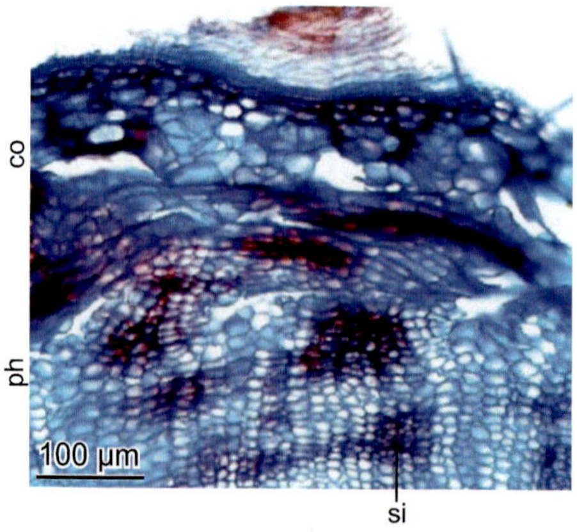

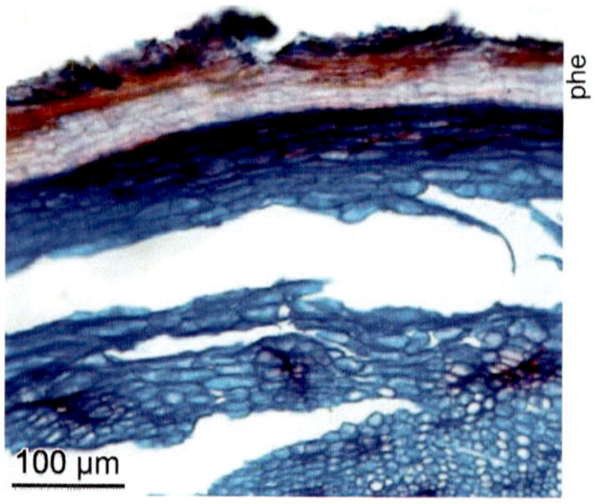

Features according to IAWA

Xylem 1 - 4 - 9.1 - 13 - 20 - 22 - 40.2 - 50.2 - 52.2 - 53.1 - 61 - 70 - 70.3 - 79 - 79.1 - 99 - 99.1 - 100.1 - 100.2 - 105

Parrya nudicaulis (L.) Boiss.

Brassicaceae

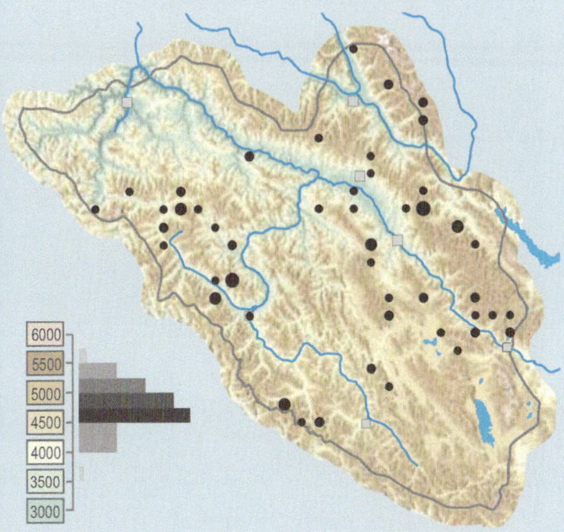

Syn.: *Cardamine nudicaulis* L., *Achoriphragma nudicaule* (L.) Soják, *Arabis nudicaulis* (L.) DC.

Description: Perennial herbs, 10–35 cm tall, cespitose, leaves narrowly spatulate, oblanceolate to lanceolate, flowers purple, ca. 1 cm across, in 3–20-flowered racemes.

Distribution: N America, Russia (Siberia, Far East), Afghanistan, Pakistan, N India, China (Qinghai, Xizang), Bhutan.

Habitat: Dry stony hillsides, stream valleys.

Stem diameter 5–10 mm. Bark/xylem proportion 1:3. Ring width 0.1–0.2 mm. Maximal 20 annual rings.

Xylem

Transverse section

Semi-ring-porous. Ring boundaries marked by groups of large earlywood vessels. Ring variability often indicated by larger (50 µm) or smaller (20 µm) vessel diameters. Vessels thick-walled (2–3 µm), lignified or unlignified. Fibers absent. Parenchyma pervasive. With remaining vascular bundles. Large ray-like parenchyma zones between vascular bundles disappear in polarized light. Rays absent within vascular bundles.

Tangential section

Multiseriate rays, confluent with axial parenchyma cells. Ray cells upright.

Radial section

Perforation plates simple. Intervessel pits scalariform to round, in alternating position. Vessel length 60–100 µm.

Bark

Phloem large, with radially or irregularly distributed groups of sieve tubes. Cortex small, cells large (15–20 µm), round to oval. Sclerenchyma and crystals absent. Phellem consists of many (up to 30) layers of collapsed cells.

Pith

Irregularly formed, consisting of large (30–40 µm), thin-walled, round cells. Some individuals contain groups of sclereids. Stem center often isolated by cork-layers.

Analyzed: Root collar of polar root of 6 individuals.

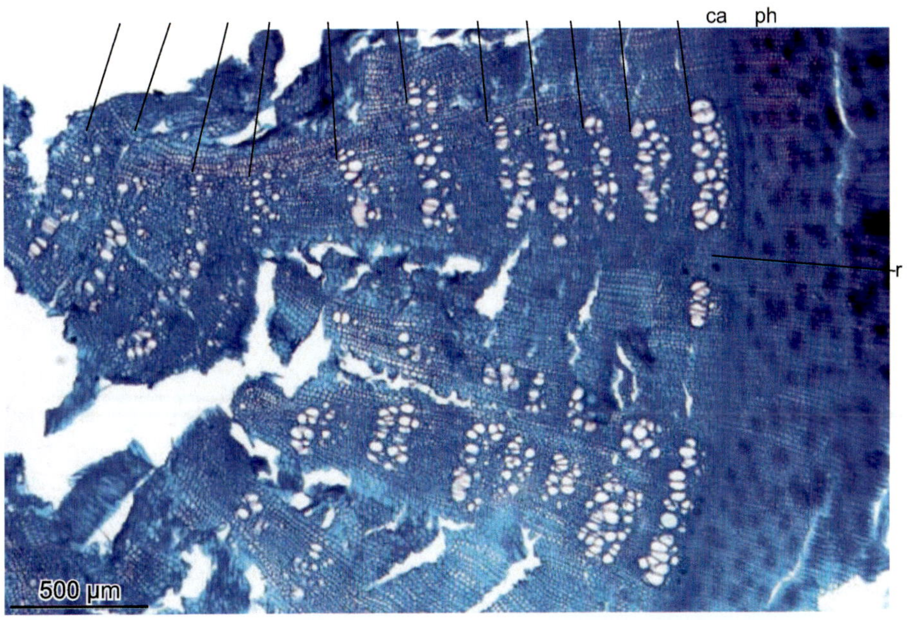

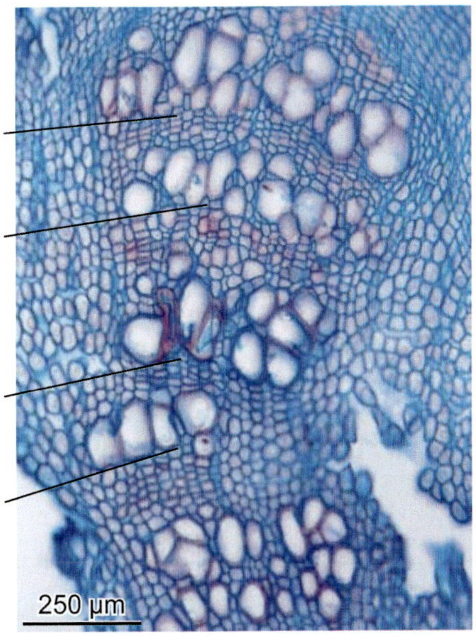

Features according to IAWA

Xylem 1 - 4 - 11 - 13 - 20 - 22 - 39.1 - 40.2 - 50.2 - 52.1 - 60.1 - 79.1 - 98 - 99.1 - 100.1 - 100.2 - 105 - 135.2

Pegaeophyton scapiflorum (Hook.f. & Thomson) C.Marquand & Airy Shaw

Brassicaceae

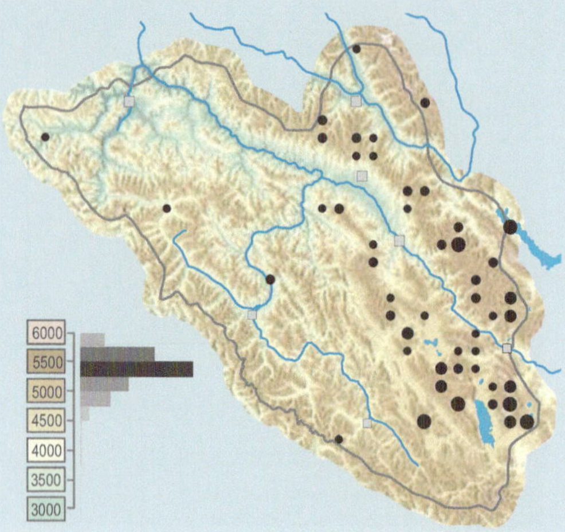

Syn.: *Cochlearia scapiflora* Hook.f. & Thomson

Description: Perennial herbs, 2–8 cm tall, stemless, leaves in basal rosette, oblong, spatulate, elliptic or linear, fleshy, flowers white, pinkish or bluish, 4–7 mm across, solitary.

Distribution: NE Pakistan, N India, Nepal, China (Gansu, Qinghai, Sichuan, Xinjiang, Xizang, Yunnan), Bhutan, N Myanmar.

Habitat: Permanently wet solifluction soils with open vegetation cover, fresh-water spring areas, stream edges.

Stem diameter 4–5 mm. Bark/xylem proportion 1:1. Ring width 0.04–0.1 mm. Maximal 8 annual rings.

Xylem

Transverse section

Semi-ring-porous. Ring boundaries marked by groups of large earlywood vessels. Ring variability often indicated by larger (60 µm) or smaller (30 µm) vessel diameters. Vessels both red-stained or unstained. Fibers absent. Parenchyma pervasive. With remaining vascular bundles. Large ray-like parenchyma zones between vascular bundles disappear in polarized light. Rays absent within vascular bundles.

Tangential section

Multiseriate rays, confluent with axial parenchyma cells. Ray cells upright.

Radial section

Perforation plates simple. Intervessel pits scalariform. Vessel length 100–150 µm.

Bark

Phloem with irregularly distributed small groups of sieve tubes. Cortex large, cells large (30–40 µm), round and irregularly formed. Sclerenchyma and crystals absent. Phellem consists of many (up to 30) layers of collapsed cells.

Pith

Irregularly formed, consisting of large (30–40µm), thin-walled, round cells. Some individuals contain groups of sclereids. Stem center often isolated by cork-layers.

Analyzed: Root collar of polar root of 5 individuals.

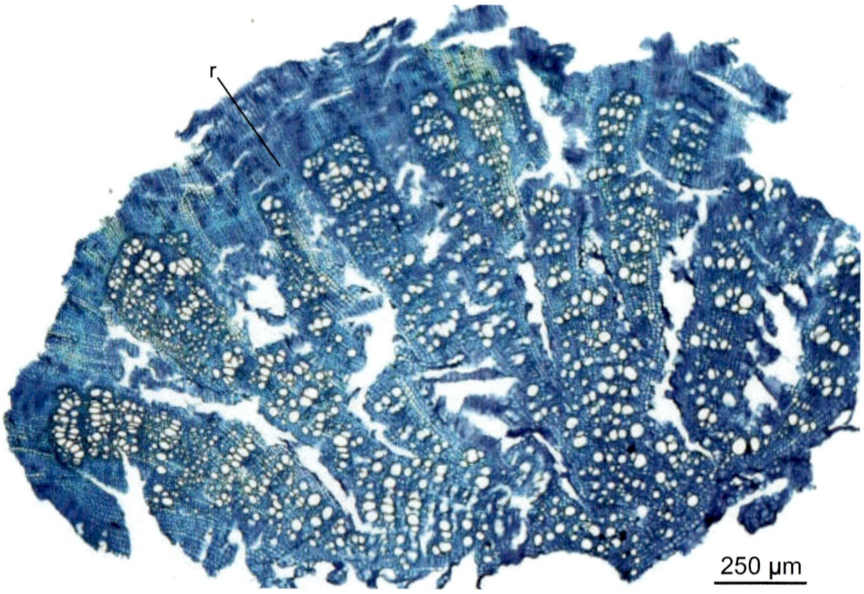

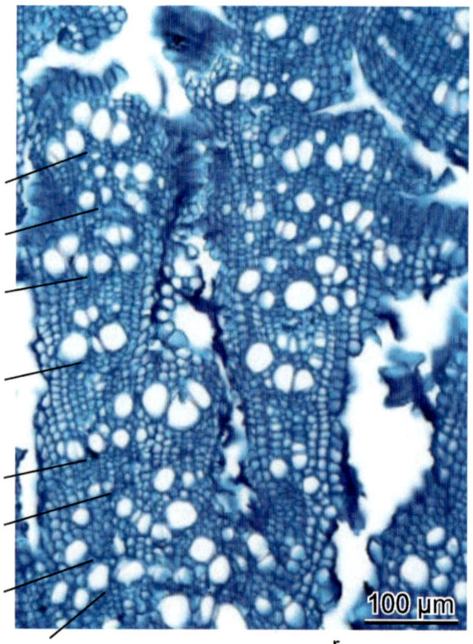

r

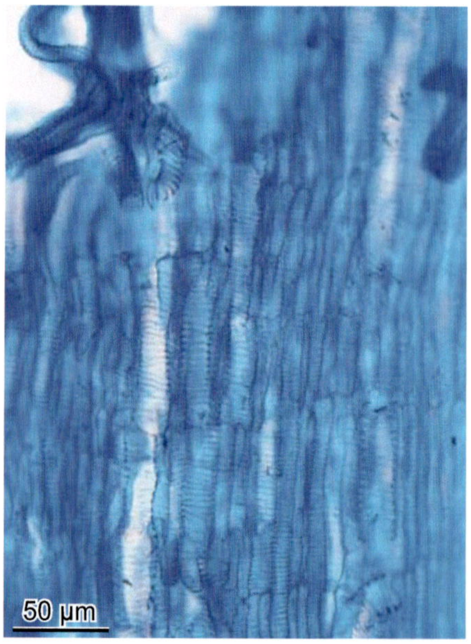

v

Features according to IAWA

Xylem 1 - 4 - 11 - 13 - 20 - 40.2 - 50.2 - 52.3 - 60.1 - 70.4 - 79.1 - 98 - 99.1 - 100.1 - 100.2 - 105

Sisymbrium brassiciforme C.A.Mey.

Brassicaceae

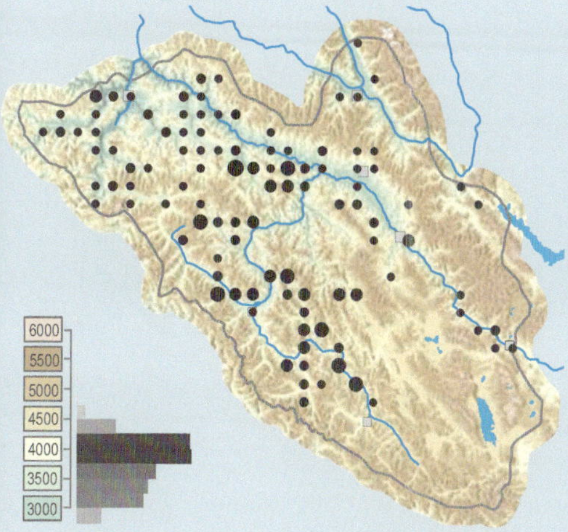

Syn.: *S. ferganense* Korsh., *S. iscandericum* Kom.

Description: Annuals, 30–90 cm tall, stems branched, leaves crowded basally, broadly oblanceolate, lyrate, margin dentate, stem leaves narrowly lanceolate, flowers yellow, ca. 5 mm across, in 20–50-flowered racemes, up to 25 cm long in fruit.

Distribution: Russia, Kazakhstan, Turkmenistan, Uzbekistan, Kyrgyzstan, Tajikistan, Afghanistan, Pakistan, India, Nepal, China (Xinjiang, Xizang), Mongolia.

Habitat: Disturbed habitats along roads, screes and stabilized loamy slopes.

Stem diameter 6–7 mm. Bark/xylem proportion 1:5. Ring width 3–4 mm. Maximal 1 annual ring.

Xylem

Transverse section

With one ring. Vessels in short radial multiples (2–4) with a diameter of 50–100 µm, arranged in tangential rows. Fibers thin- to thick-walled. Parenchyma paratracheal.

Tangential section

Rays 3–6-seriate.

Radial section

Perforation plates simple. Intervessel pits small and round in alternating position. Vessel length 70–100 µm.

Bark

Phloem consists of radial groups of sieve elements and dilated rays. Fibers and sclerenchyma cells in irregular groups in the cortex. Phellem consists of few layers of thin-walled, irregularly formed cells.

Pith

Round, filled with thin-walled cells.

Analyzed: Root collar of polar roots of 2 individuals.

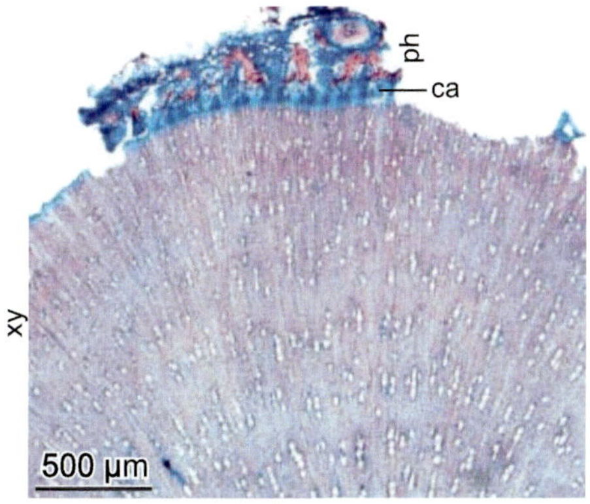

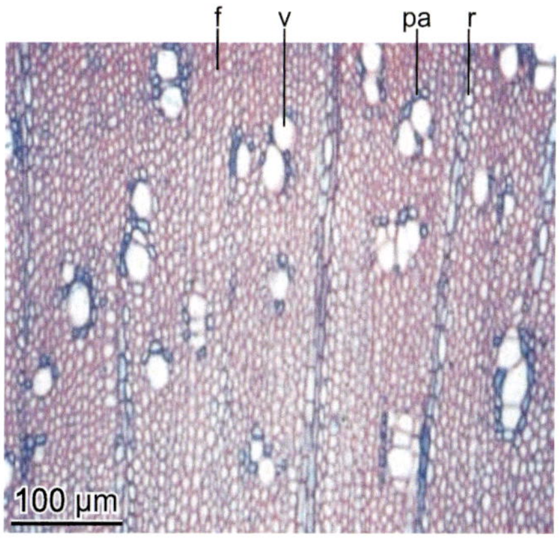

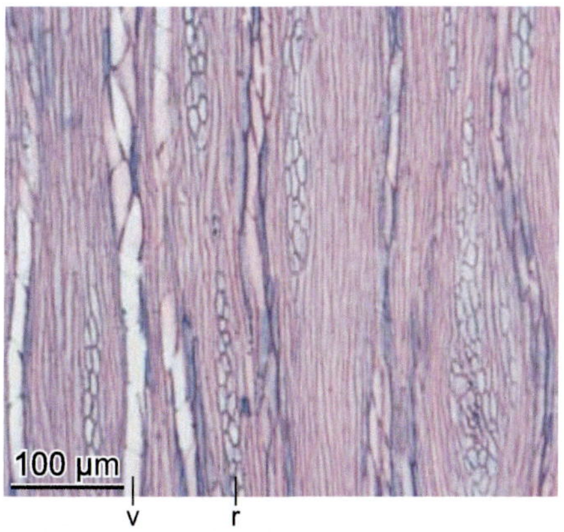

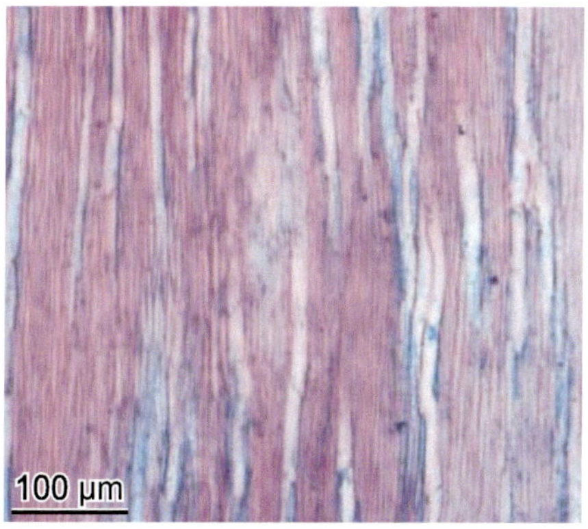

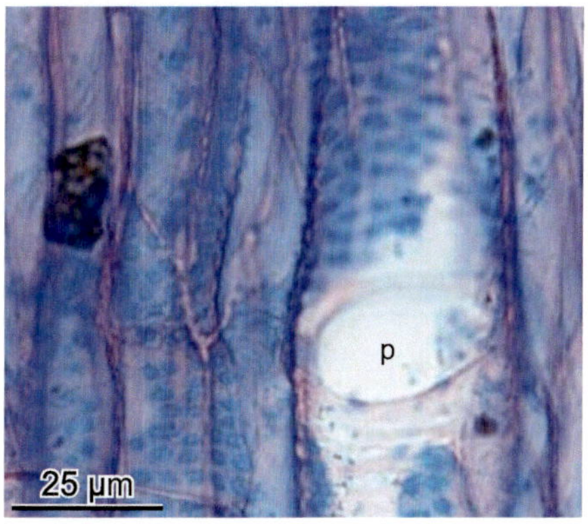

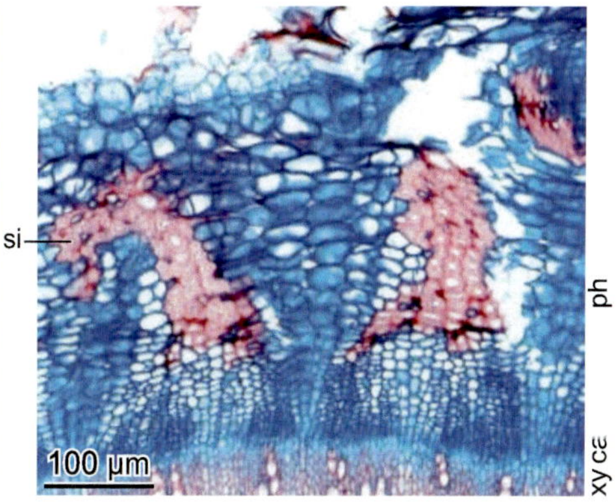

Features according to IAWA

Xylem 2.1 - 6 - 9.1 - 13 - 22 - 41 - 50.2 - 52.2 - 61 - 70 - 79 - 98 - 105

Tetracme pamirica Vassilcz.

Brassicaceae

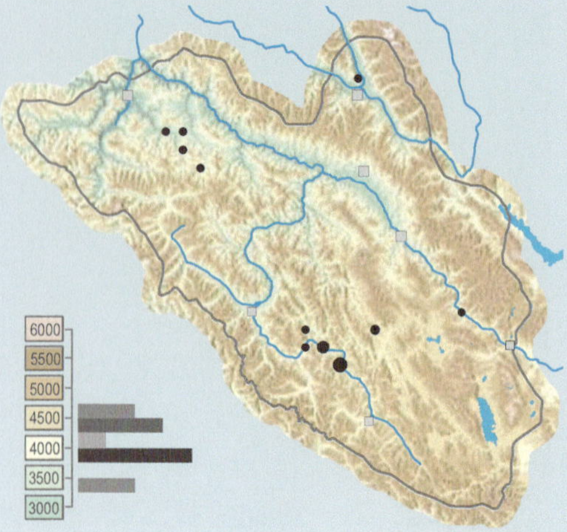

Syn.: *T. adpressa* Rech.f.

Description: Annuals, 10–30 cm tall, densely hairy, stems branched from the base, leaves oblong, flowers white, ca. 1.5 mm across, racemes elongated in fruit.

Distribution: Afghanistan, Pakistan, NW India.

Habitat: Stabilized dry stony slopes.

Stem diameter 0.8–1.6 mm. Ring width 0.25–0.7 mm. Maximal 1 annual ring.

Xylem

Transverse section

With one ring. Vessels in long radial multiples (4–8) and small groups. Vessel diameter 20–30 µm. Fibers thin- to thick-walled. Parenchyma paratracheal.

Tangential section

Rays absent.

Radial section

Perforation plates simple. Intervessel pits small and round, in alternating position. Vessel length 70–100 µm.

Bark

No slide available.

Pith

No slide available.

Analyzed: Root collar of polar roots of 3 individuals.

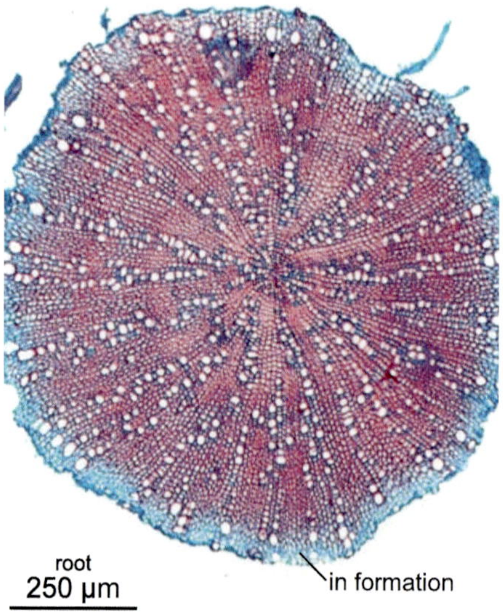

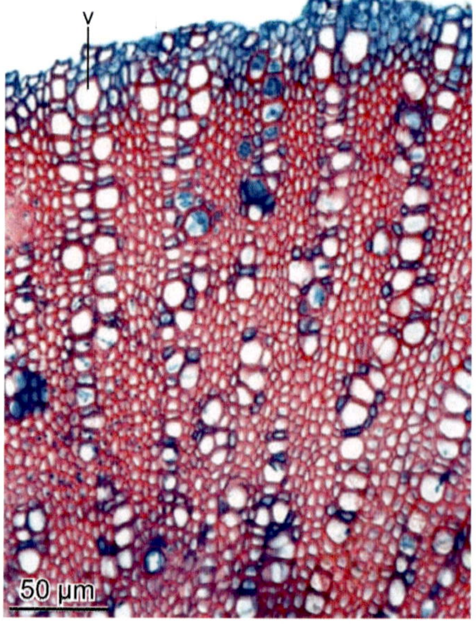

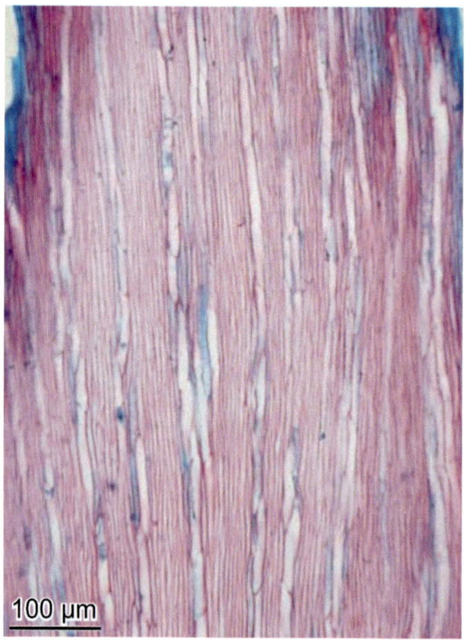

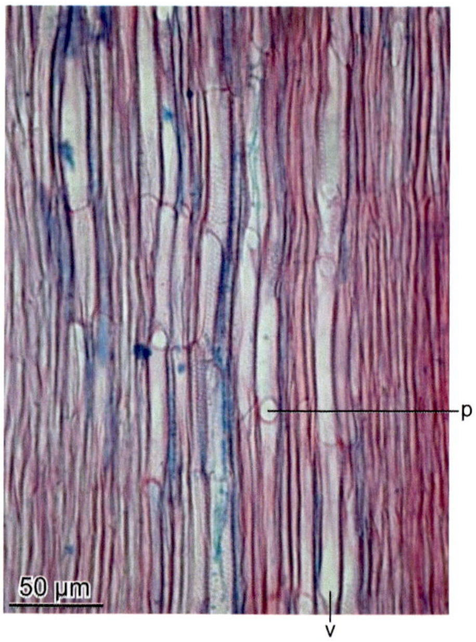

Features according to IAWA

Xylem 2.1 - 10 - 13 - 22 - 40.2 - 50.2 - 52.3 - 61 - 70 - 79 - 117

Campanula alsinoides Hook.f. & Thomson

Campanulaceae

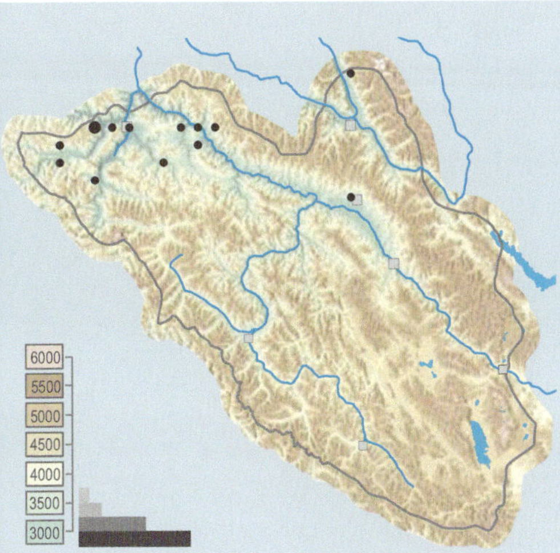

Description: Perennial herbs, 10–30 cm tall, hirsute, stems procumbent, leaves oblong to ovate, flowers ca. 3 mm long, petals absent.

Distribution: Pakistan, NW India, W China.

Habitat: Stabilized stony slopes.

Stem diameter 4–6 mm. Bark/xylem proportion 1:2. Mean ring width 0.3 mm. Maximal 5 annual rings.

Xylem

Transverse section

Semi-ring-porous. Rings distinctly marked by differences between earlywood (20–30 µm) and latewood vessel diameters (<10 µm). Vessels solitary and in groups. Fibers absent. Parenchyma pervasive. Vascular bundles remain. Crystals absent.

Tangential section

Rays >10-seriate. With 2–4 lateral rows of sheath cells.

Radial section

Perforation plates simple. Intervessel pits round and scalariform. Vessel length 100–150 µm. Ray cells upright.

Bark

Phloem with small groups of sieve elements. Cortex cells large (80–100 µm), round to oval. Phellem consists of several layers of rectangular cells.

Pith

No slide available.

Analyzed: Root collar of polar root of 1 individual.

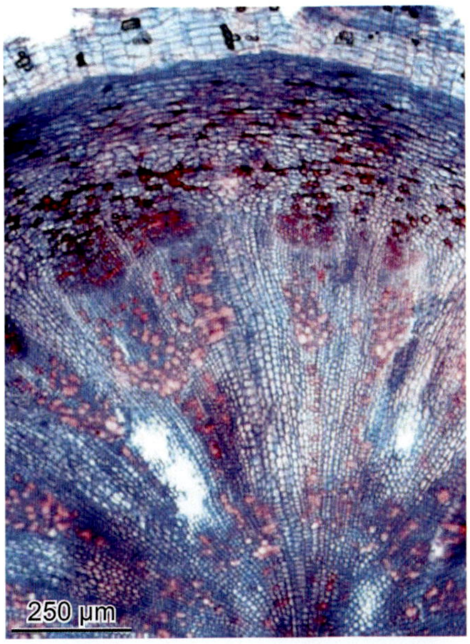

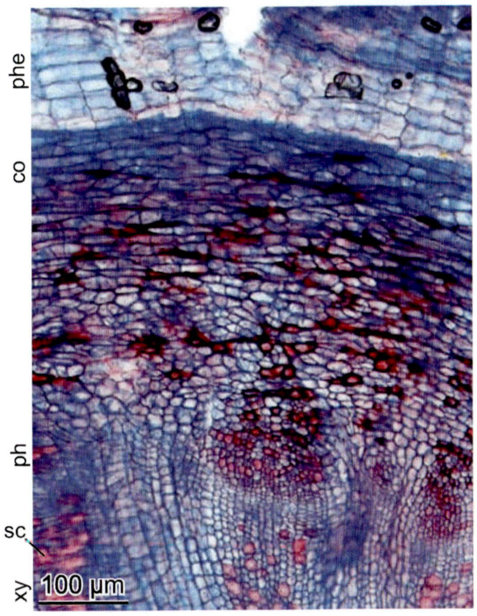

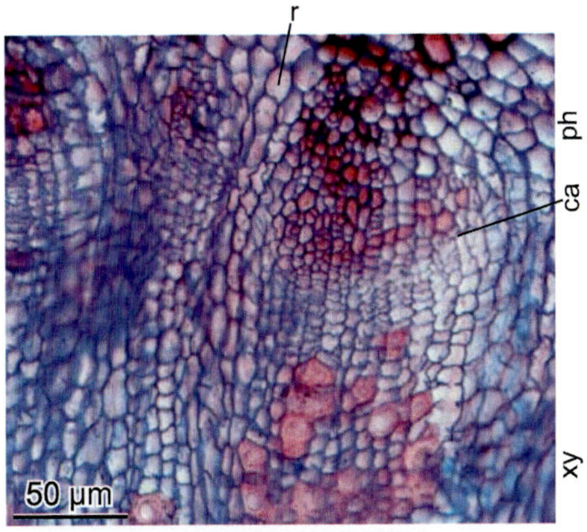

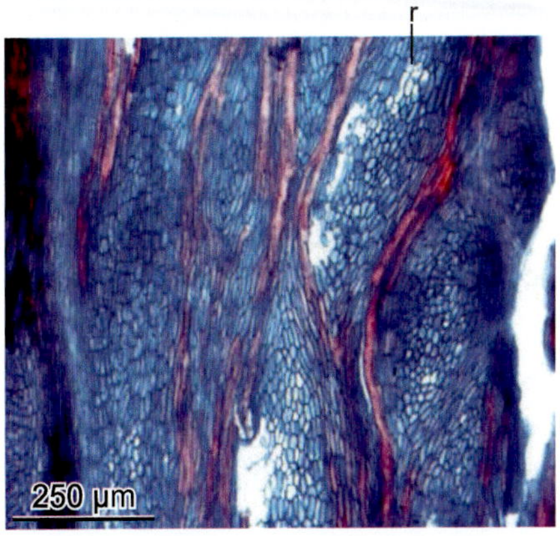

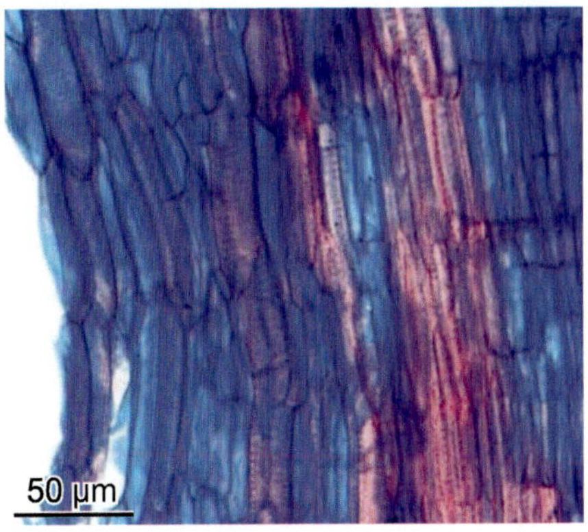

Features according to IAWA

Xylem 1 - 4 - 9 - 10 - 13 - 20 - 22 - 40.2 - 50.2 - 52.3 - 61 - 79.1 - 98 - 99 - 99.1 - 100.1 - 100.2 - 105 - 110

Campanula pallida var. *tibetica* (Hook.f. & Thomson) H.Hara
Campanulaceae

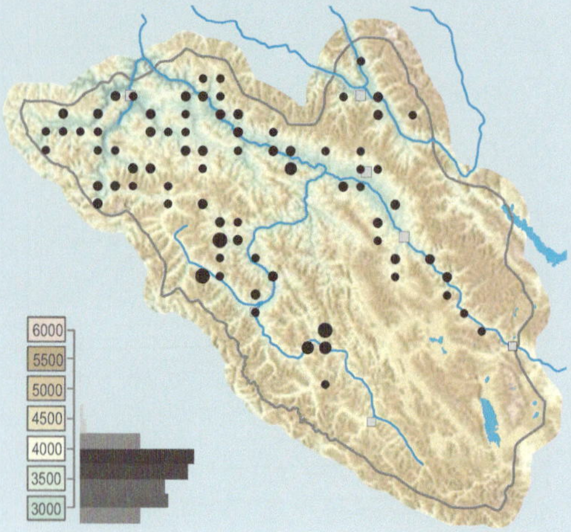

Syn.: *C. colorata* var. *tibetica* Hook.f. & Thomson

Description: Perennial herbs, 10–35 cm tall, stems ascending to erect, leaves elliptic to rhombic, flowers bluish-white or pale blue, ca. 1 cm long.

Distribution: N Pakistan, India (Jammu and Kashmir, Himachal Pradesh), China (Xizang).

Habitat: Stabilized stony slopes.

Stem diameter 4–6 mm. Bark/xylem proportion 1:3. Ring width 0.25–0.3 mm. Maximal 10 annual rings.

Xylem

Transverse section

Semi-ring-porous. Rings distinctly marked by differences between earlywood (40–50 µm) and latewood vessel diameters (10–20 µm). Vessels solitary. Fibers rare, grouped, same diameter as latewood vessels. Parenchyma pervasive. Vascular bundles remain. Crystals absent.

Tangential section

Rays >10-seriate.

Radial section

Perforation plates simple. Intervessel pits round. Vessel length 80–120 µm. Ray cells upright.

Bark

Phloem uniform. Crystal druses arranged in tangential rows. Cortex cells large, round to oval (80–100 µm), few contain crystal druses. Phellem layered, consisting of living cells with blue-stained contents (phelloids) and dead, brown-stained cells.

Pith

Consists of very thin-walled, large parenchyma cells.

Analyzed: Root collar of polar root of 3 individuals.

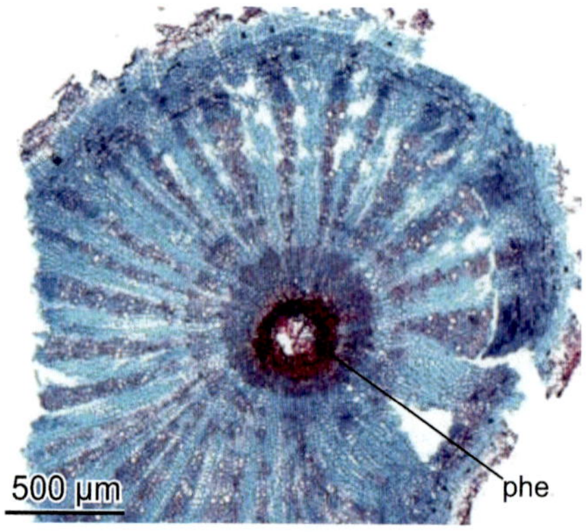

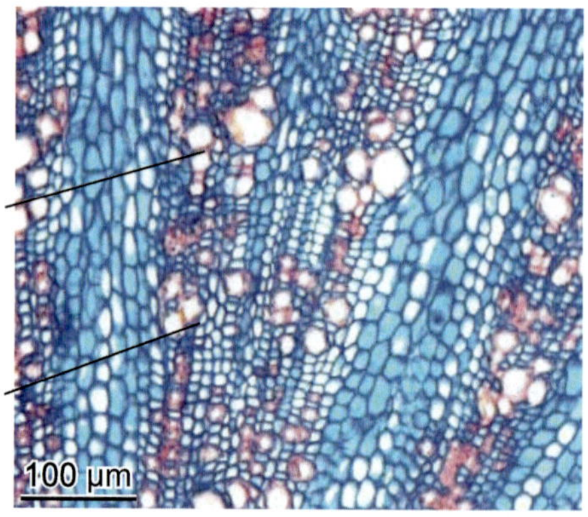

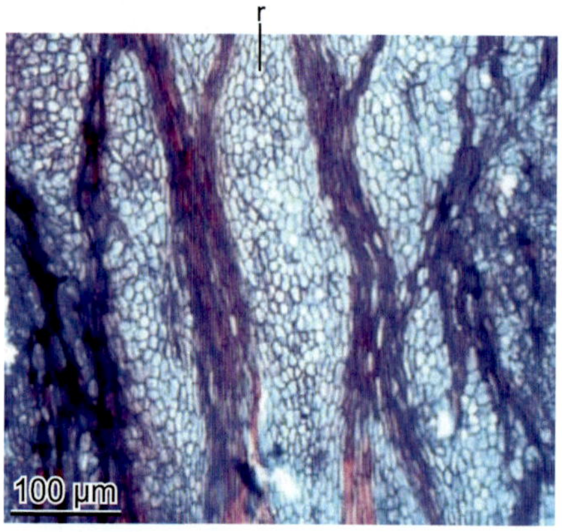

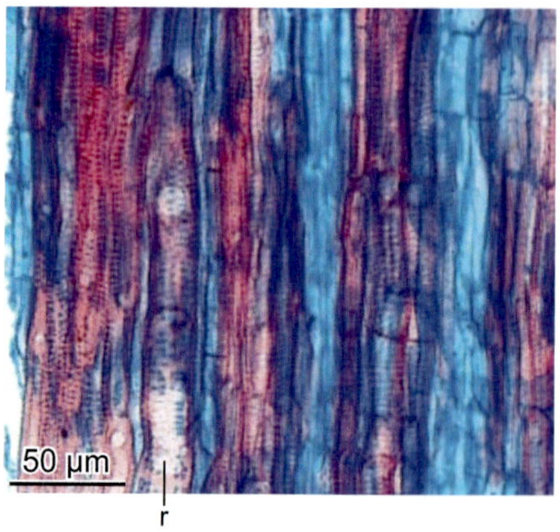

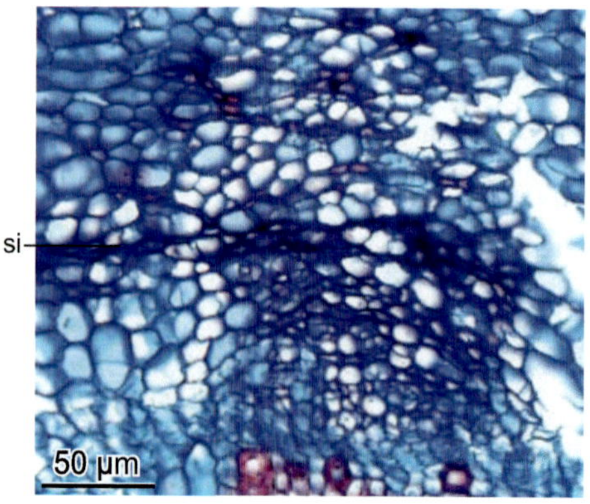

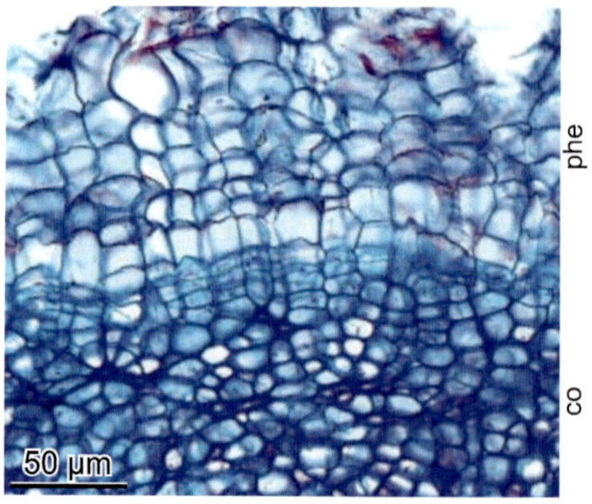

Features according to IAWA

Xylem 1 - 4 - 9 - 13 - 22 - 40.2 - 50.2 - 52.3 - 61 - 79.1 - 98 - 99 - 99.1 - 100.2 - 105 - 110

Capparis spinosa L.

Capparidaceae

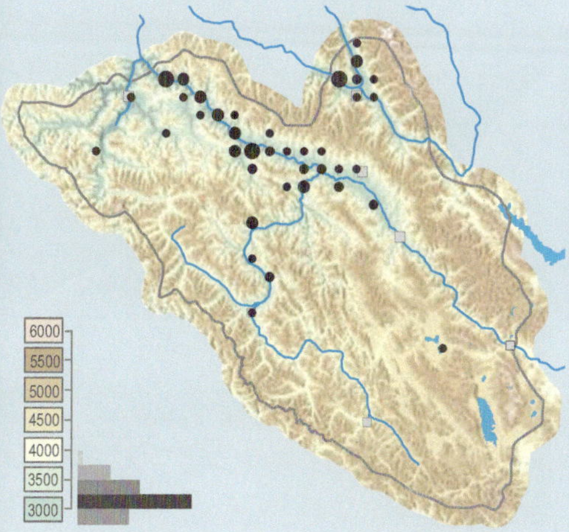

Description: Shrubs, branches prostrate, thorny, 50–100 cm long, leaves ovate, flowers whitish, 2–5 cm across, solitary in axils.

Distribution: N Africa, S Europe, SW Asia, Afghanistan, Pakistan, India, Nepal, China, (Xinjiang, Xizang), Indonesia, Australia.

Habitat: Disturbed sites along roads and dumps in settlements.

Stem diameter 30 mm. Bark/xylem proportion 1:20. Mean ring width 0.6 mm. Maximal 19 annual rings.

Xylem

Transverse section

Ring-porous. Rings distinctly marked by differences between earlywood (90–120 µm) and latewood vessel diameters (30 µm) and latewood fiber zones. Vessels generally solitary or in small groups. Vessel walls fairly thick. Fibers thin- to thick-walled. Parenchyma paratracheal and marginal. Crystals absent.

Tangential section

Rays 3–6-seriate, with sheath cells.

Radial section

Perforation plates simple. Intervessel pits round. Vessel length 80–120 µm. Ray cells procumbent.

Bark

Phloem with tangentially arranged groups of sieve tubes, which collapse in the second layer. With a few round groups of sclereids. Crystals absent.

Pith

Round, consisting of thin-walled parenchyma cells.

Analyzed: Stems of 2 individuals.

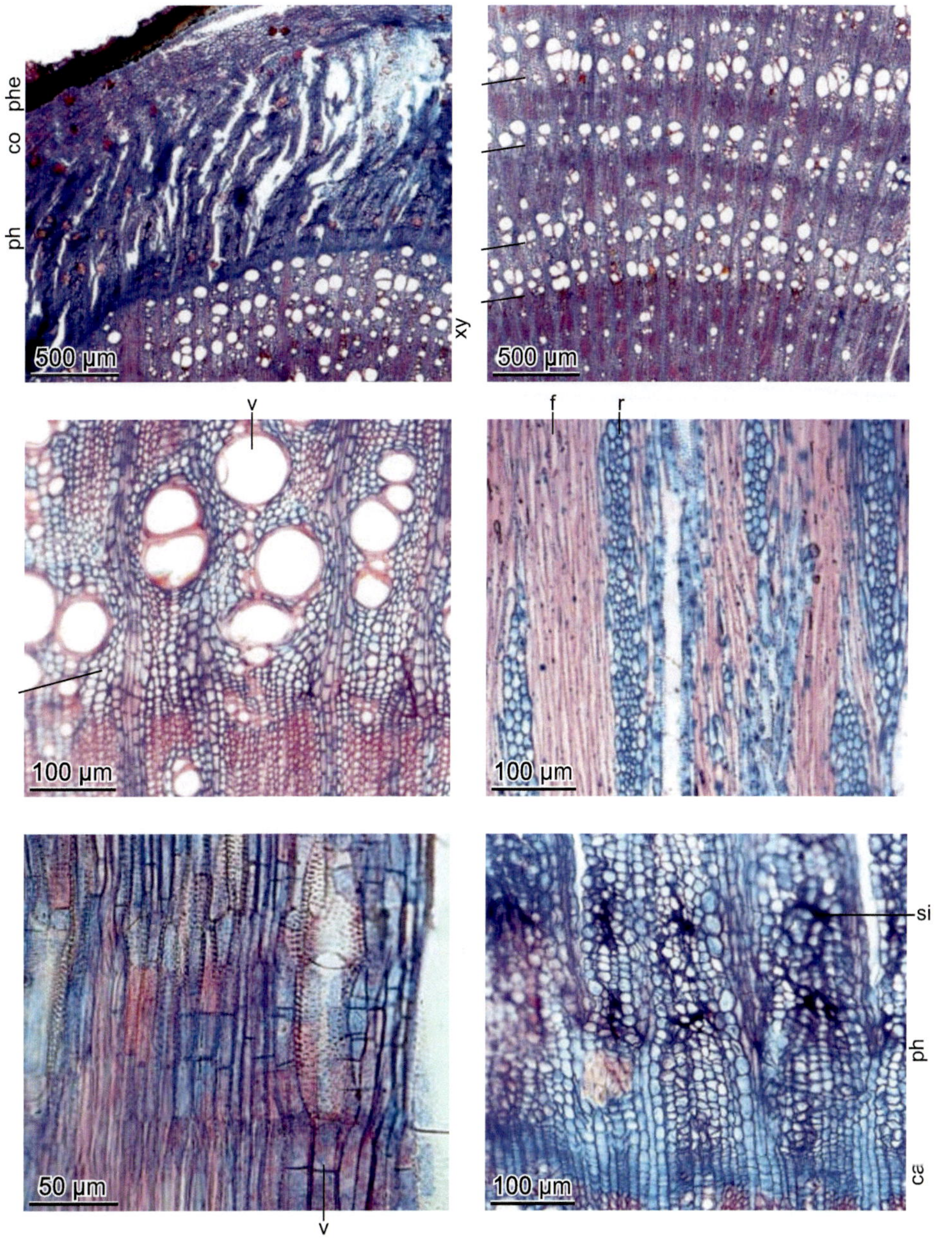

Features according to IAWA

Xylem *1 - 3 - 9 - 10 - 13 - 22 - 42 - 50 - 52.3 - 61 - 70 - 79 - 89 - 97 - 104*

Lonicera asperifolia (Decne.) Hook.f. & Thomson

Caprifoliaceae

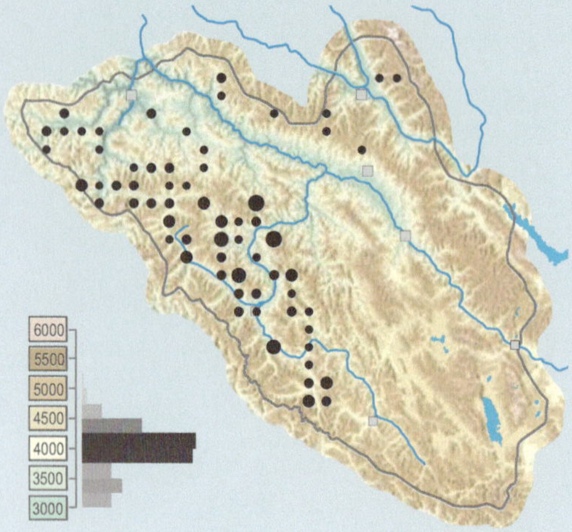

Syn.: *Xylosteon asperifolium* Decne.

Description: Shrubs, up to 1.5 m tall, hispid, leaves ovate, bristly hairy, opposite, flowers 1.5 cm long, berries in pairs, 1 cm across, orange-red.

Distribution: Iran, Tajikistan, Afghanistan, N Pakistan, India (Jammu and Kashmir, Himachal Pradesh).

Habitat: Along streams, among stones.

Stem diameter 7 mm. Bark/xylem proportion 1:10. Mean ring width 0.22 mm. Maximal 19 annual rings.

Xylem

Transverse section

Semi-ring-porous. Rings distinctly marked by differences between earlywood (50–70 µm) and latewood vessel diameters (20 µm) and a few flat latewood fibers. Vessels solitary. Fibers thick-walled. Parenchyma paratracheal. Parenchyma in the center with dark-staining substances. Crystals absent.

Tangential section

Rays of two distinct sizes: Uni- and 2–3-seriate.

Radial section

Perforation plates simple. Intervessel pits round. Vessel length 80–120 µm. With thin helical thickenings. Uniseriate rays with upright cells, 2–3-seriate rays heterogeneous, with >4 upright cells.

Bark

Phloem uniform. With a large, continuous belt of fibers.

Pith

Round, consisting of thin-walled parenchyma cells.

Analyzed: Stem of 1 individual.

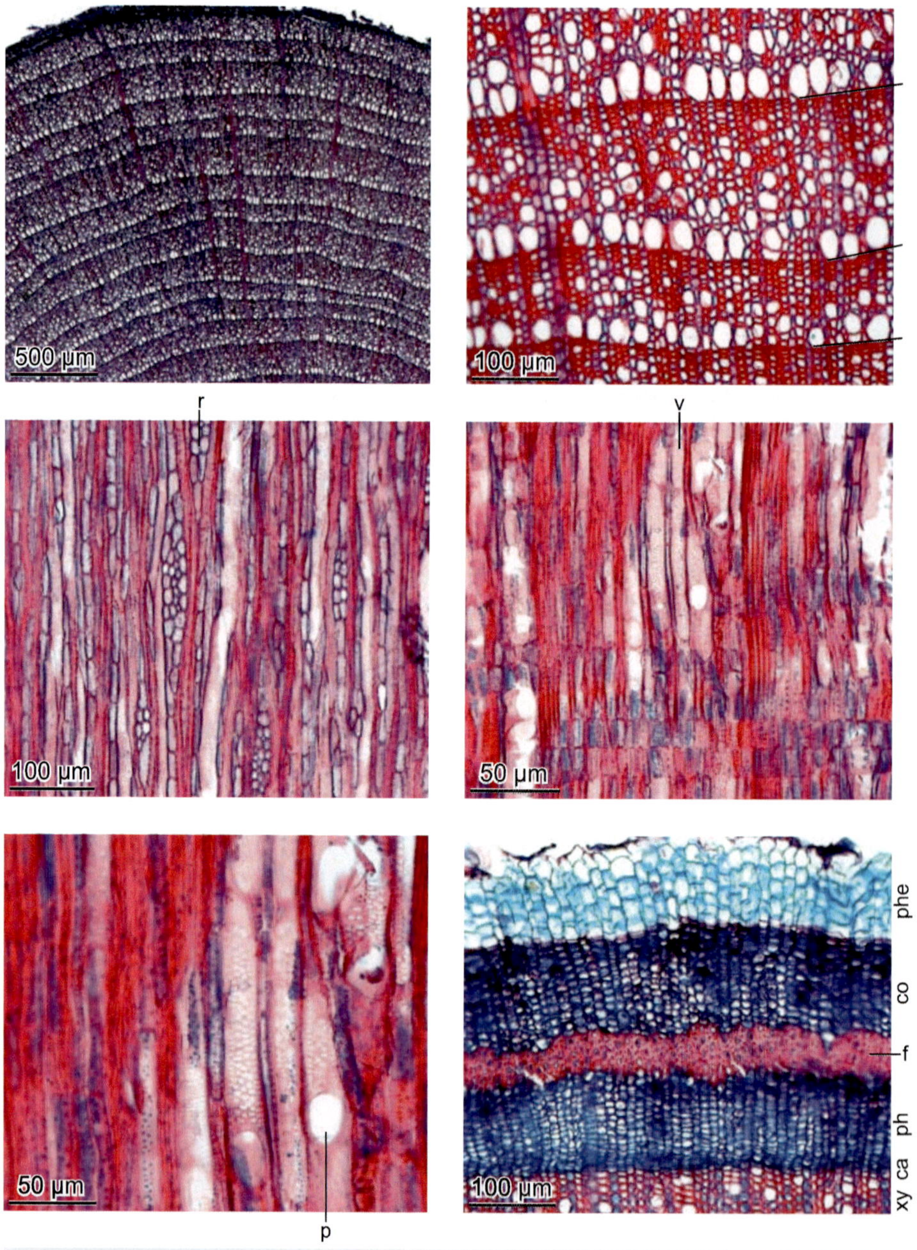

Features according to IAWA

Xylem 1 - 4 - 9 - 13 - 22 - 36 - 41 - 50.2 - 52.3 - 58 - 61 - 69 - 79 - 96 - 97 - 103 - 108

Lonicera microphylla Willd. ex Schult.

Caprifoliaceae

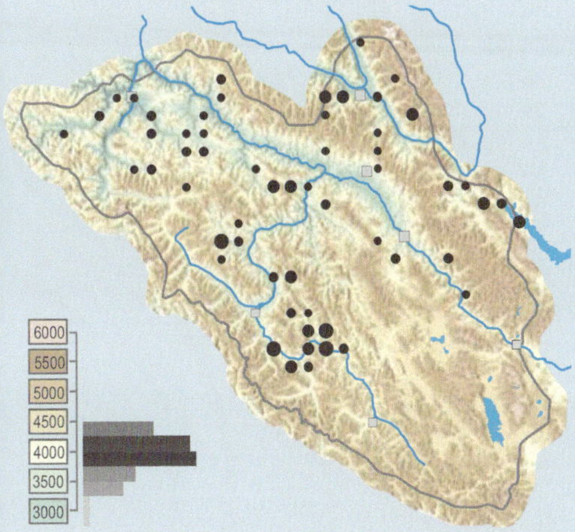

Description: Shrubs, 1–3 m tall, leaves obovate to elliptic, gray-white below, flowers white to yellow, ca. 1 cm long, arranged in pairs in axils, berries red or orange-yellow, ca. 5 mm across.

Distribution: Russia (Siberia), Iran, E Kazakhstan, Kyrgyzstan, Tajikistan, Afghanistan, N Pakistan, NW India, China (Gansu, Hebei, Nei Mongol, Ningxia, Qinghai, Shanxi, Taiwan, Xinjiang, Xizang), Mongolia.

Habitat: Dry rocky slopes.

Stem diameter 15–30 mm. Bark/xylem proportion 1:10. Ring width 0.3–0.5 mm. Maximal 27 annual rings.

Xylem

Transverse section

Semi-ring-porous. Rings distinctly marked by differences between earlywood (30–40 µm) and latewood vessel diameters (20 µm) and a few flat latewood fibers. Vessels solitary. Fibers thick-walled. Parenchyma paratracheal. Parenchyma in the center with dark-staining substances. Crystals absent.

Tangential section

Rays of two distinct size: Uni- and 2–3-seriate.

Radial section

Perforation plates simple. Intervessel pits round. Vessel length 80–120 µm. With thin helical thickenings. Uniseriate rays with upright cells, 2–3-seriate rays heterogeneous, with >4 upright cells.

Bark

Phloem uniform. With a large, continuous belt of fibers.

Pith

No slide available.

Analyzed: Stems of 2 individuals.

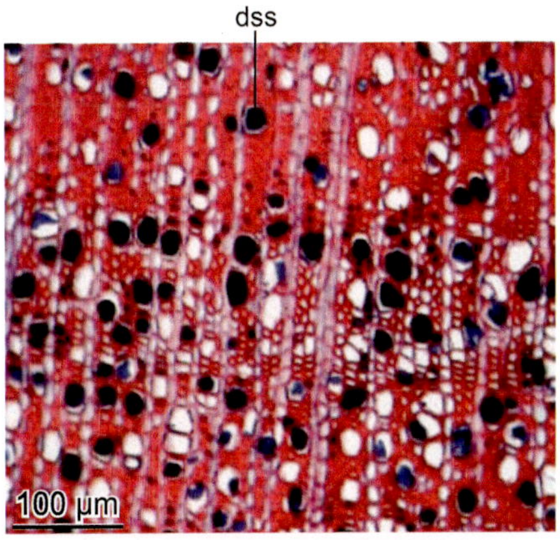

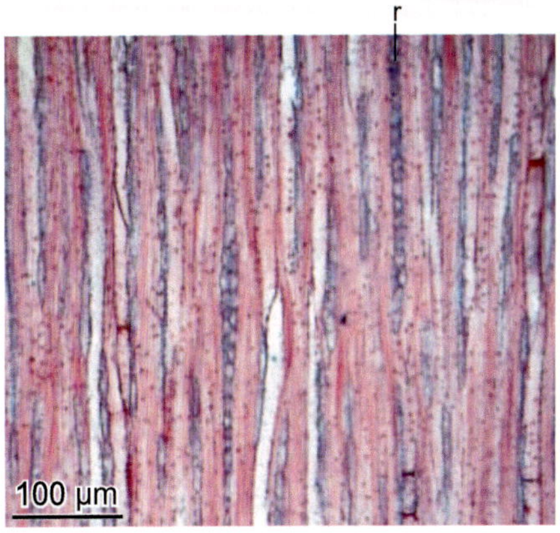

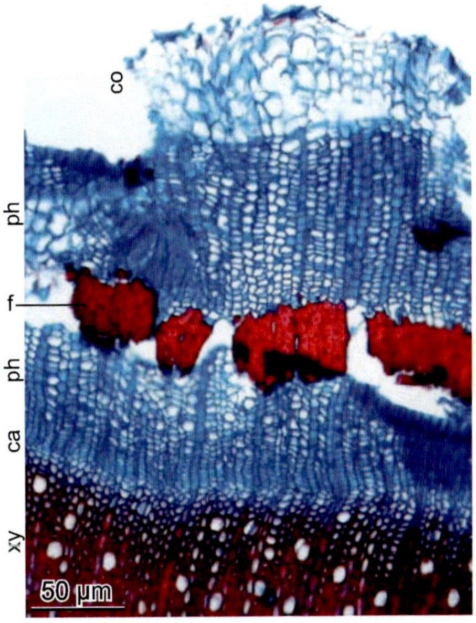

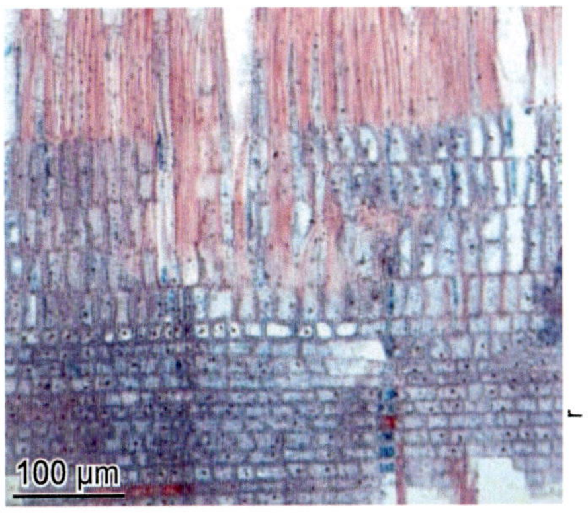

Features according to IAWA

Xylem 1 - 4 - 9 - 13 - 22 - 36 - 40.2 - 50.2 - 52.3 - 58 - 61 - 69 - 79 - 96 - 97 - 103 - 108

Lonicera semenovii Regel
Caprifoliaceae

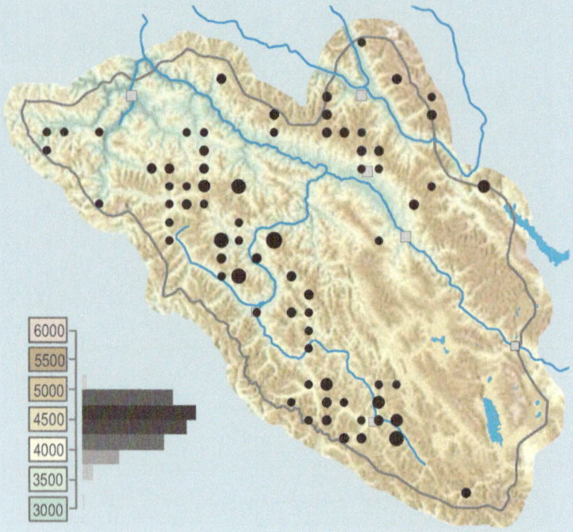

Syn.: *L. glauca* Hook.f. & Thomson

Description: Shrubs, 15–40 cm tall, leaves oblong-lanceolate, margins denticulate, somewhat revolute, flowers yellow, 1.5–3 cm long, arranged in pairs in axils, berries red.

Distribution: Iran, Kazakhstan, Kyrgyzstan, Afghanistan, Pakistan, India, China (Xinjiang, Xizang).

Habitat: Rocky crevices.

Stem diameter 7 mm. Bark/xylem proportion 1:10. Mean ring width 0.08 mm. Maximal 40 annual rings.

Xylem

Transverse section

Semi-ring-porous. Rings distinctly marked by differences between earlywood (30–50 µm) and latewood vessel diameters (20 µm) and a few flat latewood fibers. Vessels solitary. Helical thickenings absent. Fibers thin- to thick-walled. Parenchyma paratracheal. Parenchyma in the center with dark-staining substances. Crystals absent.

Tangential section

Rays of two distinct sizes: Uni- and 3-seriate.

Radial section

Perforation plates simple. Intervessel pits round. Vessel length 80–120 µm. Uniseriate rays with upright cells and 3-seriate rays with square cells in the center and upright marginal cells.

Bark

Phloem uniform. Solitary sieve tubes slightly larger than the parenchyma cells.

Pith

Round, consisting of thin-walled parenchyma cells.

Analyzed: Stem of 1 individual.

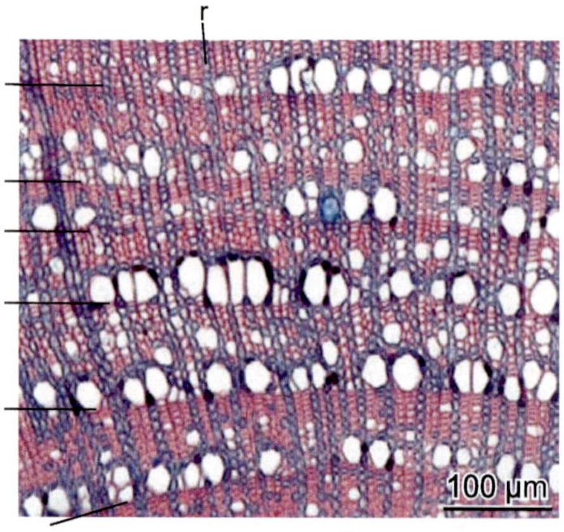

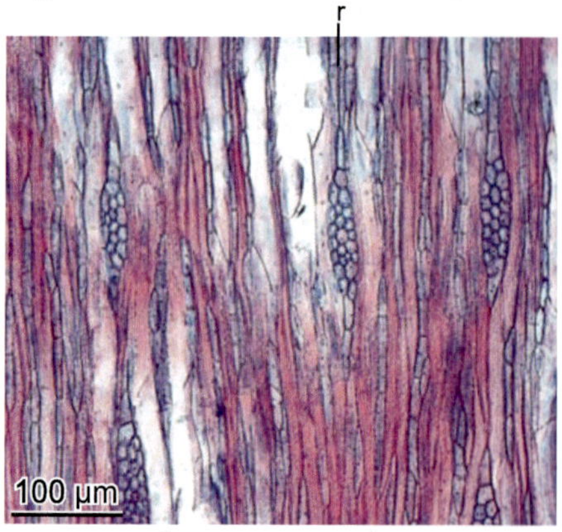

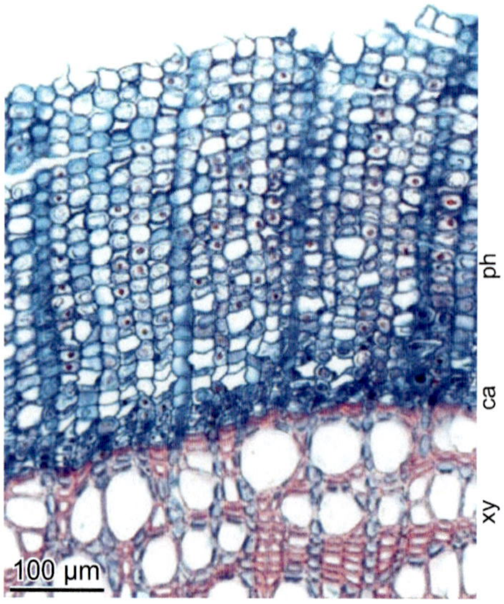

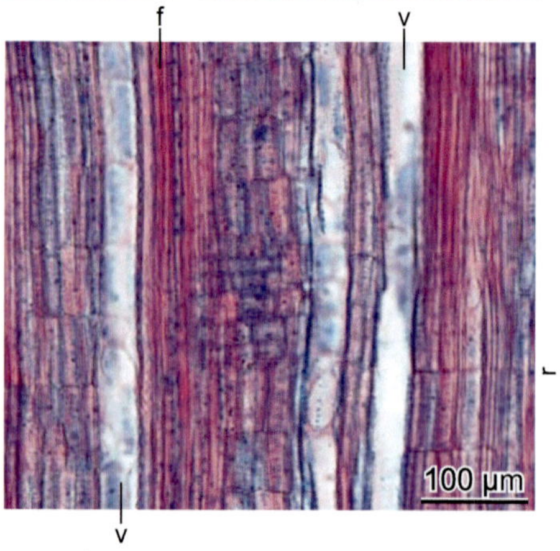

Features according to IAWA

Xylem 1 - 4 - 9 - 13 - 22 - 40.2 - 50.2 - 52.3 - 58 - 61 - 69 - 70 - 79 - 96 - 97 - 103 - 105

Lonicera spinosa (Decne.) Jacquem. ex Walp.

Caprifoliaceae

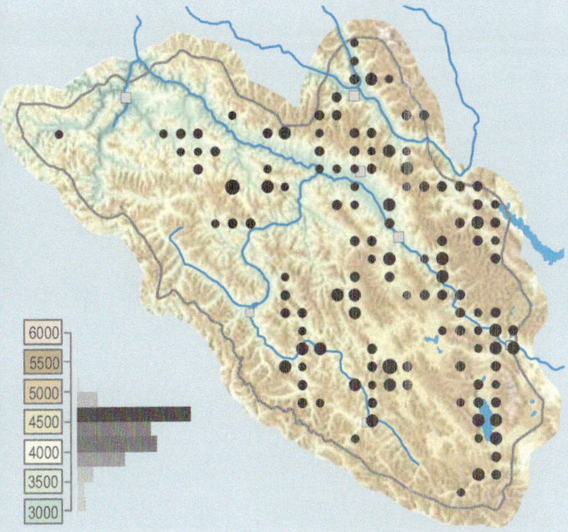

Syn.: *Xylosteon spinosum* Decne.

Description: Shrubs, up to 60 cm tall, often with spine-like leafless branchlets, leaves linear to oblong, margin revolute, flowers purplish red, later white, ca. 1 cm long, berry yellowish, ca. 5 mm across.

Distribution: Kazakhstan, Kyrgyzstan, Tajikistan, Afghanistan, India China (Xinjiang, Xizang).

Habitat: River sediments, dry stony, gravel and rocky slopes.

Stem diameter 7 mm. Bark/xylem proportion 1:10. Mean ring width 0.23 mm. Maximal 38 annual rings.

Xylem

Transverse section

Semi-ring-porous. Rings distinctly marked by differences between earlywood (30–50 µm) and latewood vessel diameters (20 µm) and a few flat latewood fibers. Vessels solitary and in radial rows. Fibers thin- to thick-walled. Parenchyma paratracheal. Parenchyma in the center with dark-staining substances. Crystals absent.

Tangential section

Rays 1–2-seriate.

Radial section

Perforation plates primarily simple, rarely scalariform with 2–4 bars. Intervessel pits round. Some vessels with helical thickenings. Vessel length 80–120 µm. Uniseriate rays with upright cells, 2-seriate rays with square cells in the center and upright marginal cells.

Bark

Phloem uniform. Solitary sieve tubes slightly larger than the parenchyma cells. With crystal druses. Phellem consisting of thin-walled, unlignified cells.

Pith

Oval, consisting of thin-walled parenchyma cells.

Analyzed: Stem of 3 individuals.

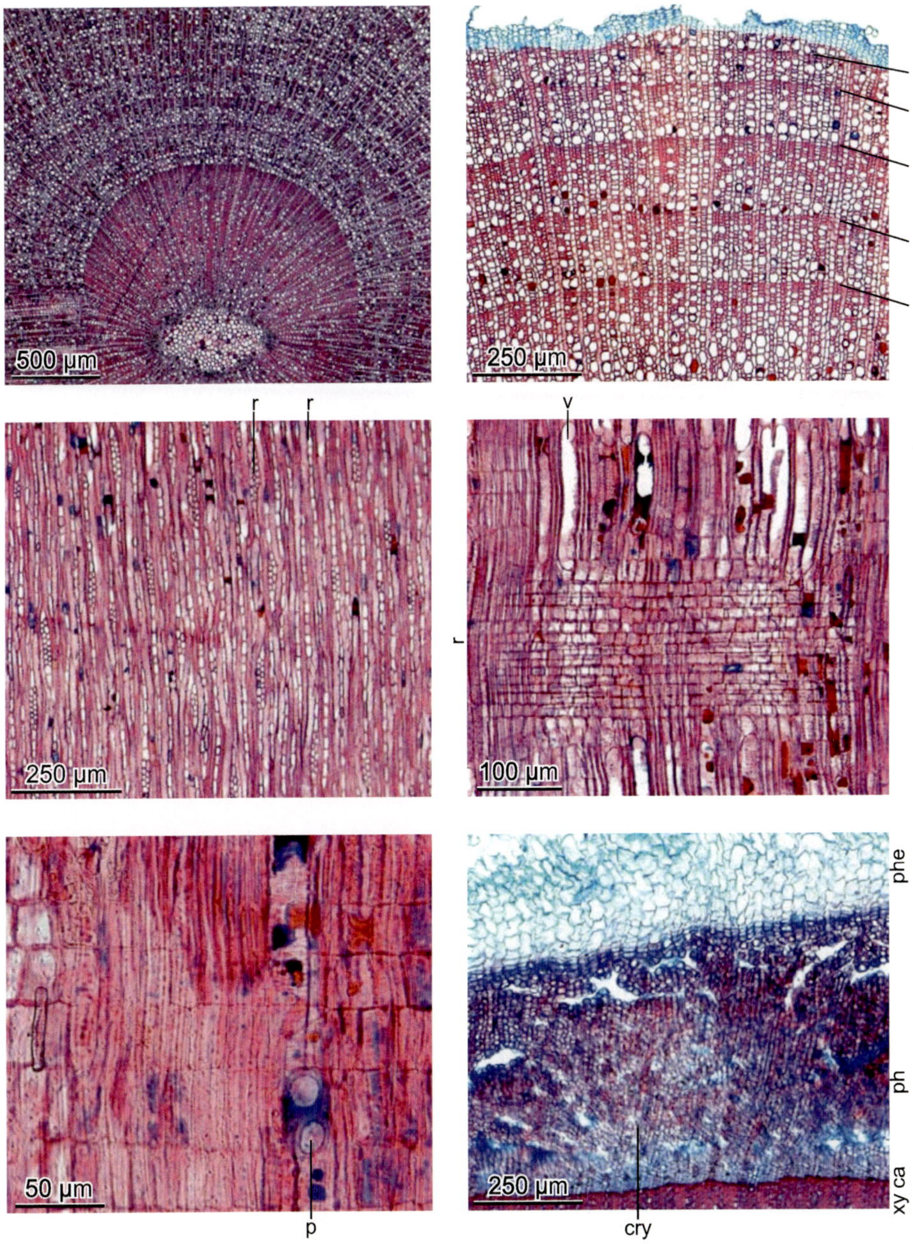

Features according to IAWA

Xylem 1 - 4 - 9 - 13 – 14 - 22 - 36 - 40.2 - 50.2 - 52.3 - 58 - 61 - 69 - 70 – 79 - 96 - 97 - 105

Arenaria bryophylla Fernald

Caryophyllaceae

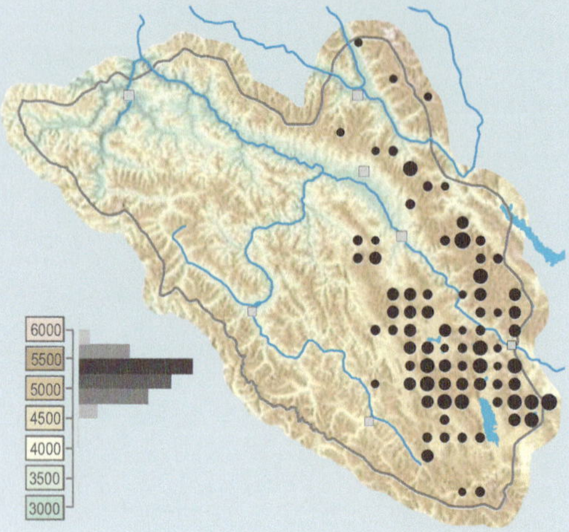

Syn.: *A. musciformis* Wall. ex Edgew. 1874 (non Triana & Planch. 1862)

Description: Perennial herbs, 1–5 cm tall, forming dense cushions up to 40 cm across, leaves linear, crowded, flowers white, ca. 1 cm across, solitary, sessile.

Distribution: India (Jammu and Kashmir, Sikkim), Nepal, China (Qinghai, Xizang).

Habitat: Alpine meadows, stony slopes, gravel sands along rivers.

Stem diameter 4–19 mm. Bark/xylem proportion 1:5. Ring width 0.12–0.4 mm. Maximal >40 annual rings. Rings partially indistinct.

Xylem

Transverse section

Semi-ring-porous. Ring boundaries often indicated by larger earlywood vessel diameters (20–30 μm) and smaller latewood vessel diameters (<10 μm). Vessels solitary, vessel walls not or poorly stained. Fibers absent. Parenchyma pervasive. Irregularly formed zones of successive cambia (Caryophyllaceae type) occur in large stems. With crystal druses.

Tangential section

Rays absent.

Radial section

Perforation plates simple. Intervessel pits scalariform. Vessel length 50–80 μm.

Bark

Phloem uniform. Phellem consists of a multilayered band of small rectangular cells. Many crystal druses in the cortex.

Pith

No slide available.

Analyzed: Root collar of polar root of 11 individuals.

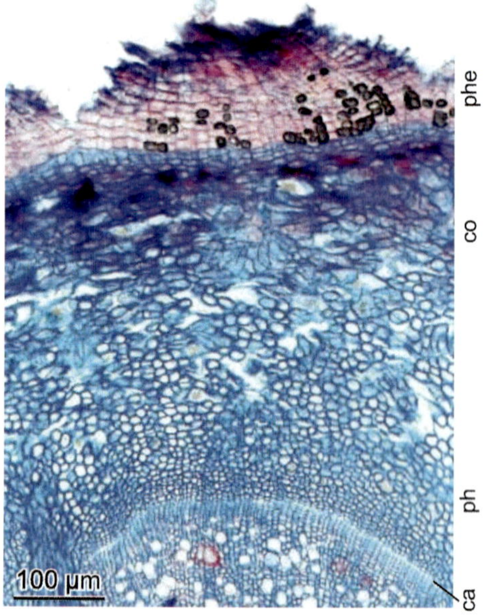

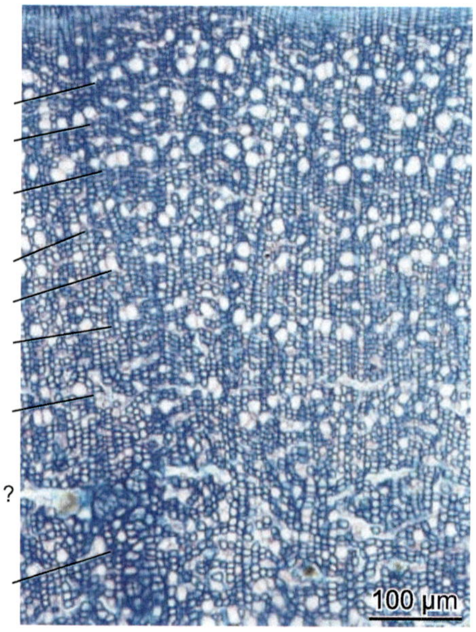

Features according to IAWA

Xylem 1 - 9 - 13 - 20 - 39.2 - 40.1 - 40.2 - 50.2 - 52.2 - 60.1 - 79.1 - 99.1 - 117 - 133 - 144

Dianthus harrissii Rech.f.

Caryophyllaceae

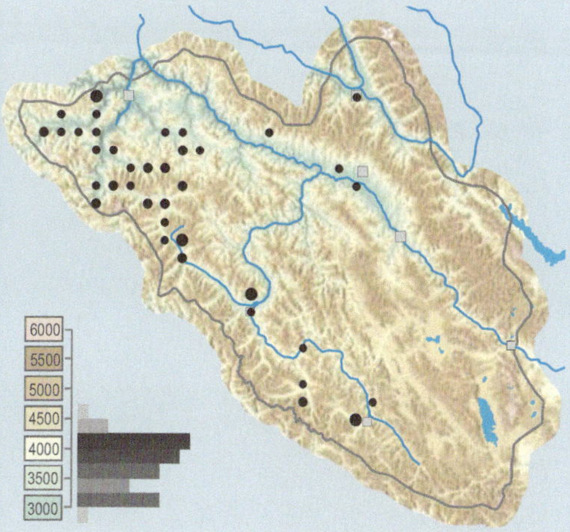

Description: Perennial herbs, 10–30 cm tall, densely cespitose, leaves linear, opposite, usually congested basally, flowers white to pink, 10–15 mm across.

Distribution: Pakistan, NW India.

Habitat: Mountain slopes.

Stem diameter 3–6 mm. Bark/xylem proportion 1:4. Ring width 0.04–0.1 mm. Maximal 12 annual rings.

Xylem

Transverse section

Semi-ring-porous. Ring boundaries indicated by larger earlywood vessel diameters (20–30 μm) and smaller latewood vessel diameters (<10 μm). Vessels solitary. Fibers absent. Parenchyma pervasive. Crystals absent.

Tangential section

Rays absent.

Radial section

Perforation plates simple. Intervessel pits scalariform. Vessel length 60–90 μm. With crystal druses.

Bark

Phloem with small groups of sieve tubes. Phellem consists of many layers of rectangular cells. Few crystal druses in the cortex.

Pith

No slide available.

Analyzed: Root collar of polar root of 4 individuals.

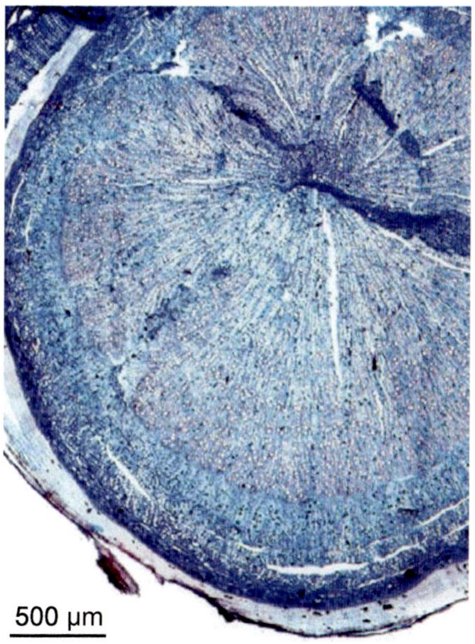

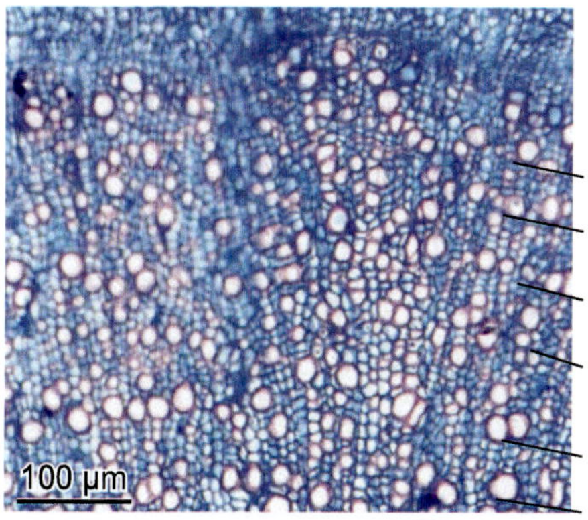

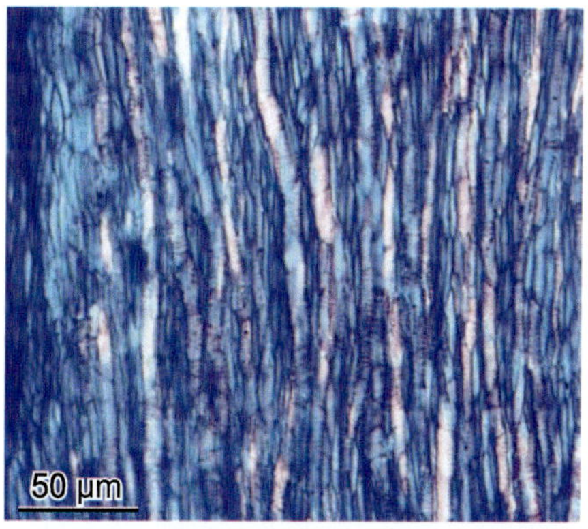

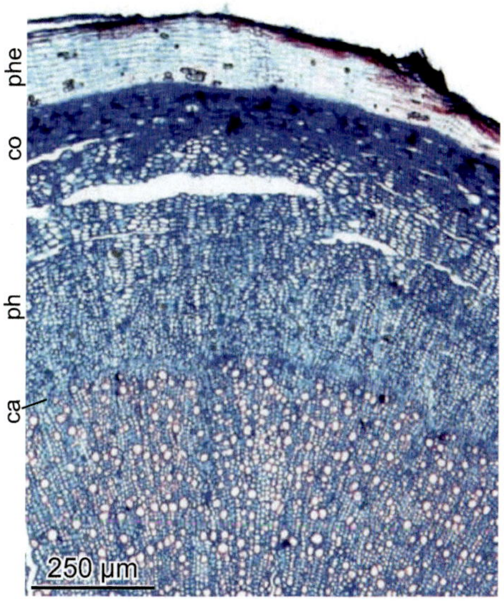

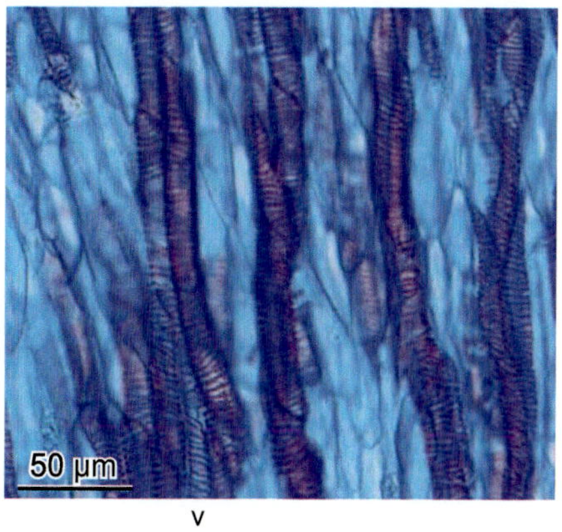

Features according to IAWA

Xylem 1 - 9 - 13 - 20 - 39.2 - 40.1 - 40.2 - 50.2 - 52.2 - 60.1 - 79.1 - 117

Gypsophila sedifolia Kurz

Caryophyllaceae

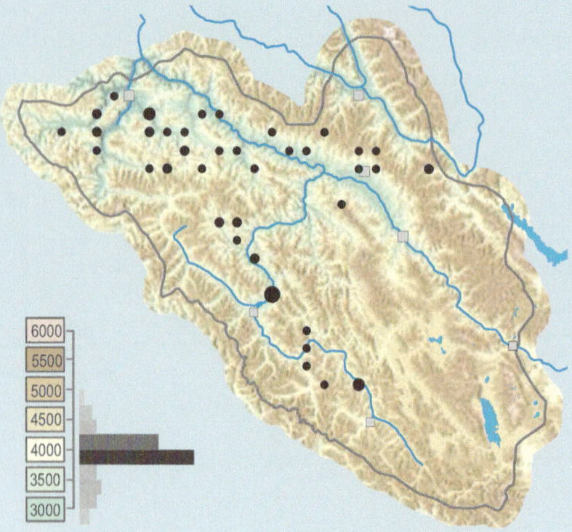

Syn.: *G. herniarioides* Boiss., *G. tibetica* Hook.f. & Thomson

Description: Perennial herbs, 3–10 cm tall, cespitose, pubescent, stems decumbent, leaves narrowly oblong, fasciculate, pubescent, flowers pink to lilac, 5–10 mm across, in terminal many-flowered cymes.

Distribution: Afghanistan, Pakistan, NW India.

Habitat: Dry gravel screes and stabilized slopes.

Stem diameter 4–6 mm. Bark/xylem proportion 1:1. Ring width 0.22–025 mm. Maximal 7 annual rings.

Xylem

Transverse section

Semi-ring-porous. Ring variability indicated by larger earlywood (30–40 μm) and smaller latewood vessel diameters (<10 μm). Vessel solitary. Fibers absent. Parenchyma pervasive. Crystal absent.

Tangential section

Rays absent.

Radial section

Perforation plates simple. Intervessel pits scalariform. Vessel length 50–80 μm.

Bark

Phloem uniform. Phellem consists of a multilayered band of small, rectangular cells. Small crystals in phellem.

Pith

Round, filled with unlignified cells.

Analyzed: Root collar of polar roots of 2 individuals.

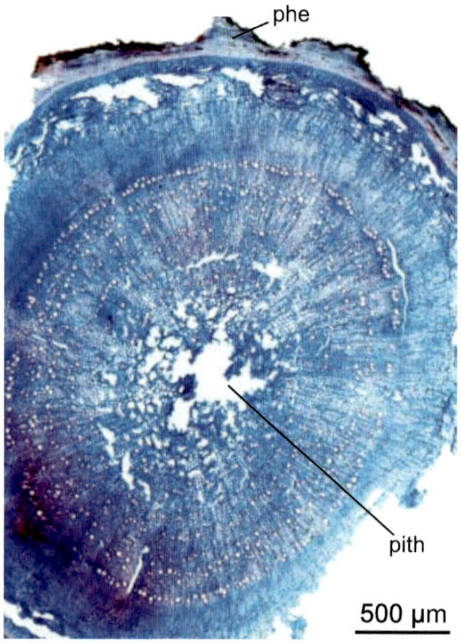

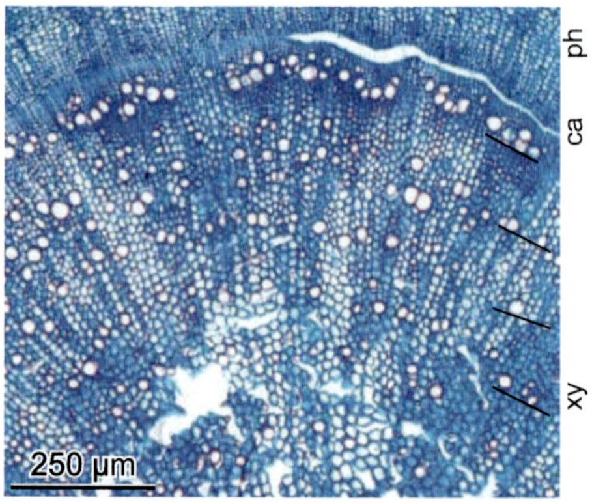

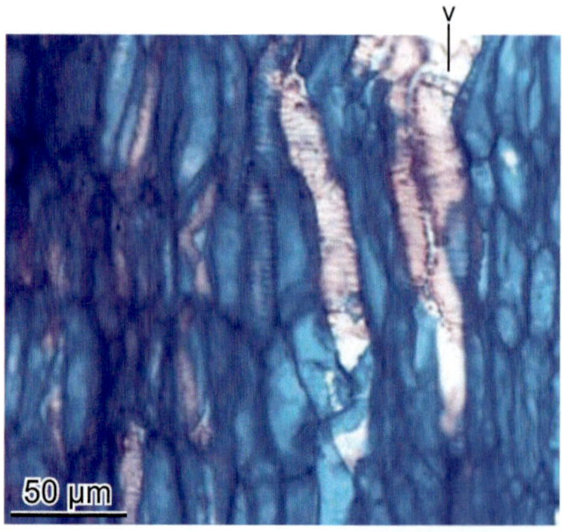

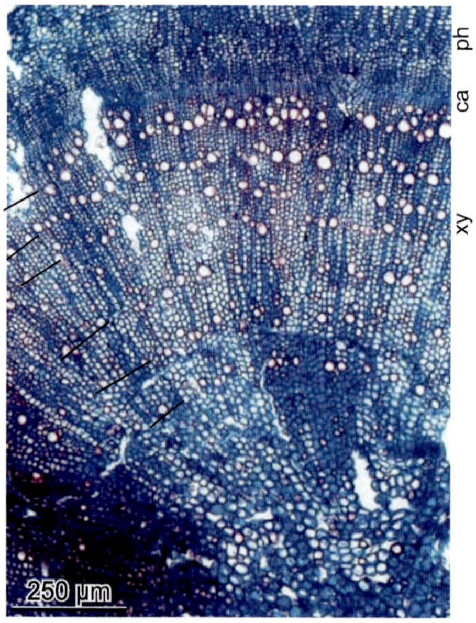

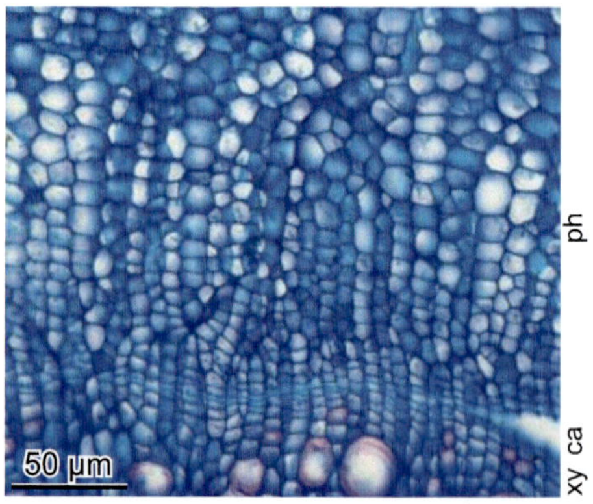

Features according to IAWA

Xylem 1 - 9 - 13 - 20 - 40.2 - 50.2 - 52.2 - 60.1 - 79.1 - 117

Minuartia foliosa (Royle ex Edgew. & Hook.f.) Majumdar

Caryophyllaceae

Syn.: *Arenaria foliosa* Royle ex Edgew. & Hook.f., *A. kashmirica* Edgew. & Hook.f., *M. kashmirica* (Edgew.) Mattf., *M. lineata* (Boiss.) Bornm., *M. ebracteolata* Majumdar & Giri *

Description: Perennial herbs, 5–25 cm tall, loosely tufted, with fasciculate sterile shoots, leaves linear, opposite, flowers white, 5–7 mm across, in lax dichasial 3–7-flowered inflorescences.

Distribution: Tajikistan, Afghanistan, N Pakistan, NW India (Jammu & Kashmir, Himachal Pradesh, Uttarakhand), W Nepal, China (Xizang).

Habitat: Rock crevices.

Stem diameter 1–2 mm. Bark/xylem proportion 1:1. Mean ring width 0.08 mm. Maximal 15 annual rings.

Xylem

Transverse section

Rings recognizable. Semi-ring-porous. Ring variability indicated by larger earlywood (20–30 µm) and smaller latewood vessel diameters (<10 µm). Vessels angular and solitary. Fibers absent. Parenchyma pervasive. Crystals absent.

Tangential section

Rays absent.

Radial section

Perforation plates simple. Intervessel pits scalariform. Vessel length 50–80 µm.

Bark

Phloem uniform. Phellem consists of a multilayered band of small, rectangular cells. Crystals absent.

Pith

No slide available.

Analyzed: Root collar of polar roots of 1 individual.

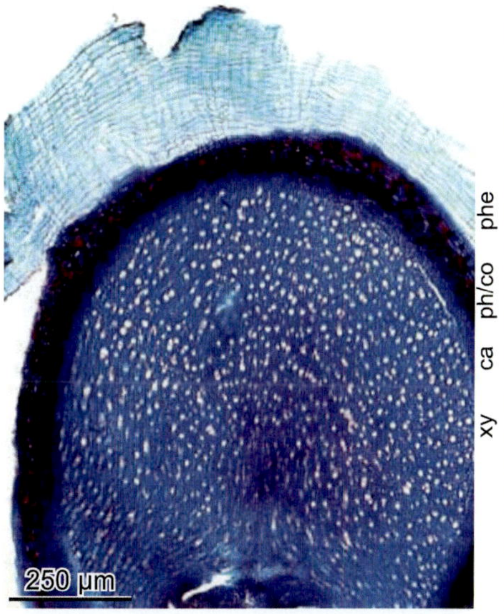

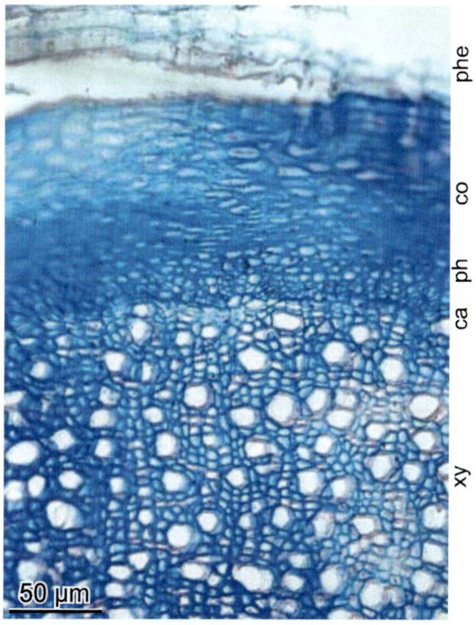

Features according to IAWA

Xylem 1 - 9 - 12 - 13 - 20 - 40.2 - 50.2 - 52.2 - 60.1 - 79.1 - 117

* Notes: A poorly understood species complex, which is possibly more wide-ranging in western Central Asia.

Sagina saginoides (L.) H.Karst.

Caryophyllaceae

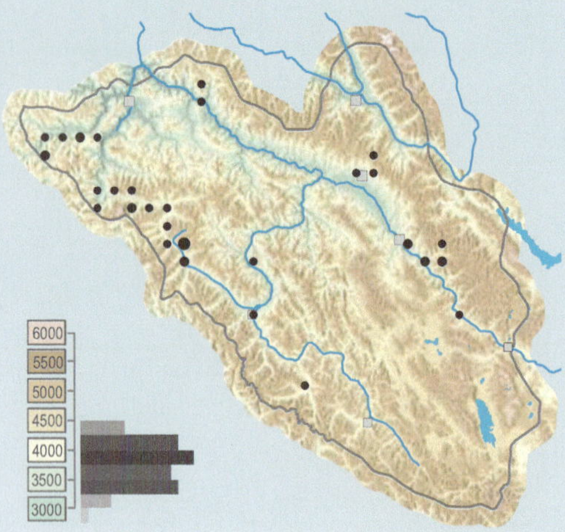

Syn.: *Spergula saginoides* L., *Sagina linnaei* C.Presl

Description: Perennial herbs, 2–10 cm tall, cespitose, glabrous, stems procumbent to ascending, leaves linear, flowers white, ca. 3 mm across, solitary, axillary or terminal.

Distribution: North America, Europe, Russia (Siberia), W and C Asia, Pakistan, India, Nepal, China, Bhutan, Vietnam, Korea, Japan.

Habitat: Mountain slopes, meadows, wet river banks, shaded trampled sites.

Stem diameter 1.5–2.5 mm. Bark/xylem proportion 1:4. Rings distinct and indistinct. Mean annual ring width 0.1 mm. Maximal 6 annual rings

Xylem

Transverse section

Semi-ring-porous. Vessel diameter 5–10 µm. Vessels hardly lignified, thick-walled, angular, solitary. Fibers absent. Parenchyma pervasive. Crystals absent.

Tangential section

Rays absent.

Radial section

Perforation plates simple. Intervessel pits scalariform with very large apertures. Vessel length 100–120 µm.

Bark

Phloem uniform. Crystals absent. Phellem consists of many layers of rectangular cells.

Pith

Pith polygonal.

Analyzed: Root collar of polar roots of 4 individuals.

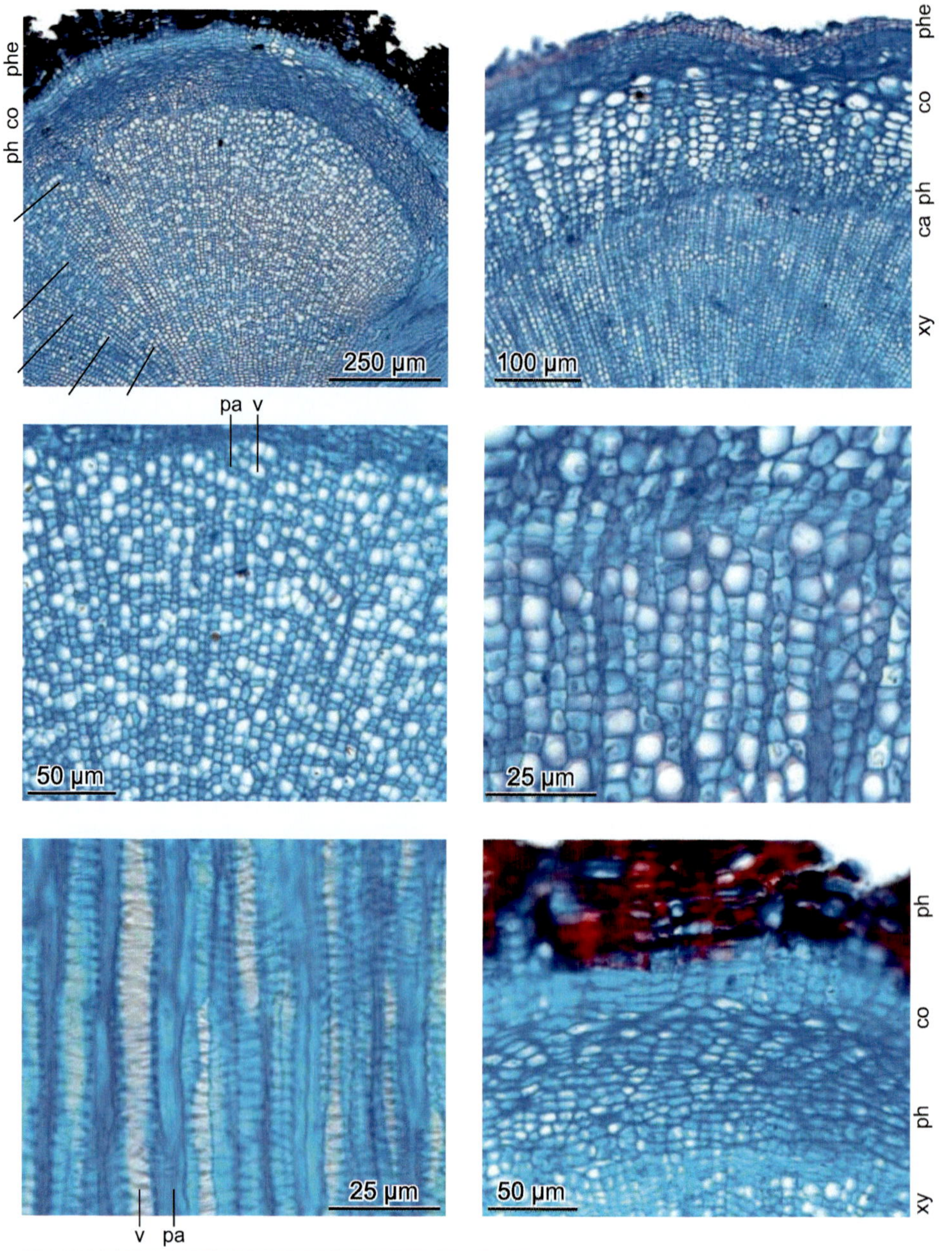

Features according to IAWA

Xylem 1 - 2 - 9 - 12 - 13 - 20 - 39.1 - 40.1 - 50.2 - 52.3 - 60.1 - 79.1 - 117

Silene caespitella F.N.Williams

Caryophyllaceae

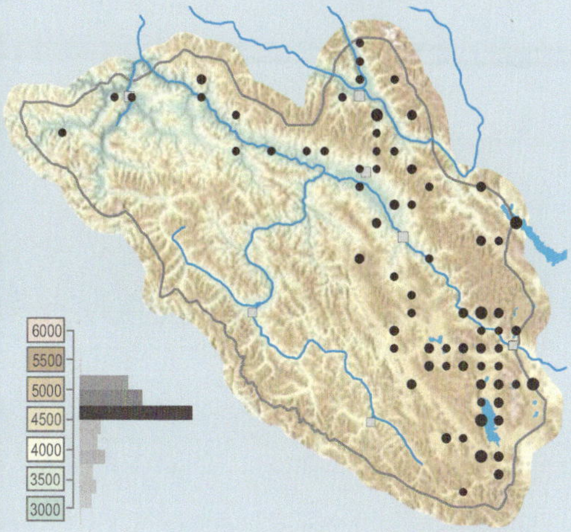

Syn.: *Melandrium neocaespitosum* Y.W.Tsui ex L.H.Zhou, *M. xainzaense* L.H.Zhou, *S. maheshwarii* Bocquet

Description: Perennial herbs, 10–40 cm tall, cespitose, leaves linear, opposite, 4–7 cm long, margin ciliate, flowers dark red, rarely greenish white, ca. 1 cm across, nodding, arranged in many-flowered inflorescences.

Distribution: N India, Nepal, Bhutan, China (Qinghai, Sichuan, Xizang).

Habitat: River banks, grazed turf, alpine meadows.

Stem diameter 3–8 mm. Bark/xylem proportion 1:4. Ring width 0.12–0.7 mm. Maximal 25 annual rings.

Xylem

Transverse section

Semi-ring-porous. Ring boundaries indicated by larger earlywood vessel diameters (20–30 μm) and smaller latewood vessel diameters (<10 μm). Vessels solitary, vessel walls not or poorly stained. Fibers absent. Parenchyma pervasive. Some individuals with remaining vascular bundles. With large crystal druses.

Tangential section

Rays mostly absent, rarely very large and confluent.

Radial section

Perforation plates simple. Intervessel pits scalariform. Vessel length 60–100 μm.

Bark

Phloem with small groups of sieve tubes. Phellem consist of many layers of square cells. Few crystal druses in the cortex.

Pith

No slide available.

Analyzed: Root collar of polar root of 4 individuals.

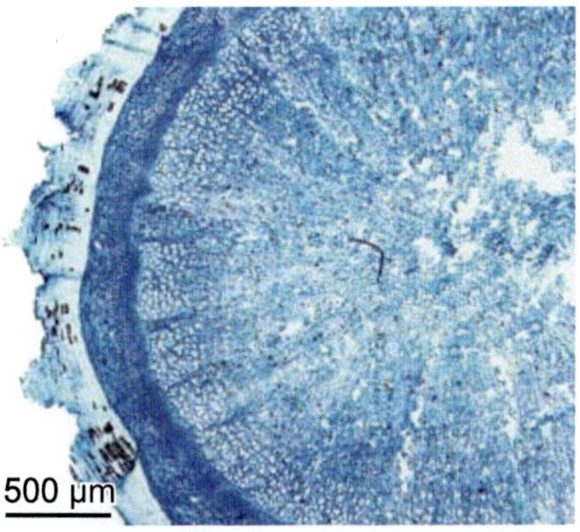

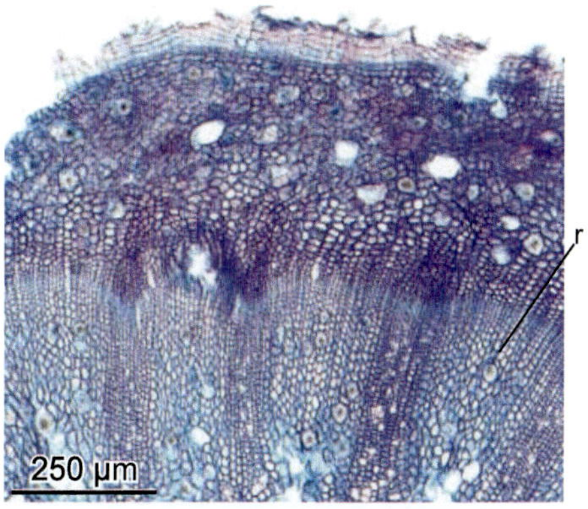

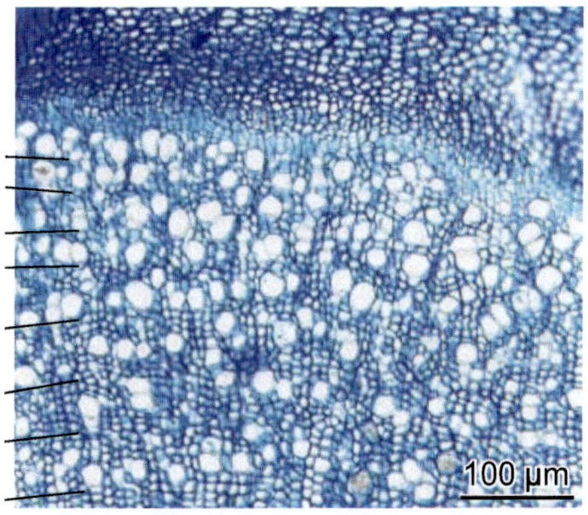

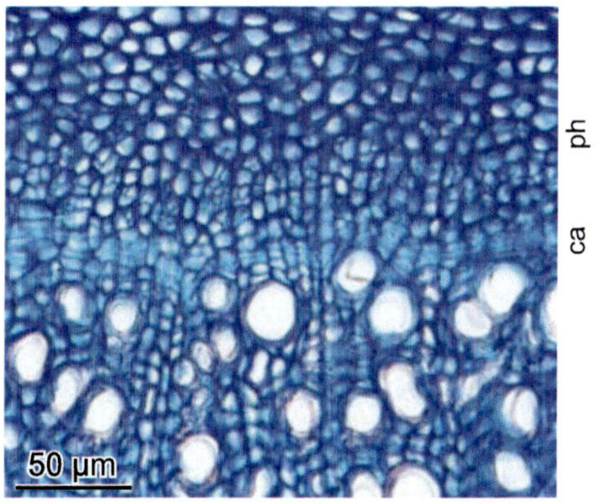

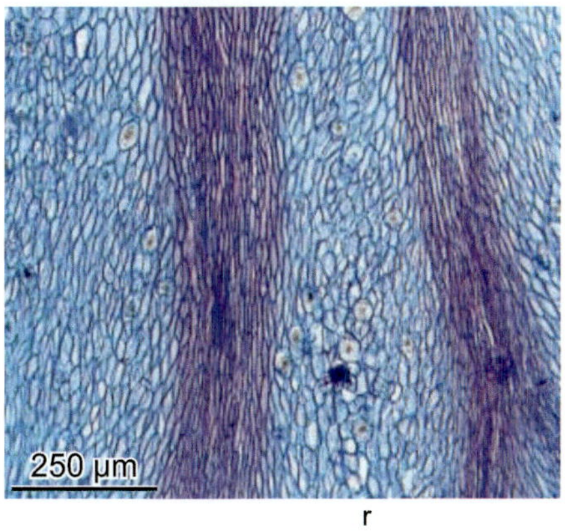

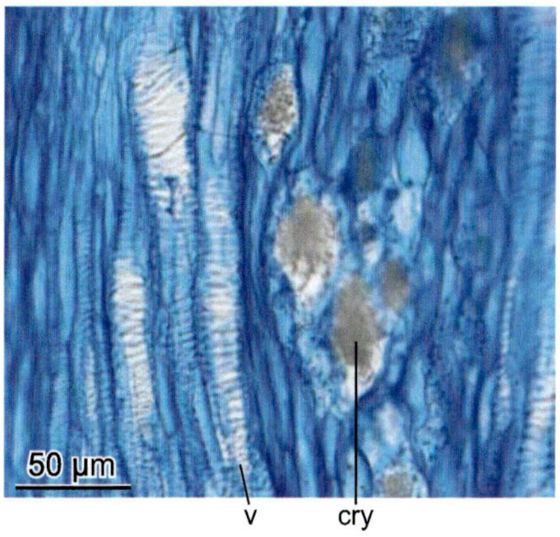

Features according to IAWA

Xylem *1 - 9 - 13 - 20 - 39.2 - 40.1 - 40.2 - 50.2 - 52.2 - 60.1 - 79.1 - 99 - 99.1 - 100.1 - 100.2 - 117 - 144*

Silene gonosperma (Rupr.) Bocquet

Caryophyllaceae

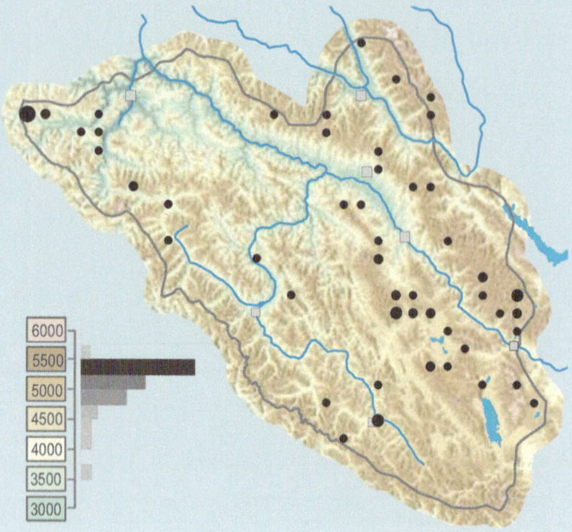

Syn.: *Physolychnis gonosperma* Rupr., *Lychnis pumila* Royle ex Benth., *Melandrium pumilum* (Benth.) Walpers

Description: Perennial herbs, 5–20 cm tall, pubescent, basal leaves linear to oblanceolate, 3–6 cm long, cauline leaves 1–3 pairs, smaller, flowers dark violet, mostly solitary or 2–3, petals usually hidden, calyx inflated, ca. 1.5 cm long, villous and glandular hairy, with dark violet veins.

Distribution: Tajikistan, Afghanistan, N Pakistan, NW India, China (Gansu, Hebei, Qinghai, Shanxi, Xinjiang, Xizang).

Habitat: Unstable slopes, moraines, scree.

Stem diameter 4–5 mm. Bark/xylem proportion 1:5. Ring width 0.1–0.3 mm. Maximal 22 annual rings.

Xylem

Transverse section

Semi-ring-porous. Ring boundaries indicated by larger earlywood vessel diameters (20–30 µm) and smaller latewood vessel diameters (<10 µm). Vessels solitary. Fibers absent. Parenchyma pervasive. With remaining vascular bundles. With large crystal druses.

Tangential section

Rays very large between vascular bundles.

Radial section

Perforation plates simple. Intervessel pits scalariform. Vessel length 70–100 µm.

Bark

Phloem uniform with isolated or paired sieve tubes. Phellem consists of a few layers of small, rectangular cells. Many large crystal druses in the cortex.

Pith

No slide available.

Analyzed: Root collar of polar root of 7 individuals.

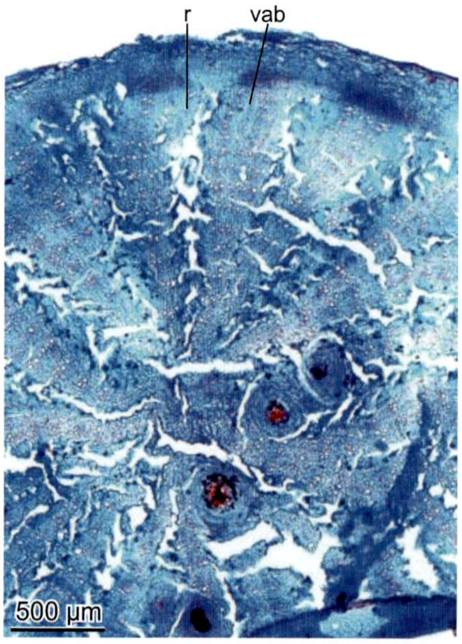

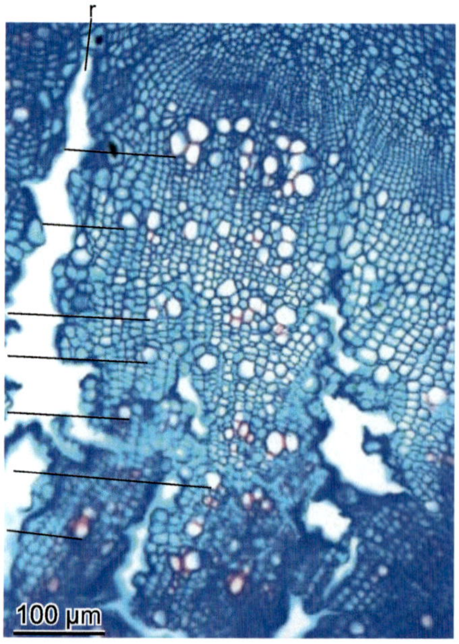

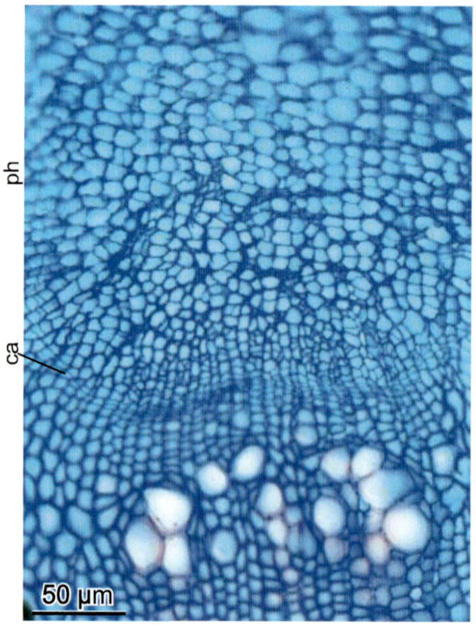

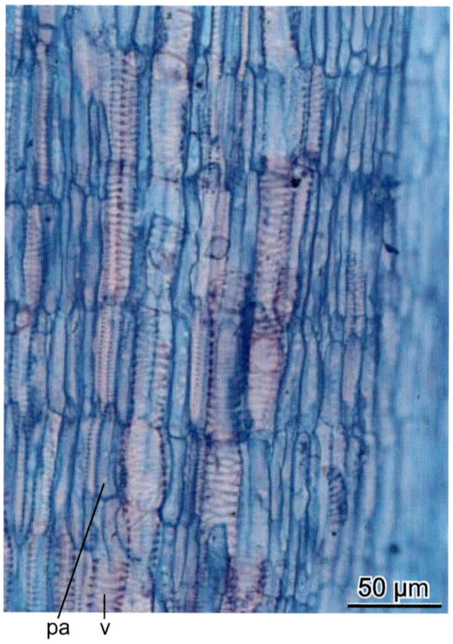

Features according to IAWA

Xylem 1 - 9 - 12 - 13 - 20 - 40.1 - 40.2 - 50.2 - 52.2 - 60.1 - 79.1 - 99 - 99.1 - 100.1 - 100.2 - 117 - 144

Silene rechingeri Bocq.

Caryophyllaceae

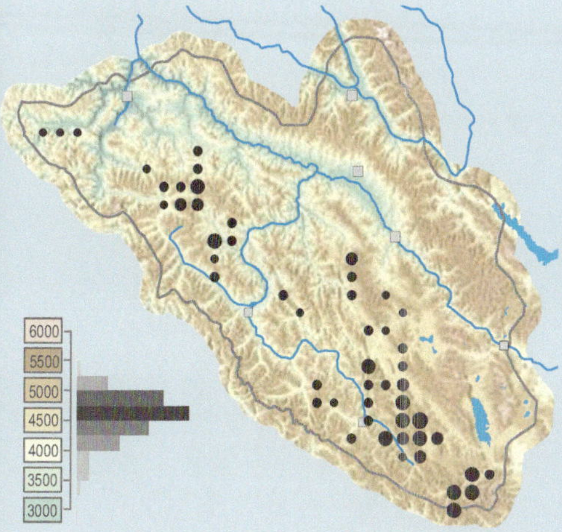

Syn.: *Lychnis cuneifolia* Royle ex Benth.

Description: Perennial herbs, 10–25 cm tall, cespitose, glandular hirsute, basal leaves oblanceolate, stem leaves narrowly elliptic, flowers purple, ca. 1.5 cm long, inclined, later erect, in 1–5-flowered cymes.

Distribution: India (Jammu and Kashmir, Himachal Pradesh), China (Xizang).

Habitat: Alpine meadows.

Stem diameter 4–4.5 mm. Bark/xylem proportion 1:6. Mean ring width 0.2 mm. Maximal 10 annual rings.

Xylem

Transverse section

Semi-ring-porous. Ring variability is indicated by larger earlywood (40–70 µm) and smaller latewood vessel diameters (<15–20 µm). Vessels hardly lignified, angular, solitary and in small groups. Fibers absent. Parenchyma pervasive. With remaining vascular bundles. With crystal druses in rays.

Tangential section

Rays very large between vascular bundles, absent within bundles.

Radial section

Perforation plates simple. Intervessel pits scalariform to reticulate. Vessel length 70–100 µm.

Bark

Phloem uniform small groups of sieve tubes. Many large crystal druses in the cortex. Phellem consists of a few layers of square cells.

Pith

No slide available.

Analyzed: Root collar of polar roots of 2 individuals.

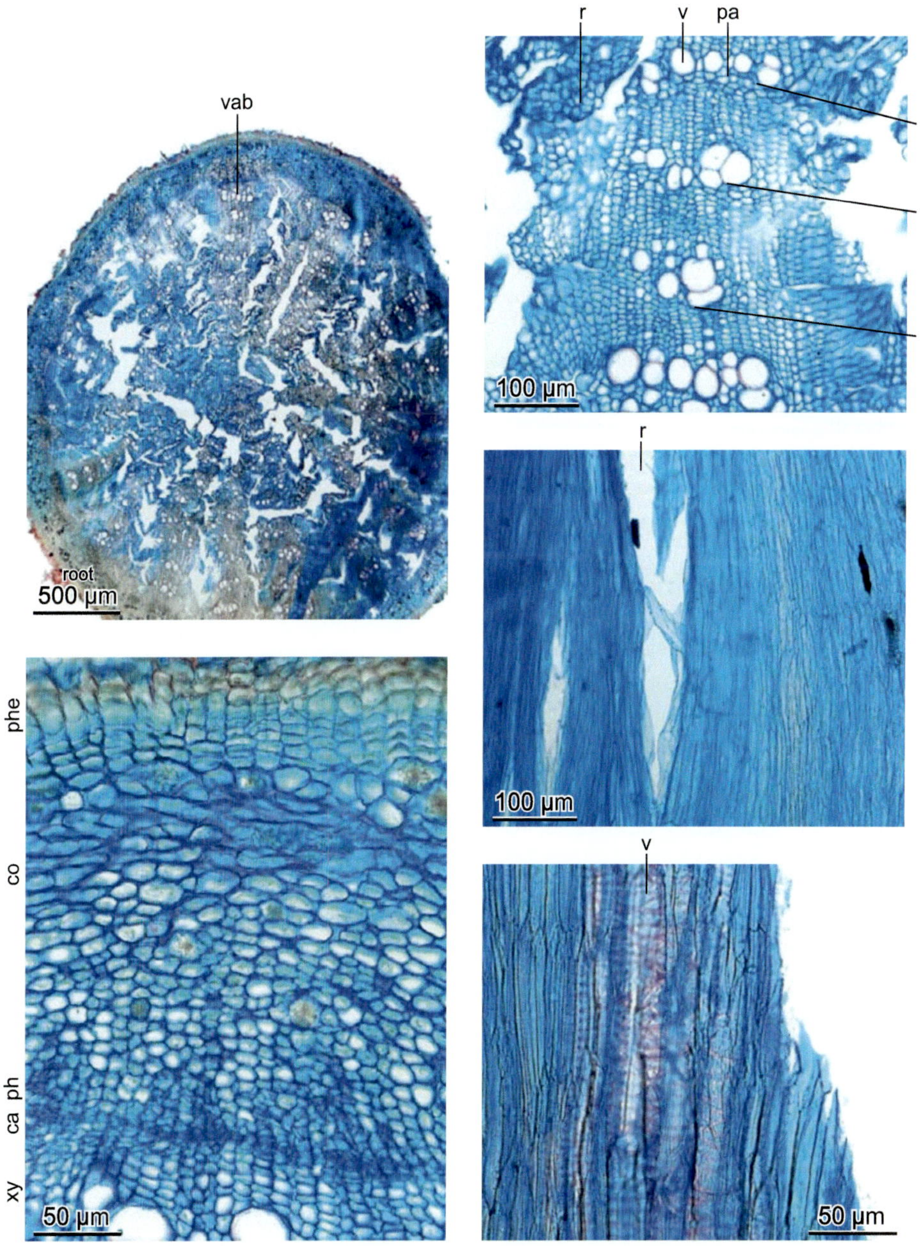

Features according to IAWA

Xylem 1 - 4 - 9 - 11 - 12 - 13 - 20 - 40.2 - 41 - 50.1 - 52.2 - 60.1 - 79.1 - 99 - 100.1 - 100.2 - 144

Silene tenuis Willd

Caryophyllaceae

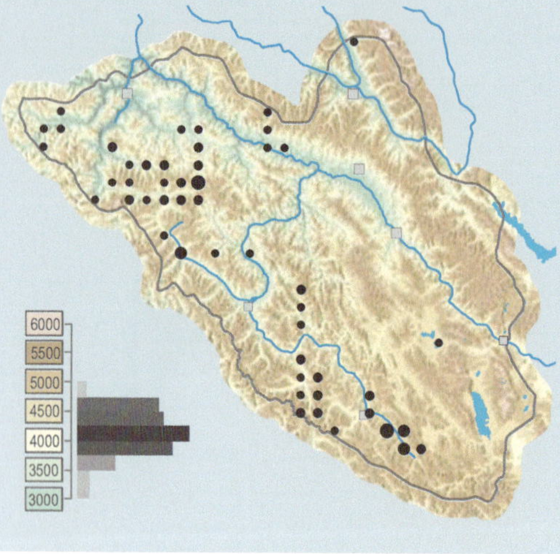

Description: Perennial herbs, 15–50 cm tall, cespitose, leaves mostly basal, narrowly linear, stem leaves 2–3 pairs, flowers whitish, 5–8 mm across, in racemes.

Distribution: Pakistan (Baltistan, Deosai, Chitral, Gilgit, Kaghan), NW India, China.

Habitat: Alpine grasslands.

Stem diameter 2–6 mm. Bark/xylem proportion 1:4. Ring width 0.12–25 mm. Maximal 19 annual rings.

Xylem

Transverse section

Semi-ring-porous. Ring variability indicated by larger earlywood (20–30 µm) and smaller latewood vessel diameters (<10 µm). Vessels solitary, vessel walls not or poorly stained. Fibers absent. Parenchyma pervasive. Crystals absent.

Tangential section

Rays absent.

Radial section

Perforation plates simple. Intervessel pits scalariform. Vessel length 60–100 µm.

Bark

Phloem with small groups of sieve tubes. Phellem consists of many layers of square cells. Crystals absent.

Pith

Round, filled with unlignified round parenchyma cells. Pith poorly delimitated from xylem.

Analyzed: Root collar of polar roots of 3 individuals.

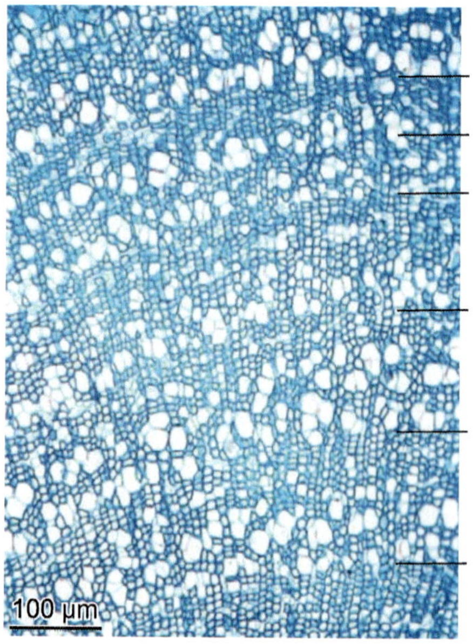

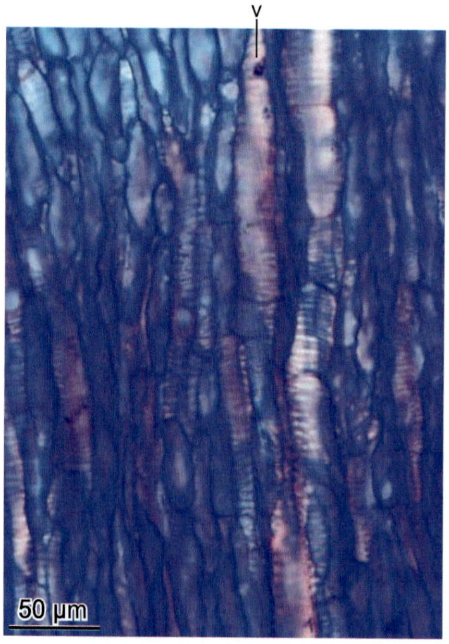

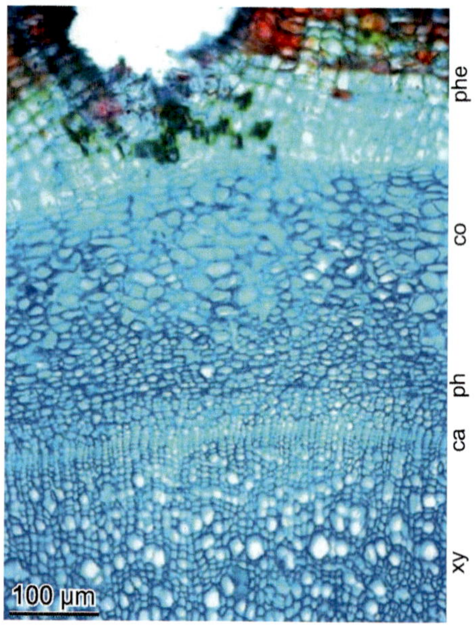

Features according to IAWA

Xylem 1 - 9 - 12 - 13 - 20 - 39.2 - 40.1 - 40.2 - 50.2 - 52.2 - 60.1 - 79.1 - 117

Silene himalayensis (Edgew.) Majumdar

Caryophyllaceae

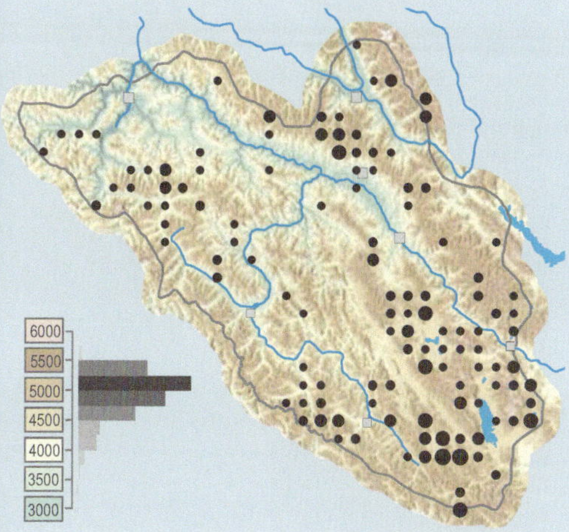

Syn.: *S. gonosperma* subsp. *himalayensis* (Rohrb.) Bocquet

Description: Perennial herbs, 15–30 cm tall, glandular pubescent, leaves narrowly oblanceolate, crowded basally, stem leaves 2–6 pairs, flowers dark red, ca. 1 cm across, slightly nodding, in terminal 3–7-flowered inflorescences.

Distribution: Afghanistan, Pakistan, N India, Nepal, Bhutan, China (Hebei, Hubei, Shaanxi, Sichuan, Xizang, Yunnan).

Habitat: Gravelly slopes, screes, alpine meadows, moraines.

Rings absent. Stem diameter 4–6 mm. Bark/xylem proportion 1:2.

Xylem

Transverse section

Rings absent. Vessel diameter 10–20 µm. Vessels thick-walled Fibers absent. Parenchyma pervasive. Vascular bundles remaining. Very large crystal druses.

Tangential section

Rays >10-seriate (between vascular bundles).

Radial section

Perforation plates simple. Intervessel pits scalariform. Vessel length 50–80 µm.

Bark

Phloem uniform. Phellem consists of a small band of small rectangular cells. Phloem with crystal druses.

Pith

Round, filled with unlignified cells.

Analyzed: Root collar of polar roots of 3 individuals.

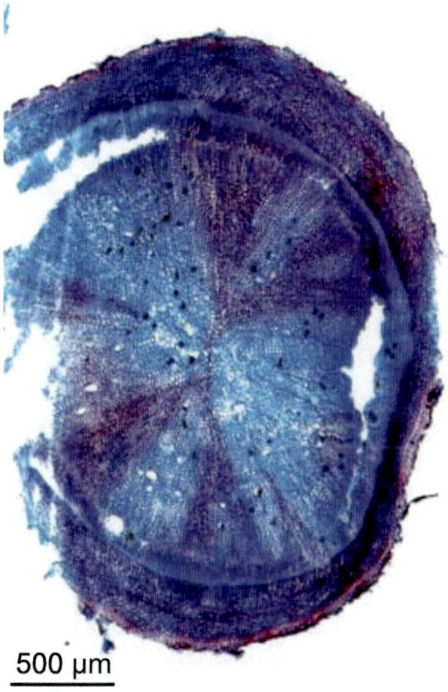

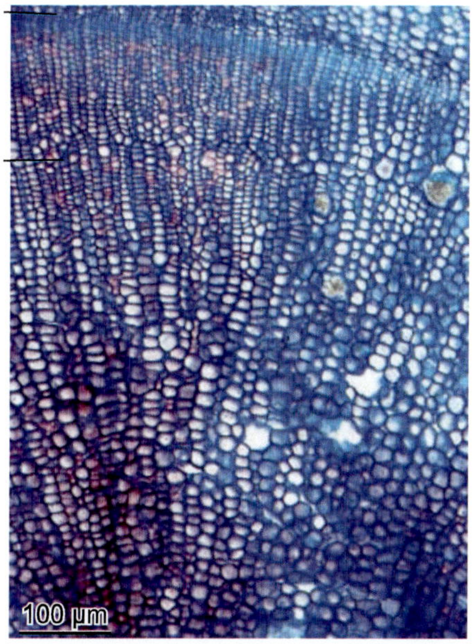

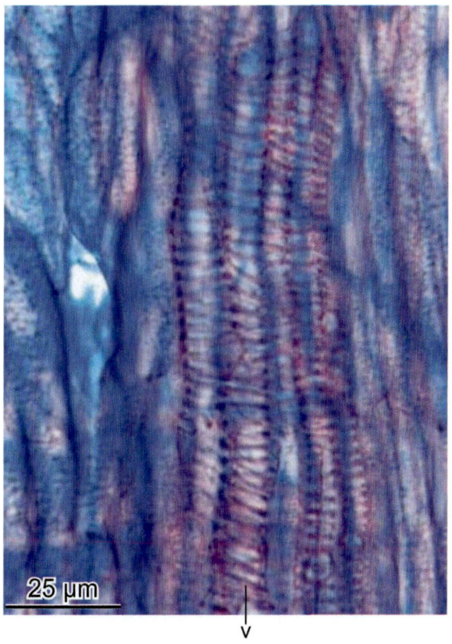

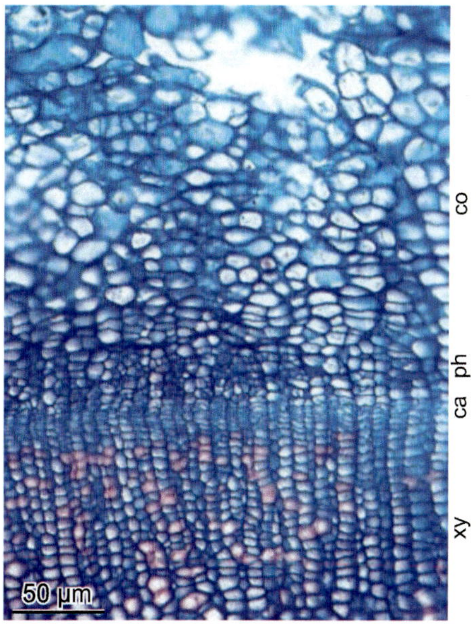

Features according to IAWA

Xylem 1 - 9 - 13 - 20 - 40.1 - 50.2 - 52.2 - 60.1 - 79.1 - 99.1 - 100.1 - 100.2 - 144

Silene moorcroftiana Wall. ex Benth.
Caryophyllaceae

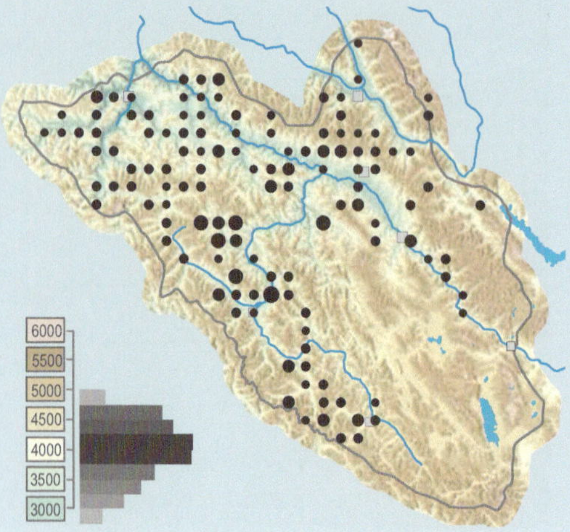

Syn.: *S. persica* subsp. *moorcroftiana* (Wall. ex Benth.) Chowdhuri

Description: Perennial herbs, 10–30 cm tall, cespitose, glandular hairy, leaves linear, lanceolate to oblanceolate, flowers white to pale red, ca. 1 cm across, mostly solitary.

Distribution: Afghanistan, N Pakistan, NW India, Nepal, China (Xizang).

Habitat: Dry gravel and stony habitats.

Stem diameter 4–5 mm. Bark/xylem proportion 1:3. Ring width 0.11–0.3 mm. Maximal 25 annual rings.

Xylem

Transverse section

Semi-ring-porous. Ring boundaries mostly indicated by larger earlywood vessel diameters (30–40 µm) and smaller latewood vessel diameters (<10 µm). Vessels solitary. Fibers absent. Parenchyma pervasive. With small crystal druses.

Tangential section

Rays generally absent, some individuals with vessel-free radial zones.

Radial section

Perforation plates simple. Intervessel pits scalariform. Vessel length 60–90 µm.

Bark

Phellem consists of many layers of rectangular cells. Few crystal druses in the cortex.

Pith

No slide available.

Analyzed: Root collar of polar root of 10 individuals.

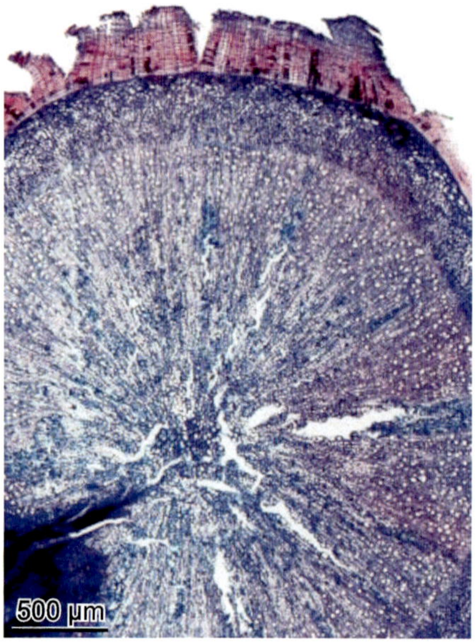

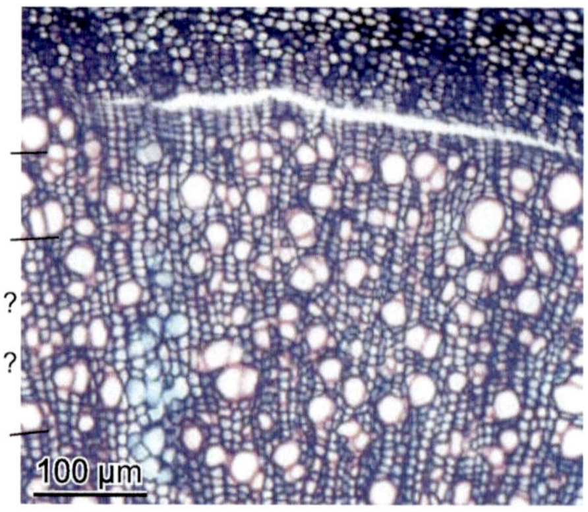

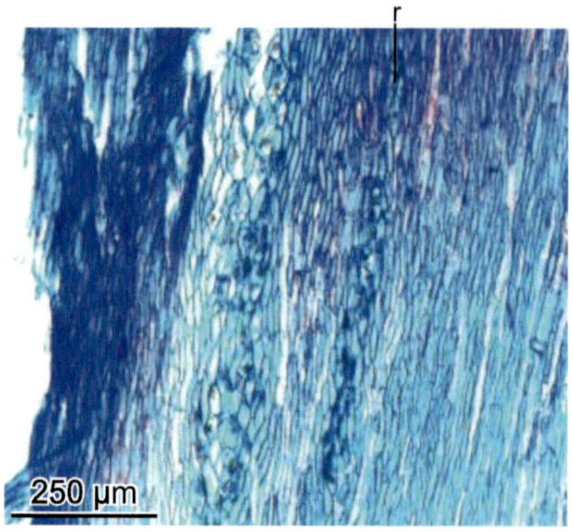

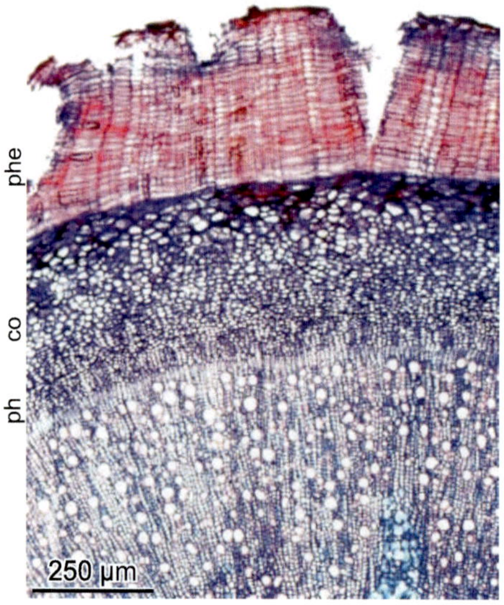

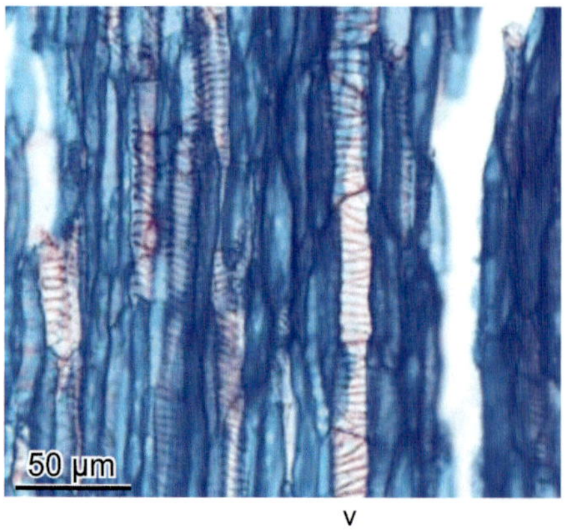

Features according to IAWA

Xylem 1 - 9 - 13 - 20 - 40.1 - 40.2 - 50.2 - 52.2 - 60.1 - 79.1 - 117 - 144

Stellaria decumbens Edgew.

Caryophyllaceae

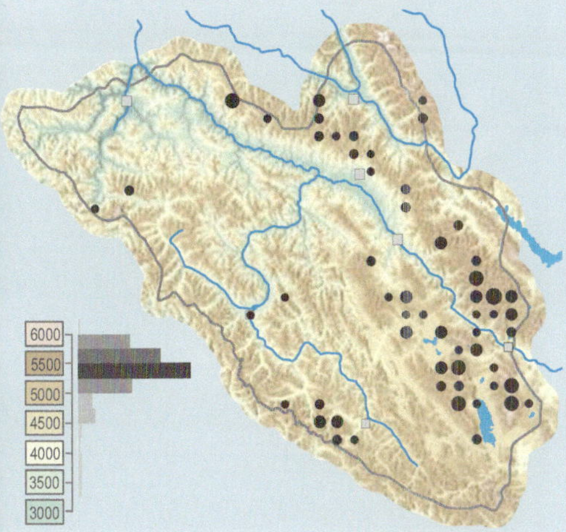

Description: Perennial herbs, 2–20 cm tall, densely cespitose, leaves sessile, ovate, margin ciliate, flowers greenish, ca. 3 mm across, solitary or few in cymes.

Distribution: Pakistan, India, Nepal, China (Qinghai, Sichuan, Xizang, Yunnan), Bhutan.

Habitat: Gravel and stony slopes.

Stem diameter 3–6 mm. Bark/xylem proportion 1:2. Mean ring width 0.09 mm. Maximal 18 annual rings.

Xylem

Transverse section

Semi-ring-porous. Ring boundaries often indicated by larger earlywood vessel diameters (30–40 μm) and smaller latewood vessel diameters (<10 μm). Vessels solitary. Vessel walls fairly thick-walled and poorly stained. Fibers absent. Parenchyma pervasive. Crystals absent.

Tangential section

Rays absent.

Radial section

Perforation plates simple. Intervessel pits scalariform. Vessel length 60–90 μm.

Bark

Phloem with small groups of sieve tubes. Phellem consists of many layers of square cells. Few crystal druses in the cortex.

Pith

No slide available.

Analyzed: Root collar of polar root of 1 individual.

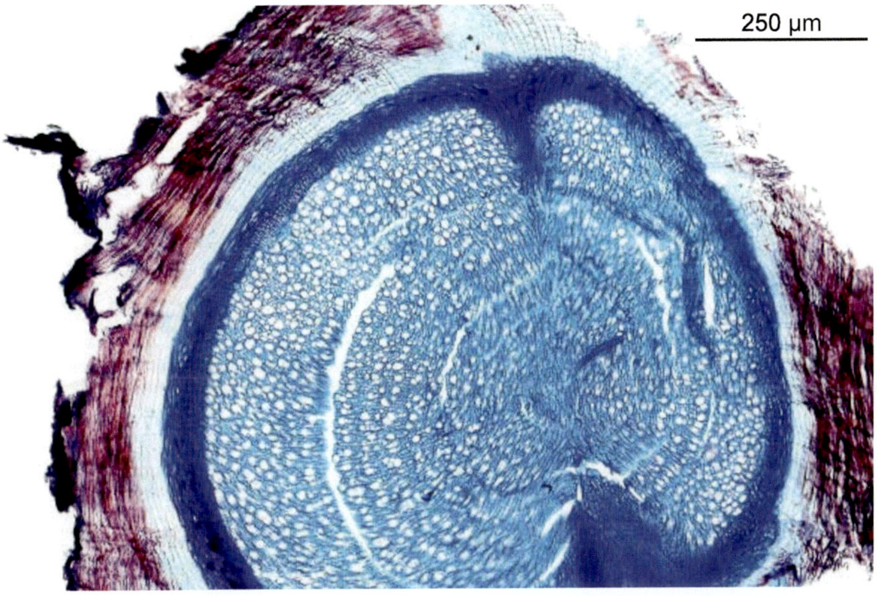

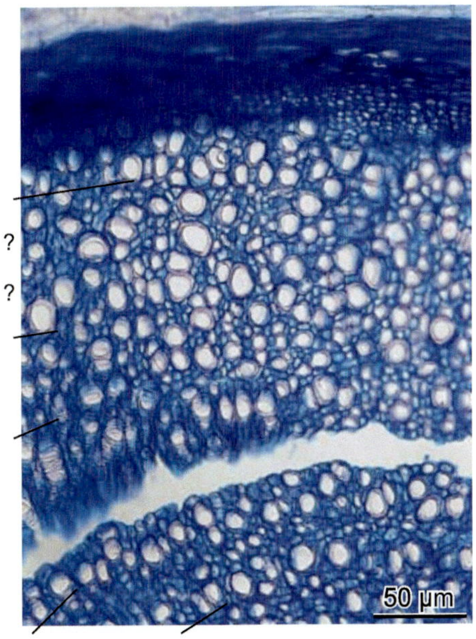

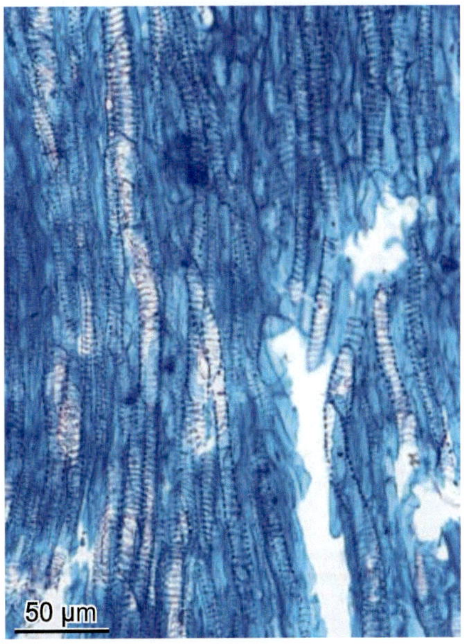

Features according to IAWA

Xylem 1 - 9 - 13 - 20 - 39.1 - 39.2 - 40.1 - 40.2 - 50.2 - 52.2 - 60.1 - 79.1 - 117

Stellaria depressa Em. Schmid
Caryophyllaceae

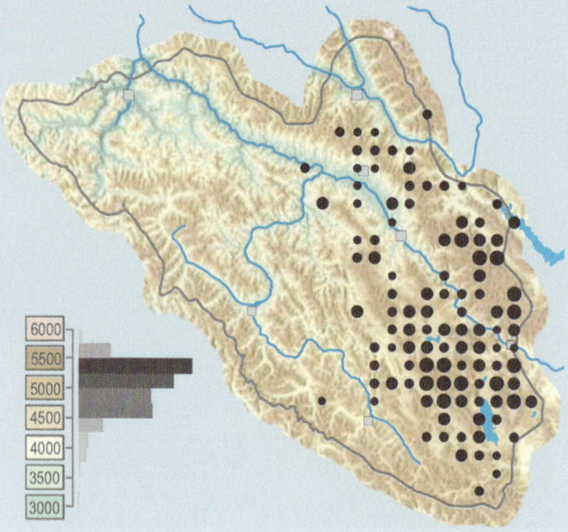

Description: Perennial herbs, 2–15 cm tall, cespitose, stems prostrate to ascending, leaves ovate to lanceolate, glabrous, flowers white, ca. 3 mm across, in 2–7-flowered cymes.

Distribution: NW India, China (?Xinjiang, Xizang).

Habitat: Open vegetation on stabilized slopes, among stones, always mesic to wet habitats.

Stem diameter 0.5 mm. Bark/xylem proportion 1:1. Mean ring width 0.1 mm. Maximal 1 annual ring.

Xylem

Transverse section
Only one ring. Vessel diameter 10–20 µm. Vessels solitary and fairly thick-walled. Fibers absent. Parenchyma pervasive. Crystals absent.

Tangential section
Rays absent.

Radial section
No slide available. (Features recognized at oblique sectioned transverse sections). Perforation plates simple. Intervessel pits scalariform.

Bark

Phloem uniform. Phellem consists of a few layers of square cells. Crystals absent.

Pith

No slide available.

Analyzed: Root collar of polar root of 1 individual.

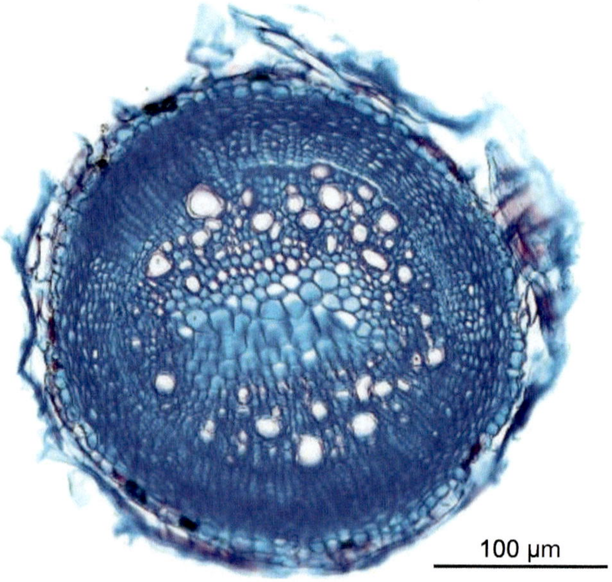

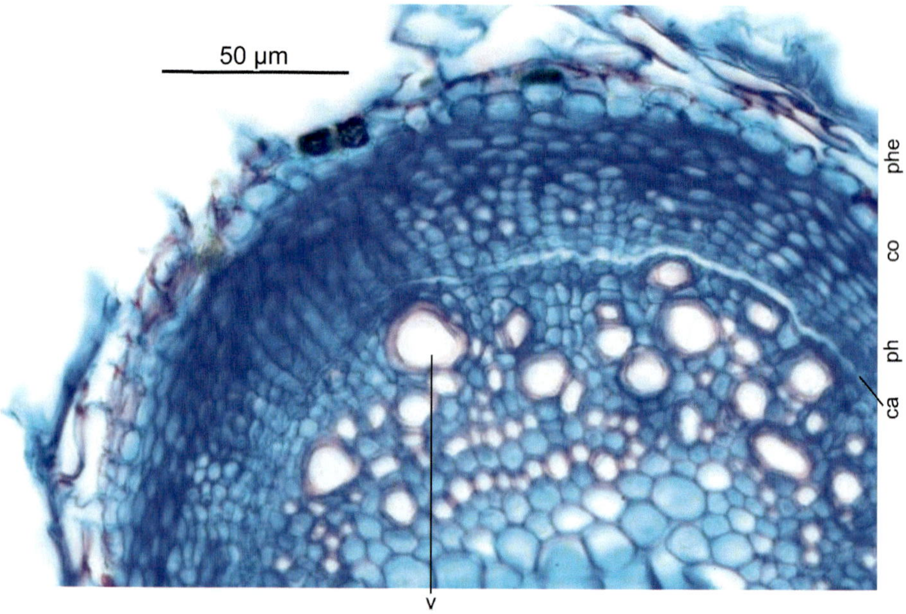

Features according to IAWA
Xylem 2.1 - 9 - 13 - 20 - 39.1 - 40.1 - 50.2 - 52.2 - 60.1 - 79.1 - 117

Stellaria tibetica Kurz

Caryophyllaceae

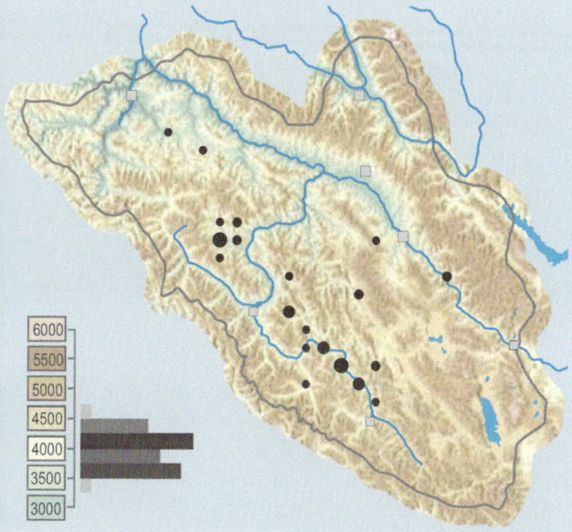

Description: Perennial herbs, 10–15 cm tall, glandular pubescent, stems ascending, leaves sessile, lanceolate, flowers white, ca. 1 cm across, in few-flowered cymes.

Distribution: N Pakistan, India (Jammu and Kashmir), China (Xizang).

Habitat: Rocky crevices, screes.

Stem diameter 2.5 mm. Bark/xylem proportion 1:4. Rings indistinct.

Xylem

Transverse section

Semi-ring-porous. Vessel diameter 25–35 μm. Vessels thick-walled, angular, solitary. Fibers absent. Parenchyma pervasive. With remaining vascular bundles. With large crystal druses in rays.

Tangential section

Rays very large between vascular bundles, absent within bundles.

Radial section

Perforation plates simple. Intervessel pits reticulate. Vessel length 140–150 μm.

Bark

Phloem uniform. Many large crystal druses in the cortex. Phellem consists of many layers of square cells.

Pith

No slide available.

Analyzed: Root collar of polar roots of 1 individual (herbarium specimen).

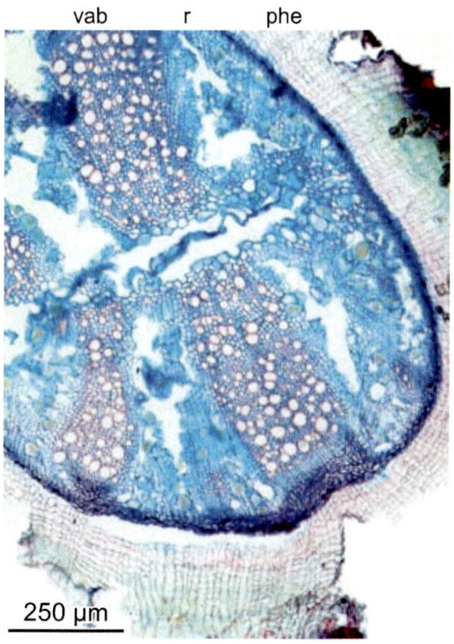

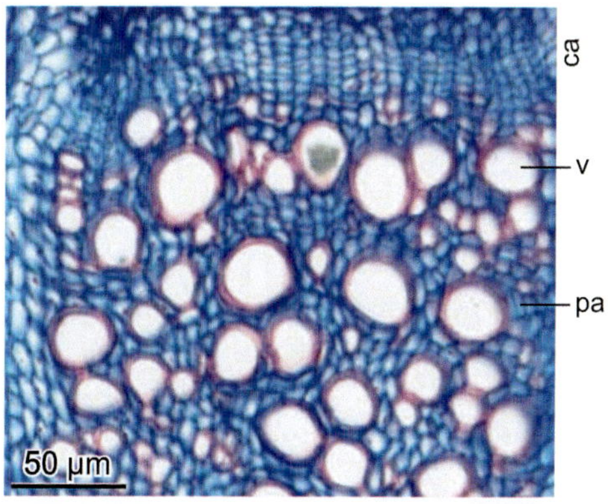

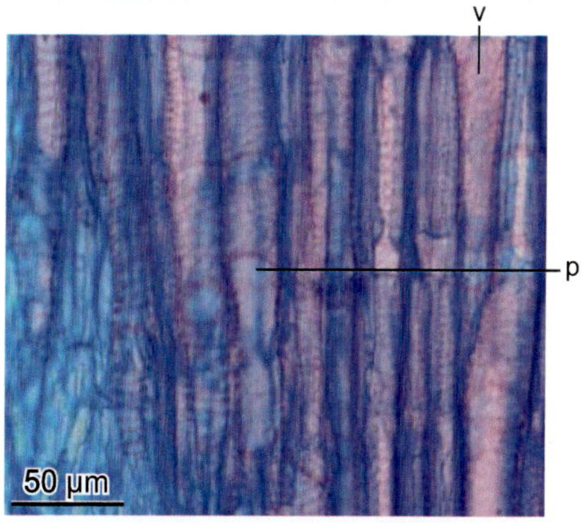

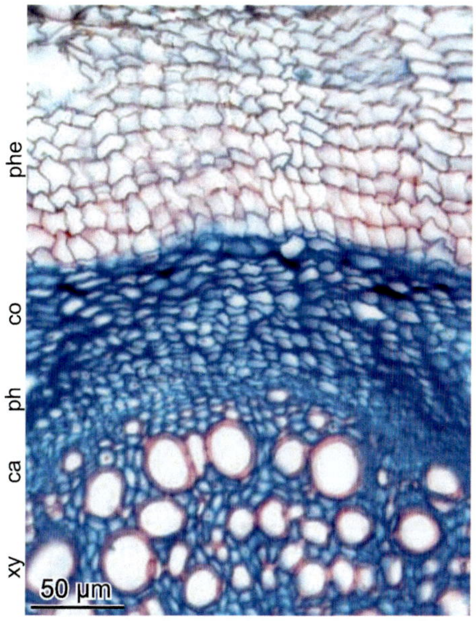

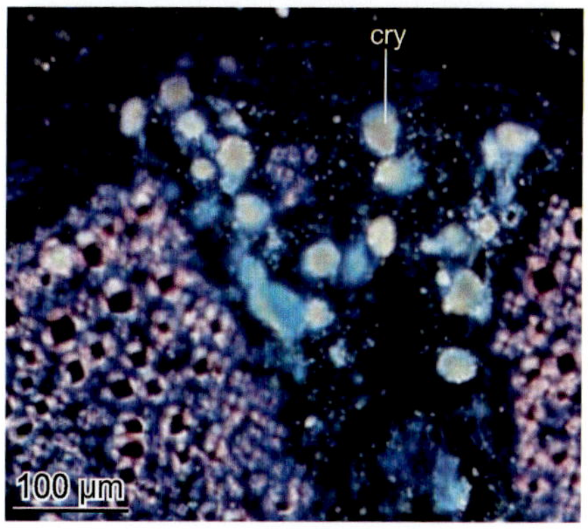

Features according to IAWA

Xylem 2 - 9 - 12 - ,13 - 20 - 39.1 - 40.2 - 50.1 - 52.2 - 60.1 - 79.1 - 99 - 100.1 - 100.2 - 144

Thylacospermum caespitosum (Cambess.) Schischk.

Caryophyllaceae

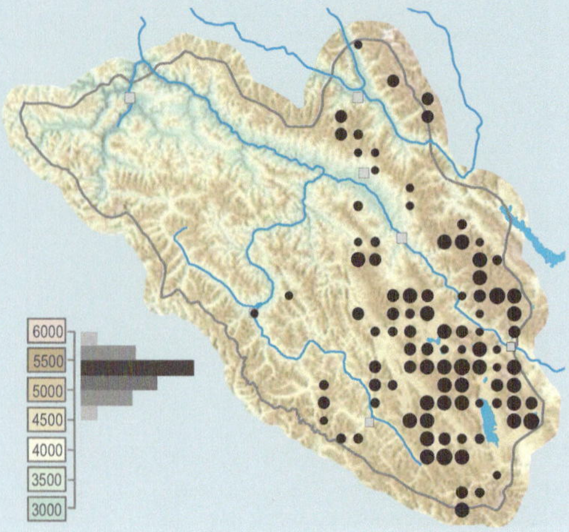

Syn.: *Periandra caespitosa* Cambess., *Arenaria caespitosa* (Cambess.) Kozhevn. 1985, non Ehrh. 1790, *Bryomorpha rupifraga* Kar. & Kir.

Description: Perennial herbs, 1–25 cm tall, forming dense semiglobose to flat cushions, 10–100 cm across, stems densely branched, leaves ovate, rigid, 2–4 × 2 mm, withered leaves persist on stems, filling room between them, flowers green to yellowish, ca. 4 mm across, solitary.

Distribution: Kazakhstan, Kyrgyzstan, NW India, Nepal, W China.

Habitat: Mesic to wet stabilized slopes, old stabilized stony river sediments, rock crevices.

Stem diameter 5–15 mm. Bark/xylem proportion 1:7. Rings absent. Maximal age unknown, probably >40 annual rings.

Xylem

Transverse section

No rings. Vessels solitary, fairly thick-walled, unstained. Vessel diameter 10–20 μm. Fibers absent. Parenchyma pervasive. Rays absent. Irregularly formed zones of successive cambia (Caryophyllaceae type). Crystals absent.

Tangential section

Rays absent.

Radial section

Perforation plates simple. Intervessel pits scalariform. Vessel length 60–80 μm.

Bark

Phloem with small groups of sieve tubes. Few crystal druses in the cortex. Phellem consists of a large, multilayered band of small, rectangular cells.

Pith

No slide available.

Analyzed: Root collar of polar root of 4 individuals.

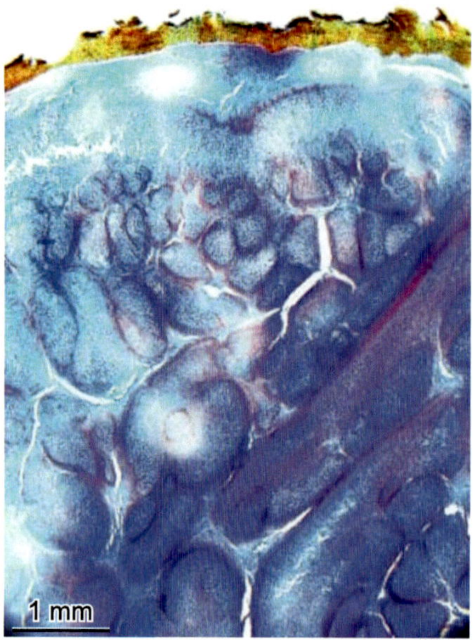

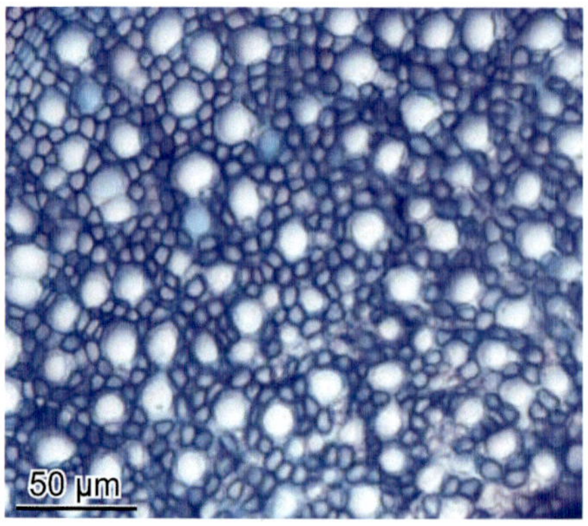

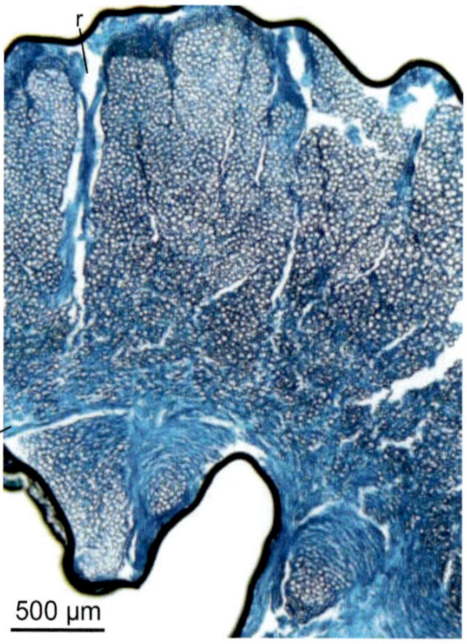

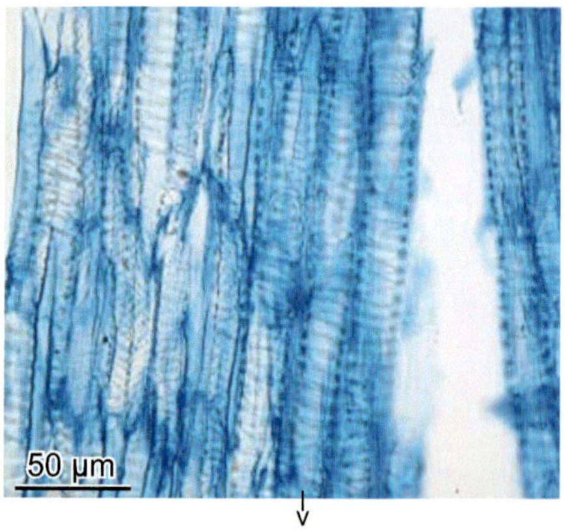

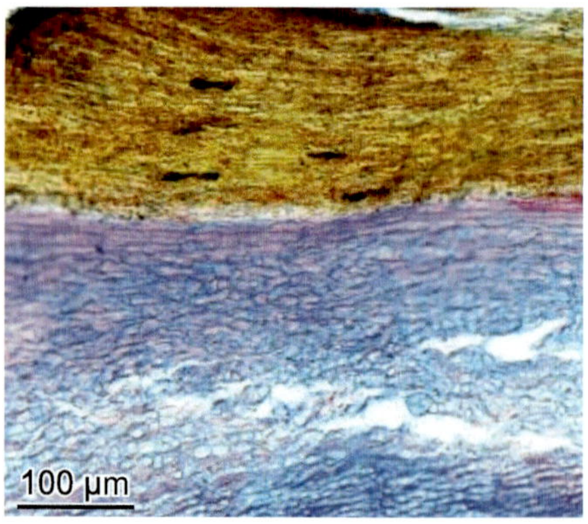

Features according to IAWA

Xylem 2 - 9 - 13 - 20 - 39.1 - 39.2 - 40.1 - 50.2 - 52.3 - 60.1 - 79.1 - 117 - 133

Parnassia pusilla Wall.

Celastraceae

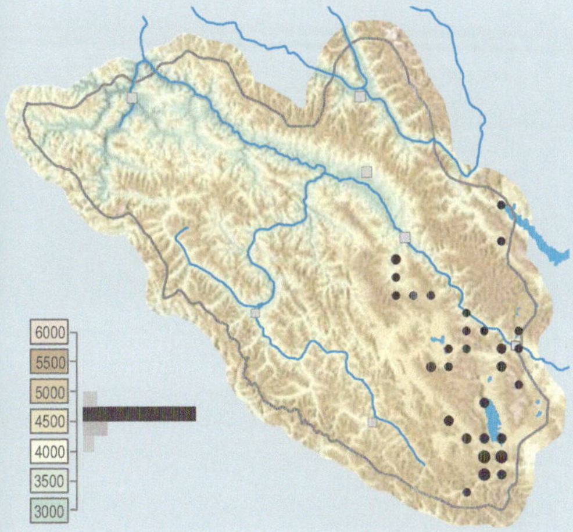

Description: Perennial herbs, 3–10 cm tall, basal leaves 2–4, reniform or ovate-cordate, stem leaf single, sessile, semi-amplexicaul, flowers white, 6–8 mm across, solitary, terminal.

Distribution: N India, Nepal, China (S Xizang), Bhutan.

Habitat: Wet meadows.

Stem diameter 3–6 mm. Bark/xylem proportion 1:0.3. Rings absent.

Xylem

Transverse section

Rings absent. Vessels angular, solitary or in small groups. Vessel diameter 10–20 µm. Fibers absent. Parenchyma pervasive. Crystals absent.

Tangential section

Rays absent.

Radial section

Perforation plates simple. Intervessel pits scalariform. Vessel length 50–70 µm.

Bark

Phloem with small groups of sieve tubes. Cortex consists of large, oval parenchyma cells. Phellem absent.

Pith

Round, filled with small, round, unlignified cells.

Analyzed: Root collar of polar root of 4 individuals.

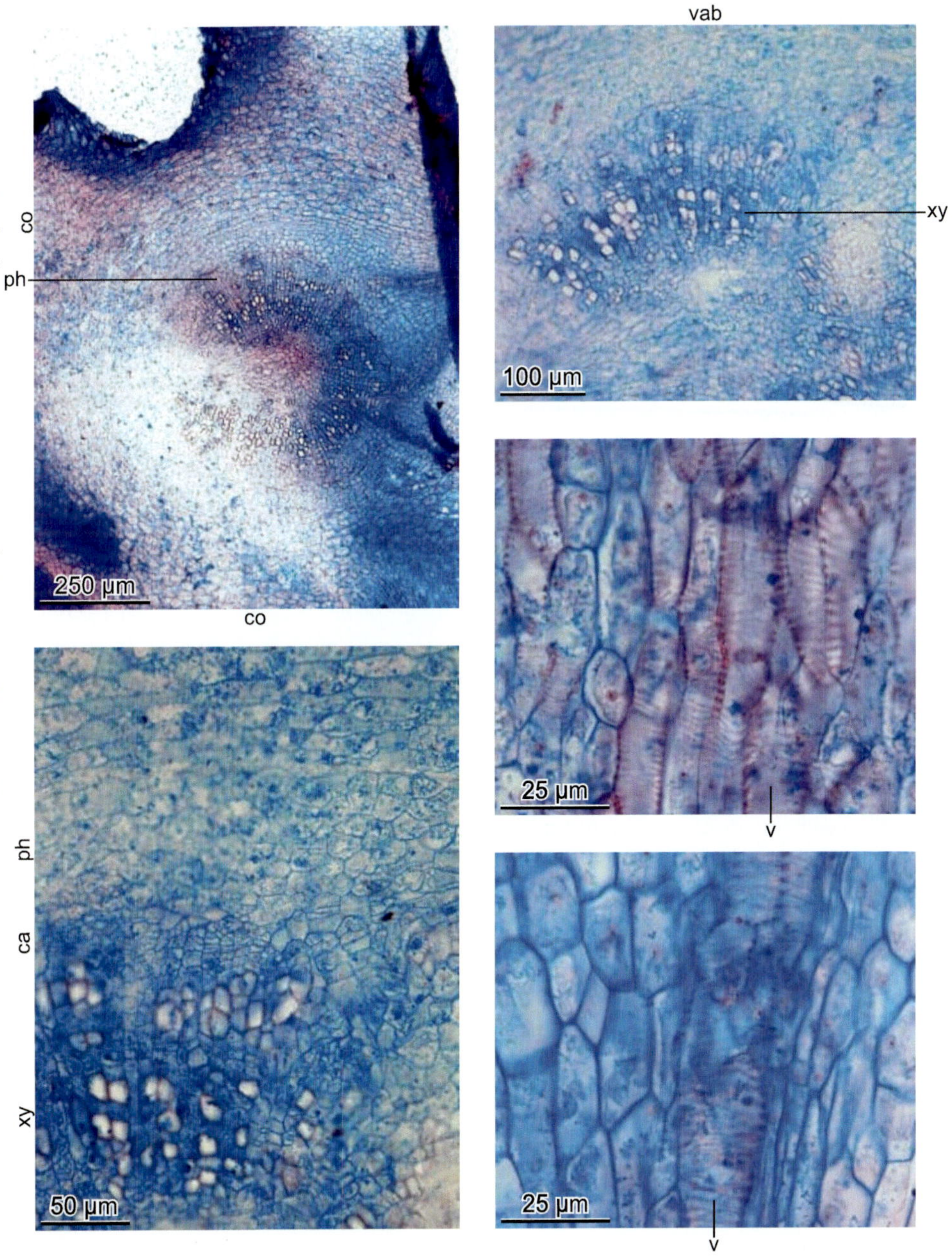

Features according to IAWA

Xylem 2 - 9 - 11 - 12 - 13 - 20 - 40.1 - 50.2 - 52.3 - 60.1 - 79.1 - 117

Hylotelephium ewersii (Ledeb.) H.Ohba

Crassulaceae

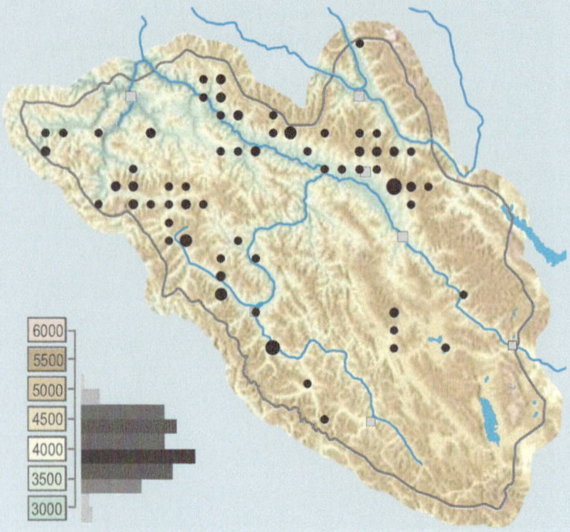

Syn.: *Sedum azureum* Royle, *S. ewersii* Ledeb., *S. pakistanicum* G.R.Sarwar

Description: Perennial herbs, 5–25 cm tall, glabrous, stems ascending, leaves broadly ovate, opposite, sessile, fleshy, flowers purplish red or pink, ca. 3 mm across, in many-flowered condensed terminal cymes, 2–3 cm across.

Distribution:
Russia, Kazakhstan, Kyrgyzstan, Tajikistan, Afghanistan, Pakistan, India, China (Nei Mongol, Xinjiang, Xizang), Mongolia.

Habitat:
Mesic to nearly wet grasslands on river banks.

Stem diameter 5–18 mm. Bark/xylem proportion 1:1. Maximal 14 annual rings. Stem segregation.

Xylem

Transverse section
Semi-ring-porous. Rings distinctly marked by differences between earlywood (40–50 µm) and latewood vessel diameters (10 µm). Angular, hardly stained vessels solitary or in groups. With a few vessel- and ray-less tangential fiber groups. Parenchyma pervasive. Rays indistinct between compartments of vessel/parenchyma zones. With intra-xylary phellem towards the pith. Crystals absent.

Tangential section
Rays indistinct and confluent with axial tissue.

Radial section
Perforation plates simple. Intervessel pits scalariform (*Crassula* type). Vessel length 70–100 µm.

Bark

Phloem with groups of sieve tubes. Ray dilatation. Phellem consists of a multilayered band of small, rectangular cells.

Pith

Round.

Analyzed: Root collar of polar root of 7 individuals.

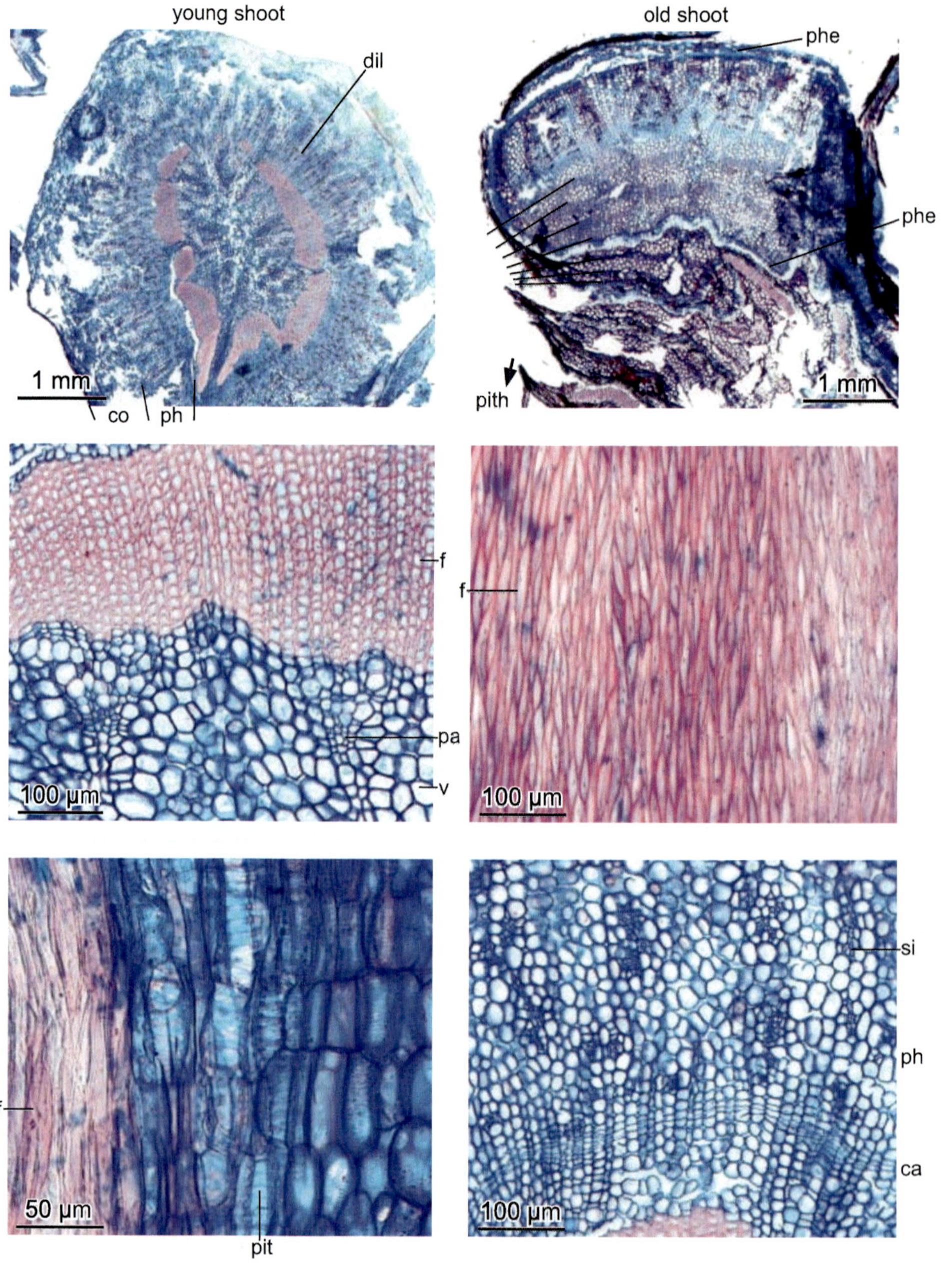

Features according to IAWA
Xylem 2 - 9 - 11 - 13 - 20 - 39.2 - 40.2 - 50.2 - 52.2 - 61 - 70 - 99.1 - 98 - 100.1 - 100.2 - 135.2

Rhodiola crenulata (Hook.f. & Thomson) H.Ohba

Crassulaceae

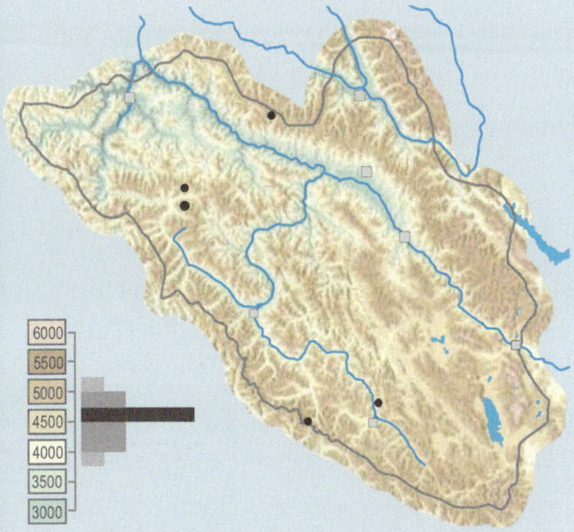

Syn.: *Sedum crenulatum* Hook.f. & Thomson

Description: Perennial herbs, 5–20 cm tall, glabrous, stems densely leafy, leaves obovate to elliptic-oblong, 1–3 cm long, margin undulate to crenate, flowers red to purplish red, inflorescence dense, more or less bracteate, 1–3 cm across.

Distribution: N India, Nepal, China (Qinghai, Sichuan, Xizang, Yunnan), Bhutan.

Habitat: Rock crevices, alpine grasslands.

Stem diameter 7–12 mm. Bark/xylem proportion 1:3. Maximum 12 annual rings. Stem segregation.

Xylem

Transverse section

Semi-ring-porous. Rings distinctly marked by differences between earlywood (30–40 µm) and latewood vessel diameters (10 µm). Angular, hardly stained vessels mainly solitary. Fibers absent. Parenchyma pervasive. Indistinct rays consisting of round, radially oriented parenchyma cells. With intra-xylary phellem towards the pith. Crystals absent.

Tangential section

Rays indistinct and confluent with axial tissue. Parenchyma cells slightly storied.

Radial section

Perforation plates simple. Intervessel pits scalariform (*Crassula* type). Vessel length 70–100 µm.

Bark

Phloem with groups of sieve tubes. Phellem consists of a multilayered band of small, rectangular, unstained cells.

Pith

No slide available.

Analyzed: Root collar of polar root of 3 individuals.

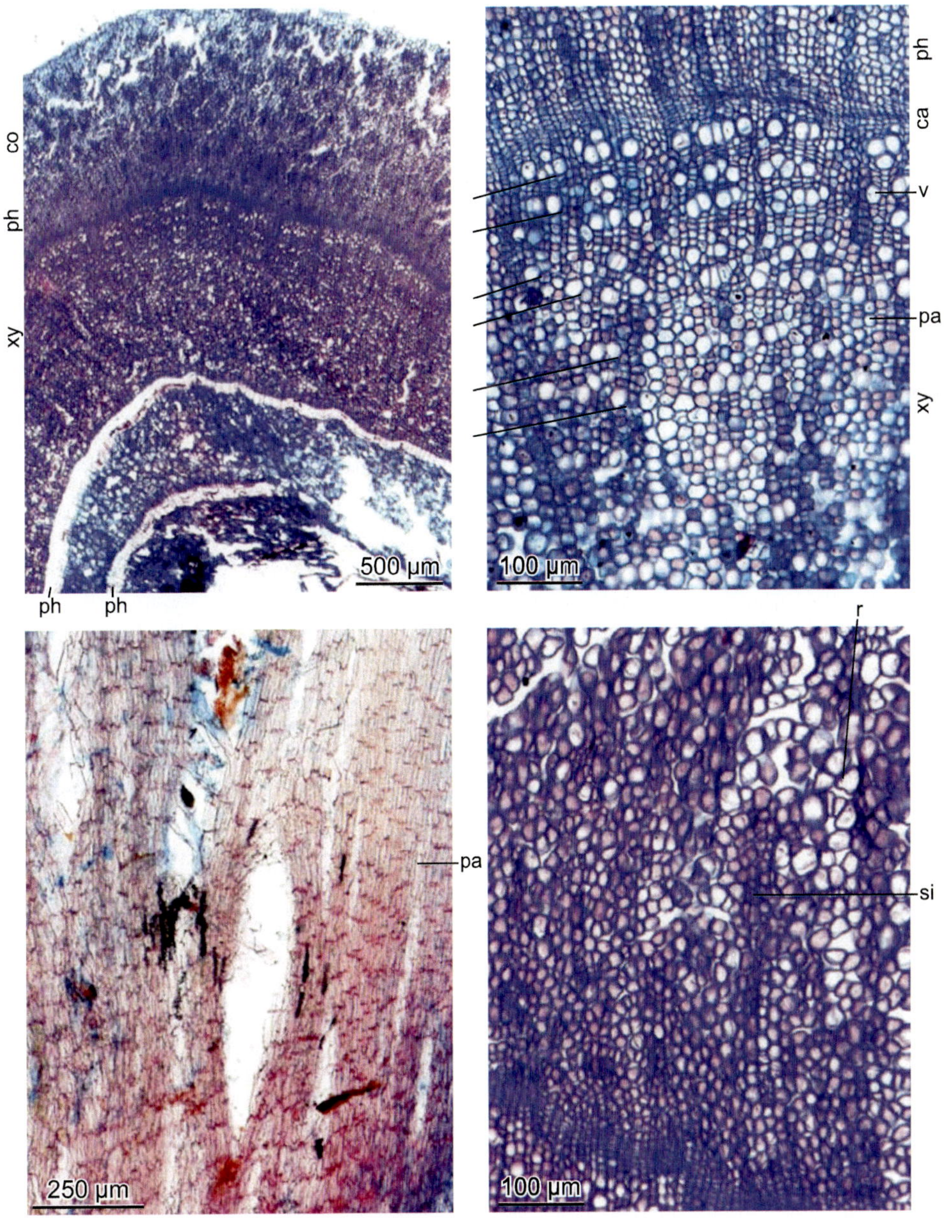

Features according to IAWA

Xylem 2 - 9 - 11 - 13 - 20 - 39.2 - 40.2 - 50.2 - 52.2 - 60.1 - 79.1 - 97 - 100.1 - 100.2 - 120 - 135.2

Rhodiola heterodonta (Hook.f. & Thomson) S.H.Fu

Crassulaceae

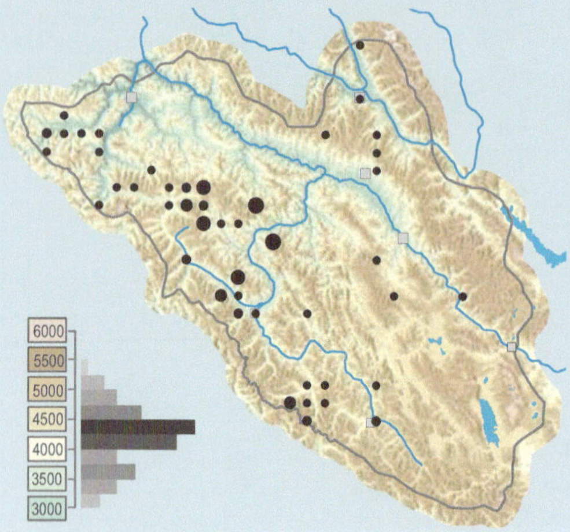

Syn.: *Sedum heterodontum* Hook.f. & Thomson

Description: Perennial herbs, 15–40 cm tall, dioecious, glabrous, leaves alternate, sessile, triangular-ovate, base cordate and amplexicaul, margin coarsely serrate, flowers greenish yellow, cymes dense, ebracteate, 1–4 cm across.

Distribution: Iran, Tajikistan, Afghanistan, Pakistan, India, Nepal, China (Xinjiang, Xizang), Mongolia.

Habitat: Dry to mesic stony slopes, river banks.

Stem diameter 5–10 mm. Bark/xylem proportion 1:2. Rings absent. Maximal 13 annual rings. Stem segregation.

Xylem

Transverse section

Semi-ring-porous. Rings marked by differences between earlywood (30–40 μm) and latewood vessel diameters (10 μm). Vessels mainly solitary. Fibers absent. Parenchyma pervasive. Indistinct rays, consisting of round, radially oriented parenchyma cells. With intra-xylary phellem towards the pith. Crystals absent.

Tangential section

Rays indistinct and confluent with axial tissue. Parenchyma cells slightly storied.

Radial section

Perforation plates simple. Intervessel pits scalariform (*Crassula* type). Vessel length 70–100 μm.

Bark

Phloem with groups of sieve tubes. Phellem consists of a multilayered band of small, rectangular, unstained cells.

Pith

No slide available.

Analyzed: Root collar of polar root of 4 individuals.

root

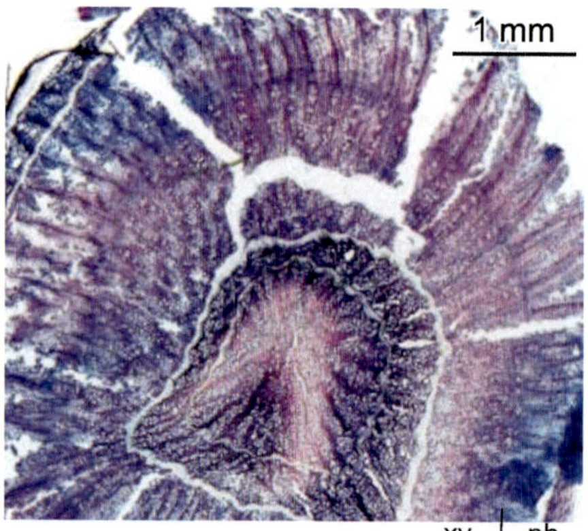

segregated stem

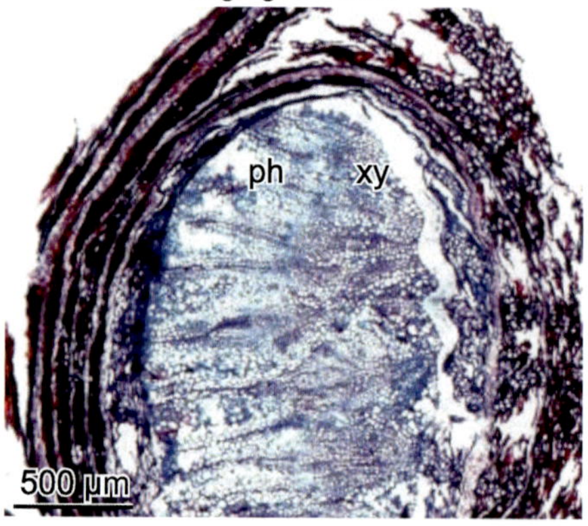

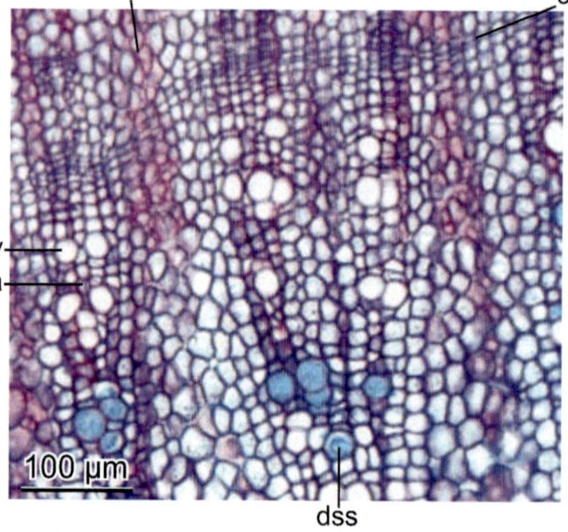

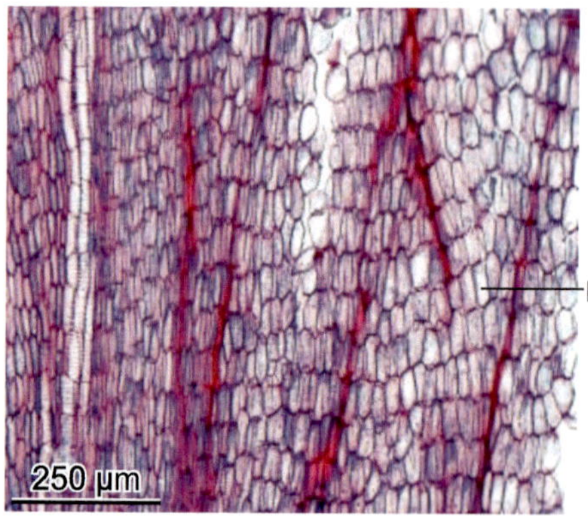

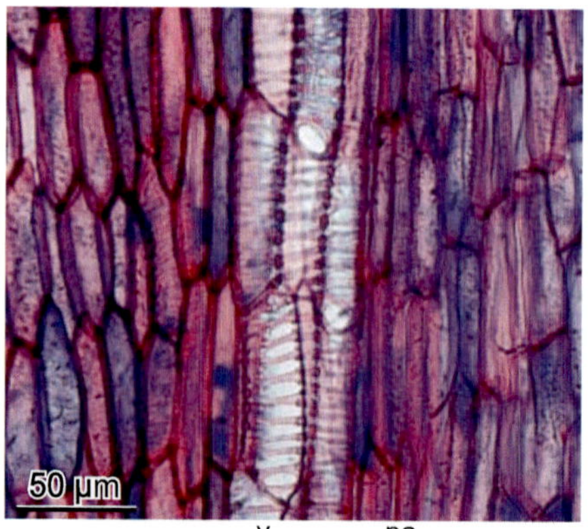

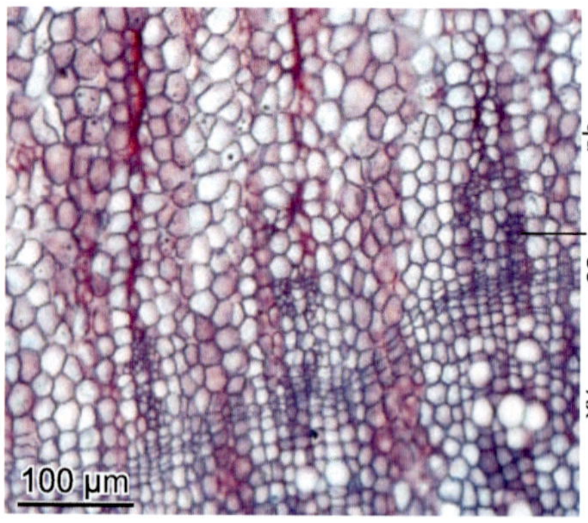

Features according to IAWA

Xylem 2 - 9 - 11 - 13 - 20 - 40.2 - 50.2 - 52.2 - 60.1 - 79.1 - 97 - 100.1 - 100.2 - 120 - 135.2

Rhodiola imbricata Edgew.

Crassulaceae

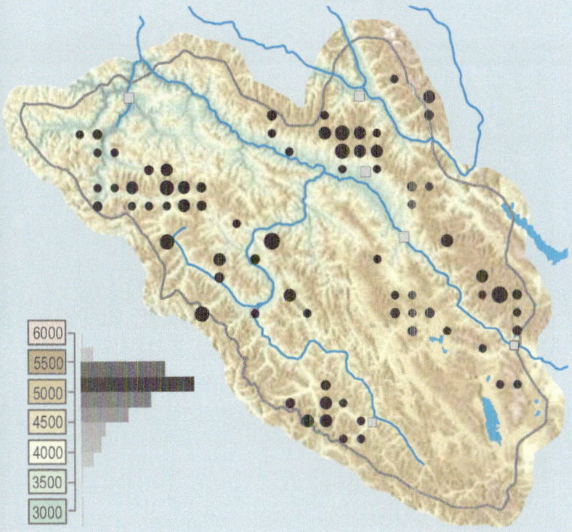

Syn.: *Sedum imbricatum* (Edgew.) Walp.

Description: Perennial herbs, 10–30 cm tall, dioecious, glabrous, stems clustered, leaves narrowly elliptic or lanceolate, sessile, imbricate, margin entire to denticulate along the upper half, flowers creamy yellow, in dense terminal compound corymbs, 1–3 cm across.

Distribution: Pakistan, NW India, Nepal.

Habitat: Stabilized stony slopes, river banks.

Stem diameter 10–15 mm. Bark/xylem proportion 1:3. Rings absent. Maximal 6 annual rings. Stem segregation. Center rotten.

Xylem

Transverse section

Semi-ring-porous. Rings slightly marked by differences between earlywood (30–40 µm) and latewood vessel diameters (20 µm). Vessels solitary and in groups. Fibers absent. Parenchyma pervasive. Indistinct rays, consisting of round, radially oriented zones of parenchyma cells. Older specimens with successive cambia. With intra-xylary phellem towards the pith. Crystals absent.

Tangential section

Rays indistinct and confluent with axial tissue.

Radial section

Perforation plates simple. Intervessel pits scalariform (*Crassula* type). Vessel length 70–100 µm.

Bark

Phloem with groups of sieve tubes. Phellem consists of a multilayered band of small, rectangular, unstained cells.

Pith

No slide available.

Analyzed: Root collar of polar root of 8 individuals.

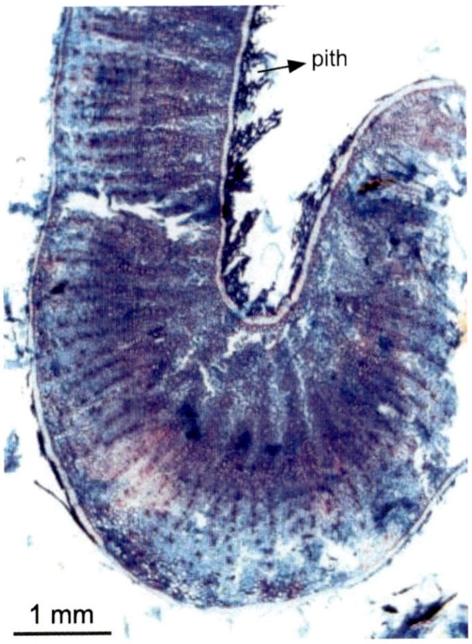

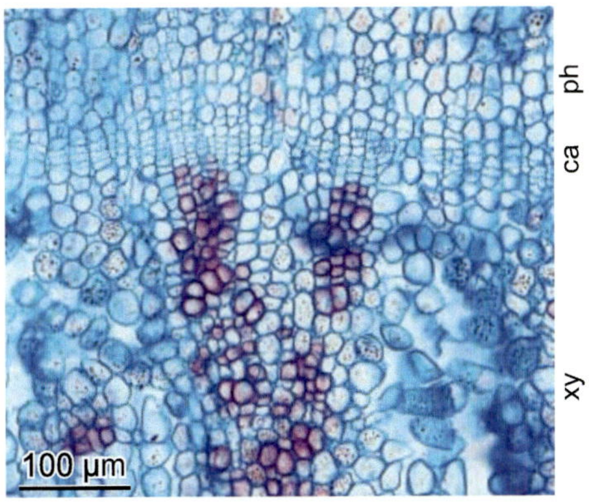

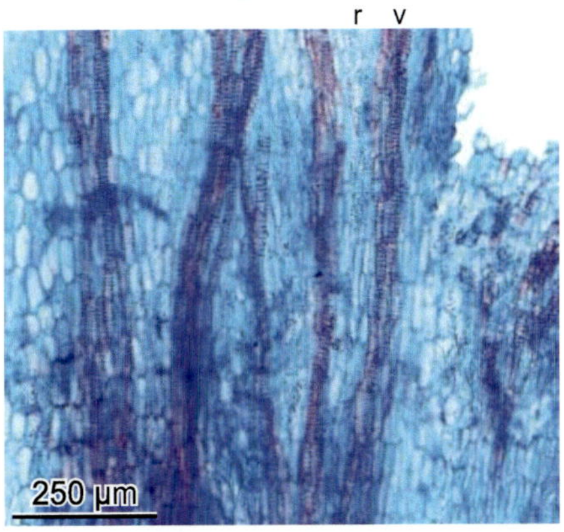

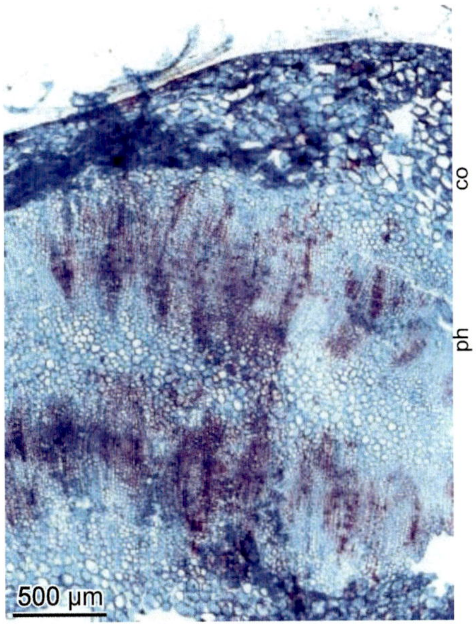

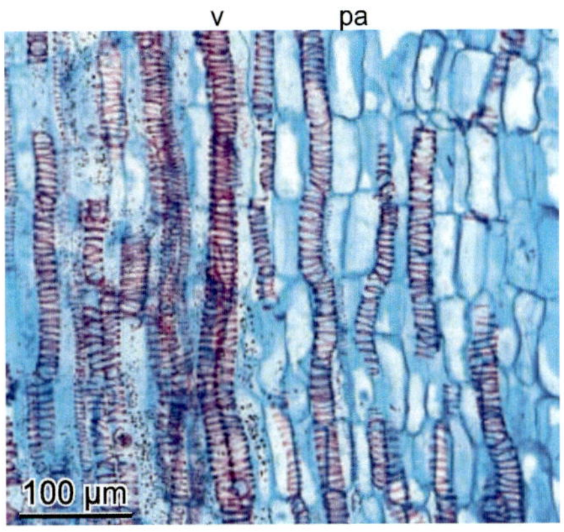

Features according to IAWA

Xylem 2 - 9 - 11 - 13 - 20 - 40.2 - 50.2 - 52.2 - 60.1 - 79.1 - 98 - 99.1 - 100.1 - 100.2 - 120 - 133.1 - 135.2

Rosularia alpestris (Kar. & Kir.) Boriss.

Crassulaceae

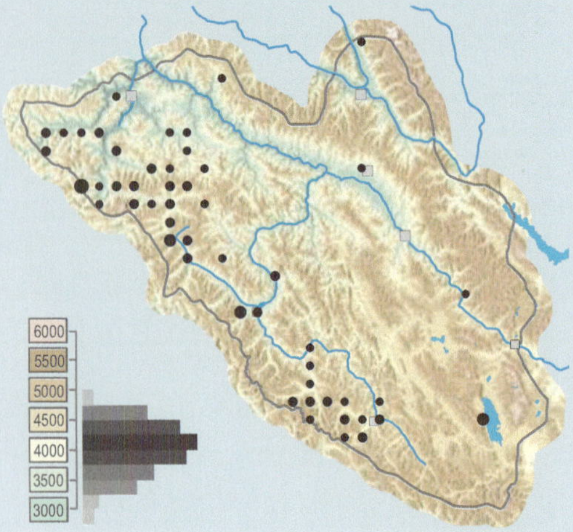

Syn.: *Umbilicus alpestris* Kar. & Kir, *Sempervivum acuminatum* Decne.

Description: Perennial herbs, 5–15 cm tall, glabrous, leaves in rosettes, obovate to lanceolate, flowers white to light purple, in many-flowered terminal corymbs, ca. 2 cm across.

Distribution: Kazakhstan, Uzbekistan, Kyrgyzstan, Afghanistan, N Pakistan, NW India, China (Xizang, Xinjiang).

Habitat: Among stones, gravel slopes.

Stem diameter 4–10 mm. Bark/xylem proportion 1:0.3. Rings absent.

Xylem

Transverse section

Rings absent. Diameter of larger vessels 15–20 µm. Vessels solitary. Fibers absent. Parenchyma pervasive. Large rays between vascular bundles. Crystals absent.

Tangential section

Rays indistinct and confluent with axial tissue.

Radial section

Perforation plates simple. Intervessel pits scalariform (*Crassula* type). Vessel length 70–100 µm.

Bark

Phloem with groups of sieve tubes in radial rows. With slime. Phellem consists of a multilayered band of square, unstained cells.

Pith

Irregularly formed, filled with thin-walled cells.

Analyzed: Root collar of polar root of 3 individuals.

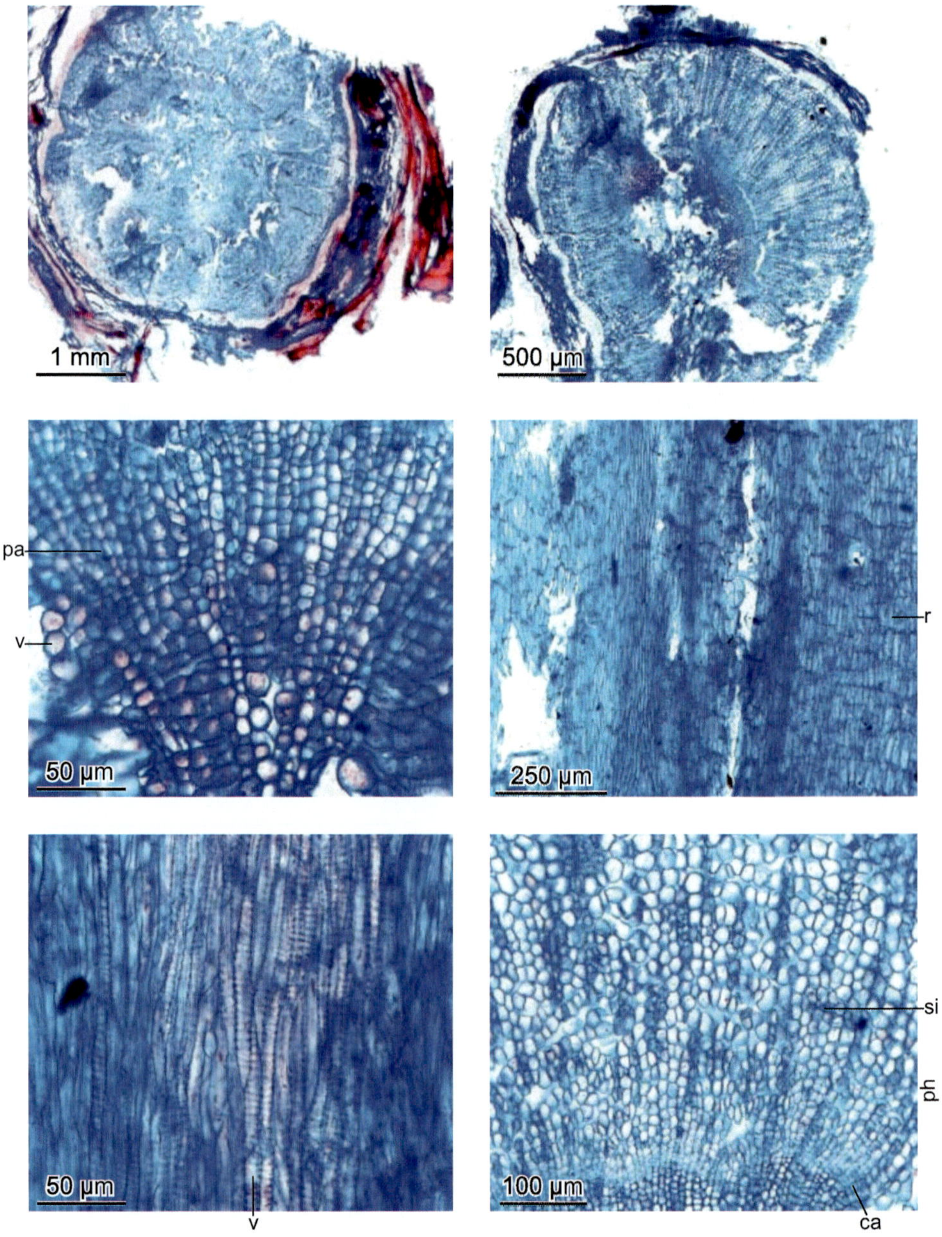

Features according to IAWA

Xylem 2 - 9 - 13 - 20 - 40.2 - 50.2 - 52.2 - 60.1 - 79.1 - 98 - 99.1 - 100.1 - 100.2

Juniperus semiglobosa Regel

Cupressaceae

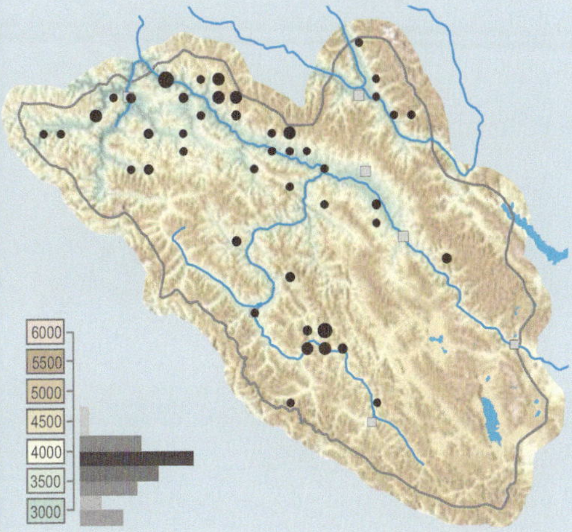

Description: Evergreen coniferous trees or shrubs, up to 12 m tall, dioecious, rarely monoecious, leaves both scale-like and needle-like, pollen cones ellipsoid, 3–5 mm long, seed cones light brown to bluish black when ripe, glaucous, globose or nearly triangular, 2- or 3-seeded.

Distribution: Kazakhstan, Uzbekistan, Kyrgyzstan, Tajikistan, Afghanistan, Pakistan, N India, China (SW Xinjiang, W Xizang).

Habitat: Dry slopes, forest margins.

Stem diameter 20 cm. Bark/xylem proportion 1:10. Ring width 0.01–2 mm. Maximal 500 annual rings.

Xylem

Transverse section

Conifer without resin ducts. Transition between earlywood and latewood fairly confluent. Latewood zone small. Often with intra-annual density variations and frost rings. Axial parenchyma cells present, filled with brown substances in the heartwood.

Tangential section

Rays uniseriate with a height of 3–8 cells.

Radial section

Ray tracheids absent. Cross field pits round, without distinct borders. Ray cells with nodular tangential walls. Horizontal walls with indentures.

Bark

Very large resin ducts in the phloem. Alternating tangential layers of rectangular sieve cells, and round parenchyma cells.

Pith

Irregularly formed.

Analyzed: Stems of 6 individuals (dry increment core).

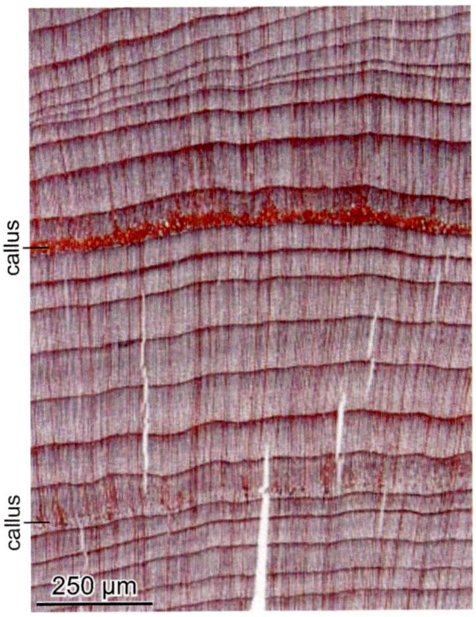

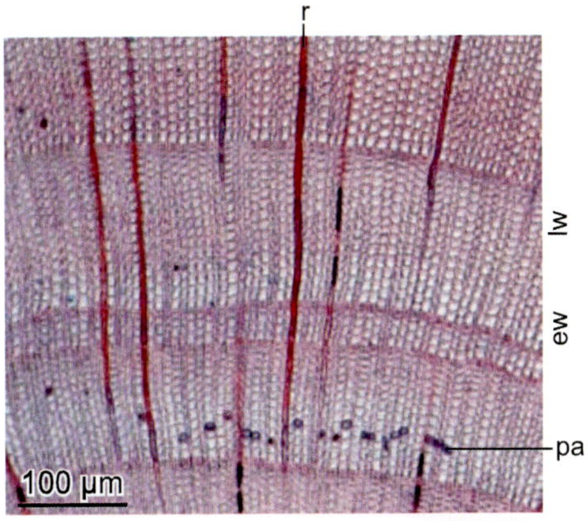

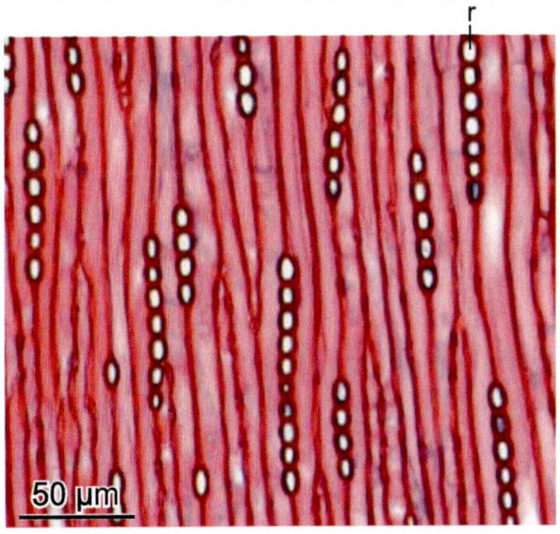

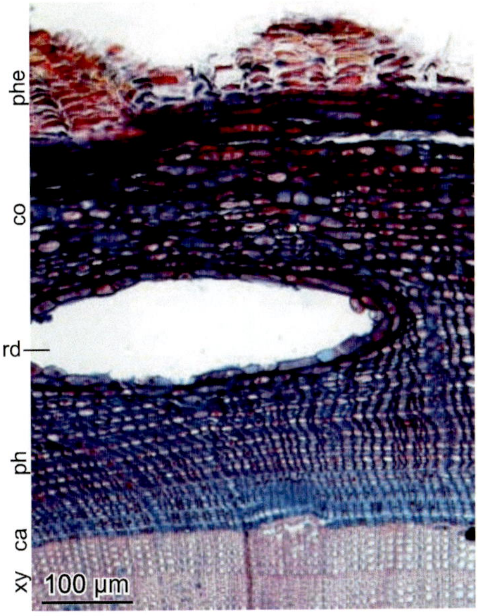

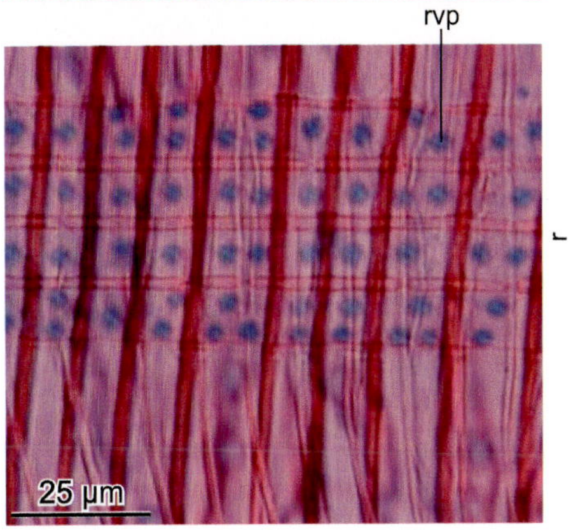

Hippophae rhamnoides subsp. *turkestanica* Rousi.

Elaeagnaceae

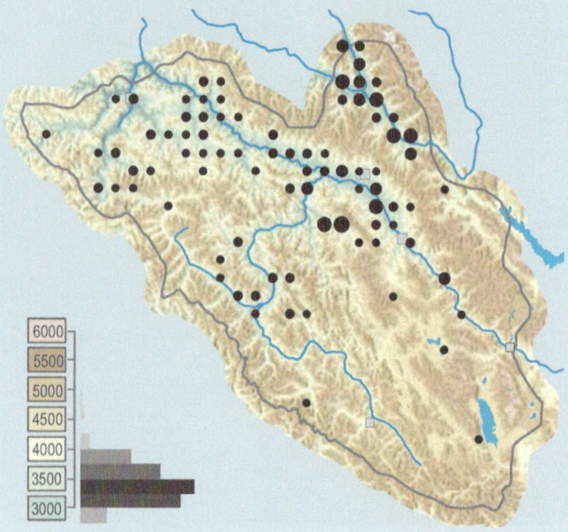

Description: Shrubs or small trees, up to 5 m tall, spiny, leaves lanceolate to linear, usually silvery on both sides, 2–6 cm long, male flowers in clusters at the base of shoots, female flowers in clusters of 2–4, fruits red or yellow.

Distribution: Kazakhstan, Turkmenistan, Uzbekistan, Kyrgyzstan, Tajikistan, Afghanistan, Pakistan, India, China (Xinjiang, N Xizang), Mongolia.

Habitat: River gravel deposits.

Stem diameter 10 cm. Bark/xylem proportion 1:10. Mean ring width 1.2 mm. More than 20 annual rings.

Xylem

Transverse section

Semi-ring-porous. Rings distinctly marked by differences between earlywood (100–150 µm) and latewood vessel diameters (20–30 µm). Vessels solitary. Smaller vessels with fine helical thickenings. Fibers thin- to tick-walled. Parenchyma paratracheal diffuse and apotracheal. Small vessels and rays in the center filled with dark-stained substances (heartwood).

Tangential section

Rays 2–3-seriate, storied.

Radial section

Perforation plates simple. Intervessel pits round in alternating position. Vessel length 150–200 µm. Fiber pits small (2–3 µm). Rays cells primarily prostrate.

Bark

Phloem with tangential to reticulate bands of sieve elements, which collapse in older parts. With distinct companion cells around single sieve elements in the young phloem. Round groups of sclereids, tangentially arranged. Older stems contain a large rhytidome. Phellem consists of radially oriented, rectangular cells. Crystals absent.

Pith

Pith more or less round. Central cells large, unlignified, thin-walled, surrounded by a few rows of smaller, lignified cells.

Analyzed: Stem of 6 individuals (dry increment core).

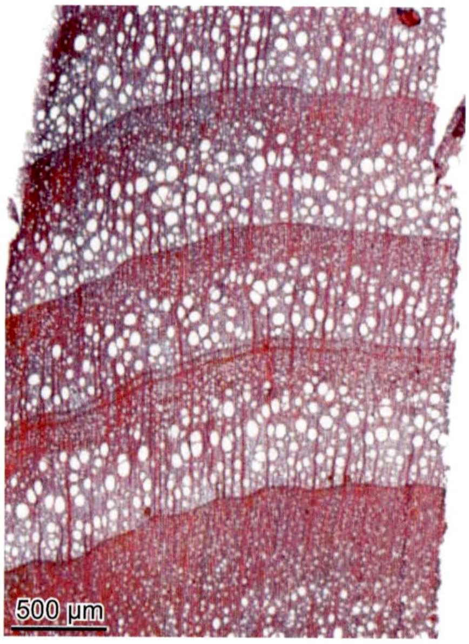

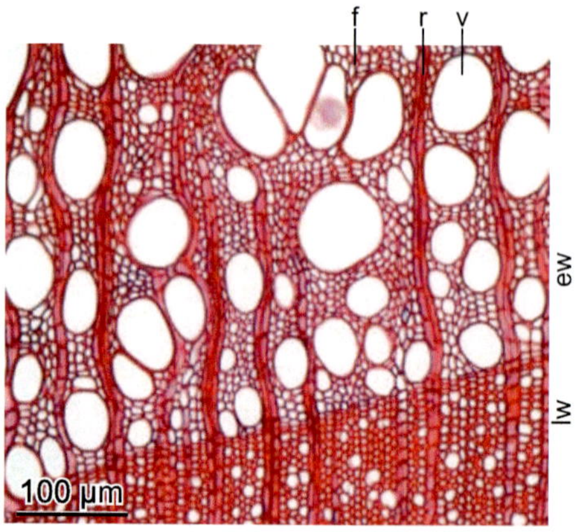

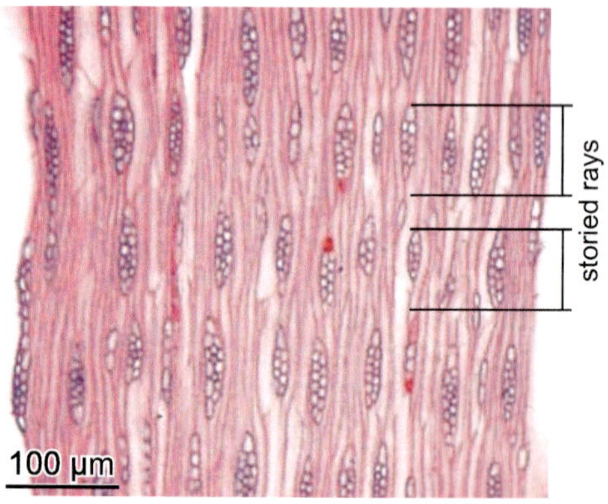

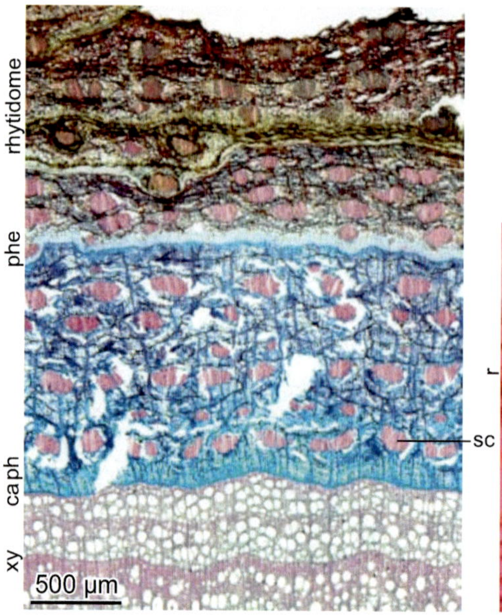

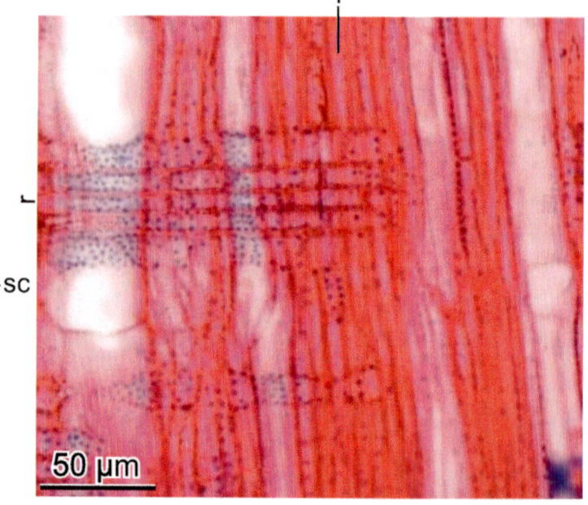

Features according to IAWA

Xylem 1 - 4 - 9 - 13 - 22 - 41 - 50.1 - 52.3 - 61 - 70 - 79 - 97 - 196

Hippophae tibetana Schltdl.

Elaeagnaceae

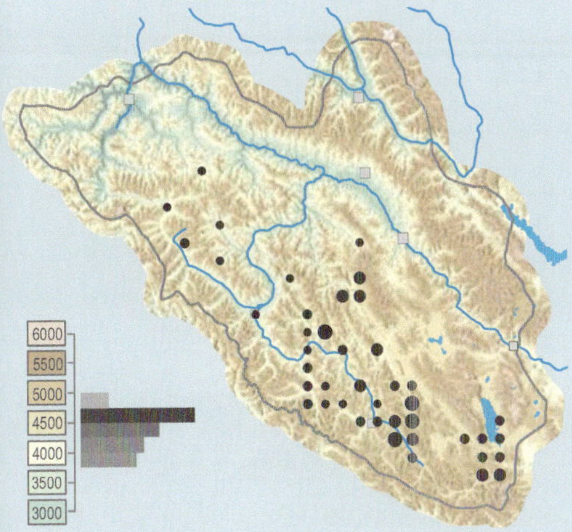

Description: Shrubs, 10–60 cm tall, leaves linear-oblong, whitish below, grayish above, mostly in whorls of 3, flowers in clusters, minute, fruits yellowish green, globose.

Distribution: N India, Nepal, China (Gansu, Qinghai, Xizang), Bhutan.

Habitat: River gravel deposits.

Stem diameter 3–4 mm. Bark/xylem proportion 1:1. Mean ring width 0.2 mm. Maximal 8 annual rings.

Xylem

Transverse section

Semi-ring-porous. Rings distinctly marked by differences between earlywood (40–50 µm) and latewood vessel diameters (10–15 µm). Vessels slightly angular, solitary. Fibers thin- to thick-walled. Parenchyma paratracheal diffuse.

Tangential section

Rays 1–2-seriate.

Radial section

Perforation plates simple. Intervessel pits round in alternating position. Vessel length 100–120 µm. Fiber pits small (2–3 µm). Rays cells primarily prostrate.

Bark

Phloem with tangential bands of sieve elements which collapse in older parts. Phellem consists of radially oriented, rectangular cells. Crystals absent.

Pith

Pith more or less round. Central cells large, unlignified, thin-walled, surrounded by a few rows of smaller lignified cells.

Analyzed: Stem of 1 individual.

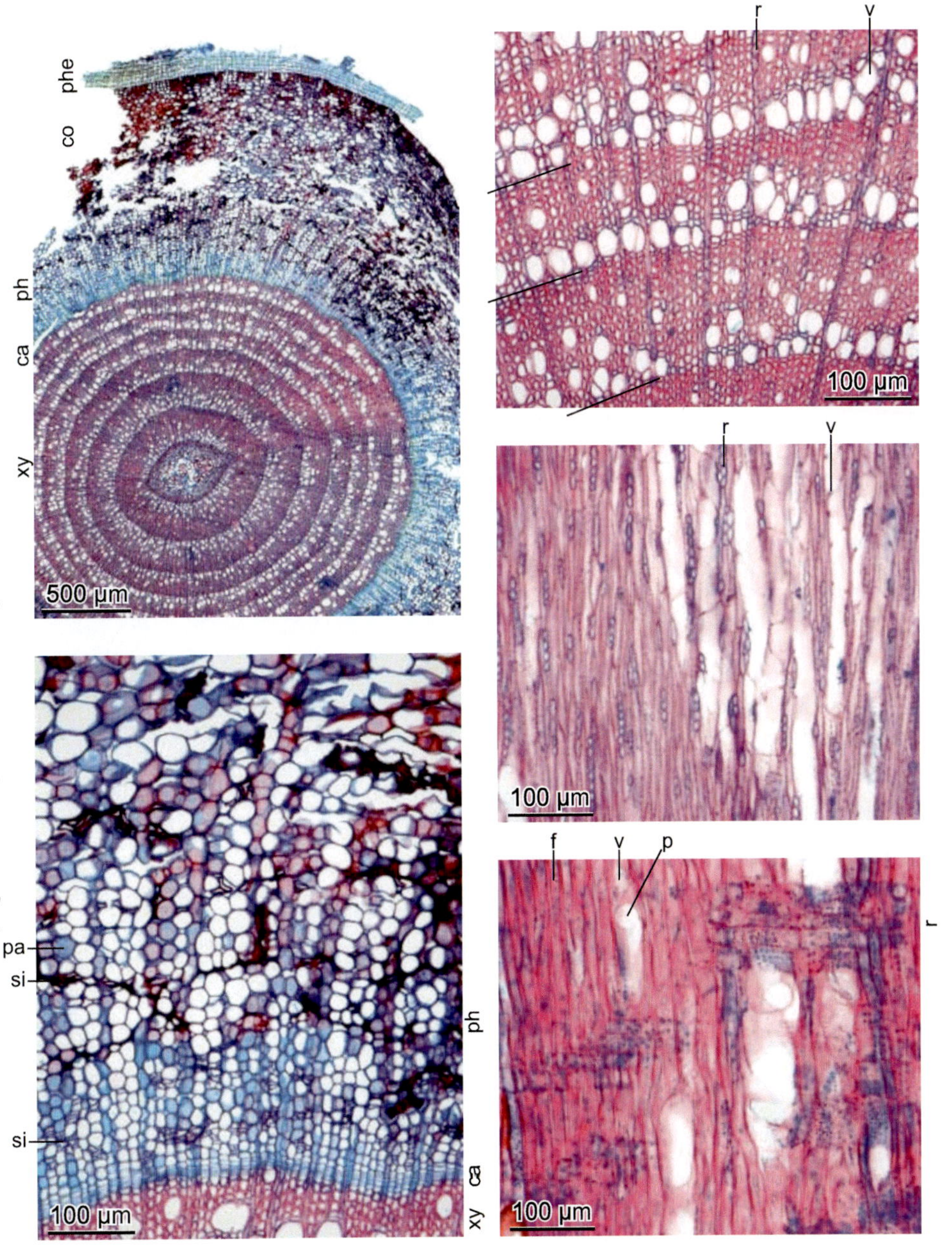

Features according to IAWA

Xylem 1 - 4 - 9 - 9.1 - 12 - 13 - 22 - 40.2 - 50.1 - 52.3 - 61 - 70 - 79 - 97 - 106

Ephedra gerardiana Wall. ex Stapf

Ephedraceae

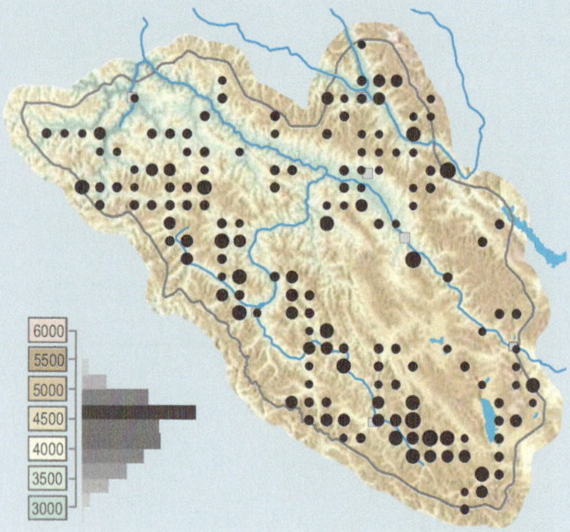

Syn.: *E. wallichii* Stapf

Description: Subshrubs, 10–20 cm tall, branchlets dark green, scabrous, stout, leaves opposite, connate, 2–3 mm long, seed cones solitary, sessile, red and fleshy at maturity, 6–7 mm across.

Distribution: Tajikistan, Afghanistan, N Pakistan, N India, Nepal, Bhutan, China (Qinghai, Xinjiang, Xizang).

Habitat: Stabilized slopes, stony and rocky habitats.

Stem diameter 10–20 mm. Bark/xylem proportion 1:10. Ring width 0.15–0.4 mm. Maximal 55 annual rings.

Xylem

Transverse section

Semi-ring-porous. Rings distinctly marked by few flat latewood fibers. Earlywood vessel diameter 30–40 µm. Vessels solitary. Fibers thin- to thick-walled. Parenchyma paratracheal. Crystal sand in the unlignified rays.

Tangential section

Rays >10-seriate.

Radial section

Foraminate perforation plates. Intervessel pits round. Vessel length 250–300 µm. Ray cells square and procumbent.

Bark

Phloem with alternating tangential layers of sieve tubes and parenchyma cells. With isolated sclereids. With ray dilatations. Phellem consists of many layers of thin-walled, rectangular cells.

Pith

Round, consisting of thin-walled parenchyma cells.

Analyzed: Stem of 6 individuals.

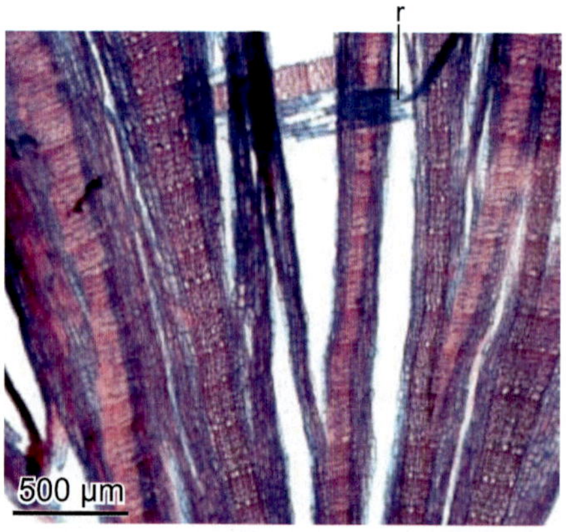

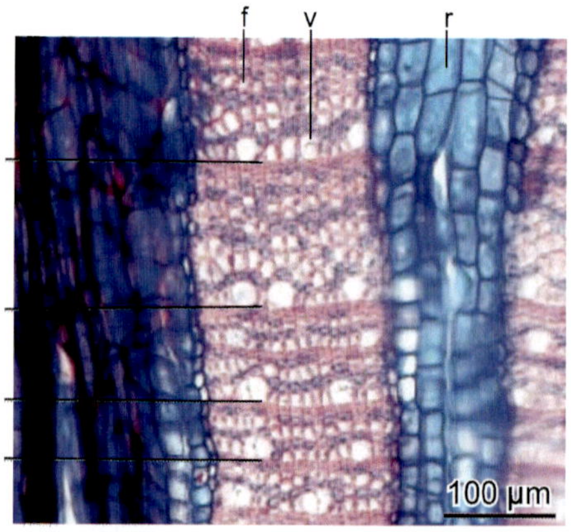

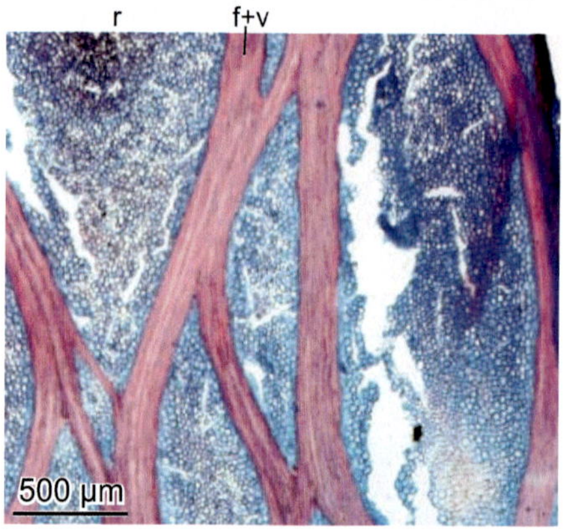

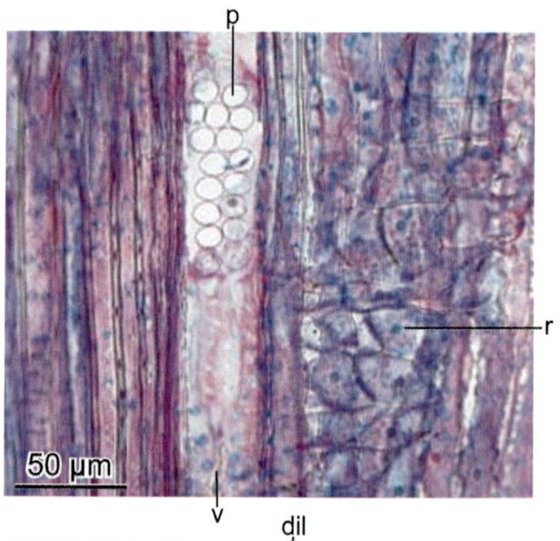

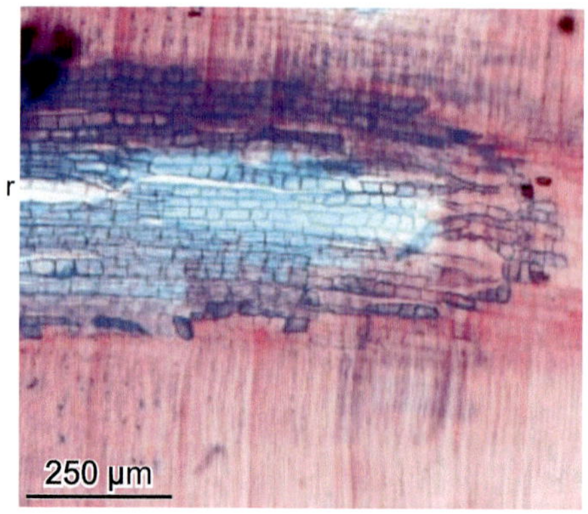

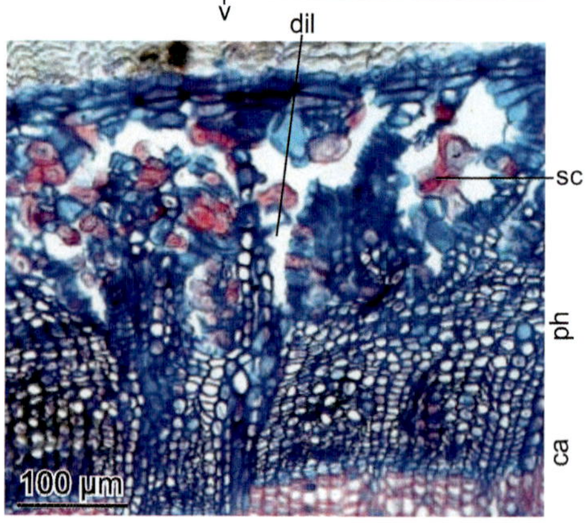

Features according to IAWA

Xylem 1 - 4 - 9 - 13a - 22 - 40.2 - 50.1 - 53.1 - 61 - 69 - 70 - 79 - 99 - 100.2 - 104

Ephedra intermedia Schrenk ex C.A.Mey.

Ephedraceae

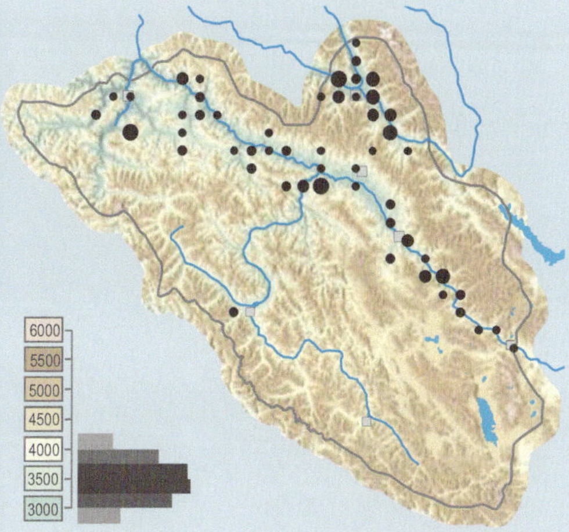

Syn.: *E. tibetica* (Stapf) V.A.Nikitin

Description: Subshrubs, 30–100 cm tall, densely branched, leaves opposite, connate for at least 2/3 their length, seed cones ovoid, red and fleshy at maturity.

Distribution: Russia (Siberia), Kazakhstan, Turkmenistan, Uzbekistan, Kyrgyzstan, Tajikistan, Afghanistan, Pakistan, India, China (Gansu, Hebei, Liaoning, Nei Mongol, Ningxia, Qinghai, Shaanxi, Shandong, Shanxi, Xinjiang, Xizang), Mongolia.

Habitat: Desert, sandy and rocky places.

Stem diameter 8–15 mm. Bark/xylem proportion 1:4. Ring width 0.3–0.4 mm. Maximal 16 annual rings.

Xylem

Transverse section

Semi-ring-porous. Rings distinctly marked by few flat latewood fibers. Earlywood vessel diameter 40–50 µm. Vessels solitary. Fibers thin- to thick-walled. Parenchyma paratracheal. Crystal sand in the rays.

Tangential section

Rays 6–8-seriate.

Radial section

Foraminate perforation plates. Intervessel pits round. Vessel length 250–300 µm. Ray cells square and procumbent.

Bark

Phloem with alternating tangential layers of sieve tubes and parenchyma cells. With solitary fibers. With ray dilatations. Phellem consists of many layers of thin-walled, rectangular cells.

Pith

Round, consisting of thin-walled parenchyma cells.

Analyzed: Stem of 3 individuals.

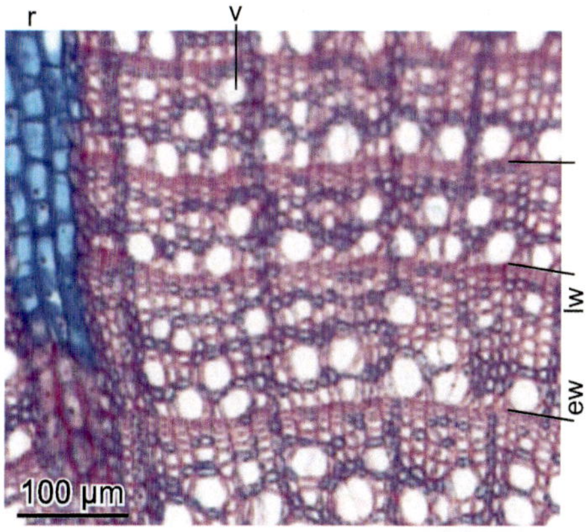

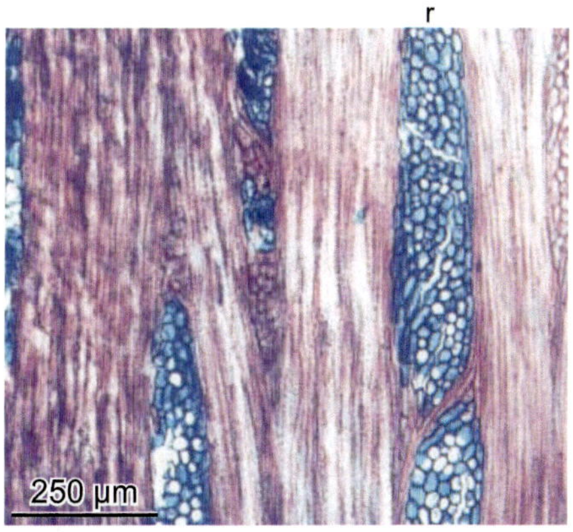

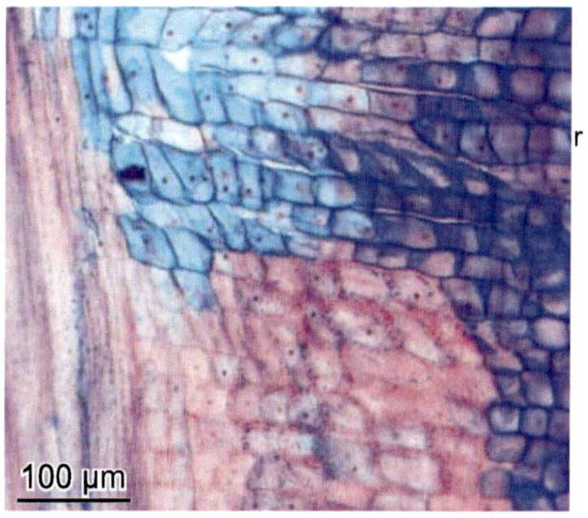

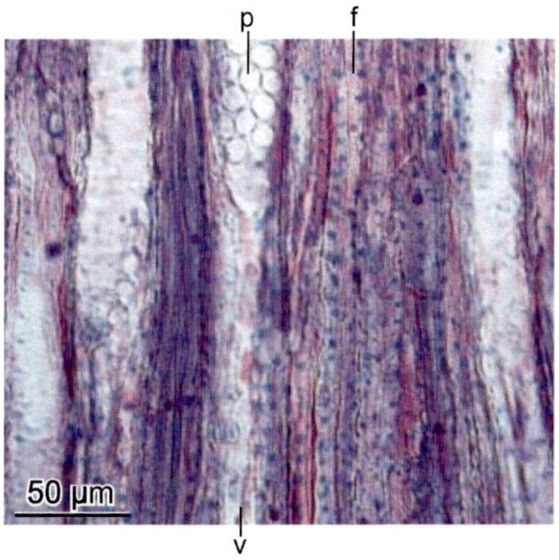

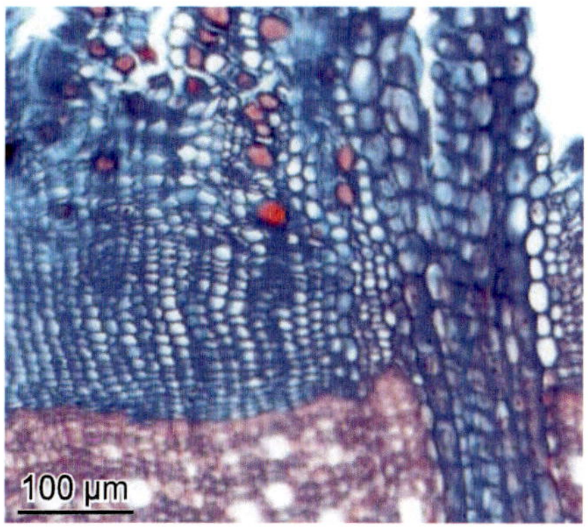

Features according to IAWA

Xylem 1 - 4 - 9 - 13 - 22 - 40.2 - 50.1 - 53.1 - 61 - 70 - 79 - 99 - 100.2 - 104 - 150

Ephedra regeliana Florin
Ephedraceae

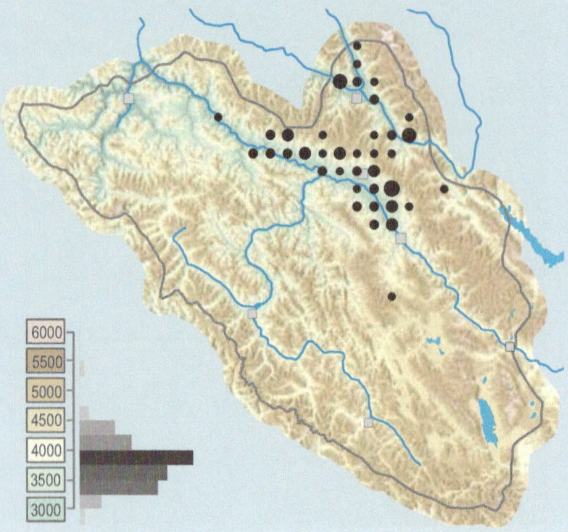

Syn.: *E. monosperma* var. *disperma* Regel, *E. pulvinaris* V.A.Nikitin

Description: Subshrubs, 8–10 cm tall, above-ground woody parts 1–2 cm short, main branches inconspicuous, internodes rough, branchlets radially arranged and clustered at nodes, leaves opposite, connate for ½–⅔ their length, seed cones ovoid, ca. 5×4 mm, red and fleshy at maturity.

Distribution: Kazakhstan, Uzbekistan, Kyrgyzstan, Tajikistan, Afghanistan, Pakistan, N India, China (Xinjiang).

Habitat: Rocks, stony semideserts.

Stem diameter 4 mm. Bark/xylem proportion 1:2. Mean ring width 0.08 mm. Maximal 12 annual rings.

Xylem

Transverse section

Semi-ring-porous. Rings distinctly marked by few flat latewood fibers. Earlywood vessel diameter 20–30 µm. Vessels solitary. Fibers thin- to thick-walled. Parenchyma paratracheal and apotracheal. Crystal sand in the rays.

Tangential section

Rays 6–8-seriate. Some rays with distinct sheath cells.

Radial section

Foraminate perforation plates. Intervessel pits round. Vessel length 250–300 µm. Rays cells square and procumbent.

Bark

Phloem with alternating tangential layers of sieve tubes and parenchyma cells. With solitary fibers. With ray dilatations. Phellem consists of many layers of thin-walled, rectangular cells. Cortex with crystal sand.

Pith

Pith round, consisting of thin-walled, lignified parenchyma cells.

Analyzed: Stem of 1 individual (herbarium specimen).

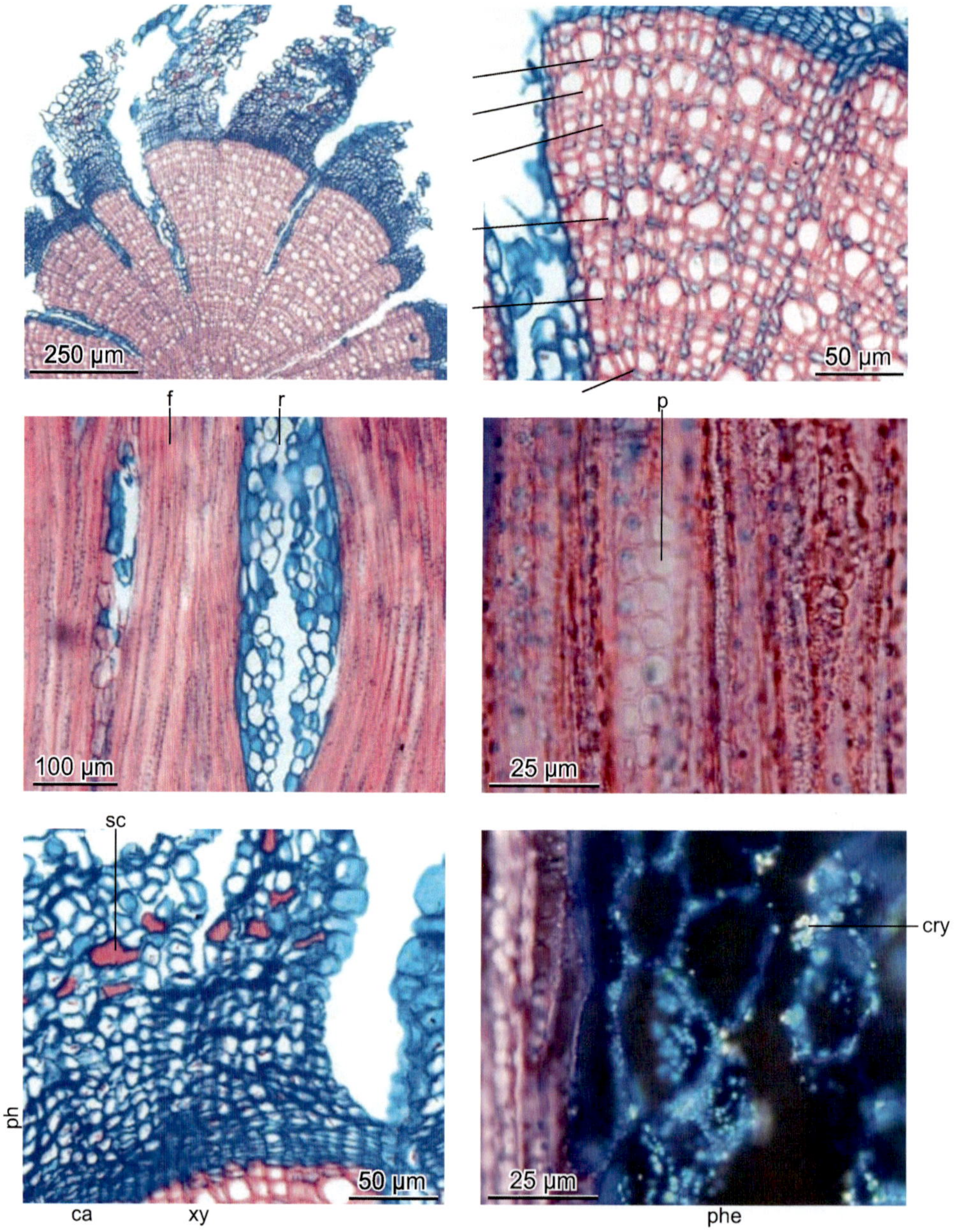

Features according to IAWA

Xylem 1 - 4 - 9 - 13 - 22 - 40.2 - 50.1 - 53.1 - 61 - 70 - 76 - 79 - 98 - 100.2 - 104 - 105 - 110 - 153

Euphorbia tibetica Boiss.

Euphorbiaceae

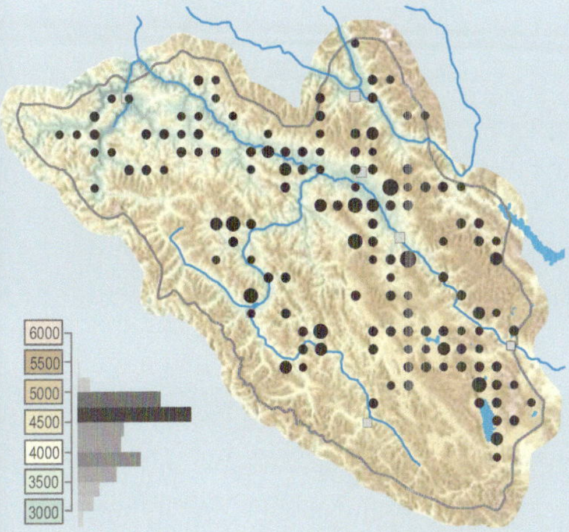

Syn.: *Tithymalus tibeticus* (Boiss.) Prokh.

Description: Perennial herbs with latex, 10–20 cm tall, stems decumbent, glaucous, leaves subsessile, flowers inconspicuous, arranged in terminal cyathia.

Distribution: C Asia, N India, China (Xizang).

Habitat: Unstable loamy, gravel and sandy slopes, sandy plains, often human-disturbed habitats along roads, stony steppes.

Stem diameter 7 mm. Bark/xylem proportion 1:1. Mean ring width 0.5 mm. Maximal 10 annual rings. Stem segmentation.

Xylem

Transverse section

Diffuse-porous and semi-ring-porous. Ring boundaries indicated by larger earlywood (40–50 µm) and smaller latewood vessel diameters (10–20 µm). Vessels stained or unstained, arranged in irregular radial rows and groups. Fibers absent. Parenchyma pervasive. Without crystals. With intraxylary periderm towards the center.

Tangential section

Rays 5–>10-seriate, partially confluent with axial tissue. Rays disappear in polarized light.

Radial section

Perforation plates simple. Intervessel pits slightly scalariform and round in alternating position. Vessel length 80–150 µm. Ray cells upright.

Bark

Phloem contains irregularly distributed groups of sieve tubes between ray dilatations. Cortex contains laticifers. Phellem consists of many layers of rectangular cells.

Pith

No slide available.

Analyzed: Root collar of polar root of 7 individuals.

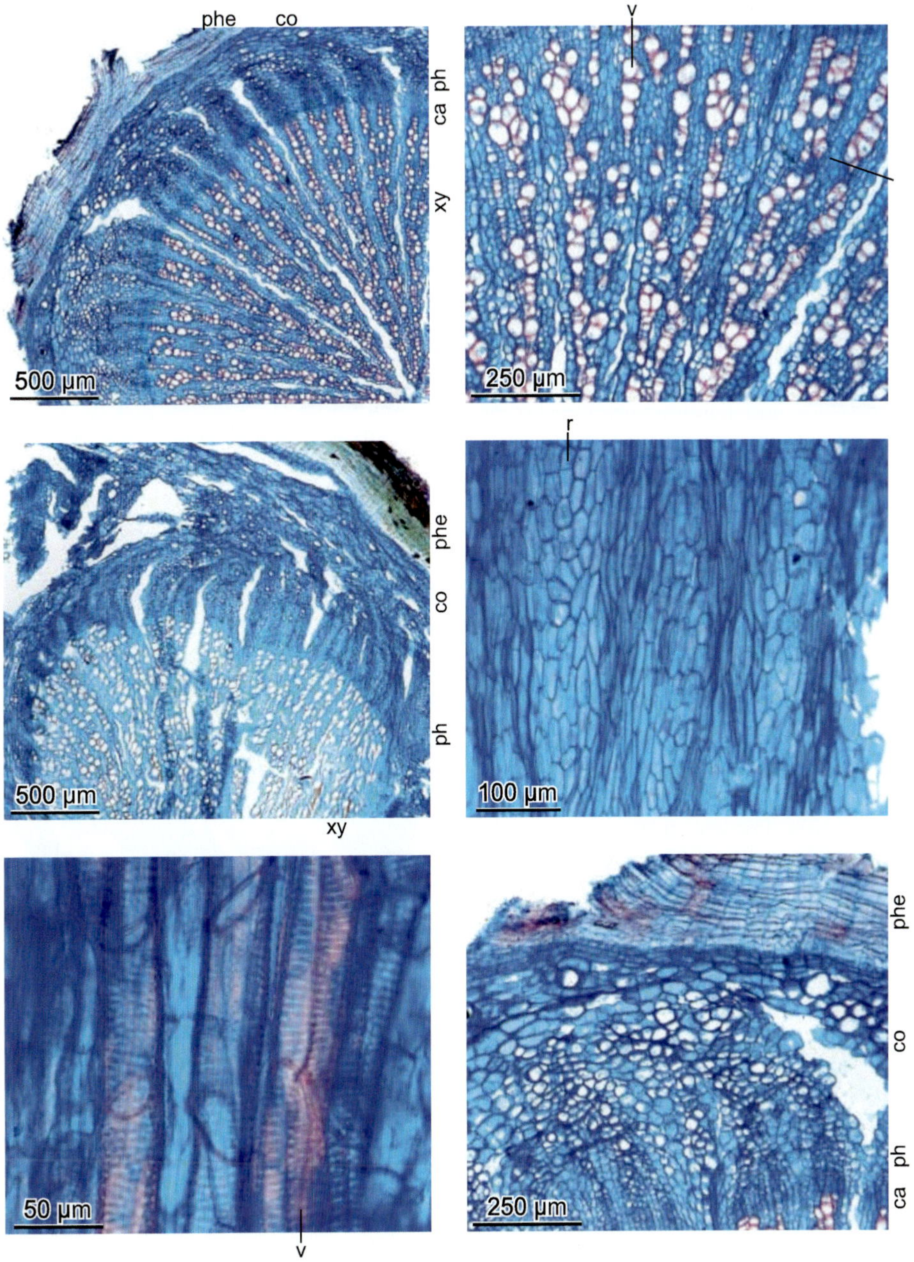

Features according to IAWA

Xylem 1 - 4 - 5 - 10 - 13 - 20 - 22 - 40.2 - 50.1 - 52.2 - 52.3 - 60.1 - 79.1 - 98 - 100.1 - 100.2 - 105

Astragalus coluteocarpus Boiss.

Fabaceae

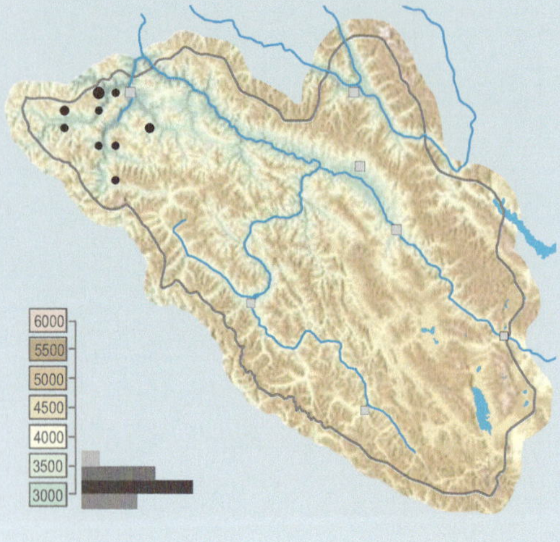

Description: Perennial herbs, 20–80 cm tall, stems ascending, glabrous, leaves imparipinnate, leaflets nearly orbicular, flowers yellowish, ca. 1.5 cm long, in 5–20-flowered racemes.

Distribution: Tajikistan, Afghanistan, N Pakistan, N India, China (Xinjiang).

Habitat: Steppe grasslands, stony slopes.

Stem diameter 17 mm. Bark/xylem proportion 1:4. Mean ring width 1 mm. Maximal 8 annual rings.

Xylem

Transverse section

Semi-ring-porous. Ring boundaries indicated by large earlywood (60–100 μm) and small latewood vessel diameters (20–30 μm) and marginal parenchyma. Vessels in small groups. Vessel walls red-stained. Fibers solitary or in groups. Parenchyma pervasive and paratracheal. Crystals absent. Without secondary intra-xylary periderm towards the pith.

Tangential section

Rays 5–6-seriate, unlignified and thin-walled.

Radial section

Perforation plates simple. Intervessel pits scalariform. Vessel length 60–120 μm. All ray cells upright.

Bark

Phloem with groups of sieve tubes. Older sieve tubes radially collapsed. Fibers in radial bands. Phellem cells tangentially flat, thin-walled, rectangular. Crystals absent.

Pith

No slide available.

Analyzed: Root collar of 1 individual.

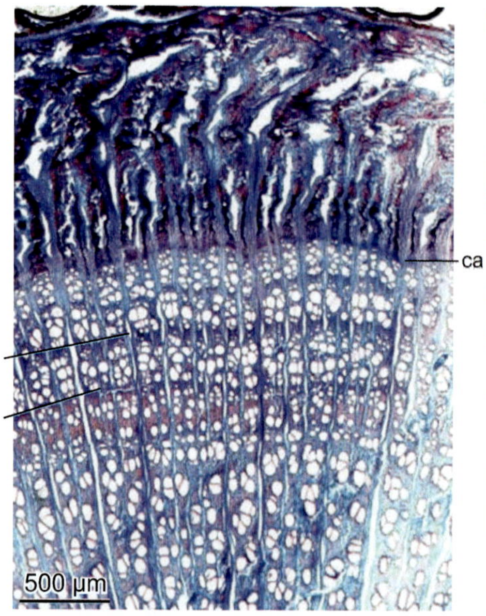

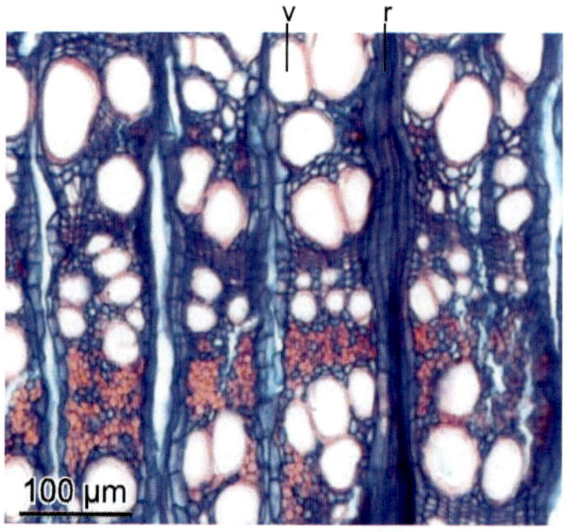

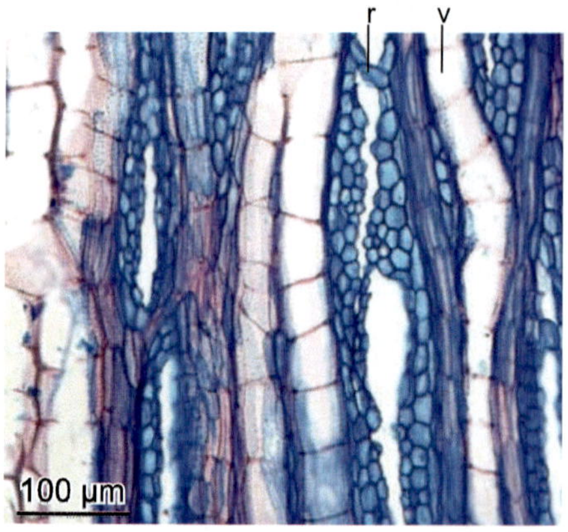

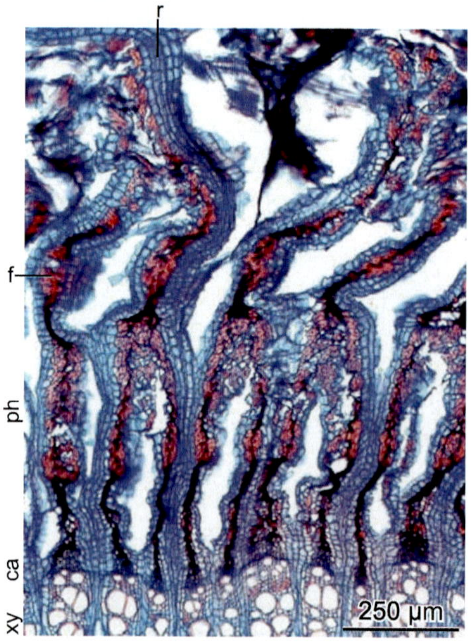

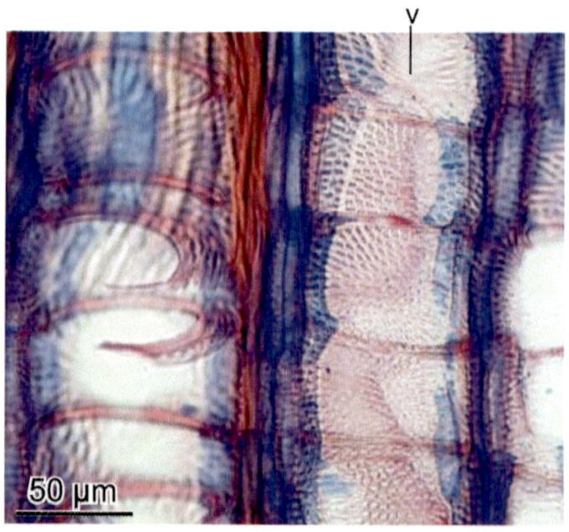

Features according to IAWA

Xylem 1 - 4 - 11 - 13 - 20 - 40.2 - 50.2 - 52.2 - 52.3 - 60.1 - 79 - 79.1 - 98 - 100.2 - 105

Astragalus confertus Benth. ex Bunge
Fabaceae

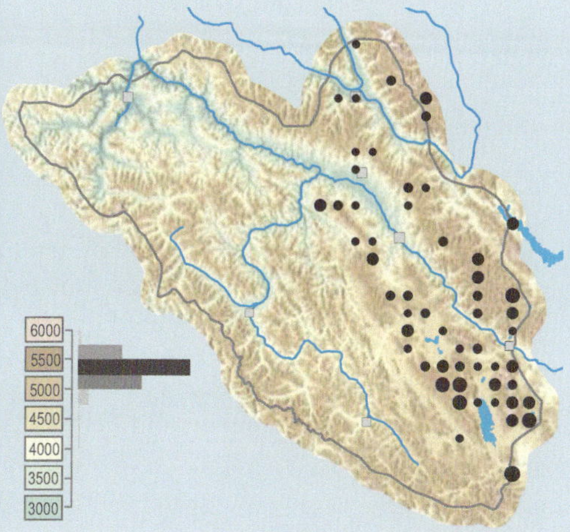

Description: Perennial herbs, 2–10 cm tall, sparsely hairy, densely cespitose with numerous stems, leaves imparipinnately compound, 1–2.5 cm long, flowers lilac to violet, ca. 1 cm across, in 1–7-flowered racemes.

Distribution: India, Nepal, China (Qinghai, Xizang).

Habitat: Open grasslands, stabilized gentle slopes.

Stem diameter 5–10 mm. Bark/xylem proportion 1:1. Ring width 0.1–0.5 mm. Maximal 17 annual rings. Stem segmentation.

Xylem

Transverse section
Semi-ring-porous. Ring boundaries indicated by large earlywood (40–50 µm) and small latewood vessel diameters (10–20 µm). Vessels solitary or in small groups. Vessel walls poorly stained. Thick-walled fibers rare, occuring in small groups. Parenchyma pervasive. Crystals absent. With secondary intra-xylary periderm towards the pith.

Tangential section
Rays 5–8-seriate, unlignified and thin-walled.

Radial section
Perforation plates simple. Intervessel pits scalariform. Vessel length 60–80 µm. All ray cells upright. With sheath cells.

Bark

Phloem with groups of sieve tubes and tangential groups of thick-walled fibers. Phellem consists of few tangential layers of cork cells, composed of tangentially flat, thin-walled, rectangular cells. Crystals absent.

Pith

No slide available.

Analyzed: Root collar of 7 individuals.

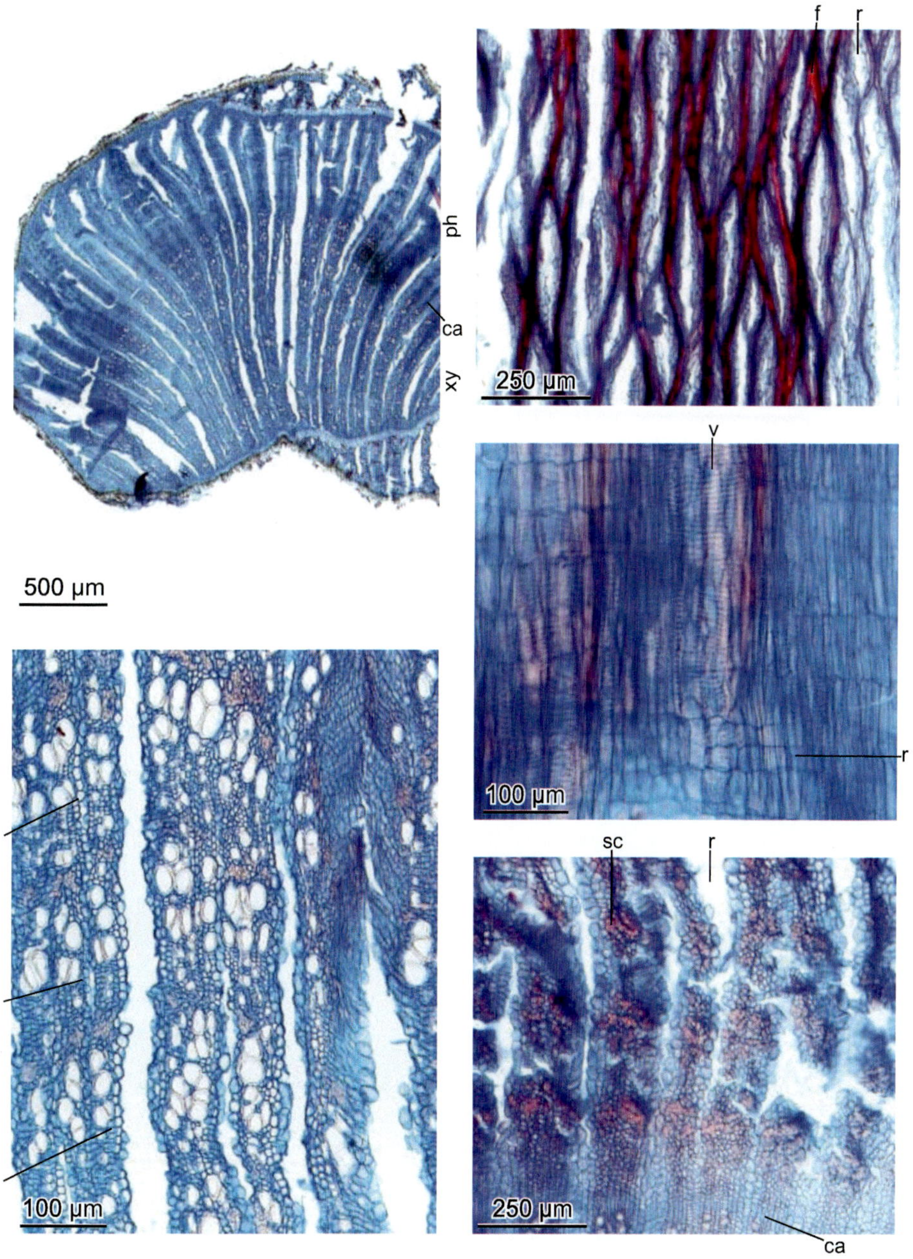

Features according to IAWA
Xylem 1 - 4 - 9 - 11 - 13 - 20 - 39.2 - 40.2 - 50.2 - 52.2 - 61 - 69 - 79.1 - 98 - 100.2 - 105 - 110 - 135.2

Astragalus densiflorus Kar. & Kir.

Fabaceae

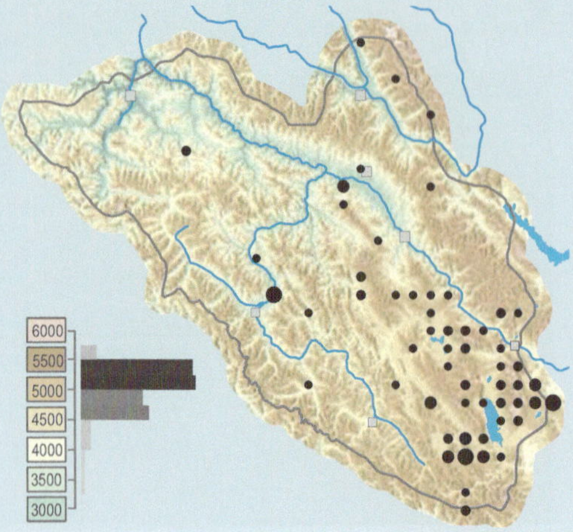

Description: Perennial herbs, 7–20 cm tall, stems ascending, hairy, leaves imparipinnately compound, leaflets narrowly elliptic, flowers violet, ca. 5 mm long, in dense many-flowered racemes.

Distribution: Kazakhstan, Kyrgyzstan, Pakistan, N India, China (Gansu, Qinghai, Sichuan, Xinjiang, Xizang).

Habitat: Grasslands, alpine meadows, along streams.

Stem diameter 15 mm. Bark/xylem proportion 1:1. Ring width 0.2–0.3 mm. Maximal 9 annual rings. Stem segmentation.

Xylem

Transverse section

Semi-ring-porous to ring-porous. Ring boundaries indicated by large earlywood (50–70 µm) and small latewood vessel diameters (10–20 µm). Vessels solitary. Vessel walls poorly stained. Thick-walled, red-stained fibers rare, occuring in small groups. Parenchyma pervasive. Crystals absent. With secondary intra-xylary periderm towards the pith.

Tangential section

Rays 5–8-seriate, unlignified and thin-walled.

Radial section

Perforation plates simple. Intervessel pits scalariform. Vessel length 60–80 µm. All ray cells upright.

Bark

Phloem with groups of sieve tubes and tangential groups of thick-walled fibers. Phellem consists of few tangential layers of cork cells, composed of tangentially flat, thin-walled, rectangular cells. Crystals absent.

Pith

No slide available.

Analyzed: Root collar of 3 individuals.

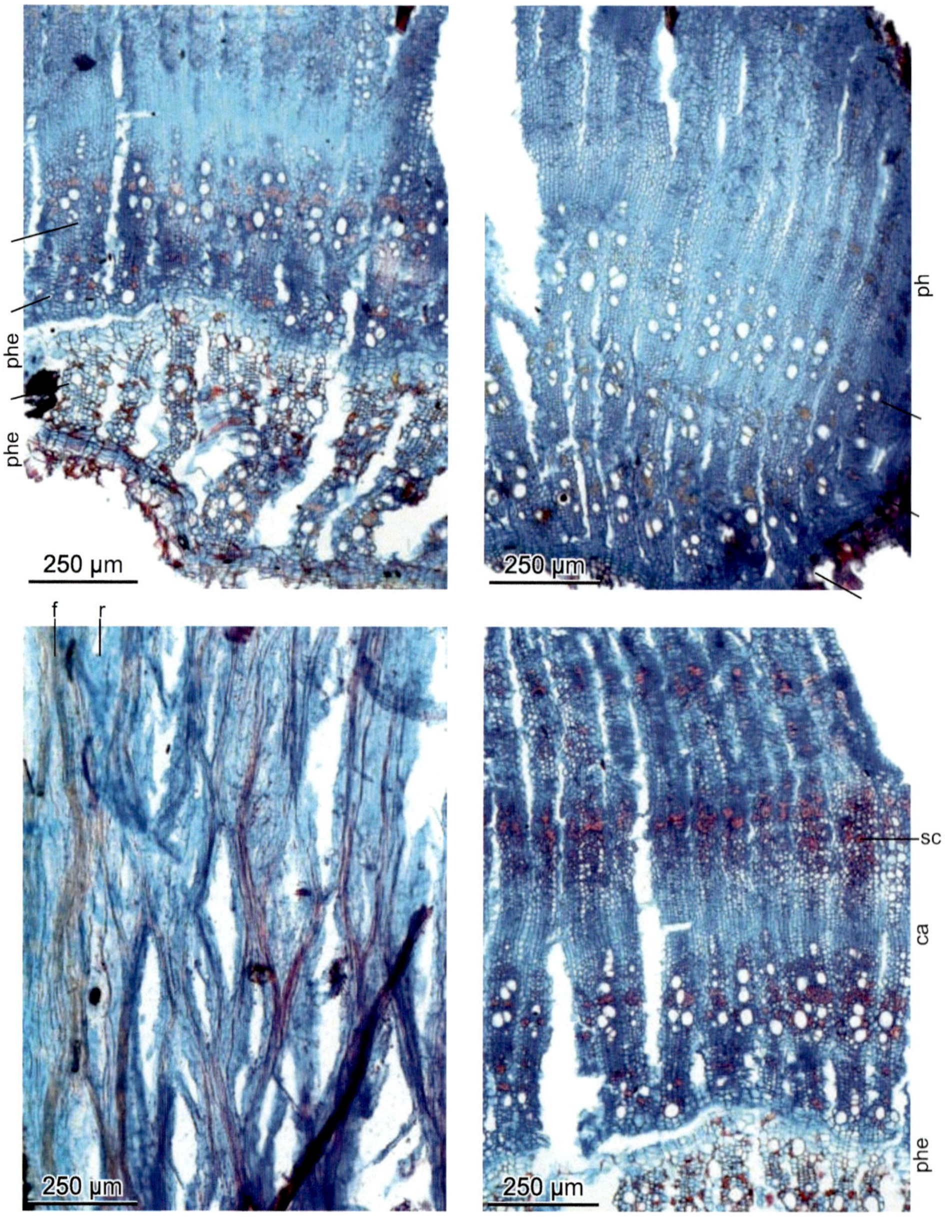

Features according to IAWA

Xylem 1 - 3 - 4 - 9 - 11 - 13 - 20 - 39.2 - 41 - 50.2 - 52.2 - 61 - 69 - 79.1 - 98 - 100.2 - 105 - 135.2

Astragalus falconeri Bunge

Fabaceae

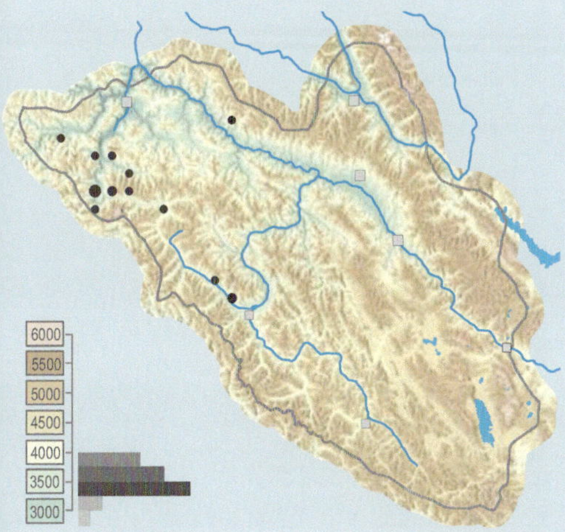

Syn.: *A. badachschanicus* Borissova

Description: Perennial herbs, 15–40 cm tall, stems prostrate to ascending, pubescent, leaves imparipinnately compound, leaflets obovate, flowers pink-purple, 10–13 mm long, in 3–15-flowered racemes.

Distribution: Tajikistan, NE Afghanistan, N Pakistan, NW India, China (Xizang).

Habitat: Gravelly areas.

Stem diameter 5–13 mm. Bark/xylem proportion 1:2. Ring width 0.15–0.3 mm. Maximal 16 annual rings.

Xylem

Transverse section

Semi-ring-porous to ring-porous. Ring boundaries indicated by large earlywood (50–70 µm) and small latewood vessel diameters (20–30 µm). Vessels solitary and in small groups. Vessel walls intensively or poorly stained. Thick-walled, poorly stained fibers occur in groups. Parenchyma pervasive and paratracheal. Crystals absent. Without secondary intra-xylary periderm towards the pith.

Tangential section

Rays 3–6-seriate, unlignified and thin-walled. With sheath cells.

Radial section

Perforation plates simple. Intervessel pits scalariform and round. Vessel length 60–80 µm. All ray cells upright.

Bark

Phloem with groups of sieve tubes and tangential groups of thick-walled fibers. With ray dilatations. Phellem consists of few tangential layers of cork cells, composed of tangentially flat, thin-walled, rectangular cells. Crystals absent.

Pith

No slide available.

Analyzed: Root collar of 4 individuals.

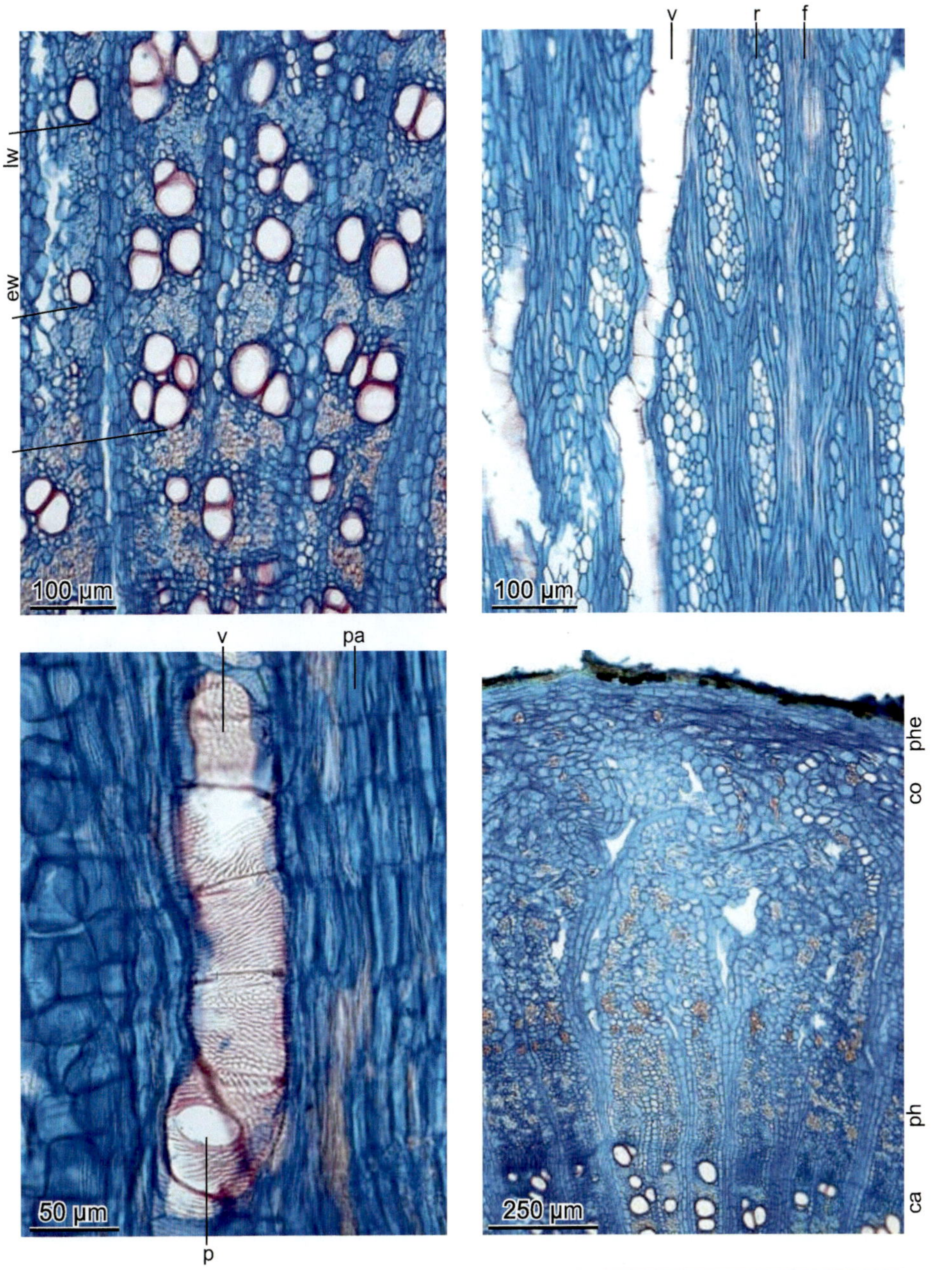

Features according to IAWA

Xylem 1 - 3 - 4 - 9 - 11 - 13 - 20 - 22 - 39.2 - 41 - 50.2 - 52.2 - 61 - 69 - 79 - 79.1 - 98 - 100.2 - 105 - 110

Astragalus melanostachys Benth. ex Bunge

Fabaceae

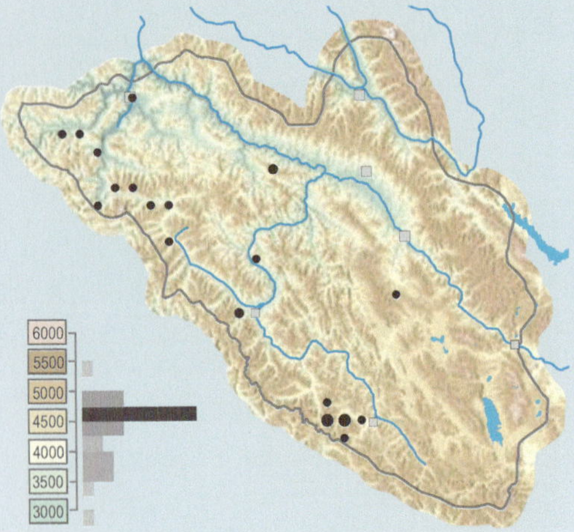

Syn.: *A. bracteosus* Klotzsch, *A. macrostegius* Rech.f., *A. supraglaber* Kitam.

Description: Perennial herbs, 10–60 cm tall, leaves imparipinnate, leaflets widely elliptic, flowers red or pale pink, ca. 7 mm long, in dense many-flowered racemes.

Distribution: Russia, Uzbekistan, Tajikistan, Afghanistan, Pakistan, NW India, China (Xinjiang).

Habitat: Alpine meadows, along streams.

Stem diameter 3–5 mm. Bark/xylem proportion 1:2. Mean ring width 0.3–0.4 mm. Maximal 4 annual rings.

Xylem

Transverse section

Diffuse- to semi-ring-porous. Ring boundaries indicated by large earlywood (30–50 μm) and small latewood vessel diameters (10–20 μm). Vessels round to angular and mainly in groups. Fibers thick-walled, in groups. Parenchyma pervasive. Crystals absent.

Tangential section

Rays 4–6-seriate, unlignified and thin-walled. With sheath cells.

Radial section

Simple perforation plates. Intervessel pits scalariform. Vessel length 50–60 μm. All ray cells upright.

Bark

Bad slide, herbarium material. Phloem with groups of sieve tubes. Fibers absent. Phellem composed of tangentially flat, thin-walled, rectangular cells. Crystals absent.

Pith

No slide available.

Analyzed: Root collar of 3 individuals.

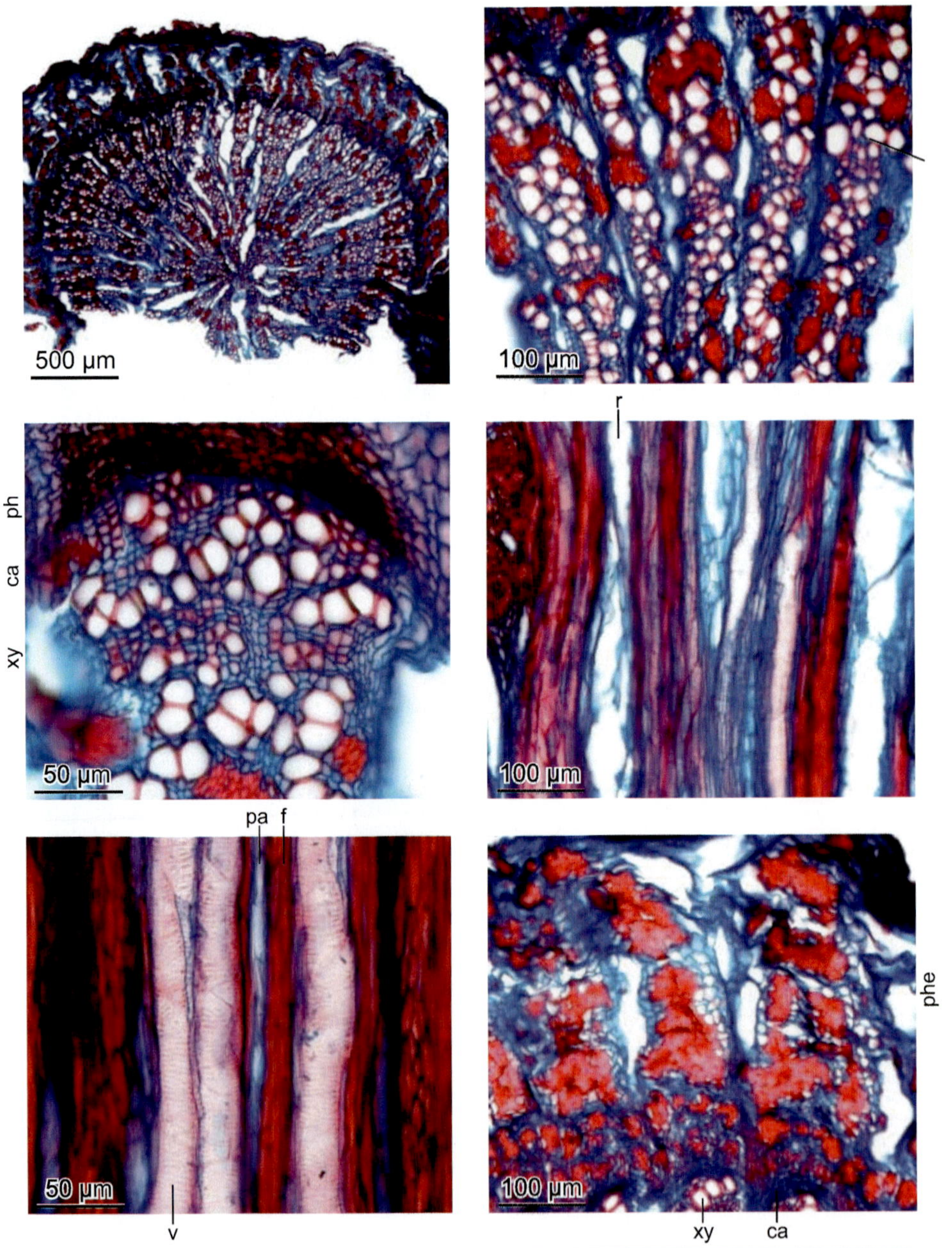

Features according to IAWA

Xylem 1 - 4 - 5 - 9 - 11 - 12 - 13 - 20 - 40.2 - 50.2 - 52.3 - 61 - 69 - 79.1 - 98 - 100.2 - 105 - 110

Astragalus munroi Benth. ex Bunge

Fabaceae

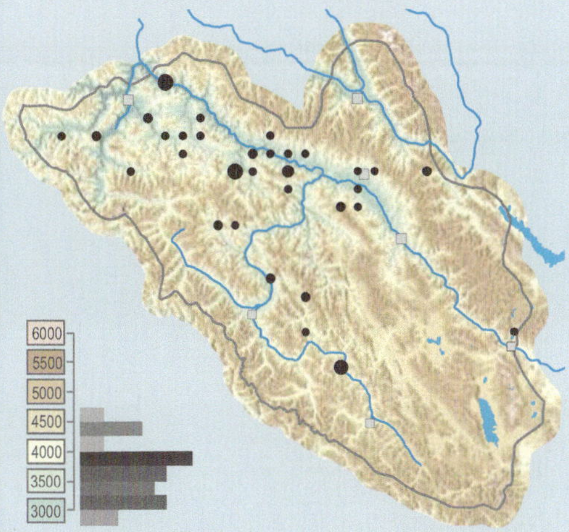

Syn.: *A. longicalyx* C.C.Ni & P.C.Li

Description: Perennial herbs, 30–70 cm tall, densely white villous, stems prostrate, leaves imparipinnately compound, leaflets oblong, flowers yellow, ca. 2 cm across, in 3–6-flowered racemes, legumes inflated, hairy, 2–3 cm long.

Distribution: Pakistan, India (Punjab, Jammu and Kashmir), China (Xinjiang, Xizang).

Habitat: Unstable screes.

Stem diameter 4–5 mm. Bark/xylem proportion 1:1. Ring width 0.05–0.06 mm. Maximal 10 annual rings. Stem segmentation.

Xylem

Transverse section

Semi-ring-porous. Ring boundaries slightly indicated by large earlywood (30–50 µm) and small latewood vessel diameters (10–20 µm). Vessels solitary. Vessel walls poorly stained. Fibers absent. Parenchyma pervasive. Crystals absent. With secondary intra-xylary periderm towards the pith.

Tangential section

Rays 4–6-seriate, unlignified and thin-walled.

Radial section

Perforation plates simple. Intervessel pits scalariform. Vessel length 60–80 µm. All ray cells upright.

Bark

Phloem with groups of sieve tubes. Older sieve tubes radially collapsed. Fibers absent. Phellem cells tangentially flat, thin-walled, rectangular. Crystals absent.

Pith

No slide available.

Analyzed: Root collar of 2 individuals.

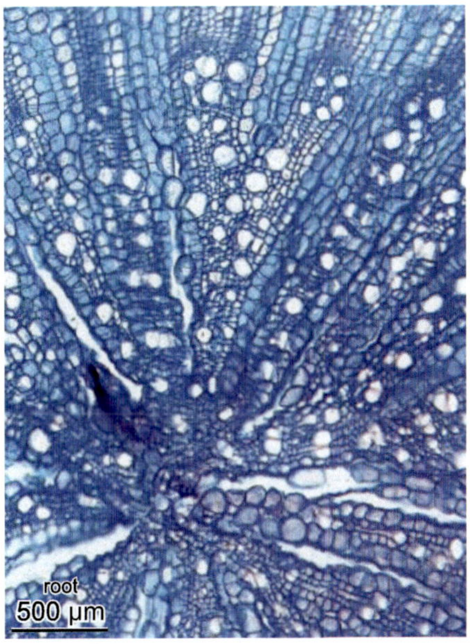

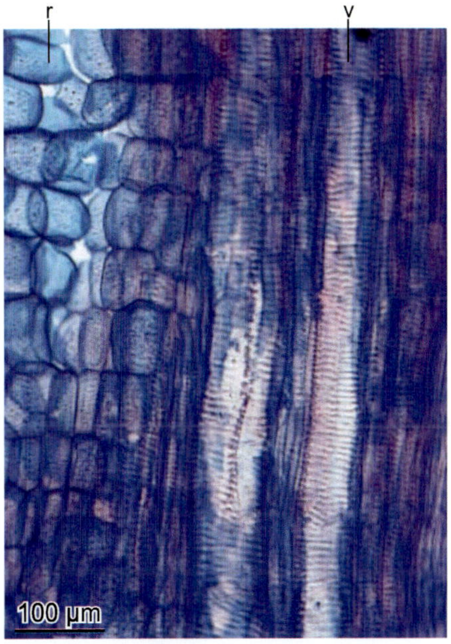

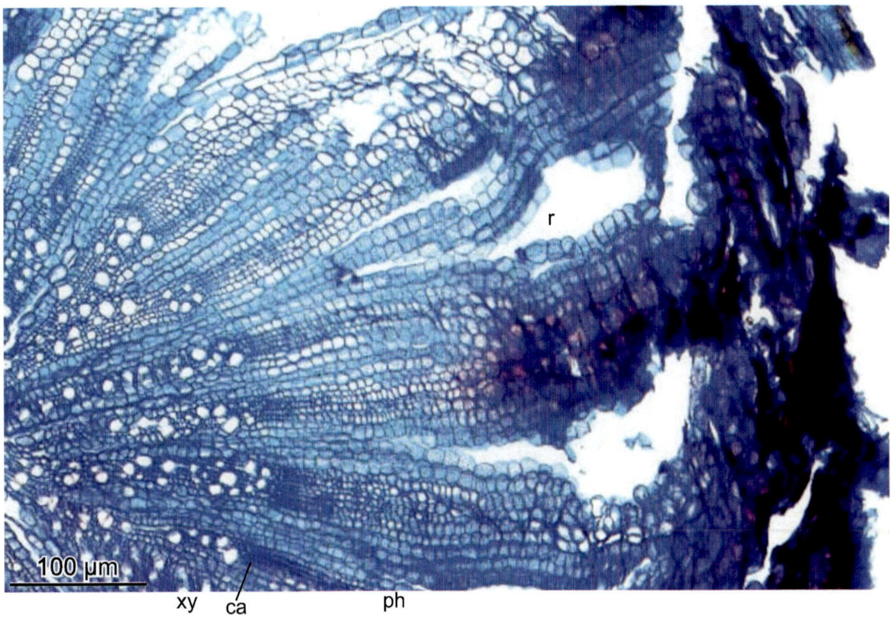

Features according to IAWA

Xylem 1 - 4 - 9 - 13 - 20 - 39.2 - 40.2 - 50.2 - 52.2 - 60.1 - 79.1 - 98 - 100.2 - 105 - 135.2

Astragalus nivalis Kar. & Kir.

Fabaceae

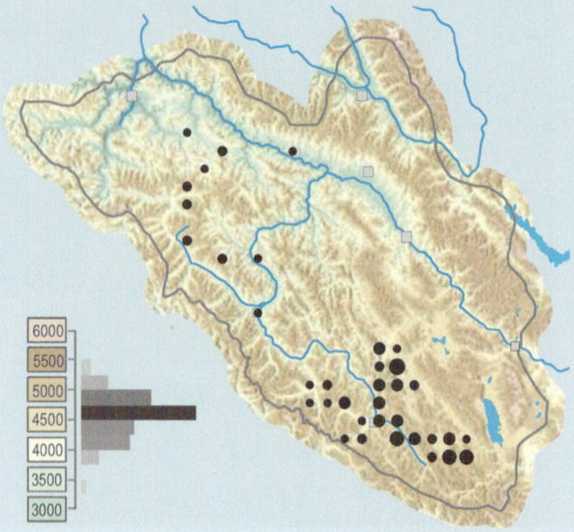

Description: Perennial herbs, 10–20 cm tall, leaves imparipinnately compound, leaflets ovate, flowers violet, ca. 1.5 cm long, in dense racemes.

Distribution: Pakistan, India (Punjab, Jammu and Kashmir), China (Xizang).

Habitat: River beds, unstable steep slopes.

Stem diameter 8–12 mm. Bark/xylem proportion 1:2. Ring width 0.2–0.3 mm. Maximal 20 annual rings. Stem segmentation.

Xylem

Transverse section

Semi-ring-porous. Ring boundaries indicated by large earlywood (50–80 µm) and small latewood vessel diameters (10–20 µm). Vessels solitary and in small groups. Vessel walls red-stained, vessels fairly thick-walled. Fibers in groups, but rare. Parenchyma pervasive. Crystals absent. With secondary intra-xylary periderm towards the pith.

Tangential section

Rays 4–5-seriate, unlignified and thin-walled.

Radial section

Perforation plates simple. Intervessel pits scalariform. Vessel length 80–120 µm. All ray cells upright.

Bark

Phloem with groups of sieve tubes. Older sieve tubes radially collapsed. Fibers isolated or in tangential groups. Phellem cells tangentially flat, thin-walled, rectangular. Crystals absent.

Pith

No slide available.

Analyzed: Root collar of 4 individuals.

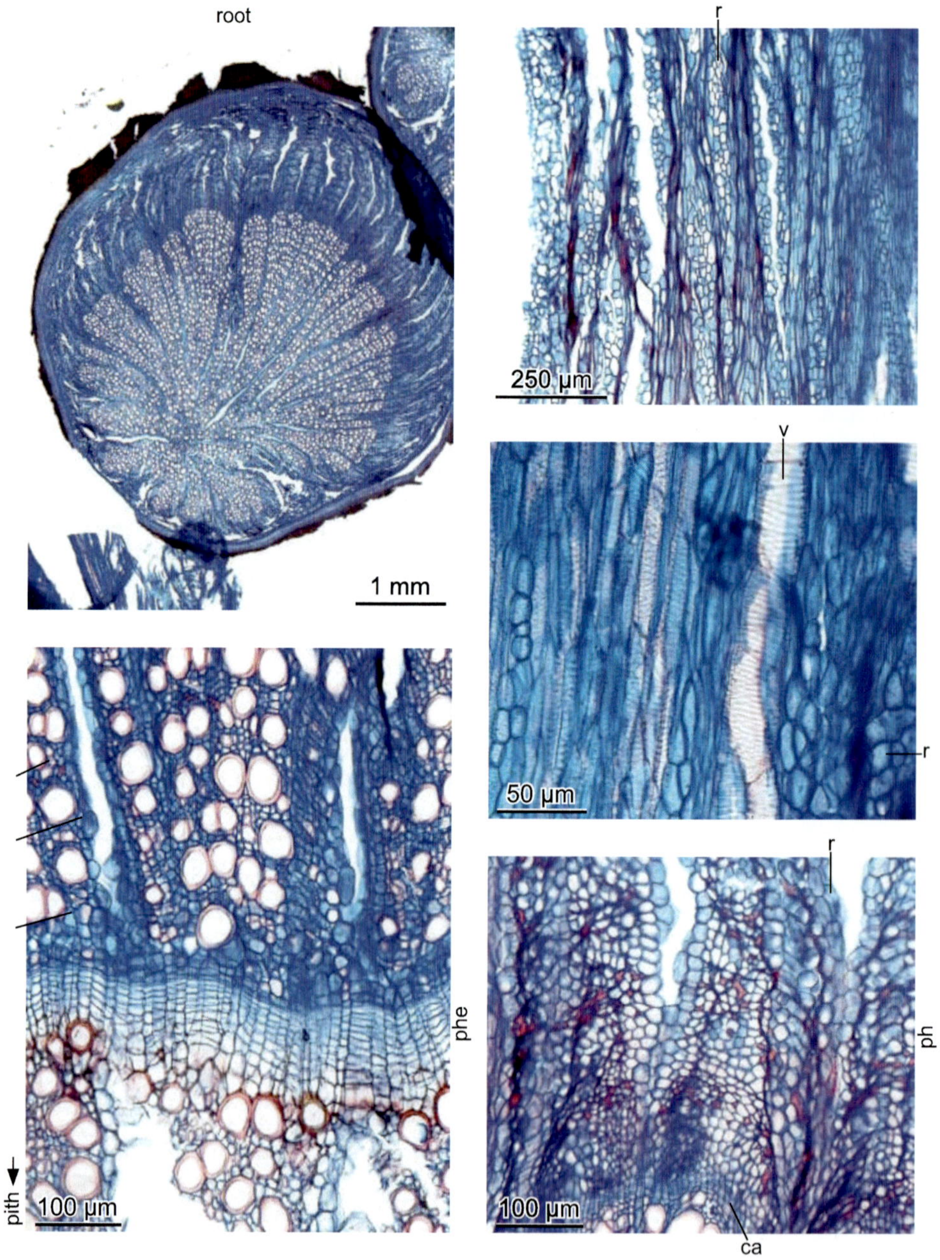

Features according to IAWA

Xylem 1 - 4 - 9 - 11 - 13 - 20 - 39.1 - 40.2 - 41 - 50.2 - 52.2 - 52.3 - 61 - 69 - 79.1 - 98 - 100.2 - 105 - 135.2

Astragalus oplites Benth. ex Parker

Fabaceae

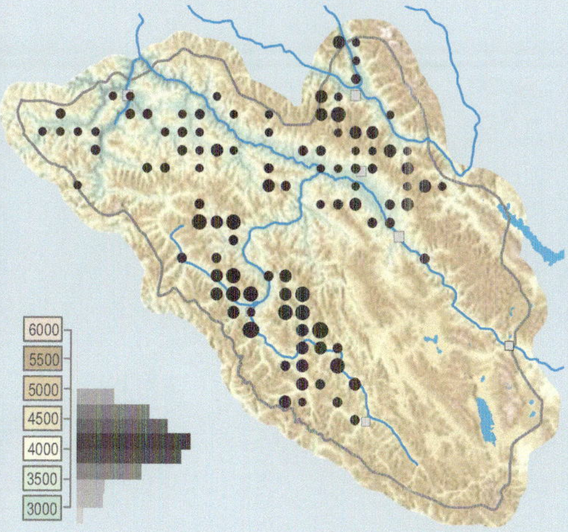

Syn.: *A. zanskarensis* subsp. *oplites* (Benth. ex Parker) Deml

Description: Subshrubs, 20–40 cm tall, leaves imparipinnately compound, rachis with pungent tip, flowers yellow, 2–3 cm long, in subsessile 2–5-flowered racemes.

Distribution: India (Jammu and Kashmir), China (Xinjiang, Xizang).

Habitat: Stabilized slopes.

Stem diameter 5–11 mm. Bark/xylem proportion 1:1. Ring width 0.15–0.6 mm. Maximal 18 annual rings.

Xylem

Transverse section

Semi-ring-porous to ring-porous. Ring boundaries indicated by large earlywood (60–90 µm) and small latewood vessel diameters (20–30 µm). Vessels solitary and in small groups. Vessel walls red-stained. Fibers in groups. Parenchyma pervasive and paratracheal. Crystals absent. With secondary intra-xylary periderm towards the pith.

Tangential section

Rays 3–5-seriate, unlignified and thin-walled.

Radial section

Perforation plates simple. Intervessel pits scalariform. Vessel length 60–100 µm. All ray cells upright.

Bark

Phloem with groups of sieve tubes. Older sieve tubes radially collapsed. Fibers isolated or in tangential groups. With ray dilatations. Phellem cells tangentially flat, thin-walled, rectangular. Crystals absent.

Pith

No slide available.

Analyzed: Root collar of 7 individuals.

root

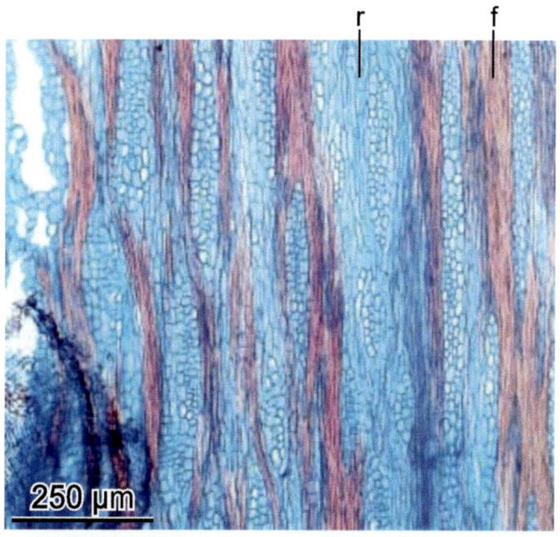

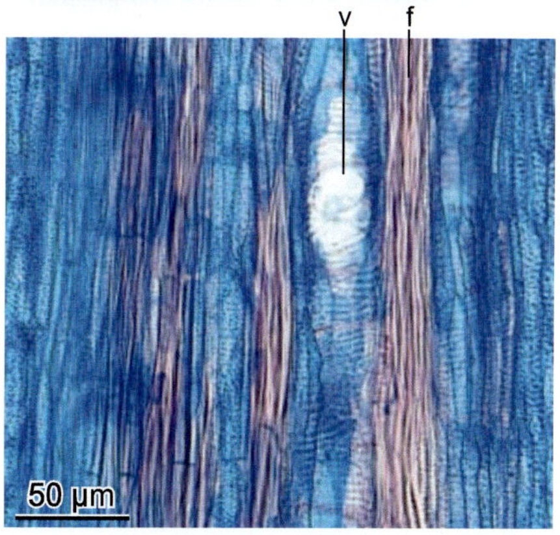

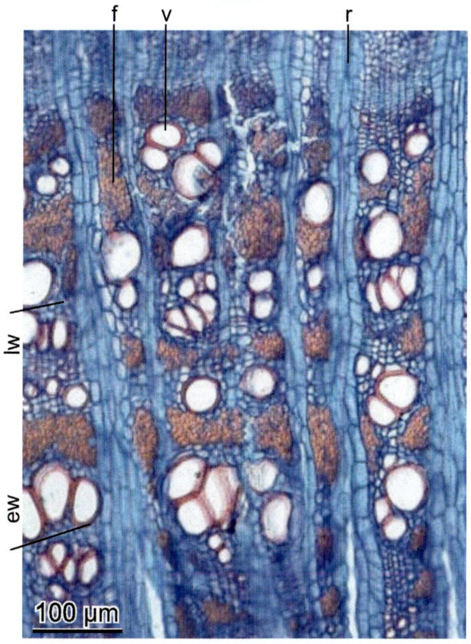

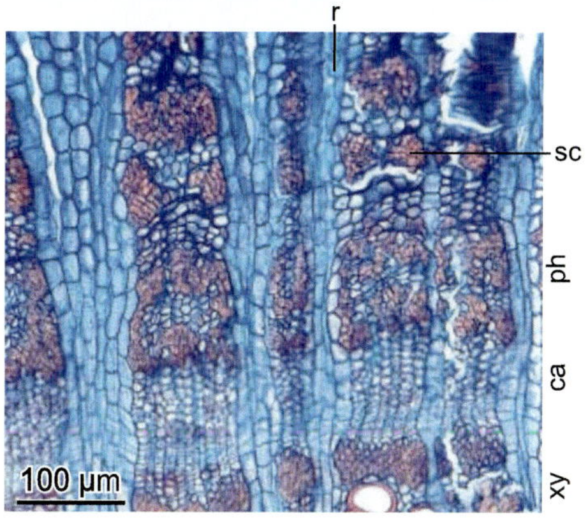

Features according to IAWA

Xylem 1 - 4 - 9 - 11 - 13 - 20 - 39.1 - 40.2 - 50.2 - 52.2 - 61 - 69 - 79 - 79.1 - 98 - 100.2 - 105 - 135.2

Astragalus peduncularis Royle ex Bth.

Fabaceae

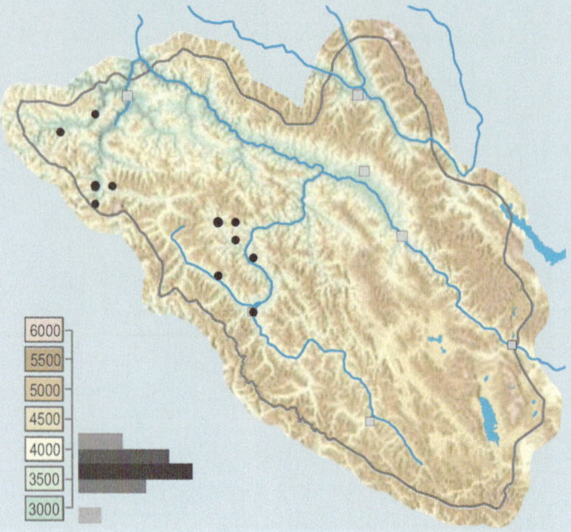

Description: Perennial herbs, 25–70 cm tall, leaves imparipinnately compound, leaflets narrowly elliptic, hairy below, subglabrous above, flowers greenish white, ca. 1.5 cm long, in many-flowered racemes.

Distribution: Kazakhstan, Uzbekistan, Kyrgyzstan, Tajikistan, Afghanistan, Pakistan, India (Jammu and Kashmir, Himachal Pradesh), China (Qinghai, Xizang).

Habitat: Steppes, open scrub.

Stem diameter 10–13 mm. Bark/xylem proportion 1:2. Ring width 1.2–1.6 mm. Maximal 3 annual rings.

Xylem

Transverse section

Semi-ring-porous to ring-porous. Ring boundaries indicated by large earlywood (70–120 µm) and small latewood vessel diameters (10–20 µm). Vessels solitary or in small groups. Fibers thick-walled, rare, occuring in groups. Parenchyma pervasive. Crystals absent. With secondary intra-xylem periderm towards the pith.

Tangential section

Rays 5–8-seriate, unlignified and thin-walled. With sheath cells.

Radial section

Perforation plates simple. Intervessel pits round to scalariform. Vessel length 60–100 µm. All ray cells upright.

Bark

Phloem with groups of sieve tubes, often radially collapsed. With tangential groups of thick-walled fibers. Rays slightly dilated. Cortex with irregular formed radial spaces. Phellem consists of few tangential layers of cork cells, composed of tangentially flat, thin-walled, rectangular cells. Crystals absent.

Pith

No slide available.

Analyzed: Root collar of 7 individuals.

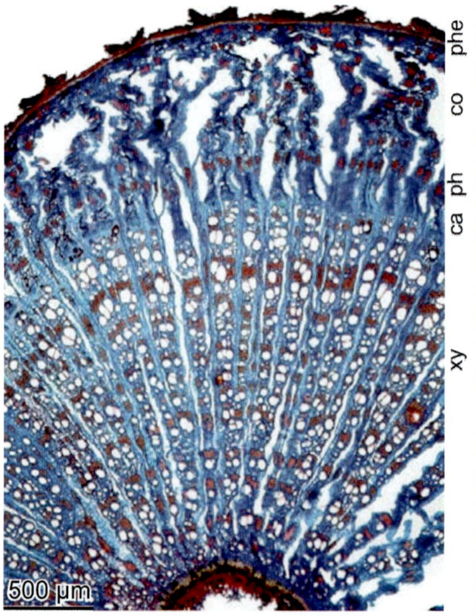

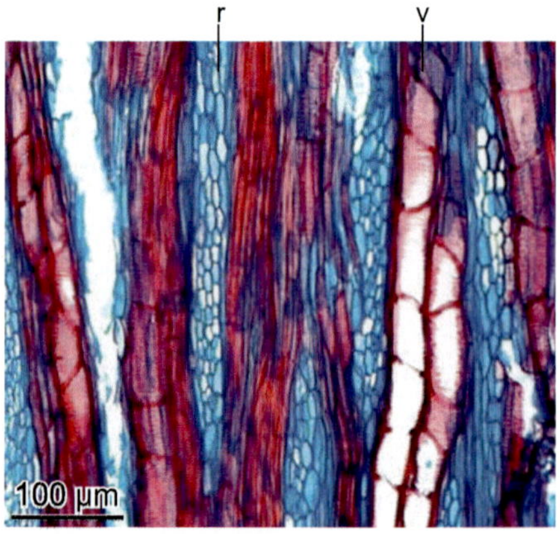

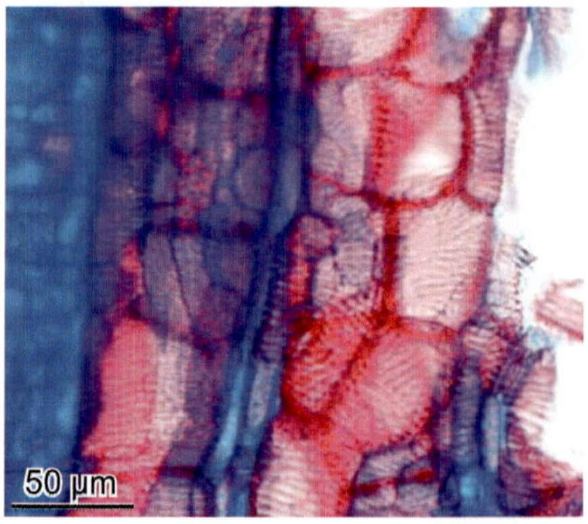

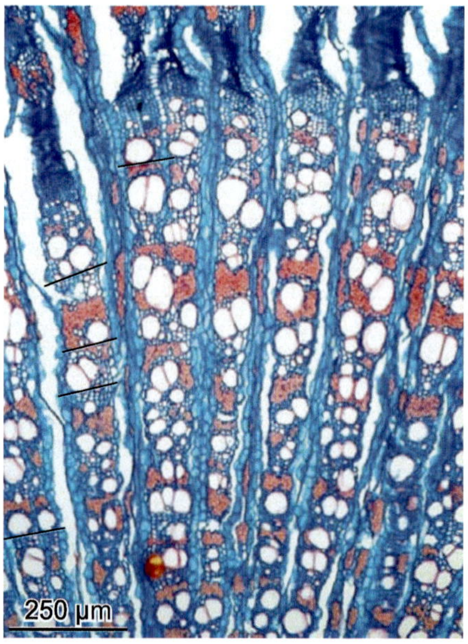

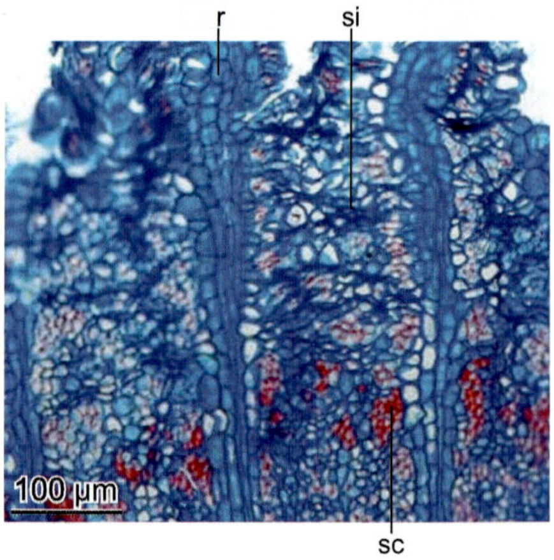

Features according to IAWA

Xylem 1 - 3 - 4 - 9 - 11 - 13 - 20 - 22 - 41 - 42 - 50.2 - 52.2 - 61 - 69 - 79.1 - 98 - 100.2 - 105 - 110 - 135.2

Astragalus rhizanthus Royle ex Benth.

Fabaceae

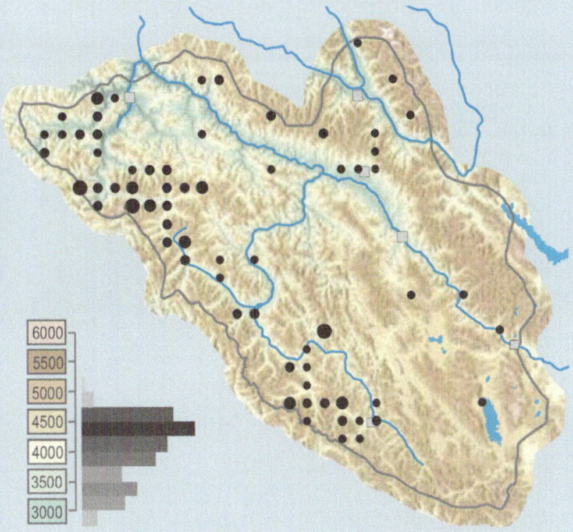

Syn.: *A. hindukushensis* Wendelbo

Description: Subshrubs, 10–25 cm tall, white hairy, acaulescent or with spreading stems, leaves imparipinnately compound, flowers yellow, ca. 2 cm long, in subsessile racemes.

Distribution: Afghanistan, Pakistan, India, Nepal, China (Xinjiang, Xizang).

Habitat: Subalpine to alpine meadows and stony slopes.

Stem diameter 16–22 mm. Bark/xylem proportion 1:1. Ring width 0.1–0.3 mm. Maximal 30 annual rings. Stem segmentation. Center rotten.

Xylem

Transverse section

Semi-ring-porous to ring-porous. Ring boundaries indicated by large earlywood (60–90 µm) and small latewood vessel diameters (20–30 µm). Vessels solitary and in small groups. Vessel walls red-stained. Fibers in groups. Parenchyma pervasive. Crystals absent. Without secondary intraxylary periderm towards the pith.

Tangential section

Rays 4–8-seriate, unlignified and thin-walled.

Radial section

Perforation plates simple. Intervessel pits scalariform. Vessel length 80–120 µm. All ray cells upright.

Bark

Phloem with groups of sieve tubes. Older sieve tubes radially collapsed. Fibers in tangential groups. Phellem cells tangentially flat, thin-walled, rectangular. Crystals absent.

Pith

No slide available.

Analyzed: Root collar of 7 individuals.

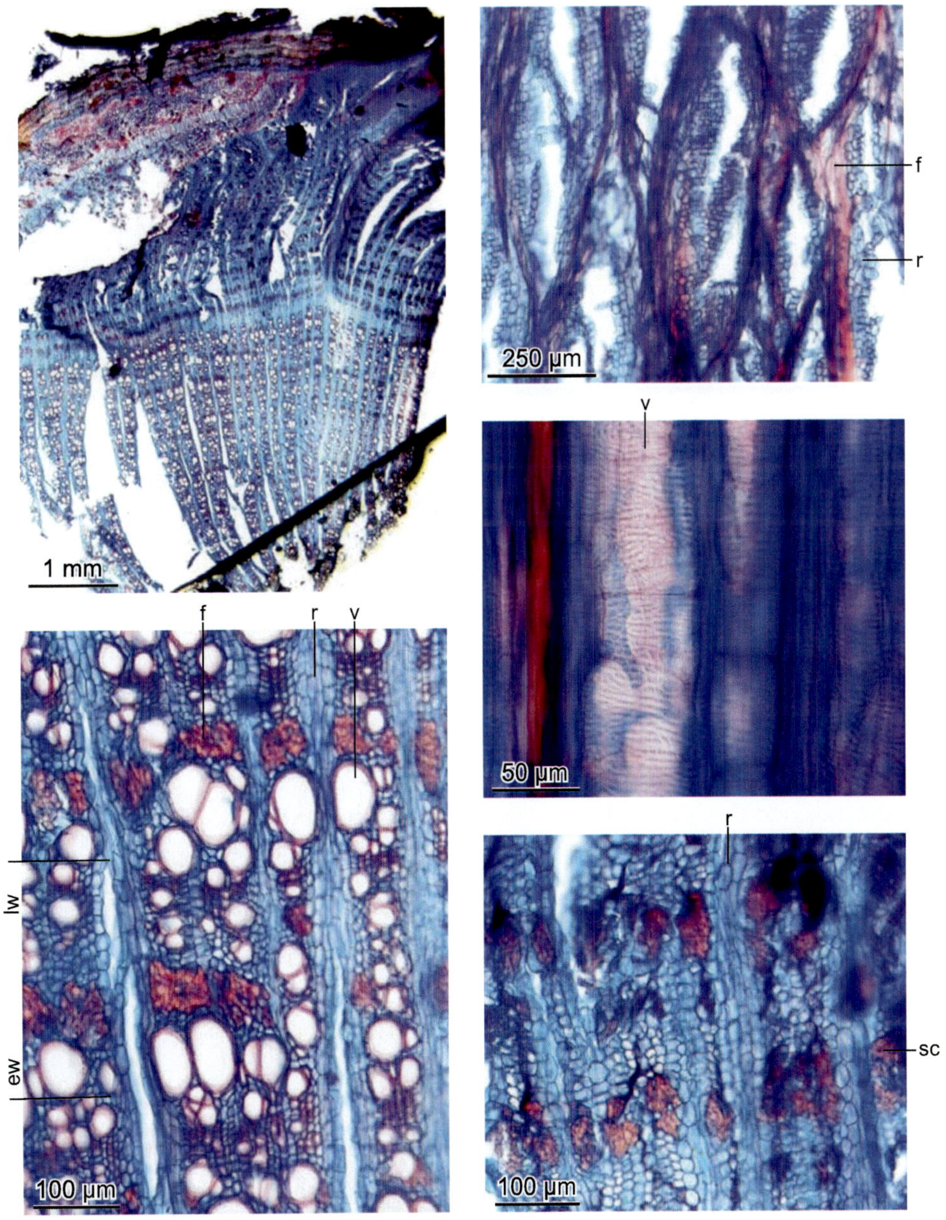

Features according to IAWA

Xylem 1 - 3 - 4 - 9 - 11 - 13 - 20 - 41 - 50.2 - 52.2 - 52.3 - 61 - 69 - 79.1 - 98 - 100.2 - 105

Astragalus strictus Grah. ex Benth.

Fabaceae

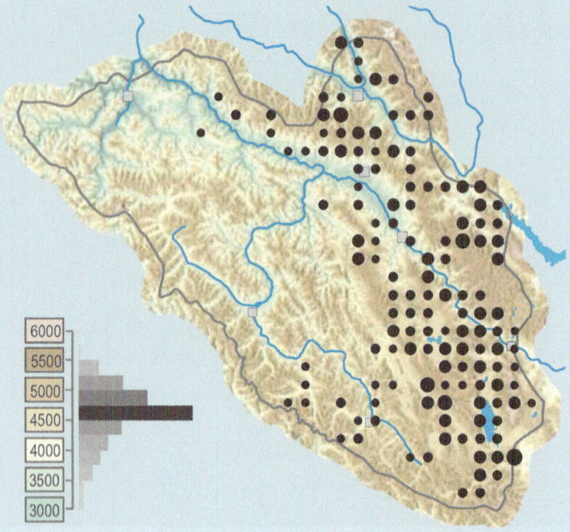

Syn.: *A. ladakensis* Balakrishnan

Description: Perennial herbs, 8–30 cm tall, cespitose, stems ascending, leaves imparipinnately compound, flowers pale pinkish to purple, ca. 1 cm long, in dense racemes.

Distribution: India, Nepal, Bhutan, China (Qinghai, Sichuan, Xinjiang, Xizang, Yunnan), Bangladesh.

Habitat: Mesic to wet loamy habitats, stony steppes, gravel river sediments, saline habitats.

Stem diameter 0.3–0.4 mm. Bark/xylem proportion 1:1. Ring width 0.16–0.5 mm. Approximately 11 annual rings. Center rotten.

Xylem

Transverse section

Semi-ring-porous. Ring boundaries indicated by large earlywood (30–50 µm) and small latewood vessel diameters (10–20 µm). Vessels solitary, rarely in small groups. Vessel walls poorly stained. Fibers in groups. Parenchyma pervasive and paratracheal. Crystals absent. Without secondary intra-xylary periderm towards the pith.

Tangential section

Rays 3–5-seriate, unlignified and thin-walled. With sheath cells.

Radial section

Perforation plates simple. Intervessel pits scalariform. Vessel length 60–100 µm. All ray cells upright.

Bark

Phloem with groups of sieve tubes. Older sieve tubes radially collapsed. Fibers in tangential groups. With ray dilatations. Phellem cells tangentially flat, thin-walled, rectangular. Crystals absent.

Pith

No slide available.

Analyzed: Root collar of 3 individuals.

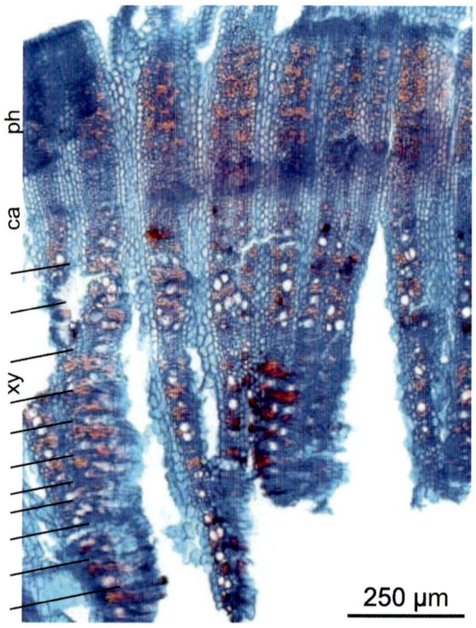

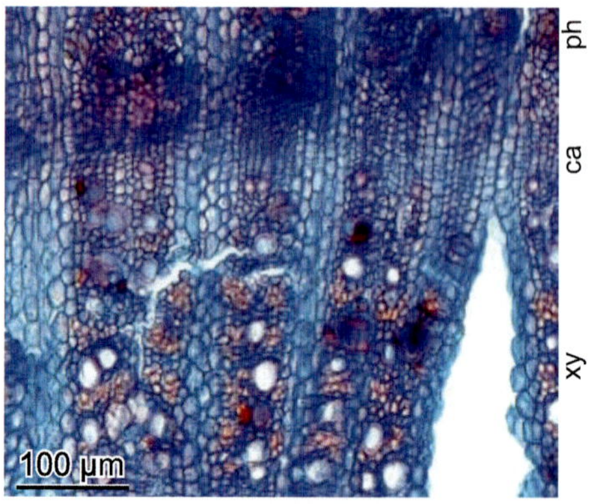

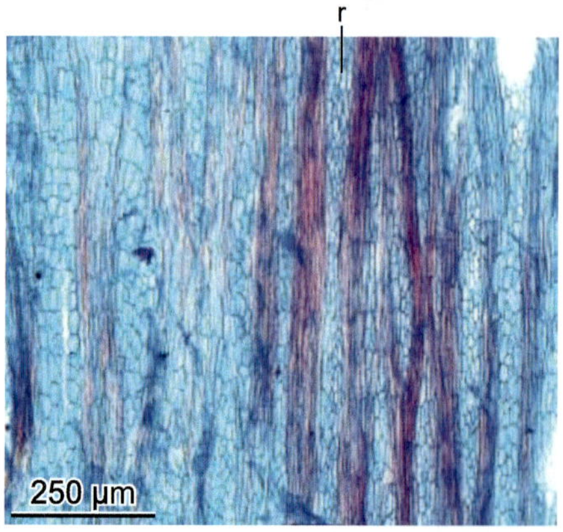

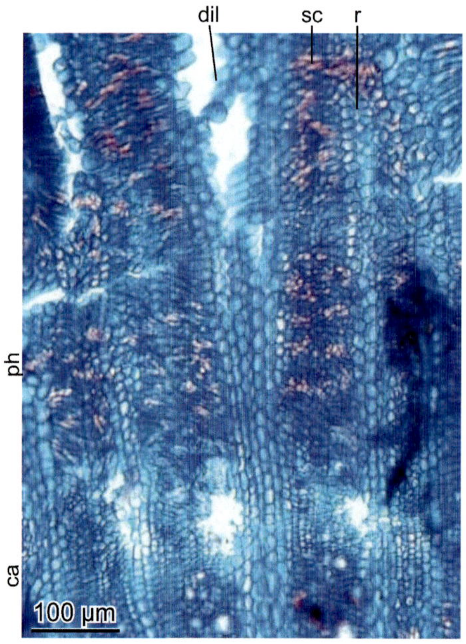

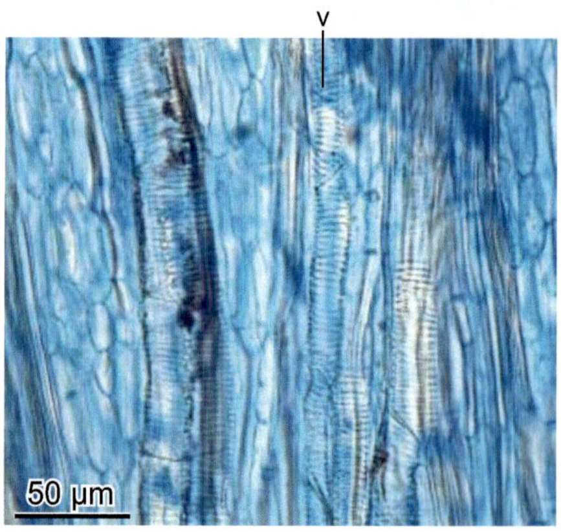

Features according to IAWA

Xylem *1 - 4 - 9 - 11 - 13 - 20 - 39.2 - 40.2 - 50.2 - 52.2 - 61 - 69 - 79 - 79.1 - 98 - 100.2 - 105 - 110*

Astragalus tecti-mundi Freyn

Fabaceae

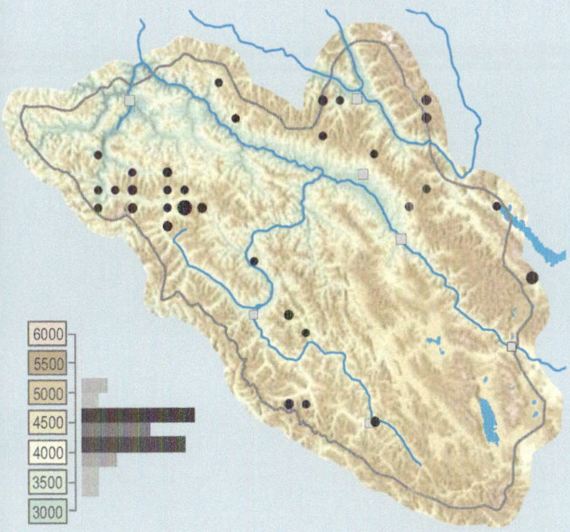

Description: Perennial herbs, 30–90 cm tall, mostly glabrous, leaves imparipinnately compound, leaflets ovate, flowers yellow, ca. 2 cm long, in many-flowered dense racemes.

Distribution: Tajikistan, Afghanistan, Pakistan, India, China (Xinjiang, Xizang).

Habitat: Subalpine and alpine wet meadows.

Stem diameter 7–12 mm. Bark/xylem proportion 1:3. Ring width 0.4–0.5 mm. Maximal 13 annual rings.

Xylem

Transverse section

Semi-ring-porous to ring-porous. Ring boundaries indicated by large earlywood (100–130 μm) and small latewood vessel diameters (10–20 μm). Vessels solitary and in small groups. Thick-walled fibers in groups in the latewood. Parenchyma pervasive. Crystals absent. With intra-xylary periderm towards the pith.

Tangential section

Rays 4–6-seriate, unlignified and thin-walled. With sheath cells. Some ray cells much smaller than the rest.

Radial section

Perforation plates simple. Intervessel pits round to scalariform. Vessel length 80–120 μm. All ray cells upright.

Bark

Phloem with small groups of sieve tubes, radially collapsed. With small irregular groups of thick-walled fibers, arranged in tangential lines. Phloem with irregular formed radial spaces. Ray course wavy. Phellem consists of few tangential layers of cork cells, composed of tangentially flat, thin-walled, rectangular cells. Crystals absent.

Pith

No slide available.

Analyzed: Root collar of 3 individuals.

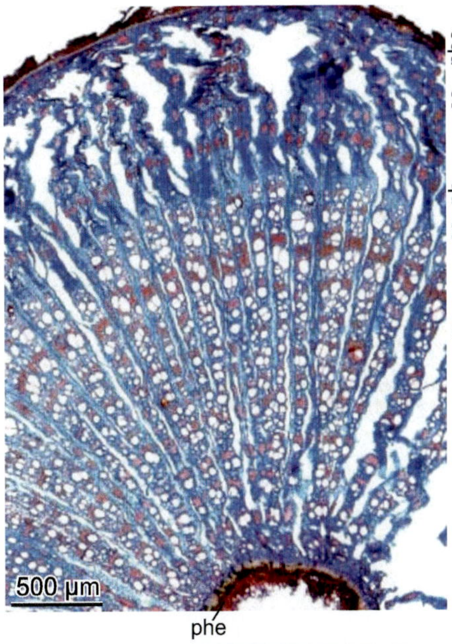

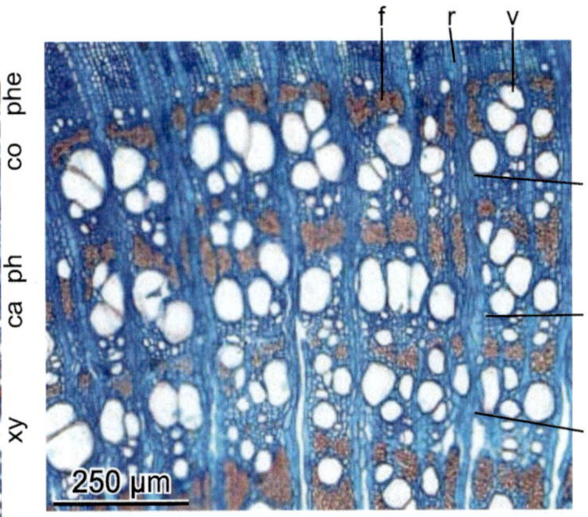

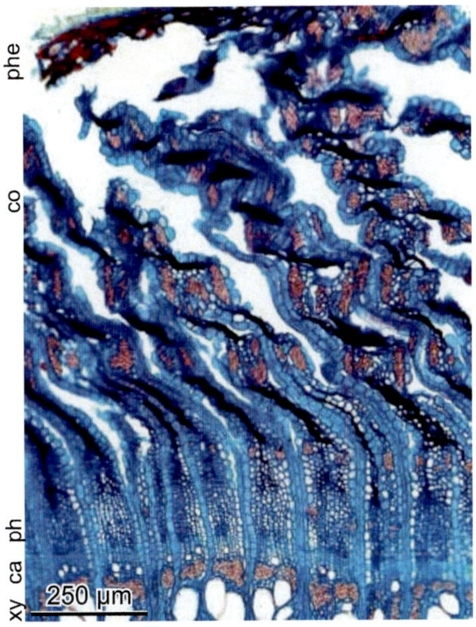

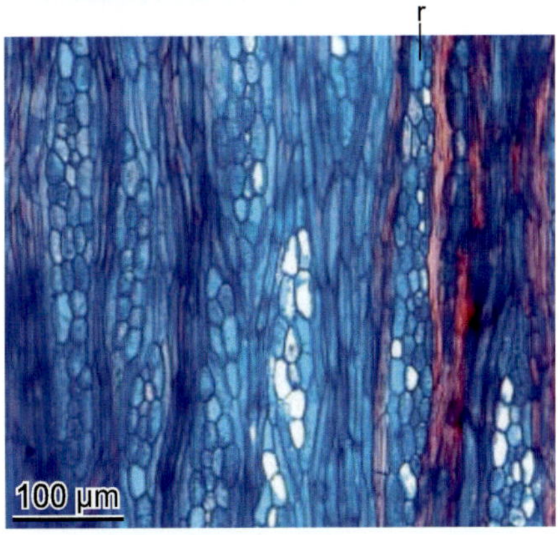

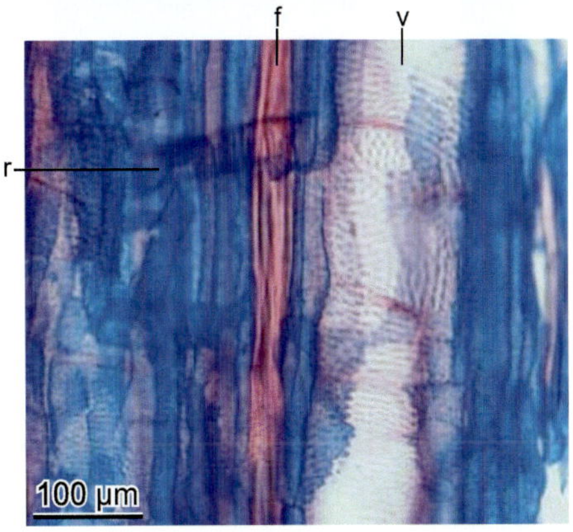

Features according to IAWA

Xylem 1 - 3 - 4 - 11 - 13 - 20 - 22 - 41 - 42 - 50.2 - 52.2 - 61 - 69 - 79.1 - 98 - 100.2 - 105 - 109 - 110 - 135.2

Astragalus thomsonii Podlech

Fabaceae

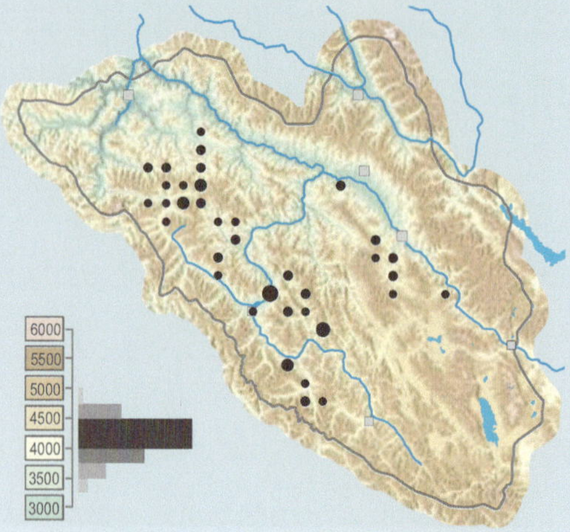

Description: Perennial herbs, 10–35 cm tall, stems erect, hairy, leaves imparipinnately compound, leaflets narrowly elliptic, flowers 6–8 mm long, in dense ovoid to oblong racemes.

Distribution: India, Nepal, China (Xizang).

Habitat: Alpine meadows.

Stem diameter 7–11 mm. Bark/xylem proportion 1:3. Ring width 0.2–0.36 mm. Maximal 12 annual rings.

Xylem

Transverse section

Semi-ring-porous to ring-porous. Ring boundaries indicated by large earlywood (50–80 μm) and small latewood vessel diameters (10–20 μm). Vessels in small groups. Thick-walled fibers rare, in groups. Parenchyma pervasive. Crystals absent. With intra-xylary periderm towards the pith.

Tangential section

Rays 5–8-seriate, unlignified and thin-walled, often confluent. With sheath cells.

Radial section

Perforation plates simple. Intervessel pits round to scalariform. Vessel length 60–100 μm. All ray cells upright.

Bark

Phloem with small groups of sieve tubes, often radially collapsed. With small irregular groups of thick-walled fibers, arranged in tangential lines. Phloem with irregular formed radial spaces. Phellem consists of few tangential layers of cork cells, composed of tangentially flat, thin-walled, rectangular cells. Crystals absent.

Pith

No slide available.

Analyzed: Root collar of 7 individuals.

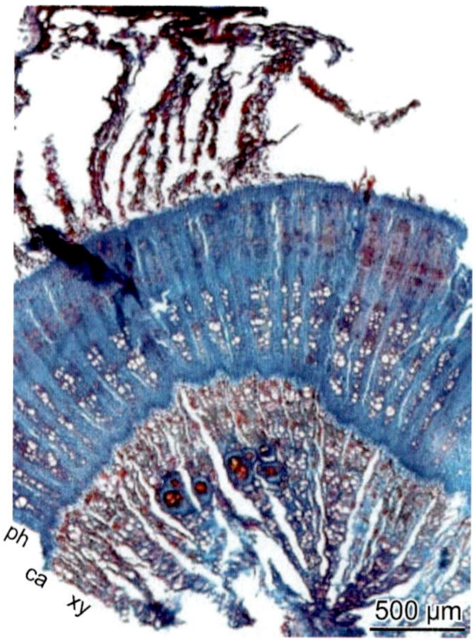

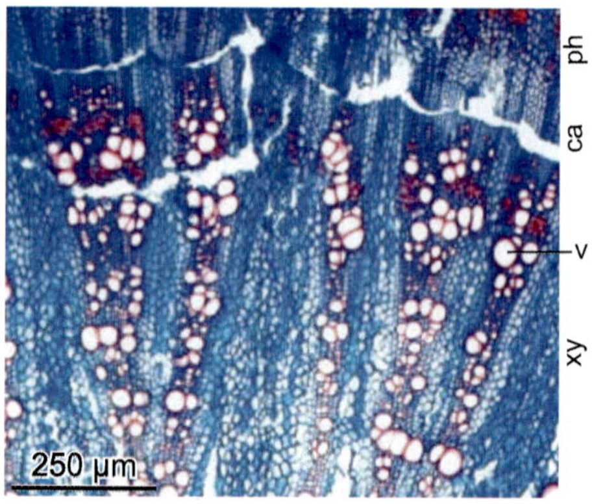

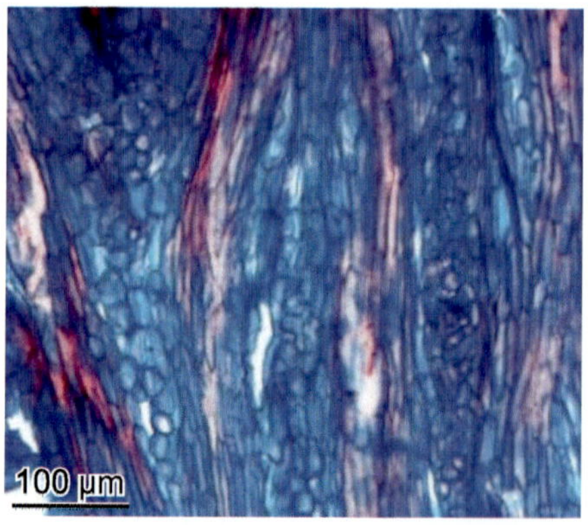

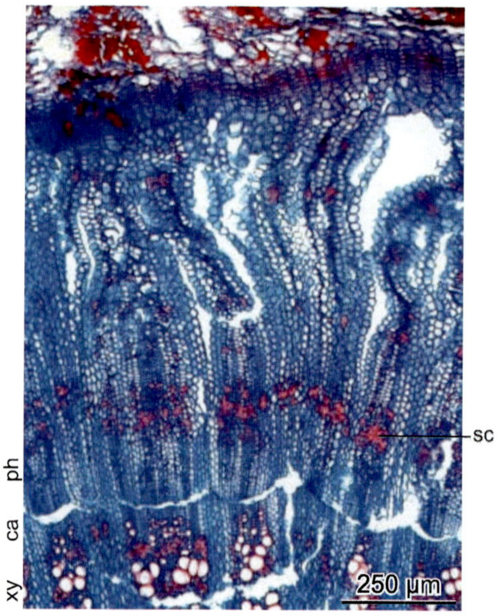

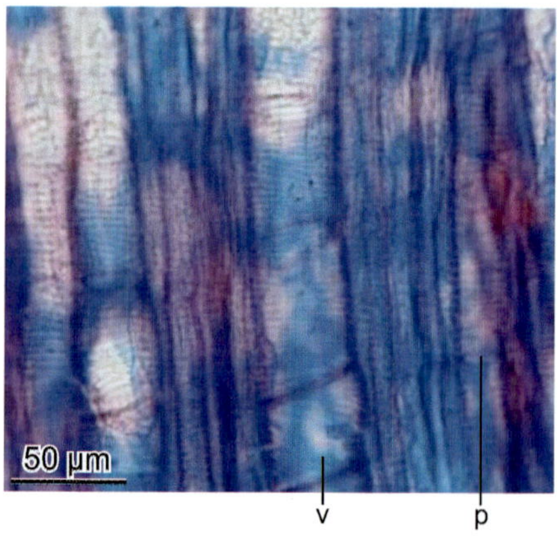

Features according to IAWA

Xylem 1 - 3 - 4 - 11 - 13 - 20 - 22 - 41 - 50.2 - 52.2 - 61 - 69 - 79.1 - 98 - 100.1 - 100.2 - 105 - 110 - 135.2

Astragalus tibetanus Benth. ex Bunge

Fabaceae

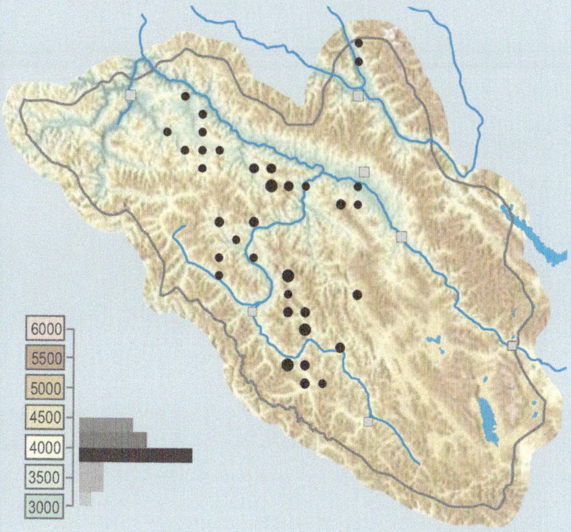

Syn.: *A. chadjanensis* Franch., *A. narynensis* Freyn, *A. olufsenii* Freyn, *A. talievii* Sirj.

Description: Perennial herbs, 4–35 cm tall, stems ascending, leaves imparipinnately compound, leaflets narrowly oblong, pubescent, flowers violet, 15–20 mm long, in dense ovoid racemes.

Distribution: S Russia, Kazakhstan, Kyrgyzstan, Tajikistan, NE Afghanistan, N Pakistan, NW India, China (Xinjiang), Mongolia.

Habitat: Ditches of irrigation canals, weedy on arable land.

Stem diameter 5 mm. Bark/xylem proportion 1:1. Mean ring width 0.2 mm. Maximal 8 annual rings.

Xylem

Transverse section

Rings absent. Diameter of earlywood vessels 40–50 µm. Vessels solitary. Thick-walled fibers rare, in groups. Parenchyma pervasive. Crystals absent.

Tangential section

Rays 5–8-seriate, unlignified and thin-walled, often confluent. With sheath cells.

Radial section

Perforation plates simple. Intervessel pits round to scalariform. Vessel length 60–80 µm. All ray cells upright.

Bark

Phloem with small groups of sieve tubes. With small irregular groups of thick-walled fibers, arranged in tangential lines. Rays dilated. Phloem with irregular formed radial spaces. Phellem consists of few tangential layers of cork cells, composed of tangentially flat, thin-walled, rectangular cells. Crystals absent.

Pith

No slide available.

Analyzed: Root collar of 1 individual.

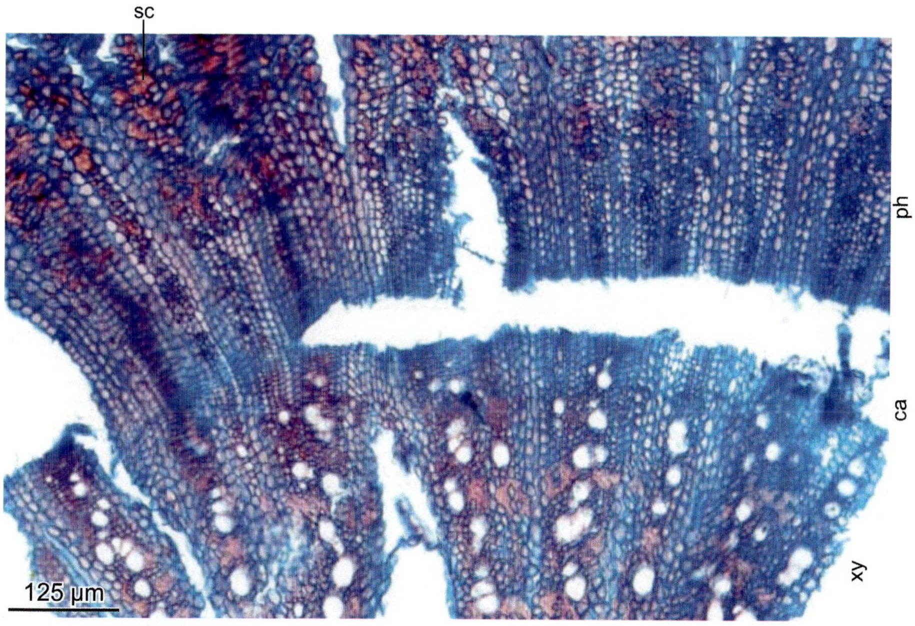

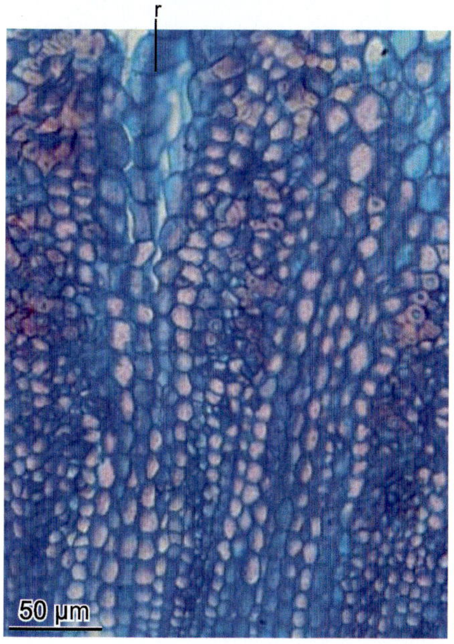

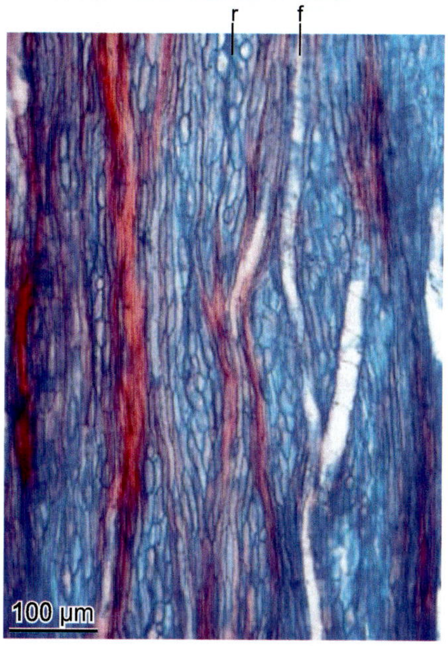

Features according to IAWA

Xylem 2 - 9 - 13 - 20 - 22 - 40.2 - 50.2 - 52.2 - 61 - 69 - 79.1 - 98 - 100.1 - 100.2 - 105 - 110

Caragana versicolor Benth.

Fabaceae

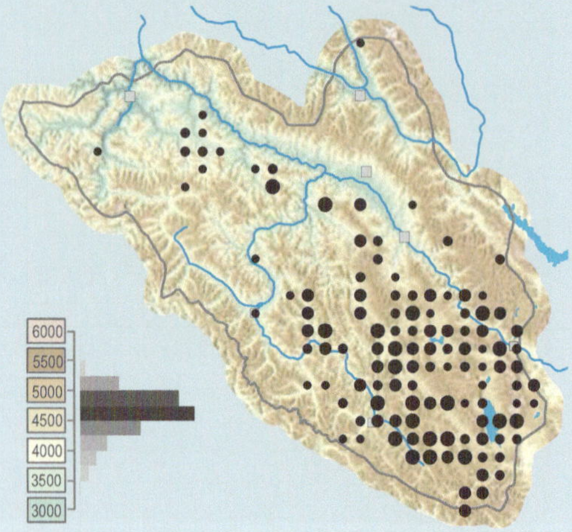

Syn.: *Genista versicolor* Wall.

Description: Shrubs, 30–80 cm tall, thorny, semiglobose in shape, leaves digitately compound, leaflets 4, flowers yellow, 1.5 cm long, solitary in axils.

Distribution: SW Asia, Afghanistan, Pakistan, NW India, Nepal, China (Qinghai, Sichuan, Xizang).

Habitat: Dry slopes, dry watersheds.

Stem diameter 15 mm. Bark/xylem proportion 1:5. Ring width 0.12–0.3 mm. Maximal 75 annual rings.

Xylem

Transverse section

Diffuse-porous. Rings indistinctly marked by a small band of marginal parenchyma. Vessel groups distributed in dendritic patterns. Vessel diameter 30–40 µm. Fibers thick-walled. Fiber groups surrounded by sheath parenchyma cells. Vessels in the stem center with dark-staining substances. With prismatic crystals in rays.

Tangential section

Rays of two distinct sizes: 1- and 2-seriate and >10-seriate. Large rays with sheath cells. Prismatic crystals in large rays.

Radial section

Perforation plates simple. Intervessel pits round. Small vessels with helical thickenings. Vessel length 60–100 µm. Uniseriate rays with procumbent cells and large rays with procumbent cells in the center and upright sheath cells. Large rays >1 mm in height.

Bark

Phloem with collapsed sieve tubes and tangential groups of thick-walled fibers. Phellem consists of tangentially layered packages (annual?) of thin-walled, rectangular cells.

Pith

No slide available.

Analyzed: Stem of 4 individuals.

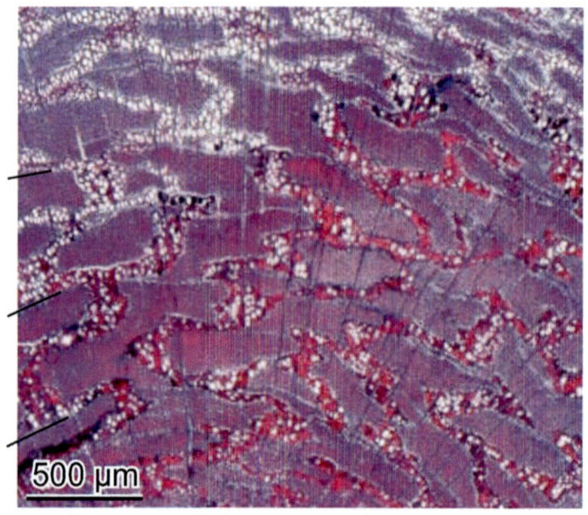

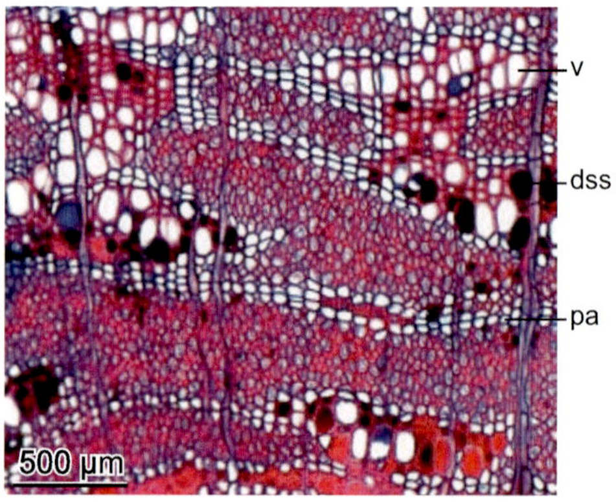

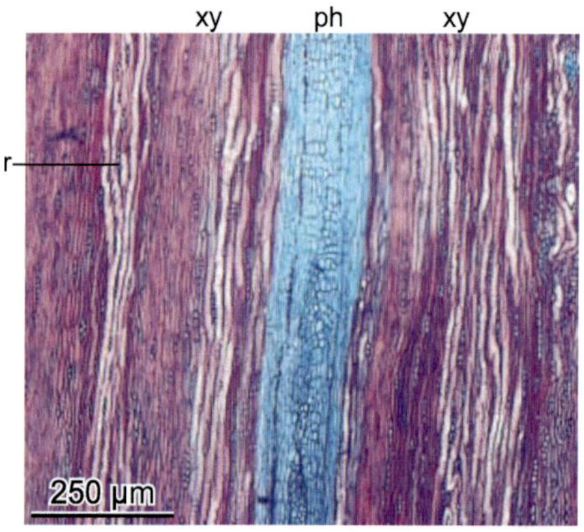

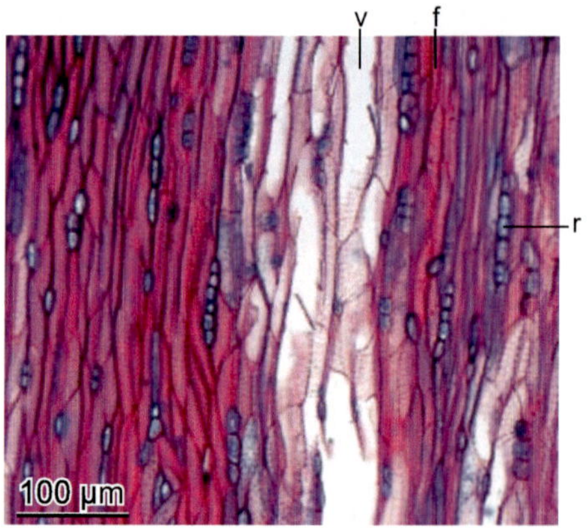

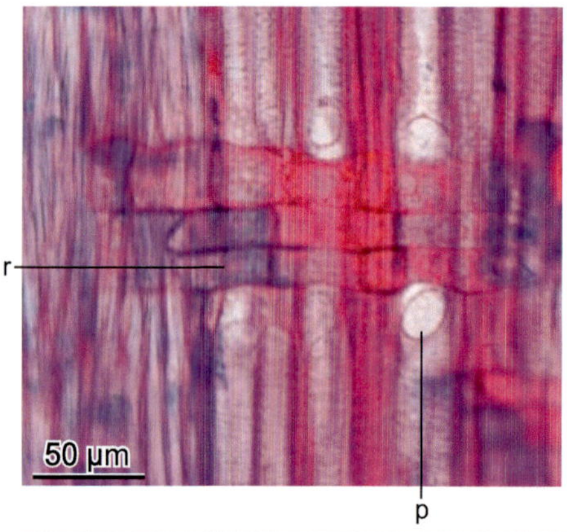

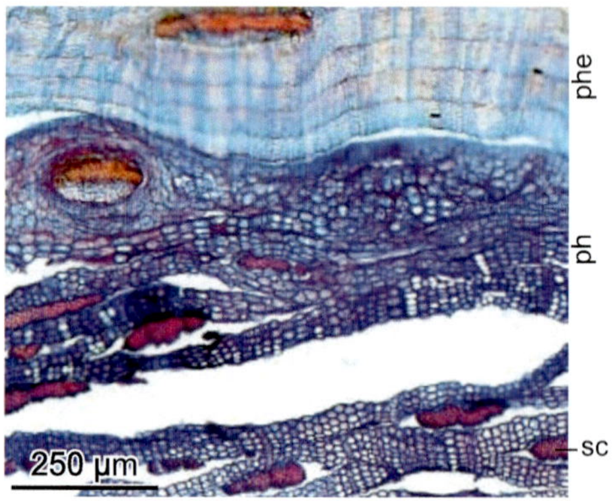

Features according to IAWA

Xylem *1 - 5 - 8 - 13 - 22 - 36 - 40.2 - 50.2 - 52.2 - 58 - 61 - 69 - 79 - 96 - 102 - 103 - 105 - 110 - 144*

Chesneya cuneata (Benth.) Ali

Fabaceae

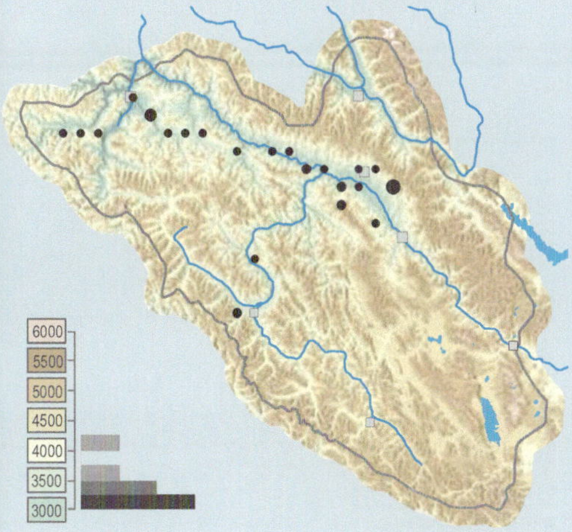

Syn.: *Gueldenstaedtia cuneata* Benth., *Calophaca cuneata* (Benth.) Kom.

Description: Perennial herbs, 5–15 cm tall, tomentose, stems prostrate or ascending, leaves imparipinnate, 2–10 cm long, leaflets obovate, flowers purple, 2–3 cm long, in 1–4-flowered axillary racemes.

Distribution: N Pakistan, N India, China (Xinjiang, Xizang).

Habitat: Semideserts, dry slopes, gravel areas.

Stem diameter 10 mm. Bark/xylem proportion 1:3. Mean ring width 0.16 mm. Maximal 15 annual rings.

Xylem

Transverse section

Semi-ring-porous. Ring boundaries indicated by large earlywood (30–50 µm) and small latewood vessel diameters (10–20 µm). Vessels in small groups. Vessel walls red-stained. Fibers absent. Parenchyma pervasive. Crystals absent. Without secondary intra-xylary periderm towards the pith.

Tangential section

Rays >10-seriate, unlignified and thin-walled.

Radial section

Perforation plates simple. Intervessel pits scalariform. Vessel length 60–120 µm. All ray cells upright.

Bark

Phloem with groups of sieve tubes. Older sieve tubes radially collapsed. Fibers in radial bands. With ray dilatations. Groups of unstained fibers in the cortex. Phellem cells tangentially flat, thin-walled, rectangular. Crystals absent.

Pith

No slide available.

Analyzed: Stem of 1 individual.

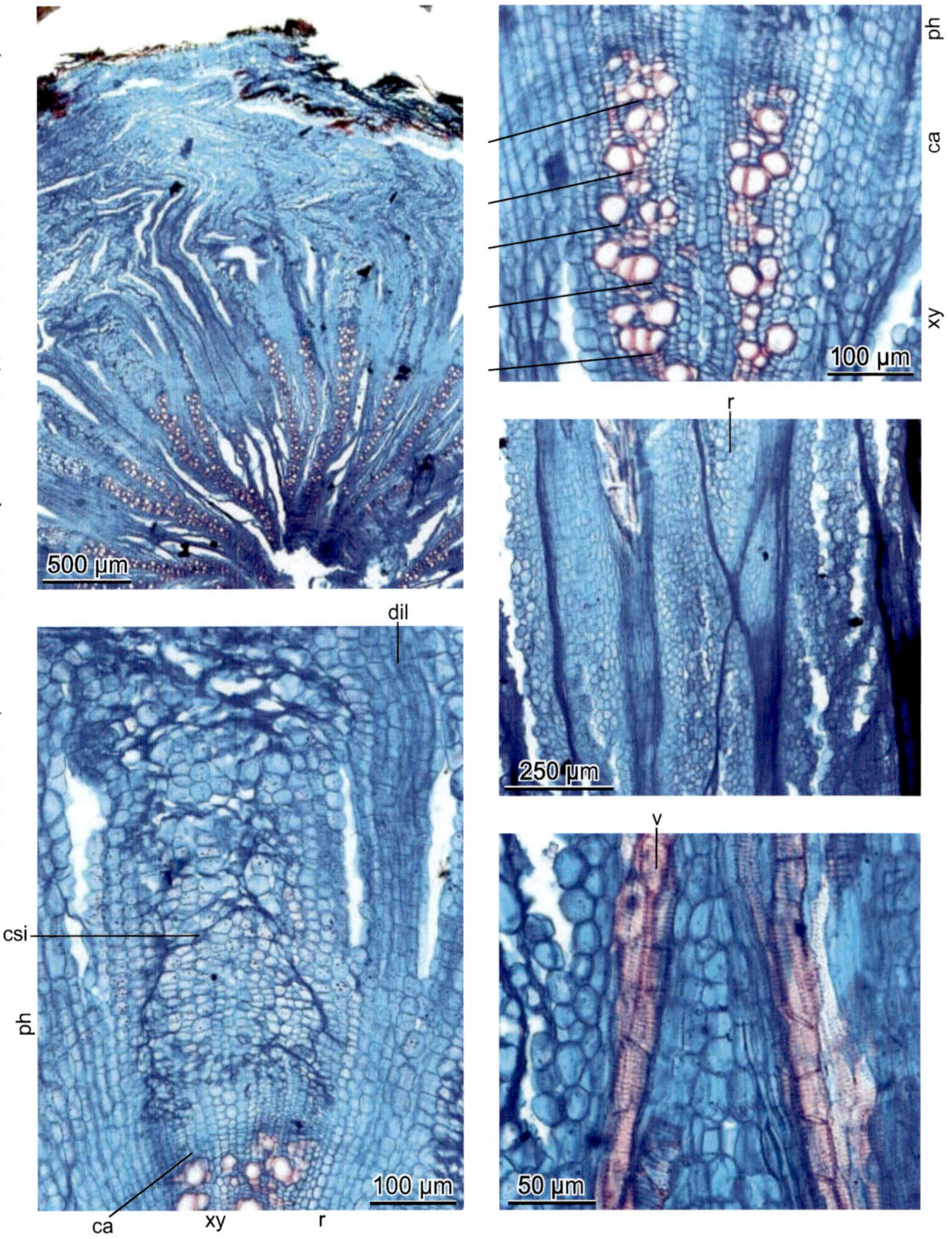

Features according to IAWA

Xylem 1 - 4 - 11 - 13 - 20 - 40.2 - 50.2 - 52.2 - 52.3 - 60.1 - 79.1 - 99 - 99.1 - 100.2 - 105

Cicer microphyllum Benth.

Fabaceae

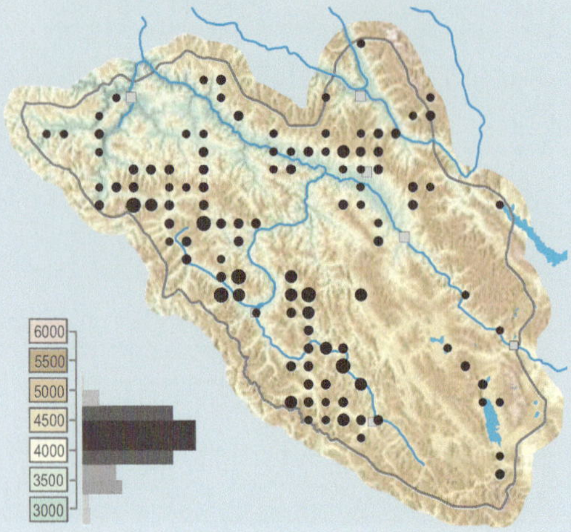

Syn.: *C. jacquemontii* Jaub. & Spach

Description: Perennial herbs, 15–40 cm tall, branched, leaves paripinnate with a terminal tendril, flowers blue-purple, ca. 2 cm long, solitary.

Distribution: E Afghanistan, N Pakistan, India (Jammu and Kashmir, Himachal Pradesh, Uttarakhand), W Nepal, China (Xinjiang, Xizang).

Habitat: Stabilized slopes, unstable gravel and stony screes.

Stem diameter 9 mm. Bark/xylem proportion 1:1. Ring width 0.16–0.5 mm. Maximal 8 annual rings. Center rotten.

Xylem

Transverse section

Semi-ring-porous to ring-porous. Ring boundaries indicated by large earlywood (60–120 µm) and small latewood vessel diameters (30–40 µm). Vessels solitary and in small groups. Vessel walls red-stained. Fibers solitary or in groups. Parenchyma pervasive. Crystals absent. With secondary intra-xylary periderm towards the pith.

Tangential section

Rays 5–6-seriate, unlignified and thin-walled.

Radial section

Perforation plates simple. Intervessel pits round. Vessel length 60–120 µm. All ray cells upright.

Bark

Phloem with groups of sieve tubes. Fibers in tangential groups. Phellem cells tangentially flat, thin-walled, rectangular. Crystals absent.

Pith

No slide available.

Analyzed: Root collar of 4 individuals.

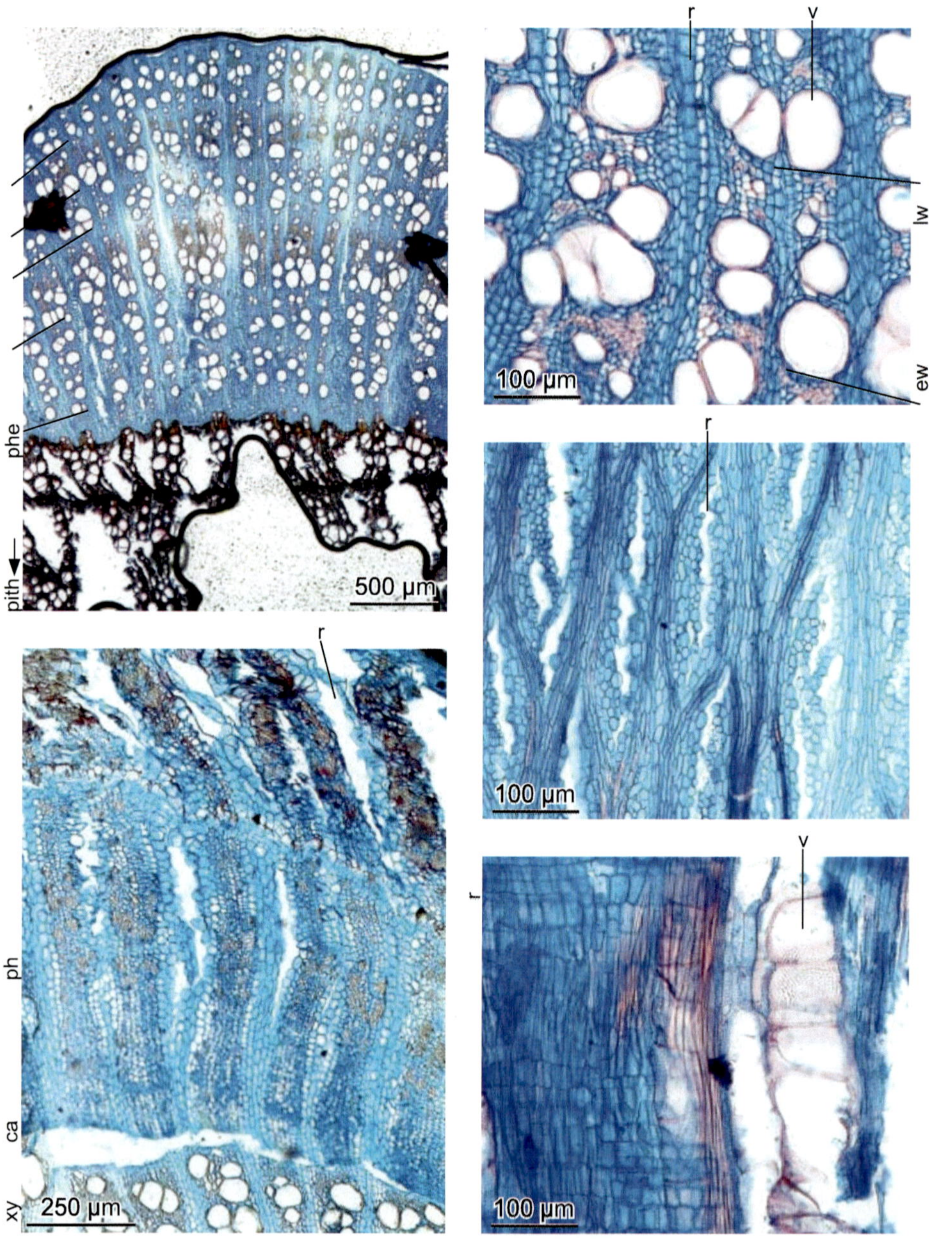

Features according to IAWA

Xylem 1 - 3 - 4 - 9 - 11 - 13 - 22 - 41 - 42 - 50.2 - 52.2 - 52.3 - 61 - 69 - 79.1 - 98 - 100.2 - 105 - 135.2

Colutea nepalensis Sims

Fabaceae

Syn.: *C. arborescens* var. *nepalensis* (Sims) Baker, *C. paulsenii* Freyn, *C. karakoramensis* Kitam., *C. mesantha* Shap. ex Ali

Description: Shrubs, 0.5–3 m tall, leaves imparipinnately compoud, leaflets elliptic to obovate, flowers yellow, 2–2.5 cm long, in 3–5-flowered racemes.

Distribution: Afghanistan, Pakistan, India, Nepal, China (Qinghai, Xizang).

Habitat: Mountain slopes, gravel places.

Stem diameter 20 mm. Bark/xylem proportion 1:5. Mean ring width 0.9 mm. Maximal 7 annual rings.

Xylem

Transverse section

Distinct rings. Ring-porous. Rings marked by large earlywood (60–100 µm) and small latewood vessel diameters (10–20 µm). Vessel groups in dendritic patterns. Vessel dimorphism in the latewood. Fibers thick-walled. Parenchyma paratracheal and marginal. Crystals absent.

Tangential section

Rays 1–6-seriate. Few large rays with sheath cells.

Radial section

Perforation plates simple. Intervessel pits round. Small vessels with helical thickenings. Vessel length 100–140 µm. Ray cells mostly square, rarely mixed with procumbent cells.

Bark

Phloem with tangential rows of collapsed sieve tubes and round groups of thick-walled fibers. No slide with phellem available.

Pith

Round, empty.

Analyzed: Stem of 1 individual.

Features according to IAWA

Xylem 1 - 5 - 8 - 13 - 22 - 36 - 41 - 45 - 50.1 - 52.3 - 61 - 69 - 79 - 89 - 98 - 102 - 103 - 105 - 109 - 110

Oxytropis chiliophylla Royle ex Benth.

Fabaceae

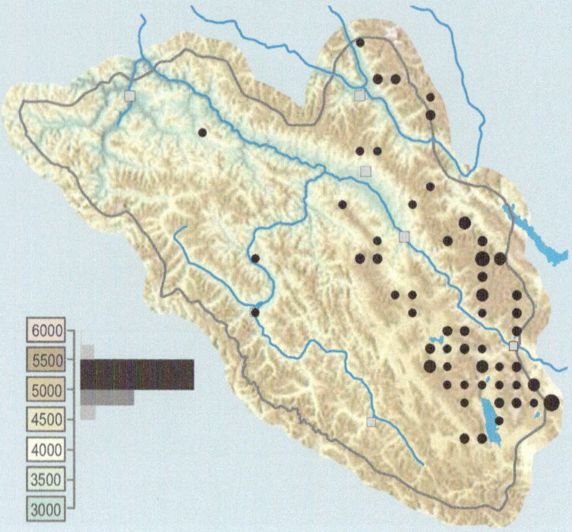

Description: Perennial herbs, 6–15 cm tall, aromatic, densely villous and viscous, stems prostrate or ascending, leaves imparipinnately compound, 4–6 cm long, leaflets verticillate, flowers purple, pink or bluish purple, 15–20 mm long, in compact racemes, legumes hairy, glands almost obscure.

Distribution: N Pakistan (Baltistan), NW India, Nepal, China (Qinghai, Xinjiang, Xizang).

Habitat: Wind-exposed stabilized slopes.

Stem diameter 8 mm. Bark/xylem proportion 1:1. Ring width 0.3–0.4 mm. Maximal 9 annual rings. Stem segmentation.

Xylem

Transverse section

Semi-ring-porous. Ring boundaries indicated by large earlywood (40–50 µm) and small latewood vessel diameters (10–20 µm). Vessels solitary or in groups. Vessel walls poorly stained. Fibers thick-walled, in small groups. Parenchyma pervasive. Crystals absent. With secondary intra-xylary periderm towards the pith.

Tangential section

Rays 4–6-seriate, unlignified and thin-walled.

Radial section

Perforation plates simple. Intervessel pits scalariform. Vessel length 60–100 µm. All ray cells upright.

Bark

Phloem with groups of sieve tubes and tangentially arranged groups of thick-walled fibers. Phellem consists of few tangential layers of cork cells, composed of tangentially flat, thin-walled, rectangular cells. Crystals absent.

Pith

No slide available.

Analyzed: Root collar of 2 individuals.

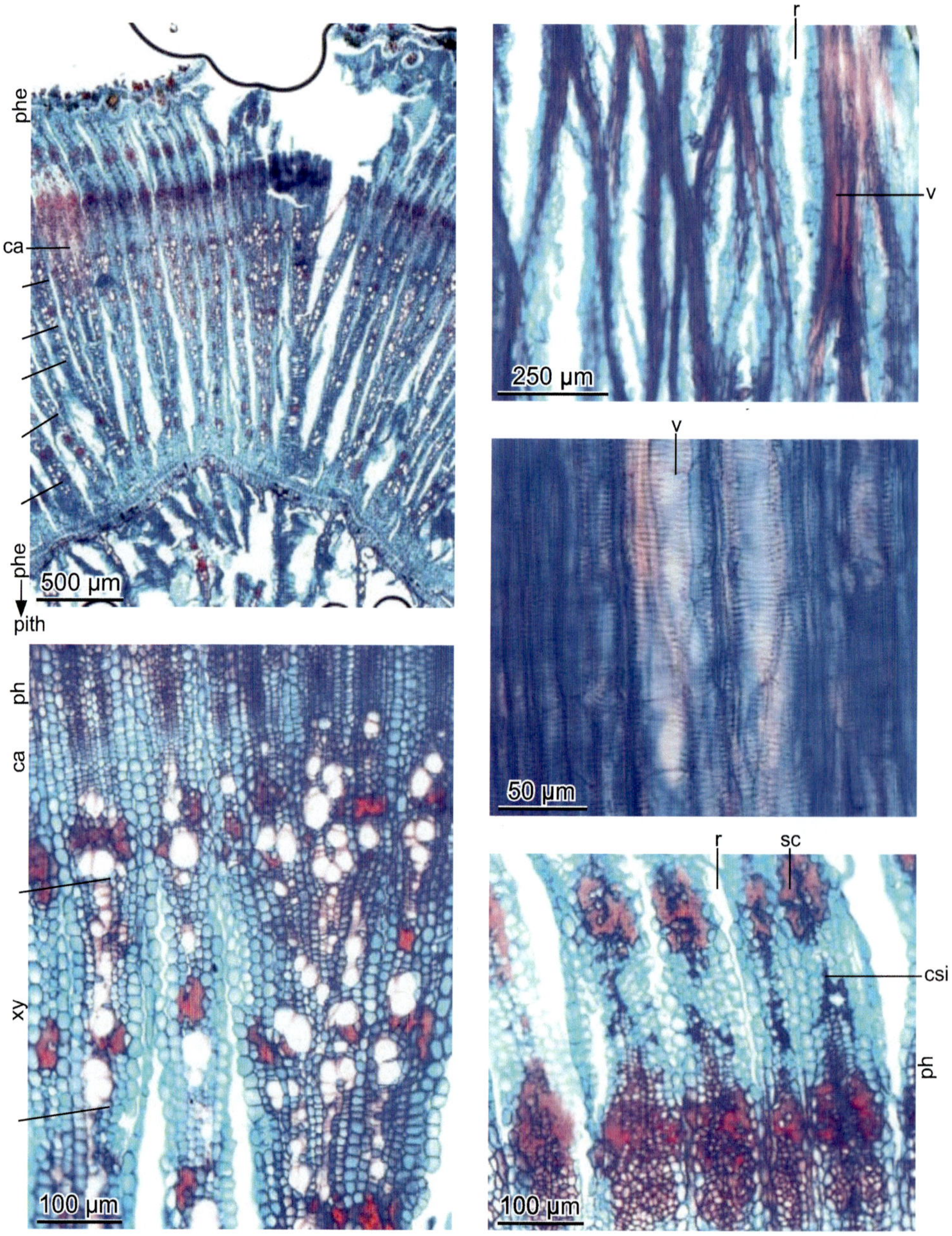

Features according to IAWA

Xylem 1 - 4 - 9 - 11 - 13 - 20 - 39.2 - 40.2 - 50.2 - 52.2 - 61 - 69 - 79.1 - 98 - 100.2 - 105 - 135.2

Oxytropis densa Benth. ex Bunge

Fabaceae

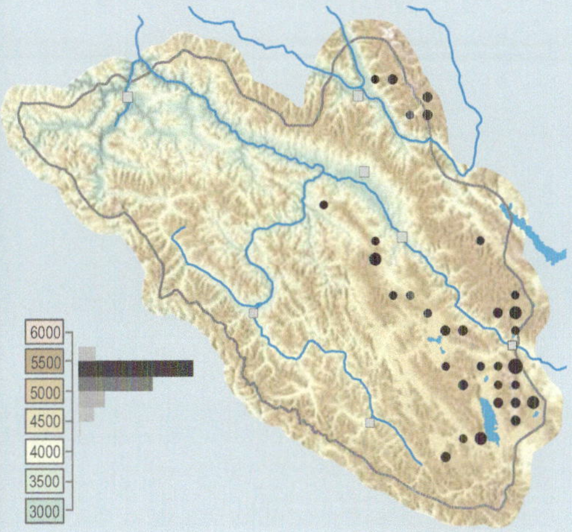

Description: Perennial herbs, 2–7 cm tall, cespitose, densely white hairy, leaves imparipinnately compound, leaflets ovate, flowers purple to bluish, 5–8 mm long, in 2–10-flowered racemes.

Distribution: Tajikistan, NE Afghanistan, N Pakistan, NW India, China (Gansu, Qinghai, Xinjiang, Xizang).

Habitat: Wind-exposed gravel slopes.

Stem diameter 8 mm. Bark/xylem proportion 1:1. Mean ring width 0.2 mm. More than 6 annual rings. Stem segmentation.

Xylem

Transverse section

Semi-ring-porous. Ring boundaries indicated by large earlywood (40–50 µm) and small latewood vessel diameters (10–20 µm). Vessels solitary or in small groups. Vessel walls poorly stained. Fibers thick-walled, in small groups. Parenchyma pervasive. Crystals absent. With secondary intra-xylary periderm towards the pith.

Tangential section

Rays 4–8-seriate, often confluent with axial tissue, cell walls unlignified and thin-walled.

Radial section

Perforation plates simple. Intervessel pits scalariform. Vessel length 60–100 µm. All ray cells upright.

Bark

Phloem with groups of sieve tubes and tangentially and radially arranged groups of thick-walled fibers. Phellem consists of few tangential layers of cork cells, composed of tangentially flat, thin-walled, rectangular cells. Crystals absent.

Pith

No slide available.

Analyzed: Root collar of 1 individual.

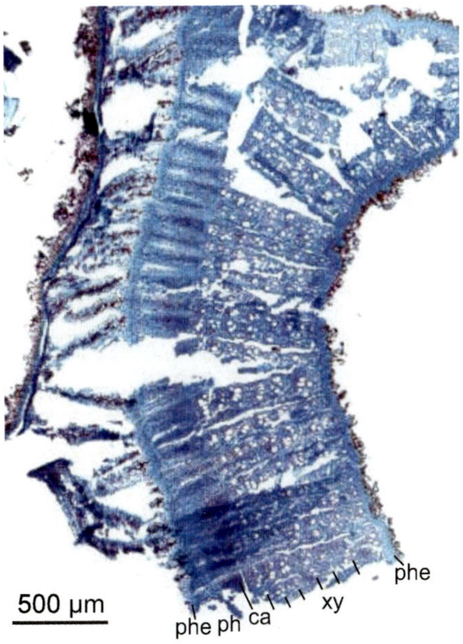

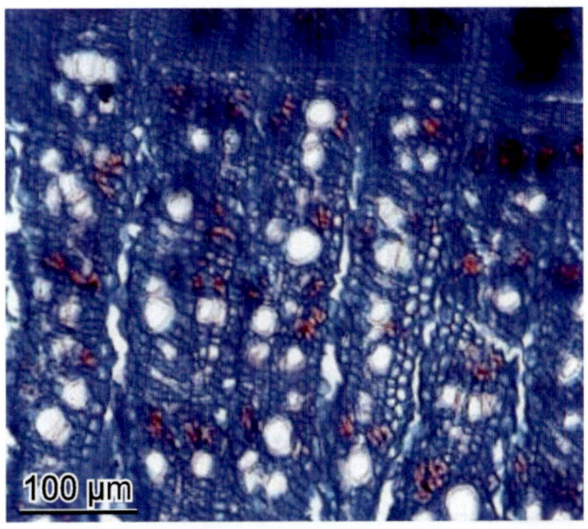

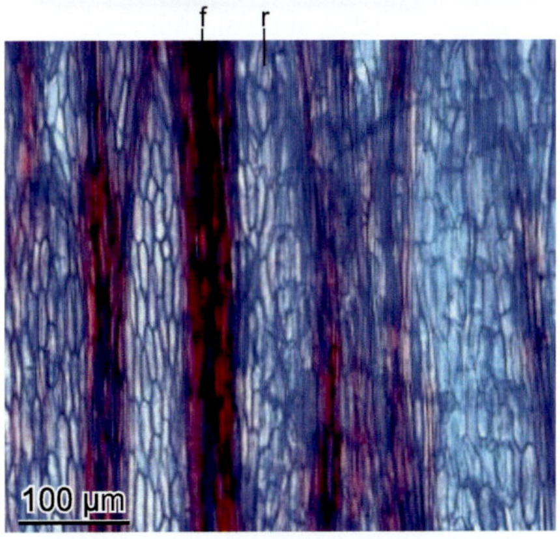

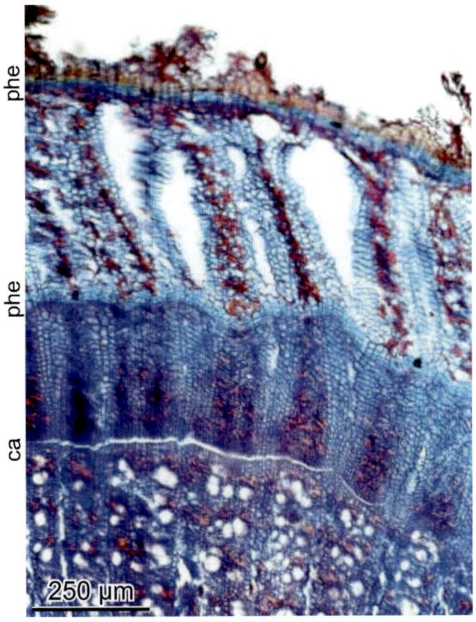

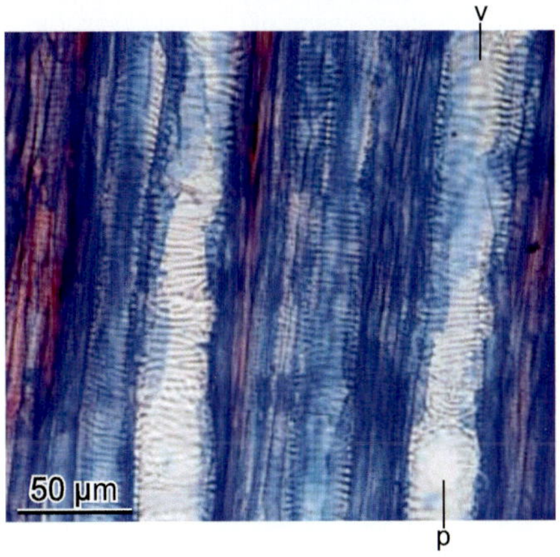

Features according to IAWA

Xylem 1 - 4 - 9 - 11 - 13 - 20 - 39.2 - 40.2 - 50.2 - 52.2 - 61 - 69 - 79.1 - 98 - 100.1 - 100.2 - 105 - 135.2

Oxytropis humifusa Kar. & Kir.

Fabaceae

Syn.: *O. lapponica* var. *humifusa* (Kar. & Kir.) Baker

Description: Perennial herbs, 2–5 cm tall, cespitose, apressed hairy, leaves imparipinnate, leaflets ovate to lanceolate, flowers purple, 8–13 mm long, in compact 6–10-flowered racemes.

Distribution: Kazakhstan, Turkmenistan, Uzbekistan, Kyrgyzstan, Tajikistan, Afghanistan, N Pakistan, NW India, Nepal, China (Xinjiang, Xizang).

Habitat: Wind-exposed slopes.

Stem diameter 3–5 mm. Bark/xylem proportion 1:1. Rings indistinct.

Xylem

Transverse section

Vessel diameter 30–40 µm. Vessels angular, solitary or in small groups. Fibers absent. Parenchyma pervasive. Crystals absent.

Tangential section

Rays 4–8-seriate, unlignified and thin-walled, confluent with the axial tissue.

Radial section

Perforation plates simple. Intervessel pits scalariform. Vessel length 100–150 µm. Structures longitudinal bent due to root contraction.

Bark

Phloem with small groups of sieve tubes. With ray dilatations. Phellem consists of few layers of rectangular cells. Crystals absent.

Pith

No slide available.

Analyzed: Root collar of 2 individuals (herbarium specimen).

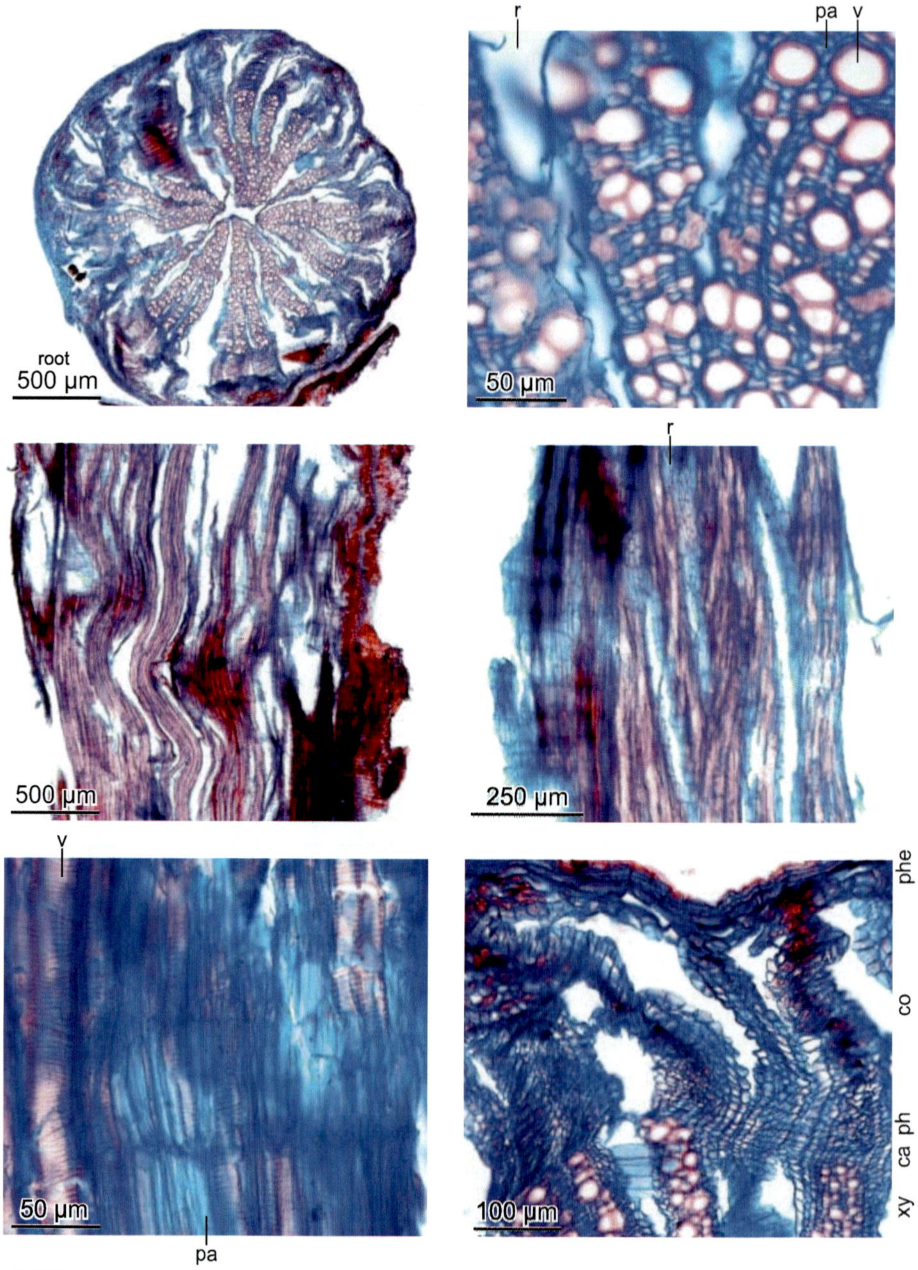

Features according to IAWA

Xylem 2.1 - 9 - 11 - 13 - 20 - 40.2 - 50.1 - 52.3 - 60.1 - 79.1 - 98 - 100.1 - 100.2 - 105

Oxytropis hypoglottoides (Baker) Ali
Fabaceae

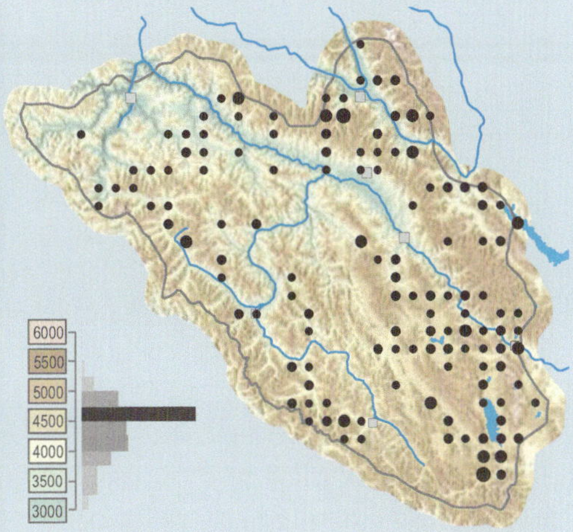

Description: Perennial herbs, 5–25 cm tall, stems suberect, leaves imparipinnately compound, 3–7 cm long, leaflets elliptic-ovate, flowers blue to violet, ca. 10 mm long, in pedunculate axillary racemes.

Distribution: N Pakistan, NW India, China (Xinjiang, Xizang).

Habitat: Mesic shores of fresh-water pools and brooks, saline grasslands.

Stem diameter 3 mm. Bark/xylem proportion 1:1. Ring width 0.13–1 mm. Maximal 6 annual rings.

Xylem

Transverse section

Semi-ring-porous. Ring boundaries indicated by large earlywood (40–50 µm) and small latewood vessel diameters (10–20 µm). Vessels solitary. Vessel walls poorly stained. Fibers absent or very rare. Parenchyma pervasive. Crystals absent. Without secondary intra-xylary periderm towards the pith.

Tangential section

Rays 8–>10-seriate, often confluent with axial tissue, cell walls unlignified and thin-walled.

Radial section

Perforation plates simple. Intervessel pits scalariform. Vessel length 80–120 µm. All ray cells upright.

Bark

Phloem with groups of sieve tubes. Fibers absent. With ray dilatations. Phellem consists of few tangential layers of cork cells, composed of tangentially flat, thin-walled, rectangular cells. Crystals absent.

Pith

No slide available.

Analyzed: Root collar of 4 individuals.

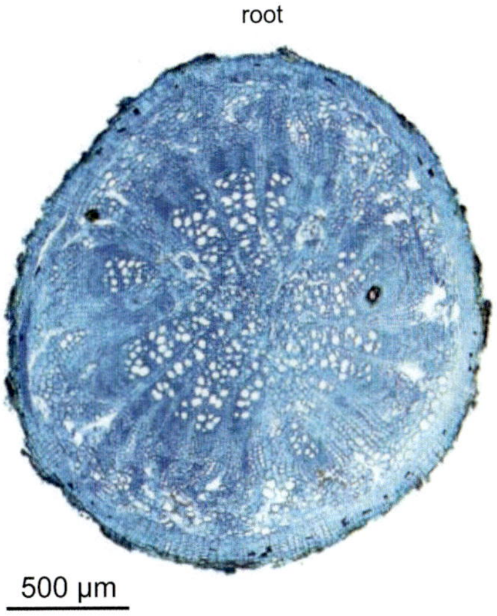

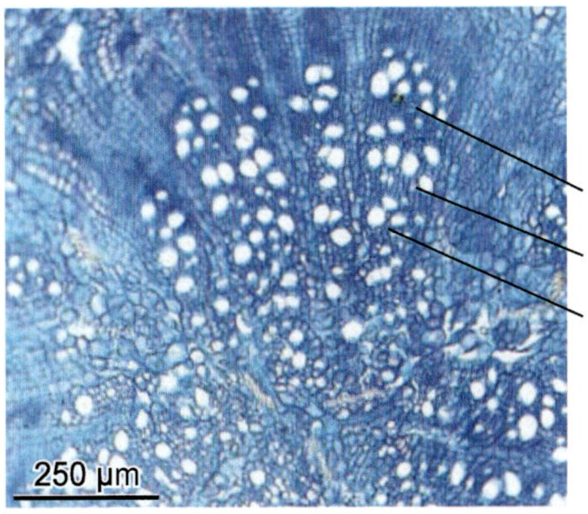

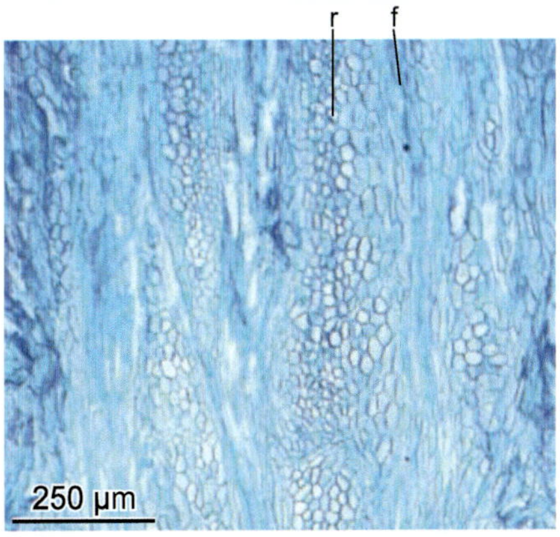

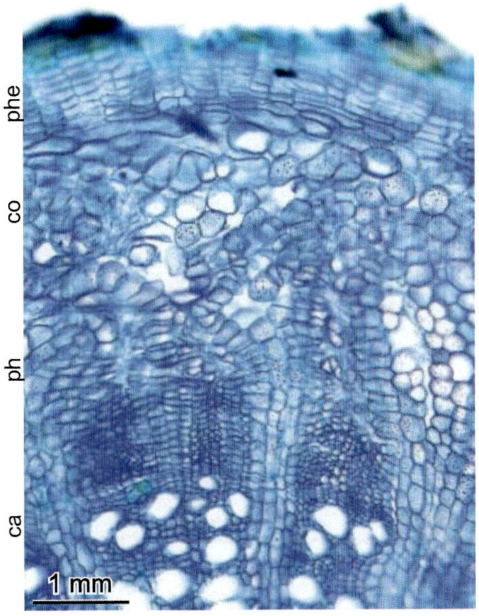

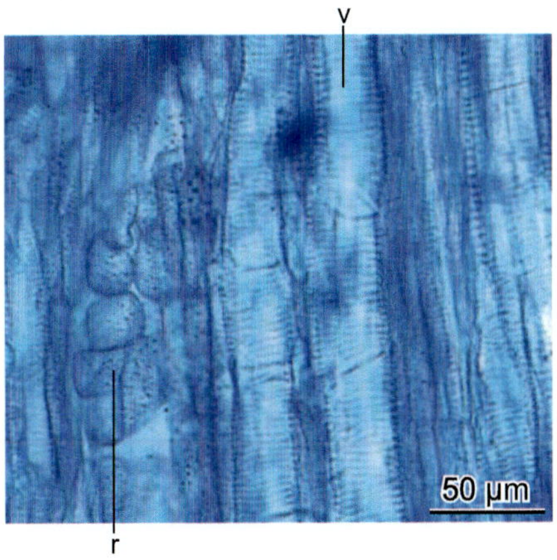

Features according to IAWA

Xylem 1 - 4 - 9 - 13 - 20 - 39.2 - 40.2 - 50.2 - 52.2 - 52.3 - 61 - 69 - 79.1 - 98 - 99 - 100.1 - 100.2 - 105

Oxytropis microphylla (Pall.) DC.

Fabaceae

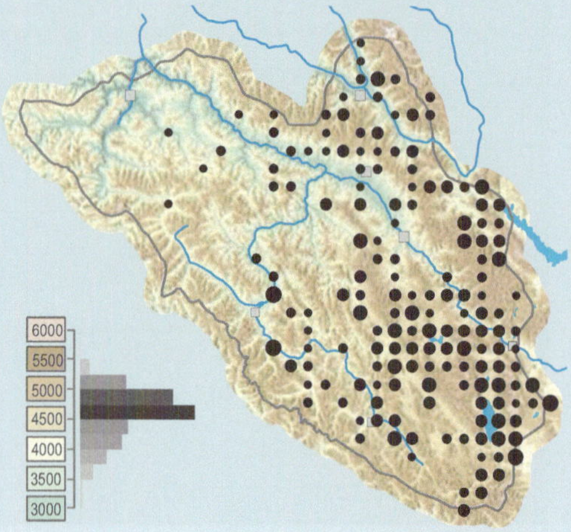

Syn.: *Phaca microphylla* Pall.

Description: Perennial herbs, 6–15 cm tall, aromatic, densely villous and viscous, stems prostrate or ascending, leaves imparipinnately compound, 4–6 cm long, leaflets verticillate, flowers purple, pink or bluish purple, 15–20 mm long, in compact racemes, legumes glabrous, tuberculate from glands.

Distribution: Russia, Kyrgyzstan, Tajikistan, Afghanistan, N Pakistan, NW India, Nepal, China (Gansu, Nei Mongol, Qinghai, Xinjiang, Xizang), Mongolia.

Habitat: Semideserts, dry stabilized slopes, steppes, sandy plains.

Rings indistinct. Stem diameter 3 mm. Bark/xylem proportion 1:2. Ring width 0.3–0.4 mm. Stem segmentation.

Xylem

Transverse section

Vessels solitary and in small groups. Diameter of large vessels 30–40 µm and of small vessels 10–20 µm.Fibers in small groups. Parenchyma pervasive. Crystals absent. With secondary intra-xylary periderm towards the pith.

Tangential section

Rays 4–6-seriate, unlignified, thin-walled.

Radial section

Perforation plates simple. Intervessel pits scalariform. Vessel length 60–100 µm. All ray cells upright.

Bark

Phloem with groups of sieve tubes and irregular groups of thick-walled fibers. Phellem cells tangentially flat, thin-walled, rectangular. Crystals absent.

Pith

No slide available.

Analyzed: Root collar of 2 individuals.

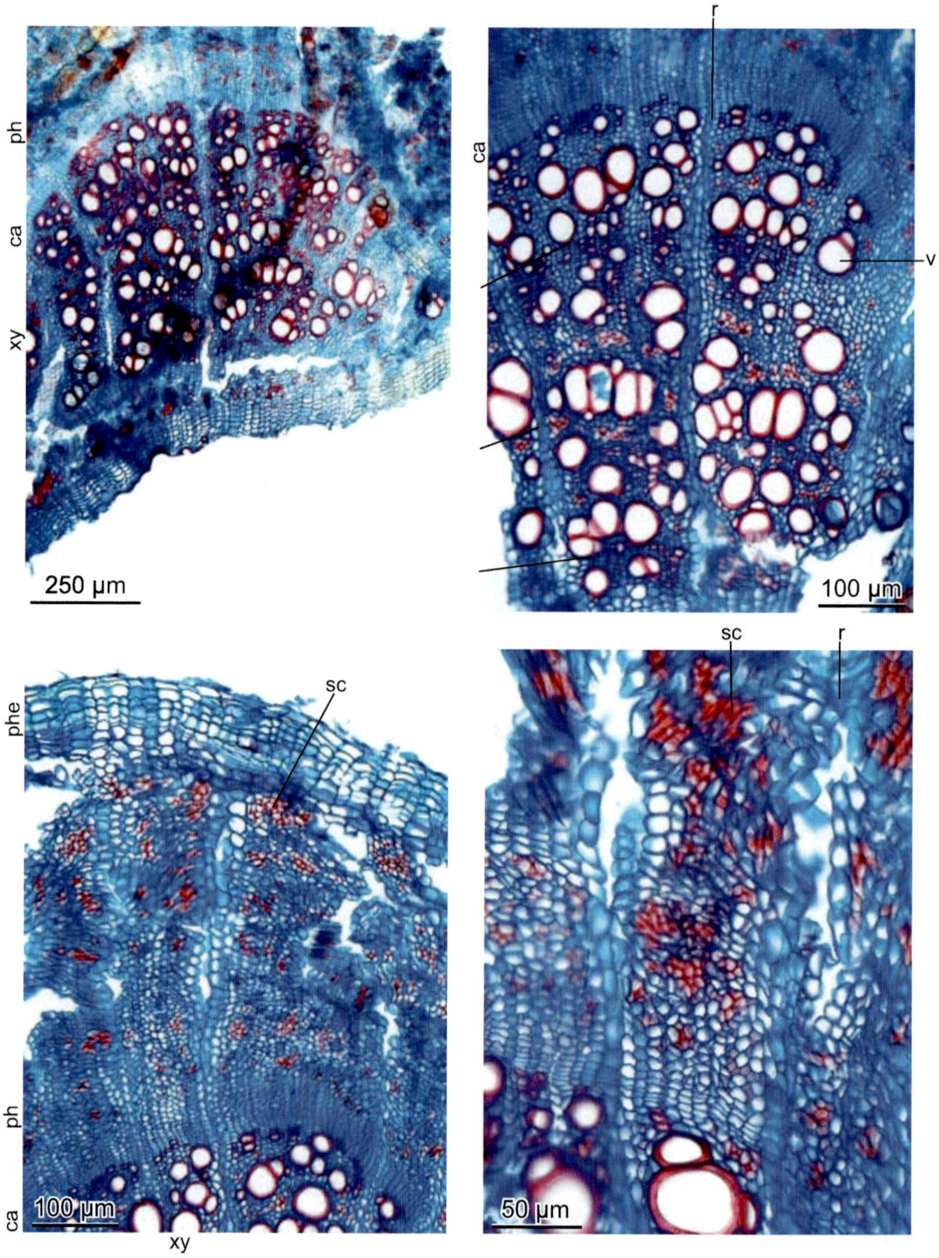

Features according to IAWA

Xylem 2 - 4 - 9 - 11 - 13 - 20 - 40.2 - 50.2 - 52.2 - 61 - 69 - 79.1 - 98 - 100.2 - 105 - 135.2

Oxytropis pauciflora Bunge

Fabaceae

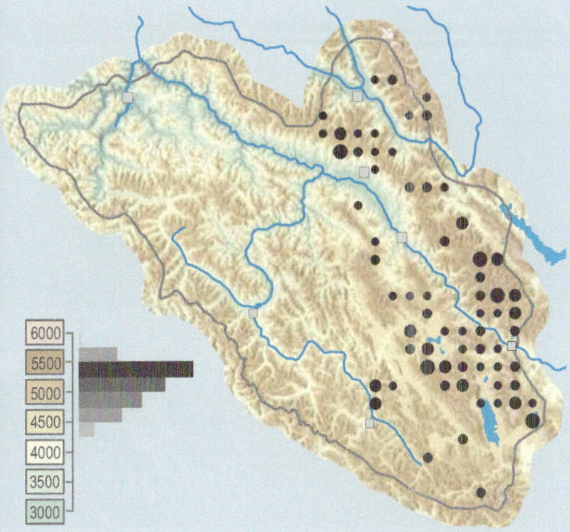

Syn.: *O. platysema* Schrenk

Description: Perennial herbs, 2–8 cm tall, almost glabrous, leaves imparipinnately compound, leaflets ovate, flowers purple, ca. 10 mm long, in compact 3–7-flowered racemes.

Distribution: Kazakhstan, Turkmenistan, Uzbekistan, Kyrgyzstan, Tajikistan, NW India, China (Qinghai, Xinjiang, Xizang).

Habitat: Short alpine grasslands.

Stem diameter 0.3–0.4 mm. Bark/xylem proportion 1:1. Ring width 0.16–0.5 mm. More than 6 annual rings. Stem segmentation.

Xylem

Transverse section

Semi-ring-porous. Ring boundaries indicated by large earlywood (30–40 μm) and small latewood vessel diameters (10 μm). Vessels solitary or in small groups. Vessel walls poorly stained. Fibers absent. Parenchyma pervasive. Crystals absent. With secondary intra-xylary periderm towards the pith.

Tangential section

Rays 8–>10-seriate, often confluent with axial tissue, cell walls unlignified and thin-walled.

Radial section

Perforation plates simple. Intervessel pits scalariform. Vessel length 80–120 μm. All ray cells upright.

Bark

Phloem with groups of sieve tubes. Fibers absent. Phellem consists of few tangential layers of cork cells, composed of tangentially flat, thin-walled, rectangular cells. Crystals absent.

Pith

No slide available.

Analyzed: Root collar of 1 individual.

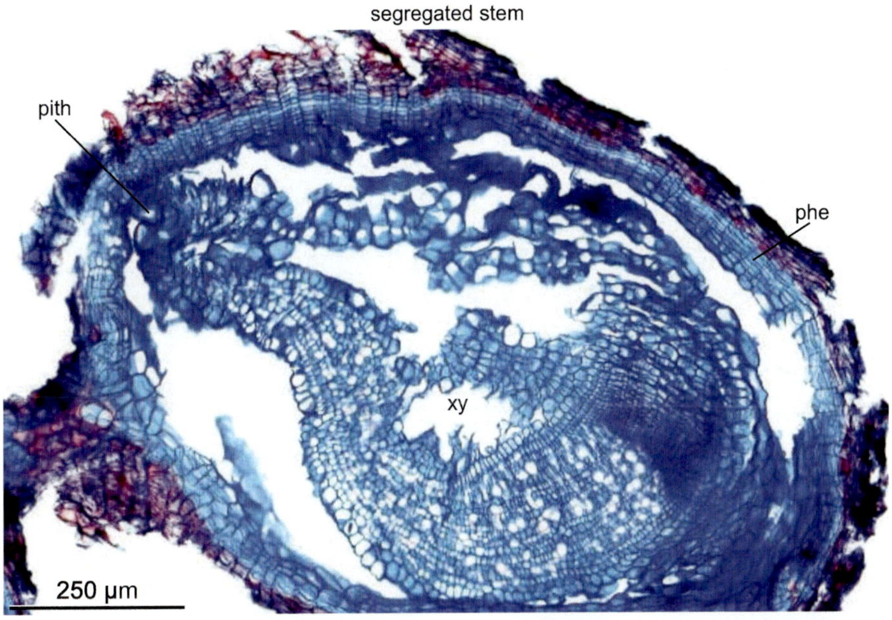

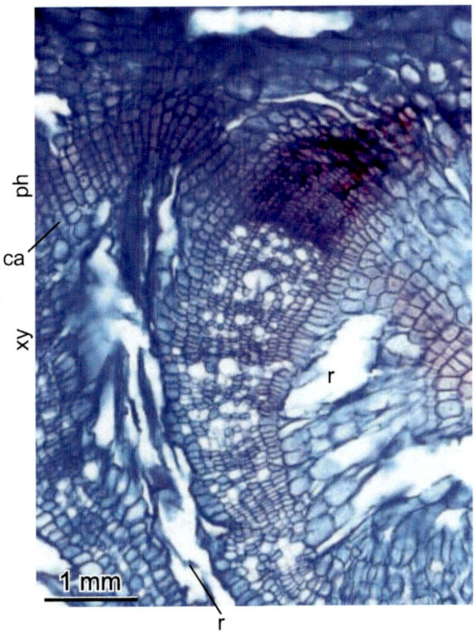

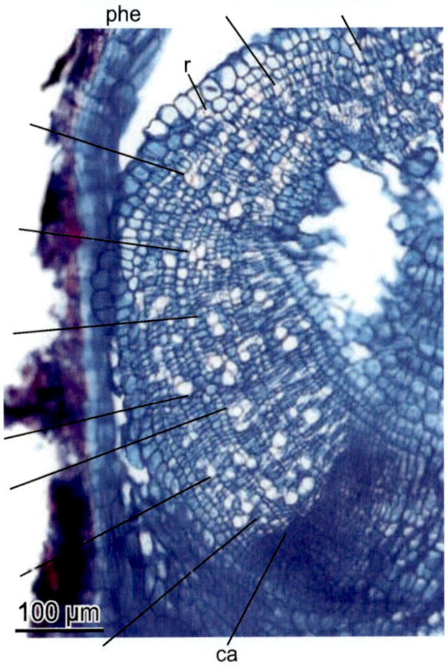

Features according to IAWA

Xylem 1 - 4 - 9 - 11 - 13 - 20 - 39.2 - 40.2 - 50.2 - 52.2 - 52.3 - 60.1 - 79.1 - 98 - 99 - 100.1 - 100.2 - 105 - 135.2

Oxytropis pusilla Bunge

Fabaceae

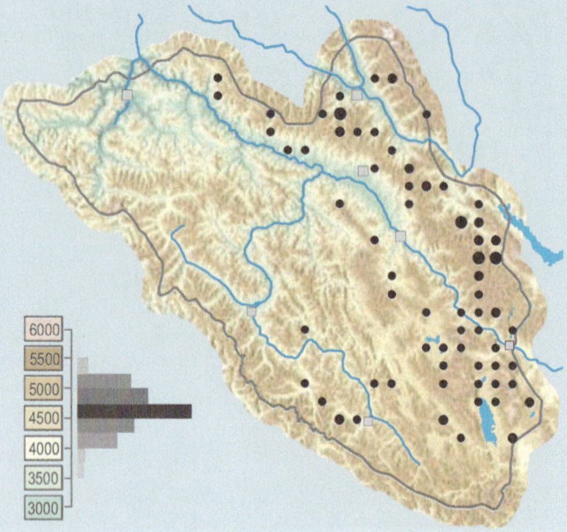

Syn.: *O. asterocarpa* Vassilcz.

Description: Perennial herbs, 2–10 cm tall, stems erect, leaves imparipinnately compound, 2–7 cm long, with 7–13 opposite leaflets, flowers purple, 6–8 mm long, in 2–5-flowered racemes.

Distribution: NE Afghanistan, N Pakistan (Baltistan) NW India, China (?Xinjiang, S Xizang).

Habitat: Short fresh or saline grasslands, floodplains.

Stem diameter 3–9 mm. Bark/xylem proportion 1:1. Ring width 0.2–0.3 mm. More than 9 annual rings. Stem segmentation.

Xylem

Transverse section

Semi-ring-porous. Ring boundaries indicated by large earlywood (40–60 µm) and small latewood vessel diameters (10 µm). Vessels solitary or in small groups. Earlywood vessel walls poorly stained. Fibers absent. Parenchyma pervasive. Crystals absent. Without secondary intra-xylary periderm towards the pith.

Tangential section

Rays 6–8-seriate, confluent with axial tissue, cell walls unlignified and thin-walled.

Radial section

Perforation plates simple. Intervessel pits scalariform. Vessel length 60–100 µm. All ray cells upright.

Bark

Phloem with groups of sieve tubes. Older sieve tubes radially collapsed. Fibers absent. Phellem consists of few tangential layers of cork cells, composed of tangentially flat, thin-walled, rectangular cells. Crystals absent.

Pith

No slide available.

Analyzed: Root collar of 3 individuals.

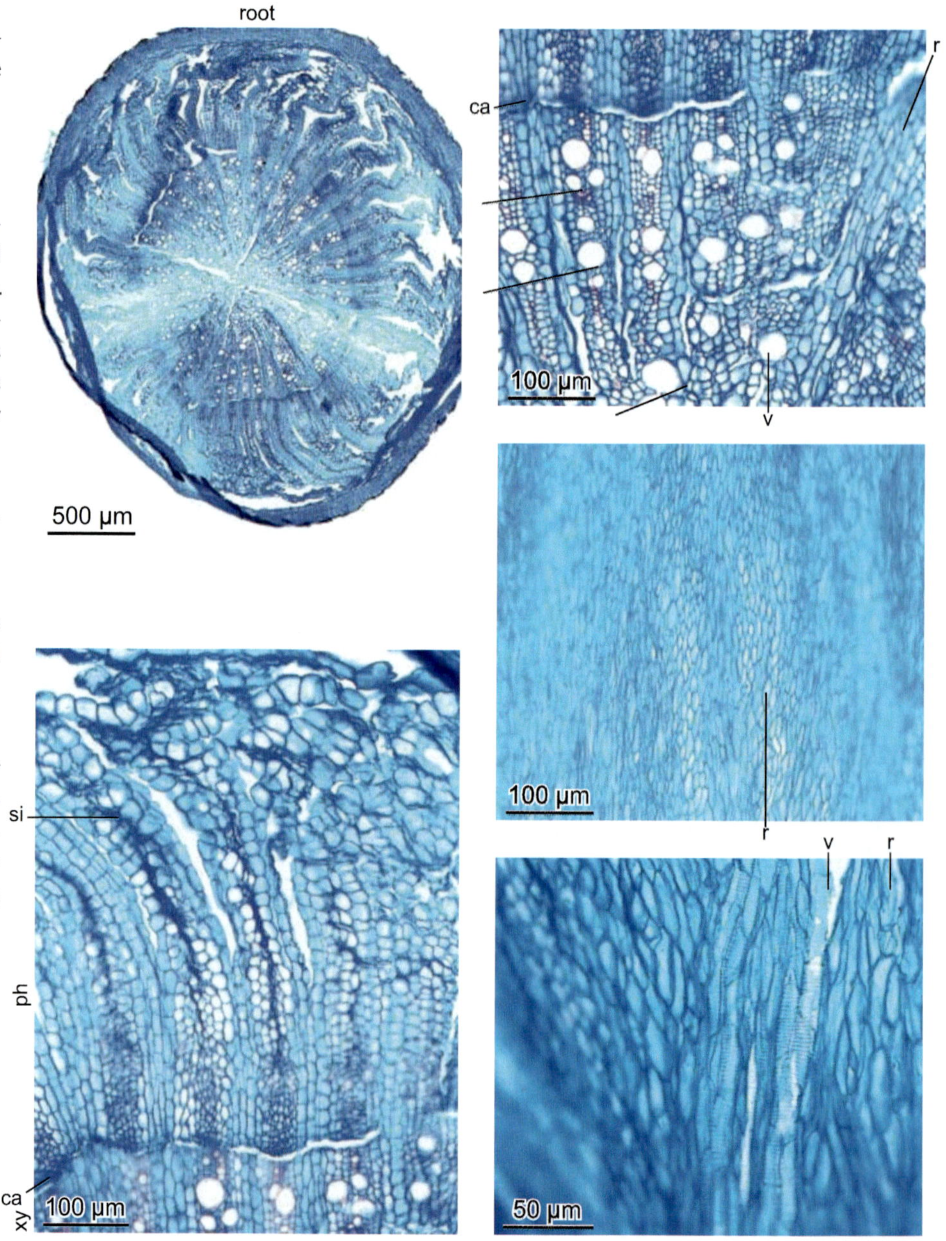

Features according to IAWA

Xylem 1 - 4 - 9 - 11 - 13 - 20 - 39.2 - 40.2 - 50.2 - 52.2 - 60.1 - 79.1 - 98 - 100.1 - 100.2 - 105

Oxytropis tatarica Cambess. ex Bunge

Fabaceae

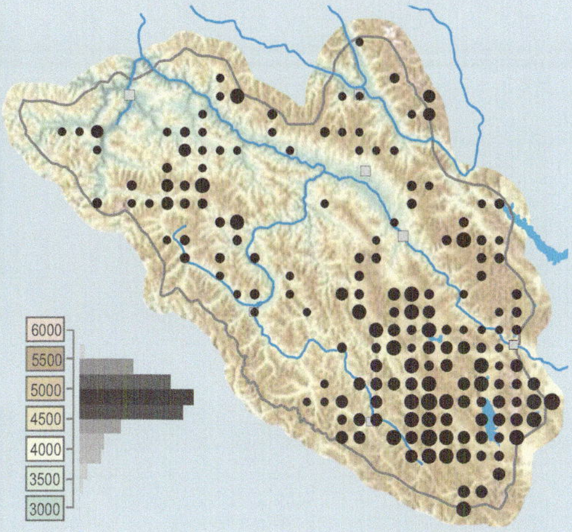

Syn.: *O. glacialis* Benth. ex Bunge, *O. proboscidea* Bunge

Description: Perennial herbs, 6–15 cm tall, canescent with gray hairs, leaves imparipinnately compound, 2–5 cm long, flowers blue or rarely purple, 7–9 mm long, in compact racemes resembling capitula, fruits inflated.

Distribution: N Pakistan, NW India, Nepal, China (Xizang).

Habitat: Dry stabilized slopes, semideserts, sandy plains, steppes.

Stem diameter 3–10 mm. Bark/xylem proportion 1:1. Ring width 0.1–0.3 mm. Maximal 20 annual rings.

Xylem

Transverse section

Semi-ring-porous. Ring boundaries indicated by large earlywood (40–70 µm) and small latewood vessel diameters (10–20 µm). Vessels solitary. Vessel walls fairly thick-walled, red-stained. Fibers thick-walled, in small groups. Parenchyma pervasive. Crystals absent. Without secondary intra-xylary periderm towards the pith.

Tangential section

Rays 4–8-seriate, unlignified and thin-walled, confluent with axial tissue.

Radial section

Perforation plates simple. Intervessel pits scalariform. Vessel length 60–100 µm. All ray cells upright.

Bark

Phloem with small groups of sieve tubes and thick-walled fibers. With ray dilatations. Phellem consists of few tangential layers of cork cells, composed of tangentially flat, thin-walled, rectangular cells. Crystals absent.

Pith

No slide available.

Analyzed: Root collar of 12 individuals.

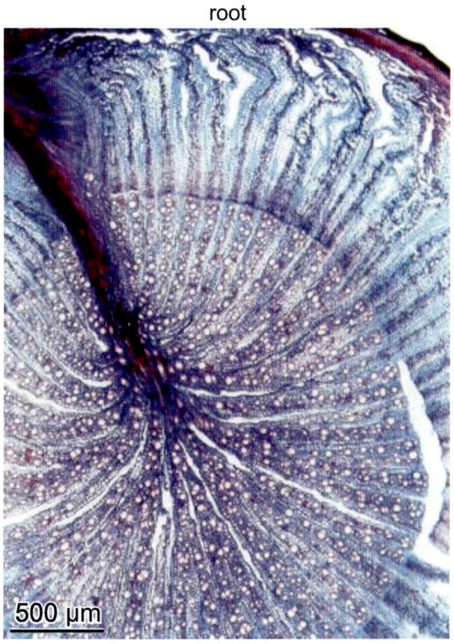

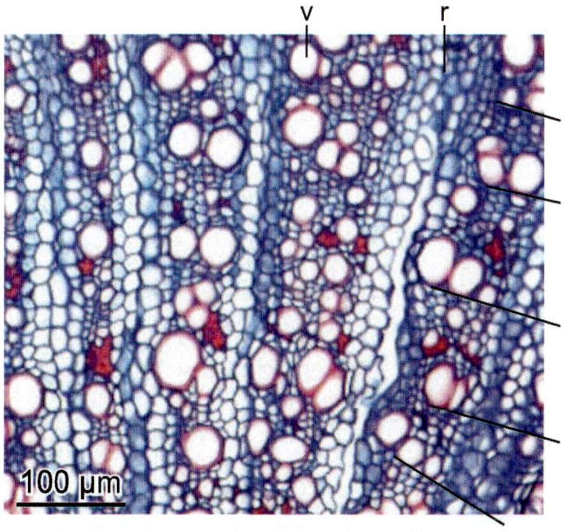

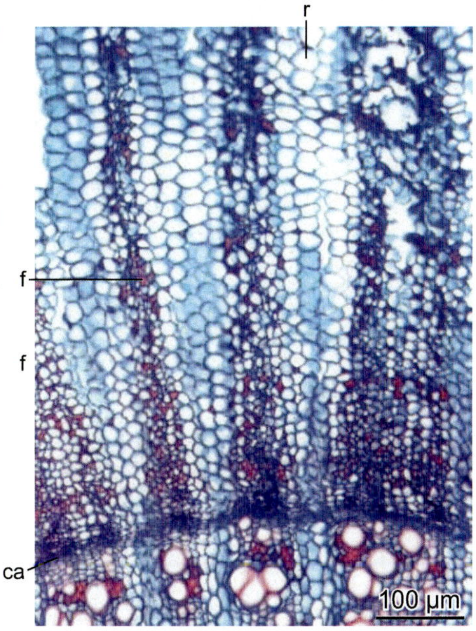

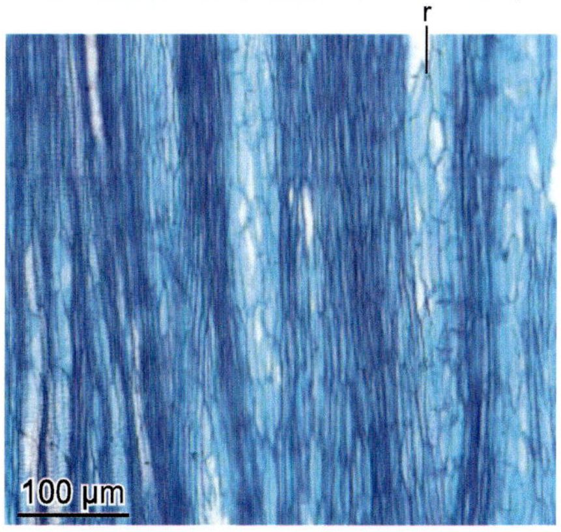

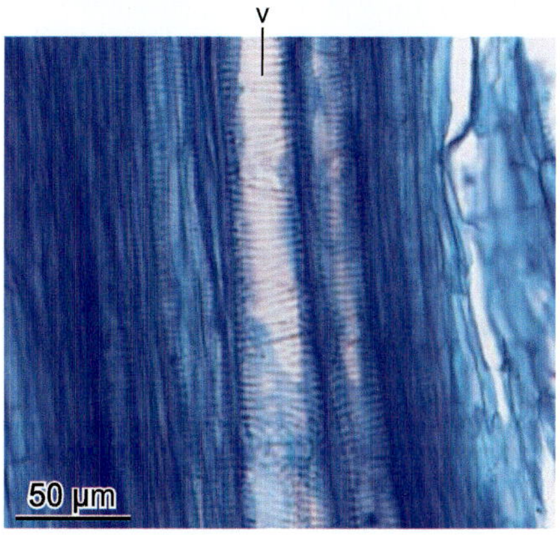

Features according to IAWA

Xylem 1 - 4 - 9 - 13 - 20 - 39.1 - 40.2 - 41 - 50.2 - 52.2 - 61 - 69 - 79.1 - 98 - 100.1 - 100.2 - 105

Phyllolobium hendersonii comb. nov. ined.

Fabaceae

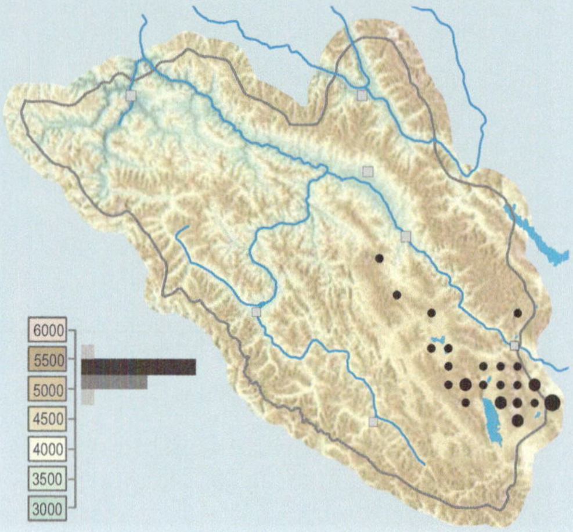

Syn.: *Astragalus hendersonii* Baker

Description: Perennial herbs, 1–5 cm tall, densely pilose, leaves imparipinnately compound, leaflets densely covered by white hairs below, subglabrous above, flowers mauve to purple, ca. 1 cm long, in 1–3-flowered racemes.

Distribution: Pakistan, N India, China (Xizang).

Habitat: Stabilized slopes.

Stem diameter 4–6 mm. Bark/xylem proportion 1:1. Ring width 0.08–0.1 mm. More than 25 annual rings. Stem segmentation.

Xylem

Transverse section

Semi-ring-porous. Ring boundaries indicated by large earlywood (30–50 µm) and small latewood vessel diameters (10–20 µm). Vessels solitary. Vessel walls poorly stained. Fibers absent. Parenchyma pervasive. Crystals absent. With secondary intra-xylary periderm towards the pith.

Tangential section

Rays 4–6-seriate, unlignified and thin-walled. Central ray cells mostly decayed.

Radial section

Perforation plates simple. Intervessel pits scalariform. Vessel length 60–80 µm. All ray cells upright.

Bark

Phloem with groups of sieve tubes. Older sieve tubes radially collapsed. Fibers absent. Phellem cells tangentially flat, thin-walled, rectangular. Crystals absent.

Pith

No slide available.

Analyzed: Root collar of 3 individuals.

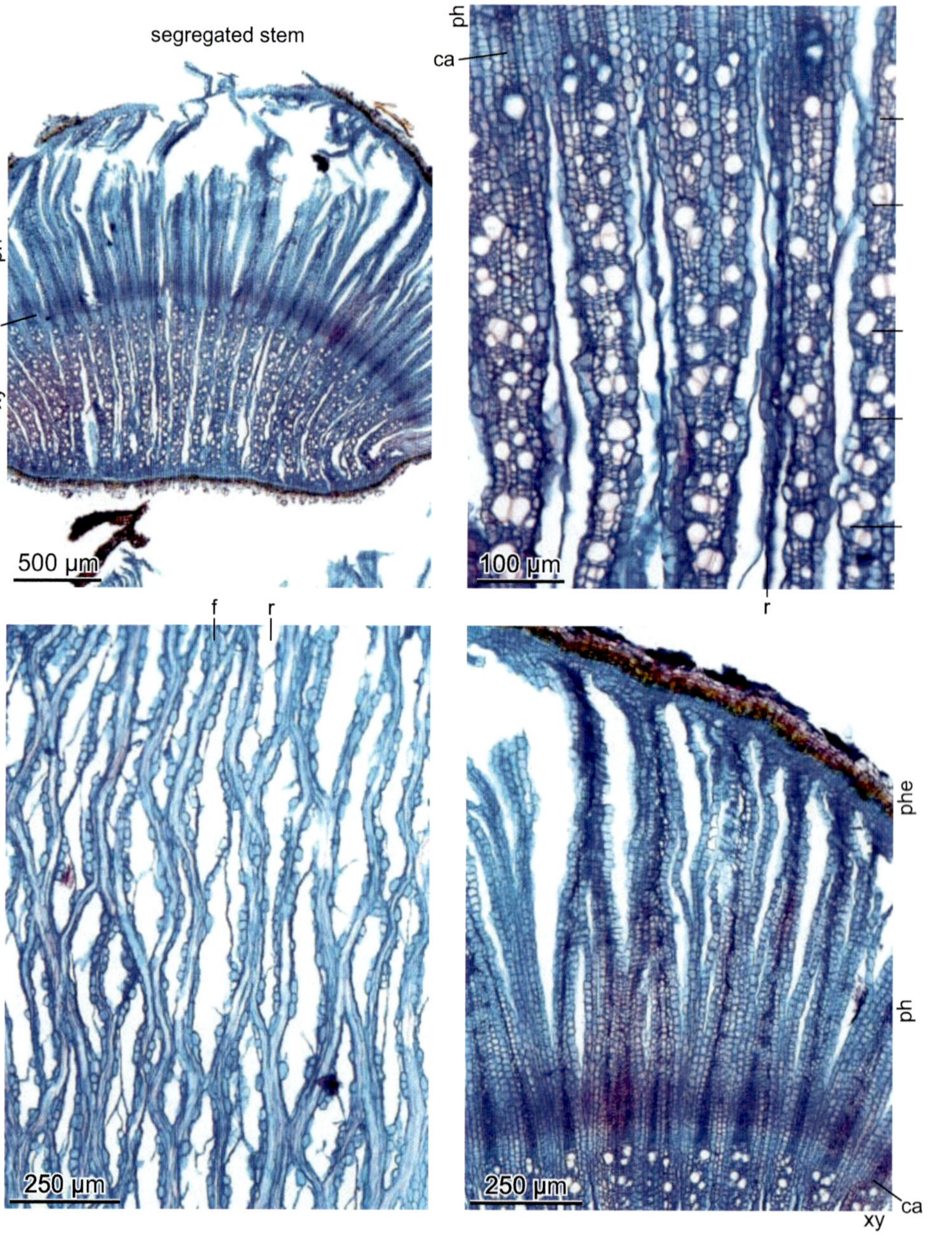

Features according to IAWA

Xylem 1 - 4 - 9 - 13 - 20 - 39.2 - 40.2 - 50.2 - 52.2 - 60.1 - 79.1 - 98 - 100.2 - 105 - 135.2

Thermopsis inflata Cambess.

Fabaceae

Description: Perennial herbs, 10–20 cm tall, stoloniferous, white villous, leaves trifoliolate with leaflet-like stipules, flowers yellow, ca. 2.5 cm long, in lax racemes, legume 3–5 cm long, inflated, villous.

Distribution: N Pakistan, N India, Nepal, China (Xinjiang, Xizang), Bhutan.

Habitat: Unstable loamy slopes, screes.

Stem diameter 3–13 mm. Bark/xylem proportion 1:1. Ring width 0.2–0.3 mm. Maximal 16 annual rings.

Xylem

Transverse section

Semi-ring-porous. Ring boundaries indicated by large earlywood (50–80 μm) and small latewood vessel diameters (20–30 μm). Vessels solitary and in small groups. Thick-walled fibers in the latewood. Parenchyma paratracheal and pervasive. Crystals absent.

Tangential section

Rays 3–7-seriate, unlignified, thin-walled.

Radial section

Perforation plates simple. Intervessel pits round. Vessel length 100–150 μm. All ray cells upright.

Bark

Phloem with small groups of sieve tubes, arranged in tangential rows. With slight ray dilatations. Ray course undulated. Phellem consists of few tangential layers of cork cells, composed of tangentially flat, rectangular cells. Crystals absent.

Pith

Triangular.

Analyzed: Root collar of 3 individuals.

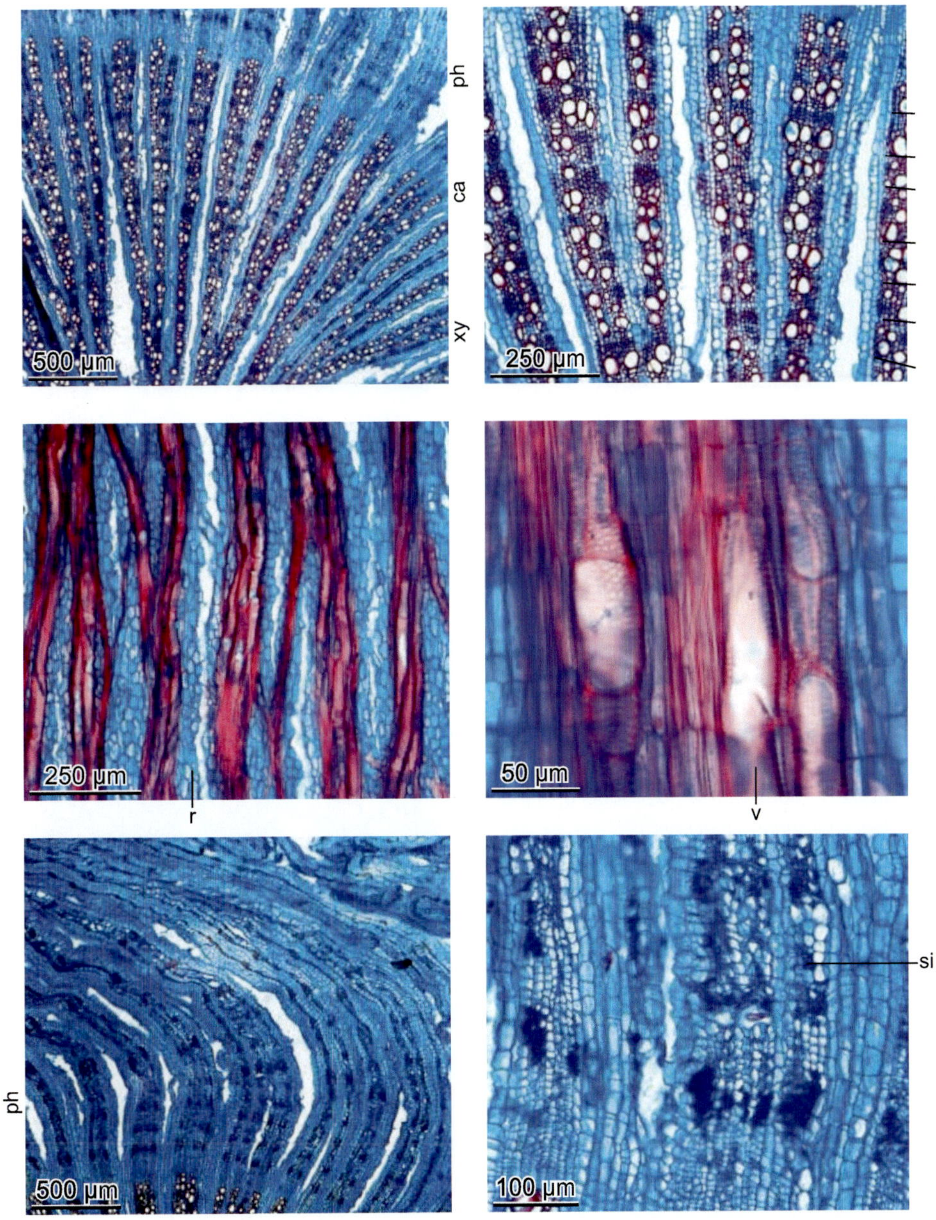

Features according to IAWA

Xylem 1 - 4 - 9 - 11 - 13 - 20 - 41 - 50.2 - 52.3 - 61 - 69 - 79 - 79.1 - 98 - 100.2 - 105

Hypecoum leptocarpum Hook.f. & Thomson

Fumariaceae

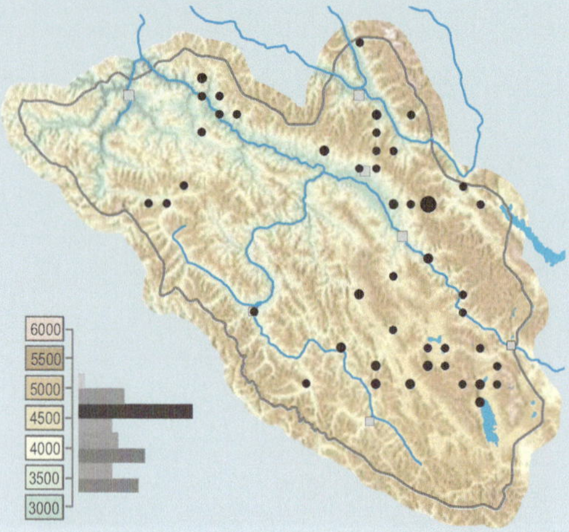

Description: Annuals, 10–30 cm tall, stems erect to procumbent, leaves pinnatisect, flowers white, 8–12 mm across, in dichasial 2–4-flowered cymes.

Distribution: Tajikistan, Afghanistan, W Pakistan, India, Nepal, China, (Gansu, Hebei, Nei Mongol, Qinghai, Shaanxi, Shanxi, Sichuan, Xinjiang, Xizang, Yunnan), Bhutan, Mongolia.

Habitat: Sandy to loamy stabilized slopes, disturbed habitats along roads.

Stem diameter 2 mm. Bark/xylem proportion 1:1. Only one ring.

Xylem

Transverse section

Vessel diameter 20–30 μm. Vessels thin- to thick-walled, angular, solitary and in short radial multiples. Fibers absent. Parenchyma pervasive. Crystals absent.

Tangential section

Rays 5–10-seriate, with sheath cells.

Radial section

Perforation plates simple. Intervessel pits round to horizontally enlarged. Vessel length 100–150 μm.

Bark

Phloem with groups of sieve tubes, radially compressed. Crystals absent. Phellem consists of some layers of thin-walled rectangular cells.

Pith

No slide available.

Analyzed: Root collar of polar roots of 1 individual (herbarium specimen).

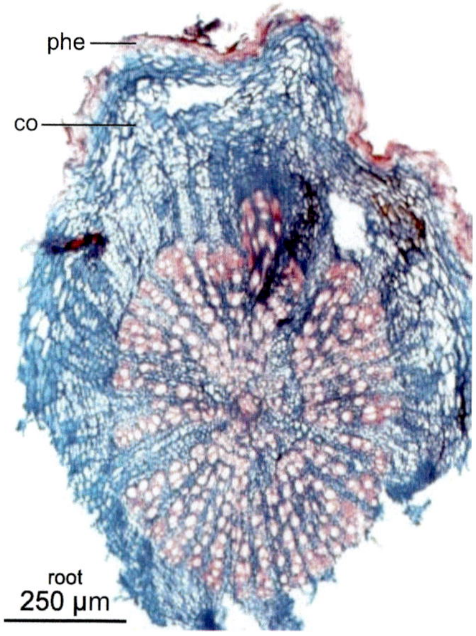

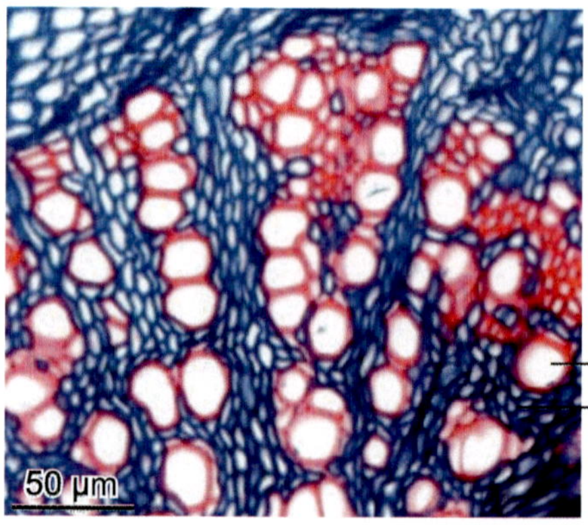

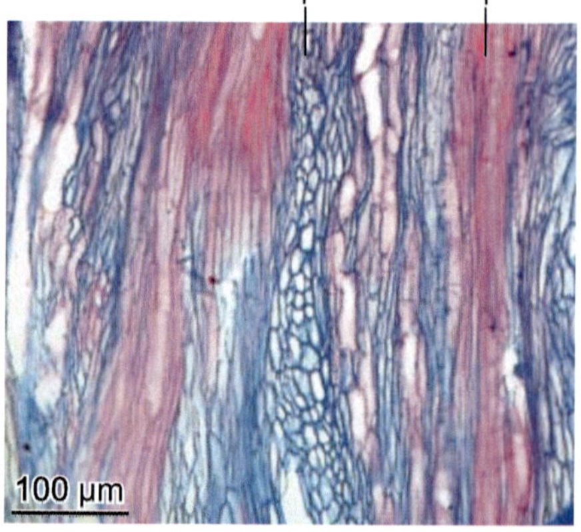

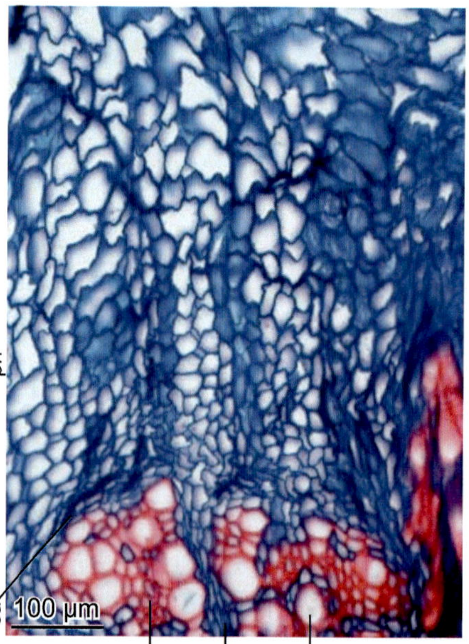

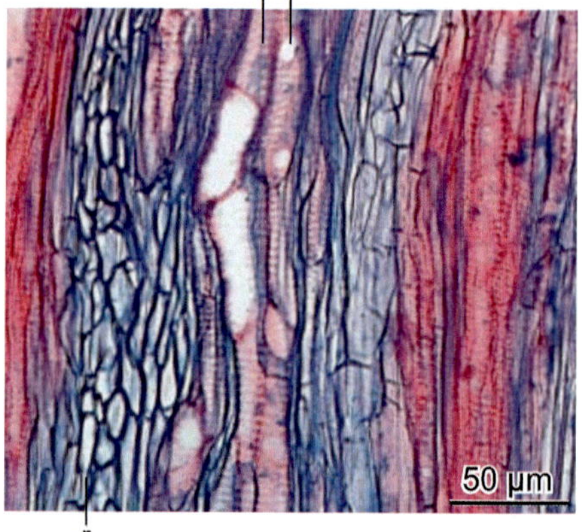

Features according to IAWA

Xylem 2.1 - 9 - 9.1 - 12 - 13 - 22 - 40.2 - 50.1 - 52.3 - 60.1 - 79.1 - 99 - 100.1 - 100.2 - 105 - 110

Comastoma azureum (Bunge) V.V.Zuev

Gentianaceae

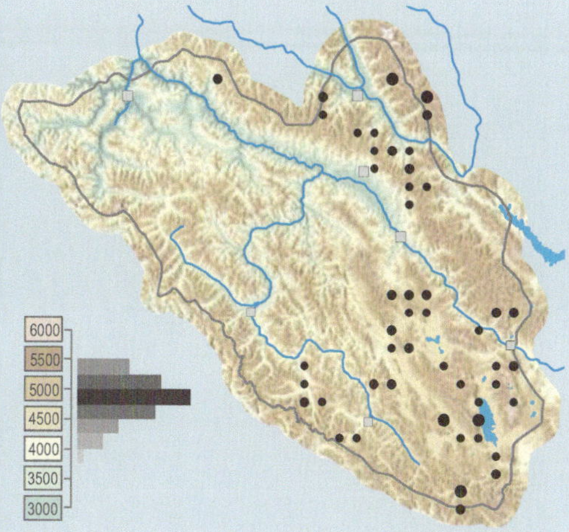

Syn.: *Gentiana azurea* Bunge, *Gentianella azurea* (Bunge) Holub, *Aloitis azurea* (Bunge) Omer

Description: Annuals, 2–15 cm tall, stems ascending to erect, branched from the base, leaves oblong to elliptic, flowers pale blue to blue, 0.5–1 cm long, mostly solitary.

Distribution: Russia (Siberia), Kazakhstan, Kyrgyzstan, Pakistan, NW India, China (Gansu, Qinghai, Sichuan, Xinjiang, Xizang, Yunnan), Bhutan, Mongolia.

Habitat: Short mesic turf of river beds.

Stem diameter 1–2 mm. Bark/xylem proportion 1:. Ring width 0.3–0.4 mm. Maximal 1 annual ring.

Xylem

Transverse section

Only one ring. Vessel diameter 10–15 µm. Vessels solitary. Fibers thin- to thick-walled. Fibers absent in the center. Parenchyma absent in the peripheral fiber zone and pervasive in the center. Crystals absent.

Tangential section

Rays absent.

Radial section

Perforation plates simple. Intervessel pits round in the fiber zone and scalariform in the center. Vessel length 80–100 µm.

Bark

Phloem with small groups of sieve tubes. Phellem absent.

Pith

No slide available.

Analyzed: Root collar of polar root of 2 individuals.

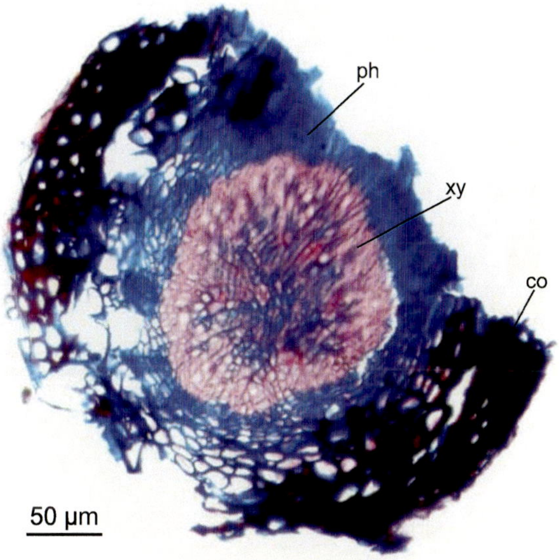

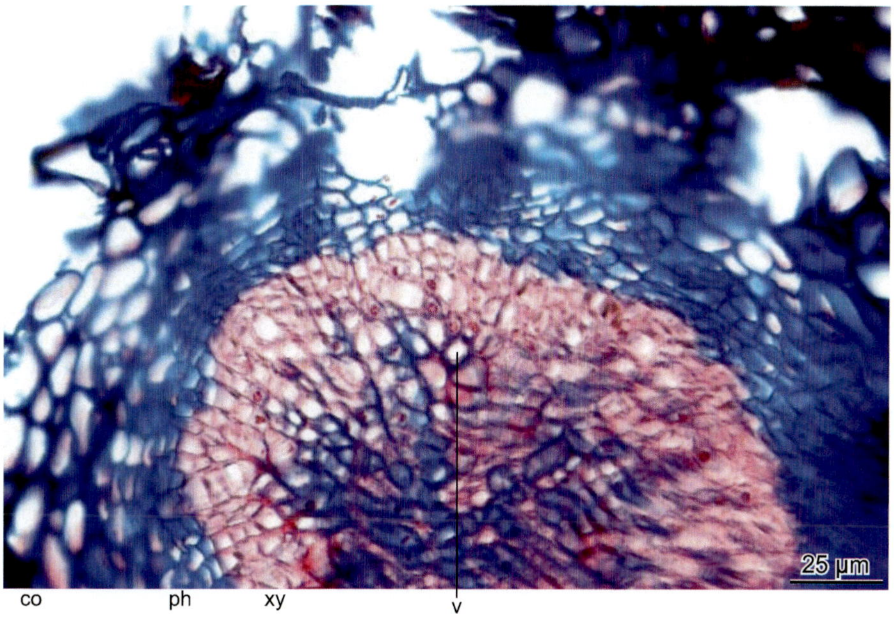

Features according to IAWA

Xylem *2.1 - 9 - 13 - 20 - 22 - 40.1 - 50.2 - 52.2 - 60.1 - 61 - 70 - 70.3 - 75 - 79.1 - 117 - P1*

Comastoma pedunculatum (Royle ex D.Don) Holub
Gentianaceae

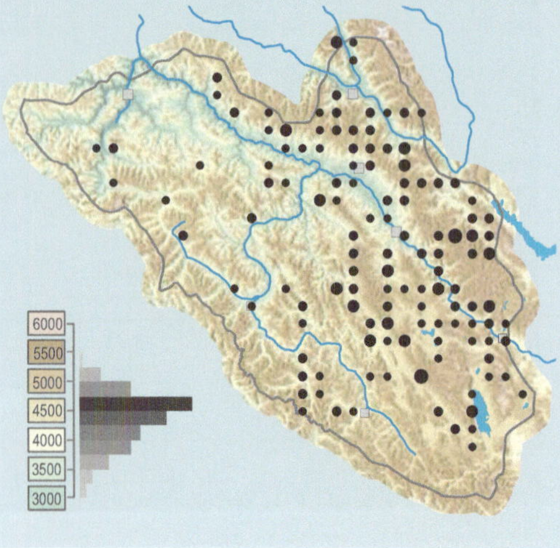

Syn.: *Eurythalia pedunculata* Royle ex D.Don, *Gentianella pedunculata* (Royle ex D.Don) Harry Sm.

Description: Annuals or biennials, 5–15 cm tall, leaves elliptic to oblong, flowers blue or purple-blue with pale yellow base and blackish veins, 6–10 mm across, terminal, solitary.

Distribution: Pakistan, N India, Nepal, China (Gansu, Qinghai, Sichuan, Xizang, Yunnan), Bhutan.

Habitat: Alpine meadows, pastures, wet river banks.

Stem diameter 1–1.5 mm. Bark/xylem proportion 1:1. Ring width 0.15–0.2 mm. Maximal 1 annual ring.

Xylem

Transverse section

Only one ring. Vessel diameter 10–15 µm. Vessels solitary. Fibers thin- to thick-walled. Fibers absent in the center. Parenchyma absent in the peripheral fiber zone and pervasive in the center. Crystals absent.

Tangential section

Rays absent.

Radial section

Perforation plates simple. Intervessel pits round in the fiber zone and scalariform in the center. Vessel length 80–120 µm.

Bark

Phloem with small groups of sieve tubes. Phellem absent.

Pith

With medullary phloem.

Analyzed: Root collar of polar root of 4 individuals.

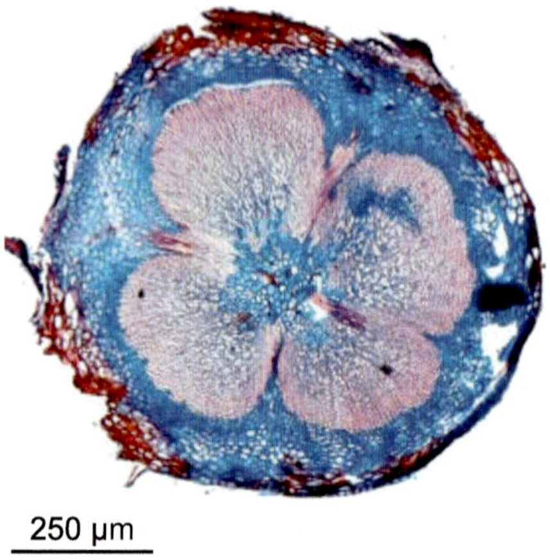

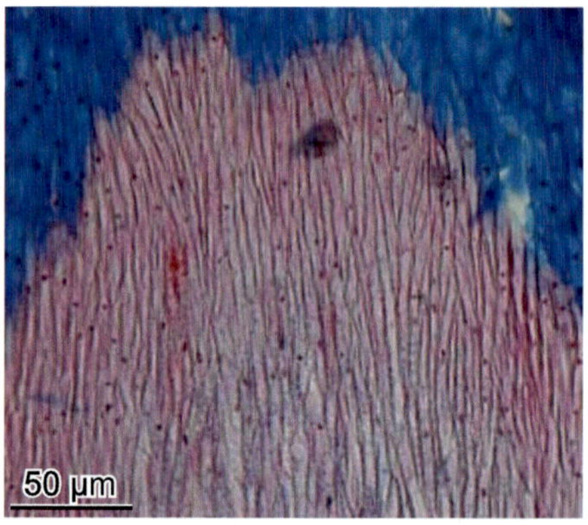

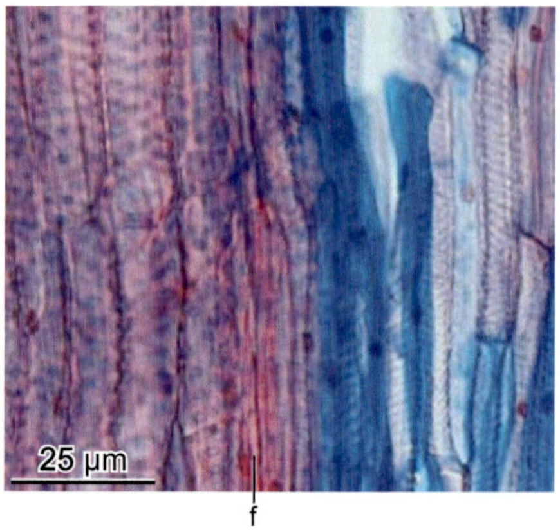

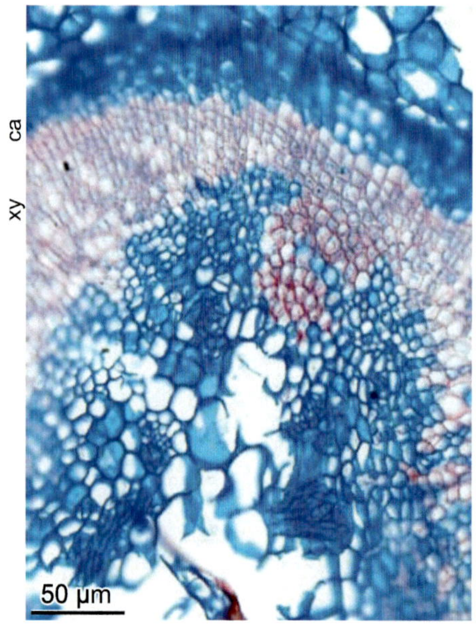

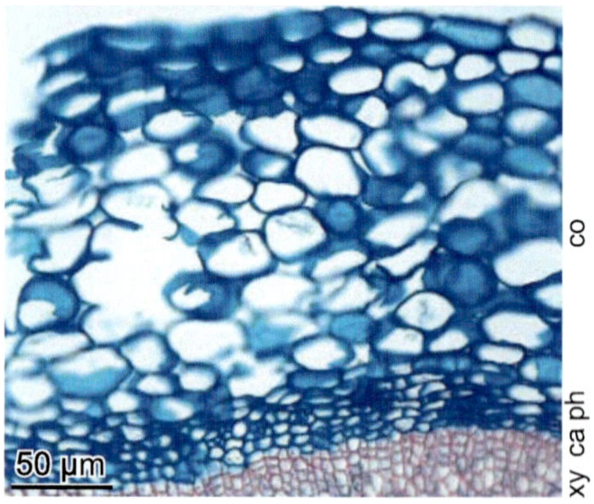

Features according to IAWA

Xylem 2.1 - 9 - 13 - 20 - 22 - 40.1 - 50.2 - 52.2 - 52.3 - 60.1 - 61 - 70 - 70.3 - 75 - 79 - 79.1 - 117 - P1

Gentiana leucomelaena Maxim.

Gentianaceae

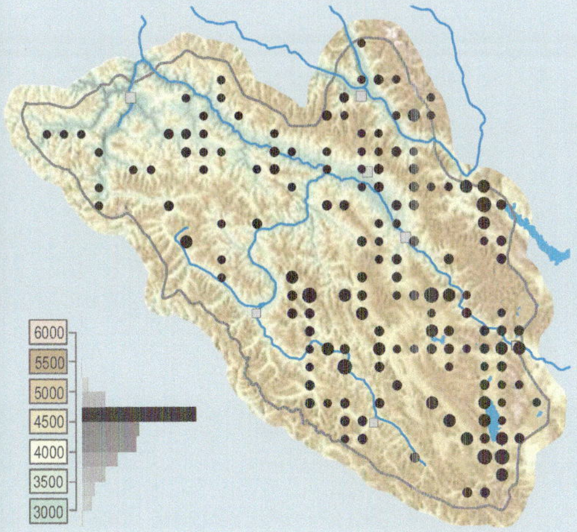

Syn.: *Ciminalis leucomelaena* (Maxim.) V.V.Zuev

Description: Annuals, 2–10 cm tall, stems prostrate to ascending, branched from the base, leaves ovate to elliptic, opposite, flowers 7–12 mm across, pale blue or white, with dark blue spots.

Distribution: Russia, Kazakhstan, Kyrgyzstan, Tajikistan, Pakistan, India, Nepal, China (Gansu, Qinghai, Sichuan, Xinjiang, Xizang), N Mongolia.

Habitat: River banks, moist saline grasslands.

Stem diameter 1 mm. Bark/xylem proportion 1:3. Ring width 0.15–0.35 mm. Maximal 3 annual rings.

Xylem

Transverse section

Semi-ring-porous. Ring boundaries indicated by larger earlywood (25–35 µm) and smaller latewood vessel diameters (<10 µm). Vessels solitary and in short radial multiples. Fibers thin- to thick-walled, absent in the center. Parenchyma paratracheal in the peripheral fiber zone and pervasive in the center. Crystals absent.

Tangential section

Rays absent.

Radial section

Perforation plates simple. Intervessel pits round in the fiber zone and scalariform in the center. Vessel length 50–70 µm.

Bark

Phloem with small groups of sieve tubes. Phellem absent.

Pith

No slide available.

Analyzed: Root collar of polar root of 2 individuals.

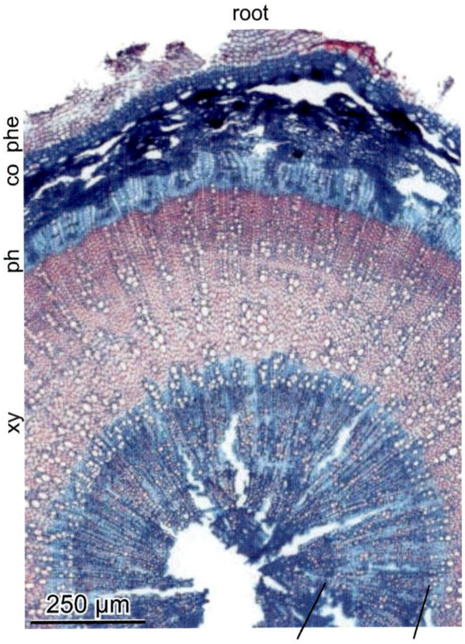

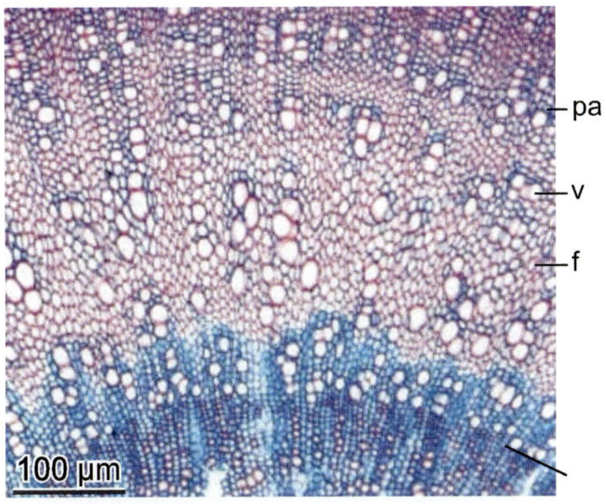

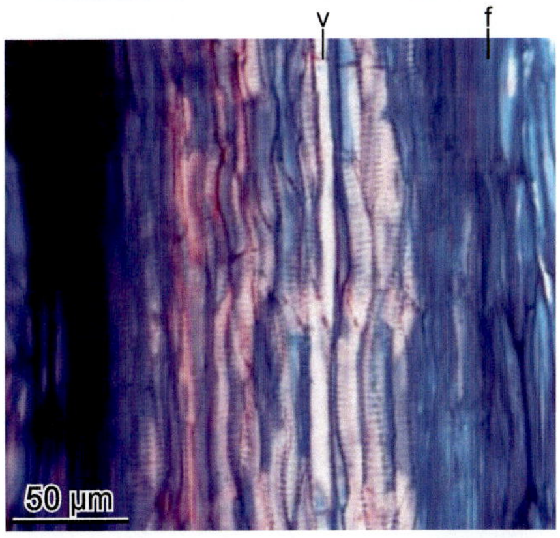

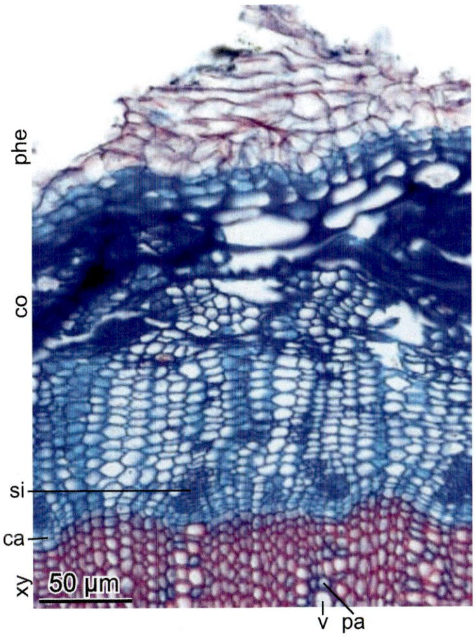

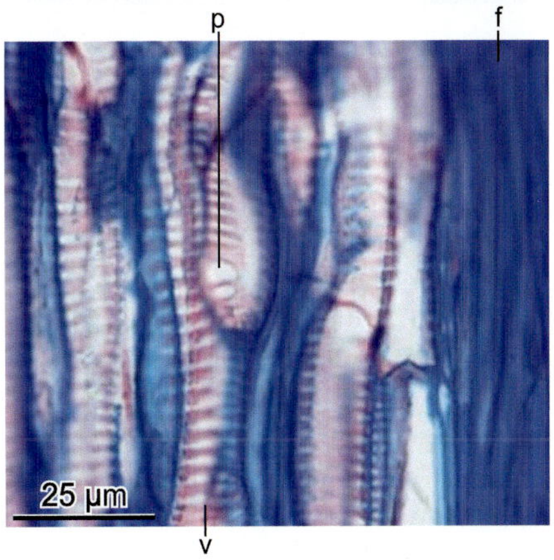

Features according to IAWA

Xylem 1 - 9 - 9.1 - 13 - 20 - 22 - 40.2 - 50.2 - 52.2 - 60.1 - 61 - 70 - 70.3 - 79 - 79.1 - 117 - P1

Gentiana nubigena Edgew.

Gentianaceae

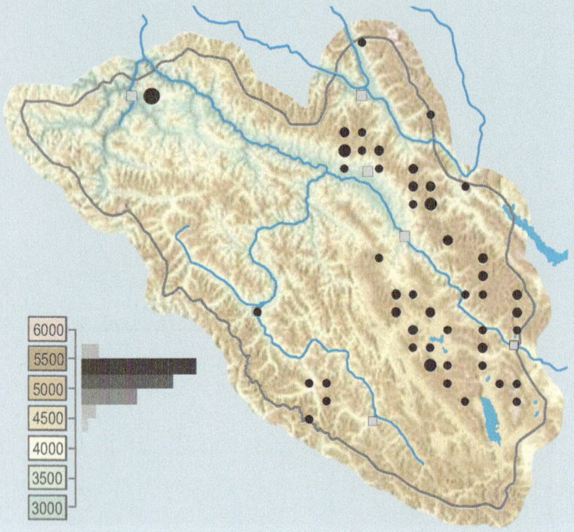

Syn.: *G. algida* var. *nubigena* (Edgew.) Kusn., *G. przewalskii* Maxim., *Gentianodes algida* var. *nubigena* (Edgew.) Omer, Ali & Qaiser

Description: Perennial herbs, 3–10 cm tall, glabrous, basal leaves in erect rosettes, flat or somewhat folded, lanceolate to narrowly elliptic, stem leaves in 1–3 pairs, flowers dark blue with pale yellow base, 4–6 cm long, in 1–3-flowered terminal clusters.

Distribution: Pakistan, NW India, Nepal, China (Gansu, Qinghai, Xizang), Bhutan.

Habitat: Wet solifluction soils, flushes, snow beds and moss-rich spring.

Rings indistinct. Stem diameter 2–5 mm. Bark/xylem proportion 1:3. Ring width 0.05 mm. Stem segregation.

Xylem

Transverse section

Semi-ring-porous. Vessel diameter 15–20 μm. Vessels solitary. Fibers absent. Parenchyma pervasive. Vascular bundles remaining. Crystals absent.

Tangential section

Rays absent.

Radial section

Perforation plates simple. Intervessel pits scalariform. Vessel length 50–120 μm.

Bark

Phloem with small groups of sieve tubes arranged in tangential rows. Phellem absent.

Pith

Round, with medullary phloem.

Analyzed: Root collar of polar root of 3 individuals.

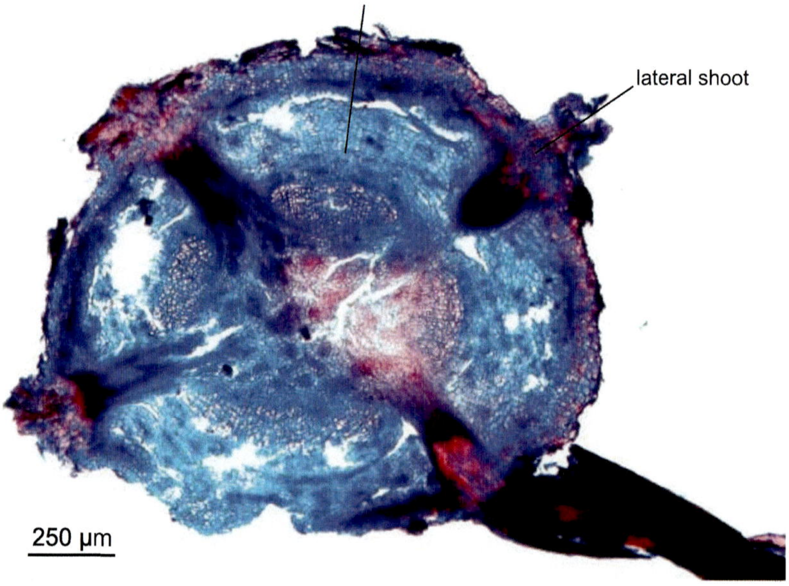

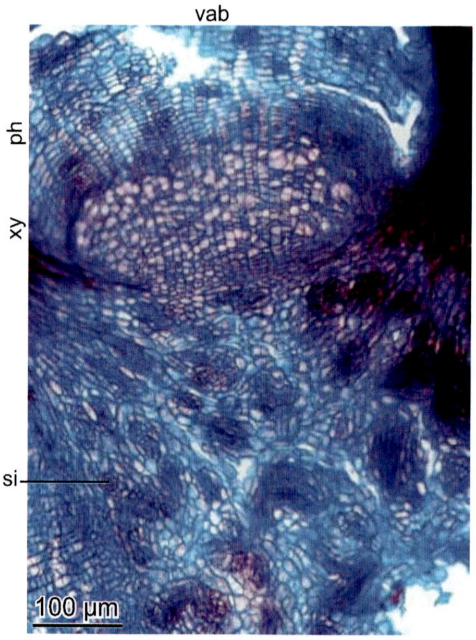

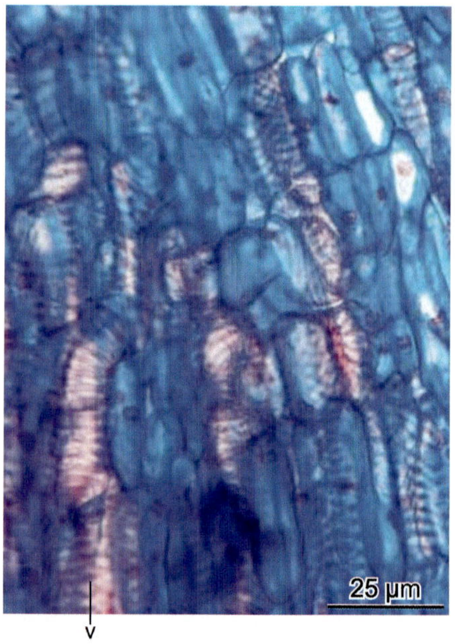

Features according to IAWA

Xylem 1 - 4 - 9 - 9.1 - 13 - 20 - 40.1 - 40.2 - 50.2 - 560.1 - 70.3 - 79.1 - 99.1 - 117 - P1

Gentiana tianschanica Rupr.

Gentianaceae

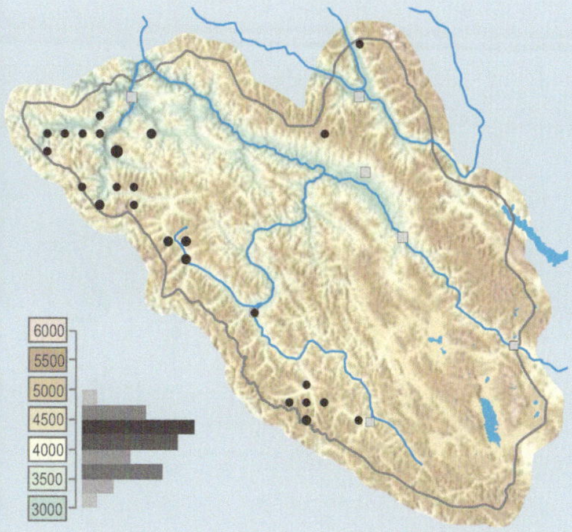

Syn.: *Gentianodes tianschanica* (Rupr.) Omer, Ali & Qaiser

Description: Perennial herbs, 15–30 cm tall, glabrous, basal leaves rosulate, lanceolate, up to 10 cm long, cauline leaves opposite, much shorter, flowers blue, 1.5–3 cm long, loosely clustered in cymes.

Distribution: Kyrgyzstan, Kazakhstan, Pakistan, NW India, Nepal, China (Xinjiang, Xizang).

Habitat: Grasslands, alpine slopes, along streams.

Rings indistinct. Stem diameter 3–5 mm. Bark/xylem proportion 1:3. Mean ring width 0.2 mm. Maximal 3 annual rings. Stem segregation.

Xylem

Transverse section

Semi-ring-porous. Ring boundaries indicated by larger earlywood (20–30 µm) and smaller latewood vessel diameters (<10 µm). Vessels solitary. Fibers absent. Parenchyma pervasive. Crystals absent.

Tangential section

Rays absent.

Radial section

Perforation plates simple. Intervessel pits scalariform. Vessel length 100–120 µm.

Bark

Phloem with small groups of sieve tubes arranged in tangential rows. Phellem absent.

Pith

No slide available.

Analyzed: Root collar of polar root of 4 individuals.

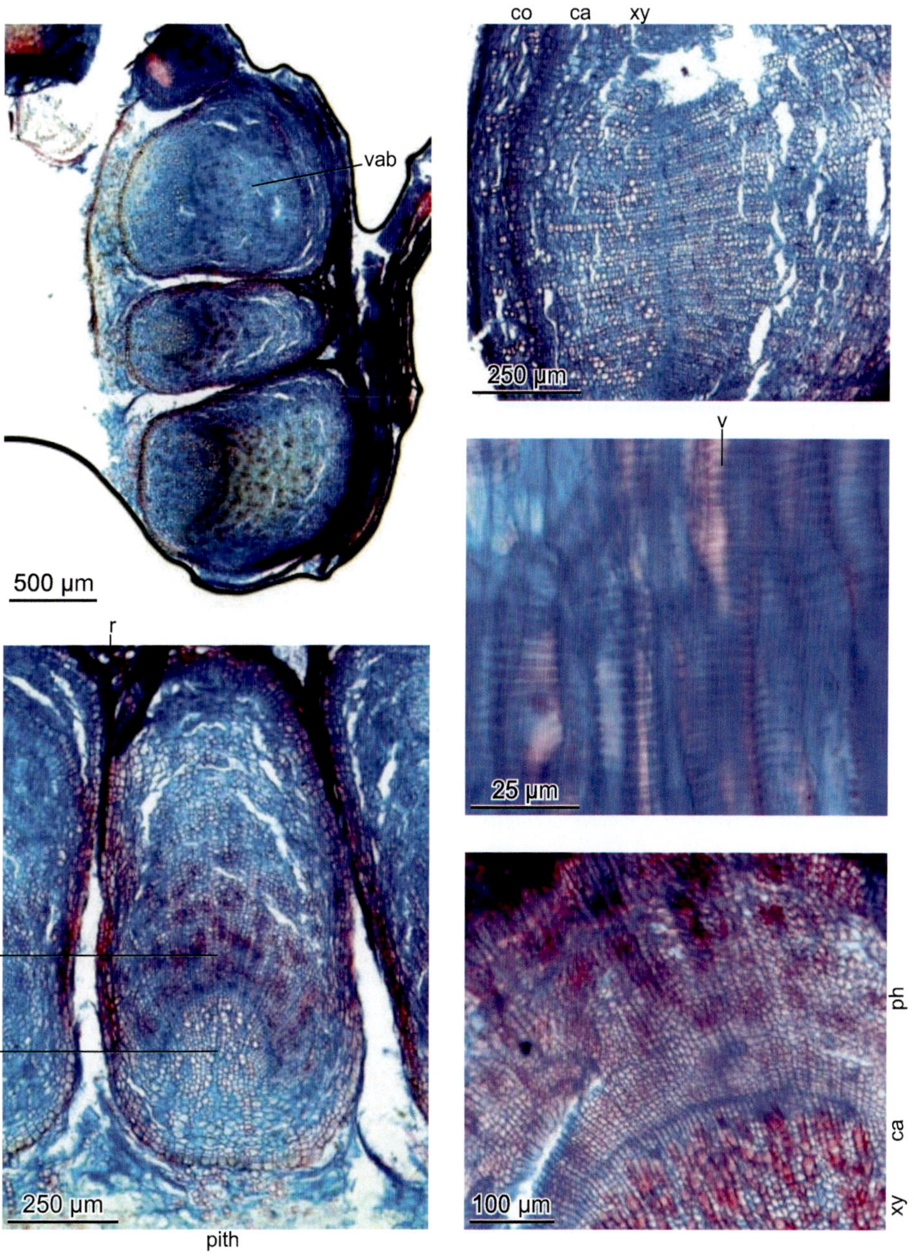

Features according to IAWA

Xylem 1 - 4 - 9 - 9.1 - 13 - 20 - 40.1 - 40.2 - 50.2 - 52.3 - 61 - 70.3 - 79.1 - 117 - P1

Gentianella arenaria (Maxim.) T.N.Ho

Gentianaceae

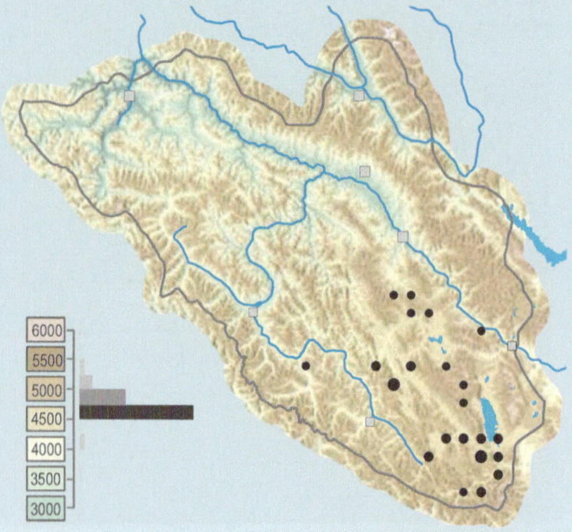

Syn.: *Gentiana arenaria* Maxim.

Description: Annuals, 2–4 cm tall, whole plants purple, stems prostrate to ascending, much branched, leaves spatulate, flowers purple, ca. 3 mm across, solitary, terminal.

Distribution: NW India, China (Gansu, Qinghai, Xizang).

Habitat: Sandy places on river banks, wet gravel sediments, alpine rocky places.

Stem diameter 0.8 mm. Bark/xylem proportion 1:1. Mean ring width 0.1 mm. Maximal 1 annual ring.

Xylem

Transverse section

Only one ring. Vessel diameter 10–15 μm. Vessels in radial multiples. Fibers thin- to thick-walled, absent in the center. Parenchyma absent in the peripheral fiber zone and pervasive in the center. Crystals absent.

Tangential section

Rays absent.

Radial section

Section missing

Bark

Phloem with small groups of sieve tubes. Phellem consists of a few rows of square and rectangular cells.

Pith

Pith polygonal. With medullary phloem.

Analyzed: Root collar of 1 individual (herbarium specimen).

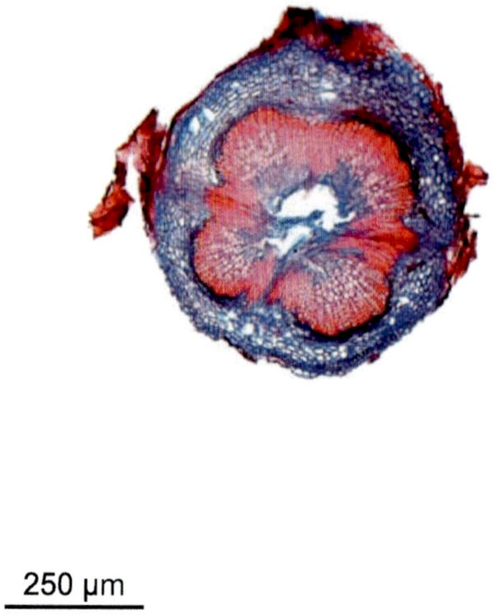

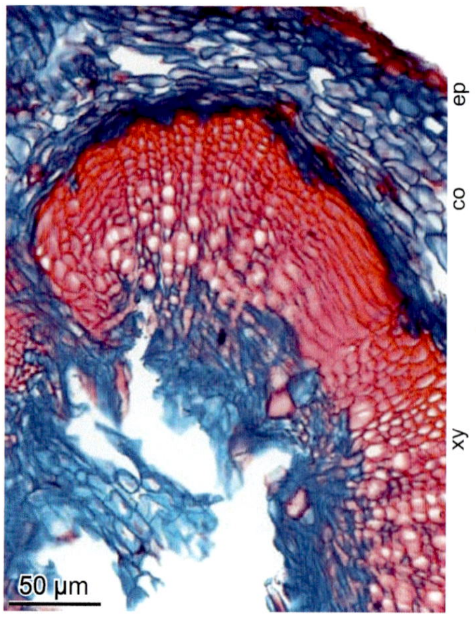

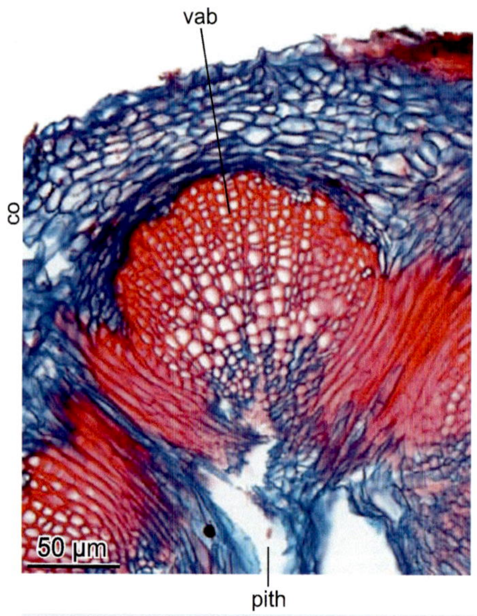

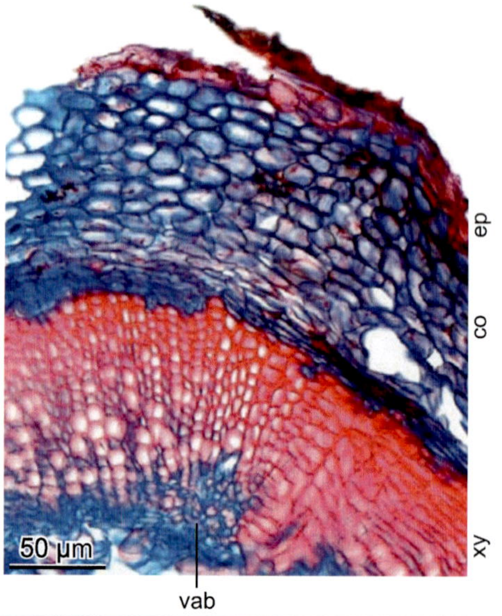

Features according to IAWA

Xylem 2.1 - 9.1 - 13 - 40.1 - 50.2 - 52.3 - 61 - 70 - 79 - 117
(incomplete due to unavailable longitudinal section)

Gentianella moorcroftiana (Wall. ex G.Don) Airy Shaw

Gentianaceae

Syn.: *Gentiana moorcroftiana* Wall. ex G.Don, *Aliopsis moorcroftiana* (Wall. ex G.Don) Omer, Qaiser, & Ali, *G. moorcroftiana* var. *maddenii* C.B.Clarke

Description: Annuals, 5–20 cm tall, stems purplish, branched from the base, leaves linear, opposite, flowers blue with pale yellow base, 1–2.5 cm long.

Distribution: Afghanistan, Pakistan, NW India, Nepal, China (Xizang).

Habitat: Dry and mesic slopes, river banks.

Stem diameter 2 mm. Bark/xylem proportion 1:3. Ring width 0.5–1 mm. Maximal 2 annual rings.

Xylem

Transverse section

Semi-ring-porous. Diameter of earlywood vessels 25–35 µm and of latewood vessels <10 µm. Vessels solitary and in short radial multiples. Fibers thin-walled, absent in the center. Parenchyma absent in the peripheral fiber zone and pervasive in the center. Crystals absent.

Tangential section

Rays absent.

Radial section

Perforation plates simple. Intervessel pits large and round in the fiber zone and scalariform in the center. Vessel length 150–200 µm.

Bark

Phloem with small groups of sieve tubes. Phellem consists of large, irregularly formed, thin-walled cells.

Pith

Round, with medullary phloem.

Analyzed: Root collar of polar root of 8 individuals.

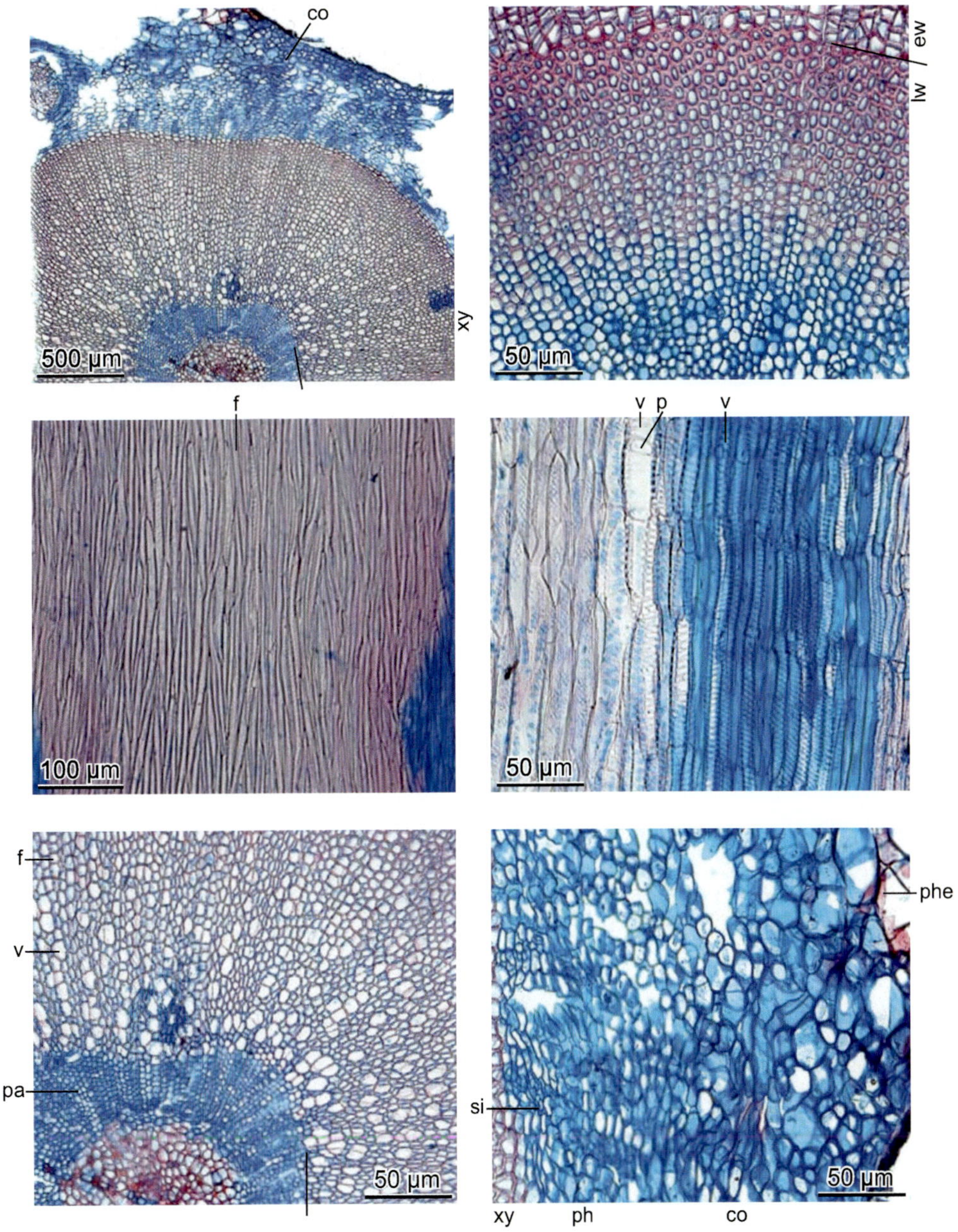

Features according to IAWA

Xylem 1 - 9 - 9.1 - 13 - 20 - 22 - 40.2 - 50.2 - 52.3 - 60.1 - 61 - 68 - 70.3 - 75 - 79.1 - 117 - P1

Gentianopsis paludosa (Munro ex Hook.f.) Ma

Gentianaceae

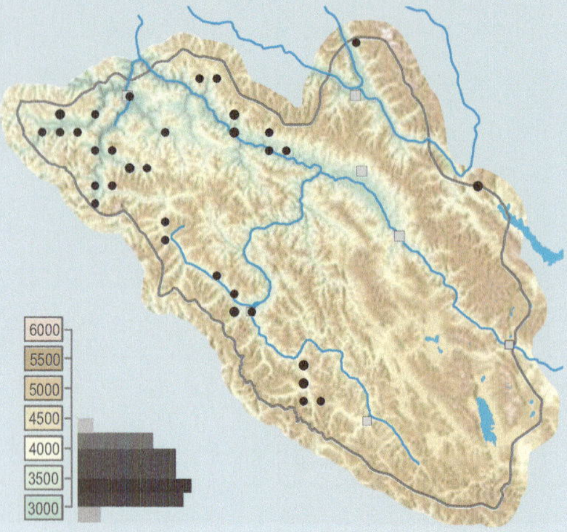

Syn.: *Gentiana paludosa* Munro ex Hook.f., *G. stracheyi* (Clarke) Kitam., *G. detonsa* var. *paludosa* Munro ex Hook.f.

Description: Annuals to biennials, 5–40 cm tall, basal leaves rosulate, spatulate, stem leaves lanceolate, opposite, flowers blue or yellowish white, 3–6 cm long, solitary, terminal.

Distribution: Pakistan, India, Nepal, China (Gansu, Hebei, Hubei, Nei Mongol, Ningxia, Qinghai, Shaanxi, Shanxi, Sichuan, Xizang, Yunnan), Bhutan.

Habitat: Wet meadows and pastures, stream banks.

Stem diameter 1–1.5 mm. Bark/xylem proportion 1:3. Ring width 0.5–0.6 mm. Maximal 1 annual ring.

Xylem

Transverse section

Only one ring. Vessel diameter 15–30 µm. Vessels solitary. Fibers thin- to thick-walled. Fibers absent in the center. Parenchyma absent in the peripheral fiber zone and pervasive in the center. Crystals absent.

Tangential section

Rays absent.

Radial section

Perforation plates simple. Intervessel pits round in the fiber zone and scalariform in the center. Vessel length 80–120 µm.

Bark

Phloem with small groups of sieve tubes. Phellem absent.

Pith

Round, with medullary phloem.

Analyzed: Root collar of polar root of 3 individuals.

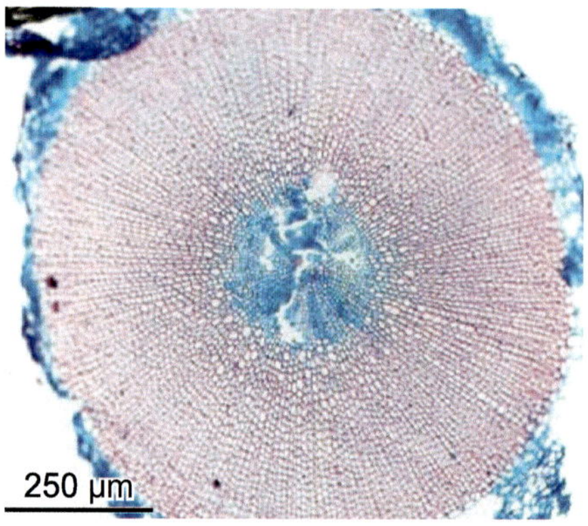

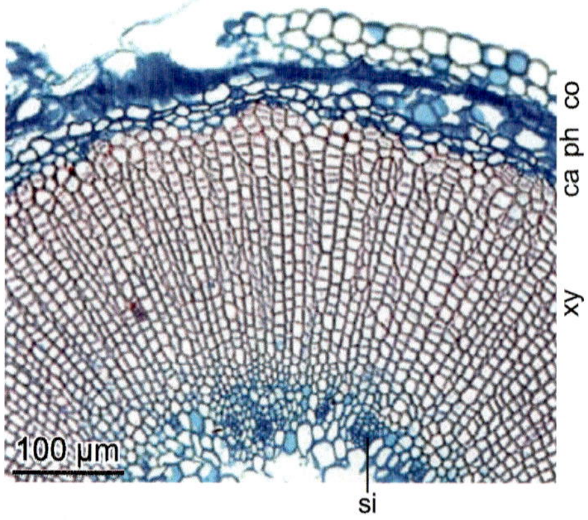

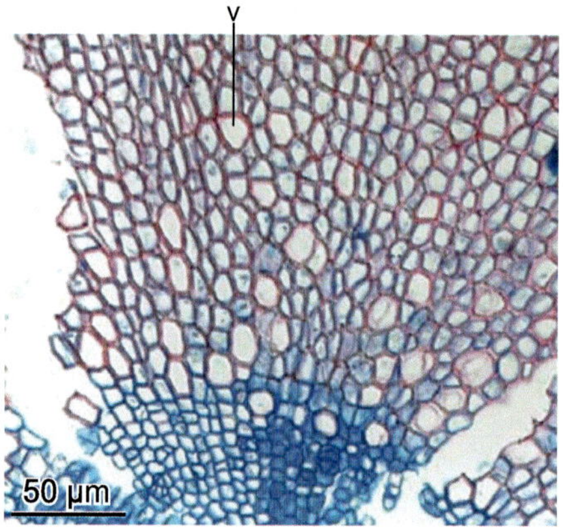

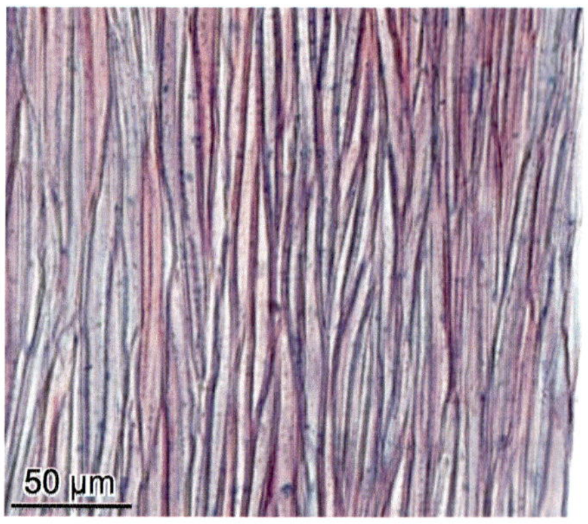

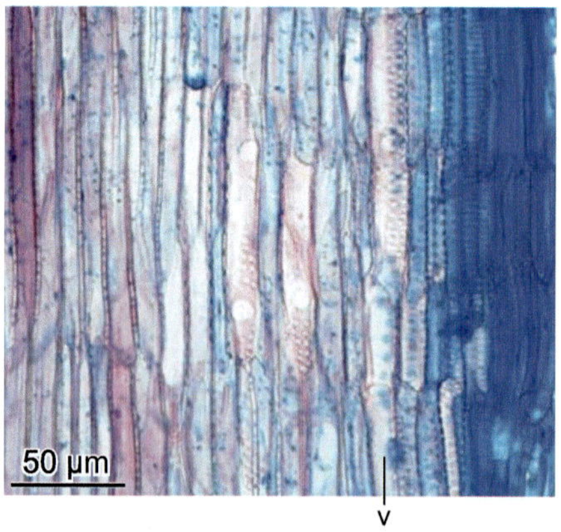

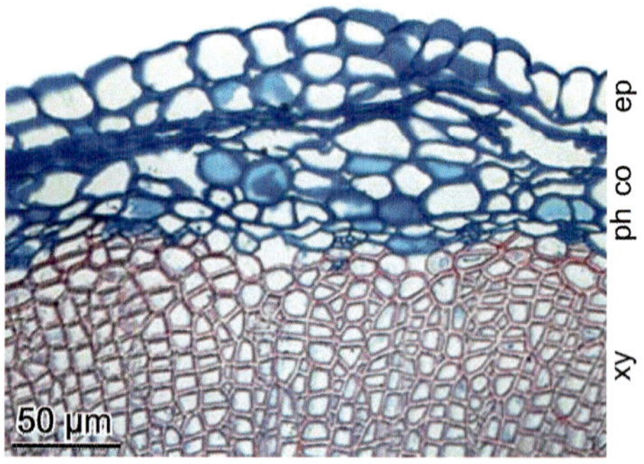

Features according to IAWA

Xylem 2.1 - 9 - 13 - 20 - 22 - 40.1 - 50.2 - 52.2 - 52.3 - 61 - 70 - 70.3 - 75 - 79 - 79.1 - 117 - P1

Gentianopsis vvedenskyi (Grossh.) Pissjauk.

Gentianaceae

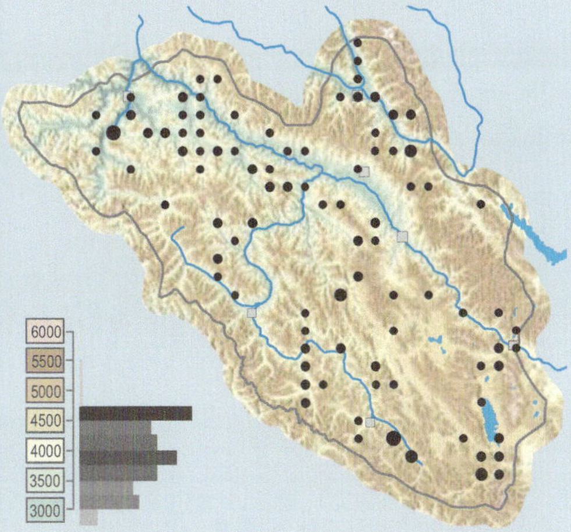

Syn.: *Gentiana vvedenskyi* Grossh., *G. stricta* Klotzsch 1862, non Willd. ex Schult. 1820

Description: Annuals, 15–40 cm tall, basal leaves rosulate, obovate to oblong, withering, stem leaves sessile, opposite, linear-lanceolate, flowers blue, up to 4.5 cm long, solitary on long peduncles.

Distribution: Russia, Afghanistan, Pakistan, India, Nepal, China, Bhutan.

Habitat: Mesic and wet grasslands on river banks, mesic slopes, sandy soils.

Stem diameter 1–2 mm. Bark/xylem proportion 1:2. Ring width 0.3–0.4 mm. Maximal 1 annual ring.

Xylem

Transverse section

Only one ring. Vessel diameter 10–15 µm. Vessels solitary, often not distinctly differenced from fibers. Fibers thin- to thick-walled, absent in the center. Parenchyma absent in the peripheral fiber zone and pervasive in the center. Crystals absent.

Tangential section

Rays absent.

Radial section

Perforation plates simple. Intervessel pits round in the fiber zone and scalariform in the center. Vessel length 120–140 µm.

Bark

Phloem with small groups of sieve tubes. Phellem absent.

Pith

No slide available.

Analyzed: Root collar of polar root of 2 individuals.

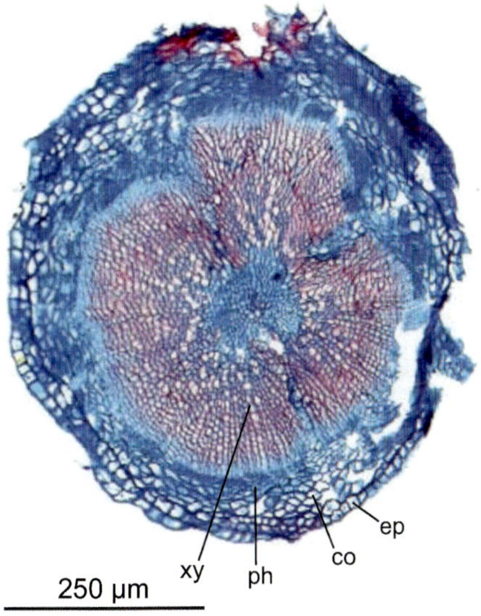

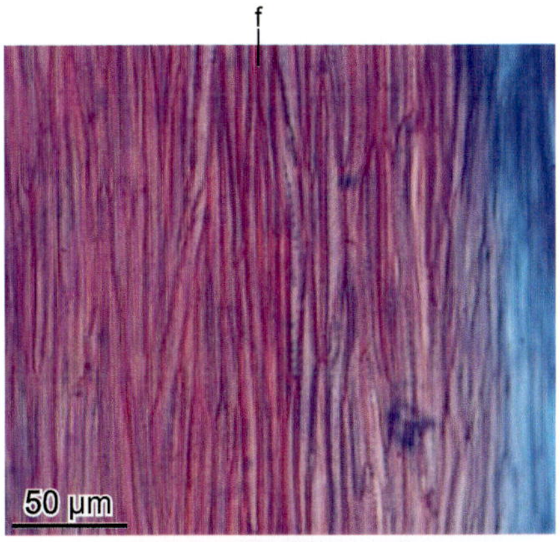

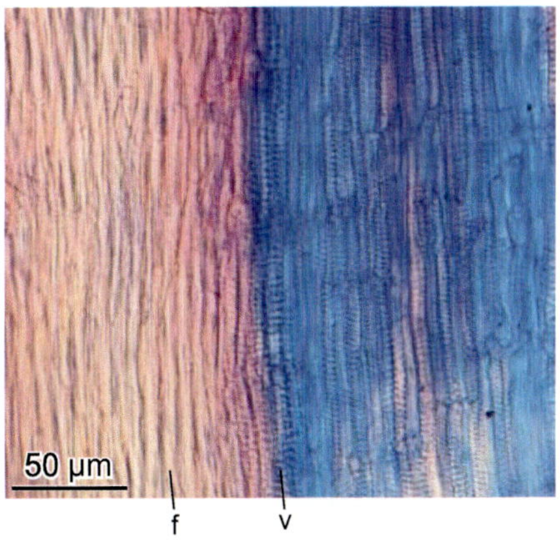

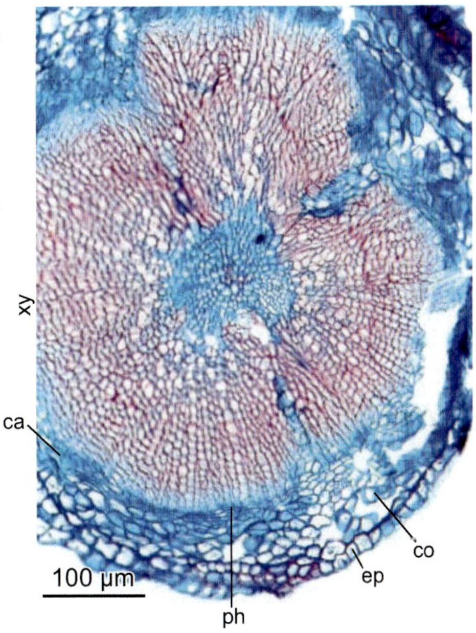

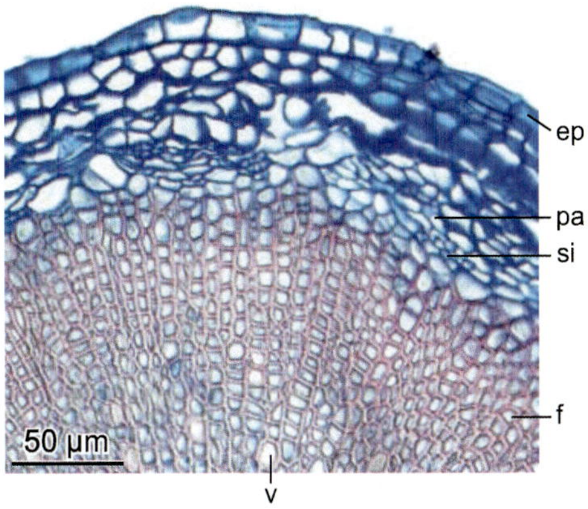

Features according to IAWA

Xylem 2.1 - 9 - 13 - 20 - 22 - 40.1 - 50.2 - 52.3 - 60.1 - 61 - 70 - 70.3 - 75 - 79.1 - 117

Lomatogonium carinthiacum (Wulf.) A.Br.

Gentianaceae

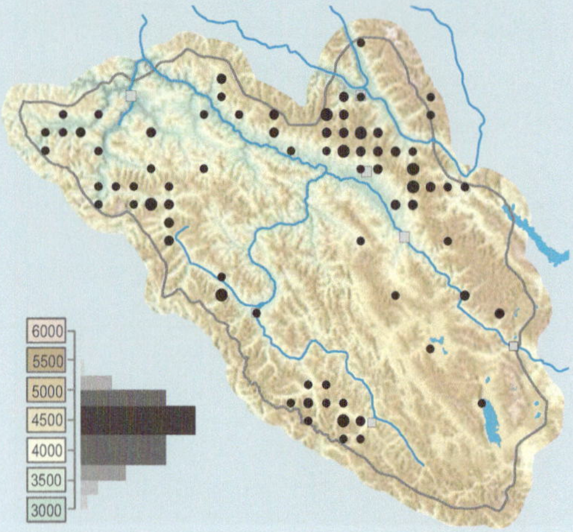

Syn.: *Swertia carinthiaca* Wulf., *Pleurogyne carinthiaca* (Wulf.) Griseb.

Description: Annuals, 3–30 cm tall, stems glabrous, ascending, branched from the base, basal leaves spatulate, stem leaves ovate, opposite, flowers blue, 0.8–2 cm across, in lax terminal and axillary cymes.

Distribution: Europe, Russia, Kyrgyzstan, Tajikistan, Afghanistan, Pakistan, India, China (Gansu, Hebei, Qinghai, Shanxi, Sichuan, Xizang, Yunnan), Bhutan, Mongolia, Japan.

Habitat: Mesic to wet river banks.

Stem diameter 1 mm. Bark/xylem proportion 1:3. Mean ring width 0.5 mm. Maximal 1 annual ring.

Xylem

Transverse section

Only one ring. Vessel diameter 10–15 µm. Vessels solitary and fairly thick-walled, often not distinctly differenced from fibers. Fibers thin- to thick-walled, absent in the center. Parenchyma pervasive in the center and absent or paratracheal in the peripheral fiber zone. Crystals absent.

Tangential section

Rays absent. Many living fibers.

Radial section

Perforation plates simple. Intervessel pits round in the fiber zone and scalariform in the center. Vessel length 50–70 µm.

Bark

Phloem with small groups of sieve tubes. Phellem absent.

Pith

No slide available.

Analyzed: Root collar of polar root of 1 individual.

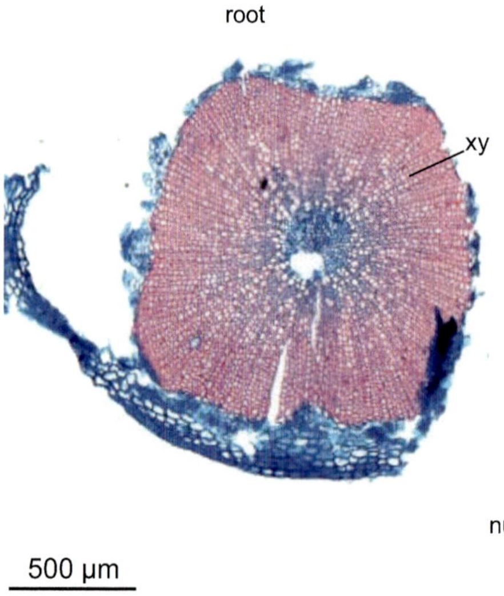

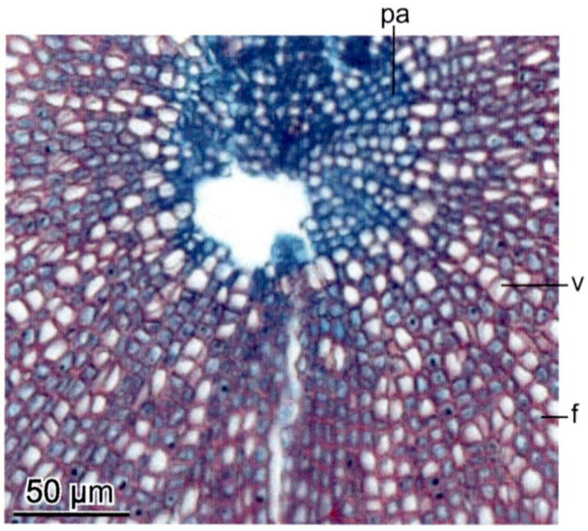

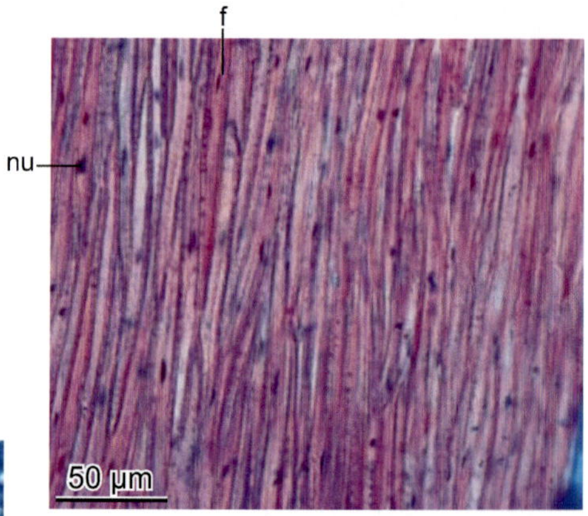

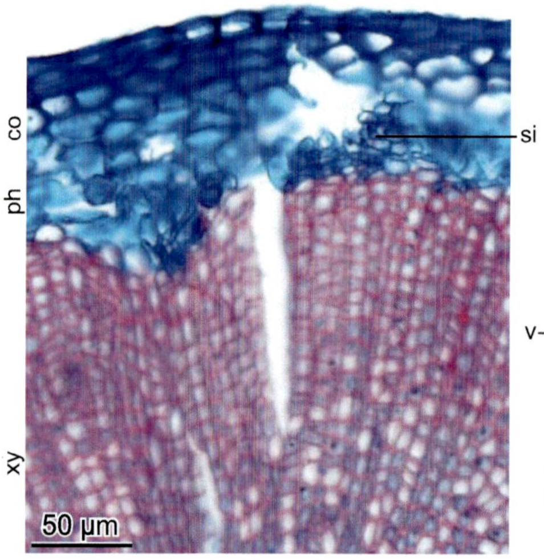

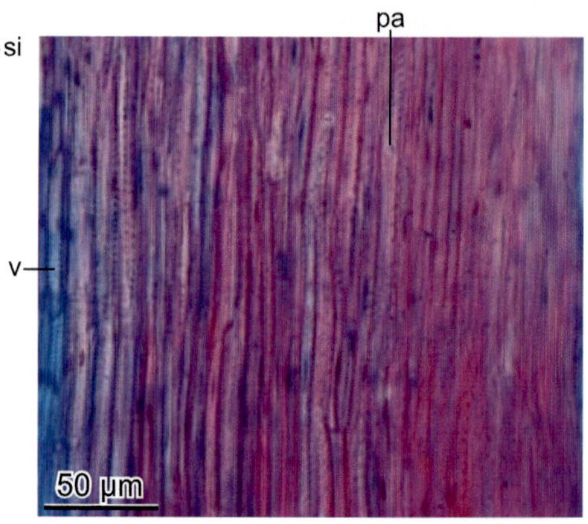

Features according to IAWA

Xylem 2.1 - 9 - 13 - 20 - 22 - 40.1 - 50.2 - 52.2 - 60.1 - 61 - 70 - 70.3 - 75 - 79 - 79.1 - 117

Lomatogonium brachyantherum (C.B.Clarke) Fernald
Gentianaceae

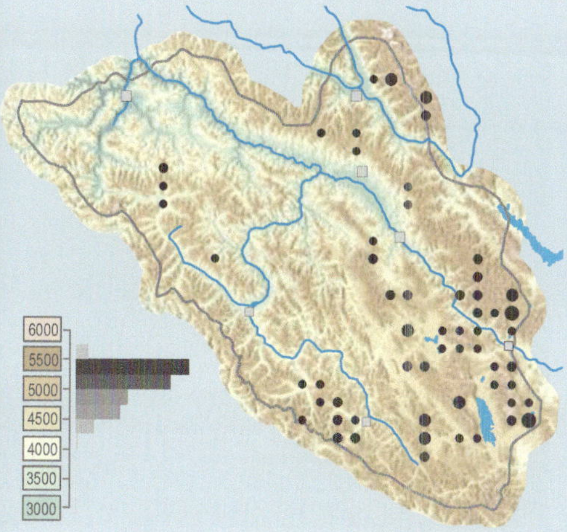

Syn.: *L. thomsonii* (C.B.Clarke) Fernald

Description: Annuals, 2–10 cm tall, glabrous, stems prostrate, angular, much branched from the base, basal leaves spatulate, stem leaves elliptic to lanceolate, flowers blue to purple, 5–8 mm across, terminal, solitary.

Distribution: Afghanistan, Pakistan, N India, Nepal, China (Gansu, Qinghai, Xizang), Bhutan.

Habitat: Alpine meadows, wet grasslands.

Stem diameter 1 mm. Bark/xylem proportion 1:3. Mean ring width 0.5 mm. Maximal 1 annual ring.

Xylem

Transverse section

Only one ring. Vessel diameter 10–15 µm. Vessels solitary and fairly thick-walled, often not distinctly differenciated from fibers. Fibers thin- to thick-walled. Fibers absent in the center. Parenchyma pervasive in the center and paratracheal in the peripheral fiber zone. Crystals absent.

Tangential section

Rays absent. Many living fibers.

Radial section

Perforation plates simple. Intervessel pits round in the fiber zone and scalariform in the center. Vessel length 50–70 µm.

Bark

Phloem with small groups of sieve tubes. Phellem absent.

Pith

No slide available.

Analyzed: Root collar of polar root of 1 individual.

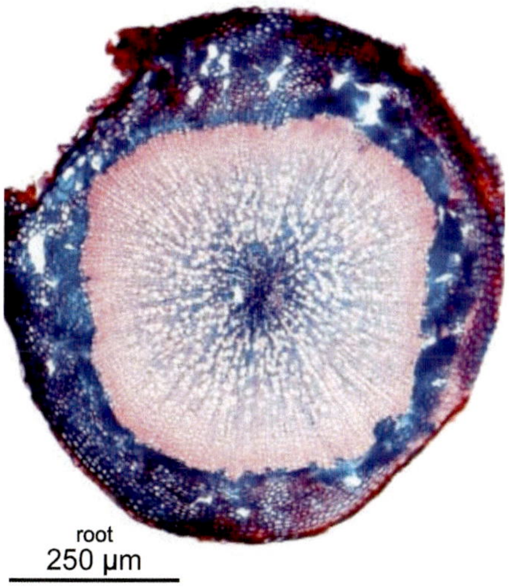

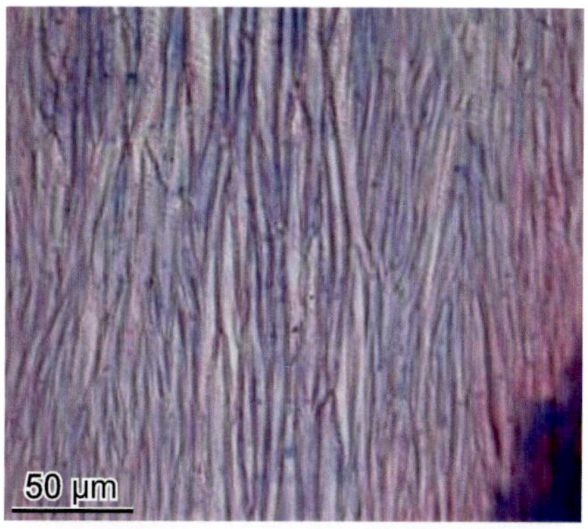

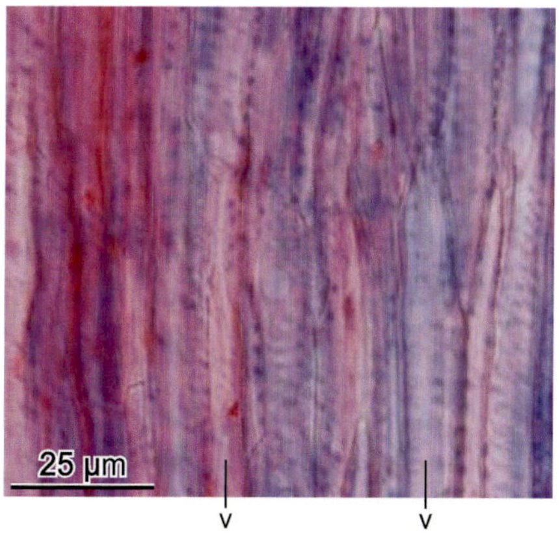

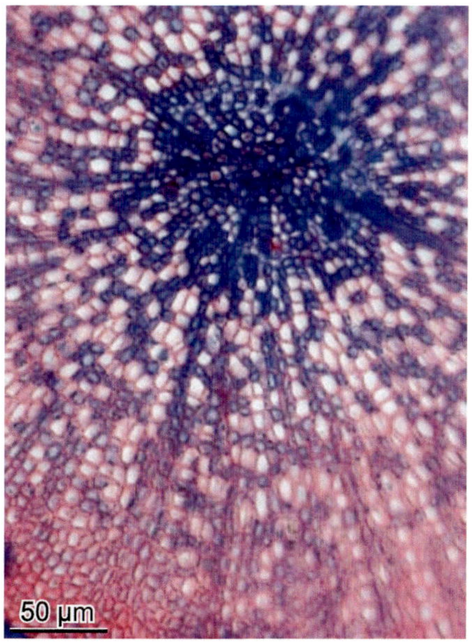

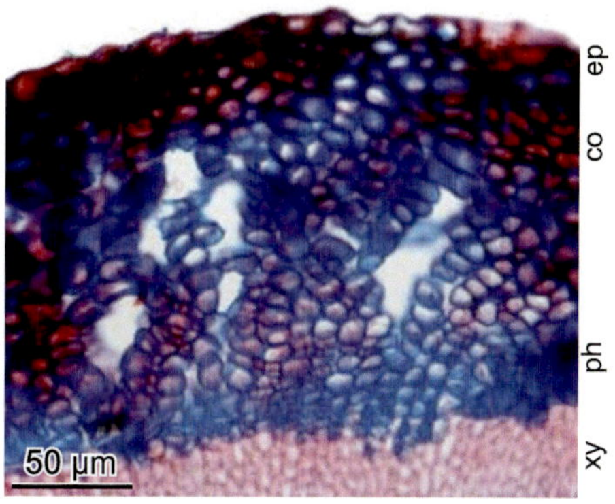

Features according to IAWA

Xylem 2.1 - 9 - 13 - 20 - 22 - 40.1 - 50.2 - 52.2 - 61 - 70 - 70.3 - 79 - 79.1 - 117 - P1

Swertia petiolata Royle ex D.Don

Gentianaceae

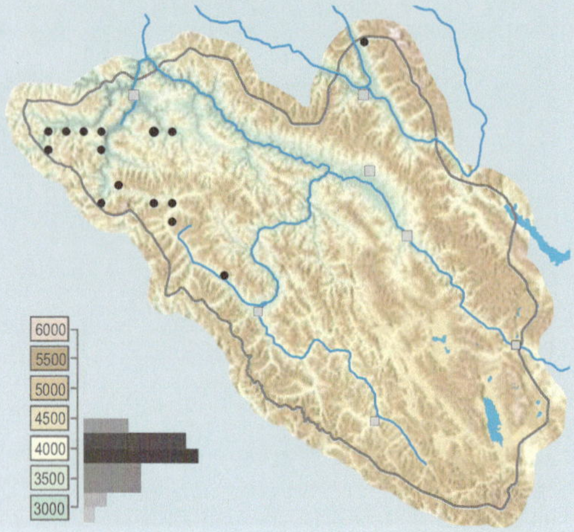

Description: Perennial herbs, 15–50 cm tall, basal leaves oblong, stem leaves ovate, opposite, connate, flowers greenish white, ca. 2 cm across, in dense panicles.

Distribution: Afghanistan, Pakistan, NW India, Nepal, China.

Habitat: Wet meadows, spring areas

Stem diameter 2.5–3.5 mm. Bark/xylem proportion 1:0.5. Annual ring absent.

Xylem

Transverse section

Vessel diameter 20–25 µm. Vessels angular, solitary and in short radial multiples. Fibers absent. Parenchyma pervasive. Crystals absent.

Tangential section

Rays absent.

Radial section

Perforation plates simple. Intervessel pits scalariform with very large apertures. Vessel length 60–70 µm.

Bark

Tangential layers of sieve elements. Phellem absent.

Pith

Pith round to polygonal, with medullary phloem.

Analyzed: Root collar of 2 individuals (herbarium specimens).

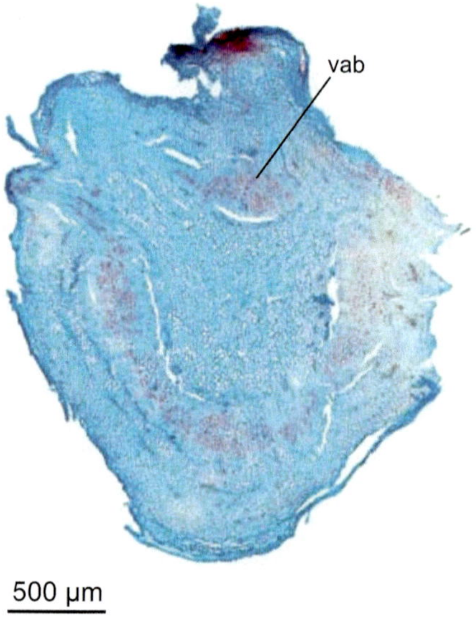

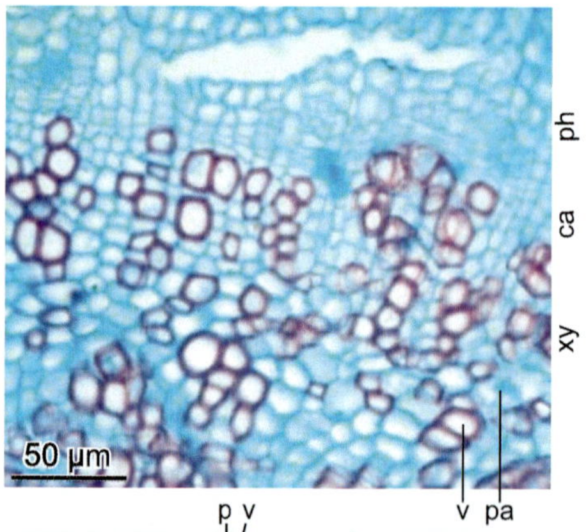

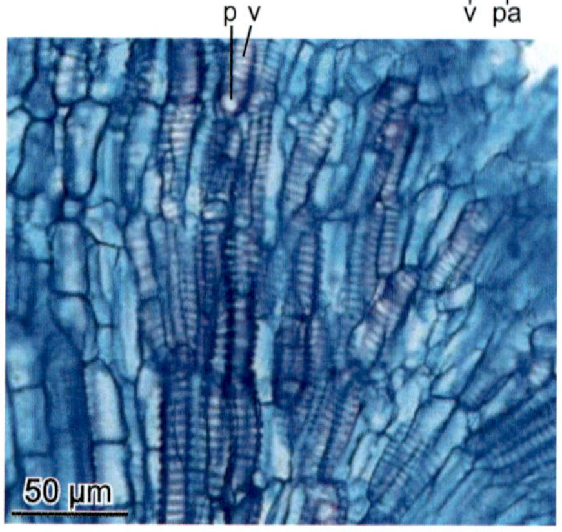

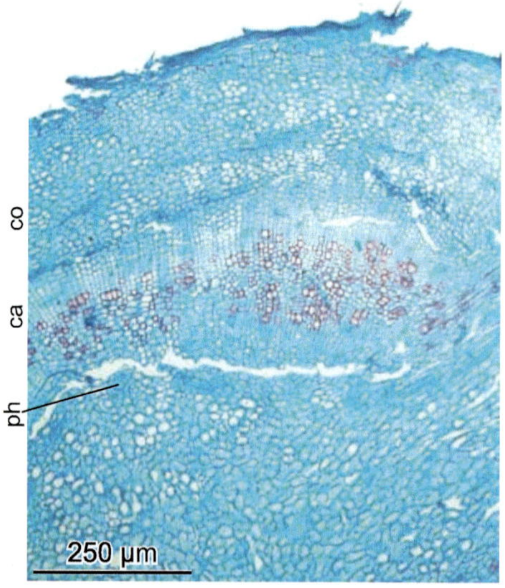

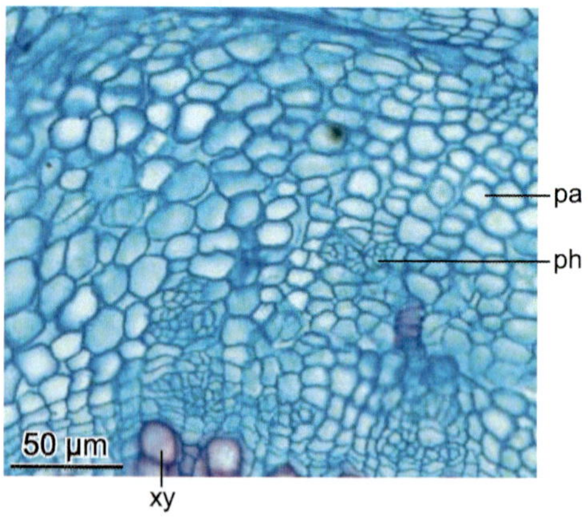

Features according to IAWA

Xylem 2 - 9 - 12 - 13 - 20 - 40.2 - 50.2 - 52.2 - 60.1 - 79.1 - 117 - P1

Erodium tibetanum Edgew. & Hook.f.

Geraniaceae

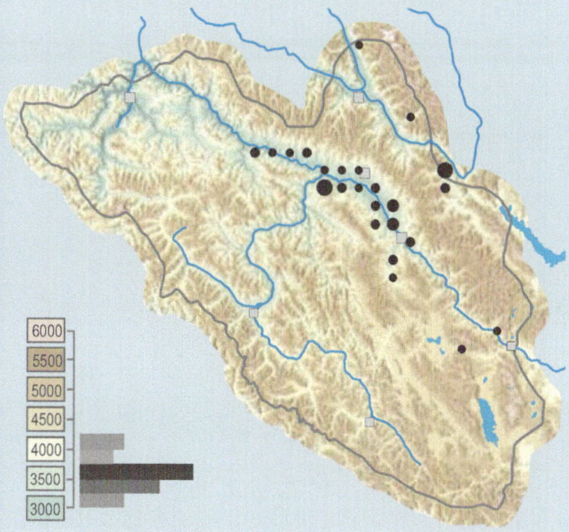

Description: Annuals, 2–10 cm tall, stems decumbent, leaves pinnately parted, in basal rosette, flowers whitish, ca. 1 cm across.

Distribution: Tajikistan, India, China (Gansu, Nei Mongol, Xinjiang, Xizang), Mongolia.

Habitat: Semideserts and deserts, human-induced places near settlements, along roads.

Stem diameter 5 mm. Bark/xylem proportion 1:1. Ring width 0.3–0.5 mm. Maximal 3 annual rings.

Xylem

Transverse section

Diffuse- and semi-ring-porous. Diameter of earlywood vessels 30–40 µm. Vessels lignified, arranged in irregular radial rows and groups. With remaining vascular bundles. Fibers absent. Parenchyma pervasive. Many crystal druses in rays.

Tangential section

Rays 5–>10-seriate, disappearing in polarized light.

Radial section

Perforation plates simple. Intervessel pits slightly scalariform and round in alternating position. Vessel length 70–100 µm. Ray cells upright.

Bark

Phloem contains tangentially arranged groups of sieve tubes in the continuation of vessels and xylem rays. Layers don't correspond with the numbers of xylem rings. Cortex consists of round to oval parenchyma cells with diameters of 40–60 µm. With crystal druses. Phellem layered, consisting of many layers of square to rectangular cells.

Pith

Pentagonal. Cells contain many crystal druses.

Analyzed: Root collar of polar root of 2 individuals.

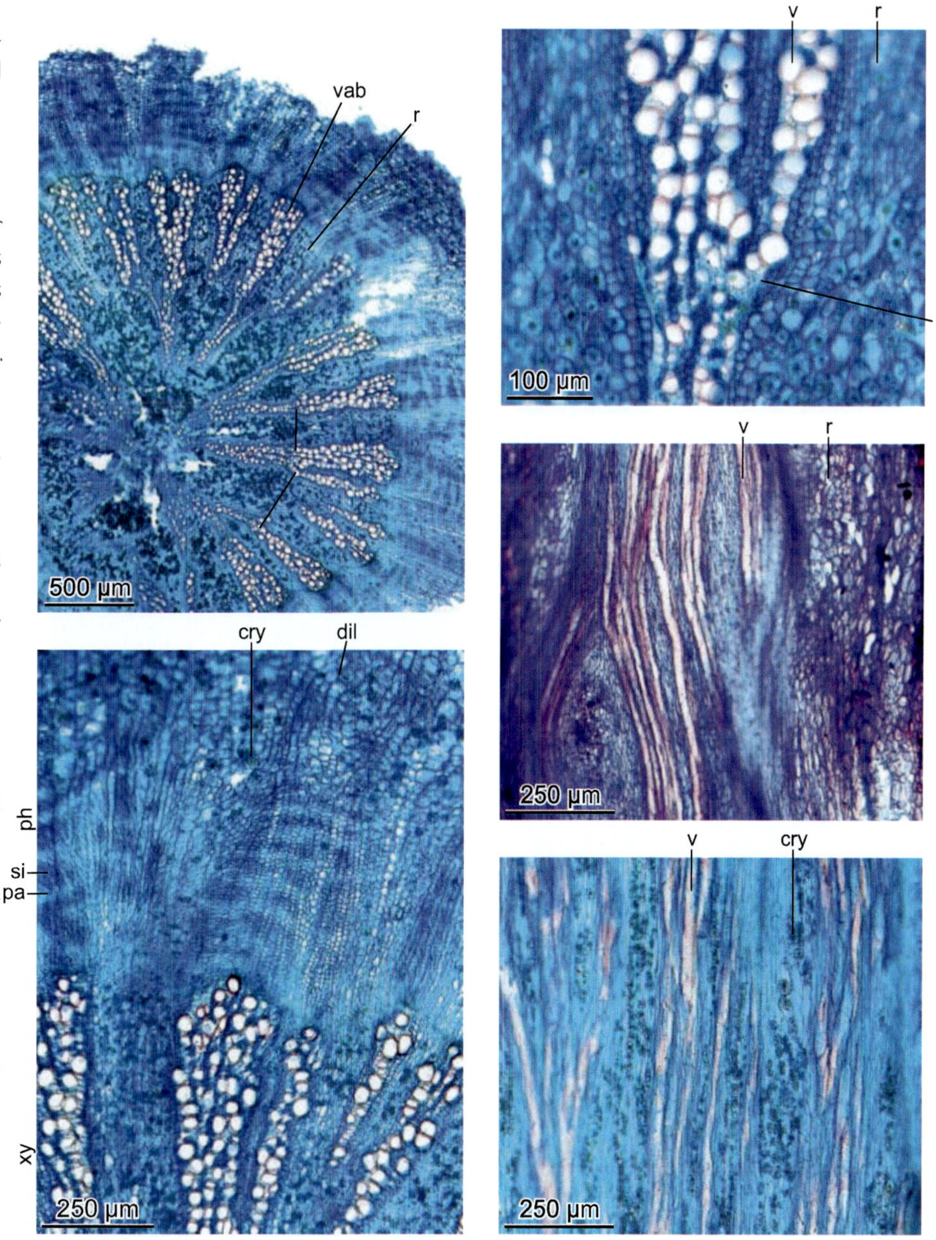

Features according to IAWA

Xylem 1 - 4 - 10 - 11 - 13 - 20 - 22 - 40.2 - 50.1 - 52.2 - 60.1 - 79.1 - 98 - 99 - 99.1 - 100.2 - 105

Geranium himalayense Klotzsch

Geraniaceae

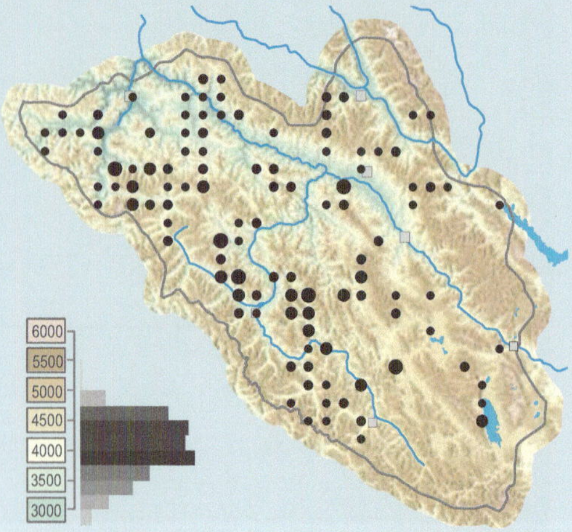

Description: Perennial herbs, 10–40 cm tall, leaves palmately 5-lobed, petioles with patent hairs, often glandular, flowers blue, ca. 2 cm across.

Distribution: Tajikistan, Afghanistan, N Pakistan, India (Jammu and Kashmir, Himachal Pradesh, Uttarakhand), Nepal, China (Xizang).

Habitat: Mesic to wet river banks, spring areas, alpine meadows.

Stem diameter 5–10 mm. Bark/xylem proportion 1:1. Ring width 0.2–0.3 mm. Maximal 10 annual rings.

Xylem

Transverse section

Semi-ring-porous. Diameter of earlywood vessels 50–70 µm. Angular lignified vessels solitary or in groups. With remaining vascular bundles. Fibers absent. Parenchyma pervasive. With crystal druses in rays.

Tangential section

Rays 5–>10-seriate, disappearing in polarized light.

Radial section

Perforation plates simple. Intervessel pits slightly scalariform and round in alternating position. Vessel length 70–100 µm. Ray cells upright.

Bark

Phloem contains indistinctly tangentially arranged small groups of sieve tubes. Cortex consists of round to oval parenchyma cells with diameters of 40–60 µm. Phellem layered, consisting of many layers of square to rectangular cells. With crystal druses.

Pith

Pentagonal. Cells contain many crystal druses.

Analyzed: Root collar of polar root of 5 individuals.

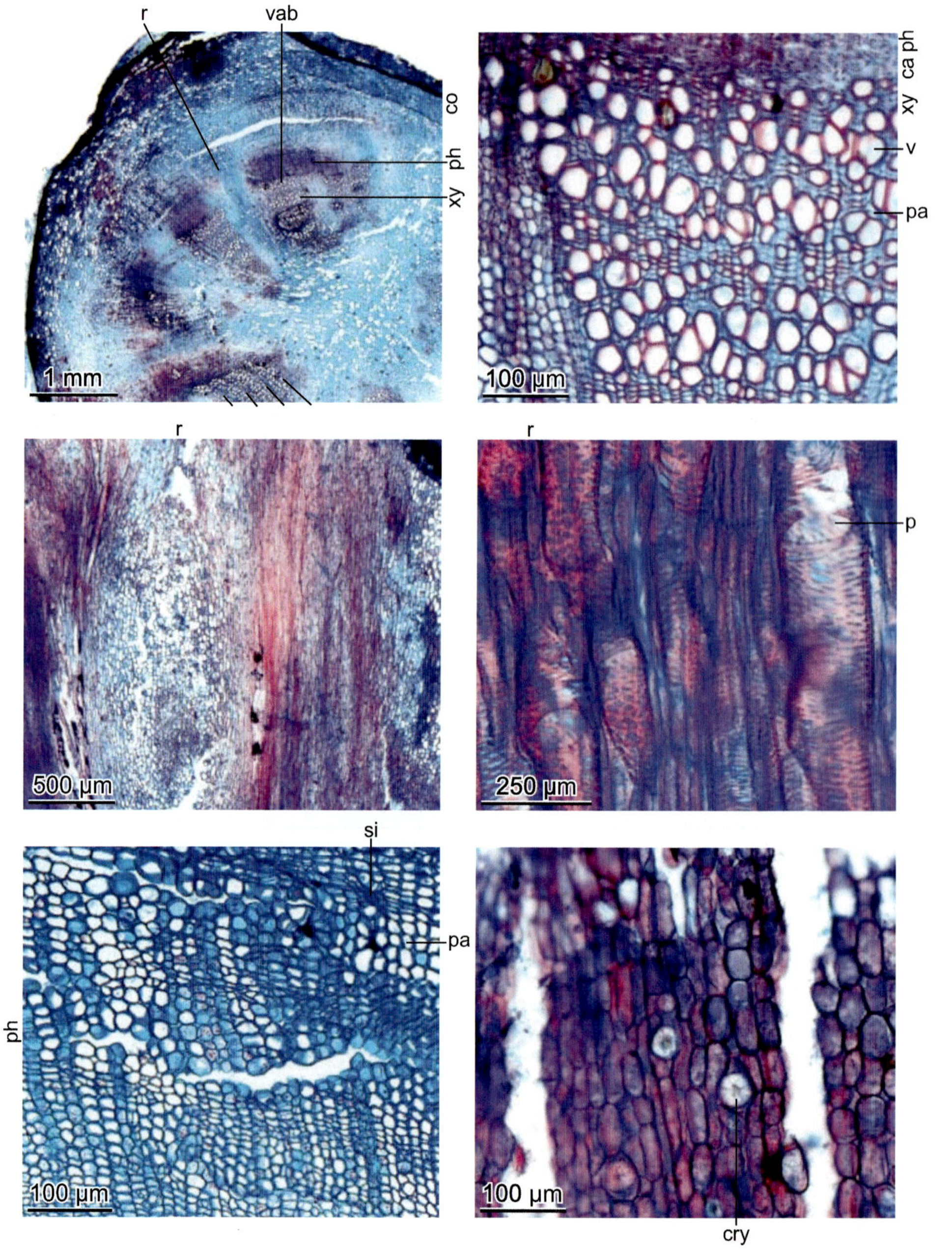

Features according to IAWA

Xylem 1 - 4 - 11 - 13 - 20 - 22 - 41 - 50.1 - 52.2 - 60.1 - 79.1 - 99 - 99.1 - 100.1 - 100.2 - 105 - 144

Geranium regelii Nevski

Geraniaceae

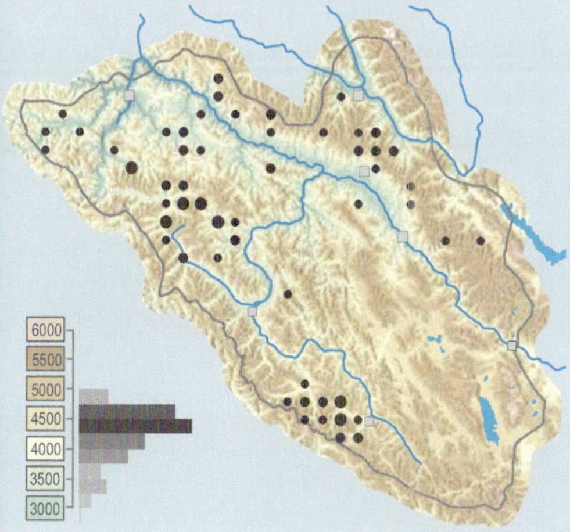

Description: Perennial herbs, 10–35 cm tall, leaves palmately 5-lobed, petioles with patent to appressed non-glandular hairs, flowers blue to whitish, ca. 2 cm across.

Distribution: Afghanistan, Pakistan, India (Jammu and Kashmir, Himachal Pradesh).

Habitat: Alpine meadows.

Stem diameter 5–10 mm. Bark/xylem proportion 1:1. Ring width 0.2–0.25 mm. Maximal 5 annual rings.

Xylem

Transverse section

Semi-ring-porous. Diameter of earlywood vessels 30–50 µm. Angular lignified vessels in short radial multipels or in groups. With remaining vascular bundles. Fibers in small groups, thin- to thick-walled. Parenchyma pervasive. With crystal druses in rays.

Tangential section

Rays 5–>10-seriate, disappearing in polarized light.

Radial section

Perforation plates simple. Intervessel pits slightly scalariform and round in alternating position. Vessel length 70–100 µm. Ray cells upright.

Bark

Phloem contains indistinctly tangentially arranged small groups of sieve tubes. Cortex consists of round to oval parenchyma cells with diameters of 40–60 µm. Phellem layered, consisting of many layers of square to rectangular cells. With crystal druses.

Pith

Pentagonal. Cells contain many crystal druses.

Analyzed: Root collar of polar root of 4 individuals.

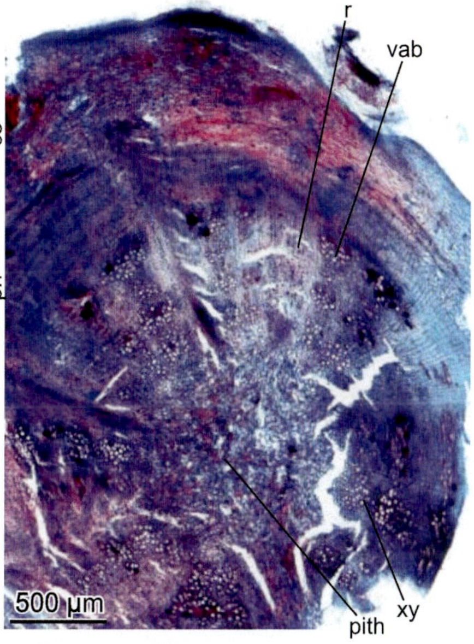

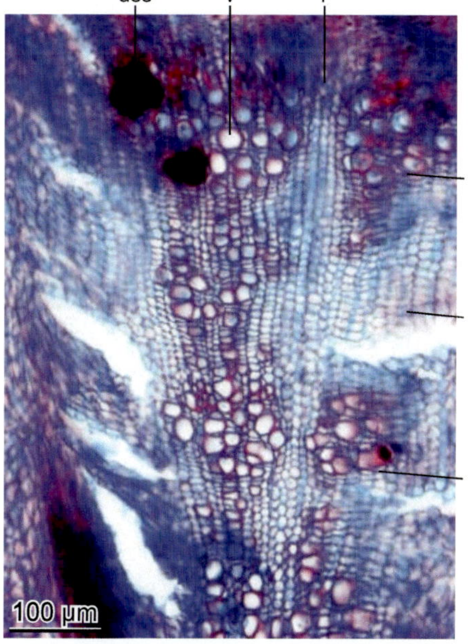

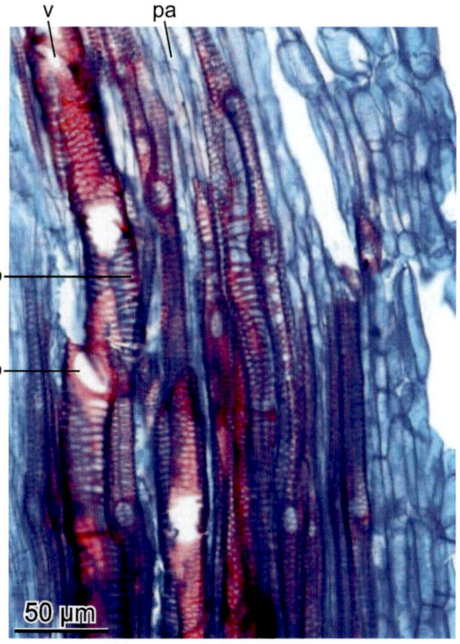

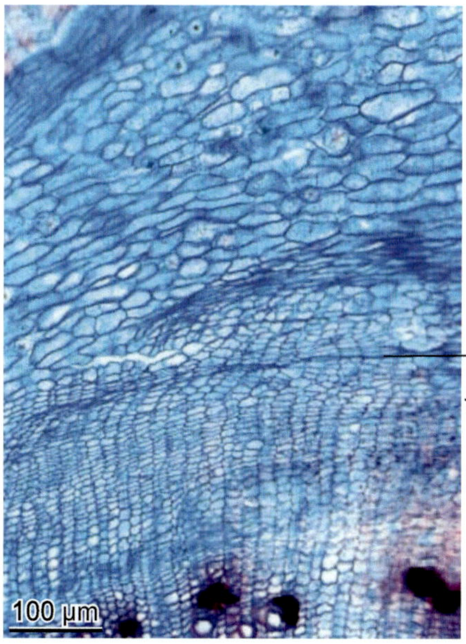

Features according to IAWA

Xylem 1 - 4 - 9.1 - 11 - 13 - 20 - 20 - 22 - 40.2 - 50.1 - 52.2 - 61 - 70 - 79.1 - 99 - 99.1 - 100.1 - 100.2 - 105 - 144

Ribes orientale Desf.

Grossulariaceae

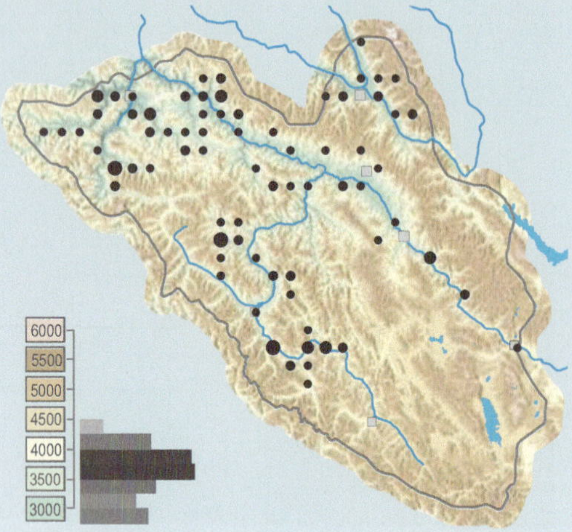

Description: Shrubs, 0.5–2 m tall, usually dioecious, pubescent, glandular and sticky, leaves suborbicular to reniform, 3–5-lobate, margin coarsely dentate, flowers greenish, 3–5 mm across, in dense racemes, berries red to purple, 7–9 mm across, pubescent and glandular.

Distribution: Greece, Turkey, Iraq, Lebanon, Syria, Iran, Armenia, Azerbaijan, Georgia, Russia, Afghanistan, Pakistan, N India, Nepal, China (Sichuan, Xizang, Yunnan), Bhutan.

Habitat: Stony river terraces, rock crevices.

Stem diameter 6–20 mm. Bark/xylem proportion 1:15. Ring width 0.12–0.21 mm. Maximal 47 annual rings.

Xylem

Transverse section

Semi-ring-porous. Rings distinctly marked by differences between earlywood (30–50 μm) and latewood vessel diameters (10–15 μm) and a few layers of tangentially flat fibers. Vessels angular, solitary or in small groups. Fibers thin- to thick-walled. Parenchyma absent.

Tangential section

Rays uniseriate and >10-seriate (ray dimorphism).

Radial section

Perforations scalariform with 5–12 bars. Intervessel pits round in alternating position. Vessel length 200–250 μm. Fiber pits round and relatively large (2–3 μm). Ray cells square and upright.

Bark

Phloem consists of tangential layers of sieve elements, parenchyma cells and fibers. Prismatic crystals adjacent to fibers. Phellem layered.

Pith

Round, filled with round, unlignified cells.

Analyzed: Root collar of polar root of 2 individuals.

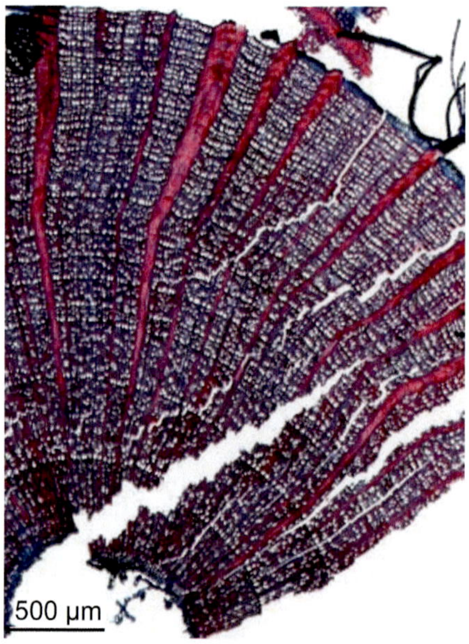

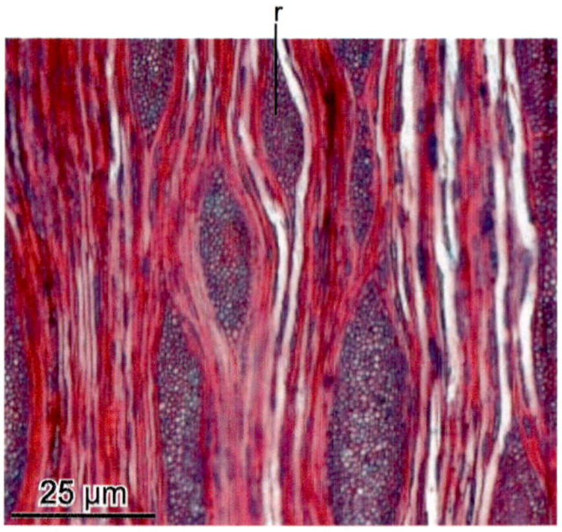

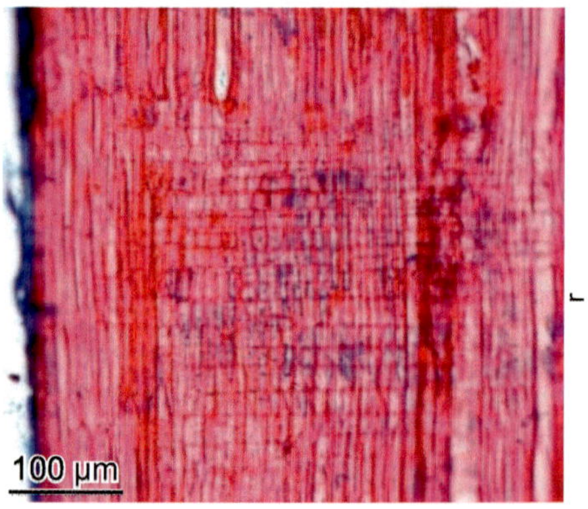

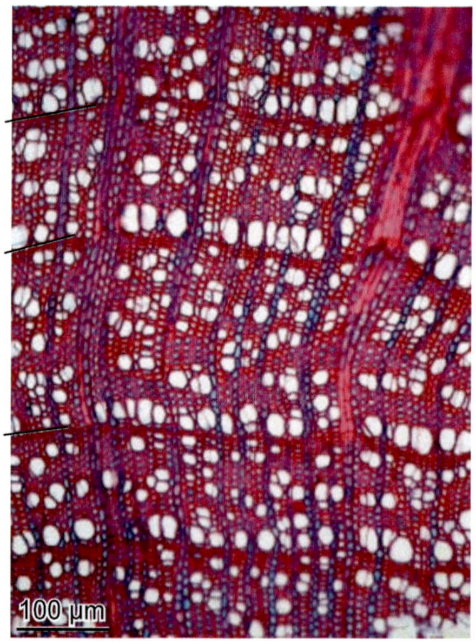

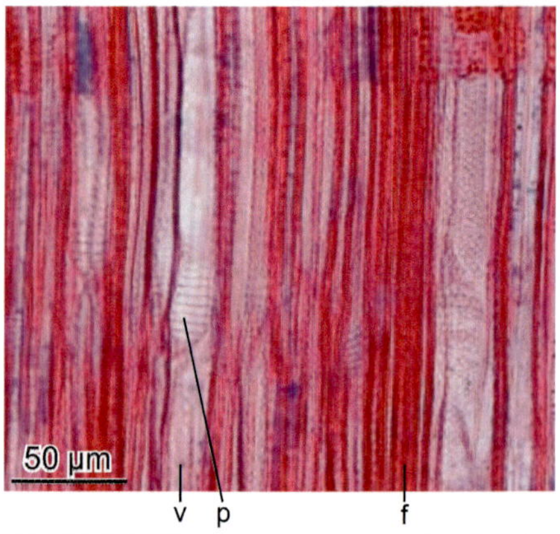

Features according to IAWA

Xylem 1 - 4 - 5 - 9 - 12 - 13 - 22 - 40.2 - 50.1 - 53.1 - 62 - 70 - 75 - 96 - 99 - 103 - 105

Hippuris vulgaris L.

Hippuridaceae

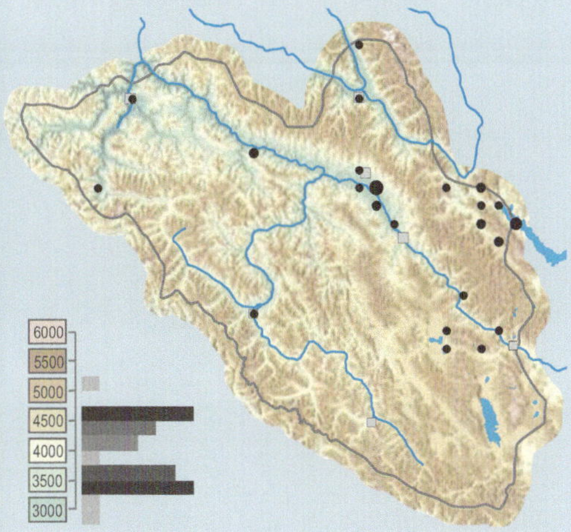

Description: Perennial herbs, 10–100 cm tall, fully or partly submerged in water, leaves linear, in whorls of 8–12, flowers minute, arranged axillary.

Distribution: Temperate regions of Europe, Asia, N Africa and N America.

Habitat: Fresh-water lakes, irrigation canals.

Diameter of rhizome 5–6 mm. Bark/xylem proportion 1:0.2. Mean ring width 0.1 mm. Maximal 1 annual ring.

Xylem

Transverse section

Only one ring. Xylem of vascular consists of angular, grouped, unlignified vessels and pervasive parenchyma. Vessel diameter 20–30 µm. Fibers and rays absent.

Tangential section

Rays absent.

Radial section

Perforations simple, intervessel pits scalariform. Vessel length 150–180 µm.

Bark

Phloem consists of large sieve tubes and groups of small companion cells. Endodermis of thin-walled cells surrounds xylem and phloem (stele). Cortex large, characterized by a netlike aerenchyma with well-delimited air cavities. Epidermis cells small, roundish, thin-walled.

Pith

Round, cells unlignified, round.

Analyzed: Rhizomes of 2 individuals.

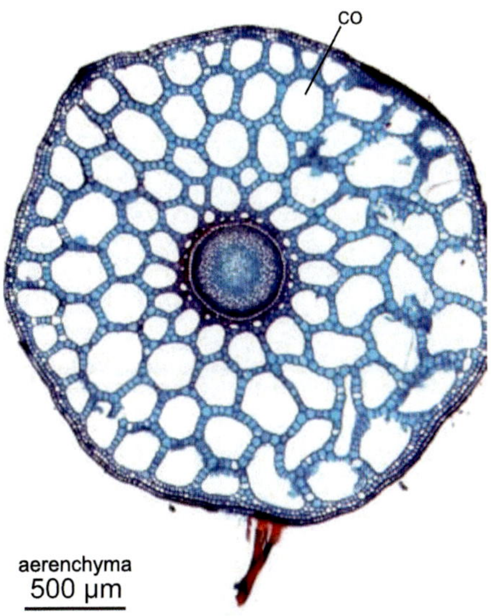

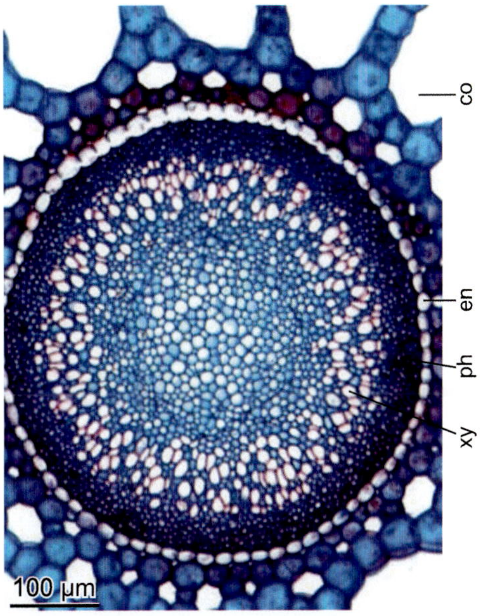

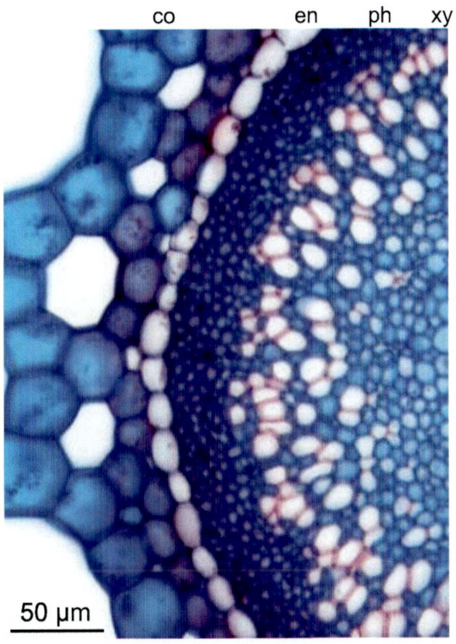

Features according to IAWA

Xylem 2.1 - 6 - 11 - 13 - 20 - 40.2 - 50.2 - 52.2 - 60.1 - 79.1 - 117

Dracocephalum heterophyllum Benth.

Lamiaceae

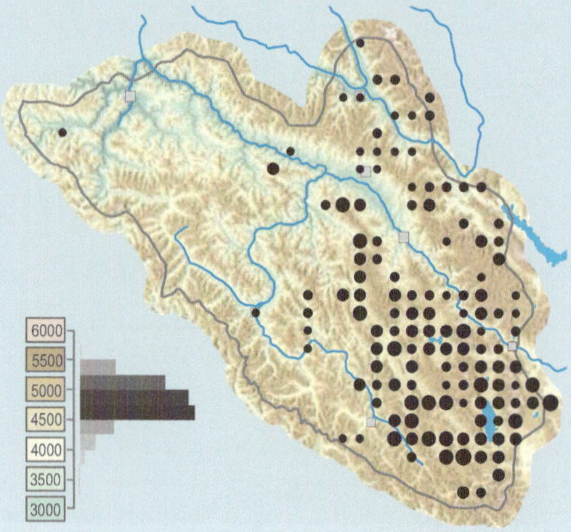

Syn.: *D. acanthoides* Edgew. ex Benth., *D. kaschgaricum* Rupr., *D. pamiricum* Briq.

Description: Perennial herbs, 10–20 cm tall, aromatic, pubescent, stem ascending or decumbent, leaves ovate, flowers white tinged with pink, ca. 25 mm long, in dense compact inflorescences.

Distribution: NE Afghanistan, N Pakistan, N India, Nepal, China (Gansu, Nei Mongol, Ningxia, Qinghai, Shanxi, Sichuan, Xinjiang, Xizang).

Habitat: Stony and gravel screes, sandy plains, stabilized slopes.

Stem diameter 5–8 mm. Bark/xylem proportion 1:2. Ring width 0.2–1.0 mm. Maximal 8 living annual rings. Stem segregation.

Xylem

Transverse section

Semi-ring-porous. Rings distinctly marked by large differences between earlywood (40–50 µm) and latewood vessel diameters (10–20 µm). Vessels solitary and in groups. Fibers absent. Parenchyma pervasive. Vascular bundles remain. Few irregular crystals in rays. With secondary intraxylary periderm towards the dead central part.

Tangential section

Rays >10-seriate, disappearing in polarized light. Rays absent within vascular bundles.

Radial section

Perforation plates simple. Intervessel pits round in alternating position or partially scalariform. Vessel length 60–100 µm. Ray cells upright.

Bark

Phloem with small groups of sieve tubes. With irregular crystals. Cortex missing. Phellem layered, consisting of a few rows of square cells.

Pith

No slide available.

Analyzed: Root collar of polar root of 7 individuals.

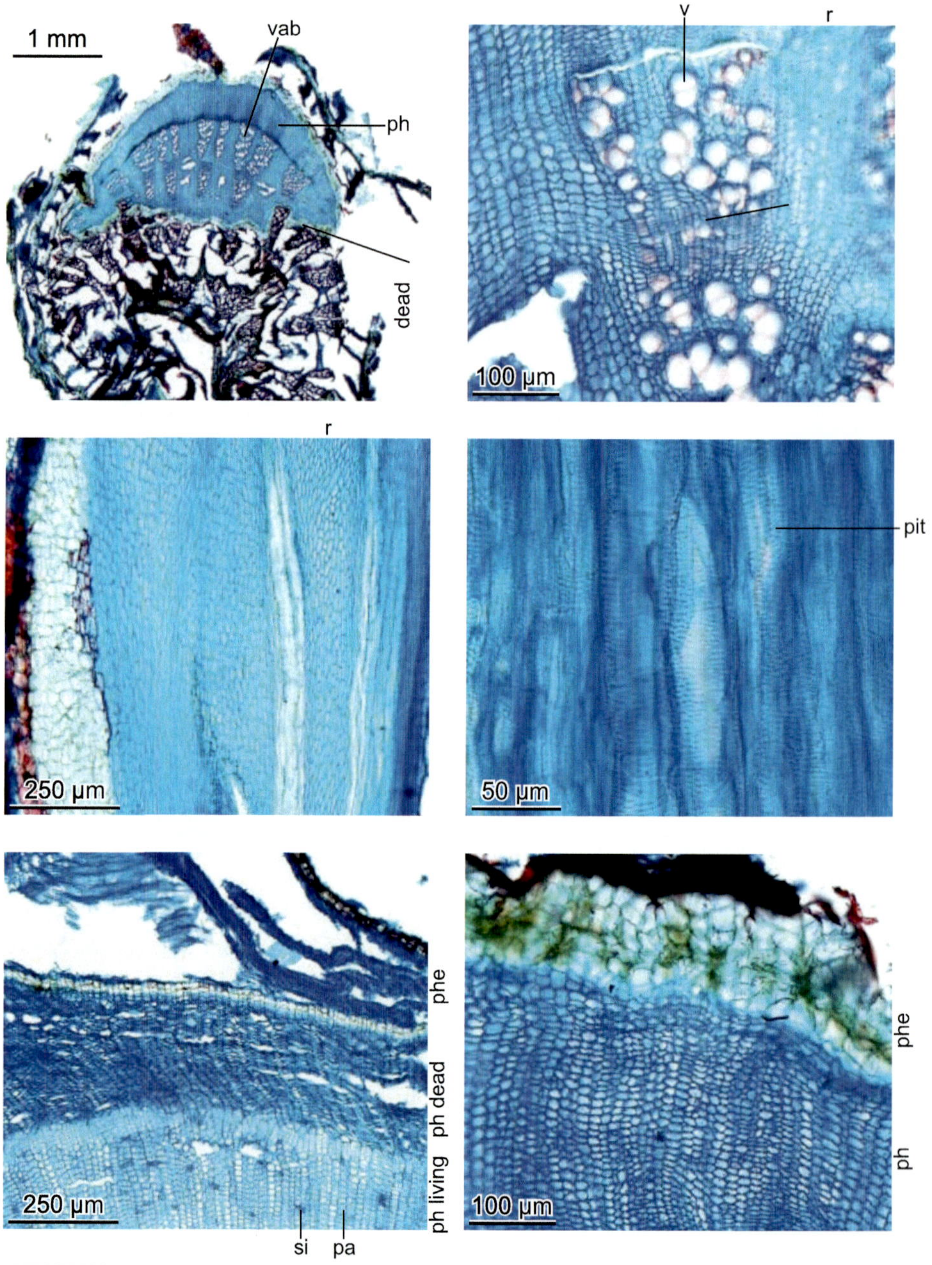

Features according to IAWA

Xylem 1 - 4 - 9 - 11 - 13 - 20 - 22 - 40.2 - 50.1 - 52.2 - 60.1 - 79.1 - 99 - 99.1 - 100.1 - 100.2 - 105 - 135.2

Dracocephalum stamineum Kar. & Kir.

Lamiaceae

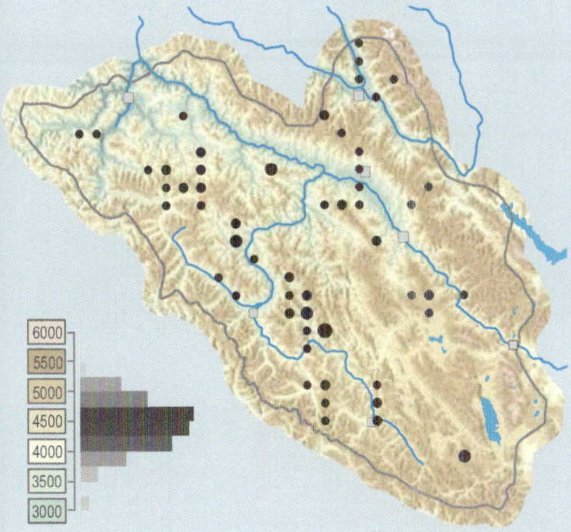

Syn.: *D. pulchellum* Briq., *Fedtschenkiella stamineа* (Kar. & Kir.) Kudr.

Description: Perennial herbs, 10–30 cm tall, stems numerous, purple, leaves broadly ovate with cordate base, flowers purple-blue with exserted stamens, 6–8 mm long, in compact inflorescences.

Distribution: Kazakhstan, Kyrgyzstan, Tajikistan, Afghanistan, Pakistan, India, China (Xinjiang, Xizang).

Habitat: Unstable screes and loamy slopes.

Stem diameter 3–4 mm. Bark/xylem proportion 1:2. Ring width 0.2–0.4 mm. Maximal 5 living annual rings. Stems lobed, segregating.

Xylem

Transverse section

Semi-ring-porous. Rings distinctly marked by large differences between earlywood (40–50 µm) and latewood vessel diameters (10–20 µm). Vessels in groups and solitar. Fibers absent. Parenchyma pervasive. Vascular bundles remain. Few irregular crystals in rays. With secondary intraxylary periderm towards the dead central part.

Tangential section

Rays >10-seriate, disappearing in polarized light. Rays absent within vascular bundles.

Radial section

Perforation plates simple. Intervessel pits round in alternating position or partially scalariform. Vessel length 60–100 µm. Ray cells upright.

Bark

Phloem with small groups of sieve tubes. Large rays dilate. With a few elongated and irregular crystals. Cortex missing. Phellem layered, consisting of a few rows of square cells.

Pith

No slide available.

Analyzed: Root collar of polar root of 4 individuals.

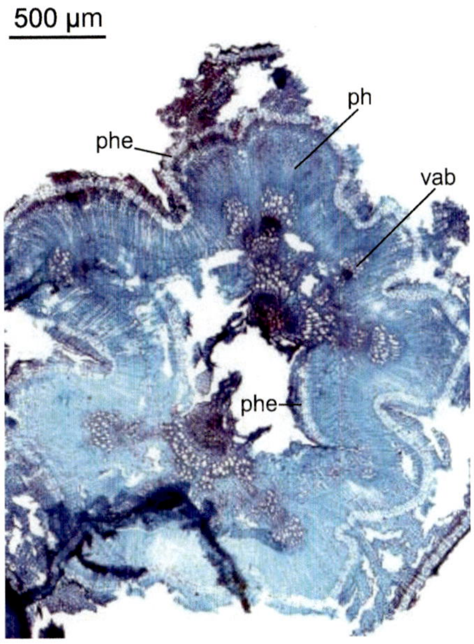

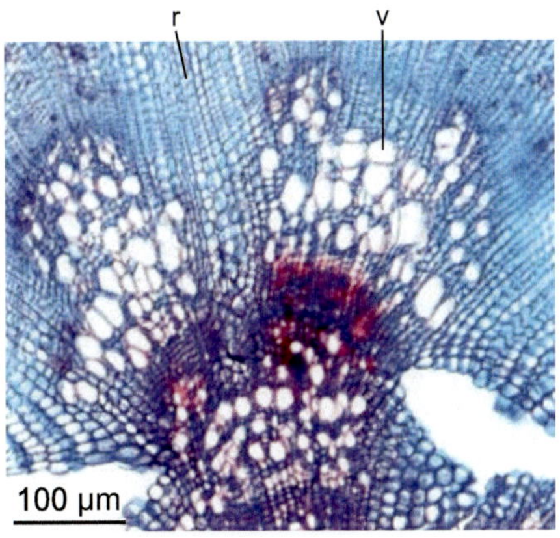

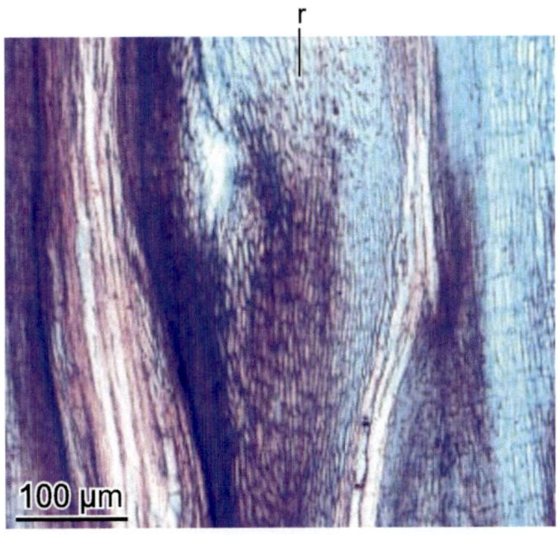

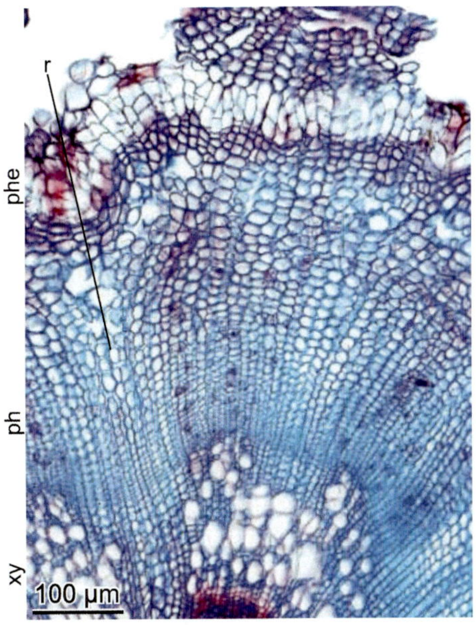

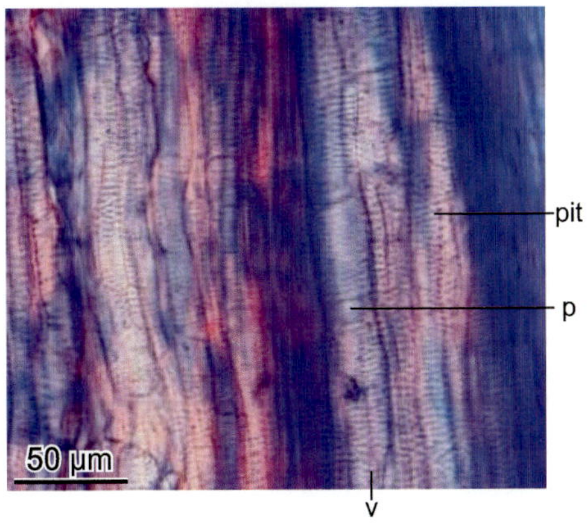

Features according to IAWA

Xylem 1 - 4 - 9 - 11 - 13 - 20 - 22 - 40.2 - 50.1 - 52.2 - 60.1 - 79.1 - 99 - 99.1 - 100.1 - 100.2 - 105 - 135.2

Elsholtzia ciliata (Thunb.) Hyl.

Lamiaceae

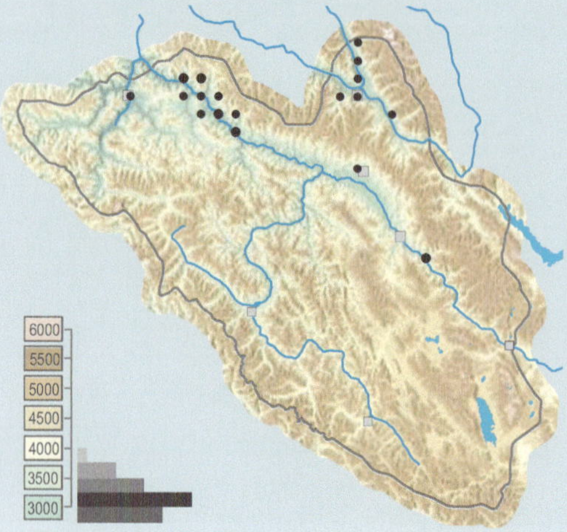

Syn.: *Sideritis ciliata* Thunb.

Description: Annuals, 30–50 cm tall, aromatic, sparsely pubescent, stems quadrangular, often purplish, leaves ovate to lanceolate, margin serrate, flowers purplish, ca. 4 mm long, in condensed verticillasters, composing terminal and axillary spikes.

Distribution: Russia, Afghanistan, N India, China, Mongolia, Korea, Japan, Cambodia, Laos, Myanmar, Thailand, Vietnam, Malaysia; naturalized in North America and Europe.

Habitat: River banks, forests.

Stem diameter 1 mm. Bark/xylem proportion 1:0.2. One annual ring.

Xylem

Transverse section

One ring. Vessel diameter 40–50 µm. Vessels mostly in short radial groups. Fibers thin- to thick-walled. Parenchyma absent. With a few acicular crystals.

Tangential section

Rays 2–3-seriate.

Radial section

Perforation plates simple. Intervessel pits round in alternating position. Vessel length 100–200 µm. Fiber pits very small.

Bark

Phloem with small groups of sieve tubes. Cortex consists of round parenchyma cells. With a few solitary fibers. Phellem absent. Crystals absent.

Pith

Pith round.

Analyzed: Root collar of 1 individual.

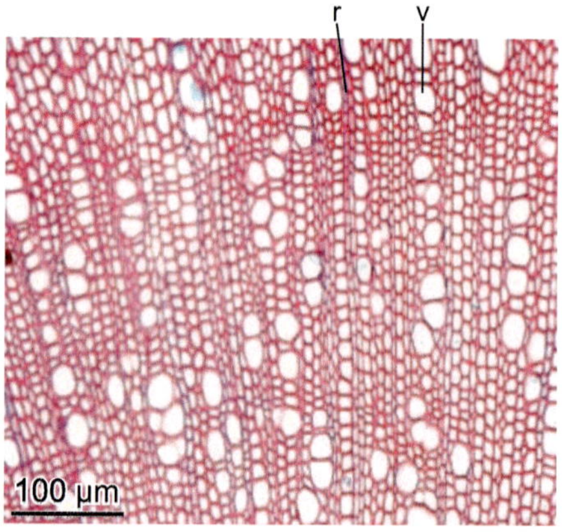

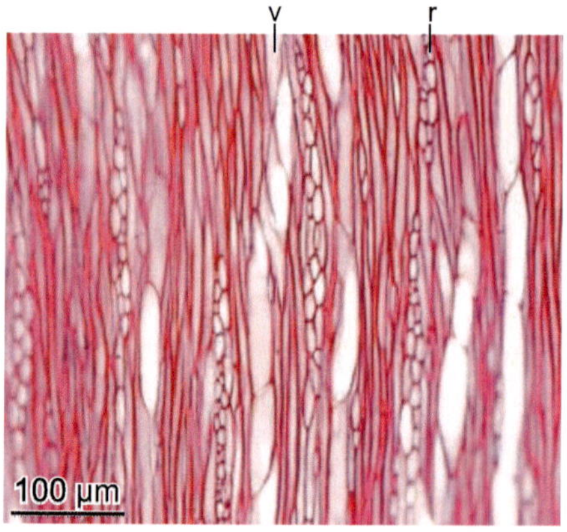

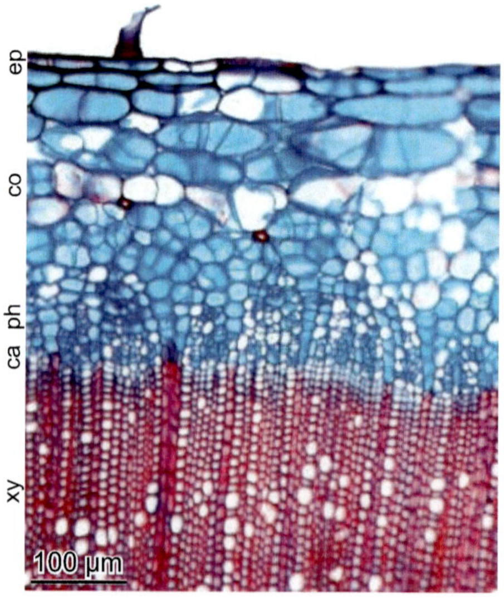

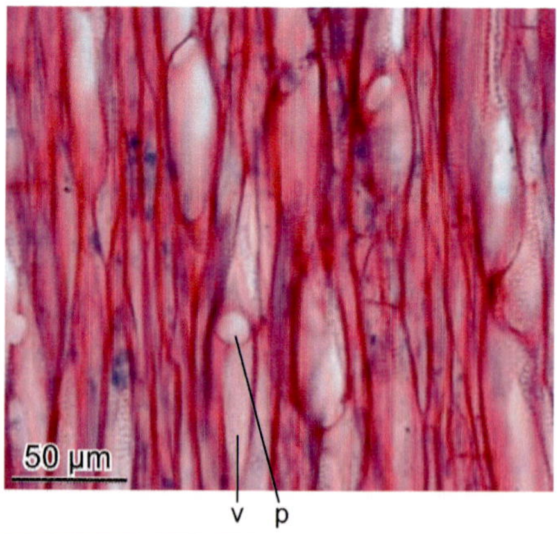

Features according to IAWA

Xylem 1 - 2.1 - 9.1 - 13 - 22 - 40.2 - 50.2 - 52.3 - 61 - 75 - 97 - 105 - 150

Elsholtzia eriostachya (Benth.) Benth.

Lamiaceae

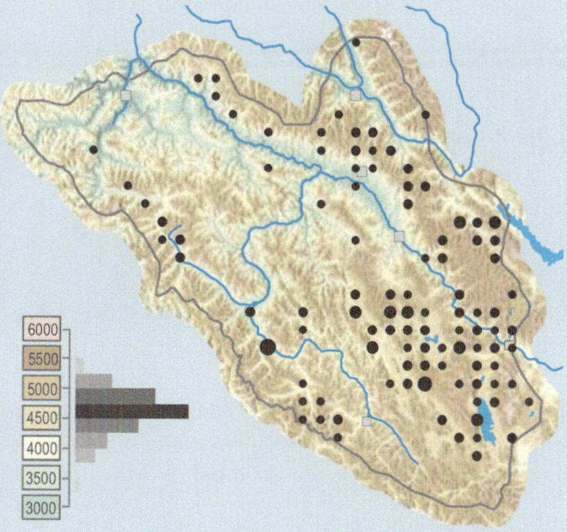

Syn.: *Aphanochilus eriostachyus* Benth.

Description: Annuals, 5–40 cm tall, aromatic, stems quadrangular, often purplish, puberulent, in larger plants branched, leaves oblong, opposite, margin crenate-serrulate, flowers yellow, ca. 2 mm long, in condensed verticillasters, composing terminal and axillary spikes.

Distribution: Pakistan, India, China (Gansu, Sichuan, Xizang, Yunnan), Bhutan.

Habitat: Mesic to dry disturbed habitats, stream gravel, sandy banks.

Stem diameter 1 mm. Bark/xylem proportion 1:0.5. Indistinct rings. Total age unknown.

Xylem

Transverse section

Rings absent. Vessels angular with a diameter of 25–40 µm, mostly in small groups. Fibers absent. Parenchyma paratracheal. Crystals absent.

Tangential section

Rays absent.

Radial section

Perforation plates simple. Intervessel pits round in alternating position. Vessel length 100–150 µm.

Bark

Phloem homogeneous. Cortex consists of round parenchyma cells and irregular large intercellulars. Phellem layered, consisting of several rows of square cells. Crystals absent.

Pith

Round.

Analyzed: Root collar of polar root of 5 individuals.

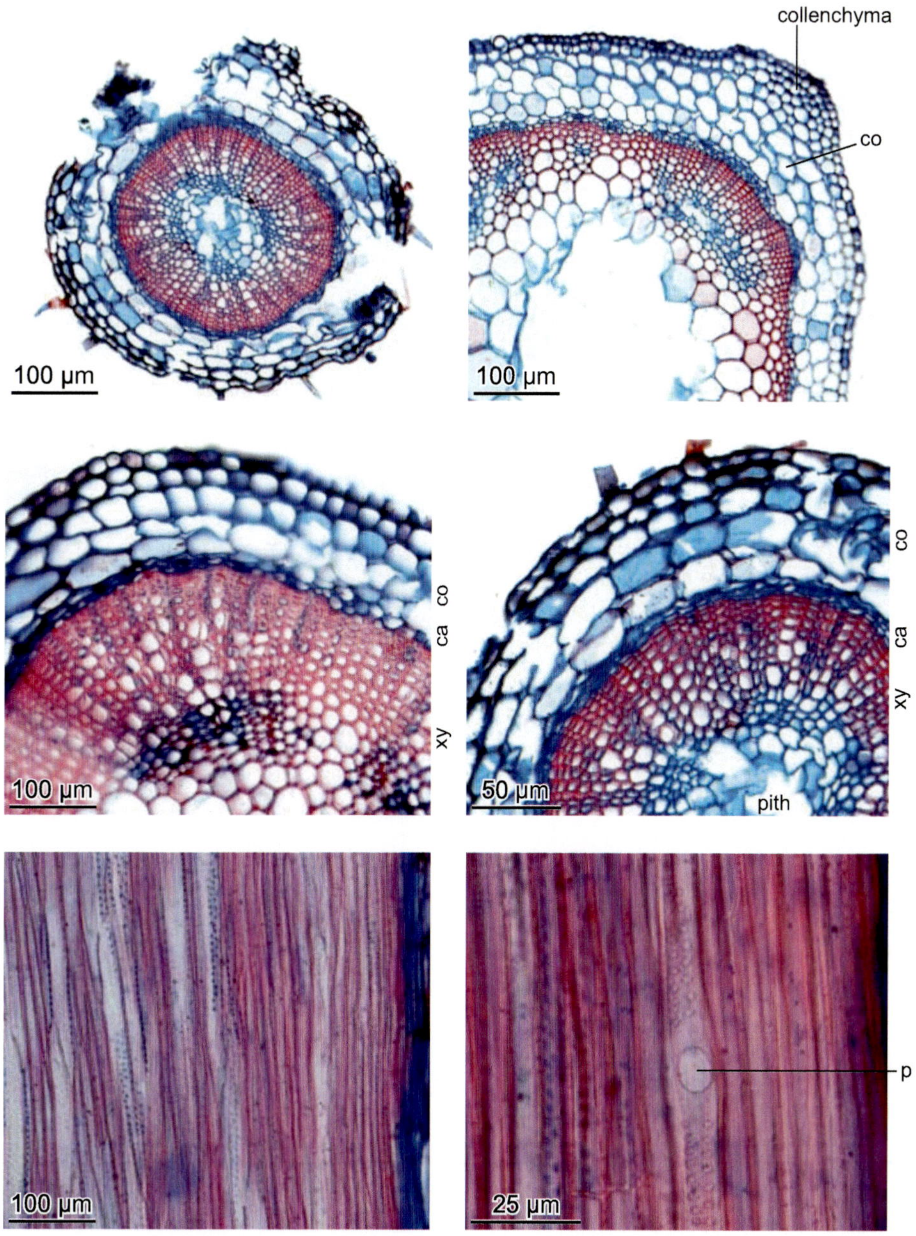

Features according to IAWA

Xylem 2 - 11 - 13 - 22 - 40.2 - 50.2 - 52.3 - 60.1 - 79 - 117

Marmoritis rotundifolia Benth.

Lamiaceae

Syn.: *Glechoma thibetica* (Jacq. ex Benth.) Briquet, *Nepeta rotundifolia* (Benth.) Benth., *N. thibetica* Jacq. ex Benth., *Phyllophyton thibeticum* (Jacq. ex Benth.) C.Y.Wu,

Description: Perennial herbs, 6–10 cm tall, stems ascending, leaves reniform to orbicular, petiolate, margin overlapping upwards, flowers white, ca. 2 cm long, usually solitary in leaf axils.

Distribution: Pakistan, N India, China (Xizang).

Habitat: Unstable gravel slopes, scree habitats.

Stem diameter 3–5 mm. Bark/xylem proportion 1:2. Ring width 0.2–1.0 mm. Maximal 6 living annual rings. Stem segregation.

Xylem

Transverse section

Semi-ring-porous. Rings distinctly marked by large differences between earlywood (40–50 µm) and latewood vessel diameters (5–10 µm). Vessels solitary and in groups. Fibers absent. Parenchyma pervasive. Vascular bundles remain. Without crystals. With secondary intra-xylary periderm towards the pith. Soon after stem and root formation radial growth occurs lobe-like, and the inner part of the root or rhizome disintegrates.

Tangential section

Rays >10-seriate, disappearing in polarized light. Rays absent within vascular bundles.

Radial section

Perforation plates simple. Intervessel pits round in alternating position or partially scalariform. Vessel length 60–100 µm. Ray cells upright.

Bark

Phloem with small groups of sieve tubes, arranged in indistinct tangential rows. Crystals absent. Phellem layered, consisting of one row of square cells (phelloids) and many round cells and large aerenchyma-like, irregular formed lacunae.

Pith

No slide available.

Analyzed: Root collar of polar root of 8 individuals.

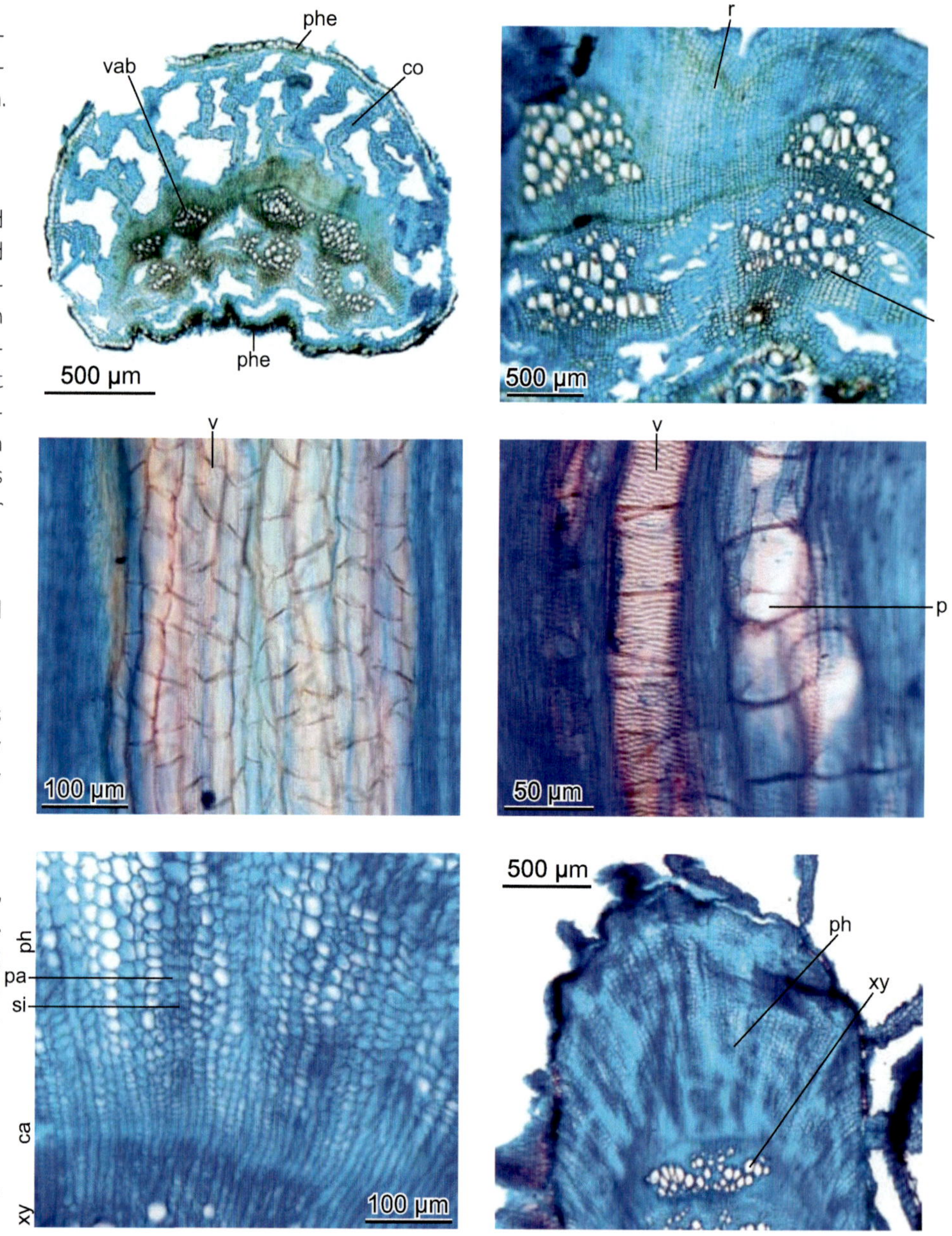

Features according to IAWA

Xylem 1 - 4 - 9 - 11 - 13 - 20 - 22 - 40.2 - 50.1 - 52.2 - 60.1 - 79.1 - 99 - 99.1 - 100.1 - 100.2 - 105 - 135.2

Mentha royleana Benth.

Lamiaceae

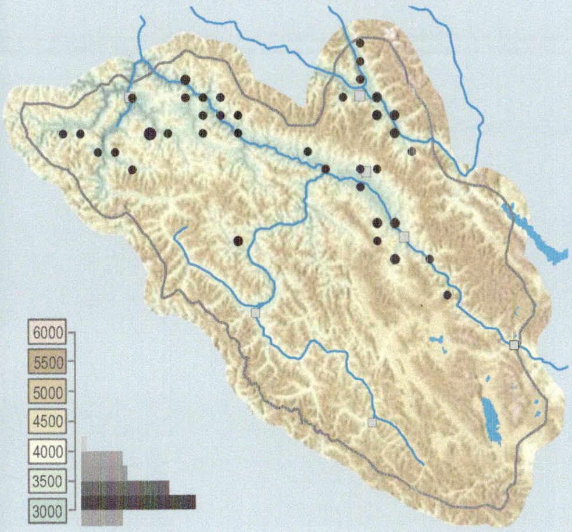

Description: Perennial herbs, 30–100 cm tall, aromatic, tomentose, leaves ovate to oblong-lanceolate, margin serrulate to coarsely dentate, flowers light purple, violet, mauve or white, ca. 2 mm across, arranged in numerous congested many-flowered verticillasters forming terminal spikes.

Distribution: Afghanistan, Pakistan, NW India, China (Xizang).

Habitat: Wet places along streams.

Stem diameter 2–2.5 mm. Bark/xylem proportion 1:3. Ring width 0.34–0.4 mm. Maximal 2 living annual rings.

Xylem

Transverse section

Rings distinct. Diffuse-porous. Rings marked by hardly lignified latewood fibers. Vessel diameter 40–50 µm. Vessels solitary and in short radial multiples. Fibers thin-walled. With septate fibers. Parenchyma absent or unrecognizable. Crystals absent.

Tangential section

Rays absent.

Radial section

Perforation plates simple. Intervessel pits round in alternating position. Vessel length 60–80 µm.

Bark

Phloem homogeneous. Cortex cells round to oval (40–50 µm). Phellem absent.

Pith

Round to oval, filled with unlignified cells.

Analyzed: Root collar of polar root of 2 individuals.

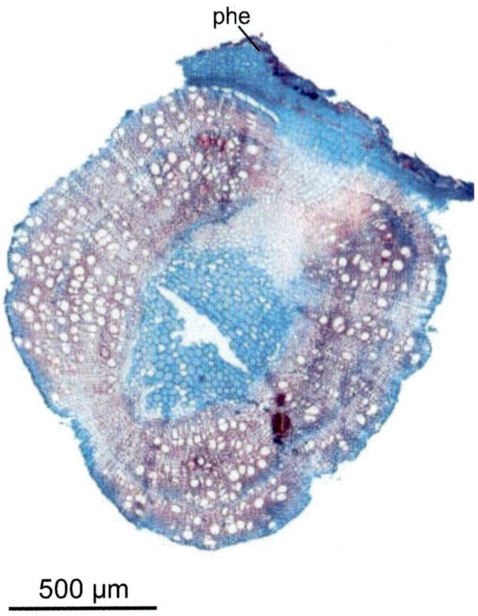

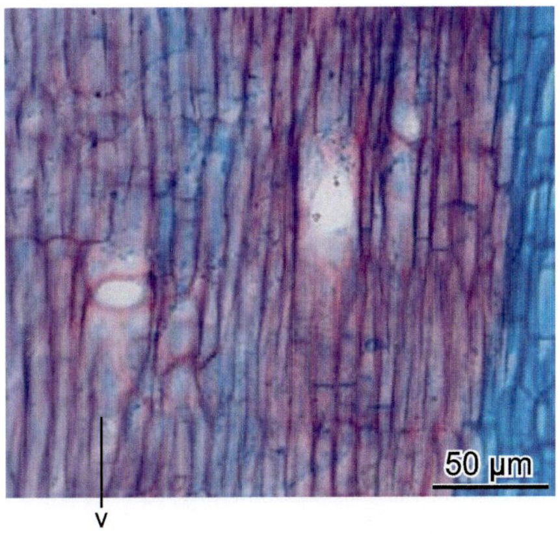

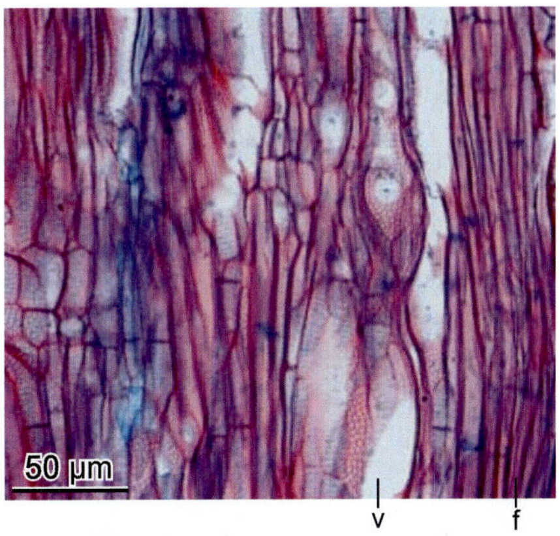

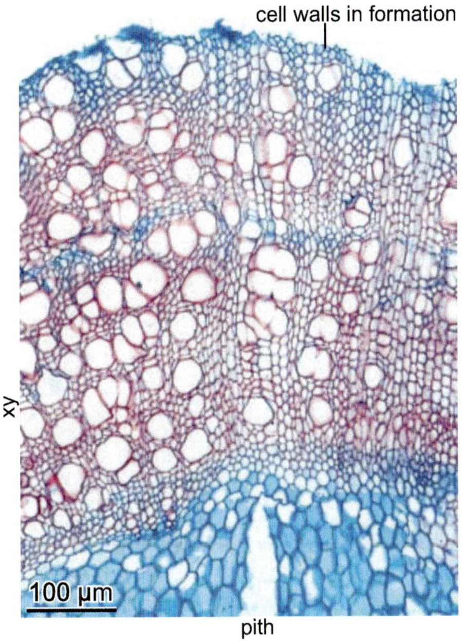

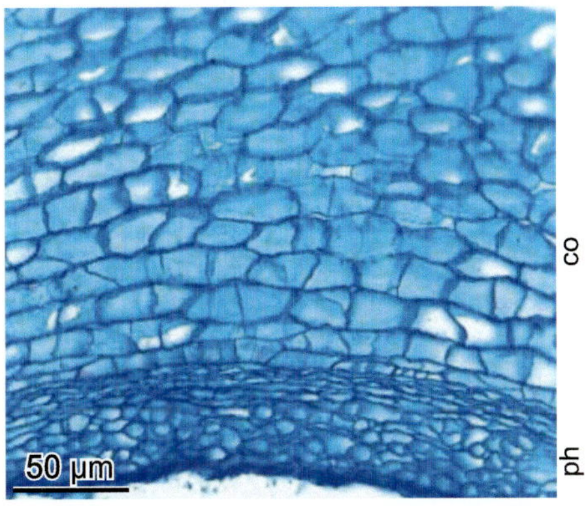

Features according to IAWA

Xylem 1 - 5 - 9.1 - 13 - 22 - 40.2 - 50.2 - 52.2 - 61 - 65 - 68 - 75 - 117

Nepeta coerulescens Maxim.

Lamiaceae

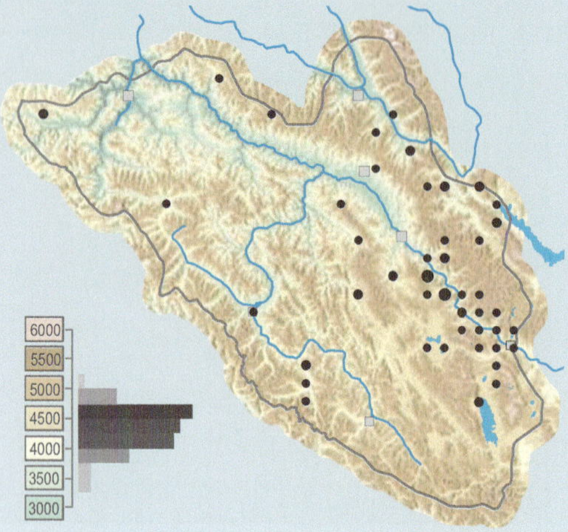

Syn.: *N. thomsonii* Benth. ex Hook.f.

Description: Perennial herbs, 15–40 cm tall, pubescent, aromatic, leaves oblong-lanceolate, margin crenate-serrate, flowers pale blue, ca. 1 cm long, arranged in verticillasters crowded in axils of uppermost leaves.

Distribution: Pakistan, NW India, China (Gansu, Qinghai, Sichuan, Xizang).

Habitat: Gravelly river beds, stony river banks, rocky slopes, field margins.

Stem diameter 4–10 mm. Bark/xylem proportion 1:1. Ring width 0.12–0.4 mm. Maximal 10 living annual rings.

Xylem

Transverse section

Semi-ring-porous. Rings marked by differences between earlywood (30–50 µm) and latewood vessel diameters (<20 µm). Vessels solitary and in small groups. Fibers absent. Parenchyma pervasive. Crystals absent. Vascular bundles remaining.

Tangential section

Rays >10-seriate between vascular bundles. Rays disappear in polarized light.

Radial section

Perforation plates simple. Intervessel pits round in alternating position. Vessel length 60–80 µm. Ray cells upright.

Bark

Phloem with small groups of sieve elements. Crystals absent. Cortex missing. Phellem layered, consisting of few rows of square cells.

Pith

No slide available.

Analyzed: Root collar of polar root of 3 individuals.

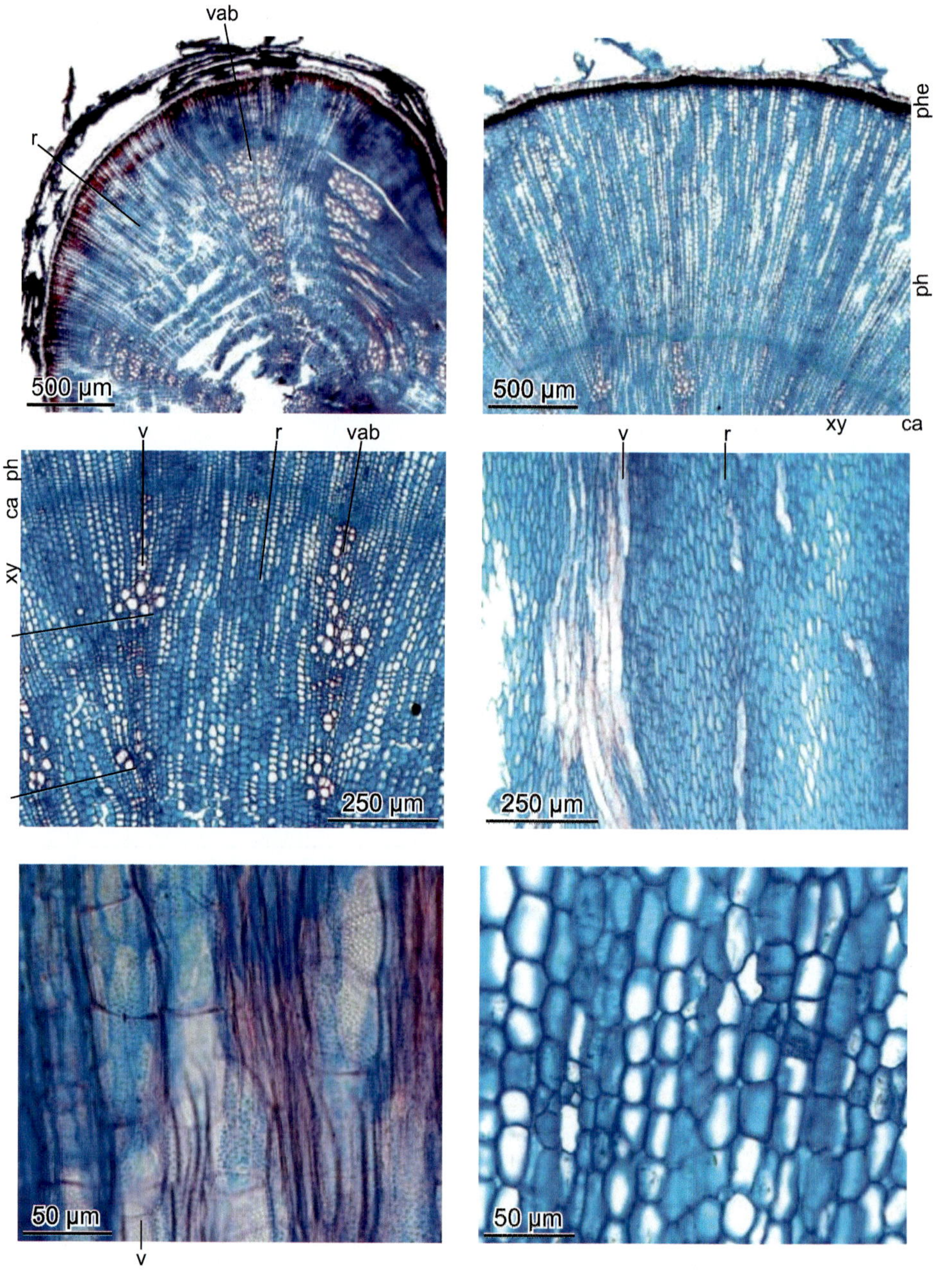

Features according to IAWA

Xylem 1 - 4 - 11 - 13 - 22 - 40.2 - 50.1 - 52.2 - 60.1 - 79.1 - 99.1 - 100.1 - 100.2 - 105

Nepeta discolor Royle ex Benth.

Lamiaceae

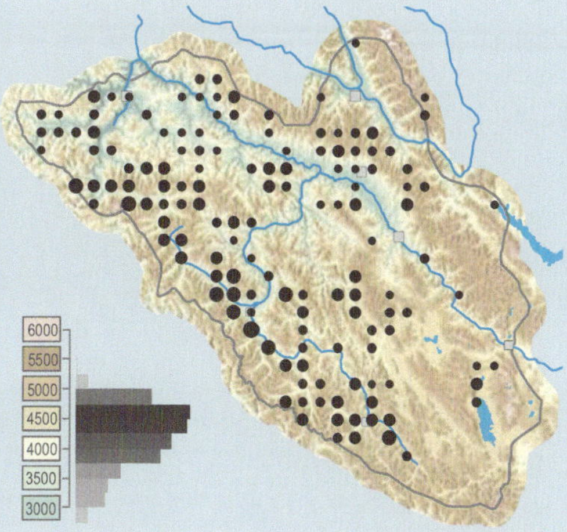

Description: Perennial herbs, 15–30 cm tall, gray-pubescent, stems ascending, leaves broad ovate, crenate or serrate, greenish above, villous or densely white tomentose below, flowers pale lilac or whitish, ca. 1 cm long, in many-flowered verticillasters.

Distribution: ?Tajikistan, E Afghanistan, N Pakistan, NW India (Jammu and Kashmir, Himachal Pradesh, Uttarakhand), Nepal, China (Xizang).

Habitat: Stabilized slopes, sandy sediments, steppes, exposed rocky slopes, often synanthropic along roads.

Stem diameter 5 mm. Bark/xylem proportion 1:2. Ring width 0.4–0.6 mm. More than 6 living annual rings. Total age unknown. Stem lobed. Stem segregation. Center rotten.

Xylem

Transverse section

Semi-ring-porous. Rings distinctly marked by large differences between earlywood (40–50 µm) and latewood vessel diameters (10–20 µm). Vessels in groups and solitary. Vascular bundles remain. Fibers thin- to thick-walled. Parenchyma paratracheal. Vascular bundles remain. Few irregular crystals in rays. With secondary intra-xylary periderm towards the dead central part.

Tangential section

Rays >10-seriate, disappearing in polarized light. Rays absent within vascular bundles.

Radial section

Perforation plates simple. Intervessel pits round in alternating position. Vessel length 150–200 µm. Ray cells upright.

Bark

Phloem with small groups of sieve tubes. With a few elongated and irregular crystals. Cortex missing. Phellem layered, consisting of very few rows of square cells.

Pith

No slide available.

Analyzed: Root collar of polar root of 2 individuals.

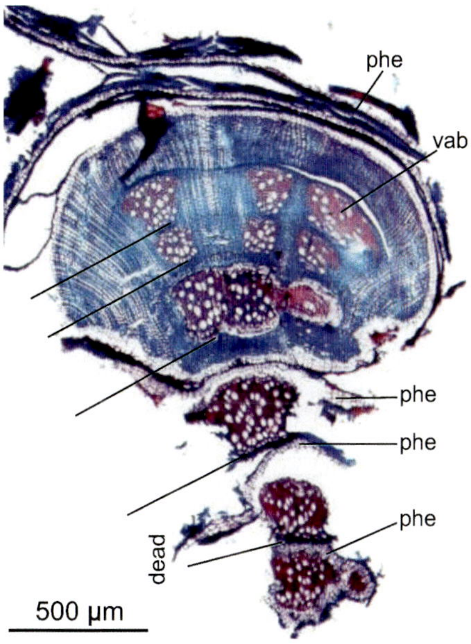

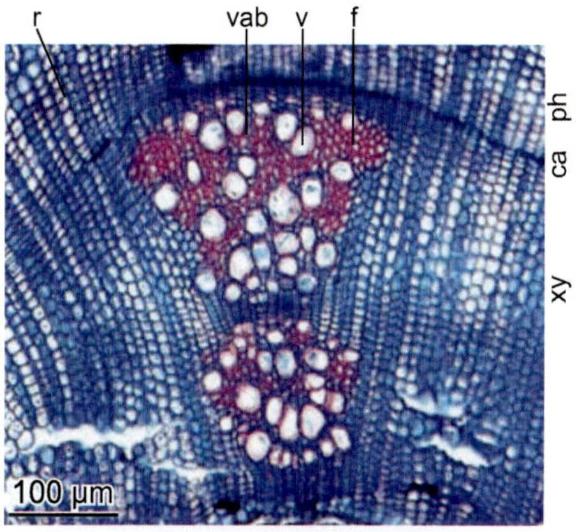

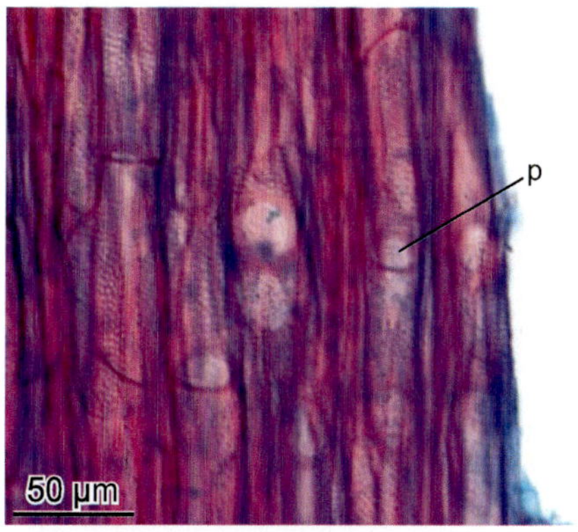

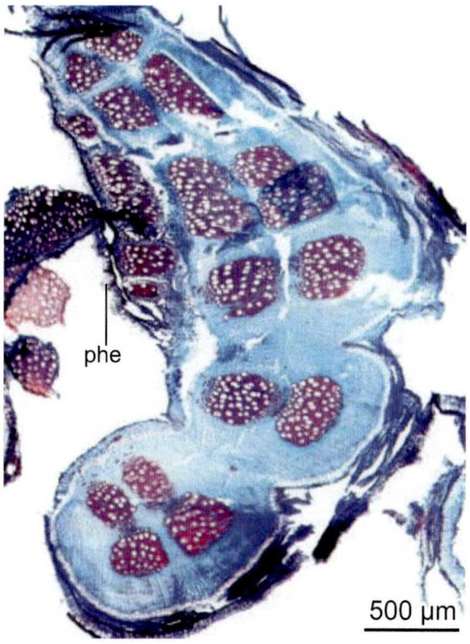

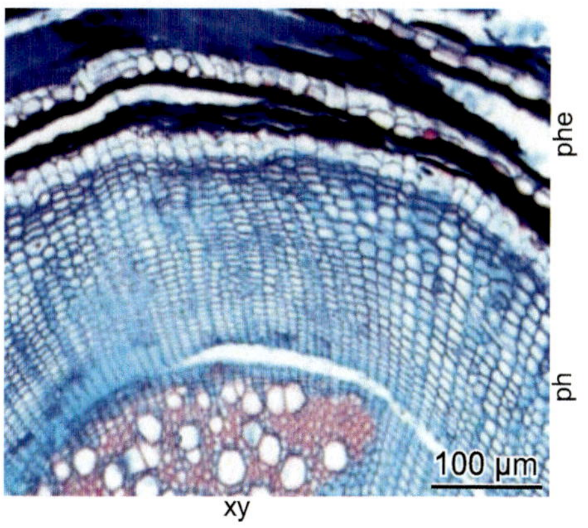

Features according to IAWA

Xylem 1 - 4 - 9 - 11 - 13 - 22 - 40.2 - 50.1 - 52.3 - 61 - 79 - 99 - 99.1 - 100.1 - 100.2 - 105 - 135.2

Nepeta floccosa Benth.

Lamiaceae

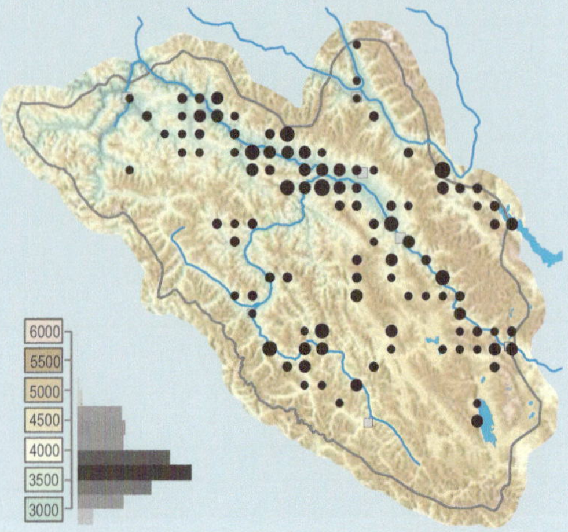

Description: Perennial herbs, 15–40 cm tall, aromatic, white tomentose, stems erect or ascending, quadrangular, leaves cordate, rigid, corrugated, margin crenulate, basal leaves clustered, long-petioled, upper leaves subsessile, flowers blue to pale violet, ca. 6 mm long, in 6–8-flowered capitate verticillasters.

Distribution: Afghanistan, N Pakistan, NW India (Jammu & Kashmir, Himachal Pradesh), China (Xinjiang, Xizang).

Habitat: Unstable loamy and sandy slopes, deserts.

Stem diameter 3–5 mm. Bark/xylem proportion 1:1. Ring width 0.2–1.0 mm. More than 7 living annual rings. Total age unknown. Stem segregation.

Xylem

Transverse section

Diffuse-porous. Rings marked by small differences between earlywood (50–80 μm) and latewood vessel diameters (40–50 μm). Vessels solitary and in groups. Vessel walls unstained. Fibers thin- to thick-walled. Parenchyma paratracheal. Few irregular crystals in rays. With secondary intra-xylary periderm towards the dead central part.

Tangential section

Rays 6–8-seriate, disappearing in polarized light.

Radial section

Perforation plates simple. Intervessel pits round in alternating position. Vessel length 80–120 μm. Ray cells upright.

Bark

Phloem homogeneous. With a few elongated and irregular crystals. Cortex missing. Phellem layered, consisting of very few rows of square cells.

Pith

No slide available.

Analyzed: Root collar of polar root of 5 individuals.

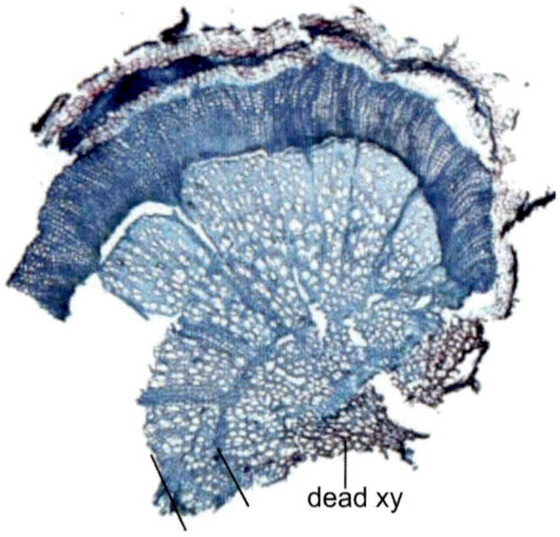

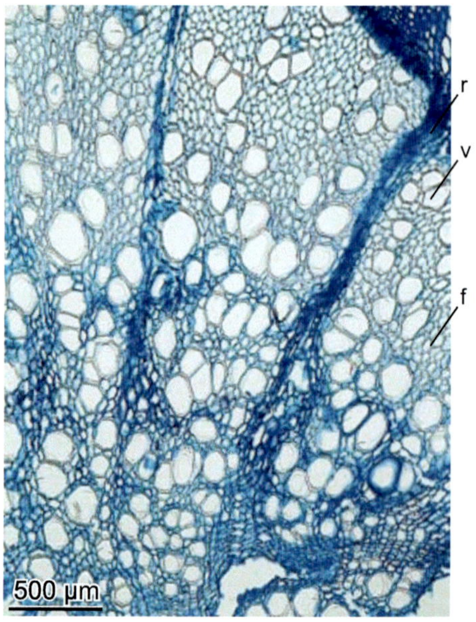

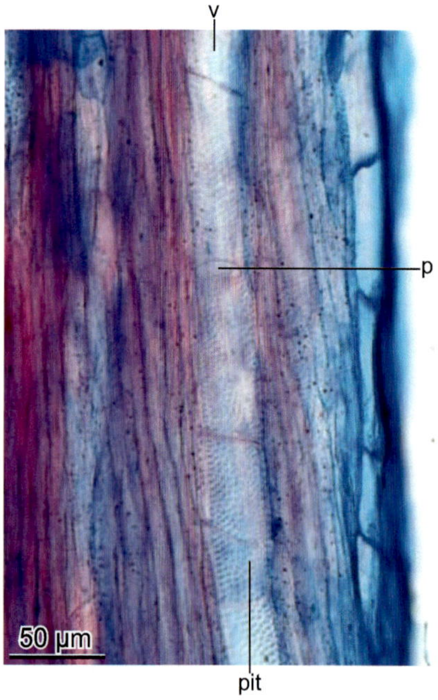

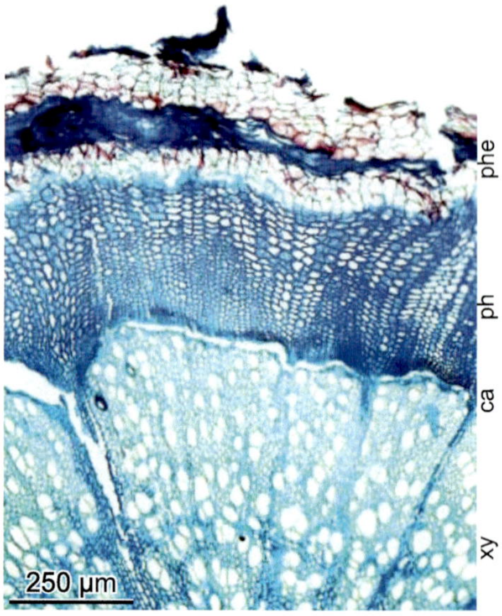

Features according to IAWA

Xylem 1 - 5 - 9 - 11 - 13 - 22 - 39.2 - 41 - 50.1 - 52.2 - 52.3 - 61 - 79 - 98 - 100.2 - 105 - 135.2

Nepeta glutinosa Benth.

Lamiaceae

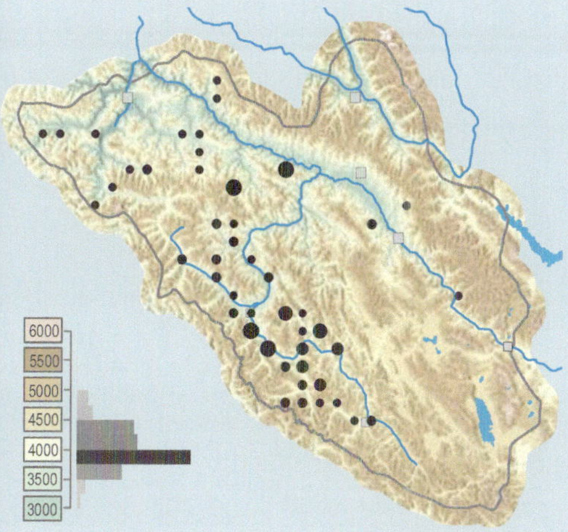

Description: Perennial herbs, 30–70 cm tall, aromatic, cespitose, densely glandular villous, leaves cordate-ovate, deeply serrate-incised, sessile and almost amplexicaul, flowers blue to purplish, ca. 2 cm long, in 4–5-flowered verticillasters.

Distribution: Tajikistan, Afghanistan, N Pakistan, NW India, China (Xinjiang).

Habitat: Stony river beds, both unstable and stabilized slopes, along roads.

Stem diameter 1–3 mm. Bark/xylem proportion 1:0.5. Rings indistinct. Total age unknown. Stem segregation.

Xylem

Transverse section

Diameter of larger vessels 50–80 µm. Vessels solitary and in groups. Secondary walls of fibers unstained, thin- to thick-walled. Vascular bundles remain. Parenchyma paratracheal (difficult to recognize). Crystals absent. With secondary intra-xylary periderm towards the central part.

Tangential section

Rays >10-seriate, disappearing in polarized light. Rays absent within the vascular bundle.

Radial section

Perforation plates simple. Intervessel pits round in alternating position. Vessel length 100–200 µm. Ray cells upright. Crystals absent.

Bark

Phloem homogeneous. Crystals absent. Cortex missing. Phellem layered, consisting of very few rows of square cells.

Pith

No slide available.

Analyzed: Root collar of polar root of 1 individual.

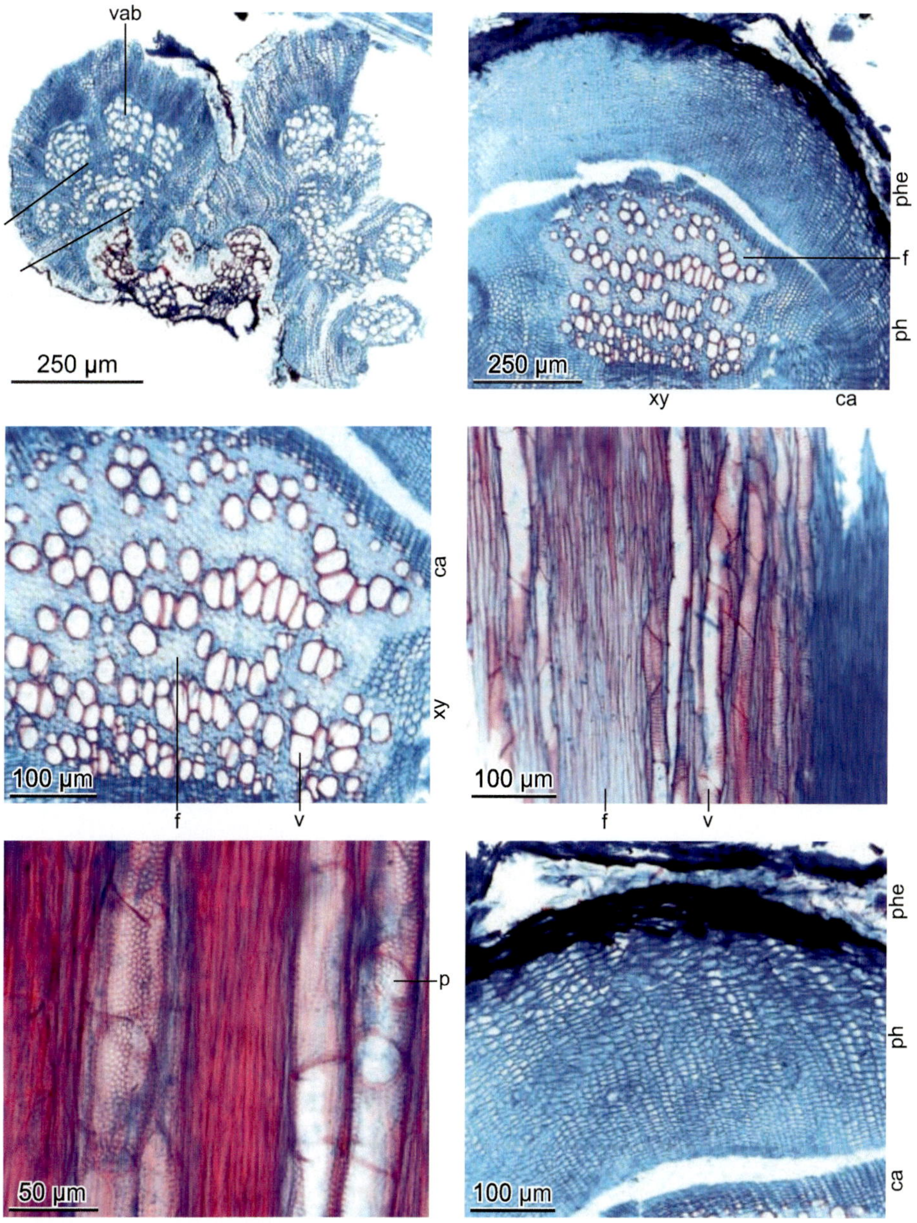

Features according to IAWA

Xylem 2 - 9 - 11 - 13 - 22 - 41 - 50.1 - 52.2 - 52.3 - 61 - 79 - 99 - 99.1 - 100.1 - 100.2 - 105 - 135.2

Nepeta longibracteata Benth.

Lamiaceae

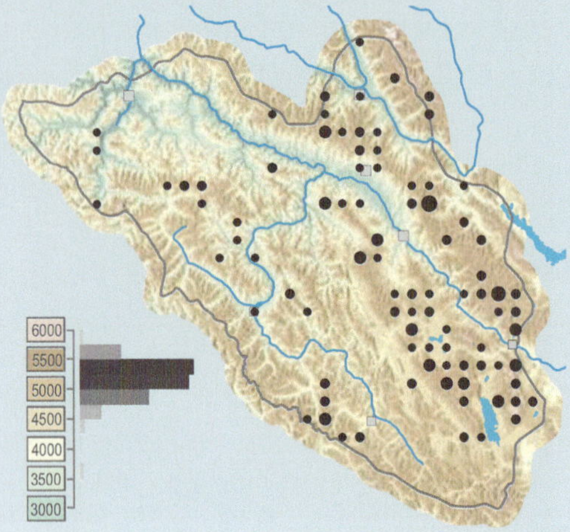

Syn.: *Glechoma longibracteata* (Benth.) Kuntze

Description: Perennial herbs, 8–15 cm tall, aromatic, tomentose, leaves ovate, margin coarsely crenate to lobed, flowers blue-violet, ca. 1.5 cm long, in a globose inflorescence, bracts longer than flowers, purplish, densely ciliate.

Distribution: Tajikistan, Afghanistan, Pakistan, India, China (Xinjiang, Xizang).

Habitat: Stony and gravel slopes, screes.

Stem diameter 1–2 mm. Bark/xylem proportion 1:2. Ring width 0.2–1.0 mm. Maximal 4 living annual rings. Total age unknown. Stem segregation. Center rotten.

Xylem

Transverse section

Diffuse-porous. Rings marked by small differences between earlywood (40–50 μm) and latewood vessel diameters (10–20 μm). Vessels solitary and in groups. Vascular bundles remain. Fibers thin- to thick-walled. Secondary walls of fibers unstained, thin- to thick-walled. Parenchyma paratracheal (difficult to recognize). Few irregular crystals in rays. With secondary intra-xylary periderm towards the dead central part.

Tangential section

Rays 6–>10-seriate, disappearing in polarized light.

Radial section

Perforation plates simple. Intervessel pits round in alternating position. Vessel length 80–120 μm. Ray cells upright.

Bark

Phloem homogeneous. Crystals absent. Cortex missing. Phellem layered, consisting of few rows of square cells.

Pith

No slide available.

Analyzed: Root collar of polar root of 2 individuals.

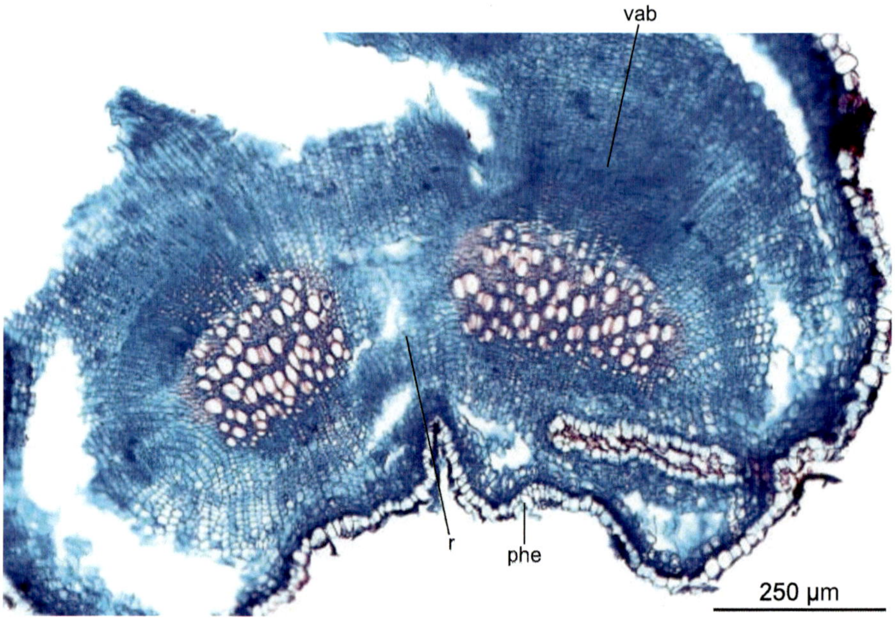

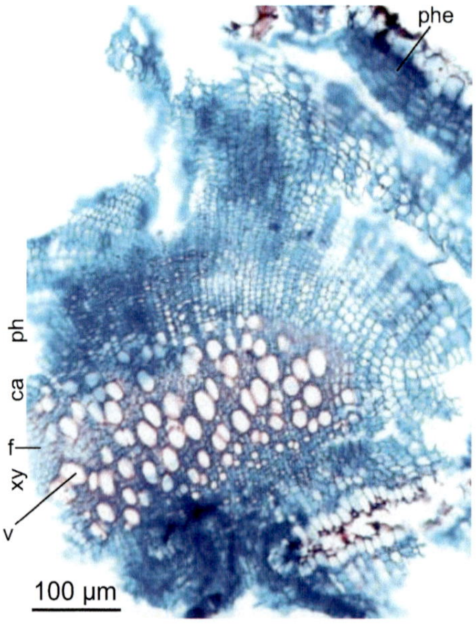

Features according to IAWA

Xylem 1 - 5 - 9 - 11 - 13 - 22 - 40.2 - 50.1 - 52.2 - 61 - 79 - 99 - 100.1 - 100.2 - 105 - 135.2

Perovskia abrotanoides Kar.

Lamiaceae

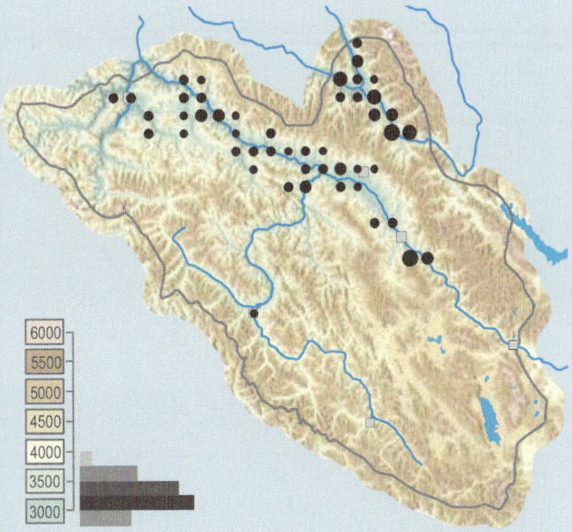

Syn.: *P. artemisioides* Boiss.

Description: Subshrubs, 0.3–1.5 m tall, aromatic, pubescent and glandular, branched from a thick woody rootstock, leaves bipinnatisect, segments oblong, below with a dense indumentum, flowers violet-blue, ca. 1 cm long, in 2–4-flowered verticillasters, composing a large branched inflorescence.

Distribution: Iran, Turkmenistan, Tajikistan, Afghanistan, Pakistan, India, China (Xizang).

Habitat: Dry riverbeds, among large stones.

Stem diameter 18 mm. Bark/xylem proportion 1:10. Ring width 0.5–1.3 mm. Maximal 17 living annual rings.

Xylem

Transverse section

Semi-ring-porous. Rings marked by differences between earlywood (40–50 µm) and latewood vessel diameters (20–30 µm), and by marginal parenchyma in the earlywood. Vessels mostly solitary or in short radial multiples. Fibers thin- to thick-walled. Parenchyma paratracheal. Crystals absent.

Tangential section

Rays 3–5-seriate.

Radial section

Perforation plates simple. Intervessel pits round in alternating position. Vessel length 100–120 µm. Ray cells square and upright.

Bark

Phloem homogeneous. Small acicular crystals in rays. Cortex missing. Phellem layered, consisting of several rows of square cells.

Pith

Round.

Analyzed: Root collar of polar root of 2 individuals.

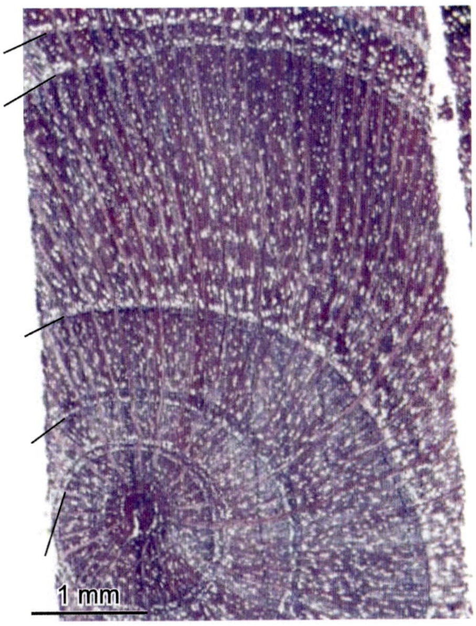

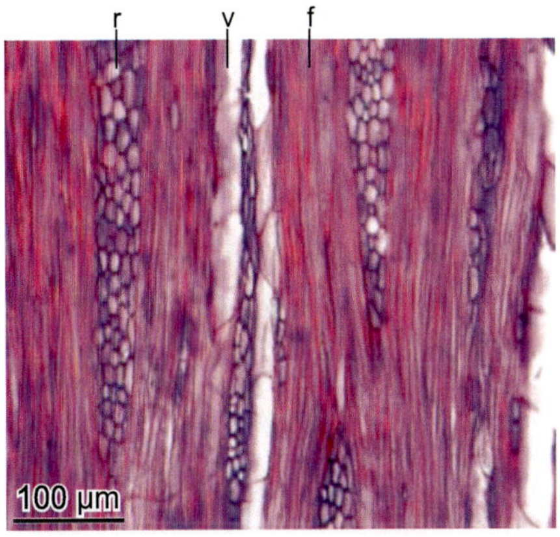

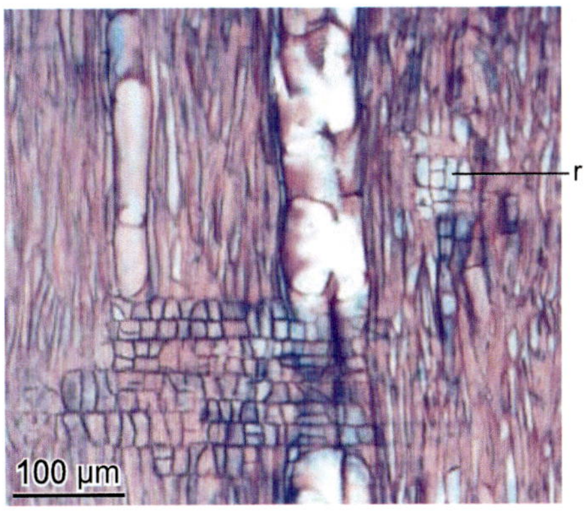

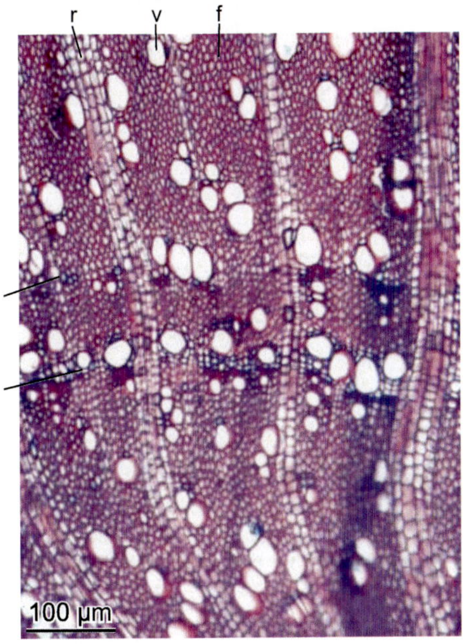

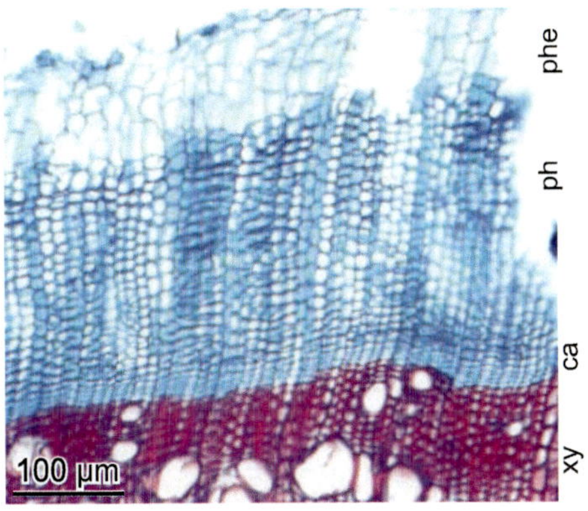

Features according to IAWA

Xylem 1 - 4 - 9.1 - 13 - 22 - 40.2 - 41 - 50.1 - 52.2 - 52.3 - 61 - 79 - 98 - 109

Scutellaria heydei Hook.f.

Lamiaceae

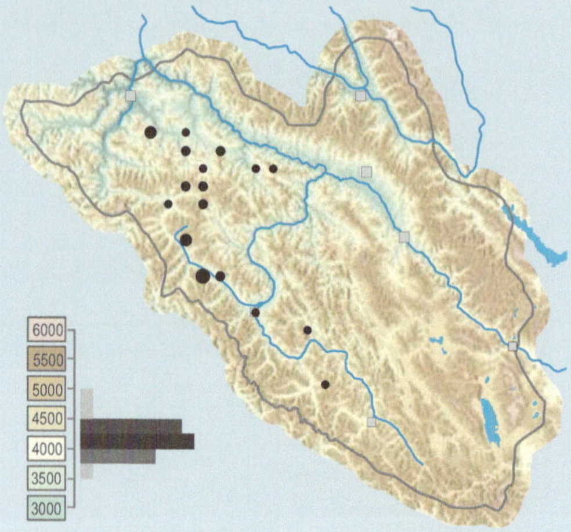

Description: Perennial herbs, 10–40 cm tall, hairy, stems prostrate to ascending, quadrangular, purplish, leaves broadly ovate or triangular, rugose, margin crenate, flowers creamy yellow with pink spots, 12–20 mm long, in a condensed 4-sided terminal inflorescence with densely villose bracts.

Distribution: NE Afghanistan, N Pakistan, NW India.

Habitat: Screes.

Stem diameter 5 mm. Bark/xylem proportion 1:2. Mean ring width 0.3 mm. Maximal 8 living annual rings.

Xylem

Transverse section

Semi-ring-porous. Rings marked by differences between earlywood (40–60 µm) and latewood vessel diameters (15–25 µm). Vessels mostly solitary or in small groups. Fibers thin- to thick-walled. Parenchyma paratracheal. Crystals absent.

Tangential section

Rays uniseriate, unlignified.

Radial section

Perforation plates simple. Intervessel pits round in alternating position. Vessel length 100–120 µm. Ray cells upright.

Bark

Phloem homogeneous. Phellem layered, consisting of several rows of square cells. Small prismatic crystals in phellem.

Pith

Round.

Analyzed: Root collar of polar root of 2 individuals.

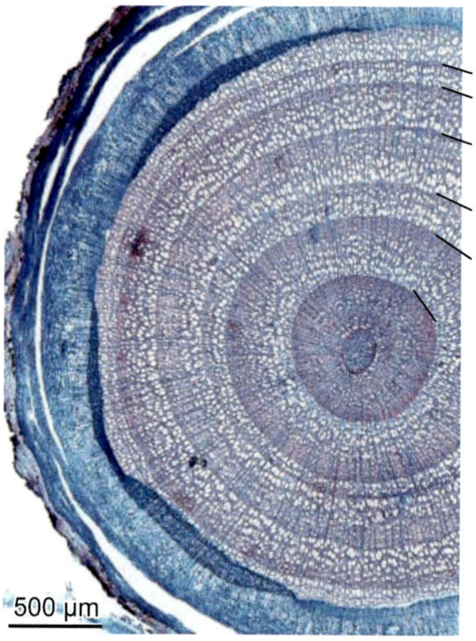

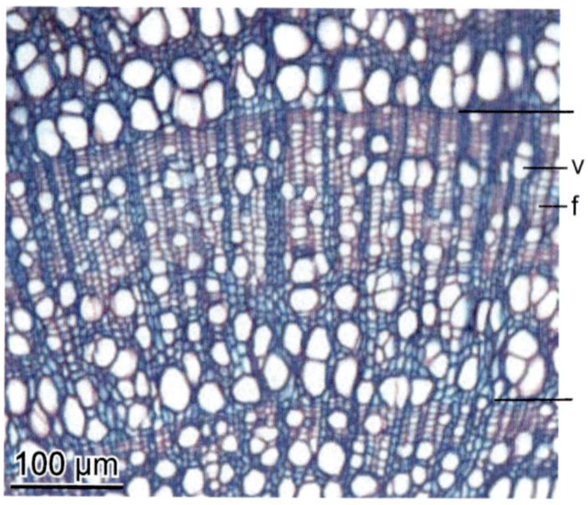

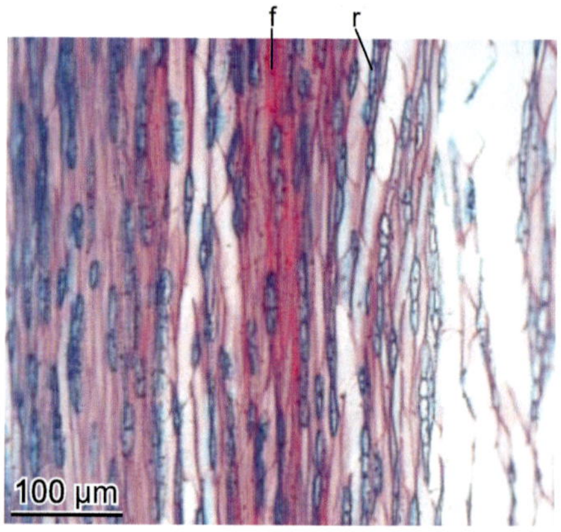

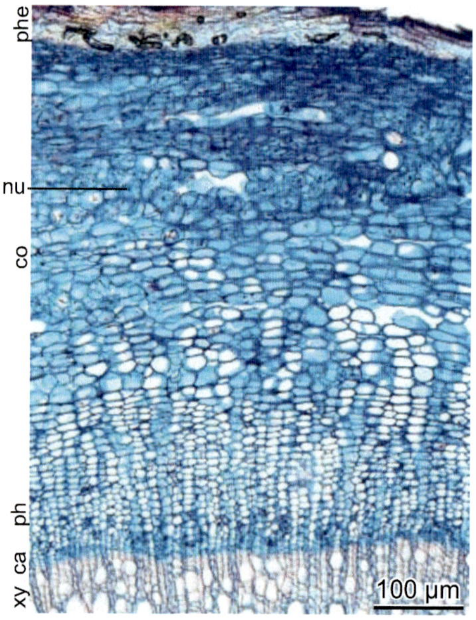

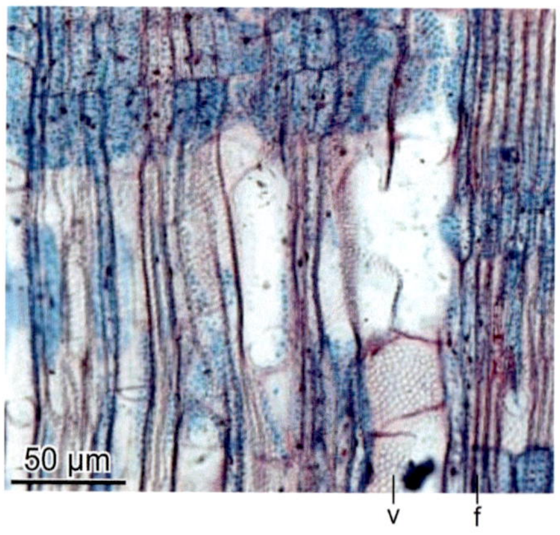

Features according to IAWA

Xylem 1 - 5 - 9 - 11 - 13 - 22 - 40.2 - 41 - 50.1 - 52.2 - 52.3 - 61 - 79 - 98 - 105

Menitskia tibetica (Vatke) Krestovsk.

Lamiaceae

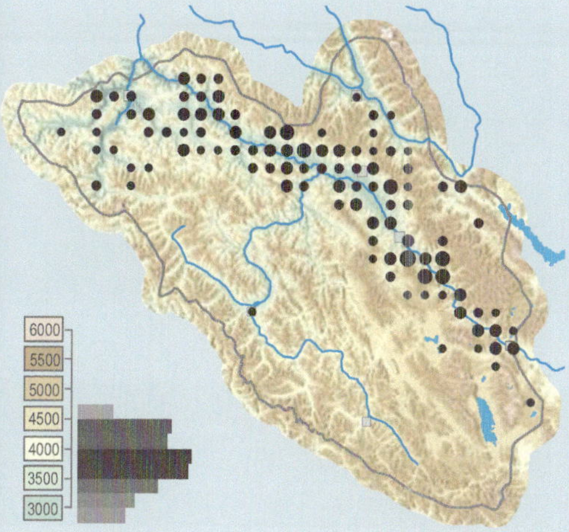

Syn.: *Stachys tibetica* Vatke

Description: Perennial herbs, 20–50 cm tall, stems quadrangular, branched, leaves oblong, opposite, flowers pink, 16–20 mm long, in distant 2–6-flowered verticillasters, bracts and calyx teeth spiny.

Distribution: N Pakistan, NW India, ?China (?W Xizang).

Habitat: Dry unstable slopes and other stony habitats.

Stem diameter 3–6 mm. Bark/xylem proportion 1:2. Mean ring width 0.8 mm. Maximal 6 living annual rings. Total age unknown.

Xylem

Transverse section

Semi-ring-porous. Rings distinctly marked by large differences between earlywood (40–50 µm) and latewood vessel diameters (5–10 µm). Vessels solitary and in groups. Fibers thin- to thick-walled. Parenchyma paratracheal and pervasive. Vascular bundles remain. With many small elongated crystals in all ray cells. With secondary intra-xylary periderm towards the pith. Soon after stem and root formation, radial growth occurs lobe-like and the inner part of the root or rhizome disintegrates.

Tangential section

Rays >10-seriate, disappearing in polarized light. Rays absent within vascular bundles.

Radial section

Perforation plates simple. Intervessel pits round in alternating position. Vessel length 60–100 µm. Ray cells upright.

Bark

Phloem with small groups of sieve tubes which are arranged in indistinct tangential rows. Cortex cells round to oval (40–50 µm). Phellem layered, consisting of one row of square cells (phelloids), many round cells and large aerenchyma-like, irregularly formed lacunae.

Pith

No slide available.

Analyzed: Root collar of polar root of 1 individual.

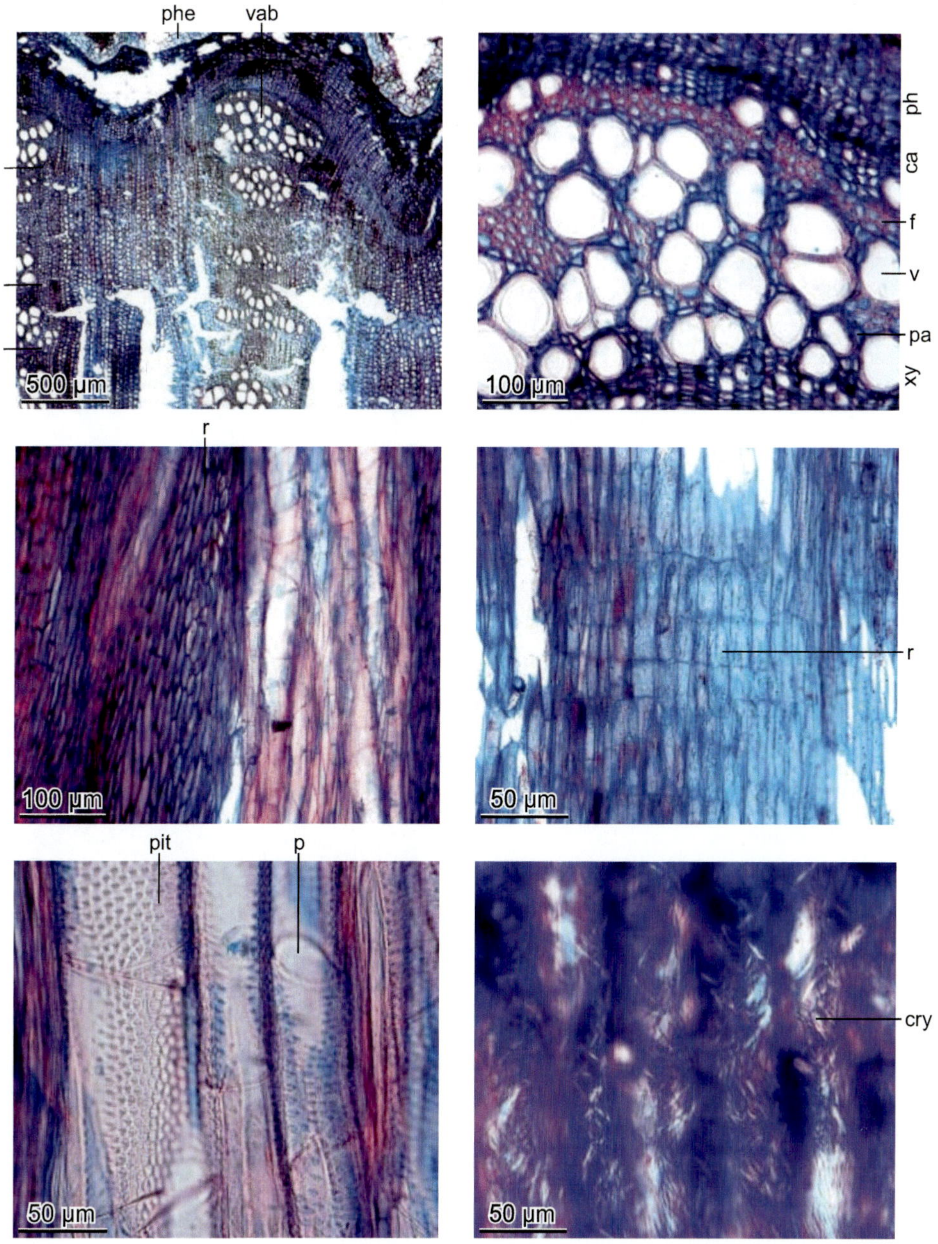

Features according to IAWA

Xylem 1 - 4 - 9 - 11 - 13 - 22 - 40.2 - 50.1 - 52.2 - 60.1 - 79.1 - 99 - 99.1 - 100.1 - 100.2 - 105

Thymus linearis Benth.
Lamiaceae

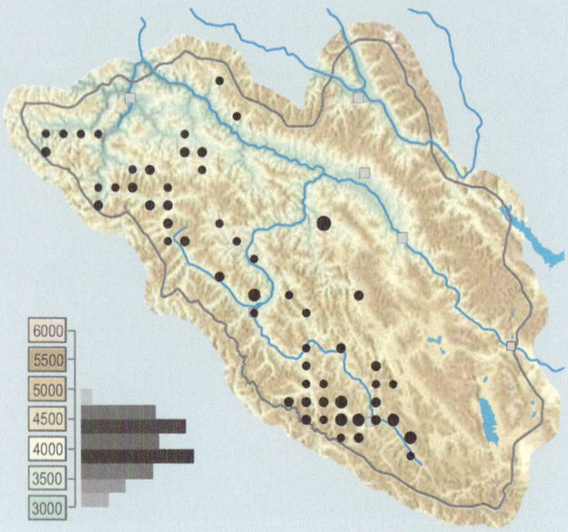

Description: Subshrubs, 2–6 cm tall, aromatic, cespitose, stems creeping, quadrangular, leaves elliptic-obovate to linear-lanceolate, flowers lilac to purplish, ca. 6 mm long, in verticillasters forming capitate inflorescences.

Distribution: NE Afghanistan, Pakistan, NW India, W Nepal.

Habitat: Dry to mesic stabilized slopes, rock crevices.

Stem diameter 1–3 mm. Bark/xylem proportion 1:2. Bark/xylem proportion 1:15. Ring width 0.1–0.25 mm. Maximal 9 annual rings.

Xylem

Transverse section

Rings very distinct. Semi-ring-porous. Rings marked by differences between earlywood (30–50 μm) and latewood vessel diameters (15–20 μm) and few rows of thick-walled latewood fibers. Vessels angular and solitary or in small groups. Fibers thin- to thick-walled. Parenchyma paratracheal. Crystals absent.

Tangential section

Rays uniseriate, cells spindle-like.

Radial section

Perforation plates simple. Intervessel pits round in alternating position. Vessel length 50–100 μm. Ray cells upright.

Bark

Phloem homogeneous. Phellem layered, consisting of a large zone of square cells. Crystals absent.

Pith

Round, empty.

Analyzed: Root collar of polar root of 4 individuals.

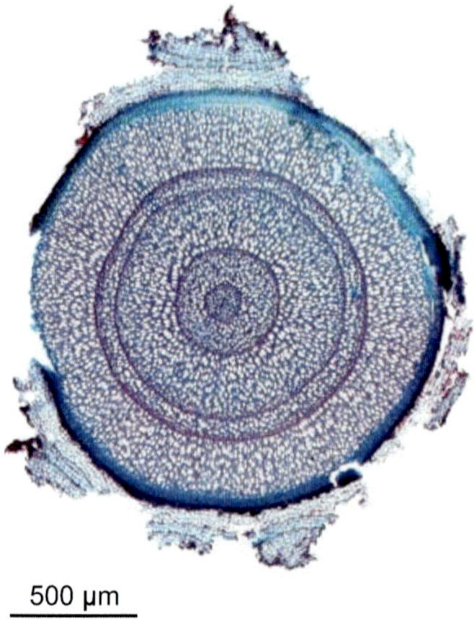

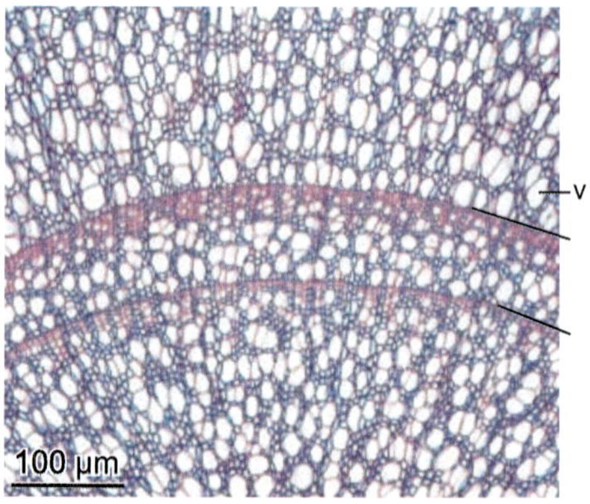

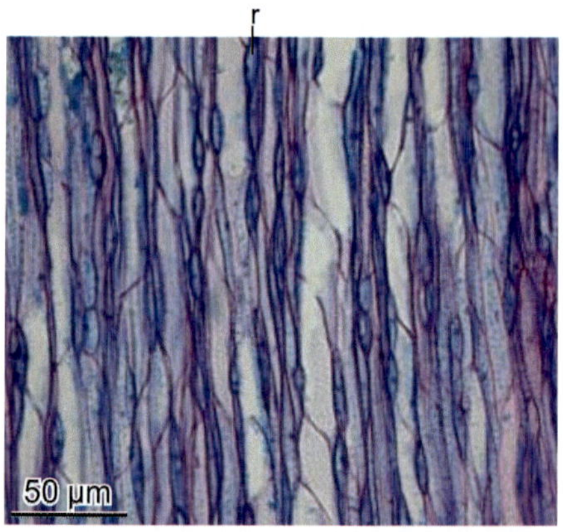

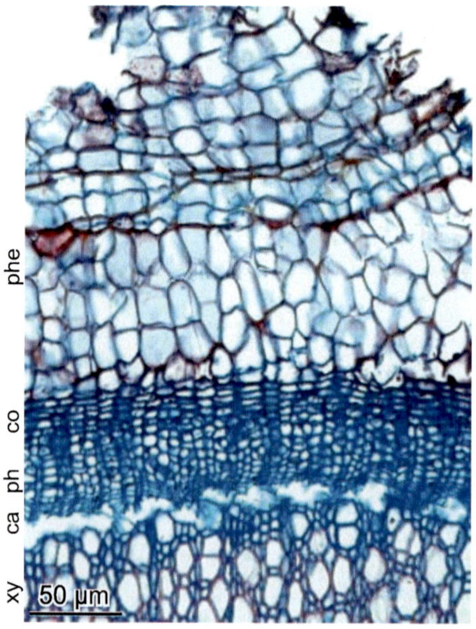

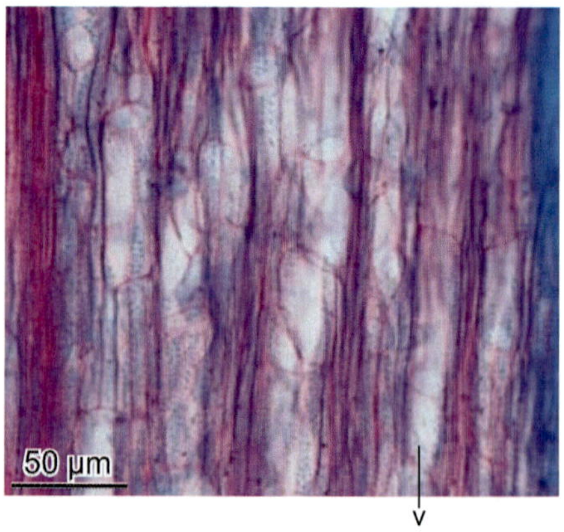

Features according to IAWA

Xylem 1 - 4 - 9 - 11 - 13 - 22 - 40.2 - 50.2 - 52.2 - 53.1 - 61 - 79 - 96 - 105

Morina coulteriana Royle

Morinaceae

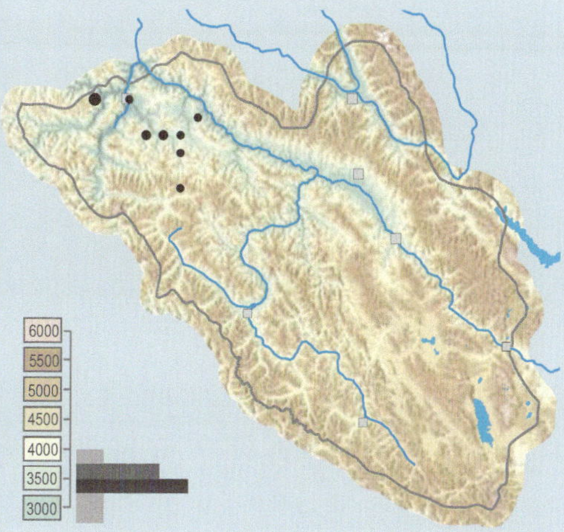

Syn.: *M. brevifolia* Edgew.

Description: Perennial herbs 30–80 cm tall, stems often pubescent and villous at nodes, leaves lanceolate, margin spiny, dentate, stem leaves in whorls, flowers yellow to greenish, ca. 3 cm long, inflorescence composed of 3–8 whorls of numerous flowers.

Distribution: Kyrgyzstan, Uzbekistan, Tajikistan, Afghanistan, N Pakistan, India (Jammu and Kashmir, Himachal Pradesh, Uttarakhand), China (Xinjiang, Xizang).

Habitat: Grassy slopes.

Stem diameter 3–6 mm. Bark/xylem proportion 1:0.3. Ring width 0.3–0.4 mm. Maximal 7 annual rings. Segmented stems.

Xylem

Transverse section

Semi-ring-porous, earlywood vessels solitary. Ring boundary marked by large earlywood (60–80 µm) and small latewood vessel diameters (10–20 µm). Intervessel pits scalariform. Fibers absent. Parenchyma pervasive. With remaining vascular bundles. Large ray-like parenchyma zones between vascular bundles. Crystal druses in rays between the vascular bundles.

Tangential section

Large, multi-seriate rays between vascular bundles. Ray cells upright. Rays disappear in polarized light.

Radial section

Perforation plates simple. Intervessel pits scalariform. Vessel length 120–150 µm.

Bark

Phloem with scattered, small groups of sieve tubes. With large crystal druses. Phellem consists of a few layers of square cells. Secondary intra-xylary phloem towards the pith separates living and dead tissues.

Pith

No slide available.

Analyzed: Root collar of polar root of 3 individuals.

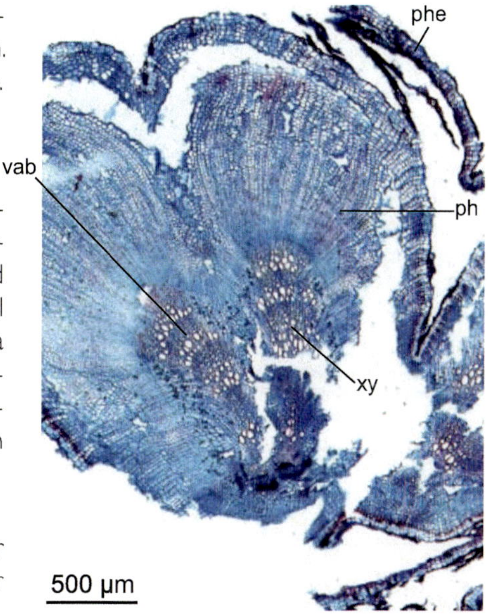

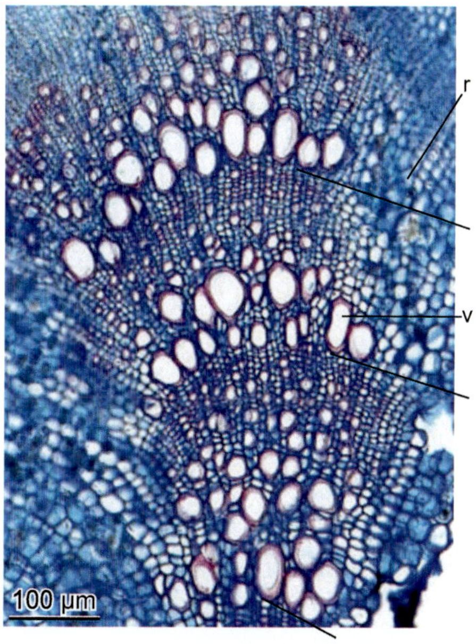

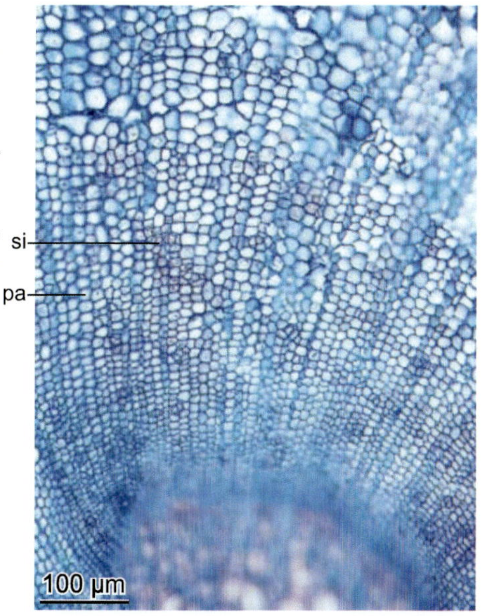

Features according to IAWA

Xylem 1 - 4 - 9 - 13 - 20 - 41 - 50.2 - 52.3 - 60.1 - 79.1 - 99 - 99.1 - 100.1 - 100.2 - 105 - 135.2

Peganum harmala L.

Nitrariaceae

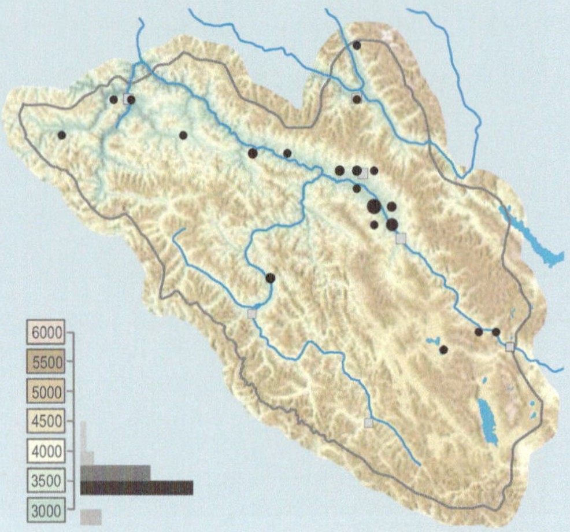

Description: Perennial herbs, 25–60 cm tall, glabrous, stem branched, leaves sessile, pinnatisect, segments linear-lanceolate, flowers white or yellowish white, 2–2.5 cm across, solitary, opposite to leaves.

Distribution: N Africa, W Asia, S Europe, Russia, Kazakhstan, Turkmenistan, Uzbekistan, Kyrgyzstan, Tajikistan, Afghanistan, Pakistan, NW India, China (Gansu, Hebei, Nei Mongol, Ningxia, Qinghai, Shanxi, Xinjiang, Xizang), Mongolia.

Habitat: Disturbed habitats along roads, often synanthropic in settlements, dry grasslands in desert areas.

Stem diameter 3–15 mm. Bark/xylem proportion 1:5. Rings absent.

Xylem

Transverse section

Structure very different between sites. Rings probably absent, if present, ring-porous. Rings marked by differences between earlywood (70–90 µm) and latewood vessel diameters (20–30 µm) and terminal bands of unlignified parenchyma. Vessels solitary. Fibers thin- to thick-walled. Parenchyma marginal. Crystals absent. Marginal (terminal) parenchyma missing in some individuals

Tangential section

Rays 1–4-seriate, larger rays with sheath cells.

Radial section

Perforation plates simple. Intervessel pits round in alternating position. Vessel length 150–180 µm. Small rays with upright cells, larger rays with few prostrate and several square and upright cells.

Bark

Phloem homogeneous. With a few small groups of sclereids. Phellem layered, consisting of a large zone of rectangular cells. Acicular crystals present in all individuals from all sites.

Pith

Round, filled with unlignified, thin-walled cells.

Analyzed: Root collar of polar root of 3 individuals.

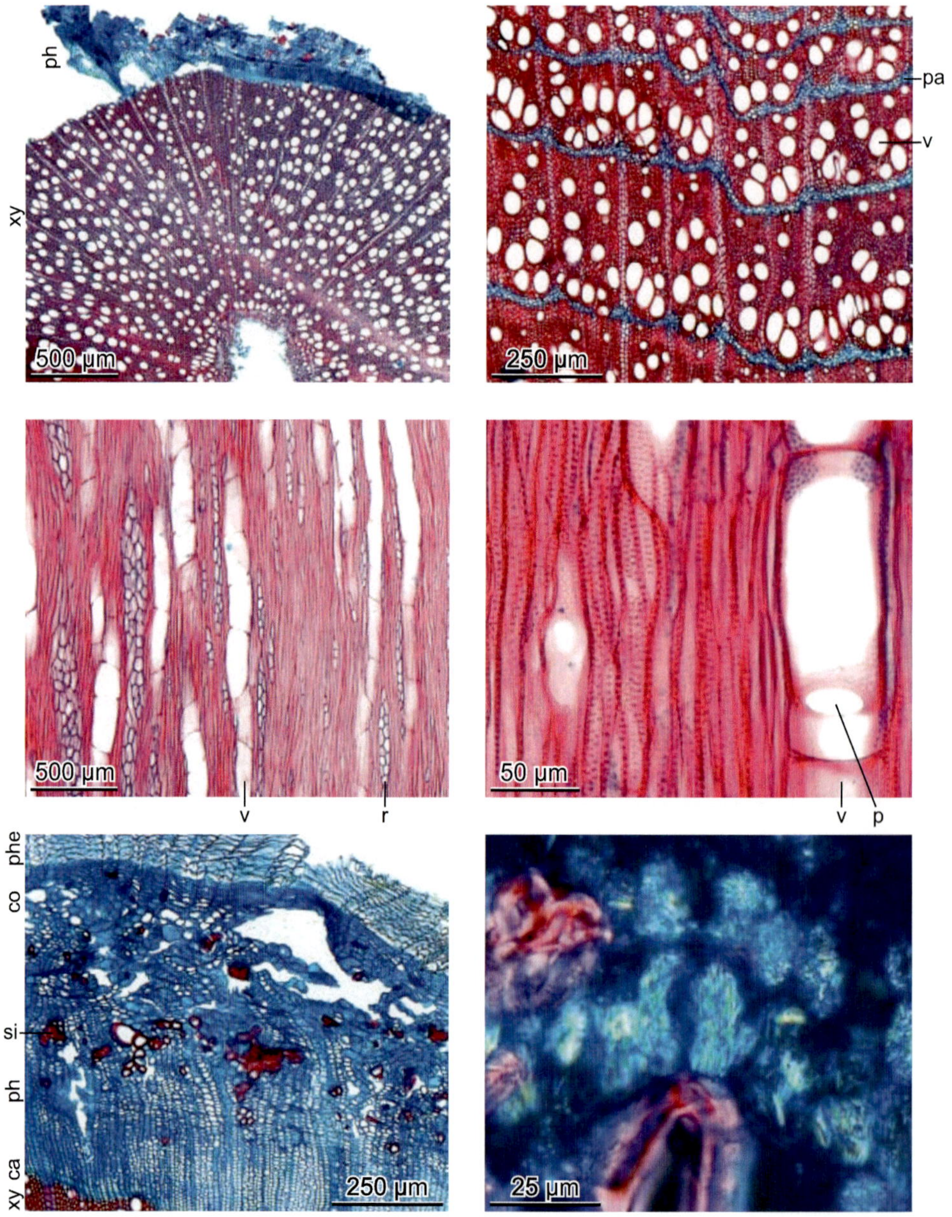

Features according to IAWA

Xylem 1 - 3 - 9 - 11 - 13 - 22 - 41 - 50.1 - 52.3 - 61 - 70 - 89 - 97 - 105 - 150

Epilobium chitralense P.H.Raven

Onagraceae

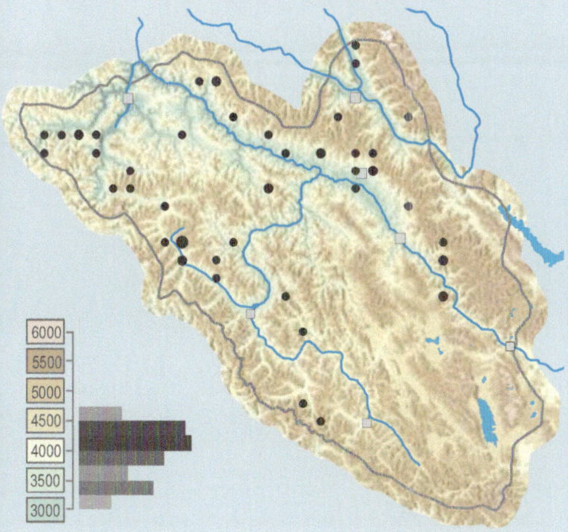

Description: Perennial herbs, 2–20 cm tall, with small fleshy basal turions from which stems arise, leaves narrowly ovate, crowded, subsessile, margin serrulate, flowers pink ot purple, 3–5 mm across, in erect racemes.

Distribution: Afghanistan, N Pakistan, NW India.

Habitat: Spring areas, wetlands.

Stem diameter 1–1.5 mm. Bark/xylem proportion 1:1. Ring width 0.3 mm. Maximal 1 annual ring.

Xylem

Transverse section

One ring present. Vessels angular, solitary. Vessel diameter 15–20 µm. Fibers thin- to thick-walled. Fibers absent in the central part of the stem. Parenchyma paratracheal. Crystals absent.

Tangential section

Rays uniseriate

Radial section

Perforation plates simple. Intervessel pits round in alternating position. Vessel length 50–80 µm. Ray cells upright. Crystals absent.

Bark

Phloem with small groups of sieve elements. With raphids in idioblasts. Phellem absent. With an endodermis.

Pith

Round, with medullary phloem.

Analyzed: Root collar of polar root of 2 individuals.

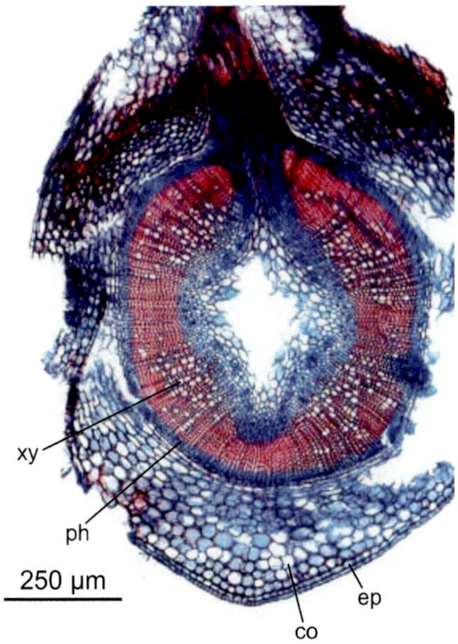

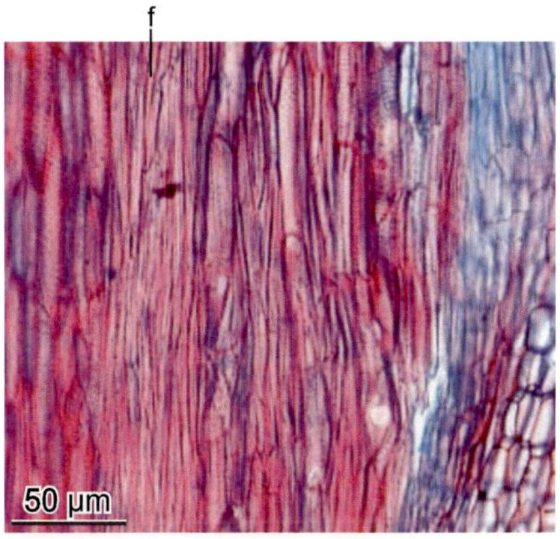

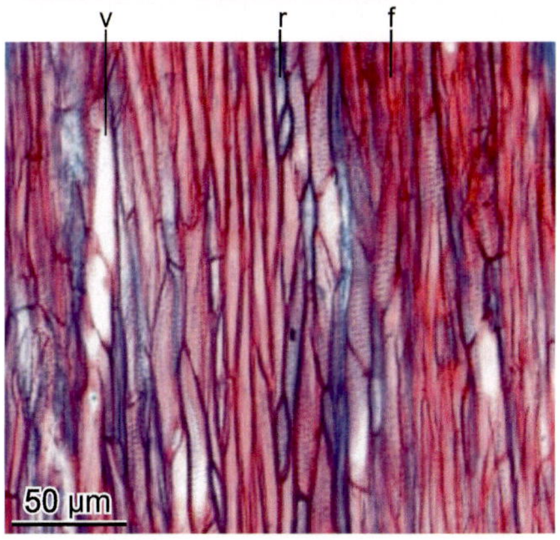

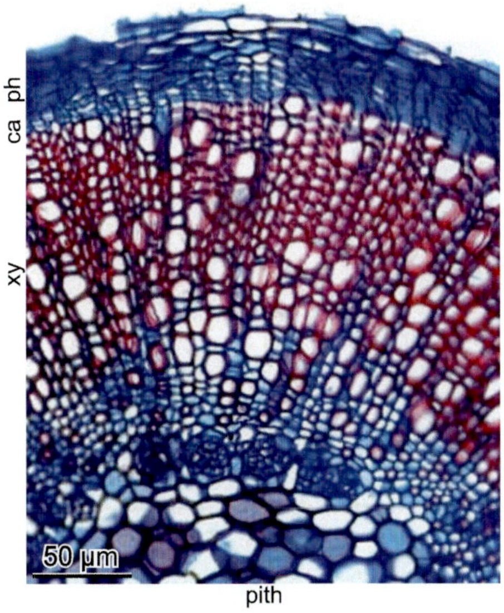

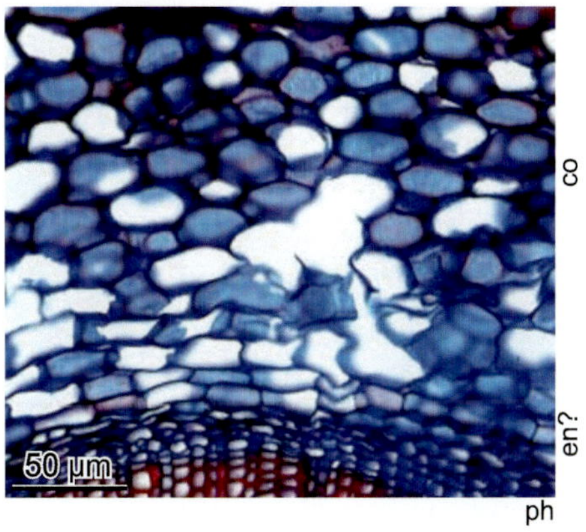

Features according to IAWA

Xylem 2.1 - 9 - 11 - 12 - 13 - 22 - 40.1 - 50.2 - 52.2 - 60.1 - 61 - 70 - 70.3 - 79 - 96 - 105 - P1

Epilobium latifolium L.

Onagraceae

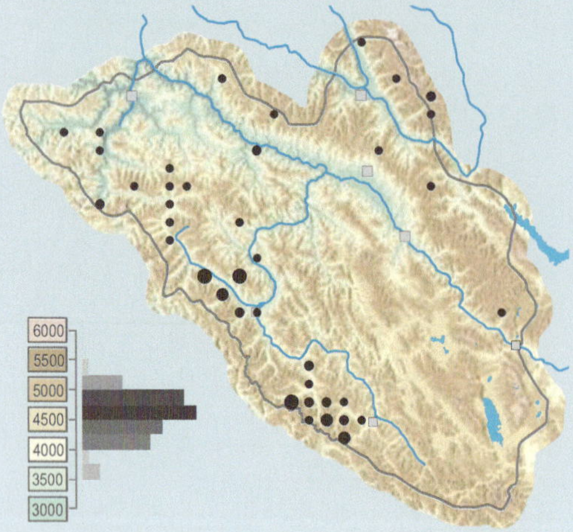

Syn.: *Chamerion latifolium* (L.) Holub, *Chamaenerion latifolium* (L.) Sweet

Description: Perennial herbs, 5–35 cm tall, stems erect or ascending, leaves ovate to elliptic, subsessile, flowers rose-purple, up to 5 cm across, in axillary racemes.

Distribution: N America, Greenland, Europe, SW Asia, Russia (Siberia, Far East), Kazakhstan, Kyrgyzstan, Tajikistan, Afghanistan, Pakistan, India (Jammu and Kashmir), Nepal, China (Qinghai, Xinjiang, Xizang, Yunnan), Bhutan, Mongolia, Japan.

Habitat: Stony river beds, near brooks.

Stem diameter 8–10 mm. Bark/xylem proportion 1:1. Ring width 0.1–0.3 mm. Maximal 2 annual rings. Total age unknown. Segmented stems. Center rotten.

Xylem

Transverse section

Semi-ring-porous, earlywood vessels solitary, angular. Ring boundary marked by large earlywood (50–80 µm) and small latewood vessel diameters (10–20 µm). Fibers absent. Parenchyma pervasive. With remaining vascular bundles. With a few isolated sieve tubes. Large ray-like parenchyma zones between vascular bundles. Bundles of raphids in large idioblasts. Secondary intra-xylary phloem towards the pith separates living and dead tissues.

Tangential section

Rays absent. With short parenchyma cells.

Radial section

Perforation plates simple. Intervessel pits scalariform. Vessel length 70–100 µm. Parenchyma cells rectangular and in-line.

Bark

Phloem homogeneous with some idioblasts with rhapids. Phellem consists of a few layers of rectangular cells. Older phellem parts flake.

Pith

No slide available.

Analyzed: Root collar of polar root of 2 individuals.

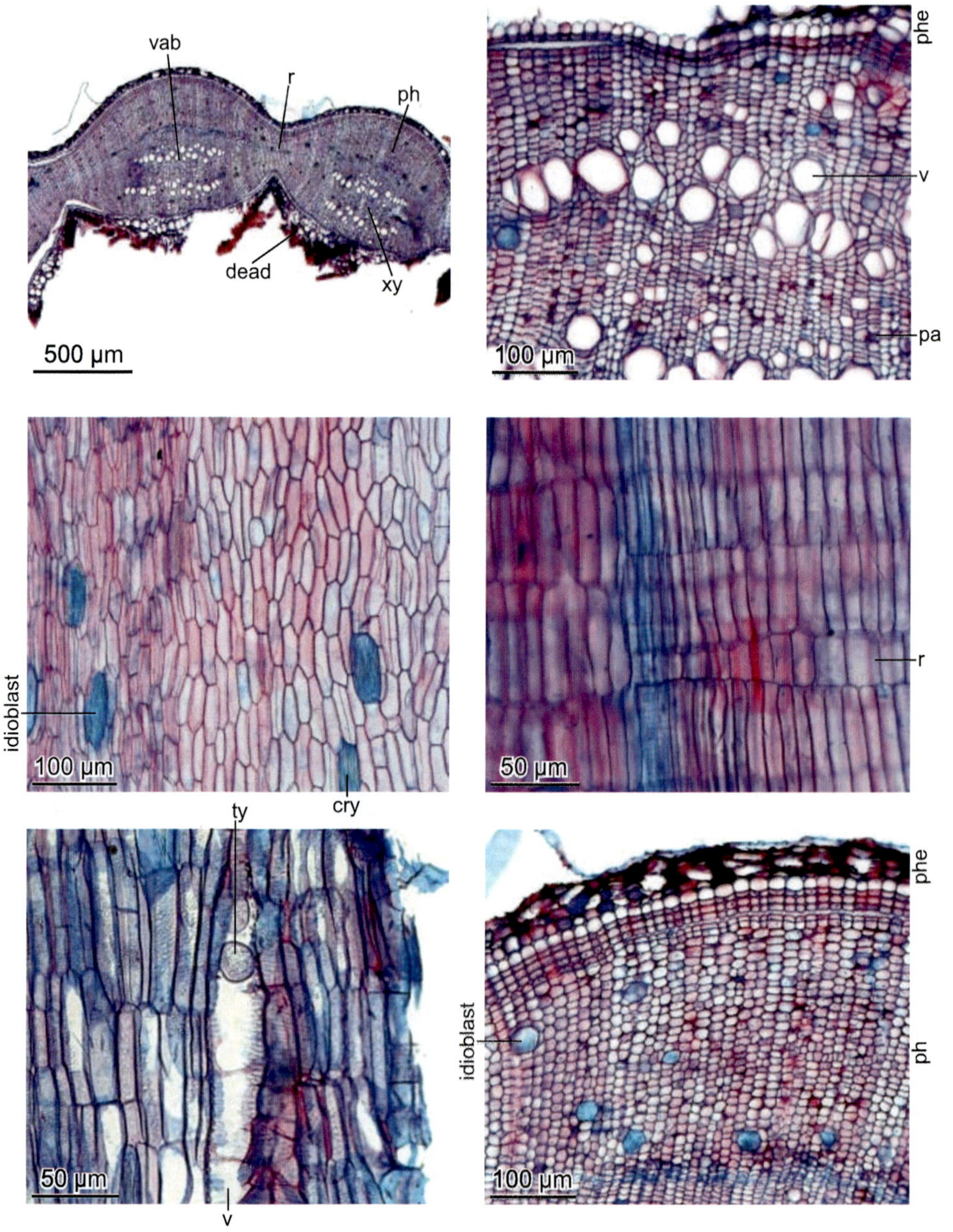

Features according to IAWA

Xylem 1 - 4 - 9 - 13 - 20 - 41 - 50.2 - 52.2 - 60.1 - 79.1 - 99 - 99.1 - 100.1 - 100.2 - 105 - 135 - 135.2

Leptorhabdos parviflora (Benth.) Benth. in DC.

Orobanchaceae

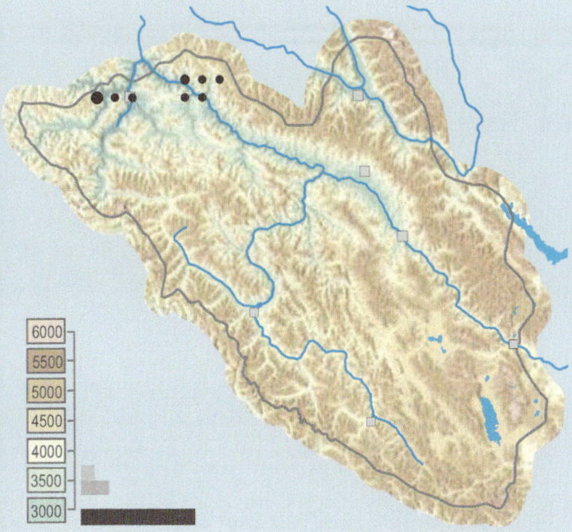

Syn.: *Gerardia parviflora* Benth.

Description: Annuals, 20–100 cm tall, glandular hairy, stems much branched, leaves pinnatisect, segments linear, flowers pink, ca. 5 mm across, in long racemes.

Distribution: SW Asia, Kazakhstan, Turkmenistan, Uzbekistan, Kyrgyzstan, Tajikistan, Afghanistan, Pakistan, NW India, China (Gansu, Xinjiang).

Habitat: Steppes, river and lake banks.

Stem diameter 1.5–4 mm. Bark/xylem proportion 1:5. Ring width 0.7–2.2 mm. Maximal 1 annual ring.

Xylem

Transverse section

One ring present. Vessels in short and long radial rows. Vessel diameter 30–50 µm. Fibers thin- to thick-walled. Fibers absent in the central part of the stem. Parenchyma paratracheal. Crystals absent.

Tangential section

Rays 1–3-seriate.

Radial section

Perforation plates simple. Intervessel pits round in alternating position. Vessel length 100–150 µm. Ray cells upright.

Bark

Phloem uniform. Cortex cells large, rectangular. Phellem absent.

Pith

No slide available.

Analyzed: Root collar of polar root of 6 individuals.

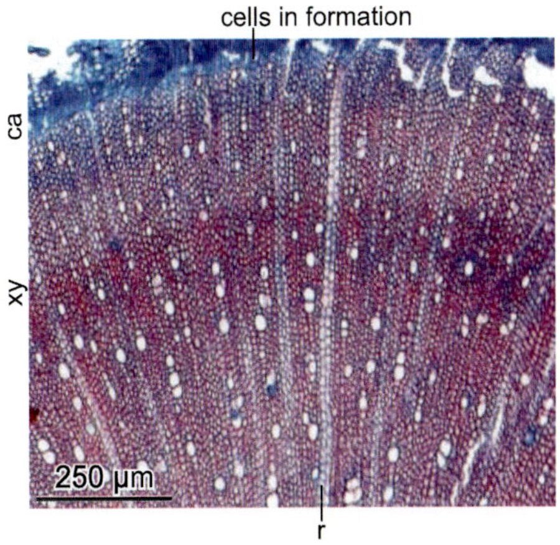

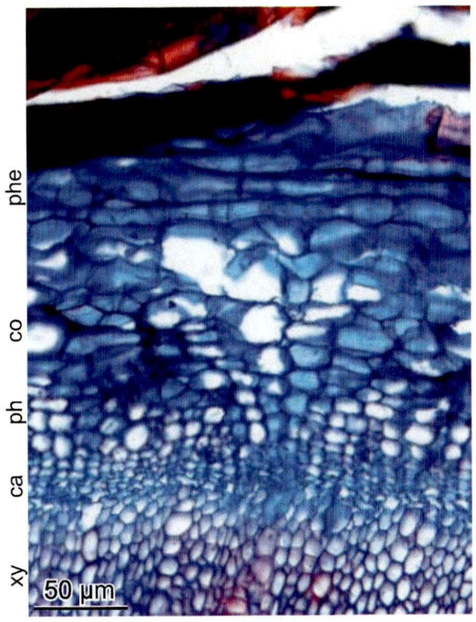

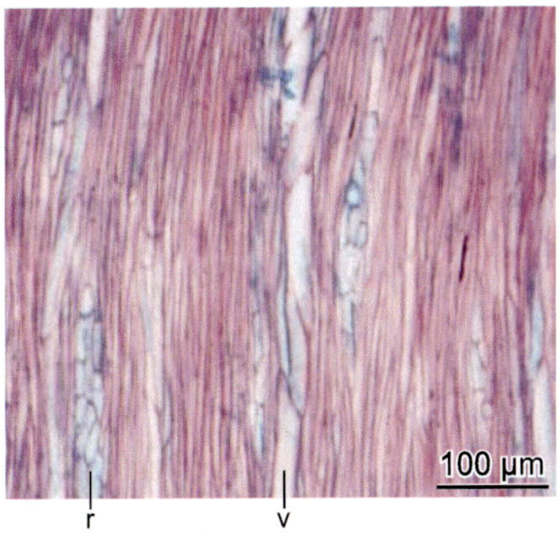

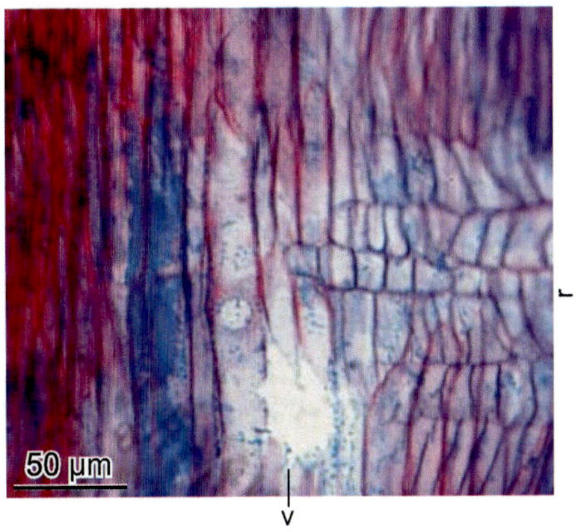

Features according to IAWA

Xylem 2.1 - 9.1 - 10 - 13 - 22 - 40.2 - 50.2 - 52.3 - 61 - 70 - 79 - 97 - 105

Corydalis crassifolia Royle

Papaveraceae

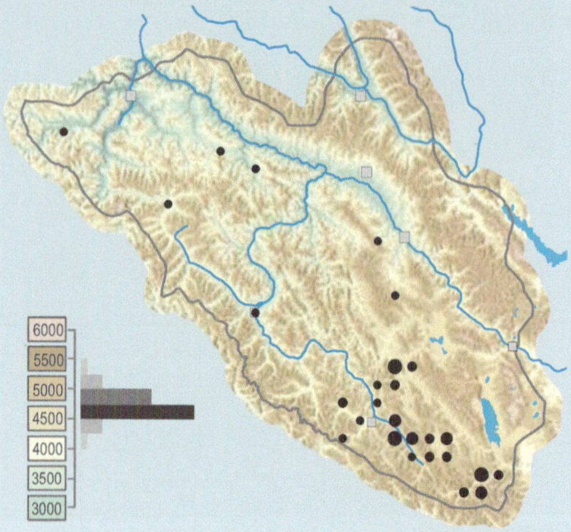

Syn.: *C. crassissima* Cambess.

Description: Perennial herbs, 10–20 cm tall, leaves trilobed, flowers white-violet, spurred, 18–25 mm long, in dense racemes.

Distribution: N Pakistan, N India, China (Xizang).

Habitat: Stony, unstable screes.

Stem diameter 8–10 mm. Bark/xylem proportion 1:1. Ring width 0.3–0.4 mm. Maximal 4 annual rings. Total age unknown. Segmented stems.

Xylem

Transverse section

Semi-ring-porous. Rings indicated by larger earlywood (90–120 µm) and smaller latewood vessel diameters (40–50 µm). Vessels solitary and in groups. Fibers absent. Parenchyma pervasive. Remaining vascular bundles. Crystals absent. Secondary intraxylary phloem towards the pith.

Tangential section

Rays >10-seriate.

Radial section

Perforation plates simple. Intervessel pits scalariform. Vessel length 100–200 µm.

Bark

Phloem homogeneous. Phellem consists of a few rows of rectangular cells. Crystals absent.

Pith

No slide available.

Analyzed: Root collar of polar root of 6 individuals.

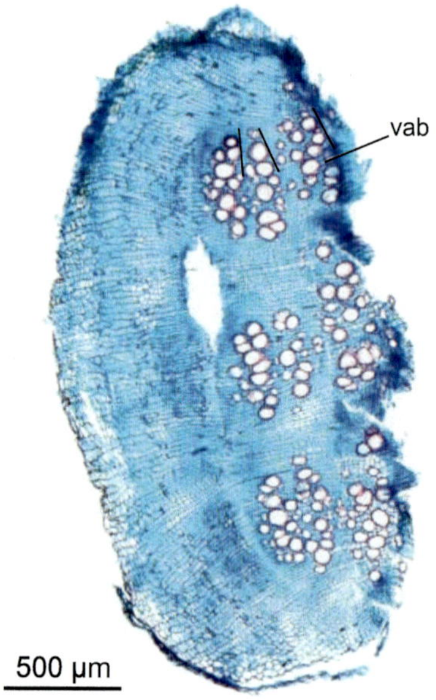

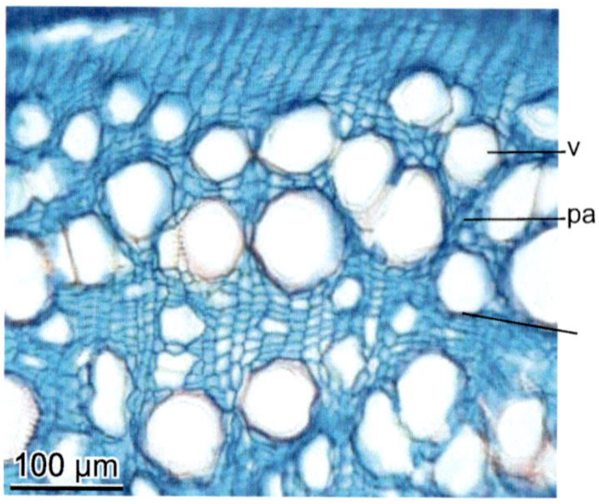

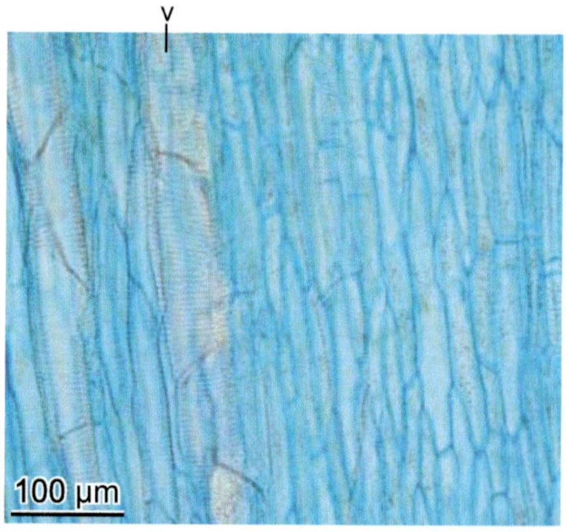

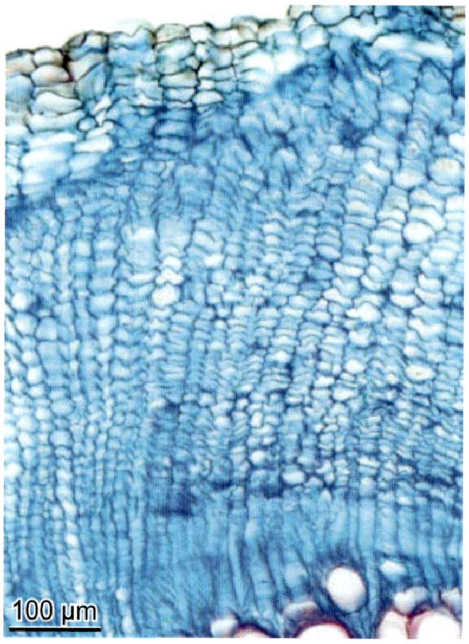

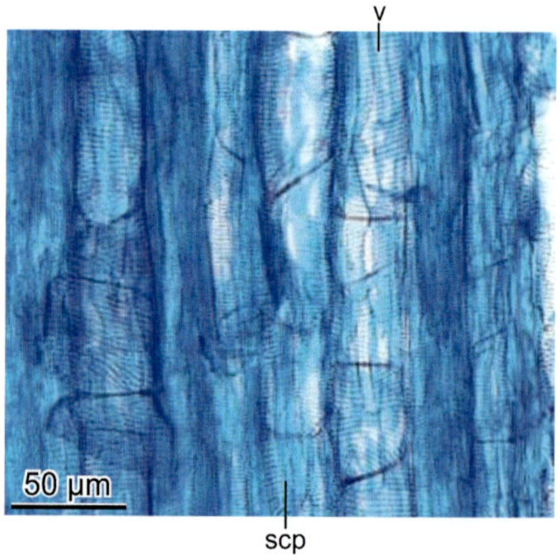

Features according to IAWA

Xylem 1 - 9 - 11 - 13 - 20 - 41 - 42 - 50.2 - 52.3 - 60.1 - 79.1 - 99 - 99.1 - 100.1 - 100.2 - 135.2

Corydalis flabellata Edgew.

Papaveraceae

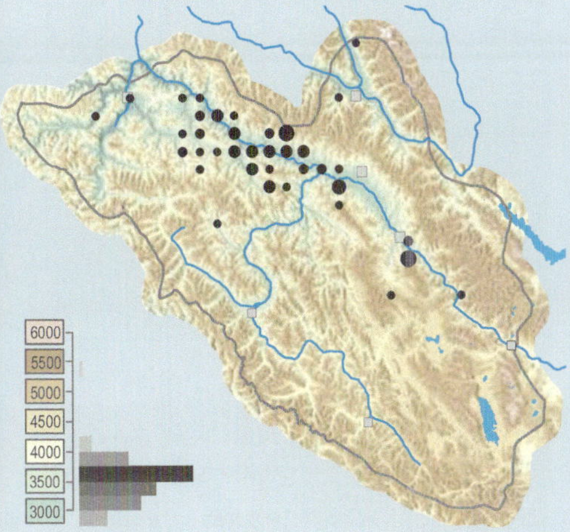

Syn.: *C. adiantifolia* Hook.f. & Thomson

Description: Perennial herbs, 20–60 cm tall, glabrous, glaucous, much branched, leaves pinnate, leaflets fan-shaped, flowers yellow, spurred, 12–18 mm long, in branched racemes.

Distribution: N Pakistan, N India, Nepal, China (Xinjiang, W Xizang).

Habitat: Dry stony areas, semideserts.

Stem diameter 8–10 mm. Bark/xylem proportion 1:1. Ring width 0.2–0.3 mm. Maximal 6 annual rings. Total age unknown. Segmented stems.

Xylem

Transverse section

Semi-ring-porous. Rings indicated by larger earlywood (80–120 μm) and smaller latewood vessel diameters (40–50 μm). Vessel solitary and in groups. Fibers absent. Parenchyma pervasive. Vascular bundles remaining. Crystals absent. Secondary intra-xylary phloem towards the pith.

Tangential section

Rays >10-seriate.

Radial section

Perforation plates simple. Intervessel pits scalariform and round. Vessel length 100–200 μm.

Bark

Phloem homogeneous. Phellem consists of a few rows of square cells. Crystals absent.

Pith

No slide available.

Analyzed: Root collar of polar root of 2 individuals.

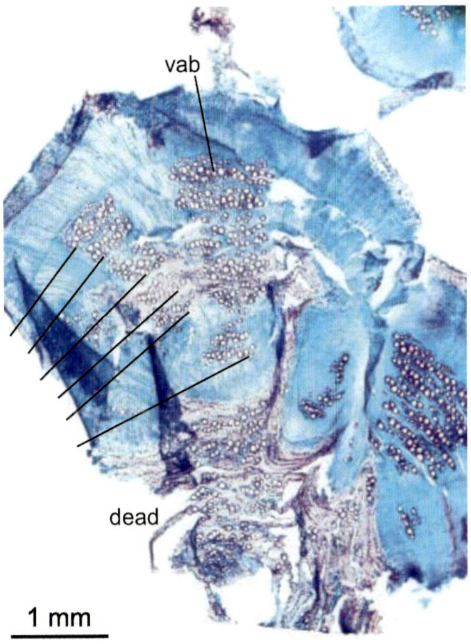

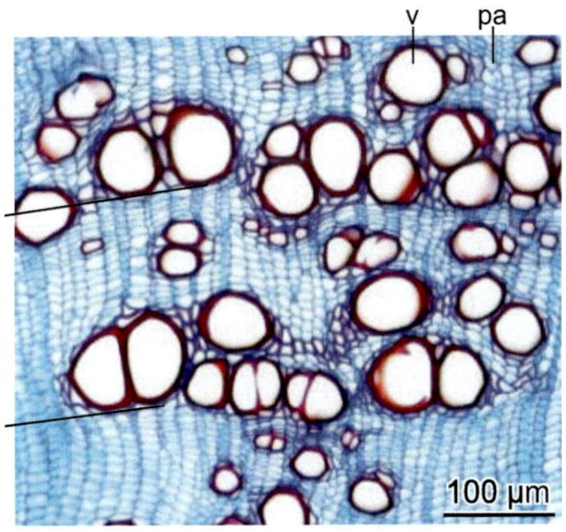

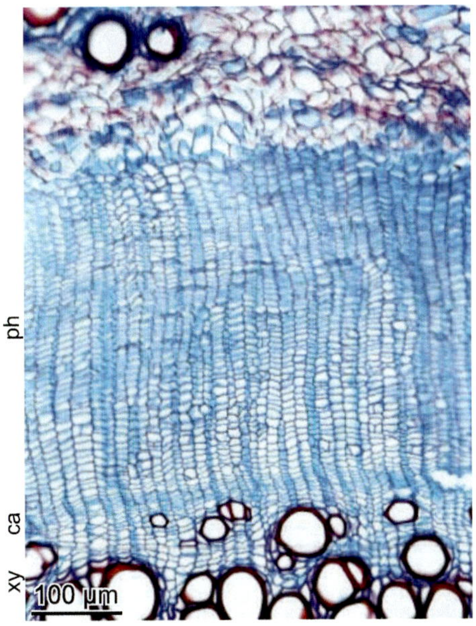

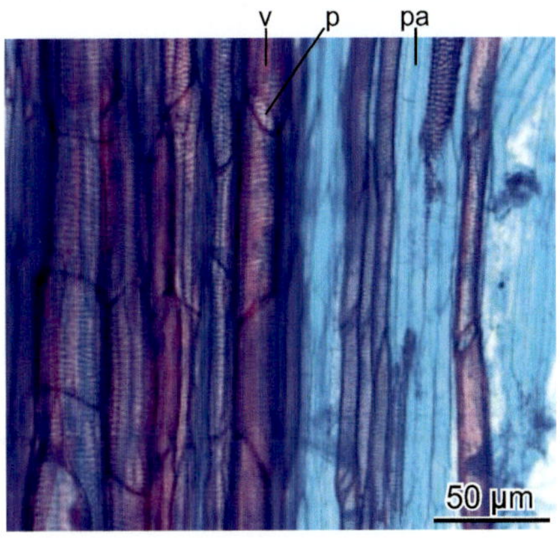

Features according to IAWA

Xylem 1 - 9 - 11 - 13 - 20 - 22 - 41 - 42 - 50.2 - 52.3 - 60.1 - 79.1 - 99 - 99.1 - 100.1 - 100.2 - 135.2

Corydalis inopinata Prain ex Fedde

Papaveraceae

Syn.: *C. fimbripetala* Ludlow & Stearn

Description: Perennial herbs, 2–7 cm tall, cespitose, stems branched, leaves 2–3-ternatisect, segments orbicular, flowers yellow, spurred, 10–15 mm long, in corymbose 2–8-flowered racemes.

Distribution: India (Jammu and Kashmir), China (Xizang).

Habitat: Alpine scree.

Stem diameter 3–6 mm. Bark/xylem proportion 1:0.3. Rings indistinct. Total age unknown.

Xylem

Transverse section

Rings absent. Larger vessels with a diameter of 30–40 µm. Vessel solitary, in groups and in short radial multiples. Fibers absent. Parenchyma pervasive. Vascular bundles remaining. Crystals absent.

Tangential section

Rays >10-seriate.

Radial section

Perforation plates simple. Intervessel pits round. Vessel length 150–200 µm.

Bark

Phloem homogeneous. Phellem consists of a few rows of large, oval cells. Crystals absent.

Pith

No slide available.

Analyzed: Root collar of polar root of 2 individuals.

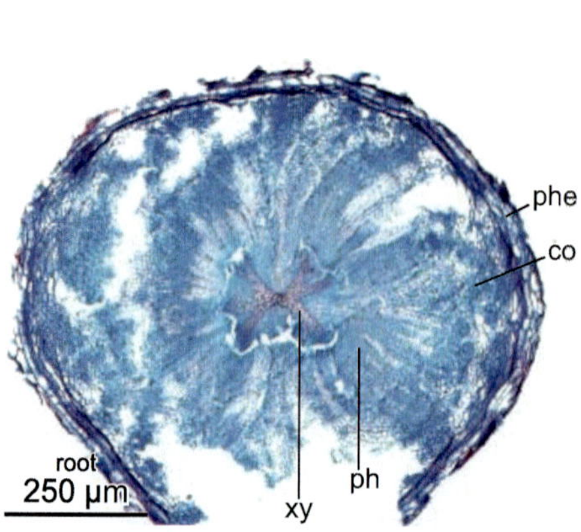

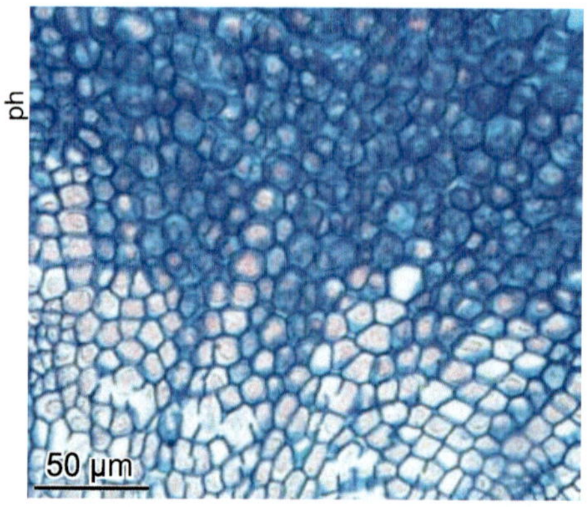

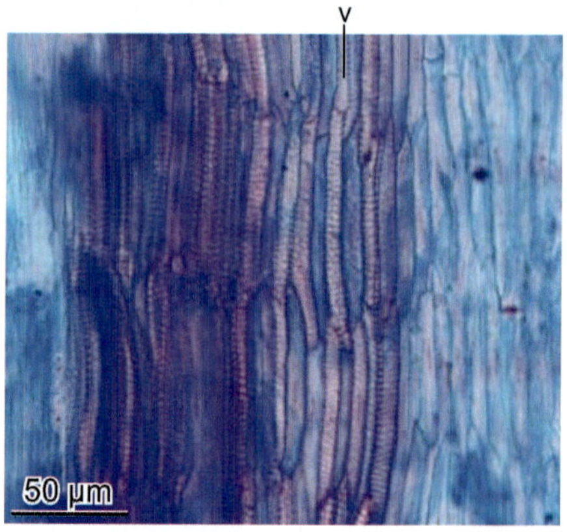

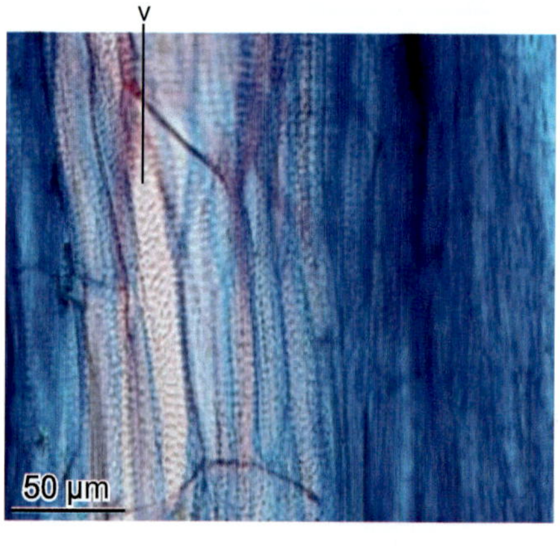

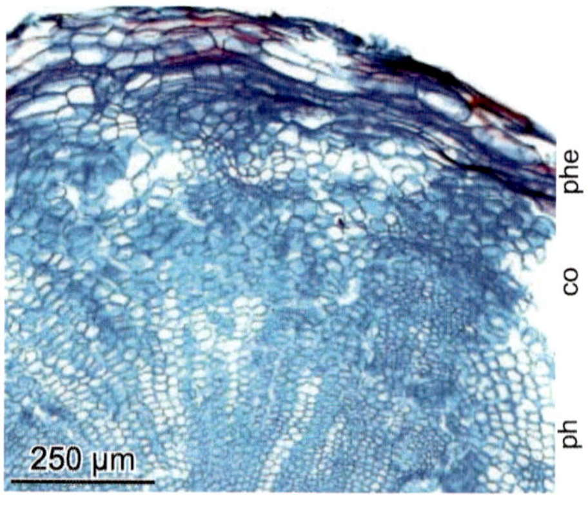

Features according to IAWA

Xylem 1 - 9 - 9.1 - 11 - 13 - 22 - 40.2 - 50.2 - 52.3 - 60.1 - 79.1 - 99 - 99.1 - 100.1 - 100.2

Corydalis meifolia Wall.

Papaveraceae

Description: Perennial herbs, 10–30 cm tall, leaves 3-pinnatisect, ultimate segments very narrow, flowers yellow, spurred, 12–18 mm long, in dense racemes.

Distribution: Pakistan, N India, Nepal, China (S Xizang), Bhutan.

Habitat: Screes, stony slopes.

Stem diameter 8–10 mm. Bark/xylem proportion 1:1. Ring width 0.3–0.5 mm. Maximal 3 annual rings. Total age unknown. Segmented stems.

Xylem

Transverse section

Semi-ring-porous. Rings indicated by larger earlywood (50–70 µm) and smaller latewood vessel diameters (30–40 µm). Vessels solitary and in small groups. Fibers absent. Parenchyma pervasive. Remaining vascular bundles. Crystals absent. With intra-xylary phloem towards the pith.

Tangential section

Rays >10-seriate between vascular bundles.

Radial section

Perforation plates simple. Intervessel pits scalariform. Vessel length 60–90 µm.

Bark

Phloem with a few small groups of sieve elements. Phellem consists of a few rows of rectangular cells. Crystals absent.

Pith

No slide available.

Analyzed: Root collar of polar root of 4 individuals.

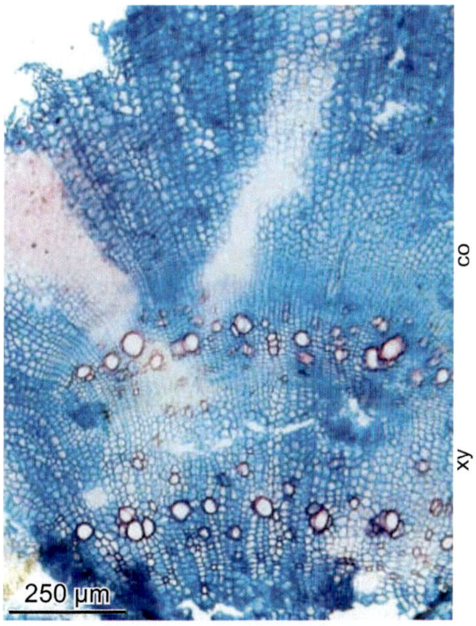

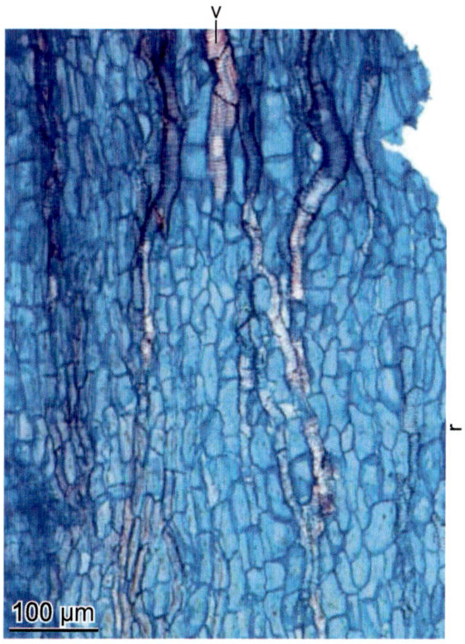

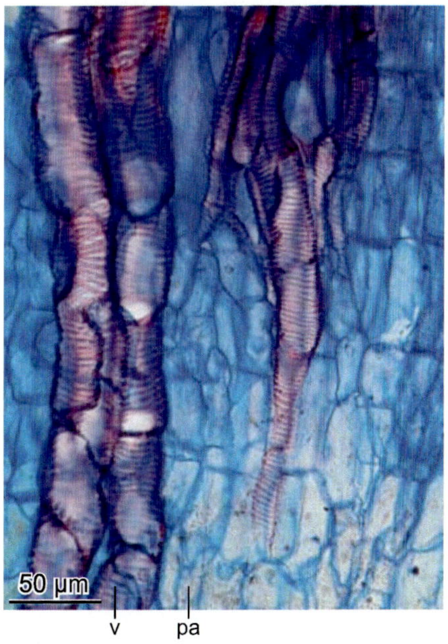

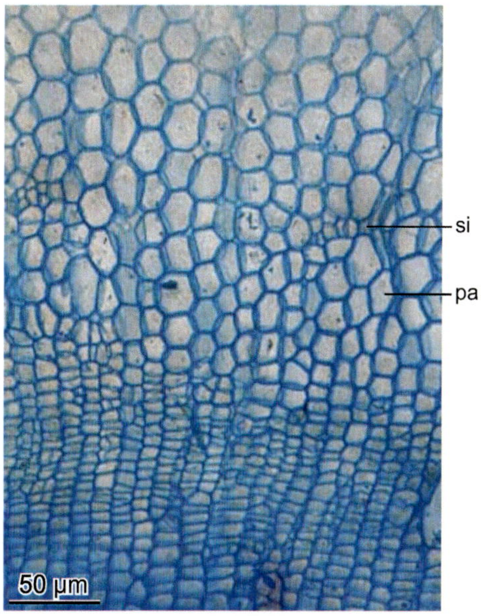

Features according to IAWA

Xylem 1 - 9 - 11 - 13 - 20 - 41 - 50.2 - 52.3 - 60.1 - 79.1 - 99 - 99.1 - 100.1 - 100.2 - 105 - 135.2

Corydalis moorcroftiana Wall. ex Hook.f. & Thomson

Papaveraceae

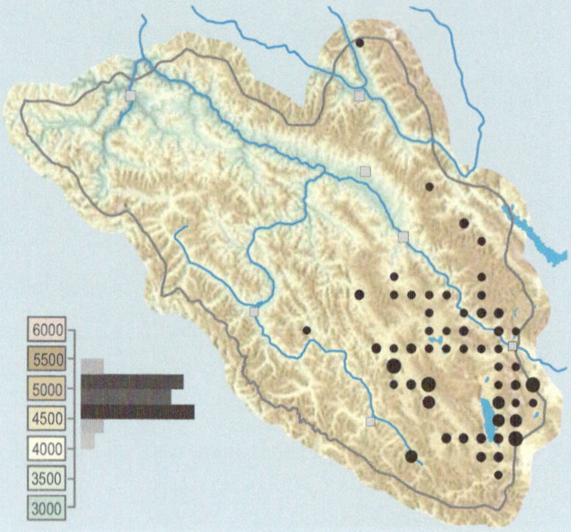

Description: Perennial herbs, 10–40 cm tall, glaucous, leaves 2-pinnate, segments deeply cut, flowers bright yellow, spurred, 19–22 mm long, in dense racemes.

Distribution: India (Jammu and Kashmir), China (Xizang).

Habitat: Scree, river sands, gravel areas.

Stem diameter 8 mm. Bark/xylem proportion 1:3. Ring width 1 mm. Maximal 2 annual rings. Total age unknown. Segmented stems. Center rotten.

Xylem

Transverse section

Semi-ring-porous. Rings indicated by larger earlywood (80–120 µm) and smaller latewood vessel diameters (40–50 µm). Vessels solitary and in groups. Fibers absent. Parenchyma pervasive. Vascular bundles remaining. Crystals absent. Secondary intra-xylary phloem towards the pith.

Tangential section

Rays >10-seriate.

Radial section

Perforation plates simple. Intervessel pits scalariform and round. Vessel length 100–150 µm.

Bark

Phloem homogeneous. Phellem consists of a few rows of square cells. Crystals absent.

Pith

No slide available.

Analyzed: Root collar of polar root of 1 individual.

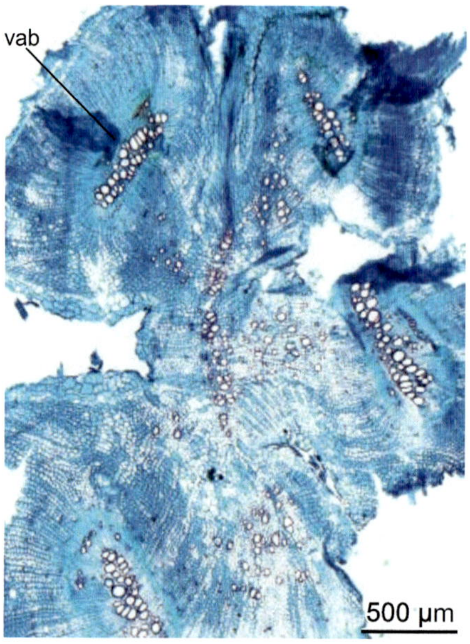

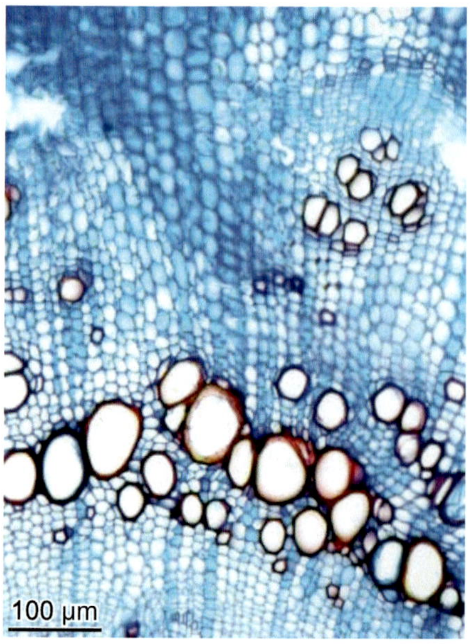

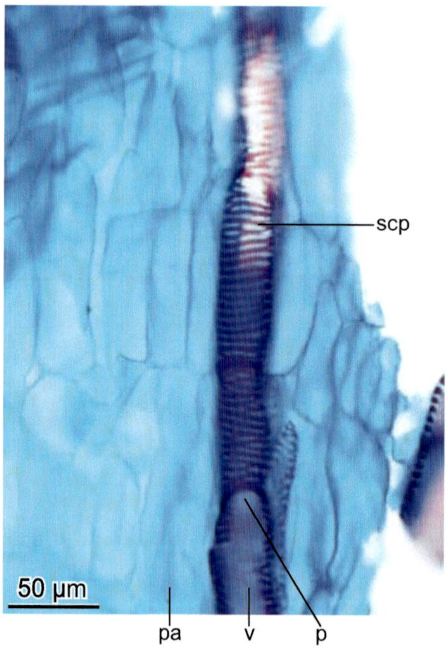

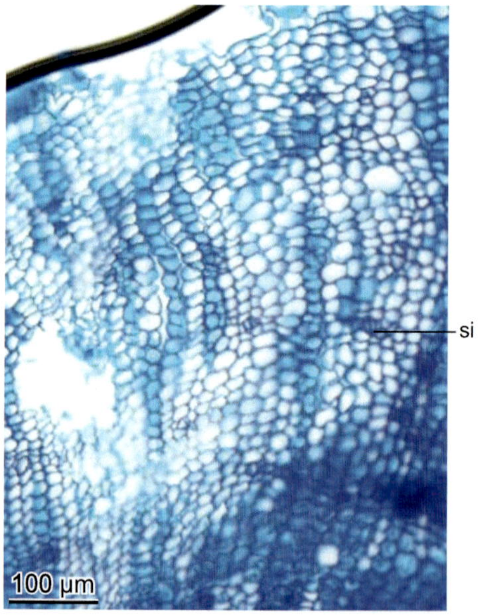

Features according to IAWA

Xylem 1 - 9 - 11 - 13 - 20 - 41 - 42 - 50.2 - 52.3 - 60.1 - 79.1 - 99 - 99.1 - 100.1 - 100.2 - 135.2

Corydalis thyrsiflora Prain

Papaveraceae

Description: Perennial herbs, 15–30 cm tall, leaves pinnatisect, ultimate segments oblong to linear, flowers yellow, spurred, 12–15 mm long, in dense thyrsiform racemes.

Distribution: Pakistan, NW India.

Habitat: Wet river banks, along streams, alpine meadows.

Stem diameter 8–10 mm. Bark/xylem proportion 1:1. Ring width 0.2–0.5 mm. Maximal 3 annual rings. Total age unknown. Segmented stems.

Xylem

Transverse section

Semi-ring-porous. Rings indicated by larger earlywood (50–70 µm) and smaller latewood vessel diameters (30–40 µm). Vessel solitary and in small groups. Fibers absent. Parenchyma pervasive. Remaining vascular bundles. Crystals absent. With intra-xylary phloem towards the pith.

Tangential section

Rays >10-seriate between vascular bundles.

Radial section

Perforation plates simple. Intervessel pits scalariform. Vessel length 150–250 µm.

Bark

Phloem with small groups of square sieve elements. Phellem consists of a few rows of rectangular cells. Crystals absent.

Pith

No slide available.

Analyzed: Root collar of polar root of 4 individuals.

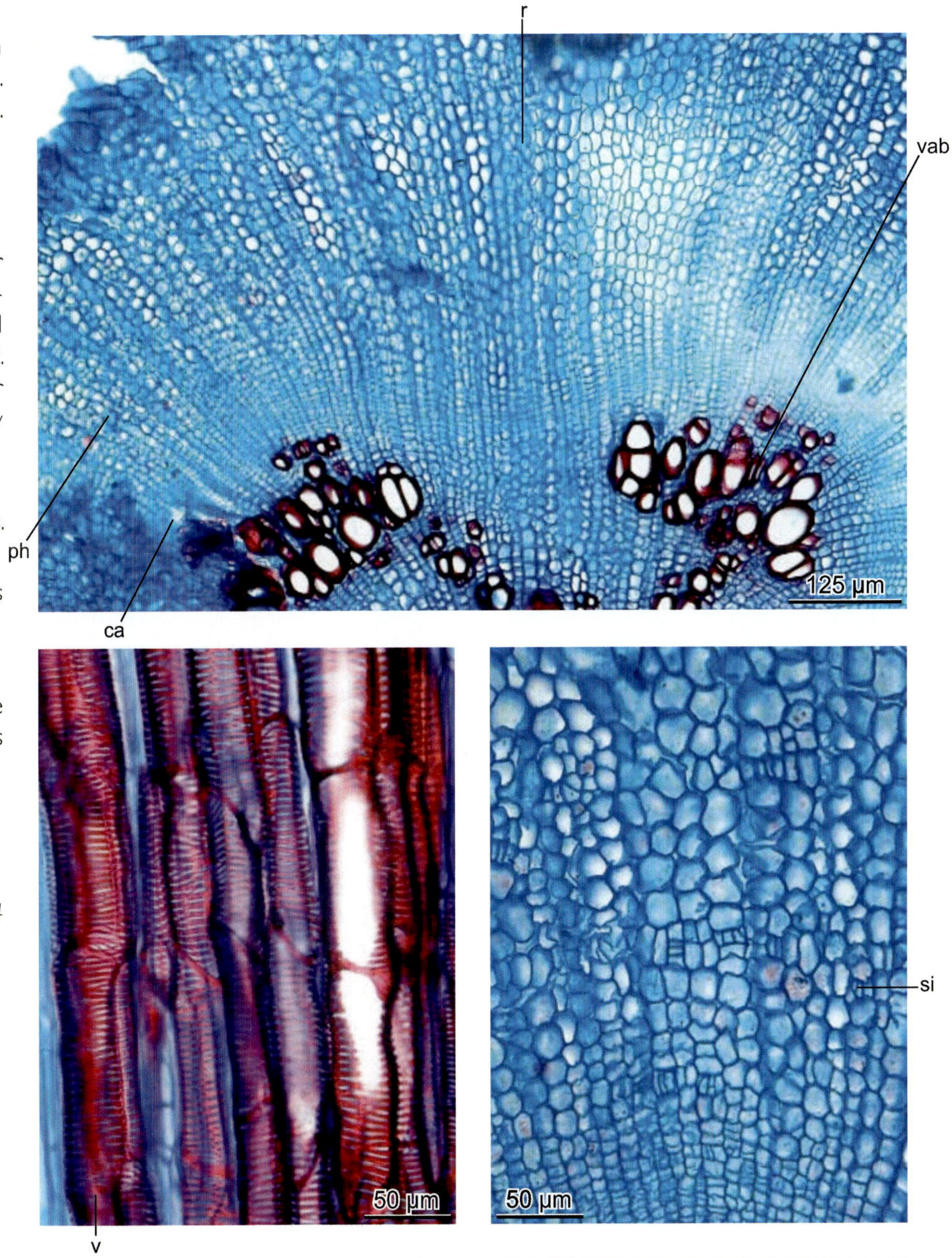

Features according to IAWA

Xylem 1 - 9 - 11 - 13 - 20 - 41 - 50.2 - 52.3 - 53.1 - 60.1 - 79.1 - 99 - 99.1 - 100.1 - 100.2 - 105 - 135.2

Corydalis tibetoalpina C.Y.Wu & Z.Y.Su

Papaveraceae

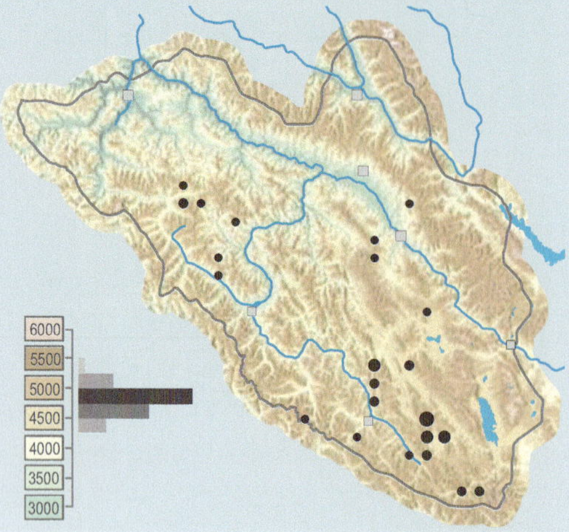

Description: Perennial herbs, 4–7 cm tall, glabrous, glaucous, leaves 2-pinnate, segments lanceolate, fleshy, flowers pale yellow, ca. 15 mm long, in dense racemes.

Distribution: India (Jammu and Kashmir), China (Xizang).

Habitat: Alpine slopes.

Stem diameter 8–10 mm. Bark/xylem proportion 1:0.3. Ring width 0.2–0.3 mm. Maximal 2 annual rings. Total age unknown.

Xylem

Transverse section

Semi-ring-porous. Rings indicated by larger earlywood (60–80 µm) and smaller latewood vessel diameters (20–40 µm). Angular vessels solitary and in groups. Fibers absent. Parenchyma pervasive. Remaining vascular bundles. Crystals absent. Secondary intra-xylary phloem towards the pith.

Tangential section

Rays >10-seriate.

Radial section

Perforation plates simple. Intervessel pits scalariform and round. Vessel length 120–200 µm.

Bark

Phloem with a few isolated sieve tubes. Phellem consists of a few rows of rectangular cells. Crystals absent.

Pith

No slide available.

Analyzed: Root collar of polar root of 2 individuals.

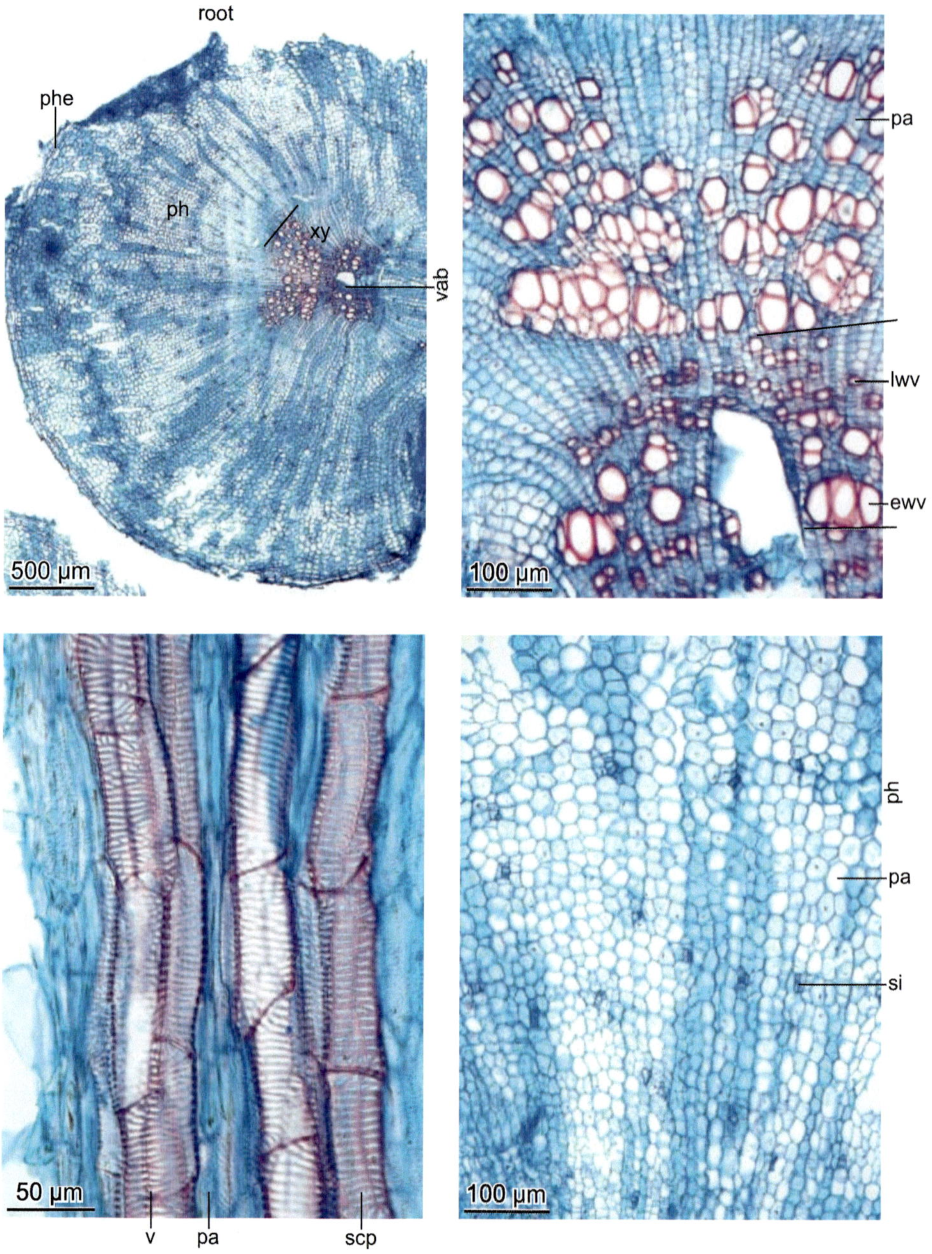

Features according to IAWA

Xylem 1 - 9 - 11 - 12 - 13 - 20 - 41 - 50.2 - 52.3 - 60.1 - 79.1 - 99 - 99.1 - 100.1 - 100.2 - 135.2

Meconopsis aculeata Royle

Papaveraceae

Description: Perennial herbs, 20–70 cm tall, prickly, leaves oblong or cordate to deltoid, pinnatifid to irregularly pinnatisect, flowers sky blue, 5–7 cm across, in long racemes, fruits densely bristly.

Distribution: Pakistan, NW India, China (SW Xizang).

Habitat: Scree, rocky slopes, moraines, alpine meadows.

Stem diameter 4–6 mm. Bark/xylem proportion 1:3. Rings mostly indistinct. Maximum age probably >5 years. Segmented stems.

Xylem

Transverse section

Rings indicated by larger vessels with a diameter of 40–50 μm. Vessel solitary and in radial groups. Vessels hardly lignified. Fibers absent. Parenchyma pervasive. Vascular bundles remaining. Rays bent and compressed. With ducts. Crystals absent. Secondary intra-xylary phloem towards the pith.

Tangential section

Rays 5–10-seriate.

Radial section

Perforation plates simple. Intervessel pits scalariform and round. Vessel length 150–180 μm. Ray cells square and upright.

Bark

Phloem homogeneous. Phellem consists of a few rows of rectangular cells. Crystals absent.

Pith

No slide available.

Analyzed: Polar root of 3 individuals.

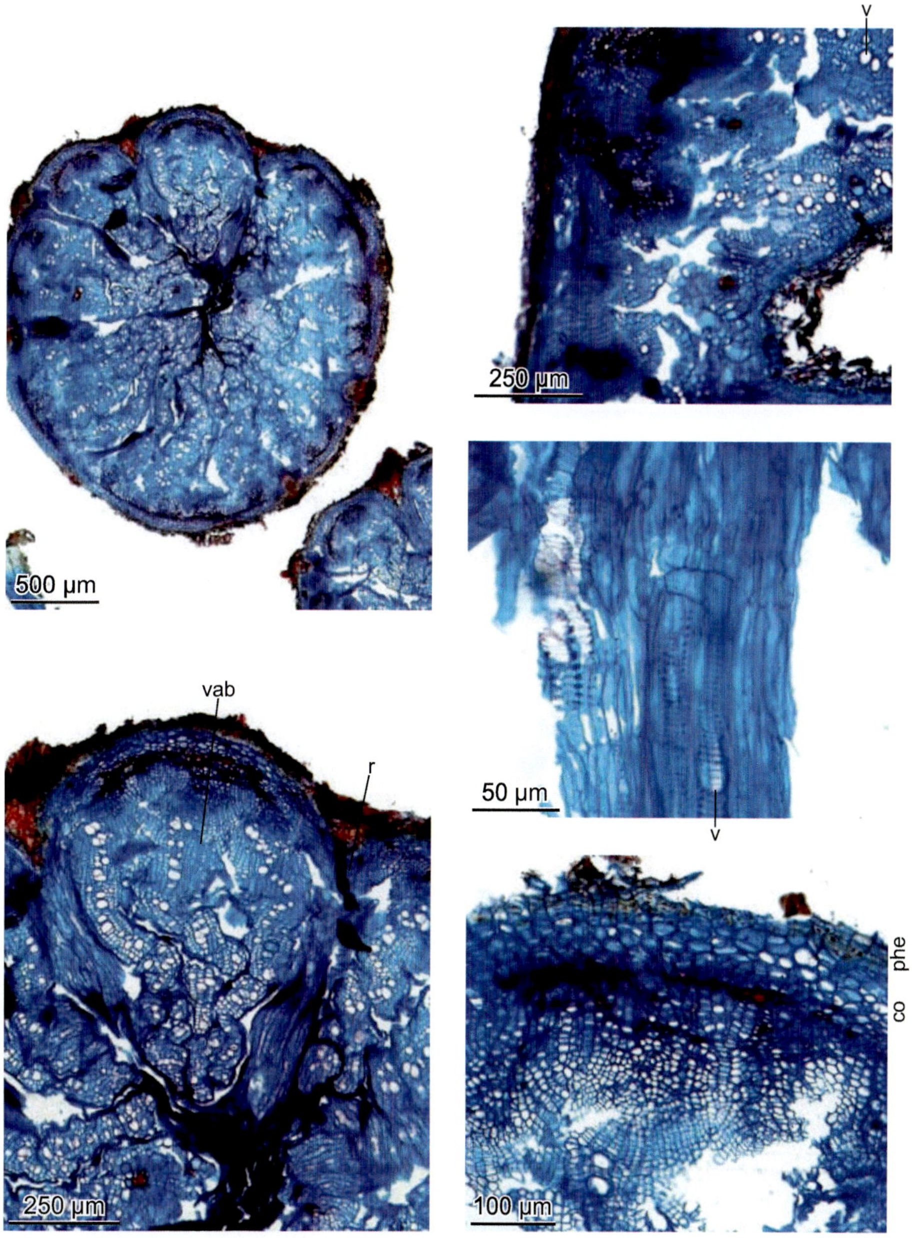

Features according to IAWA

Xylem *1 - 2 - 9 - 9.1 - 11 - 13 - 20 - 40.2 - 50.1 - 52.3 - 60.1 - 79.1 - 99 - 99.1 - 100.1 - 100.2 - 105 - 129*

Papaver nudicaule L.

Papaveraceae

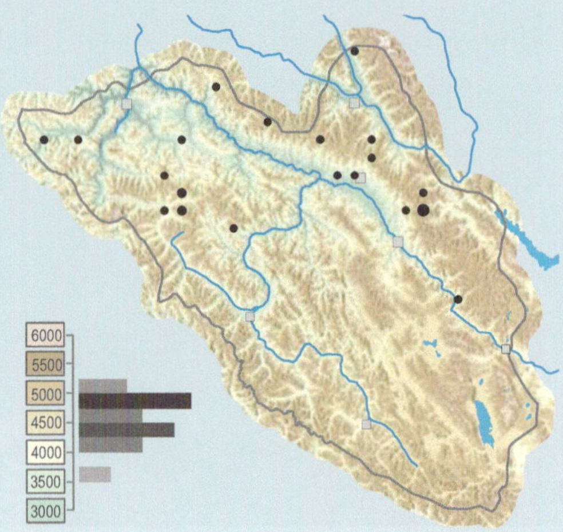

Syn.: *P. croceum* Ledeb.

Description: Perennial herbs, 20–60 cm tall, gray setose, leaves all basal, ovate to lanceolate, pinnatilobate, pinnatipartite, or pinnatisect, segments entire or futher divided, flowers yellowish, yellow, or orange, rarely red, 4–6 cm across, solitary.

Distribution: Subpolar regions of Europe, Asia and North America, and the mountains of C Asia.

Habitat: Grassy steppes, slopes, meadows, valleys, river gravel, moraines.

Stem diameter 3 mm. Bark/xylem proportion 1:4. Mean ring width 0.3 mm. Maximal 4 annual rings.

Xylem

Transverse section

Semi-ring-porous. Rings indicated by larger earlywood (20–30 µm) and smaller latewood vessel diameters (10–15 µm). Vessels angular, solitary and in small groups. Fibers absent. Parenchyma pervasive. Remaining vascular bundles. Crystals absent.

Tangential section

Rays >10-seriate between vascular bundles.

Radial section

Perforation plates simple. Intervessel pits scalariform. Vessel length 120–140 µm.

Bark

Phloem with small groups of sieve tubes. Phellem consists of a few rows of square cells. Crystals absent.

Pith

No slide available.

Analyzed: Root collar of 1 individual (herbarium specimen).

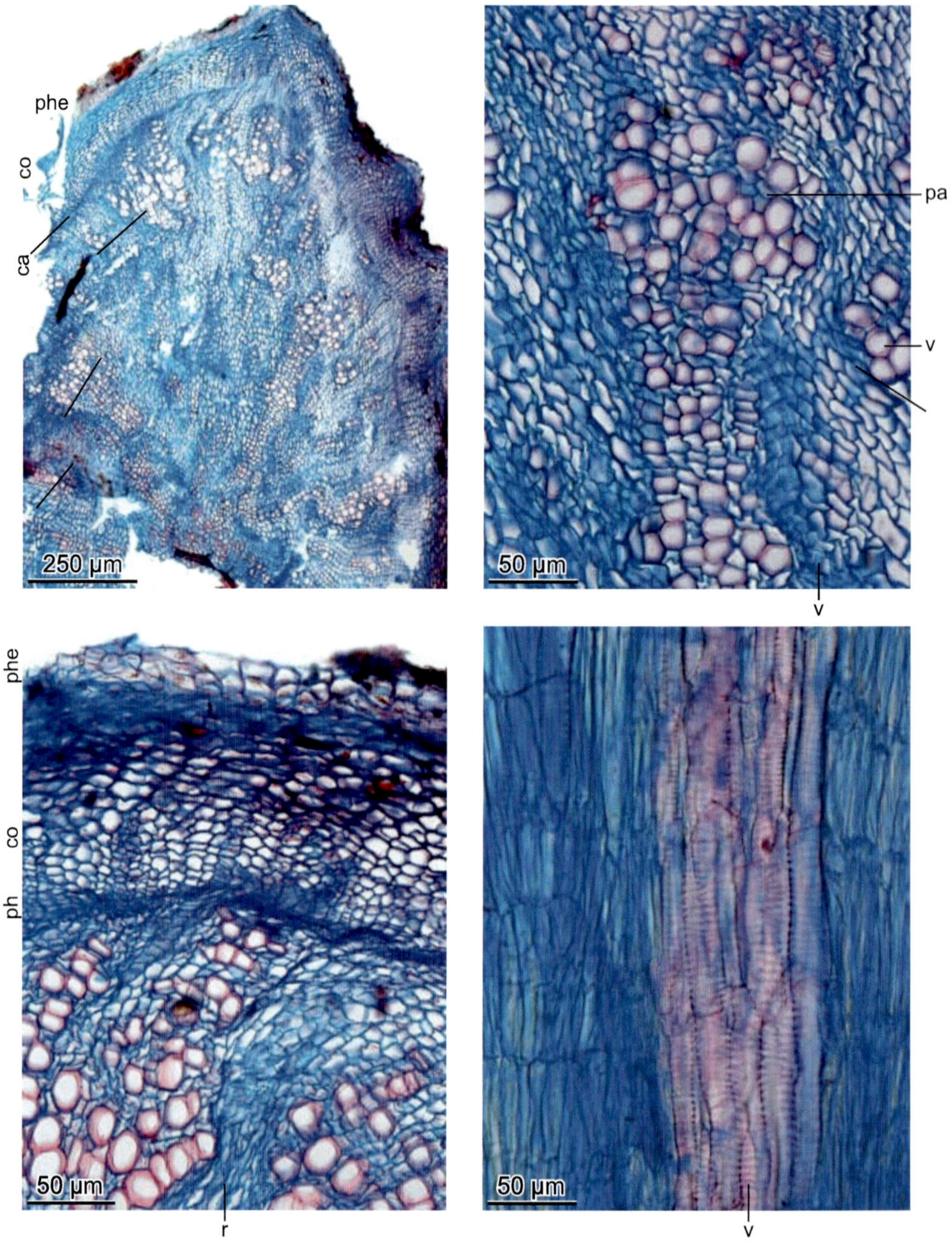

Features according to IAWA

Xylem 1 - 9 - 11 - 13 - 20 - 41 - 50.1 - 52.3 - 60.1 - 79.1 - 99 - 99.1 - 100.1 - 100.2 - 105

Lancea tibetica Hook.f. & Thomson

Phrymaceae

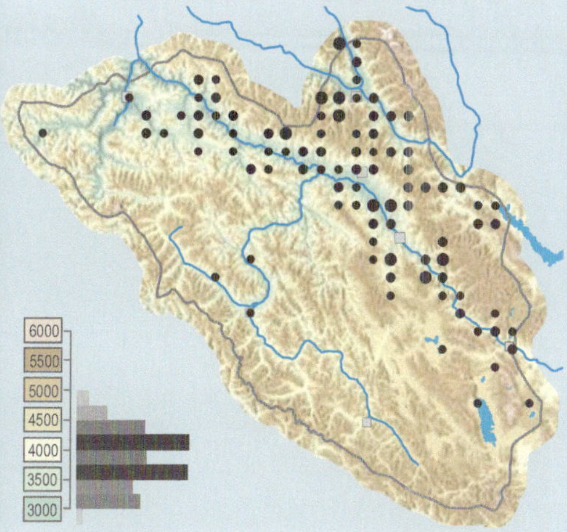

Description: Perennial herbs, 3–10 cm tall, leaves in rosette, obovate, flowers dark blue to purple, ca. 2 cm long, arranged in fascicles by 3–5.

Distribution: N India, China (Gansu, Qinghai, Sichuan, Xizang, Yunnan), Bhutan, Mongolia.

Habitat: Wet pastures, grasslands, along streams.

Stem diameter 1.2 mm. Bark/xylem proportion 1:0.5. Mean ring width 0.18 mm. Maximal 1 annual ring.

Xylem

Transverse section

Only one ring. Diameter of larger the angular vessels 20–50 µm. Vessel solitary, in short radial multiples and in groups. Fibers thin-walled. Parenchyma paratracheal. Crystals absent.

Tangential section

Rays absent.

Radial section

Perforation plates simple. Intervessel pits round. Vessel length 50–100 µm.

Bark

Phloem with small groups of sieve tubes and a few lignified fibers. Phellem absent. Epidermis cells round, thin-walled, with a red-stained periphery. Crystals absent.

Pith

Round, filled with unstained, round cells.

Analyzed: Root collar of polar root of 1 individual.

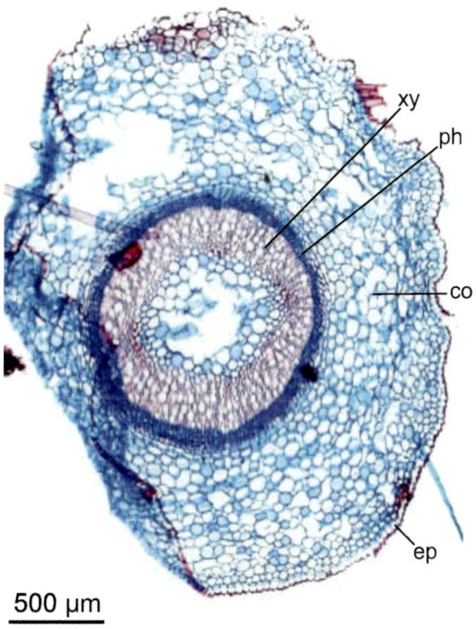

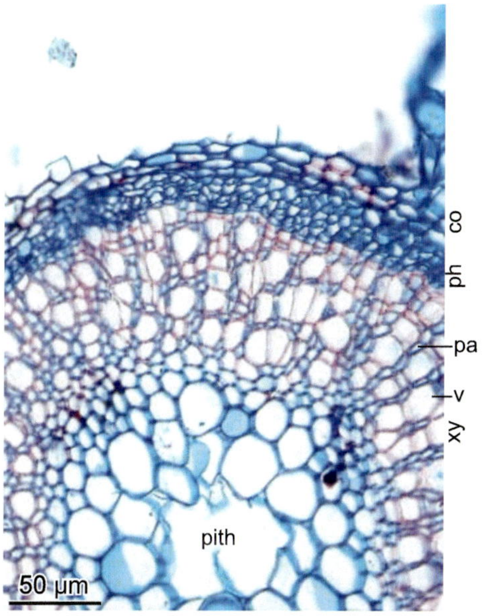

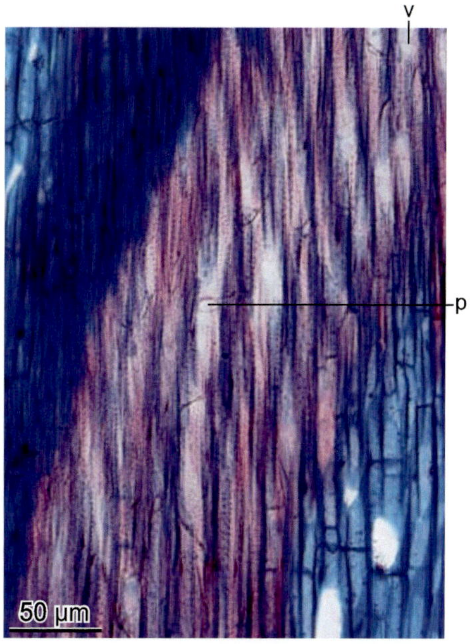

Features according to IAWA

Xylem 2.1 - 9 - 9.1 - 11 - 13 - 22 - 40.2 - 50.2 - 52.2 - 61 - 79 - 117

Picrorhiza kurroa Royle ex Benth.

Plantaginaceae

Description: Perennial herbs, 5–20 cm tall, with long woody rhizomes, leaves almost all basal, oblanceolate to obovate, margin dentate, flowers blue, ca. 5 mm across, with much longer stamens, in cylindric spikes.

Distribution: N India, Nepal, Bhutan

Habitat: Alpine meadows.

Stem diameter 3–5 mm. Bark/xylem proportion 1:0.7. Ring width 0.05–0.1 mm. Maximal 24 annual rings.

Xylem

Transverse section

Semi-ring-porous. Diameter of earlywood vessels 15–20 µm and of latewood vessels <10 µm. Vessels angular, solitary. Vascular bundles remaining. Fibers absent. Parenchyma pervasive. Crystals absent.

Tangential section

Rays between vascular bundles 8–>10-seriate, confluent with axial tissue.

Radial section

Perforation plates simple. Intervessel pits round and scalariform. Vessel length 80–100 µm.

Bark

Phloem homogeneous. Phellem consists of many layers of thin-walled, rectangular cells. Crystals absent.

Pith

Round to star-shaped.

Analyzed: Root collar of polar roots of 5 individuals.

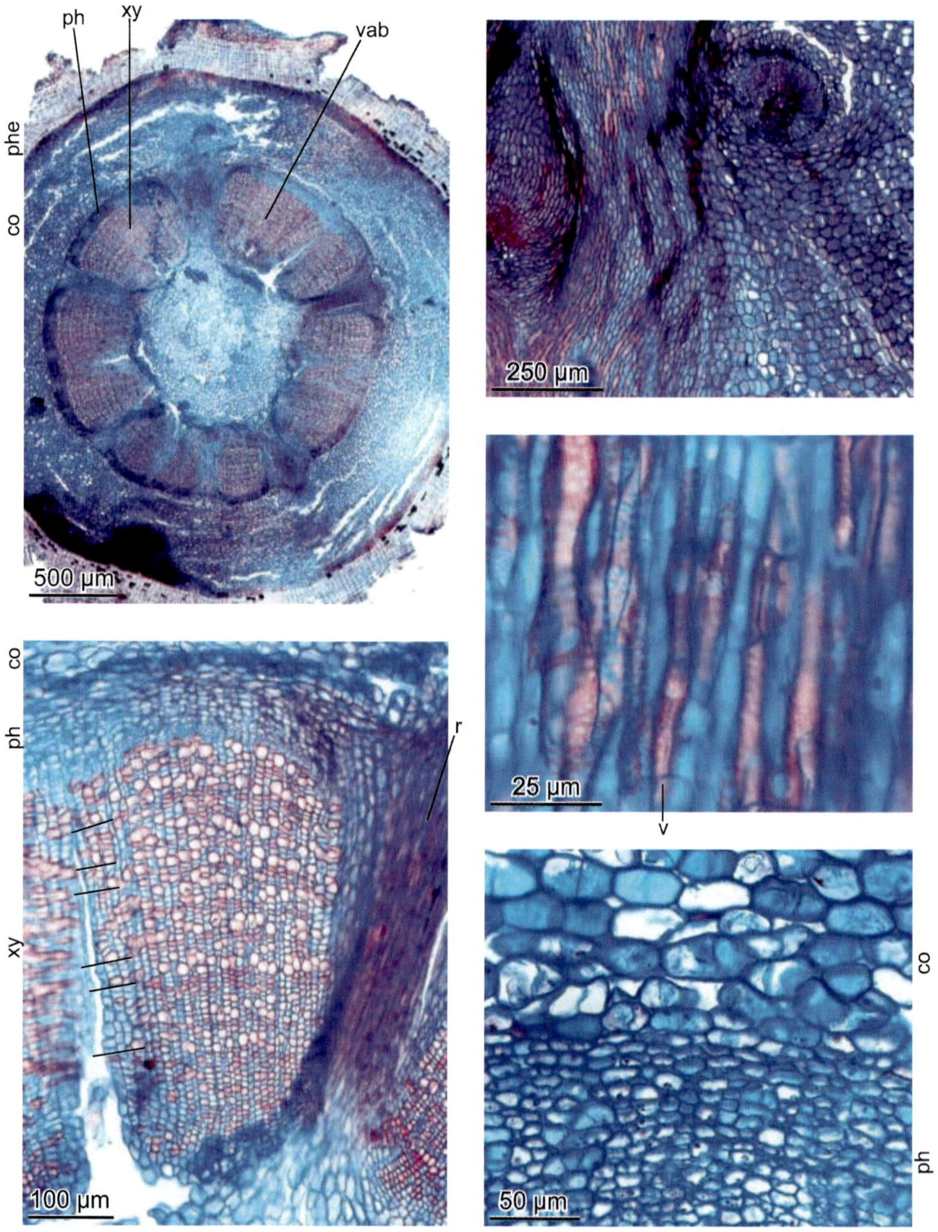

Features according to IAWA

Xylem 1 - 9 - 12 - 13 - 20 - 22 - 40.2 - 50.2 - 52.2 - 60.1 - 79.1 - 99 - 99.1 - 100.1 - 100.2 - 105

Plantago depressa Willd.

Plantaginaceae

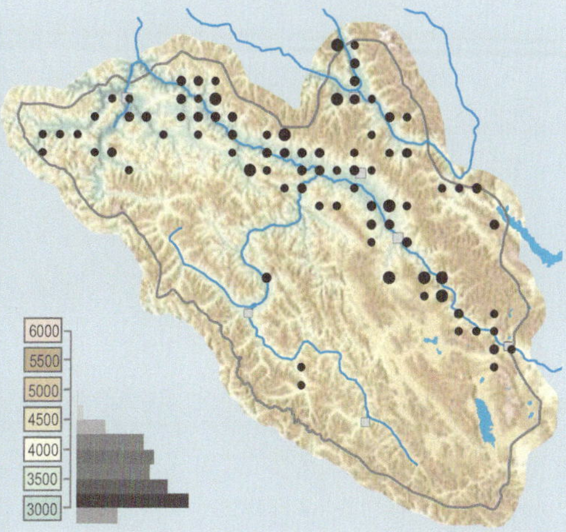

Syn.: *P. leptostachys* Ledeb., *P. sibirica* Poir., *P. tibetica* Hook.f. & Thomson

Description: Annuals or short-lived perennials, 5–20 cm tall, thinly villous, leaves only basal, elliptic to lanceolate, flowers white, minute, in dense narrowly cylindric spikes, 6–12 cm long.

Distribution: Russia (Siberia, Far East), Kazakhstan, Kyrgyzstan, Afghanistan, N Pakistan, India (Jammu and Kashmir, Himachal Pradesh), N and W China, Bhutan, Mongolia, Korea.

Habitat: Gardens, field margins, loamy and often eutrophicated soils.

Stem diameter 5 mm. Bark/xylem proportion 1:0.7. Ring width 0.3–0.6 mm. Maximal 3 annual rings.

Xylem

Transverse section

Semi-ring-porous. Ring variability indicated by larger earlywood (20–30 µm) and smaller latewood vessel diameters (<10 µm). Angular vessels solitary in short radial multiples and in groups. Vascular bundles remain partially. Fibers absent. Parenchyma pervasive. Crystals absent.

Tangential section

Rays 8–>10-seriate, confluent with the axial tissue.

Radial section

Perforation plates simple. Intervessel pits round and scalariform. Vessel length 70–100 µm.

Bark

Phloem with small groups of sieve tubes. Phellem consists of a few layers of thin-walled, rectangular cells. Crystals absent.

Pith

Star-shaped.

Analyzed: Root collar of polar root of 4 individuals.

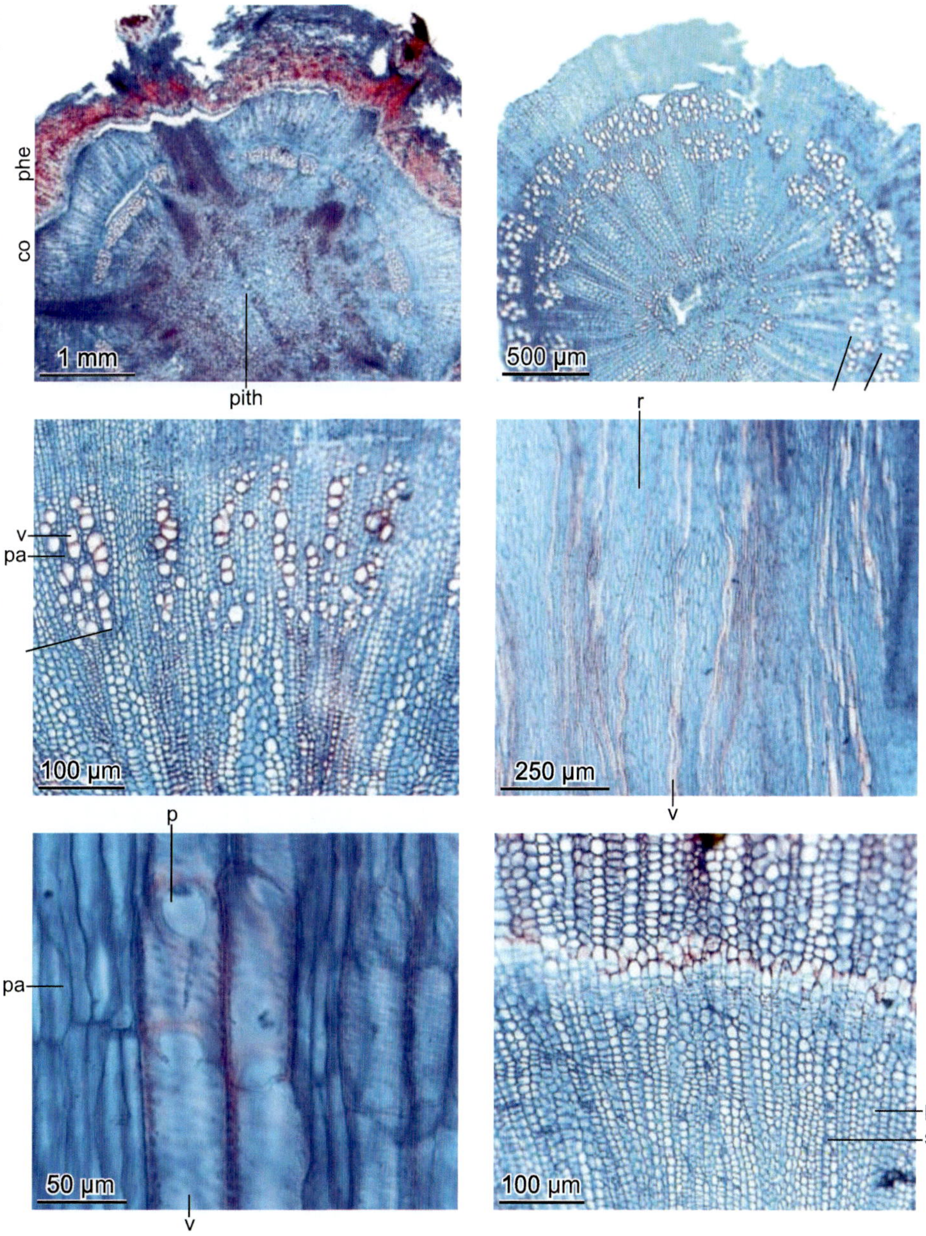

Features according to IAWA

Xylem *1 - 9 - 9.1 - 11 - 12 - 13 - 22 - 40.2 - 50.2 - 52.2 - 60.1 - 79.1 - 98 - 99 - 99.1 - 100.1 - 100.2*

Plantago himalaica Pilg.

Plantaginaceae

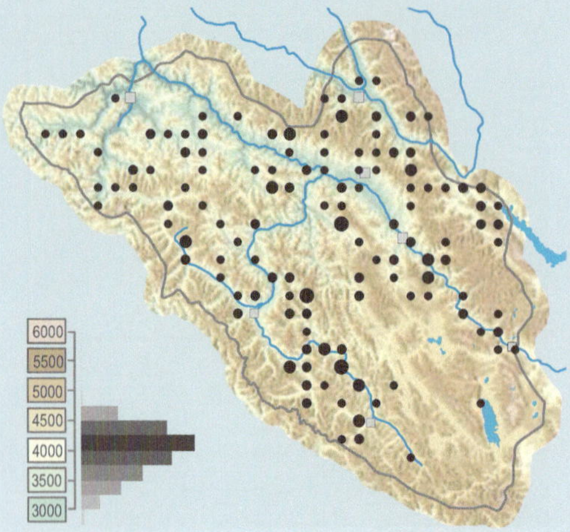

Syn.: *P. brachyphylla* Edgew. ex Decne.

Description: Perennial herbs, 5–20 cm tall, leaves only basal, ovate to broad ovate, hairy when young, flowers minute, in dense narrowly cylindric spikes, 2–5 cm long.

Distribution: Pakistan, NW India, Nepal.

Habitat: River banks, settlements, gardens, field margins, along irrigation canals.

Stem diameter 5 mm. Bark/xylem proportion 1:1. Mean ring width 1 mm. Maximal 1 annual ring.

Xylem

Transverse section

Only one ring present. Diameter of larger vessels 20–25 µm. Angular vessels solitary and fairly thick-walled. Vascular bundles remain partially. Fibers absent. Parenchyma pervasive. Crystals absent.

Tangential section

Rays 8–>10-seriate, confluent with the axial tissue.

Radial section

Perforation plates simple. Intervessel pits scalariform. Vessel length 50–80 µm.

Bark

Phloem with small groups of sieve tubes. Phellem not observed. Crystals absent.

Pith

Large, round and star-shaped.

Analyzed: Root collar of polar root of 3 individuals.

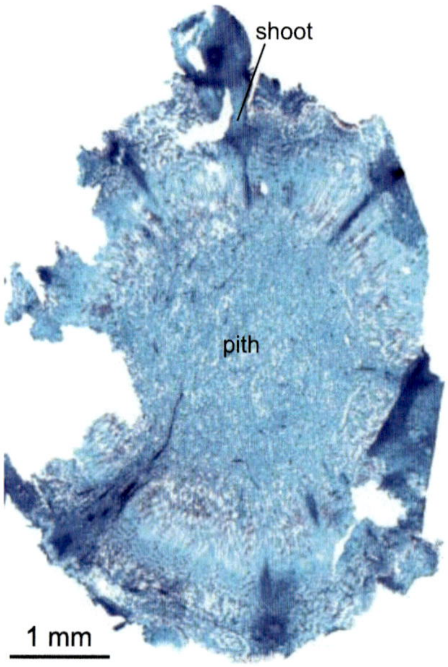

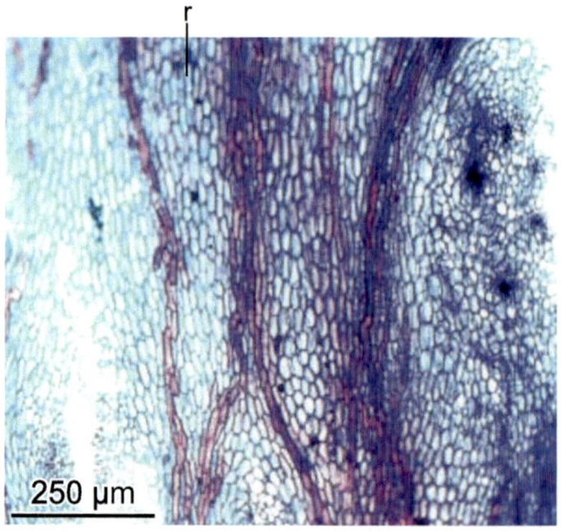

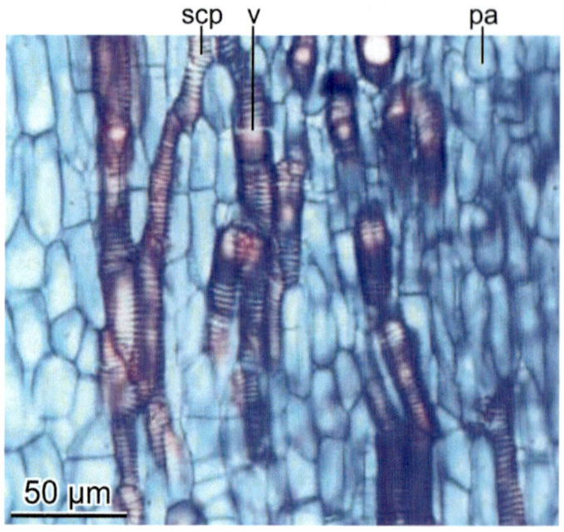

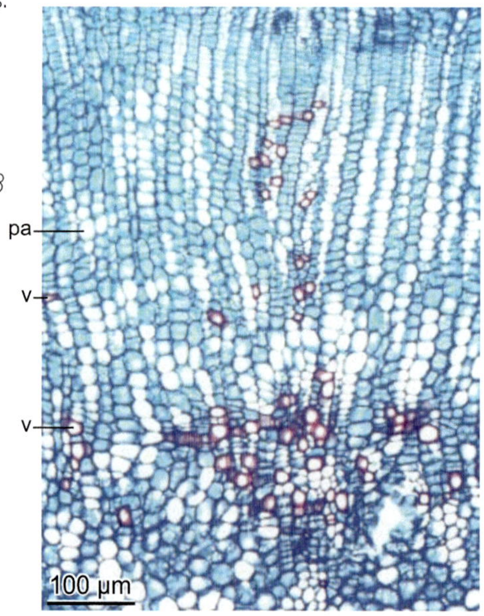

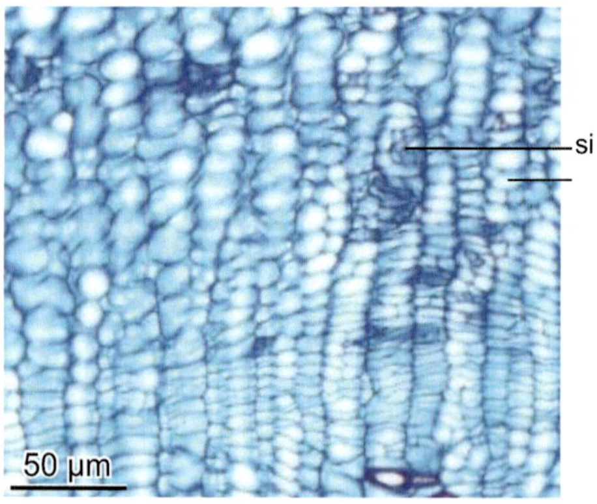

Features according to IAWA

Xylem 1 - 9 - 13 - 20 - 40.2 - 50.2 - 52.2 - 60.1 - 79.1 - 98 - 99 - 99.1 - 100.1 - 100.2

Acantholimon lycopodioides (Girard) Boiss.

Plumbaginaceae

Syn.: *Statice lycopodioides* Girard, *A. tibeticum* Hook.f. & Thomson ex C.B.Clarke

Description: Subshrubs, forming dense semiglobose or flat cushions 30–100 cm across, leaves linear and pungent, flowers white to pink, arranged in spikelets.

Distribution: Tajikistan, Afghanistan, Pakistan, India (Jammu and Kashmir), China (NW Xinjiang).

Habitat: Dry stable slopes and rocks.

Stem diameter 5–6 mm. Bark/xylem proportion 1:4. Mean ring with 0.2–0.6 mm. Maximal 50 annual rings.

Xylem

Transverse section

Semi-ring-porous. Ring variability indicated by larger earlywood (30–40 µm) and smaller latewood vessel diameters (<10 µm) and a small band of marginal parenchyma cells. Vessel solitary and fairly thick-walled. Fibers thick-walled in groups. Parenchyma paratracheal and pervasive. Crystals absent.

Tangential section

Rays 5–8-seriate.

Radial section

Perforation plates simple. Intervessel pits round. Vessel length 80–100 µm.

Bark

Phloem with large groups of fibers. Phellem consists of a band of upright, thick-walled cells. Crystals absent.

Pith

No slide available.

Analyzed: Root collar of polar root of 4 individuals.

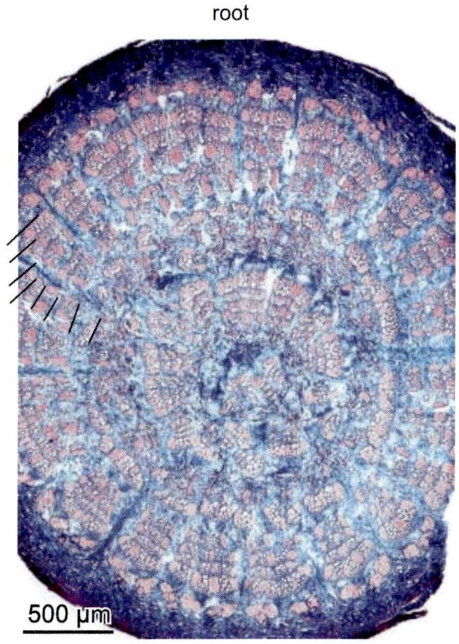

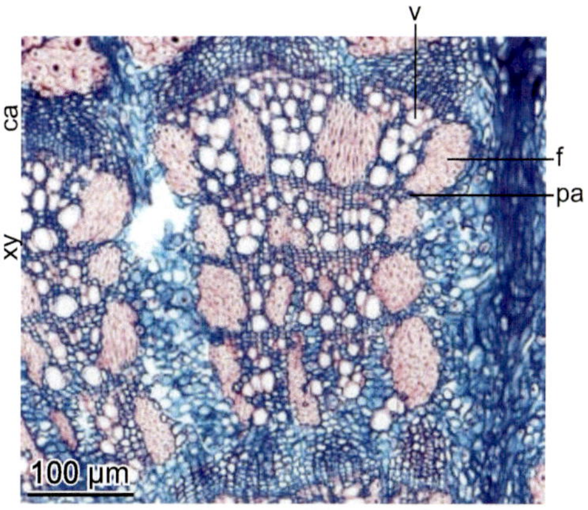

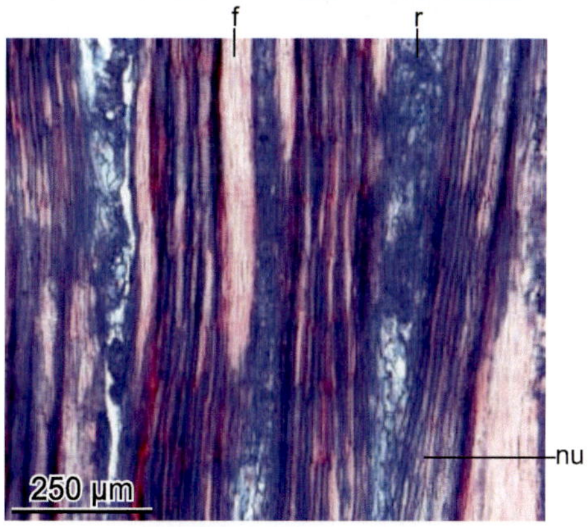

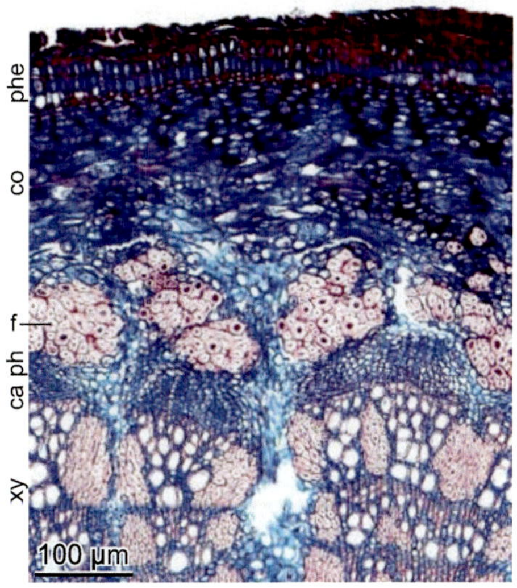

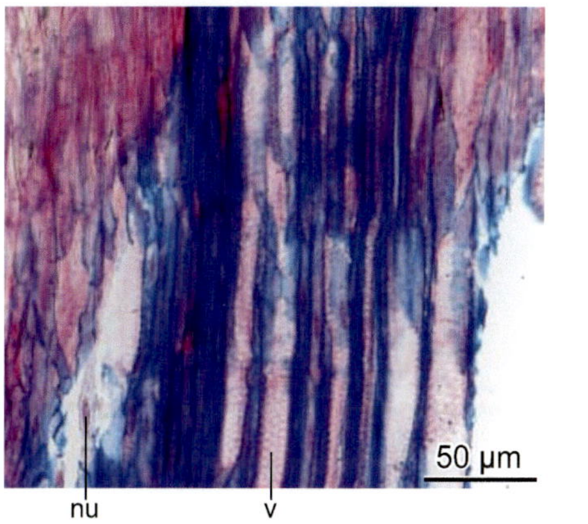

Features according to IAWA

Xylem 1 - 9 - 13 - 22 - 39.1 - 40.2 - 50.2 - 52.2 - 61 - 69 - 79 - 79.1 - 98 - 100.1 - 100.2

Bukiniczia cabulica (Boiss.) Lincz.

Plumbaginaceae

Syn.: *Statice cabulica* Boiss., *S. macrorrhabdos* Boiss., *Limonium cabulicum* (Boiss.) Kuntze, *Dictyolimon macrorrhabdos* (Boiss.) Rech.f.

Description: Biennial herbs, 10–30 cm tall, leaves in basal rosette, obovate to spatulate, fleshy, glaucous with whitish blurs along veins, flowers whitish, arranged in 3–4-flowered spikelets in clusters along the scape branches, bracts ovate, with a broad hyaline margin.

Distribution: Afghanistan, Pakistan, India (Jammu and Kashmir).

Habitat: Dry habitats, stony steppe.

Stem diameter 5 mm. Bark/xylem proportion 1:1.5. Ring width 0.25–0.7 mm. Maximal 4 annual rings.

Xylem

Transverse section

Diffuse-porous. Ring variability indicated by larger earlywood (20–30 µm) and smaller latewood vessel diameters (10 µm). Vessels solitary and in short radial multiples. Fibers absent. Parenchyma paratracheal and pervasive. Crystals absent.

Tangential section

Rays 5–8-seriate, confluent, not lignified.

Radial section

Perforation plates simple. Intervessel pits round and scalariform. Vessel length 80–100 µm. Ray cells upright.

Bark

Phloem with small groups of sieve tubes. Phellem consists of a band of upright, thick-walled cells.

Pith

Star-shaped.

Analyzed: Root collar of polar root of 2 individuals.

Features according to IAWA

Xylem 1 - 5 - 9 - 9.1 - 13 - 20 - 22 - 40.2 - 50.2 - 52.2 - 60.1 - 79 - 79.1 - 98 - 100.1 - 100.2

Aconogonon rumicifolium (Royle ex Bab.) H.Hara

Polygonaceae

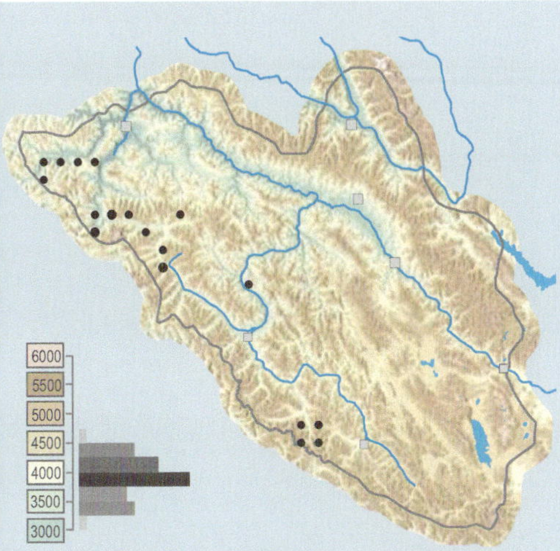

Syn.: *Polygonum rumicifolium* Royle ex Bab., *P. glaberrimum* Kitam., *P. somdevae* Aswal & Mehrotra

Description: Perennial herbs, 15–120 cm tall, leaves broadly ovate, flowers greenish, 4–6 mm across, in dense axillary and terminal clusters.

Distribution: Afghanistan, Pakistan, N India, Nepal, China (Xizang).

Habitat: Dry stony slopes.

Stem diameter 15–20 mm. Bark/xylem proportion 1:8. Ring width 0.5–2 mm. More than 7 annual rings. Stem center rotten.

Xylem

Transverse section

Semi-ring-porous. Vessels solitary. Rings indicated by larger earlywood (60–80 μm) and smaller latewood vessel diameters (20–30 μm). Fibers absent. Parenchyma pervasive. With crystal druses in idioblasts. With inter-xylary phloem.

Tangential section

Rays >10-seriate, unlignified, confluent.

Radial section

Perforation plates simple. Intervessel pits scalariform. Vessel length 100–120 μm.

Bark

Phloem with a few small groups of sieve elements. With large, irregularly formed intercellulars. Phellem consists of few rows of rectangular, thin-walled cells.

Pith

No slide available.

Analyzed: Root collar of polar roots of 3 individuals.

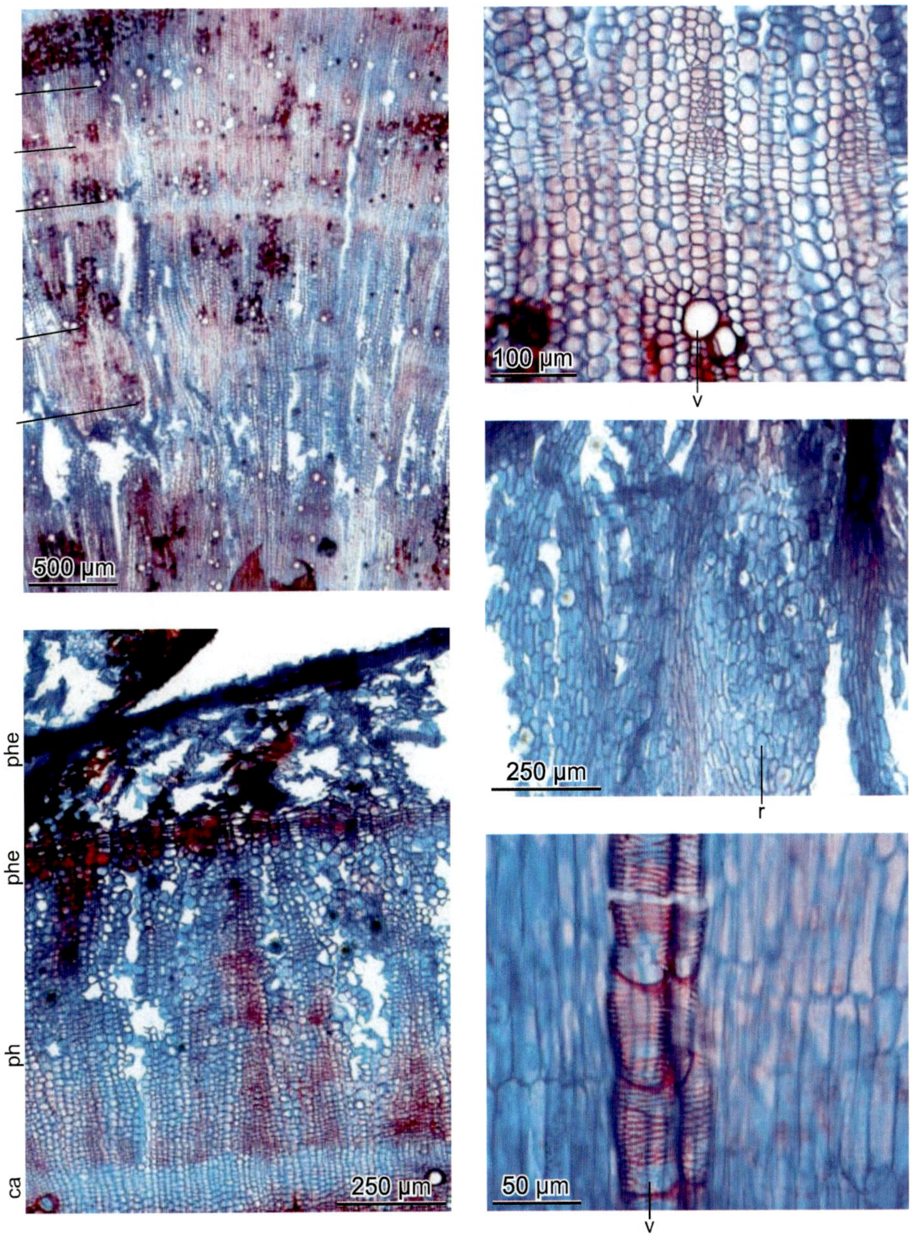

Features according to IAWA

Xylem 1 - 4 - 9 - 13 - 20 - 41 - 50.1 - 52.3 - 60.1 - 79.1 - 89 - 100.1 - 100.2 - 135 - 144

Aconogonon tortuosum (D.Don) H.Hara

Polygonaceae

Syn.: *Polygonum tortuosum* D.Don, *Pleuropteropyrum tortuosum* (D.Don) Munshi & Javeid

Description: Perennial herbs, 30–150 cm tall, forming dense groups of shoots, stems dichotomously branched, leaves elliptic to lanceolate-ovate, flowers creamy white, 2–3.5 mm across, in terminal or axillary panicles 1–5 cm long.

Distribution: Afghanistan, N Pakistan, N India (Jammu and Kashmir, Himachal Pradesh, Uttarakhand, Sikkim), Nepal, China (Xizang), Bhutan.

Habitat: Sandy slopes with stones, unstable loamy slopes, screes.

Stem diameter 10–25 mm. Bark/xylem proportion 1:10. Ring width 1–1.5 m. More than 6 annual rings. Center rotten.

Xylem

Transverse section

Semi-ring-porous. Vessels solitary. Rings indicated by larger earlywood (90–100 µm) or smaller latewood vessel diameters (20–30 µm). Fibers absent. Parenchyma pervasive. With crystal druses. With inter-xylary periderm towards the pith.

Tangential section

Rays confluent or absent.

Radial section

Perforation plates simple. Intervessel pits scalariform. Vessel length 100–120 µm.

Bark

Phloem homogeneous. With large, irregularly formed intercellulars. Phellem consists of a few rows of rectangular, thin-walled cells.

Pith

No slide available.

Analyzed: Root collar of polar root of 2 individuals.

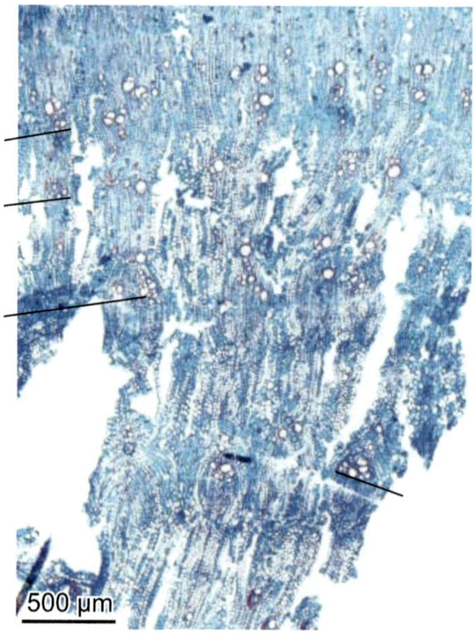

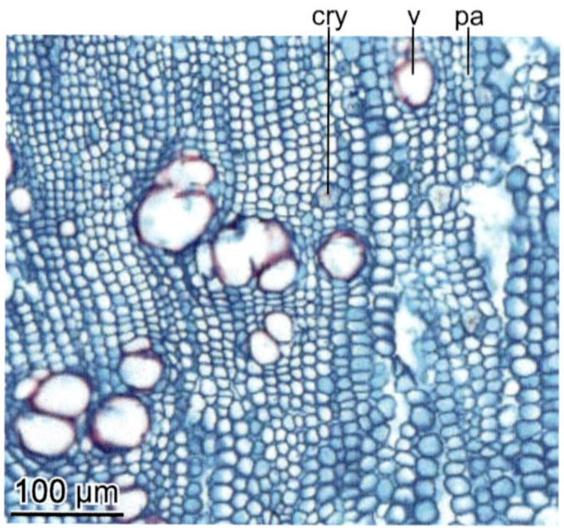

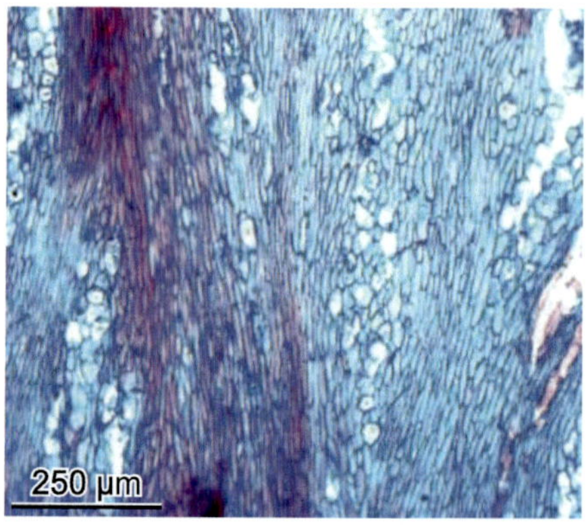

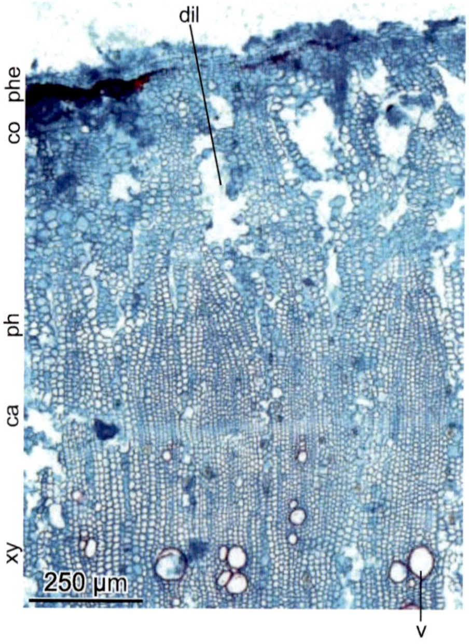

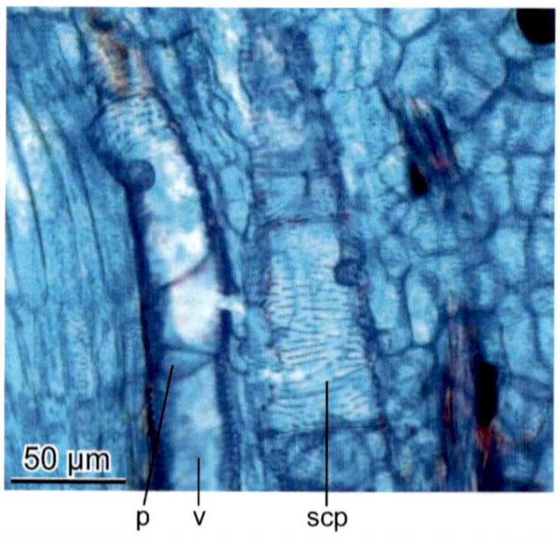

Features according to IAWA

Xylem 1 - 4 - 9 - 13 - 20 - 41 - 50.1 - 52.3 - 60.1 - 79.1 - 117 - 135.2 - 144

Bistorta affinis (D.Don) Greene

Polygonaceae

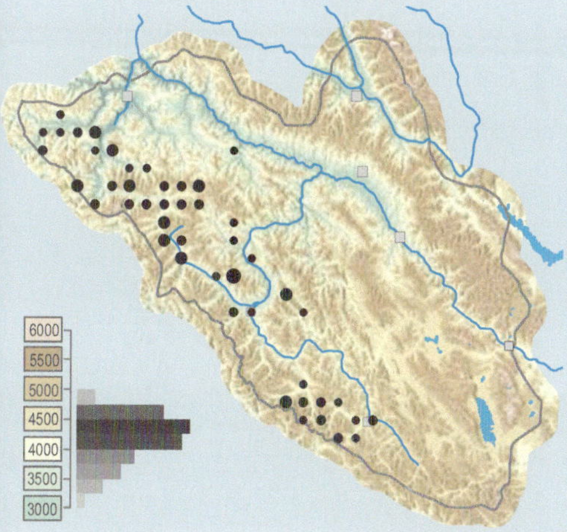

Syn.: *Polygonum affine* D.Don, *P. donianum* Spreng., *Persicaria affinis* (D.Don) Ronse Decr.

Description: Perennial herbs, 10–20 cm tall, glabrous, densely cespitose, leaves lanceolate or oblanceolate, 5–10 cm long, flowers pink, in dense erect spikes, 3–6 cm long.

Distribution: Pakistan, NW India, Nepal, W Myanmar.

Habitat: Open slopes, rocky places, river beds.

Stem diameter 3–5 mm. Bark/xylem proportion 1:4. Ring width 0.06–0.08 mm. Maximal 14 annual rings.

Xylem

Transverse section

Semi-ring-porous. Vessels solitary and in small groups. Rings indicated by larger earlywood (60–70 µm) or smaller latewood vessel diameters (20–30 µm). Fibers thin- to thick-walled. Parenchyma absent. With few large crystal druses.

Tangential section

Rays absent.

Radial section

Perforation plates simple. Intervessel pits round. Vessel length 120–200 µm.

Bark

Phloem homogeneous. With a few large crystal druses. Cortex cells large, oval, thin-walled. Phellem and cortex separated by an endodermis. Phellem consists of few rows of rectangular, thin-walled cells.

Pith

Round, surrounded by distinct vascular bundles. With a few dark-stained cells.

Analyzed: Root collar of polar root of 6 individuals.

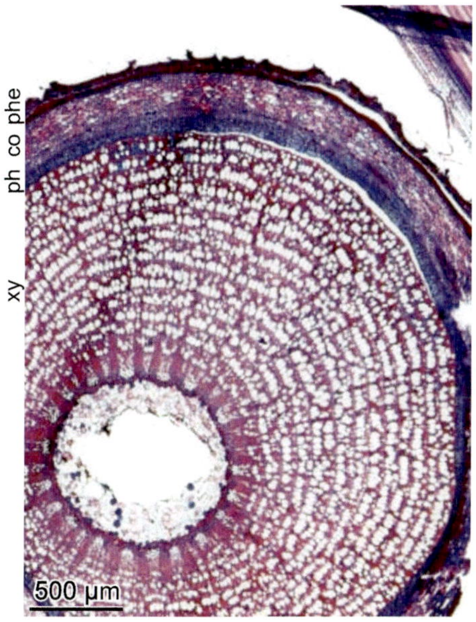

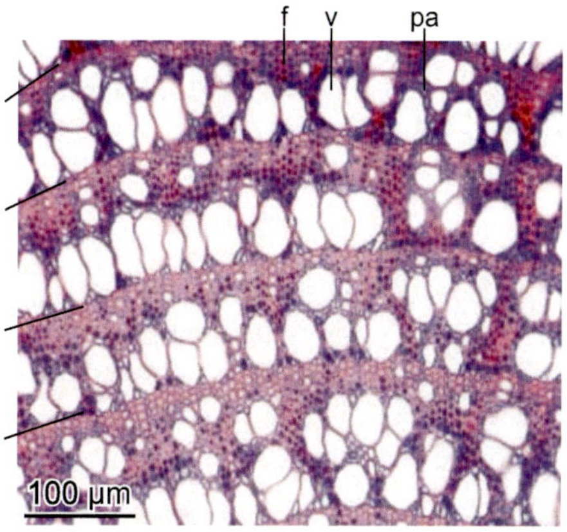

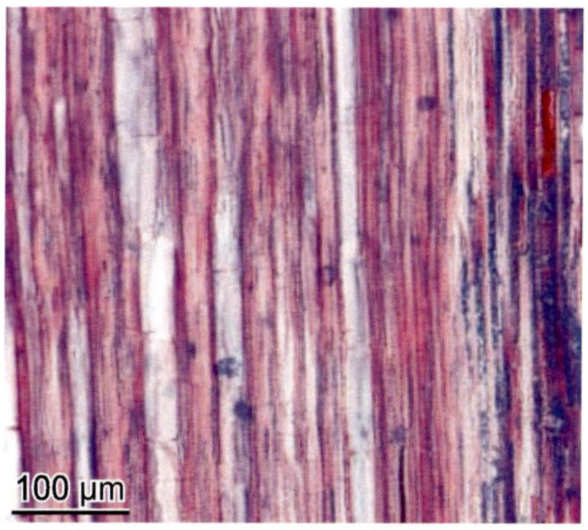

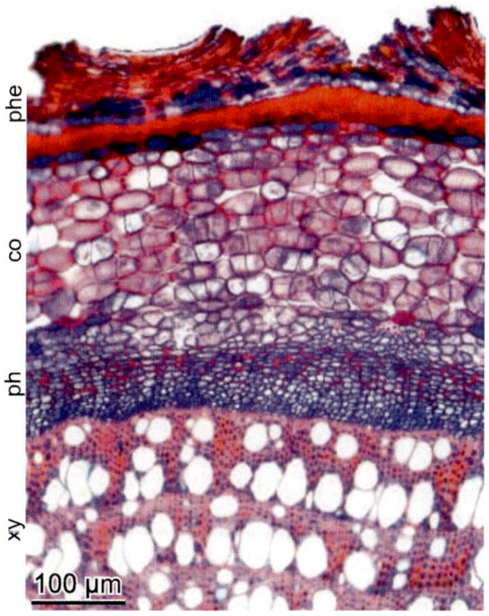

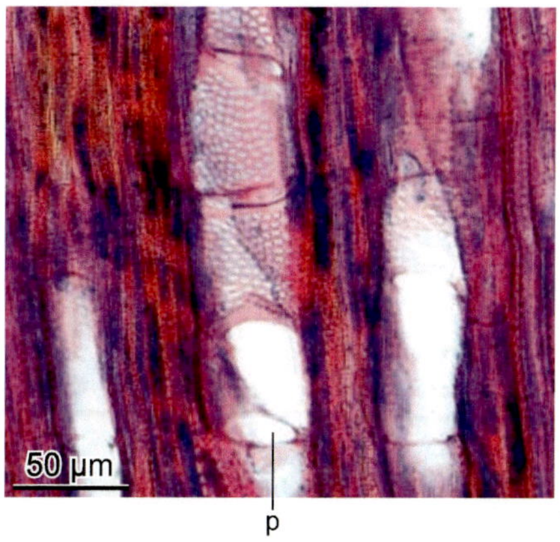

Features according to IAWA

Xylem 1 - 4 - 9 - 11 - 13 - 22 - 41 - 50.2 - 52.3 - 61 - 70 - 75 - 117 - 144

Bistorta vivipara (L.) S.F.Gray

Polygonaceae

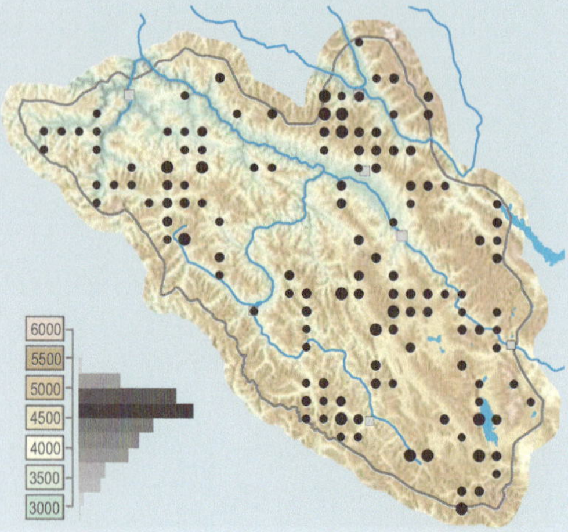

Syn.: *Polygonum viviparum* L.

Description: Perennial herbs, 10–35 cm tall, leaves linear to lanceolate, flowers whitish or pink, 2–4 mm across, in narrow dense spikes, usually with brownish bulbils of vegetative propagation.

Distribution: N America, Greenland, Europe, Russia (Siberia), W and C Asia, N India, Nepal, China (Xizang), Bhutan, Japan.

Habitat: Mesic and wet turf, grasslands.

Bulb diameter 4–8 mm. Bark/xylem proportion 1:0.3. Rings absent. Parenchymatic center oft bulb is aerenchymatic.

Xylem

Transverse section

Diameter of vessels 10–20 µm. Vessels mostly solitary. Fibers absent. Parenchyma pervasive. Vascular bundles remain. Many large crystal druses (50 µm) occur in large, unlignified parenchyma cells.

Tangential section

Rays consist of large, inter-vascular parenchyma zones. Rays disappear in polarized light.

Radial section

Perforation plates simple. Intervessel pits round in alternating position. Vessel length 50–70 µm.

Bark

Phloem of some vessels contains groups of sieve tubes. Cortex consists of round to oval parenchyma cells with diameters of 80–120 µm. Individuals on wet sites with aerenchyma. Many lacunas in the cortex. Phellem layered, consisting of few layers of rectangular cells.

Pith

No slide available.

Analyzed: Bulb of 5 individuals.

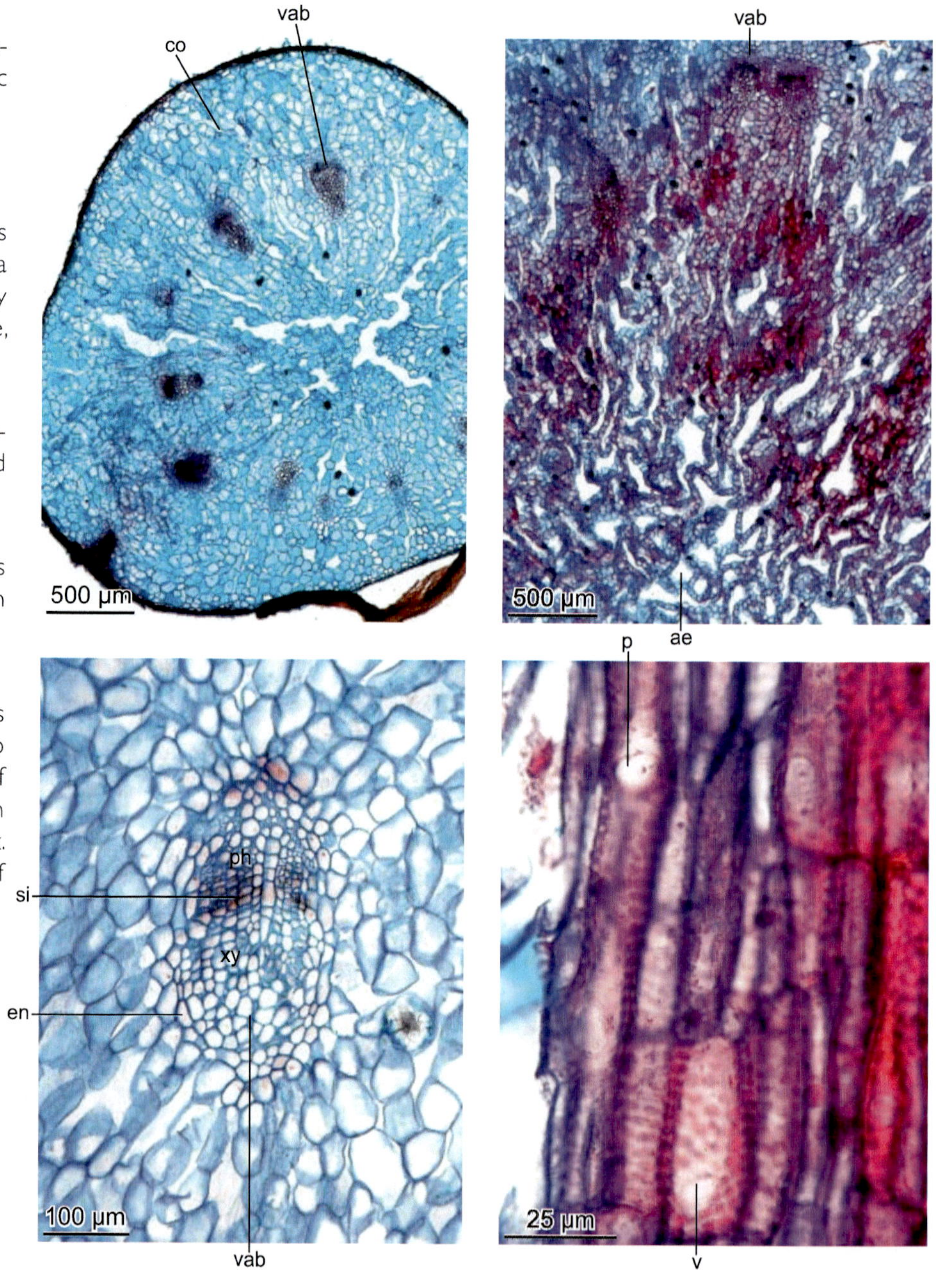

Features according to IAWA

Xylem 2 - 2.1 - 9 - 13 - 22 - 40.1 - 50.1 - 52.1 - 60.1 - 79 - 99 - 99.1 - 100.2 - 105 - 144

Fagopyrum esculentum Moench

Polygonaceae

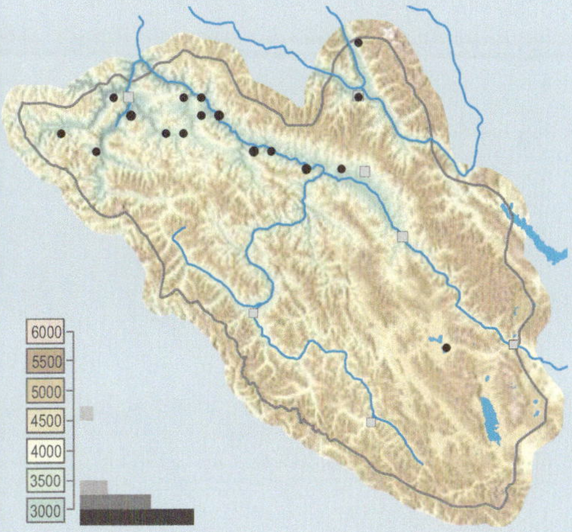

Syn.: *Polygonum fagopyrum* L.

Description: Annuals, 30–90 cm tall, stems branched, leaves triangular with cordate base, flowers white or pink, 3–4 mm across, in terminal or axillary racemes or corymbs.

Distribution: Probably native to China; widely cultivated and introduced.

Habitat: Cultivated, sometimes escaping.

Stem diameter 3–5 mm. Bark/xylem proportion 1:1. Ring width 0.5–0.7 mm. Maximal 1 annual ring.

Xylem

Transverse section

Vessels in short radial rows and groups. Diameter of larger vessels 30–40 µm. Fibers thin- to thick-walled. Parenchyma paratracheal. Crystal absent.

Tangential section

Rays absent or indistinct, 1–3-seriate.

Radial section

Perforation plates simple. Intervessel pits round. Vessel length 80–100 µm.

Bark

Phloem with groups of sieve elements. Cortex cells large, thin-walled. Phellem absent. With crystal druses.

Pith

No slide available.

Analyzed: Root collar of polar roots of 2 individuals.

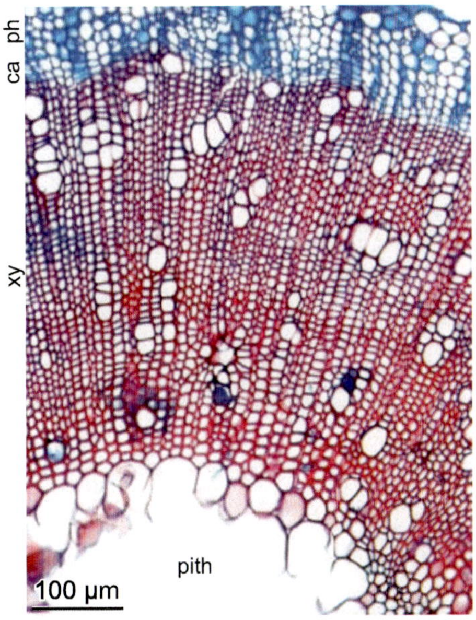

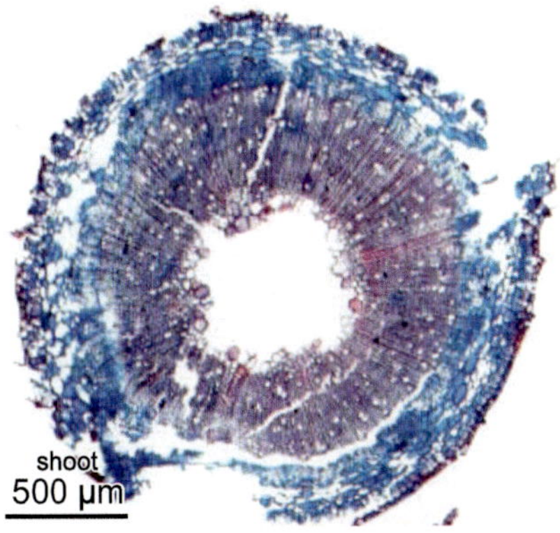

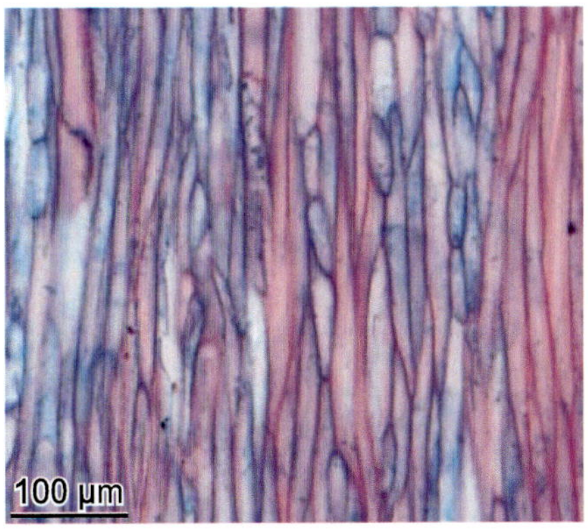

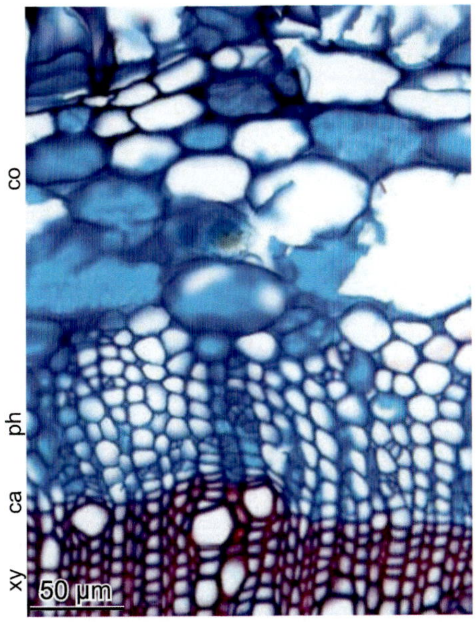

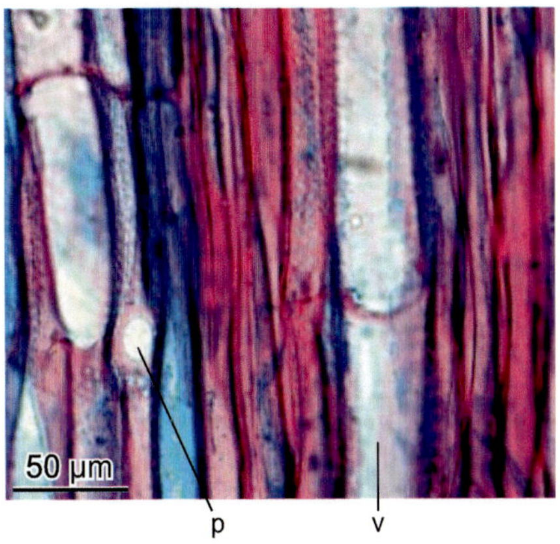

Features according to IAWA

Xylem 2.1 - 9.1 - 10 - 13 - 22 - 40.2 - 50.1 - 52.2 - 61 - 70 - 79 - 117

Knorringia pamirica (Korsh.) Tzvelev

Polygonaceae

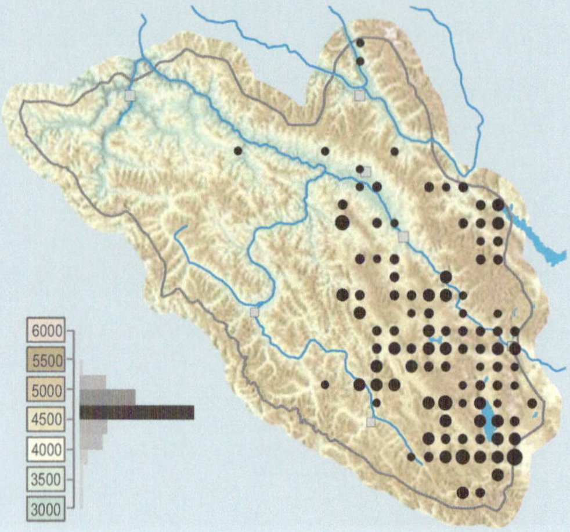

Syn.: *Polygonum sibiricum* var. *thomsonii* Meissn., *P. pamiricum* Korsh.

Description: Perennial herbs, 2–5 cm tall, rhizomatous, stems prostrate to ascending, leaves linear, sometimes with hastate base, flowers whitish, ca. 2 mm across, in capitate panicles.

Distribution: Russia, Kyrgyzstan, Tajikistan, Afghanistan, Pakistan, India, Nepal, China (Qinghai, Xizang).

Habitat: Disturbed loamy and sandy sediments, road margins, salty wetlands and grasslands, weedy in barley fields.

Stem diameter 1–1.5 mm. Bark/xylem proportion 2:1. Mean ring width 0.3 mm. Maximal 1 annual rings.

Xylem

Transverse section

Only one ring. Diameter of earlywood vessels 40–50 µm. Vessels solitary or in small groups. Fibers thin- to thick-walled. Parenchyma paratracheal and pervasive. Vascular bundles remain, circular arranged. Crystals absent.

Tangential section

Rays 4–6-seriate between vascular bundles. Rays unlignified and confluent with the axial tissue.

Radial section

Perforation plates simple. Intervessel pits round to scalariform. Vessel length 50–80 µm.

Bark

Phloem contains groups of sieve tubes and fibers. Ray dilatation between vascular bundles. Cortex contains irregular intracellular spaces. Phellem not observed.

Pith

Polygonal.

Analyzed: Root collar of polar roots of 2 individuals.

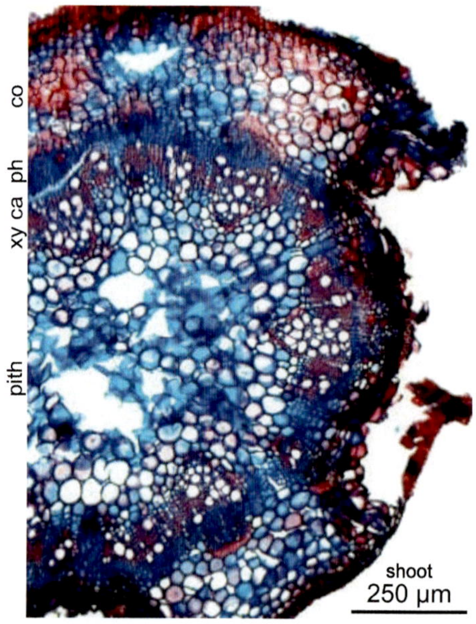

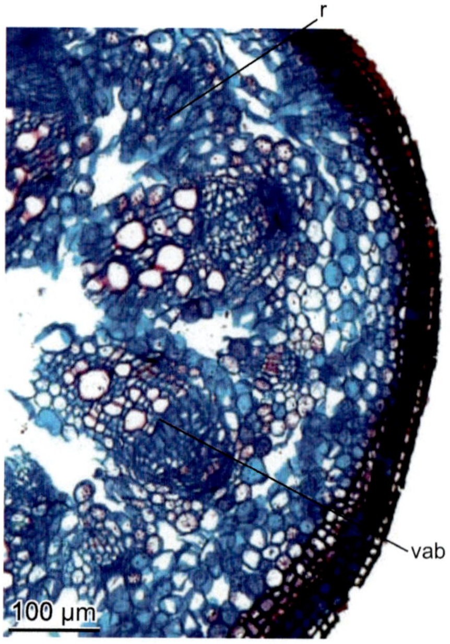

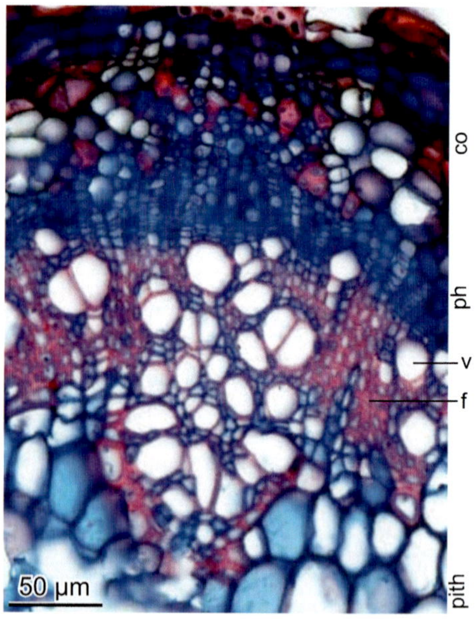

Features according to IAWA

Xylem 2.1 - 9 - 11 - 13 - 20 - 22 - 40.2 - 50.1 - 52.2 - 61 - 70 - 79 - 79.1 - 98 - 99.1 - 100.1 - 100.2 - 105

Koenigia islandica L.

Polygonaceae

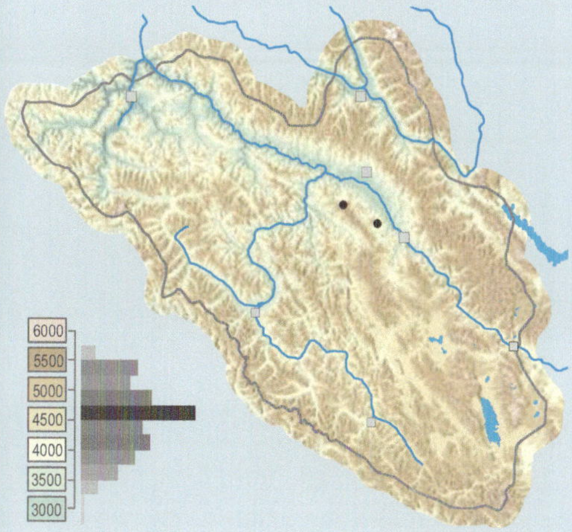

Syn.: *Polygonum islandicum* (L.) Hook.f.

Description: Annuals, 1–8 cm tall, cespitose, stems reddish, branched or simple, rooting adventitiously from proximal nodes, leaves obovate to broadly elliptic, flowers creamy white, ca. 1 mm across, in terminal 10–15-flowered corymbs.

Distribution: N America, Greenland, South America (Argentina, Chile), N Europe, Russia, Kazakhstan, Kyrgyzstan, Pakistan, NW India, Nepal, China, Bhutan, Mongolia.

Habitat: Wet and waterlogged habitats close to streams and lakes, especially on loamy soils.

Stem diameter 0.8 mm. Bark/xylem proportion 1:0.5. Ring width 0.1 mm. Maximal 1 annual ring.

Xylem

Transverse section

Vessels solitary and in groups. Diameter of larger vessels 20–30 µm. Fibers thin- to thick-walled. Parenchyma paratracheal, rare. Lateral parenchymatic zones between vascular bundles With a few isolated vascular bundles. Crystals absent.

Tangential section missing (features described from transverse section)

Rays absent within vascular bundles. Parenchyma zones between vascular bundles.

Radial section missing (features described from thick, oblique transverse section)

Perforation plates simple. Intervessel pits scalariform.

Bark

Phloem with groups of sieve elements. Cortex cells large, thin-walled. Phellem absent. Epidermis cells rectangular.

Pith

Star-shaped, filled with thin-walled, unlignified cells.

Analyzed: Root collar and stem of 1 individual.

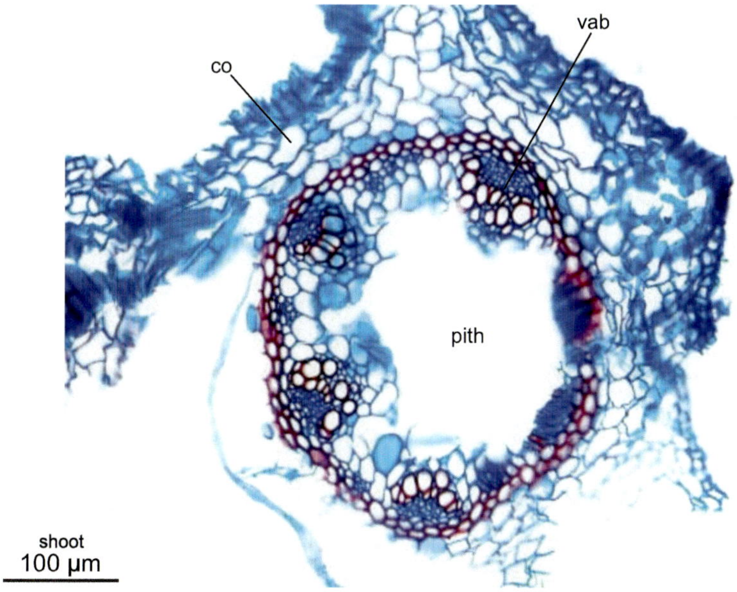

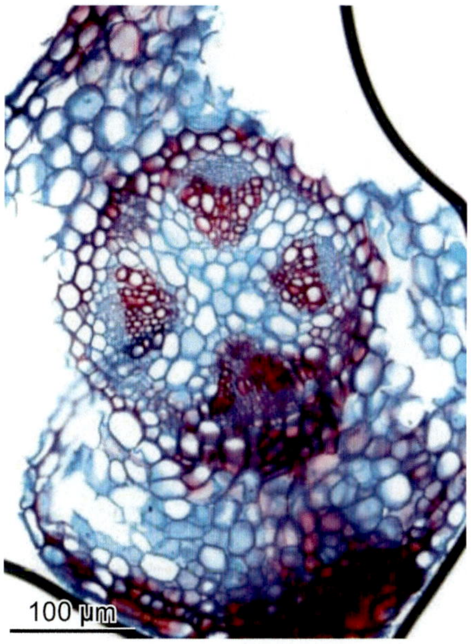

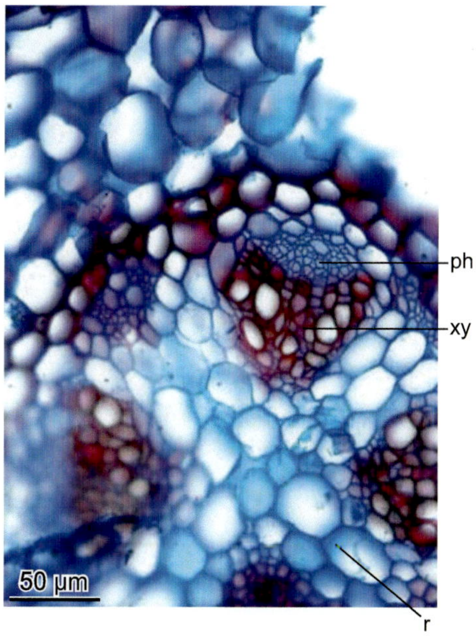

Features according to IAWA

Xylem 2.1 - 9.1 - 10 - 13 - 22 - 40.2 - 50.1 - 52.2 - 61 - 70 - 79 - 99.1 - 100.1 - 100.2 - 117

Oxyria digyna (L.) Hill

Polygonaceae

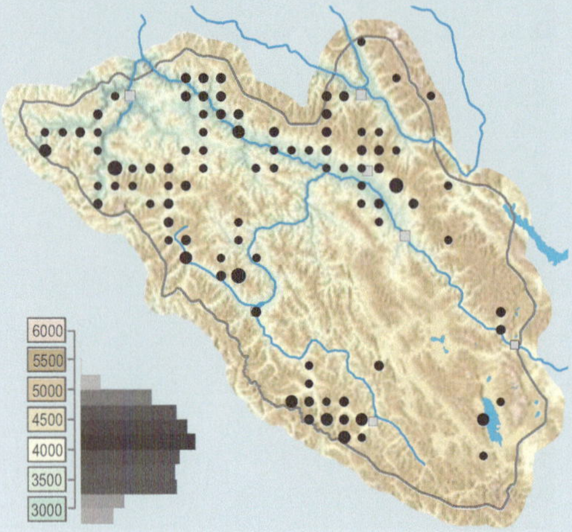

Syn.: *Rumex digynus* L.

Description: Perennial herbs, 5–30 cm tall, stems erect, leaves reniform to orbicular, somewhat fleshy, mostly basal, long-petioled, 2–5 cm across, flowers in terminal panicle.

Distribution: Circumpolar (America, Europe, Asia).

Habitat: Mesic to wet spring areas and snow beds, along streams.

Stem diameter 6 mm. Bark/xylem proportion 1:1.5. Ring width 0.25–1 mm. Maximal 7 annual rings.

Xylem

Transverse section

Semi-ring-porous. Diameter of earlywood vessels 40–50 μm, diameter of latewood vessels 20 μm. Vessels mostly solitary. Fibers absent. Parenchyma pervasive. Vascular bundles remain. Many large crystal druses (50 μm) in chambered cells in rays.

Tangential section

Rays 5–>10-seriate, confluent with the axial tissue.

Radial section

Perforation plates simple. Intervessel pits slightly scalariform and round in alternating position. Vessel length 70–100 μm. Ray cells upright.

Bark

Phloem contains groups of sieve tubes in continuation of vessels and xylem rays. Rays dilatate. Cortex consists of round to oval parenchyma cells with diameters of 80–120 μm. Many large, irregulary formed intercellulars in the cortex. Phellem layered, consisting of few layers of rectangular cells.

Pith

Round, with large, thin-walled cells, containing many crystal druses.

Analyzed: Root collar of polar root of 8 individuals.

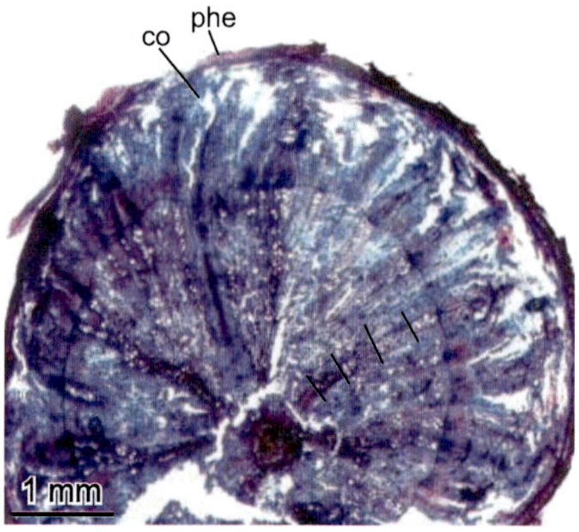

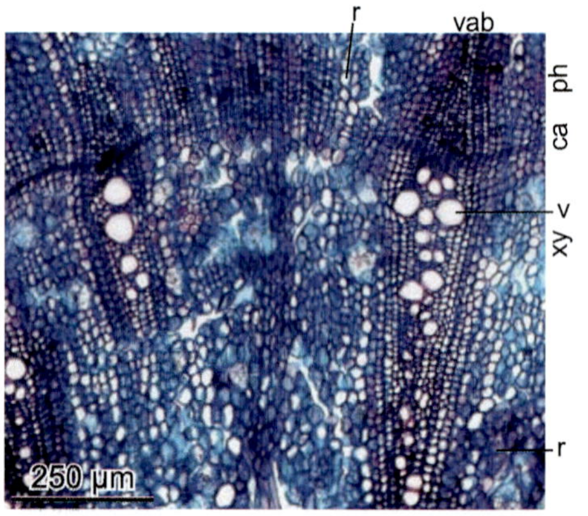

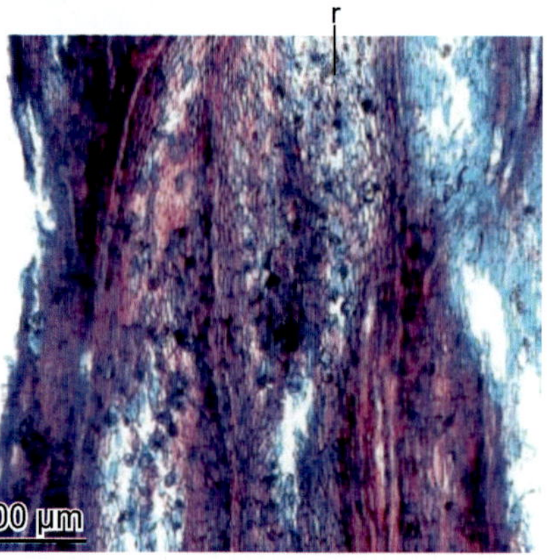

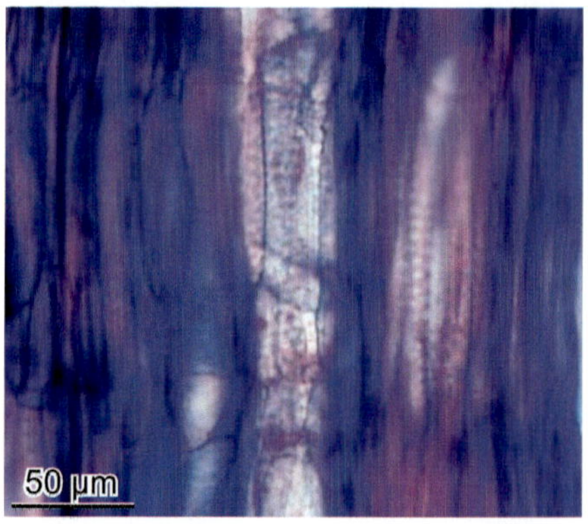

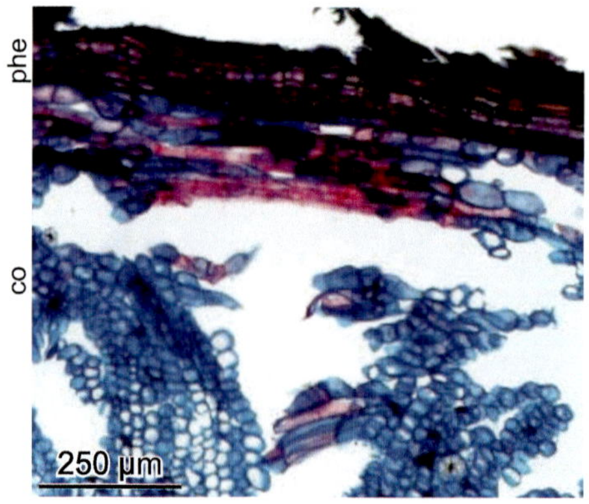

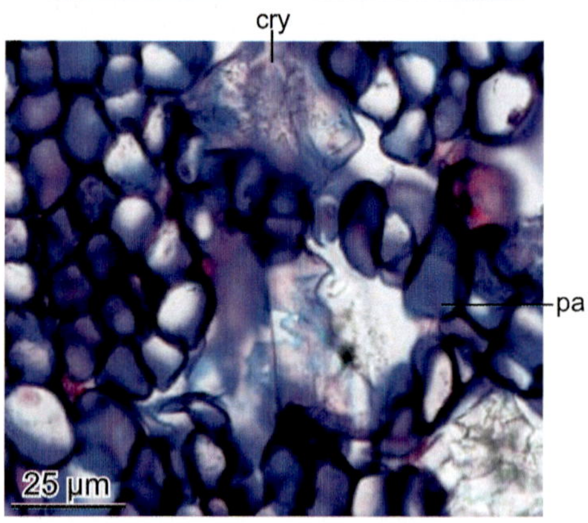

Features according to IAWA

Xylem 1 - 4 - 9 - 13 - 20 - 22 - 40.2 - 50.2 - 52.2 - 60.1 - 79.1 - 99 - 99.1 - 100.1 - 100.2 - 105 - 144

Persicaria glacialis (Meisn.) H.Hara
Polygonaceae

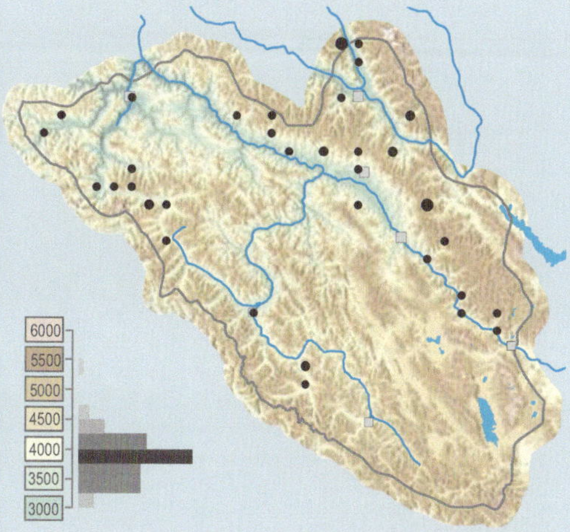

Syn.: *Polygonum glaciale* (Meisn.) Hook.f.

Description: Annuals, 3–15 cm tall, glabrous, stems prostrate to suberect, branched, leaves ovate, acute, entire, flowers white or pinkish, ca. 1 mm across, in terminal or axillary capitate panicles.

Distribution: Afghanistan, Pakistan, N India, Nepal, China (Gansu, Hebei, Qinghai, Shaanxi, Shanxi, Sichuan, Xizang, Yunnan).

Habitat: Grassy slopes, moist and shady places.

Stem diameter 1 mm. Bark/xylem proportion 1:2. Mean ring width 0.2 mm. Maximal 1 annual ring.

Xylem

Transverse section
Vessels solitary. Diameter of larger vessels 20–30 µm. Fibers thin- to thick-walled. Parenchyma paratracheal. Crystals absent.

Tangential section
Rays absent.

Radial section
Perforation plates simple. Intervessel pits round. Vessel length 100–150 µm.

Bark

Phloem homogeneous. Cortex cells large, thin-walled. With a few irregular formed crystals.

Pith

Round, filled with large unlignified cells.

Analyzed: Root collar of polar root of 1 individual.

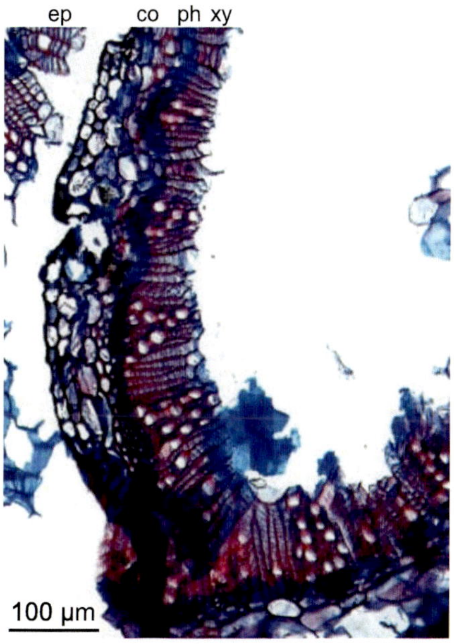

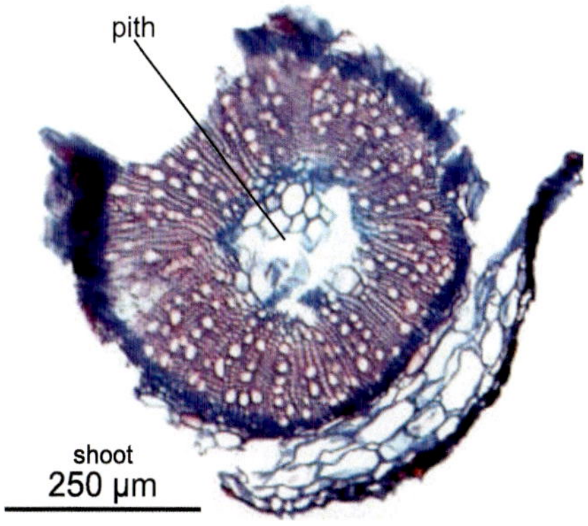

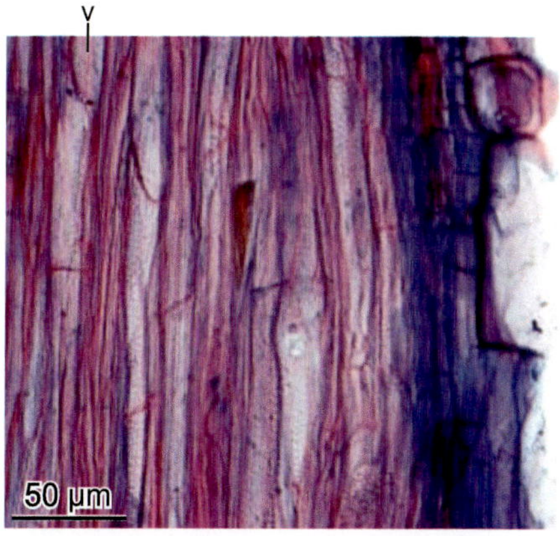

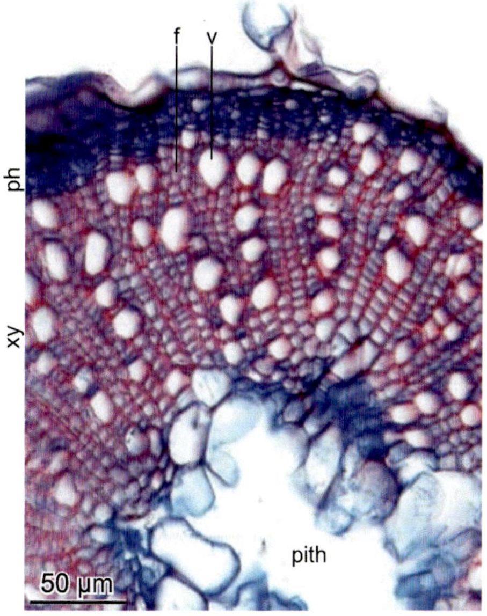

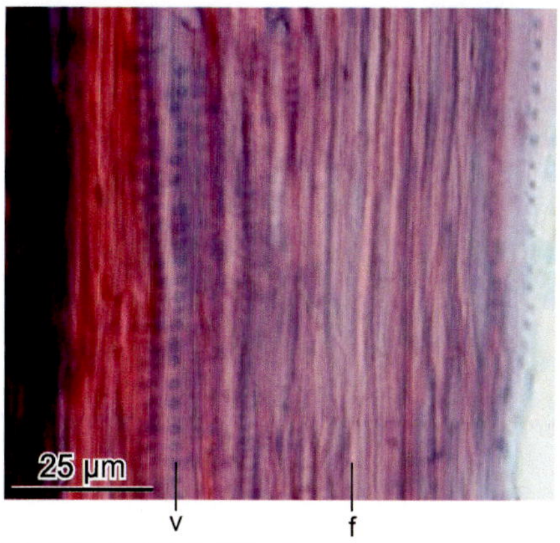

Features according to IAWA

Xylem *2.1 - 9 - 13 - 22 - 40.2 - 50.2 - 52.3 - 61 - 70 - 79 - 117*

Polygonum arenastrum Boreau

Polygonaceae

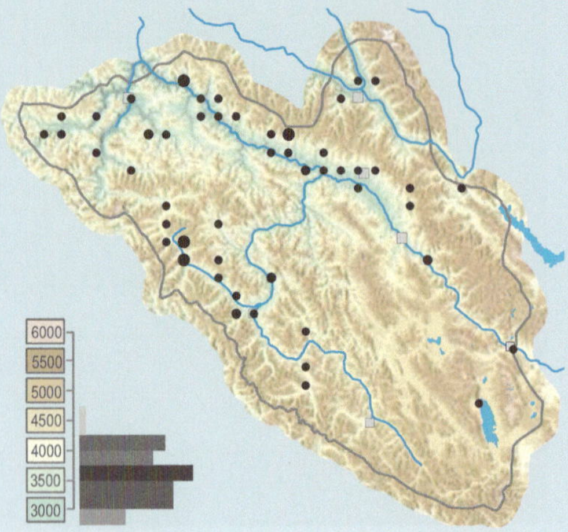

Syn.: *P. calcatum* Lindm., *P. aequale* Lindm., *P. aviculare* subsp. *aequale* (Lindm.) Asch. & Graebn., *P. aviculare* subsp. *calcatum* (Lindm.) Thell.

Description: Annuals, 15–30 cm tall, stems ascending to procumbent, branched from base, leaves elliptic to oblanceolate, flowers green with white margin, 1–2 mm across, arranged by 1–5 in axils.

Distribution: Nearly cosmopolitan.

Habitat: Field margins, dumps, eutrophicated wetlands, trampled places in settlements.

Stem diameter 2–5 mm. Bark/xylem proportion 1:4. Ring width 0.4–1 mm. Maximal 1 annual ring.

Xylem

Transverse section

Vessels in long radial rows. Diameter of larger vessels 20–30 µm. Fibers thin- to thick-walled. Parenchyma paratracheal. Crystals absent.

Tangential section

Rays 1–2-seriate.

Radial section

Perforation plates simple. Intervessel pits round. Vessel length 120–180 µm.

Bark

Phloem homogeneous. Cortex cells large, oval, thin-walled. With small groups of sclereids. Phellem consists of a few layers of rectangular cells.

Pith

No slide available.

Analyzed: Root collar of polar roots of 6 individuals.

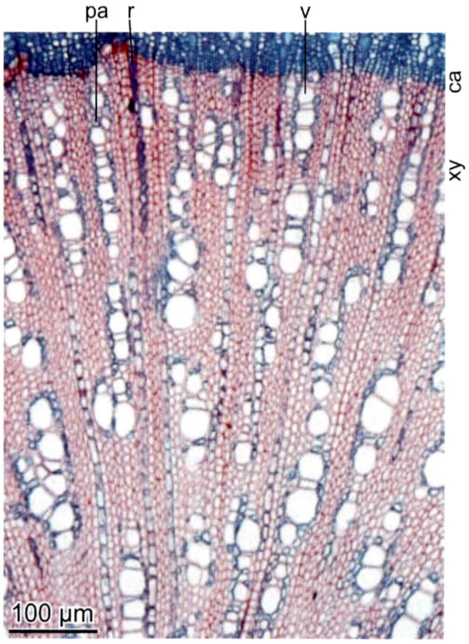

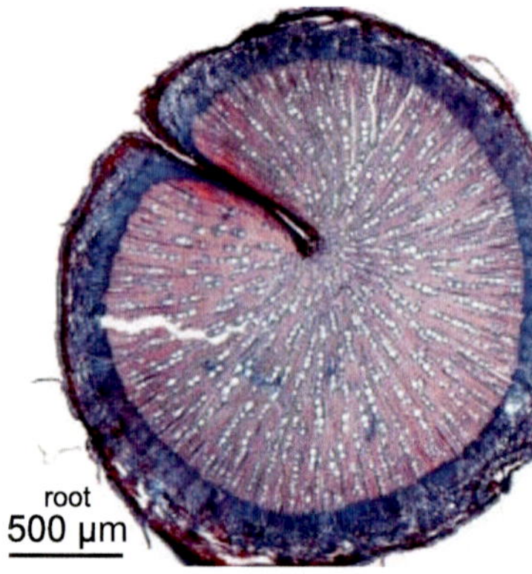

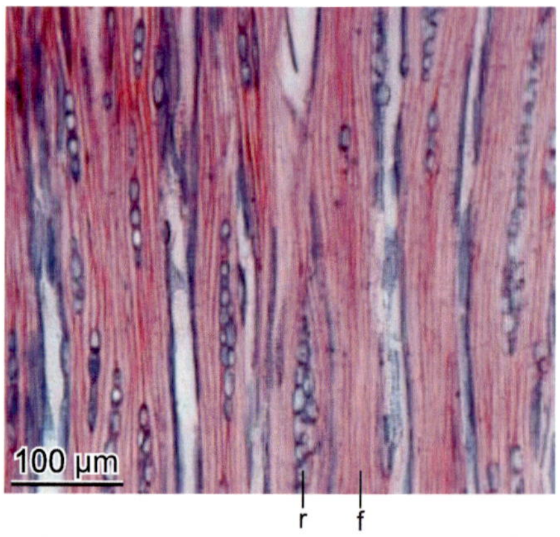

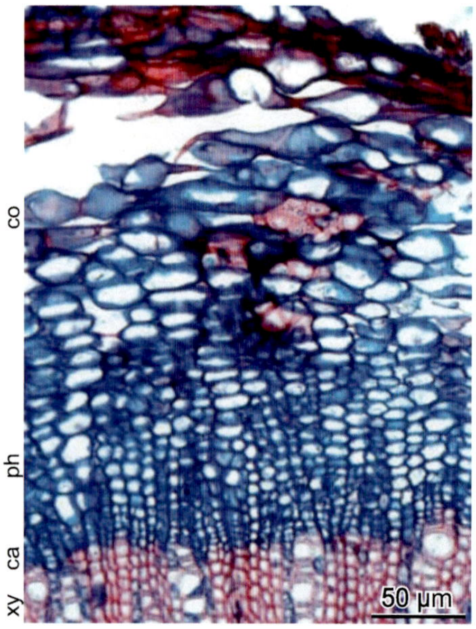

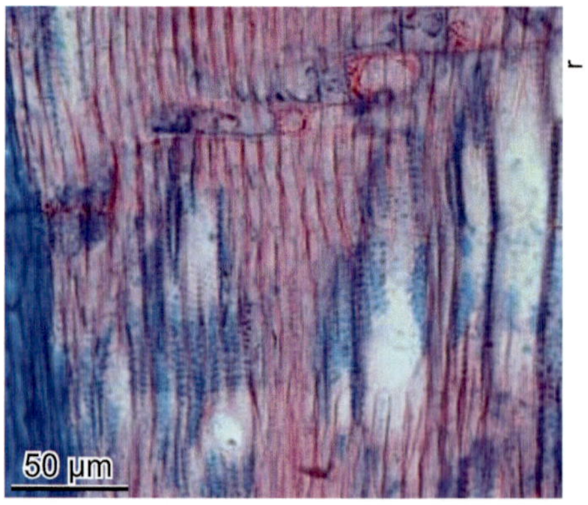

Features according to IAWA

Xylem *2.1 - 10 - 13 - 22 - 40.2 - 50.2 - 52.3 - 61 - 70 - 79 - 97 - 105*

Polygonum biaristatum Aitch. & Hemsl.

Polygonaceae

Description: Subshrubs, 5–20 cm tall, stems prostrate, leaves elliptic-ovate, sessile, margins revolute, flowers red, 1–2 mm across, by 1–2 in axils.

Distribution: C Asia, Afghanistan, Pakistan, India (Jammu and Kashmir).

Habitat: Dry stony slopes.

Stem diameter 2–3 mm. Bark/xylem proportion 1:5. Mean ring width 0.1 mm. Maximal 14 annual rings.

Xylem

Transverse section

Semi-ring-porous. Vessels angular and solitary. Rings indicated by larger earlywood (25–30 μm) and very small latewood vessel diameters (<10 μm). Fibers thick-walled. Parenchyma absent. Crystal absent.

Tangential section

Rays uniseriate.

Radial section

Perforation plates simple. Intervessel pits round. Vessel length 70–90 μm.

Bark

Phloem homogeneous. With a few large crystal druses. Cortex consists of large, oval, thin-walled cells and a few small fiber groups. Phellem very small, old parts flaking off.

Pith

No slide available.

Analyzed: Root collar of polar root of 2 individuals.

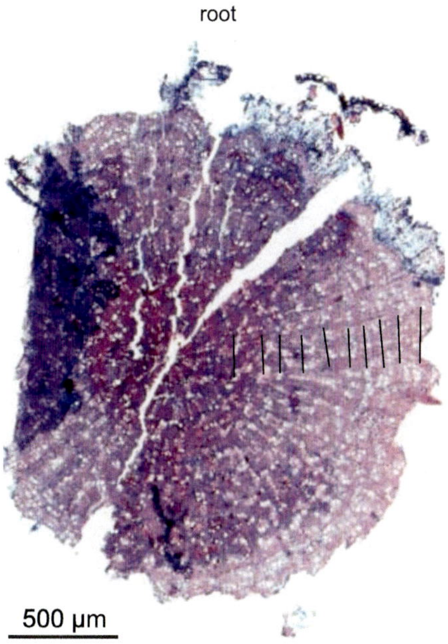

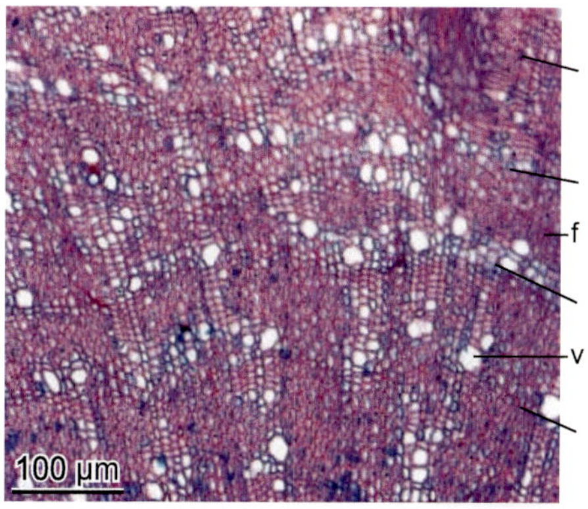

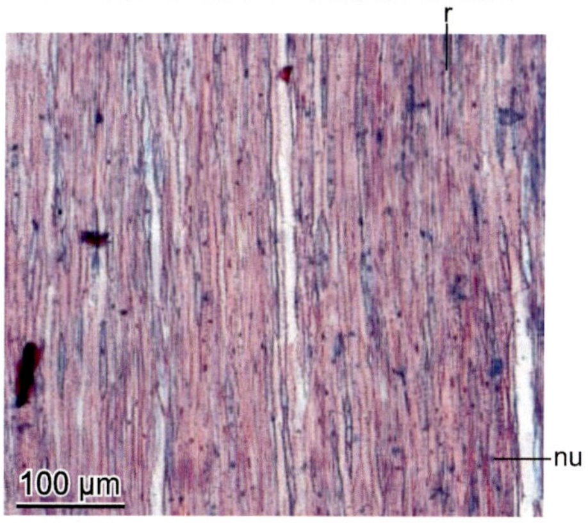

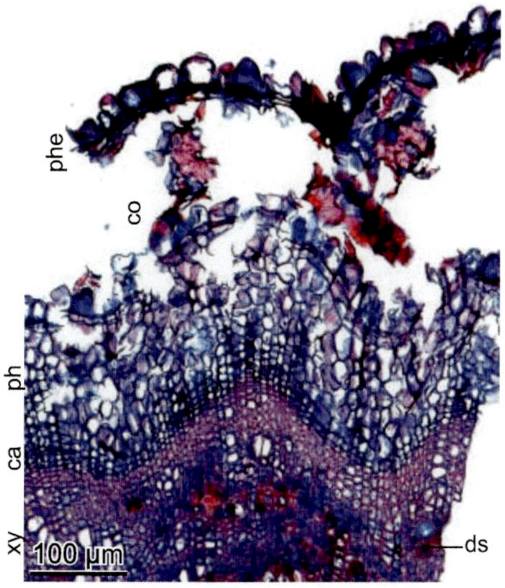

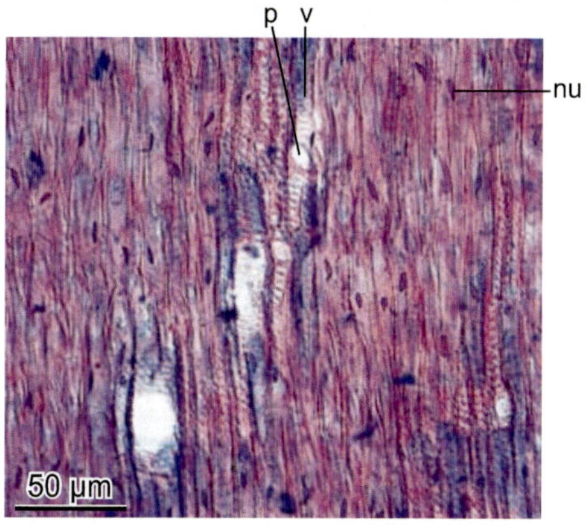

Features according to IAWA

Xylem 1 - 4 - 9 - 12 - 13 - 22 - 40.2 - 50.1 - 52.1 - 61 - 69 - 75 - 96 - 105

Polygonum cognatum Meisn.

Polygonaceae

Syn.: *P. alpestre* C.A.Mey, *P. pamiroalaicum* Kom., *P. rupestre* Kar. & Kir.

Description: Perennial herbs, 15–30 cm tall, stems prostrate to ascending, leaves elliptic-oblong, subsessile, flowers white, ca. 1.5 mm across, axillary.

Distribution: Russia (Siberia), Kazakhstan, Kyrgyzstan, Tajikistan, India (Jammu and Kashmir), China (Nei Mongol, Xinjiang), Mongolia.

Habitat: Mesic to wet habitats, often disturbed by trampling, gravel mountain slopes, dry riverbeds.

Stem diameter 7–1.1 mm. Bark/xylem proportion 1:4. Ring width 0.5–1.5 mm. More than 7 annual rings.

Xylem

Transverse section

Semi-ring-porous. Vessels solitary and in small groups. Rings indicated by larger earlywood (50 µm) or smaller latewood vessel diameters (20 µm). Vascular bundles remain. Fibers absent. Parenchyma pervasive. Very large rays between radial vessel/parenchyma strips. With large crystal druses. With inter-xylary periderm towards the pith.

Tangential section

Very large, unlignified rays between radial vessel/parenchyma strips. Rays confluent with the axial tissue.

Radial section

Perforation plates simple. Intervessel pits scalariform. Vessel length 120–200 µm.

Bark

Phloem with small groups of sieve tubes. With large crystal druses. Phellem consists of few rows of rectangular, thin-walled cells.

Pith

No slide available.

Analyzed: Root collar of polar root of 2 individuals.

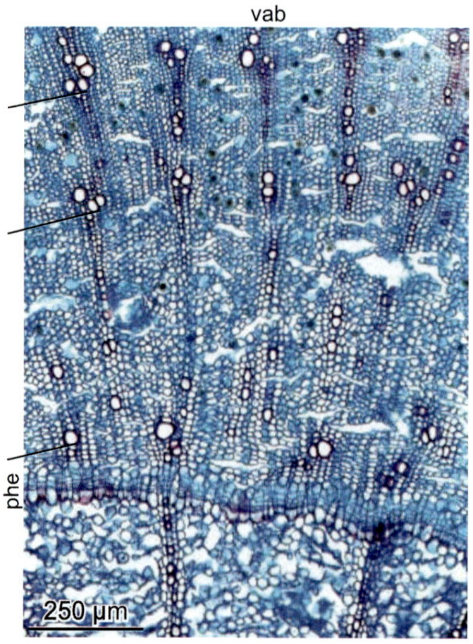

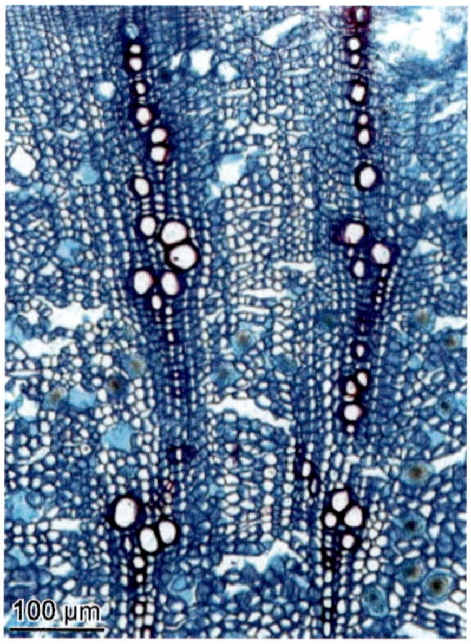

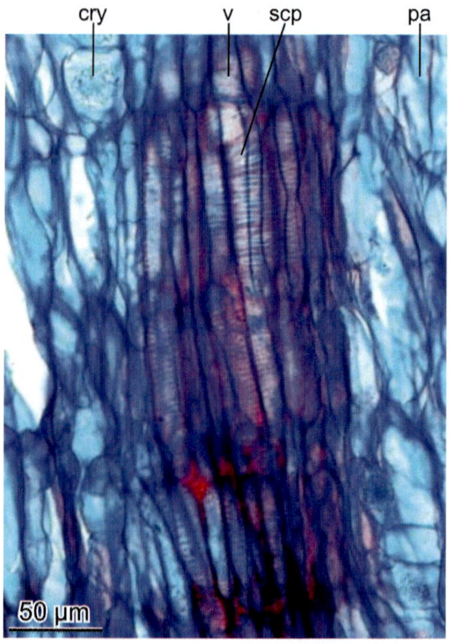

Features according to IAWA

Xylem 1 - 4 - 9 - 11 - 13 - 20 - 41 - 50.1 - 52.3 - 60.1 - 79.1 - 99 - 99.1 - 100.1 - 100.2 - 135.2

Polygonum paronychioides C.A.Mey.

Polygonaceae

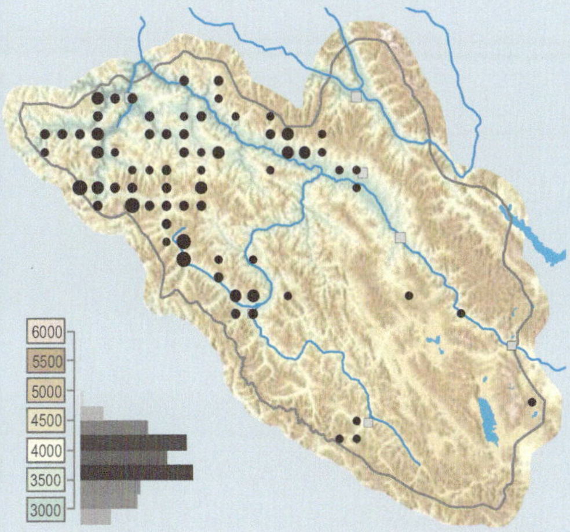

Syn.: *P. englerianum* H.Gross, *P. himalayense* H.Gross

Description: Subshrubs, 5–15 cm tall, stems prostrate to ascending, leaves linear, margins revolute, flowers pinkish, ca. 2 mm across, solitary in axils.

Distribution: Georgia, Kazakhstan, Turkmenistan, Uzbekistan, Kyrgyzstan, Tajikistan, Afghanistan, Pakistan, India (Jammu and Kashmir), China (Xizang).

Habitat: Gravel mountain slopes, dry rocky soil.

Stem diameter 3 mm. Bark/xylem proportion 1:2. Mean ring width 0.2 mm. Maximal 12 annual rings.

Xylem

Transverse section

Semi-ring-porous. Angular vessels solitary and in short radial rows. Rings indicated by larger earlywood (30–40 µm) and very small latewood vessel diameters (<10 µm). Fibers thick-walled. Parenchyma paratracheal. Dark-stained substances in vessels of the stem center (heartwood). Crystals absent.

Tangential section

Rays uniseriate.

Radial section

Perforation plates simple. Intervessel pits round. Vessel length 50–80 µm.

Bark

Phloem homogeneous. Cortex consists of large, oval, thin-walled cells and irregular fiber groups. Phellem consists of a few layers of rectangular cells. Stem is surrounded by an epidermis containing dark-stained substances.

Pith

No slide available.

Analyzed: Root collar of polar root of 5 individuals.

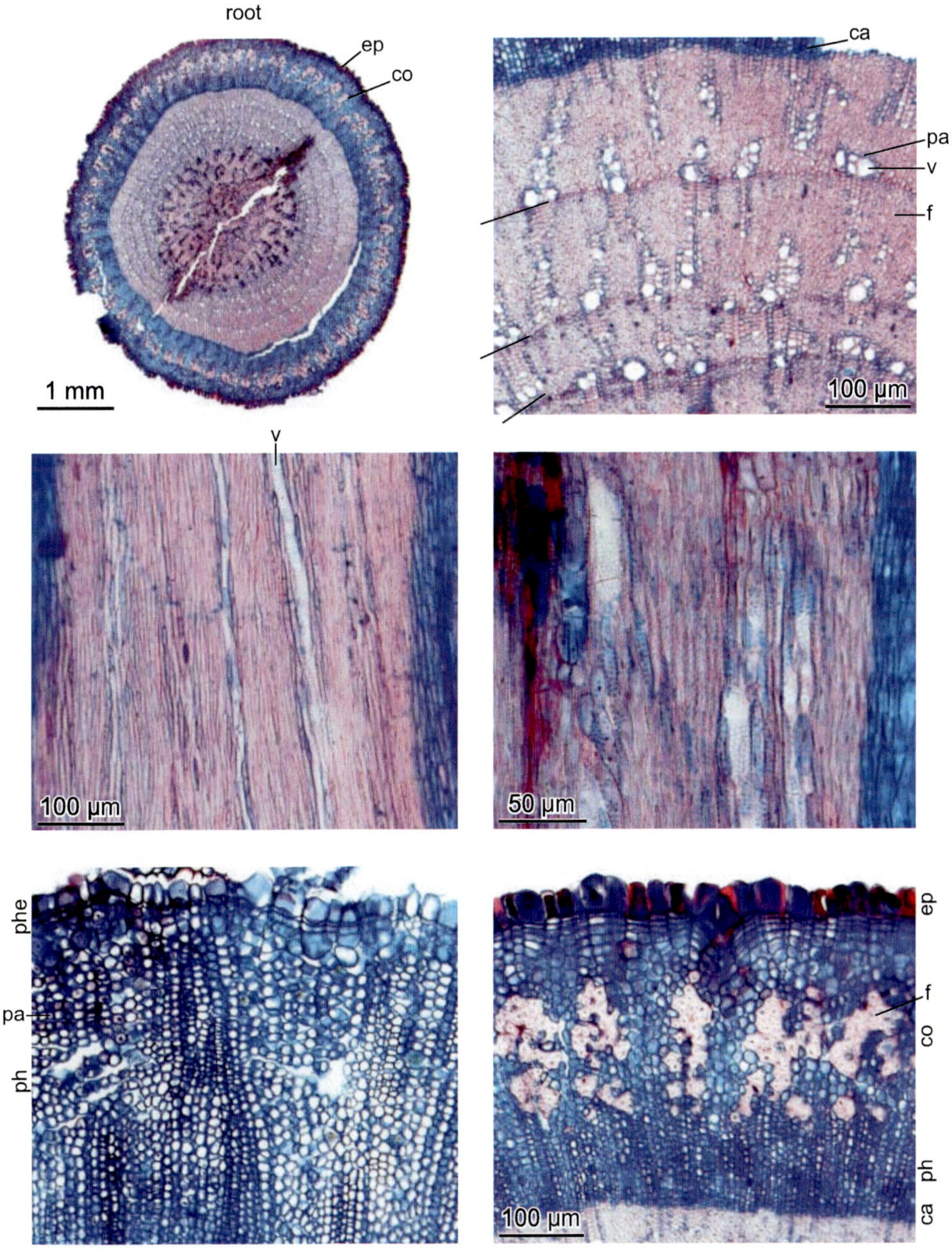

Features according to IAWA

Xylem 1 - 4 - 9 - 12 - 13 - 22 - 40.2 - 50.1 - 52.1 - 61 - 69 - 75 - 96 - 105

Polygonum plebeium R.Br.

Polygonaceae

Syn.: *Avicularia indica* Didr., *P. parviflorum* Y.L. Chang & S.H. Li, *P. changii* Kitag.

Description: Annuals, 15–30 cm tall, stems ascending to procumbent, branched from base, leaves elliptic to oblanceolate, flowers pink, 1–2 mm across, arranged by 1–5 in axils.

Distribution: E Africa, S Asia, Australia; introduced to North America.

Habitat: Disturbed sites.

Stem diameter 1–2 mm. Bark/xylem proportion 1:1. Ring width 0.4–0.8 mm. Maximal 1 annual ring.

Xylem

Transverse section

Vessels in long radial rows. Diameter of larger vessels 20–30 µm. Fibers thin- to thick-walled. Parenchyma paratracheal. Crystals absent.

Tangential section

Rays absent.

Radial section

Perforation plates simple. Intervessel pits round. Vessel length 120–180 µm.

Bark

Phloem homogeneous. Cortex with small irregular groups of sclereids. Periderm with one row of rectangular cells containing blue-stained substances.

Pith

No slide available.

Analyzed: Root collar of polar roots of 6 individuals.

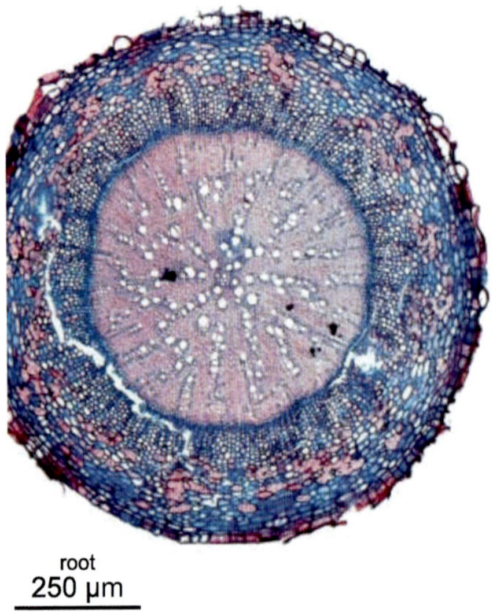

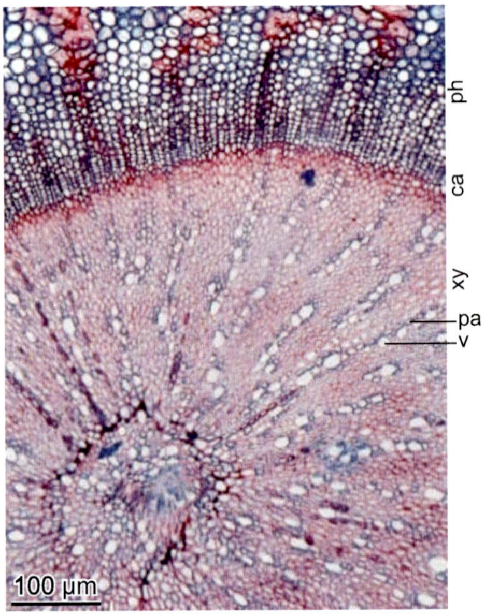

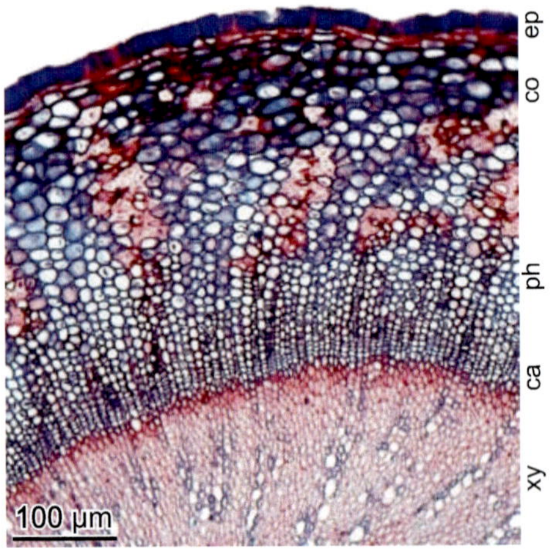

Features according to IAWA

Xylem 2.1 - 10 - 13 - 22 - 40.2 - 50.2 - 52.3 - 61 - 70 - 79 - 117

Polygonum polycnemoides Jaub. & Spach

Polygonaceae

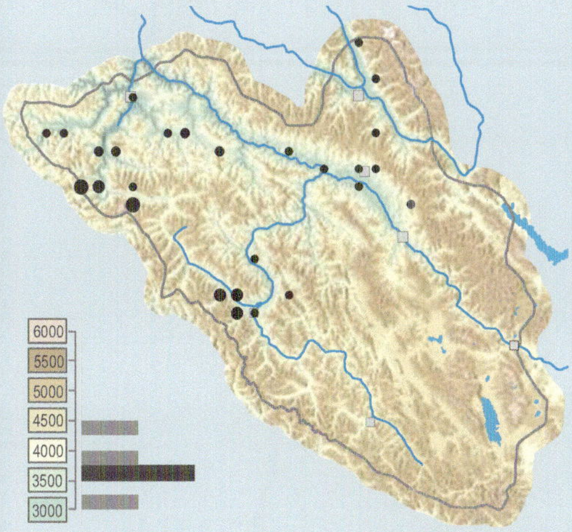

Description: Annuals, 5–20 cm tall, stems prostrate, leaves linear, flowers green with white or pinkish margin, arranged by 1–2 in axils.

Distribution: Turkey, Iraq, Iran, Georgia, Kazakhstan, Turkmenistan, Uzbekistan, Kyrgyzstan, Tajikistan, Afghanistan, Pakistan, NW India, China (Xinjiang), Mongolia.

Habitat: Dry slopes, rocky deserts, gravel grasslands.

Stem diameter 1–2 mm. Bark/xylem proportion 1:3. Mean ring width 0.3 mm. Maximal 2 annual rings.

Xylem

Transverse section

Vessels in long radial rows. Diameter of larger vessels 20–30 µm. Fibers thin- to thick-walled. Parenchyma paratracheal. Crystals absent.

Tangential section

Rays absent.

Radial section

Perforation plates simple. Intervessel pits round. Vessel length 120–150 µm.

Bark

Phloem homogeneous. Cortex with small irregular groups of sclereids. Periderm with one row of rectangular cells which contain blue-stained substances.

Pith

No slide available.

Analyzed: Root collar of polar roots of 6 individuals.

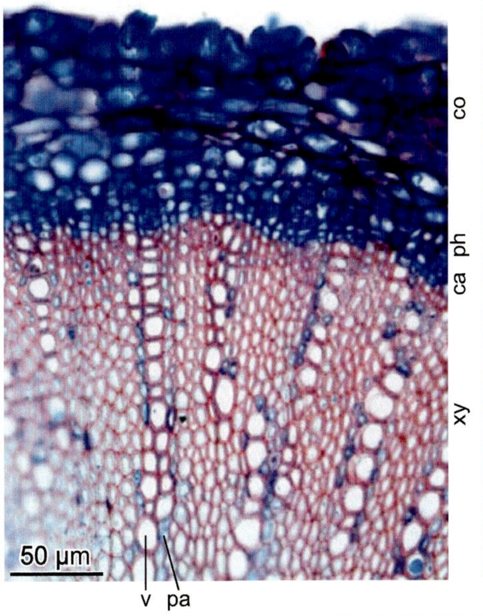

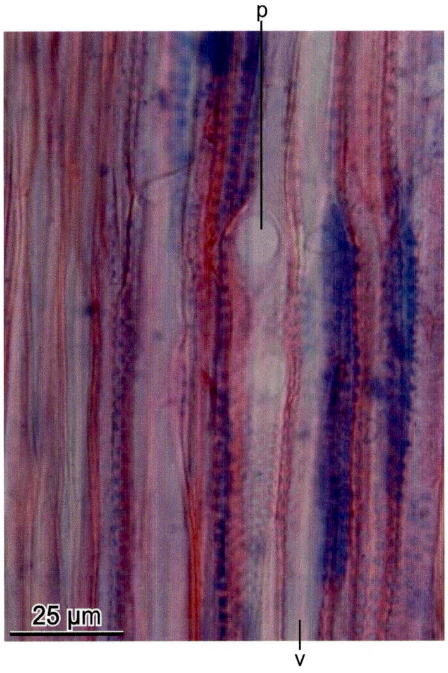

Features according to IAWA

Xylem 1 - 10 - 13 - 22 - 40.1 - 40.2 - 50.2 - 52.3 - 61 - 70 - 79 - 117

Polygonum rottboellioides Jaub. & Spach

Polygonaceae

Syn.: *P. durum* Brenckle, *P. tubulosum* Boiss.

Description: Annuals, 5–25 cm tall, stems prostrate, leaves linear, margins revolute, flowers pink or white, 1.5–2 mm across, in axillary clusters of 2–3.

Distribution: Iraq, Turkey, Iran, Afghanistan, Pakistan, NW India.

Habitat: Dry disturbed pastures, weedy in barley fields, synanthropic along roads.

Stem diameter 1–2 mm. Bark/xylem proportion 1:4. Ring width 0.2–0.5 mm. Maximal 1 annual ring.

Xylem

Transverse section

Vessels in long radial rows. Diameter of larger vessels 20–30 µm. Fibers thin- to thick-walled. Parenchyma paratracheal. Crystal absent.

Tangential section

Rays absent.

Radial section

Perforation plates simple. Intervessel pits round. Vessel length 80–100 µm.

Bark

Phloem homogeneous. Cortex cells large, oval, thin-walled. Phellem consists of a few layers of rectangular cells. Stem surrounded by an epidermis containing dark-stained substances.

Pith

No slide available.

Analyzed: Root collar of polar root of 5 individuals.

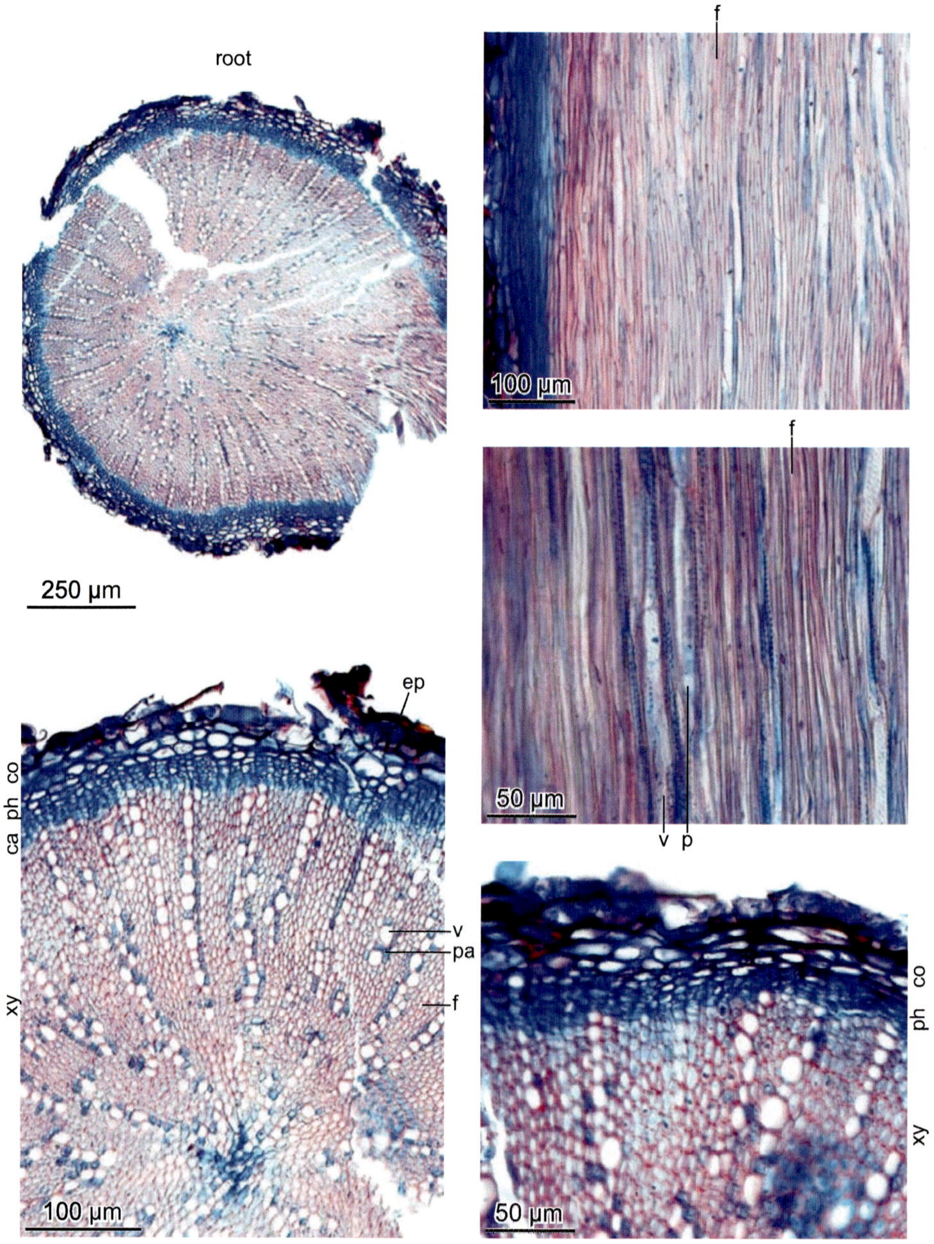

Features according to IAWA

Xylem 2.1 - 10 - 13 - 22 - 40.2 - 50.1 - 52.2 - 61 - 70 - 79 - 117

Rheum spiciforme Royle

Polygonaceae

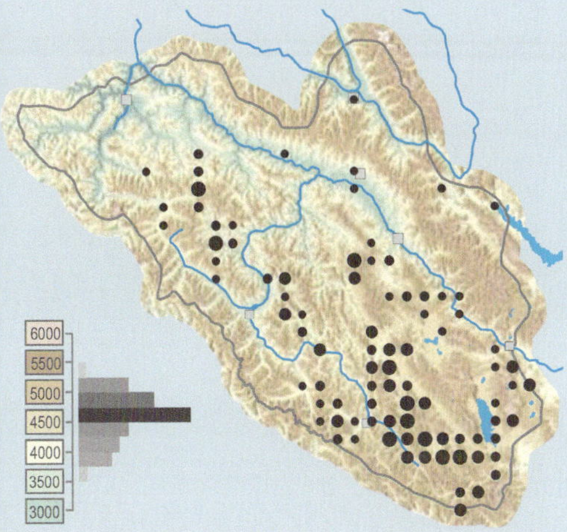

Syn.: *R. przewalskyi* Losinsk., *R. rhizostachyum* Schrenk, *R. rhomboideum* Losinsk., *R. scaberrimum* Lingelsh. ex Limpr.

Description: Perennial herbs, 10–30 cm tall, leaves all basal, ovate, 10–20 × 8–15 cm, leathery, flowers small, greenish, in many-flowered spike-like racemes.

Distribution: SE Kazakhstan, Kyrgyzstan, Tajikistan, Afghanistan, Pakistan, India (Jammu and Kashmir, Himachal Pradesh, Uttarakhand, Sikkim), China (Gansu, Qinghai, Sichuan, Xinjiang, Xizang), Bhutan.

Habitat: Gentle stabilized slopes, gravel areas, barren river beds.

Stem diameter 10 mm. Bark/xylem proportion 1:2. Mean ring width 0.3 mm. Maximal 16 annual rings.

Xylem

Transverse section

With distinct rings. Semi-ring-porous. Diameter of earlywood vessels 40–50 μm, diameter of latewood vessels 10–20 μm. Vessels in groups. Fibers absent. Parenchyma pervasive. Vascular bundles remain. Extremely many large crystal druses (up to 70 μm) in rays. Much slime is produced by parenchyma cells, anatomically not different from normal parenchyma cells. With small groups of intra-xylary phloem.

Tangential section

Very large rays between radial vessel/parenchyma strips. Rays confluent with the axial tissue.

Radial section

Perforation plates simple. Intervessel pits scalariform. Vessel length 100–120 μm.

Bark

Phloem contains groups of sieve tubes in continuation of vessels and xylem rays. Cortex consists of round to oval parenchyma cells with diameters of 80–120 μm. Phellem layered, consisting of few layers of rectangular cells.

Pith

No slide available.

Analyzed: Root collar of polar root of 2 individuals.

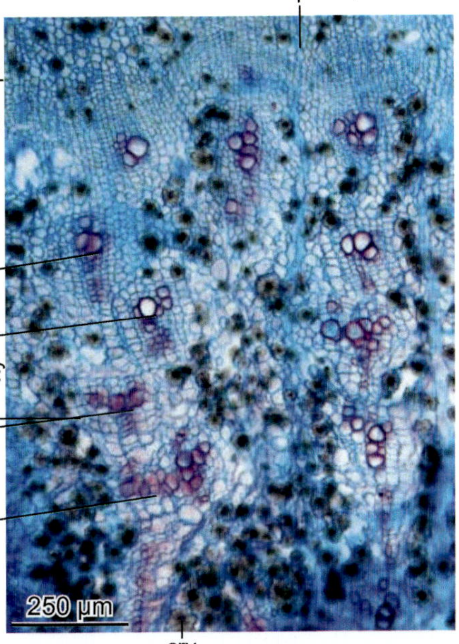

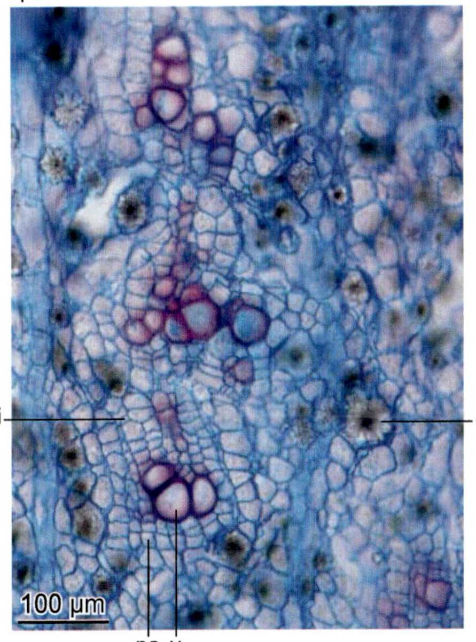

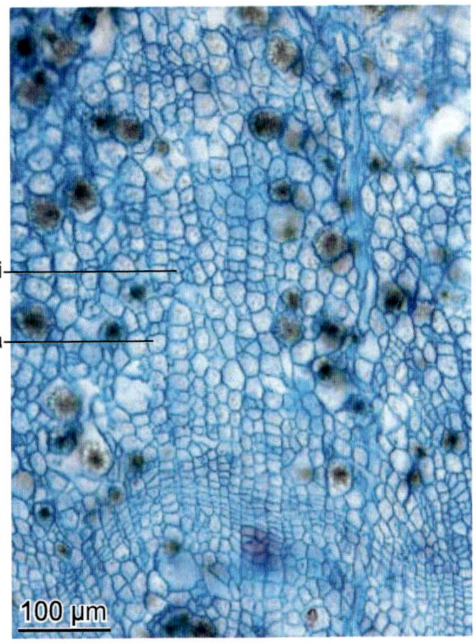

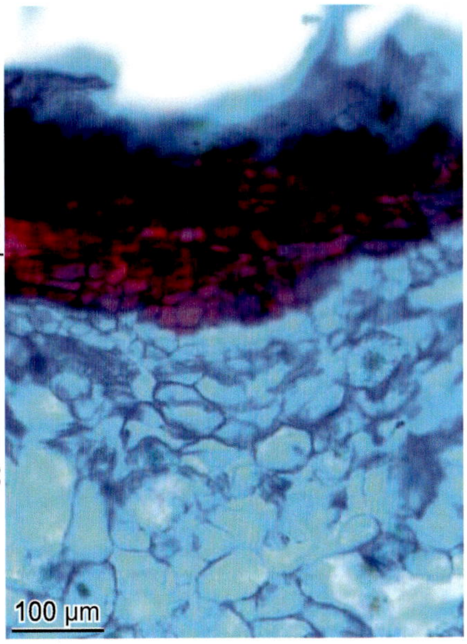

Features according to IAWA

Xylem 1 - 4 - 11 - 13 - 20 - 40.2 - 50.1 - 52.2 - 52.3 - 60.1 - 79.1 - 99 - 99.1 - 100.1 - 100.2 - 105 - 135 - 144

Rheum tibeticum Maxim. ex Hook.f.

Polygonaceae

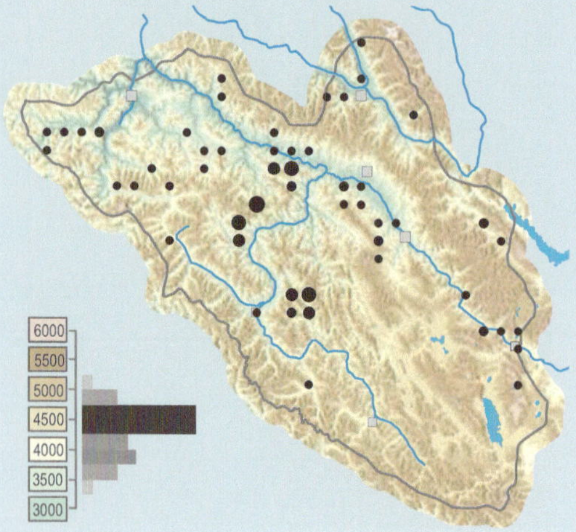

Description: Perennial herbs, 20–80 cm tall, leaves all basal, orbicular, entire or crenulate, leathery, 10–20 cm long, flowers minute, greenish, in leafless panicles, fruits 8–10 mm across, brownish red, orbicular with broad wings.

Distribution: Afghanistan, Pakistan, India, China (Xizang).

Habitat: Dry slopes.

Stem diameter 2 mm. Bark/xylem proportion 1:3. Ring width 0.4–0.7 mm. Maximal 8 annual rings. Center rotten.

Xylem

Transverse section

Diameter of larger vessels 30–40 µm. Vessels solitary. Fibers absent. Parenchyma pervasive. Vascular bundles remain. Many large crystal druses (up to 70 µm) in rays. Much slime is produced by parenchyma cells, anatomically not different from normal parenchyma cells. With small groups of intra-xylary phloem.

Tangential section

Very large rays between radial vessel/parenchyma strips. Rays confluent with the axial tissue.

Radial section

Perforation plates simple. Intervessel pits scalariform. Vessel length 150–180 µm.

Bark

Phloem contains groups of sieve tubes in continuation of vessels and xylem rays. Cortex consists of round to oval parenchyma cells with diameters of 80–120 µm. Phellem layered, consisting of few layers of rectangular cells.

Pith

No slide available.

Analyzed: Root collar of polar root of 4 individuals.

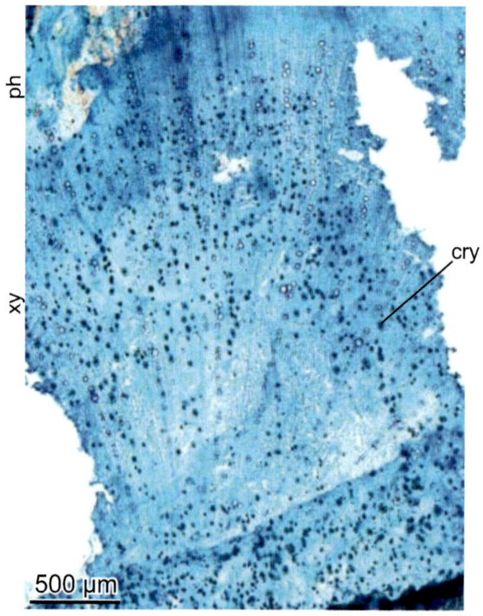

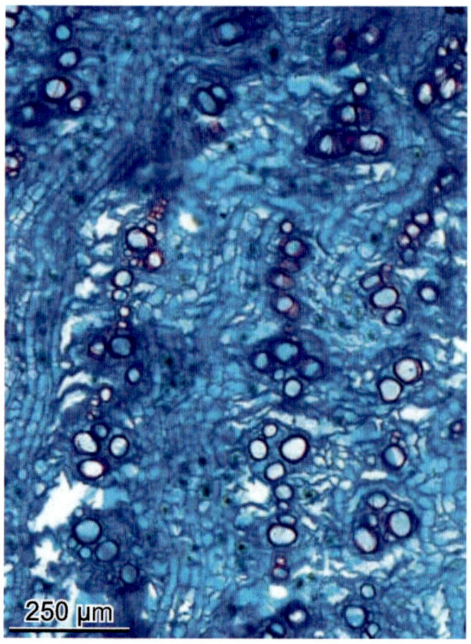

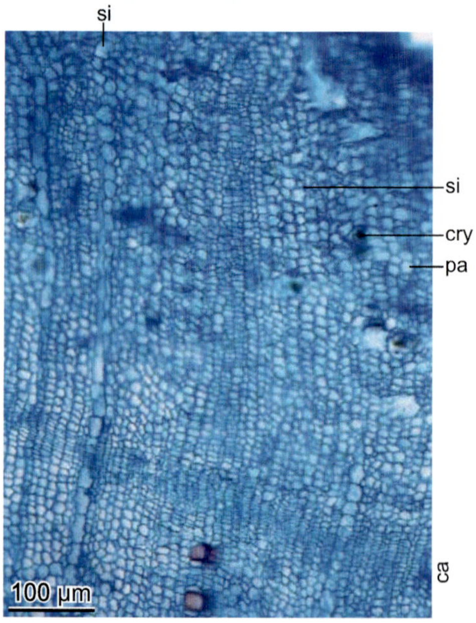

Features according to IAWA

Xylem 1 - 4 - 11 - 13 - 20 - 40.2 - 50.1 - 52.3 - 60.1 - 79.1 - 99 - 99.1 - 100.1 - 100.2 - 105 - 135 - 144

Rheum webbianum Royle

Polygonaceae

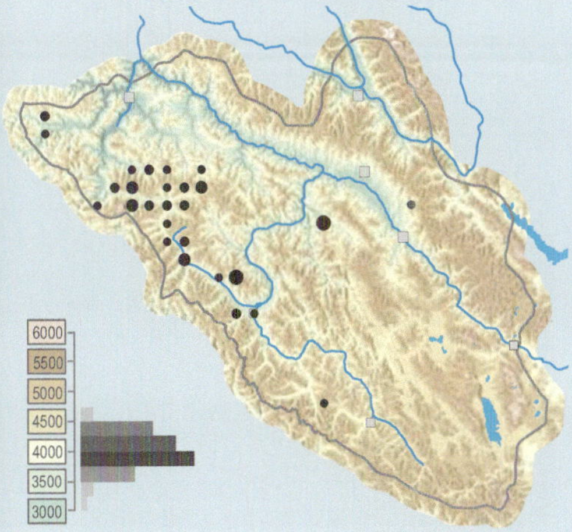

Description: Perennial herbs, 30–200 cm tall, leaves orbicular to reniform, flowers pale yellow, 2–2.5 mm across, in a branched leafy panicle.

Distribution: Pakistan, NW India, Nepal, China (Xizang).

Habitat: Dry slopes.

Stem diameter 3 mm. Bark/xylem proportion 1:0.5. Mean ring width 1.2 mm. Maximal 1 annual ring.

Xylem

Transverse section

Diameter of larger vessels 50–60 µm. Vessels solitary. Fibers thin- to thick-walled. Parenchyma pervasive. Vascular bundles remain. Crystal druses in rays.

Tangential section

Rays 4–7-seriate, confluent with the axial tissue.

Radial section

Perforation plates simple. Intervessel pits round. Vessel length 150–200 µm.

Bark

Phloem contains groups of sieve tubes in continuation of vessels and xylem rays. Cortex consists of round to oval parenchyma cells. Phellem absent.

Pith

No slide available.

Analyzed: Root collar of polar root of 1 individual.

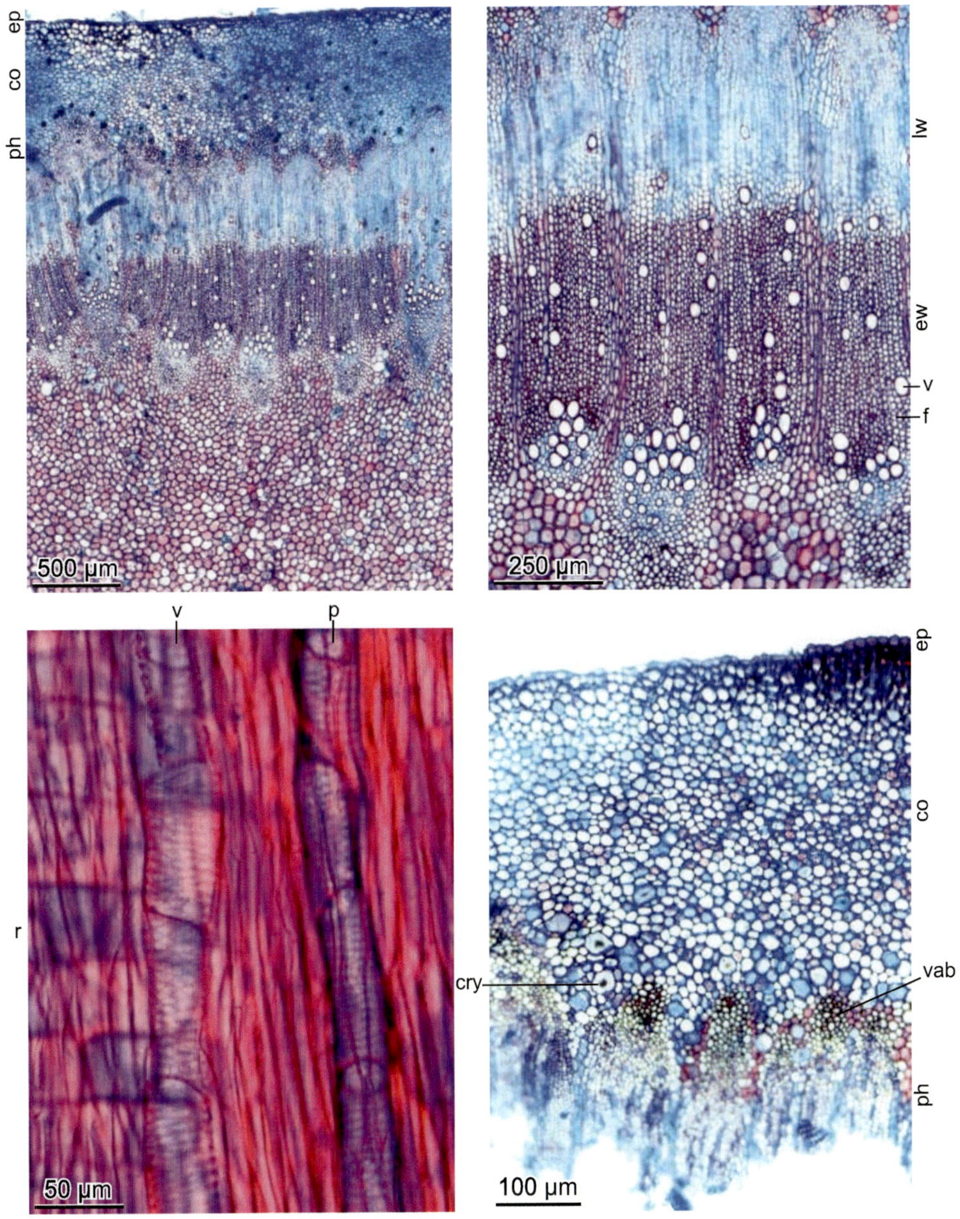

Features according to IAWA

Xylem 2.1 - 9 - 13 - 22 - 40.2 - 50.1 - 52.3 - 61 - 75 - 98 - 99.1 - 100.1 - 105 - 144

Rumex angulatus Rech.f.

Polygonaceae

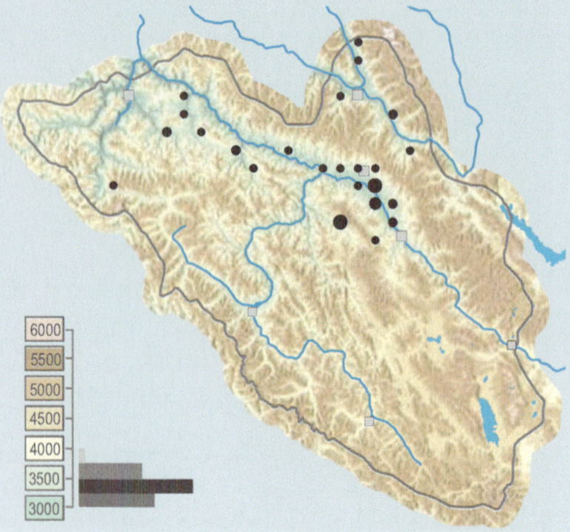

Description: Perennial herbs, 40–80 cm tall, rhizomatous, glabrous, leaves oblong-lanceolate, flowers minute, forming terminal dense panicles.

Distribution: Afghanistan, Pakistan, NW India, China (Xizang).

Habitat: Irrigation canals, wet ditches, watersides.

Stem diameter 7–12 mm. Bark/xylem proportion 1:1. Ring width 0.12–0.3 mm. Maximal 3 annual rings.

Xylem

Transverse section

Diffuse-porous. Diameter of larger vessels 40–50 µm. Vessels solitary, with angular outline. Fibers absent. Parenchyma pervasive. Vascular bundles remain. With intraxylary phloem Crystal druses in rays.

Tangential section

Rays 4–7-seriate. Rays confluent with axial tissue.

Radial section

Perforation plates simple. Intervessel pits scalariform. Vessel length 150–200 µm.

Bark

Phloem contains groups of sieve tubes. Ray dilatation. Cortex contains irregular intracellular spaces. Phellem with several layers of rectangular cells.

Pith

Round, with a netlike aerenchyma.

Analyzed: Root collar of polar roots of 3 individuals.

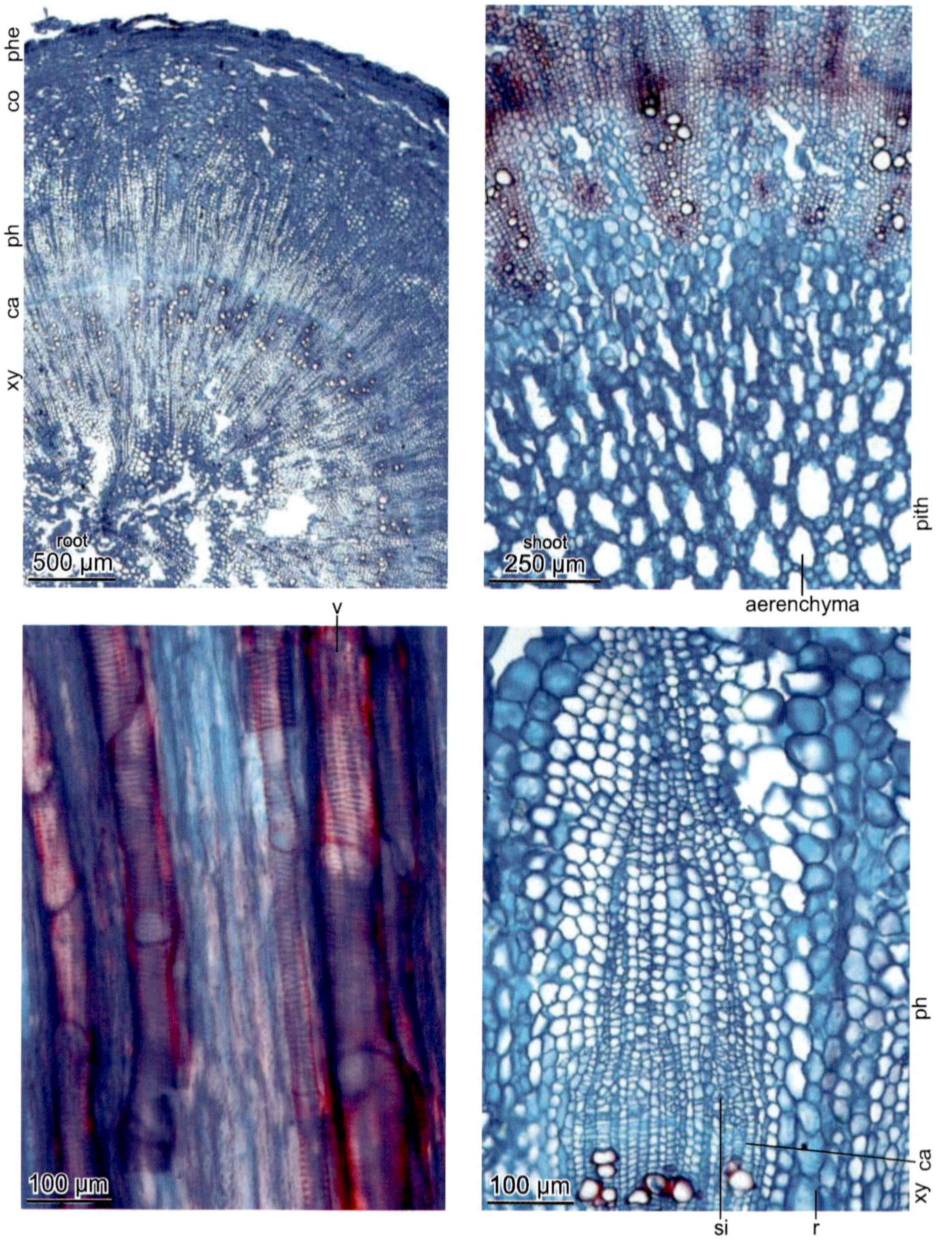

Features according to IAWA

Xylem 1 - 5 - 9 - 13 - 20 - 40.2 - 50.1 - 52.3 - 60.1 - 79.1 - 98 - 99.1 - 100.1 - 100.2 - 105 - 144

Rumex patientia L.

Polygonaceae

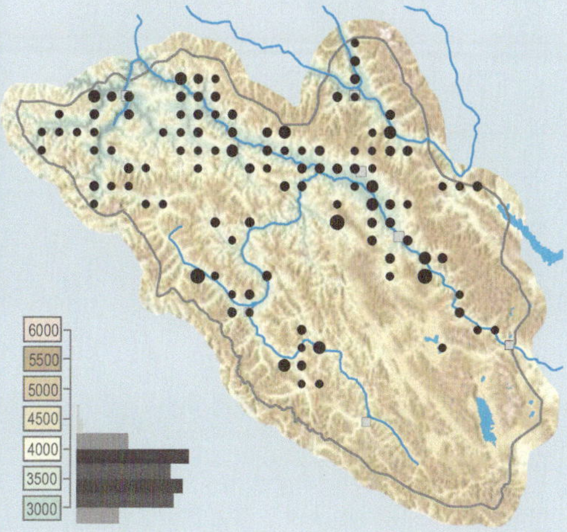

Syn.: *R. patientia* subsp. *tibeticus* (Rech.f.) Rech. f.

Description: Perennial herbs, 60–150 cm tall, leaves oblong, margin undulate, flowers green, in large paniculate inflorescences.

Distribution: Europe, Russia, W Asia, Caucasus, Kazakhstan, Kyrgyzstan Tajikistan, Pakistan, N India, Nepal, China (Gansu, Hebei, Heilongjiang, Henan, Hubei, Hunan, Jilin, Liaoning, Nei Mongol, Ningxia, Qinghai, Shaanxi, Shandong, Shanxi, Sichuan, Xinjiang, Xizang), Mongolia; naturalized in North America.

Habitat: Abandoned fields, field margins, along roads and ditches.

Stem diameter 20 mm. Bark/xylem proportion 1:3. Mean ring width 1 mm. Maximal 7 annual rings.

Xylem

Transverse section

Semi-ring-porous. Diameter of vessels 60–80 µm. Vessels solitary or in short radial rows. Thin- to thick-walled fibers in a belt at the stem periphery and around vessels in the stem center. Parenchyma paratracheal and pervasive. Vascular bundles remain. With intra-xylary phloem. Crystal druses in rays.

Tangential section

Rays 4–>10-seriate. Rays unlignified and confluent with axial tissue.

Radial section

Perforation plates simple. Intervessel pits scalariform. Vessel length 80–120 µm.

Bark

Phloem contains groups of sieve tubes and fibers. Ray dilatation. Cortex contains irregular intracellular spaces. Phellem not observed.

Pith

No slide available.

Analyzed: Root collar of polar root of 1 individual.

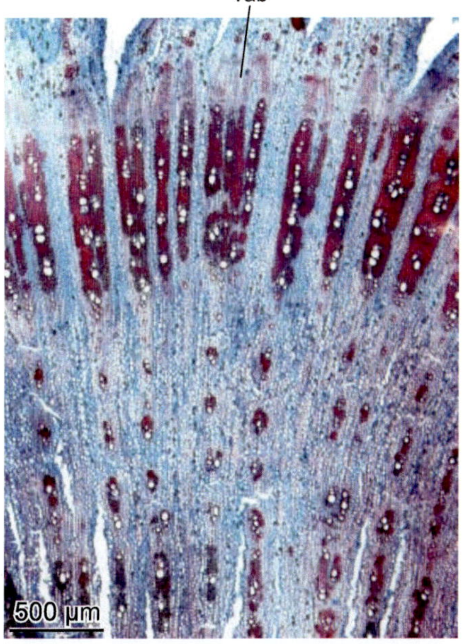

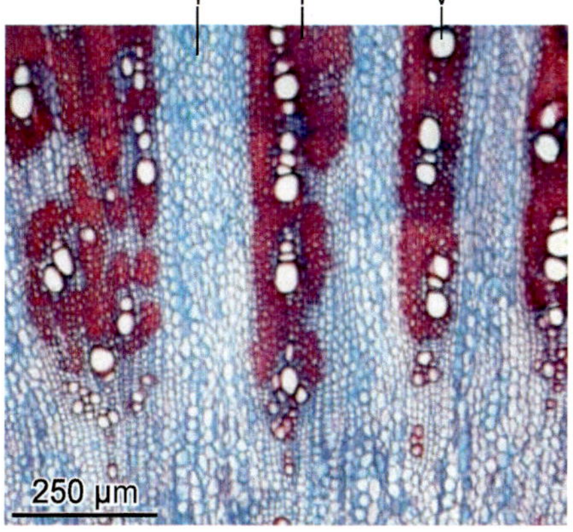

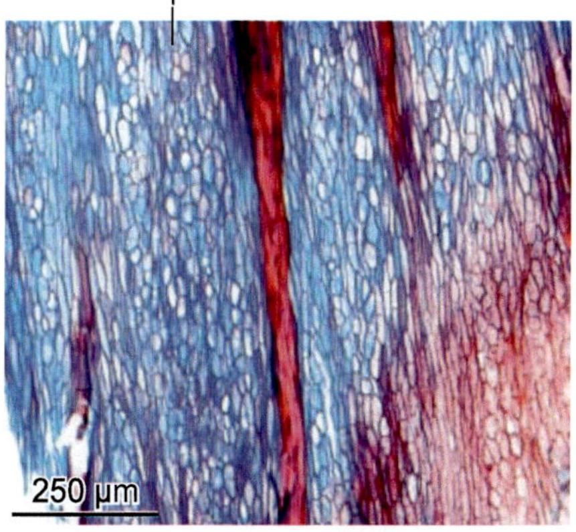

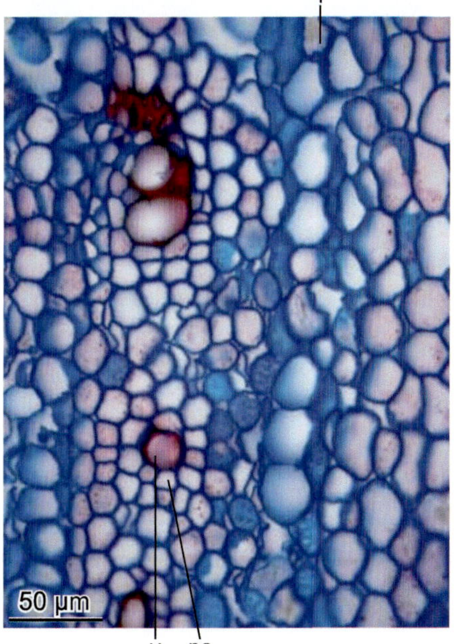

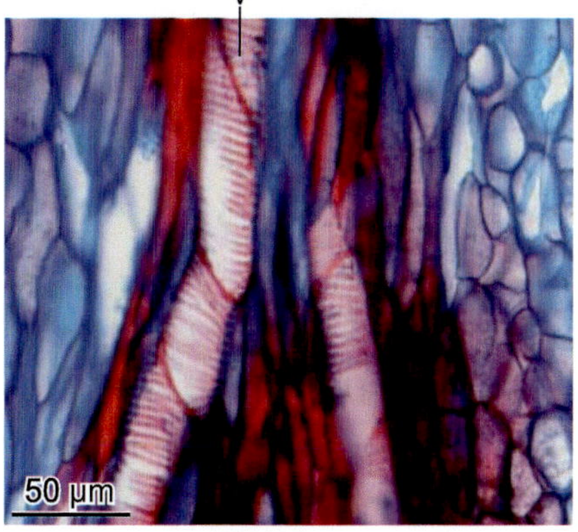

Features according to IAWA

Xylem 1 - 4 - 9.1 - 13 - 20 - 41 - 50.1 - 52.2 - 61 - 70 - 70.3 - 79 - 79.1 - 98 - 99 - 99.1 - 100.1 - 100.2 - 105 - 135 - 144

Androsace mucronifolia Watt

Primulaceae

Syn.: *A. globifera* Klatt, *A. microphylla* Hook.f.

Description: Perennial herbs, 1–3 cm tall, forming small cushions of globular rosettes up to 1 cm across, leaves obovate, flowers white to deep pink, fragrant, 6–8 mm across, in short-stalked umbels.

Distribution: E Afghanistan, NW Pakistan, India (Jammu and Kashmir), China (W Xizang).

Habitat: Alpine meadows, screes, rocky slopes.

Stem diameter 1 mm. Bark/xylem proportion 1:1. Mean ring width 0.07 mm. Maximal 5 annual rings.

Xylem

Transverse section

Semi-ring-porous. Vessels thick-walled, angular, solitary and in short radial multiples. Diameter of larger vessels 5–10 μm. Fibers absent. Parenchyma pervasive. Crystals absent.

Tangential section

Rays absent.

Radial section

Perforation plates simple. Intervessel pits scalariform. Vessel length 100–120 μm.

Bark

Phloem contains small groups of sieve tubes. Cortex consists of thick-walled parenchyma cells. Sclerenchyma and crystals absent. Epidermis consists of large cells, with a thick cuticula.

Pith

No slide available.

Analyzed: Root collar of polar roots of 1 individual (herbarium specimen).

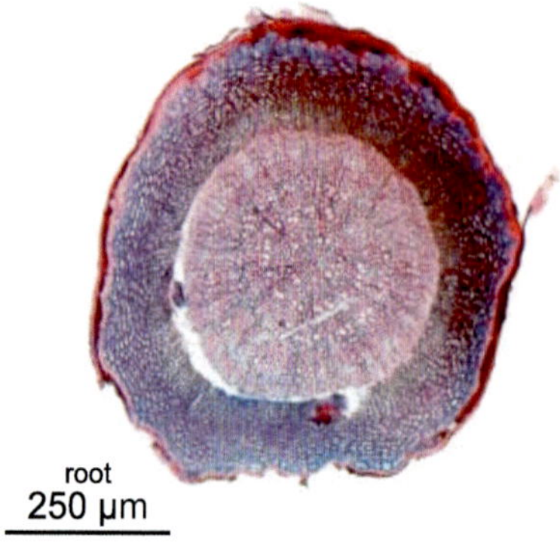

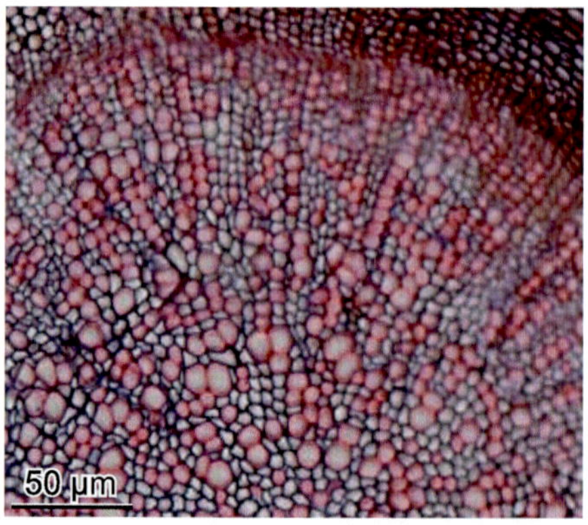

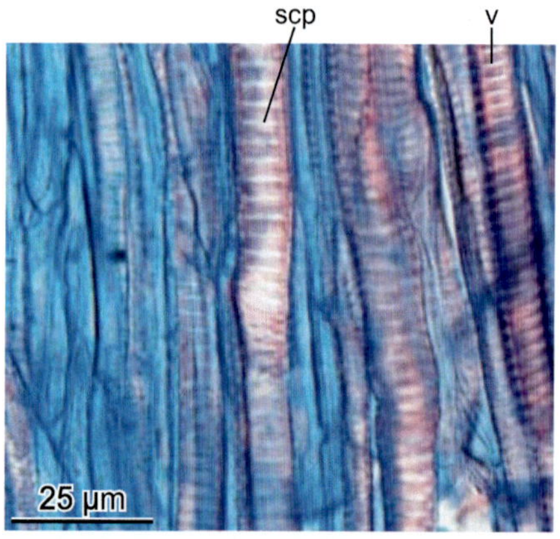

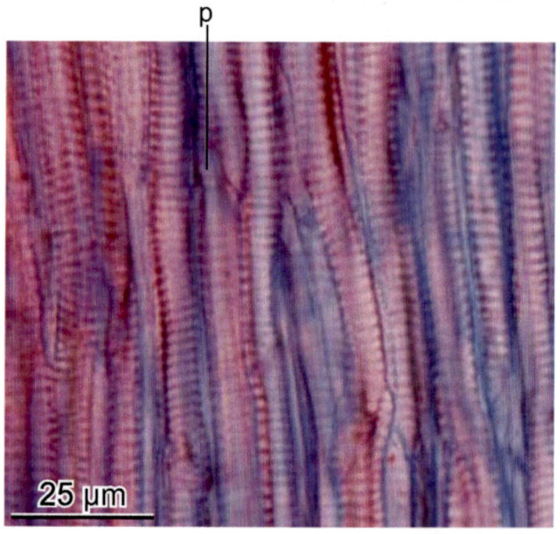

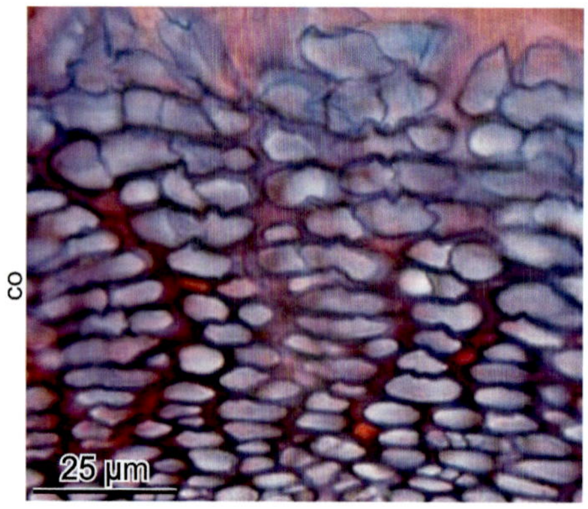

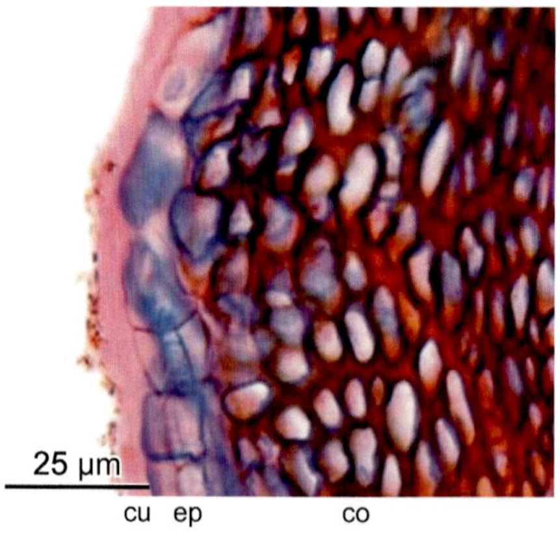

Features according to IAWA

Xylem 1 - 4 - 9 - 9.1 - 13 - 20 - 40.1 - 50.2 - 52.3 - 60.1 - 79.1 - 117

Androsace robusta (R.Knuth) Hand.-Mazz.

Primulaceae

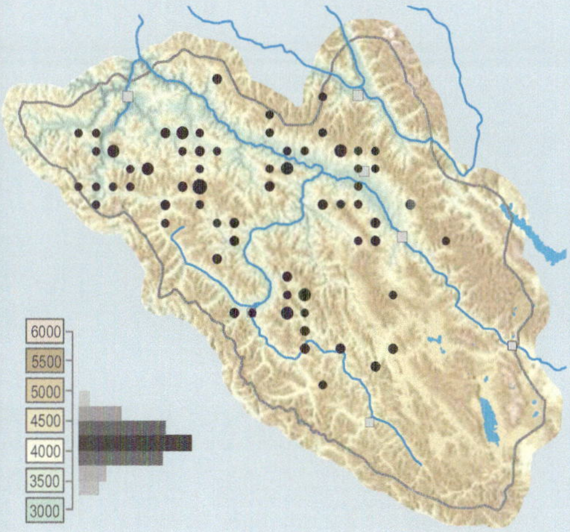

Syn.: *A. villosa* var. *robusta* R.Knuth

Description: Perennial herbs, 3–10 cm tall, villous, forming dense cushions, leaves oblong to elliptic, sessile, overlapping in globose rosettes, 0.8–1.5 cm across, flowers purple with a yellow centre, 6–8 mm across, arranged in 4–8-flowered umbels.

Distribution: Pakistan, N India, Nepal, China (S Xizang).

Habitat: Rock crevices.

Stem diameter 2 mm. Bark/xylem proportion 1:1. Ring width 0.05–0.2 mm. Maximal 8 annual rings.

Xylem

Transverse section

Rings recognizable, semi-ring-porous. Vessels solitary and in short radial multiples. Diameter of larger vessels 20–25 µm. Fibers absent. Parenchyma pervasive. Crystals absent.

Tangential section

Rays absent.

Radial section

Perforation plates simple. Intervessel pits scalariform. Vessel length 70–100 µm.

Bark

Phloem contains small groups of sieve tubes. Cortex consists of fairly thick-walled parenchyma cells. Sclerenchyma and crystals absent. Phellem absent.

Pith

No slide available.

Analyzed: Root collar of polar root of 8 individuals.

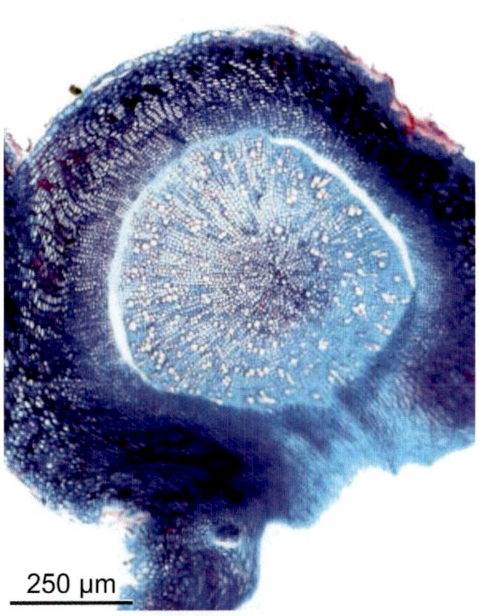

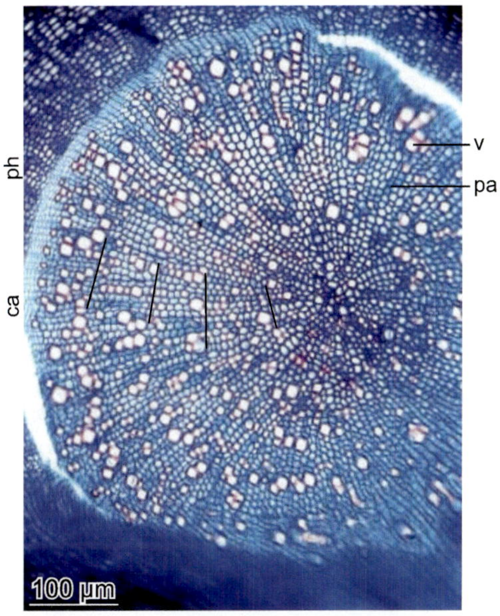

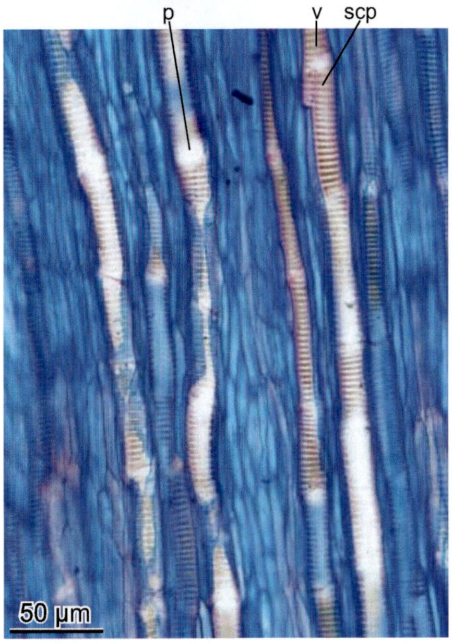

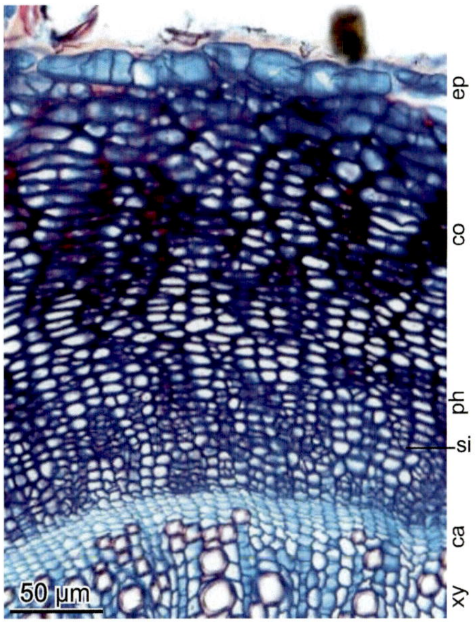

Features according to IAWA

Xylem 1 - 4 - 9 - 9.1 - 13 - 22 - 40.1 - 40.2 - 50.2 - 52.2 - 60.1 - 79.1 - 117

Glaux maritima L.

Primulaceae

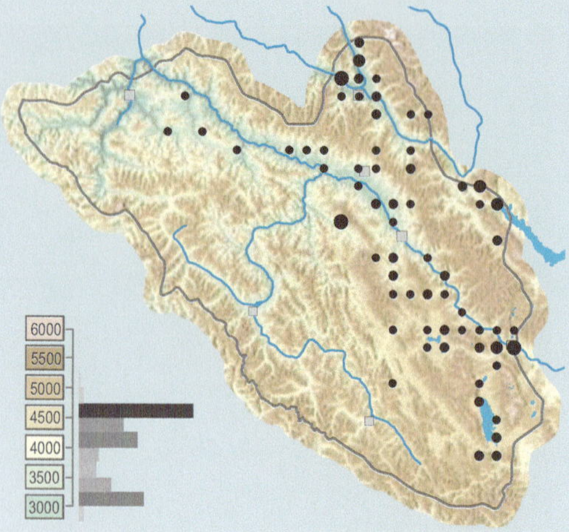

Description: Perennial herbs, 3–25 cm tall, stems prostrate to erect, rooting at nodes, leaves opposite, elliptic-oblong, fleshy, flowers pink, 3–5 mm across, solitary, axillary.

Distribution: Temperate and arctic regions of the Northern hemisphere.

Habitat: Marshes and wet sandy sediments, alkaline swamps, saline areas.

Stem diameter 1–2.5 mm. Bark/xylem proportion 1:0.3. Ring width 0.1–0.25 mm. Maximal 1 annual ring.

Xylem

Transverse section

Only one ring. Vessels solitary. Diameter of larger vessels 15–20 µm. Fibers absent. Parenchyma paratracheal. Crystals absent.

Tangential section

Rays absent.

Radial section

Perforation plates simple. Intervessel pits round. Vessel length 50–70 µm.

Bark

Phloem contains small groups of sieve tubes and companion cells. Large, round thin-walled cells are probably laticifers. Cortex consists of a belt of fibers. Large aerenchymatic tissue surrounds stele. Endodermis with distinct Caparian strips separates stele from cortex. (Grigore et al. 2014)

Pith

Round, filled with unlignified, round cells.

Analyzed: Root collar of polar roots of 3 individuals.

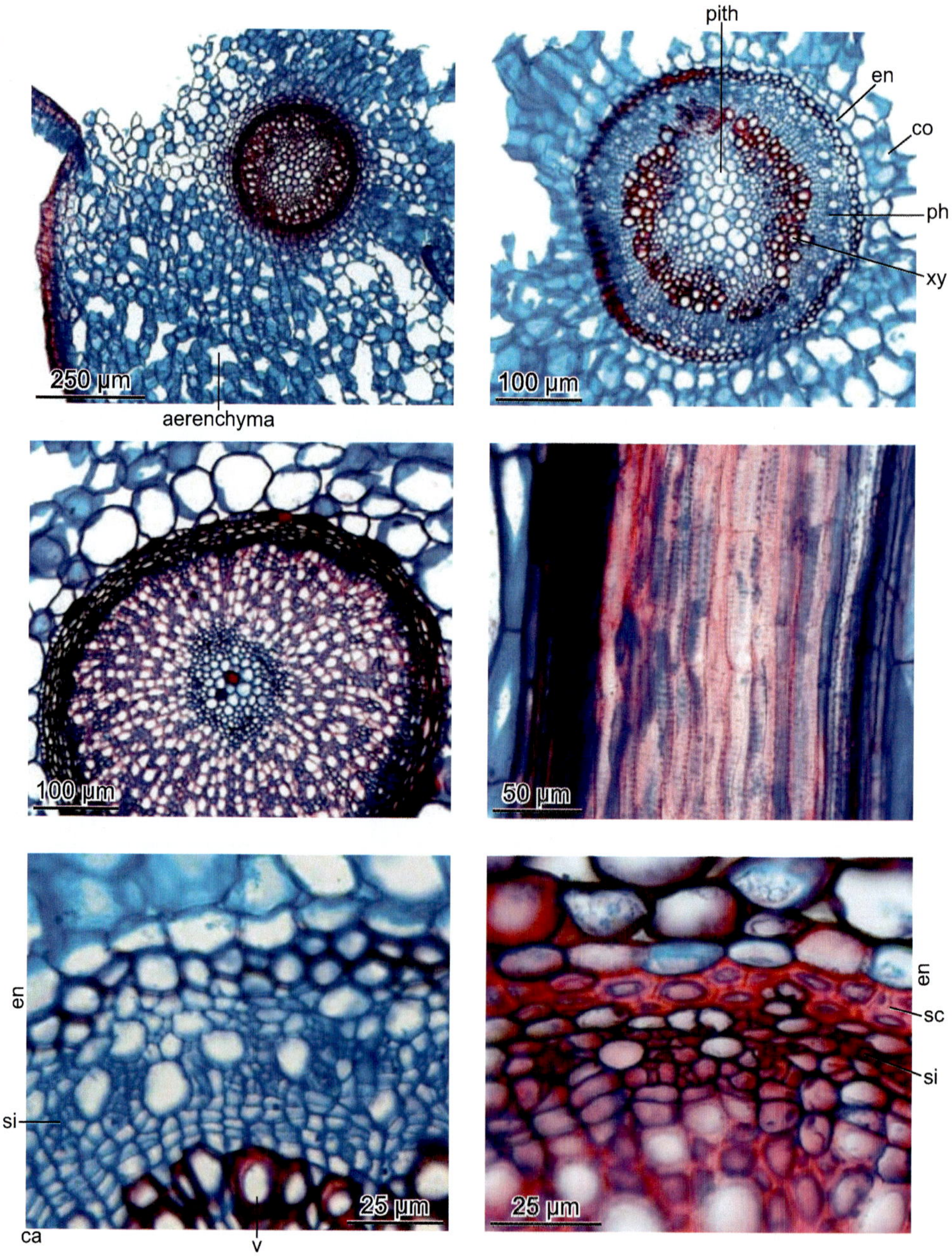

Features according to IAWA

Xylem 2.1 - 9 - 13 - 22 - 40.1 - 50.2 - 52.2 - 60.1 - 79 - 117

Primula macrophylla D.Don

Primulaceae

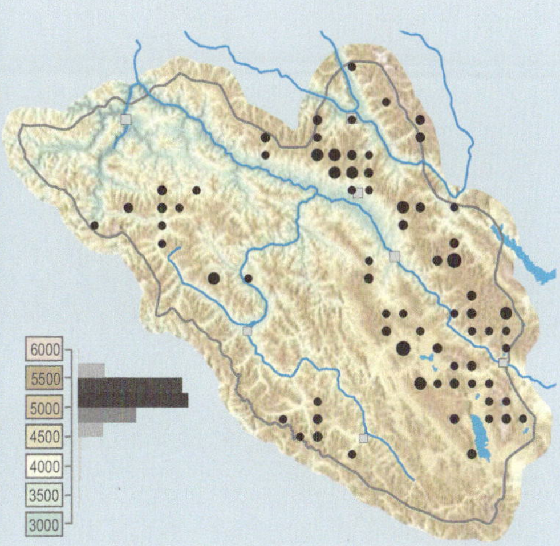

Description: Perennial herbs, 10–25 cm tall, leaves in basal rosette, lanceolate to oblanceolate with revolute margin, farinose below, flowers violet to purple, ca. 2 cm across, in many-flowered umbels.

Distribution: Tajikistan, Afghanistan, N Pakistan, India (Jammu and Kashmir, Himachal Pradesh, Uttarakhand, Sikkim), Nepal, China (Xinjiang, Xizang), Bhutan.

Habitat: Snow beds, solifluction soils, flushes and wet spring areas.

Stem diameter 5–11 mm. Bark/xylem proportion 1:0.5. Rings absent.

Xylem

Transverse section

Xylem small, consisting of irregularly distributed and directed parenchyma cells and vessels. Diameter of larger vessels 20–30 µm. Irregularly formed vascular bundles remain.

Tangential section

Rays absent.

Radial section

Perforation plates simple. Intervessel pits scalariform. Vessel length 80–100 µm.

Bark

Construction of the phloem indistinct. Cortex and phloem separated by a thin-walled, lignified endodermis. Cortex consists of round parenchyma cells. Sclerenchyma and crystals absent. Phellem generally absent.

Pith

Consists of uniform parenchyma cells.

Analyzed: Root collar of polar root of 8 individuals.

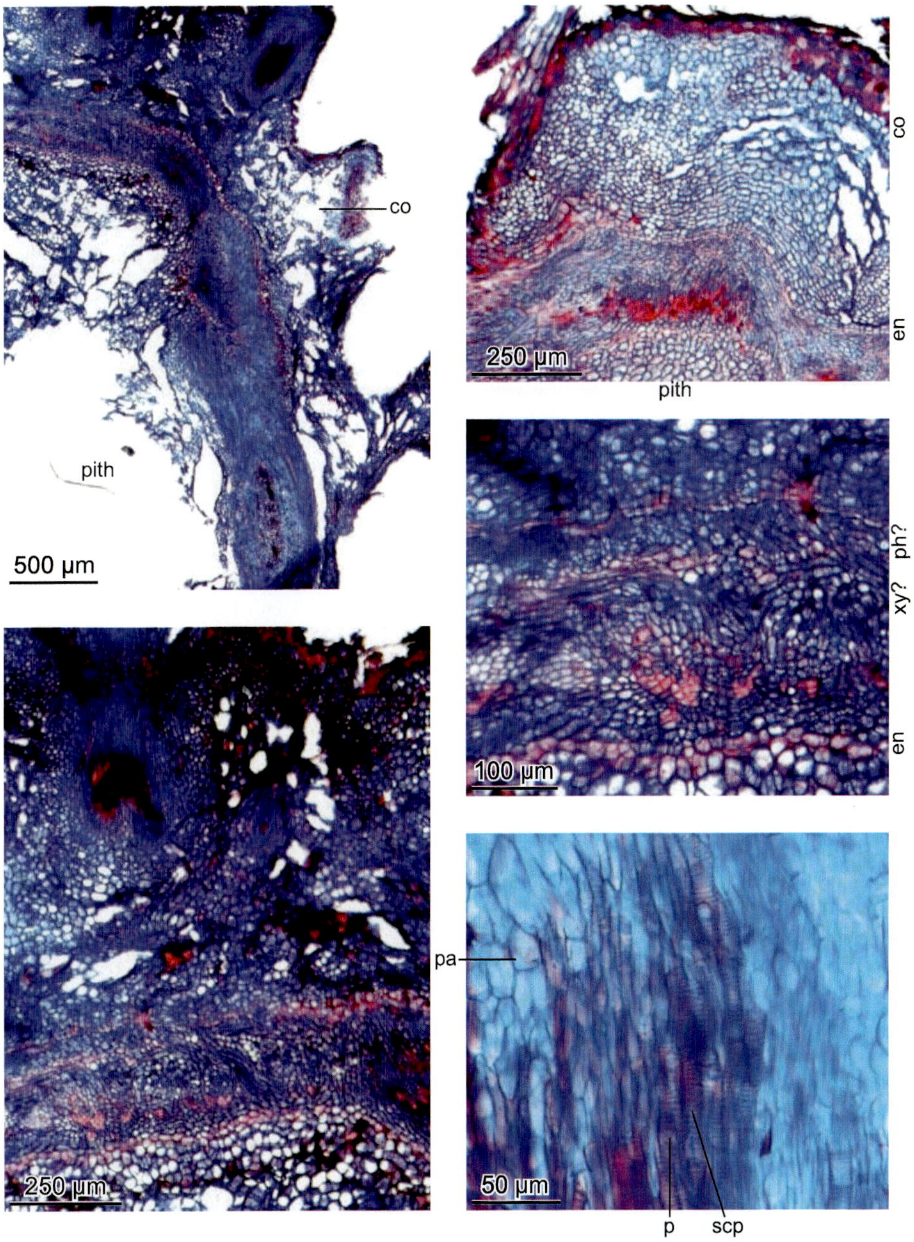

Features according to IAWA

Xylem 2.1 - 9 - 13 - 22 - 40.2 - 50.1 - 52.2 - 60.1 - 79.1 - 117

Aconitum rotundifolium Kar. & Kir.

Ranunculaceae

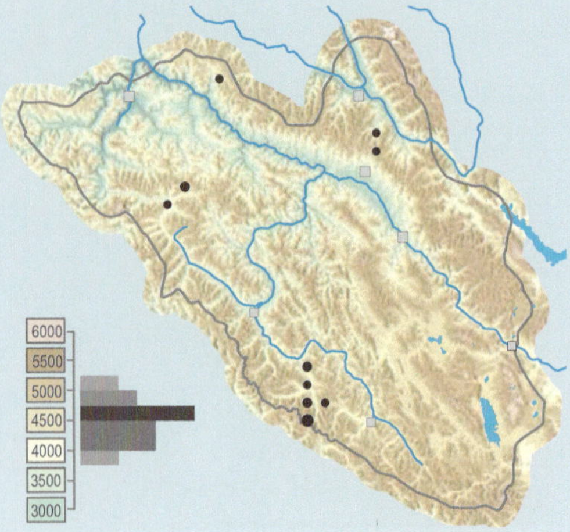

Syn.: *A. napellus* var. *rotundifolium* (Kar. & Kir.) Hook.f. & Thomson

Description: Biennial herbs, 20–40 cm tall, pubescent, leaves mostly basal, orbicular in shape, 3-parted with 3-fid lobes, flowers purplish, ca. 2 cm long, in lax racemes.

Distribution: Russia, Kyrgyztan, Tajikistan, Afghanistan, N Pakistan, NW India, China (Xinjiang).

Habitat: Subalpine and alpine grasslands.

Diameter 5–6 mm. Bark/xylem proportion 1:0.1. Rings indistinct.

Xylem

Transverse section

Very small xylem without annual rings in the center of the bulb. Vessels angular, diameter 30–40 µm. Fibers and rays absent. Parenchyma pervasive.

Tangential section

Parenchyma between vascular bundles.

Radial section

Perforation plates simple. Intervessel pits scalariform. Vessel length 100–150 µm.

Bark

Large phloem with isolated groups of sieve tubes. Cortex consists of round and angular parenchyma cells. Phellem consists of several layers of rectangular cells.

Pith

No slide available.

Analyzed: Basis of bulb of 3 individuals.

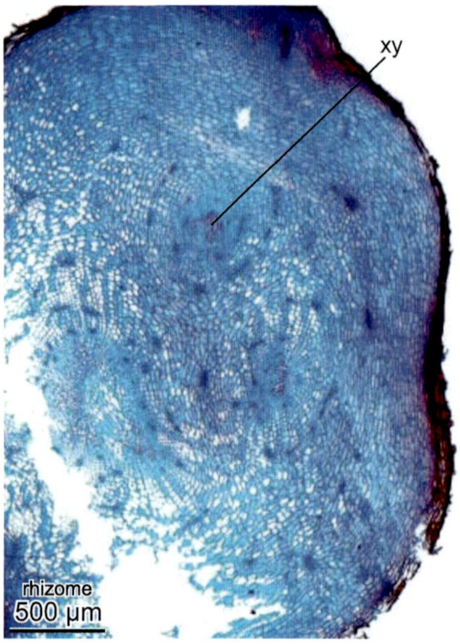

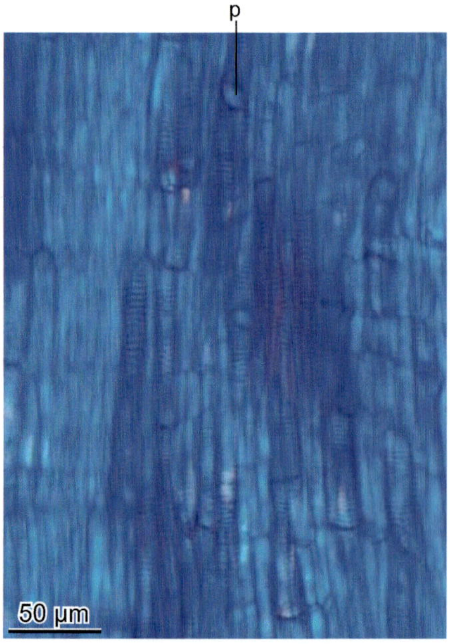

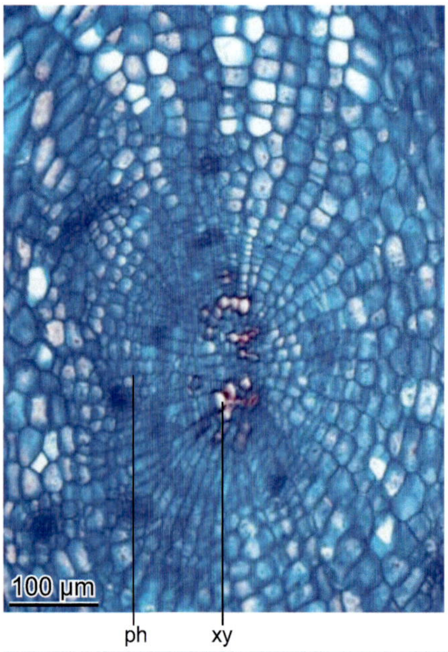

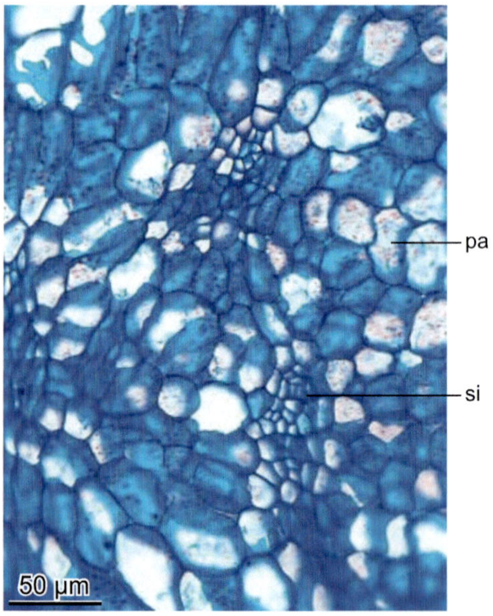

Features according to IAWA

Xylem 2 - 9 - 12 - 13 - 20 - 40.2 - 50.1 - 52.3 - 60.1 - 79.1 - 117

Aconitum violaceum Jacquem. ex Stapf

Ranunculaceae

Syn.: *A. multifidum* Royle, *A. violaceum* var. *multifidum* (Royle) Qureshi & Chaudhri

Description: Biennial herbs, 10–30 cm tall, leaves mostly basal, orbicular in shape, 5-palmatipartite, segments deeply pinnatisect, flowers violet, rarely blue or yellowish, ca. 2 cm long, in lax racemes.

Distribution: N Pakistan, NW India, Nepal.

Habitat: Montane and subalpine grasslands.

Diameter 5 mm. Bark/xylem proportion 1:ca. 2. Rings absent.

Xylem

Transverse section

Isolated vascular bundles in circles (annual rings?). Vascular bundles in the center consist of xylem with vessels, those at the periphery of phloem. Fibers within the bundles absent. Bundles surrounded by sclerenchyma.

Tangential section

Parenchyma between vascular bundles.

Radial section

Perforation plates simple. Intervessel pits round. Vessel length 100–150 µm.

Bark

Phloem absent. Cortex consists of round and angular parenchyma cells, phellem of several layers of rectangular cells.

Pith

Round.

Analyzed: Basis of bulb of 1 individual.

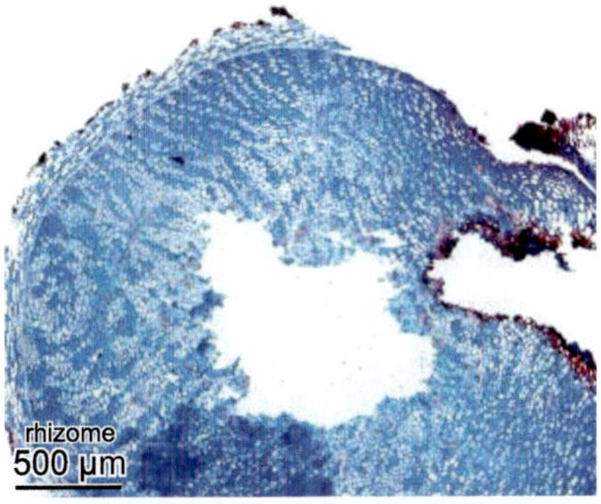

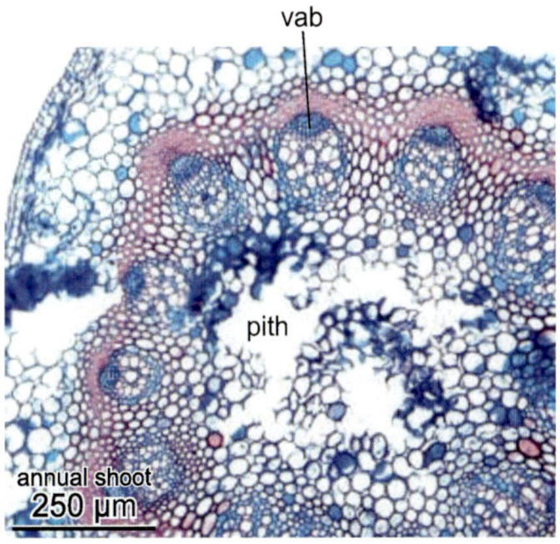

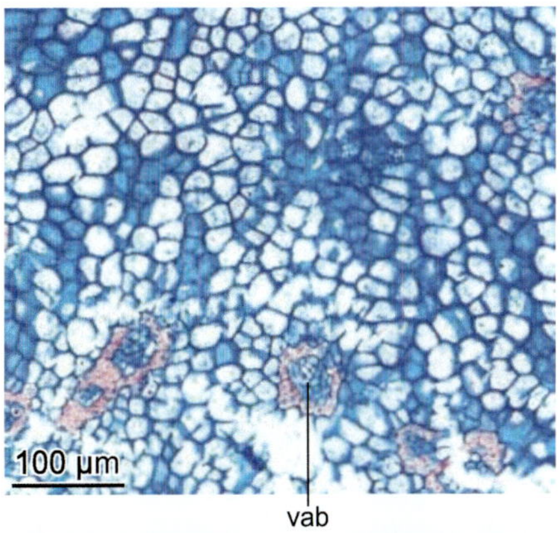

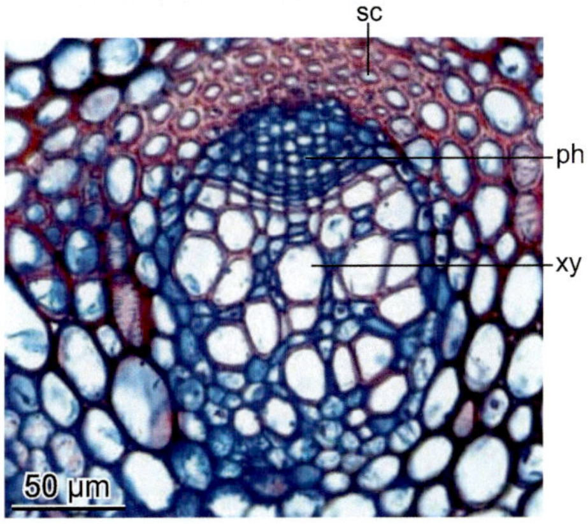

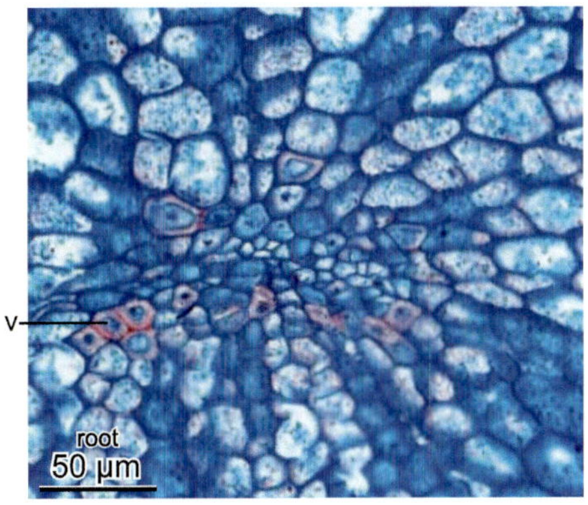

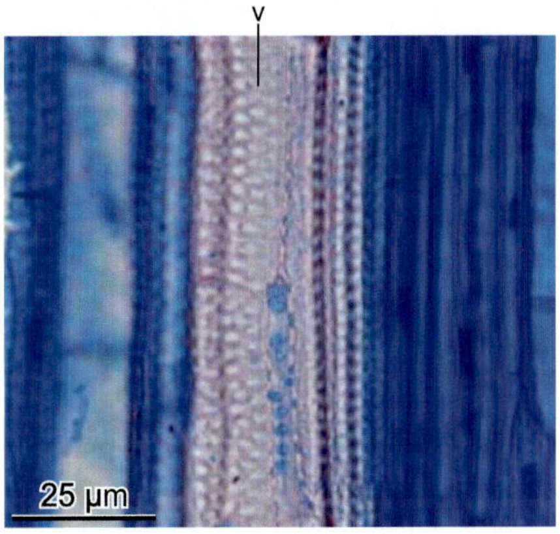

Features according to IAWA

Xylem 2 - 11 - 13 - 22 - 40.2 - 50.1 - 52.3 - 60.1 - 79.1 - 99 - 99.1 - 100.2

Anemone rupicola Cambess.

Ranunculaceae

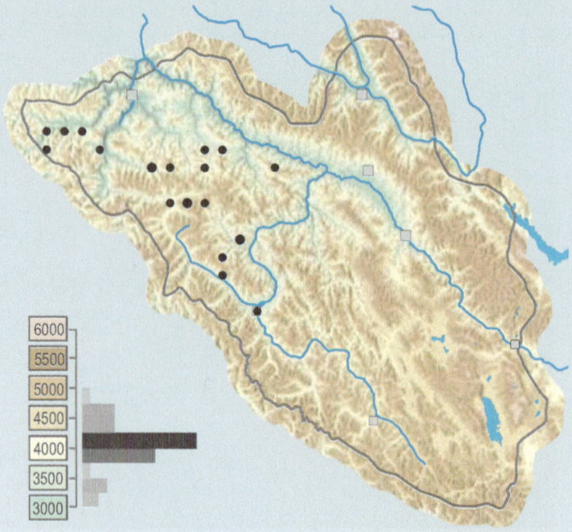

Syn.: *A. batangensis* Finet, *A. rupestris* Jacquem.

Description: Perennial herbs, 5–30 cm tall, with long rhizomes, softly hairy, leaves 3-sect, segments deeply lobed and sharply toothed, shiny above, flowers white, yellowish or purplish, 4–7 cm across, usually solitary, achenes densely woolly.

Distribution: E Afghanistan, Pakistan, N India, Nepal, China (W Sichuan, S Xizang, NW Yunnan), Bhutan.

Habitat: Rocks, moraines, cliffs, gravelly slopes.

Stem diameter 3–4 mm. Bark/xylem proportion 1:3. Ring width 0.3–0.4 mm. Maximal 6 annual rings.

Xylem

Transverse section

Diffuse- to semi-ring-porous. Diameter of earlywood vessels 40–70 µm. Diameter of latewood vessels 20–30 µm. Vessels angular, solitary and in small irregular groups. Fibers rare, in small bands in the latewood. Parenchyma pervasive. Vascular bundles remain. Crystals absent. With periderm towards the pith.

Tangential section

Rays >10-seriate, unlignified, confluent with the axial tissue.

Radial section

Perforation plates simple. Intervessel pits round to scalariform. Vessel length 100–150 µm.

Bark

Phloem consists primarily of parenchyma. Large sieve tubes in tangential rows. Phellem consists of several layers of rectangular cells.

Pith

No slide available.

Analyzed: Rhizome of 2 individuals.

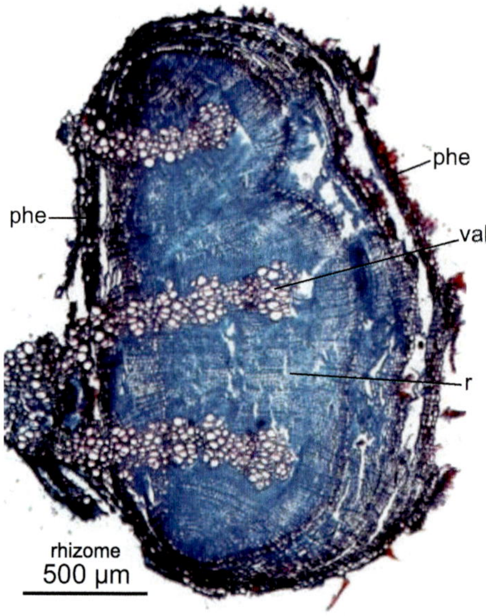

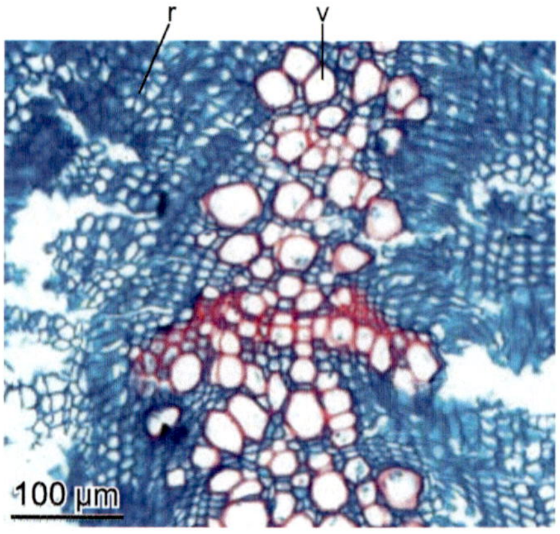

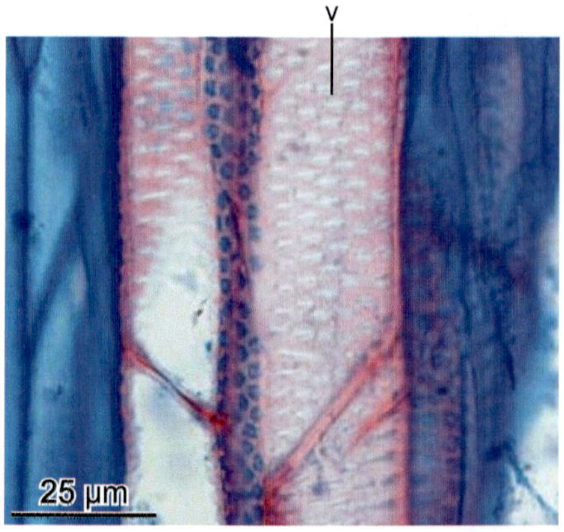

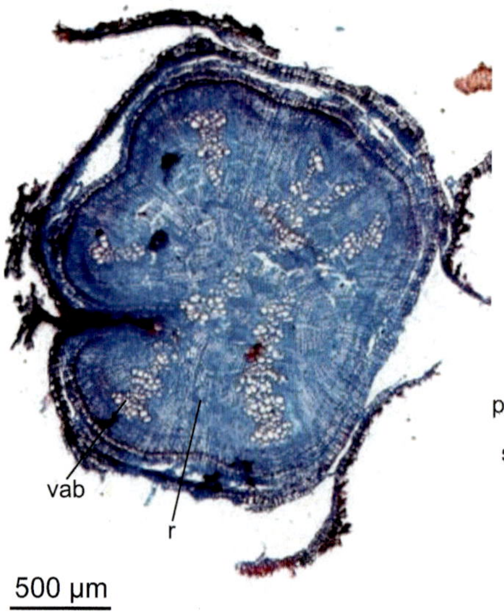

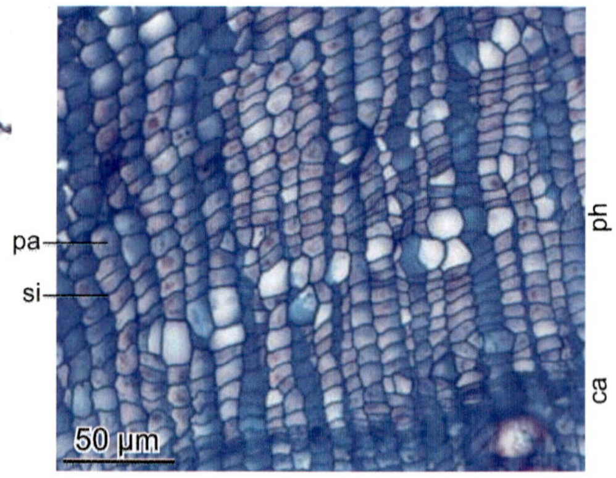

Features according to IAWA

Xylem 1 - 4 - 5 - 9 - 11 - 12 - 13 - 20 - 22 - 41 - 42 - 50.2 - 52.3 - 61 - 79.1 - 99.1 - 100.1 - 100.2 - 105 - 135.2

Aquilegia fragrans Benth.

Ranunculaceae

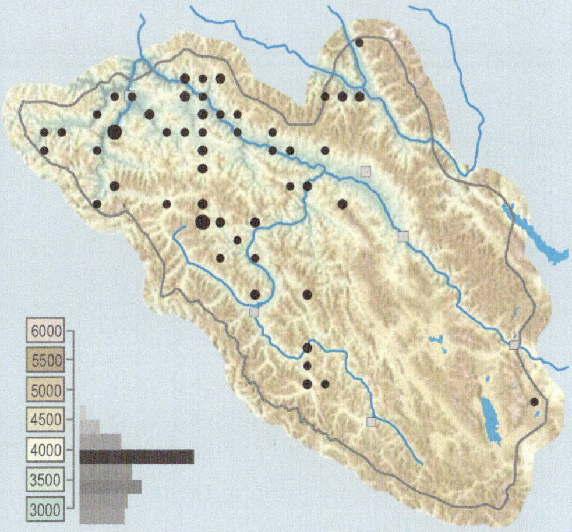

Description: Perennial herbs, 30–80 cm tall, leaves 2–3-ternate, flowers whitish to purple, spurred, usually nodding, 3–4 cm across.

Distribution: NE Pakistan, NW India.

Habitat: Stabilized slopes, screes, rock crevices.

Stem diameter 6 mm. Bark/xylem proportion 1:1. Mean ring width 0.3 mm. More than 3 annual rings.

Xylem

Transverse section

Semi-ring-porous. Diameter of earlywood vessels 50–60 µm. Diameter of latewood vessels 10–20 µm. Vessels mostly solitary. Fibers absent. Parenchyma pervasive. Vascular bundles remain. Crystals absent. With periderm towards the pith.

Tangential section

Rays >10-seriate, unlignified, confluent with the axial tissue.

Radial section

Perforation plates simple. Intervessel pits round to scalariform. Vessel length 120–150 µm.

Bark

Phloem contains groups of sieve elements in tangential layers. Cortex consists of round to oval parenchyma cells, phellem of several layers of rectangular cells.

Pith

No slide available.

Analyzed: Rhizome of 1 individual.

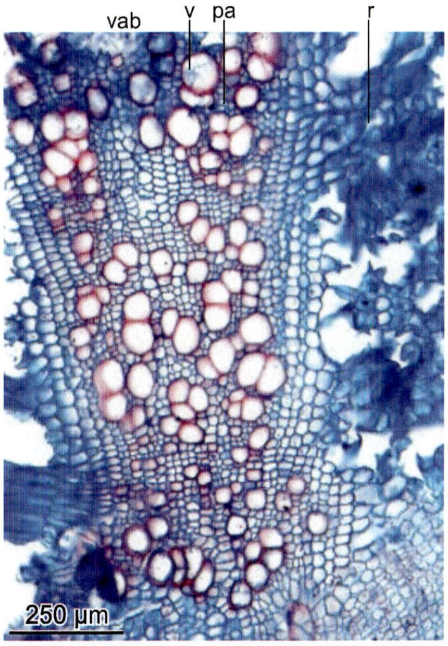

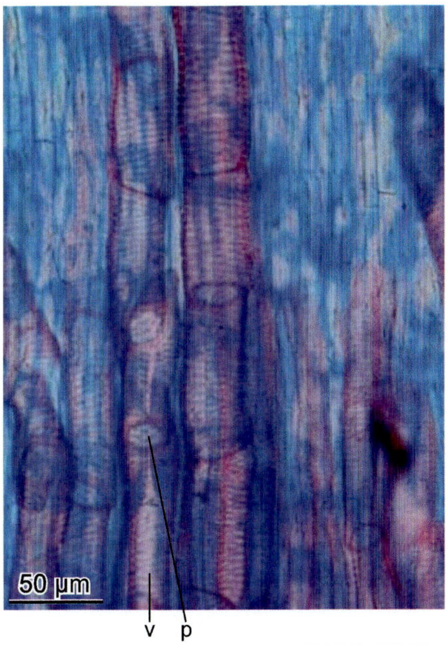

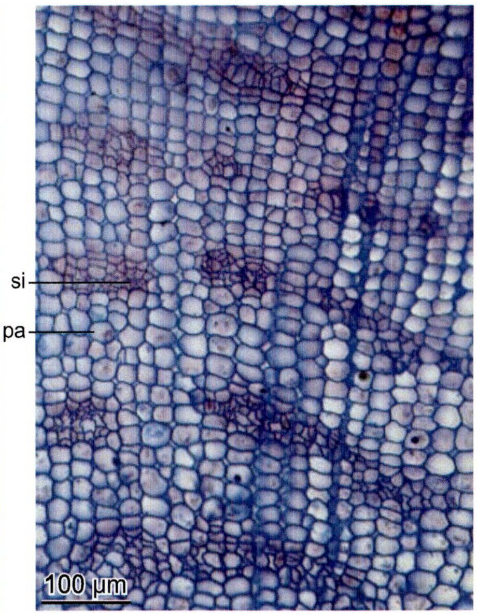

Features according to IAWA

Xylem 1 - 4 - 9 - 13 - 20 - 22 - 42 - 50.2 - 52.3 - 60.1 - 79.1 - 99.1 - 100.1 - 100.2 - 105 - 135.2

Aquilegia moorcroftiana Wall. ex Royle

Ranunculaceae

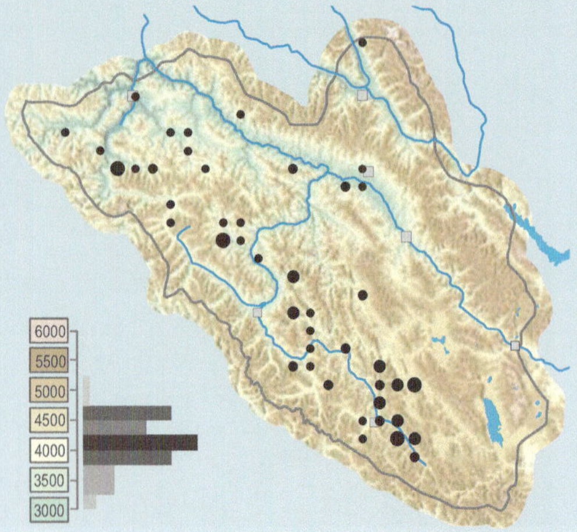

Syn: *A. vulgaris* subsp. *moorcroftiana* (Wall.) Bruehl

Description: Perennial herbs, 20–40 cm tall, glandular pubescent, leaves biternate, leaflets 2–3-fid, flowers yellow to yellow tinged with blue, spurred, usually nodding, ca. 3 cm across.

Distribution: Afghanistan, Pakistan, India (Jammu and Kashmir), China (Xizang).

Habitat: Stabilized slopes, screes, rock crevices, shady places near rivers.

Stem diameter 5 mm. Bark/xylem proportion 1:2. Ring width 0.5–0.6 mm. Maximal 5 annual rings.

Xylem

Transverse section

Semi-ring-porous. Diameter of earlywood vessels 30–40 μm. Vessels in groups. Fibers absent. Parenchyma pervasive. Vascular bundles remain. Crystals absent.

Tangential section

Rays 8–>10-seriate, confluent with the axial tissue.

Radial section

Perforation plates simple. Intervessel pits scalariform. Vessel length 100–150 μm.

Bark

Phloem contains groups of sieve tubes in tangential layers. Cortex consists of round to oval parenchyma cells, phellem of several layers of rectangular cells.

Pith

No slide available.

Analyzed: Rhizome of 2 individuals.

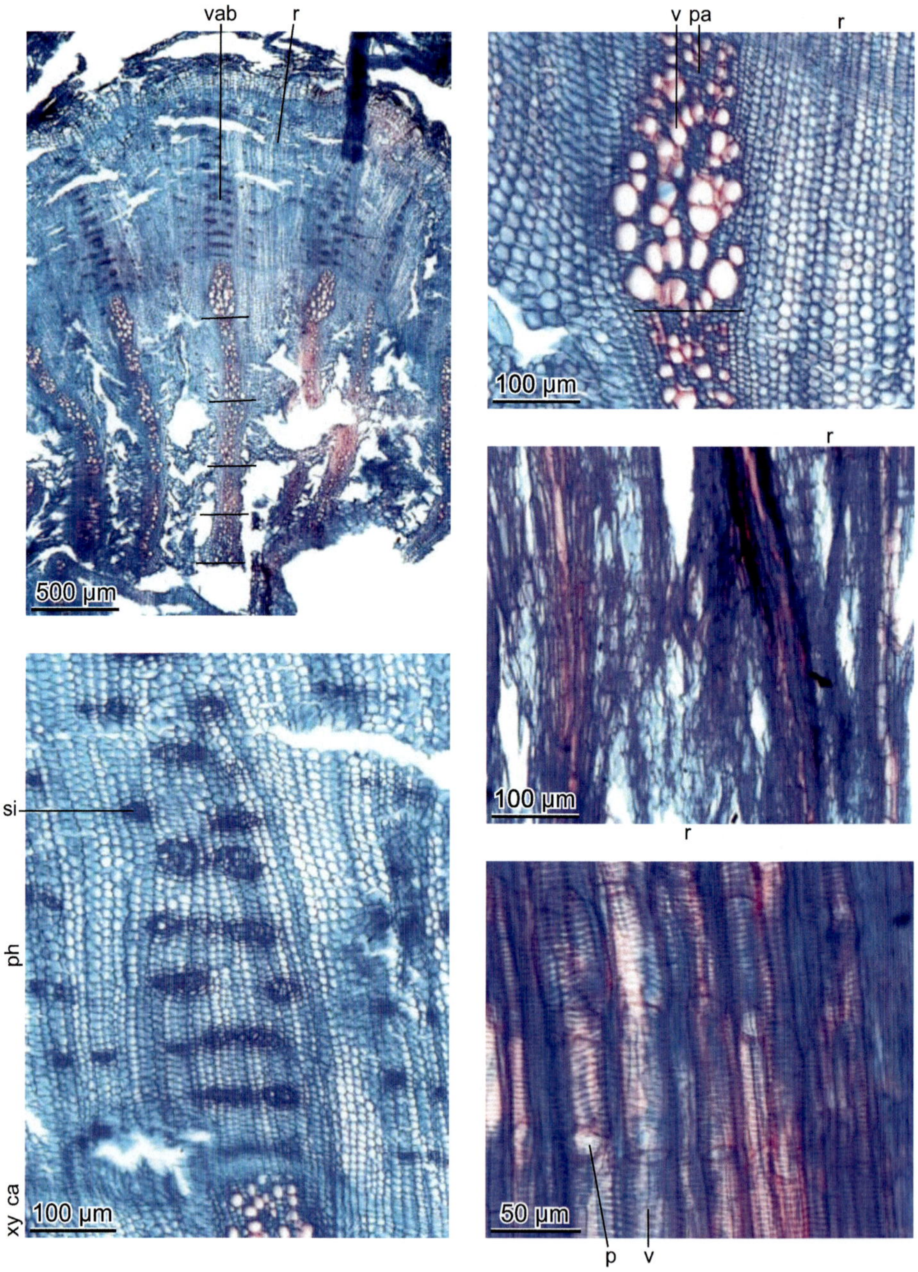

Features according to IAWA

Xylem 1 - 4 - 9 - 13 - 20 - 40.2 - 50.1 - 52.3 - 60.1 - 79.1 - 98 - 99.1 - 100.1 - 100.2 - 105 - 135.2

Clematis orientalis L.

Ranunculaceae

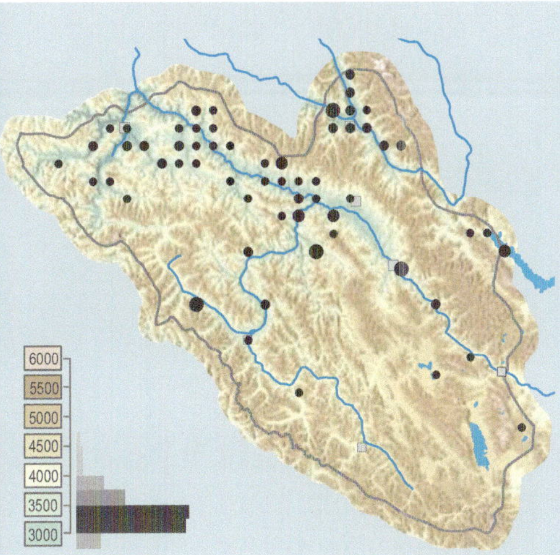

Syn.: *C. tibetana* Kuntze, *C. ladakhiana* Grey-Wilson

Description: Woody climbers, leaves variable, pinnate to 2-pinnate, leaflets ovate to narrowly oblong, margin lobed or incised, flowers yellow or purplish in parts, 2–4 cm across, in lax panicles.

Distribution: SE Europe, SW Asia, Russia (Siberia), Afghanistan, Pakistan, India, China (Xizang) Mongolia.

Habitat: Stabilized stony slopes, human-made habitats, such as stony walls, dumps, road banks.

Stem diameter 14 mm. Bark/xylem proportion 1:ca.3. Ring width 0.4–1 mm. Maximal 33 annual rings.

Xylem

Transverse section

Ring-porous. Conducting tissue consists of large and small vessels (vessel dimorphism). Diameter of large earlywood vessels 80–120 µm. Diameter of small earlywood vessels 20–30 µm. Large vessels primarily solitary, small vessels in groups. Fibers thin- to thick-walled. Parenchyma paratracheal and marginal. Vascular bundles remain. Crystals absent.

Tangential section

Rays 8–>10-seriate, with distinct sheath cells.

Radial section

Perforation plates simple. Intervessel pits round. Vessel length 150–200 µm.

Bark

Phloem layered, consisting of a fairly homogeneous, parenchymateous, unlignified tissue and a small layer of thin- to thick-walled fibers. Uniseriate layer of larger cells (phelloids) between the layers. Cortex absent.

Pith

No slide available.

Analyzed: Rhizome of 3 individuals.

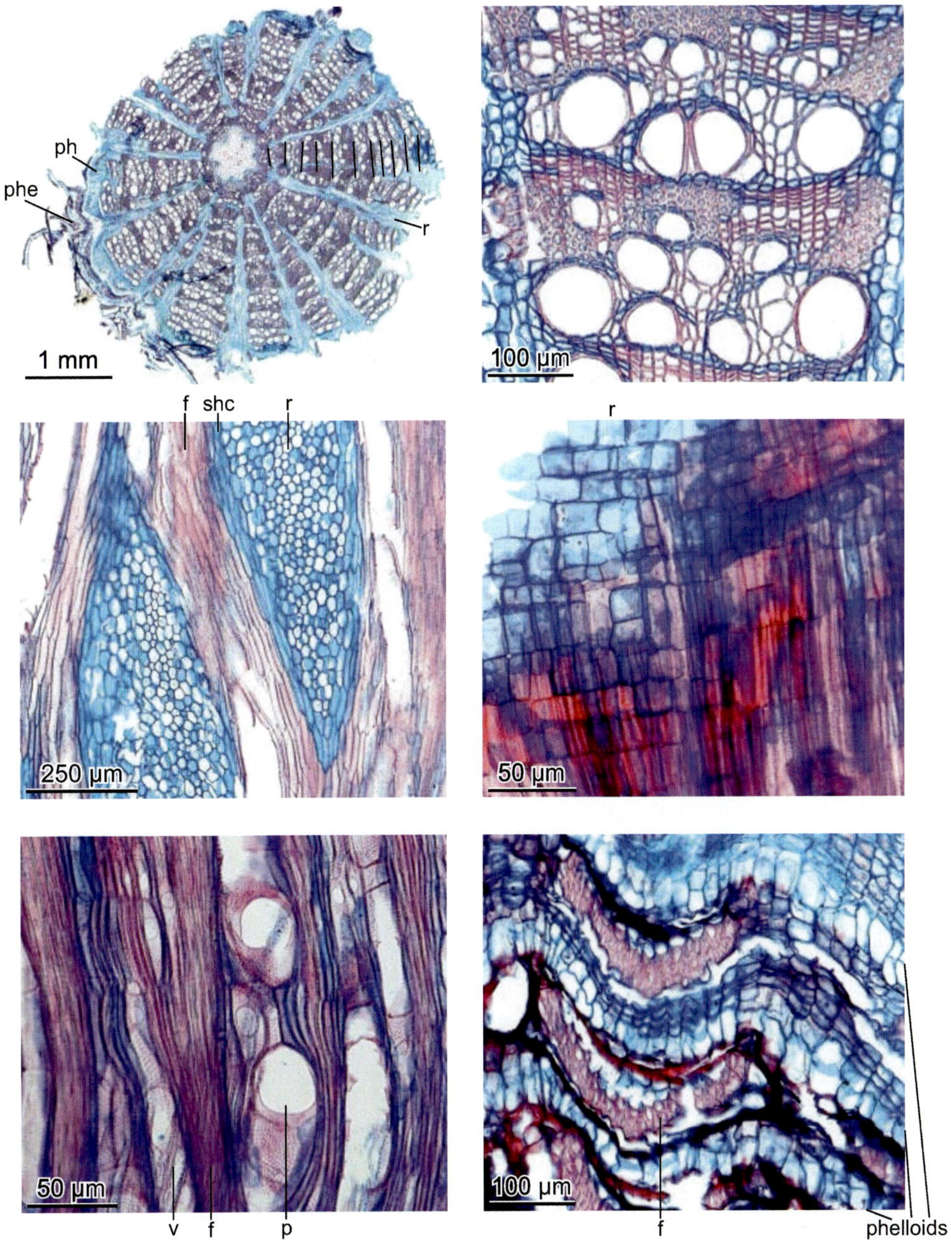

Note: Several species and subordinate taxa described from the broader area seem to be impossible to delimit and fall into the wide range of variation of *Clematis orientalis*.

Features according to IAWA

Xylem 1 - 3 - 9 - 11 - 13 - 22 - 40.2 - 50.1 - 52.2 - 61 - 79 - 89 - 99 - 100.2 - 104 - 108 - 110

Clematis tangutica (Maxim.) Korsh.

Ranunculaceae

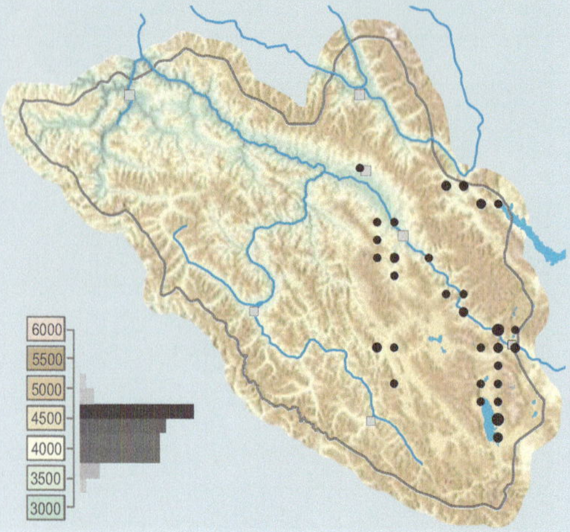

Syn.: *C. orientalis* var. *tangutica* Maxim.

Description: Woody climbers or stout semi-erect shrubs when growing in dry regions, leaves 1–2-pinnate, flowers yellow, 4–6 cm across, solitary or 2–3 in cymes.

Distribution: Kazakhstan, N India, China (Gansu, Qinghai, Shaanxi, Sichuan, Xinjiang, Xizang), Mongolia.

Habitat: Gravel along rivers, unstable screes, stony walls in settlements.

Stem diameter 9–20 mm. Bark/xylem proportion 1 : 5. Ring width 0.2–0.4 mm. Maximal 30 annual rings.

Xylem

Transverse section

Ring porous. Vessel dimorphism. Diameter of large earlywood vessels 200–250 µm. Diameter of small earlywood vessels 20–30 µm. Large and small vessels solitary and in groups. Fibers thick-walled, fiber tracheids thin-walled. Parenchyma paratracheal and marginal. Vascular bundles remain. Crystals absent.

Tangential section

Rays 8–>10-seriate, unlignified with distinct sheath cells.

Radial section

Perforation plates simple. Intervessel pits round. Vessel length 150–200 µm.

Bark

Phloem layered, consisting of fairly homogeneous, parenchymateous, unlignified tissue and a small layer of thin- to thick-walled fibers. Uniseriate layer of larger cells (phelloids) between the layers. Cortex absent.

Pith

Pith round.

Analyzed: Rhizome of 4 individuals.

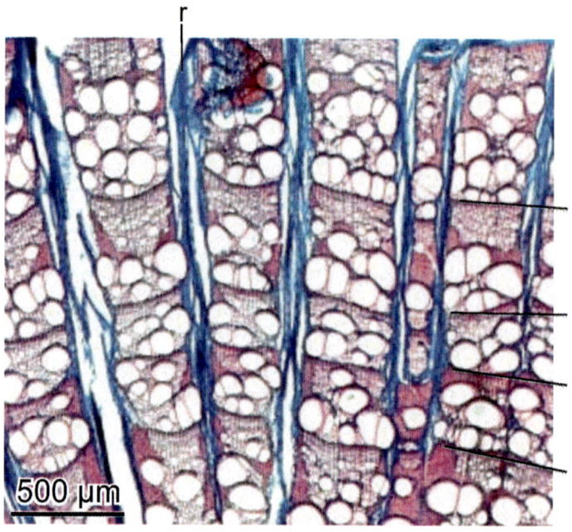

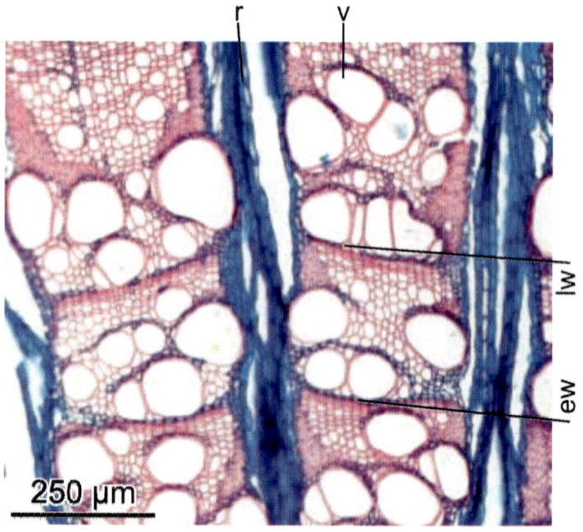

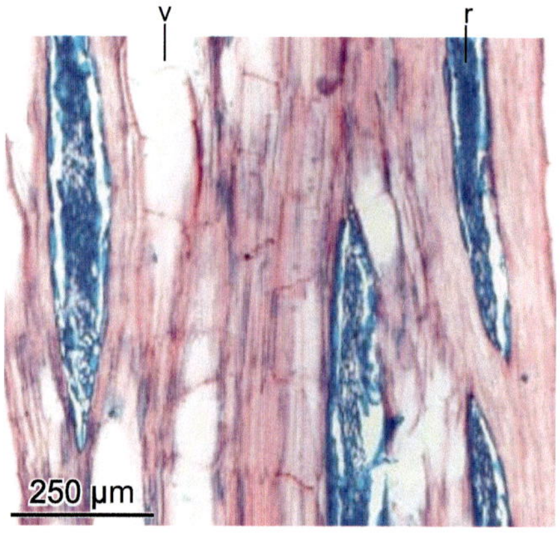

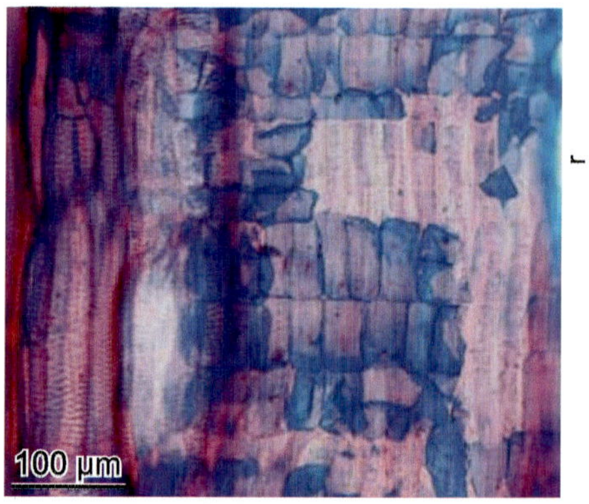

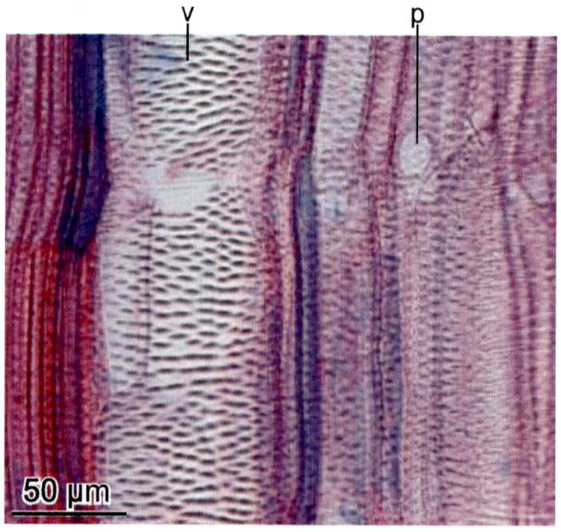

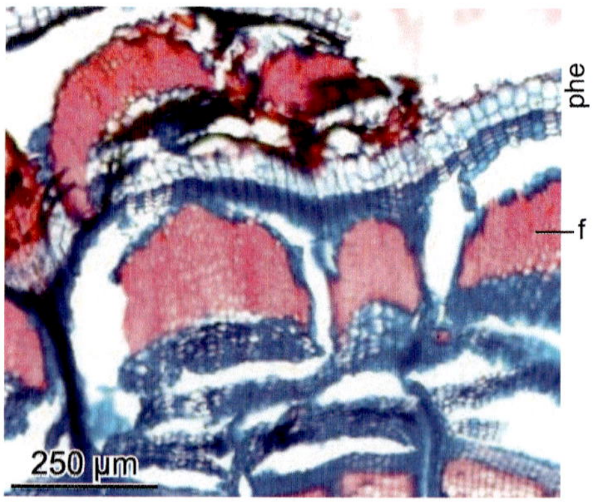

Features according to IAWA

Xylem 1 - 3 - 9 - 11 - 13 - 22 - 40.2 - 45 - 50.1 - 52.2 - 61 - 79 - 89 - 99 - 100.2 - 105 - 110

Delphinium brunonianum Royle

Ranunculaceae

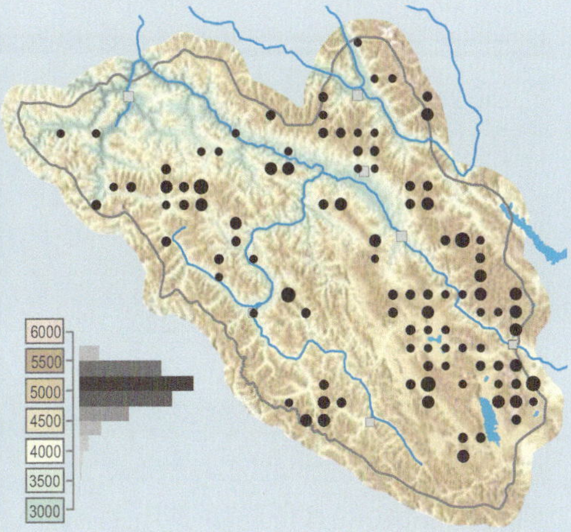

Syn.: *D. nordhagenii* Wendelbo

Description: Perennial herbs, 10–100 cm tall, yellow glandular hairy, leaves palmately 6–7-lobed to the middle, flowers dark purple to light blue, 2–3 cm long, spurred, arranged in few-flowered corymbs.

Distribution: Tajikistan, Afghanistan, N Pakistan, NW India (Jammu and Kashmir, Himachal Pradesh, Uttarakhand), Nepal, China (Xinjiang, Xizang).

Habitat: Stony and gravel slopes, both stabilized and unstable.

Stem diameter 14 mm. Bark/xylem proportion 1:ca. 3. Mean ring width 0.3 mm. Maximal 11 annual rings.

Xylem

Transverse section

With distinct rings. Since the center starts decaying after ca. 5 years, the total age cannot be determined. Tangential splits occur along ring boundaries.

Semi-ring-porous. Diameter of earlywood vessels 60–80 µm. Diameter of latewood vessels 30–40 µm. Vessels in groups. Fibers absent and parenchyma pervasive. Remaining vascular bundles separated by extremely large rays. Crystals absent.

Tangential section

Rays >10-seriate, confluent with the axial tissue. Rays disappear in polarized light.

Radial section

Perforation plates simple. Inter vessel pits round. Vessel length 100–150 µm. Ray construction not observed.

Bark

Phloem fairly homogeneous, with sieve tubes and a few isolated fibers in continuation of vessels and xylem rays. Cortex consists of round to oval parenchyma cells with diameters of 80–120 µm. Phellem not observed.

Pith

No slide available.

Analyzed: Rhizome of 2 individuals.

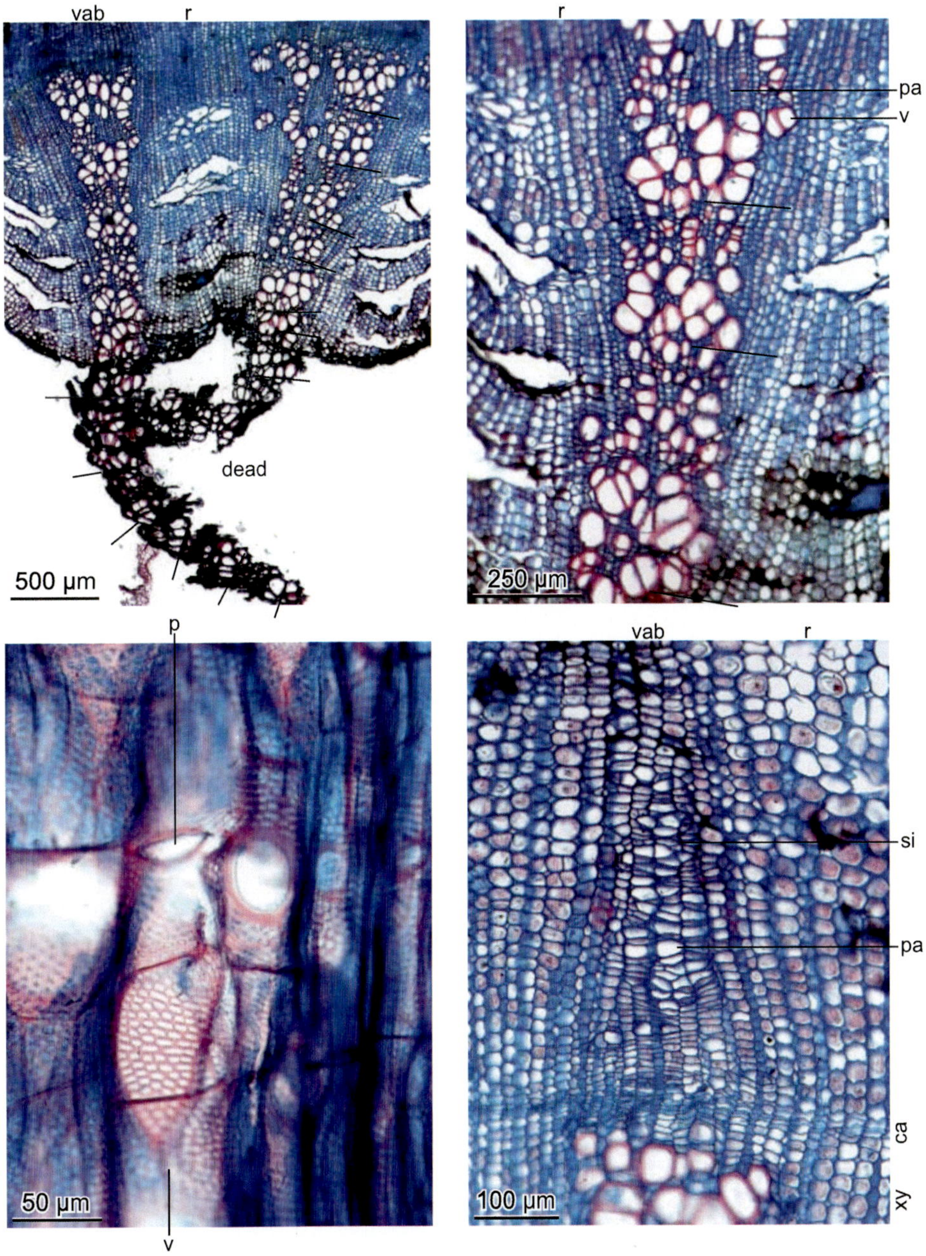

Features according to IAWA

Xylem 1 - 4 - 11 - 13 - 22 - 41 - 50.1 - 52.3 - 60.1 - 79.1 - 99 - 99.1 - 100.1 - 100.2 - 105 - 135.2

Delphinium cashmerianum Royle

Ranunculaceae

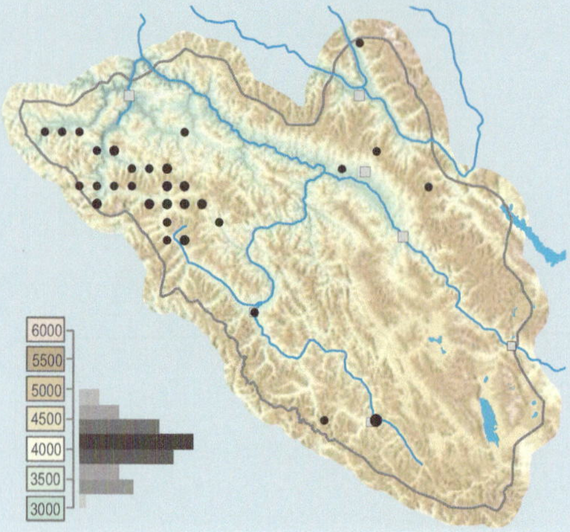

Syn.: *D. aitchisonii* Huth

Description: Perennial herbs, 10–50 cm tall, without glandular hairs, leaves palmately 6–7-lobed to the middle, flowers bluish-purple, 2–3 cm long, spurred, arranged in few-flowered corymbs.

Distribution: N Pakistan, NW India (Jammu and Kashmir, Himachal Pradesh, Uttarakhand), China (Xizang).

Habitat: Alpine meadows, wet streams banks.

Stem diameter 2–3 mm. Bark/xylem proportion 1:1. Mean ring width 0.2 mm. Maximal 2 annual rings.

Xylem

Transverse section

Semi-ring-porous. Diameter of earlywood vessels 30–40 µm. Vessels, angular groups. Fibers absent. Parenchyma pervasive. Vascular bundles remain. Crystals absent.

Tangential section

Rays >10-seriate, confluent between vascular bundles, absent within the bundles.

Radial section

Perforation plates simple. Intervessel pits round to scalariform. Vessel length 140–170 µm. Parenchyma paratracheal and pervasive

Bark

Phloem uniform. Cortex consists of round to oval parenchyma cells. With secondary phellem towards the pith.

Pith

Pith decayed.

Analyzed: Rhizome of 2 individuals (herbarium specimens).

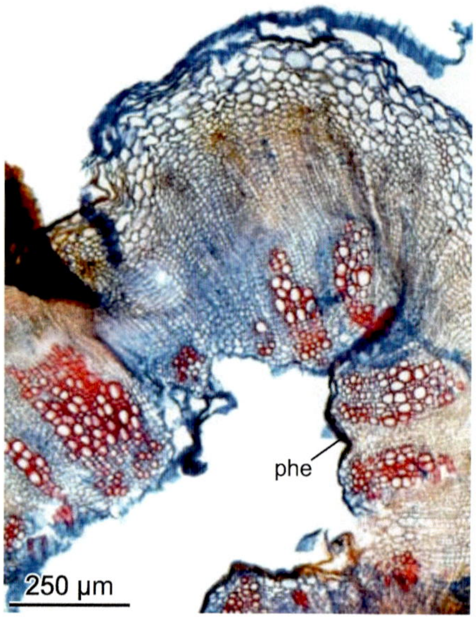

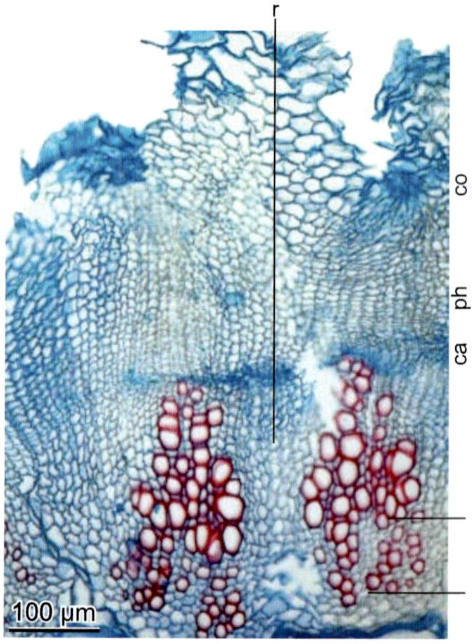

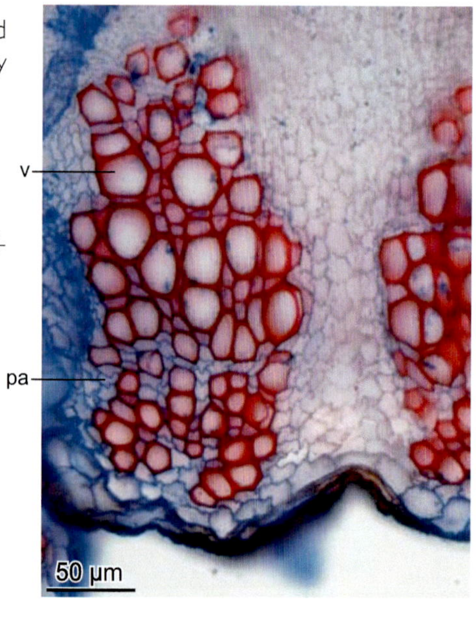

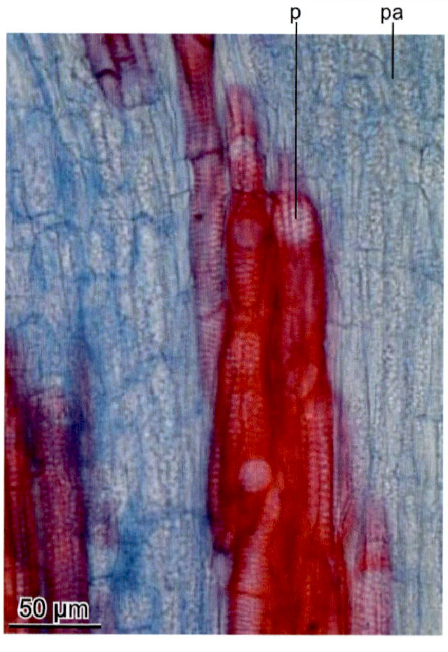

Features according to IAWA

Xylem 1 - 4 - 11 - 13 - 20 - 22 - 40.2 - 50.1 - 52.3 - 60.1 - 79 - 79.1 - 99 - 99.1 - 100.1 - 100.2 - 105 - 135.2

Halerpestes sarmentosa (Adams) Kom.

Ranunculaceae

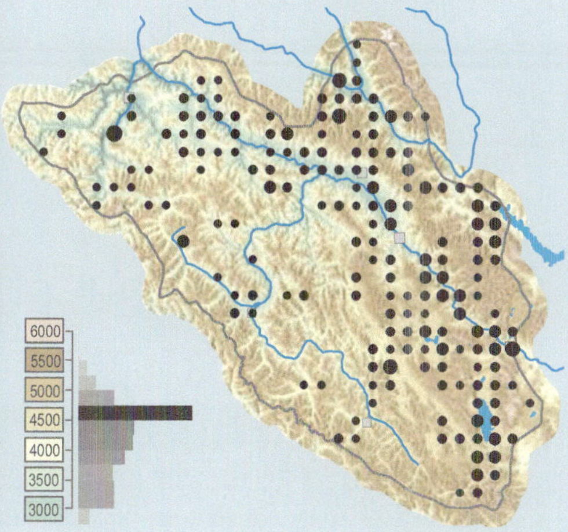

Syn.: *Ranunculus sarmentosus* Adams

Description: Perennial herbs, 1–5 cm tall, with creeping stolons 10–20 cm long, leaves variable from undivided to tripartite, flowers yellow, usually single, ca. 5 mm across.

Distribution: Russia (Siberia), Kazakhstan, N Pakistan, India, China (Gansu, Hebei, Heilongjiang, Jilin, Liaoning, Nei Mongol, Ningxia, Qinghai, Shaanxi, Shanxi, Sichuan, Xinjiang, Xizang), Mongolia, N Korea.

Habitat: Emerged sandy and loamy bottoms, saline habitats, irrigation canals, mesic eutrophicated grasslands.

Stem diameter 6 mm. Bark/xylem proportion 1:0.5. Rings absent.

Xylem

Transverse section

Rings absent, but secondary growth occurs. Vessel diameter 20–30 µm. Vessels mostly solitary in small irregular groups. Fibers absent. Parenchyma pervasive. Irregular formed vascular bundles remain. Ray-like parenchymatic zones between vascular bundles and fibers.

Tangential section

Rays 2–8-seriate.

Radial section

Perforation plates simple. Intervessel pits round in alternating position to scalariform. Vessel length 50–80 µm.

Bark

Phloem more or less uniform. Cortex consists of round to oval parenchyma cells and a few groups of thick-walled cells. Phellem layered, consisting of irregularly formed cells.

Pith

Round, consisting of very thin-walled, large parenchyma cells.

Analyzed: Root collar of polar root of 1 individual.

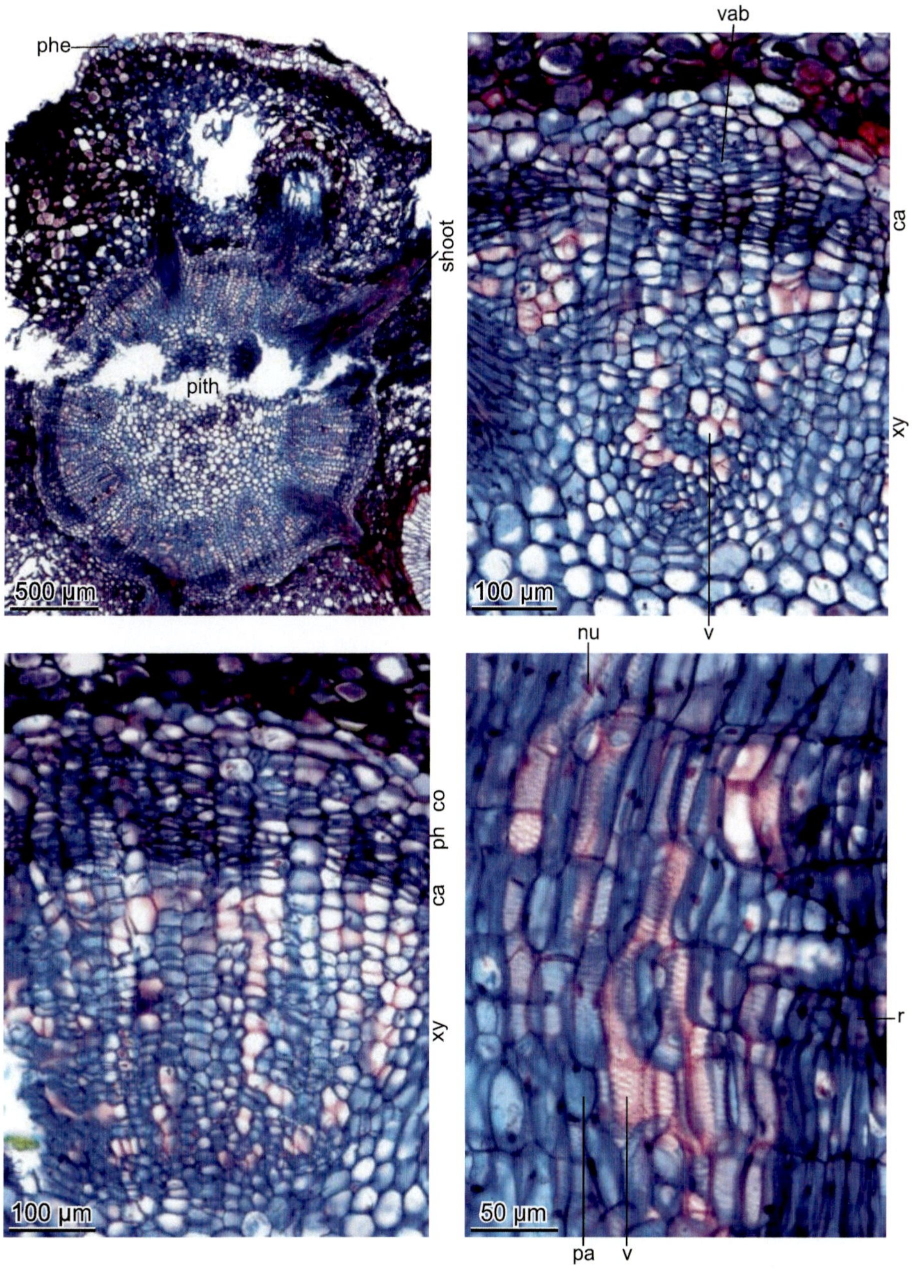

Features according to IAWA

Xylem 2 - 9 - 11 - 13 - 20 - 22 - 40.1 - 50.1 - 52.1 - 60.1 - 79.1 - 98 - 99 - 99.1 - 100.2 - 105

Paraquilegia anemonoides (Willd.) Ulbr.

Ranunculaceae

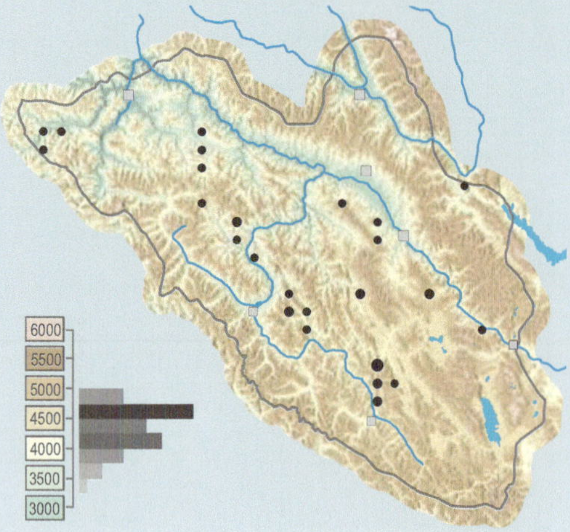

Syn.: *Aquilegia anemonoides* Willd., *Isopyrum grandiflorum* Fisch. ex DC.

Description: Perennial herbs, 5–10 cm tall, cespitose, glabrous, leaves biternate, flowers white, sometimes tinged with violet on the underside, ca. 2 cm across.

Distribution: Russia, Kazakhstan, Afghanistan, Pakistan, N India, China (Gansu, Ningxia, Qinghai, Xinjiang, Xizang), Bhutan, Mongolia.

Habitat: Rock crevices, gravel screes.

Stem diameter 5–7 mm. Bark/xylem proportion 1:2. Rings distinct and indistinct. ring width 0.2 mm. Maximal 3 annual rings.

Xylem

Transverse section

Diffuse-porous. Diameter of earlywood vessels 20–25 µm. Vessels round to angular, solitary. Fibers absent. Parenchyma pervasive. Vascular bundles remain. Crystals absent.

Tangential section

Rays >10-seriate, confluent between vascular bundles. Rays absent within the bundles.

Radial section

Simple perforation plates. Intervessel pits scalariform. Vessel length 100–130 µm.

Bark

Phloem contains groups of sieve tubes, tangentially arranged. Cortex consists of round to oval parenchyma cells. Phellem consists of many layers of square cells. With secondary phellem towards the pith.

Pith

Pith decayed.

Analyzed: Rhizome of 3 individuals (herbarium specimens).

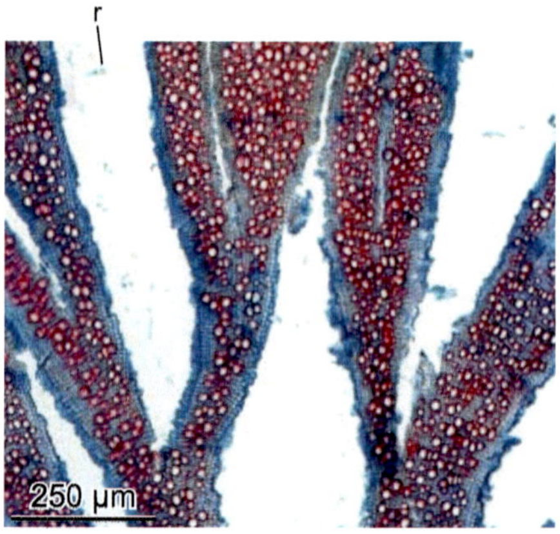

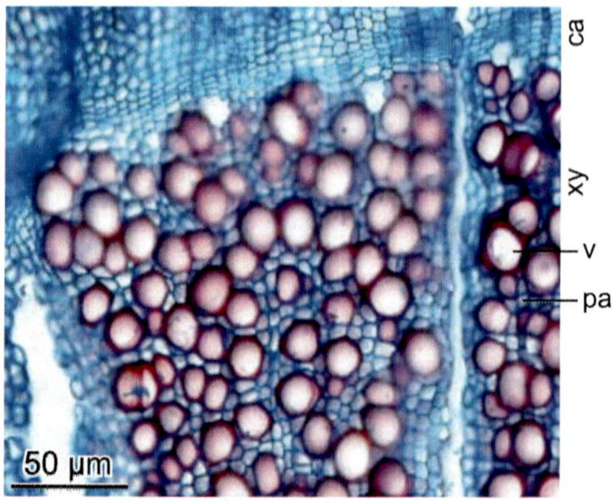

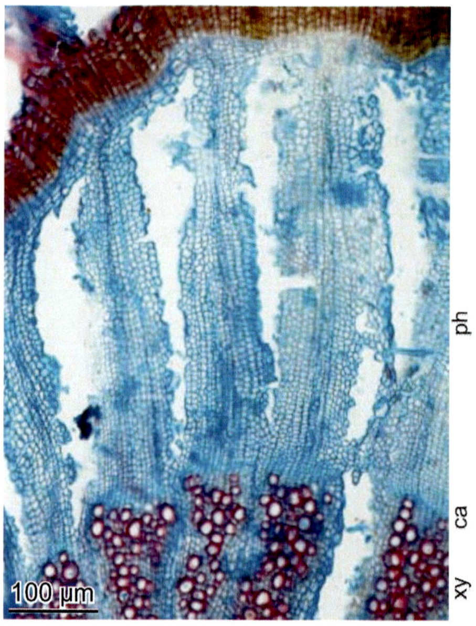

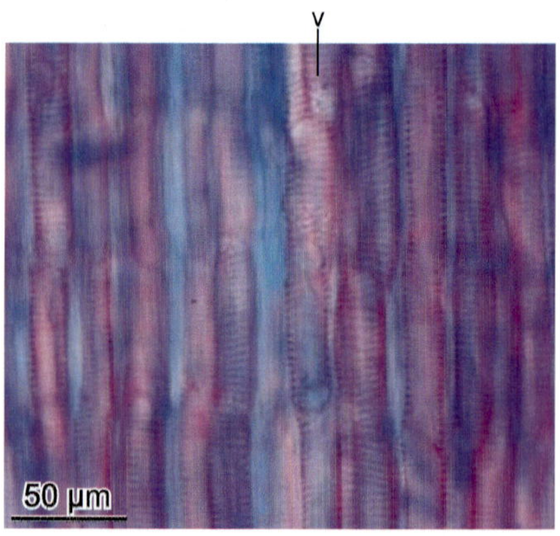

Features according to IAWA

Xylem 1 - 4 - 11 - 13 - 20 - 40.2 - 50.1 - 52.3 - 60.1 - 79.1 - 99 - 99.1 - 100.1 - 100.2 - 105 - 135.2

Pulsatilla wallichiana (Royle) Ulbr.

Ranunculaceae

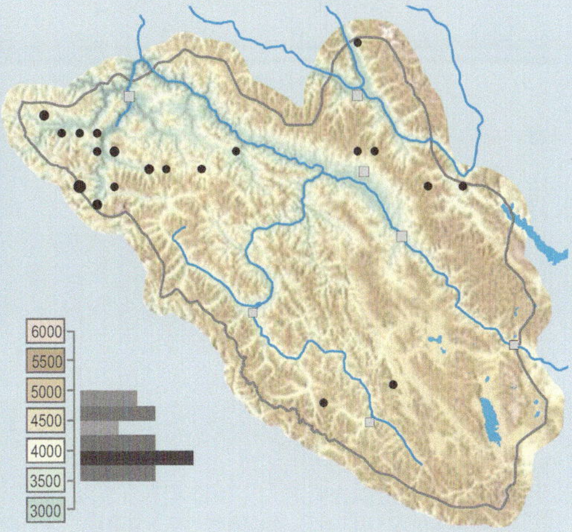

Syn.: *Anemone wallichiana* Royle, *A. albana* Hook.f. & Thoms., *P. duthiei* Gand.

Description: Perennial herbs, 10–15 cm tall, cespitose, hairy, leaves 3–5-pinnatisect, ultimate segments cuneate, tips 3-dentate, flowers reddish-brown, ca. 2 cm across, solitary, nodding, scapes elongating after flowering.

Distribution: Pakistan, NW India.

Habitat: Stabilized slopes, among stones.

Stem diameter 14 mm. Bark/xylem proportion 1:ca.3. Ring width 0.4–1 mm. Maximal 11 annual rings, not distinct. Center rotten.

Xylem

Transverse section

Semi-ring-porous. Diameter of earlywood vessels 30–40 µm. Vessels in radial groups. Vessels not red-stained. Fibers absent. Parenchyma pervasive. Vascular bundles remain. Crystals absent.

Tangential section

Rays >10-seriate, confluent with the axial tissue.

Radial section

Perforation plates simple. Intervessel pits round. Vessel length 100–150 µm.

Bark

Groups of sieve tubes tangentially arranged in phloem. Cortex consists of round to oval parenchyma cells. Phellem flaked.

Pith

No slide available.

Analyzed: Rhizome of 6 individuals.

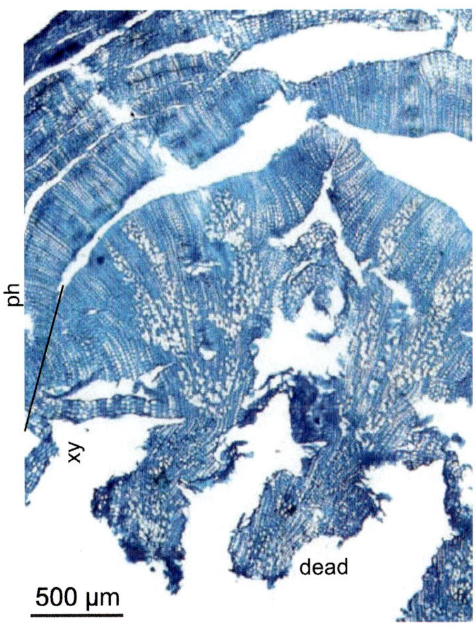

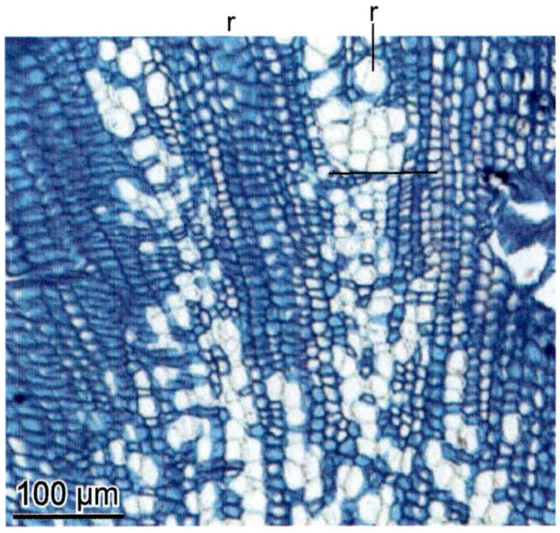

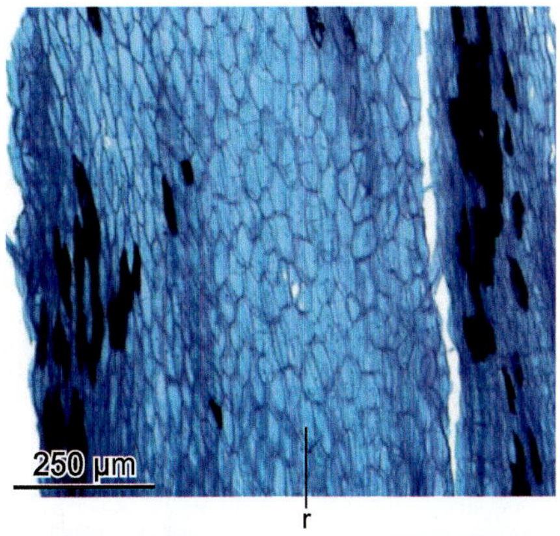

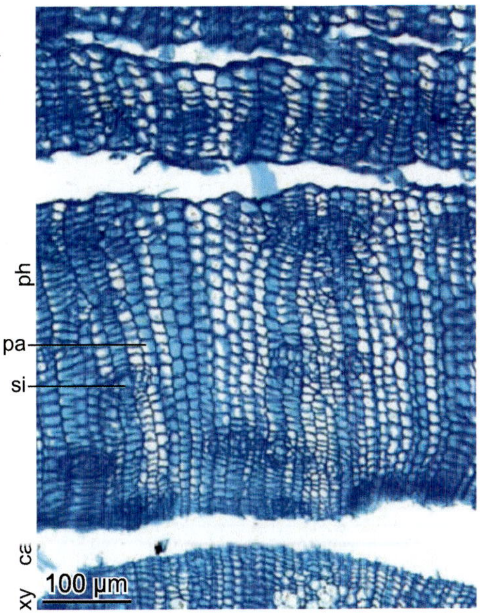

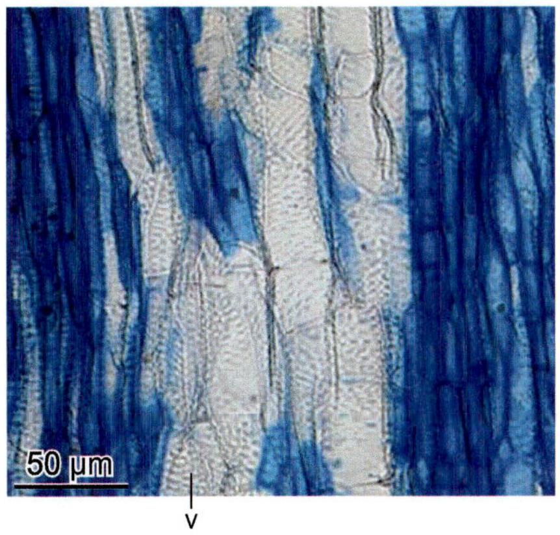

Features according to IAWA

Xylem 1 - 4 - 11 - 13 - 22 - 39.2 - 40.2 - 50.1 - 52.2 - 60.1 - 79.1 - 99 - 99.1 - 100.1 - 100.2 - 105 - 135.2

Ranunculus jacquemontii Riedl

Ranunculaceae

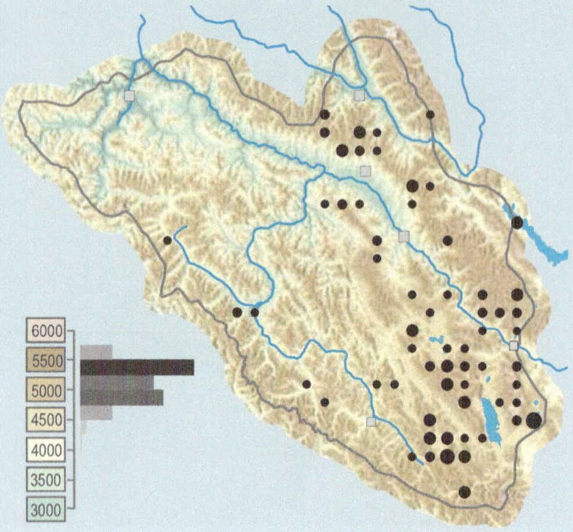

Syn.: *R. lobatus* Jacquem. ex Cambess.

Description: Perennial herbs, 2–8 cm tall, glabrous, stems decumbent, leaves suborbicular, margin crenate with broadly rounded teeth, flowers yellow, 1.5–2 cm across, solitary.

Distribution: N Pakistan, NW India, China (Xizang).

Habitat: Flushes, permanently wet brook banks, snow beds.

Stem diameter 3 mm. Bark/xylem proportion 1:ca.3. Annual rings absent

Xylem

Transverse section

Center surrounded by a few circular arranged, irregularly formed, open collateral vascular bundles, embedded in a thin-walled parenchymatic tissue. Xylem is composed of a group of vessels with diameters of 20–25 µm. Fibers absent.

Tangential section

Rays consist of large parenchymatic zones.

Radial section

Perforation plates simple. Intervessel pits round in alternating position. Vessel length 50–70 µm.

Bark

Phloem is composed of small groups of sieve tubes between parenchyma cells. Cortex consists of unlignified, thin-walled parenchyma cells and small intercellulars.

Pith

Irregularly formed, not well separated from primary xylem.

Analyzed: Rhizome of 2 individuals.

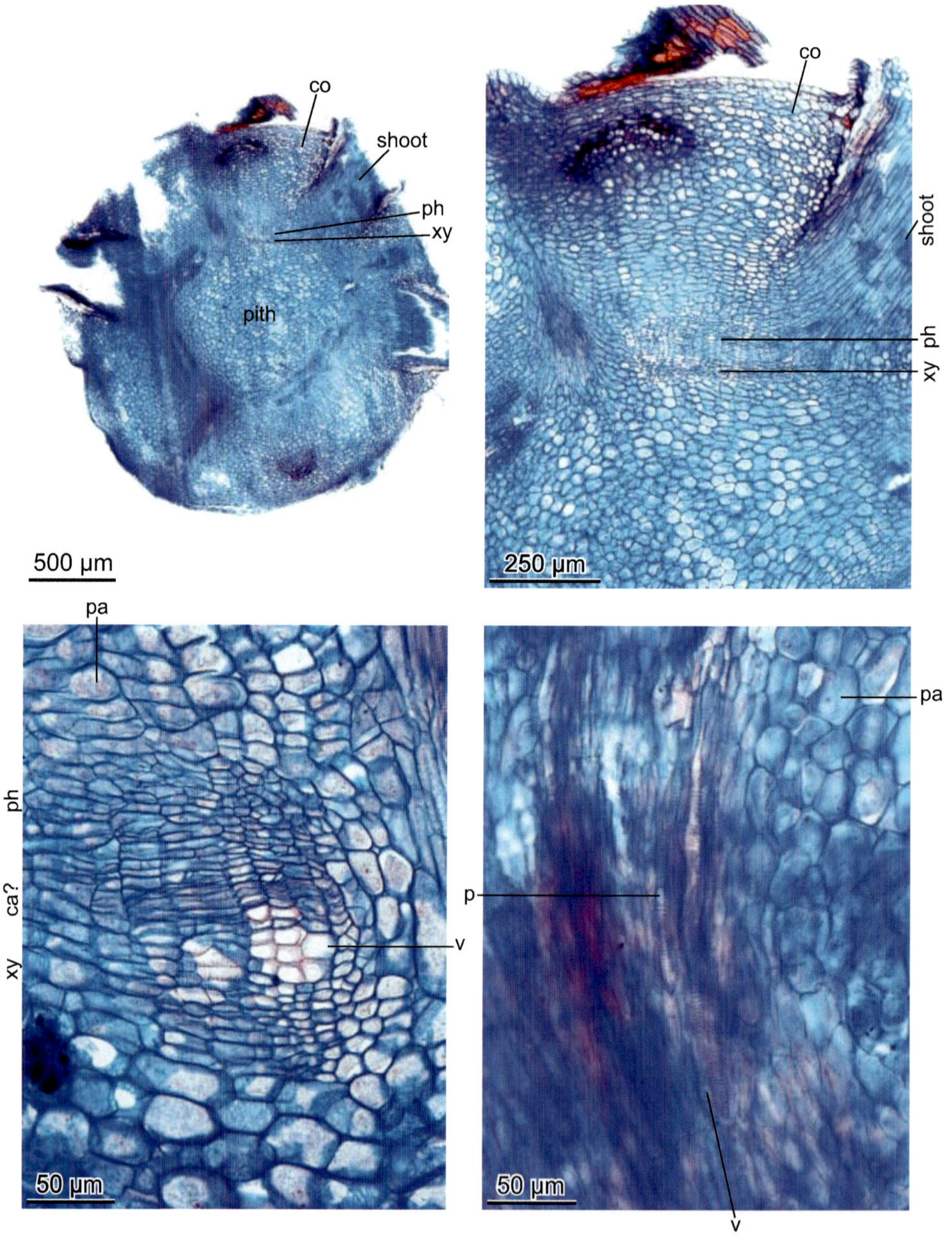

Features according to IAWA

Xylem 2.1 - 9.1 - 13 - 22 - 40.2 - 50.2 - 52.1 - 60.1 - 79.1 - 99 - 99.1 - 100.1 - 100.2

Ranunculus longicaulis C.A.Mey.

Ranunculaceae

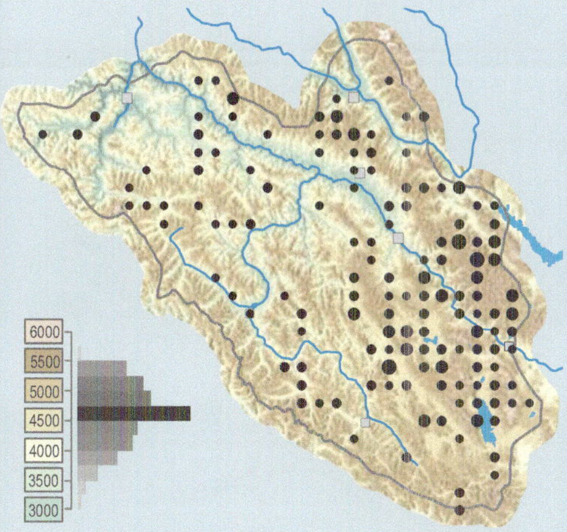

Syn.: *R. pulchellus* var. *longicaulis* Trautv., *R. nephelogenes* var. *longicaulis* (Trautv.) W.T. Wang

Description: Perennial herbs, 5–30 cm tall, glabrous, leaves ovate, narrowly elliptic, oblong, or linear-lanceolate, flowers yellow, 8–12 mm across, solitary.

Distribution: Russia (Siberia), Kazakhstan, Pakistan, NW India, China (Gansu, Qinghai, Shanxi, C and S Xinjiang, W Xizang), Mongolia.

Habitat: Wet meadows, along streams.

Stem diameter 2–2.5 mm. Bark/xylem proportion 1:0.5. Rings absent.

Xylem

Transverse section

Vessel diameter 15–20 µm. Vessels solitary or in small irregular groups. Fibers absent. Parenchyma pervasive. Irregular formed vascular bundles remain. Ray-like parenchymatic zones between vascular bundles and fibers.

Tangential section

Rays 4–>10-seriate, between vascular bundles.

Radial section

Perforation plates simple. Intervessel pits scalariform and round, in alternating position. Vessel length 80–100 µm.

Bark

Phloem generally uniform. Cortex consists of round to oval parenchyma cells and a few groups of thick-walled cells. Phellem absent. Vascular bundles surrounded by an endodermis.

Pith

Round.

Analyzed: Root collar of polar root of 4 individuals.

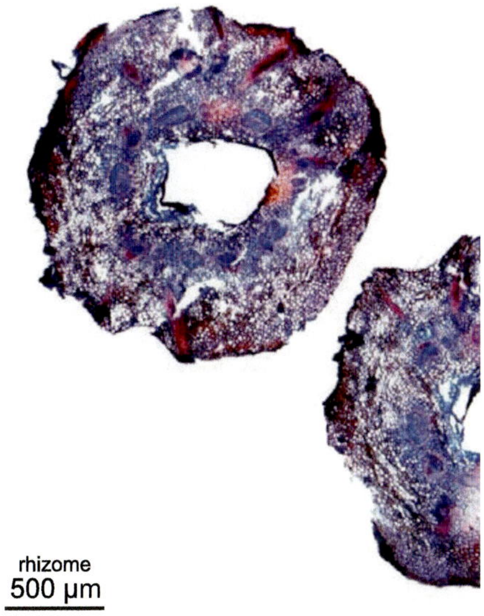

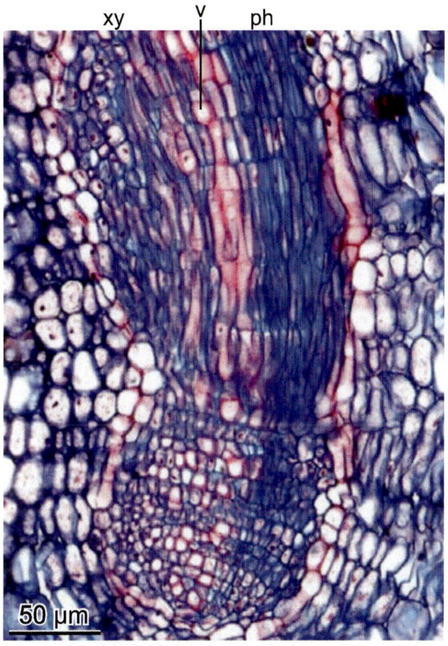

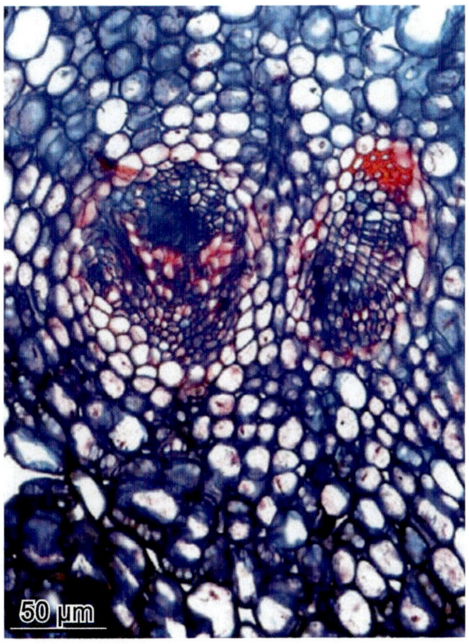

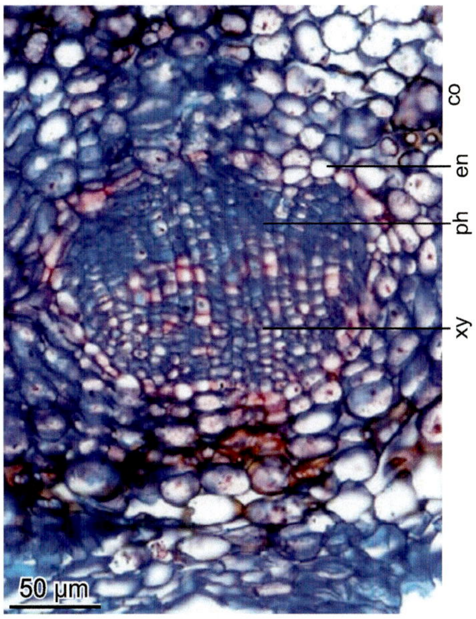

Features according to IAWA

Xylem 2 - 9 - 11 - 13 - 20 - 22 - 40.1 - 50.1 - 52.2 - 60.1 - 79.1 - 98 - 99 - 99.1 - 100.1 - 100.2 - 105

Ranunculus membranaceus Royle

Ranunculaceae

Description: Perennial herbs, 5–25 cm tall, leaves lanceolate, undivided or 3-lobed, flowers yellow, 1–2 cm across, usually solitary.

Distribution: Pakistan, N India, Nepal, China (Gansu, SW Nei Mongol, Ningxia, Qinghai, Sichuan, Xinjiang, Xizang).

Habitat: Wetlands in spring areas, alpine meadows.

Stem diameter 1 mm. Bark/xylem proportion 1:0.5. Rings absent.

Xylem

Transverse section

Vessel diameter 15–20 µm. Vessels in groups. Fibers absent. Parenchyma pervasive. Vascular bundles remain. Ray-like parenchymatic zones between vascular bundles and fibers.

Tangential section

Rays 4–>10-seriate between vascular bundles.

Radial section

No slide available, features observed on thick transverse section. Perforation plates simple. Intervessel pits round in alternating position to scalariform.

Bark

Phloem uniform. Cortex consists of thin-walled, round to oval parenchyma cells. Phellem absent. Vascular bundles surrounded by an endodermis

Pith

Round, filled with large, thin-walled, unlignified cells.

Analyzed: Root collar of polar root of 1 individual.

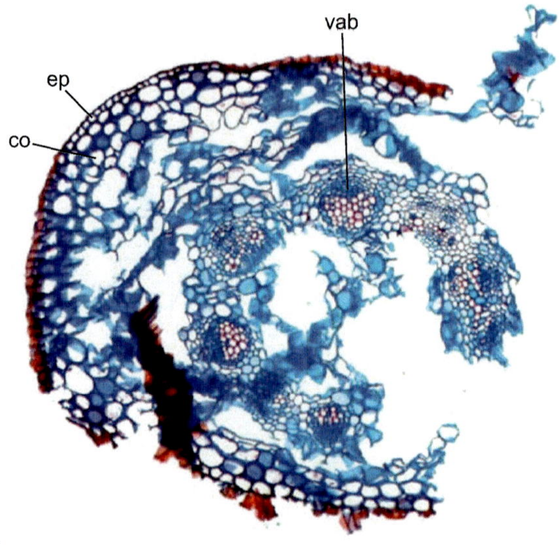

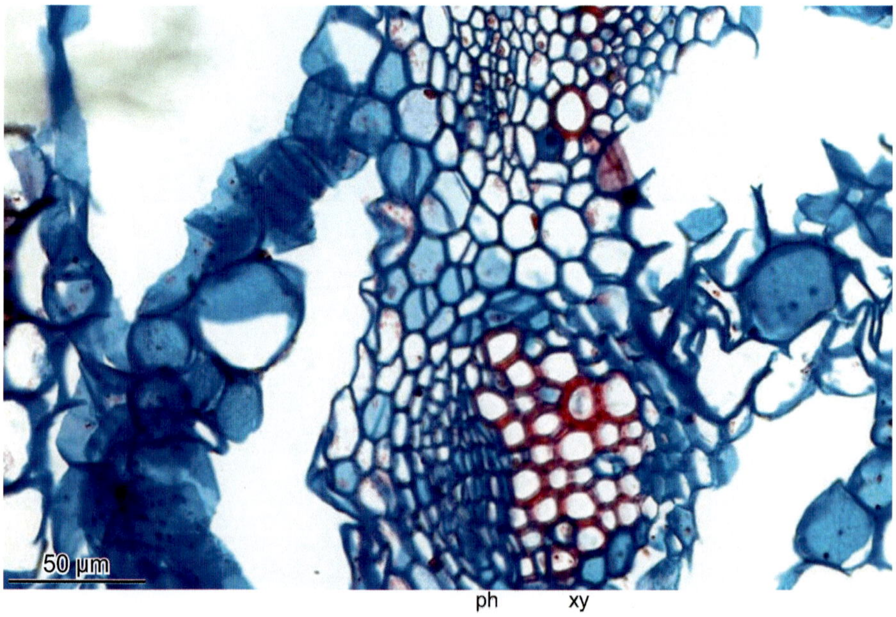

Features according to IAWA

Xylem 2 - 11 - 13 - 20 - 22 - 40.1 - 50.1 - 60.1 - 79.1 - 98 - 99 - 99.1 - 100.1 - 100.2 - 105

Thalictrum alpinum L.

Ranunculaceae

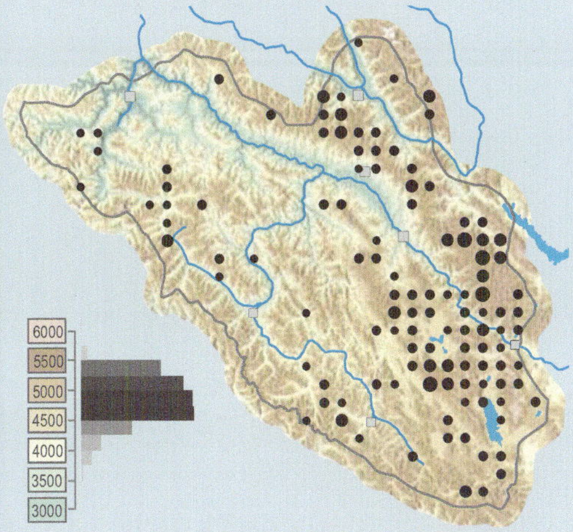

Description: Perennial herbs, 5–20 cm tall, leaves pinnately 2-ternate, all basal, leaflets cuneate to ovate, apex 3–5-lobed, flowers purplish, 3–5 mm across, in few-flowered racemes.

Distribution: North America, Europe, Russia, Kazakhstan, Kyrgyzstan, Tajikistan, Afghanistan, Pakistan, N India, Nepal, China, Bhutan, Mongolia, N Myanmar, Vietnam.

Habitat: Wet meadows and rocks, grasslands along streams.

Root collar with a diameter of 10 mm. Xylem to bark ratio 1:1. Rings distinct and indistinct. Maximal 2 rings. Mean ring width 0.1 mm.

Xylem

Transverse section

Diffuse-porous. Earlywood diameter 10–15 µm. Rings, if present, marked by a zone of missing vessels in the latewood. Vessels angular, solitary and in small groups. Fibers within vascular bundles absent. Parenchyma pervasive. Vascular bundles remaining. Crystals absent. Groups of sclereids in the large rays and the pith.

Tangential section

Rays very large, >10-seriate and confluent, between vascular bundles. Rays absent within vascular bundles.

Radial section

Perforation plates simple. Intervessel pits scalariform. Vessel length not measured.

Bark

Phloem with large parenchyma cells and irregularly distributed small sieve tubes or companion cells. Phellem consists of a 3–4-seriate band of unlignified, round cells. With groups of sclereids in the phellem. Without crystals.

Pith

Pith large and polygonal.

Analyzed: Root collar of 3 individuals.

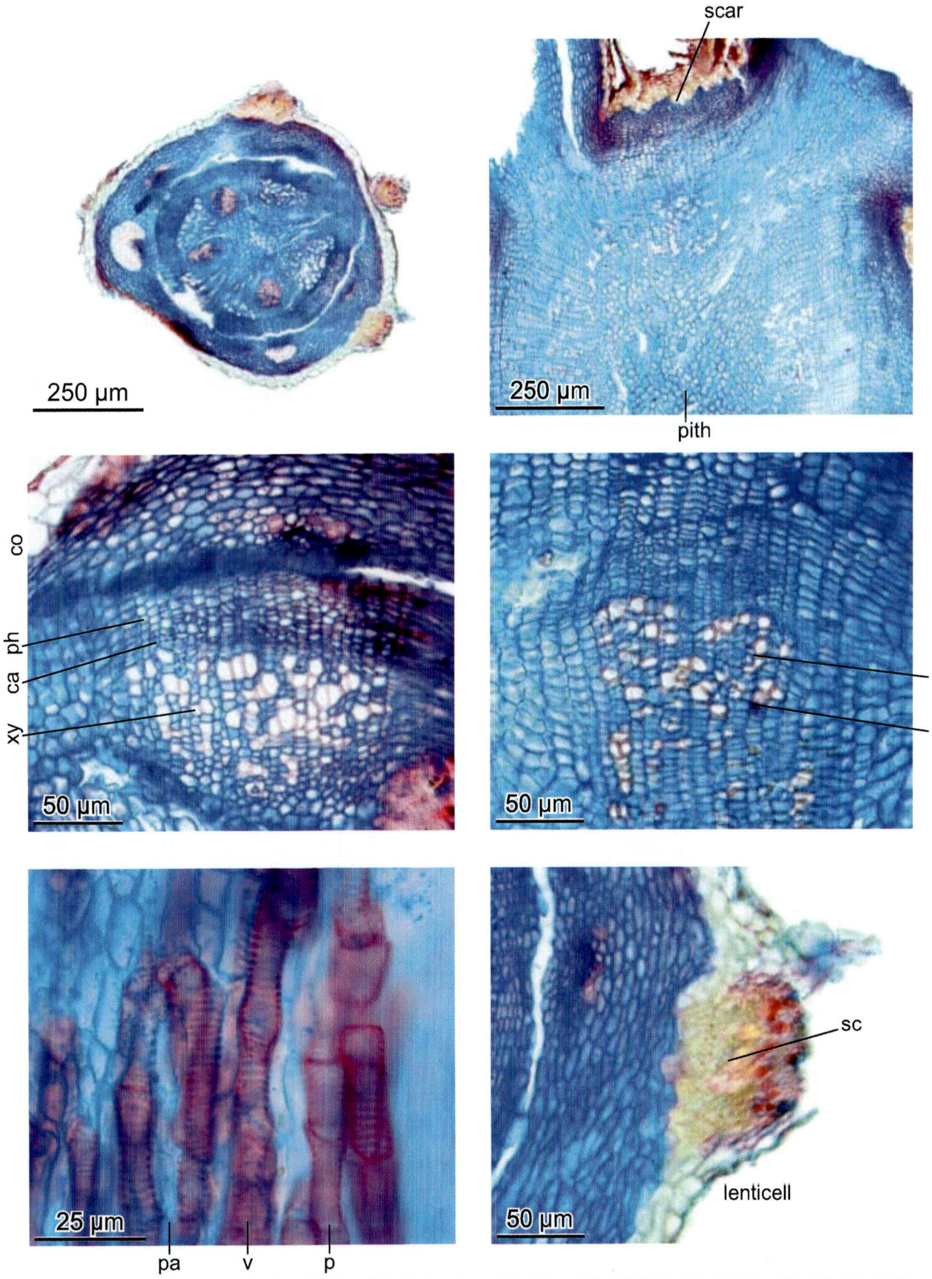

Features according to IAWA

Xylem 1 - 2 - 9 - 11 - 12 - 13 - 20 - 40.1 - 50.2 - 52.2 - 60.1 - 79.1 - 99 - 99.1 - 100.1 - 100.2 - 105

Thalictrum foetidum L.

Ranunculaceae

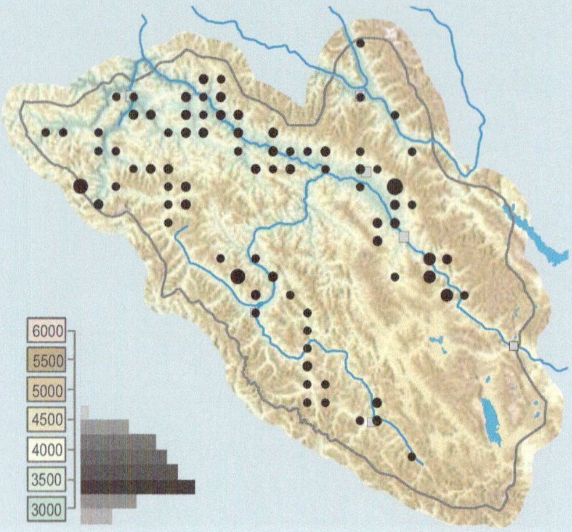

Description: Perennial herbs, 15–60 cm, leaves 3-pinnate, glandular hairy, flowers yellow to greenish, ca. 5 mm across, in panicles.

Distribution: Europe, Iran, Turkey, Caucasus, Russia (Siberia, Far East), SE Kazakhstan, Kyrgyzstan, Tajikistan, Afghanistan, N Pakistan, India (Jammu and Kashmir, Himachal Pradesh, Uttarakhand, Sikkim), Nepal, China (Gansu, Hebei, Nei Mongol, Qinghai, Shaanxi, Shanxi, Sichuan, Xinjiang, Xizang), Bhutan, Mongolia, Japan.

Habitat: Gravel and stony sites, often human-induced habitats, in villages, along stony walls.

Stem diameter 7 mm. Bark/xylem proportion 1:0.2. Rings absent.

Xylem

Transverse section

Tissue distorted due to the occurrence of many secondary roots. Rings only partially distinct. Diffuse-porous. Vessel diameter 20–30 µm. Vessels thick-walled, mostly solitary or in short radial rows. Some vessels with dark-stained substances. Fibers very thick-walled, in irregularly formed tangential bands or groups. Parenchyma pervasive. Irregularly formed vascular bundles remain. Ray-like parenchymatic zones between the vascular bundles and the fibers.

Tangential section

Rays not observed due to distortion of the tissue.

Radial section

Perforation plates simple. Intervessel pits round in alternating position. Vessel length 80–100 µm.

Bark

Phloem more or less uniform. Cortex consists of round to oval parenchyma cells and a few groups of thick-walled cells. Phellem layered, consisting of irregularly formed cells.

Pith

Irregularly formed, not well separated from the primary xylem.

Analyzed: Rhizome of 4 individuals.

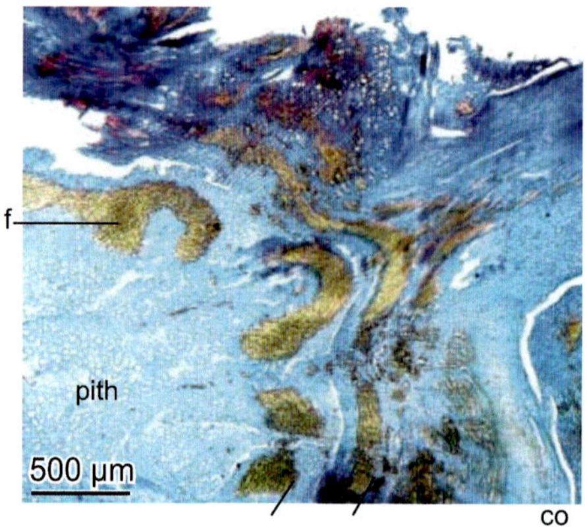

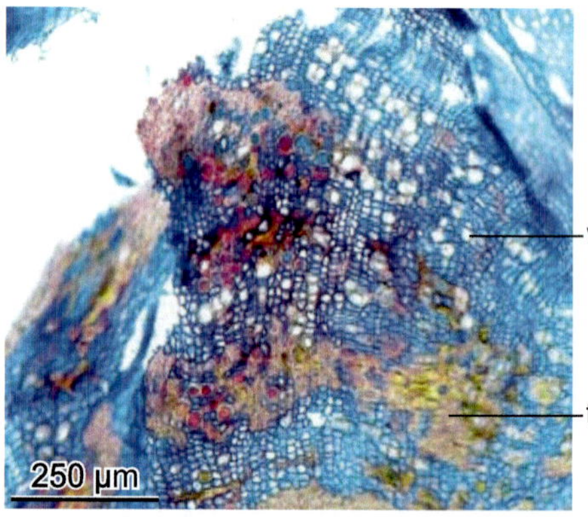

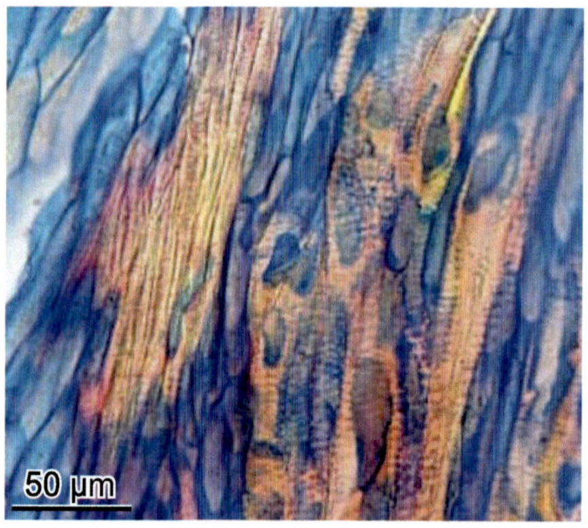

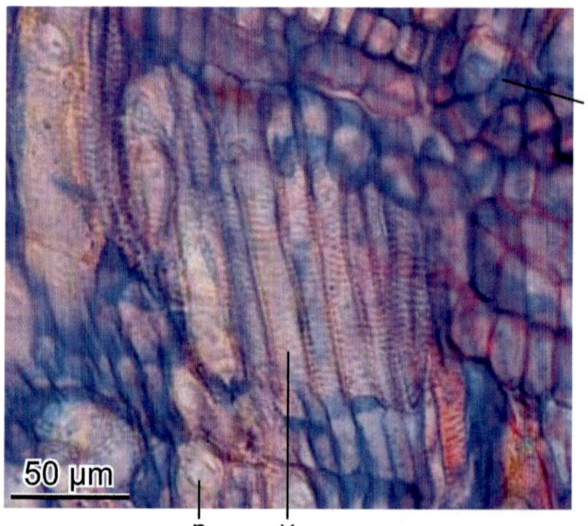

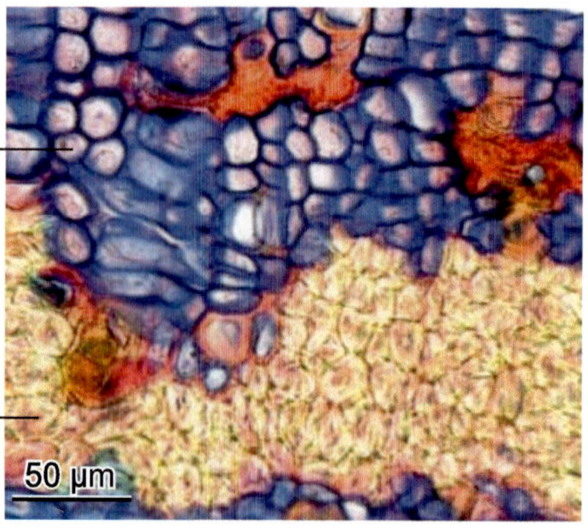

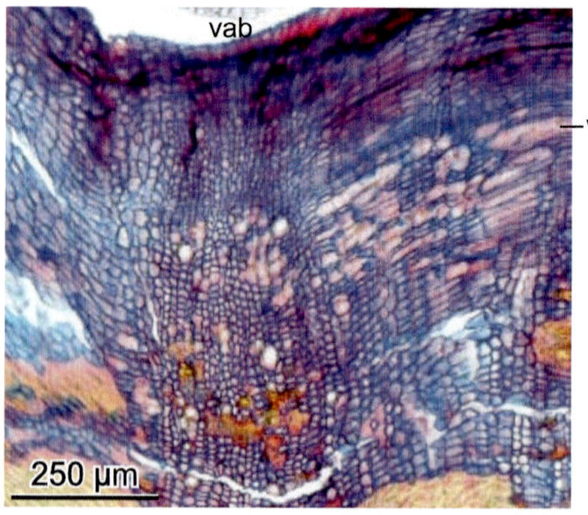

Features according to IAWA

Xylem 1 - 5 - 9 - 9.1 - 13 - 22 - 40.2 - 50.2 - 52.1 - 61 - 69 - 79.1 - 99 - 99.1 - 100.2

Rhamnus prostrata Jacquem. ex R.Parker

Rhamnaceae

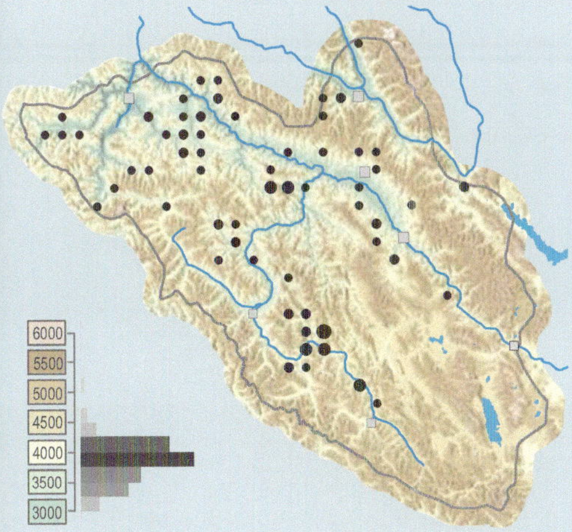

Description: Shrubs, up to 1 m tall, branches prostrate, thorny, leaves elliptic, obscurely serrate, clustered, flowers yellow-green, minute, solitary, axillary, fruits purple-red, turning black at maturity, obovoid-globose, ca. 5 mm across.

Distribution: Afghanistan, Pakistan, W and E India, China (W Xizang).

Habitat: Dry stony slopes, screes, rock crevices.

Stem diameter 5–10 mm. Bark/xylem proportion 1:10. Mean ring width 0.2 mm. Maximal 26 annual rings.

Xylem

Transverse section

Semi-ring-porous. Rings distinctly marked by differences between earlywood (30–35 µm) and latewood vessel diameters (15–20 µm). Vessels thin- to thick-walled, in a flame-like arrangement. Heartwood vessels with dark-staining substances. Fibers thick-walled, often with gelatinous fibers. With fiber tracheids. Parenchyma paratracheal, often absent.

Tangential section

Rays 1–2-seriate.

Radial section

Simple perforation plates. Intervessel pits round in alternating position. Vessels with fine helical thickenings. Vessel length 130–200 µm. Fiber pits round and relatively large (2–3 µm). Ray cells prostrate and square.

Bark

Phloem contains large, oval groups of fibers. Sieve tubes and parenchyma cells undifferentiated. Phellem cells rectangular.

Pith

Pith round to pentagonal.

Analyzed: Stem of 2 individuals.

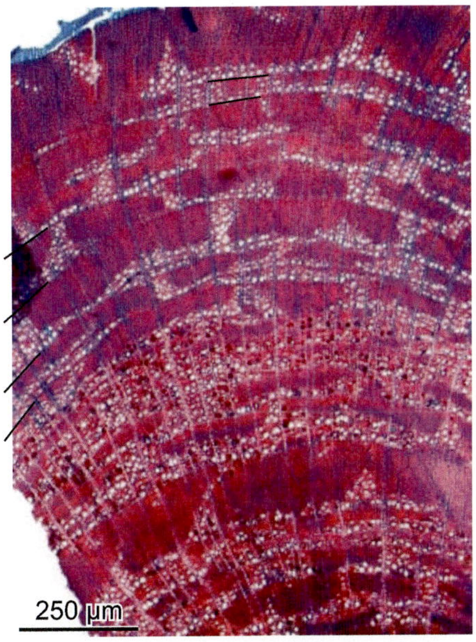

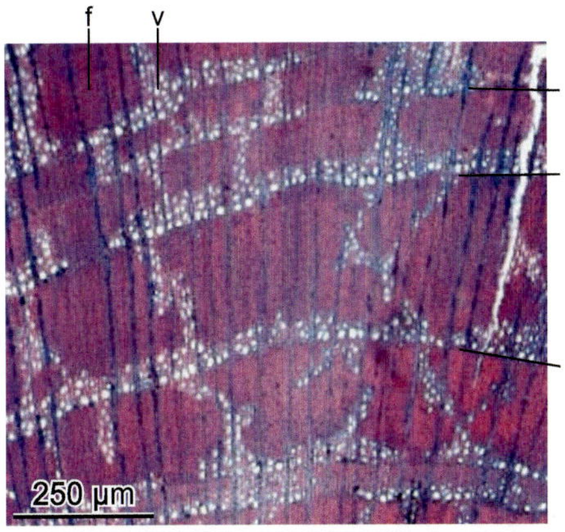

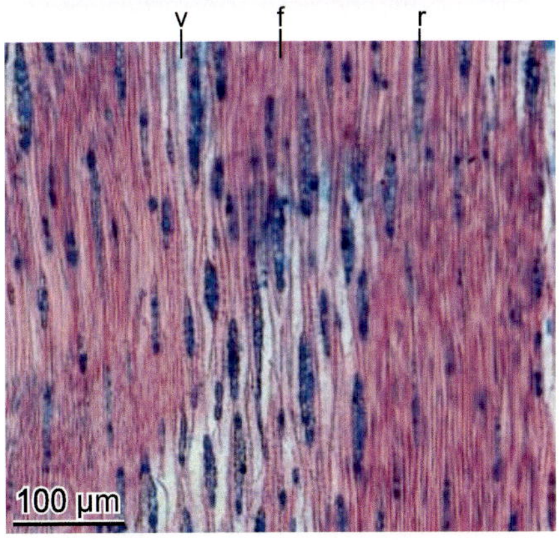

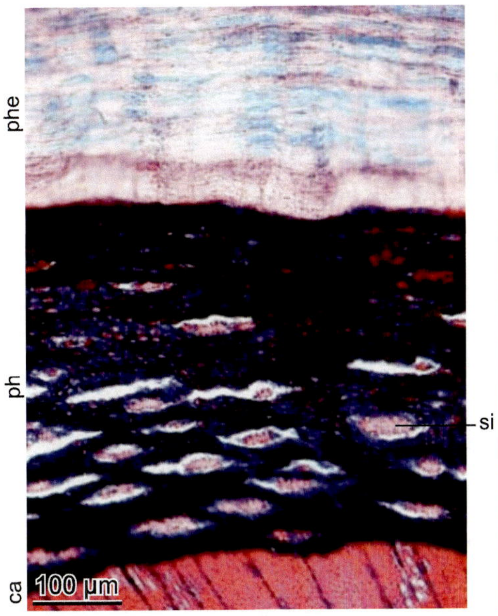

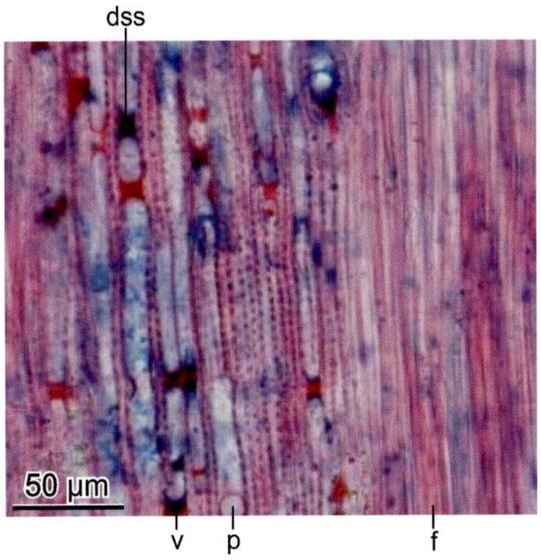

Features according to IAWA

Xylem 1 - 4 - 7 - 9 - 9.1 - 13 - 22 - 36 - 40.2 - 50.2 - 52.3 - 62 - 70 - 70.2 - 75 - 79 - 97 - 104

Chamaerhodos sabulosa Bunge

Rosaceae

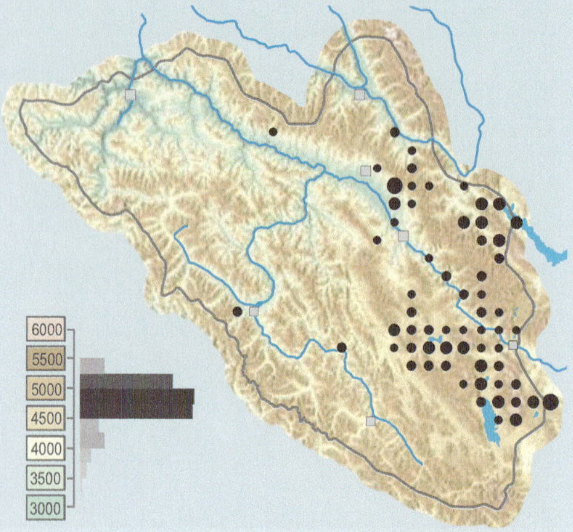

Syn.: *Sibbaldia sabulosa* (Bunge) Steud.

Description: Perennial herbs, 5–10 cm tall, glandular hairy, leaves 3-pinnatisect, flowers white or pink, 3–5 mm across, in lax terminal many-flowered inflorescences.

Distribution: Russia, C Asia, Pakistan, NW India, China (Nei Mongol, Xinjiang, Xizang), Mongolia.

Habitat: Sandy plains, sandy steppes, river beds.

Stem diameter 1.8–2 mm. Bark/xylem proportion 1:6. Mean ring width 0.16 mm. Maximal 7 annual rings.

Xylem

Transverse section

Semi-ring-porous. Rings distinctly marked by differences between earlywood (20–30 µm) and latewood vessel diameters (5–10 µm) and an unlignified parenchymateous zone in the earlywood. Vessels thick-walled, solitary. Fibers in the center deformed. Parenchyma apotracheal or absent in fiber zones and pervasive in parenchymateous zones. Vascular bundles remain in the rhizome, absent in the primary root.

Tangential section

Rays 1–3-seriate and >10-seriate (ray dimorphism) in the rhizome, 1–2-seriate in the root. Rays unlignified, disappear in polarized light.

Radial section

Perforation plates simple. Intervessel pits round in alternating position, slit-like. Vessel length 50–70 µm. Fiber pits round, relatively large (2–3 µm). Ray cells square and upright.

Bark

Older parts of bark flake. Phloem very small, fairly homogeneous. Very few sieve tubes slightly smaller than parenchyma cells. Phellem layered. Between each layer is a row of very thin-walled, rectangular cells (phelloids), which appear light in polarized light.

Pith

More or less round. Large, unlignified, thin-walled central cells disappear in polarized light.

Analyzed: Root collar of polar root of 2 individuals.

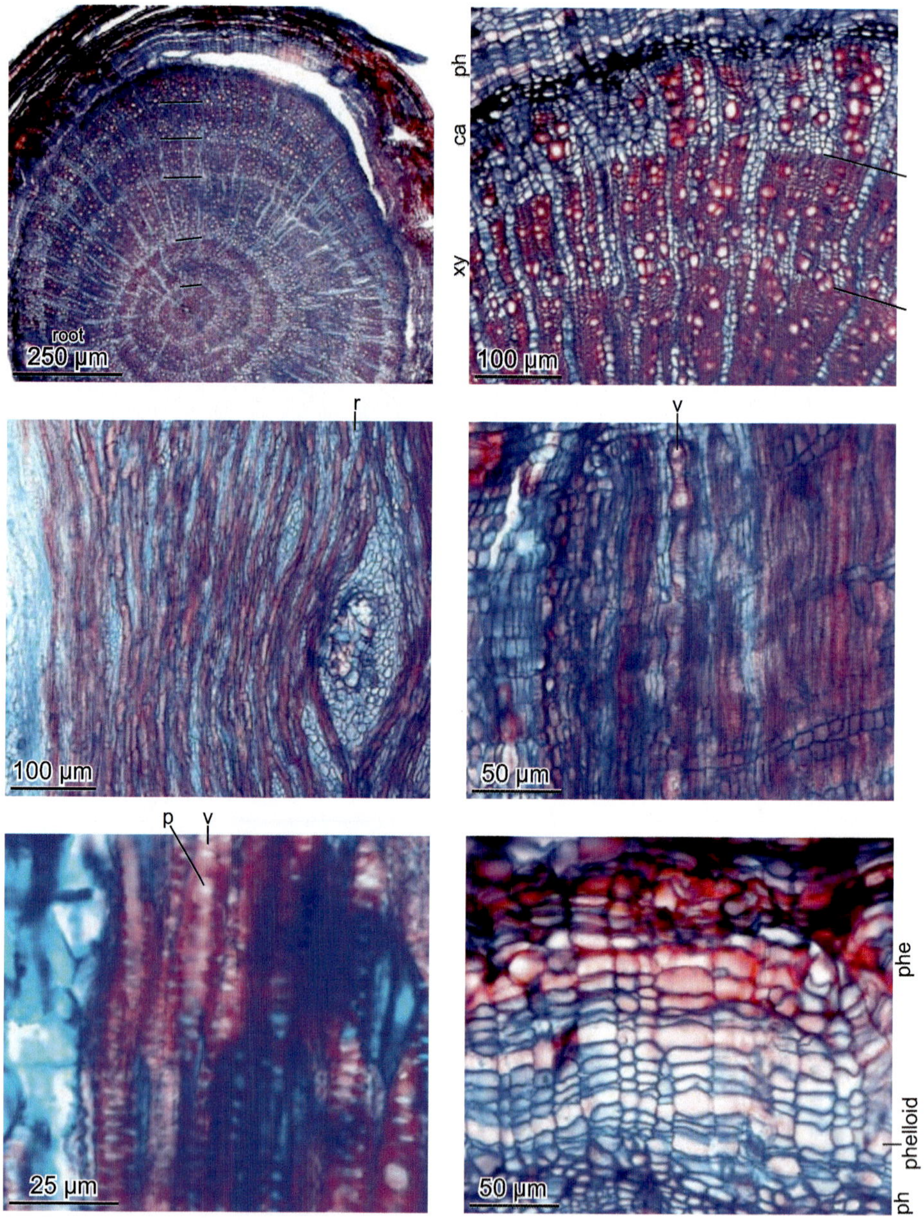

Features according to IAWA

Xylem 1 - 4 - 9 - 13 - 20 - 22 - 39.1 - 40.1 - 50.2 - 52.2 - 62 - 70 - 75 - 76 - 79.1 - 97 - 99.1 - 100.2 - 105

Comarum salesovianum (Steph.) Asch. & Graebn.

Rosaceae

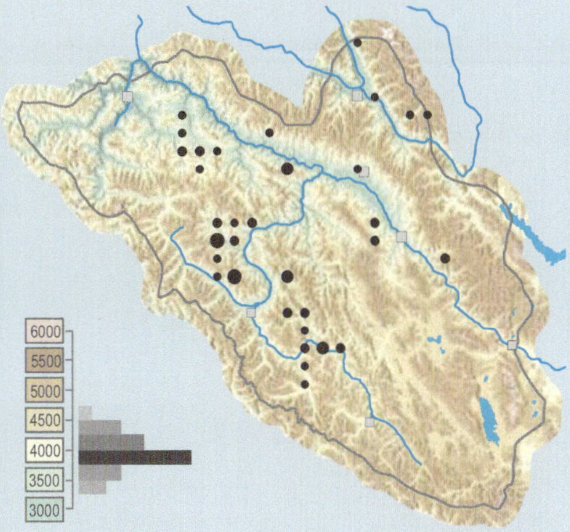

Syn.: *Potentilla salesoviana* Steph.

Description: Perennial herbs, 15–50 cm tall, villous, stem branched, prostrate, leaves imparipinnately compound, flowers white with yellowish centre, 2.5–3 cm across, in lax cymes.

Distribution: Russia, Kyrgyzstan, Tajikistan, Afghanistan, Pakistan, NW India, China (Gansu, Nei Mongol, Ningxia, Qinghai, Xinjiang, Xizang), Mongolia.

Habitat: Stony slopes near rivers and streams.

Stem diameter 3–6 mm. Bark/xylem proportion 1:10. Mean ring width 0.16 mm. Maximal 25 annual rings.

Xylem

Transverse section

Semi-ring-porous. Rings distinctly marked by differences between earlywood (40–50 μm) and latewood vessel diameters (15–20 μm). Vessels angular, thin- to thick-walled, solitary. Parenchyma paratracheal or absent.

Tangential section

Rays 1–2-seriate.

Radial section

Perforation plates simple. Intervessel pits round in alternating position. Vessel length 130–200 μm. Fiber pits round, relatively large (2–3 μm). Ray cells square and upright.

Bark

Older parts of the bark flake. Phloem very small, contains small groups of sieve tubes. Phellem layered. Between a few layers of rectangular, thin-walled cells is a row of round, dark-stained cells (phelloids).

Pith

Round to pentagonal. Large, unlignified, thin-walled central cells disappear in polarized light.

Analyzed: Root collar of polar root of 13 individuals.

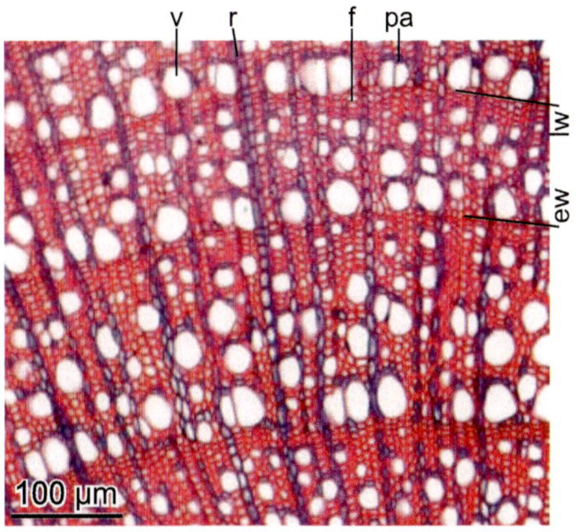

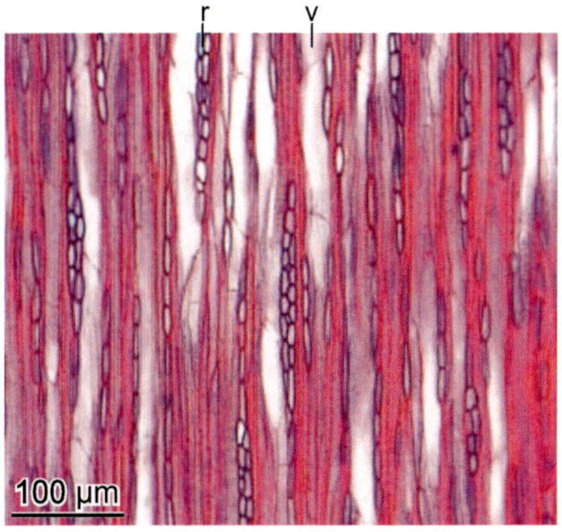

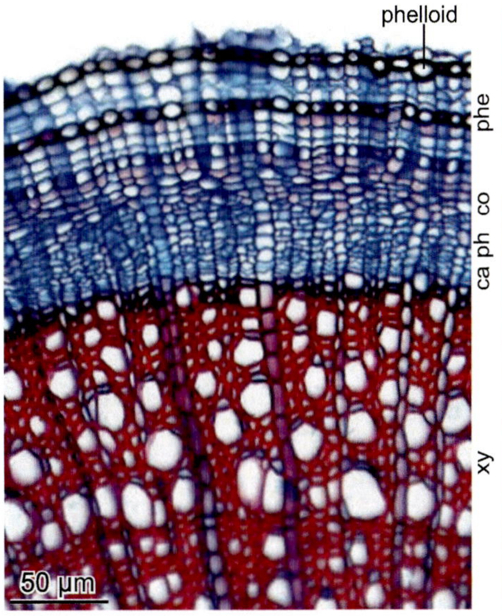

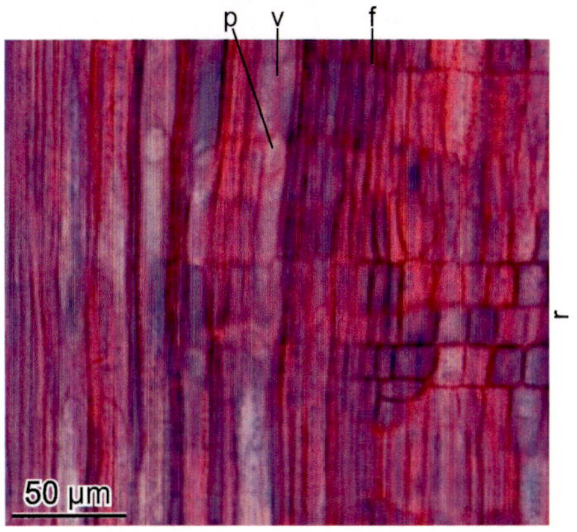

Features according to IAWA

Xylem 1 - 4 - 9 - 12 - 13 - 22 - 40.2 - 50.2 - 52.3 - 62 - 70 - 75 - 79 - 97 - 105

Cotoneaster affinis Lindl.

Rosaceae

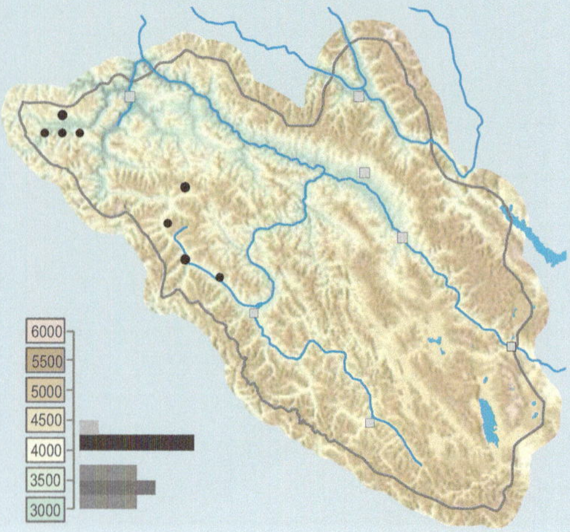

Description: Shrubs, 1–5 m tall, leaves ovate, densely tomentose below, flowers white, ca. 1 cm across, in composed 15–30-flowered corymbs.

Distribution: Pakistan, India, Nepal, China (Sichuan, Xizang, Yunnan), Bhutan.

Habitat: Slopes, forest margins.

Stem diameter 3–4 mm. Bark/xylem proportion 1:0.5. Mean ring width 0.15 mm. Maximal 8 annual rings.

Xylem

Transverse section

Diffuse- to semi-ring-porous. Rings distinctly marked by differences between earlywood (30–40 µm) and latewood vessel diameters (15–20 µm). Vessels angular, solitary. Fibers thin- to thick-walled. Parenchyma apotracheal and paratracheal. Parenchyma cells contain dark-stained substances.

Tangential section

Rays 1–2-seriate.

Radial section

Perforation plates simple. Intervessel pits round in alternating position. Vessel length 100–120 µm. Fiber pits small (2–3 µm). Ray cells primarily upright.

Bark

Phloem homogeneous. Sieve tubes and parenchyma indistinguishable. Phellem small, consisting of radially oriented, rectangular cells. Crystals absent.

Pith

Pith round to slightly square. Central cells large, unlignified, thin-walled, surrounded by a few rows of smaller, lignified cells

Analyzed: Stem of 1 individual.

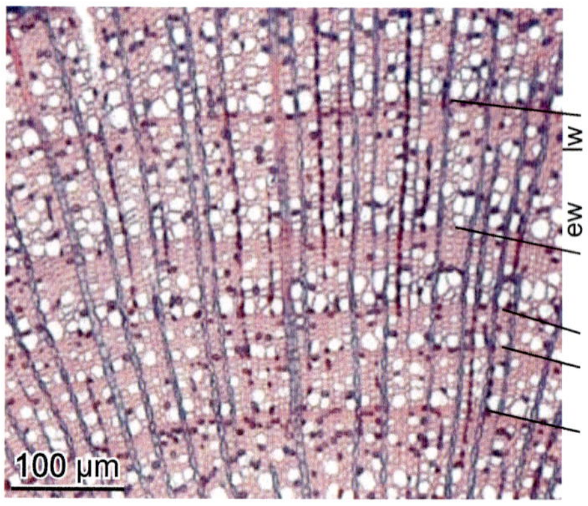

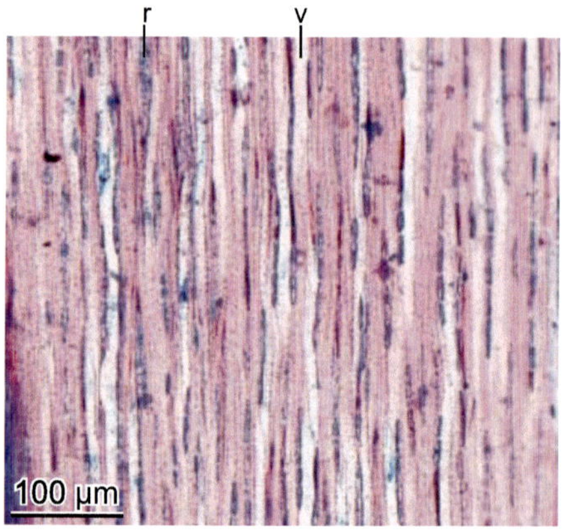

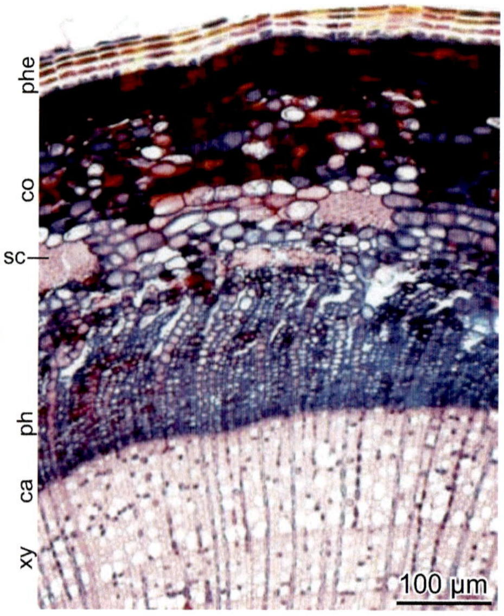

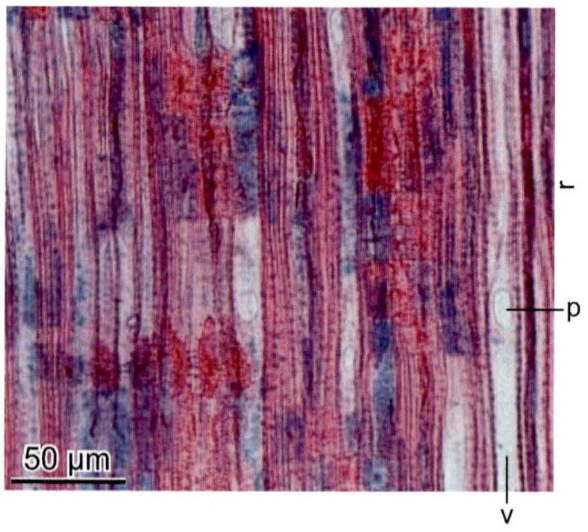

Features according to IAWA

Xylem 1 - 4 - 9 - 9.1 - 12 - 13 - 22 - 40.2 - 50.1 - 52.3 - 58 - 61 - 70 - 97 - 105

Dasiphora dryadanthoides Juz.

Rosaceae

Syn.: *Potentilla dryadanthoides* (Juz.) Vorosh., *P. arbuscula* var. *pumila* (Hook.f.) Hand.-Mazz., *P. fruticosa* var. *pumila* Hook.f.

Description: Shrubs, 6–10 cm tall, villous, branches prostrate, ascending, leaves pinnatisect, flowers yellow, ca. 1.5 cm across, solitary.

Distribution: Tajikistan, NE Afghanistan, N Pakistan, N India, Nepal, Bhutan, China (Qinghai, Xinjiang, Xizang).

Habitat: Stony places in the dry subalpine zone.

Stem diameter 3–4 mm. Bark/xylem proportion 1:>10. Mean ring width 0.1 mm. Maximal 29 annual rings. This species can anatomically not be distinguished from *Dasiphora parviflora*.

Xylem

Transverse section

Semi-ring-porous. Rings distinctly marked by differences between earlywood (25–45 µm) and latewood vessel diameters (10–15 µm). Vessels slightly angular, solitary. Fibers thick-walled. Parenchyma apotracheal and paratracheal diffuse.

Tangential section

Rays 1–2-seriate.

Radial section

Perforation plates simple. Intervessel pits round in alternating position. Vessel length 100–120 µm. Fiber pits round, relatively large (2–3 µm). Ray cells upright.

Bark

Older parts flake. Phloem small, fairly homogeneous. Few sieve cells slightly smaller than parenchyma cells. Phellem layered. Between each layer is a row of very thin-walled, rectangular cells (phelloids), which appear light in polarized light. With crystal druses in the phellem.

Pith

More or less round. Large, unlingnified, thin-walled central cells surrounded by a few rows of smaller, lignified cells.

Analyzed: Root collar of polar root of 3 individuals.

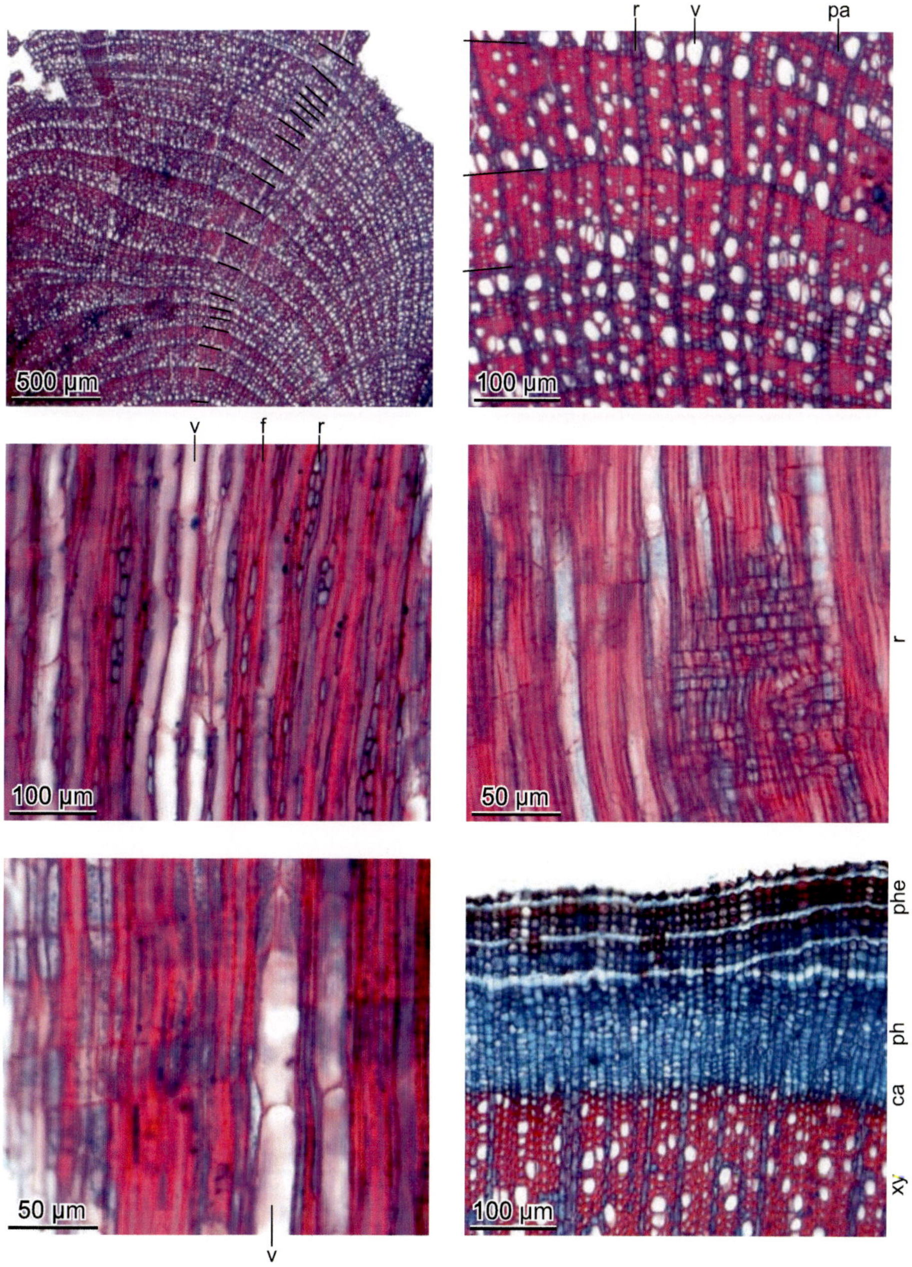

Features according to IAWA

Xylem 1 - 4 - 9 - 12 - 13 - 22 - 40.2 - 50.1 - 52.3 - 62 - 69 - 76 - 96 - 97 - 105

Dasiphora arbuscula (D.Don) Soják

Rosaceae

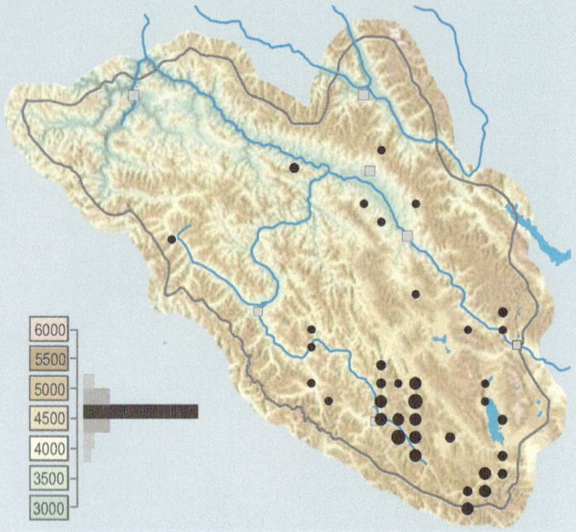

Syn.: *Potentilla arbuscula* D.Don, *P. fruticosa* var. *arbuscula* (D.Don) Maxim.

Description: Shrubs, 0.5–2 m tall, leaves pinnate, margin of leaflets revolute, flowers yellow, 1.5–2 cm across, solitary.

Distribution: N Pakistan, N India, Nepal, China (Sichuan, Xizang, Yunnan), Bhutan.

Habitat: Rocky slopes.

Stem diameter 12–14 mm. Bark/xylem proportion 1:40. Mean ring width 0.2 mm. Maximal 38 living annual rings.

Xylem

Transverse section

Semi-ring-porous. Rings distinctly marked by differences between earlywood (25–45 µm) and latewood vessel diameters (10–15 µm). Vessels angular and solitary. Fibers thick-walled. Parenchyma mainly paratracheal. Dark-staining substances in vessels of the stem center (heartwood).

Tangential section

Rays 1–2-seriate.

Radial section

Perforation plates simple. Intervessel pits round, in alternating position. Vessel length 100–120 µm. Fiber pits round and relatively large (2–3 µm). Ray cells upright.

Bark

Phloem small. Older parts flake. Phloem fairly homogeneous. A few sieve cells slightly smaller than parenchyma cells. Phellem layered. Between layers are rows of very thin-walled cells (phelloids), consisting of rectangular cells which appear light in polarized light. Prismatic crystals in the phellem.

Pith

No slide available.

Analyzed: Root collar of polar root of 3 individuals.

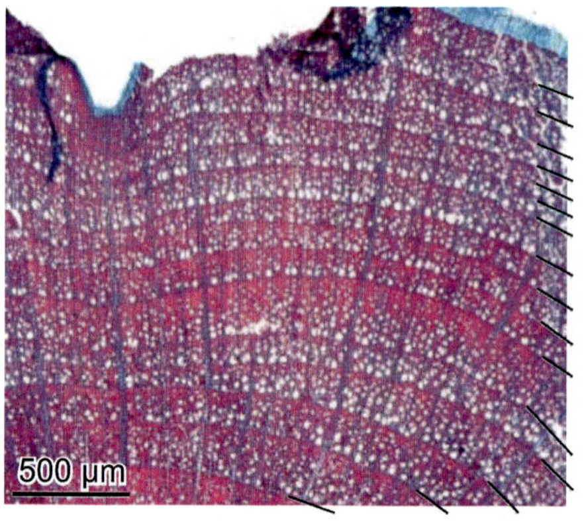

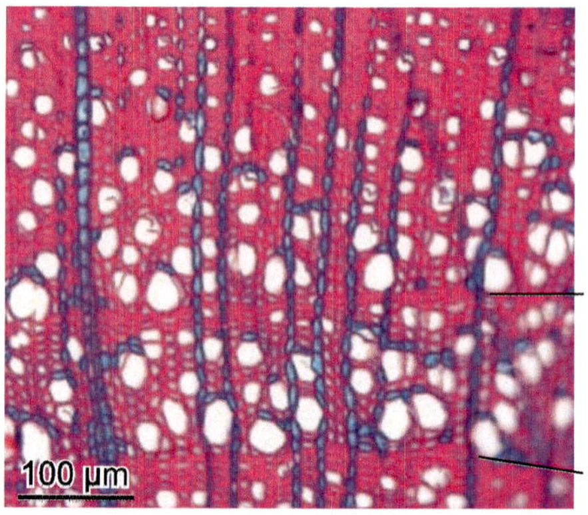

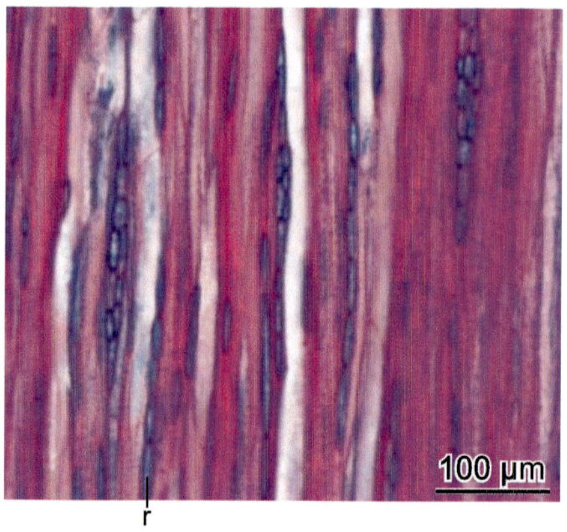

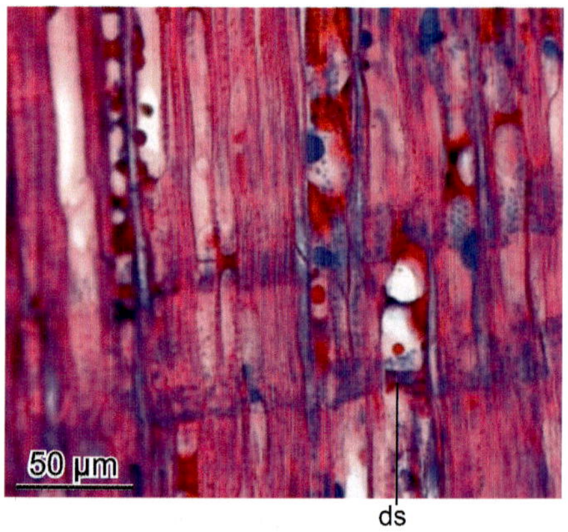

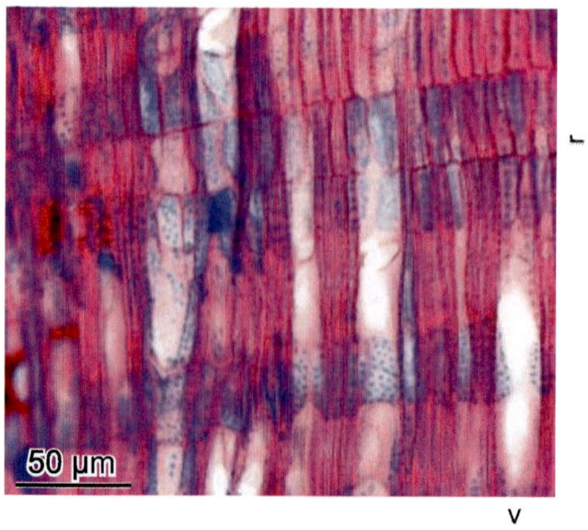

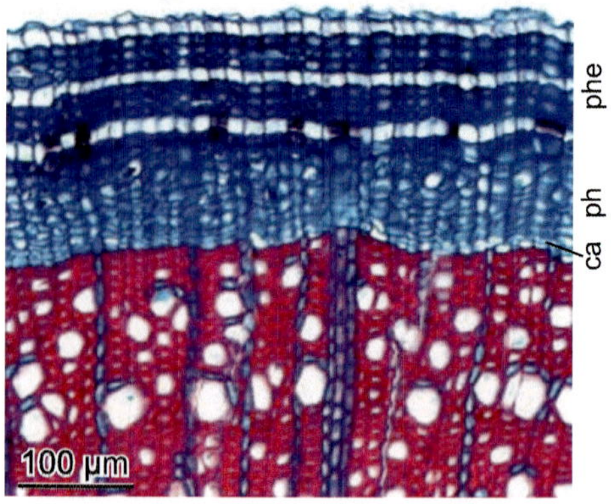

Features according to IAWA

Xylem 1 - 4 - 9 - 12 - 13 - 22 - 40.2 - 50.1 - 52.3 - 58 - 62 - 69 - 79 - 96 - 97 - 105

Potentilla agrimonioides M.Bieb.

Rosaceae

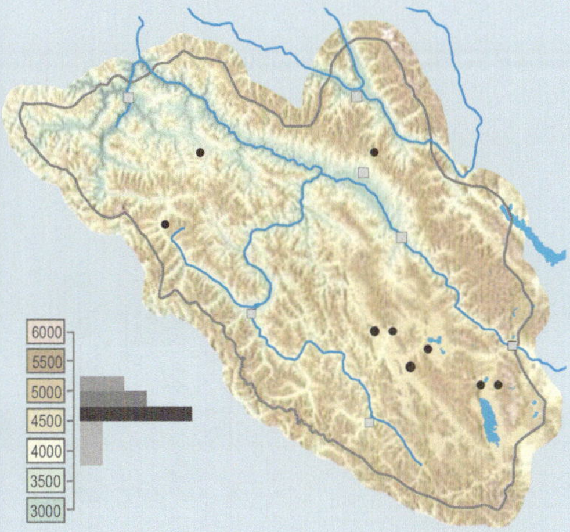

Syn.: *P. cinerascens* Bertol.

Description: Perennial herbs, 10–20 cm tall, tomentose, pilose and glandular hairy, leaves imparipinnately compound, leaflets 3–5 pairs, obovate, deeply incised, flowers yellow, 1–1.5 cm across, in lax corymbs.

Distribution: Russia (Caucasus), Iran, Afghanistan, Pakistan, India (Jammu and Kashmir).

Habitat: Dry scree, stony places in alpine pastures.

Stem diameter 5–12 mm. Bark/xylem proportion 1:5. Ring width 0.1–0.2 mm. Maximal 23 annual rings.

Xylem

Transverse section

Semi-ring-porous. Rings distinctly marked by differences between earlywood (30–45 µm) and latewood vessel diameters (10–20 µm). Vessels round, fairly thick-walled, solitary, hardly stained. Fibers absent. Parenchyma pervasive. With remaining vascular bundles. With crystal druses in large rays. With intra-xylary periderm towards the pith.

Tangential section

Rays 1–2-seriate.

Radial section

Perforation plates simple. Intervessel pits round in alternating position. Vessel length 100–120 µm. Fiber pits round, relatively large (2–3 µm). Ray cells upright.

Bark

Phloem small, fairly homogeneous. Few sieve cells slightly smaller than parenchyma cells. With crystal druses. Phellem layered. Between each layer is a row of very thin-walled, rectangular cells (phelloids), which appear light in polarized light. With prismatic crystals in the phellem.

Pith

No slide available.

Analyzed: Root collar of polar root of 3 individuals.

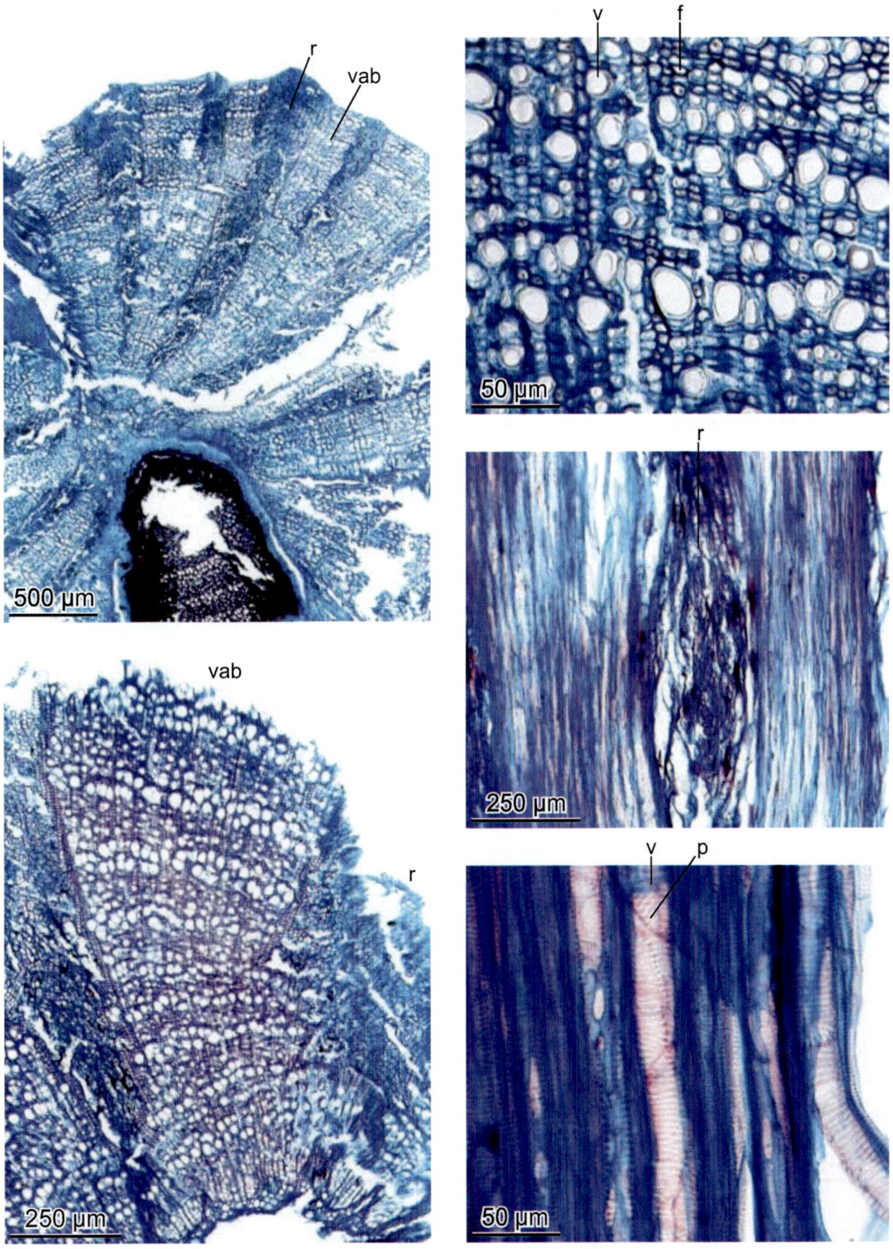

Features according to IAWA

Xylem *1 - 4 - 9 - 13 - 22 - 39.1 - 39.2 - 40.2 - 50.1 - 52.2 - 60.1 - 79.1 - 99 - 99.1 - 100.1 - 100.2 - 105 - 135.2 - 144*

Potentilla evestita var. *orthotricha* Soják

Rosaceae

Description: Perennial herbs, 15–30 cm tall, sparsely pilose, stems ascending to erect, leaves 3-foliolate, leaflets broadly obovate, green above, canescent below, later glabrescent, margin obtusely 3–6-incised serrate, flowers yellow, ca. 1.5 cm across, in lax corymbs.

Distribution: Russia (Siberia), Kazakhstan, Kyrgyzstan, Tajikistan, Pakistan, NW India, China (Xinjiang), Mongolia.

Habitat: Stream sides, rock crevices.

Stem diameter 10–14 mm. Bark/xylem proportion 1:5. Mean ring width 0.2 mm. Maximal 46 annual rings. Stem segregation.

Xylem

Transverse section

Semi-ring-porous. Rings marked by differences between earlywood (30–40 µm) and latewood vessel diameters (10–15 µm). Thick-walled vessels solitary. Fibers absent. Parenchyma pervasive. Remaining vascular bundles. Crystal druses in large rays. Intra-xylary periderm towards the pith.

Tangential section

Rays very large (>10-seriate), between vessel/parenchyma strips.

Radial section

Perforation plates simple. Intervessel pits scalariform. Vessel length 60–100 µm. Ray cells upright.

Bark

Phloem small, fairly homogeneous. A few sieve cells slightly smaller than parenchyma cells. With crystal druses. Phellem layered. Between layers are rows of very thin-walled cells (phelloids), consisting of rectangular cells which appear light in polarized light. Prismatic crystals in the phellem.

Pith

No slide available.

Analyzed: Root collar of polar root of 7 individuals.

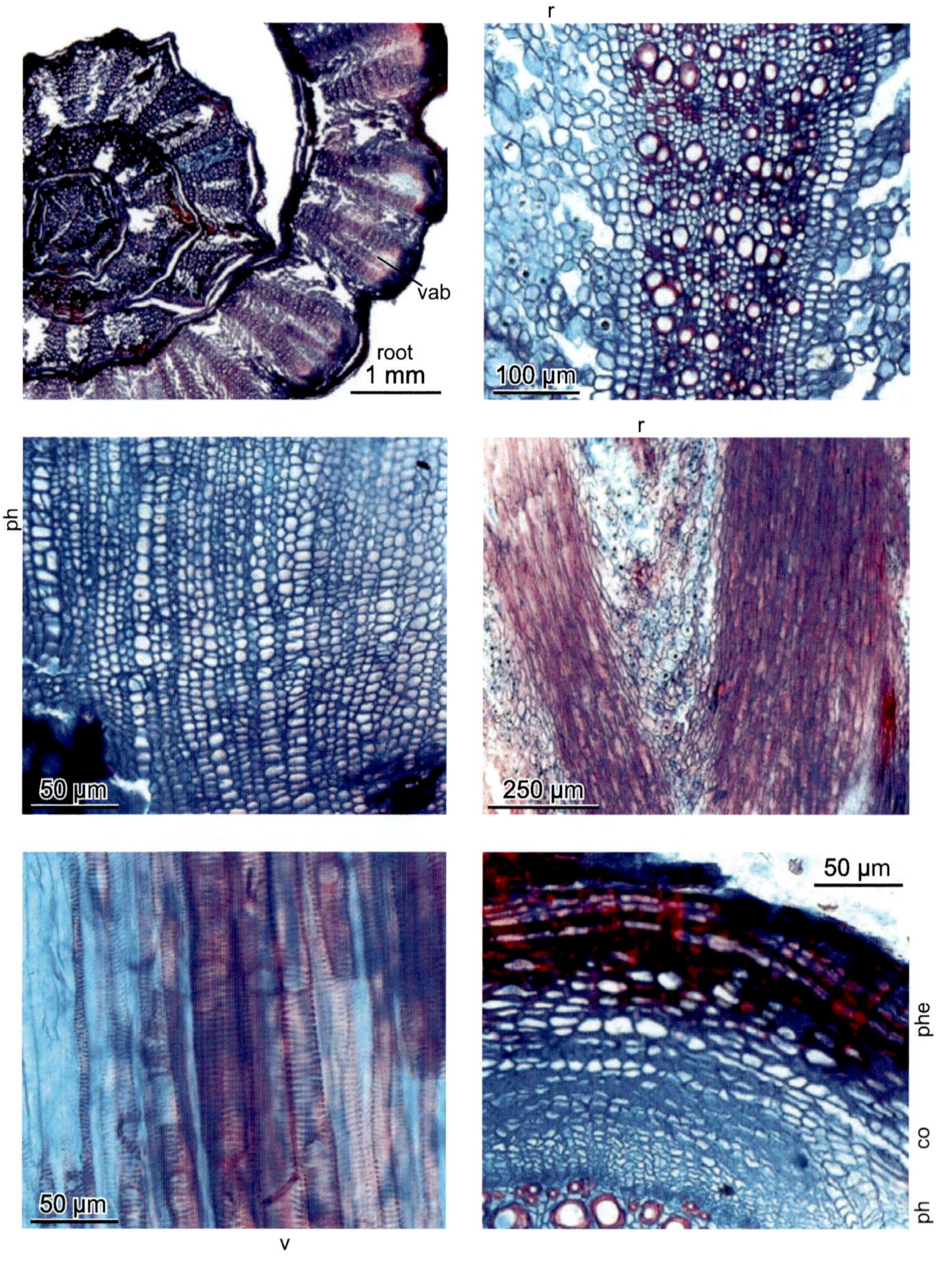

Features according to IAWA

Xylem 1 - 4 - 9 - 13 - 22 - 39.1 - 40.2 - 50.1 - 52.2 - 60.1 - 79.1 - 99 - 99.1 - 100.1 - 100.2 - 105 - 135.2 - 144

Potentilla multifida L.

Rosaceae

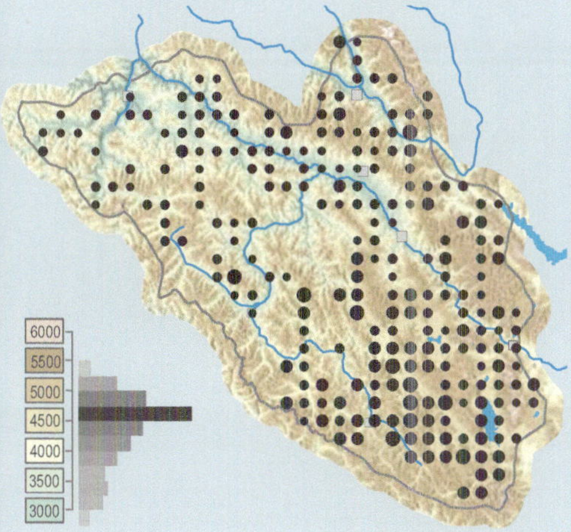

Syn.: *P. angustiloba* T.T.Yu & C.L.Li

Description: Perennial herbs, 6–40 cm tall, stems prostrate to ascending, leaves 2–3 pinnate, usually white tomentose below, flowers yellow, ca. 1.5 cm across, in lax corymbs.

Distribution: N America, Europe, Russia (Siberia), W and C Asia, Pakistan, India, China (Gansu, Hebei, Heilongjiang, Jilin, Liaoning, Nei Mongol, Qinghai, Shaanxi, Shanxi, Sichuan, Xinjiang, Xizang, Yunnan).

Habitat: Semi-deserts, mesic grasslands on various substrates, sandy river banks.

Stem diameter 5–12 mm. Bark/xylem proportion 1:4. Mean ring width 0.2 mm. Maximal 38 annual rings. Stem segregation

Xylem

Transverse section

Semi-ring-porous. Rings marked by differences between earlywood (40–50 µm) and latewood vessel diameters (10–15 µm). Vessels solitary. Fibers absent. Parenchyma pervasive. Remaining vascular bundles. Crystal druses in large rays. Intraxylary periderm towards the pith.

Tangential section

Rays very large (>10-seriate) and confluent, between vessel/parenchyma strips.

Radial section

Perforation plates simple. Intervessel pits round to scalariform. Vessel length 60–100 µm. Ray cells upright.

Bark

Phloem small, fairly homogeneous. A few sieve tubes slightly smaller than parenchyma cells. Older parts of phloem collapsed. With crystal druses. Phellem layered, cells rectangular.

Pith

No slide available.

Analyzed: Root collar of polar root of 7 individuals.

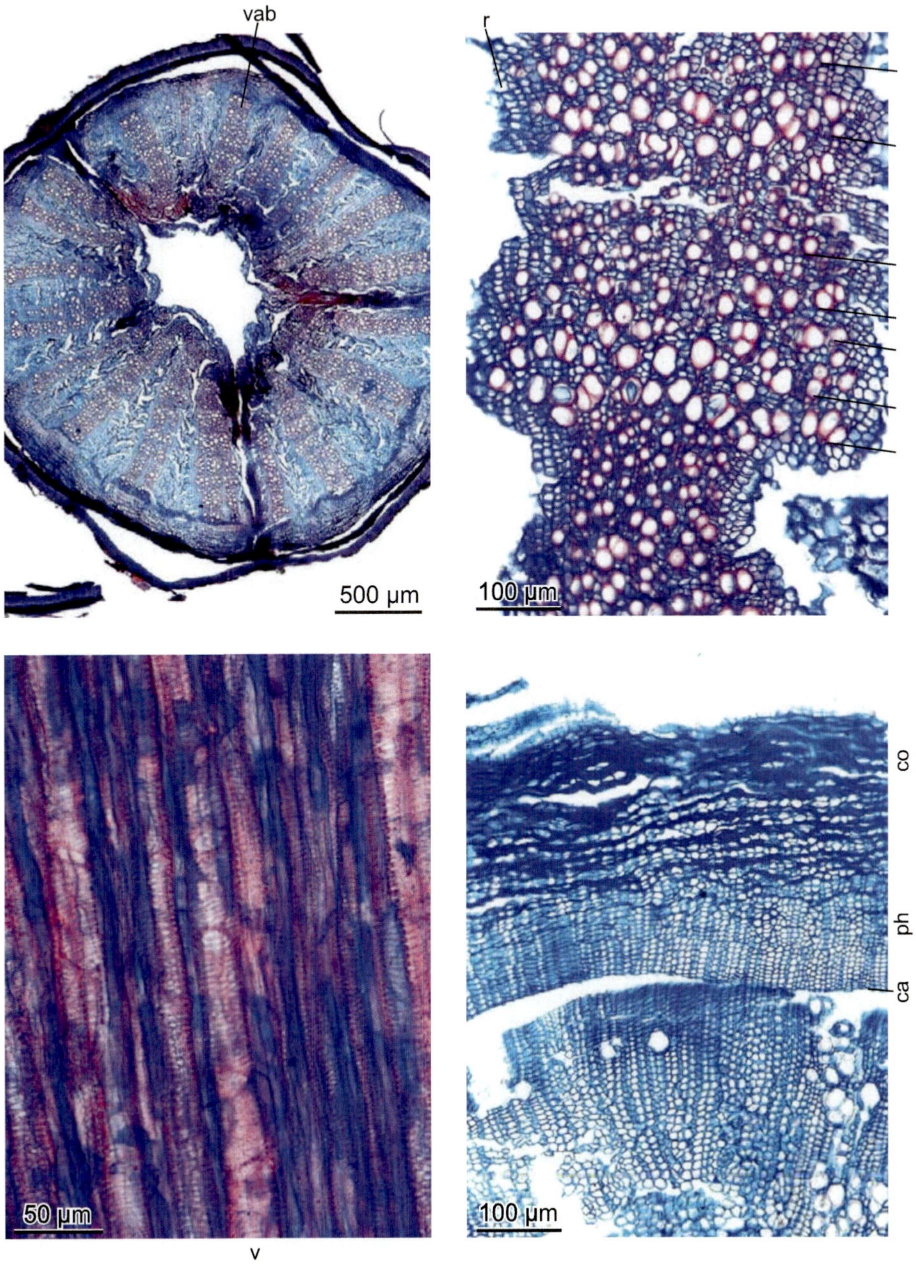

Features according to IAWA

Xylem 1 - 4 - 9 - 13 - 22 - 40.2 - 50.1 - 52.2 - 60.1 - 79.1 - 99 - 99.1 - 100.1 - 100.2 - 105 - 135.2 - 144

Potentilla pamirica Wolf

Rosaceae

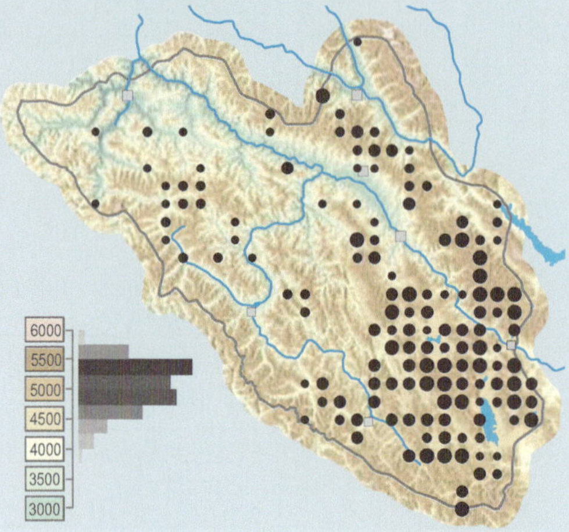

Description: Perennial herbs, 5–15 cm tall, stems prostrate, silky hairy, leaves pinnate with 4–5 pairs of pinnatisect leaflets, gray below, flowers yellow, ca. 1 cm across, in lax corymbs.

Distribution: ?Iran, Tajikistan, Afghanistan, N Pakistan, NW India (Jammu & Kashmir), China (Xizang), SW Mongolia.

Habitat: Mesic stony slopes, mesic to dry gravel sediments.

Stem diameter 5–14 mm. Bark/xylem proportion 1:5. Ring width 0.06–0.15 mm. Maximal 70 annual rings.

Xylem

Transverse section

Semi-ring-porous. Rings marked by differences between earlywood (40–50 µm) and latewood vessel diameters (10–15 µm). Vessels solitary. Fibers absent. Parenchyma pervasive. Remaining vascular bundles. Crystal druses in large rays. Intraxylary periderm towards the pith.

Tangential section

Rays very large (>10-seriate), between vessel/parenchyma strips.

Radial section

Perforation plates simple. Intervessel pits scalariform. Vessel length 100–150 µm. Ray cells upright.

Bark

Phloem small, fairly homogeneous. A few sieve tubes slightly smaller than parenchyma cells. Older parts of phloem collapsed. With crystal druses. Phellem layered, cells rectangular.

Pith

No slide available.

Analyzed: Root collar of polar root of 24 individuals.

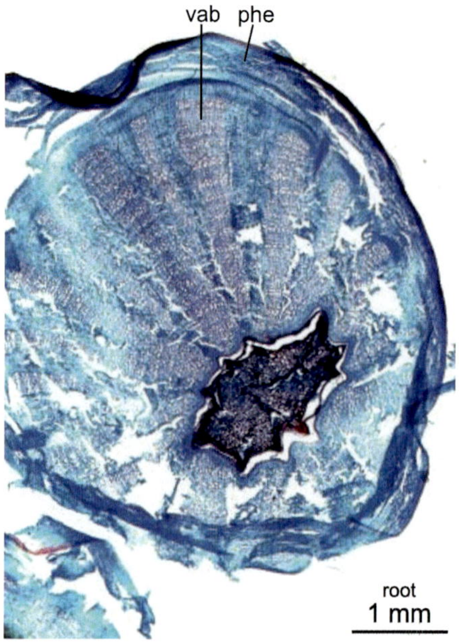

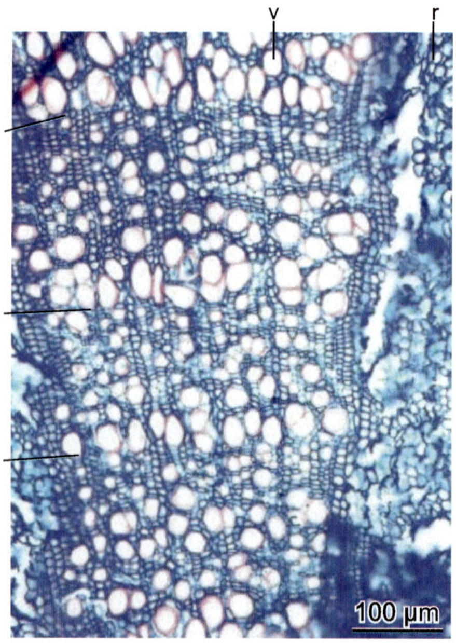

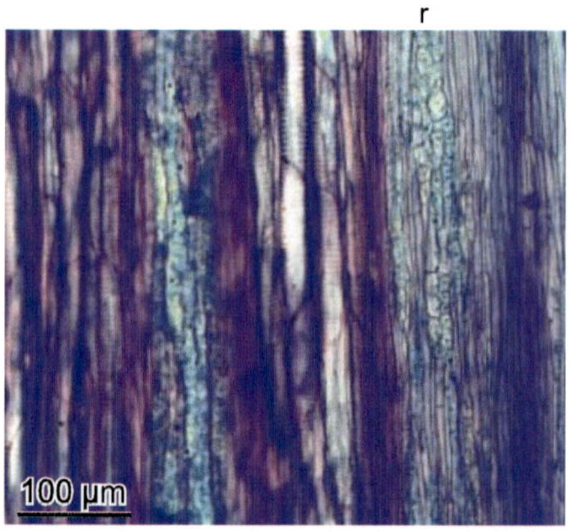

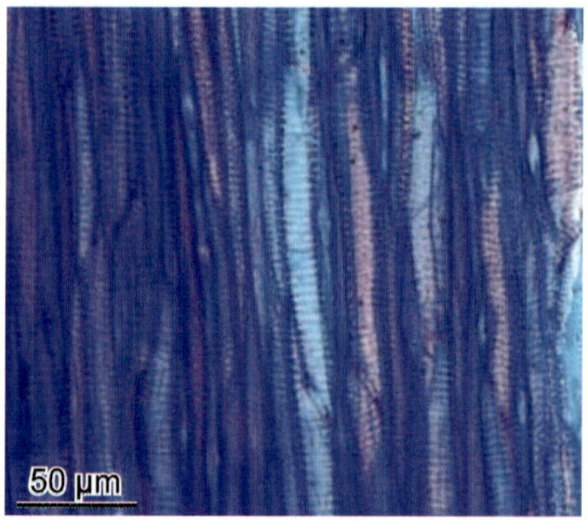

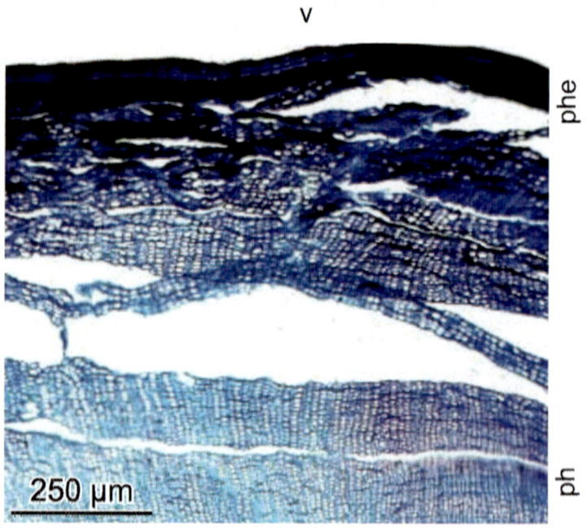

Features according to IAWA

Xylem 1 - 4 - 9 - 13 - 22 - 39.2 - 40.2 - 50.1 - 52.3 - 60.1 - 79.1 - 99 - 99.1 - 100.1 - 100.2 - 105 - 135.2 - 144

Potentilla sojakii Panigrahi & Dixit

Rosaceae

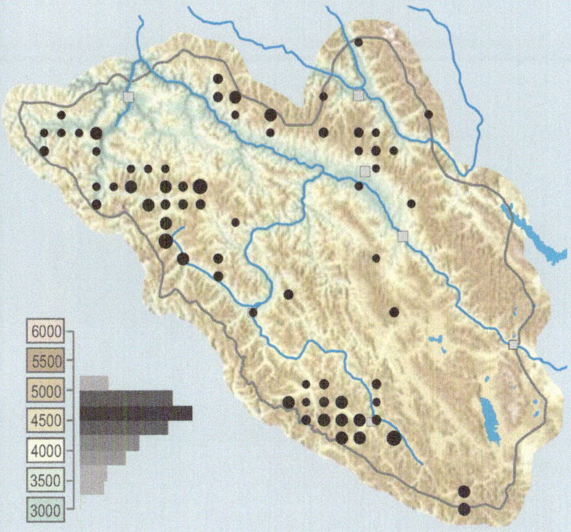

Syn.: *P. venusta* Soják

Description: Perennial herbs, 5–20 cm tall, sericeous, leaves ternate, leaflets elliptic-ovate, dull white tomentose below, margin serrate, flowers yellow or red, 1.5–2 cm across, in corymbs.

Distribution: Afghanistan, Pakistan, India, Nepal, Bhutan.

Habitat: Dry slopes, alpine meadows.

Stem diameter 12 mm. Bark/xylem proportion 1:4. Mean ring width 0.2 mm. Maximal 26 annual rings. Stem segregation. Center rotten.

Xylem

Transverse section

Semi-ring-porous. Rings marked by differences between earlywood (40–50 µm) and latewood vessel diameters (10–15 µm). Vessels solitary. Fibers absent. Parenchyma pervasive. Remaining vascular bundles. Crystal druses in large rays. Intraxylary periderm towards the pith.

Tangential section

Rays very large (>10-seriate), between vessel/parenchyma strips.

Radial section

Perforation plates simple. Intervessel pits scalariform. Vessel length 100–150 µm. Ray cells upright.

Bark

Phloem small, fairly homogeneous. A few sieve tubes slightly smaller than the parenchyma cells. Older parts of phloem collapsed. With crystal druses. Phellem layered, cells rectangular.

Pith

No slide available.

Analyzed: Root collar of polar root of 4 individuals.

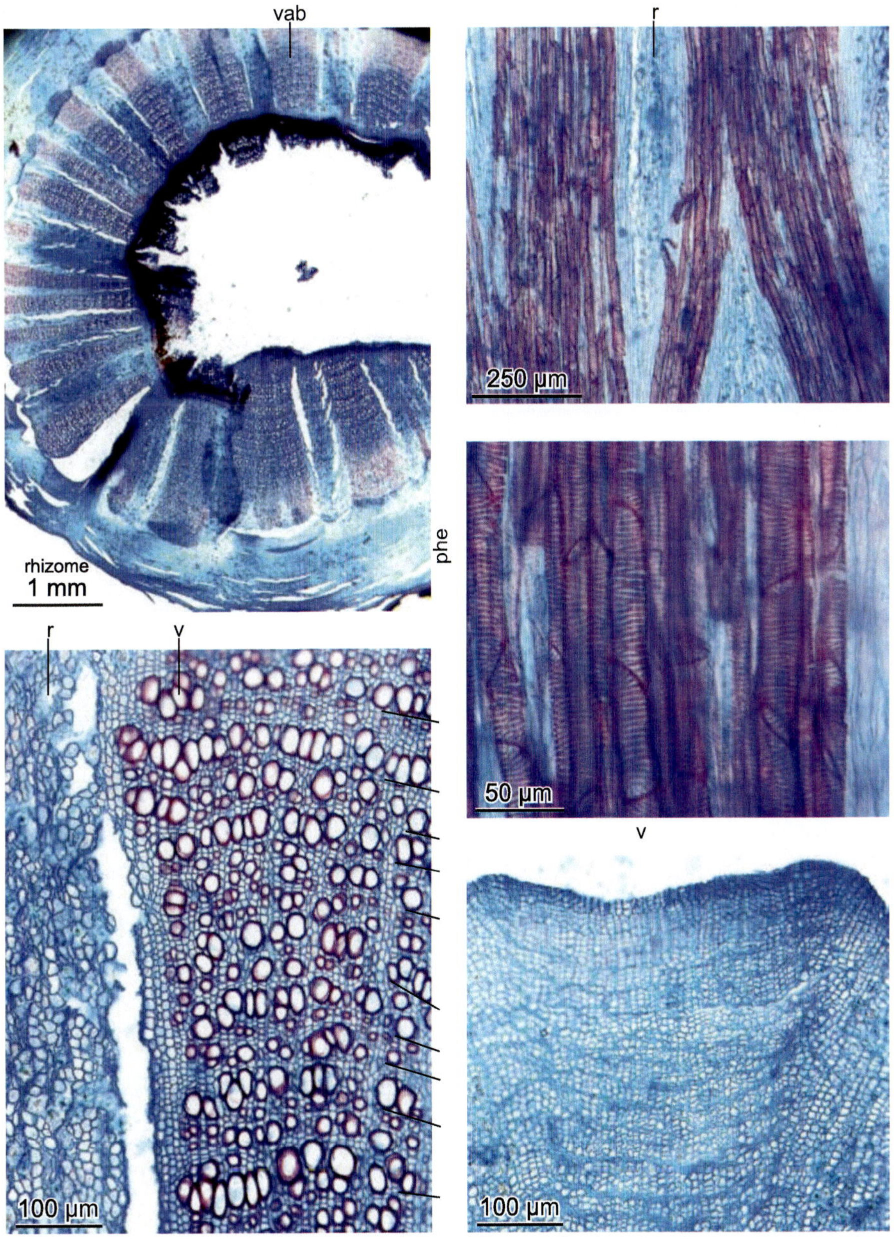

Features according to IAWA

Xylem 1 - 4 - 9 - 13 - 22 - 40.2 - 50.1 - 52.3 - 60.1 - 79.1 - 99 - 99.1 - 100.1 - 100.2 - 105 - 135.2 - 144

Potentilla turczaninowiana Stschegl.

Rosaceae

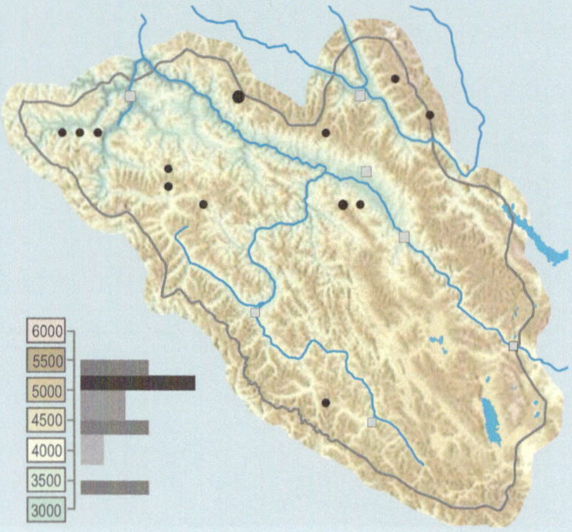

Syn.: *P. gelida* var. *turczaninowiana* Th.Wolf

Description: Perennial herbs, 2–10 cm tall, glabrous, stems prostrate, leaves ternate, leaflets broadly obovate-cuneate, margin serrate, flowers pale yellow, 6–15 mm across, in 2–4-flowered corymbs.

Distribution: Kazakhstan, Kyrgyzstan, Tajikistan, N Pakistan, India (Himachal Pradesh, Jammu and Kashmir).

Habitat: Stony slopes, along streams.

Stem diameter 4 mm. Bark/xylem proportion 1:5. Rings indistinct.

Xylem

Transverse section

Semi-ring-porous. Rings marked by differences between earlywood (30–40 µm) and latewood vessel diameters (10–15 µm). Vessels round to angular, solitary. Fibers absent. Parenchyma pervasive. Remaining vascular bundles. Crystal druses in large rays.

Tangential section

Rays very large (>10-seriate), between vessel/parenchyma strips.

Radial section

Perforation plates simple. Intervessel pits scalariform. Vessel length 100–150 µm. Ray cells upright.

Bark

Phloem small, fairly homogeneous. A few sieve tubes slightly smaller than the parenchyma cells. Older parts of phloem collapsed. With crystal druses. Phellem layered, cells rectangular.

Pith

No slide available.

Analyzed: Root collar of polar root of 3 individuals.

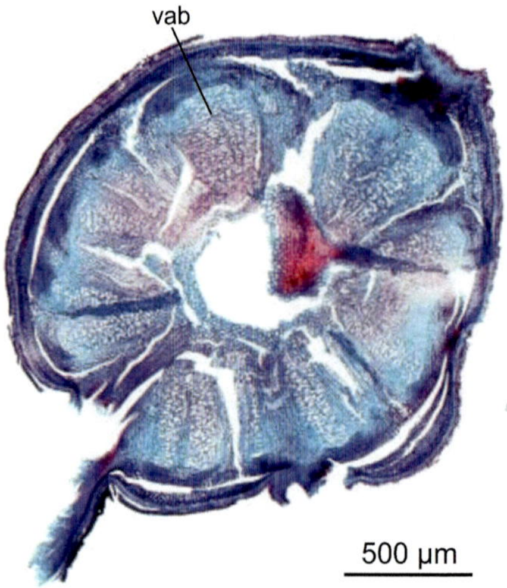

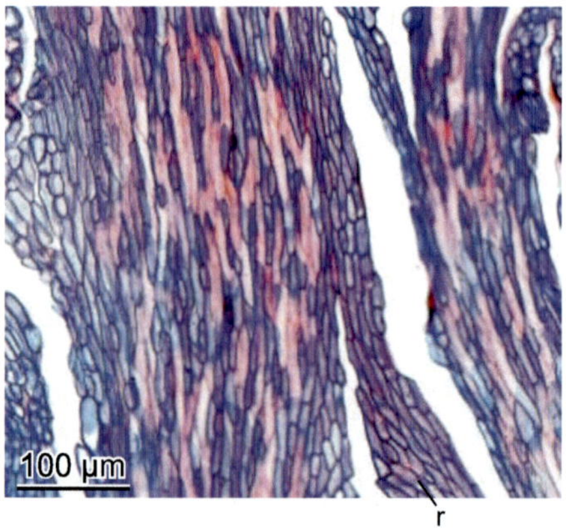

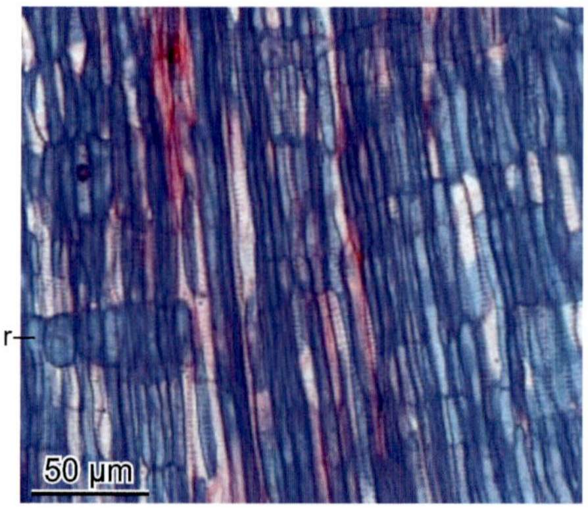

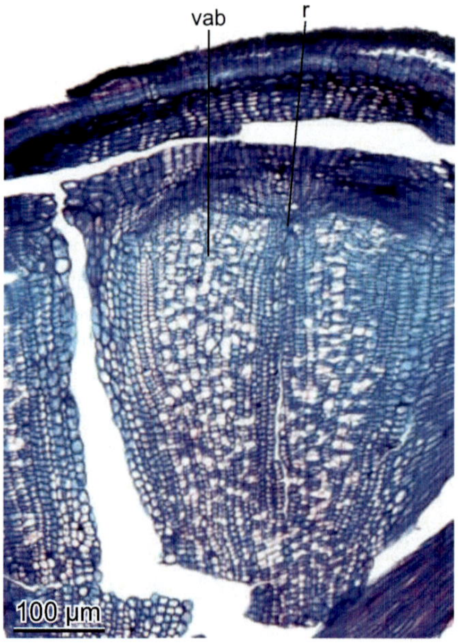

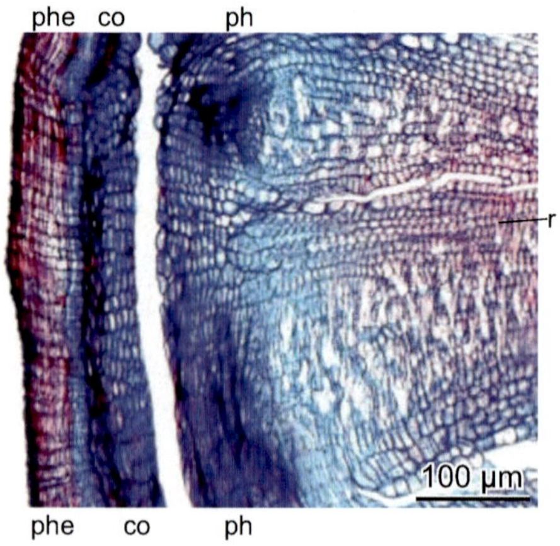

Features according to IAWA

Xylem 1 - 4 - 9 - 12 - 13 - 22 - 40.2 - 50.1 - 52.3 - 60.1 - 79.1 - 99 - 99.1 - 100.1 - 100.2 - 105 - 144

Rosa webbiana Wall. ex Royle

Rosaceae

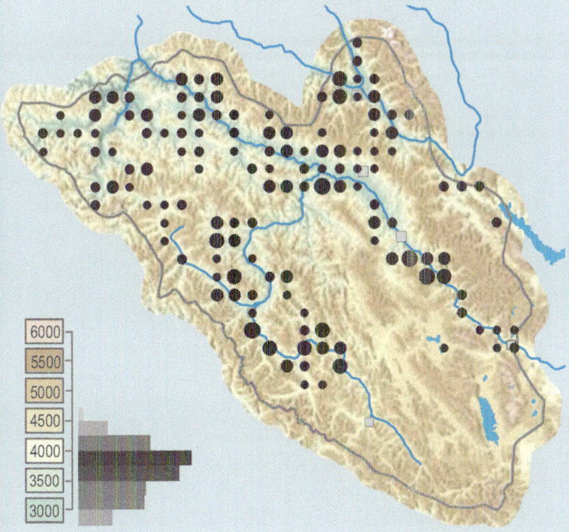

Syn.: *R. fedtschenkoana* Regel, *R. hissarica* Slobod., *R. korschinskyana* Bouleng., *R. maracandica* Bunge, *R. nanothamnus* Bouleng.

Description: Shrubs, 1–3 m tall, thorny with prickles, flowers rose, showy, mostly solitary, 3.6–5 cm across, hip nodding, bright red, 1.5–2 cm across.

Distribution: Tajikistan, Afghanistan, N Pakistan, NW India, W Nepal, China (Xizang).

Habitat: Stony river terraces, dry stony slopes, rock crevices.

Stem diameter 50 mm. Bark/xylem proportion 1:10. Mean ring width 1.1 mm. Maximal 7 living annual rings.

Xylem

Transverse section

Ring-porous. Rings distinctly marked by large differences between earlywood (60–100 µm) and latewood diameters (10–20 µm). Earlywood vessels arranged in a belt, latewood vessels mostly solitary. Fibers thick-walled. Parenchyma apotracheal diffuse.

Tangential section

Rays uniseriate or >10-seriate (ray dimorphism).

Radial section

Perforation plates simple. Intervessel pits round in alternating position. Vessel length 100–250 µm. Fiber pits round and large (2–3 µm). Ray cells prostrate.

Bark

Phloem with bands collapsed groups of sieve tubes. Fibers in small tangentially enlarged groups. Cortex not observed. Phellem layered, cells rectangular. Prismatic crystals in parenchyma cells.

Pith

Heterocellular: Small cells groups with diameters of 29–30 µm occur between very large, thin-walled cells (100–200 µm).

Analyzed: Root collar of polar root of 1 individual.

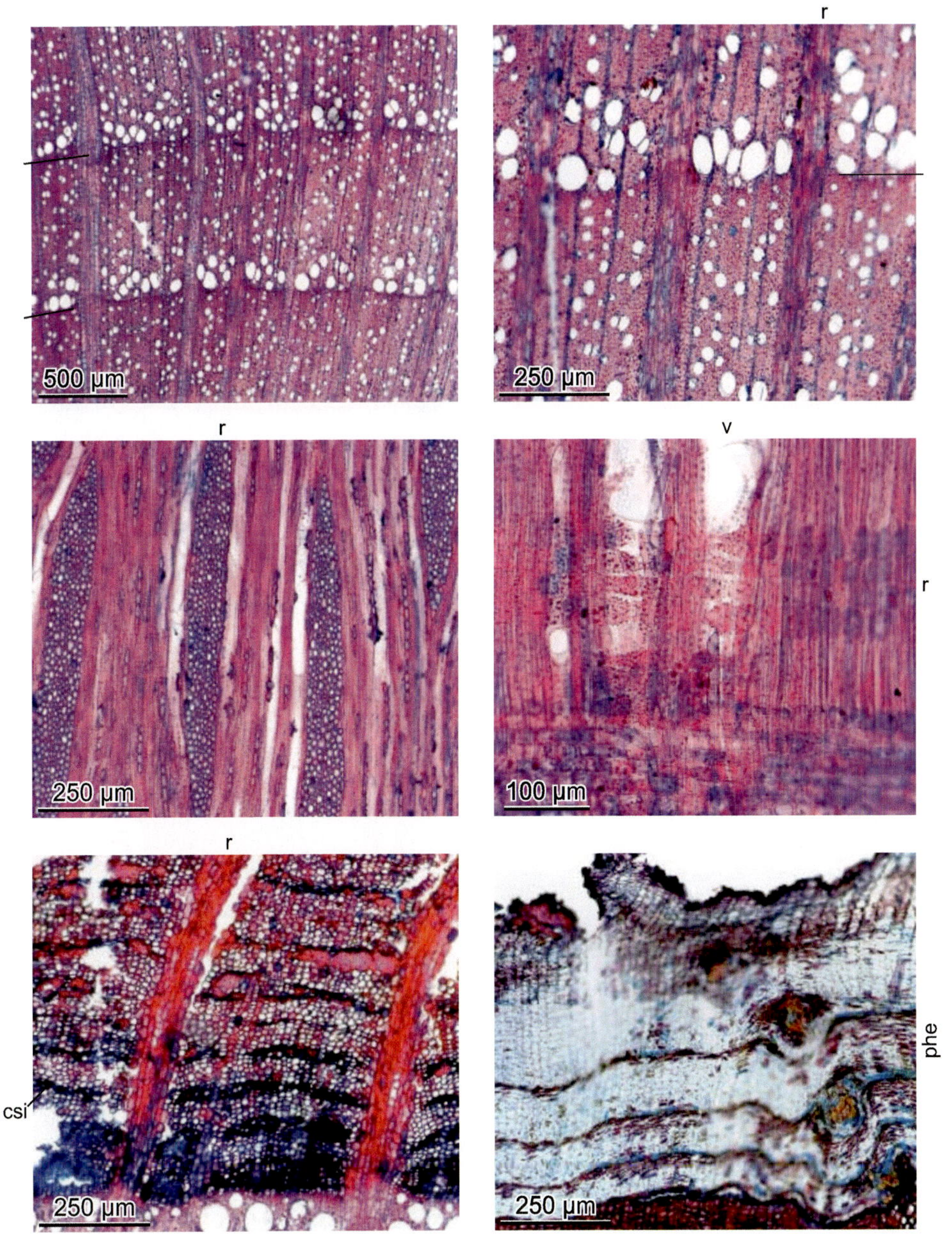

Features according to IAWA

Xylem 1 - 3 - 9 - 13 - 22 - 40.1 - 50.1 - 52.3 - 62 - 69 - 76 - 96 - 99 - 103 - 104

Schistophyllidium bifurcum (L.) Ikonn.

Rosaceae

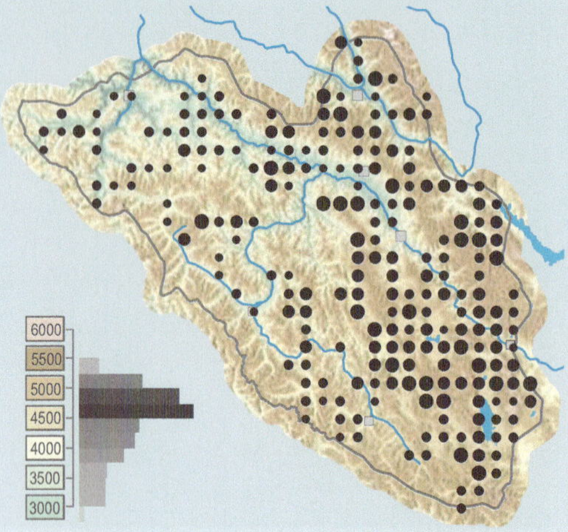

Syn.: *Potentilla bifurca* subsp. *moorcroftii* (Wall. ex Lehm.) Soják ex Panigrahi

Description: Perennial herbs, 5–20 cm tall, stems prostrate, leaves imparipinnately compound, leaflets 3–8 pairs, entire or 2–3-partite, flowers yellow, 0.7–1.5 cm across.

Distribution: SE Europe (Romania), Ukraine, Russia, Turkey, Caucasus, Iran, Afghanistan, Pakistan, India, Nepal, China, Bhutan, S Mongolia.

Habitat: Dry steppes and semideserts, sandy and loamy deposits, abandoned fields, sandy road margins.

Stem diameter 1 mm. Bark/xylem proportion 1:4. Mean ring width 0.08 mm. Maximal 6 annual rings.

Xylem

Transverse section

Semi-ring-porous. Rings often marked by differences between earlywood (40–60 µm) and latewood vessel diameters (20–25 µm) in large rings. Vessels solitary or in groups. Fibers thin- to thick-walled. Parenchyma paratracheal.

Tangential section

Rays uniseriate.

Radial section

Perforation plates simple. Intervessel pits scalariform. Vessel length 100–150 µm. Ray cells upright.

Bark

Phloem small, homogeneous. Crystals absent. Phellem layered, cells rectangular.

Pith

Round, filled with unlignified, round cells.

Analyzed: Root collar of polar root of 3 individuals.

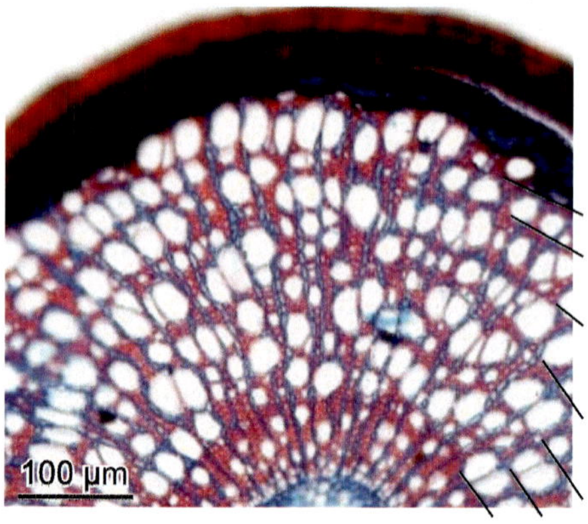

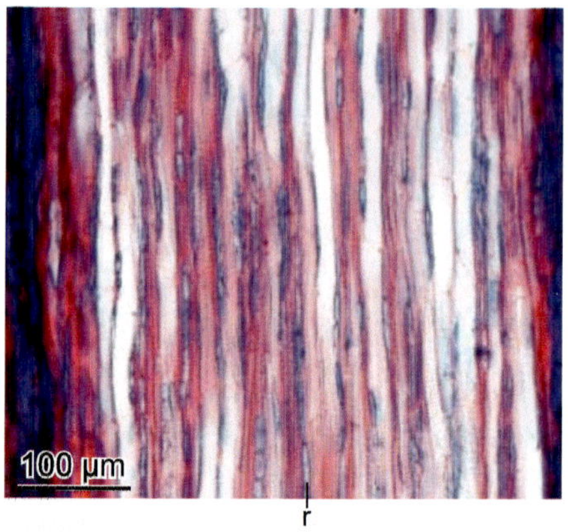

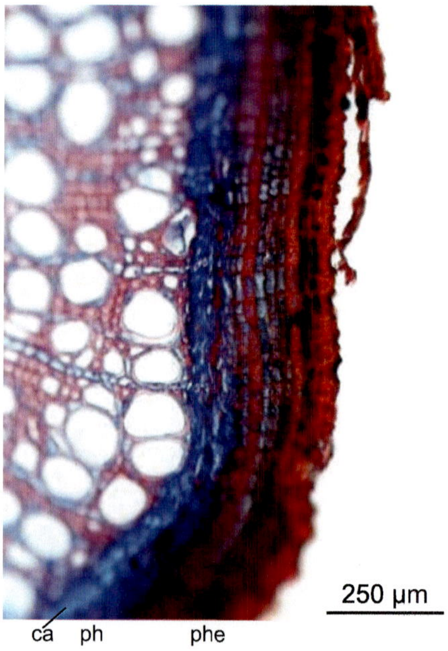

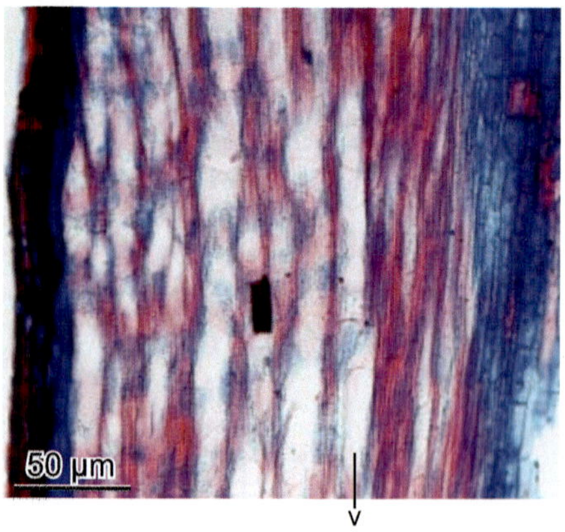

Features according to IAWA

Xylem 1 - 4 - 9 - 11 - 13 - 22 - 40.2 - 50.1 - 52.3 - 61 - 70 - 79 - 96 - 105

Sibbaldia cuneata Kunze

Rosaceae

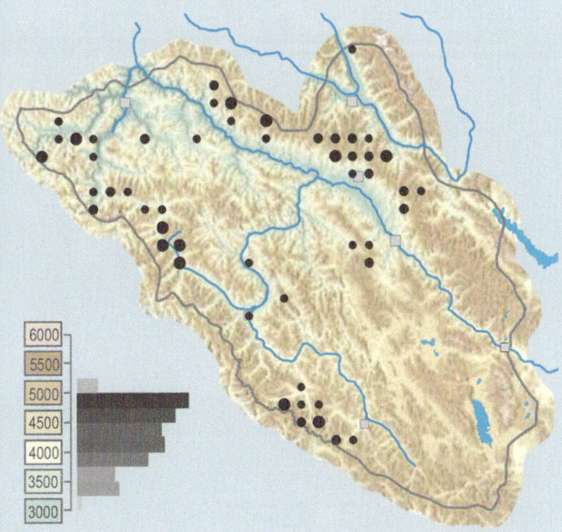

Description: Perennial herbs, 5–15 cm tall, pilose, stems ascending, leaves ternate, leaflets obovate, margin 3–5-dentate, flowers yellow, 5–7 mm across, in compact terminal inflorescences.

Distribution: Russia, Afghanistan, Pakistan, NW India, Nepal, China (Qinghai, Sichuan, Xizang, Yunnan), Bhutan.

Habitat: Alpine meadows, rock crevices.

Stem diameter 5–6 mm. Bark/xylem proportion 1:4. Ring width 0.1–0.2 mm. Maximal 20 rings.

Xylem

Transverse section

Semi-ring-porous. Rings distinctly marked by differences between earlywood (25–45 µm) and latewood vessel diameters (10–15 µm). Vessels solitary. Fibers thick-walled. Parenchyma apotracheal diffuse and paratracheal.

Tangential section

Rays 1–2-seriate or >10-seriate (ray dimorphism).

Radial section

Perforation plates simple. Intervessel pits round, in alternating position. Vessel length 100–120 µm. Fiber pits round. Ray cells upright.

Bark

Phloem small, fairly homogeneous. A few sieve cells slightly smaller than the parenchyma cells. Phellem layered. Between layers are rows of very thin-walled cells (phelloids), consisting of rectangular cells which appear light in polarized light. Crystals absent in phloem and phellem.

Pith

More or less round and hollow. Unlignified, thin-walled cells towards the xylem. With a few crystal druses.

Analyzed: Root collar of polar root of 5 individuals.

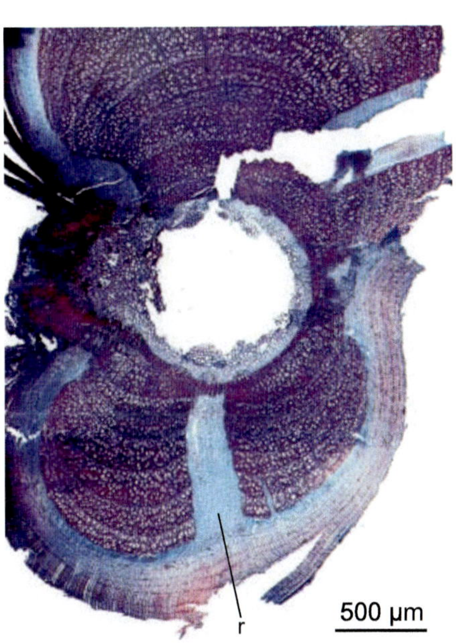

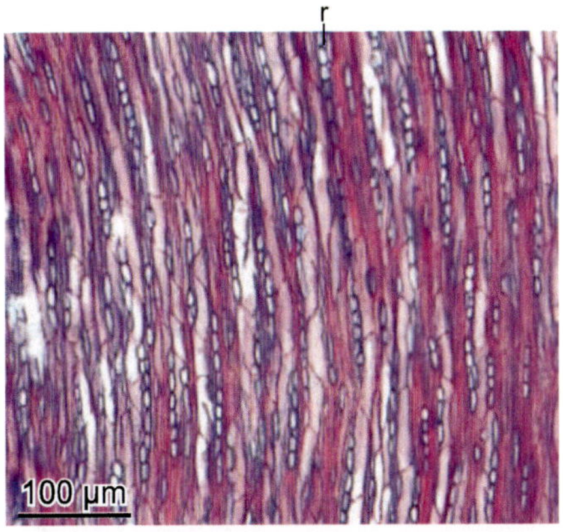

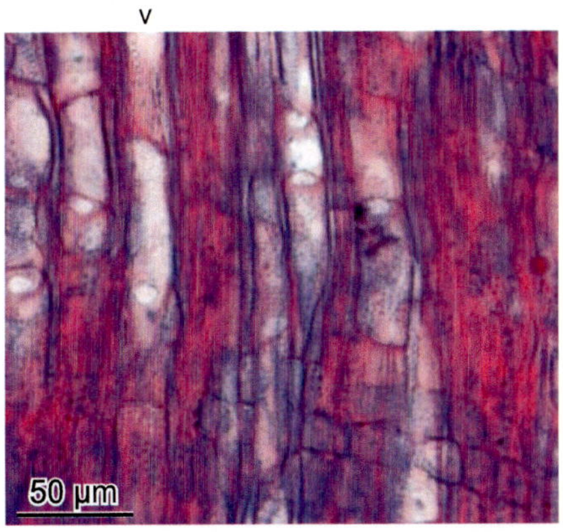

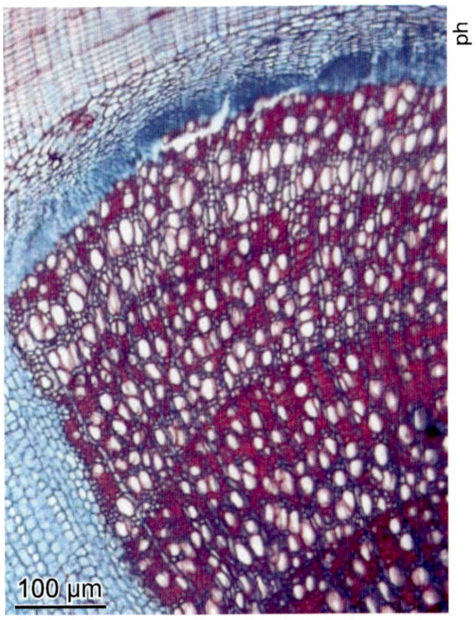

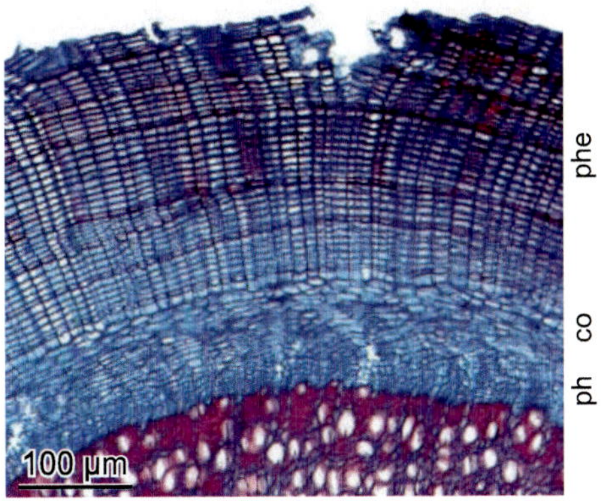

Features according to IAWA

Xylem 1 - 4 - 9 - 13 - 22 - 40.2 - 50.1 - 52.3 - 61 - 69 - 76 - 79 - 96 - 99 - 103 - 105

Sibbaldia tetrandra Bunge

Rosaceae

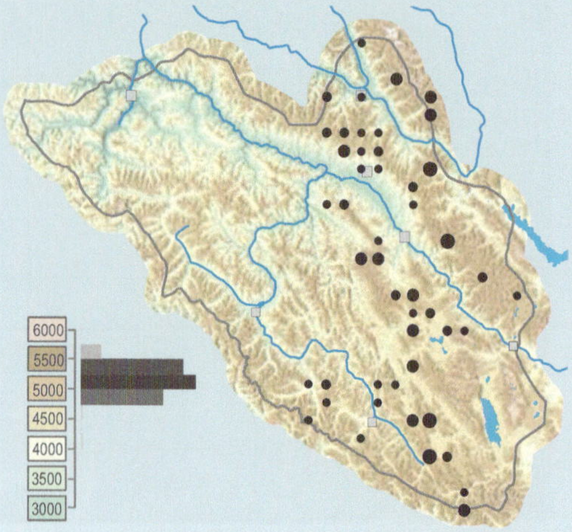

Syn.: *Dryadanthe bungeana* Ledeb., *D. tetrandra* (Bunge) Juz., *Potentilla tetrandra* (Bunge) Hook.f.

Description: Perennial herbs, 5–15 cm tall, cespitose, leaves ternate, tomentose, middle leaflets with 3 apical teeth, lateral leaflets with 2 apical teeth, flowers pale yellow, 4–8 mm across, solitary or by 2–3 at the ends of short branches.

Distribution: C Asia, Russia (W Siberia), NE Afghanistan, N Pakistan, N India, Nepal, China (Qinghai, Xinjiang, Xizang), Mongolia.

Habitat: Mountain slopes, rock crevices.

Stem diameter 1–2 mm. Bark/xylem proportion 1:3. Mean ring width 0.05–0.07 mm. Maximal 20 rings.

Xylem

Transverse section

Semi-ring-porous. Rings distinct or indistinct. Ring boundaries marked, if present, by differences between earlywood (15–20 μm) and latewood vessel diameters (5–10 μm). Vessels round, solitary. Vessel walls thick (2 μm). Fibers absent. Parenchyma pervasive and paratracheal. Vascular bundles remaining.

Tangential section

Rays within vascular bundles not recognizable in tangential sections, uniseriate and unlignified in transverse sections.

Radial section

Simple perforation plates. Intervessel pits scalariform. Vessel length 100–120 μm.

Bark

Phloem small and fairly homogeneous. A few sieve cells slightly smaller than the parenchyma cells. Phellem layered. Between layers are rows of very thin-walled cells (phelloids), consisting of rectangular cells which appear bright in polarized light. A few crystal druses in the pith.

Pith

Pith round to polygonal, unliginifed. With a few crystal druses in the pith.

Analyzed: Polar root of 3 individuals.

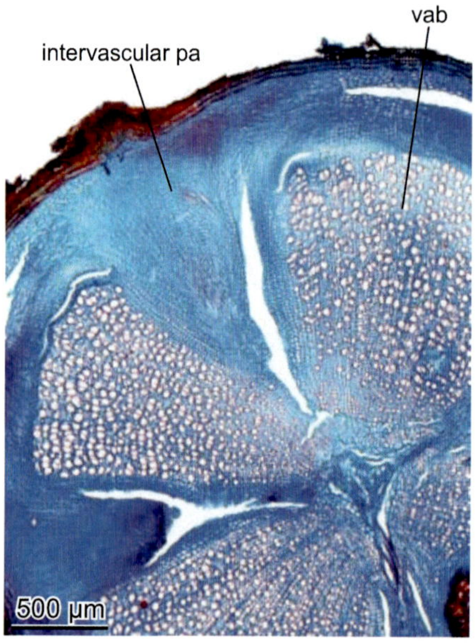

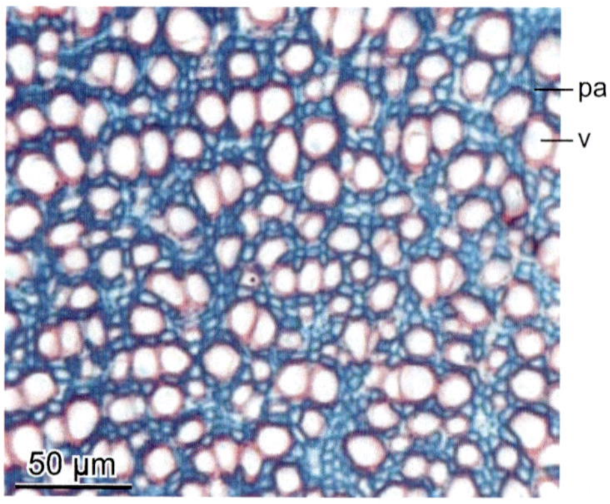

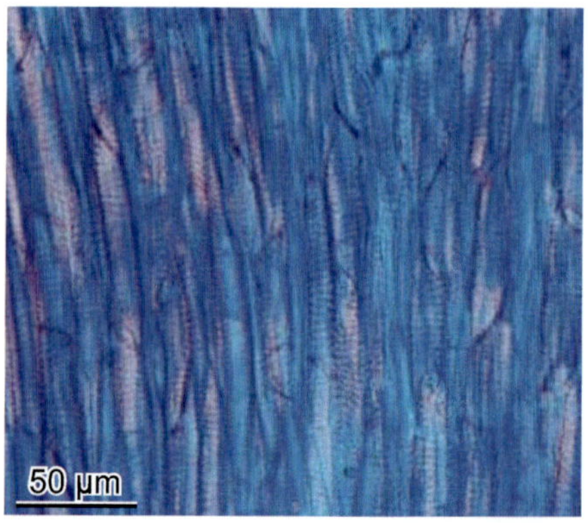

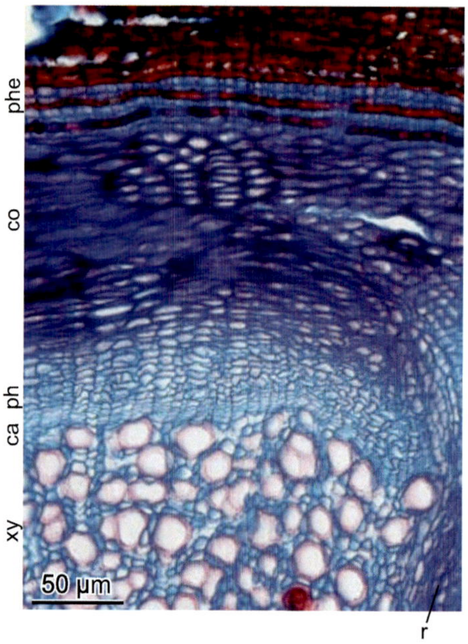

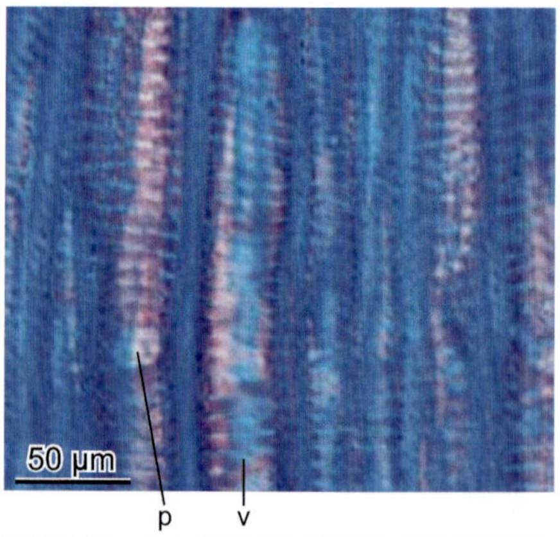

Features according to IAWA

Xylem 1 - 4 - 9 - 13 - 20 - 39.1 - 40.1 - 50.1 - 52.3 - 60.1 - 79 - 79.1 - 96 - 99 - 99.1 - 105

Sibbaldianthe adpressa (Bunge) Juz.

Rosaceae

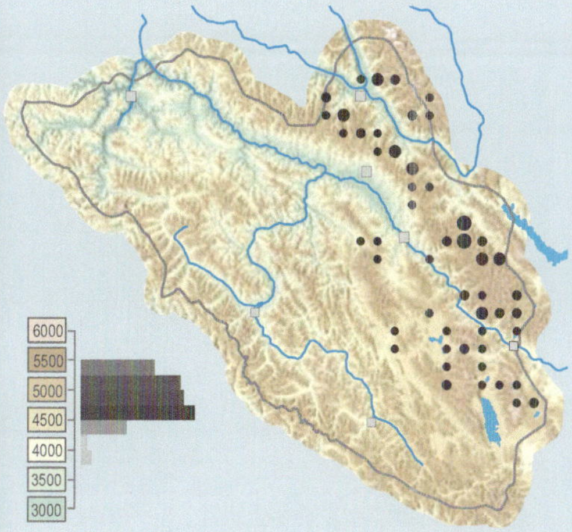

Syn.: *Sibbaldia adpressa* Bunge

Description: Perennial herbs, 1–10 cm tall, cespitose, leaves pinnate with 2 pairs of leaflets, sometimes 3-foliolate, terminal leaflet oblanceolate or obovate, apex 3-serrate, lateral leaflets lanceolate, entire, flowers pale yellow or whitish, 6–10 mm across, solitary or in several-flowered inflorescences.

Distribution: Russia, N Pakistan, NW India, Nepal, China (Gansu, Hebei, Heilongjiang, Nei Mongol, Qinghai, Xinjiang, Xizang), Mongolia.

Habitat: Mountain slopes, sandy river banks, gravel areas.

Stem diameter 2–2.5 mm. Bark/xylem proportion 1:6. Mean ring width 0.05 mm. Maximal 25 annual rings.

Xylem

Transverse section

Diffuse-porous. Rings distinctly marked by differences between earlywood (2–25 µm) and latewood vessel diameters (10–15 µm). Vessels round, solitary. Fibers thick-walled. Parenchyma mainly paratracheal.

Tangential section

Rays 1–2-seriate.

Radial section

Perforation plates simple. Intervessel pits round in alternating position. Vessel length 120–150 µm. Fiber pits round and relatively large (2–3 µm). Ray cells upright.

Bark

Phloem small. Older parts flake. Phloem fairly homogeneous. A few sieve cells slightly smaller than the parenchyma cells. Phellem layered. Between layers are rows of very thin-walled cells (phelloids), consisting of thin-walled, rectangular cells. Crystals absent.

Pith

No slide available.

Analyzed: Stem of 2 individuals (herbarium specimen).

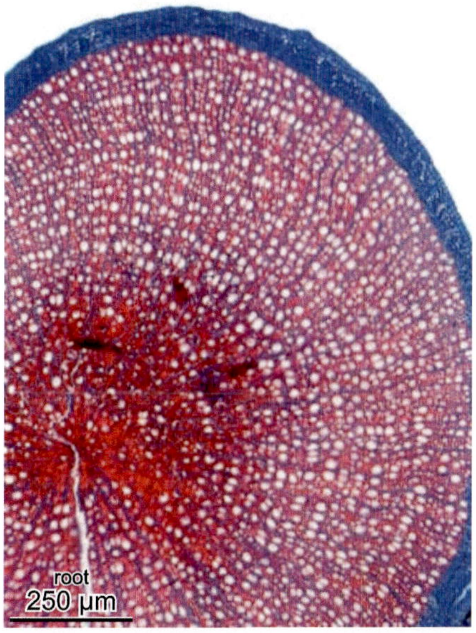

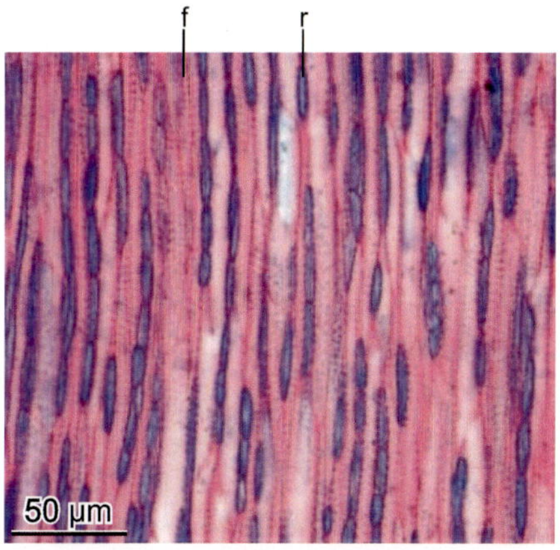

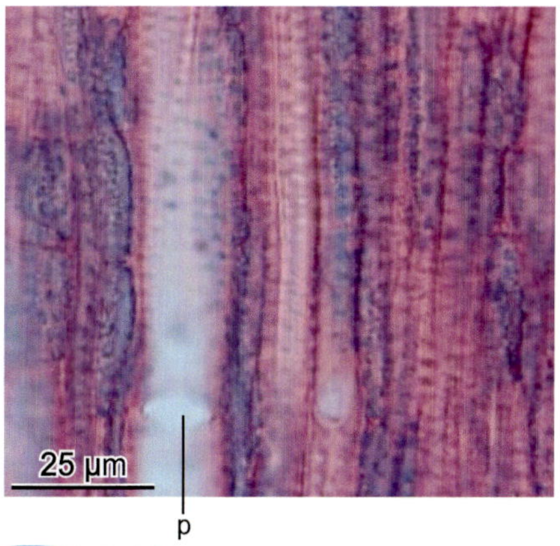

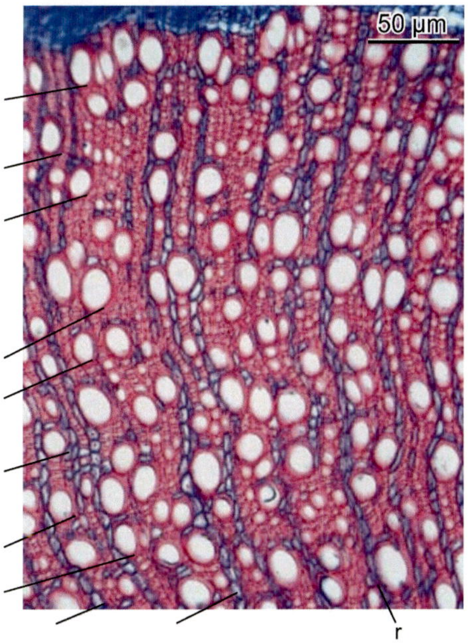

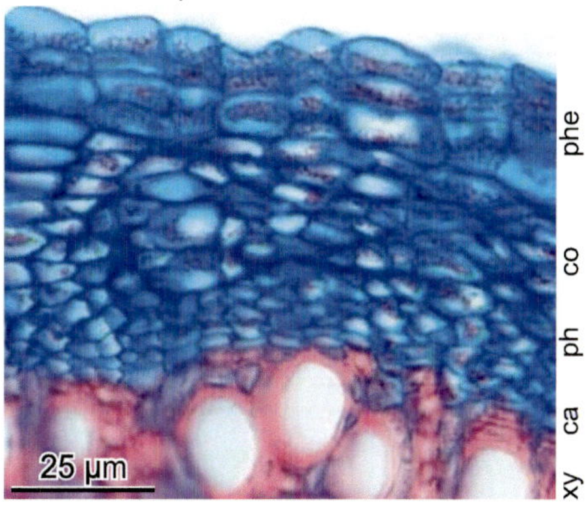

Features according to IAWA

Xylem 1 - 5 - 9 - 13 - 22 - 40.2 - 50.2 - 52.3 - 62 - 69 - 79 - 96 - 97 - 106

Sorbaria tomentosa (Lindl.) Rehder

Rosaceae

Description: Shrubs, up to 1 m tall, branches prostrate, thorny, leaves elliptic, obscurely serrate, clustered, flowers yellow-green, minute, solitary, axillary, fruits purple-red, turning black at maturity, obovoid-globose, ca. 5 mm across.

Distribution: Afghanistan, Pakistan, W and E India, China (W Xizang).

Habitat: Dry stony slopes, screes, rock crevices.

Stem diameter 3–4 mm. Bark/xylem proportion 1:4. Mean ring width 1.2 mm. Maximal 19 annual rings.

Xylem

Transverse section

Semi-ring-porous. Rings distinctly marked by differences between earlywood (40–50 µm) and latewood vessel diameters (15–20 µm). Vessels angular, solitary. Vessels in the heartwood contain dark-stained substances. Fibers thin- to thick-walled. Parenchyma apotracheal and paratracheal. Parenchyma cells contain dark-stained substances.

Tangential section

Rays 2–3-seriate.

Radial section

Perforation plates simple. Intervessel pits round in alternating position. Vessel length 100–120 µm. Locally with fine helical thickenings. Fiber pits fairly large (2–3 µm). Ray cells prostrate.

Bark

With small groups of sieve tubes. Sieve tubes collapsed in older phloem. Groups of sclereids in tangential rows. Phellem small, consisting of radially oriented, rectangular cells. Many prismatic crystals and crystal druses.

Pith

Pith round. Central cells large, unlignified, thin-walled, accompanied by single cells with cell contents.

Analyzed: Stem of 1 individual.

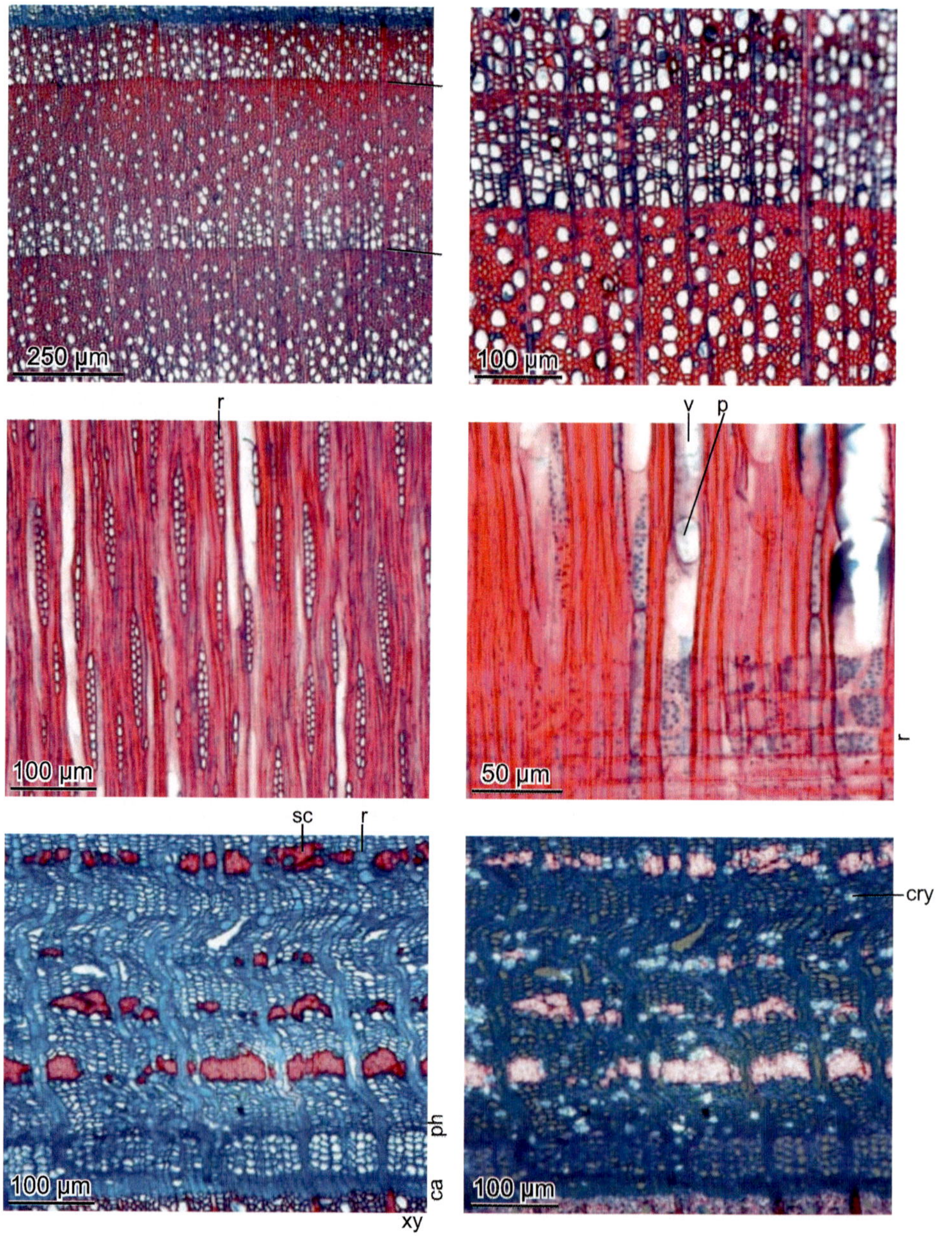

Features according to IAWA

Xylem 1 - 4 - 9 - 9.1 - 12 - 13 - 22 - 36 - 40.2 - 50.1 - 52.3 - 58 - 61 - 70 - 97 - 105

Galium spurium L.

Rubiaceae

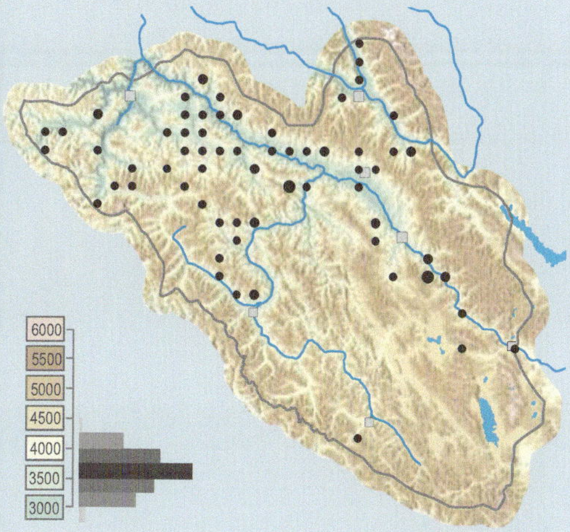

Syn.: *G. aparine* var. *pauciflorum* (Bunge) Maxim., *G. pauciflorum* Bunge, *G. vaillantii* DC.

Description: Annuals, 30–50 cm tall, stems procumbent or climbing, quadrangular, leaves oblanceolate, in whorls of 4–6, flowers white to greenish, 1–1.5 mm across, in terminal and axillary inflorescences, fruits with uncinate trichomes.

Distribution: N Africa, temperate Eurasia; naturalized elsewhere.

Habitat: Weedy in barley fields, on field margins.

Stem diameter 0.25 mm. Bark/xylem proportion 1:4. Mean ring width 0.08–0.12 mm. Maximal 1 annual ring.

Xylem

Transverse section

Only one ring. Vessel diameter 25–35 µm. Concentration of fibers in the latewood. Vessels angular, solitary. Fibers thin- to thick-walled. Parenchyma paratracheal.

Tangential section

Rays absent.

Radial section

No slide available.

Bark

Phloem uniform. With bundles of raphids in the cortex.

Pith

Pith more or less round. Central cells large, unlignified, thin-walled, disappearing in polarized light.

Analyzed: Polar roots of 2 individuals.

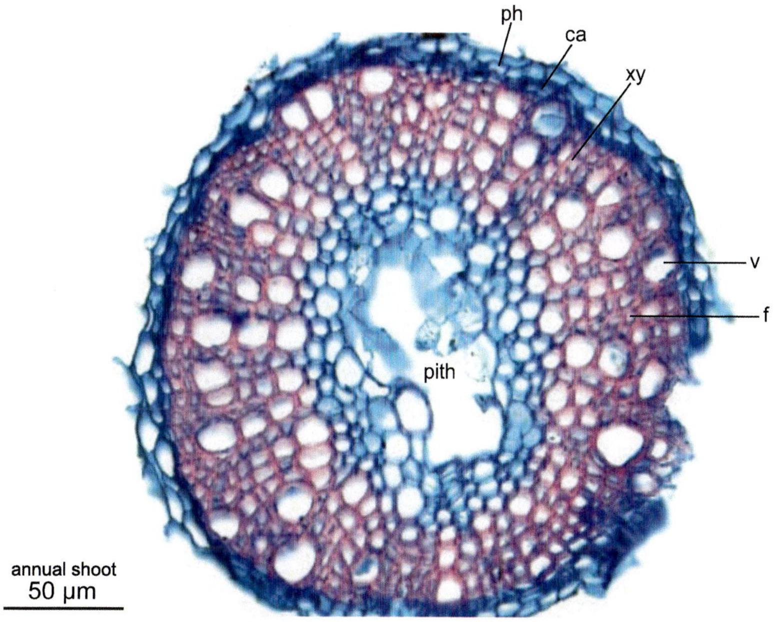

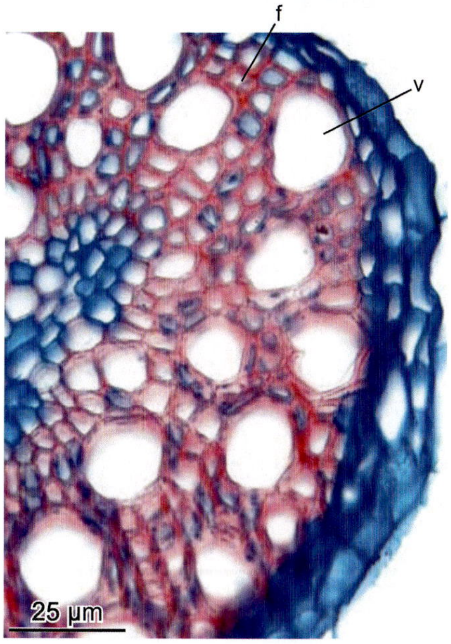

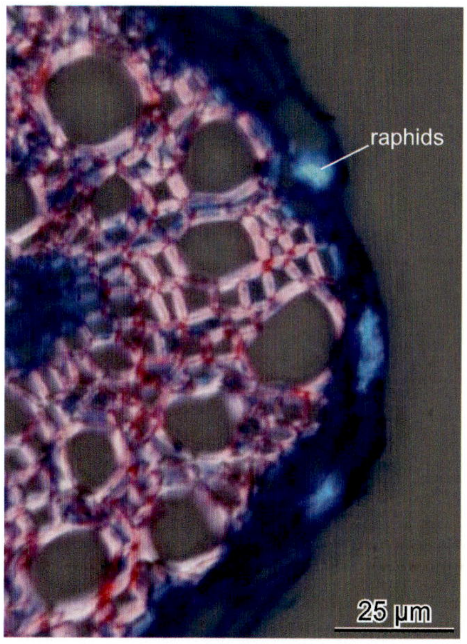

Features according to IAWA

No codes due to missing longitudinal sections.

Rubia tibetica Hook.f.

Rubiaceae

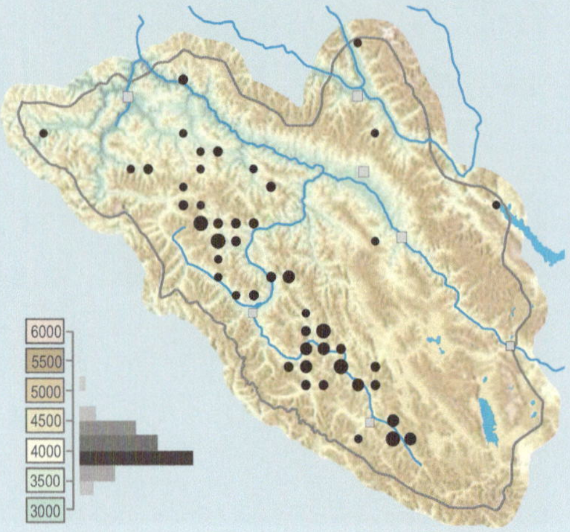

Description: Perennial herbs, 10–30 cm tall, stems quadrangular, leaves opposite, ovate to elliptic, in whorls of 4 (–6) with leaf-like stipules, flowers yellow, 6–8 mm across, in few-flowered axillary or terminal cymes.

Distribution: Kyrgyzstan, Tajikistan, Afghanistan, Pakistan, India, China (Xinjiang, Xizang).

Habitat: Dry stony slopes, screes, rock crevices.

Stem diameter 3–5 mm. Bark/xylem proportion 1:4. Ring width 0.15–0.22 mm. Maximal 25 annual rings.

Xylem

Transverse section

Semi-ring-porous. Rings distinctly marked by differences between earlywood (80–100 μm) and latewood vessel diameters (15–20 μm), and a concentration of fibers in the latewood. Angular vessels solitary. Fibers thin- to thick-walled. Parenchyma paratracheal.

Tangential section

Rays 1–2-seriate.

Radial section

Perforation plates simple. Intervessel pits round in alternating position. Fiber pits round (2–3 μm). Vessel length 160–200 μm. Ray cells upright. Bundles of raphids in enlarged parenchyma cells.

Bark

Phloem uniform. Older parts flake. A few rows of very thin-walled cells (phelloids) between each layer, consisting of rectangular cells. Many parenchyma cells with raphids.

Pith

More ore less round. Large, unlingnified, thin-walled central cells disappear in polarized light.

Analyzed: Root collar of polar root of 8 individuals.

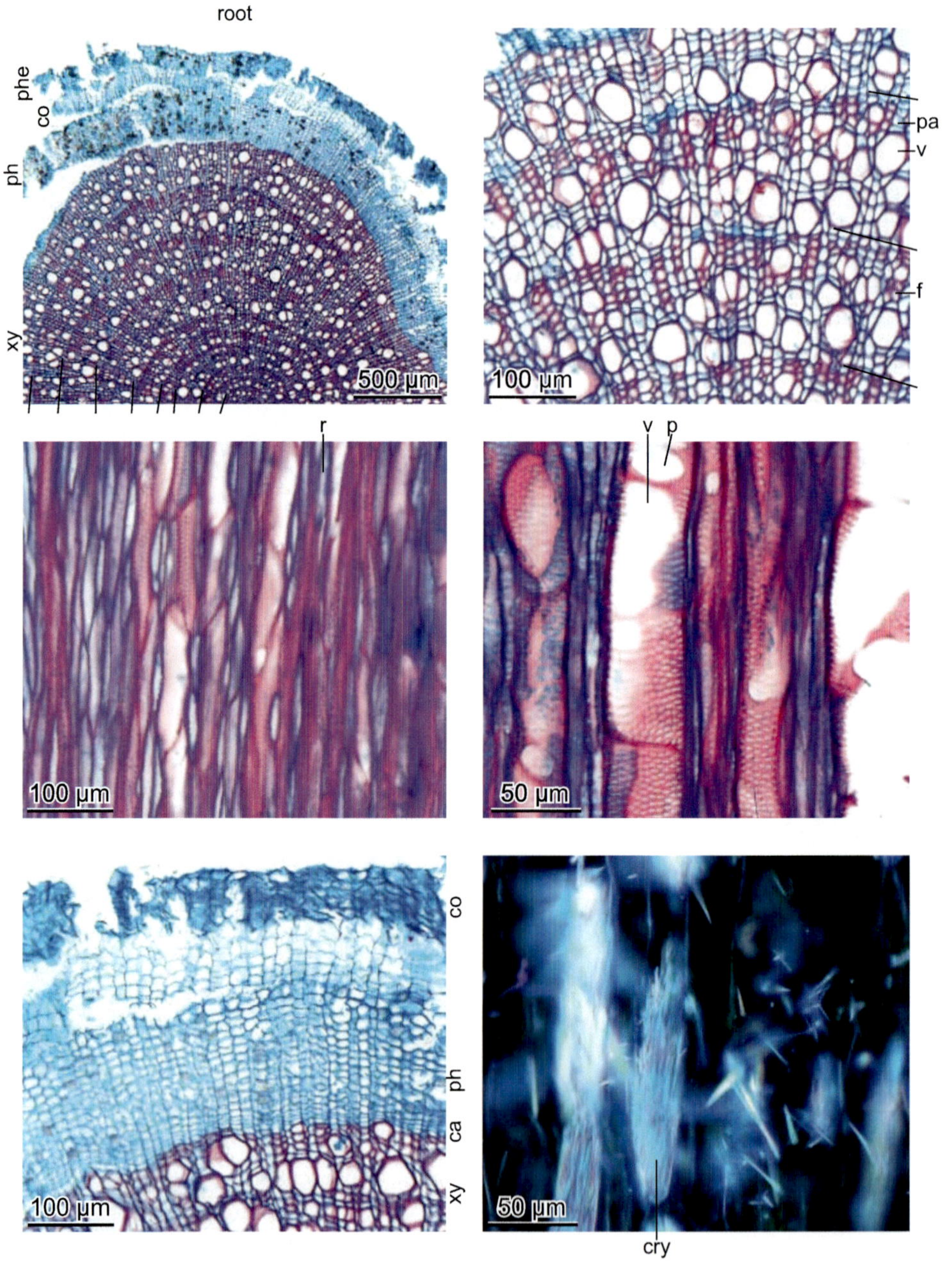

Features according to IAWA

Xylem 1 - 4 - 9 - 12 - 13 - 22 - 41 - 50.2 - 52.3 - 61 - 70 - 79 - 96 - 105 - 149

Populus pamirica Kom.

Salicaceae

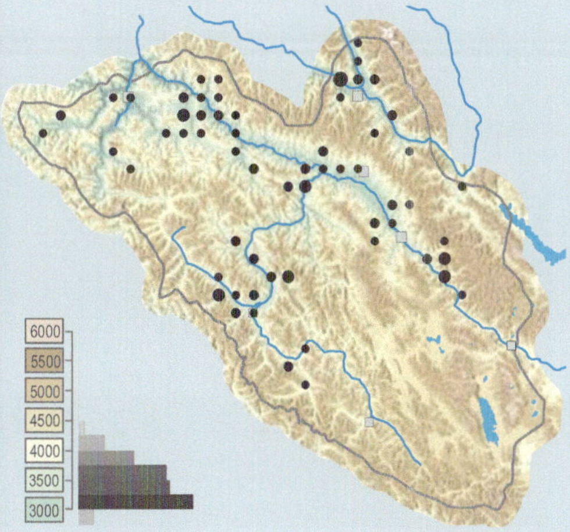

Description: Stout trees, leaves orbicular, 5–8 cm, greenish above, slightly pubescent along veins, green below, base rounded or broadly cuneate, margin coarsely sinuate dentate, ciliate, apex apiculate, petioles terete.

Distribution: Tajikistan, Afghanistan, Pakistan (Chitral, Gilgit), China (Xinjiang).

Habitat: Along rivers; in Ladakh cultivated only.

Stem diameter >30 cm. Bark/xylem proportion 1:10. Mean ring width 0.5 mm. More than 14 annual rings.

Xylem

Transverse section

Distinct rings. Semi-ring-porous. Vessels thin- to thick-walled, and solitary and in short radial multiples. Earlywood vessel diameter 60–75 µm. Latewood vessel diameter 30–50 µm. Fibers thin- to thick-walled. With gelatinous fibers (tension wood). Parenchyma paratracheal.

Tangential section

Rays uniseriate.

Radial section

Perforation plates simple. Intervessel pits round in alternating position. Vessel-ray pits with large apertures. Vessel length 300–400 µm. Rays homogenous, cells prostrate.

Bark

Phloem with tangentially alternating rows of sieve tubes (rectangular), parenchyma cells with crystal druses (round) and fibers. Groups of sclereids in older parts of the phloem. Phellem consists of tangential layers of thin-walled cork cells.

Pith

No slide available.

Analyzed: Stem of 2 individuals (dry increment cores).

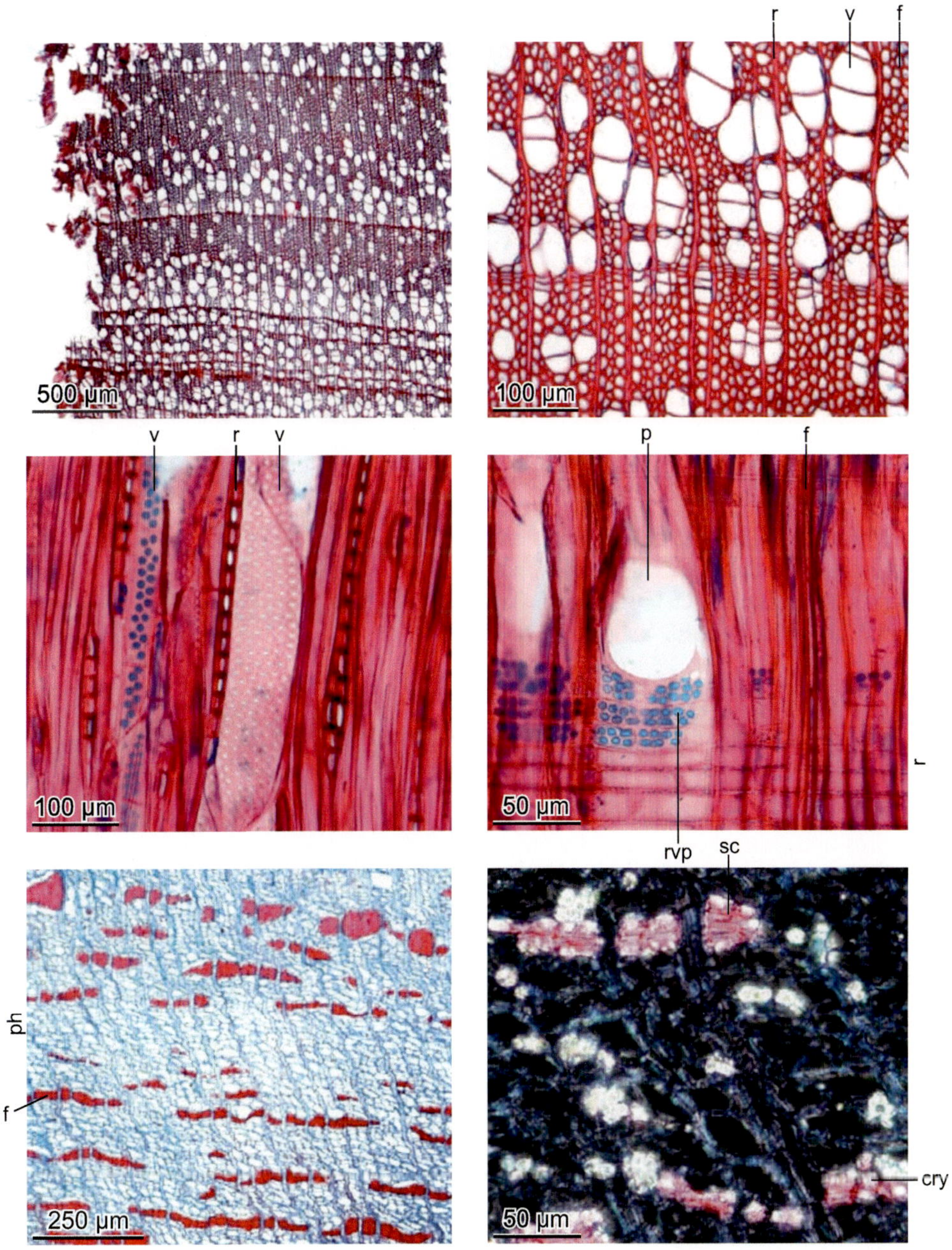

Features according to IAWA

Xylem 1 - 9 - 9.1 - 13 - 22 - 31 - 41 - 50.1 - 53.1 - 61 - 70 - 70.2 - 79 - 96 - 104

Salix caesia Vill.

Salicaceae

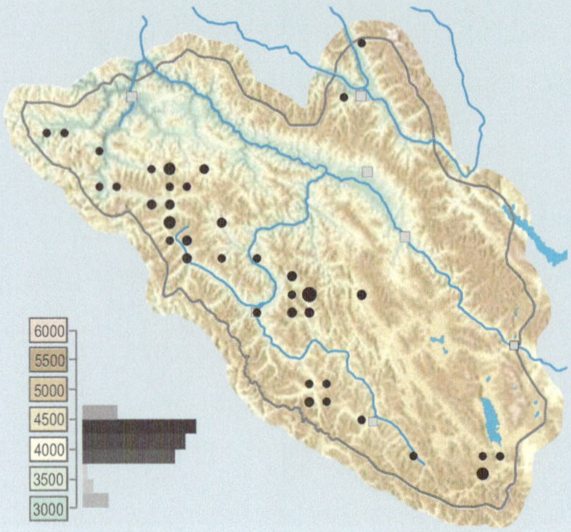

Syn.: *S. minutiflora* Turcz. ex E.L.Wolf, *S. myricifolia* Andersson

Description: Shrubs, 40–70 cm tall, leaves oblong, elliptic or lanceolate, 0.5–3 × 0.3–1 cm, grayish green, glabrous, catkins 5–20 mm long.

Distribution: Europe, Russia (Siberia), Kyrgyzstan, Tajikistan, Afghanistan, Pakistan (Chitral, Gilgit), NW India, China (Xinjiang, Xizang), Mongolia.

Habitat: Gravelly river beds.

Stem diameter 3–5 mm. Bark/xylem proportion 1:3. Mean ring width 0.1 mm. Maximal 19 annual rings.

Xylem

Transverse section

Diffuse- to semi-ring-porous. Diameter of large vessels 25–30 µm. Rings distinctly marked by a uniseriate layer of marginal parenchyma. Angular vessels solitary. Fibers thin- to thick-walled. Often with tension wood. Parenchyma paratracheal.

Tangential section

Rays uniseriate.

Radial section

Perforation plates simple. Intervessel pits round in alternating position. Fiber pits very small. Vessel length 150–200 µm. Ray cells upright. Large vessel-ray pits. Crystals absent.

Bark

Phloem consists of small groups of tangentially arranged fibers. Parenchyma cells and sieve tubes indistinguishable. Crystal druses and prismatic crystals frequent in the phloem. Cortex composed of round parenchyma cells, often separated by intercellulars. Contains a few groups of sclereids. Phellem cells rectangular.

Pith

Round.

Analyzed: Root collar of polar root of 2 individuals.

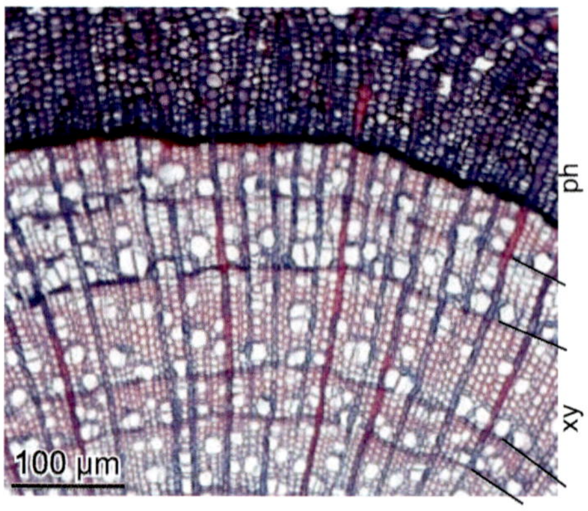

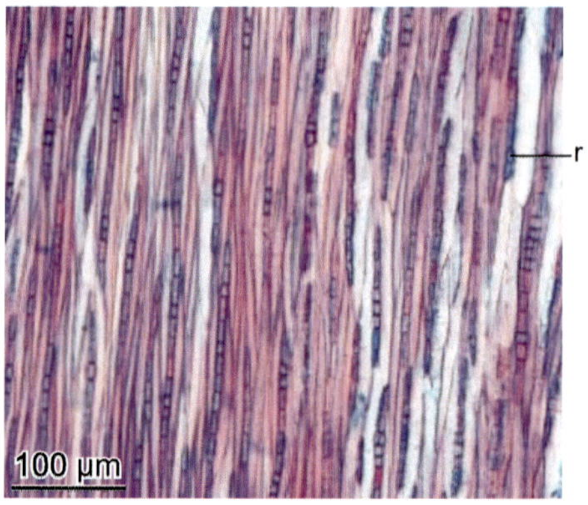

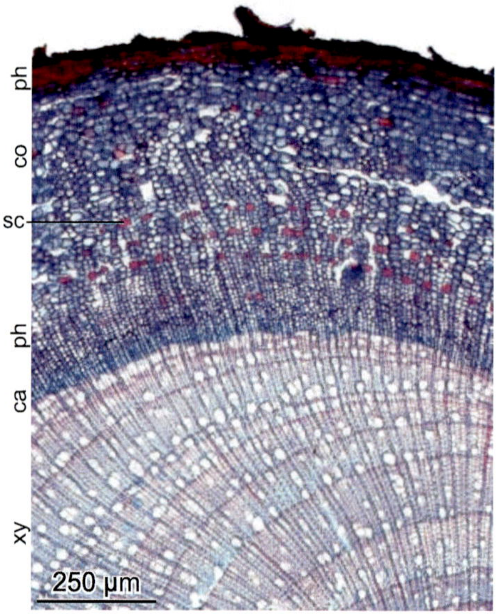

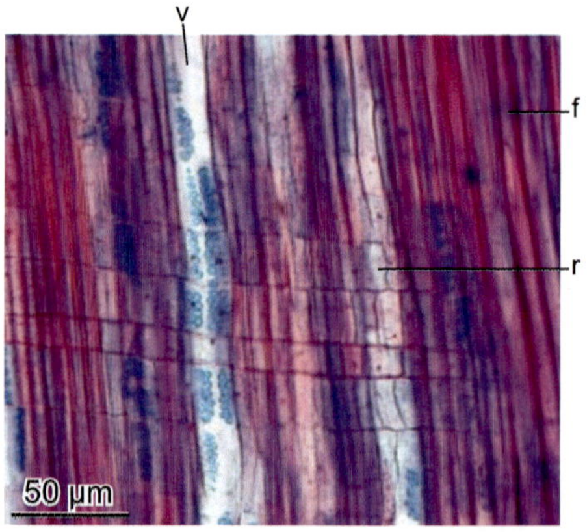

Features according to IAWA

Xylem 1 - 4 - 5 - 9 - 12 - 13 - 22 - 40.2 - 50.2 - 52.3 - 61 - 70 - 79 - 89 - 96 - 105

Salix flabellaris Andersson
Salicaceae

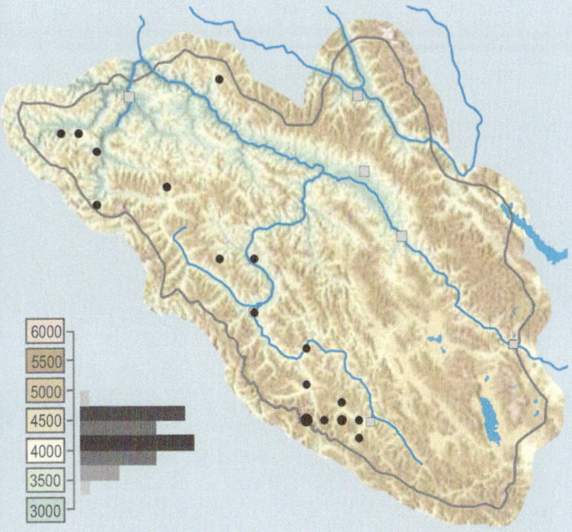

Description: Shrubs, up 20 cm tall, branches procumbent, leaves obovate to elliptic, 0.9–1.5 × 0.5–1 cm, margin glandular serrate, catkins ca. 15 mm long.

Distribution: Pakistan, N India, Nepal, China (Sichuan, Xizang, Yunnan), Bhutan.

Habitat: Alpine slopes and moraines, river beds.

Stem diameter 5–6 mm. Bark/xylem proportion 1:3. Ring width 0.2–0.3 mm. Maximal 22 annual rings

Xylem

Transverse section

Diffuse-porous. Vessels angular, generally in short radial groups. Diameter of large vessels 35–50 µm. Rings distinctly marked by a uniseriate layer of marginal parenchyma. Fibers thin- to thick-walled. Often with tension wood. Parenchyma paratracheal.

Tangential section

Rays generally uniseriate, rarely biseriate.

Radial section

Perforation plates simple. Intervessel pits large, round and angular in alternating position. Fiber pits very small. Vessel length 150–250 µm. Rays consist of a few prostrate and some upright cells. Large vessel-ray pits. Crystals absent.

Bark

Phloem consists of small, tangentially arranged fiber groups. Parenchyma cells and sieve tubes indistinguishable. Crystal druses and prismatic crystals frequent in the phloem. Slight ray dilatations. Cortex with tangentially arranged intercellulars. Phellem cells rectangular.

Pith

Round, filled with round cells.

Analyzed: Stem of 2 individuals.

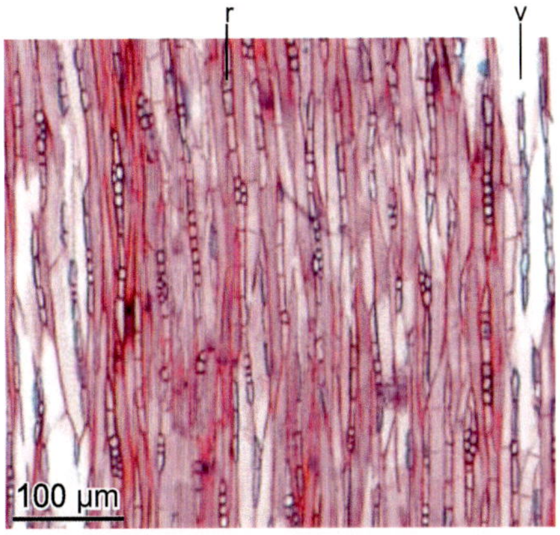

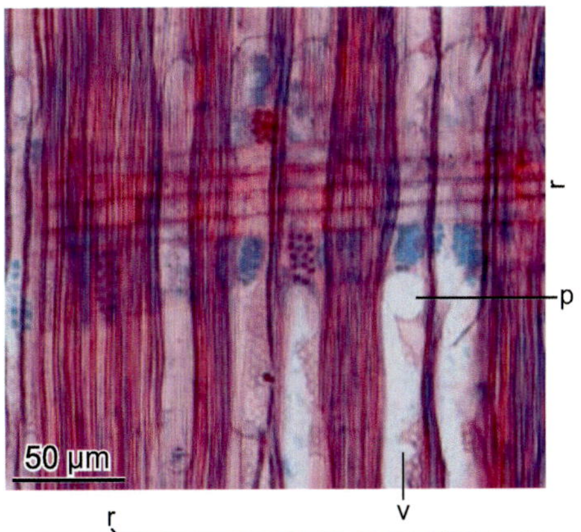

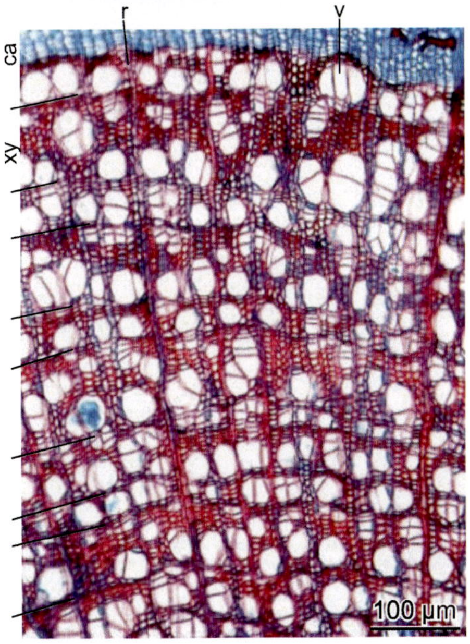

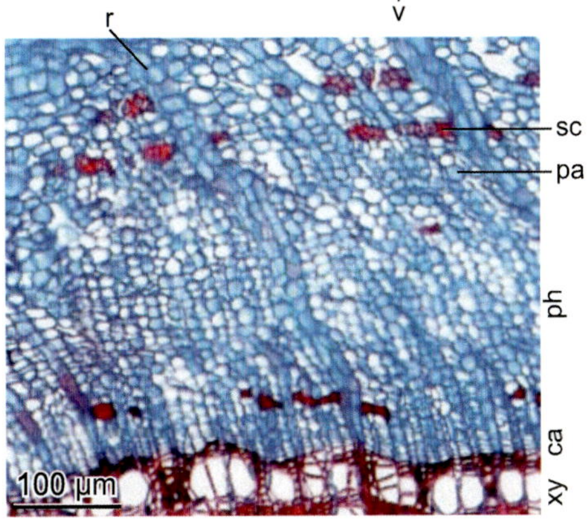

Features according to IAWA

Xylem 1 - 5 - 9.1 - 12 - 13 - 22 - 31 - 40.2 - 50.2 - 52.3 - 61 - 70 - 79 - 89 - 96 - 97 - 107

Salix karelinii Turcz. ex Stschegl.

Salicaceae

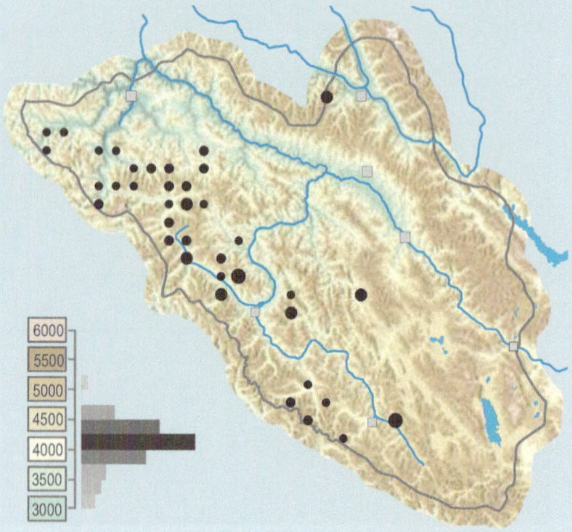

Syn.: *S. hastata* var. *himalayensis* Andersson

Description: Shrubs, 15–150 cm tall, branched from the base, leaves ovate, lanceolate to elliptic, 2.5–12 × 1–5 cm, green above, pale below, margin denticulate, catkins 20–30 mm long.

Distribution: Kyrgyzstan, Tajikistan, Afghanistan, N Pakistan, NW India, Nepal, China (W Xinjiang).

Habitat: Among stones and gravel in river beds, mountain slopes.

Stem diameter 10 mm. Bark/xylem proportion 1:4. Mean ring width 0.12 mm. Maximal 22 annual rings.

Xylem

Transverse section

Diffuse-porous. Vessels generally solitary or in short radial groups. Diameter of large vessels 35–50 µm. Rings distinctly marked by a uniseriate layer of marginal parenchyma. Fibers thin- to thick-walled. Parenchyma paratracheal.

Tangential section

Rays uniseriate.

Radial section

Perforation plates simple. Intervessel pits large, round and angular in alternating position. Fiber pits very small. Vessel length 200–250 µm. Rays consist of a few prostrate and some upright cells. Large vessel-ray pits. Crystals absent.

Bark

Phloem consists of small, tangentially arranged fiber groups. Parenchyma cells and sieve tubes indistinguishable. Crystal druses and prismatic crystals frequent in the phloem. Few rays with dilatations. Phellem cells rectangular. With lenticels.

Pith

Round, empty.

Analyzed: Stem of 1 individual.

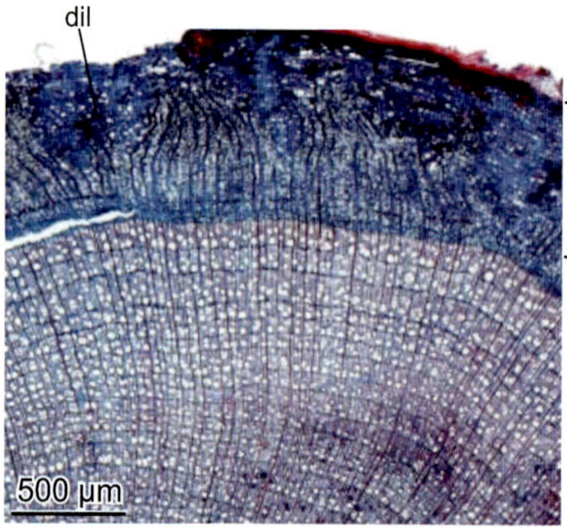

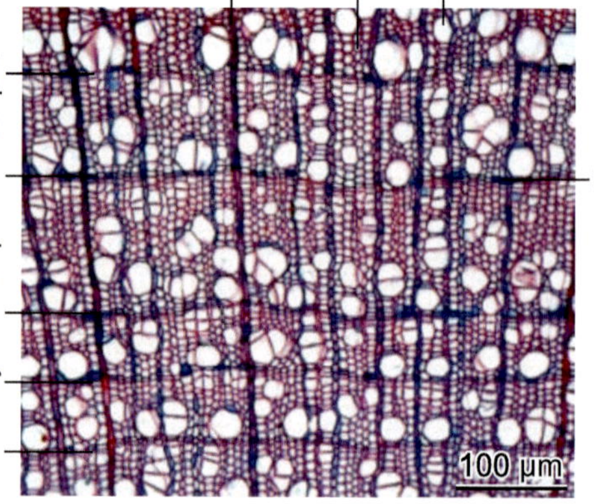

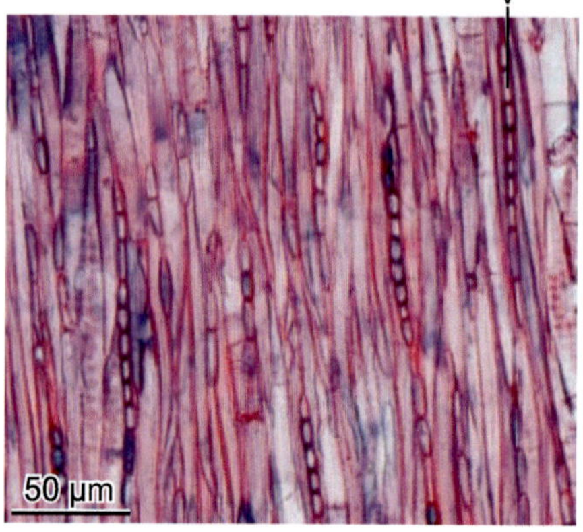

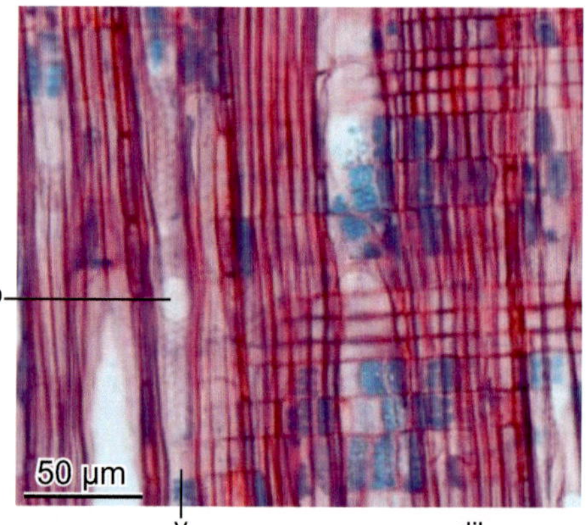

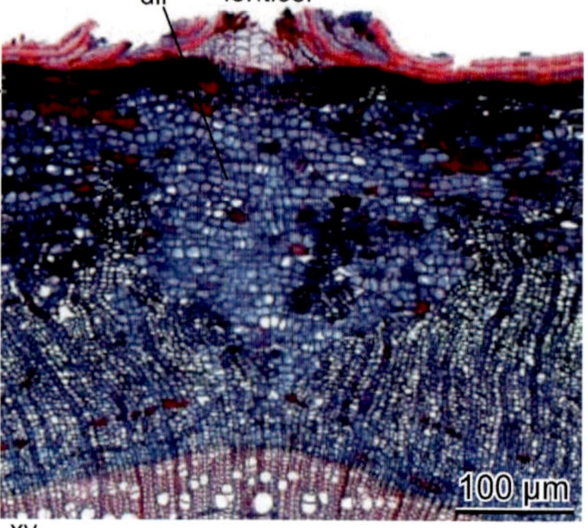

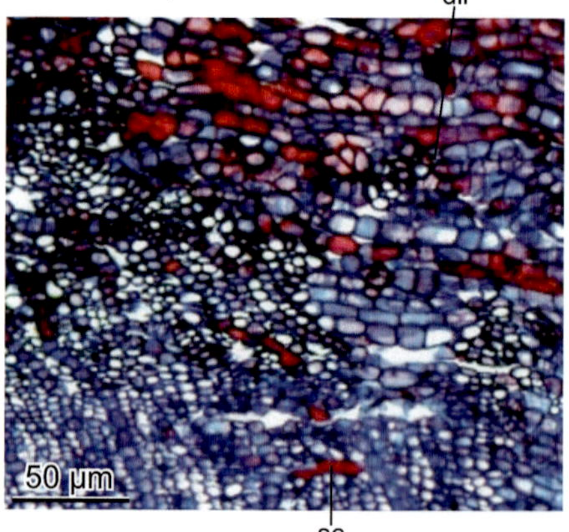

Features according to IAWA

Xylem 1 - 5 - 9.1 - 12 - 13 - 22 - 31 - 40.2 - 50.2 - 52.3 - 61 - 70 - 79 - 89 - 96 - 107

Salix sericocarpa Andersson

Salicaceae

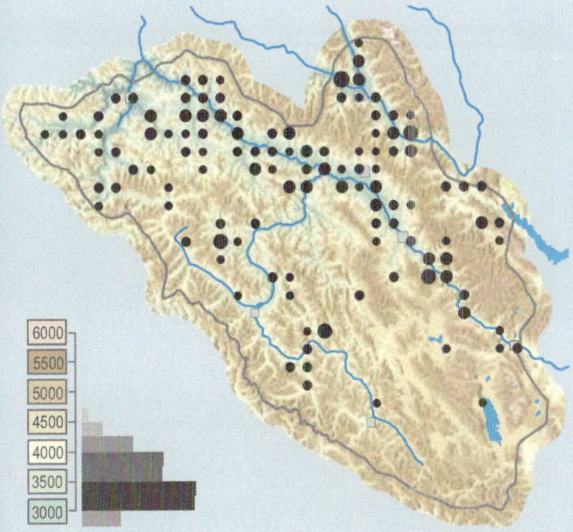

Description: Shrubs or trees, up to 10 m tall, leaves lanceolate, 5–8 × 1.2–1.8 cm, grayish white below, usually pilose along veins, green above, margin serrulate, catkins 1.5–3 cm long

Distribution: Afghanistan, Pakistan (Chitral, Gilgit, Hunza), NW India, Nepal, China (Xizang, Yunnan).

Habitat: River valleys, along irrigation canals, planted in gardens.

Stem diameter more than 2.5 cm. Bark/xylem proportion 1:10. Mean ring width 0.6 mm. More than 12 annual rings.

Xylem

Transverse section

Distinct rings. Diffuse-porous. Vessels thin- to thick-walled, primarily solitary and in short radial multiples. Earlywood vessel diameters 60–80 µm. Latewood vessel diameters 40–50 µm. Fibers thin- to thick-walled. With gelatinous fibers (tension wood). Parenchyma absent or rarely apotracheal.

Tangential section

Rays uniseriate.

Radial section

Perforation plates simple. Intervessel pits round in alternating position. Vessel-ray pits with large apertures. Vessel length 300–500 µm. Rays heterogeneous with one or two square or upright cells.

Bark

Phloem with large polyangular sieve tubes, small rectangular parenchyma cells and tangential bands of fibers. Prismatic crystals, and, rarely, crystal druses mainly around the fiber bands in older parts of the phloem. Groups of phellem consists of tangential layers of thin-walled cork cells. Often with a large rhytidome.

Pith

No slide available.

Analyzed: Stem of 2 individuals (dry increment cores).

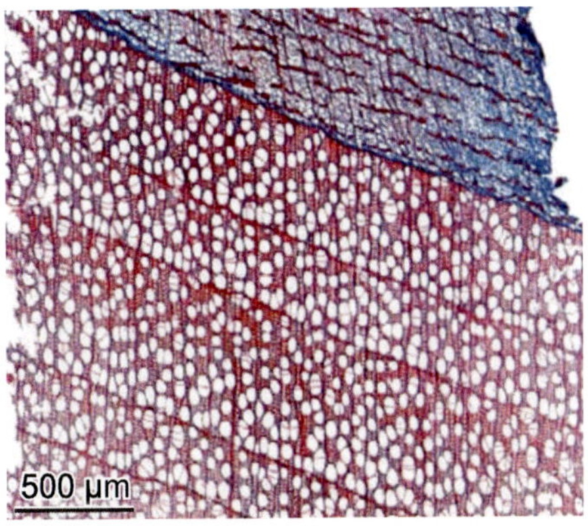

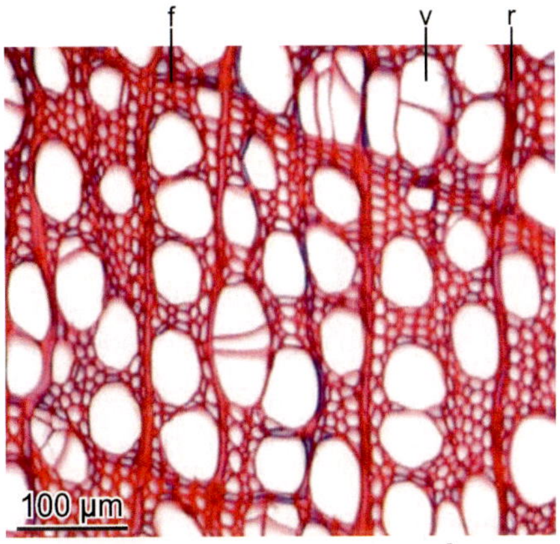

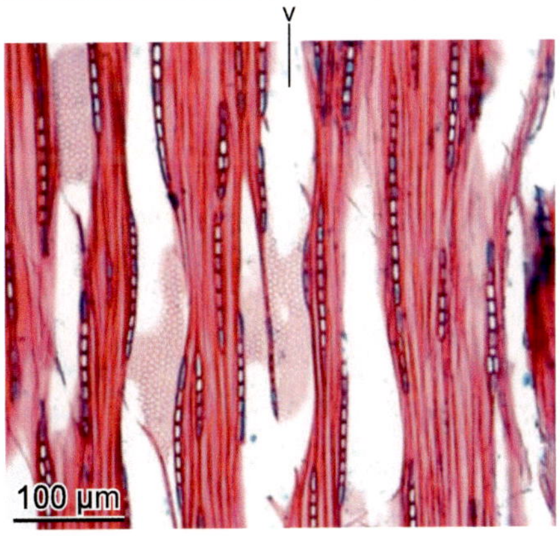

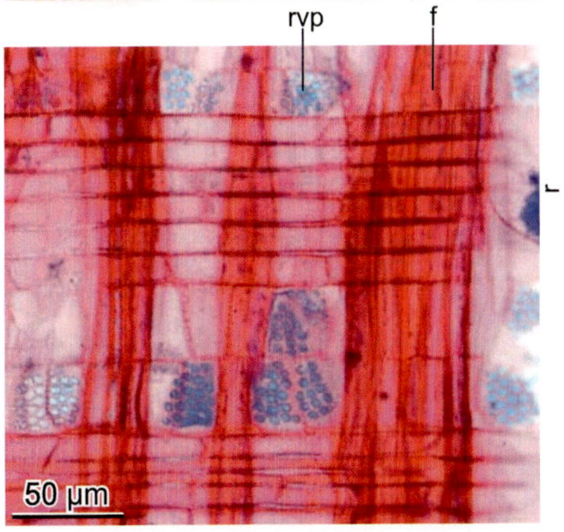

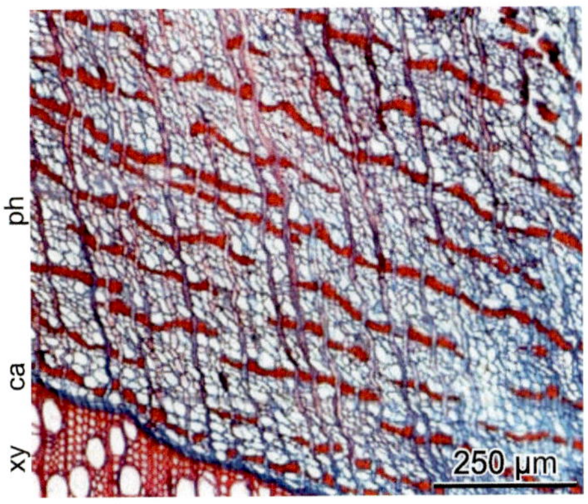

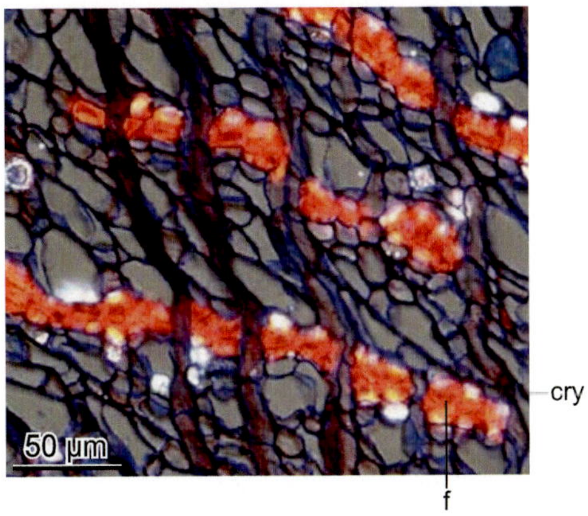

Features according to IAWA

Xylem 1 - 9 - 9.1 - 13 - 22 - 31 - 41 - 50.1 - 53.1 - 61 - 70.2 - 75 - 76 - 96 - 106

Thesium hookeri Hendrych

Santalaceae

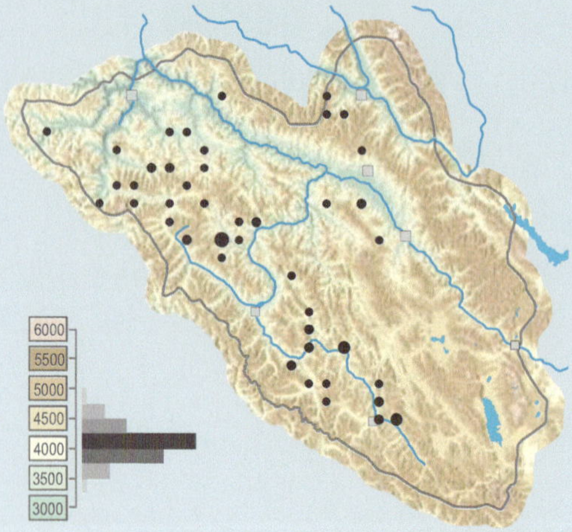

Description: Perennial root hemiparasite, 15–30 cm tall, stoloniferous, stems ascending or decumbent, leaves linear-lanceolate, flowers greenish, 3.5–5 mm long, campanulate, in many-flowered racemes.

Distribution: Pakistan, NW India.

Habitat: Stony slopes, gravel, screes.

Stem diameter 5–20 mm. Bark/xylem proportion 1:2. Ring width 0.2–0.7 mm. Maximal 14 rings.

Xylem

Transverse section

Semi-ring-porous. Rings distinctly marked by differences between earlywood (40–50 µm) and latewood vessel diameters (10–20 µm). Vessels solitary and in groups. Fibers absent. Parenchyma pervasive.

Tangential section

Rays >10-seriate, often with sheath cells.

Radial section

Perforation plates simple. Intervessel pits generally scalariform, rarely round in alternating position. Vessel length 50–80 µm. Ray cells upright. Crystals absent.

Bark

Phloem consists of parenchyma cells, small groups of sieve tubes, fiber groups and a few sclereids or fibers. Fine-grained crystal sand in parenchyma cells of phloem and cortex. All rays dilated. Cortex is composed of small and large, round to oval parenchyma cells, often separated by intercellulars. Phellem cells irregularly formed.

Pith

No slide available.

Analyzed: Root collar of polar root of 5 individuals.

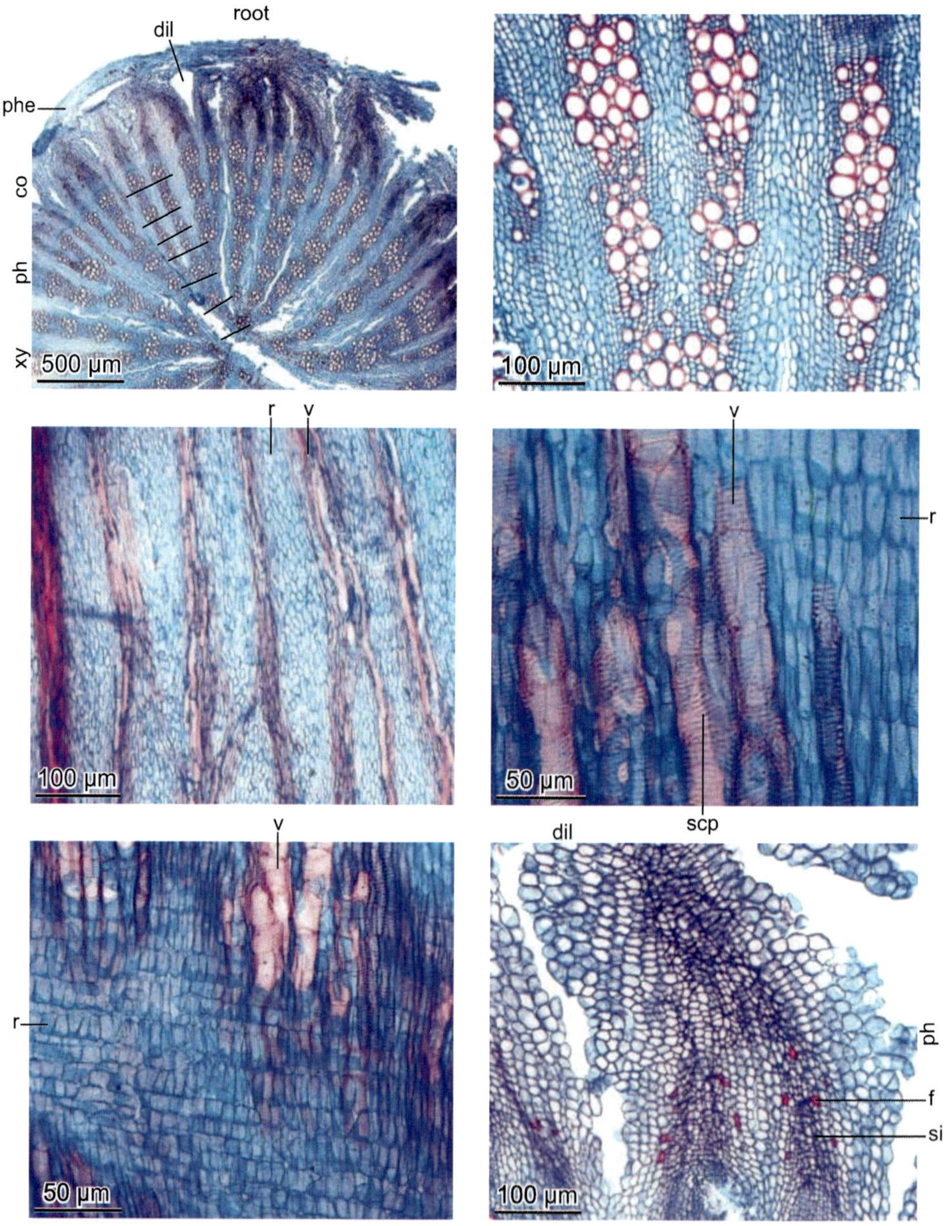

Features according to IAWA

Xylem 1 - 4 - 5 - 9 - 11 - 13 - 20 - 40.2 - 50.2 - 52.2 - 60.1 - 79.1 - 99 - 105 - 110

Bergenia ciliata (Haw.) Sternb.
Saxifragaceae

Syn.: *Megasea ciliata* Haw., ?*Bergenia pacumbis* (Buch.-Ham. ex D.Don) C.Y.Wu & J.T.Pan)

Description: Perennial herbs, 15–30 cm tall, leaves orbicular or broadly obovate, 5–12 × 3–10 cm, glabrous or hirsute, margin entire or obscurely crenate, ciliate, flowers white or pink, ca. 8 mm across, in dense cymes.

Distribution: Afghanistan, Pakistan, N India, Nepal, W China, Bhutan.

Habitat: Rock crevices, forest margins.

Stem diameter 5–6 mm. Bark/xylem proportion 1:1. Mean ring width 0.5 mm. Maximal 3 rings.

Xylem

Transverse section

Diffuse-porous. Rings marked by slight differences between earlywood (30–40 µm) and latewood vessel diameters (20 µm) and often by the occurrence of stained substances in vessels. Angular vessels primarily solitary. Vascular bundles remain. Fibers absent. Parenchyma pervasive.

Tangential section

Rays >10-seriate, often with sheath cells.

Radial section

Perforation plates simple. Intervessel pits scalariform. Vessel length 150–180 µm. Ray cells upright. Crystal druses frequent in rays.

Bark

Phloem consists of parenchyma cells and small groups of sieve tubes. Cortex is composed of round to oval parenchyma cells, often separated by intercellulars. Phellem cells irregularly formed. Crystal druses absent in phloem but frequent in cortex.

Pith

In the rhizome round and very large.

Analyzed: Root collar of polar root of 1 individual.

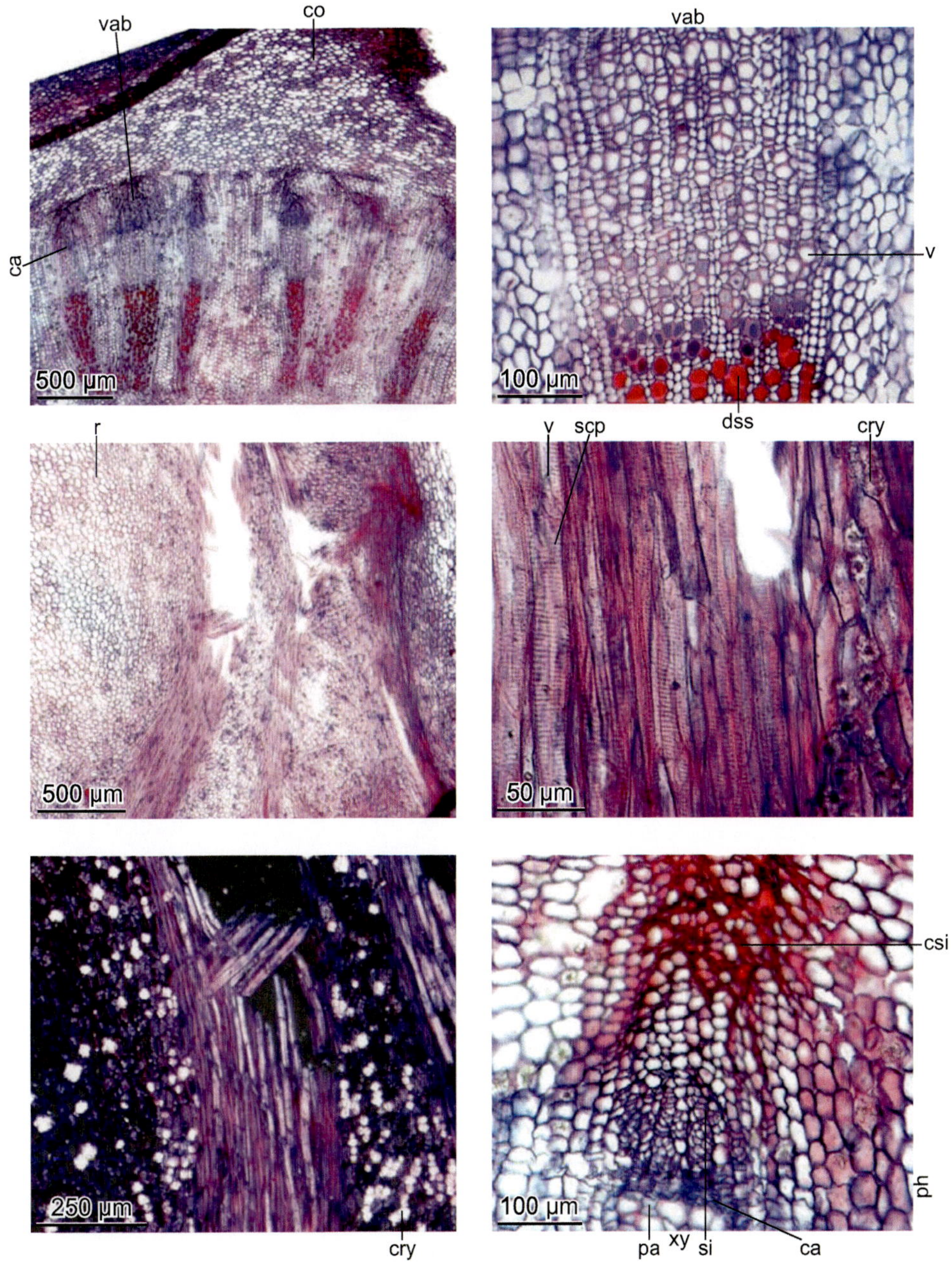

Features according to IAWA

Xylem 1 - 4 - 5 - 9 - 11 - 13 - 20 - 40.2 - 50.2 - 52.3 - 60.1 - 79.1 - 99 - 99.1 - 105 - 110 - 144

Bergenia stracheyi (Hook.f. & Thomson) Engl.

Saxifragaceae

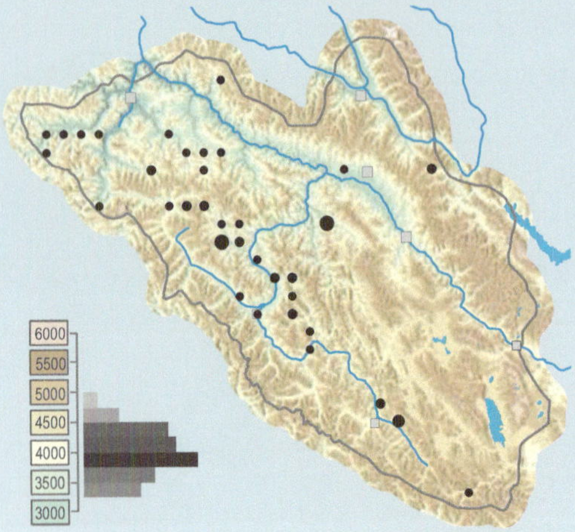

Syn.: *Saxifraga stracheyi* Hook.f. & Thomson

Description: Perennial herbs, 15–30 cm tall, leaves obovate, 6–15 × 4–8 cm, margin dentate and ciliate, flowers white-pink or yellowish, ca. 8 mm across, in dense cymes.

Distribution: Tajikistan, Afghanistan, Pakistan (Kashmir, Baltistan, Gilgit, Chitral, Upper Kaghan), India (Jammu and Kashmir), China (Xizang).

Habitat: Moist rocky slopes.

Stem diameter 5–6 mm. Bark/xylem proportion 1:2. Mean ring width 1 mm. Maximal 6 indistinct rings.

Xylem

Transverse section

Diffuse-porous. Rings poorly marked by slight differences between earlywood (30–40 μm) and latewood vessel diameters (15–20 μm). Angular vessels primarily solitary. Vascular bundles remain. Fibers absent. Parenchyma pervasive.

Tangential section

Rays >10-seriate, often with sheath cells.

Radial section

Perforation plates simple. Intervessel pits scalariform. Vessel length 150–180 μm. Ray cells upright. Crystal druses frequent in rays.

Bark

Phloem consists of parenchyma cells and small groups of sieve tubes. Cortex is composed of round to oval parenchyma cells, often separated by intercellulars. Phellem cells irregularly formed. Crystal druses absent in the phloem but frequent in the cortex.

Pith

In the rhizome round and very large.

Analyzed: Root collar of polar root of 2 individuals.

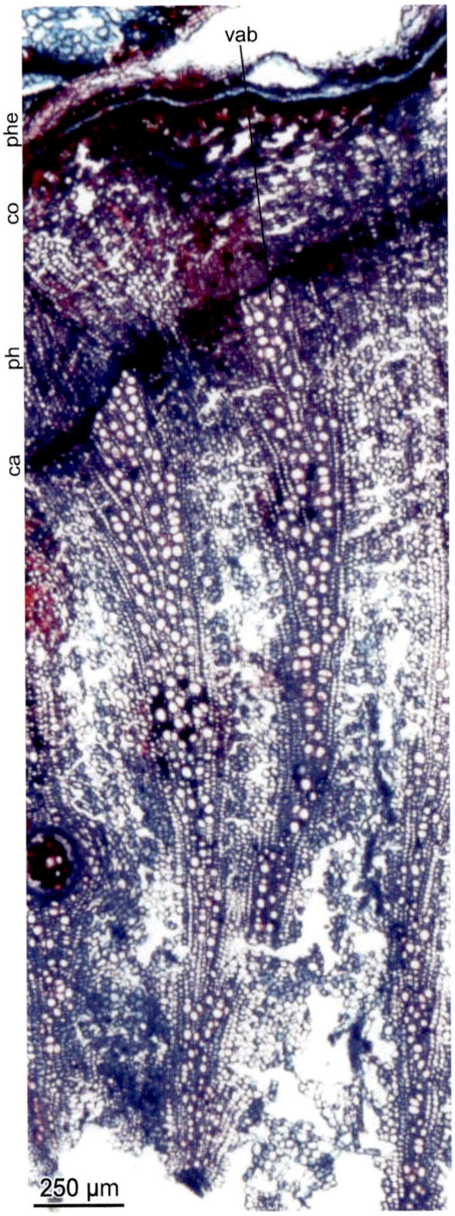

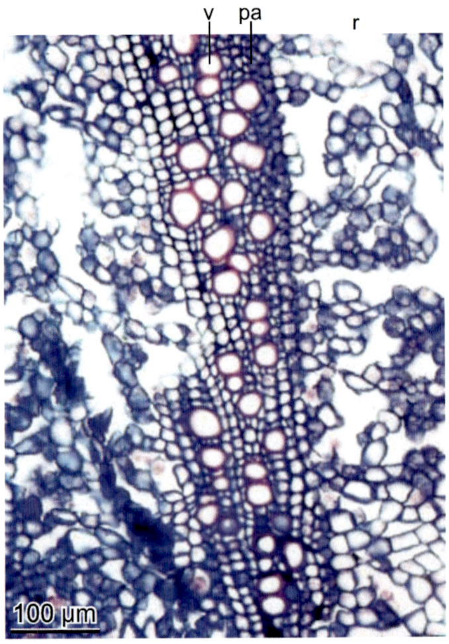

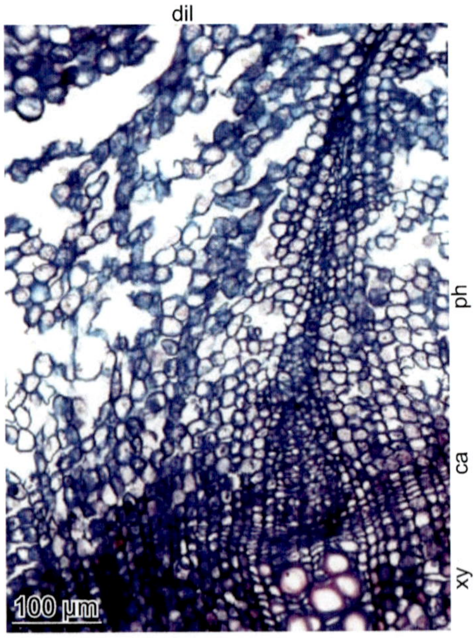

Features according to IAWA

Xylem 1 - 4 - 5 - 9 - 11 - 13 - 20 - 40.2 - 50.2 - 52.3 - 60.1 - 79.1 - 99 - 99.1 - 105 - 110 - 144

Saxifraga cernua L.

Saxifragaceae

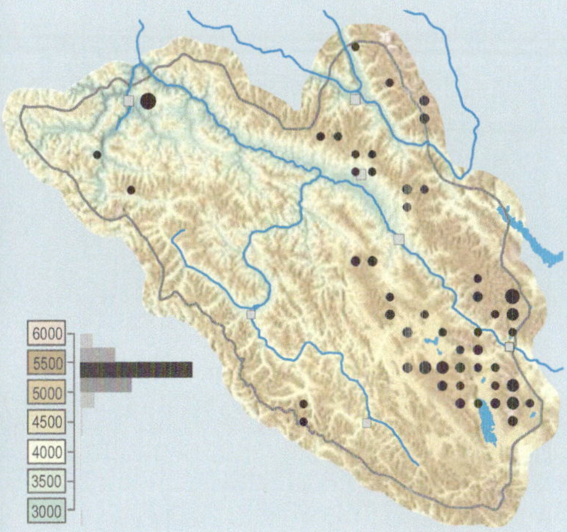

Description: Perennial herbs, 3–20 cm tall, glandular pubescent, with red or brown bulbils in axils of cauline leaves, leaves reniform, margin 5–7-lobed, flowers white or yellowish, 5–10 mm across, solitary or in 2–5-flowered cymes.

Distribution: N America, N and C Europe, N and C Asia.

Habitat: Solifluction soils, brook banks, mesic to wet rock crevices.

Stem diameter 5–6 mm. Bark/xylem proportion 1:0.5. Rings absent.

Xylem

Transverse section

Rings absent. Vessel diameter 10–20 µm. Vessels primarily solitary. Fibers absent. Parenchyma pervasive. Xylem structure indistinct due to intensive spiral grain.

Tangential section

Rays absent.

Radial section

Perforation plates simple. Intervessel pits scalariform. Vessel length <50 µm. Crystals absent.

Bark

Phloem uniform. Cortex is composed of round parenchyma cells. Phellem consists of a few rows of tangentially enlarged, thick-walled cells. Crystals absent.

Pith

In the rhizome round and very large.

Analyzed: Root collar of 2 individuals.

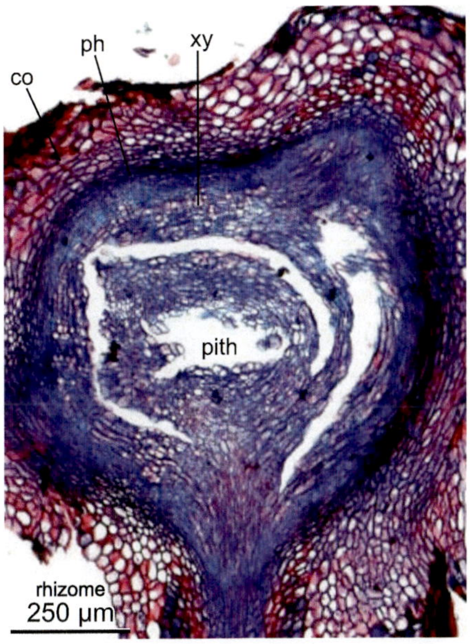

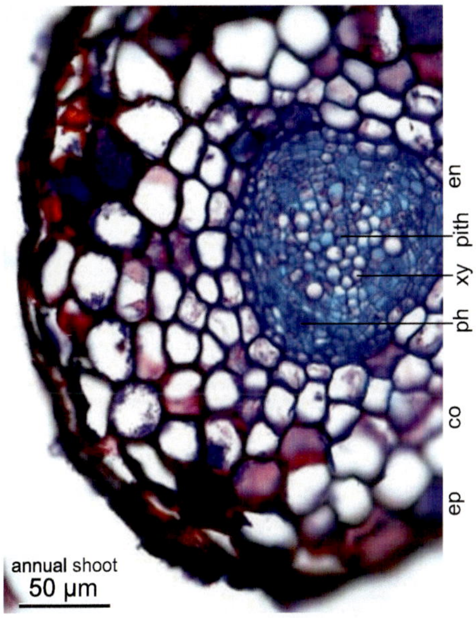

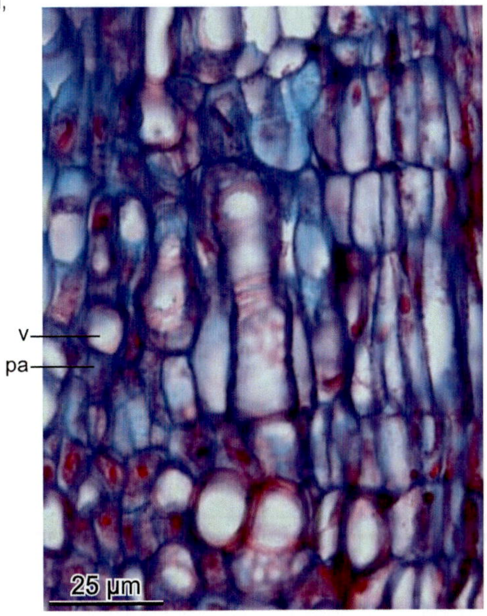

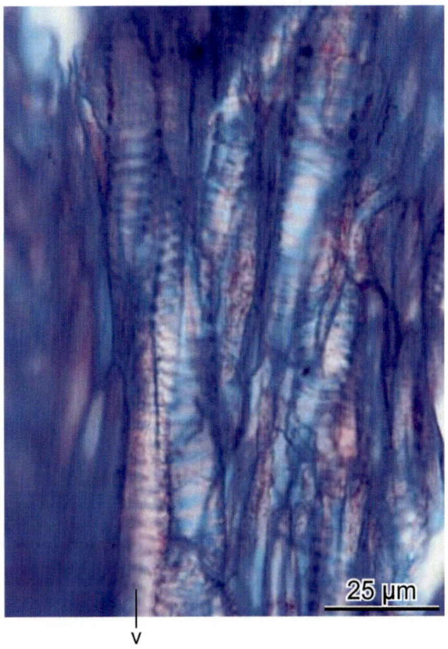

Features according to IAWA

Xylem 2.1 - 9 - 13 - 22 - 40.1 - 50.1 - 52.1 - 60.1 - 79.1 - 117

Saxifraga flagellaris Willd. ex Sternb.
Saxifragaceae

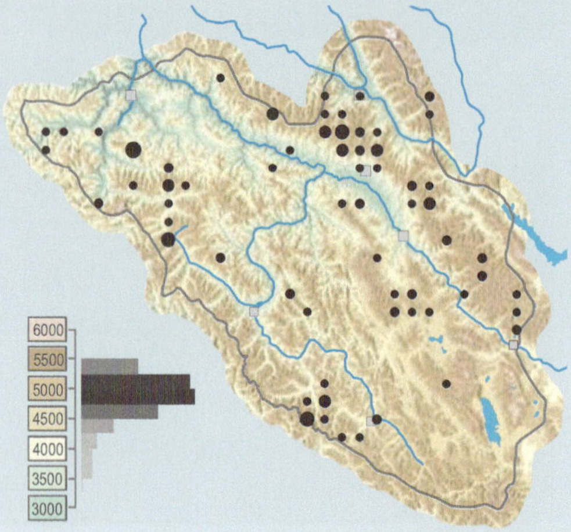

Description: Perennial herbs, 4–12 cm tall, glandular hairy, stems with stolons, leaves crowded basally, obovate to ovate-lanceolate, flowers yellow, 1–1.5 cm across, in 1–5-flowered cymes.

Distribution: N America, N Europe (Svalbard), Caucasus, Russia (Siberia, Far East), N Pakistan, NW India.

Habitat: Wet rock crevices.

Stem diameter 0.5–0.66 mm. Bark/xylem proportion 1:3. Ring width 0.10–0.15 mm. Maximal 2 annual rings.

Xylem

Transverse section
Semi-ring-porous. Vessels solitary and in small groups. Diameter of large vessels 15–20 µm. Rings distinctly marked. Fibers and rays absent. Parenchyma pervasive.

Tangential section
Rays absent.

Radial section
No slide available, features observed on thick transversal sections. Perforation plates simple. Intervessel pits scalariform. Vessel length not observed. Crystals absent.

Bark

Phloem uniform. Phellem absent. Epidermis cells thin-walled, round.

Pith

Pith round to polygonal

Analyzed: Stem of 3 individuals.

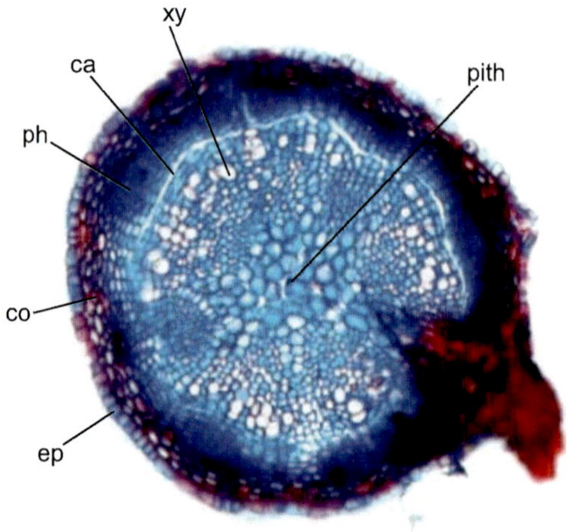

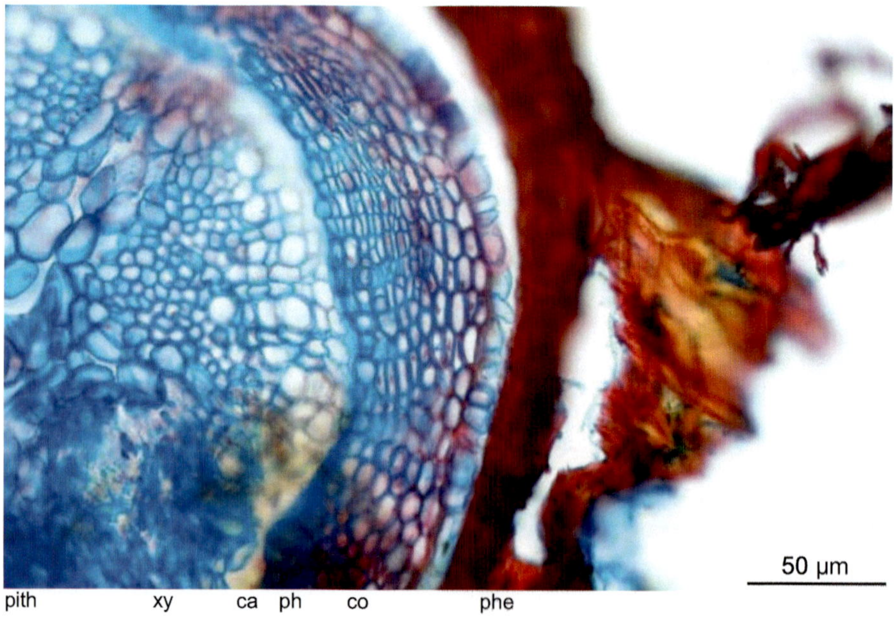

Features according to IAWA
Xylem 1 - 4 - 9 - 13 - 20 - 40.1 - 50.2 - 60.1 - 79.1 - 117

Saxifraga jacquemontiana Decne.

Saxifragaceae

Description: Perennial herbs, 2–3 cm tall, glandular hairy, forming dense cushions, leaves mostly in dense basal rosettes, spatulate to broadly oblong, flowering stems very short, flowers yellow, ca. 5 mm across, solitary.

Distribution: N Pakistan, N India, Nepal, China (SE Xizang), Bhutan.

Habitat: Rock crevices.

Stem diameter 0.6–1 mm. Bark/xylem proportion 1:2. Rings indistinct or absent.

Xylem

Transverse section

Vessel diameter 10–15 μm. Vessels solitary or in groups. Fibers absent. Parenchyma pervasive. Vessels hardly stained.

Tangential section

Rays absent.

Radial section

Perforation plates simple. Intervessel pits scalariform. Vessel length <50 μm. Crystals absent.

Bark

Phloem uniform. Cortex composed of round parenchyma cells. Epidermis cells large.

Pith

Round to polygonal.

Analyzed: Root collar of polar root of 3 individuals.

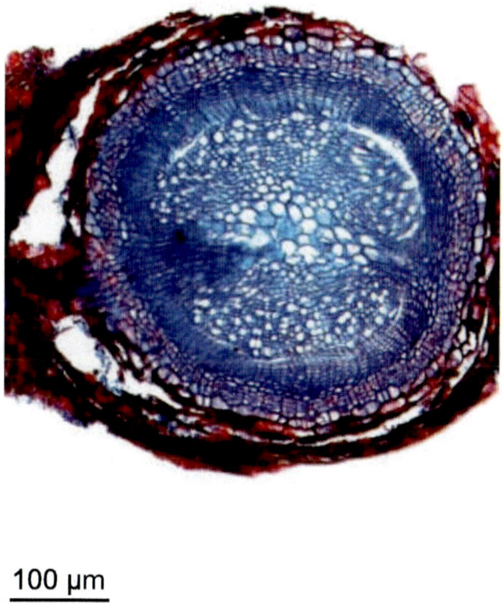

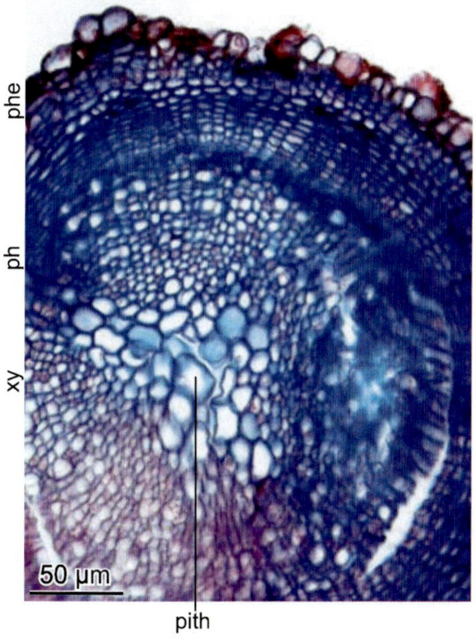

Features according to IAWA

Xylem 2.1 - 9 - 13 - 22 - 39.2 - 40.1 - 50.2 - 52.1 - 60.1 - 79.1 - 117

Saxifraga prorepens Fisch. ex Sternb.

Saxifragaceae

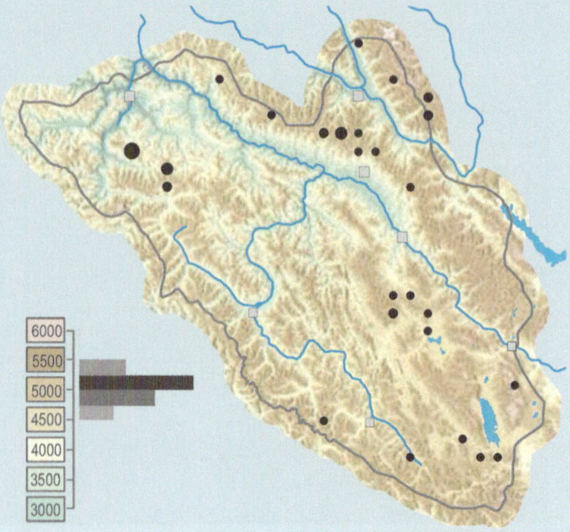

Syn.: *S. hirculus* var. *alpina* Engl.

Description: Perennial herbs, 5–20 cm tall, leaves elliptic, lanceolate or oblong, margin villous, flowers yellow, ca. 1 cm across, solitary or by 2–4 in a cyme.

Distribution: Kazakhstan, Tajikistan, N Pakistan, NW India, China (Xinjiang, Xizang).

Habitat: River banks, wet rock crevices, snow fields.

Stem diameter 5–6 mm. Bark/xylem proportion 1:0.5. Rings absent.

Xylem

Transverse section

Rings absent. Vessel diameter 10–20 µm. Vessels solitary or in groups. Fibers absent. Parenchyma pervasive. Vascular bundles remain.

Tangential section

Rays 5–10-seriate.

Radial section

Perforation plates simple. Intervessel pits scalariform. Vessel length <50 µm. Ray cells upright. Crystals absent.

Bark

Phloem consists of small groups of sieve tubes and parenchyma cells. Cortex is composed of round parenchyma cells. Phellem consists of a thick layer of irregularly formed, brown cells.

Pith

In the rhizome round and very large.

Analyzed: Root collar of polar root of 3 individuals.

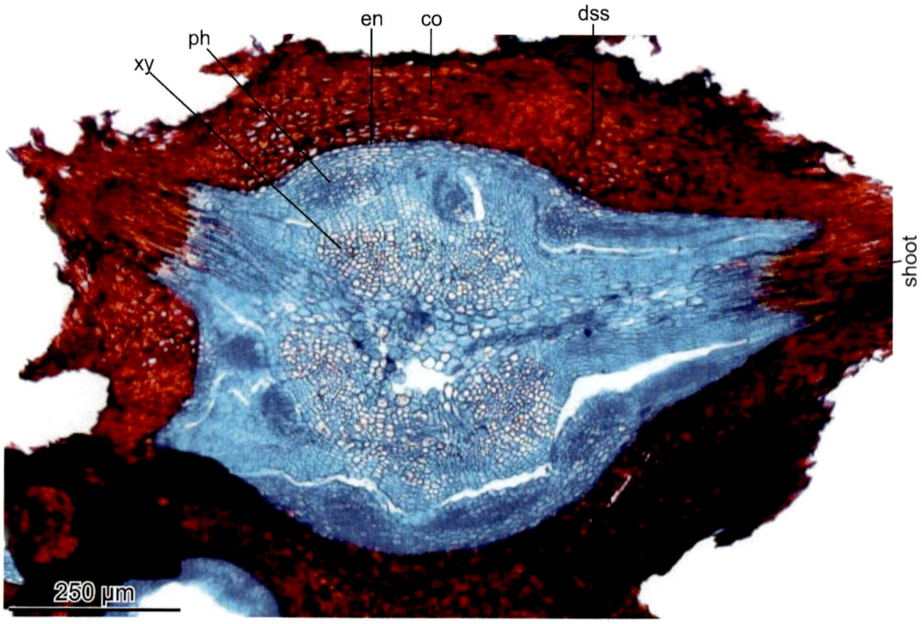

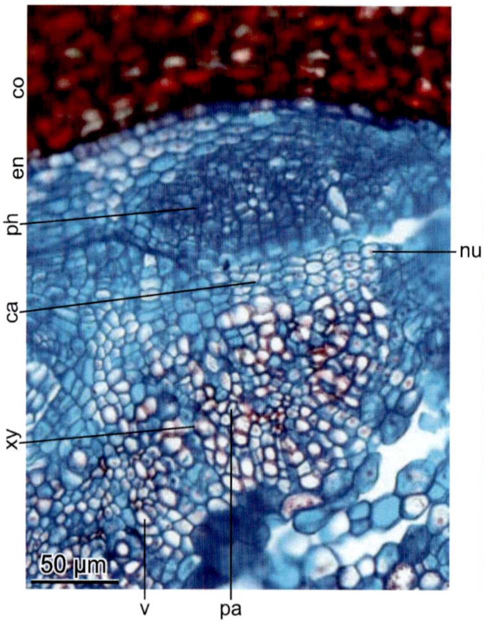

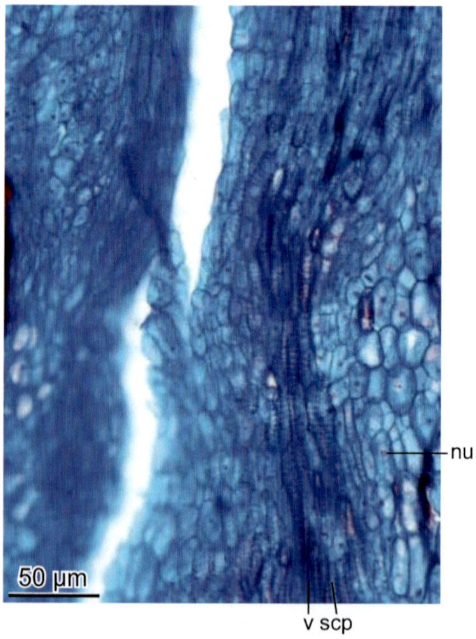

Features according to IAWA

Xylem 2.1 - 9 - 11 - 13 - 22 - 40.1 - 50.1 - 52.1 - 60.1 - 79.1 - 98 - 99.1 - 105

Saxifraga sibirica L.
Saxifragaceae

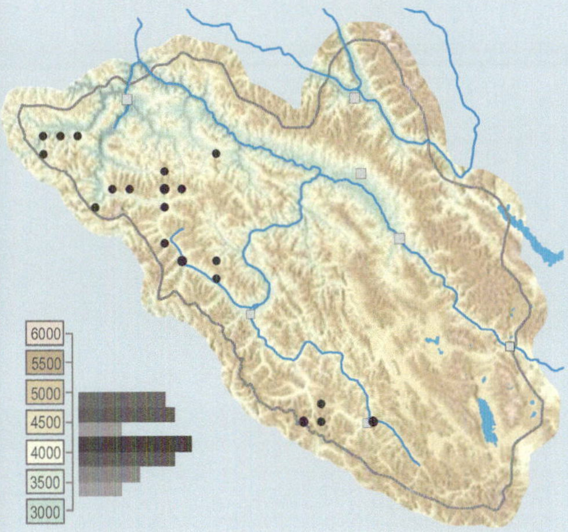

Description: Perennial herbs, 5–25 cm tall, densely glandular pilose, leaves reniform, margin 7–9-lobed, flowers white, 8–12 mm across, in 1–6-flowered cymes.

Distribution: E Europe, Russia, Pakistan, India, Nepal, China (Gansu, Hebei, Heilongjiang, Hubei, Hunan, Nei Mongol, Shaanxi, Shandong, Shanxi, Sichuan, Xinjiang, Xizang, Yunnan), Mongolia.

Habitat: Alpine meadows, moist rocky slopes.

Stem diameter 0.3 mm. Bark/xylem proportion 1:0.5. Ring width 0.1 mm. Maximal 1 annual ring

Xylem

Transverse section
Only one ring. Vessels solitary and in small groups. Diameter of large vessels 10–15 µm. Fibers and rays absent. Parenchyma pervasive.

Tangential section
Rays absent.

Radial section
No slide available, features observed on thick transversal sections. Simple perforation plates. Intervessel pits scalariform. Vessel length not observed. Crystals absent.

Bark

Cortex and phloem separated by an endodermis. Phloem contains small groups of sieve tubes. Cortex cells large, thin-walled. Phellem absent.

Pith

Pith round to oval.

Analyzed: Stem of 1 individual.

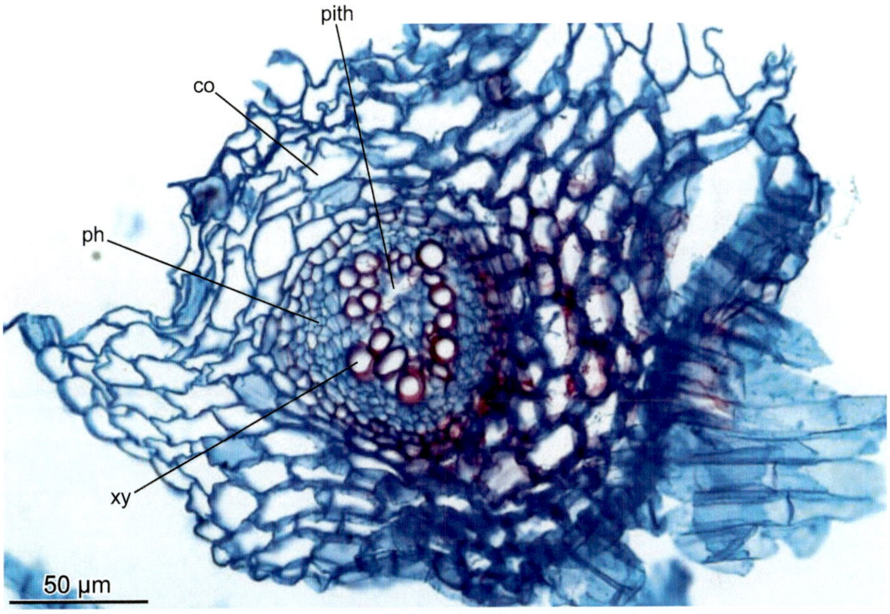

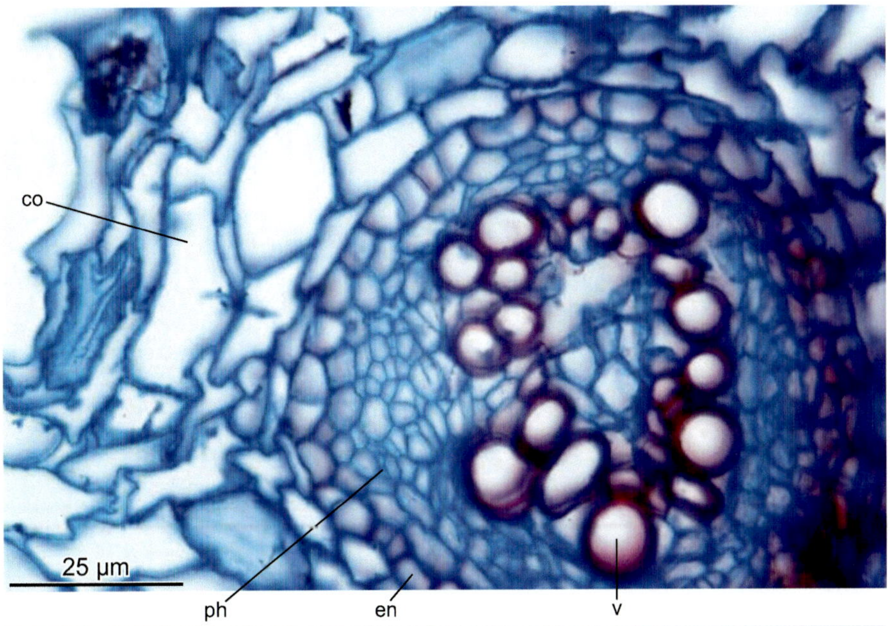

Features according to IAWA
Xylem 2.1 - 9 - 13 - 20 - 40.1 - 50.2 - 60.1 - 79.1 - 117

Pedicularis bicornuta Klotzsch

Scrophulariaceae

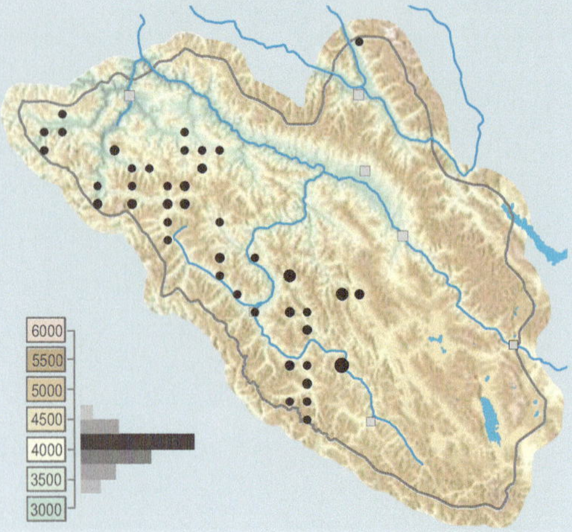

Description: Perennial herbs, 20–60 cm tall, leaves pinnately lobed, flowers yellow, ca. 2 cm long, with a long hairy tube, in dense spikes.

Distribution: Afghanistan, Pakistan, N India, China.

Habitat: Mesic and often shaded river banks with loamy soils, alpine meadows.

Stem diameter 5–8 mm. Bark/xylem proportion 1:0.5. Ring width 0.6-1 mm. Maximal 3 rings.

Xylem

Transverse section

Semi-ring-porous. Rings marked by differences between earlywood (40–60 µm) and latewood vessel diameters (15–20 µm) and the absence of fibers in the first ring. Vessels solitary, in short radial multiples and in small groups. Vascular bundles remain. Fibers absent in the central part and present in various amounts at the periphery. Parenchyma paratracheal and pervasive.

Tangential section

Rays 5–10-seriate.

Radial section

Perforation plates simple. Intervessel pits scalariform. Vessel length 50–100 µm. Ray cells upright. Crystal absent.

Bark

Phloem consists of parenchyma cells and small groups of sieve tubes. Rays slightly dilate. Cortex is composed of oval, often compressed parenchyma cells, often separated by intercellulars. Older parts of phloem undulated. Phellem cells irregularly formed. Crystals absent.

Pith

No slide available.

Analyzed: Root collar of polar root of 4 individuals.

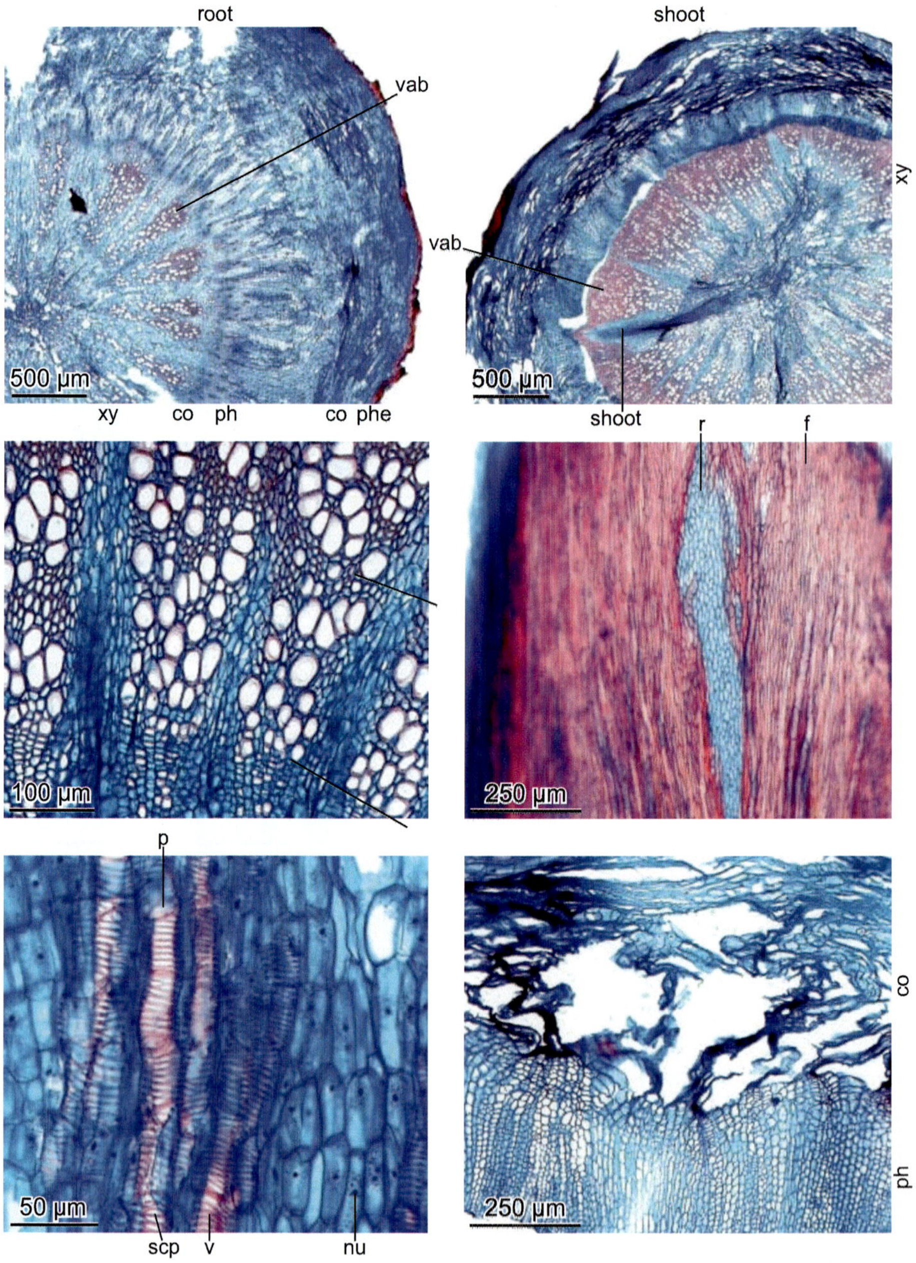

Features according to IAWA

Xylem 1 - 4 - 9 - 9.1 - 11 - 13 - 22 - 40.2 - 50.2 - 52.2 - 60.1 - 61 - 70 - 79.1 - 98 - 99.1 - 100.2 - 105

Pedicularis cheilanthifolia Schrenk

Scrophulariaceae

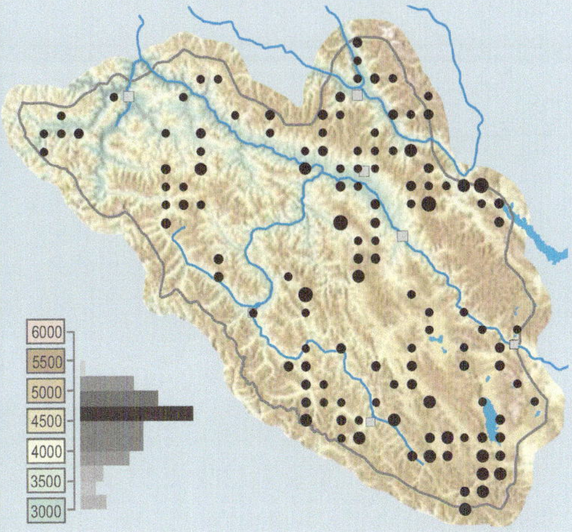

Description: Perennial herbs, 6–30 cm tall, basal leaves with petioles 3–4 cm long, stem leaves verticillate by 4, pinnatisect, segments pinnatifid or dentate, flowers white or purple-red, ca. 1 cm long, in spikes, sometimes basally interrupted.

Distribution: Kazakhstan, Kyrgyzstan, Tajikistan, Afghanistan, India, China (Gansu, Qinghai, Xinjiang, Xizang), Mongolia.

Habitat: Turf wetlands and marshes.

Stem diameter 5–9 mm. Bark/xylem proportion 1:2. Ring width 0.3–1 mm. Maximal 5 rings.

Xylem

Transverse section

Semi-ring-porous. Rings marked by differences between earlywood (40–50 μm) and latewood vessel diameters (10–20 μm) and the absence of fibers in the first ring. Vessels solitary, in short radial multiples and in small groups. Vascular bundles remain. Fibers absent. Parenchyma pervasive.

Tangential section

Rays 5–>10-seriate, often confluent with the axial tissue.

Radial section

Perforation plates simple. Intervessel pits scalariform. Vessel length 100–150 μm. Ray cells upright. Crystals absent.

Bark

Phloem consists of parenchyma cells and small groups of sieve tubes. Rays slightly dilate. Cortex composed of oval, often compressed parenchyma cells, often separated by intercellulars. Phellem cells irregularly formed. Crystals absent.

Pith

No slide available.

Analyzed: Root collar of polar root of 10 individuals.

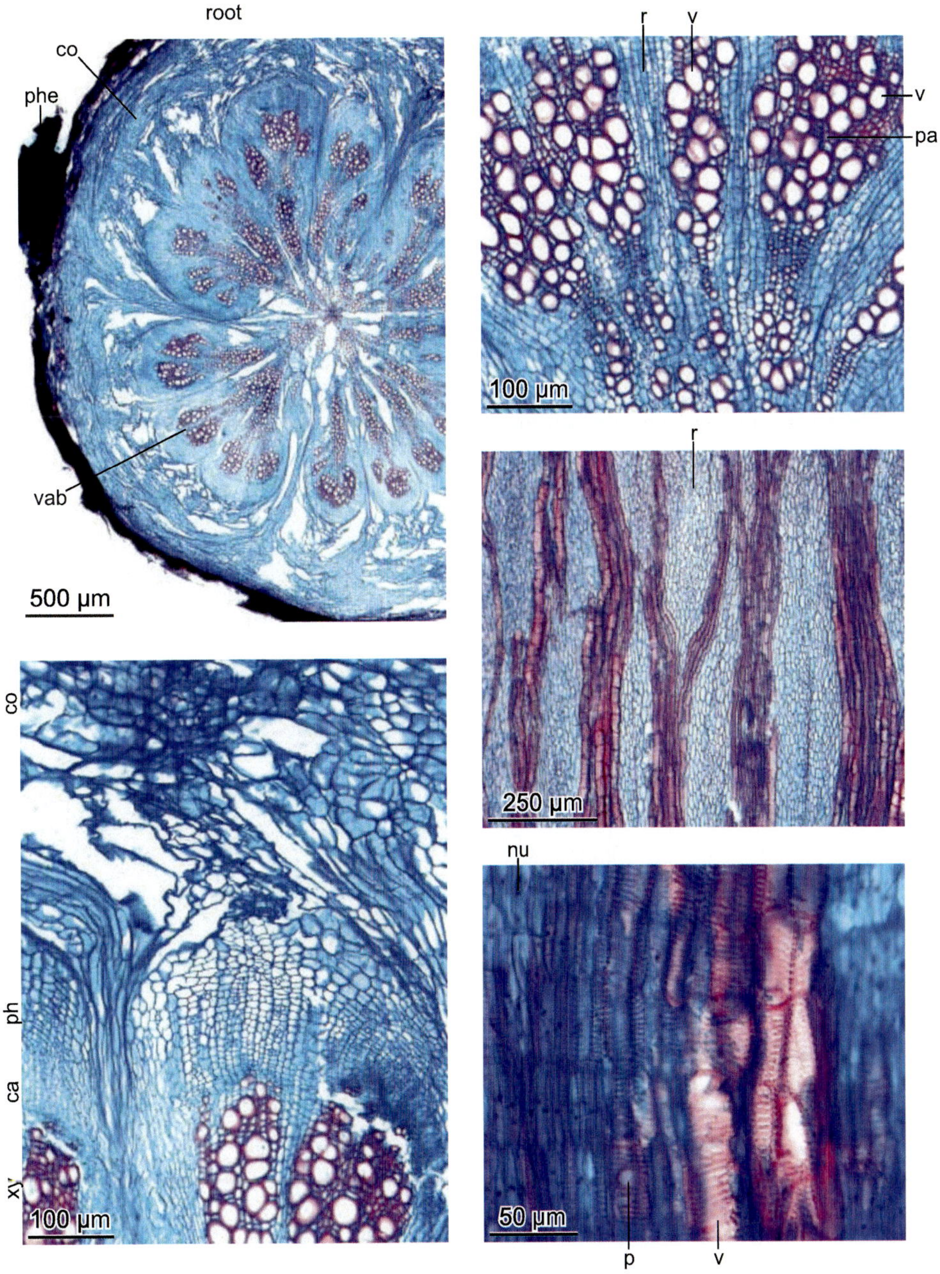

Features according to IAWA

Xylem *1 - 4 - 9 - 9.1 - 11 - 13 - 22 - 40.2 - 50.2 - 52.3 - 60.1 - 61 - 70 - 79.1 - 98 - 99 - 99.1 - 100.1 - 100.2 - 105 - 110*

Pedicularis heydei Prain

Scrophulariaceae

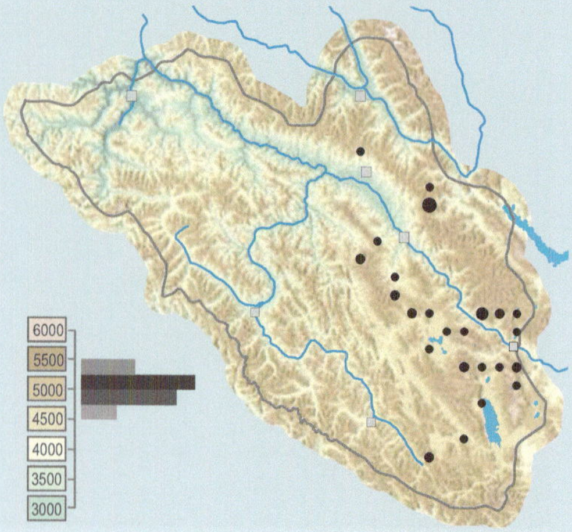

Description: Perennial herbs, 5–15 cm tall, hairy, leaves opposite, lanceolate, pinnatipartite to pinnatisect, segments dentate, flowers purple, ca. 1 cm long, in terminal spikes.

Distribution: NW India, Nepal.

Habitat: Wetlands, turf by springs, along brooks, in marshes.

Stem diameter 3–5 mm. Bark/xylem proportion 1:1. Mean ring width 0.6 mm. Maximal 3 rings.

Xylem

Transverse section

Semi-ring-porous. Rings marked by differences between earlywood (40–50 µm) and latewood vessel diameters (10–20 µm) and the absence of fibers in the first ring. Vessels solitary, in short radial multiples and in small groups. Vascular bundles remain. Fibers absent. Parenchyma pervasive.

Tangential section

Rays 5–>10-seriate, often confluent with the axial tissue.

Radial section

Perforation plates simple. Intervessel pits scalariform. Vessel length 50–110 µm. Ray cells upright. Crystals absent.

Bark

Phloem consists of parenchyma cells and small groups of sieve tubes. Rays slightly dilate. Cortex composed of oval, often compressed parenchyma cells, often separated by intercellulars. Older parts of the phloem undulated. Phellem cells irregularly formed. Crystals absent.

Pith

No slide available.

Analyzed: Root collar of polar root of 2 individuals.

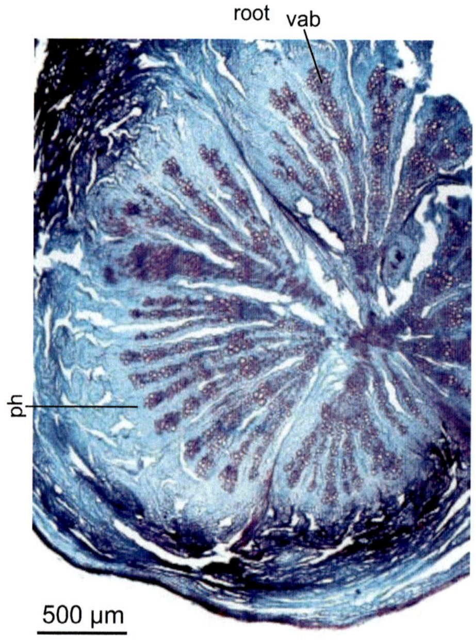

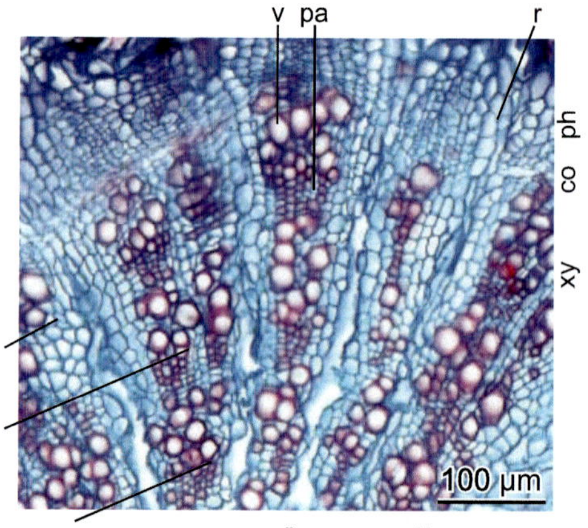

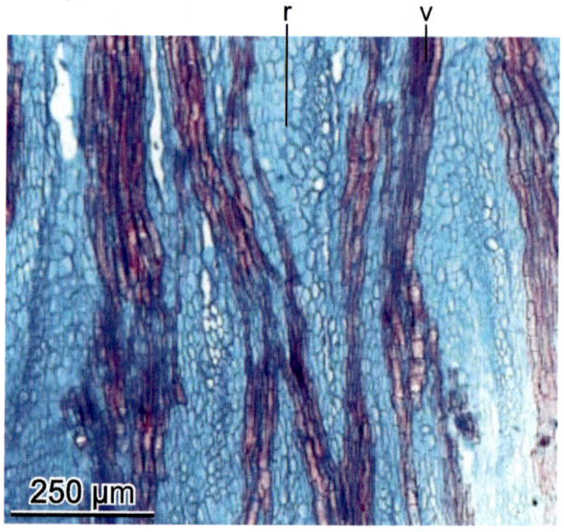

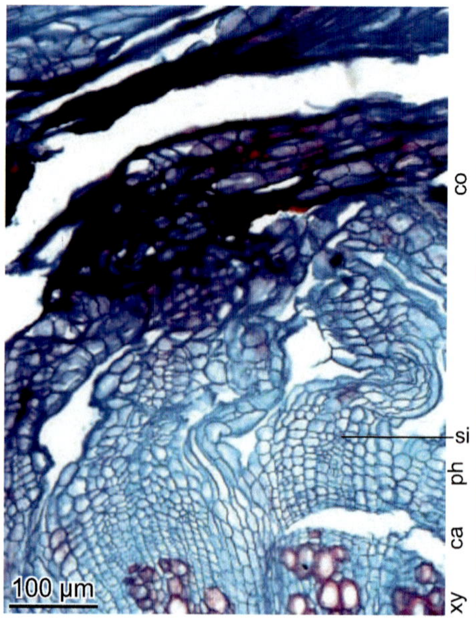

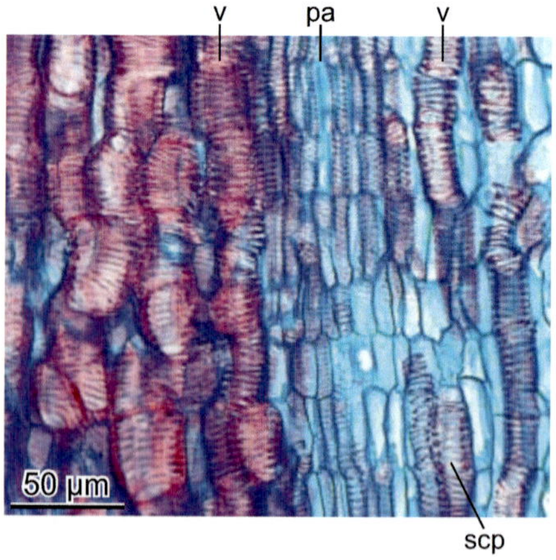

Features according to IAWA

Xylem 1 - 4 - 9 - 9.1 - 11 - 13 - 22 - 40.2 - 50.2 - 52.2 - 60.1 - 79.1 - 98 - 99 - 99.1 - 100.1 - 100.2 - 105

Pedicularis longiflora var. tubiformis (Klotzsch) P.C.Tsoong

Scrophulariaceae

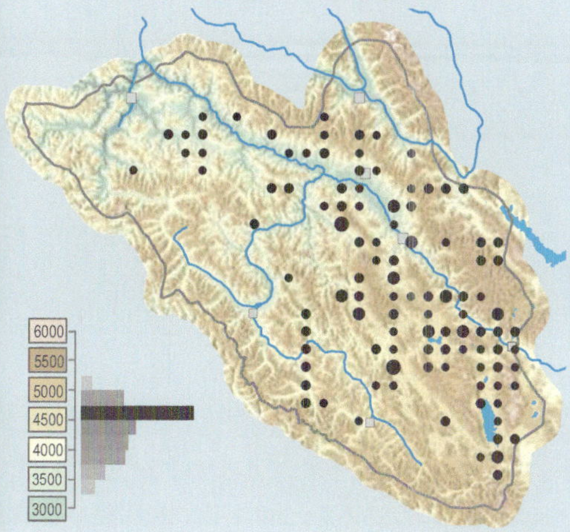

Description: Perennial herbs, 3–15 cm tall, leaves lanceolate, pinnatipartite, alternate, margin double dentate, flowers yellow, tube 5–8 cm long, situated in axils.

Distribution: Pakistan, N India, Nepal, China (Sichuan, Xizang, Yunnan).

Habitat: Marshes, spring areas, mesic to wet river banks.

Stem diameter 3–5 mm. Bark/xylem proportion 1:0.3. Rings absent.

Xylem

Transverse section

Semi-ring-porous. Rings marked by differences between earlywood (40–50 µm) and latewood vessel diameters (10–20 µm). Vessels mainly solitary. Vascular bundles remain. Fibers absent. Parenchyma pervasive.

Tangential section

Rays 4–6-seriate.

Radial section

Perforation plates simple. Intervessel pits scalariform. Vessel length 70–100 µm. Ray cells upright. Crystals absent.

Bark

Phloem consists of parenchyma cells and small groups of sieve tubes. Some rays dilate. Cortex composed of oval, parenchyma cells, often separated by intercellulars. Crystals and phellem absent.

Pith

No slide available.

Analyzed: Root collar of polar root of 2 individuals.

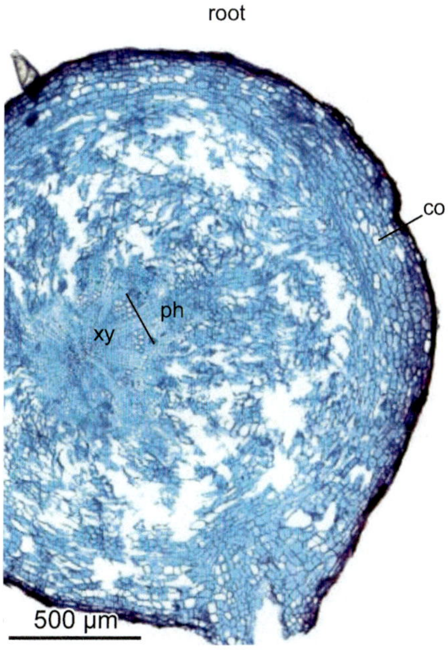

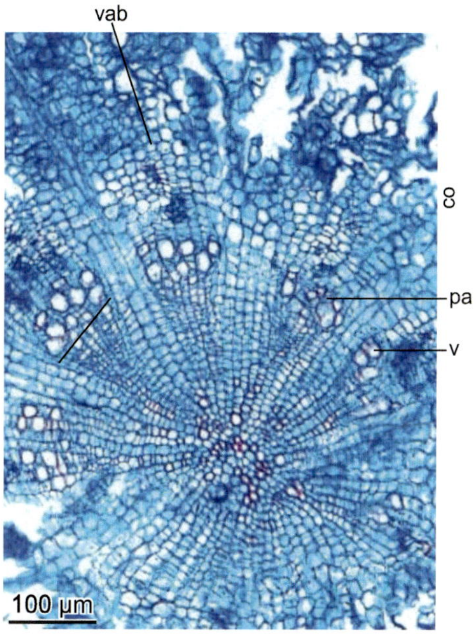

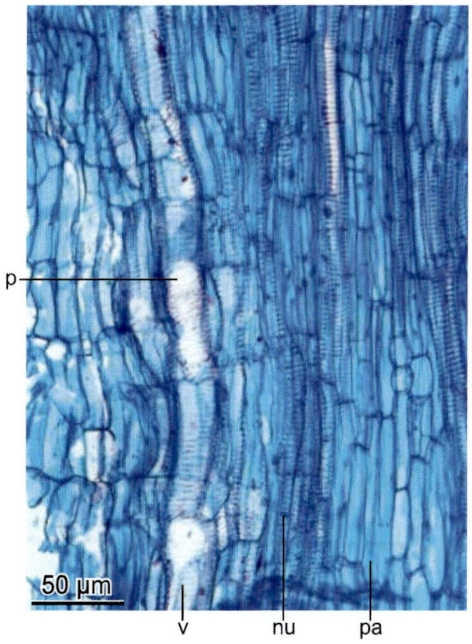

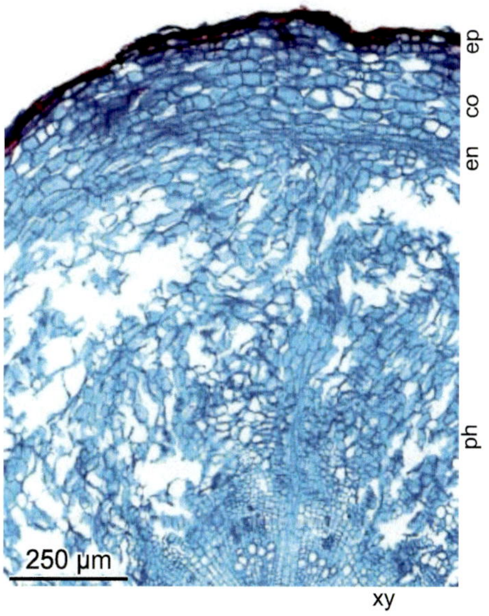

Features according to IAWA

Xylem 1 - 4 - 9 - 13 - 22 - 40.2 - 50.1 - 52.2 - 60.1 - 79.1 - 98 - 99.1 - 100.2 - 105

Pedicularis rhinanthoides Schrenk ex Fisch. & C.A.Mey.

Scrophulariaceae

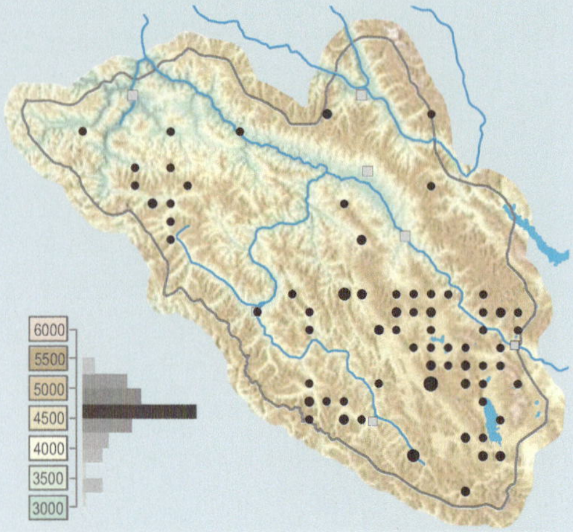

Description: Perennial herbs, 4–30 cm tall, glabrous, leaves mostly basal, oblong, pinnatisect, flowers pink to violet, 1.5–2.5 cm long, in short racemes.

Distribution: Russia, Kazakhstan, Kyrgyzstan, Tajikistan, Pakistan, India, China (Gansu, Hebei, Qinghai, Shaanxi, Shanxi, Sichuan, Xinjiang, Xizang, Yunnan), Mongolia.

Habitat: Wetlands, mesic to wet river banks, marshes.

Stem diameter 3–7 mm. Bark/xylem proportion 1:1. Ring width 0.3–0.7 mm. Maximal 4 rings.

Xylem

Transverse section

Semi-ring-porous. Rings marked by differences between earlywood (40–50 µm) and latewood vessel diameters (10–20 µm) and the absence of fibers in the first ring. Vessels solitary, in short radial multiples and in small groups. Vascular bundles remain. Thin- to thick-walled fibers present in the peripheral ring and absent in the center. Parenchyma pervasive.

Tangential section

Rays 5–>10-seriate.

Radial section

Perforation plates simple. Intervessel pits scalariform. Vessel length 70–100 µm. Ray cells upright. Crystals absent.

Bark

Phloem consists of parenchyma cells and small groups of sieve tubes. Rays dilate. Cortex composed of oval, often compressed parenchyma cells, often separated by intercellulars. Phellem cells irregularly formed. Crystals absent.

Pith

No slide available.

Analyzed: Root collar of polar root of 6 individuals.

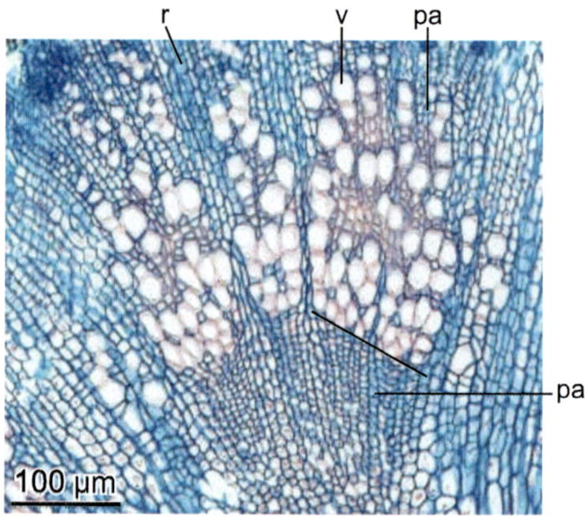

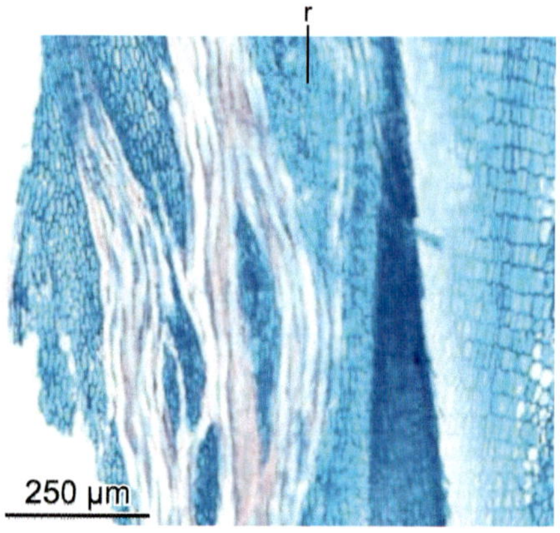

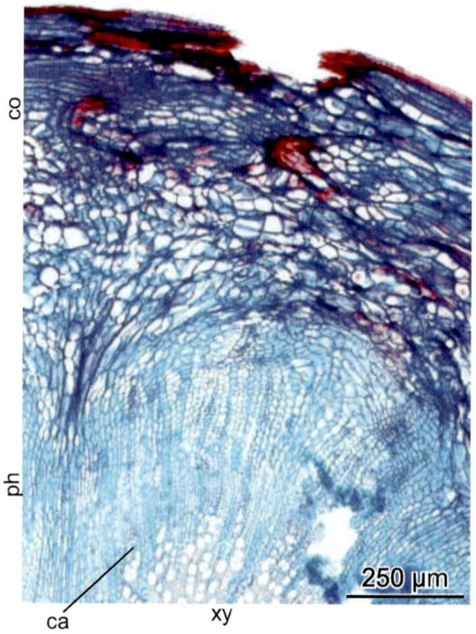

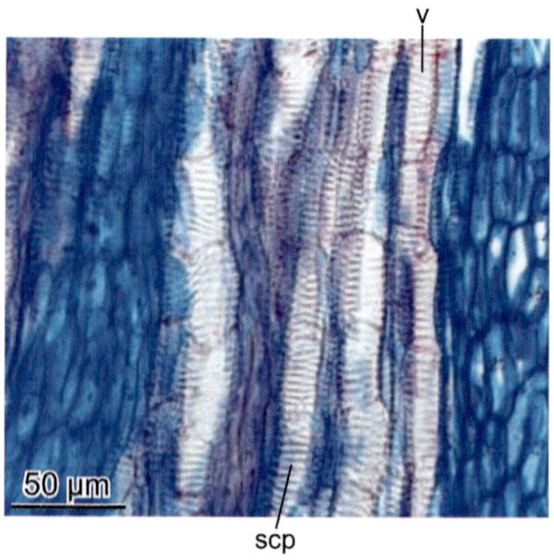

Features according to IAWA

Xylem 1 - 4 - 9 - 9.1 - 11 - 13 - 22 - 40.2 - 50.2 - 52.2 - 60.1 - 61 - 70 - 79.1 - 98 - 99 - 99.1 - 100.2 - 105

Scrophularia dentata Royle ex Benth.

Scrophulariaceae

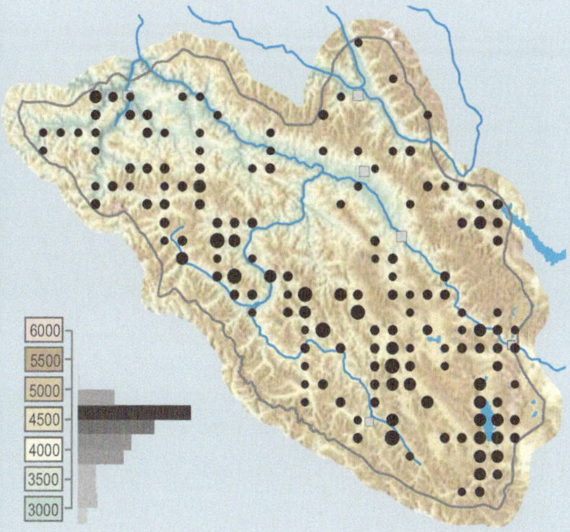

Description: Perennial herbs, 20–40 cm tall, branched from the base, leaves oblong, subsessile, 1.5–5 cm long, margin dentate or pinnately parted, flowers purple-red, ca. 6 mm long, in lax terminal panicles.

Distribution: N Pakistan, NW India, China (Xizang).

Habitat: Semi-deserts, steppes, disturbed sites, dry screes, floodplains, along roads.

Stem diameter 5–6 mm. Bark/xylem proportion 1:4. Ring width 0.1–0.6 mm. Maximal 3 rings.

Xylem

Transverse section

Semi-ring-porous. Earlywood vessels solitary and in groups. Ring boundary marked by large earlywood (60–80 µm) and small latewood vessel diameters (10–20 µm), and a band of intra-xylary periderm. Fibers thin- to thick-walled. Parenchyma unrecognizeable. Secondary intra-xylary phloem towards the pith separates living and dead tissues. Canals (laticifers?) in the intra-xylary phloem zone. Crystals absent.

Tangential section

Rays uniseriate, with 1–4 cells in height.

Radial section

Perforation plates simple. Intervessel pits round to oval. Vessel length 150–200 µm. Crystals absent.

Bark

Phloem consists of parenchyma cells, and groups of sieve tubes and laticifer-like cells. Phellem consist of a few layers of square cells. Older phellem parts flake.

Pith

No slide available.

Analyzed: Root collar of polar root of 4 individuals.

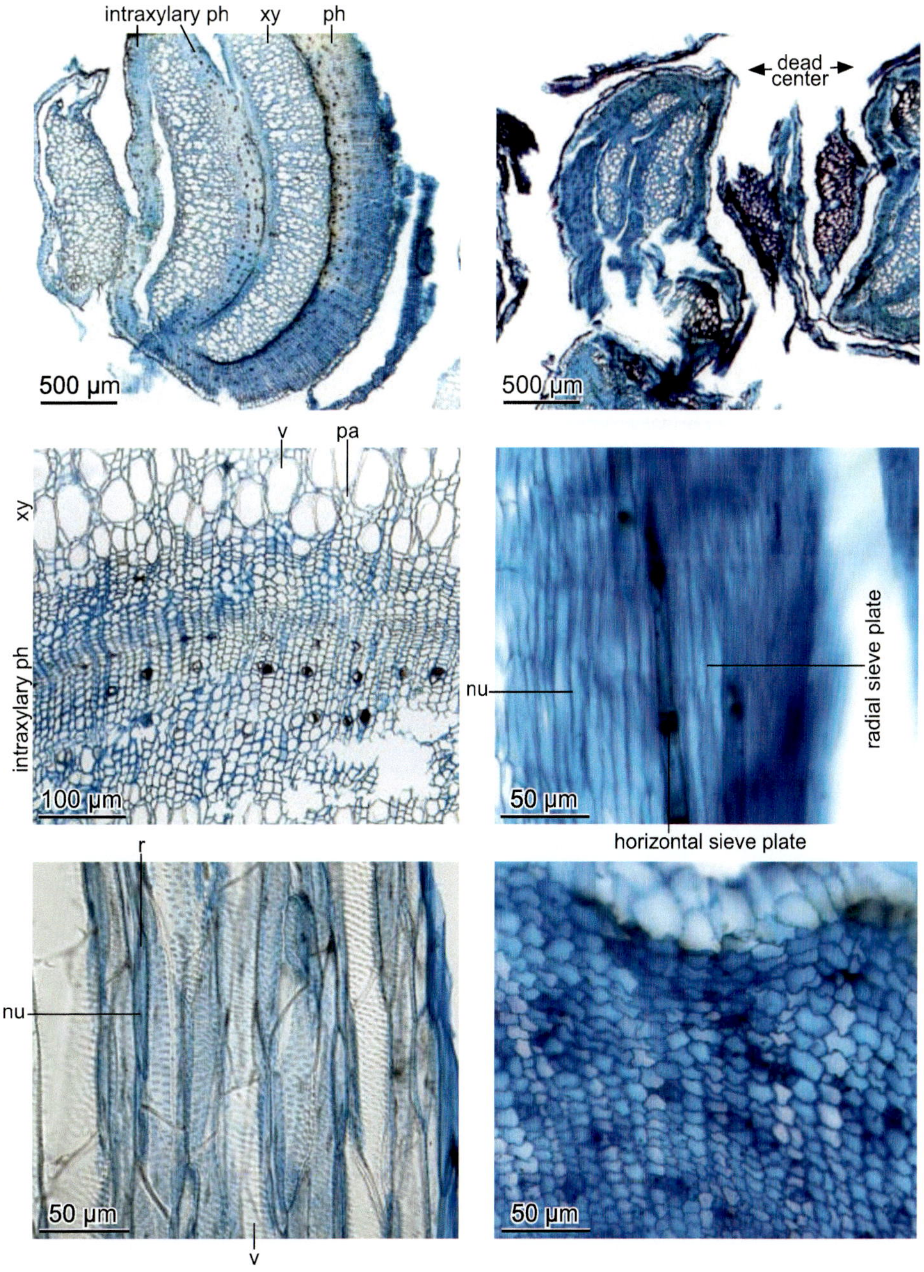

Features according to IAWA

Xylem 1 - 4 - 9 - 11 - 13 - 20 - 41 - 50.2 - 52.3 - 61 - 70 - 75 - 96 - 105 - 135.2

Scrophularia koelzii Pennell
Scrophulariaceae

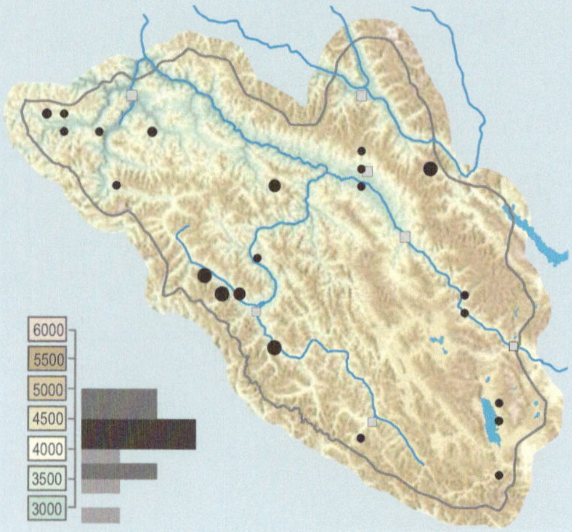

Description: Perennial herbs, 20–40 cm tall, glandular puberulent, leaves oblong, pinnately cut towards base, flowers purple-brown, 5–7 mm long, in lax terminal panicles.

Distribution: Afghanistan, Pakistan, NW India, Nepal.

Habitat: Semideserts, steppes.

Stem diameter 5–6 mm. Bark/xylem proportion 1:4. Mean ring width 0.4 mm. Maximal 7 rings.

Xylem

Transverse section

Semi-ring-porous. Earlywood vessels solitary and in groups. Ring boundary marked by large earlywood (50–60 µm) and small latewood vessel diameters (10–20 µm), and a band of intra-xylary periderm. Fibers thin- to thick-walled. Parenchyma unrecognizeable. Secondary intra-xylary phloem towards the pith separates living and dead tissues. Canals (laticifers?) in the intra-xylary phloem zone. Crystals absent.

Tangential section

Rays uniseriate, with 1–4 cells in height.

Radial section

Perforation plates simple. Intervessel pits round. Vessel length 150–200 µm. Crystals absent.

Bark

Phloem consists of parenchyma cells, groups of sieve tubes and laticifer-like cells. Phellem consists of a few layers of square cells. Older phellem parts flake.

Pith

No slide available.

Analyzed: Root collar of polar root of 1 individual.

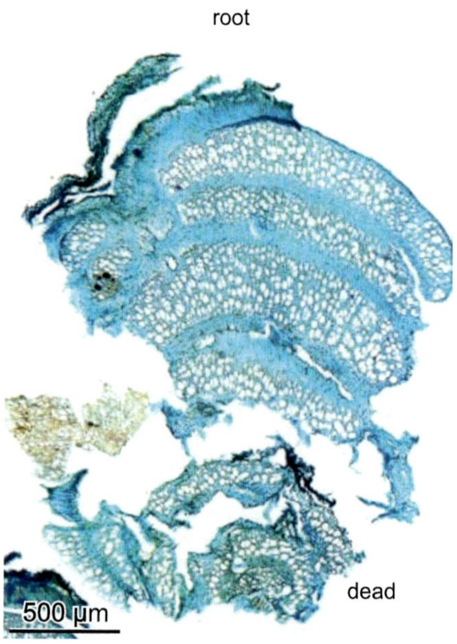

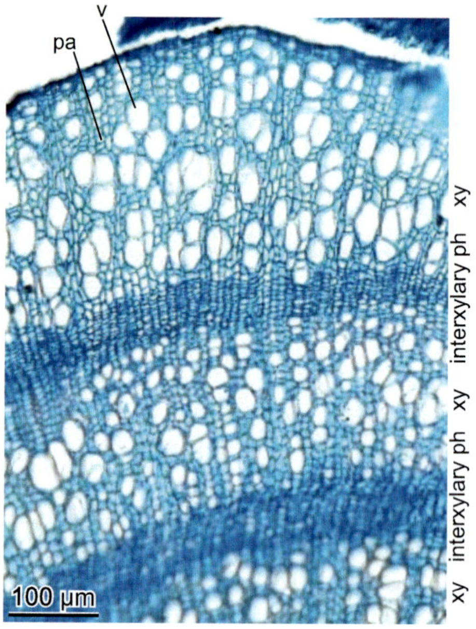

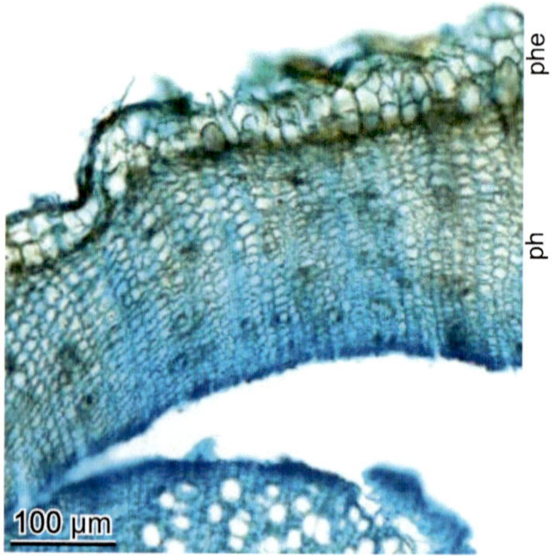

Features according to IAWA

Xylem 1 - 4 - 9 - 11 - 13 - 20 - 41 - 50.2 - 52.3 - 61 - 70 - 75 - 96 - 105 - 135.2

Veronica anagalloides Guss.

Scrophulariaceae

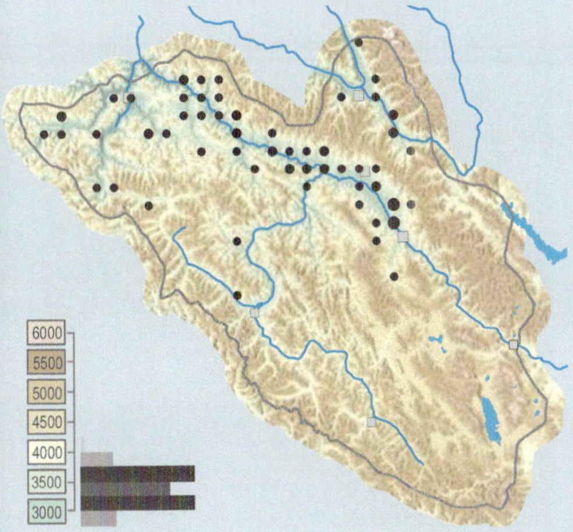

Syn.: *V. anagallis-aquatica* var. *anagalloides* (Guss.) K.Koch

Description: Annuals, 15–50 cm tall, sparsely glandular hairy, leaves lanceolate, sessile, semiamplexicaul, margin subentire to serrate, flowers pale blue, ca. 3 mm across, in many-flowered axillary racemes.

Distribution: C and S Europe, Russia, Iran, Kazakhstan, Turkmenistan, Kyrgyzstan, Tajikistan, Afghanistan, Pakistan, NW India, China (Gansu, Heilongjiang, Nei Mongol, Qinghai, Shaanxi, Shanxi, Xizang), Mongolia, Korea, Japan.

Habitat: Ditches, river sides, wet places.

Stem diameter 4 mm. Bark/xylem proportion 1:0.3. Ring width 0.2 mm. Maximal 1 ring.

Xylem

Transverse section

Vessels angular and solitary. Diameter of larger vessels 30–40 µm. Fibers thin- to thic-walled. Parenchyma absent. Crystals absent.

Tangential section

Rays absent.

Radial section

Perforation plates simple. Intervessel pits round. Vessel length 150–200 µm. Crystals absent.

Bark

Phloem contains small groups of sieve tubes. Cortex consists of large, thin-walled, unlignified parenchyma cells. With a distinct endodermis. Crystals absent. With irregularly formed intercellulares. Phellem absent.

Pith

Round and very large, filled with round, unlignified cells.

Analyzed: Root collar of polar root of 1 individual.

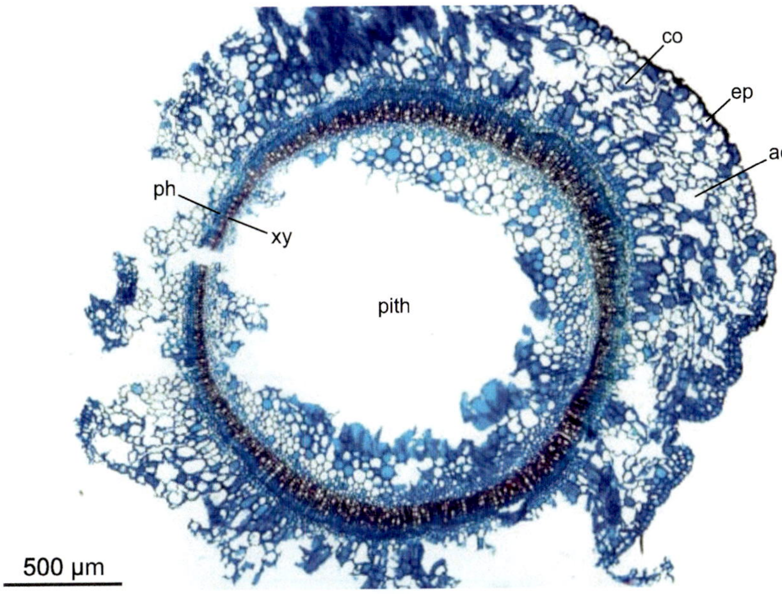

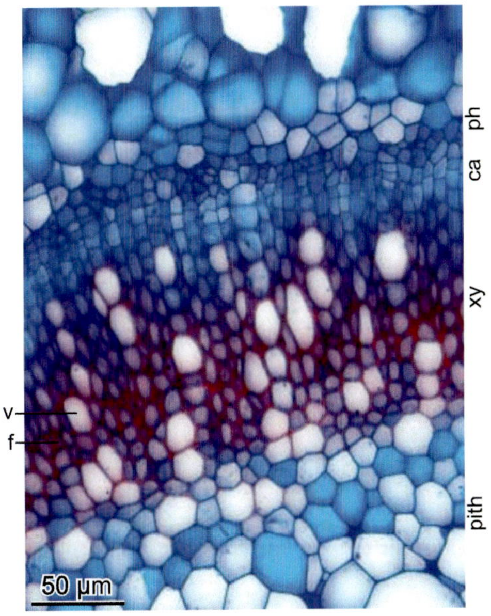

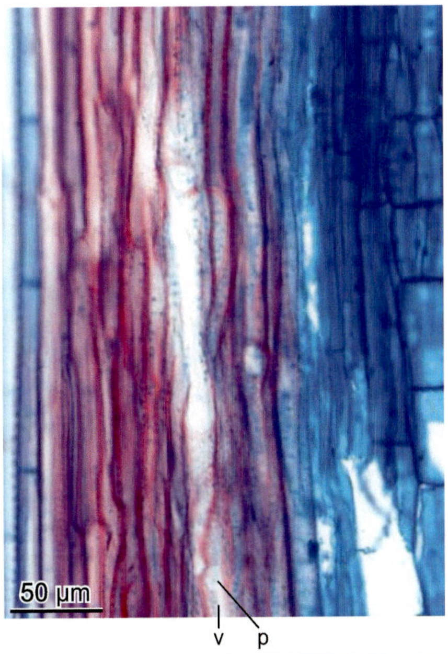

Features according to IAWA

Xylem 1 - 9 - 9.1 - 12 - 13 - 22 - 40.2 - 50.2 - 52.3 - 61 - 75 - 117

Veronica beccabunga subsp. *muscosa* (Korsh.) Elenevsky
Scrophulariaceae

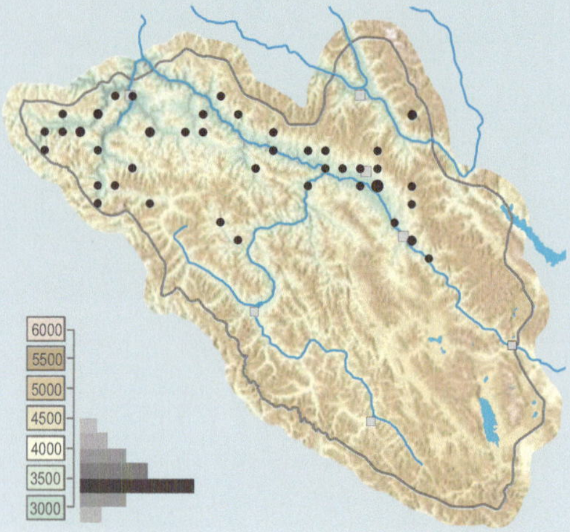

Description: Perennial herbs, 10–30 cm tall, glabrous, stems prostrate or ascending, leaves suborbicular, ovate, or oblong, margin shallowly serrulate to crenulate, flowers pale purple to pale blue, ca. 5 mm across, in 3–15-flowered axillary racemes.

Distribution: E Iran, Kazakhstan, Turkmenistan, Uzbekistan, Kyrgyzstan, Tajikistan, Afghanistan, Pakistan, N India, Nepal, China (Sichuan, Xinjiang, Yunnan).

Habitat: Irrigation canals and ditches, shallow streams.

Stem diameter 1.5–2 mm. Bark/xylem proportion 1:0.2. Ring width 0.1–0.1.5 mm. Maximal 1 ring.

Xylem

Transverse section

Vessels angular and in groups. Diameter of larger vessels 15–30 µm. Fibers absent. Parenchyma pervasive. Crystals absent.

Tangential section

Rays absent.

Radial section

Perforation plates simple. Intervessel pits round to scalariform. Vessel length 50–70 µm. Crystals absent.

Bark

Phloem contains small groups of sieve tubes. Cortex consists of large, thin-walled, unlignified parenchyma cells. With a distinct endodermis. Crystals absent. With irregularly formed intercellulares. Phellem absent.

Pith

Round, filled with round, unlignified cells.

Analyzed: Root collar of polar root of 2 individuals.

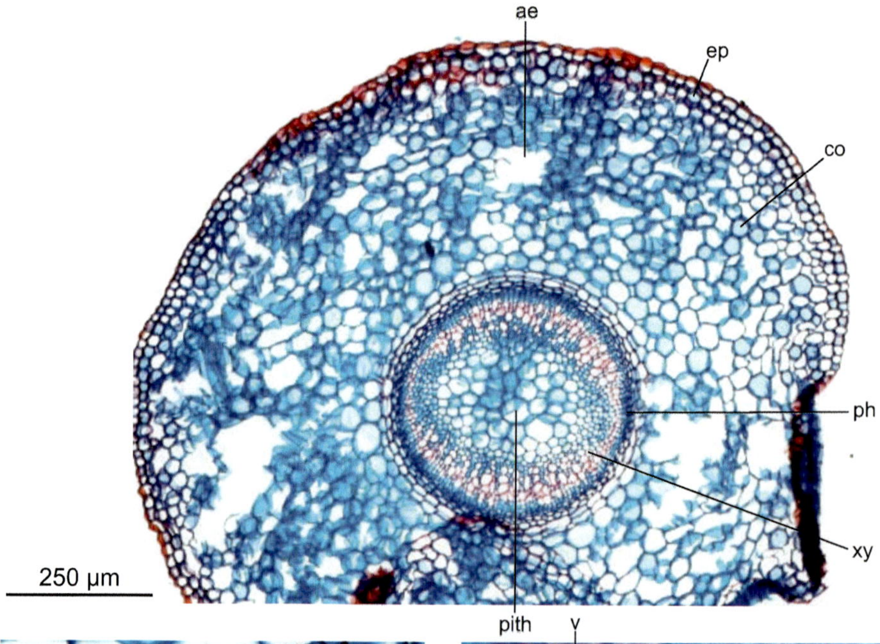

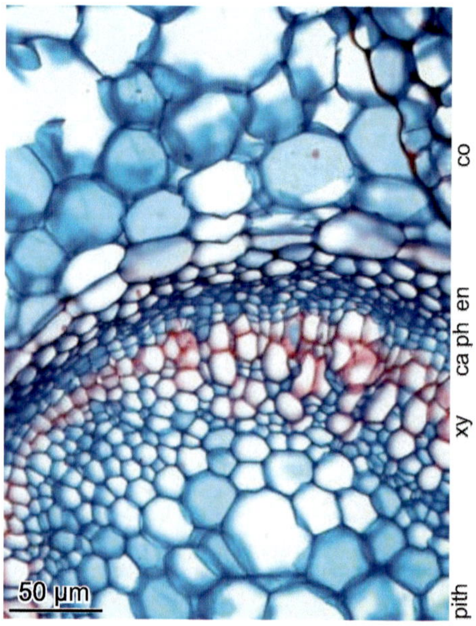

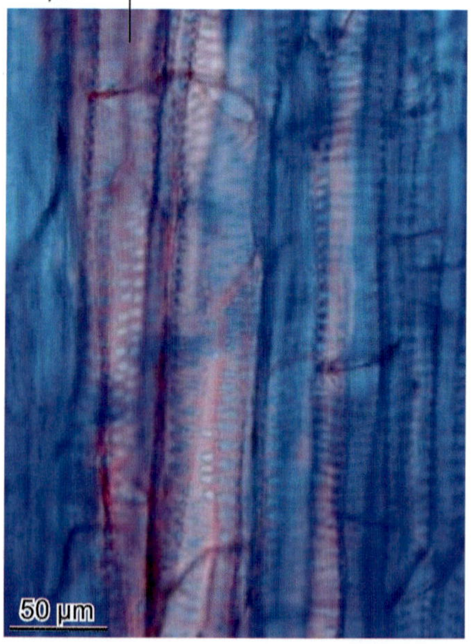

Features according to IAWA

Xylem 2 - 9 - 9.1 - 12 - 13 - 20 - 22 - 40.1 - 40.2 - 50.2 - 52.2 - 60.1 - 61 - 70 - 79.1 - 117

Veronica biloba L.

Scrophulariaceae

Description: Annuals, 5–20 cm tall, sparsely pubescent and glandular, leaves oblong to ovate-lanceolate, margin sparsely shallowly dentate, flowers white, blue, or purple, 3–4 mm across, in many-flowered terminal racemes.

Distribution: Russia, Iran, Kazakhstan, Turkmenistan, Uzbekistan, Kyrgyzstan, Tajikistan, Afghanistan, Pakistan, India, Nepal, China (Gansu, Nei Mongol, Ningxia, Qinghai, Shaanxi, Sichuan, Xinjiang, Xizang), Mongolia.

Habitat: Dry pastures, disturbed sites, unstable slopes, eroded dry stream banks.

Stem diameter 0.8–1.0 mm. Bark/xylem proportion 1:4. Mean ring width 0.2–0.4 mm. Maximal 2 rings.

Xylem

Transverse section

Rings distinctly marked by a few rows of flat latewood cells. Diffuse- to semi-ring-porous. Vessels in long, radial rows. Diameter of vessels 20–30 µm. Fibers thin- to thick-walled. Parenchyma and rays absent. Crystals absent.

Tangential section

Rays absent.

Radial section

Perforation plates simple. Intervessel pits small, round. Vessel length 150–200 µm.

Bark

Phloem contains small groups of sieve tubes. Cortex consists of thin-walled, large parenchyma cells. Crystals absent. Phellem absent.

Pith

Pith round.

Analyzed: Polar root of 3 individuals.

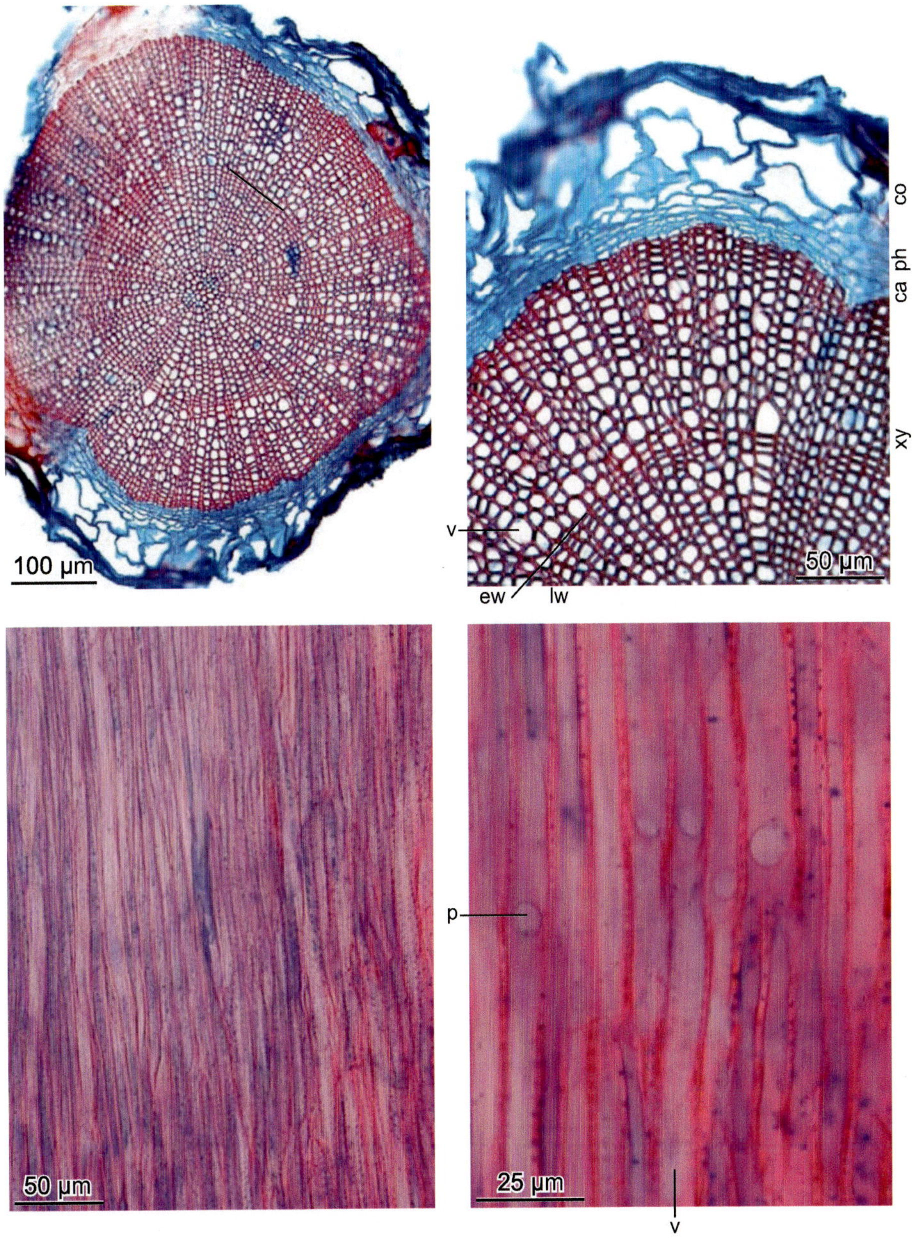

Features according to IAWA

Xylem 1 - 4 - 5 - 9 - 10 - 13 - 22 - 40.1 - 40.2 - 50.2 - 52.3 - 61 - 70 - 75 - 117

Veronica ciliata Fisch.

Scrophulariaceae

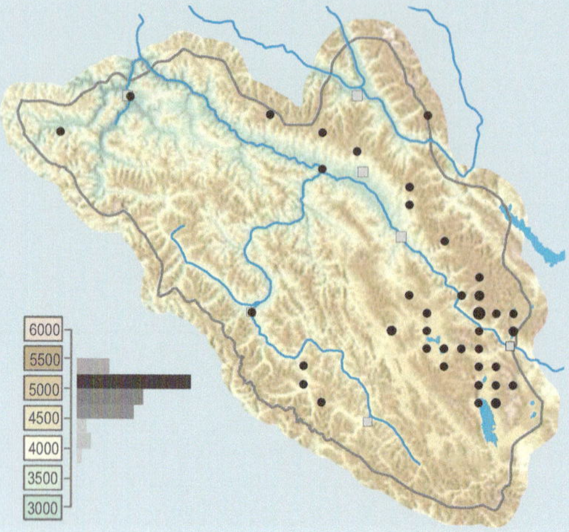

Syn.: *V. koelzii* Pennell

Description: Perennial herbs, 5–25 cm tall, hairy, stems ascending, leaves oblong to ovate-lanceolate, sessile, margin entire or serrate, flowers blue, ca. 3 mm across, in axillary racemes.

Distribution: Russia (Siberia), Kazakhstan, Kyrgyzstan, Tajikistan, Pakistan, NW India, Nepal, China (Gansu, Nei Monggol, Ningxia, Qinghai, Shaanxi, Sichuan, Xinjiang, Xizang, Yunnan), Mongolia.

Habitat: Alpine meadows.

Stem diameter 2 mm. Bark/xylem proportion 1:1. Ring width 0.15–0.2 mm. Maximal 2 rings.

Xylem

Transverse section

Semi-ring-porous. Vessels solitary and in short radial rows. Diameter of larger vessels 15–20 µm. Fibers absent in the rhizome and thin- to thick-walled in the shoot. Parenchyma pervasive in the rhizome and absent in the shoot. Crystals absent.

Tangential section

Rays absent.

Radial section

Perforation plates simple. Intervessel pits scalariform. Vessel length 80–100 µm.

Bark

Phloem contains small groups of sieve tubes. Cortex consists of thin-walled parenchyma cells. Crystals absent. Phellem absent.

Pith

Round.

Analyzed: Root collar of polar root and shoot of 3 individuals.

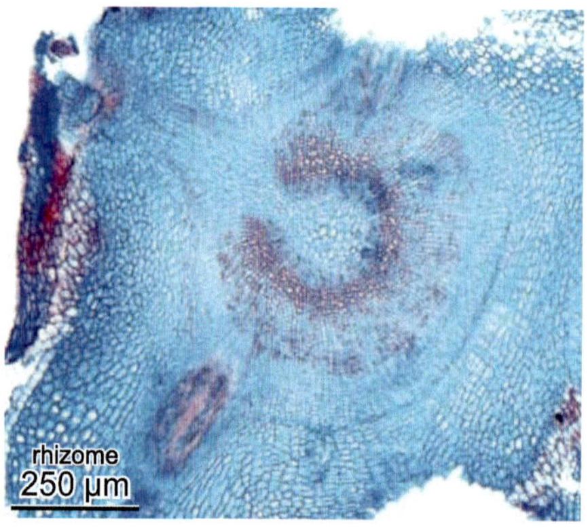

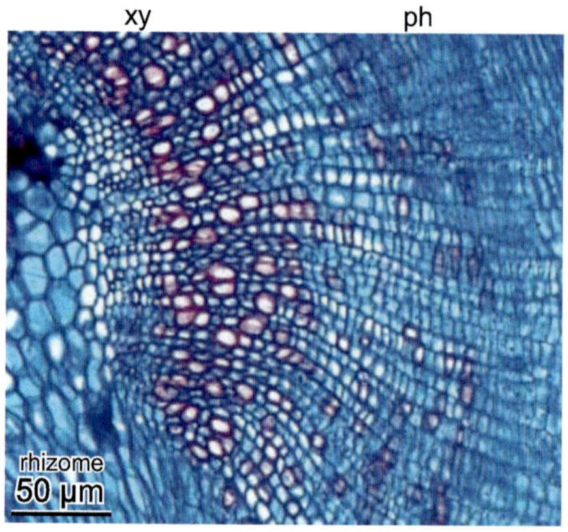

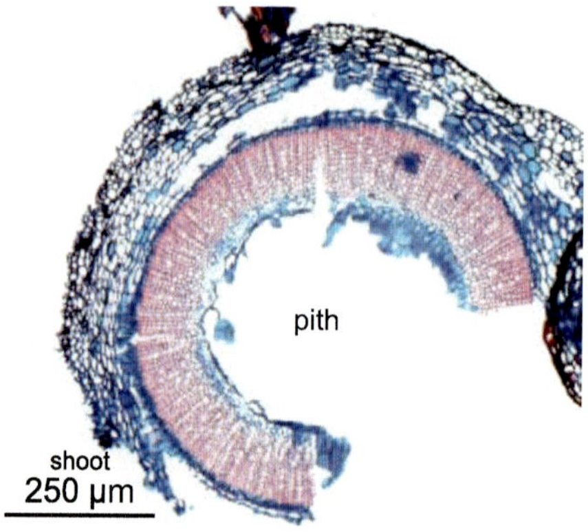

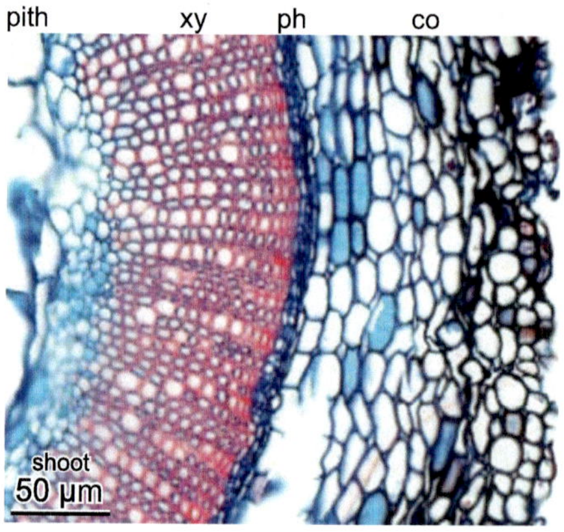

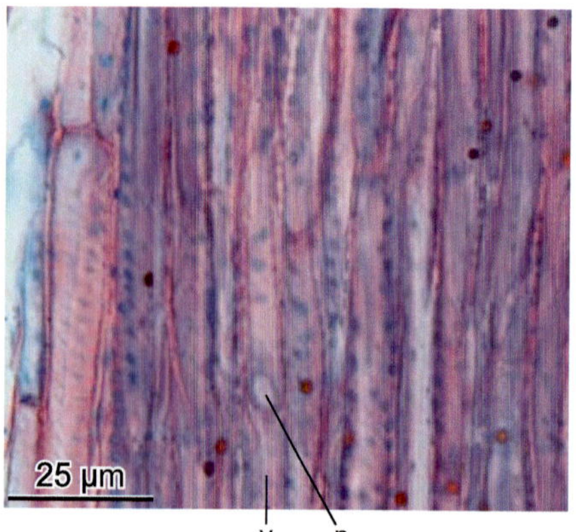

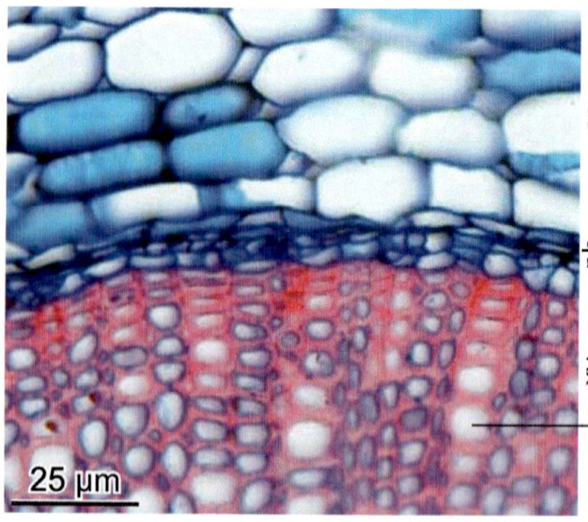

Features according to IAWA

Xylem 1 - 4 - 9 - 9.1 - 13 - 22 - 40.1 - 50.2 - 52.1 - 60.1 - 61 - 70 - 75 - 79.1 - 117

Hyoscyamus pusillus L.

Solanaceae

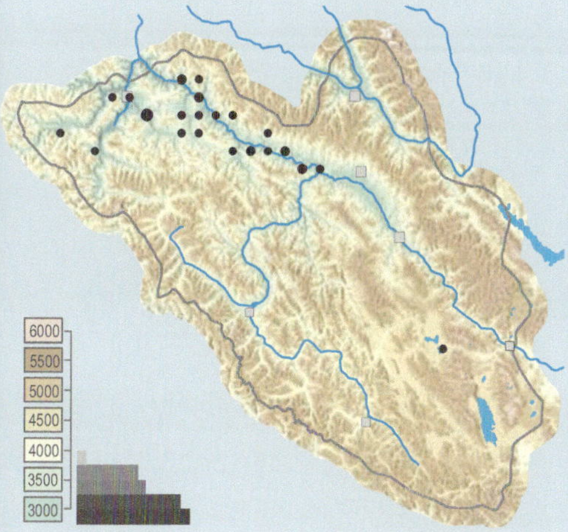

Description: Annuals, 5–30 cm tall, glandular pubescent, stem branched from the base, leaves mostly basal, ovate to oblong, margin entire or dentate, sometimes pinnatifid or pinnatisect, flowers yellow with a dark purple throat, 1–1.5 cm long, solitary, axillary.

Distribution: Arabia, Israel, Egypt, Turkey, Iran, Caucasus, Russia, Kazakhstan, Turkmenistan, Uzbekistan, Kyrgyzstan, Tajikistan, Afghanistan, Pakistan, India, China, Mongolia.

Habitat: Steppes, deserts, wastelands.

Stem diameter 1.2–1.0 mm. Bark/xylem proportion 1:5. Ring width 0.4–0.6 mm. Only one ring.

Xylem

Transverse section

Vessels annular, solitary. Vessel diameter 20–25 µm. Fibers thin- to thick-walled. Parenchyma absent or rarely paratracheal.

Tangential section

Rays uniseriate and similar to axial parenchyma cells.

Radial section

Simple perforation plates. Intervessel pits round in alternating position. Vessel length unrecognizable. Fibers pits rare, very small, round. Rays cells prostrate. Crystals absent.

Bark

Phloem with a few sieve tubes which collapse soon after formation. Dark-stained substances in undulated, radial lines. With crystal sand in parenchyma cells. Phellem consists of some layers of square to rectangular cells.

Pith

Pith round, with a few medullary sieve tubes.

Analyzed: Stem of 2 individuals (herbarium specimens).

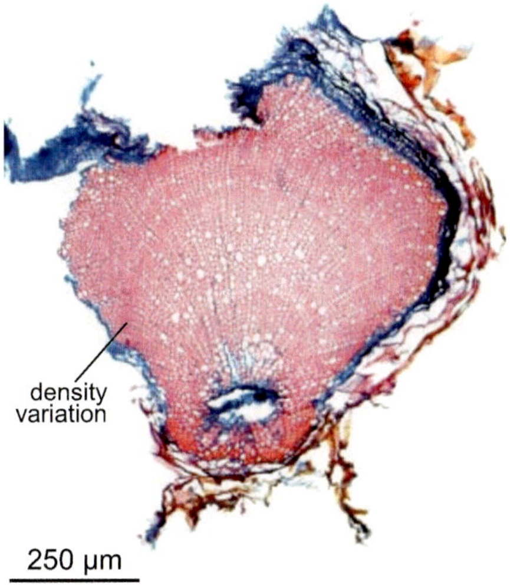

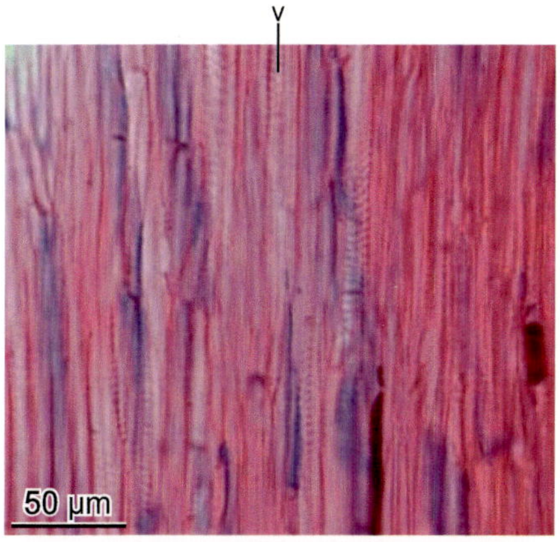

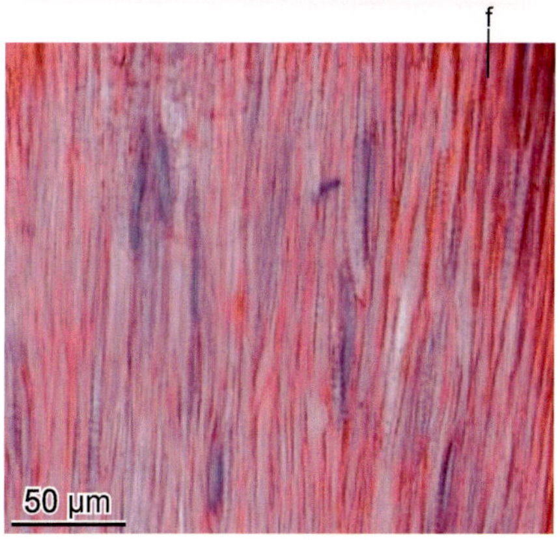

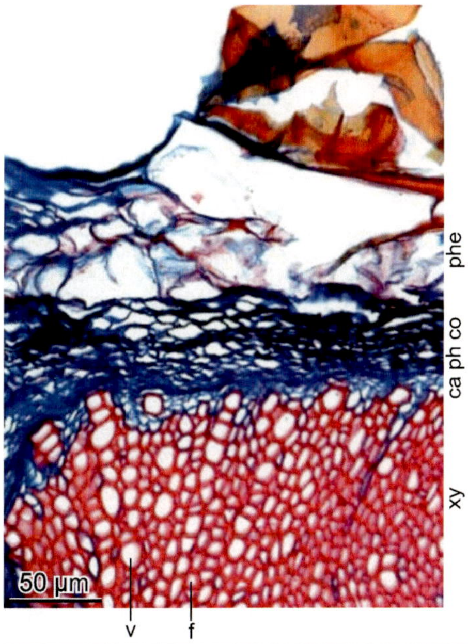

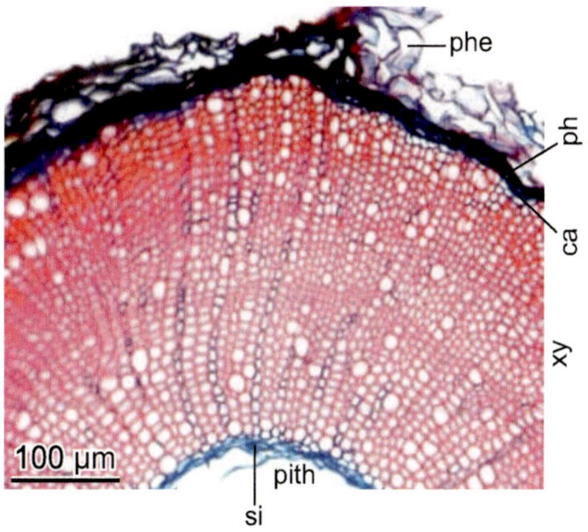

Features according to IAWA

Xylem 2.1 - 9 - 13 - 22 - 40.2 - 50.2 - 52.3 - 61 - 70 - 79 - 96 - 105

Lycium ruthenicum Murray

Solanaceae

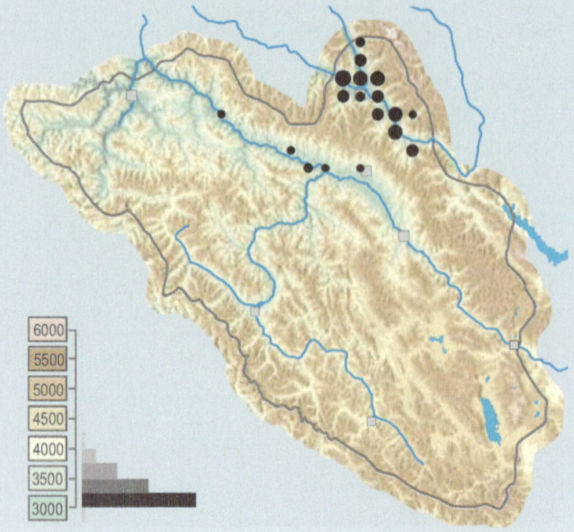

Description: Shrubs, 20–100 cm tall, spiny, glabrous, leaves linear, spatulate to aciculate, arranged in fascicles, flowers purple, 5–8 mm long, solitary, axillary, berries black, 5–8 mm across.

Distribution: Russia (European part), E Turkey, Iraq, Iran, Kazakhstan, Turkmenistan, Uzbekistan, Kyrgyzstan, Tajikistan, Afghanistan, Pakistan (Baltistan, Baluchistan), NW India, China (Gansu, Nei Mongol, Ningxia, Qinghai, Shaanxi, Xinjiang, Xizang), Mongolia.

Habitat: Disturbed places along roads, semideserts.

Stem diameter 15–50 mm. Bark/xylem proportion 1:20. Mean ring width 0.2 mm. Maximal 28 rings.

Xylem

Transverse section

Undulated ring boundaries. Ring-porous, vessels solitary. Ring boundary marked by large earlywood (150–200 μm) and smaller latewood vessel diameters (20–40 μm, and a uniseriate layer of marginal parenchyma cells. Fibers thick-walled. Parenchyma rare, paratracheal.

Tangential section

Rays mainly uniseriate, rarely biseriate.

Radial section

Perforation plates simple. Intervessel pits round in alternating position. Vessel length 150–200 μm. Fibers with round to oval pits. Ray cells prostrate. Crystals absent.

Bark

Phloem with a few sieve tubes which collapse soon after formation. Dark-stained substances in undulated, radial lines. With crystal sand in parenchyma cells. Phellem consists of some layers of square to rectangular cells.

Pith

Round, with crystal sand and dark-stained substances.

Analyzed: Stem of 3 individuals.

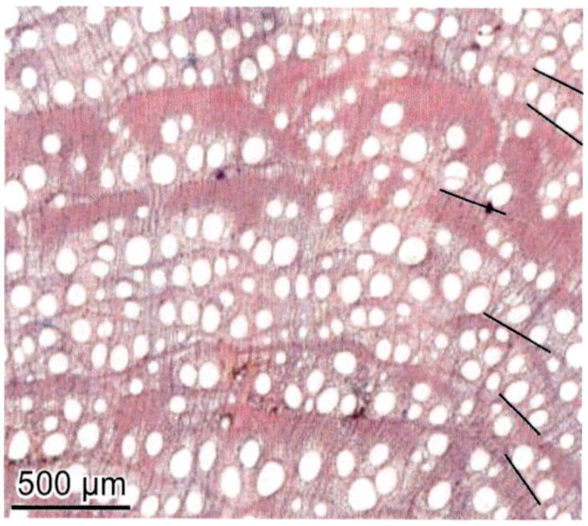

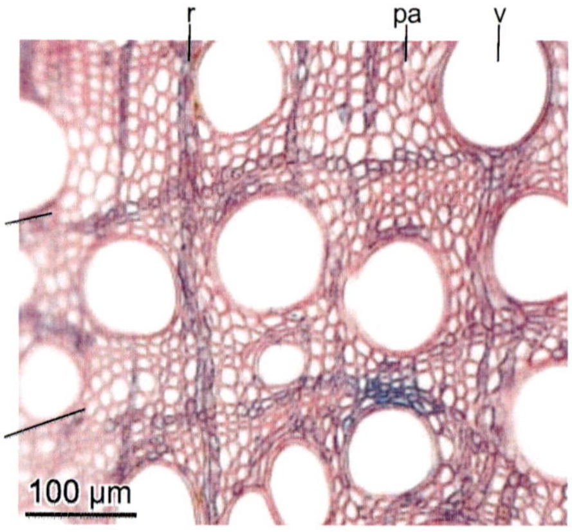

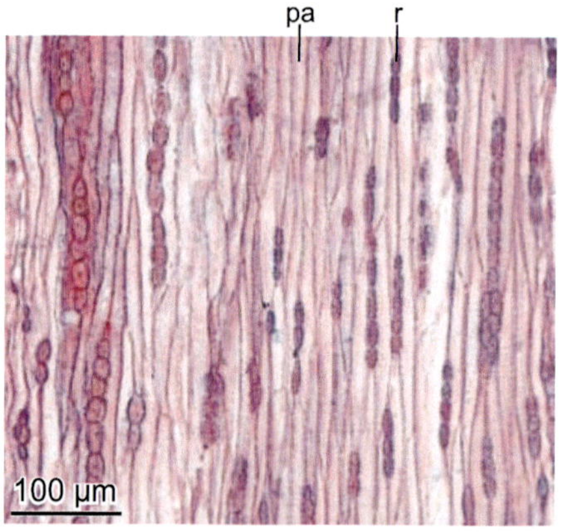

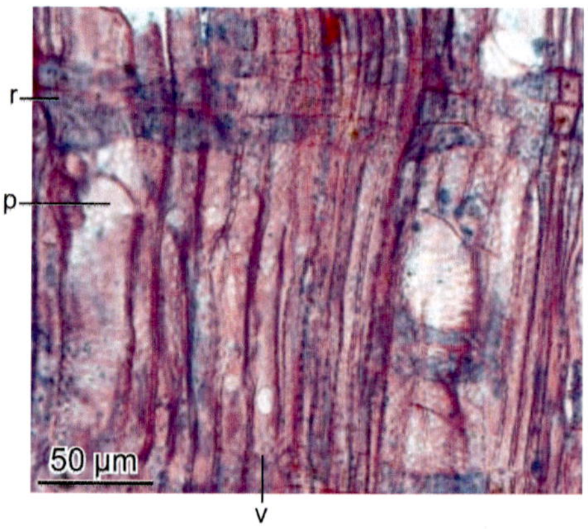

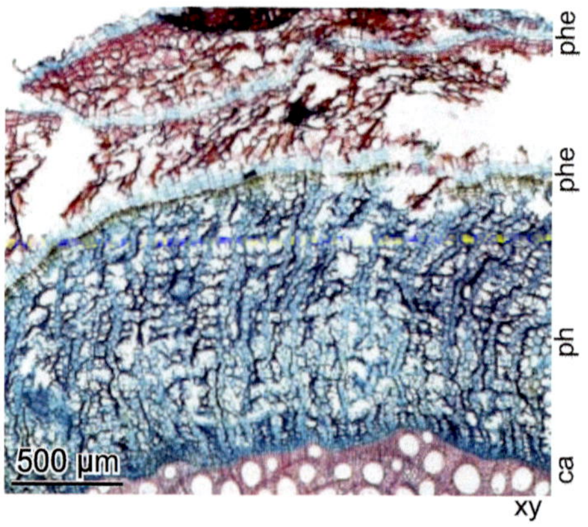

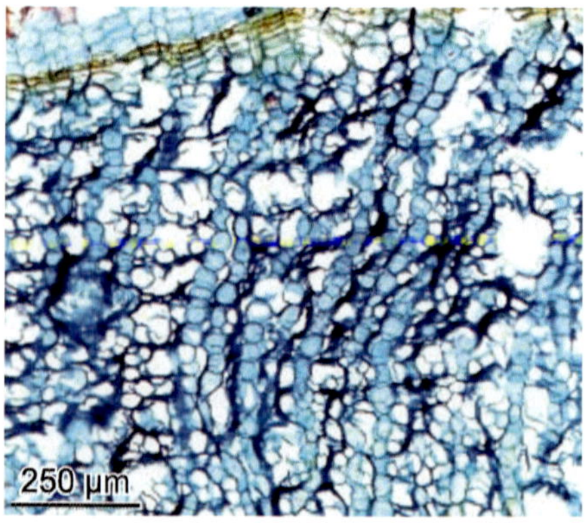

Features according to IAWA

Xylem 1 - 3 - 9 - 13 - 20 - 42 - 50.1 - 52.3 - 61 - 70 - 79 - 89 - 96 - 97 - 104

Physochlaina praealta (Decne.) Miers

Solanaceae

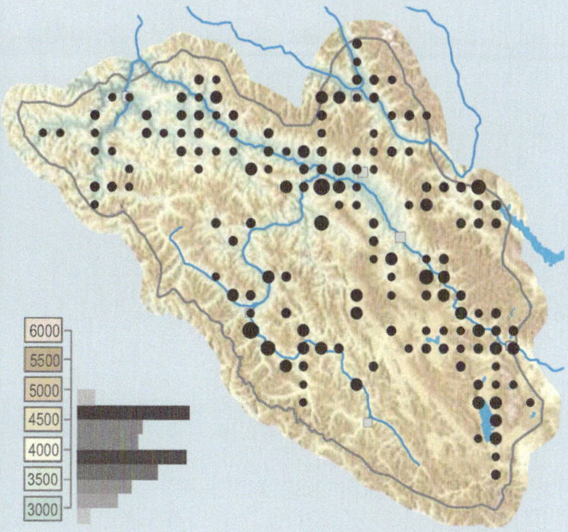

Syn.: *Belenia praealta* Decne., *Hyoscyamus praealtus* (Decne.) Walp., *Scopolia praealta* (Decne.) Dunal

Description: Perennial herbs, 30–50 cm tall, glandular pubescent, stems much branched, leaves ovate to oblong, flowers yellow, 2–3 cm long, funnel-shaped, in few-flowered inflorescences.

Distribution: Pakistan, N India, China (Xizang), Bhutan.

Habitat: Stony deserts, loamy slopes, always in dry habitats, often synanthropic in settlements, weedy in barley fields, sometimes cultivated.

Stem diameter 5–15 mm. Bark/xylem proportion 1:5. Ring width 0.4–0.6 mm. Maximal 9 rings.

Xylem

Transverse section

Semi-ring-porous. Rings marked by differences between earlywood (100–140 μm) and latewood vessel diameters (40–50 μm). Vessels primarily solitary, rarely in small groups. Fibers absent. Vascular bundles indistinctly remaining. Parenchyma pervasive. With small groups of sieve tubes within the axial parenchyma. A lot of crystal sand in all parenchyma cells.

Tangential section

Rays 4–>8-seriate, often confluent with the axial tissue.

Radial section

Perforation plates simple. Intervessel pits round to scalariform. Vessel length 100–150 μm. Ray cells upright. Crystals absent.

Bark

Phloem consists of large parenchyma cells, sieve tubes and small companion cells. Rays dilate. Cortex composed of oval parenchyma cells, which are often separated by small intercellulars. Phellem consists of some layers of thin-walled, rectangular cells. A lot of crystal sand in all parenchyma cells.

Pith

No slide available.

Analyzed: Root collar of polar root of 3 individuals.

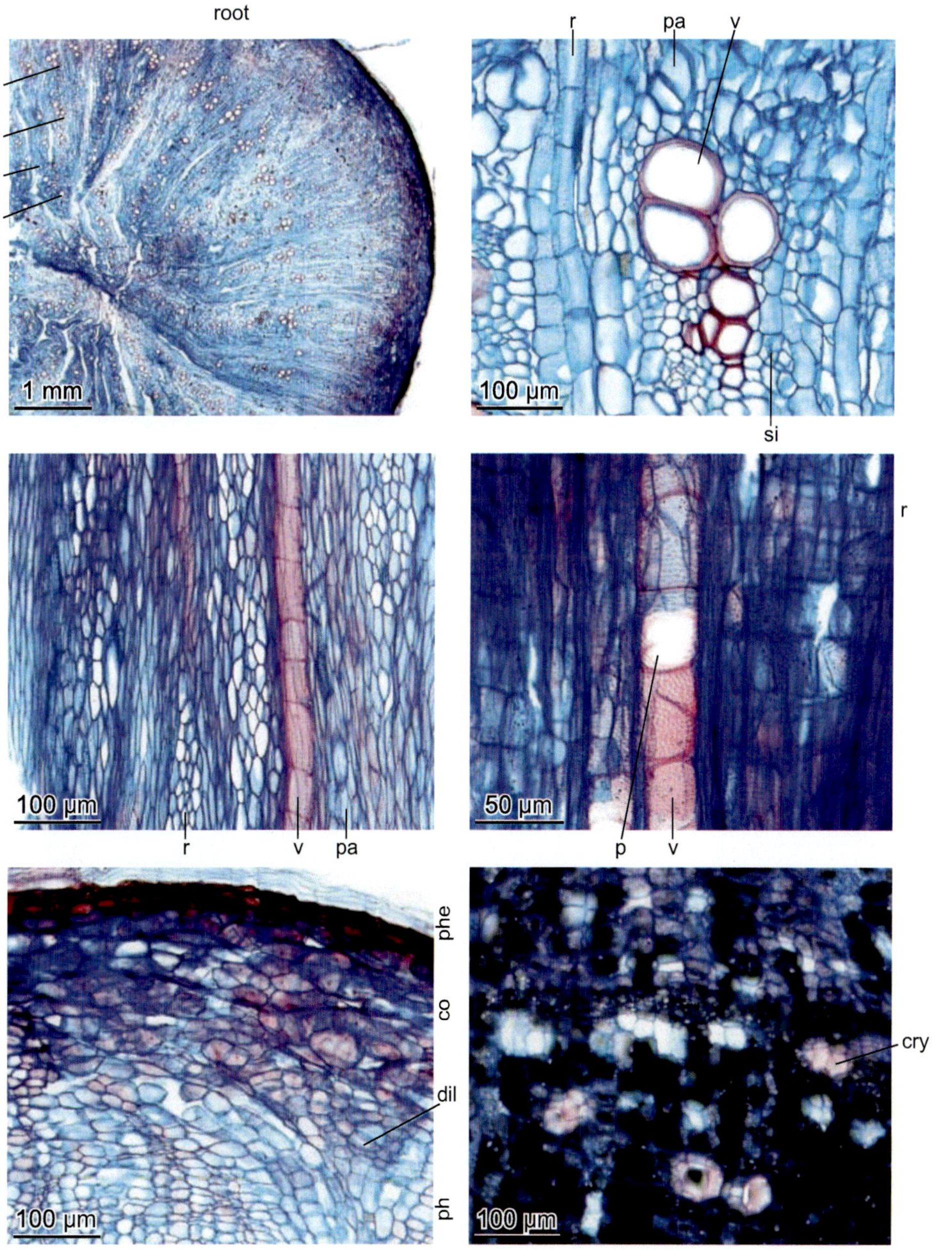

Features according to IAWA

Xylem 1 - 4 - 9 - 11 - 13 - 20 - 22 - 42 - 50.1 - 52.3 - 60.1 - 79.1 - 98 - 99.1 - 100.1 - 100.2 - 105 - 135 - 153

Myricaria elegans Royle
Tamaricaceae

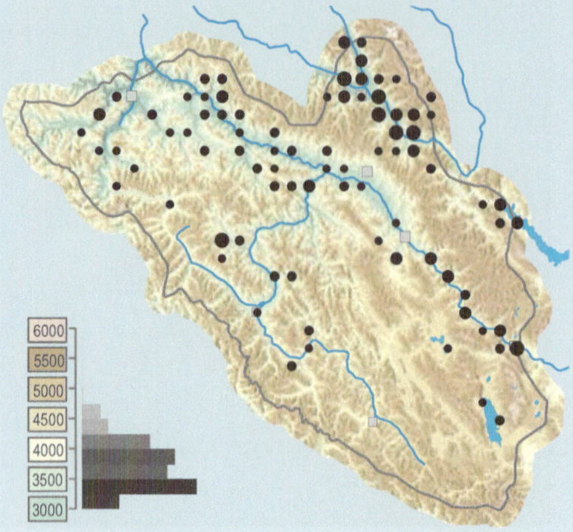

Syn.: *Tamaricaria elegans* (Royle) Qaiser & Ali, *Tamarix ladachensis* B.R.Baum

Description: Shrubs or small trees, up to 5 m tall, leaves elliptic or lanceolate, sessile, 6–15 × 2–3 mm, flowers white or pink, ca. 1.5 cm across, in lateral racemes.

Distribution: N Pakistan, NW India, China (Xinjiang, Xizang).

Habitat: River banks, stream sides.

Twig diameter 6–8 mm. Bark/xylem proportion 1:>10. Mean ring width 1 mm. Maximal 170 rings.

Xylem

Transverse section

Ring-porous. Angular vessels in groups. Ring boundary marked by large earlywood (80–100 µm) and smaller latewood vessel diameters (20–30 µm), and thickened large rays. Latewood fibers thick-walled. Parenchyma paratracheal. Crystals absent.

Tangential section

Rays 5–8-seriate.

Radial section

Perforation plates simple. Intervessel pits round in alternating position. Vessel length 50–100 µm. Fibers with small pits. Ray cells prostrate. Crystals absent.

Bark

Phloem structure indistinct, due to compressed cells. With groups of thick-walled fibers. Some sclerenchyma cells in groups. Slight ray dilatation. Crystals absent. Phellem cells square and rectangular.

Pith

Round, with crystal sand and dark-stained substances.

Analyzed: Stem and twig of 3 individuals.

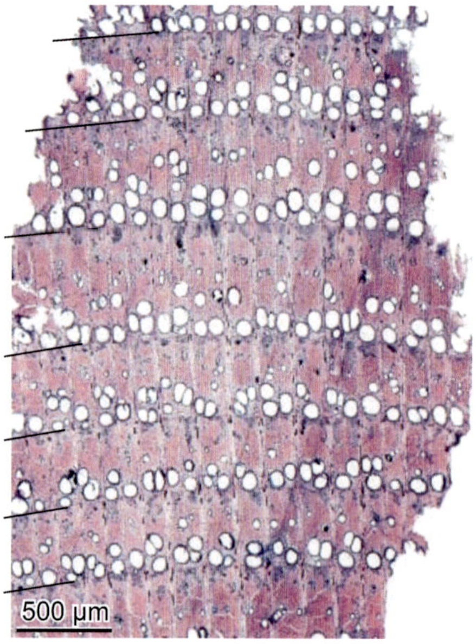

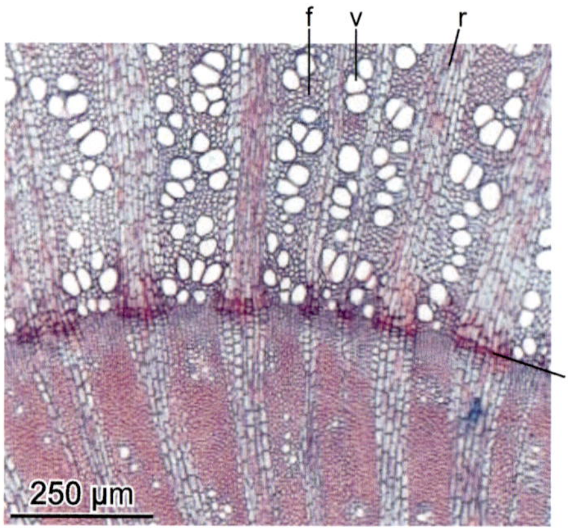

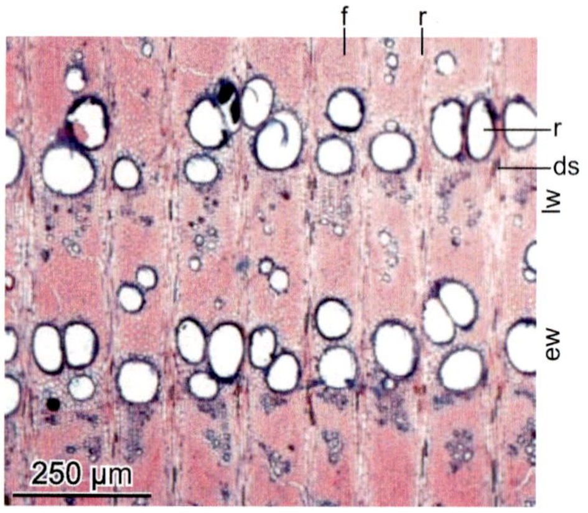

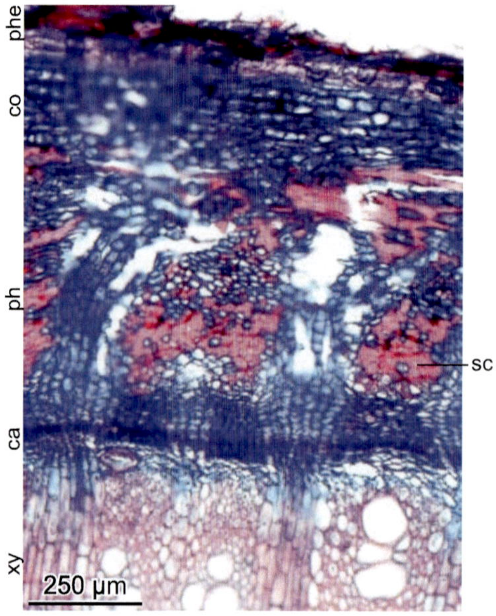

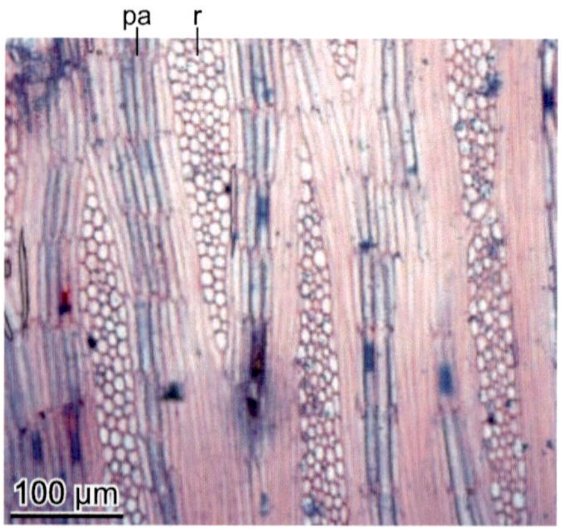

Features according to IAWA

Xylem 1 - 3 - 11 - 22 - 42 - 50.1 - 52.2 - 61 - 69 - 79 - 98 - 104

Urtica dioica L.

Urticaceae

Description: Perennial herbs, 30–120 cm tall, dioecious, covered with stinging hairs, leaves ovate, margin serrate, flowers greenish, in paniculate inflorescences.

Distribution: North America, N Africa, Europe, temperate Asia, Russia (Siberia), Pakistan, N India, Nepal, China, Bhutan; naturalized elsewhere.

Habitat: Eutrophicated sites, moist places in forests.

Stem diameter 3–6 mm. Bark/xylem proportion 1:4. Ring width 0.2–0.4 mm. Maximal 25 rings.

Xylem

Transverse section

Diffuse- to semi-ring-porous. Rings marked by differences between earlywood (50–90 μm) and latewood diameters (20–40 μm) and the absence of fibers in the latewood. Vessels solitary, in short radial multiples and in small groups. With unlignified tyloses. Vascular bundles remain. Fibers thin- to thick-walled. Parenchyma paratracheal and marginal. Bands of fibers and parenchyma cells alternate in the lateral parenchymatic zones of vascular bundles.

Tangential section

Rays dilating, >10-seriate, often confluent with the axial tissue.

Radial section

Perforation plates simple. Intervessel pits round to scalariform. Vessel length 100–150 μm. Ray cells upright. Crystals absent.

Bark

Phloem uniform, sieve tubes collapse. Rays dilate. Cortex composed of oval, often compressed parenchyma cells. With small groups of sclereides. Phellem cells rectangular. Crystals absent.

Pith

Round, filled with unlignified, round cells.

Analyzed: Rhizoms of 10 individuals.

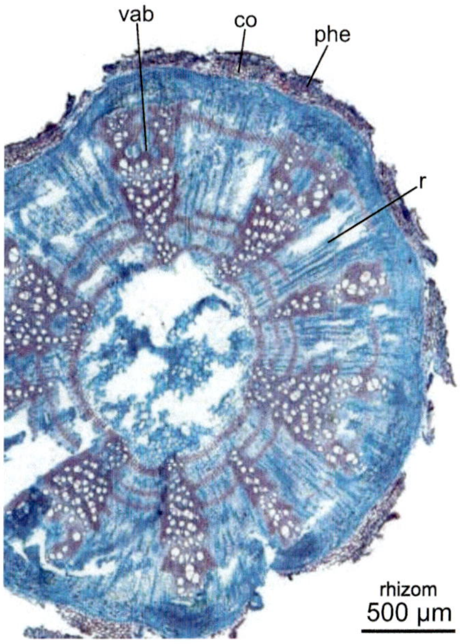

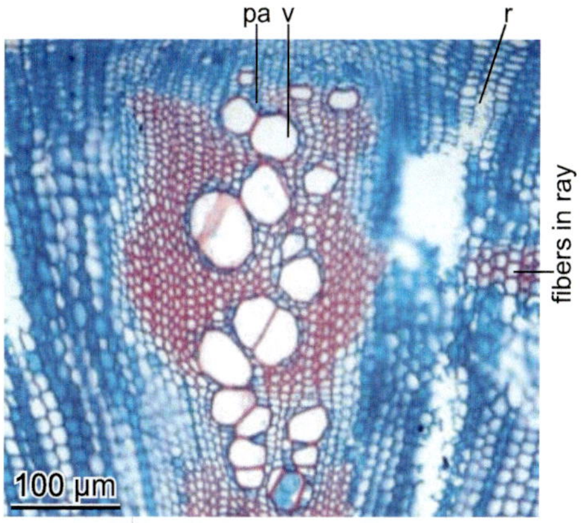

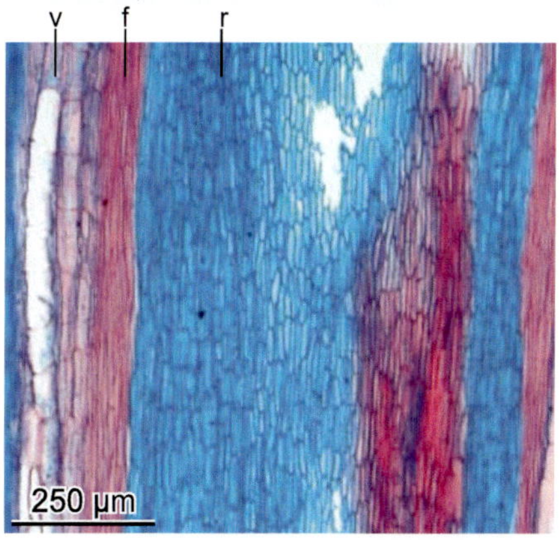

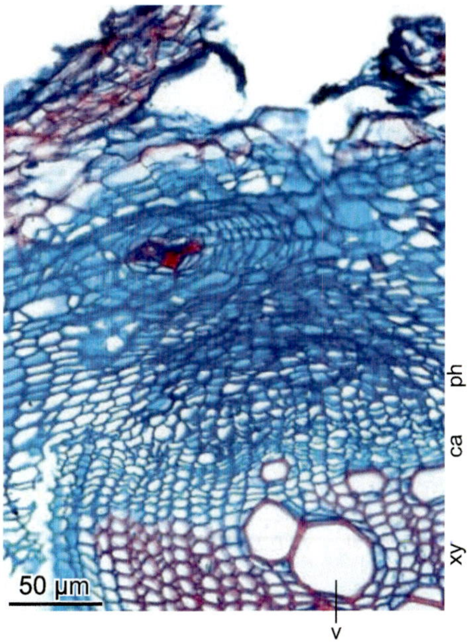

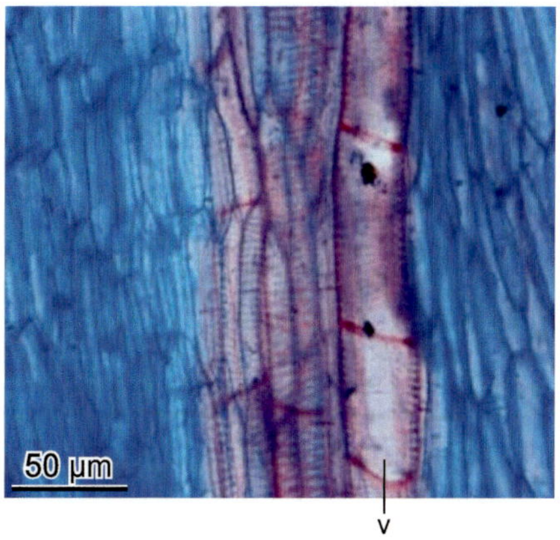

Features according to IAWA

Xylem 1 - 4 - 5 - 9 - 9.1 - 11 - 13 - 20 - 22 - 41 - 50.1 - 52.3 - 56 - 61 - 70 - 79 - 89 - 99 - 99.1 - 100.1 - 100.2 - 105

Urtica hyperborea Jacquem. ex Wedd.
Urticaceae

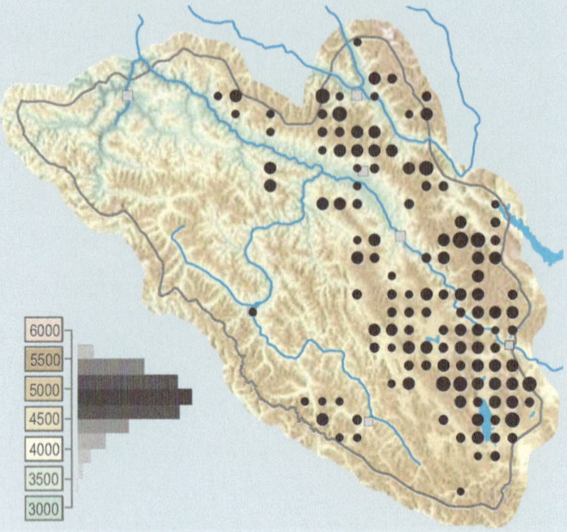

Description: Perennial herbs, 10–50 cm tall, cespitose, covered with stinging hairs, leaves ovate or cordate, floweres greenish-purple, in axillary clusters.

Distribution: Pakistan, N India, China (Gansu, Qinghai, Sichuan, Xinjiang, Xizang).

Habitat: Eutrophicated and disturbed shaded sites, dry stony stream bottoms, stabilized loamy and stony slopes.

Stem diameter 5–11 mm. Mean ring width 0.9 mm. More than 5 rings. Center rotten.

Xylem

Transverse section

Rings absent or indistinct. Diffuse-porous. Diameter of large vessels 50–70 µm. Vessels in long radial multiples and in small groups. With unlignified tyloses. Vascular bundles remain. Fibers thick-walled. Parenchyma paratracheal. Bands of thin- to thick-walled fibers and parenchyma cells tangentially alternating in the lateral parenchymatic zones of vascular bundles.

Tangential section

Rays >10-seriate, often confluent with axial tissue.

Radial section

Perforation plates simple. Intervessel pits round to scalariform. Vessel length 100–120 µm. Ray cells upright. Crystals absent.

Bark

No slide available.

Pith

No slide available.

Analyzed: Rhizomes of 3 individuals.

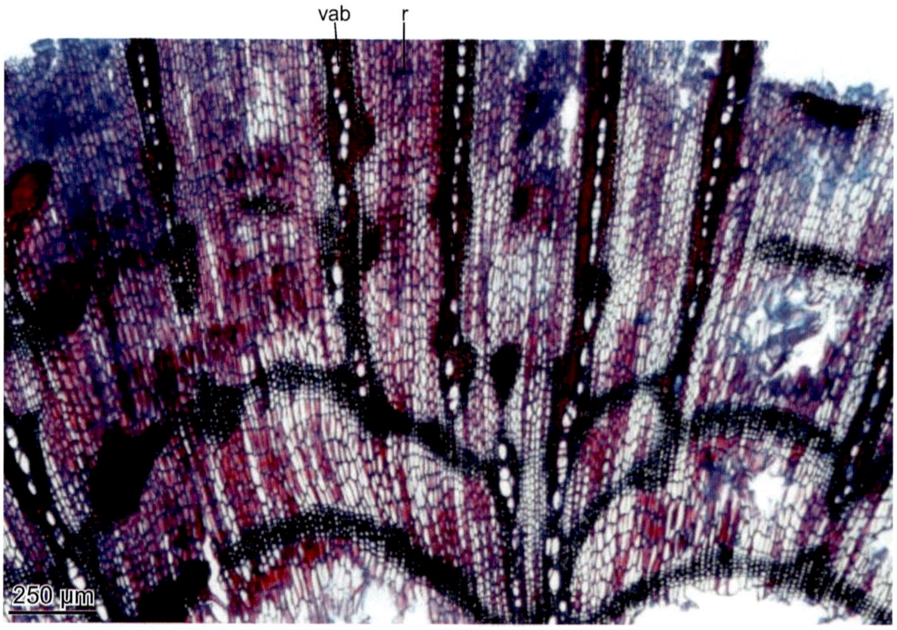

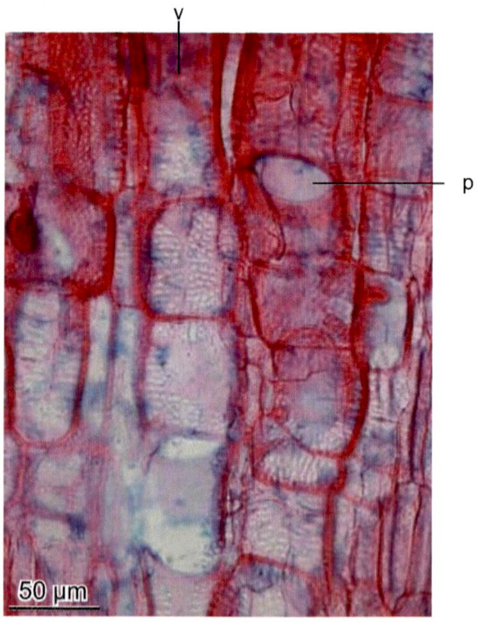

Features according to IAWA

Xylem 1 - 5 - 10 - 11 - 13 - 20 - 22 - 41 - 50.1 - 52.3 - 56 - 61 - 70 - 79 - 99 - 99.1 - 100.1 - 100.2 - 105

Valeriana himalayana Grubov

Valerianaceae

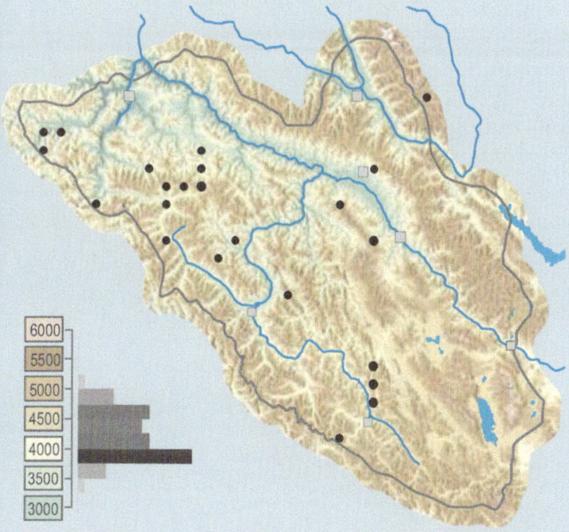

Syn.: *V. pusilla* Royle nom. nud., *V. dioica* auct. non L.: C.B.Clarke

Description: Perennial herbs, usually dioecious, 10–40 cm tall, radical leaves entire, cauline leave pinnatipartite with 1–3 lateral lobes, flowers white or pink, in dichotomous corymbs or compact capitula.

Distribution: NW Asia, Pakistan, India.

Habitat: Wet rock crevices, marshes, wet places.

Stem diameter 1 mm. Bark/xylem proportion 1:0.3. Mean ring width 0.15 mm. Maximal 1 ring.

Xylem

Transverse section

Only one ring. Angular vessels in groups. Vessel diameter 15–20 µm. Vascular bundles remain. Fibers absent. Parenchyma paratracheal and pervasive.

Tangential section

Rays 5–10-seriate between vascular bundles.

Radial section

Perforation plates simple. Intervessel pits scalariform.

Bark

Phloem uniform. Cortex cells large (50–60 µm), oval. Phellem absent.

Pith

Round.

Analyzed: Root collar of polar root of 1 individual.

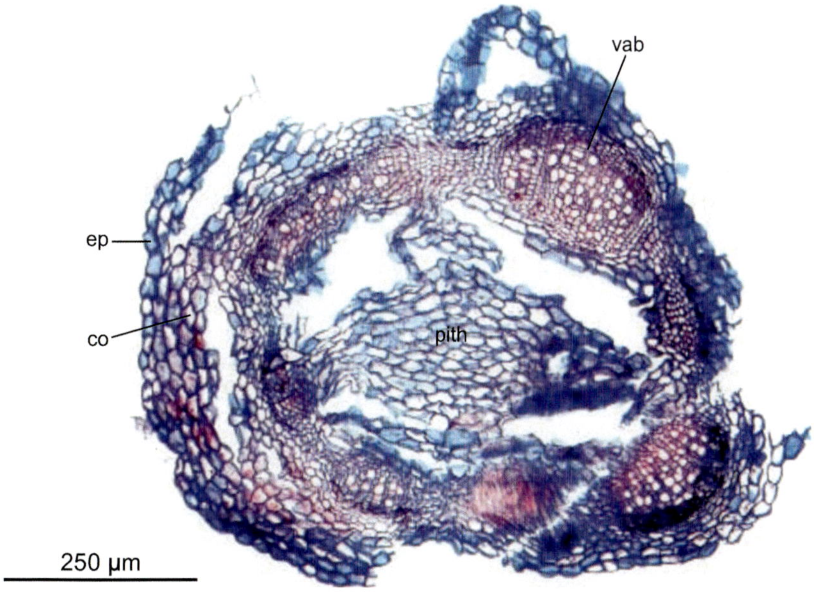

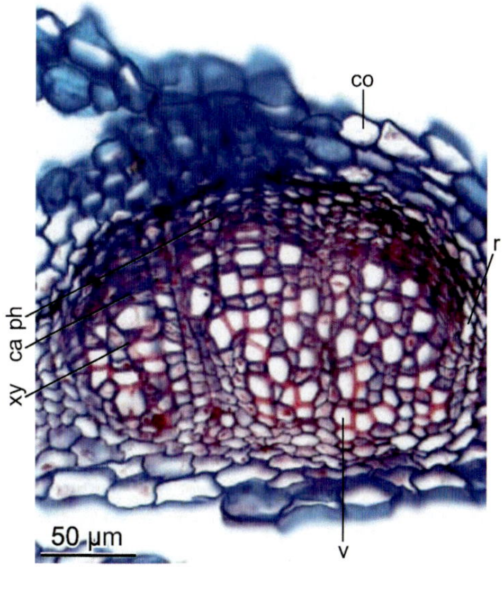

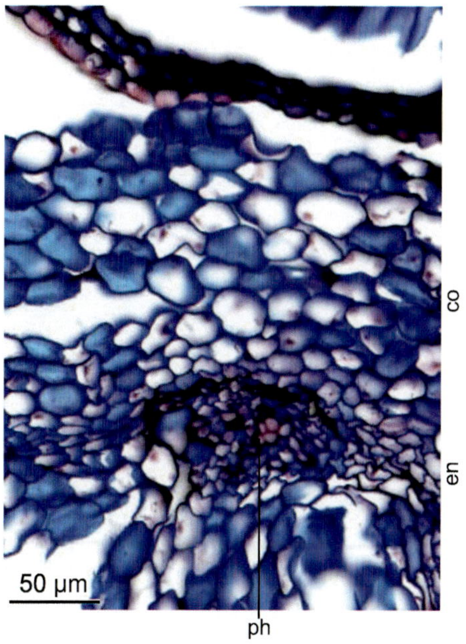

Features according to IAWA

Xylem 2.1 - 11 - 12 - 13 - 20 - 40.1 - 50.2 - 60.1 - 79 - 79.1 - 98 - 105

Viola kunawurensis Royle

Violaceae

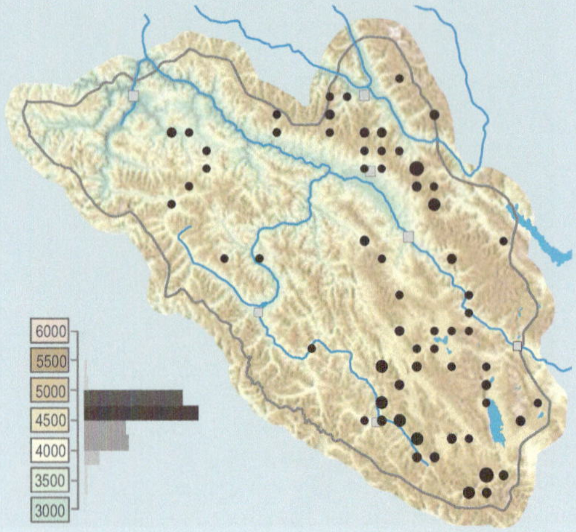

Syn.: *V. thianschanica* Maxim.

Description: Perennial herbs, 2–6 cm tall, stemless, leaves rosulate, ovate, or oblong, margin entire or remotely and shallowly crenate, flowers white, liliac or purplish-blue, 7–10 mm long, solitary on pedicels longer than leaves.

Distribution: Russia, Kazakhstan, Kyrgyzstan, Tajikistan, Afghanistan, Pakistan, India, N Nepal, China, W Mongolia.

Habitat: Grasslands of river banks, wet screes, among stones in steppes, alpine meadows.

Stem diameter 1–1.5 mm. Bark/xylem proportion 1:1. Ring width 0.2 mm. Maximal 1 ring.

Xylem

Transverse section

Diameter of vessels 15–20 µm. Vessels solitary or in small groups. Fibers absent. Vascular bundles remain. Parenchyma pervasive. Large crystal druses in rays.

Tangential section

Rays >10-seriate between vascular bundles, often confluent with axial tissue.

Radial section

Perforation plates simple. Intervessel pits scalariform. Vessel length 50–100 µm. Ray cells upright.

Bark

Phloem uniform. Phellem consists of some layers of thin-walled, rectangular cells. Large crystal druses in rays between vascular bundles.

Pith

Polygonal.

Analyzed: Rhizome of 2 individuals.

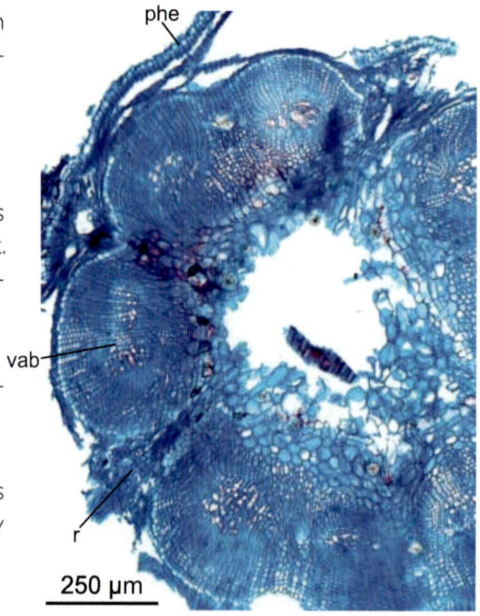

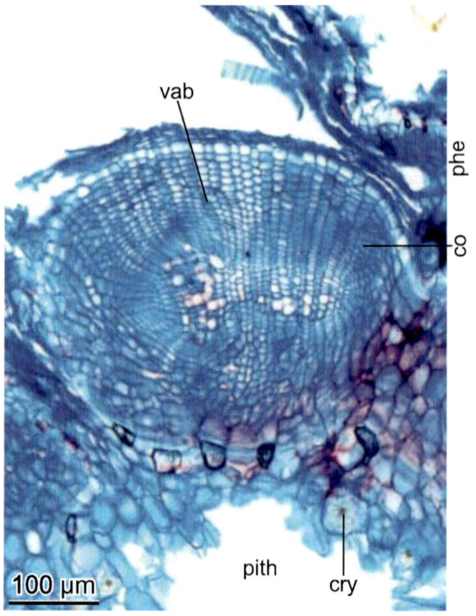

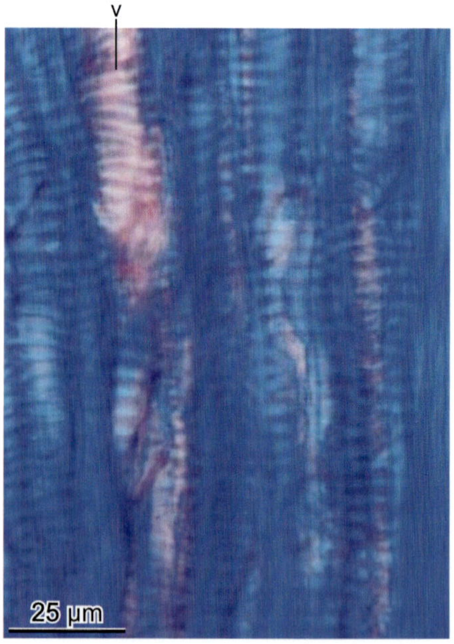

Features according to IAWA

Xylem 2.1 - 9 - 11 - 13 - 20 - 40.1 - 50.1 - 52.2 - 60.1 - 79.1 - 99.1 - 100.1 - 100.2 - 105 - 144

Tribulus terrestris L.

Zygophyllaceae

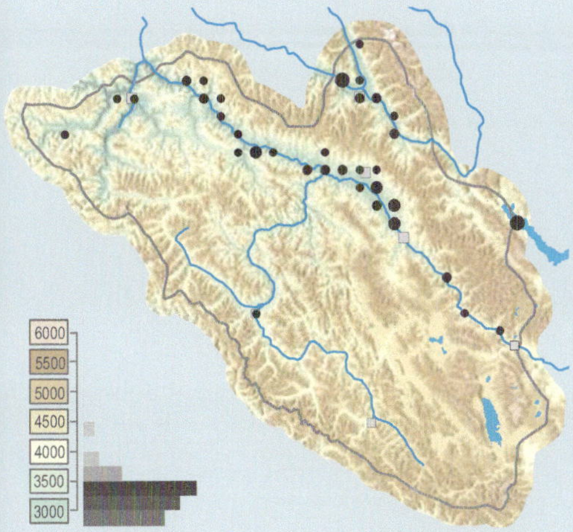

Description: Annuals or biennials, 1–5 cm tall, pubescent, branches prostrate, 5–40 cm long, leaves paripinnate, leaflets ovate to elliptic-oblong, flowers yellow, 1–1.5 cm across, solitary, fruits with spines.

Distribution: Africa, S Europe, temperate and tropical Asia, N Australia; widely naturalized.

Habitat: Wastelands, sandy areas.

Stem diameter 4 mm. Bark/xylem proportion 1:3. Ring width 1.5 mm. Maximal 1 annual ring.

Xylem

Transverse section

Vessels in short radial rows and groups, often arranged in tangential bands. Diameter of larger vessels 50–70 µm. Fibers thin- to thick-walled. Parenchyma paratracheal, rare. Crystals absent.

Tangential section

Rays 1–3-seriate, not lignified.

Radial section

Perforation plates simple. Intervessel pits round. Vessel length 70–100 µm.

Bark

Phloem with single sieve tubes and thin-walled fibers with irregular outline. Fibers in large radial groups. With few ray dialatations. Cortex cells square and rectangular, thin walled. Phellem cells rectangular. Crystals absent.

Pith

Round, filled with unlignified, round cells.

Analyzed: Root collar of polar root of 1 individual.

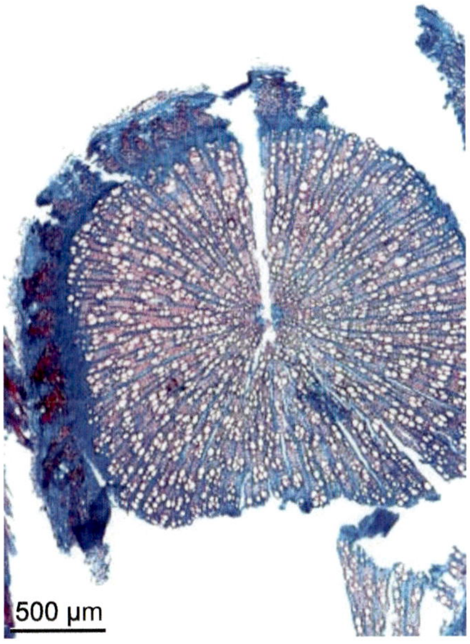

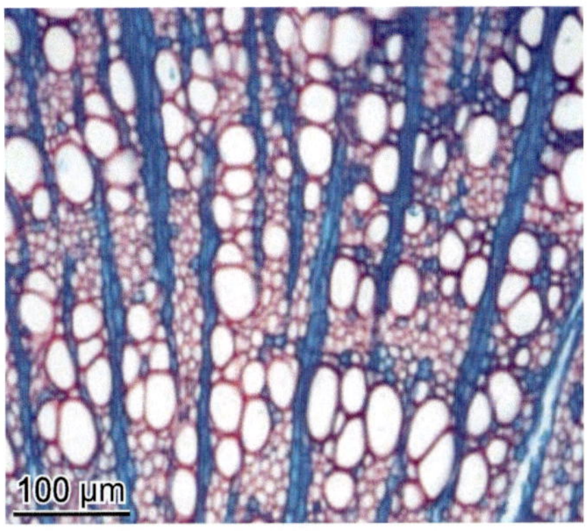

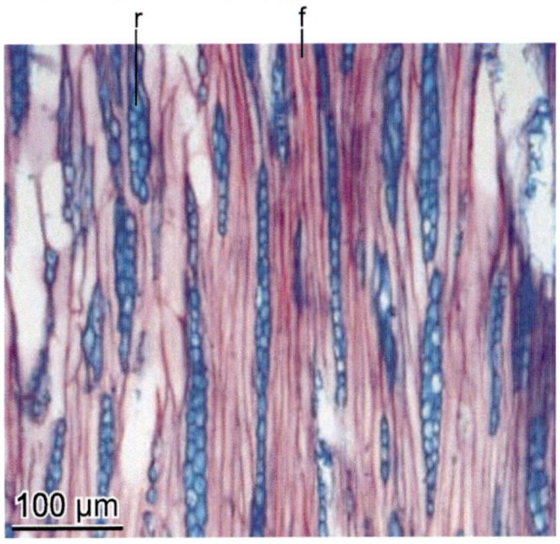

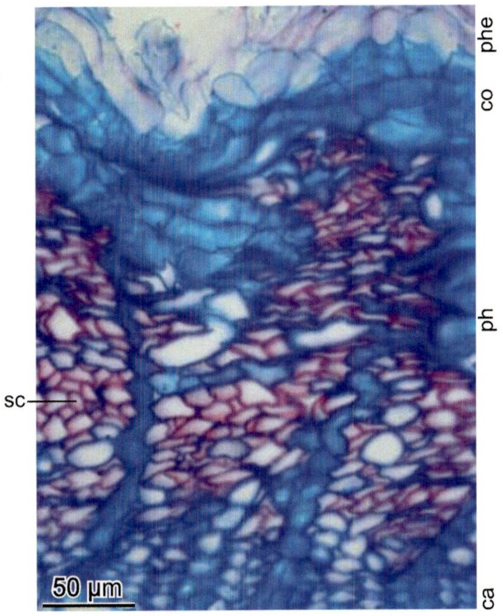

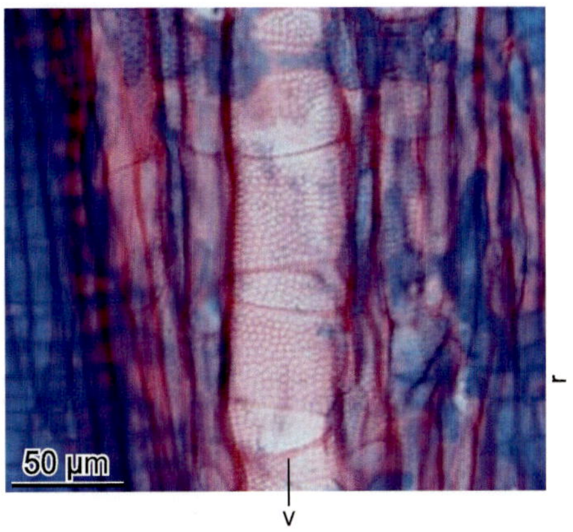

Features according to IAWA

Xylem 2.1 - 6 - 9.1 - 10 - 11 - 13 - 22 - 41 - 50.2 - 52.2 - 61 - 70 - 79 - 97 - 100.2 - 105

Anatomical descriptions of monocotyledons

J. Doležal et al., *Anatomy, Age and Ecology of High Mountain Plants in Ladakh, the Western Himalaya*, https://doi.org/10.1007/978-3-319-78699-5_4

Definition of culm features

Described are cross-section of culms primarily of Cyperaceae and Poaceae. The internal individual anatomical variability was reduced by sectioning middle parts of culms. Since culm structures are different in nodes, culm bases and rhizomes we consequently excluded them.

The following characteristics are principally based on

Metcalfe C.R. 1960. Anatomy of the Monocotyledons I. Gramineae. Oxford Clarendron Press. 731 pg.

Metcalfe C.R. 1971. Anatomy of the Monocotyledons V. Cyperaceae. Oxford Clarendron Press. 597 pg.

Characteristic features are defined by the outline of the culm, the structure of the epidermis, the arrangement of vascular bundles, the form and size of chlorenchyma (assimilatory tissue), the peripheral sclerenchymatic belt, the sclerenchymatic girders, the sclerenchymatic sheath around vascular bundles and cavities (intercellulars) in the parenchymatic tissues and vessel size and the presence of cavities in the protophloem of vascular bundles.

Outline of culms

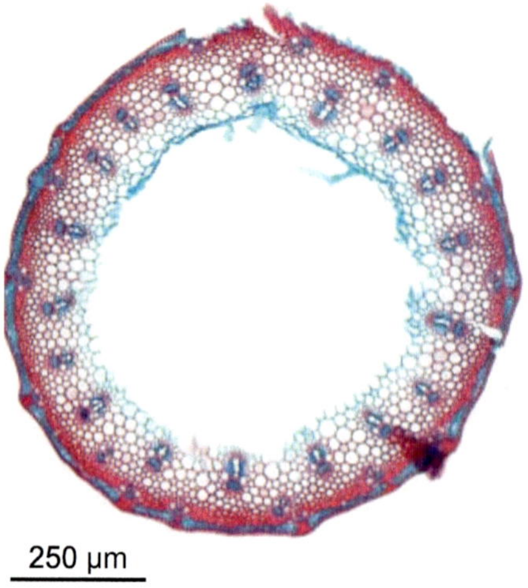

Outline circular smooth. *Koeleria macrantha*, Poaceae.

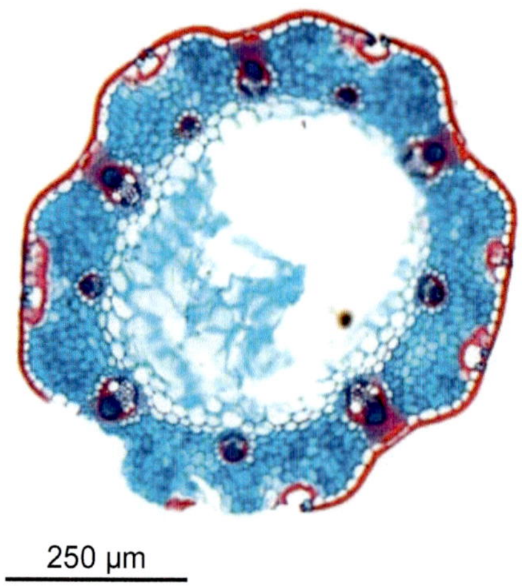

Outline circular wavy. *Trichophorum caespitosum*, Cyperaceae.

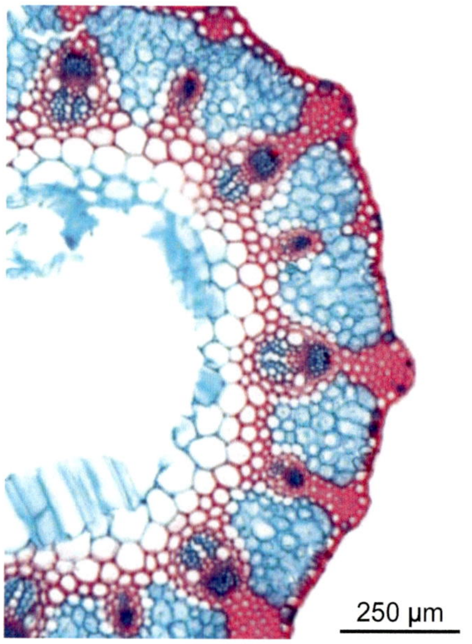

Outline circular with rips. *Sesleria caerulea*, Poaceae.

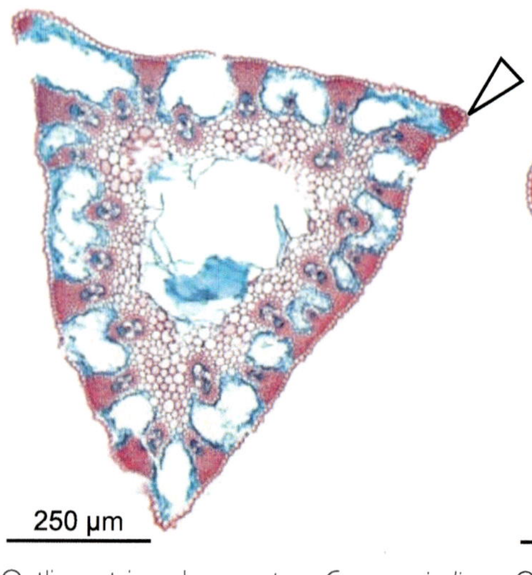

Outline triangular acute. *Carex nivalis*, Cyperaceae.

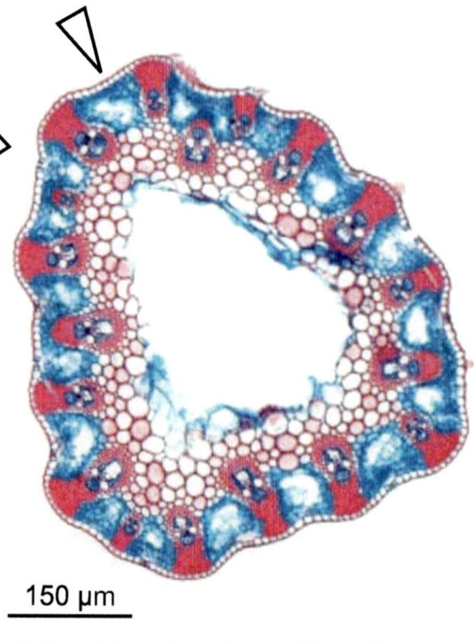

Outline triangular obtuse. *Carex haematostoma*, Cyperaceae.

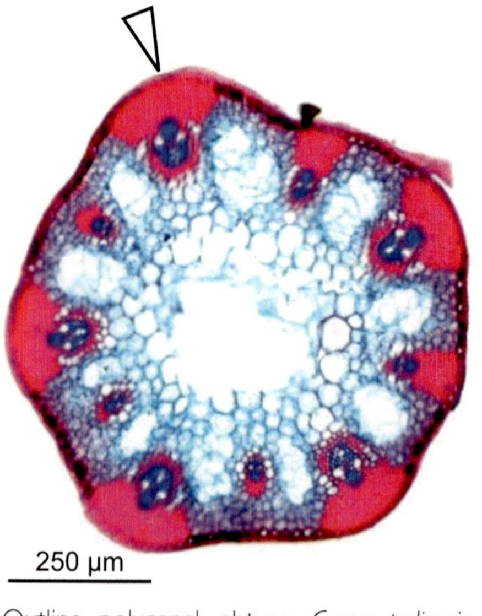

Outline polygonal obtuse. *Carex pulicaris*, Cyperaceae.

Epidermis cell walls

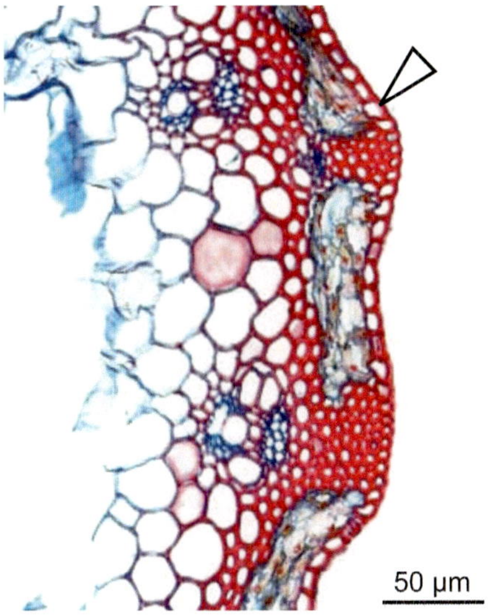

Epidermis smooth, cells thin- to thick-walled. *Alopecurus arundinaceus*, Poaceae.

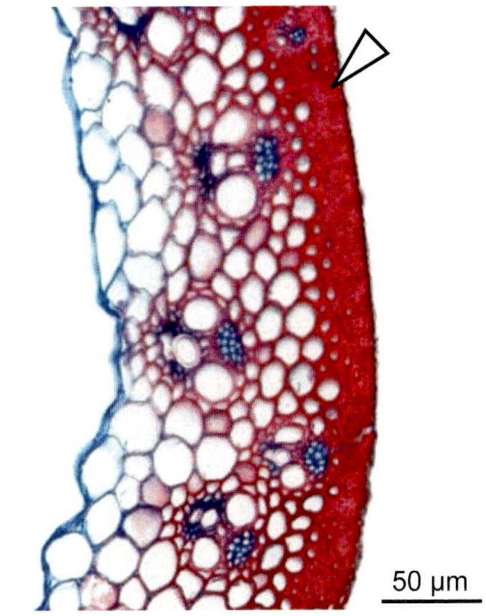

Epidermis smooth, cells thick-walled. *Muhlenbergia huegelii*, Poaceae.

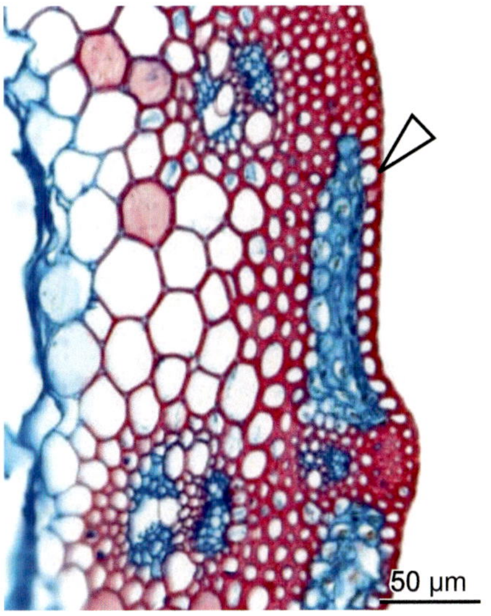

Epidermis smooth, cells thin-walled inside, thicker-walled outside. *Hierochloe laxa*, Poaceae.

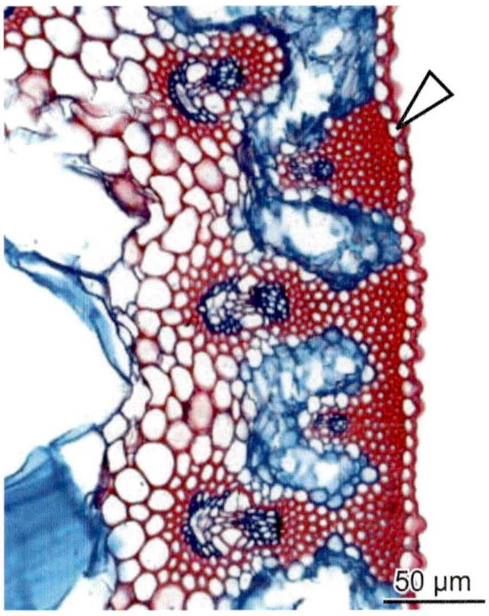

Epidermis bulbiform (papillous). *Carex nivalis*, Cyperaceae.

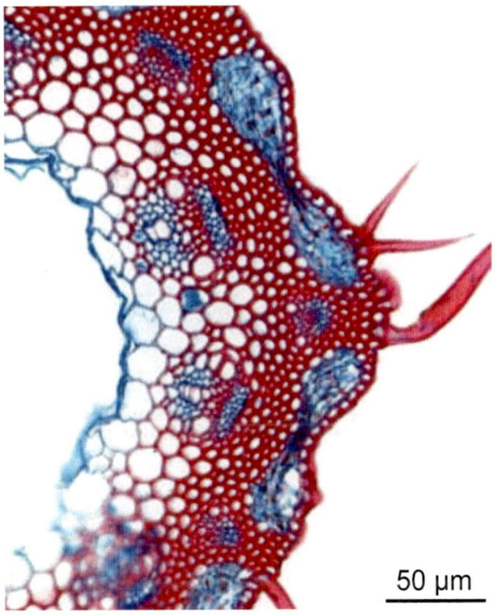

Epidermis with hairs. *Trisetum spicatum*, Poaceae.

Arrangement of vascular bundles

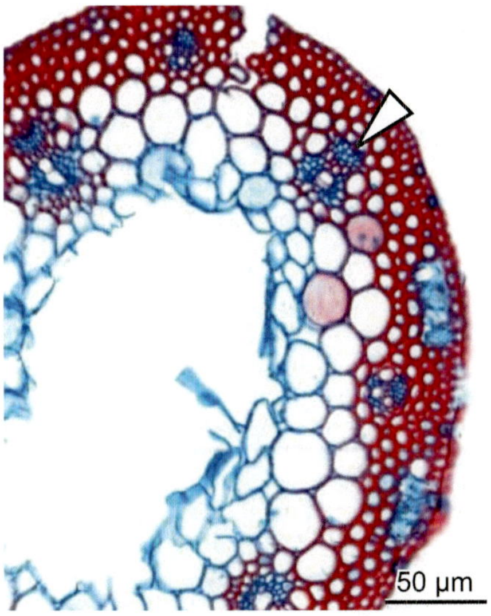

Vascular bundles arranged in one circular row. *Puccinellia ladakhensis*, Poaceae.

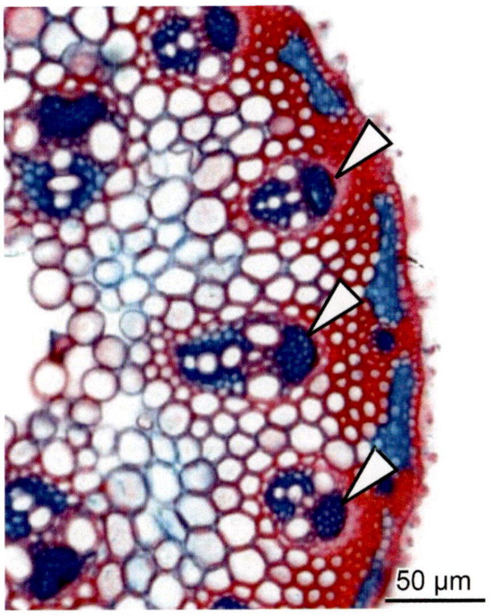

Vascular bundles arranged in two to three circular rows. *Stipa caucasica*, Poaceae.

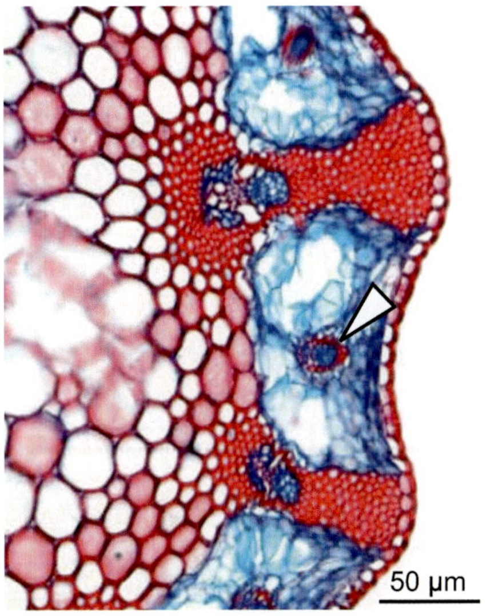

Vascular bundles arranged in peripheral bundles within the chlorenchyma. *Kobresia nitens*, Cyperaceae.

Form of chlorenchyma

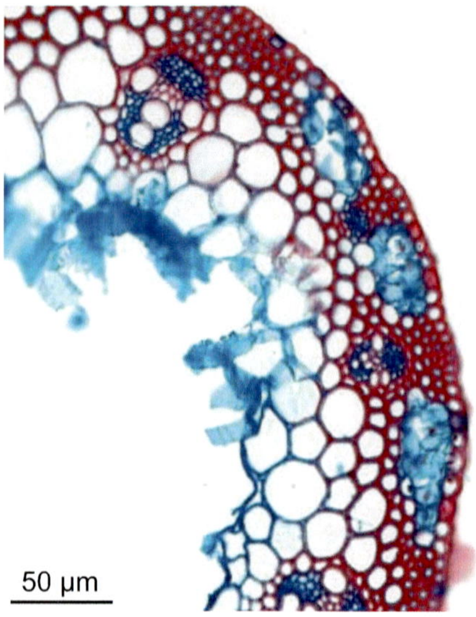

Chlorenchyma round to oval, often slightly irregular. *Puccinellia thomsonii*, Poaceae.

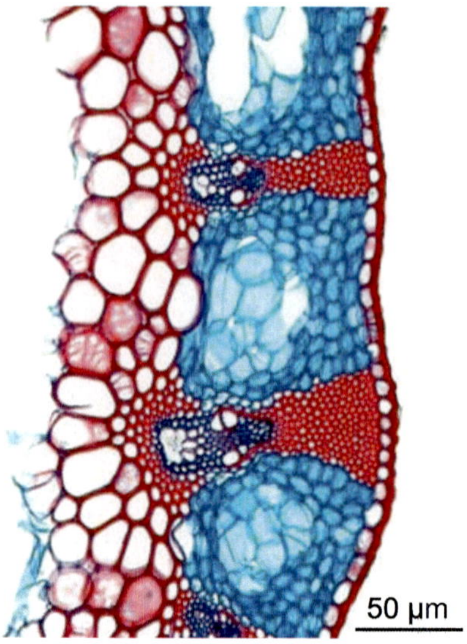

Chlorenchyma square to rectangular. *Carex montis-everesti*, Cyperaceae.

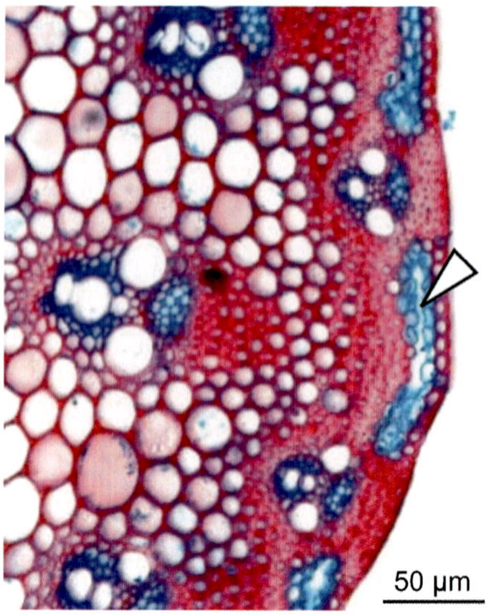

Chlorenchyma tangentially enlarged. *Bromus pectinatus*, Poaceae.

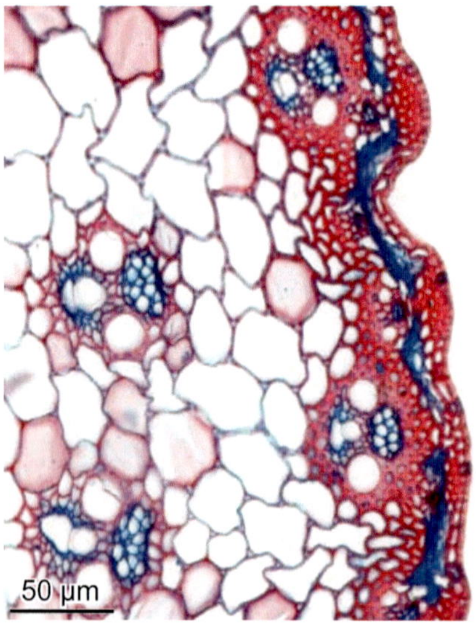

Chlorenchyma a 1–2-seriate line. *Pennisetum flaccidum*, Poaceae.

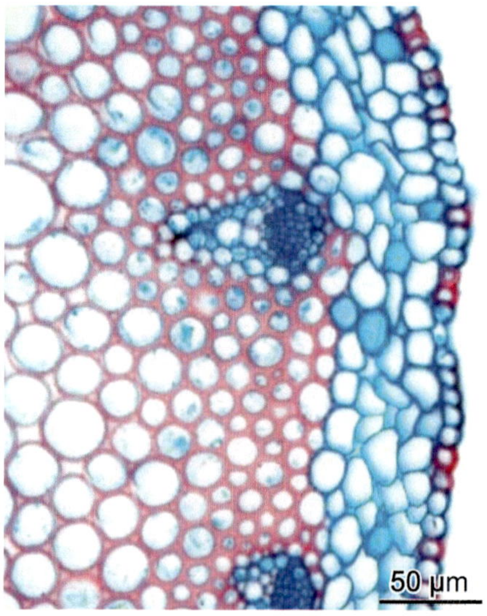

Chlorenchyma a 1–2-seriate line. *Allium* sp., Amaryllidaceae.

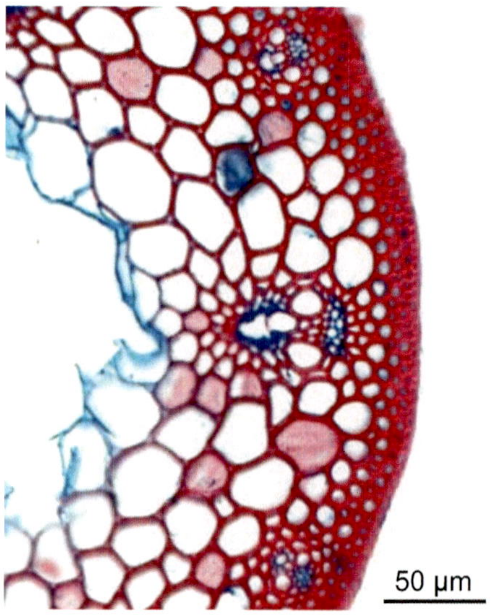

Chlorenchyma absent. *Vulpia myuros*, Poaceae.

Girders

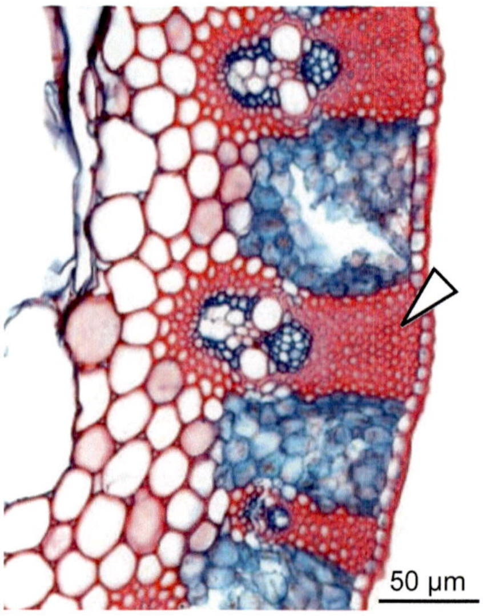

Girders square. *Carex plectobasis*, Cyperaceae.

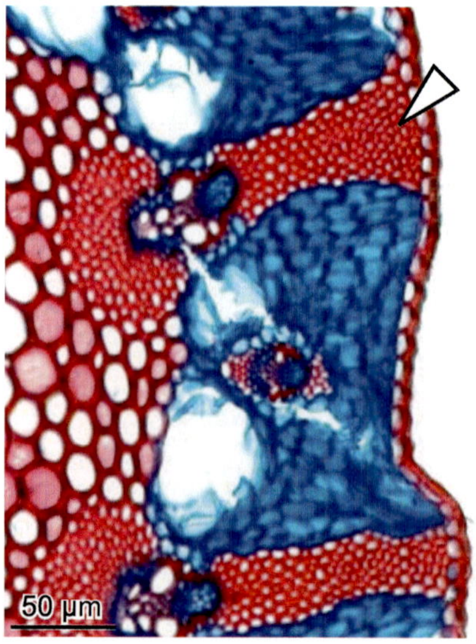

Girders radially rectangular. *Kobresia robusta*, Cyperaceae.

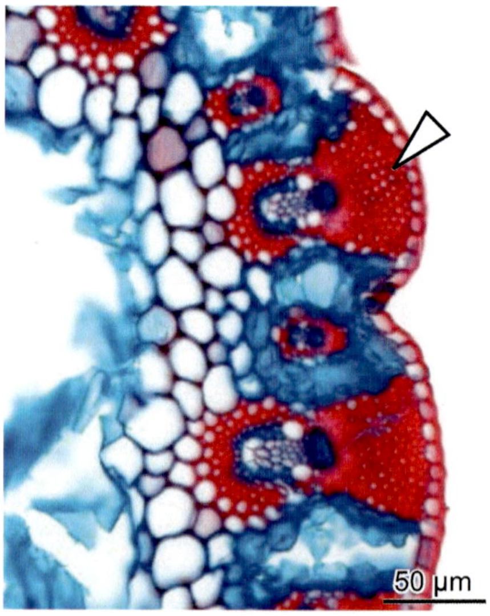

Girders conical. *Carex karoi*, Cyperaceae.

Peripheral belt of sclerenchyma

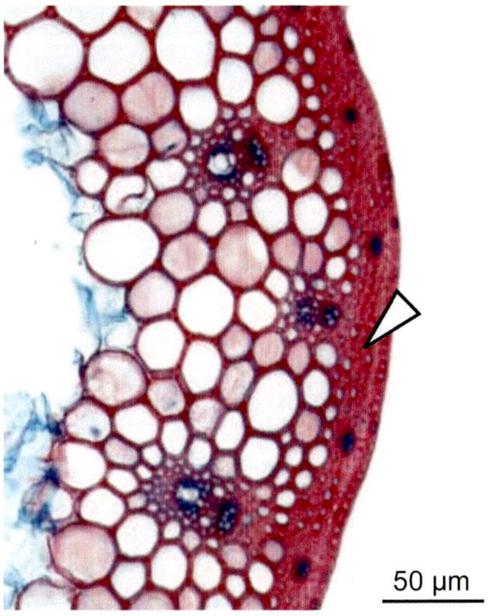

Peripheral belt of sclerenchyma >3-seriate. *Stipa mongholica*, Poaceae.

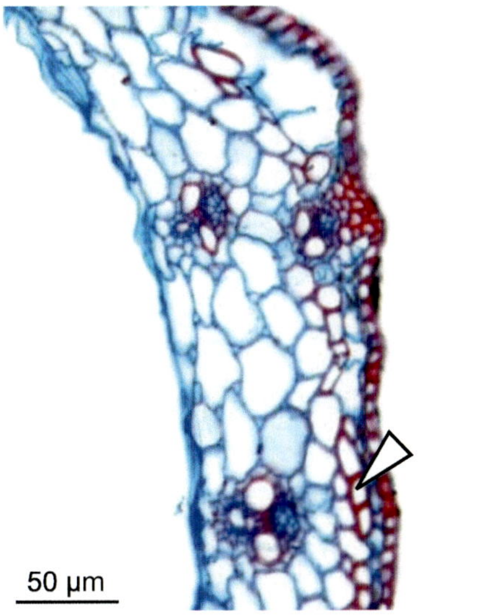

Peripheral belt of sclerenchyma 1–3-seriate. *Agrostis aequalis*, Poaceae.

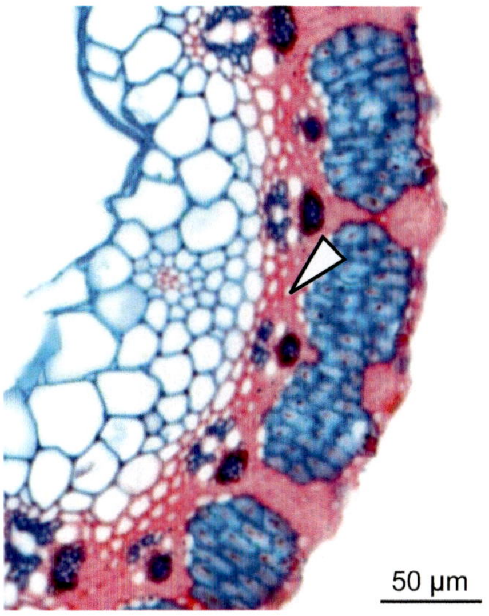

Peripheral belt of sclerenchyma inside of the chlorenchyma; smooth or wavy. *Poa sterilis*, Poaceae.

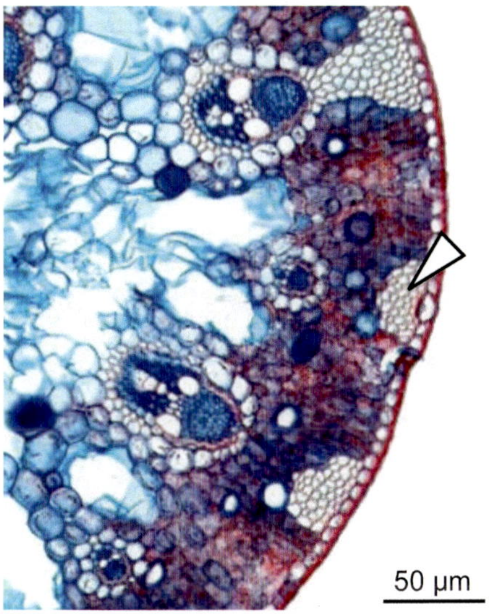

Isolated peripheral sclerenchymatic groups. *Blysmus compressus*, Cyperaceae.

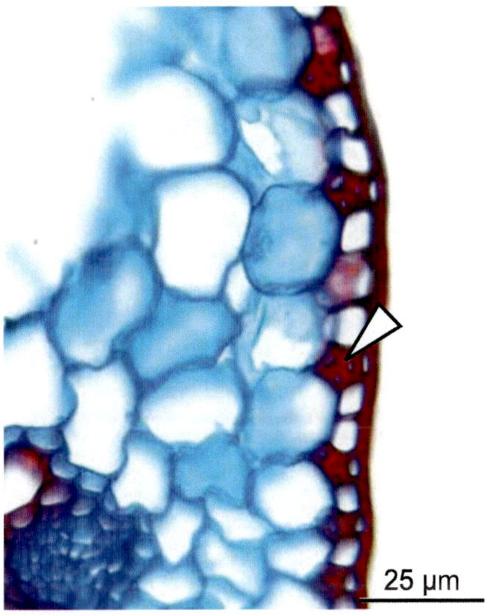

Isolated peripheral sclerenchymatic groups. *Eleocharis uniglumis*, Cyperaceae.

Vessel size

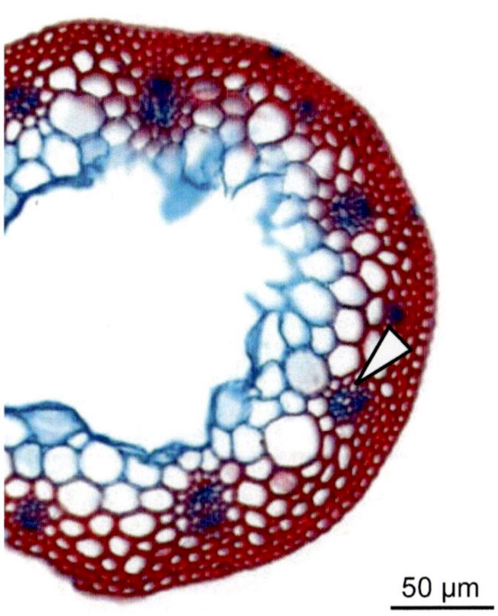

Vessel diameter <20 µm. *Festuca hartmannii*, Poaceae.

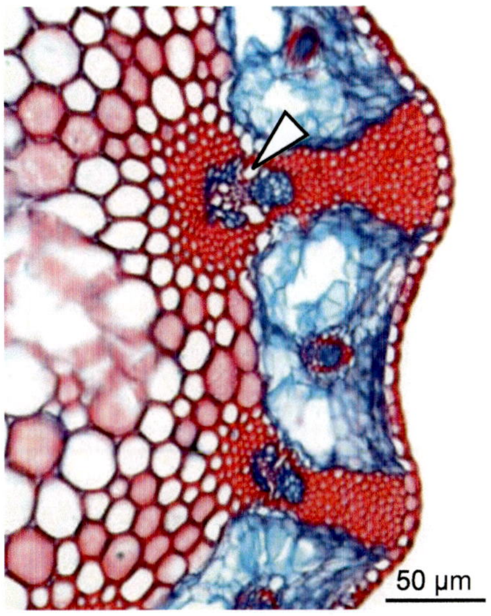

Vessel diameter 20–50 µm. *Kobresia nitens*, Cyperaceae.

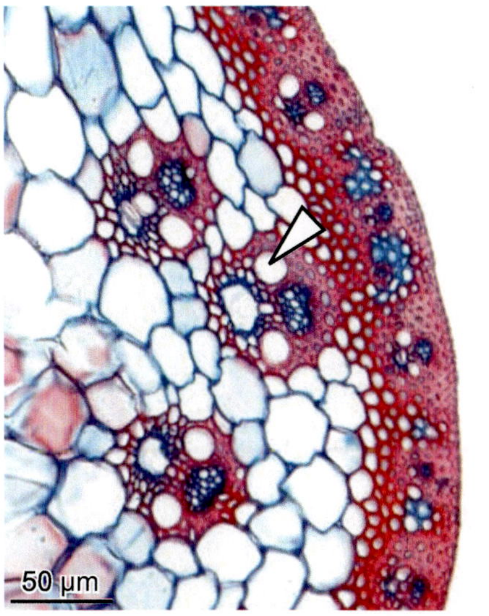

Vessel diameter 30–50 µm. *Digitaria sanguinalis*, Poaceae.

Sclerenchymatic sheath around vascular bundles

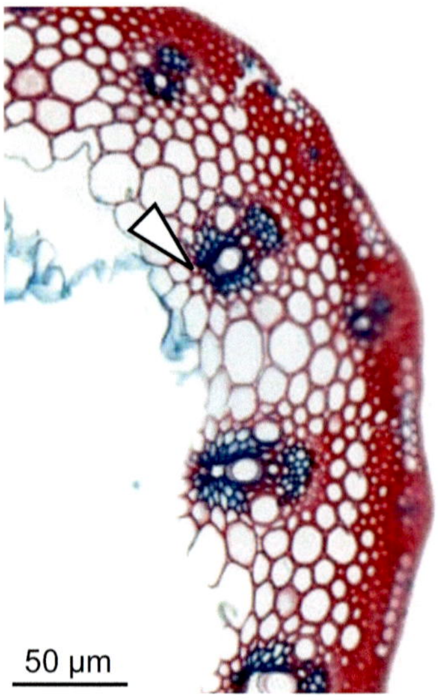

Sclerenchymatic sheath small; 1-2 cells around the vascular bundle. *Stipa himalaica*, Poaceae.

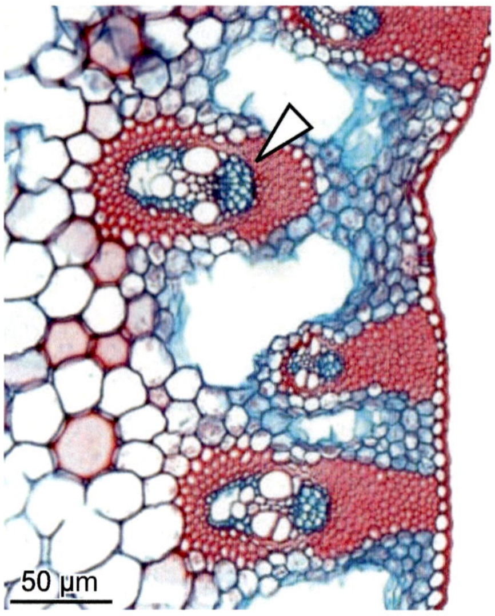

Sclerenchymatic sheath large ; 3 or more cells around or at least at the distal ends oft the vascular bundle. *Carex infuscata*, Cyperaceae.

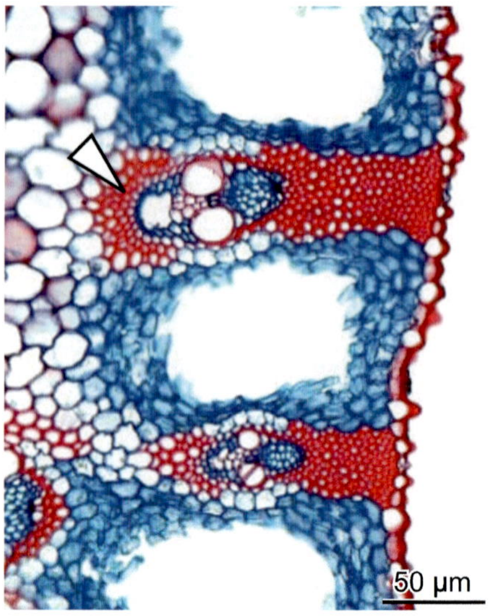

Sclerenchymatic sheath eccentric centripetally. *Carex moorcroftii*, Cyperaceae.

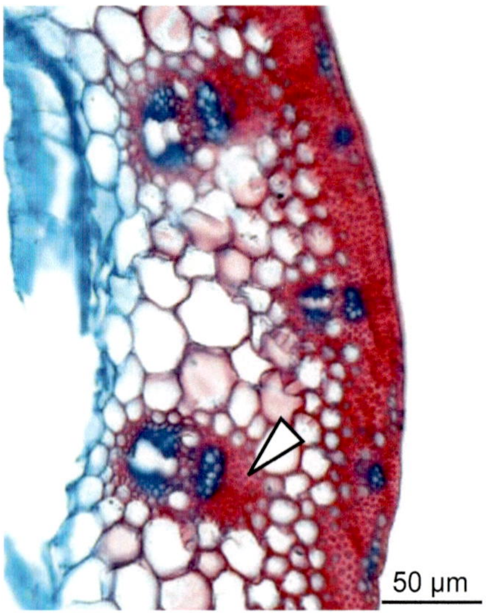

Sclerenchymatic sheath eccentric centrifugally. *Stipa brandisii*, Poaceae.

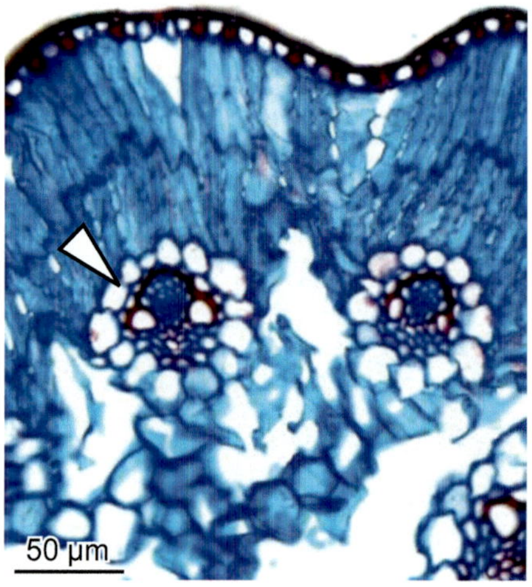

Sheath around vascular bundle not lignified. *Eleocharis uniglumis*, Cyperaceae.

Intercellulars in parenchymatic tissue

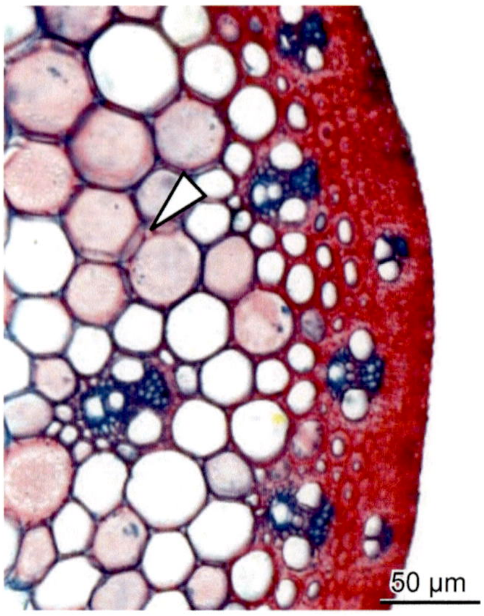

Cavities in parenchymatic zones small triangular. *Cymbopogon jwarancusa*, Poaceae.

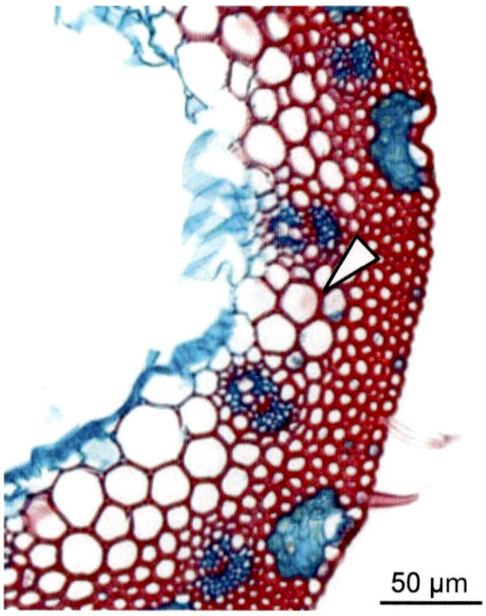

Cavities in parenchymatic zones absent. *Festuca coelestis*, Poaceae.

Intercellulars in the protoxylem

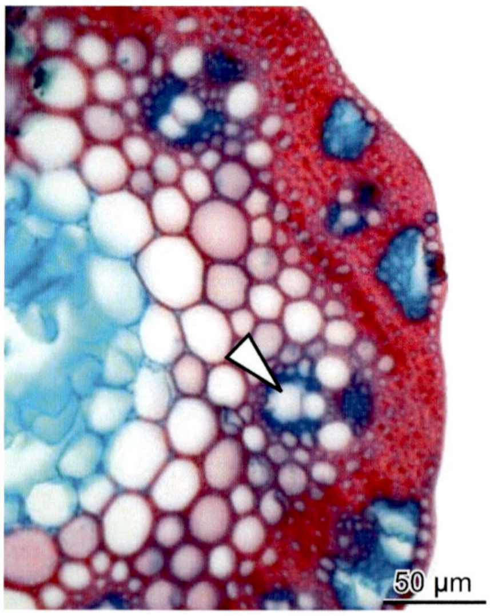

With cavities on the protoxylem of vascular bundles. *Bromus japonicus*, Poaceae.

Allium oreoprasum Schrenk

Amaryllidaceae

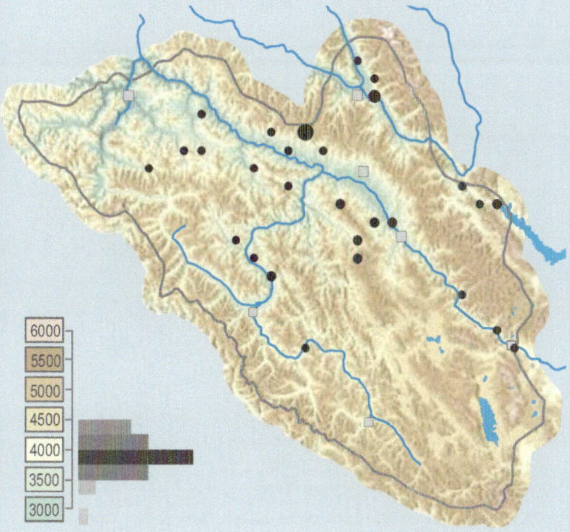

Description: Geophytes with clustered bulbs, tunic of bulbs yellowish brown and reticulate, scapes 20–40 cm tall, leaves 1–3 mm wide, shorter than scape, flowers pale red or pink, in lax umbels.

Distribution: Kazakhstan, Uzbekistan, Kyrgyzstan, Tajikistan, Afghanistan, Pakistan, India, China (Xinjiang, Xizang).

Habitat: Stony river banks, dry slopes.

Flower stalk round, filled, with a diameter of 2–2.5 mm. Center filled with large, thin-walled, polygonal, lignified cells and a few vascular bundles. Epidermis cells unlignified, thin-walled on the inside, thicker-walled and minutely dentate peripherally. Continuous, 5–7-seriate, thin- to thick-walled peripheral sclerenchymatic belt centripetal of the chlorenchyma. Vascular bundles arranged in two offset circular rows. Chlorenchyma consists of a continuous, 3–5-seriate peripheral band. Lateral vessels large, diameter 30–40 µm. Phloem consists of sieve tubes and companion cells. Intercellulars in the protophloem absent. Vascular bundle sheaths unlignified. With a few crystal druses.

Analyzed: 2 individuals

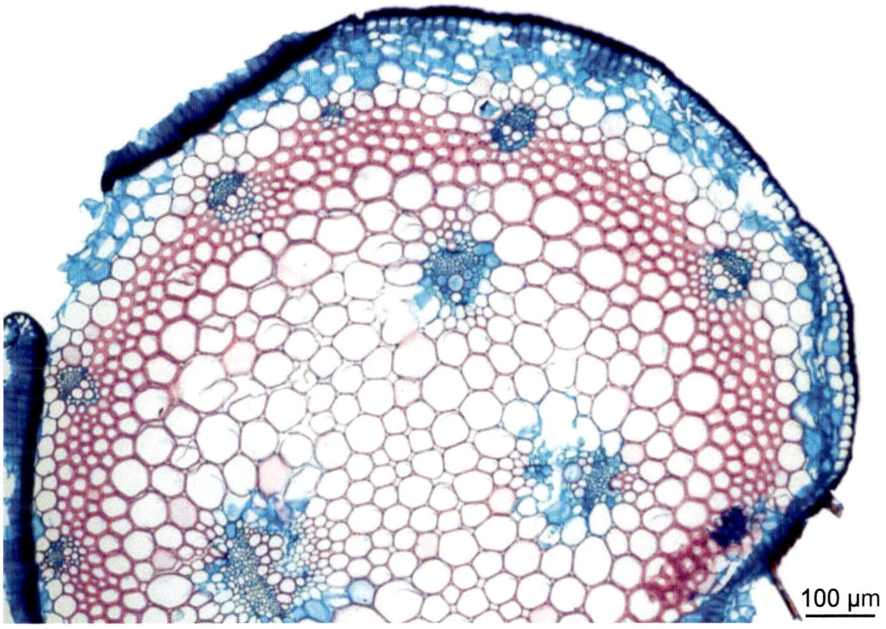

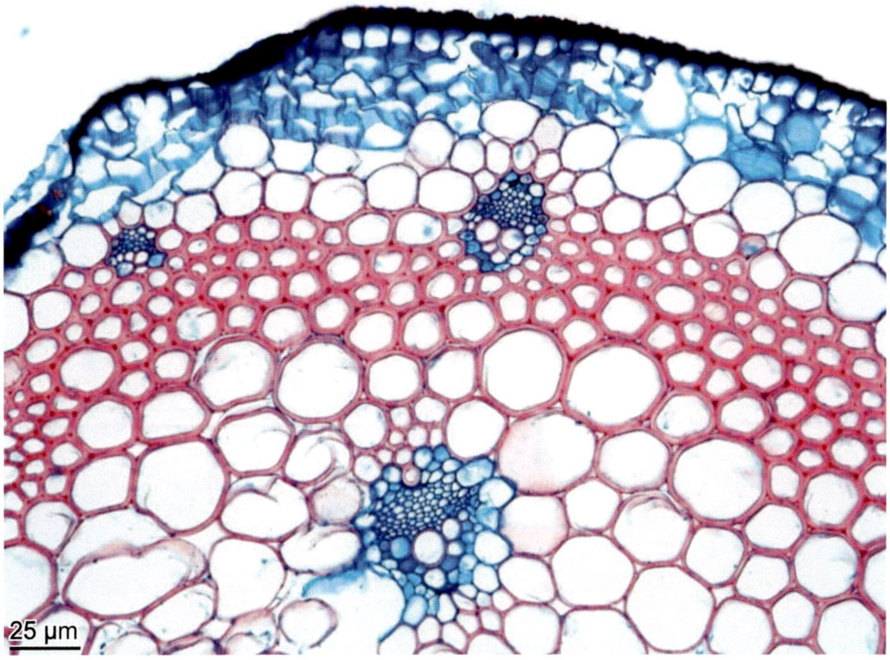

Allium przewalskianum Regel

Amaryllidaceae

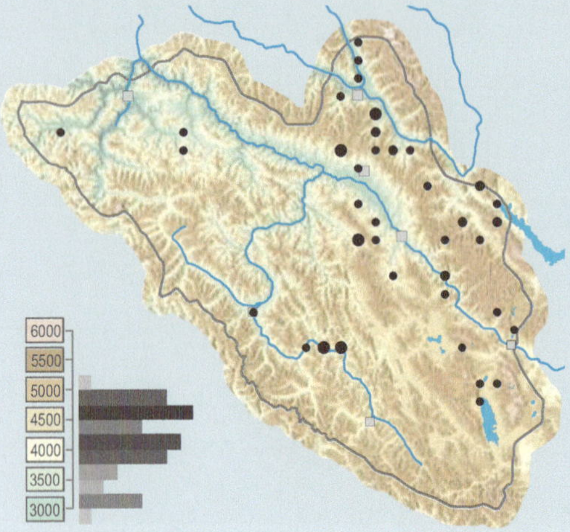

Description: Geophytes with clustered bulbs, tunic of bulbs red and reticulate, scape 10–40 cm tall, leaves 0.5–1.5 mm wide, flowers pale red to dark purple, in dense umbels.

Distribution: Pakistan, India, Nepal, China (Gansu, Nei Mongol, Ningxia, Qinghai, Shaanxi, Sichuan, Xinjiang, Xizang, Yunnan).

Habitat: Scrub, dry slopes, plains, rock crevices.

Flower stalk round, filled, with a diameter of 2–2.5 mm. Center filled with lignified, large, thin-walled, polygonal cells and poorly lignified vascular bundles. Epidermis cells unlignified, thin-walled on the inside, thicker-walled and minutely dentate peripherally. Continuous, 5–7-seriate, thin- to thick-walled peripheral sclerenchymatic belt centripetal of the chlorenchyma. Vascular bundles arranged in two offset circular rows. Chlorenchyma consists of a continuous, 5–6-seriate peripheral band. Lateral vessels large, diameter 20–30 µm. Phloem consists of sieve tubes and companion cells. Intercellulars in the protophloem absent. Vascular bundle sheaths unlignified. With a few grains of crystal sand.

Analyzed: 3 individuals

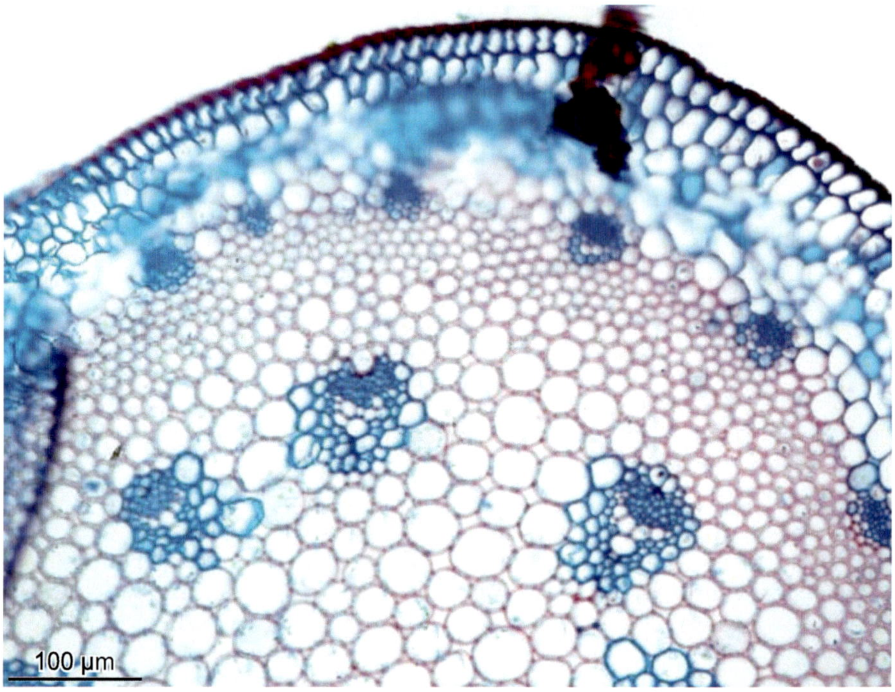

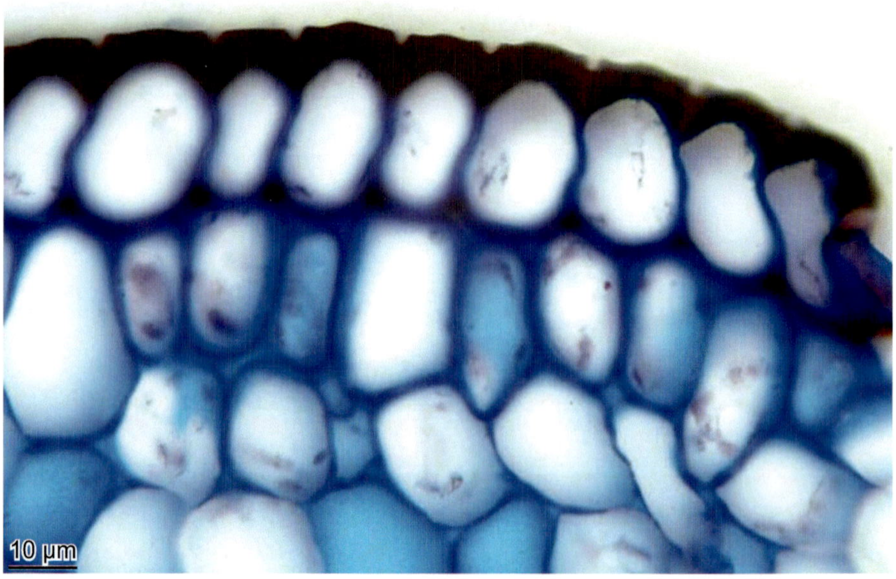

Blysmus compressus (L.) Panzer ex Link
Cyperaceae

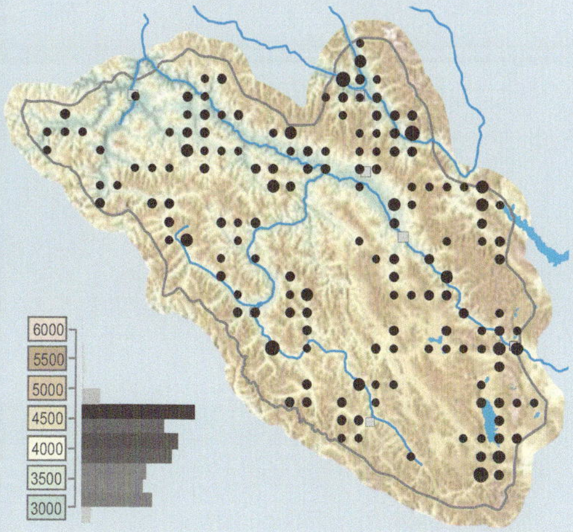

Syn.: *Schoenus compressus* L., *Scirpus compressus* (L.) Pers.

Description: Perennial herbs, 5–40 cm tall, loosely cespitose, leaves linear, shorter than stems, spikes ovoid, 10–35 × 5–12 mm, composed of 5–12 spikelets.

Distribution: NW Africa, Europe, temperate Asia, Mongolia, China.

Habitat: Wetland vegetation, tolerates salty or brackish substrates.

Culm circular to oval, hollow, with a diameter of 1–1.2 mm. Center hollow, surrounded by thin-walled unlignified cells. Epidermis cells thin-walled on the inside, thicker-walled peripherally (lignified). Large vascular bundles arranged in two peripheral rows. Chlorenchyma belt composed of round, palisade-like cells and with small vascular bundles. Large radial intercellulars inside the chlorenchyma belt. Conical girders and round groups of fibers. Vessels within the radial oval vascular bundles arranged in a T-shape or arc. Two large lateral vessels with a diameter of 50–70 µm, larger than those in the center of the bundle. Vessels of the protophloem arranged in radial groups and accompanied by centripetal intercellulars. Phloem unicellular, forming round groups. A sheath of thin-walled unlignified large round cells surrounds all bundles externally. In addition they are peripherally surrounded by an endodermis-like sheath of U-shaped lignified cells and towards the center by two layers of fibers. Laticifers or round cells occur in the whole stem and contain dark-staining substances.

Analyzed: 2 individuals.

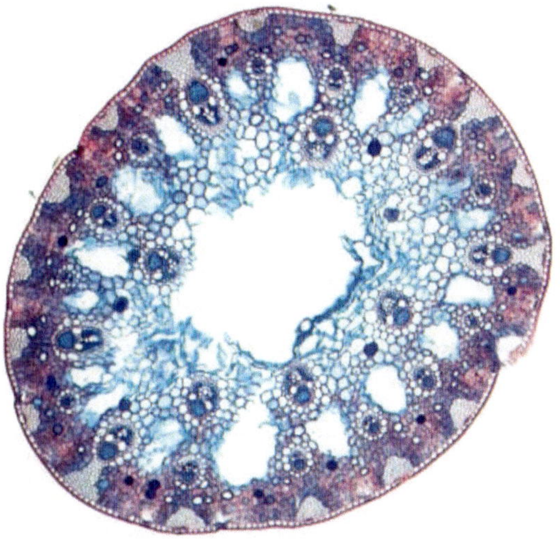

100 µm

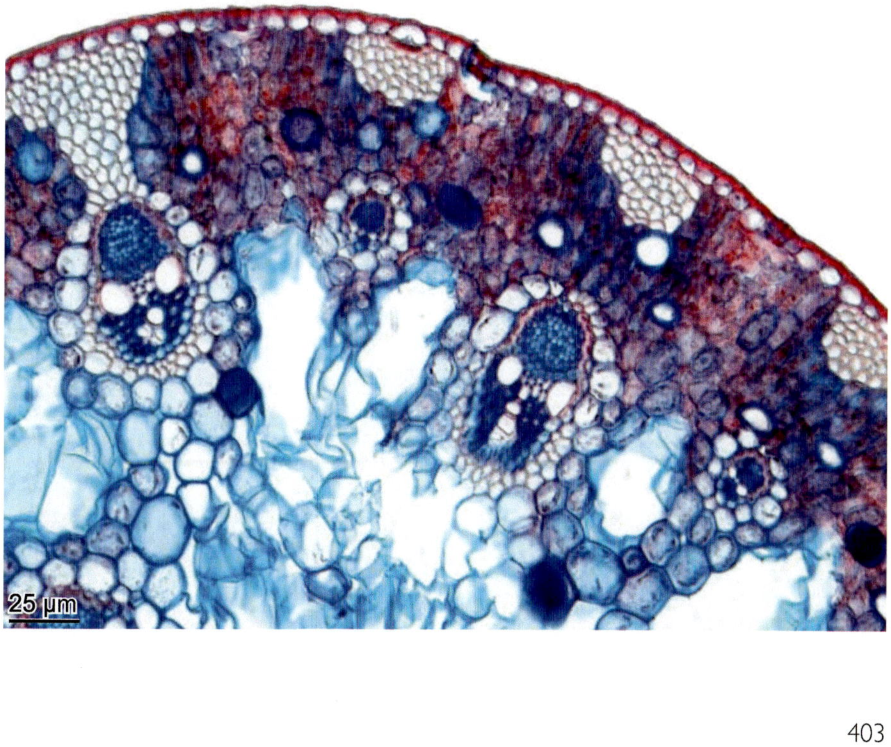

Carex atrofusca Schkuhr

Cyperaceae

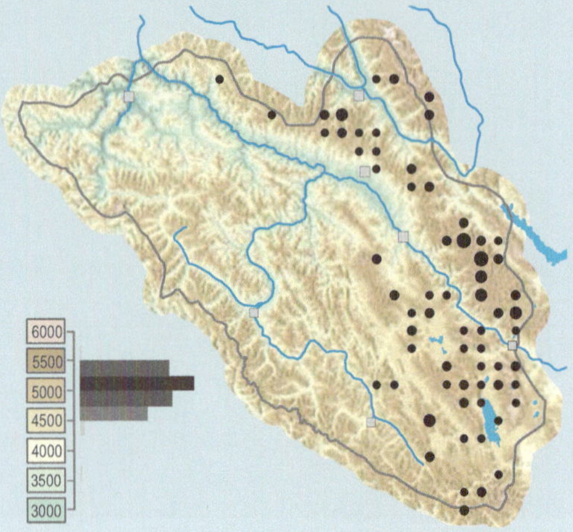

Description: Perennial herbs, 10–70 cm tall, rhizomatous, stems trigonous, leaves shorter than culms, spikes 2–5, upper 1–2 spikes male, lower ones female.

Distribution: Subarctic to temperate regions of Asia, Europe, and North America.

Habitat: Wet turf and spring wetlands, alpine meadows.

Culm obtusely triangular to obtusely pentagonal, hollow, with a diameter of 1.2 mm. Center hollow, surrounded by 2–4 rows of large, thin-walled, polygonal, lignified cells. Epidermis cells lignified, bulliform, thin-walled on the inside, thicker-walled peripherally. Vascular bundles arranged in two slightly offset peripheral rows within girders. Chlorenchyma consists of roundish groups, generally containing large, irregularly formed intercellulars. With conical to rectangular girders. Vessels within larger vascular bundles arranged in an arc. Lateral vessels large with a diameter of 15–20 µm. Vessels of the protophloem accompanied by small centripetal intercellulars. Phloem consists of round groups, containing sieve tubes and companion cells. Sclerenchymatic vessel sheaths eccentric, with thick-walled fibers in one row centrifugally, and groups in a crescent centripetally.

Analyzed: 2 individuals

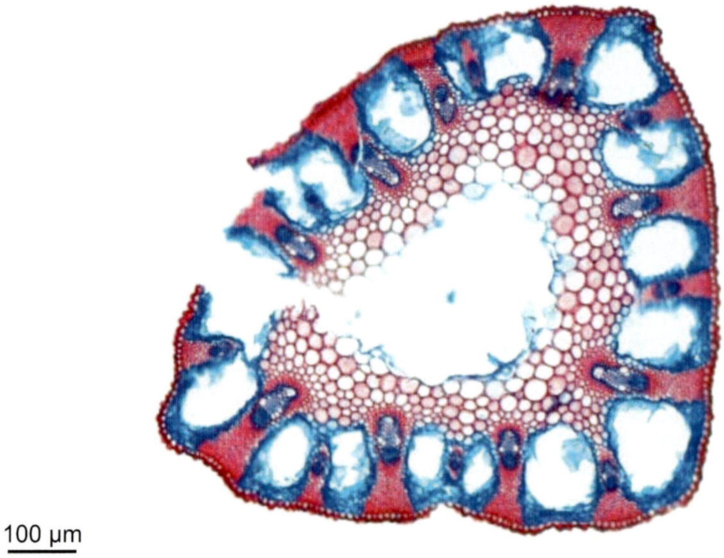

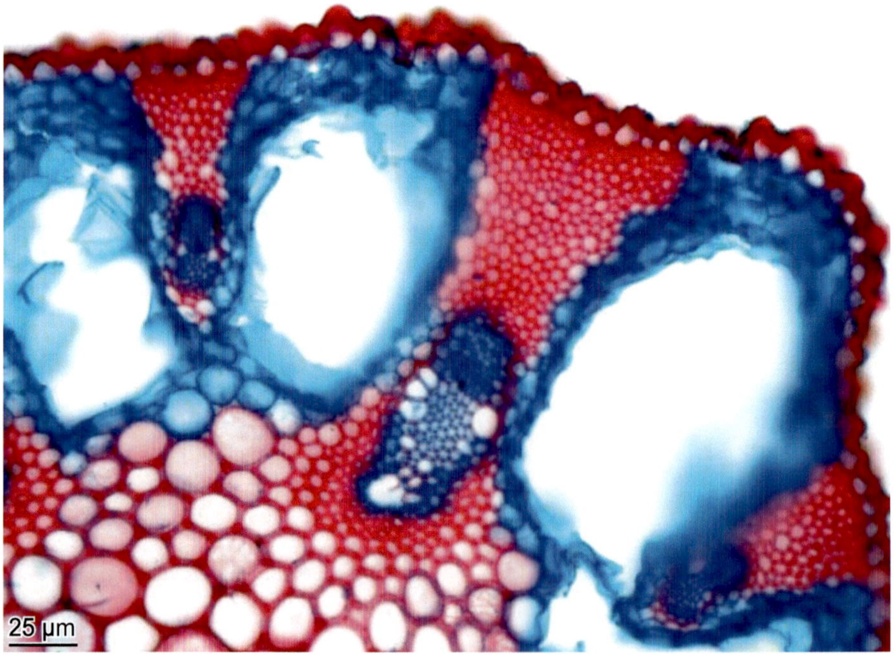

Carex borii Nelmes

Cyperaceae

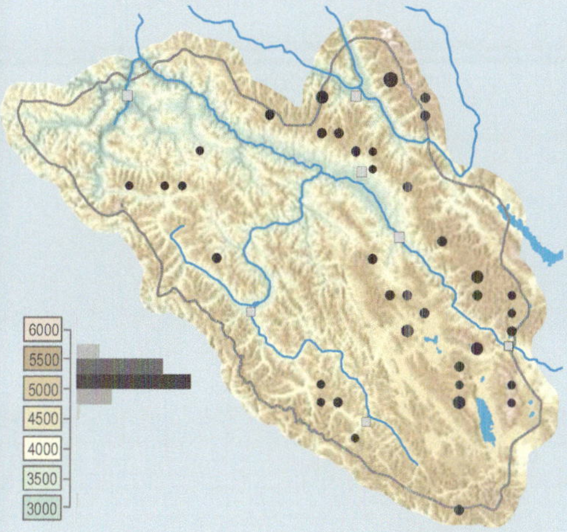

Description: Perennial herbs, 3–10 cm tall, cespitose, leaves as short as stems or shorter, inflorescence of 1–3 androgynous spikes, spikes 5–12 × 3–5 mm, ovoid to obovoid, sessile or subsessile.

Distribution: SE Tajikistan, N Pakistan, NW India (Jammu and Kashmir, Himachal Pradesh), China (Xizang).

Habitat: Dry to humid turf, water surplus habitats, scree.

Culm round, hollow, with a diameter of 0.8 mm. Center hollow, surrounded by 2–3 rows of large, thin-walled, polygonal, lignified cells. Epidermis cells lignified, smooth, thin-walled on the inside, thicker-walled peripherally. Vascular bundles arranged in two slightly offset rows within girders. Chlorenchyma consists of square to polygonal groups, generally containing large, irregularly formed intercellulars. Girders rectangular. Vessels within larger vascular bundles arranged in a T-shape. Lateral vessels large with a diameter of 5–8 µm. Phloem consists of round groups of uniform cells. Vessels of the protophloem accompanied by small centripetal intercellulars. Sclerenchymatic vessel sheaths eccentric, with thick-walled fibers in one row centrifugally, and multicellular in a U-shape centripetally.

Analyzed: 2 individuals

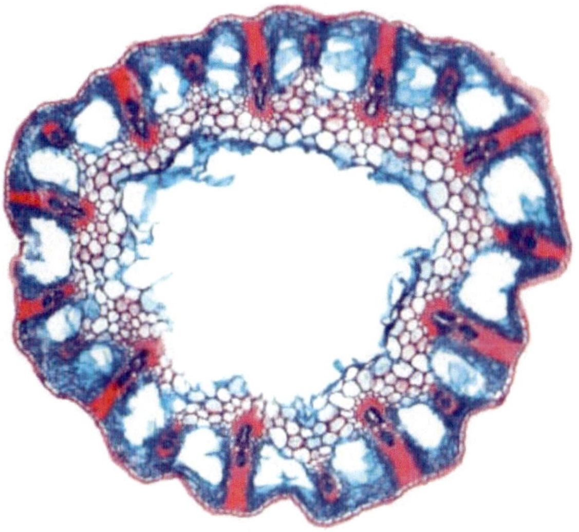

100 µm

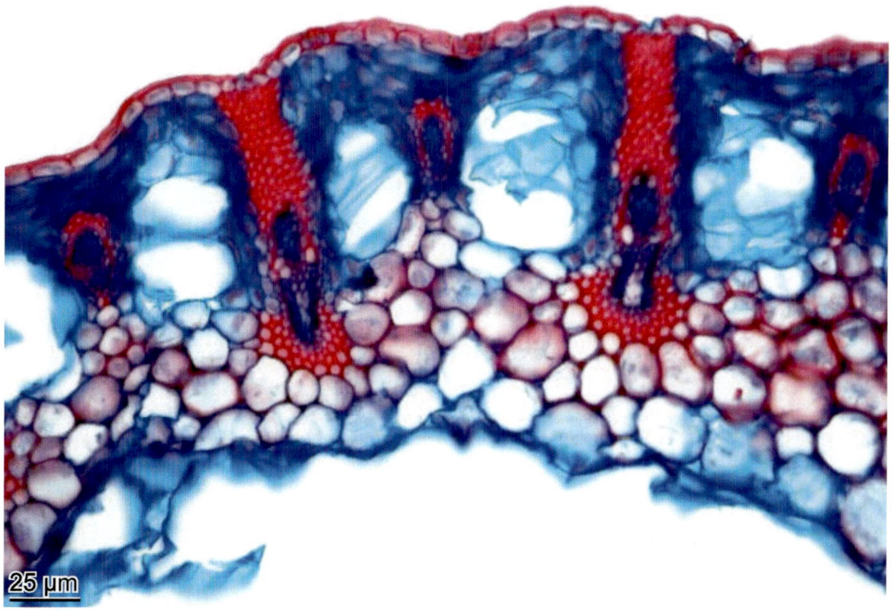

Carex haematostoma Nees

Cyperaceae

Description: Perennial herbs, 25–50 cm tall, rhizomatous, leaves shorter than stems, blades gray-green, linear, 1.5–3 mm wide, upper 2–4 spikes male, approximate, clavate, ca. 1.5 cm long, lower spikes female, narrowly cylindric, 1–3 cm long, distant, pedunculate.

Distribution: NE Afghanistan, Pakistan, N India, Nepal, China (Qinghai, SW Sichuan, Xizang, NW Yunnan), Bhutan.

Habitat: Alpine meadows.

Culm round, hollow, with a diameter of 0.8–0.9 mm. Center hollow, surrounded by 2–3 rows of large, thin-walled, polygonal, lignified cells. Epidermis cells lignified, smooth, thin-walled on the inside, thicker-walled peripherally. Vascular bundles arranged in two slightly offset peripheral rows within girders. Chlorenchyma consists of conical to rectangular groups, generally containing large irregularly formed intercellulars. With conical to rectangular girders. Vessels within larger vascular bundles arranged in a T-shape. Lateral vessels large, diameter 20 μm. Vessels of the protophloem accompanied by small centripetal intercellulars. Phloem consists of round groups, containing sieve tubes and companion cells. Sclerenchymatic vessel sheaths eccentric, with thick-walled fibers in one row centrifugally, and in multicellular U-shaped groups centripetally.

Analyzed: 2 individuals

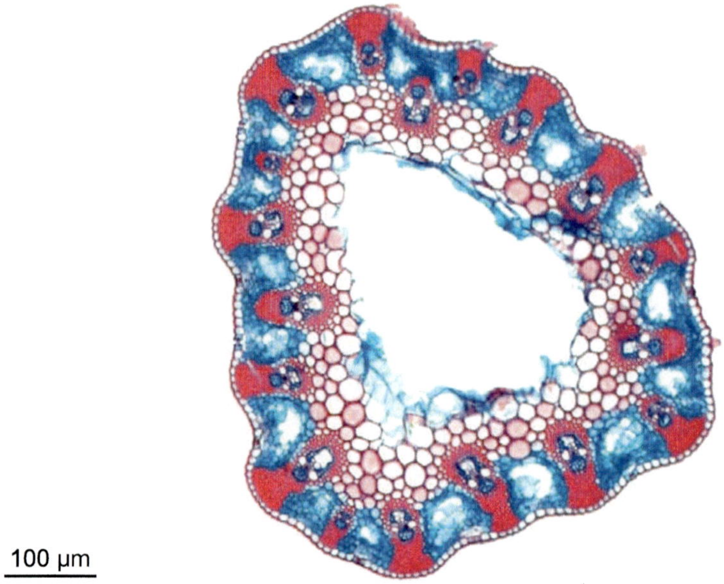

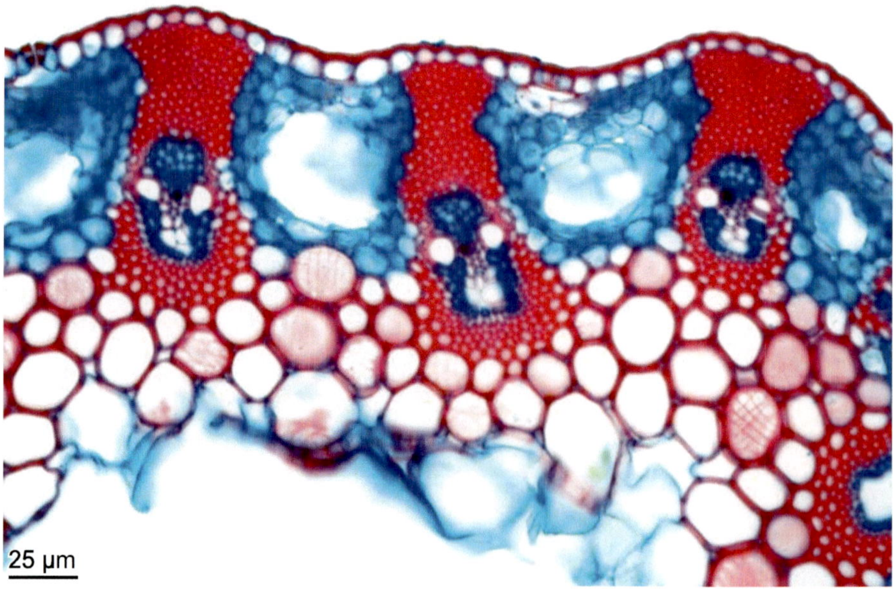

Carex heterostachya Bunge
Cyperaceae

Syn.: *C. songorica* Kar. & Kir.

Description: Perennial herbs, 20–50 cm tall, stoloniferous, leaves shorter than stems, flat, 3–5 mm wide, terminal 1 or 2 spikes male, narrowly clavate, sessile, lower spikes female, oblong, 1–3 cm long, sessile or very shortly pedunculate.

Distribution: Turkey, Iran, Russia, Kazakhstan, Afghanistan, Pakistan, NW India, China (Xinjiang), NW Mongolia.

Habitat: Moist meadows, salt marshes, lake shores.

Culm obtusely triangular, hollow, with a diameter of 1.0–1.1 mm. Center hollow, surrounded by 1–3 rows of polygonal unlignified cells. Epidermis cells smooth, lignified, thin-walled on the inside, thicker-walled peripherally. Vascular bundles arranged in two offset peripheral rows. Chlorenchyma consists of irregularly formed square groups, containing irregularly formed intercellulars. Girders square. Vessels within larger vascular bundles arranged in a T-shape. Lateral vessels large, diameter 20–25 µm. Phloem consists of round groups, containing sieve tubes and companion cells. Vessels of the protophloem often accompanied by centripetal intercellulars. Sclerenchymatic vessel sheaths eccentric, with one row of T-shaped, thin- to thick-walled fibers centrifugally, and 2–3 layers centripetally.

Analyzed: 2 individuals

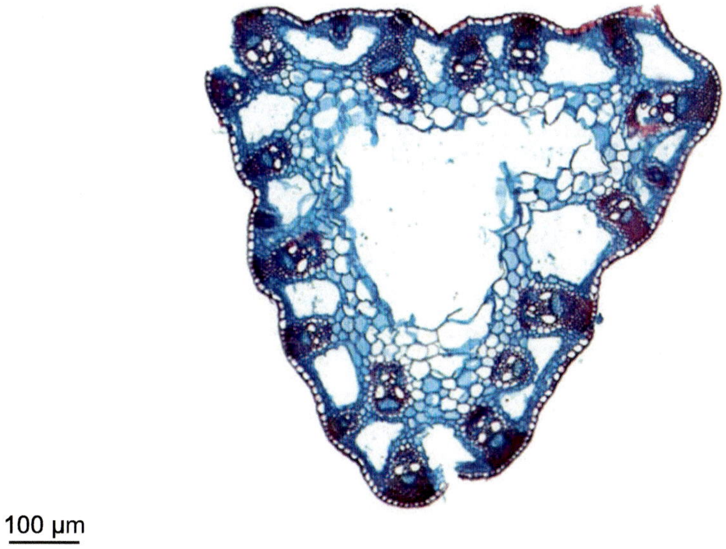

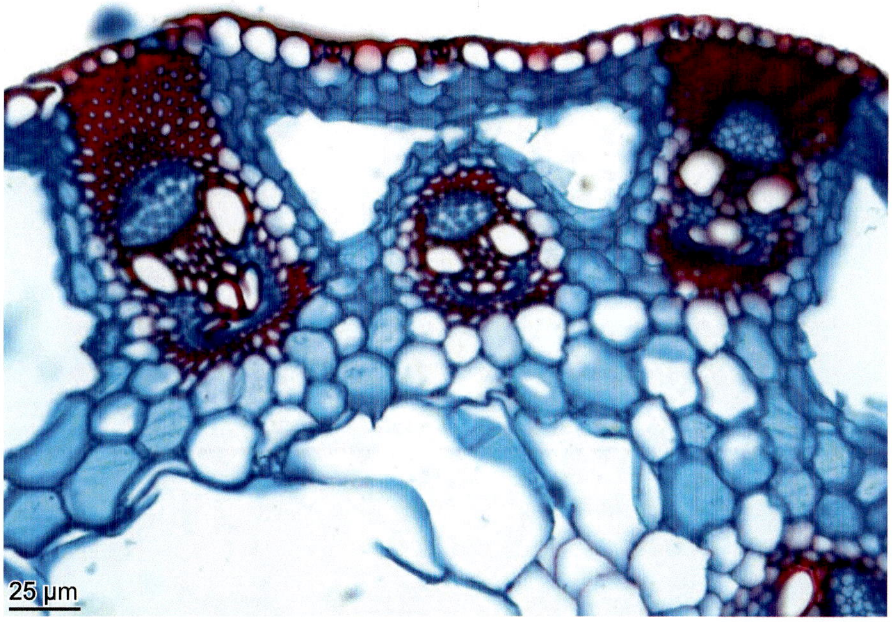

Carex infuscata Nees

Cyperaceae

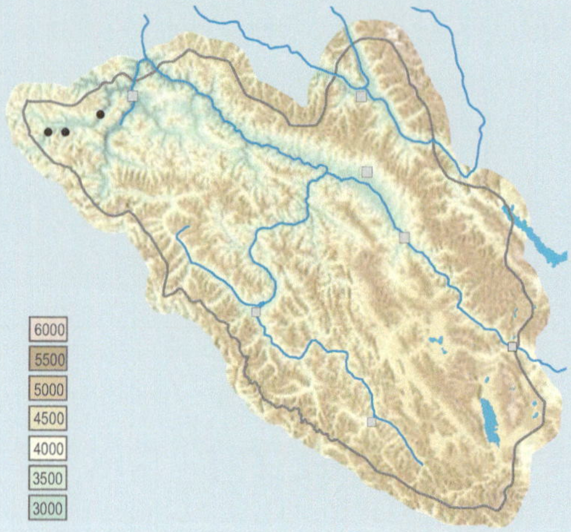

Syn.: *C. alpina* subsp. *infuscata* (Nees) Kük.

Description: Perennial herbs, 10–50 cm tall, cespitose, leaves shorter than stems, 1.5–3.5 mm wide, lower side gray, spikes 2–4, grouped closely together, 4–8 mm long, globose, male flowers variously arranged, lower spikes usually entirely female.

Distribution: Afghanistan, N Pakistan, N India, Nepal.

Habitat: Moist meadows, along streams.

Culm obtusely triangular, filled, with a diameter of 1.5 mm. Center hollow, surrounded by 2–4 rows of large, thin-walled, polygonal, lignified cells. Epidermis cells lignified, smooth, thin-walled on the inside, thicker-walled peripherally. Vascular bundles arranged in two slightly offset peripheral rows within girders. Chlorenchyma consists of rectangular groups, containing large, irregularly formed intercellulars. Girders rectangular. Vessels within larger vascular bundles arranged in a T-shape. Lateral vessels large, diameter 20–30 µm. Vessels of the protophloem accompanied by large centripetal intercellulars. Phloem consists of round groups containing sieve tubes and companion cells. Sclerenchymatic vessel sheaths eccentric, with thick-walled fibers in one row centrifugally, and in multicellular U-shaped groups centripetally.

Analyzed: 2 individuals

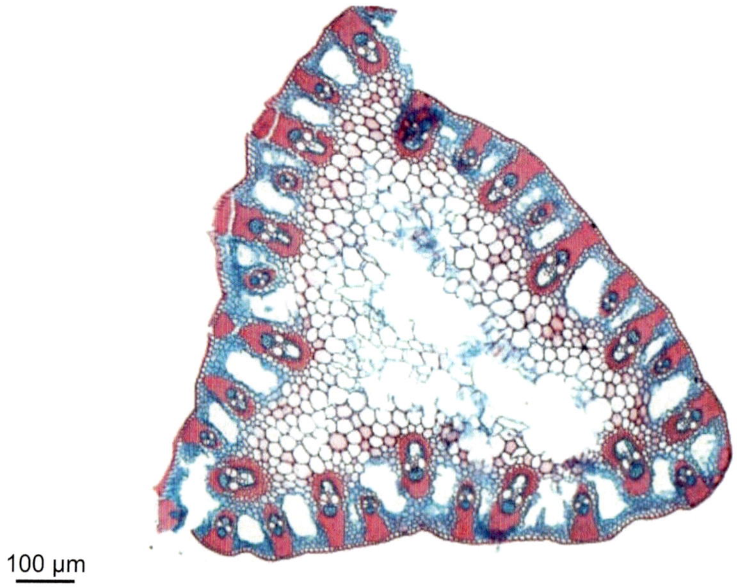

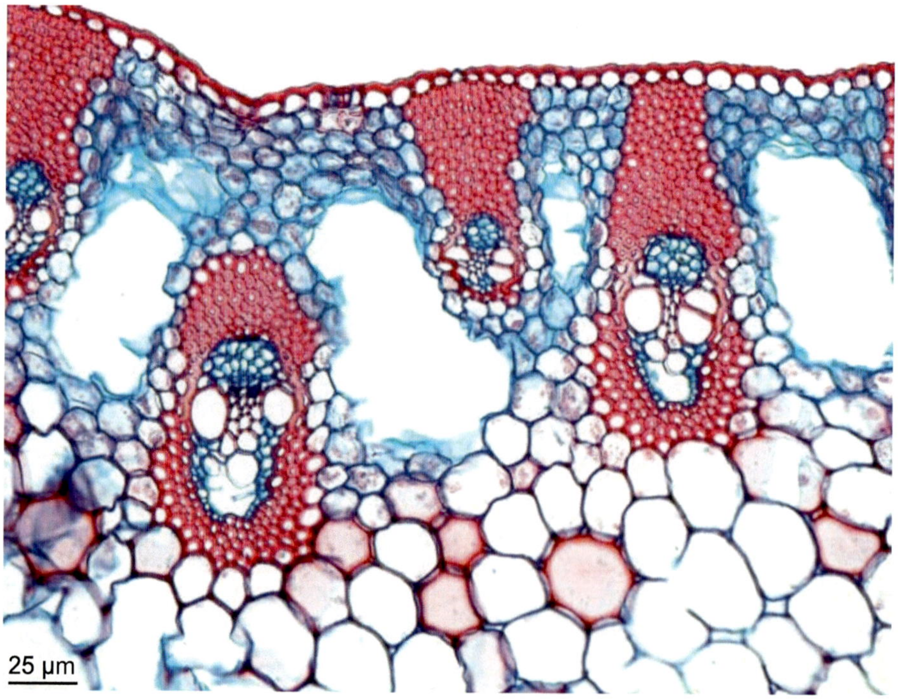

Carex karoi Freyn
Cyperaceae

Syn.: *C. capillaris* subsp. *karoi* Freyn

Description: Perennial herbs, 15–60 cm tall, densely cespitose, leaves shorter than stems, 1.5–2.5 mm wide, spikes 3–4, 4–15 mm long, upper one androgynous, clavate, lower ones female, oblong, pedunculate.

Distribution: Russia (Siberia), Pakistan, N India, China (Hebei, Liaoning, Nei Mongol, Shanxi), Mongolia.

Habitat: Riversides, moist slopes.

Culm round, mostly hollow, with a diameter of 0.9–1.2 mm. Center hollow, surrounded by 2–4 rows of large, thin-walled, polygonal, lignified cells. Epidermis cells lignified, smooth, thin-walled on the inside, thicker-walled peripherally. Vascular bundles arranged in two slightly offset peripheral rows within girders. Chlorenchyma consists of square to rectangular groups, generally containing large, irregularly formed intercellulars. With conical to rectangular girders. Vessels within larger vascular bundles arranged in a T-shape. Lateral vessels large, diameter 15 μm. Vessels of the protophloem accompanied by small centripetal intercellulars. Phloem consists of round groups, containing sieve tubes and companion cells. Sclerenchymatic vessel sheaths eccentric, with thick-walled fibers in one row centrifugally, and groups in a crescent centripetally.

Analyzed: 1 individual

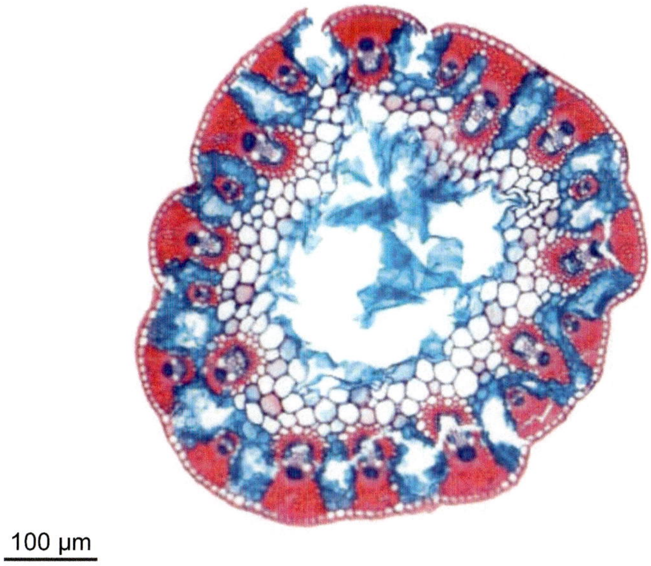

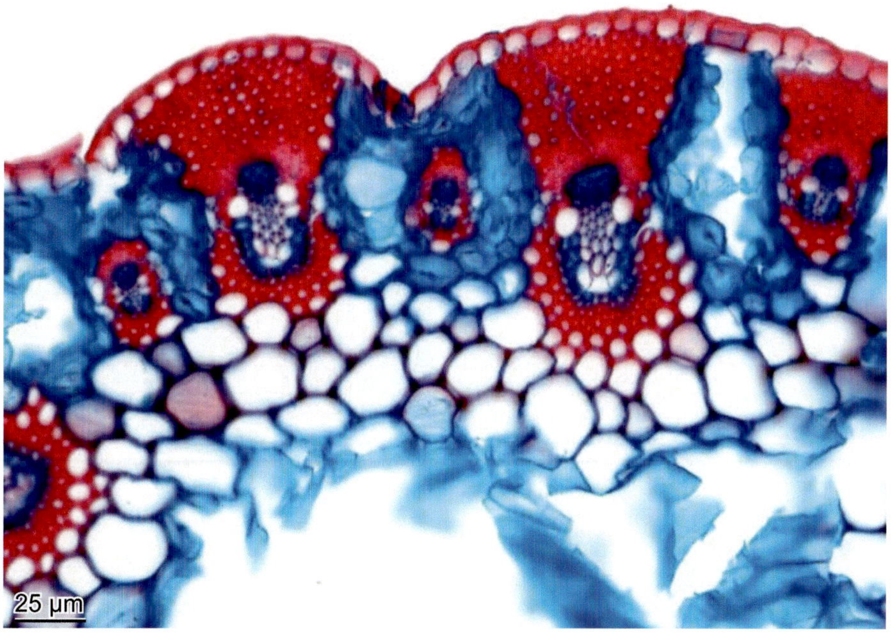

Carex melanantha C.A.Mey.

Cyperaceae

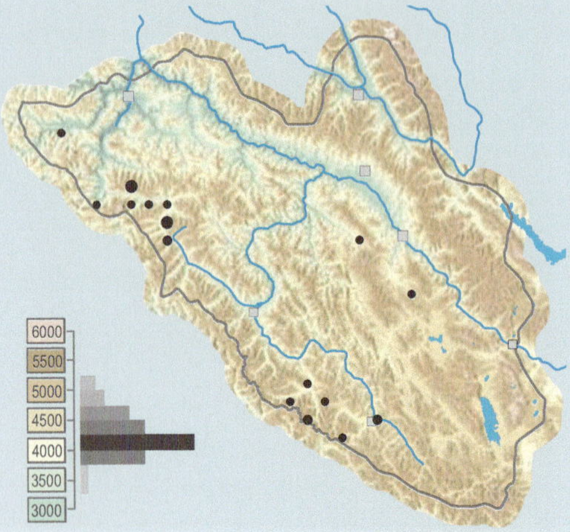

Description: Perennial herbs, 10–50 cm tall, stoloniferous, stems trigonous, firm, longer than leaves, spikes 3–6, in capitate inflorescences.

Distribution: C Asia, Russia (Siberia), Afghanistan, Pakistan, India, Nepal, China (Xinjiang), Mongolia.

Habitat: River banks, alpine meadows.

Culm obtusely triangular, hollow, with a diameter of 1.5 mm. Center hollow or with remains of thin-walled, unlignified cells, surrounded by 2–4 rows of large, thin-walled, polygonal, lignified cells. Epidermis cells lignified, smooth, thin-walled on the inside, thicker-walled peripherally. Vascular bundles arranged in two slightly offset peripheral rows within girders. Chlorenchyma consists of square to rectangular groups, containing large irregularly formed intercellulars. Girders rectangular. Vessels within larger vascular bundles arranged in a T-shape. Lateral vessels large, diameter 25–30 µm. Vessels of the protophloem accompanied by small centripetal intercellulars. Phloem consists of round groups, containing sieve tubes and companion cells. Sclerenchymatic vessel sheaths eccentric, with thick-walled fibers in one row centrifugally, and in U-shaped groups centripetally.

Analyzed: 2 individuals

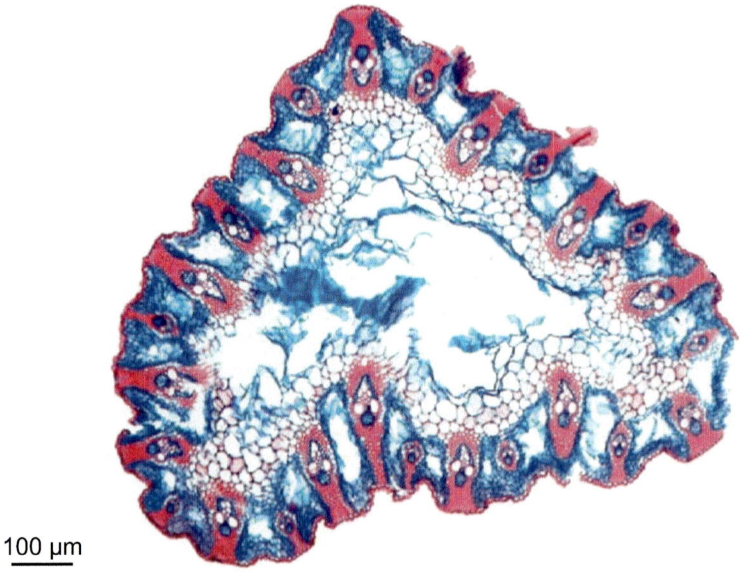

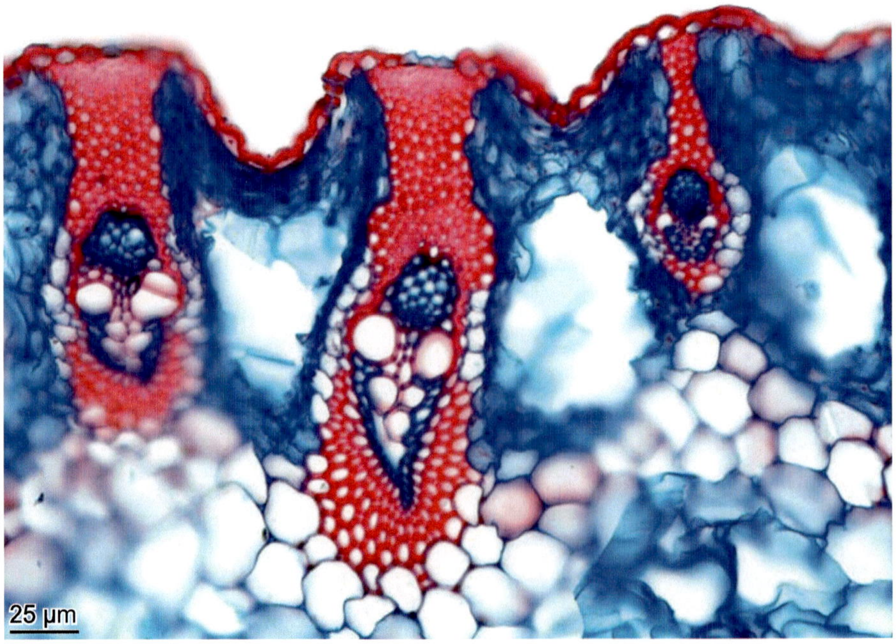

Carex microglochin Wahlenb.

Cyperaceae

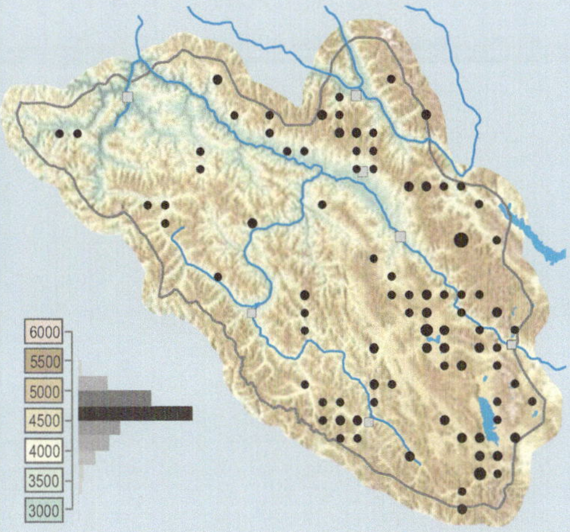

Syn.: *Uncina microglochin* (Wahlenb.) Spreng.

Description: Perennial herbs, 5–20 cm tall, densely cespitose, stems trigonous, longer than leaves, inflorescence terminal, composed of one androgynous spike.

Distribution: North America, Greenland, South America, Europe (Scandinavia, Alps), Russia (Siberia, Caucasus), NW Iran, Pakistan, NW India, Nepal, China (Qinghai, Sichuan, Xinjiang, Xizang), Bhutan, Mongolia.

Habitat: Saline or less often fresh water wetlands.

Culm round, hollow, with a diameter of 0.8–0. mm. Center hollow, surrounded by 1–3 rows of large, thin-walled, polygonal, unlignified cells. Epidermis cells lignified, smooth, thin-walled on the inside, thicker-walled peripherally. Vascular bundles arranged in one peripheral row. Chlorenchyma consists of square to polygonal groups, generally containing large, irregularly formed intercellulars. Girders conical. Vessels within larger vascular bundles arranged in a T-shape. Lateral vessels large, diameter 10–12 µm. Phloem consists of round groups of uniform cells. Vessels of the protophloem accompanied by small centripetal intercellulars. Sclerenchymatic vessels sheaths eccentric, with thick-walled fibers in one row centrifugally, and in multicellular U-shaped groups centripetally.

Analyzed: 2 individuals

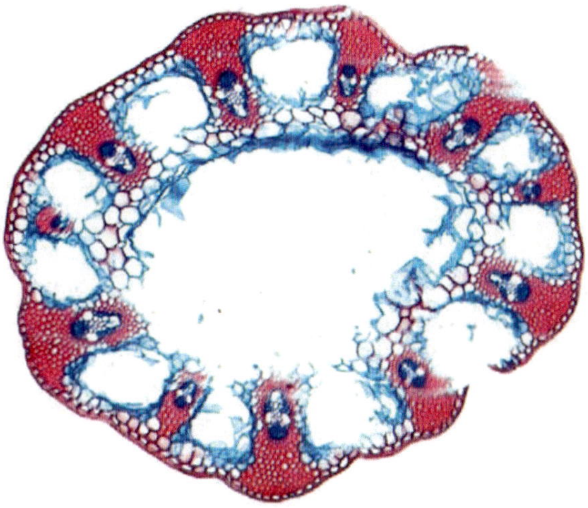

100 µm

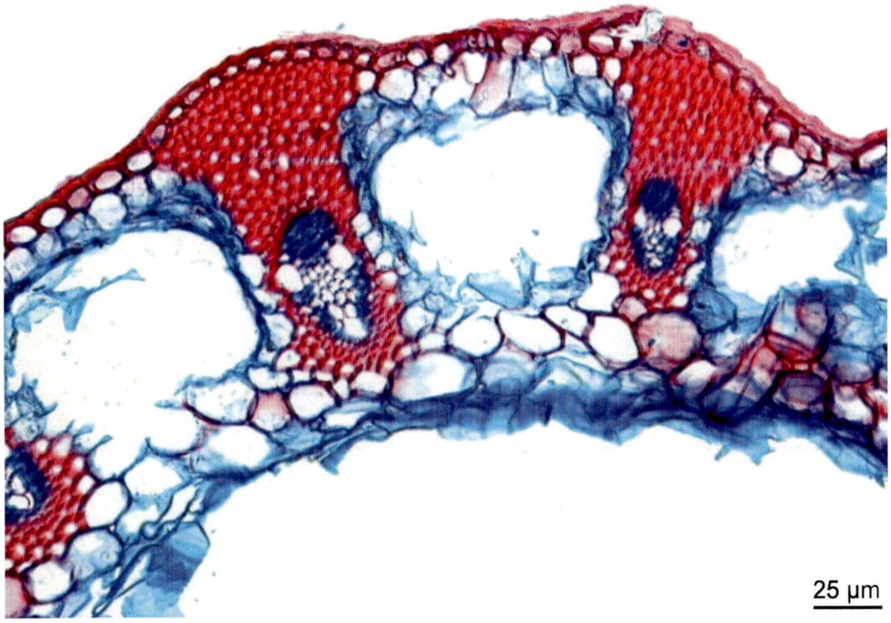

25 µm

Carex montis-everesti Kük.

Cyperaceae

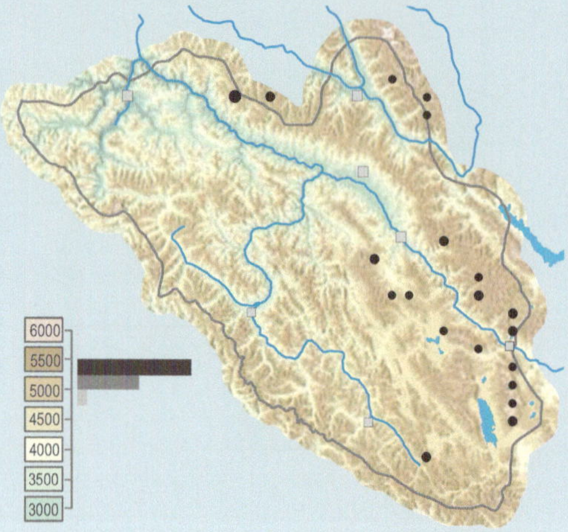

Description: Perennial herbs, 5–15 cm tall, stoloniferous, stems tufted, leaves linear, shorter than stems, spikes usually 2, rarely 3 or 4, terminal spike male, lateral spikes female.

Distribution: India (Jammu and Kashmir, Sikkim), Nepal, China (Xizang).

Habitat: High-alpine steppes and grasslands, humid alpine turf.

Culm round to obtusely triangular, hollow, with a diameter of 1.2–1.4 mm. Center hollow, surrounded by 3–4 rows of large, thin-walled, polygonal, lignified cells. Epidermis cells lignified, smooth, thin-walled on the inside, thicker-walled peripherally. Vascular bundles arranged in one peripheral row. Chlorenchyma consists of square groups, containing large, irregularly formed intercellulars. With conical to rectangular girders. Vessels within larger vascular bundles arranged in a T-shape. Lateral vessels large, diameter 12–15 µm. Phloem consists of round groups of uniform cells. Vessels of the protophloem accompanied by small centripetal intercellulars. Sclerenchymatic vessel sheaths eccentric, with thick-walled fibers in one row centrifugally, and in multicellular U-shaped groups centripetally.

Analyzed: 2 individuals

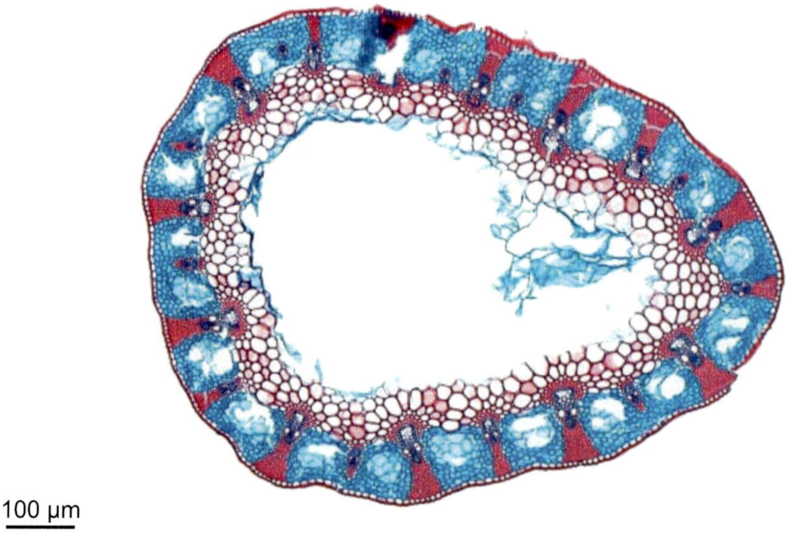

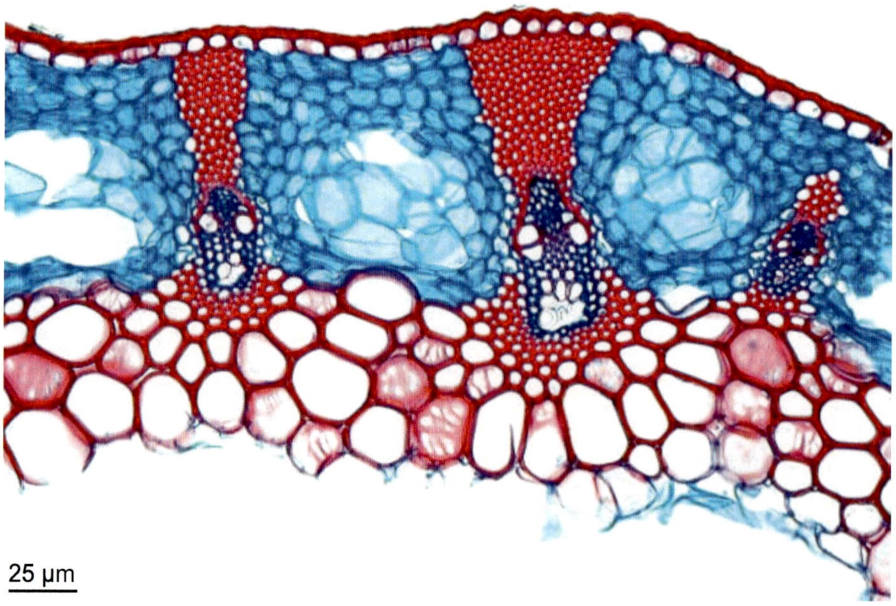

Carex moorcroftii Falc. ex Boott
Cyperaceae

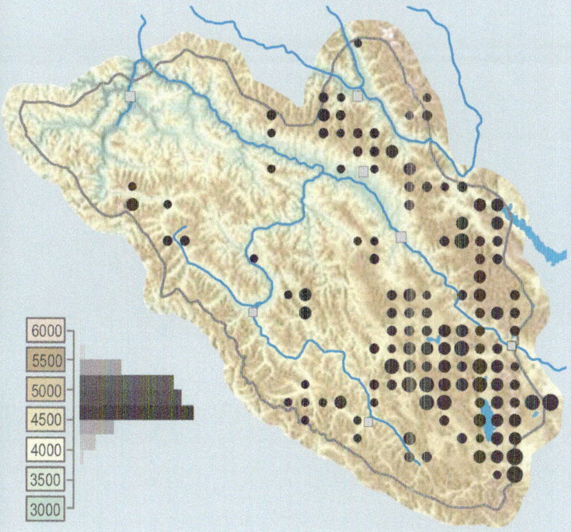

Syn.: *C. melanantha* var. *moorcroftii* (Falconer ex Boott) Kük.

Description: Perennial herbs, 10–50 cm tall, stoloniferous, stems trigonous, leaves with pungent tips, shorter than stems, inflorescence composed of 4–5 spikes, terminal spike male, lateral ones female.

Distribution: N India, China (Qinghai, Sichuan, Xizang), Bhutan.

Habitat: Sandy deposits with unstable soil surface, stabilized slopes.

Culm obtusely triangular, wavy, hollow, with a diameter of 1.5–2.0 mm. Center hollow, surrounded by 3–4 wavy rows of large thin-walled polygonal unlignified cells. Epidermis cells lignified, smooth, or papillose (bulliform), thin-walled on the inside, thicker-walled peripherally. Vascular bundles arranged in two, slightly offset peripheral rows. Chlorenchyma consists of square to triangular groups, containing large, irregularly formed intercellulars. Girders conical. Vessels within larger vascular bundles arranged in a U-shape. Lateral vessels large, diameter 30–35 µm. Phloem consists of round groups, containing sieve tubes and companion cells. Vessels of the protophloem accompanied by large centripetal intercellulars. Sclerenchymatic vessel sheaths eccentric, with thin- to thick-walled fibers in one row centrifugally, and in multicellular U-shaped groups centripetally.

Analyzed: 2 individuals

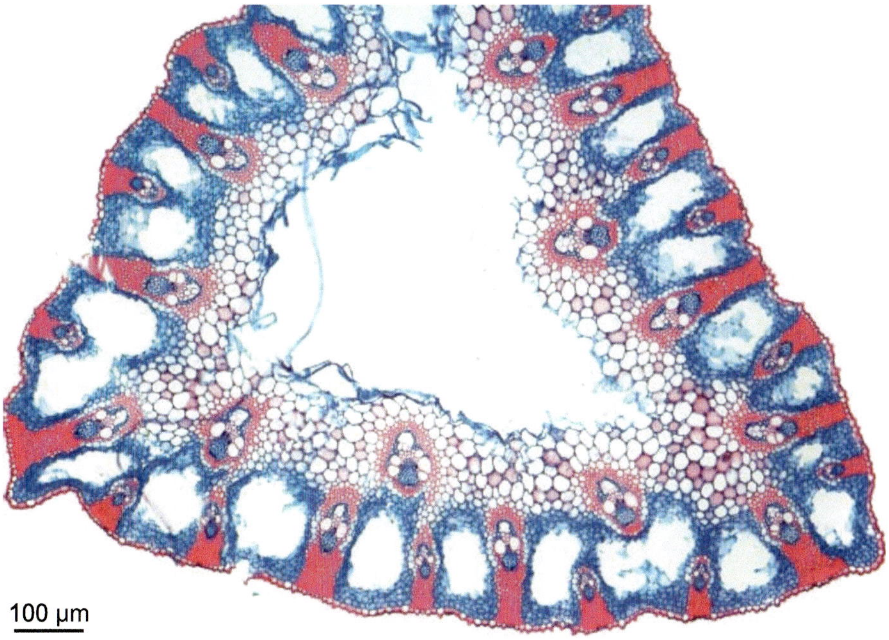

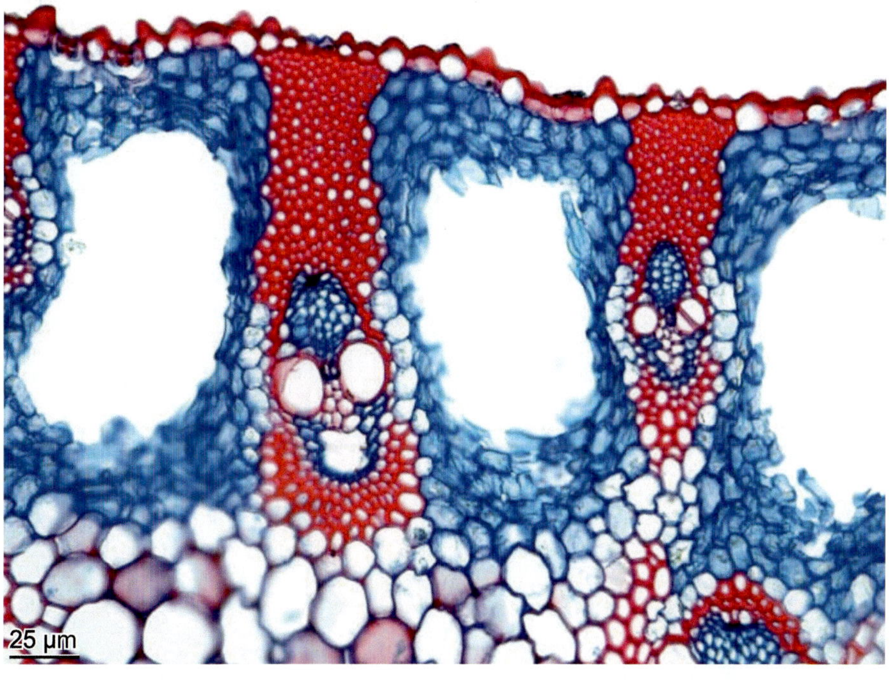

Carex nivalis Boott

Cyperaceae

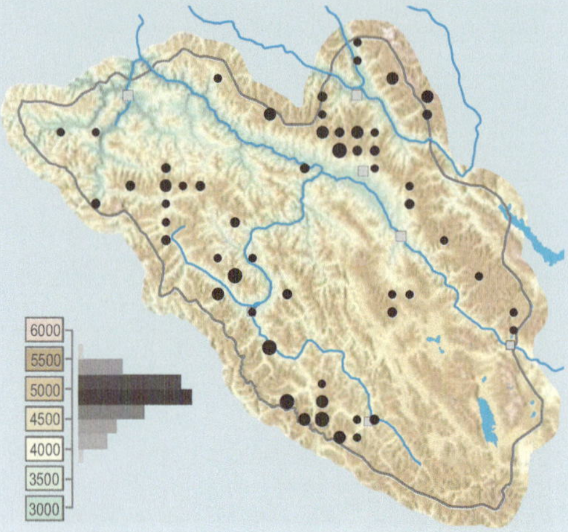

Syn.: *C. griffithii* Boott

Description: Perennial herbs, 30–60 cm tall, stems sharply trigonous, leaves half the length of stems, inflorescence composed of 2–4 nodding peduncled spikes, upper spikes of mixed sexes, lower spikes female.

Distribution: Uzbekistan, Kyrgyzstan, Afghanistan, Pakistan, India, Nepal, China (Sichuan, Xizang, Yunnan).

Habitat: Mesic turf on slopes and river banks, glacier moraines.

Culm acutely triangular, hollow, with a diameter of 0.9–1.1 mm. Center hollow, surrounded by 3–4 wavy rows of large, thin-walled, polygonal, lignified cells. Epidermis cells lignified, slightly papillose, thin-walled on the inside, thicker-walled peripherally. Vascular bundles arranged in two offset peripheral rows. Chlorenchyma consists of irregularly formed, square to rectangular groups, containing large, irregularly formed intercellulars. Girders conical. Vessels within larger vascular bundles arranged in a T-shape. Lateral vessels large, diameter 12–15 µm. Phloem consists of round groups, containing sieve tubes and companion cells. Vessels of the protophloem rarely accompanied by small centripetal intercellulars. Sclerenchymatic vessel sheaths eccentric, with thin- to thick-walled fibers in one row centrifugally, and in multicellular U-shaped groups centripetally.

Analyzed: 2 individuals

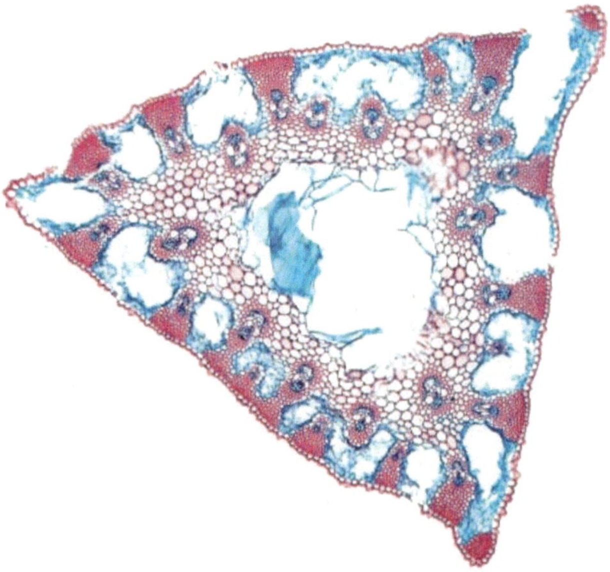

100 µm

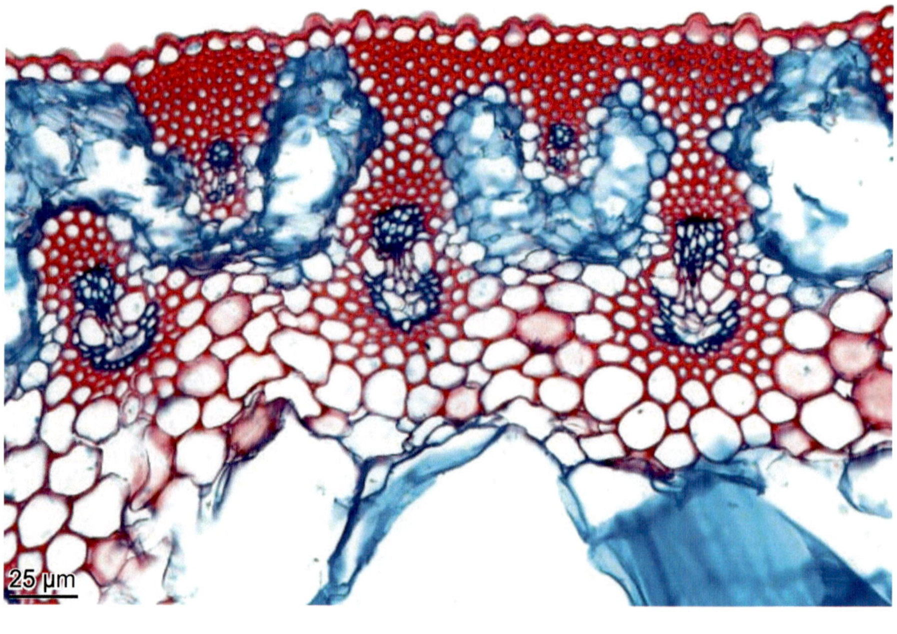

Carex orbicularis Boott

Cyperaceae

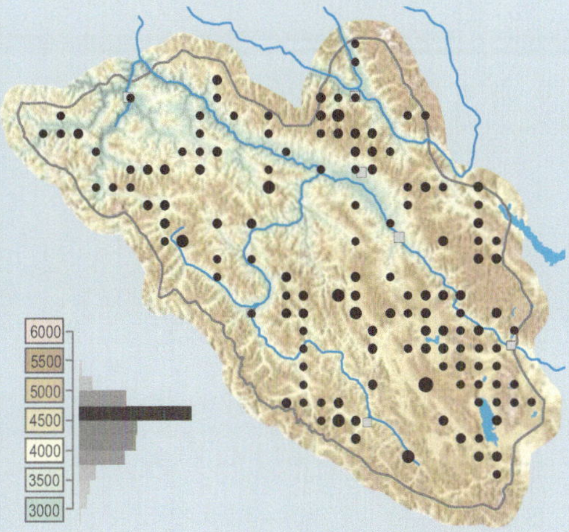

Description: Perennial herbs, 10–25 cm tall, rhizomatous, culms tufted, trigonous, leaves shorter than stems, spikes 2–3, terminal spike male, lateral spikes female, upper sessile, lowest with peduncle.

Distribution: Russia, Afghanistan, Pakistan, NW India, Nepal, China (Gansu, Qinghai, Xinjiang, Xizang).

Habitat: Saline and fresh wetlands, river banks, springs and flushes.

Culm acutely triangular, filled, with a diameter of 0.9–1.1 mm. Center filled, surrounded by 3–4 wavy rows of large, thin-walled, polygonal, lignified cells. Epidermis cells lignified, bent, thin-walled on the inside, thicker-walled peripherally. Vascular bundles arranged in two offset peripheral rows. Chlorenchyma consists of irregularly formed, square to rectangular groups, containing large, irregularly formed intercellulars. Girders rectangular. Vessels within larger vascular bundles arranged in a T-shape. Lateral vessels large, diameter 15–20 µm. Phloem consists of round groups, containing sieve tubes and companion cells. Vessels of the protophloem accompanied by large centripetal intercellulars. Sclerenchymatic vessel sheaths eccentric, with thick-walled fibers in one row centrifugally, and in multicellular U-shaped groups centripetally.

Analyzed: 3 individuals

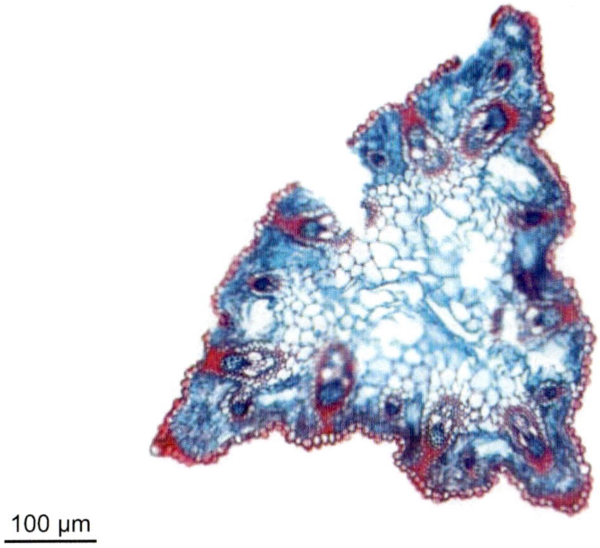

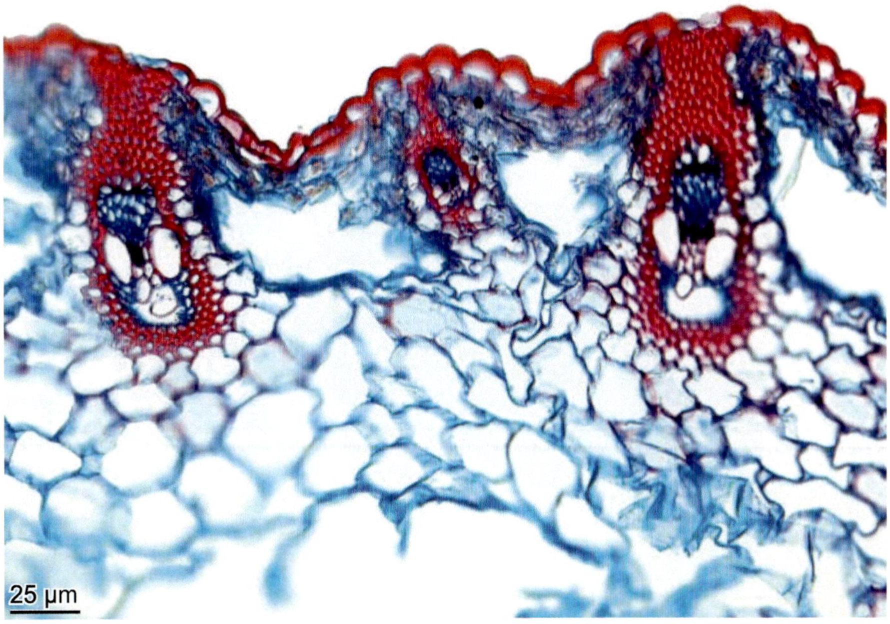

Carex pamirensis C.B.Clarke

Cyperaceae

Syn.: *C. vesicaria* var. *pamirica* O.Fedtsch., *C. pamirica* (O.Fedtsch.) O.Fedtsch. & B.Fedtsch.

Description: Perennial herbs, 30–70 cm tall, loosely cespitose, leaves up to the length of stems, inflorescence composed of 1–3 upper sessile male spikes and 1–3 lower female spikes, shortly pedunculate.

Distribution: Russia, Kazakhstan, Afghanistan, Pakistan, N India, China (Gansu, Sichuan, Xinjiang).

Habitat: Moist alpine meadows, wetlands.

Culm acutely to obtusely triangular, hollow, with a diameter of 2.0–2.2 mm. Center hollow, surrounded by 3–4 wavy rows of large, thin-walled, polygonal, lignified cells. Epidermis cells lignified, smooth, thin-walled on the inside, thicker-walled peripherally. Vascular bundles arranged in two offset peripheral rows. Chlorenchyma consists of square groups, containing large irregularly formed intercellulars. With rectangular to conical girders. Vessels within larger vascular bundles arranged in a T-shape. Lateral vessels large, diameter 30 µm. Phloem consists of round groups, containing sieve tubes and companion cells. Vessels of the protophloem accompanied by large centripetal intercellulars. Sclerenchymatic vessel sheaths eccentric, with thick-walled fibers in one row centrifugally, and in multicellular U-shaped groups centripetally.

Analyzed: 1 individual

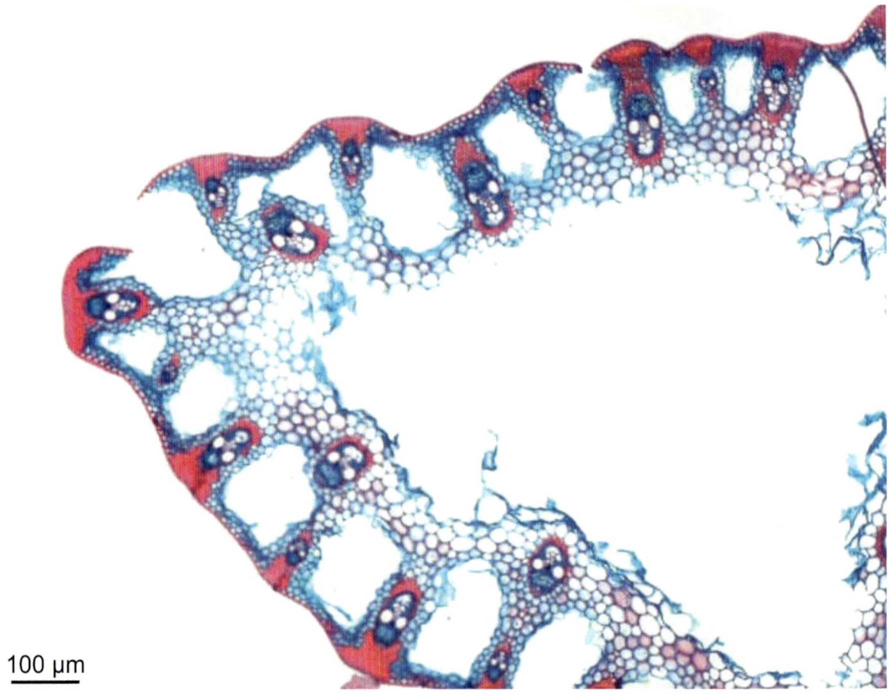

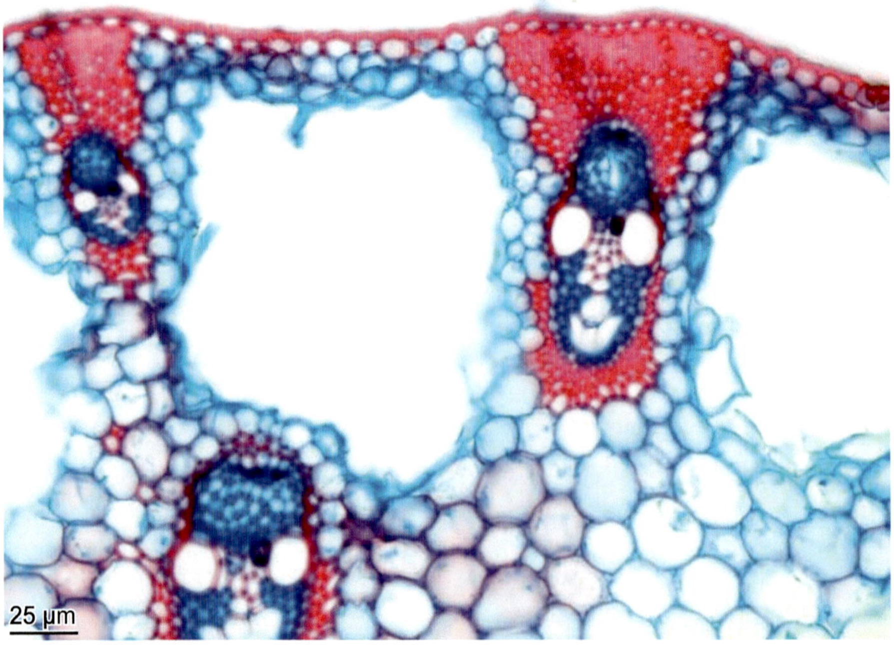

Carex plectobasis V.I.Krecz.

Cyperaceae

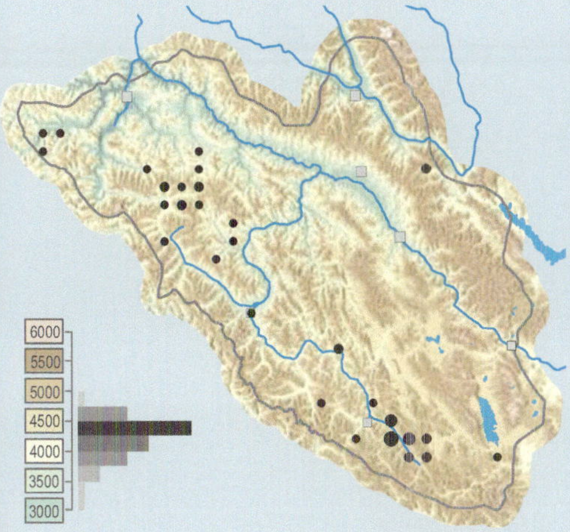

Description: Perennial herbs, 25–70 cm tall, cespitose, leaves shorter than stems, 1–1.5 mm wide, gray green, carinate, spikes 15–25 mm long, male ones 1–2, clavate, female ones 2–3, oblong, erect, lowest pedunculate.

Distribution: Afghanistan, Pakistan, N India, Nepal, China (W Sichuan, Xizang, Yunnan).

Habitat: Mountain meadows, rocky slopes.

Culm round to obtusely triangular, hollow, with a diameter of 1.0–1.1 mm. Center hollow, surrounded by 3–4 rows of large, thin-walled, polygonal, lignified cells. Epidermis cells lignified, smooth, thin-walled on the inside, thicker-walled peripherally. Vascular bundles arranged in one peripheral row. Chlorenchyma consists of square groups, often containing irregularly formed intercellulars. Girders rectangular. Vessels within larger vascular bundles arranged in a T-shape. Lateral vessels large, diameter 20–30 µm. Phloem consists of round groups, containing sieve tubes and companion cells. Vessels of the protophloem accompanied by small centripetal intercellulars. Sclerenchymatic vessel sheaths eccentric, with thick-walled fibers in one row centrifugally, and in multicellular U-shaped groups centripetally.

Analyzed: 2 individuals

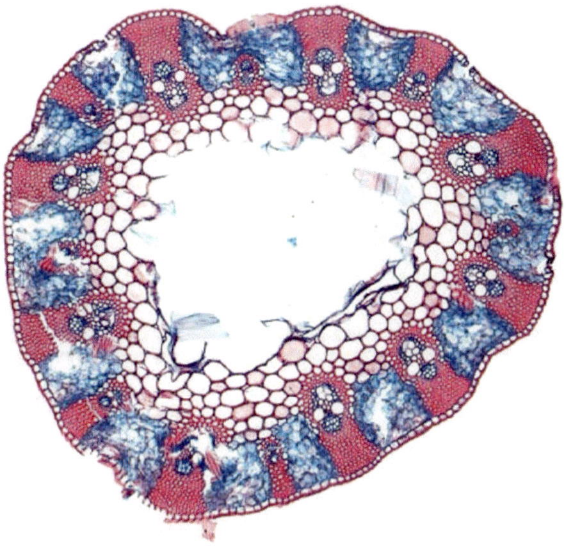

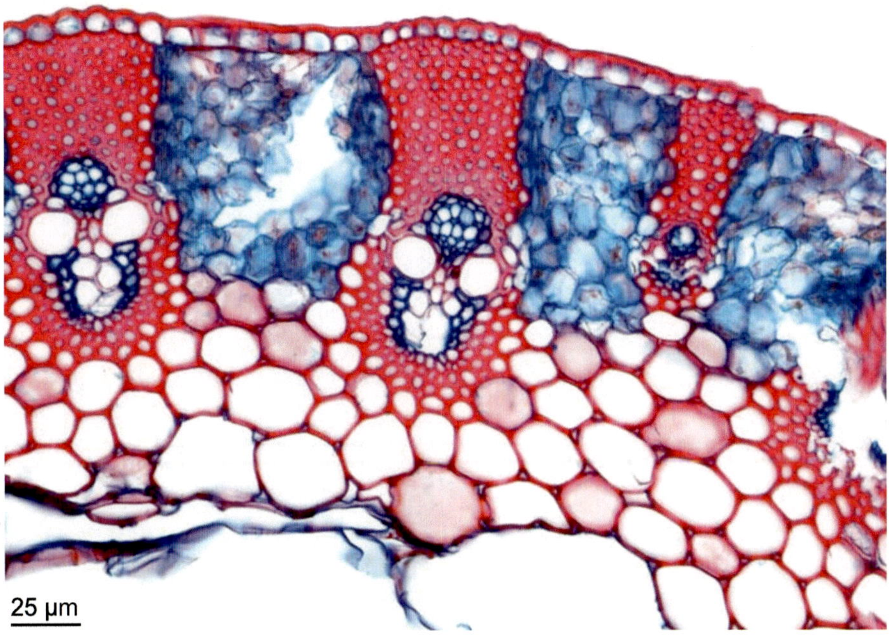

Carex pseudobicolor Boeckeler

Cyperaceae

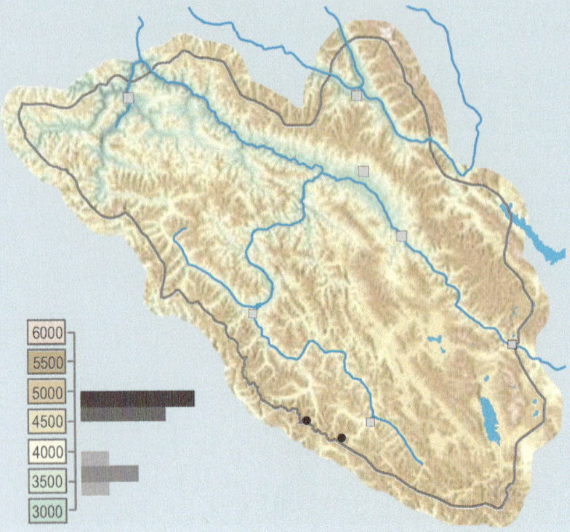

Description: Perennial herbs, 5–15 cm tall, rhizomatous, leaves nearly as long as stems, 1.5–2.5 mm wide, gray green, carinate, spikes 5–10 mm long, ovate, grouped closely together, upper one androgynous, lower ones female.

Distribution: N Pakistan, N India, China (Xizang).

Habitat: Grasslands along streams, mountain meadows.

Culm acutely triangular, wavy, filled, with a diameter of 1.0–1.1 mm. Center filled with unlignified, very large, irregularly formed cells. Epidermis cells papillose, lignified, thin-walled on the inside, thicker-walled peripherally. Vascular bundles arranged in one peripheral row. Chlorenchyma consists of irregularly formed square groups, containing irregularly formed intercellulars. Girders rectangular. Vessels within larger vascular bundles arranged in a T-shape. Lateral vessels large, diameter 12–15 µm. Phloem consists of round groups, containing sieve tubes and companion cells. Vessels of the protophloem rarely accompanied by small centripetal intercellulars. Sclerenchymatic vessel sheaths eccentric, with thick-walled fibers in one row centrifugally, and in multicellular U-shaped groups centripetally.

Analyzed: 2 individuals

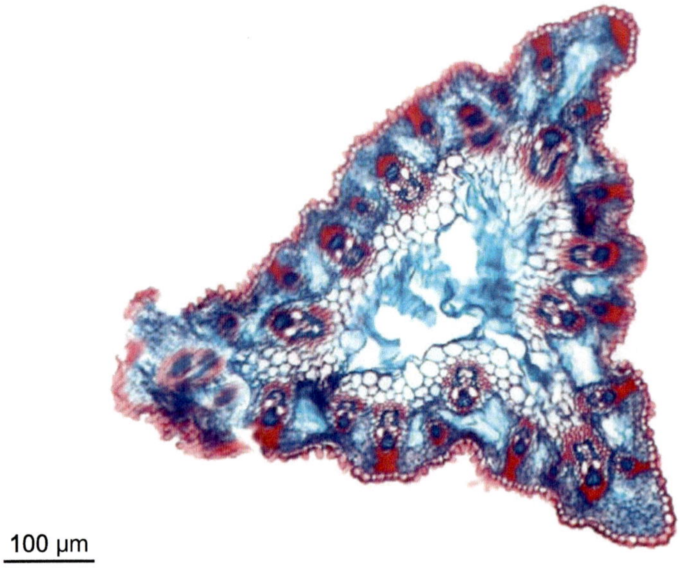

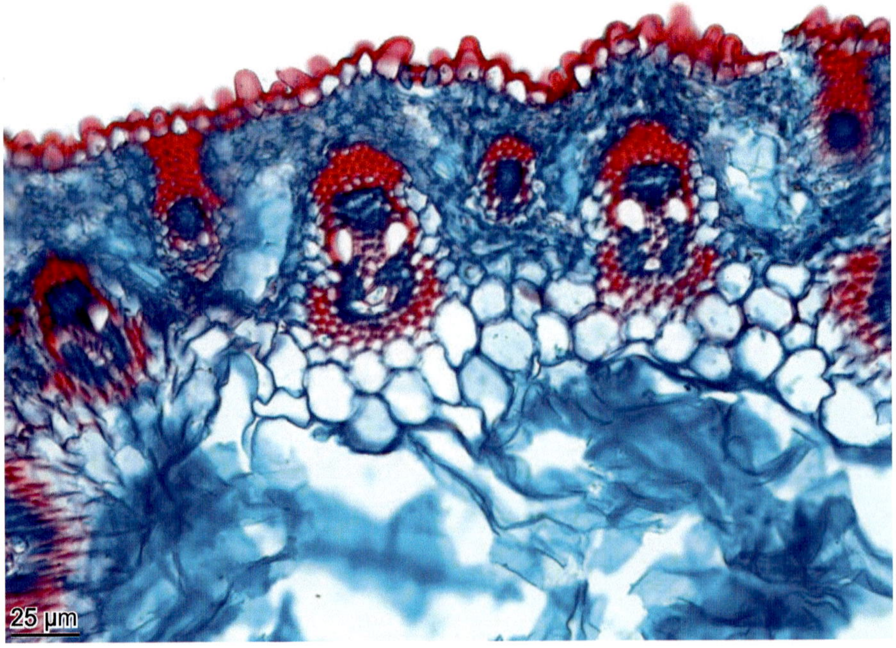

Carex pseudofoetida Kük.

Cyperaceae

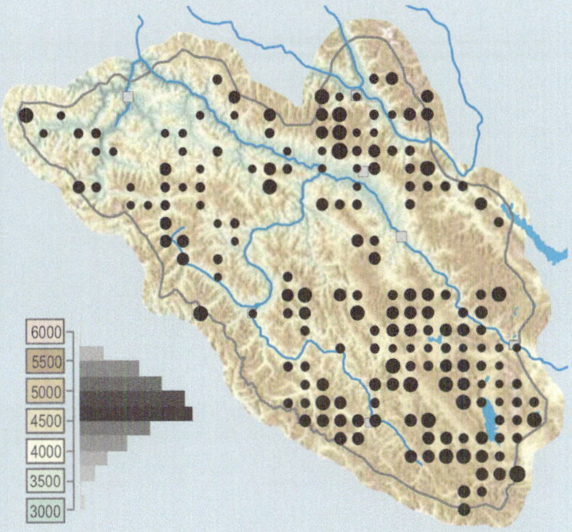

Description: Perennial herbs, 3–15 cm tall, rhizomatous, leaves rigid, spreading, shorter than stems, inflorescence ovoid, compact, composed of 3–5 androgynous spikes.

Distribution: Russia (Siberia), Iran, C Asia, Afghanistan, Pakistan, India, Nepal, China (Qinghai, Xizang), Bhutan, Mongolia.

Habitat: Forming turfs and large mats on river banks, waterlogged and flooded soils, locally in pools, freshwater or saline wetlands.

Culm obtusely triangular, wavy, filled, with a diameter of 0.8–1.1 mm. Center filled with polygonal, unlignified cells. Epidermis cells smooth, lignified, thin-walled on the inside, thicker-walled peripherally. Vascular bundles arranged in one peripheral row. Chlorenchyma consists of irregularly formed, square groups, containing irregularly formed intercellulars. Girders rectangular. Vessels within larger vascular bundles arranged in a T-shape. Lateral vessels large, diameter 8–10 µm. Phloem consists of round groups, containing sieve tubes and companion cells. Vessels of the protophloem rarely accompanied by small centripetal intercellulars. Sclerenchymatic vessel sheaths eccentric, with thin- to thick-walled fibers in one row centrifugally, and in multicellular U-shaped groups centripetally.

Analyzed: 2 individuals

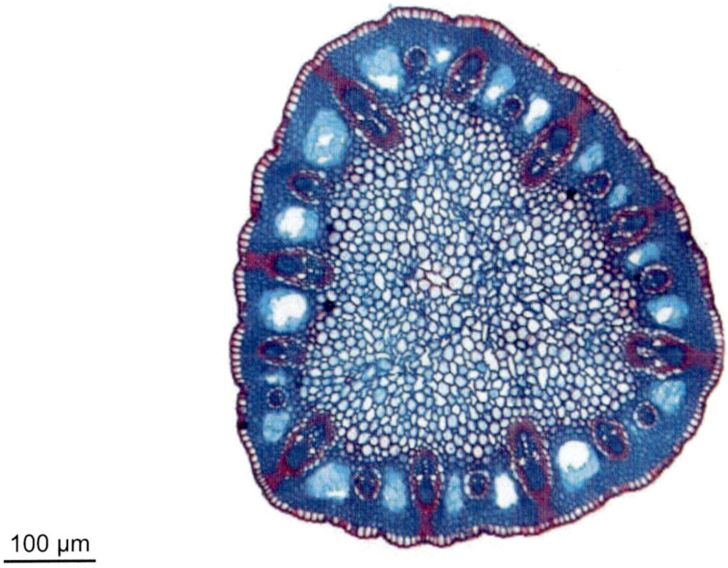

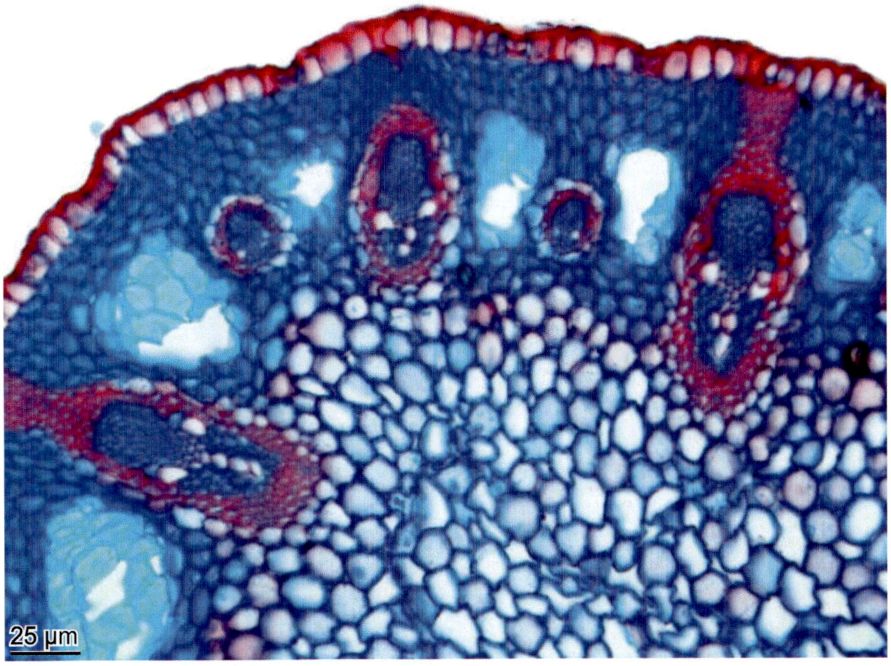

Carex sagaensis Y.C.Yang

Cyperaceae

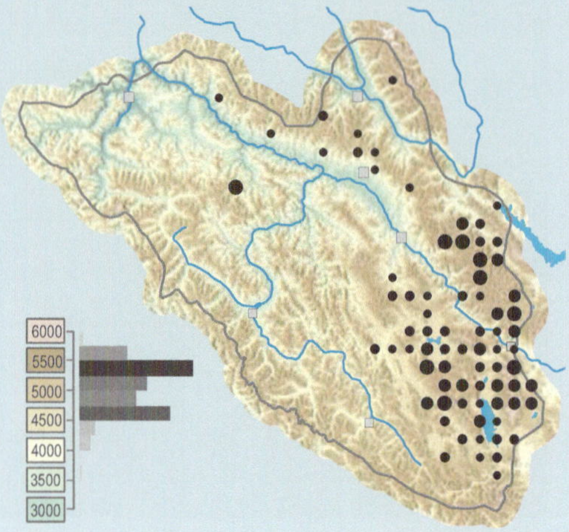

Description: Perennial herbs, 2–10 cm tall, rhizomatous, stems tufted, leaves longer or shorter than stems, inflorescence ovoid, compact, composed of 2–3 androgynous spikes.

Distribution: India (Jammu and Kashmir), China (Xizang).

Habitat: River banks, alpine turf.

Culm obtusely triangular to polygonal, hollow, with a diameter of 0.7–0.9 mm. Center hollow, surrounded by 1–2 rows of polygonal, lignified cells. Epidermis cells smooth, lignified, thin-walled on the inside, thicker-walled peripherally. Vascular bundles arranged in one peripheral row. Chlorenchyma consists of irregularly formed, roundish groups, containing irregularly formed intercellulars. With a few rectangular to conical girders. Vessels within larger vascular bundles arranged in a T-shape. Lateral vessels large, diameter 8–10 µm. Phloem consists of round groups of uniform cells. Vessels of the protophloem not accompanied by centripetal intercellulars. Sclerenchymatic vessel sheaths eccentric, with one row of T-shaped, thin- to thick-walled fibers centrifugally, and 2–3 layers centripetally.

Analyzed: 4 individuals.

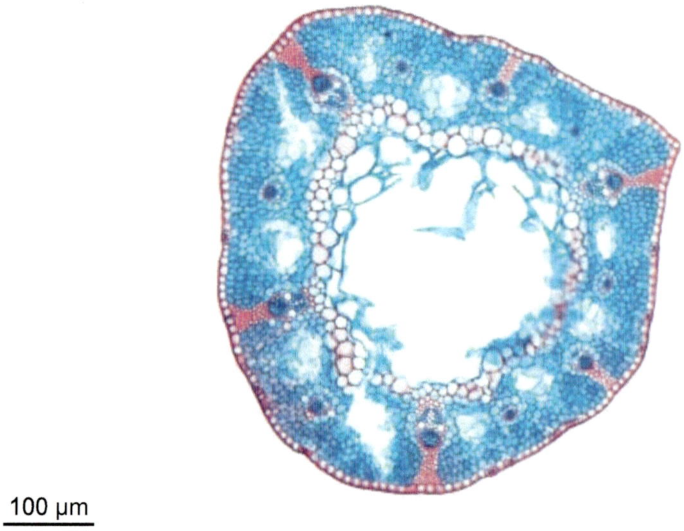

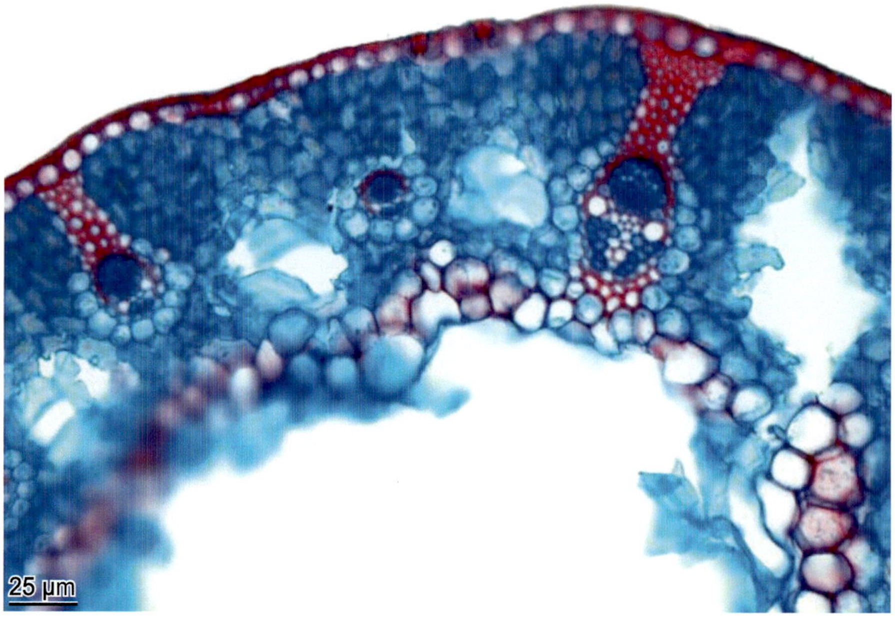

Carex setosa Boott

Cyperaceae

Description: Perennial herbs, 15–45 cm tall, rhizomatous, leaves shorter than stems, 2–4 mm wide, spikes 3–7, distant, terminal spike male, cylindric, 3–4 cm long, lateral spikes female, pedunculate.

Distribution: Pakistan, N India, Nepal, China (Gansu, Guangxi, Guizhou, Hunan, Qinghai, Shaanxi, Sichuan, Xizang, Yunnan), Bhutan.

Habitat: Alpine meadows, grassy slopes.

Culm obtusely triangular to polygonal, hollow, with a diameter of 1.0–1.1 mm. Center hollow, surrounded by 1–3 rows of polygonal unlignified cells. Epidermis cells smooth, lignified, thin-walled on the inside, thicker-walled peripherally. Vascular bundles arranged in one peripheral row. Chlorenchyma consists of irregularly formed, roundish groups, containing irregularly formed intercellulars. Girders rectangular. Vessels within larger vascular bundles arranged in a T-shape. Lateral vessels large, diameter 15–20 µm. Phloem consists of round groups of uniform cells. Vessels of the protophloem often accompanied by centripetal intercellulars. Sclerenchymatic vessel sheaths eccentric, with one row of T-shaped, thin- to thick-walled fibers centrifugally, and 2–3 layers centripetally.

Analyzed: 4 individuals

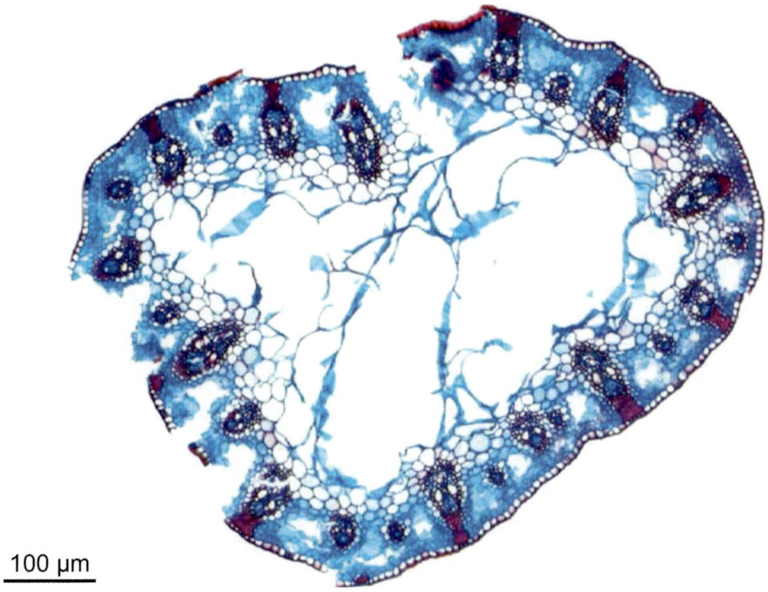

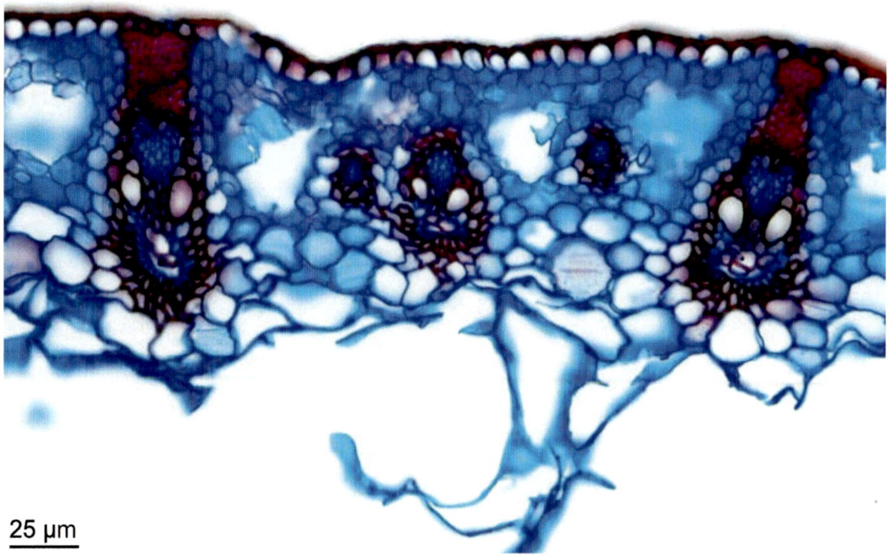

Carex stenocarpa Turcz. ex V.Krecz.

Cyperaceae

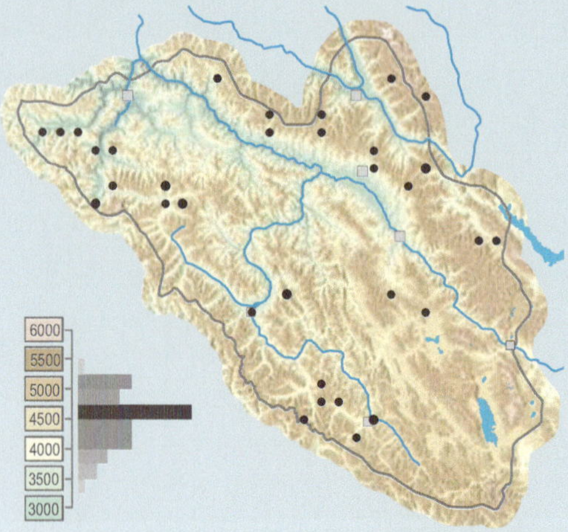

Description: Perennial herbs, 30–70 cm tall, cespitose, stems trigonous, leaves ca. ¼ of stem length, up to 6 mm wide, keeled, inflorescence lax, composed of 1–4 sessile male spikes and 1–4 female spikes on long peduncles.

Distribution: Russia, NE Afghanistan, N Pakistan, NW India, China (Gansu, Xinjiang), N Mongolia.

Habitat: Alpine meadows.

Culm obtusely triangular, hollow, with a diameter of 1.8–2.0 mm. Center hollow, surrounded by 2–4 rows of polygonal, unlignified cells. Epidermis cells smooth, lignified, thin-walled on the inside, thicker-walled peripherally. Vascular bundles arranged in two offset peripheral rows. Chlorenchyma consists of irregularly formed, square to rectangular groups, containing irregularly formed intercellulars. Girders conical and rectangular. Vessels within larger vascular bundles arranged in a T-shape. Lateral vessels large, diameter 25–30 µm. Phloem consists of round groups, containing sieve tubes and companion cells. Vessels of the protophloem often accompanied by centripetal intercellulars. Sclerenchymatic vessel sheaths eccentric, with one row of T-shaped, thin- to thick-walled fibers centrifugally, and 2–3 layers centripetally.

Analyzed: 2 individuals

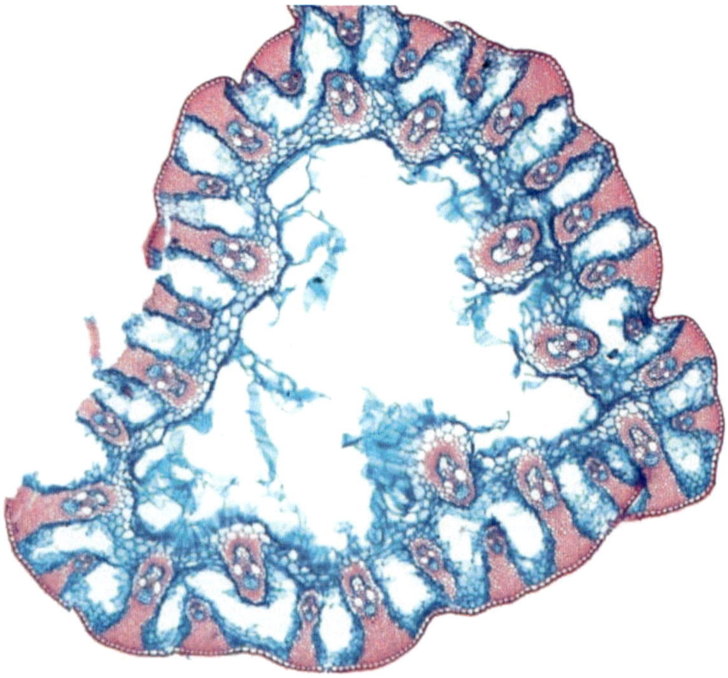

100 µm

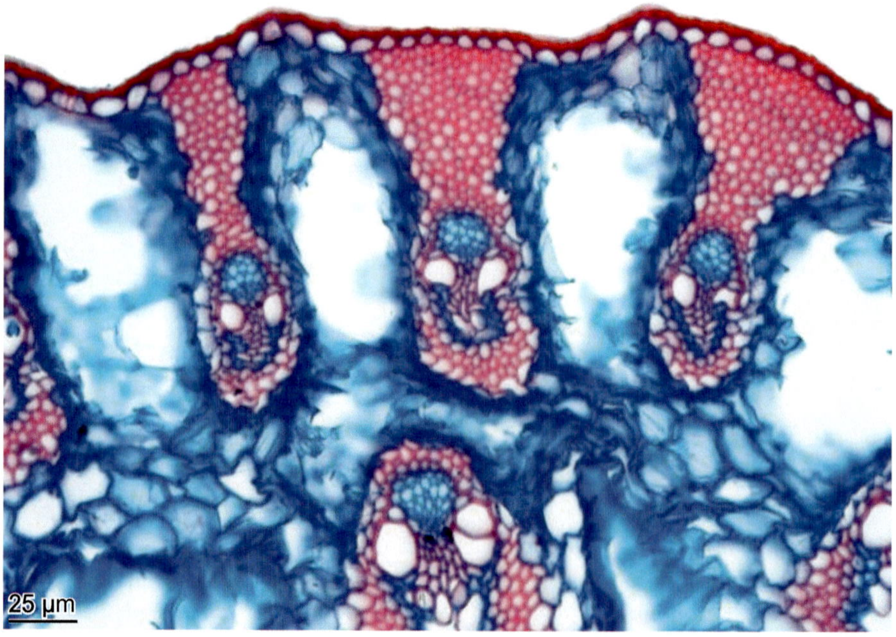

Carex stenophylla Wahlenb.

Cyperaceae

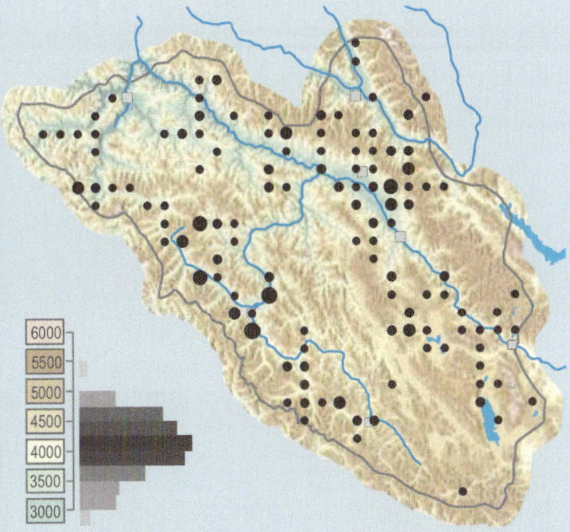

Syn.: *Vignea stenophylla* (Wahlenb.) Rchb.

Description: Perennial herbs, 10–40 cm tall, loosely cespitose, leaves shorter or as long as stems, inflorescence ovoid, compact, composed of androgynous spikes.

Distribution: Europe, Caucasus, Russia (Siberia), Iran, Kazakhstan, Turkmenistan, Uzbekistan, Kyrgyzstan, Tajikistan, Afghanistan, Pakistan, NW India, China (Gansu, Nei Monggol, Shaanxi, Xinjiang, Xizang), Mongolia.

Habitat: Dry sandy and loamy pastures, sandy plains.

Culm round, filled, with a diameter of 0.8–1.0 mm. Center contains thin-walled remains of large, unlignified cells, surrounded by 2–4 rows of unlignified cells. Epidermis cells smooth, lignified, thin-walled on the inside, thicker-walled peripherally. Vascular bundles arranged in one peripheral row. Chlorenchyma consists of irregularly formed square groups, containing irregularly formed intercellulars. Girders poorly lignified, square. Vessels within larger vascular bundles arranged in a T-shape. Lateral vessels large, diameter 25–30 µm. Phloem consists of round groups, containing sieve tubes and companion cells. Vessels of the protophloem rarely accompanied by centripetal intercellulars. Sclerenchymatic vessel sheaths eccentric, with one row of T-shaped, thin- to thick-walled fibers centrifugally, and 2–3 layers centripetally.

Analyzed: 2 individuals

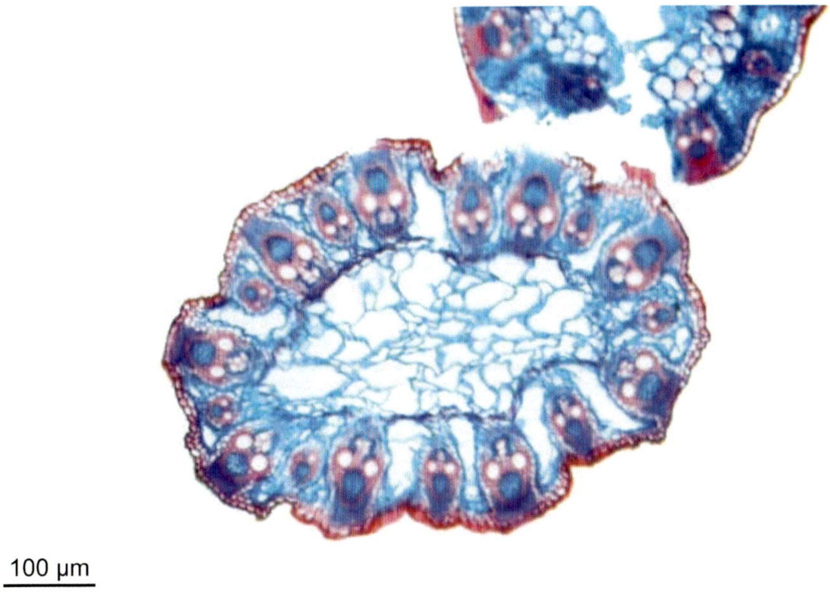

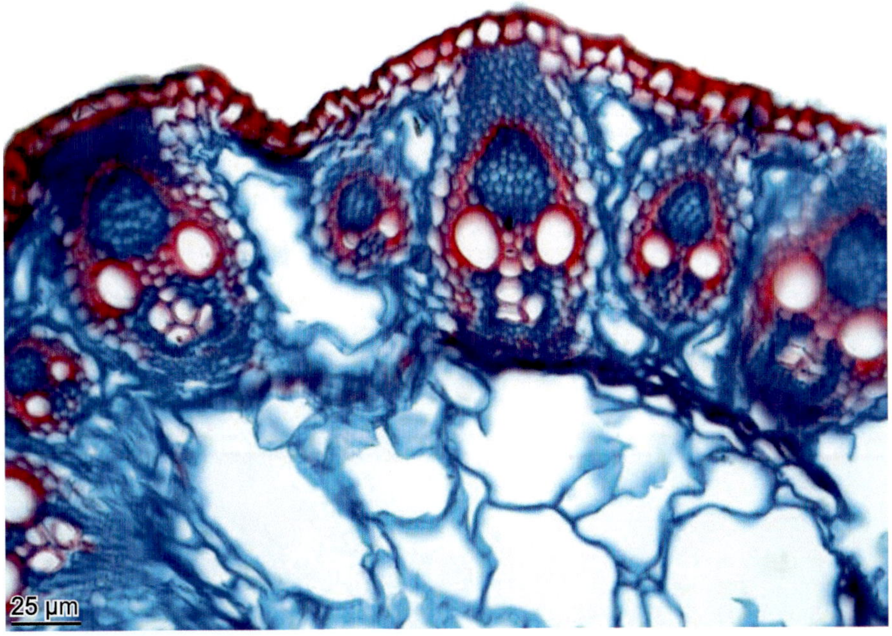

Eleocharis quinqueflora (Hartmann) O.Schwarz

Cyperaceae

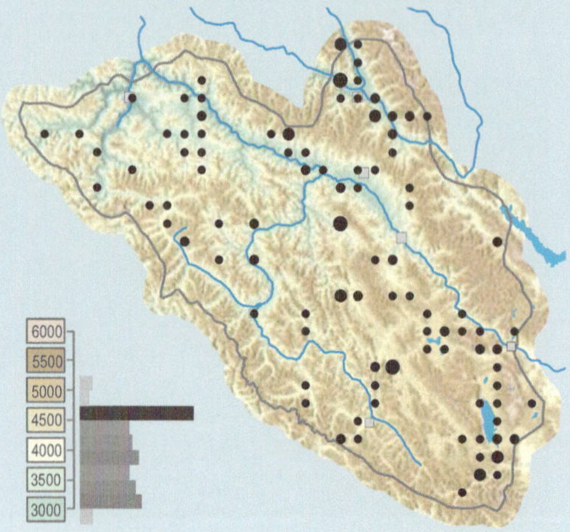

Syn.: *Scirpus quinqueflorus* Hartmann

Description: Perennial herbs, 5–25 cm tall, cespitose, spikelet brownish, ovoid to globose, 4–7 mm long, 2–7-flowered.

Distribution: North and South America, N Africa, Europe, SW Asia, Russia, Kazakhstan, Uzbekistan, Kyrgyzstan, Tajikistan, Afghanistan, Pakistan, NW India, Nepal, China, Mongolia.

Habitat: Wet peaty mires, particularly in base-rich habitats, lake margins.

Culm round, wavy, filled, with a diameter of 0.5 mm. Center contains a small group of polygonal cells and some very large intercellulars (air spaces). Epidermis cells thin-walled on the inside, thicker-walled (lignified) peripherally. Small fiber groups of thick-walled lignified fibers embedded in the epidermis. Chlorenchyma consists of palisade-like cells forming a continuous belt. A few large vascular bundles located just inside the chlorenchyma. Vessels within radial oval vascular bundles arranged in an arc. Two large lateral vessels with a diameter of 10–15 µm. Vessels of the protophloem accompanied by centripetal intercellulars. Phloem unicellular, forming round groups. Vascular bundles surrounded by a sheath of thin-walled, large, round cells and a belt of unlignified, small fibers.

Analyzed: 2 individuals

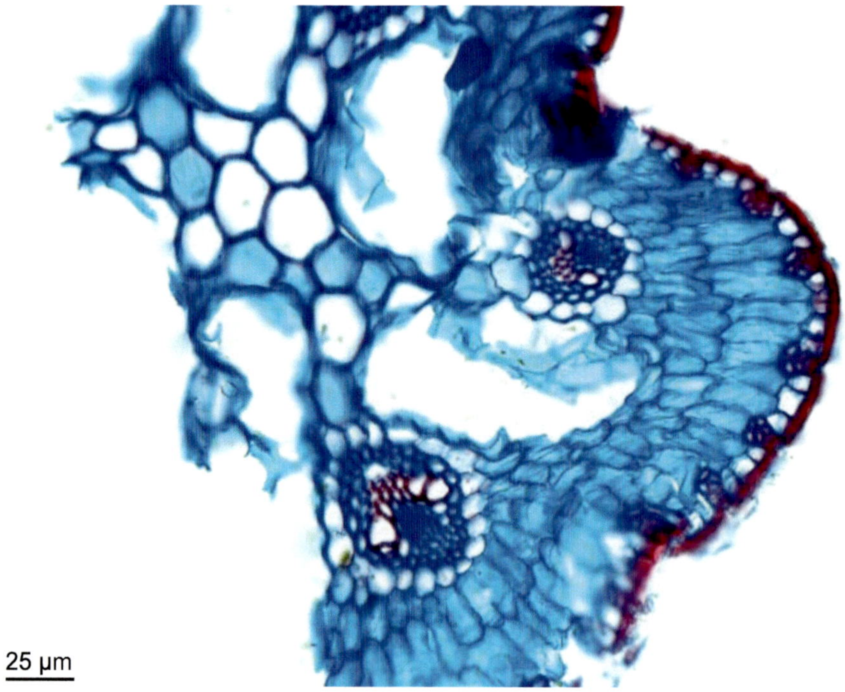

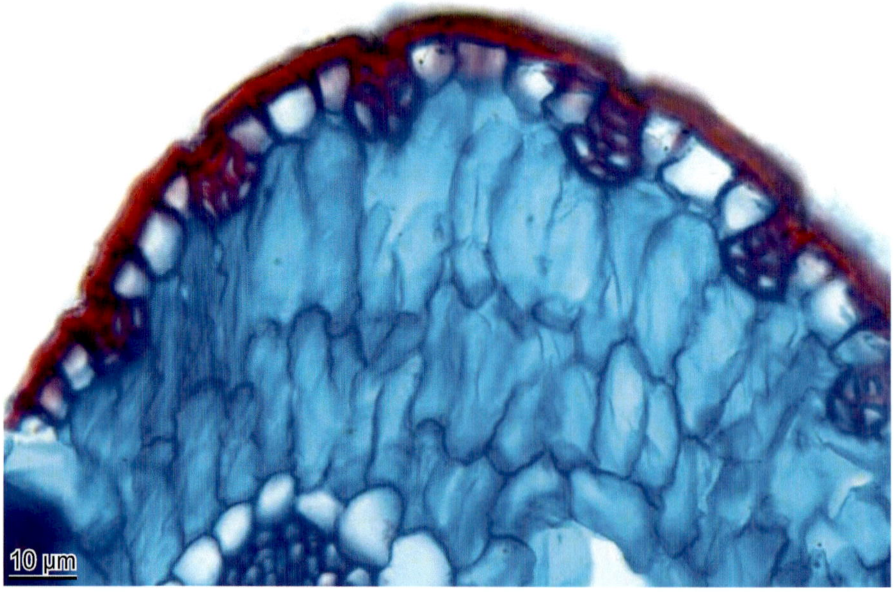

Eleocharis uniglumis (Link) Schult.

Cyperaceae

Syn.: *Scirpus uniglumis* Link

Description: Perennial herbs, 10–50 cm tall, cespitose, spikelet brown, narrowly ellipsoid or ovoid, 7–15 mm long, 3–4 mm wide, margin of basal glume widely scarious.

Distribution: North America, N Africa, Europe, SW Asia, Russia, Kazakhstan, Uzbekistan, Kyrgyzstan, Afghanistan, Pakistan, India, China, Mongolia.

Habitat: Wet meadows, salt marshes.

Culm round, wavy, filled, with a diameter of 1 mm. Center contains large, thin-walled cells and some very large intercellulars (air spaces). Epidermis cells thin-walled on the inside, thicker-walled (lignified) peripherally. Small fiber groups of thick-walled lignified fibers embedded in the epidermis. Chlorenchyma consists of palisade-like cells, forming a continuous belt. A few large vascular bundles located just inside the chlorenchyma. Vessels within the radial oval vascular bundles arranged in an arc. Two large lateral vessels with a diameter of 10–15 µm. Vessels of the protophloem frequently accompanied by centripetal intercellulars. Phloem unicellular, forming round groups. Vascular bundles surrounded by a sheath of thin-walled, unlignified, large, round cells and a discontinuous layer of lignified, small fibers.

Analyzed: 2 individuals

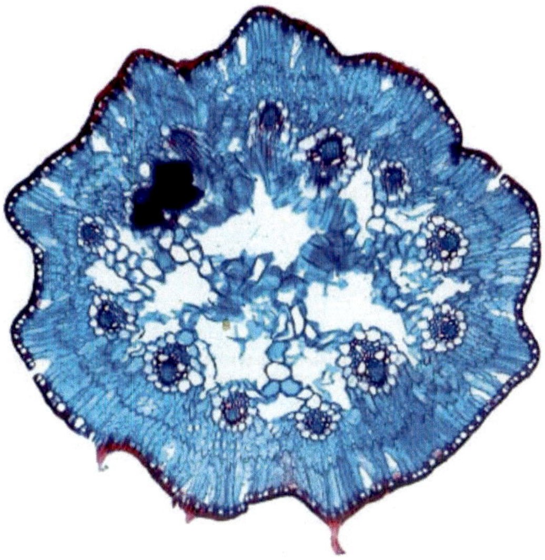

100 µm

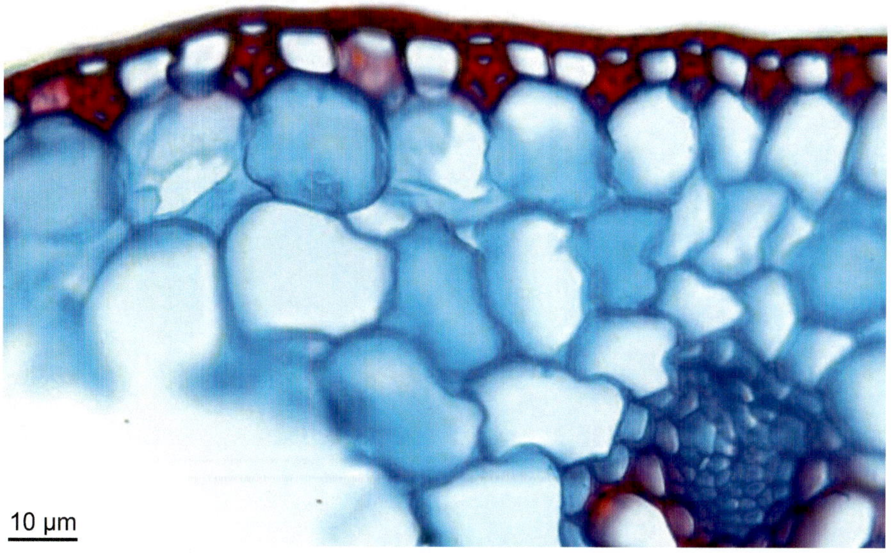

10 µm

Kobresia capillifolia (Decne.) C.B.Clarke
Cyperaceae

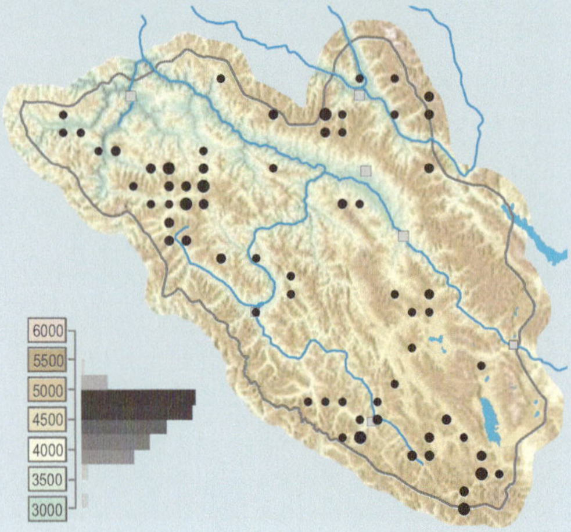

Syn.: *Elyna capillifolia* Decne.

Description: Perennial herbs, 10–45 cm tall, densely cespitose, stems terete, leaves linear, shorter than stems, often flexuous above, spikes cylindrical or clavate, 10–50 mm long, 2–4 mm wide.

Distribution: Kazakhstan, Kyrgyzstan, Tajikistan, Afghanistan, Pakistan, India (Jammu and Kashmir, Himachal Pradesh, Sikkim), Nepal, China (Gansu, Qinghai, Sichuan, Xinjiang, Xizang), Bhutan, Mongolia.

Habitat: River banks, alpine meadows.

Culm round, wavy, hollow, with a diameter of 0.9–1.1 mm. Center surrounded by 2–4 rows of large, thin-walled, lignified cells. Epidermis cells thin-walled on the inside, thicker-walled (lignified) peripherally. Large vascular bundles arranged in one peripheral row. A few small vascular bundles exist within the chlorenchyma. Chlorenchyma consists of tangentially enlarged, rectangular groups, often containing large, irregularly formed intercellulars. Girders conical. Vessels within the radial oval vascular bundles arranged in an arc. Two large lateral vessels with a diameter of 20 µm, larger than those in the center of the bundle. A few vessels of the protophloem arranged radially. Vessels of the protophloem accompanied by very small centripetal intercellulars. Sclerenchymatic vessel sheaths eccentric, with thick-walled fibers in one row centrifugally, and groups in a crescent centripetally.

Analyzed: 2 individuals

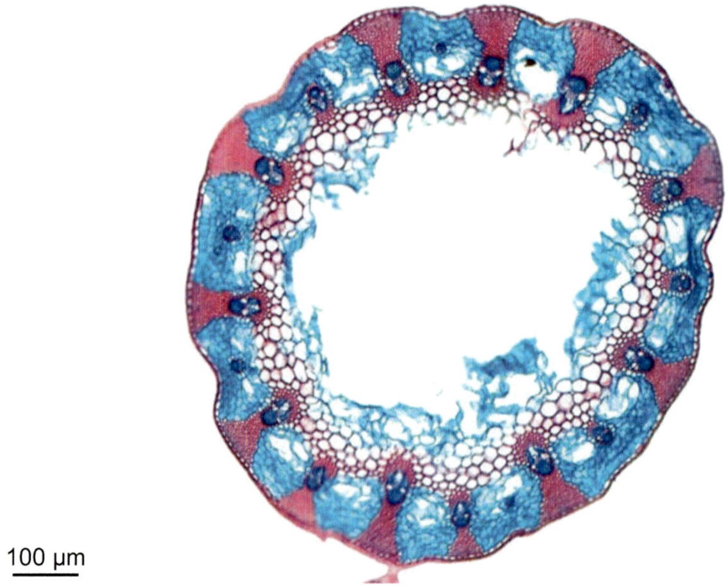

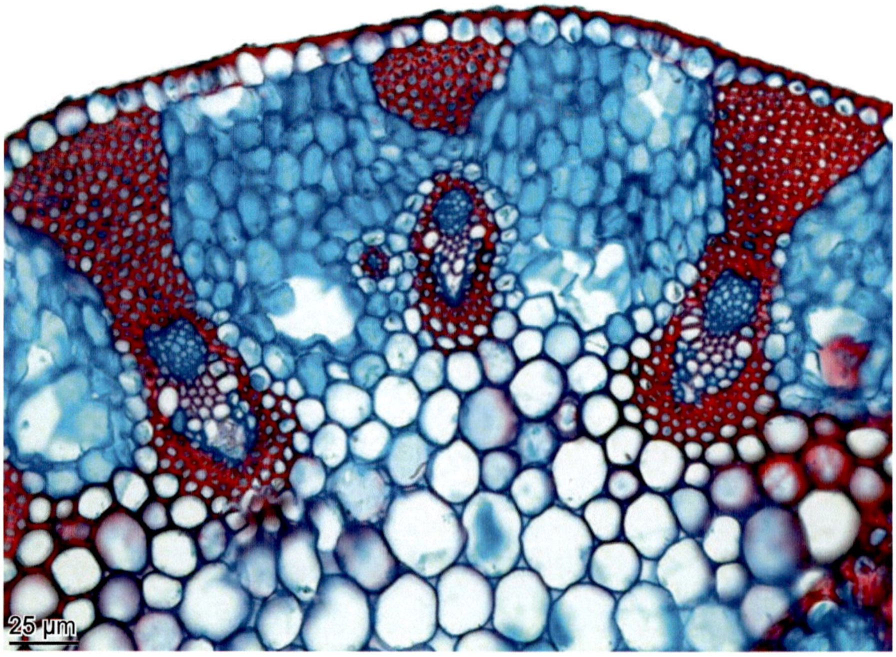

Kobresia hohxilensis R.F.Huang

Cyperaceae

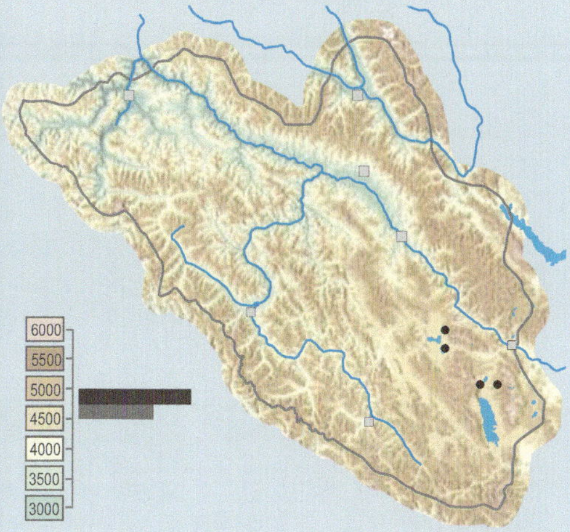

Syn.: *K. stolonifera* Y.C.Tang ex P.C.Li

Description: Perennial herbs, 5–20 cm tall, stoloniferous, stems slender, 0.5–1 mm across, leaves involute, filiform, shorter than stems, spikes narrowly ovoid, 1–2 cm long, reddish brown.

Distribution: NW India (Jammu and Kashmir), China (Gansu, Qinghai, Xizang).

Habitat: Sandy plains.

Culm round, wavy, hollow, with a diameter of 0.5–0.7 mm. Center surrounded by 2–3 rows of large, thin-walled, unlignified cells. Epidermis cells thin-walled on the inside, thicker-walled (lignified) peripherally. Large vascular bundles arranged in one peripheral row. Small vascular bundles exist within the chlorenchyma. Chlorenchyma dominating the cross-section, forming an almost continuous belt. Intercellulars absent. Girders long, rectangular to conical. Vessels within the radial oval vascular bundles arranged in a T-shape or arc. Two large lateral vessels with a diameter of 10–15 µm, larger than those in the center of the bundle. Vessels of the protophloem accompanied by centripetal intercellulars. Sclerenchymatic vessel sheaths eccentric, with thick-walled fibers in one row centrifugally, and groups in a crescent centripetally.

Analyzed: 1 individual

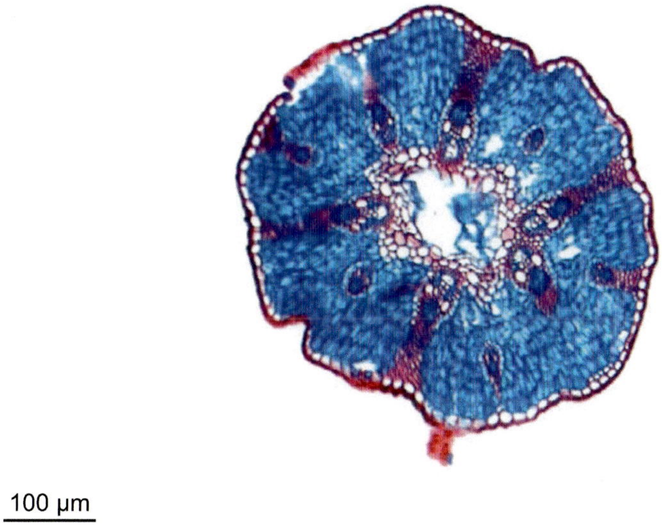

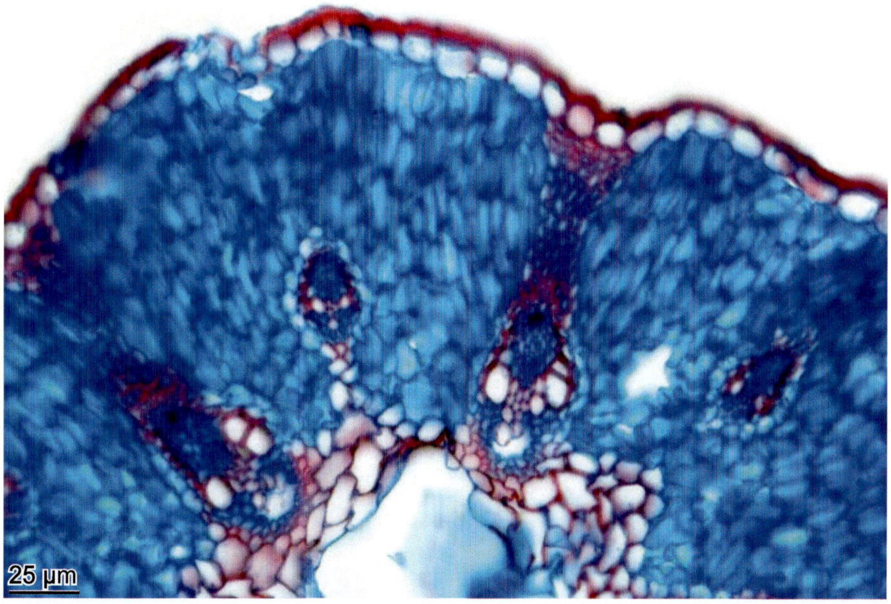

Kobresia humilis (C.A.Mey. ex Trautv.) Serg.
Cyperaceae

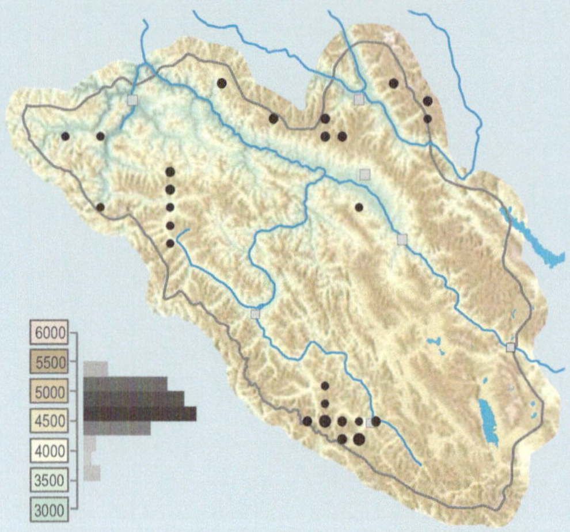

Syn.: *Elyna humilis* C.A.Mey. ex Trautv., *K. royleana* var. *humilis* (C.A.Mey. ex Trautv.) Kük.

Description: Perennial herbs, 5–15 cm tall, cespitose, stems trigonous, leaves basal, flat, 1–2 mm wide, shorter than stems, spikes compact, 7–12 mm long, 3–4 mm wide.

Distribution: SW Asia, Kazakhstan, Uzbekistan, Kyrgyzstan, Tajikistan, Afghanistan, Pakistan, India (Jammu and Kashmir, Sikkim), Nepal, China (Xinjiang, Xizang), Mongolia.

Habitat: Alpine meadows.

Culm round, wavy, hollow, with a diameter of 0.9–1.1 mm. Center surrounded by 2–4 rows of large, thin-walled, lignified cells. Epidermis cells thin-walled on the inside, thicker-walled (lignified) peripherally. Large vascular bundles arranged in one peripheral row within girders. Small vascular bundles exist within the chlorenchyma. Chlorenchyma consists of square to rectangular groups, generally containing large, irregularly formed intercellulars. Girders rectangular. Vessels within the radial oval vascular bundles arranged in a T-shape. Two large lateral vessels with a diameter of 10–15 μm, larger than those in the center of the bundle. Vessels of the protophloem accompanied by very small centripetal intercellulars. Sclerenchymatic vessel sheaths eccentric, with thick-walled fibers in one row centrifugally, and groups in a crescent centripetally.

Analyzed: 2 individuals

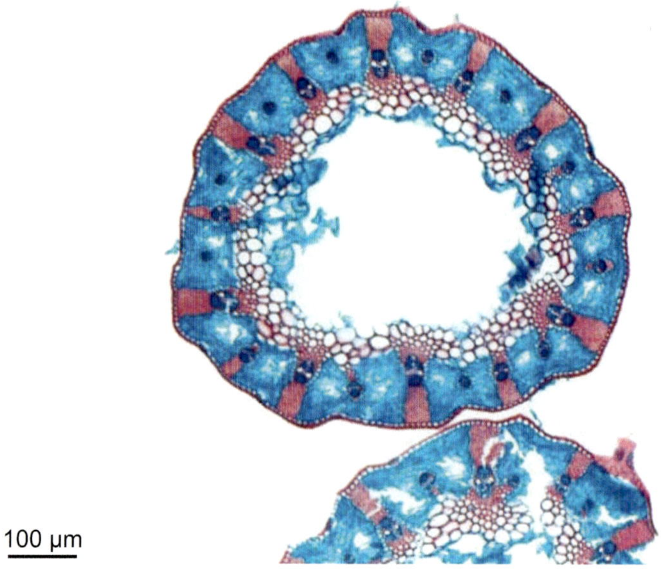

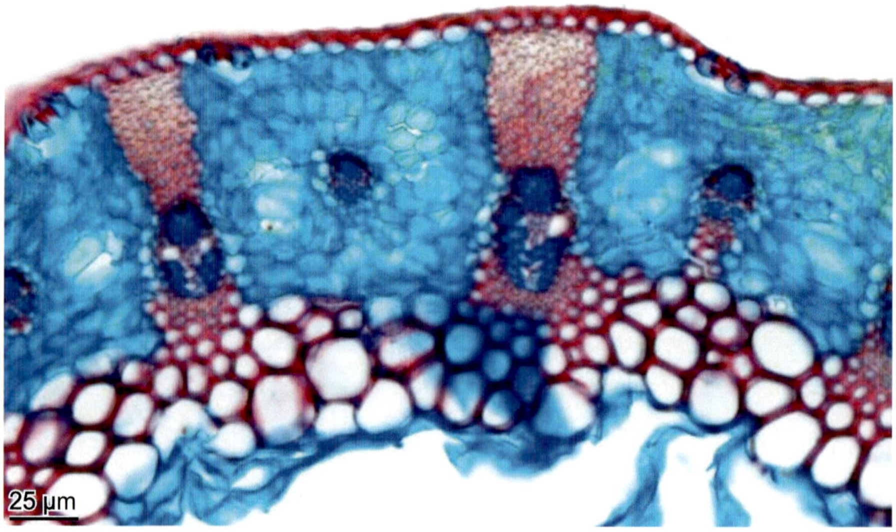

Kobresia macrantha Boeckeler

Cyperaceae

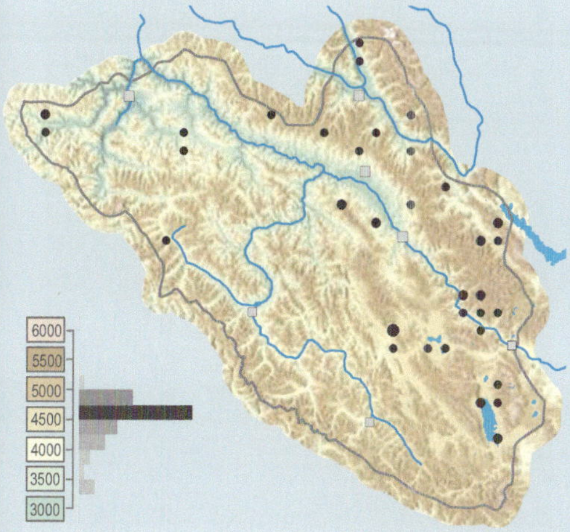

Description: Perennial herbs, 5–15 cm tall, stoloniferous, leaves flat, basal, shorter than stems, panicles cylindric to ovoid, 1–2 cm long, 0.6–1 cm wide.

Distribution: Pakistan, NW India, Nepal, China (Gansu, Qinghai, Sichuan, Xinjiang, Xizang).

Habitat: Wet turf, sandy soils, meadows in river valleys.

Culm obtusely triangular, hollow, with a diameter of 0.9–1.2 mm. Center surrounded by 2–4 rows of large, thin-walled, lignified cells. Epidermis cells thin-walled on the inside, thicker-walled (lignified) peripherally. Large vascular bundles arranged in one peripheral row. Small vascular bundles exist within the chlorenchyma. Chlorenchyma dominating the cross-section, consisting of square to slightly rectangular groups, occasionally containing large, irregularly formed intercellulars. Girders rectangular. Vessels within the radial oval vascular bundles arranged in a T-shape. Two large lateral vessels with a diameter of 10–15 µm, larger than those in the center of the bundle. A few vessels of the protophloem arranged radially. Vessels of the protophloem accompanied by very small centripetal intercellulars. Sclerenchymatic vessel sheaths eccentric, with thick-walled fibers in one row centrifugally, and groups in a crescent centripetally.

Analyzed: 2 individuals

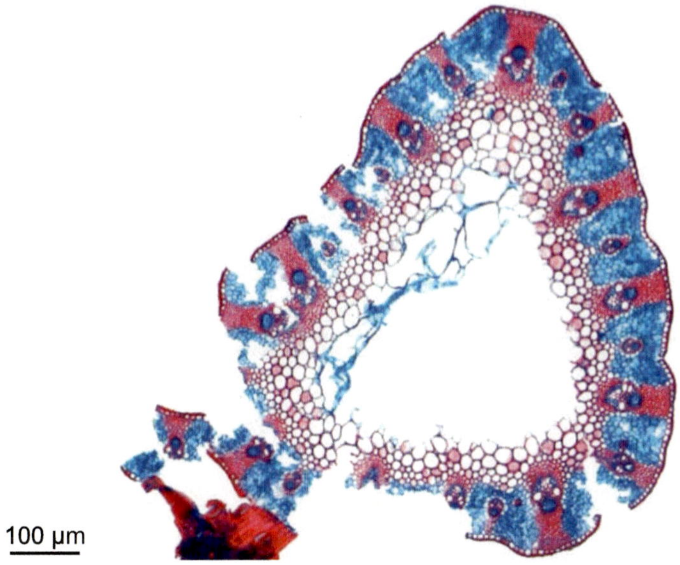

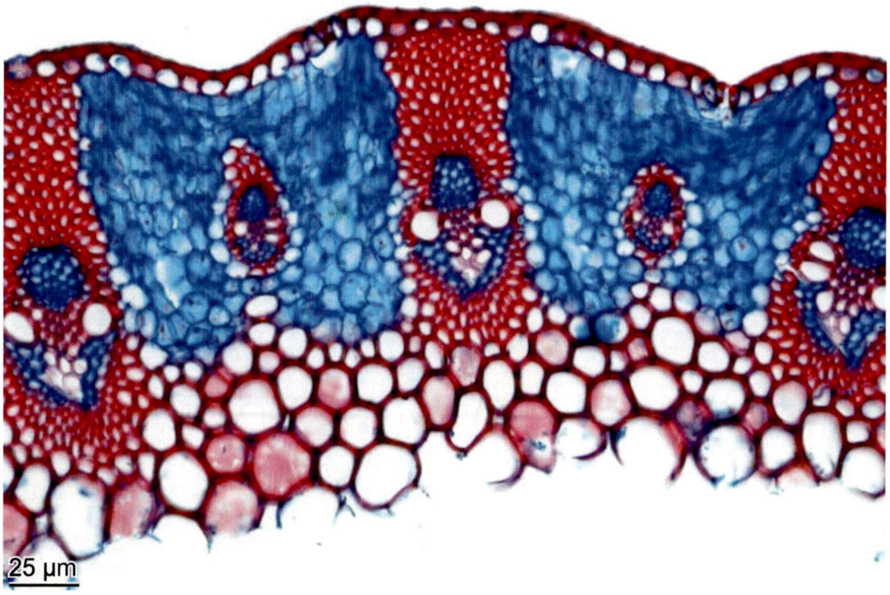

Kobresia myosuroides subsp. *bistaminata*

Cyperaceae (W.Z.Di & M.J.Zhong) S.R.Zhang

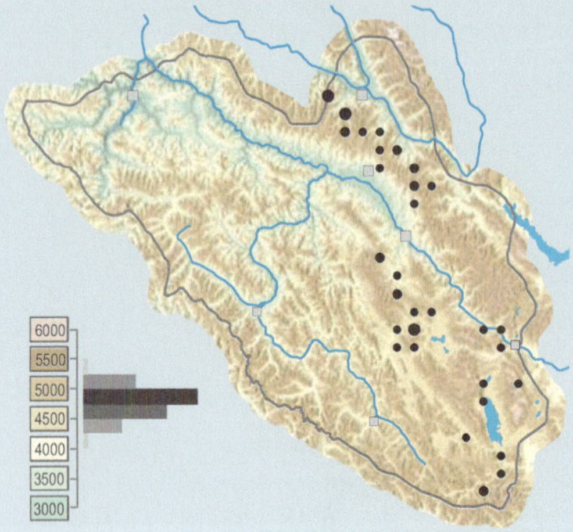

Syn.: *K. bistaminata* W.Z.Di & M.J.Zhong, *K. kashgarica* Dickoré

Description: Perennial herbs, 10–30 cm tall, densely cespitose, leaves shorter than stems, panicles narrowly cylindrical, 1–2.5 cm long, 0.2 cm wide.

Distribution: NW India, China (Gansu, Nei Mongol, Ningxia, Qinghai, Sichuan, Xinjiang, Xizang).

Habitat: Alpine meadows, grassy slopes, among shrubs.

Culm round, wavy, filled or hollow, with a diameter of 0.4–0.6 xmm. Center surrounded by 2–3 rows of large, thin-walled, unlignified cells. Epidermis cells thin-walled on the inside, thicker-walled (lignified) peripherally. Large vascular bundles arranged in one peripheral row. Small vascular bundles exist within the chlorenchyma. Chlorenchyma dominating the cross-section, consisting of large, square to slightly conical groups. Intercellulars absent. Girders rectangular. Vessels within the radial oval vascular bundles arranged in a T-shape. Two large lateral vessels with a diameter of 10 µm, larger than those in the center of the bundle. A few vessels of the protophloem arranged radially. Intercellulars within the bundles absent. Sclerenchymatic vessel sheaths eccentric, with thick-walled fibers in one row centrifugally, and groups in a crescent centripetally.

Analyzed: 2 individuals

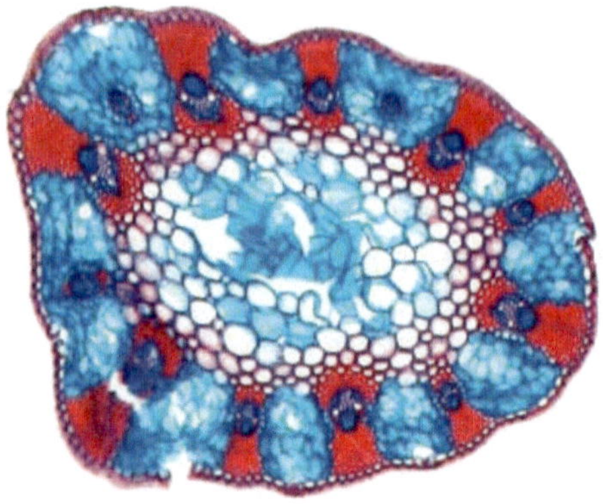

100 µm

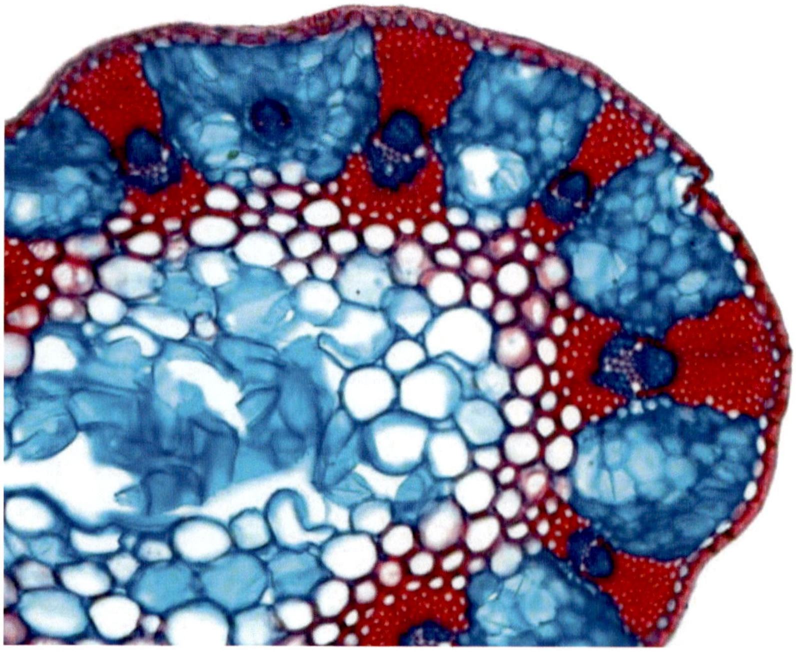

25 µm

Kobresia nepalensis (Nees) Kük.

Cyperaceae

Syn.: *Carex essenbeckii* Boott, *Hemicarex linearis* Benth., *Uncinia nepalensis* Nees

Description: Perennial herbs, 10–30 cm tall, cespitose, leaves filiform, 0.5–1 mm wide, margin involute, spike linear, 1.5–8 cm long, 2–5 mm wide, spikelets 1-flowered.

Distribution: Pakistan, India (Jammu and Kashmir, Sikkim, Uttarakhand), Nepal, China (Sichuan, Xizang, Yunnan), Bhutan.

Habitat: Dry grassy slopes, screes.

Culm round, wavy, hollow, with a diameter of 0.8–1 mm. Center surrounded by 2–4 rows of large, thin-walled, lignified, cells. Epidermis cells thin-walled on the inside, thicker-walled (lignified) peripherally. Large vascular bundles arranged in one peripheral row. Small vascular bundles within the chlorenchyma. Chlorenchyma consists of square groups, containing large irregularly formed intercellulars. Girders conical. Vessels within the radial oval vascular bundles arranged in a tangential line. Two large lateral vessels, diameter 20 μm, much larger than those in the center of the bundle. A few vessels of the protophloem accompanied by small centripetal intercellulars. Sclerenchymatic vessel sheaths eccentric, with thick-walled fibers in one row centrifugally, and groups in a crescent centripetally.

Analyzed: 2 individuals

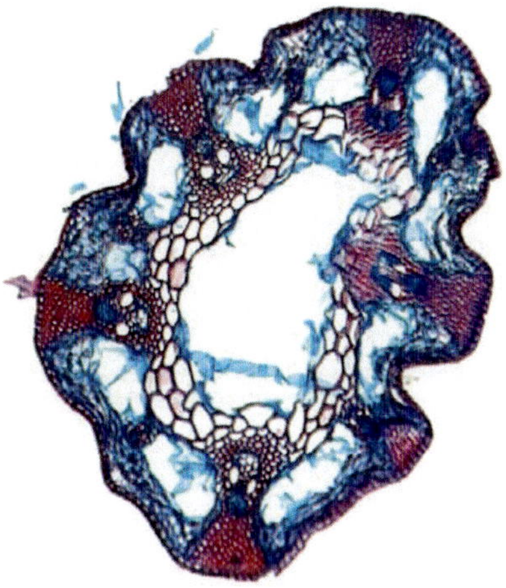

100 μm

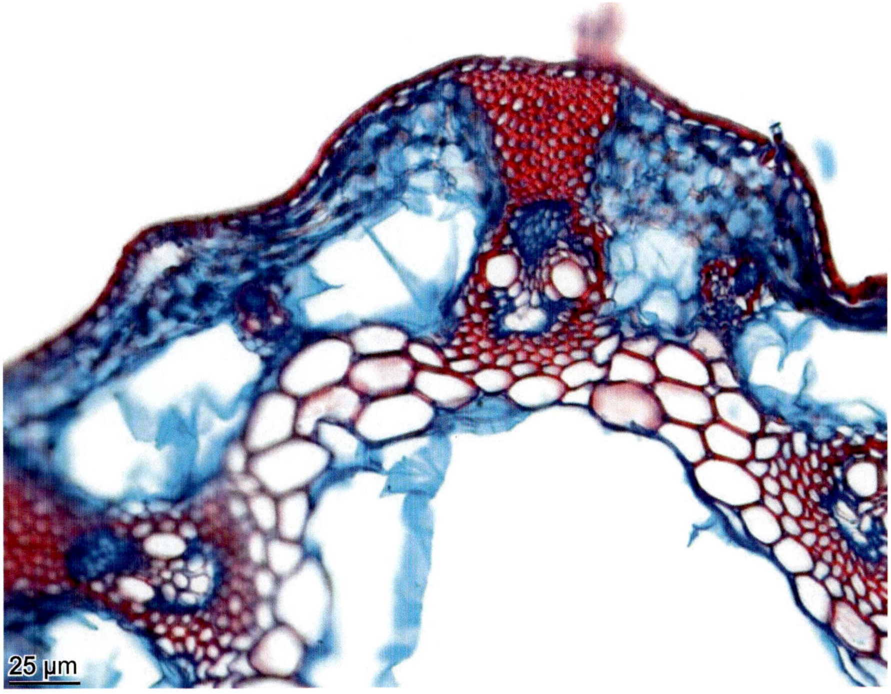

Kobresia nitens C.B.Clarke

Cyperaceae

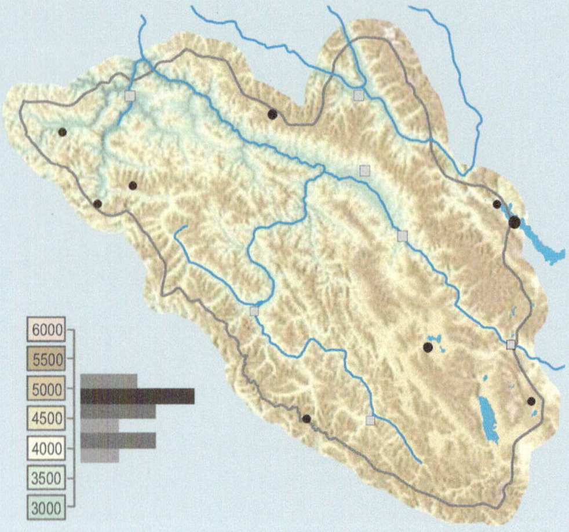

Description: Perennial herbs, 5–25 cm tall, cespitose, leaves basal, flat, 1–1.5 mm wide, shorter than stems, spike globose, 1–2 cm long, 4–8 mm wide.

Distribution: Pakistan, NW India, Nepal.

Habitat: Alpine meadows.

Culm round, wavy, filled, with a diameter of 0.9–1.2 mm. Center filled with large, thin-walled, lignified cells. Epidermis cells thin-walled on the inside, thicker-walled (lignified) peripherally. Large vascular bundles arranged in one peripheral row. A few small vascular bundles exist within the chlorenchyma. Chlorenchyma consists of tangentially enlarged, rectangular groups, often containing large, irregularly formed intercellulars. Girders conical. Vessels within the radial oval vascular bundles arranged in a T-shape. Two large lateral vessels with a diameter of 10–15 µm, larger than those in the center of the bundle. A few vessels of the protophloem arranged radially. Vessels of the protophloem not accompanied by centripetal intercellulars. Sclerenchymatic vessel sheaths eccentric, with thick-walled fibers in one row centrifugally, and groups in a crescent centripetally.

Analyzed: 3 individuals

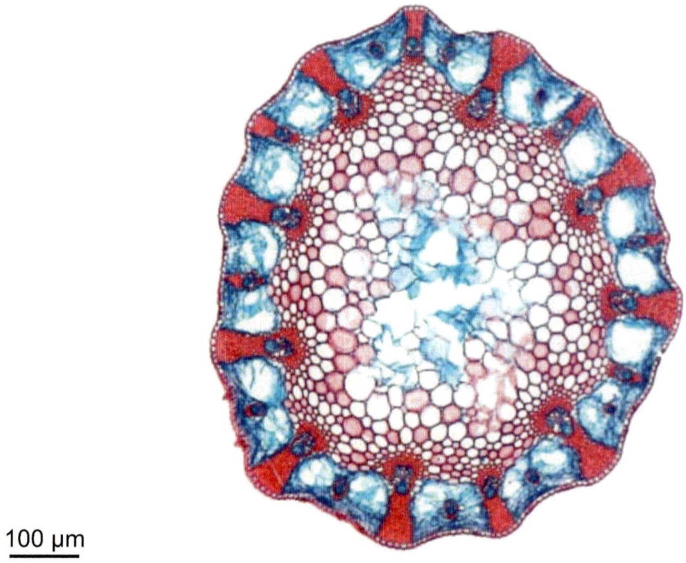

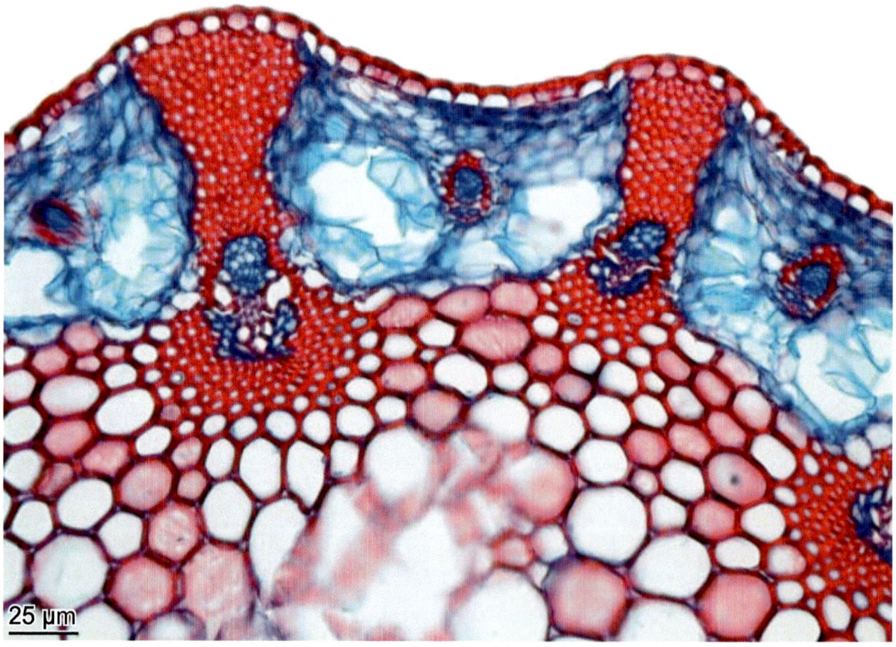

Kobresia pygmaea (C.B.Clarke) C.B.Clarke

Cyperaceae

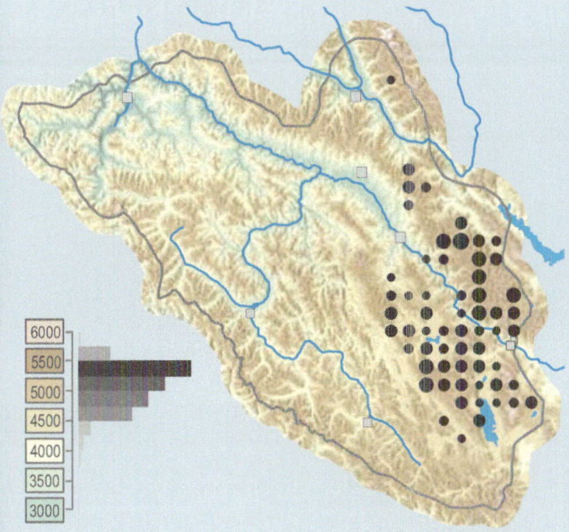

Syn.: *Hemicarex pygmaea* C.B.Clarke

Description: Perennial herbs, 1–4 cm tall, densely cespitose, leaves filiform, as long as stems, spike ovoid, dense, 2–6 mm long, 1–2 mm wide.

Distribution: Pakistan, India (Jammu and Kashmir, Himachal Pradesh, Sikkim, Uttarakhand), Nepal, China (Gansu, Hebei, Nei Mongol, Qinghai, Shanxi, Sichuan, ?Xinjiang, Xizang, Yunnan), Bhutan, Myanmar.

Habitat: Wet alpine meadows, stream sides, open grassy places.

Culm acutely triangular, filled, with a diameter of 0.5–0.7 mm. Center filled with large, thin-walled, unlignified cells. Epidermis cells thin-walled on the inside, thicker-walled (lignified) peripherally. Large vascular bundles arranged in one peripheral row. A few small vascular bundles exist within the chlorenchyma. Chlorenchyma consists of tangentially enlarged, rectangular groups, often containing large, irregularly formed intercellulars. Girders rectangular. Vessels within the radial oval vascular bundles arranged in a T-shape. Two large lateral vessels with a diameter of 5 µm, larger than those in the center of the bundle. A few vessels of the protophloem arranged radially. Vessels of the protophloem not accompanied by centripetal intercellulars. Sclerenchymatic vessel sheaths eccentric, with thick-walled fibers in one row centrifugally, and groups in a crescent centripetally.

Analyzed: 2 individuals

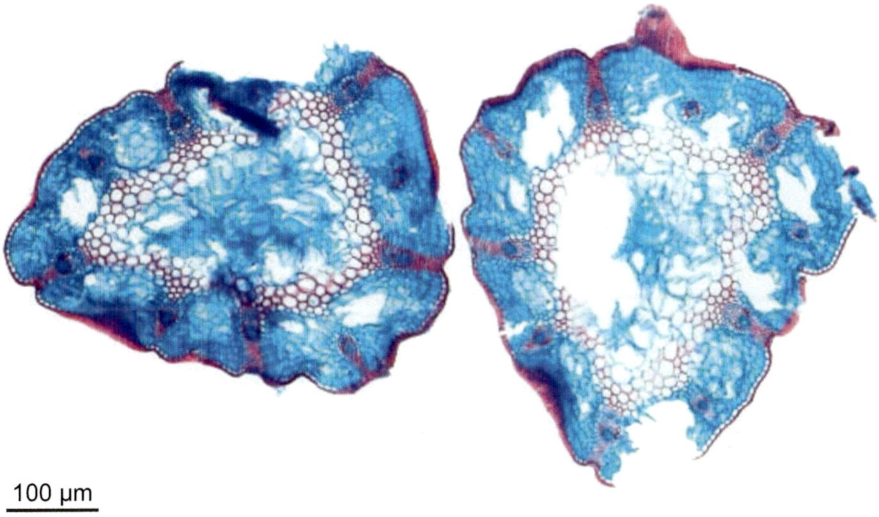

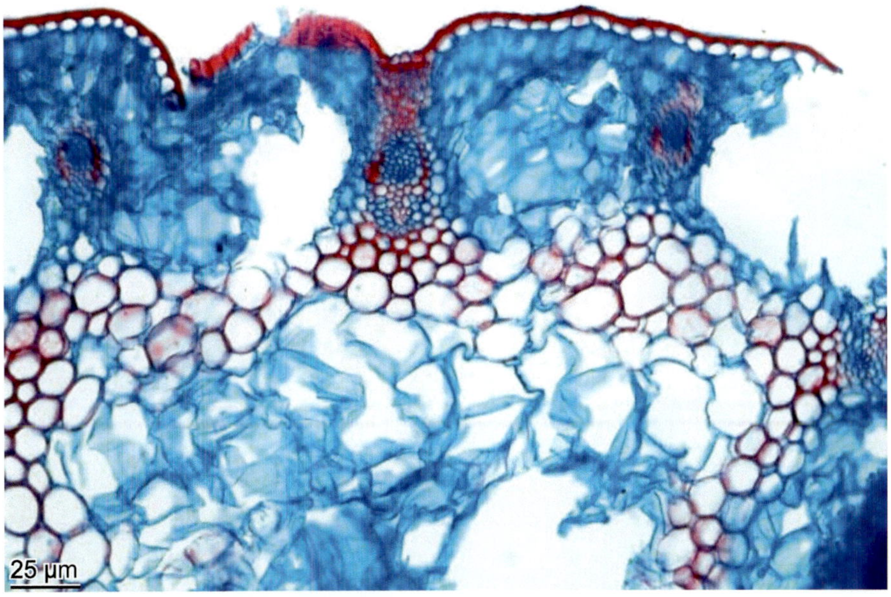

Kobresia robusta Maxim.

Cyperaceae

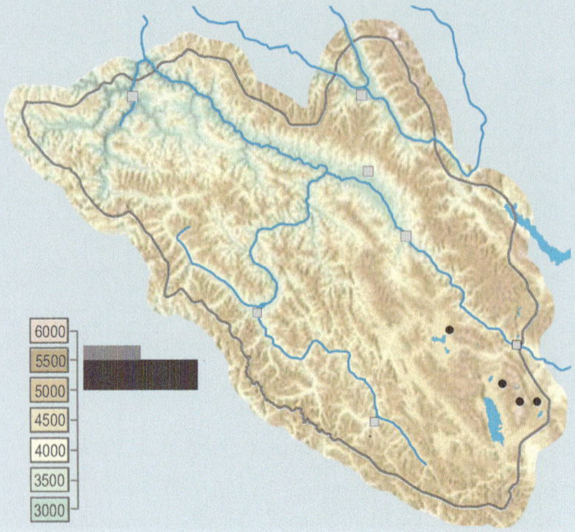

Description: Perennial herbs, 5–50 cm tall, cespitose, leaves basal, rigid, erect or curved, filiform, 0.7–2 mm wide, margin involute, spike narrowly ovoid to cylindric, glumes yellowish brown, margin very broadly hyaline.

Distribution: India (Jammu and Kashmir), China (Gansu, Qinghai, Xinjiang, Xizang), Mongolia.

Habitat: Alpine meadows.

Culm round, wavy, filled, with a diameter of 0.9–1.1 mm. Center filled with round, lignified cells. Epidermis cells thin-walled on the inside, thicker-walled (lignified) peripherally. Large vascular bundles arranged in one peripheral row. Smaller vascular bundles exist within the chlorenchyma. Chlorenchyma consists of tangentially enlarged, rectangular groups and large, round intercellulars. Girders long, rectangular. Vessels within the radial oval vascular bundles arranged in a T-shape. Two large lateral vessels with a diameter of 20 µm, larger than those in the center of the bundle. A few vessels of the protophloem arranged radially. Centripetal intercellulars within the bundle absent. Sclerenchymatic vessel sheaths eccentric, with thick-walled fibers in one row centrifugally, and groups in a crescent centripetally.

Analyzed: 2 individuals

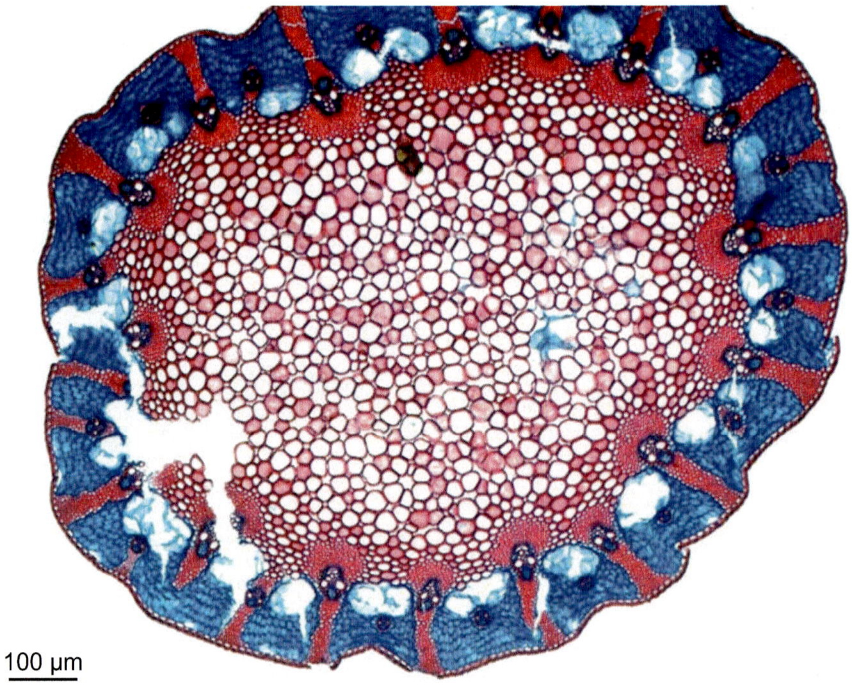

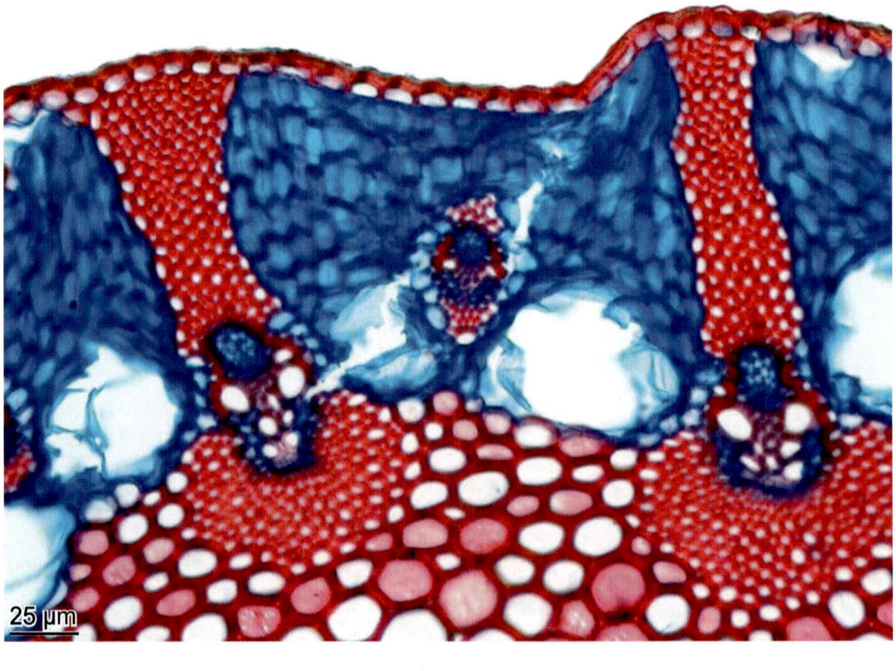

Kobresia royleana (Nees) Boeck.

Cyperaceae

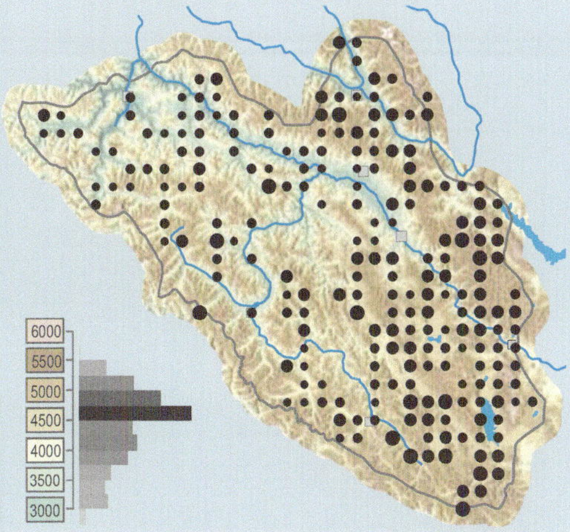

Syn.: *K. stenocarpa* (Kar. & Kir.) Steud.

Description: Perennial herbs, 10–50 cm tall, cespitose, leaves linear, flat, basal, shorter than stems, panicles dense or slightly loose, 1–3.5 cm long, 0.6–1.2 cm wide.

Distribution: Kyrgyzstan, Tajikistan, Afghanistan, Pakistan, India (Jammu and Kashmir, Himachal Pradesh, Uttarakhand), Nepal, China (Gansu, Qinghai, Sichuan, Xinjiang, Xizang, Yunnan), Bhutan.

Habitat: Mesic to dry grasslands along streams.

Culm round to acutely triangular, hollow, with a diameter of 1.5–2 mm. Center surrounded by 2–4 rows of large, thin-walled, lignified cells. Epidermis cells thin-walled on the inside, thicker-walled (lignified) peripherally. Large vascular bundles arranged in one peripheral row within girders. A few small vascular bundles exist within the chlorenchyma. Chlorenchyma consists of large, square to rectangular groups, generally containing large, irregularly formed intercellulars. Girders long, rectangular. Vessels within the radial oval vascular bundles arranged in a T-shape. Two large lateral vessels with a diameter of 15–20 µm, larger than those in the center of the bundle. Vessels of the protophloem accompanied by centripetal intercellulars. Sclerenchymatic vessel sheaths eccentric, with thick-walled fibers in one row centrifugally, and groups in a crescent centripetally.

Analyzed: 3 individuals

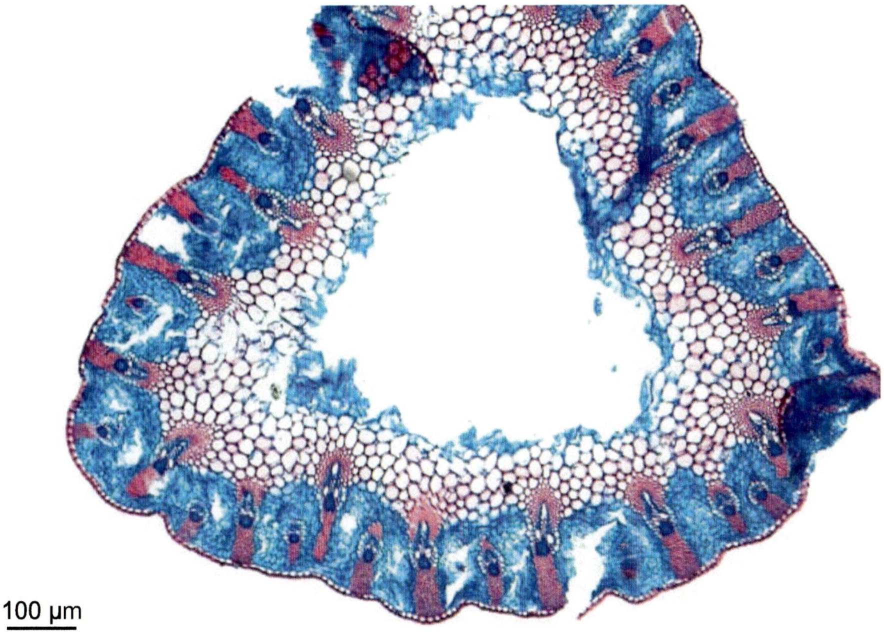

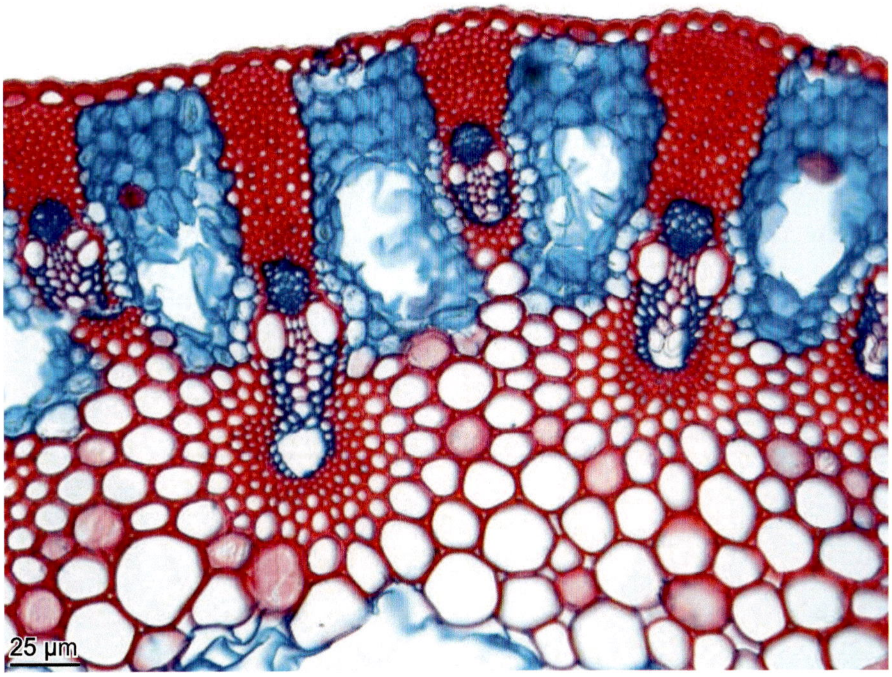

Kobresia schoenoides (C.A.Mey.) Steud.

Cyperaceae

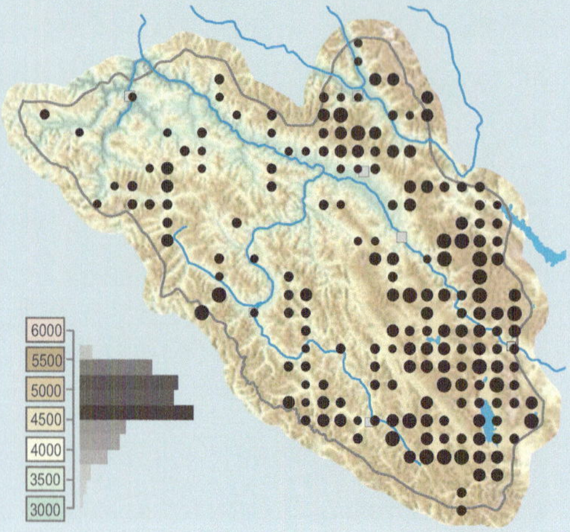

Syn.: *Elyna schoenoides* C.A.Mey., *K. deasyi* C.B.Clarke, *K. pamiroalaica* N.A.Ivanova

Description: Perennial herbs, 10–60 cm tall, densely cespitose, stems stout, 1–3 mm across, leaves involute, shorter than stems, spikes dense, oblong, 1–2 cm long, dark to yellowish brown.

Distribution: Russia, Iran, Kazakhstan, Uzbekistan, Kyrgyzstan, Tajikistan, Afghanistan, Pakistan, India (Jammu and Kashmir, Himachal Pradesh, Sikkim), Nepal, China (Gansu, Qinghai, Sichuan, Xinjiang, Xizang, Yunnan), Bhutan, Mongolia.

Habitat: Mesic to wet river banks, stabilized stony mesic slopes, saline pastures.

Culm round to acutely triangular, hollow, with a diameter of 1.2–1.5 mm. Center surrounded by 2–4 rows of large, thin-walled, lignified cells. Epidermis cells thin-walled on the inside, thicker-walled (lignified) peripherally. Large vascular bundles arranged in one peripheral row within girders. Small vascular bundles exist within the chlorenchyma. Chlorenchyma consists of large, square to rectangular groups, generally containing large, irregularly formed intercellulars. Girders long, rectangular. Vessels within the radial oval vascular bundles arranged in a T-shape. Two large lateral vessels with a diameter of 15–20 µm, larger than those in the center of the bundle. Vessels of the protophloem accompanied by centripetal intercellulars. Sclerenchymatic vessel sheaths eccentric, with thick-walled fibers in one row centrifugally, and groups in a crescent centripetally.

Analyzed: 4 individuals

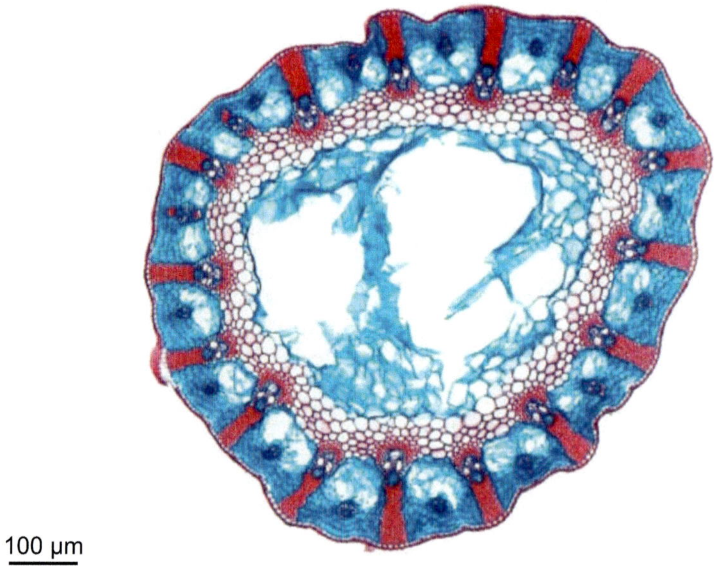

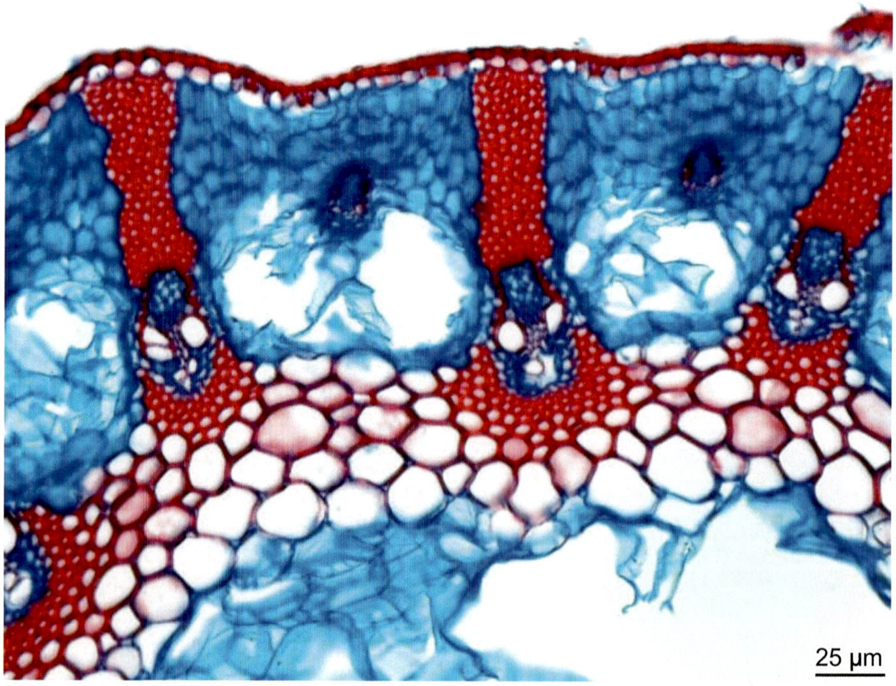

Trichophorum pumilum (Vahl) Schinz & Thell.

Cyperaceae

Syn.: *Scirpus pumilus* Vahl, *Baeothryon pumilum* (Vahl) Á.Löve & D.Löve, *Isolepis pumila* (Vahl) Roem. & Schult., *Eriophorella pumila* (Vahl) Kit Tan

Description: Perennial herbs, 4–25 cm tall, cespitose, leaves linear, mostly shorter than 2 cm, spike fusiform, 3–4 mm long.

Distribution: North America, Europe, temperate Asia.

Habitat: Marshes, springs, shallow wetlands.

Culm uniform, round, wavy, hollow or filled, with a diameter of 0.5–0.7 mm. Epidermis smooth, thin-walled on the inside, thicker-walled (lignified) peripherally. Large vascular bundles arranged in one peripheral row. Small vascular bundles exist within the chlorenchyma belt. Intercellulars absent. Girders rectangular. Vessels within the radial oval vascular bundles arranged in a T-shape or arc. Two large lateral vessels with a diameter of 10 µm, larger than those in the center of the bundle. Vessels of the protophloem accompanied by centripetal intercellulars. Phloem unicellular, forming round groups. Vascular bundles surrounded by an endodermis-like sheath of U-shaped, lignified cells.

Analyzed: 2 individuals

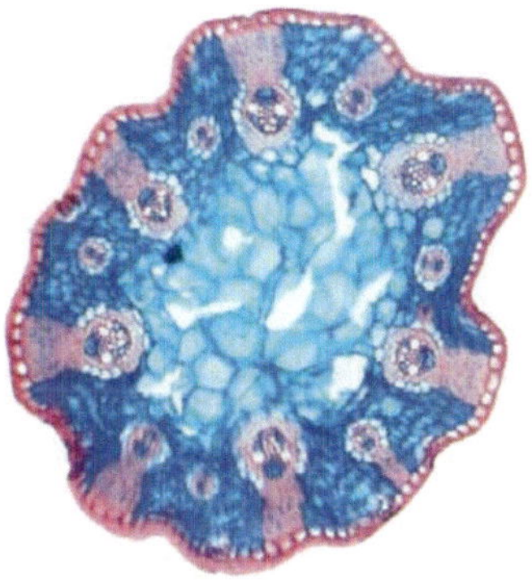

100 µm

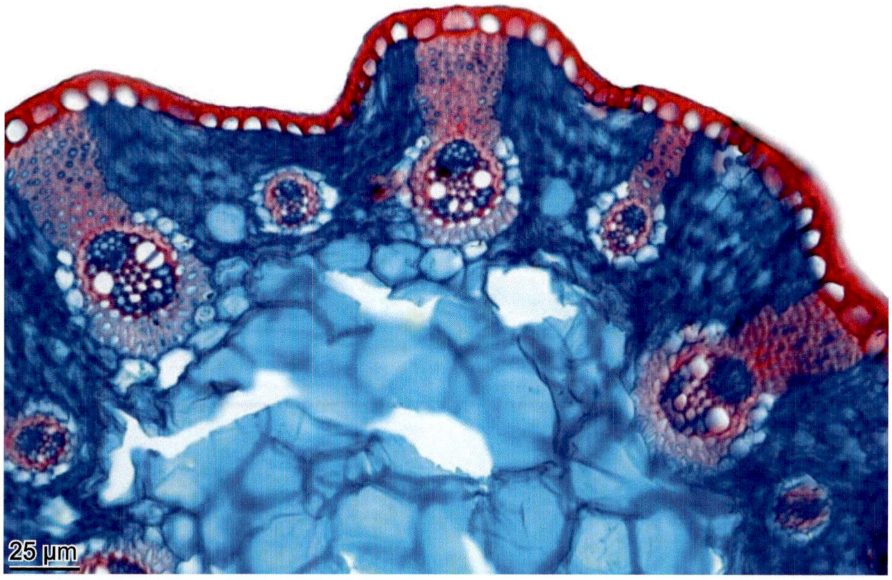

Juncus himalensis Klotzsch
Juncaceae

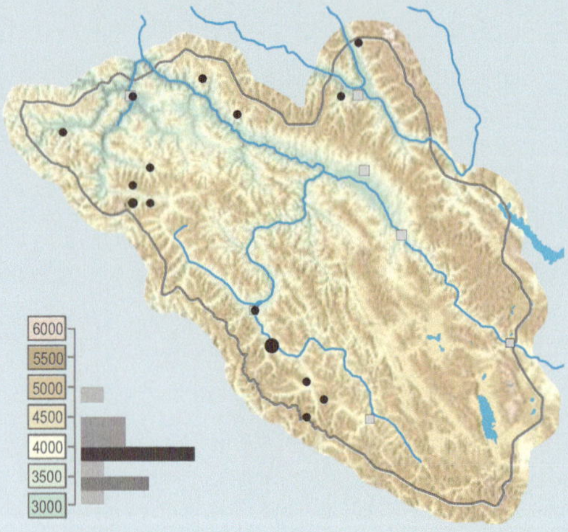

Syn.: *J. sphacelatus* var. *himalensis* (Klotzsch) Jafri, *J. afghanicus* Snog., *J. schlagintweitii* Buchenau

Description: Perennial herbs, 30–70 cm tall, stoloniferous, basal leaves 3–4, linear, channelled above, inflorescence terminal, 3–5-branched, 5–10 cm long, capitula mostly congested, 6–10 mm across, corymbosely arranged.

Distribution: Tajikistan, Pakistan, India, Nepal, China (Gansu, Qinghai, Sichuan, Xizang, Yunnan), Bhutan.

Habitat: Wet places in river valleys.

Culm round, filled, with a diameter of 1.0–1.5 mm. Center filled with large, thin-walled, polygonal, unlignified cells and a few vascular bundles. Epidermis cells lignified, smooth, thin-walled on the inside, thicker-walled peripherally. Continuous, wavy, 4–6-seriate, thin- to thick-walled peripheral sclerenchymatic belt centripetal of the chlorenchyma. Large and small vascular bundles principally arranged in two slightly offset rows. Vascular bundles laterally connected by a thin- to thick-walled zone of lignified cells. Chlorenchyma consists of a large, continuous peripheral band. Intercellulars between the unlignified cells small and triangular. Lateral vessels large, diameter 20–25 µm. Phloem consists of sieve tubes and companion cells. Vessels of the protophloem accompanied by a centripetal intercellular. Vascular bundles surrounded by an eccentric sheath of sclerenchyma, with a 3–4-seriate zone of thin- to thick-walled cells centripetally and laterally, and a large, >10-seriate group of small, thick-walled cells centrifugally. Crystal sand in the chlorenchyma.

Analyzed: 3 individuals

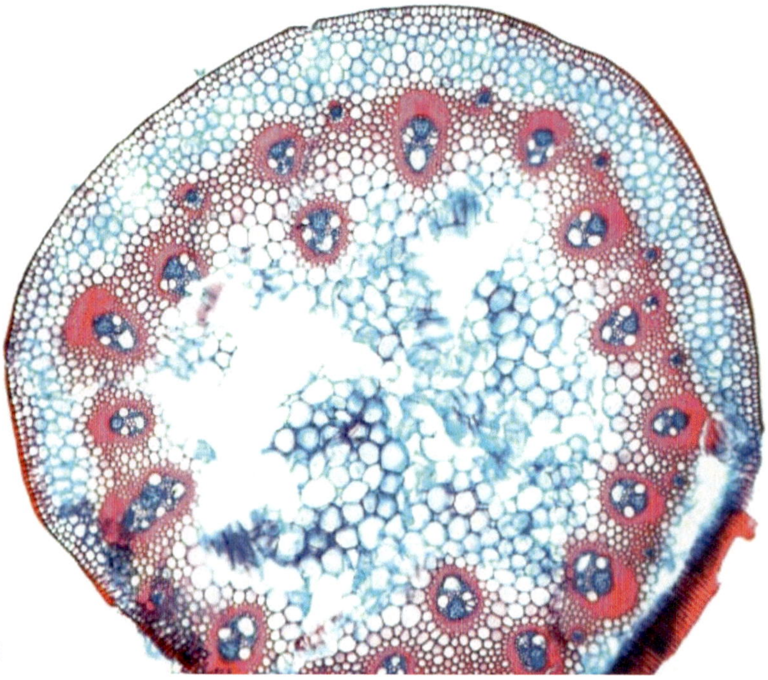

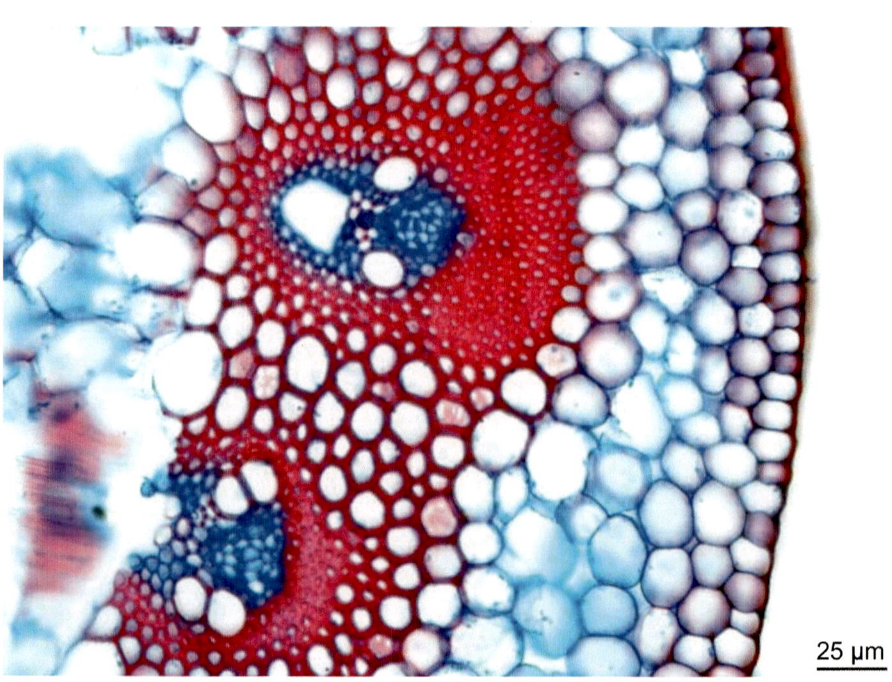

Juncus membranaceus Royle
Juncaceae

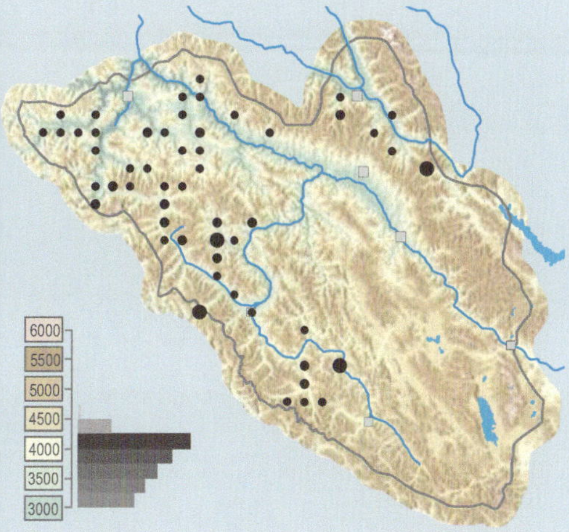

Syn.: *J. hoffmeisteri* Klotzsch

Description: Perennial herbs, 10–40 cm tall, cespitose, stoloniferous, basal leaves 1–2, up to 10 cm long, stem leaves 1–3, linear, channelled, inflorescence terminal, whitish, head solitary, 4–5 mm across, 8–20-flowered, lowest bract usually exceeding the head.

Distribution: Afghanistan, Pakistan, India, Nepal, China (Xizang, Yunnan).

Habitat: Wet fresh-water turf along brooks.

Culm round, filled, with a diameter of 2.0–2.5 mm. Center filled with large, thin-walled, polygonal, unlignified cells. Epidermis cells lignified, smooth, thin-walled on the inside, thicker-walled peripherally. Continuous, 6–8-seriate, thin- to thick-walled peripheral sclerenchymatic belt centripetal of the chlorenchyma. Large and small vascular bundles arranged in two slightly offset rows. Large vascular bundles laterally connected by a small zone of thin- to thick-walled, lignified cells. Some small bundles in the chlorenchyma zone. Chlorenchyma consists of a large continuous peripheral band. With smaller and larger irregularly formed intercellulars between the unlignified cells. Lateral vessels large, diameter 25–30 µm. Phloem consists of sieve tubes and companion cells. Vessels of the protophloem accompanied by a large centripetal intercellular. Vascular bundles surrounded by an eccentric sheath of sclerenchyma, with a 3–4-seriate zone of thin- to thick-walled cells centripetally and laterally, and a 5–7-seriate group of small, thick-walled cells centrifugally. Crystal sand in the chlorenchyma.

Analyzed: 3 individuals

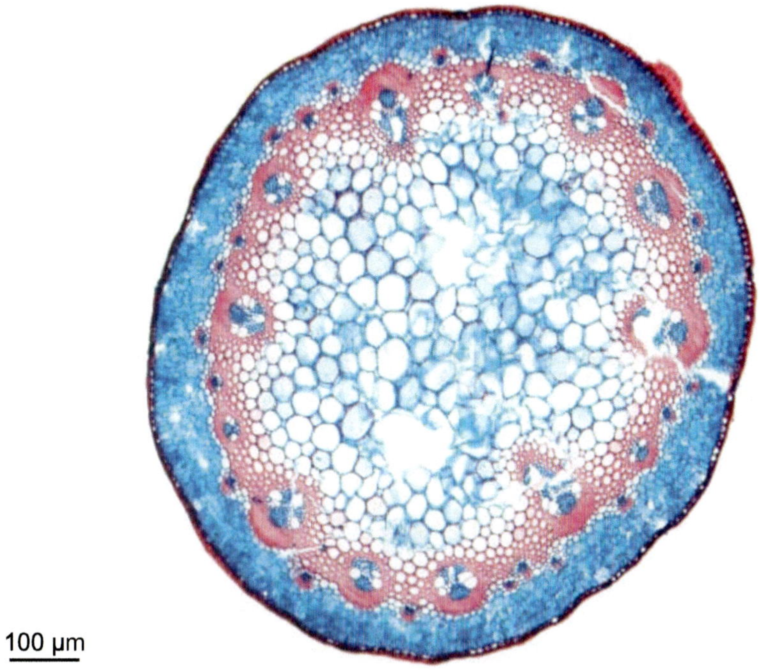

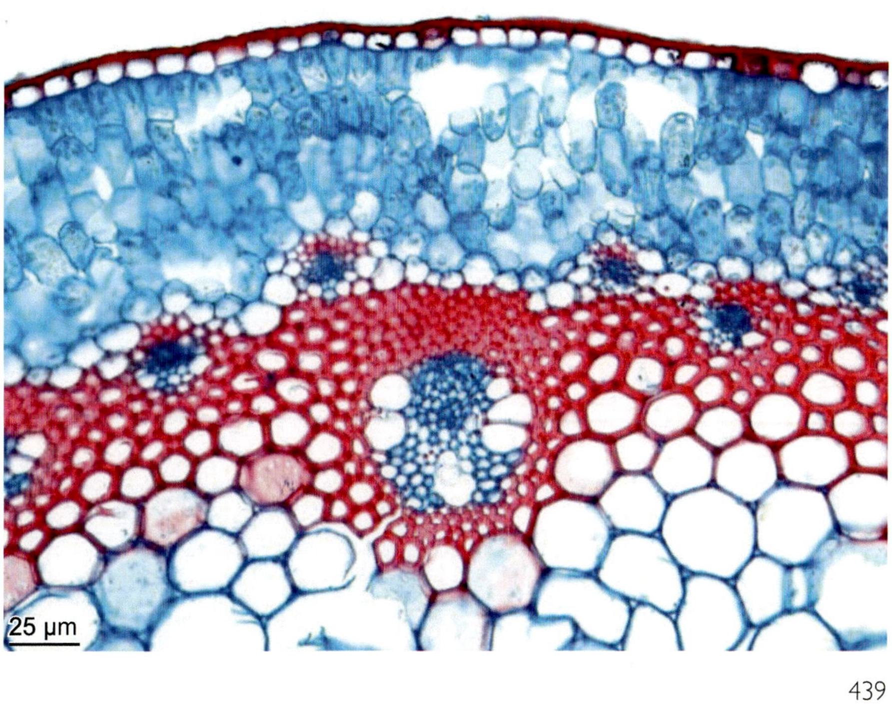

Juncus sphacelatus Decne.

Juncaceae

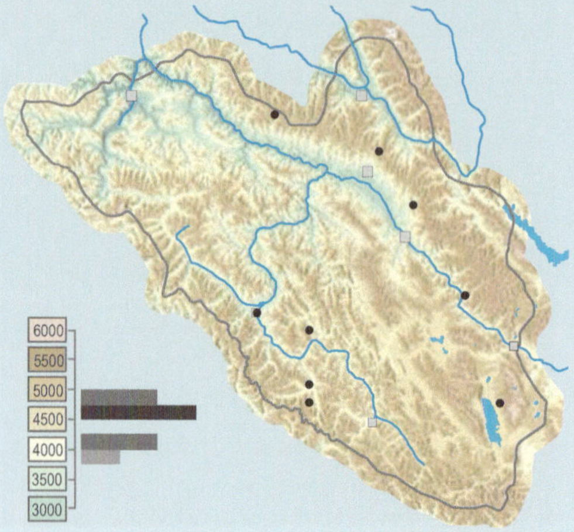

Description: Perennial herbs, 15–40 cm tall, stoloniferous, stems terete, with basal leaves and 1–2 cauline leaves, leaves linear, canaliculate, inflorescence terminal, 2–3-branched, heads 2–5, hemispheric, 1.5–2 cm across, 5–7-flowered, brown to blackish.

Distribution: Afghanistan, Pakistan, India, Nepal, China, Bhutan.

Habitat: Rocky slopes, gravelly and stony beds along streams.

Culm round, hollow, with a diameter of 1.2–1.5 mm. Center filled with large, thin-walled, polygonal, unlignified cells. Epidermis cells lignified, smooth, thin-walled on the inside, thicker-walled peripherally. Indistinct, continuous, 2–4-seriate, thin- to thick-walled peripheral sclerenchymatic belt centripetal of the chlorenchyma. Large and small vascular bundles arranged in two slightly offset rows. Large vascular bundles laterally connected by a zone of thin- to thick-walled, lignified cells. Some small bundles in the chlorenchyma zone. Chlorenchyma consists of a large, continuous peripheral band. Intercellulars between the unlignified cells irregularly formed. Lateral vessels large, diameter 15–25 µm. Phloem consists of sieve tubes and companion cells. Vessels of the protophloem accompanied by a centripetal intercellular. Vascular bundles surrounded by an eccentric sheath of sclerenchyma, with a 3–4-seriate zone of thin- to thick-walled cells centripetally and laterally, and a large, >10-seriate group of small, thick-walled cells centrifugally. Acicular crystals sand in the chlorenchyma.

Analyzed: 1 individual

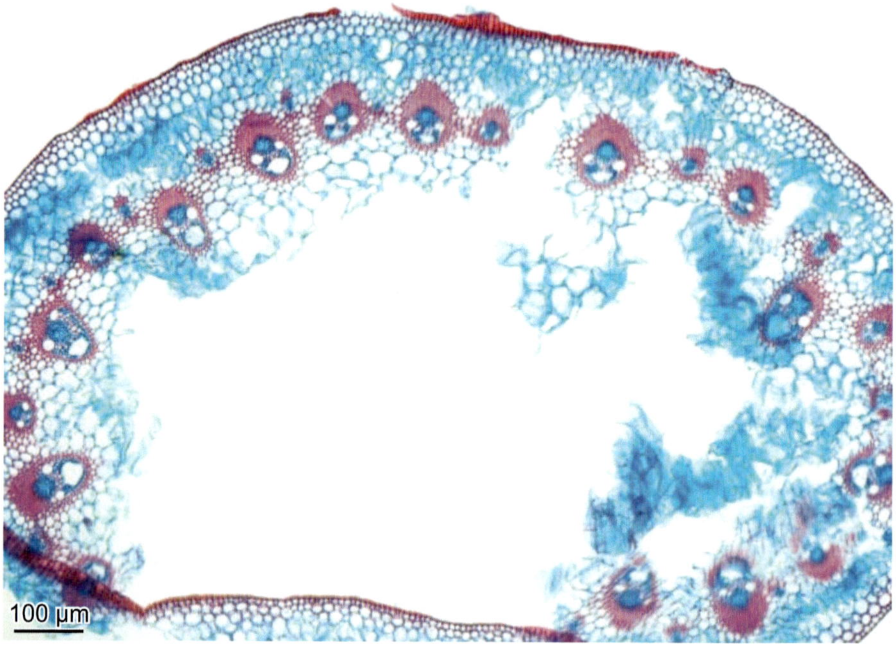

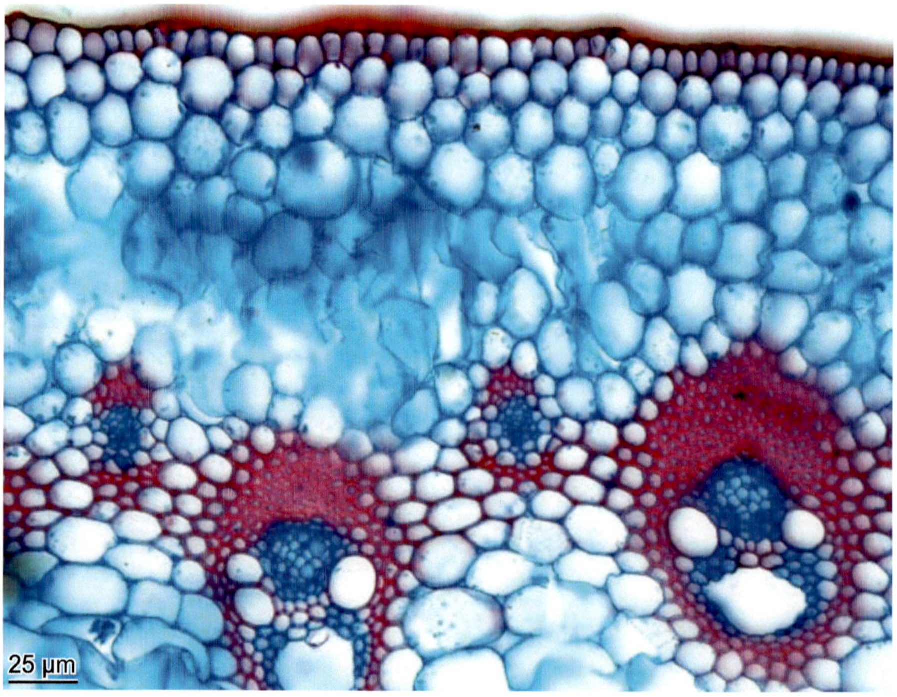

Juncus thomsonii Buchenau
Juncaceae

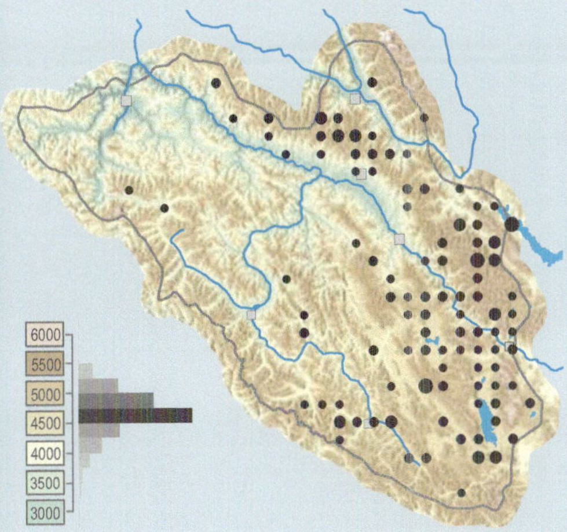

Description: Perennial herbs, 5–20 cm tall, densely cespitose, leaves 2–3, basal or subbasal, linear, 1–5 cm long, inflorescences terminal, head solitary, 5–10 mm across, 2–8-flowered, whitish.

Distribution: Kyrgyzstan, Tajikistan, Pakistan, India, Nepal, China (Gansu, Qinghai, Shaanxi, Sichuan, Xizang, Yunnan), Bhutan, Myanmar.

Habitat: Mesic to wet habitats close to watersheds, salty wetlands, alpine meadows.

Culm round, hollow, with a diameter of 0.6–0.8 mm. Center partially filled with large, thin-walled, polygonal, unlignified cells. Epidermis cells lignified, smooth, thin-walled on the inside, thicker-walled peripherally. Continuous, 6–8-seriate, thick-walled peripheral sclerenchymatic belt centripetal of the chlorenchyma. Large and small vascular bundles principally arranged in two slightly offset rows. Vascular bundles located in the sclerenchymatic belt. Chlorenchyma consists of a continuous, 2–4-seriate peripheral band. Lateral vessels large, diameter 5–8 µm. Phloem consists of sieve tubes and companion cells. Vessels of the protophloem accompanied by a centripetal intercellular. Vascular bundles surrounded by an eccentric sheath of sclerenchyma, with a 3–4-seriate zone of thin- to thick-walled cells centripetally and laterally, and a 6–8-seriate group of small, thick-walled cells centrifugally.

Analyzed: 1 individual

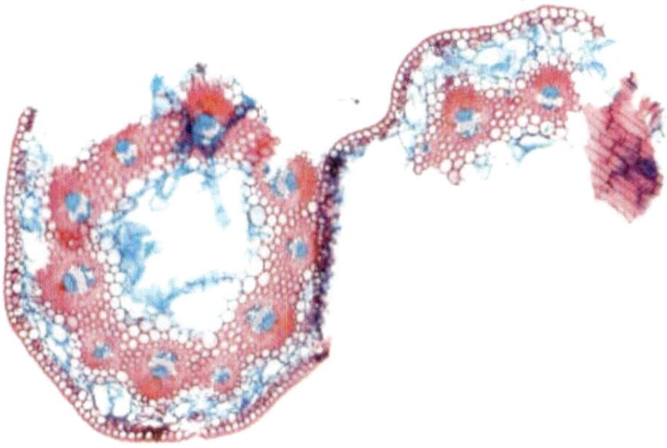

100 µm

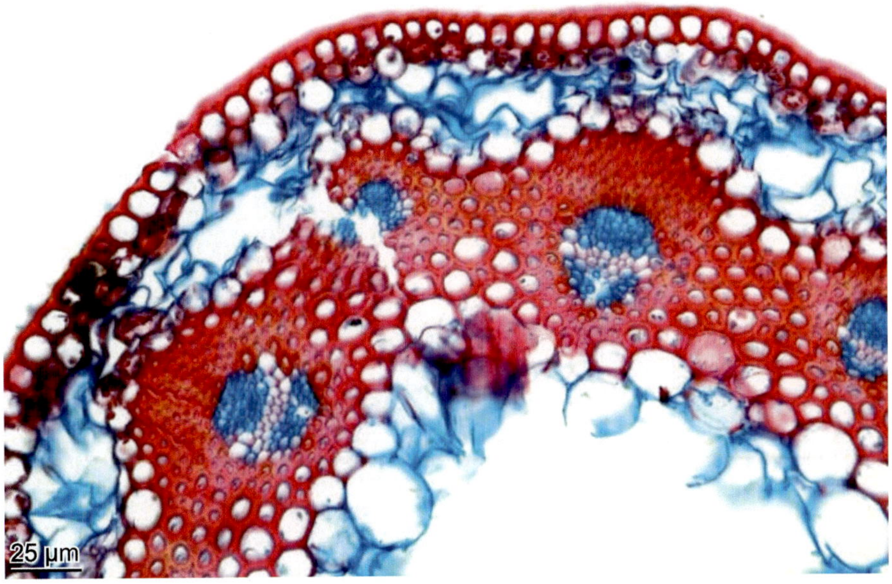

Juncus turkestanicus V.Krecz. ex Gontsch.

Juncaceae

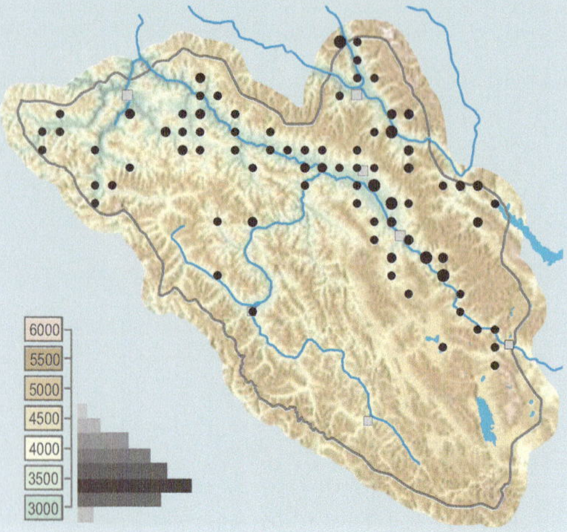

Syn.: *J. bufonius* subsp. *turkestanicus* (V.Krecz. ex Gontsch.) Novikov

Description: Annuals, 3–15 cm tall, cespitose, leaves 2–4, sub-basal, 1–5 cm long, flat, flowers single near plant base and distally in 2–5-flowered clusters.

Distribution: E Turkey, SE Russia, Iraq, Iran, Kazakhstan, Turkmenistan, Uzbekistan, Kyrgyzstan, Tajikistan, Afghanistan, Pakistan, N India, China (Gansu, Jiangsu, Nei Mongol, Qinghai, Yunnan, Xinjiang), Mongolia.

Habitat: Wet habitats, irrigation canals.

Culm round, filled, with a diameter of 0.6–0.8 mm. Center filled with large, thin-walled, polygonal, unlignified cells. Epidermis cells lignified, smooth, thin-walled on the inside, thicker-walled peripherally. Continuous, 2–3-seriate, thin- to thick-walled peripheral sclerenchymatic belt centripetal of the chlorenchyma. Large and small vascular bundles arranged in two slightly offset rows. Large vascular bundles laterally connected by a small zone of thin- to thick-walled, lignified cells. Rudimentary bundles small, in the chlorenchyma zone. Chlorenchyma consists of a large, continuous peripheral band. Intercellulars between the unlignified cells small. Lateral vessels large, diameter 12–15 µm. Phloem consists of sieve tubes and companion cells. Vessels of the protophloem accompanied by a centripetal intercellular. Vascular bundles surrounded by an eccentric sheath of sclerenchyma, with a 2–3-seriate zone of thin- to thick-walled cells centripetally and laterally, and a 3–4-seriate group of small, thick-walled cells centrifugally. Crystal sand in the chlorenchyma.

Analyzed: 1 individual

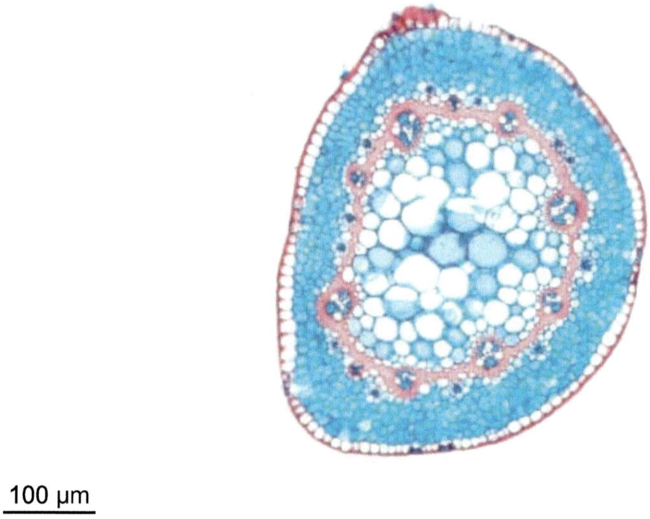

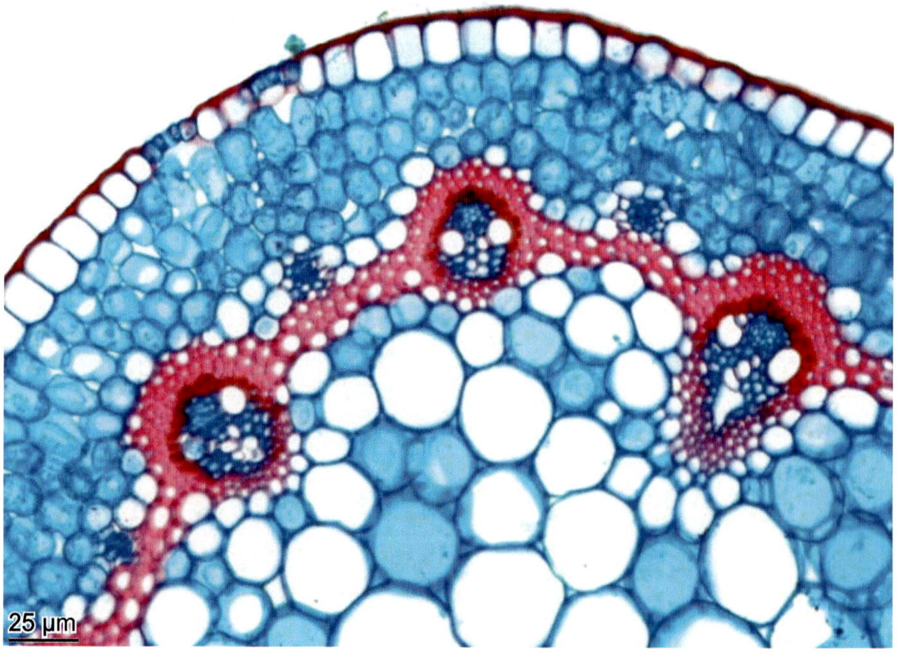

Triglochin palustris L.

Juncaginaceae

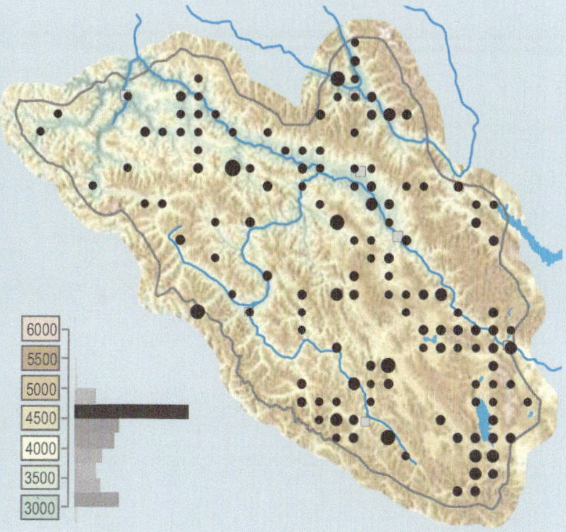

Description: Perennial herbs, 10–40 cm tall, glabrous, leaves linear, shorter than scapes, flowers greenish, ca. 3 mm across, in many-flowered racemes, fruits 8–10 mm long.

Distribution: Cosmopolitan in temperate regions (North and South America, N Africa, Europe, Asia).

Habitat: Saline marshes, springs, shallow wetlands.

Flower stalk round, filled, with a diameter of 1–2 mm. Center filled with large, thin-walled, polygonal, unlignified cells. Epidermis cells unlignified, thin-walled on the inside, thicker-walled and minutely dentate peripherally. Continuous, 6–8-seriate thin- to thick-walled peripheral sclerenchymatic belt centripetal of the chlorenchyma. Vascular bundles arranged in one circular row. Chlorenchyma consists of a continuous, 3–5-seriate peripheral band. Lateral vessels large, diameter 30 μm. Phloem consists of sieve tubes and companion cells. Intercellulars in the protophloem absent. Vascular bundles surrounded by an eccentric sheath of sclerenchyma, with a 1–4-seriate zone of thin- to thick-walled cells centripetally and laterally, and a large, 2–3-seriate group of small, thick-walled cells centrifugally. Crystals are absent.

Analyzed: 1 individual

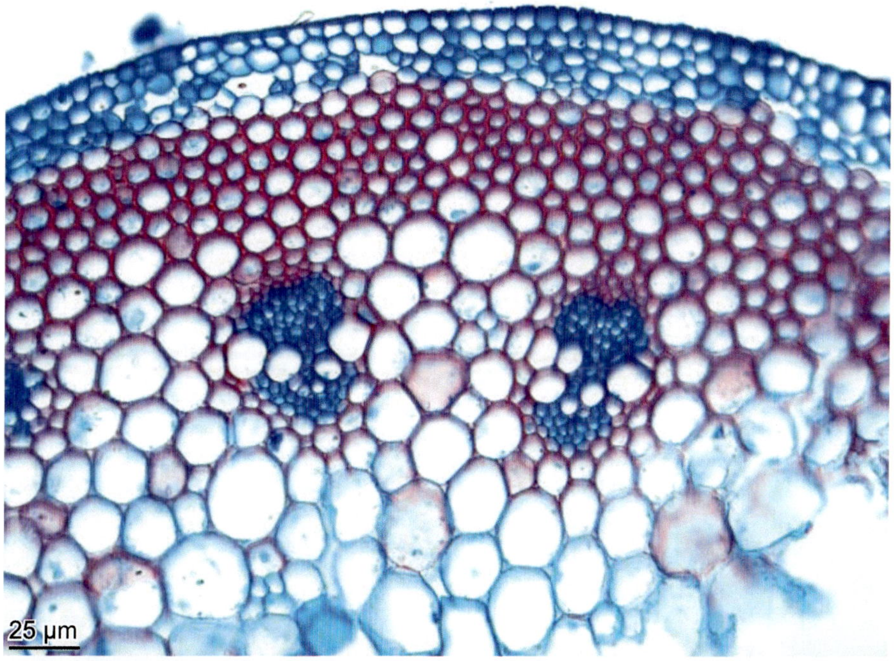

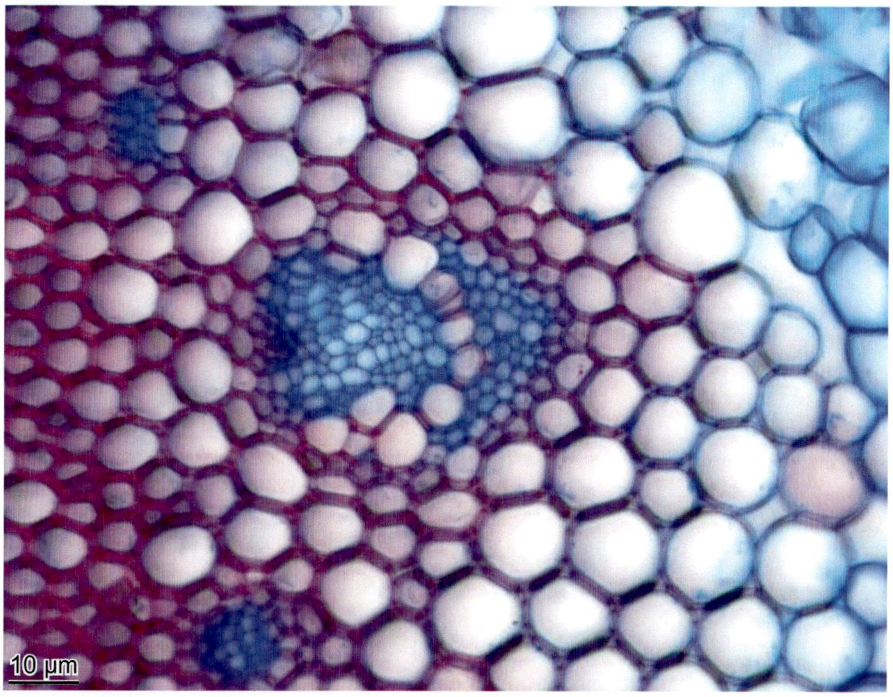

Lloydia serotina (L.) Rchb.

Liliaceae

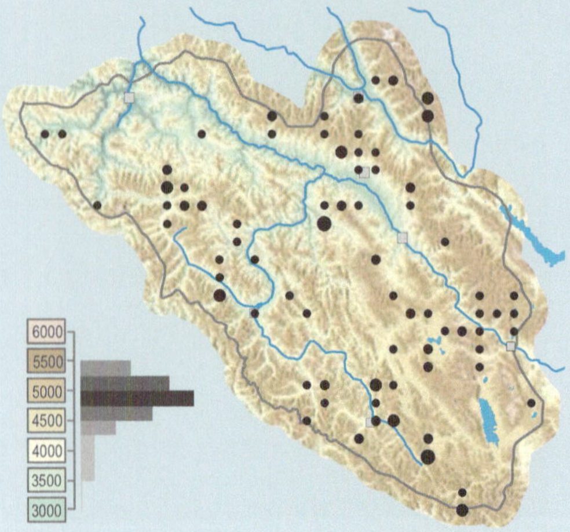

Syn.: *Bulbocodium serotinum* L., *Gagea serotina* (L.) Ker Gawl.

Description: Geophytes with a bulb, stems 5–15 cm tall, basal leaves usually 2, filiform, stem leaves 2–4, lanceolate, flowers white to yellowish with purplish veins, green at the base, ca. 1 cm across, solitary.

Distribution: North America, Europe, Russia, Kazakhstan, Kyrgyzstan, Tajikistan, Pakistan, India, Nepal, China, Bhutan, Mongolia, Korea, Japan.

Habitat: Alpine turfs, meadow-steppes, open screes, among rocks.

Flower stalk circular, with a diameter of 2.5 mm. Epidermis cells unlignified, thin-walled on the inside, thicker-walled peripherally. Continuous, 5–7-seriate, thin- to thick-walled peripheral sclerenchymatic belt centripetal of the chlorenchyma. Vascular bundles arranged in one circular row. Chlorenchyma consists of a continuous 3–5-seriate peripheral unlignified zone. Vessels poorly lignified, angular, with a diameter of 5 µm, occuring in groups. Phloem consists of sieve tubes and companion cells. Vascular bundle sheaths do not exist. Crystals absent.

Analyzed: Flower stalk of 1 individual.

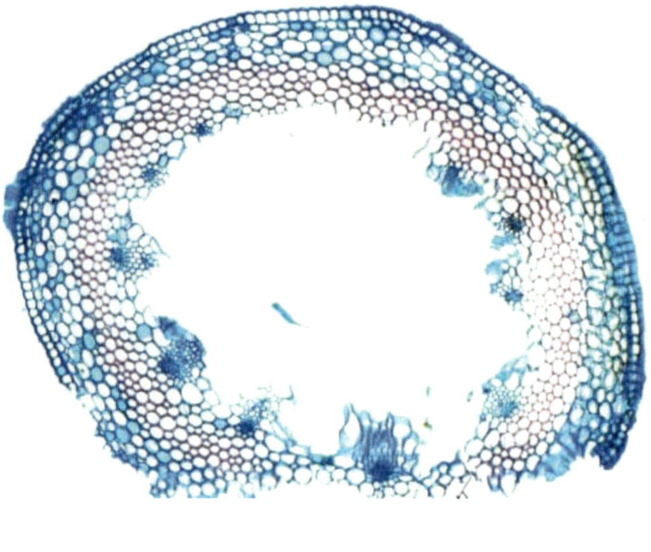

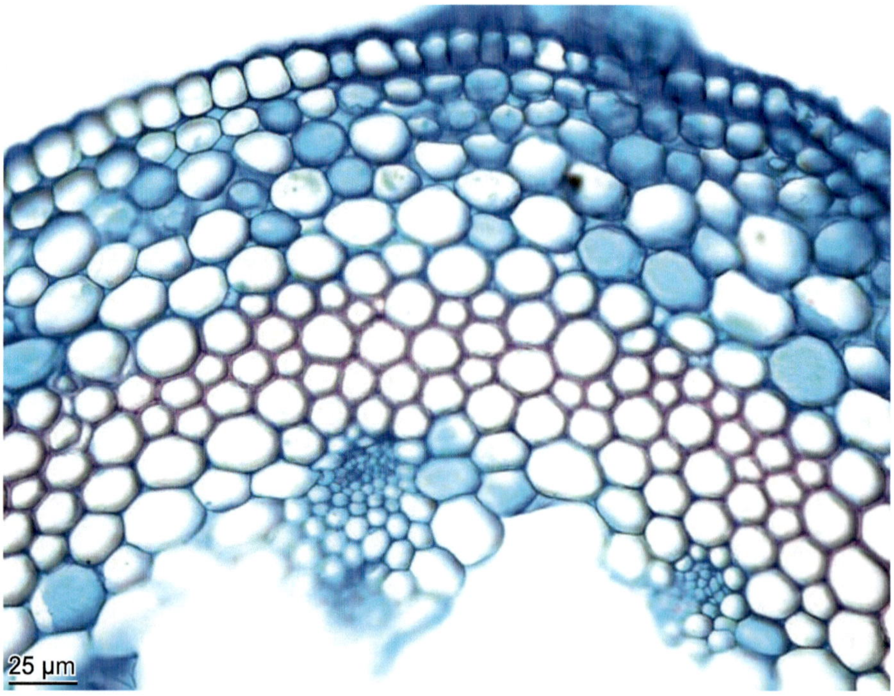

Agrostis filipes Hook.f.

Poaceae

Description: Perennial herbs, 15–35 cm tall, cespitose, leaves flat, 3–8 cm long, 1–2.5 mm wide, panicle very narrow, interrupted, 5–12 cm long, brownish or tinged with purple.

Distribution: N Pakistan, India (Jammu and Kashmir, Assam).

Habitat: Alpine meadows.

Culm uniform, round, hollow, with a diameter of 0.7–0.9 mm. Epidermis thick-walled. Continuous, 1–3-seriate peripheral sclerenchymatic belt of very thick-walled fibers. Girders absent. Small, tangentially enlarged groups of chlorenchyma within the sclerenchymatic belt. Vascular bundles arranged in two offset circular rows. Large vascular bundles within thin- to thick-walled parenchymatic tissue, small bundles within the sclerenchymatic belt. Intercellulars absent. Vascular bundles with two large lateral vessels, diameter 10–15 µm. Intercellulars within vascular bundles absent. Phloem small, oval, consisting of uniform cells. Vascular bundles surrounded by a uniseriate sheath of polygonal, thin- to thick-walled cells.

Analyzed: 1 individual

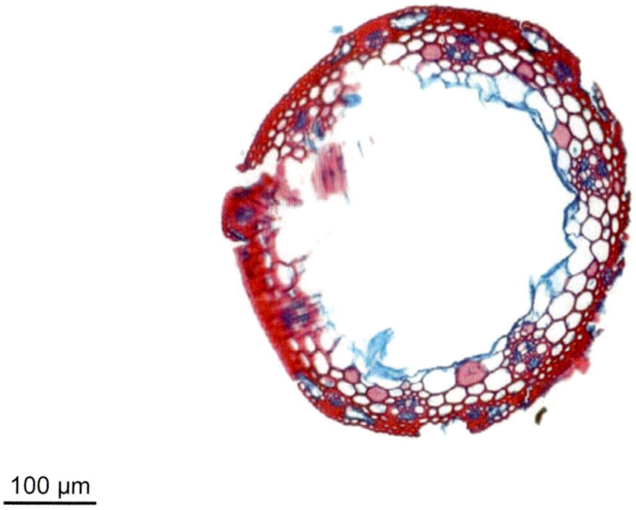

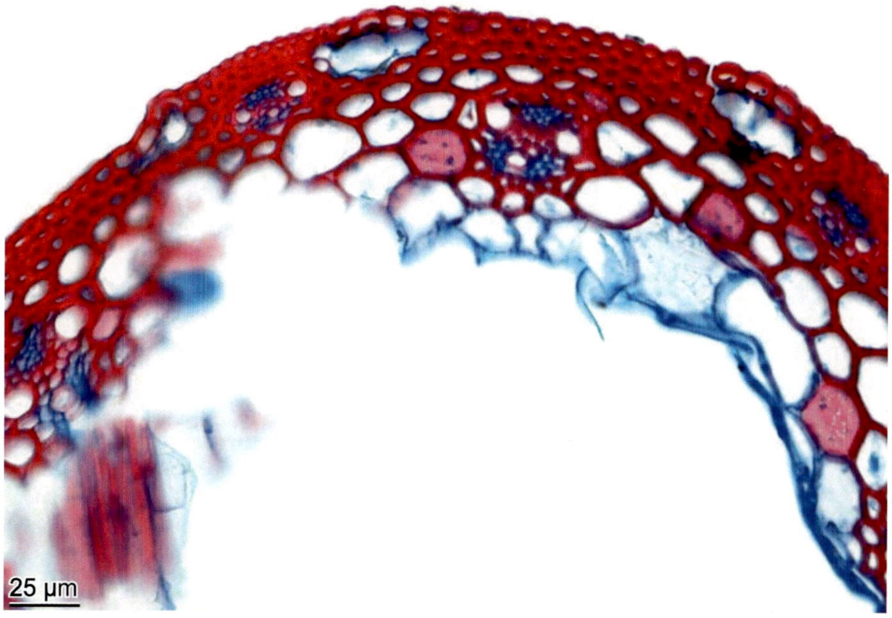

Agrostis gigantea Roth

Poaceae

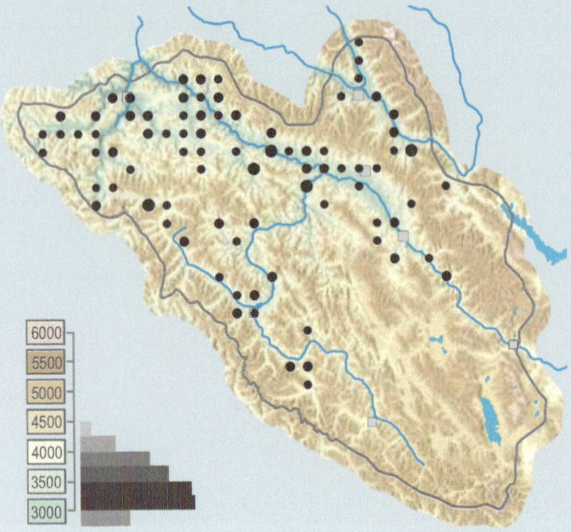

Syn.: *A. stolonifera* subsp. *gigantea* (Roth.) Maire & Weiller

Description: Perennial herbs, 40–120 cm tall, rhizomatous, culms erect or geniculately ascending, rooting and branching from the lower nodes, leaves flat, 5–20 cm long, 2–8 mm wide, panicle oblong to ovate in shape, 8–25 cm long, loose and spreading at flowering time, branches clustered.

Distribution: Europe and temperate Asia.

Habitat: Montane grasslands, mesic to moist pastures, shaded field margins, irrigation canals, ditches.

Culm uniform, round, hollow, with a diameter of 1.3–1.5 mm. Epidermis thin- to thick-walled. Continuous, 1–3-seriate peripheral sclerenchymatic belt of thin- to thick-walled fibers. Girders represented by square groups of thin- to thick-walled fibers. Bands or groups of chlorenchyma located between girders. Vascular bundles arranged in two offset circular rows. Large bundles within thin-walled parenchymatic tissue, small bundles at the centripetal side of girders. Vascular bundles with large vessels, diameter 25–35 µm. Intercellulars within vascular bundles frequent. Phloem small, oval, consisting of sieve tubes and companion cellss. Vascular bundles surrounded by a uniseriate sheath of polygonal, thin- to thick-walled cells.

Analyzed: 2 individuals

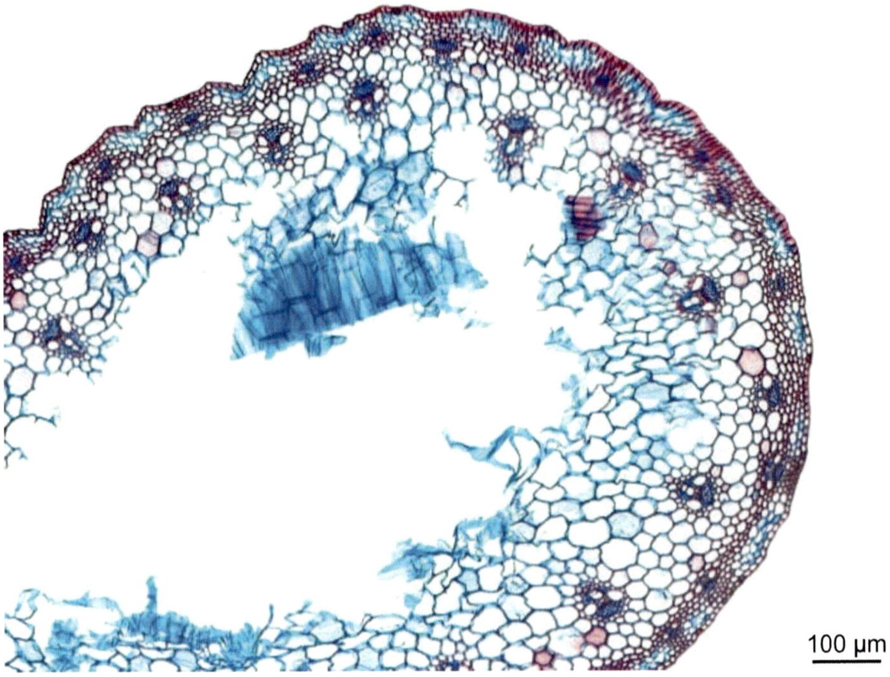

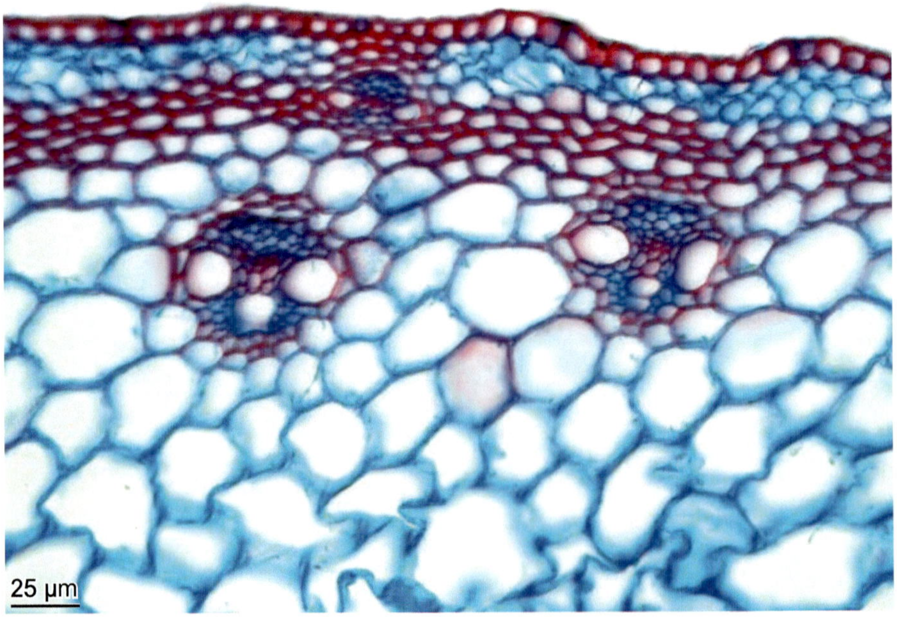

Agrostis stolonifera L.

Poaceae

Syn.: *A. diluta* Kurczenko

Description: Perennial herbs, 20–60 cm tall, stoloniferous, culms erect or ascending, leaves flat, 4–10 cm long, 2–5 mm wide, panicle linear to lanceolate or oblong, 5–20 cm long, open at anthesis, contracted at maturity, branches ascending.

Distribution: N Africa, Europe, temperate Asia; widely introduced.

Habitat: Moist meadows, lakesides.

Culm uniform, round, hollow, with a diameter of 0.9–1.0 mm Epidermis thick-walled. Continuous, 1–3-seriate peripheral sclerenchymatic belt of thick-walled fibers. Girders absent. Small, tangentially enlarged groups of chlorenchyma located within the sclerenchymatic belt. Vascular bundles arranged in two offset circular rows. Large bundles within thin-walled parenchymatic tissue, small bundles within the sclerenchymatic belt. Vascular bundles with large vessels, diameter 15–20 µm. Vessels of the protophloem hardly lignified. Intercellulars within vascular bundles absent. Phloem small, oval, consisting of sieve tubes and companion cellss. Vascular bundles surrounded by a uniseriate sheath of polygonal, thin- to thick-walled cells.

Analyzed: 2 individuals

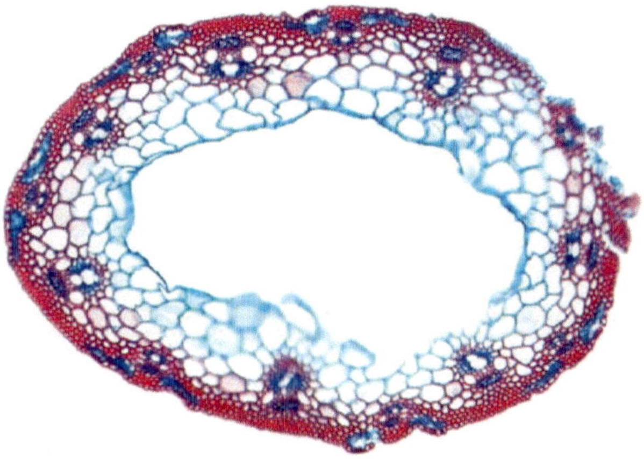

100 µm

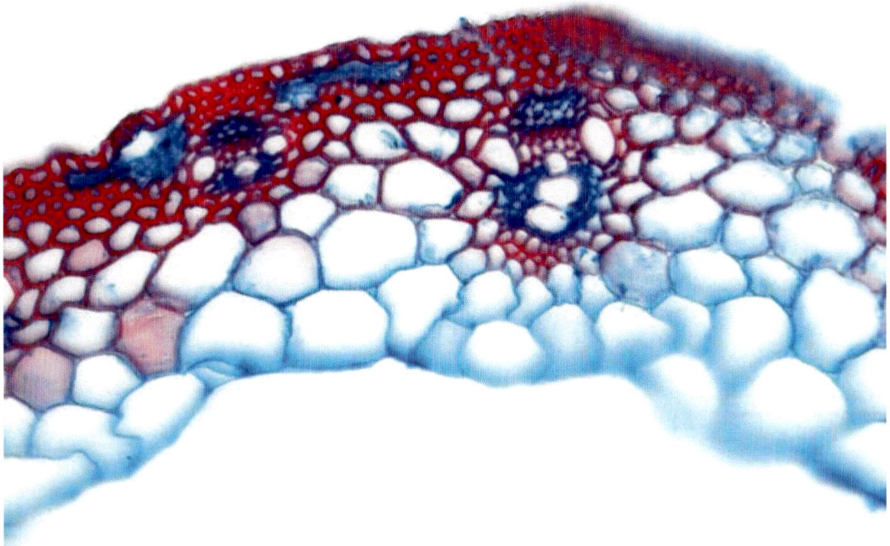

25 µm

Agrostis vinealis Schreb.

Poaceae

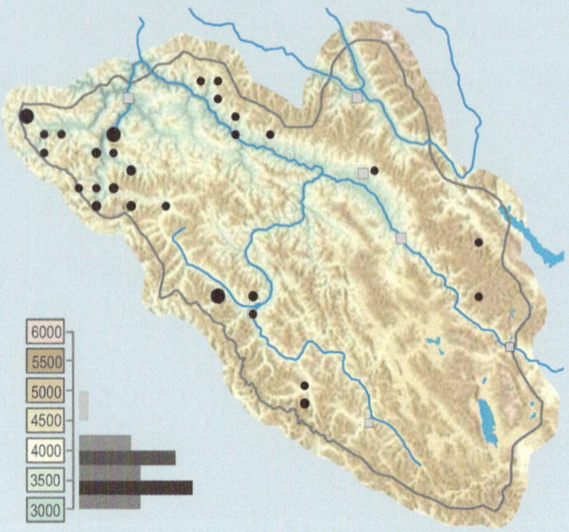

Syn.: *A. canina* subsp. *montana* (Hartm.) Hartm.

Description: Perennial herbs, 10–50 cm tall, cespitose, with rhizomes, leaves flat or convolute, 2–15 cm long, 1–3 mm wide, panicle open or contracted, lanceolate or ovate, 2–20 cm long.

Distribution: Temperate Eurasia.

Habitat: Dry grasslands.

Culm uniform, round, hollow, with a diameter of 0.3–0.4 mm. Epidermis thick-walled. Continuous, 2–4-seriate peripheral sclerenchymatic belt of thick-walled fibers. Girders absent. Round to oval groups of chlorenchyma located within the sclerenchymatic belt. Vascular bundles arranged in a circular row. Vascular bundles within the thin-walled parenchymatic tissue and the sclerenchymatic belt. Intercellulars absent. Vascular bundles with two large lateral vessels, diameter 10–15 μm. Vessels of the protophloem hardly lignified. Intercellulars within vascular bundles absent. Phloem small, oval, consisting of sieve tubes and companion cellss. Vascular bundles surrounded by a uniseriate sheath of polygonal, thin- to thick-walled cells.

Analyzed: 2 individuals

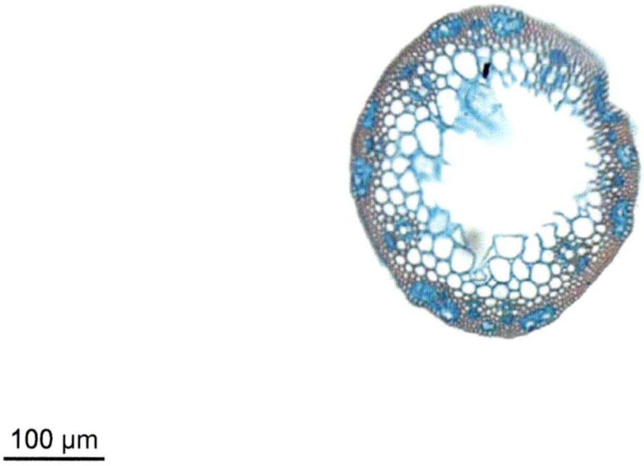

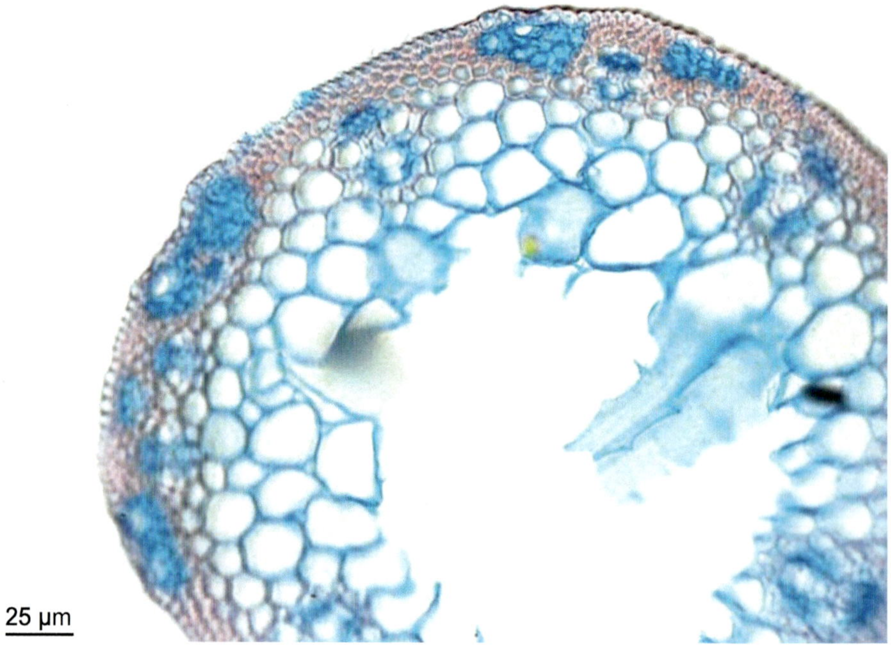

Alopecurus aequalis Sobol.

Poaceae

Syn.: *A. amurensis* Kom., *A. caespitosus* Trin., *A. fulvus* Sm.

Description: Annuals or short-lived perennials, 10–35 cm tall, culms decumbent, leaves flat, 2–10 cm long, 2–5 mm wide, panicle spike-like, linear, 1–5 cm long, 0.3–0.6 cm wide.

Distribution: North America, N Africa, Europe, temperate Asia; naturalized in South America and Australia.

Habitat: Banks of irrigation canals, moist open grasslands, flood-meadows.

Culm uniform, round, hollow, with a diameter of 1.4–1.6 mm. Epidermis thin- to thick-walled. Uniseriate peripheral belt of lignified thin- to thick-walled fibers. Small, uniseriate, tangentially enlarged groups of chlorenchyma located between epidermis and sclerenchymatic belt. Vascular bundles arranged in two offset circular rows. Small vascular bundles within thin-walled parenchymatic tissue. Girders small, conical. Vascular bundles with two large lateral vessels, diameter 20 μm. Small triangular intercellulars in the thin-walled parenchymatic tissue. Vessels of the protophloem unlignified and rarely accompanied by centripetal intercellulars. Phloem small, round, consisting of sieve tubes and companion cellss. Vascular bundles not surrounded by a sheath of lignified cell.

Analyzed: 1 individual

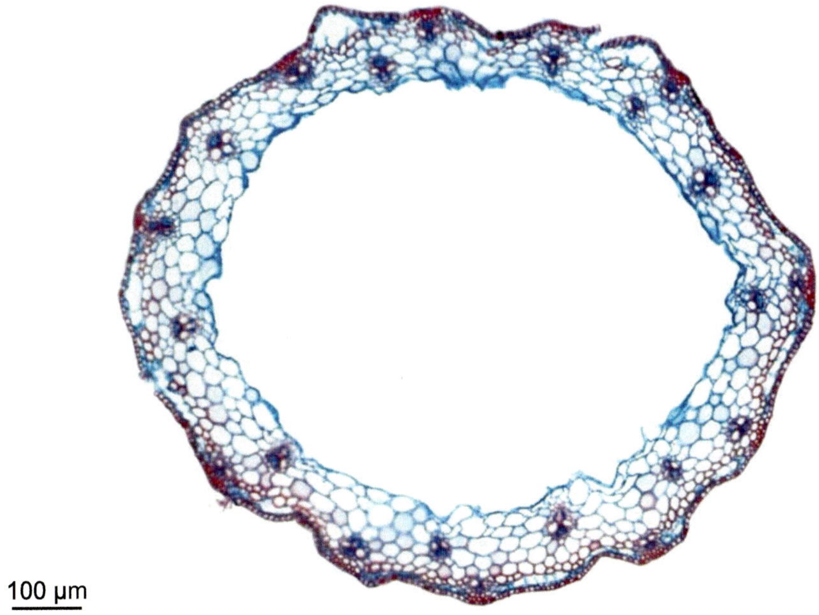

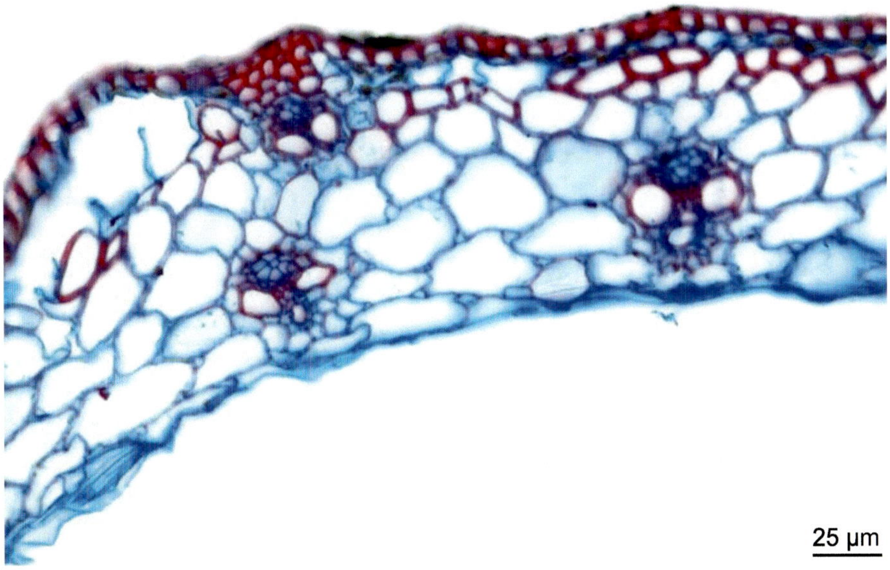

Alopecurus arundinaceus Poir.

Poaceae

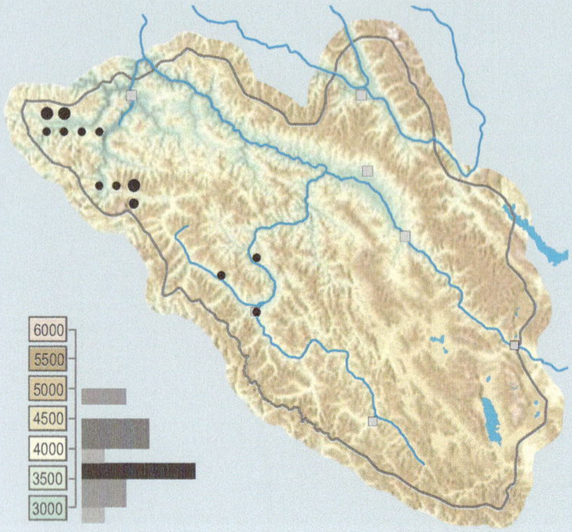

Description: Perennial herbs, 20–100 cm tall, rhizomatous, leaves flat, 6–40 cm long, 3–10 mm wide, panicle spike-like, linear, 1.5–7 cm long, 0.6–1.4 cm wide.

Distribution: Europe, SW Asia, Russia, Kazakhstan, Turkmenistan, Uzbekistan, Kyrgyzstan, Tajikistan, N Pakistan, N India, China, Mongolia; introduced in North America.

Habitat: Moist meadows.

Culm uniform, round, hollow, with a diameter of 1–1.1 mm. Epidermis thin- to thick-walled. Continuous, 1–3-seriate peripheral sclerenchymatic belt of thick-walled fibers centripetal of the chlorenchyma. Girders represented by square groups of thin- to thick-walled fibers. Square to rectangular chlorenchyma groups located between girders. Vascular bundles arranged in two offset circular rows. Large bundles within the thin-walled parenchymatic tissue, small bundles at the centripetal side of girders. Vascular bundles with large vessels, diameter 20 μm. Intercellulars within vascular bundles frequent. Phloem small, oval, consisting of sieve tubes and companion cellss. Vascular bundles surrounded by a uniseriate sheath of polygonal, thin- to thick-walled cells.

Analyzed: 2 individuals

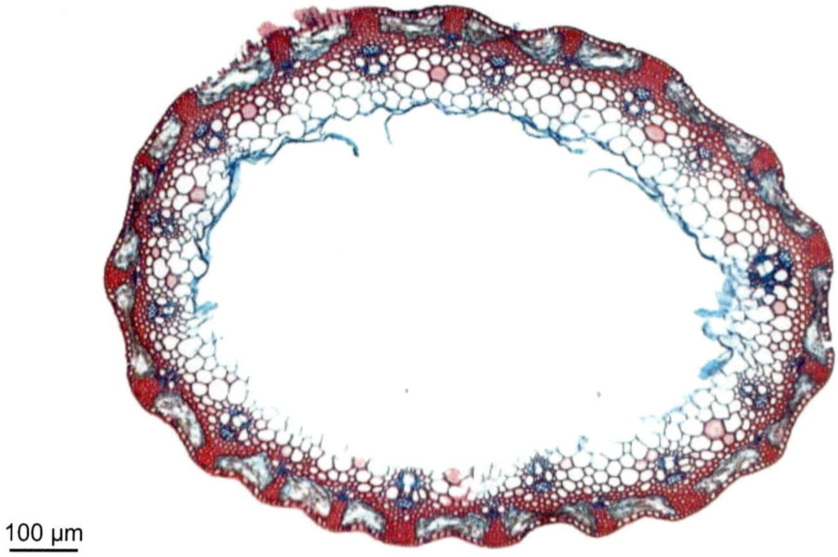

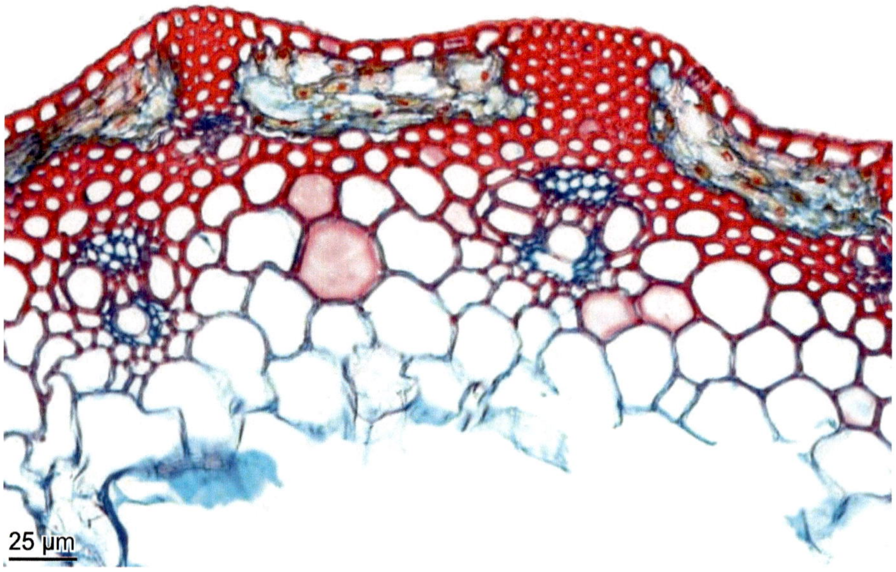

Aristida adscensionis L.

Poaceae

Description: Annuals, or short-lived perennials, 10–100 cm tall, cespitose, culms erect, ascending or decumbent, leaves flat or conduplicate, 5–20 cm long, 1–3 mm wide, panicle open or contracted, elliptic, 10–30 cm long.

Distribution: Worldwide in tropical and warm-temperate regions.

Habitat: Dry mountain slopes.

Culm uniform, round, hollow, with a diameter of 0.5–0.6 mm. Epidermis thin- to thick-walled. Continuous, 1–3-seriate peripheral sclerenchymatic belt of thick-walled fibers centripetal of the chlorenchyma. Girders represented by square to rectangular groups of thick-walled fibers. Square to rectangular chlorenchyma groups located between girders. Vascular bundles arranged in two offset circular rows. Large bundles within the thin-walled parenchymatic tissue, small bundles at the centripetal side of girders. Vascular bundles with large vessels, diameter 15 µm. Intercellulars within vascular bundles frequent. Phloem small, oval, consisting of sieve tubes and companion cellss. Vascular bundles surrounded by a 1–3-seriate sheath of polygonal, thin- to thick-walled cells.

Analyzed: 1 individual

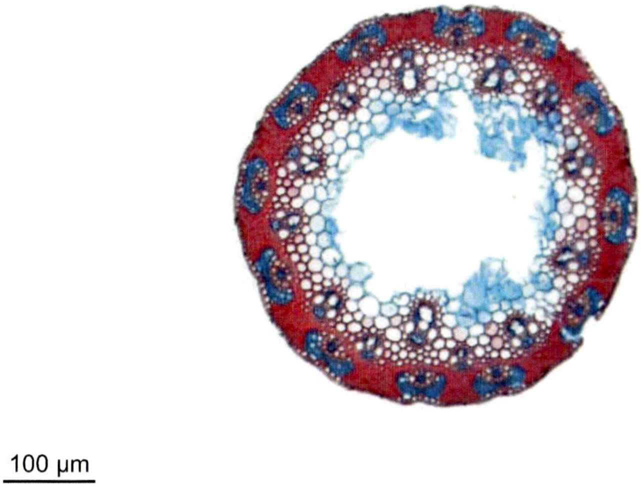

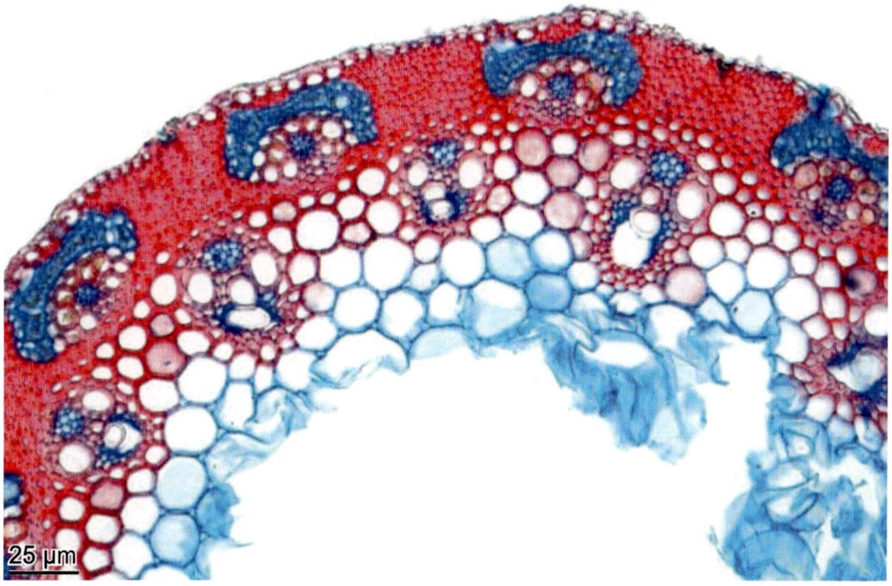

Bothriochloa ischaemum (L.) Keng

Poaceae

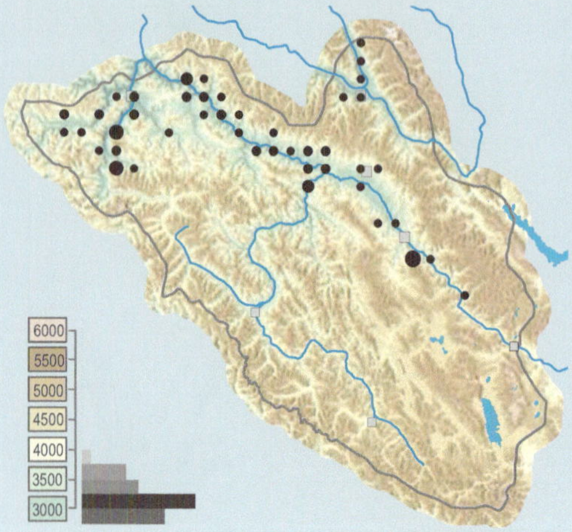

Description: Perennial herbs, 25–70 cm tall, cespitose, leaves linear, flat, inflorescence composed of 5–15 subdigitate racemes, 3–7 cm long.

Distribution: N Africa, Europe, Russia, Kazakhstan, Uzbekistan, Kyrgystan, Turkmenistan, Tajikistan, Afghanistan, Pakistan, N India, Nepal, China, Bhutan, Mongolia, Korea, Taiwan; introduced in the United States.

Habitat: Disturbed sandy places, roadsides.

Culm uniform, round, filled, with a diameter of 0.9–1.2 mm. Epidermis thick-walled. Continuous, 3- to >4-seriate peripheral sclerenchymatic belt of thin- to thick-walled fibers. Girders absent. Chlorenchyma groups absent. Intercellulars in the parenchymatic zone absent. Vascular bundles arranged in two offset circular rows. Large bundles within the thin-walled parenchymatic tissue, small bundles centripetal of the sclerenchymatic belt. Vascular bundles with large vessels, diameter 20–25 µm. Intercellulars centripetal of the protoxylem absent. Phloem small, oval, consisting of sieve tubes and companion cellss. Vascular bundles surrounded by a 1–2-seriate sheath of polygonal, thick-walled cells.

Analyzed: 2 individuals

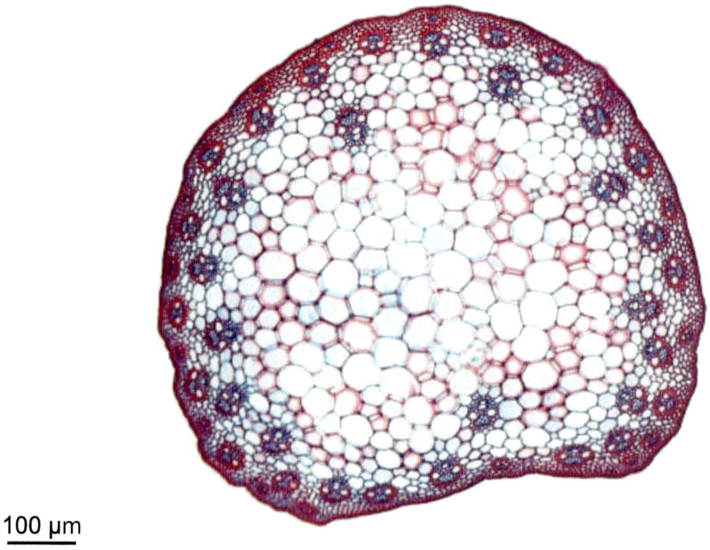

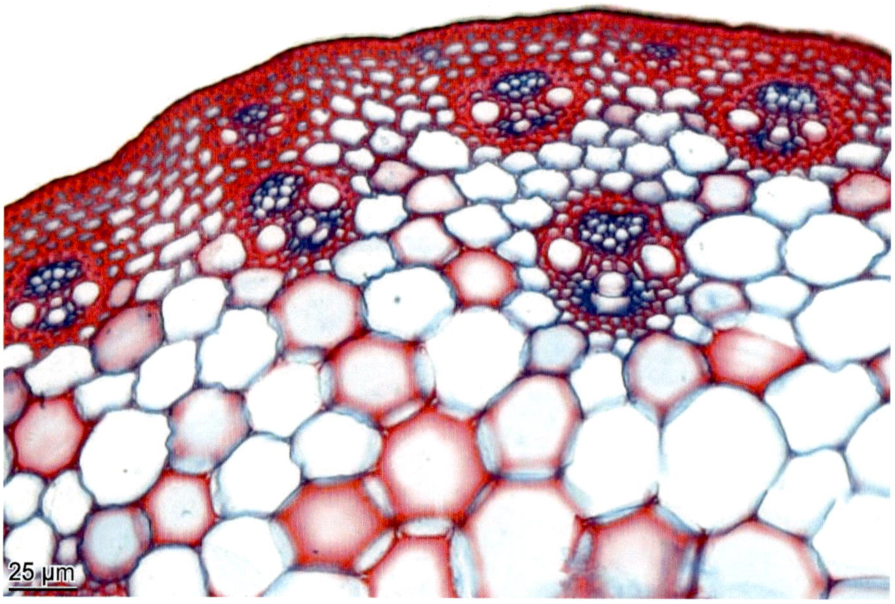

Bromus alopecuros Poir.

Poaceae

Syn.: *B. sericeus* Drobow, *Anisantha sericea* (Drobow) Nevski

Description: Annuals, 10–30 cm tall, cespitose, culms erect or ascending, leaves hairy, 5–20 cm long, 2–4 mm wide, panicle contracted, oblong, nodding, 5–10 cm long, branches drooping, awns 15–45 mm long.

Distribution: Turkey, Izrael, Iran, Iraq, Turkmenistan, Kyrgyzstan, Tajikistan, Pakistan, India (Jammu and Kashmir), China (Xizang, Yunnan).

Habitat: Steppes, dry sandy places.

Culm uniform, round, hollow, with a diameter of 1.0–1.2 mm. Epidermis thick-walled. Continuous, >4-seriate peripheral sclerenchymatic belt of thick-walled fibers. Girders absent. Round to irregular formed chlorenchyma groups located within the sclerenchymatic belt. Intercellulars triangular and very small. Vascular bundles arranged in two offset circular rows. Large bundles within the thin-walled parenchymatic tissue, small bundles within the sclerenchymatic belt. Vascular bundles with large vessels, diameter 35–40 µm. With very small triangular intercellulars within the parenchyma. Phloem small, oval, consisting of uniform cells. With small intercellulars within vascular centripetal of the protoxylem. Vascular bundles surrounded by an eccentric, 1–3-seriate sheath of polygonal thin- to thick-walled cells.

Analyzed: 2 individuals

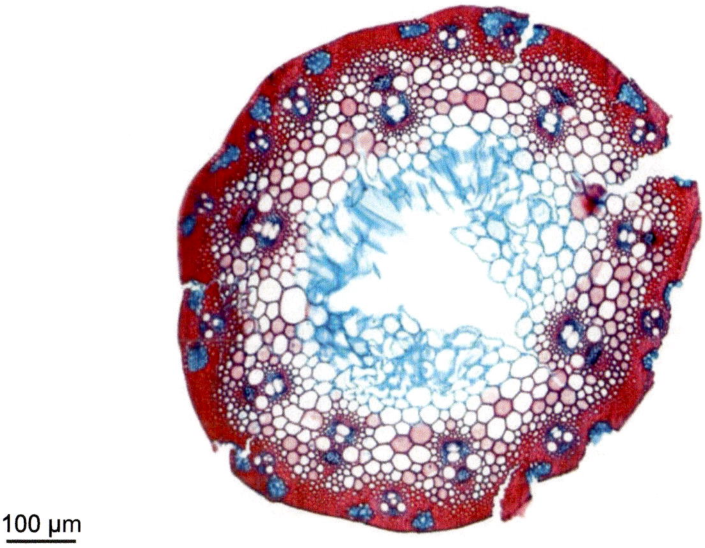

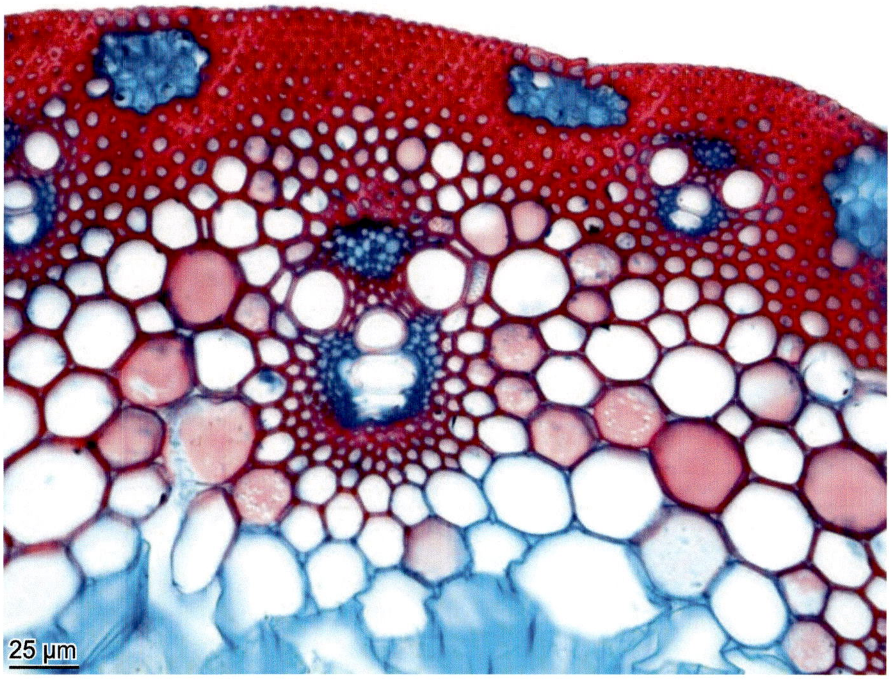

Bromus confinis Nees ex Steud.
Poaceae

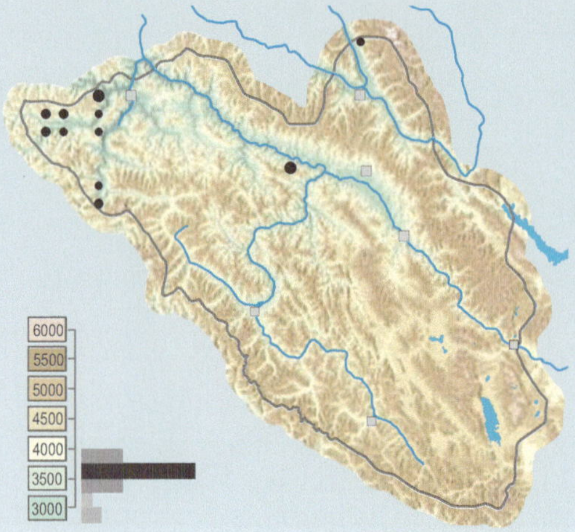

Syn.: *B. inermis* var. *confinis* (Nees) Stapf, *B. munroi* Boiss., *Bromopsis confinis* (Steud.) Holub

Description: Perennial herbs, 50–100 cm tall, rhizomatous, leaves hairy on both sides, 10–25 cm long, 3–6 mm wide, panicle open, linear, 10–18 cm long, awns 2–6 mm long.

Distribution: SW Asia, Pakistan, NW India, China (Gansu).

Habitat: Open grassy places.

Culm uniform, round, hollow, with a diameter of 1.4–1.6 mm. Epidermis thin- to thick-walled. Continuous, 1–3-seriate peripheral sclerenchymatic belt of thick-walled fibers centripetal of the chlorenchyma. Girders represented by roundish groups of thick-walled fibers. Tangentially enlarged, small rectangular chlorenchyma groups located between girders. Intercellulars triangular and very small. Vascular bundles arranged in two offset circular rows. Large bundles within the thin-walled parenchymatic tissue, small bundles at the centripetal side of girders. Vascular bundles with large vessels, diameter 35–45 μm. Intercellulars within vascular bundles absent. Phloem small, oval, consisting of sieve tubes and companion cellss. Vascular bundles surrounded by a uniseriate sheath of polygonal, thin- to thick-walled cells.

Analyzed: 2 individuals

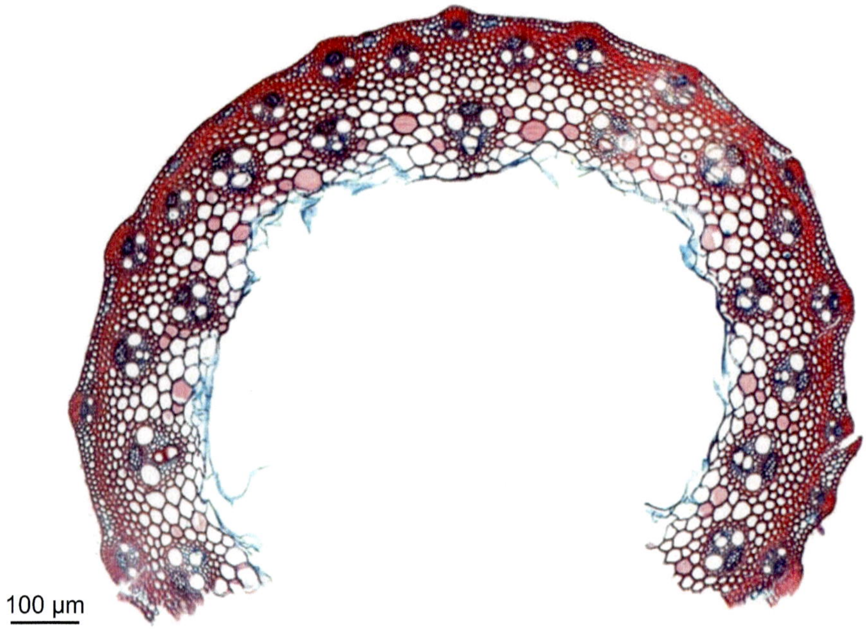

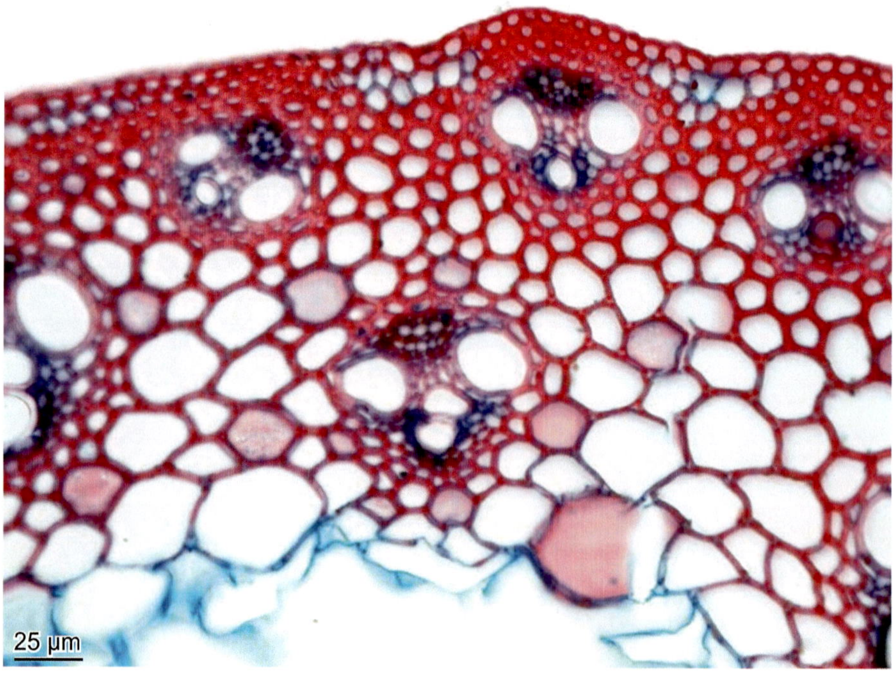

Bromus danthoniae Trin.

Poaceae

Syn.: *Boissiera danthoniae* (Trin.) A.Braun, *Triniusa danthoniae* (Trin.) Steud.

Description: Annuals, 5–40 cm tall, culms erect or ascending, leaves hairy on both sides, 2–10 cm long, 2–5 mm wide, panicle open, ovate, dense, 2–10 cm long, 1–2 cm wide, awns 5–15 mm long, curved.

Distribution: Cyprus, SW Asia, Russia, Kazakhstan, Turkmenistan, Uzbekistan, Kyrgyzstan, Tajikistan, Afghanistan, Pakistan, NW India, China (W Xizang).

Habitat: Dry grassy slopes.

Culm uniform, round, ripped, hollow, with a diameter of 0.5–0.9 mm. Epidermis thin- to thick-walled, with hairs. Continuous, 2–4-seriate peripheral sclerenchymatic belt of thick-walled fibers centripetal of the chlorenchyma. Girders represented by roundish groups of thick-walled fibers. Roundish chlorenchyma groups located between girders. Intercellulars triangular and very small. Vascular bundles arranged in two offset circular rows. Large bundles within the thin-walled parenchymatic tissue, small bundles at the centripetal side of girders. Vascular bundles with large vessels, diameter 28–35 µm. Intercellulars within vascular bundles absent. Phloem small, oval, consisting of uniform cells. Vascular bundles surrounded by a 1–2-seriate sheath of polygonal, thin- to thick-walled cells.

Analyzed: 2 individuals

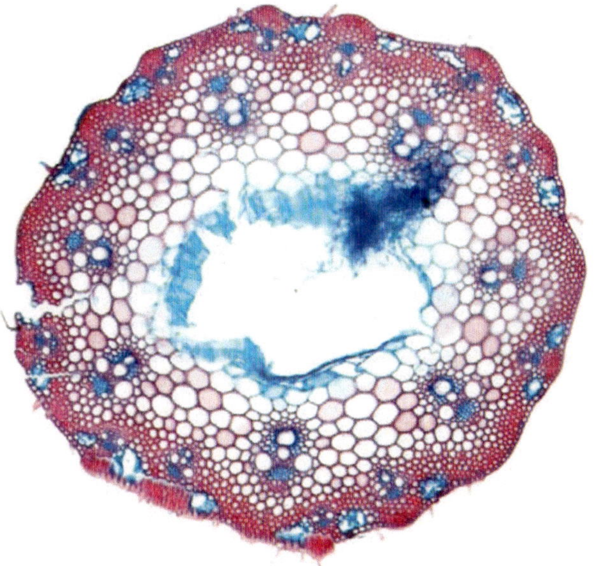

100 µm

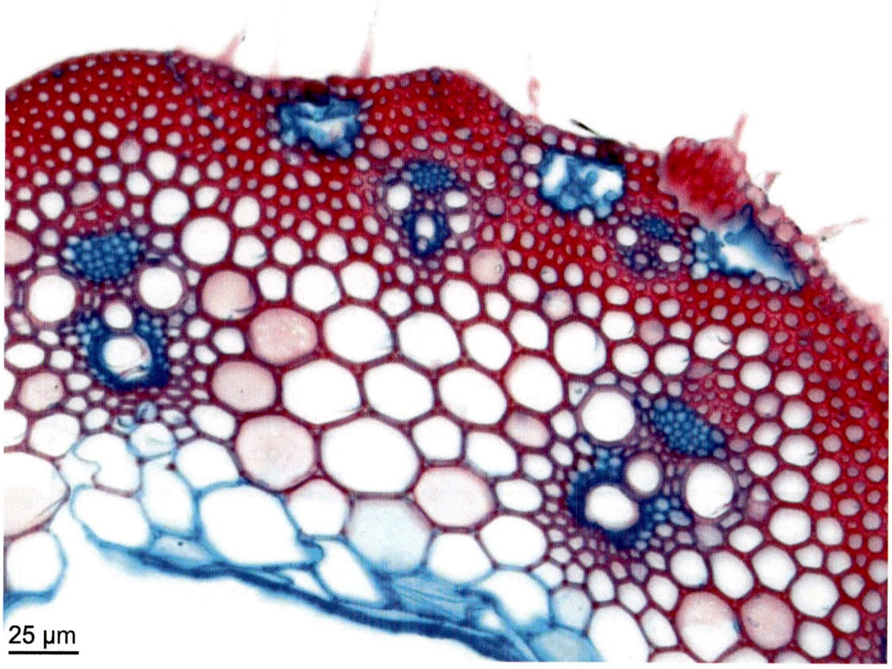

Bromus gracillimus Bunge

Poaceae

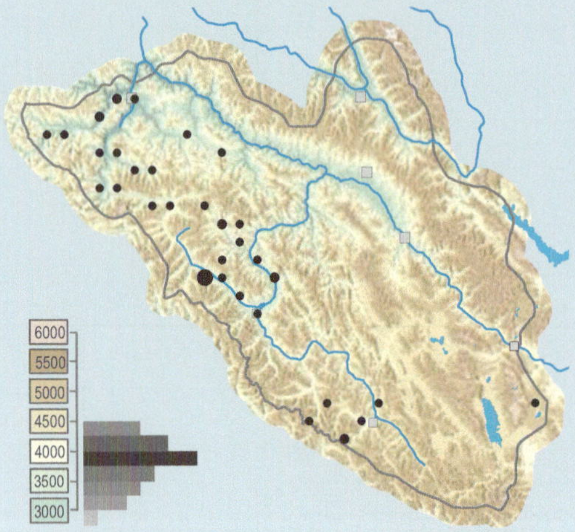

Syn.: *Deschampsia aralensis* Regel, *Nevskiella gracillima* (Bunge) Krecz. & Vved.

Description: Annuals, 10–40 cm tall, leaves hairy on both sides, 3–12 cm long, 1.5–4 mm wide, panicle open, ovate, 3–8 cm long, 1–4 cm wide, awns 12–20 mm long.

Distribution: Iran, Kazakhstan, Turkmenistan, Uzbekistan, Tajikistan, Afghanistan, Pakistan, India (Jammu and Kashmir), China (Xinjiang, W Xizang).

Habitat: Dry mountain slopes.

Culm uniform, round, hollow, with a diameter of 1.2–1.8 mm. Epidermis thick-walled. Continuous, 2–4-seriate peripheral sclerenchymatic belt of thick-walled fibers. Girders absent. With tangential Chlorenchyma groups oval. Intercellulars triangular and very small. Vascular bundles arranged in two offset circular rows. Large bundles within the thin-walled parenchymatic tissue, small bundles at the peripheral sclerenchymatic belt. Vascular bundles with large vessels, diameter 20–25 µm. Intercellulars within vascular bundles absent. Phloem small, oval, consisting of uniform cells. Vascular bundles surrounded by a uniseriate sheath of polygonal, thin- to thick-walled cells.

Analyzed: 1 individual

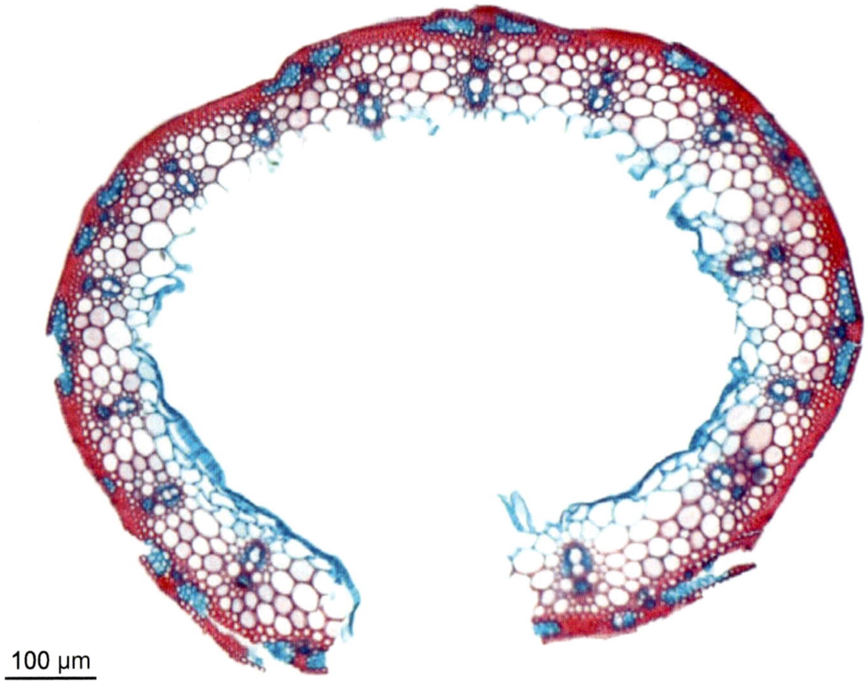

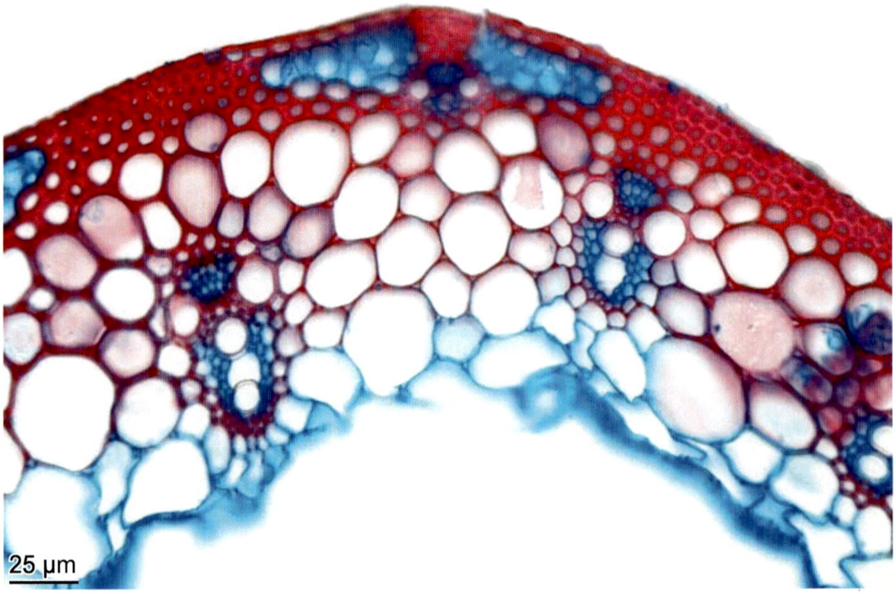

Bromus japonicus Thunb.

Poaceae

Description: Annuals, 30–70 cm tall, culms erect or ascending, leaves hairy on both sides, 5–15 cm long, 3–6 mm wide, panicle open, ovate, effuse, 10–20 cm long, 5–10 cm wide, branches spreading or drooping, awns 5–15 mm long, curved.

Distribution: Egypt, Europe, temperate Asia; introduced in North America and Australia.

Habitat: Dry grasslands, roadsides.

Culm uniform, round, mostly hollow, with a diameter of 0.9–1.0 mm. Epidermis thin- to thick-walled. Continuous, >4-seriate peripheral sclerenchymatic belt of thick-walled fibers centripetal of the chlorenchyma. Girders represented by indistinct roundish groups of thick-walled fibers. Round to rectangular chlorenchyma groups located between girders. Intercellulars triangular and very small. Vascular bundles arranged in two offset circular rows. Large bundles within the thin-walled parenchymatic tissue, small bundles at the centripetal side of girders. Vascular bundles with large vessels, diameter 15–20 µm. Intercellulars within vascular bundles absent. Phloem small, oval, consisting of uniform cells. Vascular bundles surrounded by a uniseriate sheath of polygonal. thin- to thick-walled cells.

Analyzed: 2 individuals

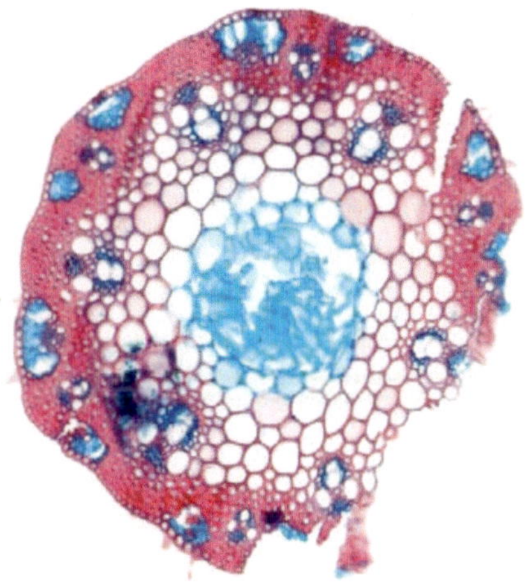

100 µm

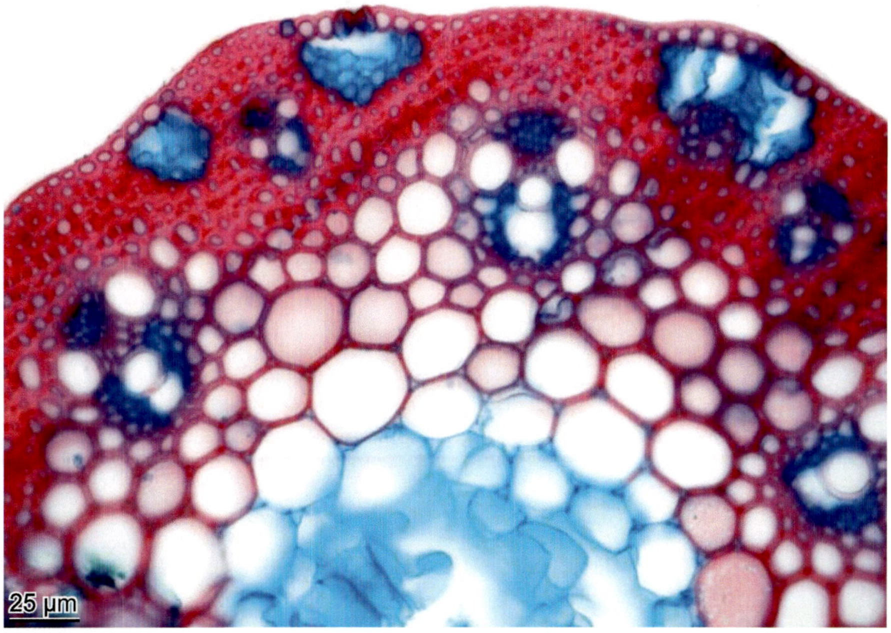

Bromus oxyodon Schrenk
Poaceae

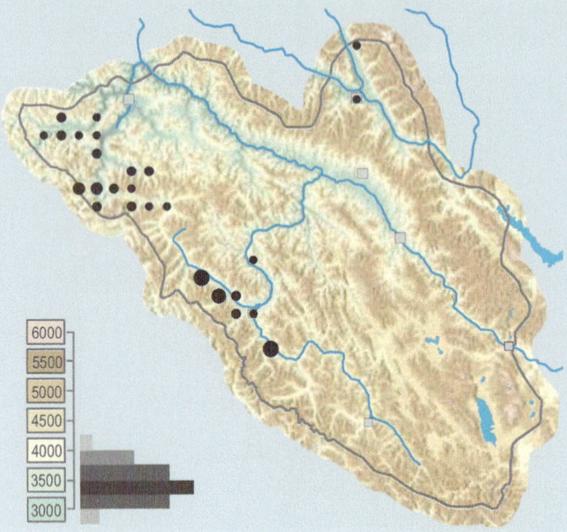

Description: Annuals, 30–60 cm tall, leaves flat, up to 20 cm long, panicle ovate, 5–20 cm long, 2–15 cm wide, awns 1.5–3 cm long.

Distribution: Kazakhstan, Kyrgyzstan, Uzbekistan, Tajikistan, Afghanistan, Pakistan, NW India, China (Xinjiang), W Mongolia.

Habitat: Semideserts, dry grasslands.

Culm uniform, round, normally hollow, with a diameter of 0.7–0.9 mm. Epidermis thick-walled. Continuous, >4-seriate peripheral sclerenchymatic belt of thick-walled fibers. Girders absent. Round to rectangular chlorenchyma groups located within the sclerenchymatic belt. Intercellulars triangular and very small. Vascular bundles arranged in two offset circular rows. Large bundles within the thin- to thick-walled walled parenchymatic tissue, small bundles within the sclerenchymatic belt. Vascular bundles with large vessels, diameter 10–15 μm. Intercellulars within vascular bundles absent. Phloem small, oval, consisting of uniform cells. Vascular bundles surrounded by an eccentric, 1–3 seriate sheath of polygonal, thin- to thick-walled cells.

Analyzed: 2 individuals

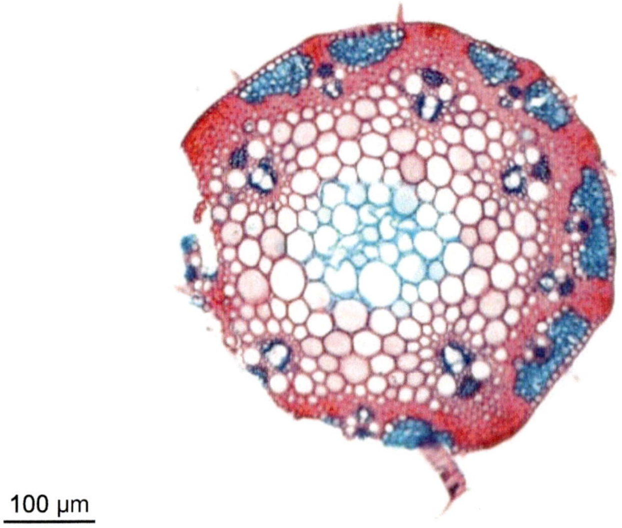

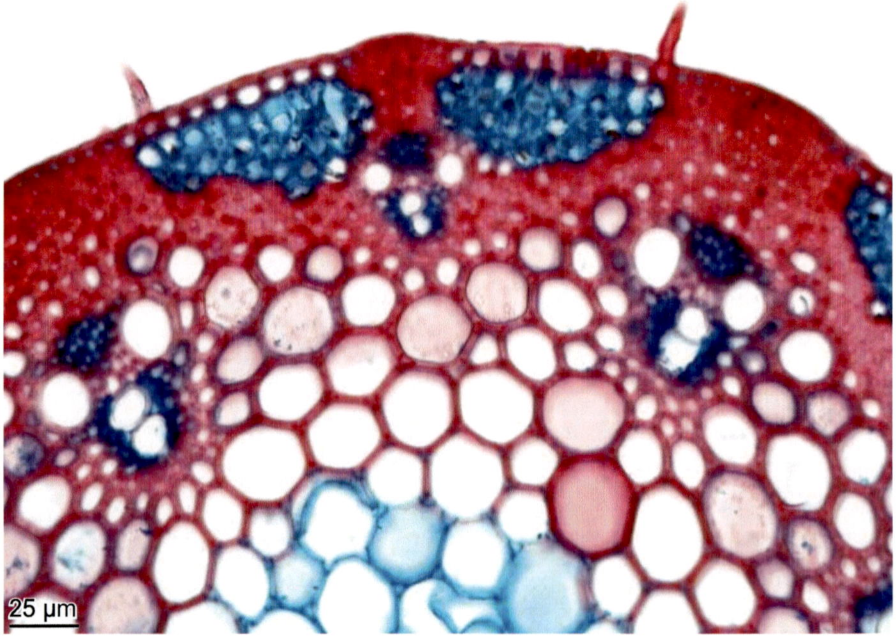

Bromus pectinatus Thunb.
Poaceae

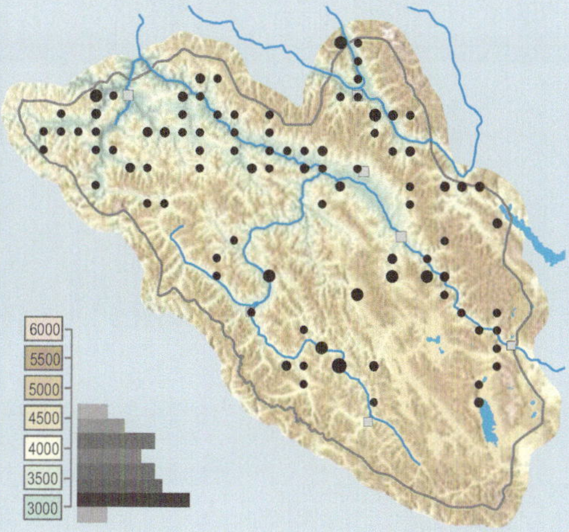

Description: Annuals, 20–60 cm tall, leaves up to 30 cm long, 2–8 mm wide, panicle 5–20 cm long, lax, branches nodding or ascending, awns 7–17 mm long.

Distribution: S and W Africa, Arabia, Iran, Tajikistan, Afghanistan, Pakistan, India, China.

Habitat: A weed in barley fields, field margins, along irrigation canals.

Culm uniform, round, hollow, with a diameter of 1.2–1.5 mm. Epidermis thin- to thick-walled. Continuous, 1–3-seriate peripheral sclerenchymatic belt of thick-walled fibers centripetal of the chlorenchyma. Girders, are represented by rectangular groups of thick-walled fibers. Tangentially enlarged chlorenchyma groups are located between girders. Intercellulars are absent. Vascular bundles are arranged in 2–3 offset circular rows. Large bundles within the thin-walled parenchymatic tissue, smaller bundles at the centripetal side of girders. Vascular bundles with large vessels, diameter 25–35 µm. With small intercellulars within vascular bundles centripetal of the protoxylem. Phloem small, oval, consisting of sieve tubes and companion cellss. Vascular bundles surrounded by an eccentric, 1–3-seriate sheath of polygonal, thin- to thick-walled cells.

Analyzed: 2 individuals

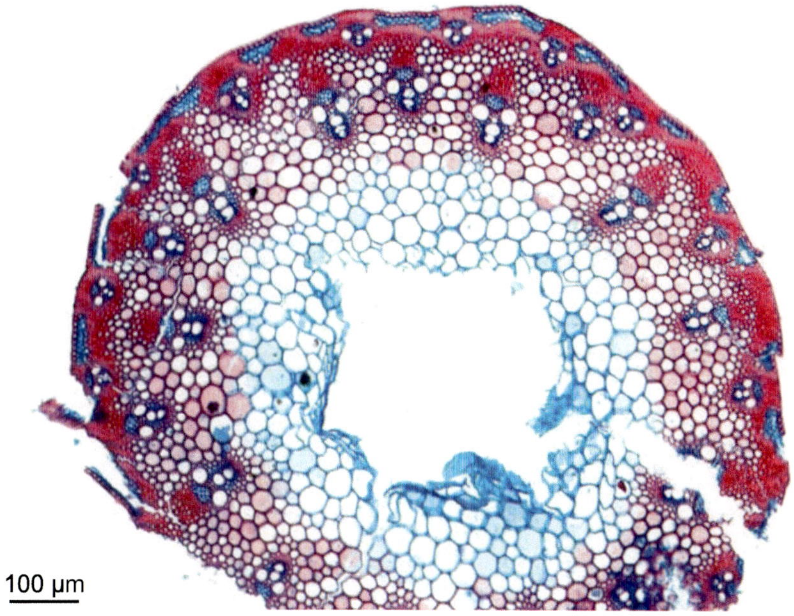

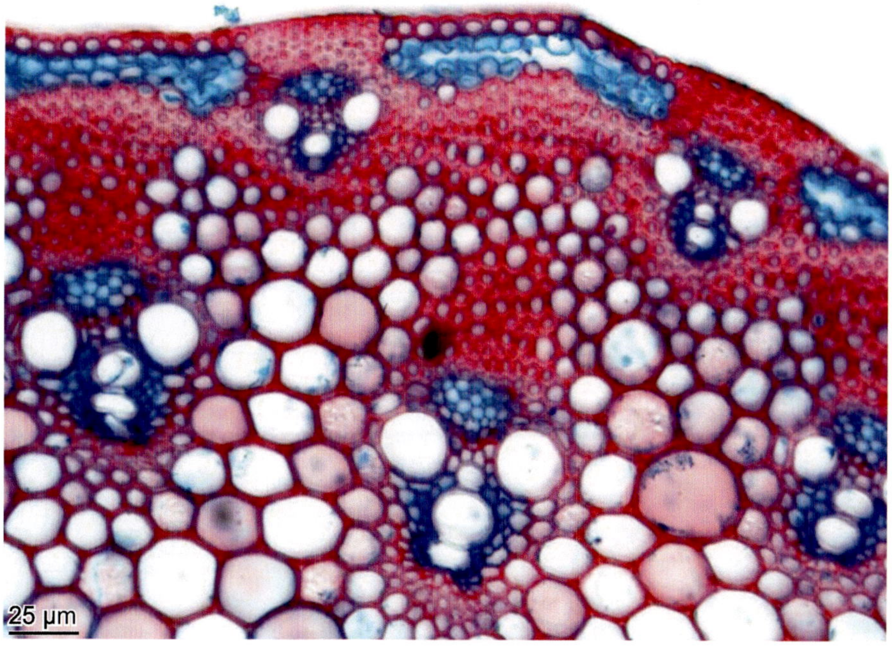

Bromus tectorum L.

Poaceae

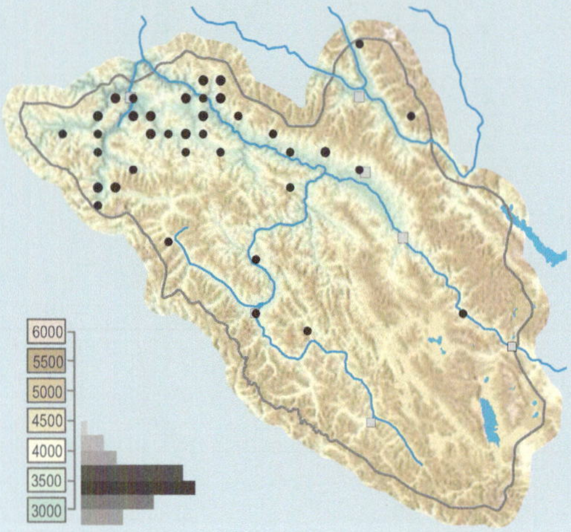

Syn.: *Anisantha tectorum* (L.) Nevski

Description: Annuals, 5–50 cm tall, leaves hairy on both sides, 2–15 cm long, 1.5–4 mm wide, panicle contracted, oblong, nodding, 4–18 cm long, branches drooping, awns 10–25 mm long.

Distribution: N Africa, Europe, SW Asia; introduced in North America and Australia, adventive in most temperate regions of the world.

Habitat: Roadsides, dry sandy places, deserts.

Culm uniform, round, hollow, with a diameter of 0.9–1.0 mm. Epidermis thick-walled. Continuous, 3- to >4-seriate peripheral sclerenchymatic belt of thick-walled fibers. Girders absent. Oval to rectangular chlorenchyma groups located within the sclerenchymatic belt. Intercellulars triangular and very small. Vascular bundles arranged in two offset circular rows. Large bundles within the thin- to thick-walled parenchymatic tissue, small bundles within the sclerenchymatic belt. Vascular bundles with large vessels, diameter 25–30 µm. Intercellulars within vascular bundles absent. Phloem small, oval, consisting of uniform cells. Vascular bundles surrounded by an eccentric, 1–3-seriate sheath of polygonal, thin- to thick-walled cells.

Analyzed: 3 individuals

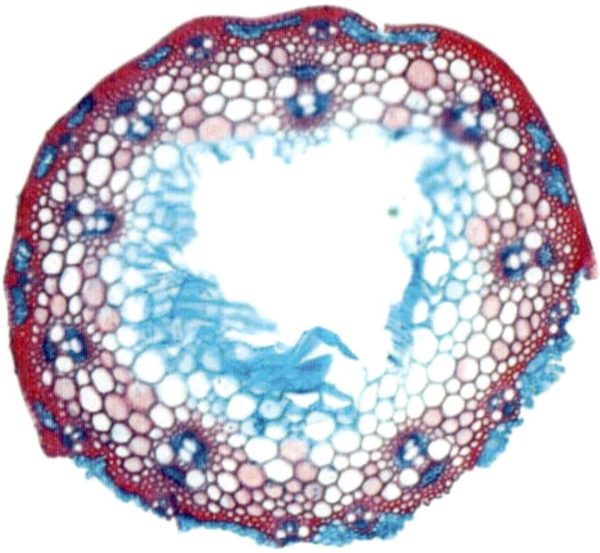

100 µm

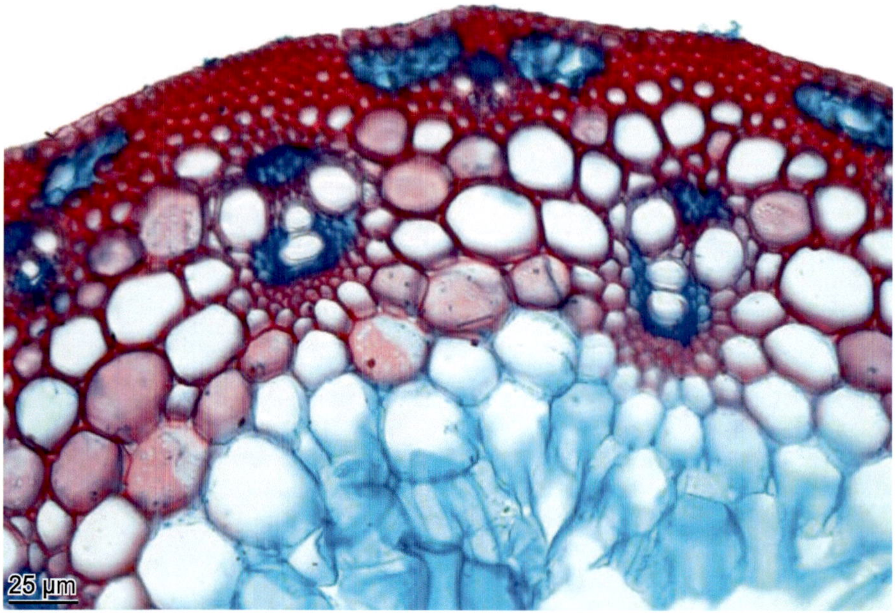

Bromus tomentosus Trin.

Poaceae

Syn.: *B. persicus* Boiss., *Zerna tomentosa* (Trin.) Nevski

Description: Perennial herbs, 10–60 cm tall, rhizomatous, culms ascending or decumbent, leaves flat or involute, hairy on both sides, 7–15 cm long, 2–4 mm wide, panicle open, linear or elliptic, 4–10 cm long, 2–4 cm wide, branches ascending, awns 2–3 mm long.

Distribution: Turkey, Caucasus, Iran, Afghanistan, Pakistan, India.

Habitat: Dry grasslands.

Culm uniform, horseshoe-shaped, hollow, with a diameter of 1–2.0 mm. Epidermis thick-walled. Continuous, 3- to >4-seriate peripheral sclerenchymatic belt of thin- to thick-walled fibers centripetal of the chlorenchyma. Girders conical and square. Rectangular chlorenchyma groups located peripheral of the sclerenchymatic belt. Intercellulars in the parenchymatic zone absent. Vascular bundles arranged in two offset circular rows. Large bundles within the thin- to thick-walled parenchymatic tissue, small bundles centripetal of girders. Vascular bundles with large vessels, diameter 30–40 µm. Intercellulars within vascular bundles centripetal of the protoxylem. Phloem small, oval, consisting of sieve tubes and companion cellss. Vascular bundles surrounded by a 1–3 seriate sheath of polygonal, thin- to thick-walled cells.

Analyzed: 1 individual

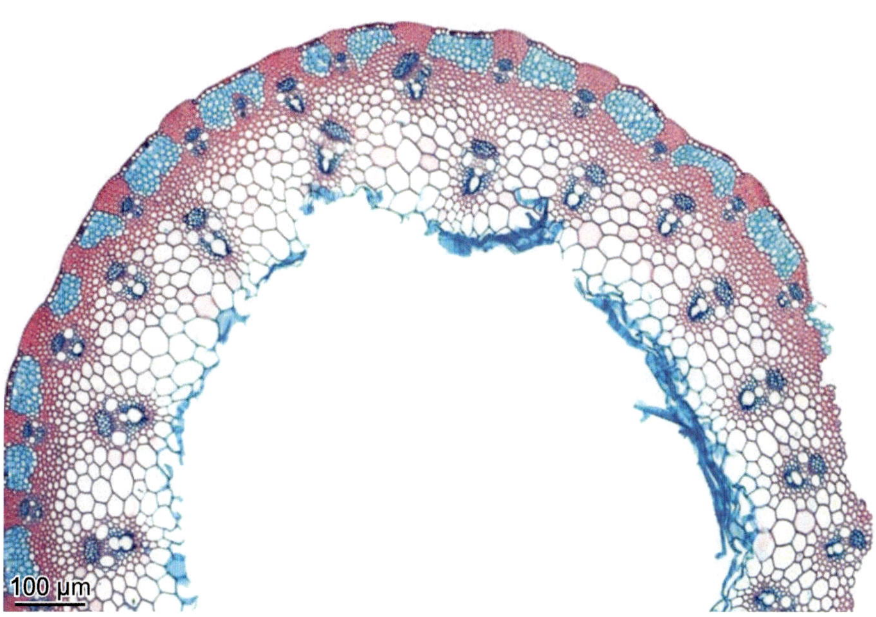

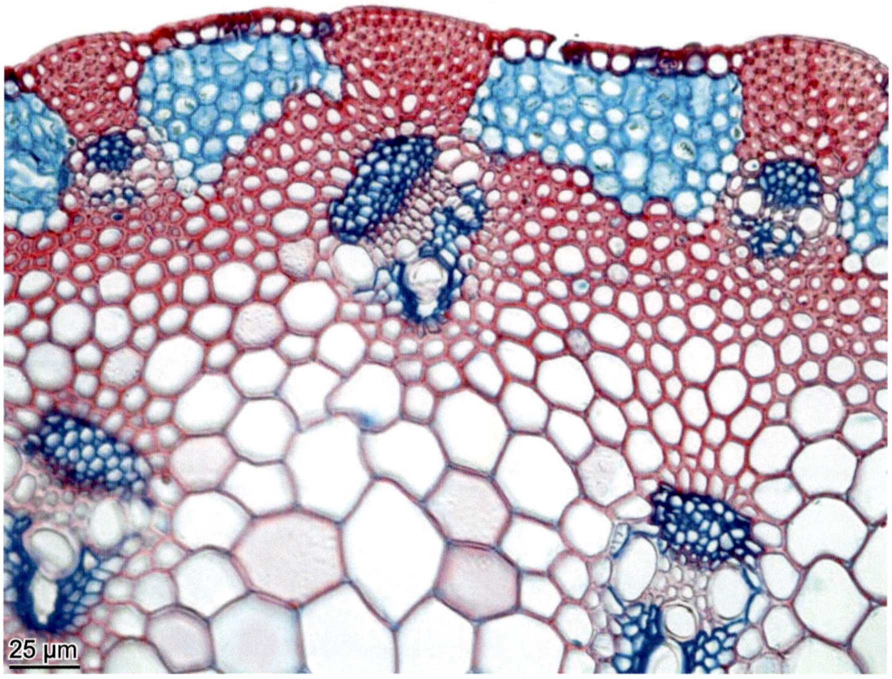

Catabrosa aquatica (L.) P.Beauv.
Poaceae

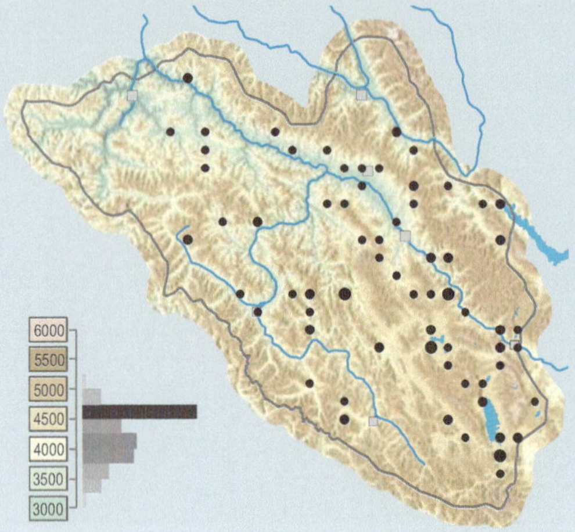

Syn.: *Aira aquatica* L.

Description: Perennial herbs, 5–70 cm tall, stoloniferous, leaves soft, panicle loosely contracted.

Distribution: North America, Europe, SW Asia, Russia, Kazakhstan, Turkmenistan, Kyrgyzstan, Tajikistan, Afghanistan, Pakistan, India, China (Gansu, Guizhou, Hebei, Hubei, Nei Mongol, Qinghai, Sichuan, Xinjiang, Xizang, Yunnan), Mongolia.

Habitat: Muddy river sediments, emerged loamy bottoms, fresh to saline habitats close to water bodies, shallow lakes and pools.

Culm uniform, round, hollow, with a diameter of 1–1.2 mm. Epidermis thin- to thick-walled. Uniseriate peripheral belt of lignified, thin- to thick-walled fibers. Tangentially enlarged groups of chlorenchyma located between the epidermis and the sclerenchymatic belt. Vascular bundles arranged in two offset circular rows. Large vascular bundles within the thin-walled parenchymatic tissue. Tangentially enlarged girders in alternate position to the chlorenchyma. Vascular bundles with two large lateral vessels, diameter 20–25 µm. Small triangular intercellulars in the thin-walled, unlignified parenchymatic tissue. Intercellulars centripetal of the protoxylem absent. Phloem small, round, consisting of uniform cells. Vascular bundles surrounded by a uniseriate sheath of polygonal, thick-walled cells.

Analyzed: 2 individuals

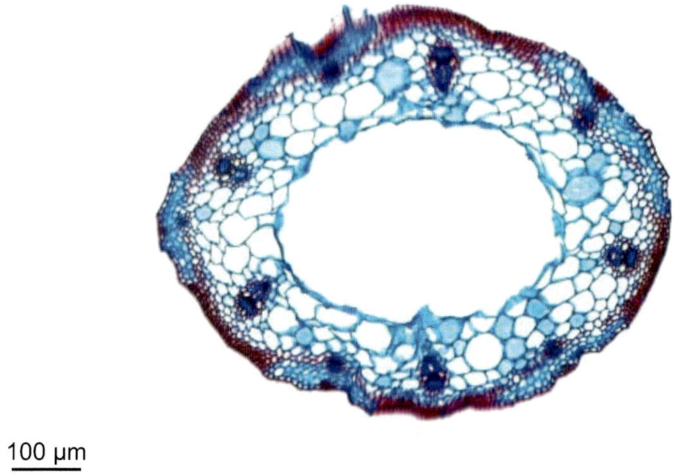

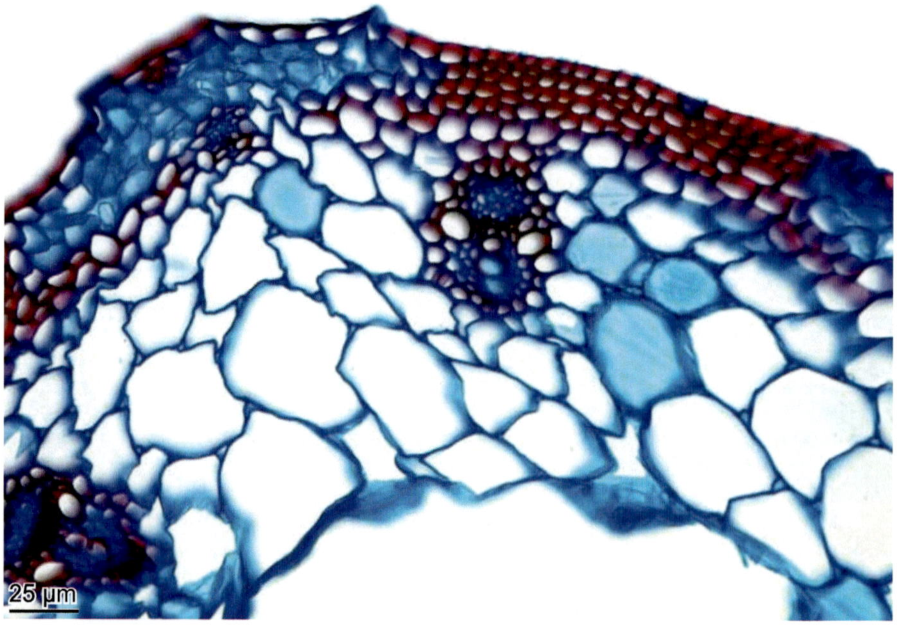

Chloris virgata Sw.

Poaceae

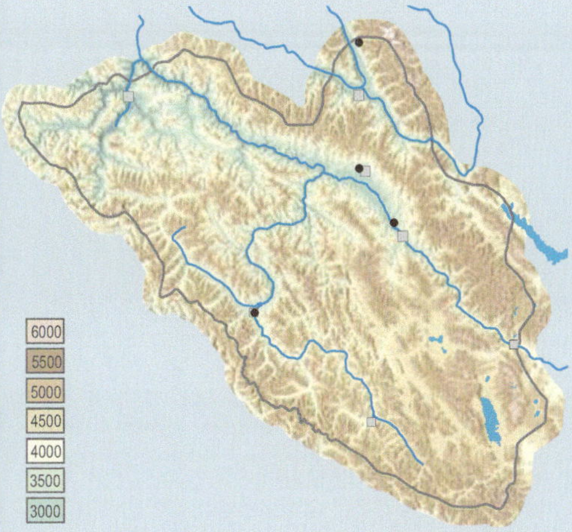

Syn.: *C. caudata* Trin.

Description: Annuals, 15–100 cm tall, cespitose, culms erect or ascending, leaf-sheaths keeled, leaves 10–30 cm long, 2–6 mm wide, racemes 4–12, digitate, spreading, 2–10 cm long, awns 5–15 mm long.

Distribution: North and South America, Africa, temperate Asia; naturalized elsewhere, also cultivated.

Habitat: Steppes, stony slopes, roadsides, sandy river sediments.

Culm uniform, round, filled, with a diameter of 0.9–1.1 mm. Epidermis thin- to thick-walled. Continuous, 2–3-seriate peripheral sclerenchymatic belt of thin- to thick-walled fibers centripetal of the chlorenchyma. Girders square to rectangular. Small, round chlorenchyma groups located within the sclerenchymatic belt or between girders. Intercellulars in the parenchymatic zone small and triangular. Vascular bundles arranged in two offset circular rows. Large bundles within the thin-walled parenchymatic tissue, small bundles centripetal of the sclerenchymatic belt. Vascular bundles with large vessels, diameter 10–15 µm. Distinct intercellulars within vascular bundles absent centripetal of the protoxylem. Phloem small, oval, consisting of uniform cells. Vascular bundles surrounded by a uniseriate sheath of polygonal, thin- to thick-walled cells.

Analyzed: 1 individual

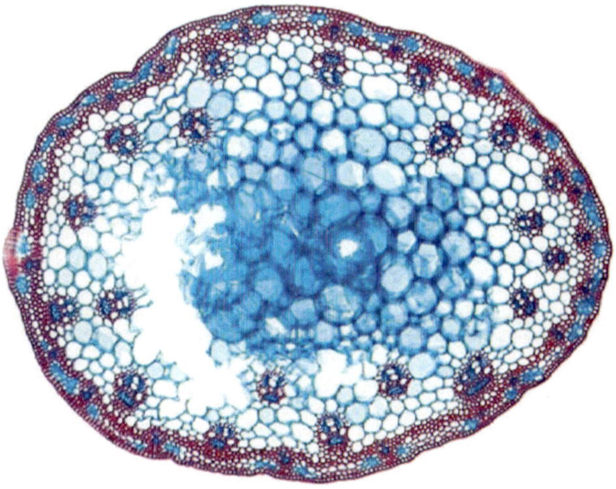

100 µm

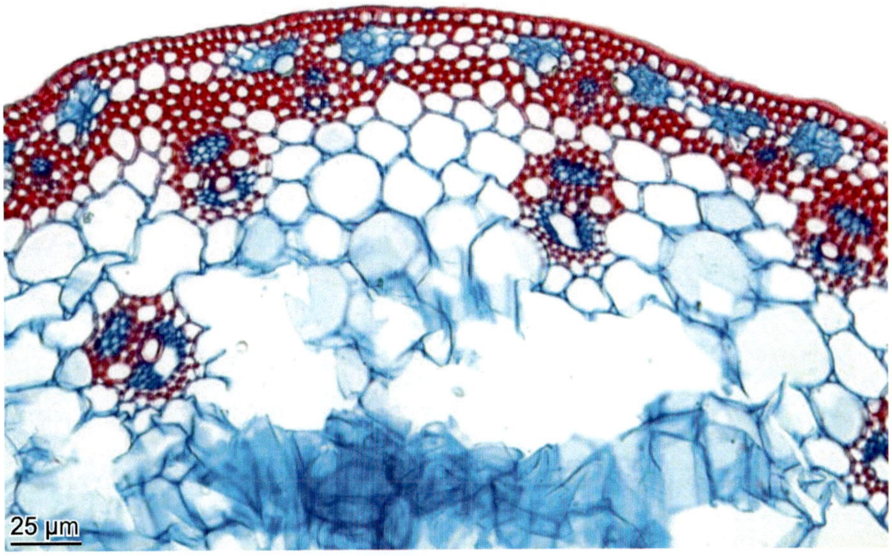

Cleistogenes songorica (Roshev.) Ohwi

Poaceae

Syn.: *Cleistogenes mutica* Keng, *Diplachne songorica* Roshev., *Kengia mutica* (Keng) Packer, *K. songorica* (Roshev.) Packer

Description: Perennial herbs, 20–30 cm tall, densely cespitose, leaves flat, rough on both sides, 2–8 cm long, 1.5–2.5 mm wide, panicle open, 2–5 cm long, usually enclosed with the uppermost leaf sheath.

Distribution: Russia, Kazakhstan, Uzbekistan, Turkmenistan, Kyrgyzstan, N India, China (Gansu, Henan, Nei Mongol, Ningxia, Qinghai, Shaanxi, Xinjiang), Mongolia.

Habitat: Desert steppes, dry grasslands.

Culm uniform, round, hollow, with a diameter of 1.2–1.7 mm. Epidermis thick-walled. Continuous, 2–5-seriate peripheral sclerenchymatic belt of thick-walled fibers. Girders absent. Chlorenchyma groups absent. Intercellulars in the parenchymatic tissue absent. Vascular bundles arranged in 2 circular offset rows. Large bundles within the thin- to thick-walled parenchymatic tissue, small bundles within the sclerenchymatic belt. Vascular bundles with large vessels, diameter 15–20 µm. Phloem oval, consisting of sieve tubes and companion cells. With intercellulars centripetal of the protoxylem. Vascular bundles surrounded by a 1–2-seriate sheath of polygonal, thin- to thick-walled cells.

Analyzed: 2 individuals

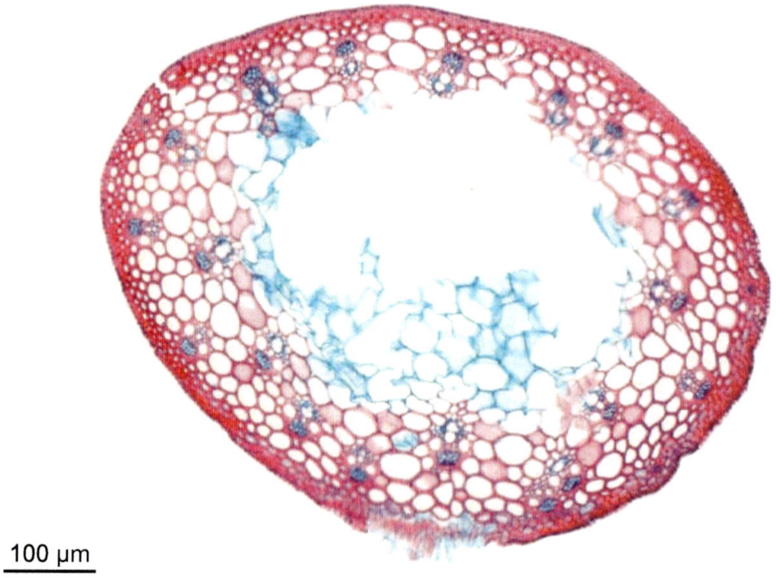

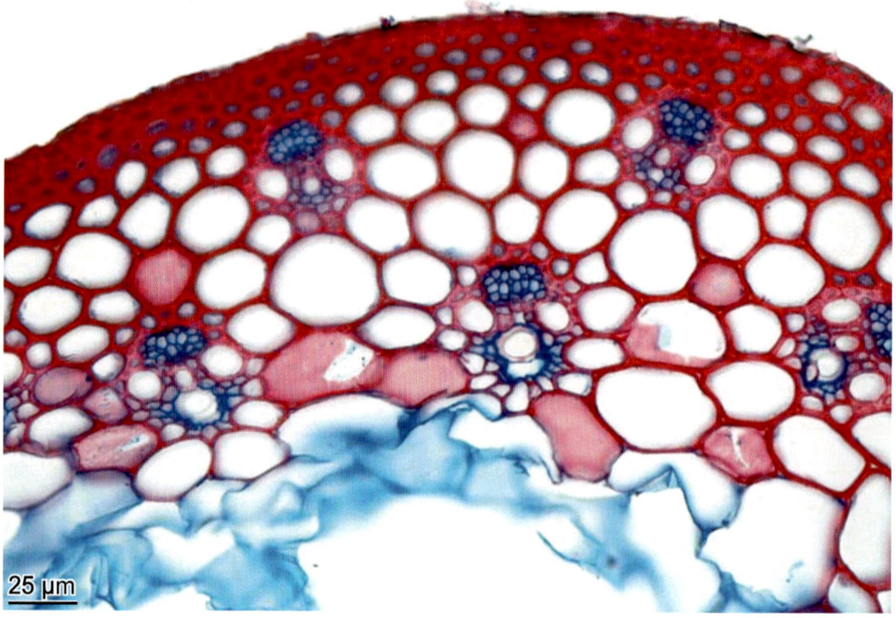

Colpodium himalaicum (Hook.f.) Bor

Poaceae

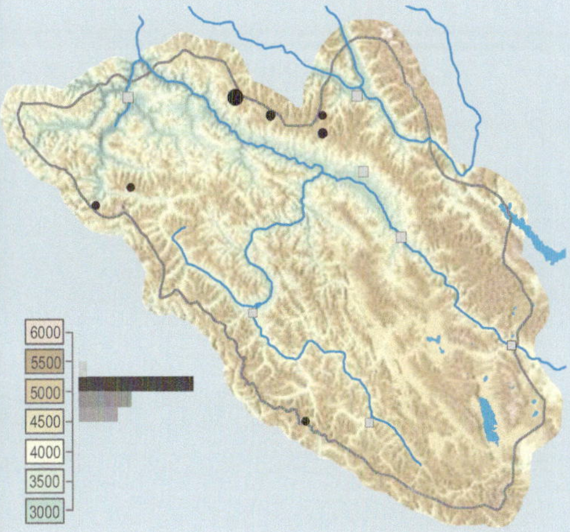

Syn.: *Catabrosa himalaica* (Hook.f.) Stapf, *Catabrosella himalaica* (Hook.f.) Tzvelev, *Phippsia himalaica* Hook.f.

Description: Perennial herbs, 5–30 cm tall, cespitose, leaves conduplicate, 1–10 cm long, 1–2 mm wide, panicle contracted, linear, 2.5–4 cm long, 0.2–0.5 cm wide

Distribution: Pakistan, India.

Habitat: Moist places in screes, open grassy places.

Culm uniform, round, wavy, hollow, with a diameter of 0.4–0.6 mm. Epidermis thin- to thick-walled. Continuous, uniseriate peripheral sclerenchymatic belt of thin- to thick-walled fibers centripetal of the chlorenchyma. Girders conical to square. Irregular formed chlorenchyma groups located between girders. Intercellulars in the parenchymatic zone small and triangular. Vascular bundles arranged in one circular row, within the thin-walled parenchymatic tissue and the sclerenchymatic belt. Vascular bundles with large vessels, diameter 5–10 µm. Intercellulars within vascular bundles absent. Phloem small, oval, consisting of uniform cells. Vascular bundles surrounded by a uniseriate sheath of polygonal, rather thick-walled cells.

Analyzed: 1 individual

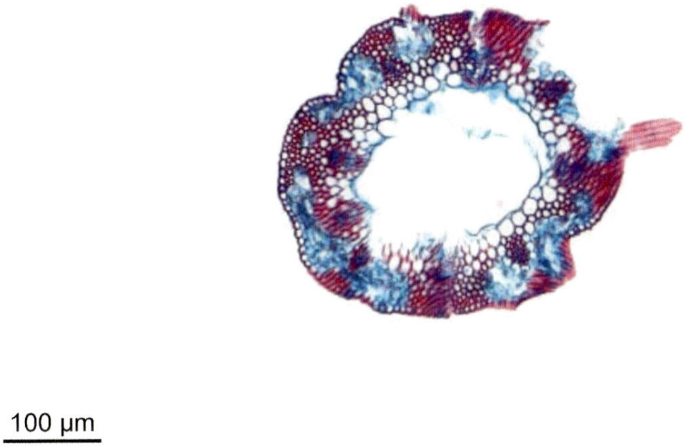

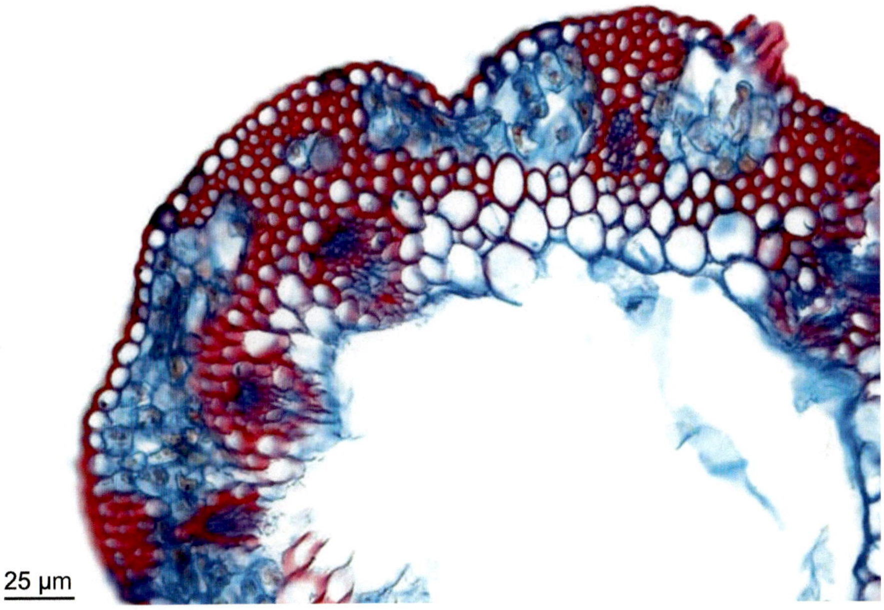

Colpodium nutans (Stapf) Bor

Poaceae

Syn.: *Catabrosa nutans* Stapf, *Hyalopoa nutans* (Stapf) E.B.Alexeev ex T.A.Cope

Description: Perennial herbs, 20–80 cm tall, cespitose, leaves flat, 5–13 cm long, 3–5 mm wide, panicle 5–15 cm long, branches long and flexuous.

Distribution: N Pakistan, NW India.

Habitat: Moist places.

Culm uniform, round, hollow, with a diameter of 0.9–1.2 mm. Epidermis thin- to thick-walled, often unlignified. Continuous, uniseriate peripheral sclerenchymatic belt of thin- to thick-walled fibers centripetal of the chlorenchyma. Girders conical. Continuous chlorenchyma band centripetal of girders. Intercellulars in the parenchymatic zone small and triangular. Vascular bundles arranged in one circular row, within the thin-walled parenchymatic tissue and the sclerenchymatic belt. Vascular bundles with large vessels, diameter 5–10 μm. Intercellulars centripetal of the protoxylem absent. Phloem small, oval, consisting of uniform cells. Vascular bundles surrounded by a uniseriate sheath of polygonal, thin- to thick-walled cells.

Analyzed: 2 individuals

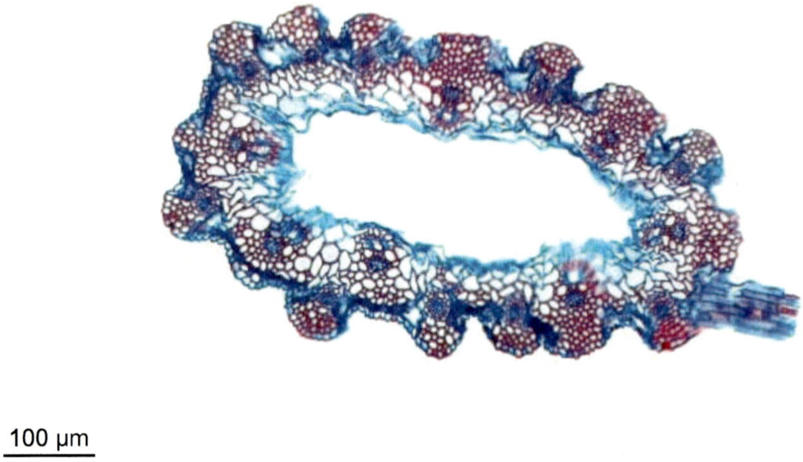

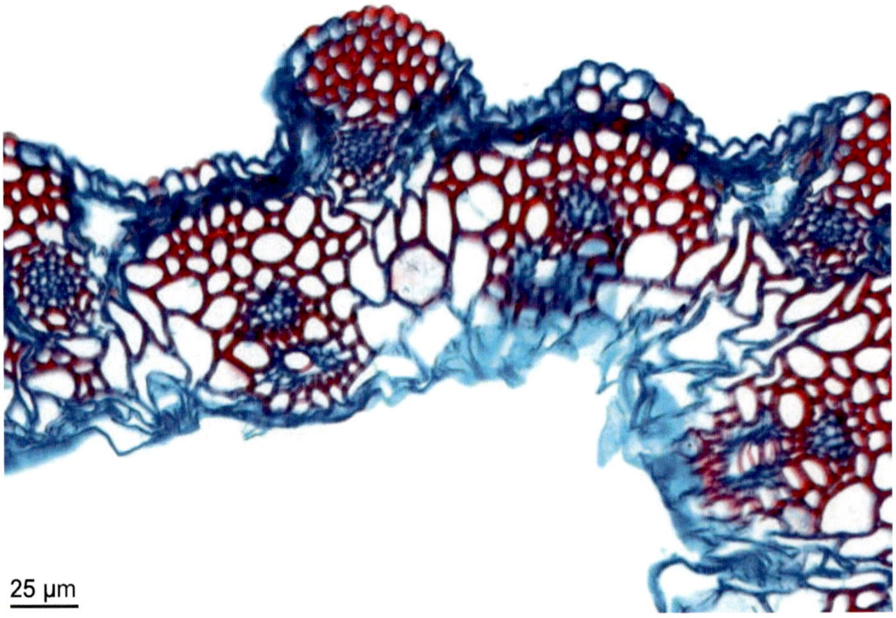

Cymbopogon jwarancusa subsp. *olivieri* (Boiss.) Soenarko

Poaceae

Syn.: *Andropogon olivieri* Boiss., *C. ladakhensis* B.K.Gupta

Description: Perennial herbs, 20–40 cm tall, cespitose, aromatic, leaves 15–30 cm long, 3–5 mm wide, inflorescence composed of terminal and axillary racemes, racemes 2, deflexed, 1.3–2.2 cm long, hairy, awns 7–10 mm long.

Distribution: Turkey, Lebanon, Syria, Yemen, Oman, Iraq, Iran, Afghanistan, Pakistan, NW India, China (W Xizang, Yunnan).

Habitat: Dry mountain slopes, stabilized screes.

Culm uniform, round, filled, with a diameter of 1.2–1.5 mm. Epidermis thick-walled. Continuous, 3- to >4-seriate peripheral sclerenchymatic belt of thick-walled fibers. Girders absent. Chlorenchyma groups absent. Intercellulars in the parenchymatic zone absent. Vascular bundles arranged in three offset circular rows. Large bundles within the thin-walled parenchymatic tissue, small bundles centripetal of the sclerenchymatic belt. Vascular bundles with large vessels, diameter 20–30 µm. Intercellulars centripetal of the protoxylem absent. Phloem small, oval, consisting of sieve tubes and companion cellss. Vascular bundles surrounded by a 1–2-seriate sheath of polygonal, rather thick-walled cells.

Analyzed: 2 individuals

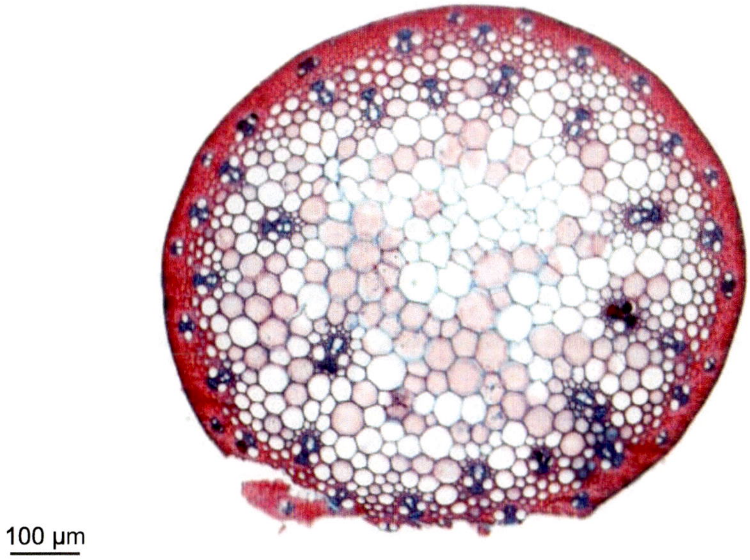

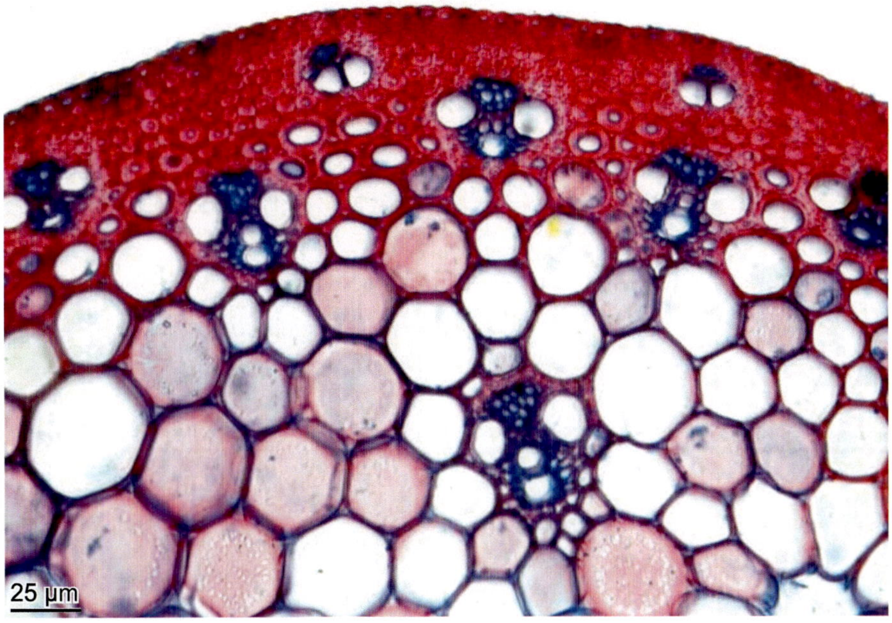

Danthonia cachemyriana Jaub. & Spach

Poaceae

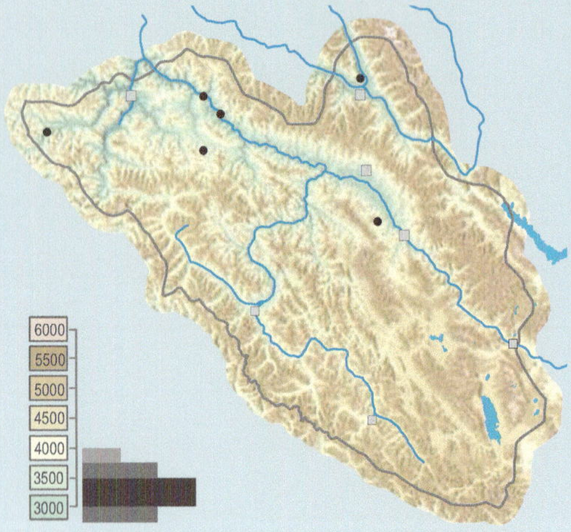

Syn.: *D. exilis* Hook.f.

Description: Perennial herbs, 10–25 cm tall, densely cespitose, leaves filiform, up to 6 cm long, ca. 1 mm wide, panicle oblong or ovate, 1–3 cm long, contracted, awns 4–5 mm long.

Distribution: E Afghanistan, N Pakistan, N India, China (Xizang).

Habitat: Dry rocky slopes.

Culm uniform, round, filled, with a diameter of 0.7–0.9 mm. Epidermis thin- to thick-walled. Continuous, 3- to >4-seriate peripheral sclerenchymatic belt of thick-walled fibers. Girders absent or conical to rectangular. Chlorenchyma groups irregularly formed. Intercellulars in the parenchymatic zone small and triangular. Vascular bundles arranged in two offset circular rows. Large bundles within the thin-walled parenchymatic tissue, small bundles centripetal of the sclerenchymatic belt. Vascular bundles with large vessels, diameter 10–15 µm. Phloem small, oval, consisting of sieve tubes and companion cellss. Distinct intercellulars centripetal of the protoxylem. Vascular bundles surrounded by a 1–2-seriate sheath of polygonal, thin- to thick-walled cells.

Analyzed: 2 individuals

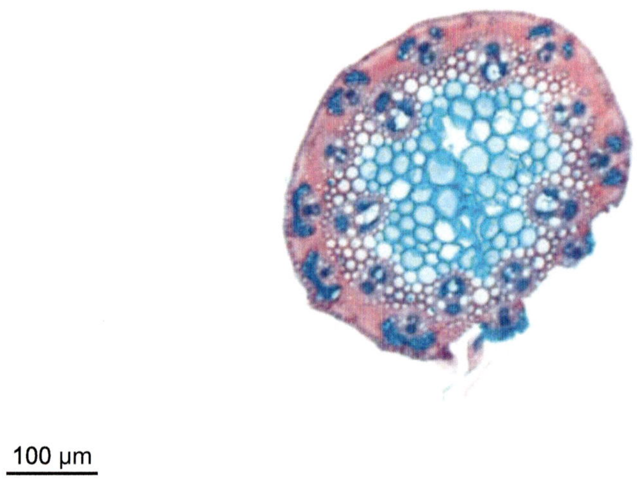

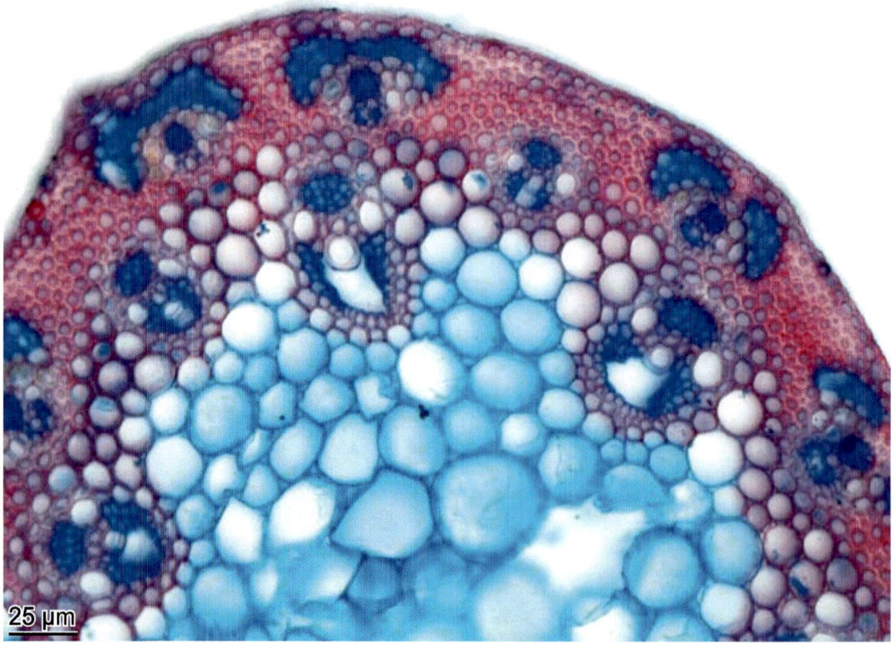

Deschampsia koelerioides Regel

Poaceae

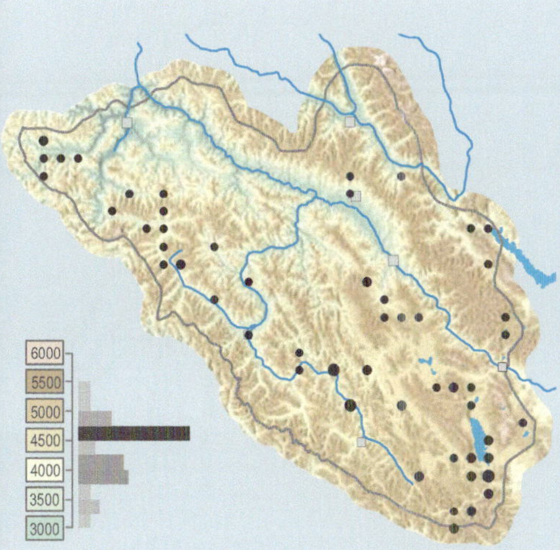

Syn.: *D. cespitosa* subsp. *koelerioides* (Regel) Tzvelev

Description: Perennial herbs, 10–40 cm tall, cespitose, leaves mostly basal, flat, 2–20 cm long, 2–3 mm wide, coarsely ridged above, panicle ovoid, purple-brown, 2–5 cm long, contracted and spike-like, awns 2–4 mm long.

Distribution: Russia (S Siberia), Kazakhstan, Uzbekistan, Kyrgyzstan, Tajikistan, E Afghanistan, N Pakistan, N India, China (Gansu, Nei Mongol, Qinghai, Xinjiang, Xizang), Mongolia.

Habitat: Alpine meadows, moist places near streams.

Culm uniform, round, hollow, with a diameter of 1.8–2.0 mm. Epidermis thin- to thick-walled. Continuous, 3–4-seriate peripheral sclerenchymatic belt of thin- to thick-walled fibers. Girders absent. Chlorenchyma groups absent. Intercellulars in the parenchymatic zone small and triangular. Vascular bundles arranged in two offset circular rows. Large bundles within the thin-walled, unlignified parenchymatic tissue, small bundles centripetal of the sclerenchymatic belt. Vascular bundles with large vessels, diameter 12–15 µm. Distinct intercellulars centripetal of the protoxylem. Phloem small, oval, consisting of sieve tubes and companion cellss. Vascular bundles surrounded by a 1–2-seriate sheath of polygonal, thin- to thick-walled cells.

Analyzed: 2 individuals

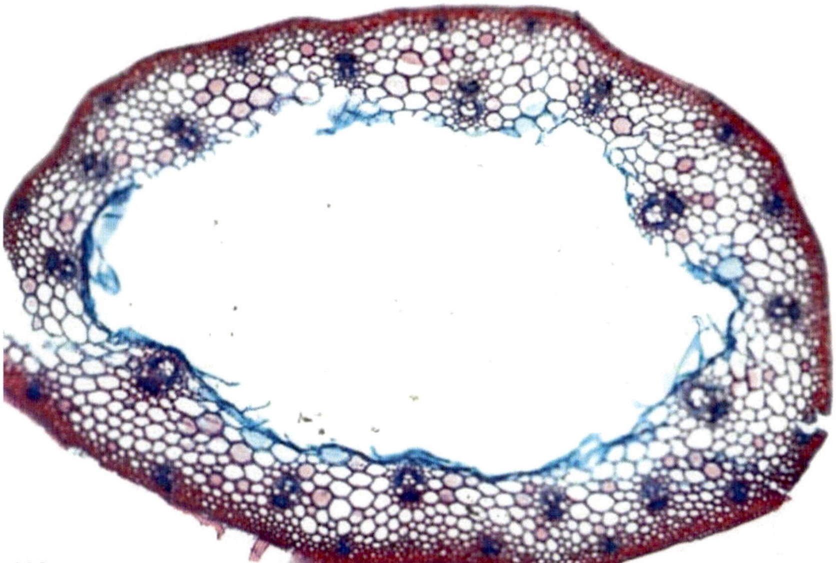

100 µm

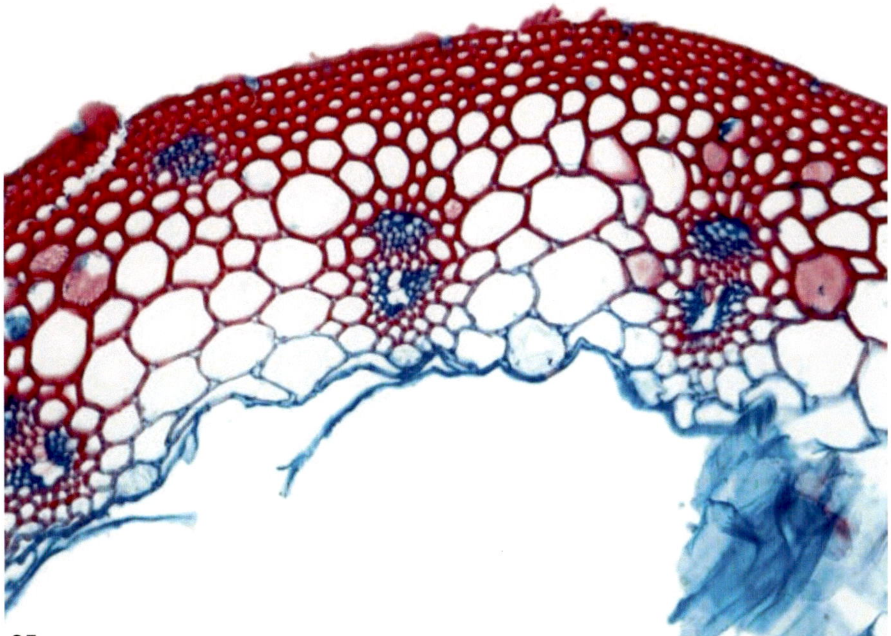

25 µm

Digitaria sanguinalis (L.) Scop.

Poaceae

Syn.: *Panicum sanguinale* L.

Description: Annuals, 10–40 cm tall, culms prostrate or ascending, leaves flat, linear-lanceolate, 5–20 cm long, 0.4–1.2 cm wide, glabrous or pilose, inflorescence composed of 4–12 digitately arranged racemes, 5–15 cm long.

Distribution: N Africa, S Europe, W and C Asia; naturalized elsewhere.

Habitat: Fields, along roads.

Culm uniform, round, hollow, with a diameter of 2–3.0 mm. Epidermis thin- to thick-walled. Continuous, 3–4-seriate peripheral sclerenchymatic belt of thin- to thick-walled fibers. Girders absent. Chlorenchyma groups absent. Intercellulars in the parenchymatic zone small and triangular. Vascular bundles arranged in two offset circular rows. Large bundles within the thin-walled, unlignified parenchymatic tissue, small bundles centripetal of the sclerenchymatic belt. Vascular bundles with large vessels, diameter 25–35 µm. Phloem small, oval, consisting of sieve tubes and companion cellss. Distinct intercellulars centripetal of the protoxylem. Vascular bundles surrounded by a uniseriate sheath of polygonal, thin- to thick-walled cells.

Analyzed: 1 individual

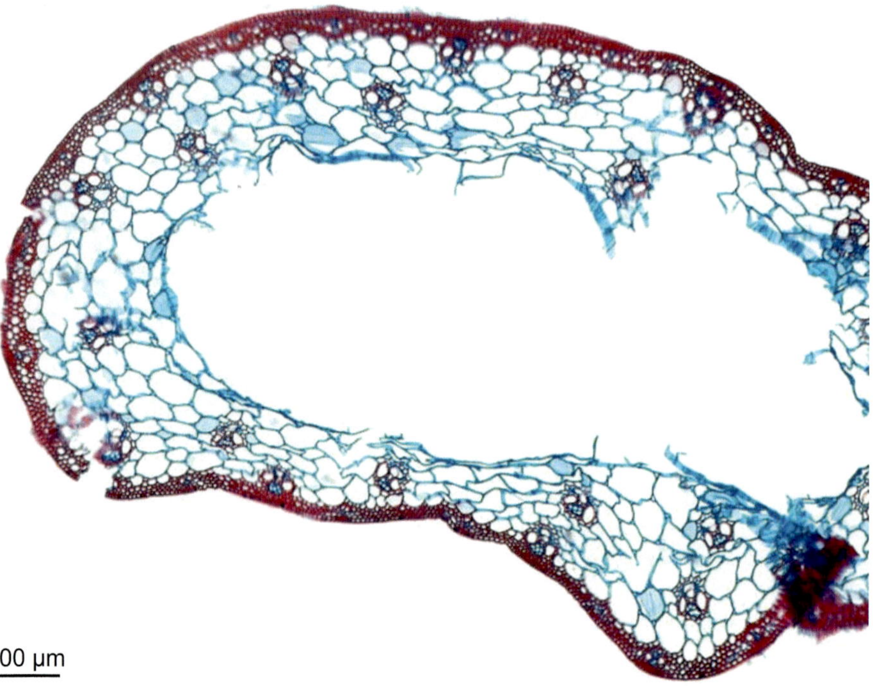

100 µm

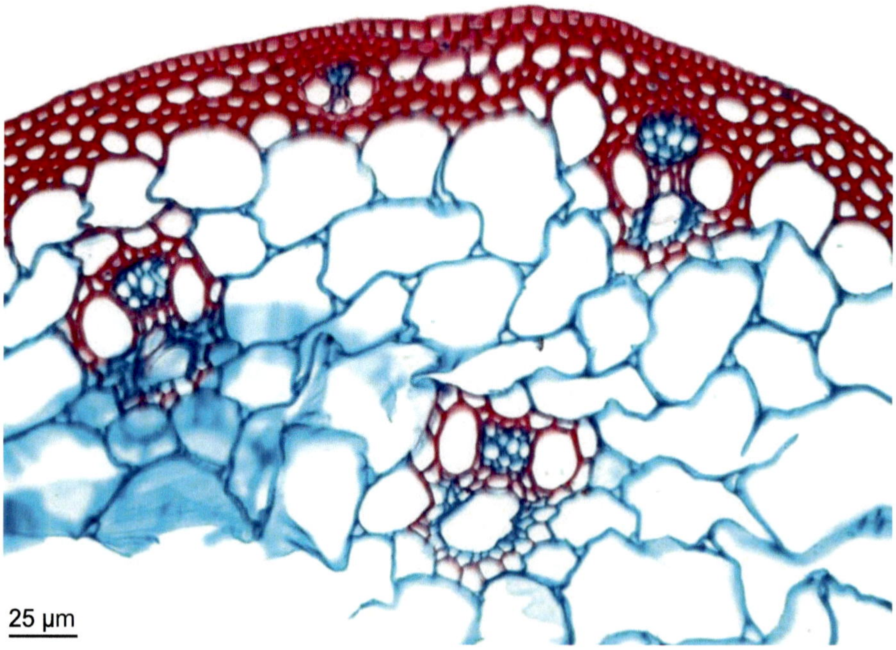

25 µm

Digitaria stewartiana Bor

Poaceae

Syn.: *D. ischaemum* subsp. *stewartiana* (Bor) Tzvelev

Description: Annuals, 10–30 cm tall, culms geniculately ascending, leaves 3–6 cm long, ca. 5 mm wide, inflorescence composed of 2–3 digitately arranged racemes, up to 5 cm long.

Distribution: Pakistan, N India.

Habitat: Open grassy places.

Culm uniform, round, filled, with a diameter of 0.6–0.8 mm. Epidermis thin- to thick-walled. Discontinuous peripheral sclerenchymatic belt of thin- to thick-walled fibers. Girders rectangular. Chlorenchyma groups small between girders. Intercellulars in the parenchymatic zone small and triangular. Vascular bundles arranged in one circular row, centripetal of girders. Vascular bundles with small vessels, diameter 10–15 µm. Phloem small, oval, consisting of uniform cells. A few intercellulars centripetal of the protoxylem. Vascular bundles surrounded by a 1–2-seriate sheath of polygonal, thin- to thick-walled cells.

Analyzed: 1 individual

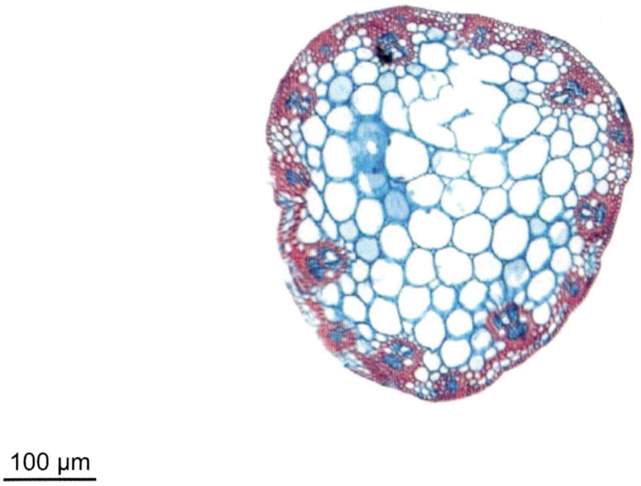

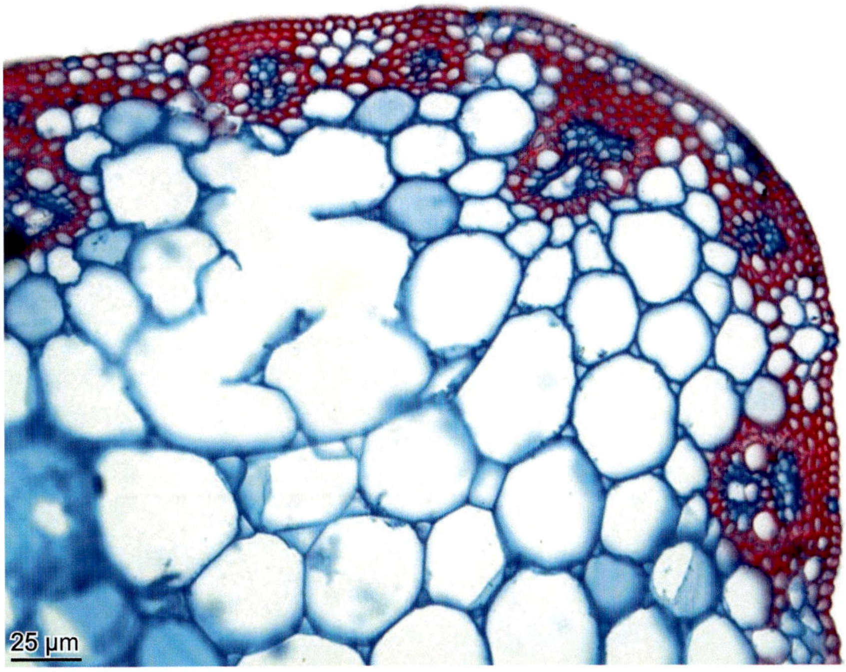

Elymus canaliculatus (Nevski) Tzvelev
Poaceae

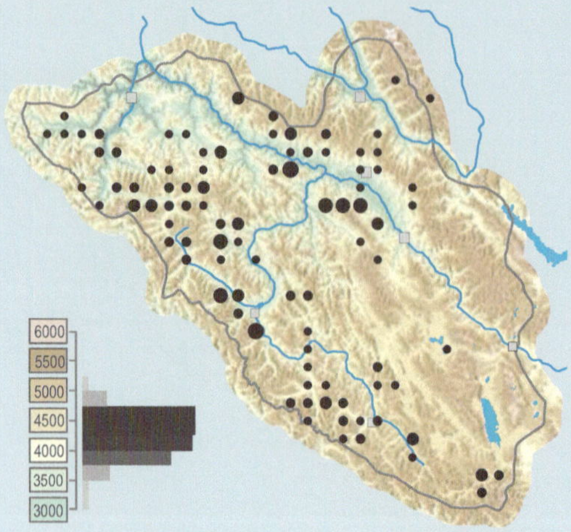

Syn.: *Agropyron canaliculatum* Nevski, *Elymus longearistatus* subsp. *canaliculatus* (Nevski) Tzvelev

Description: Perennial herbs, 25–70 cm tall, cespitose, leaves flat or loosely rolled, 5–15 cm long, 2–3 mm wide, puberulous above, spike lax, 8–15 cm long, erect or nodding, awns strongly curved, 3–6 cm long.

Distribution: N Iran, Tajikistan, Afghanistan, N Pakistan, NW India, Nepal, China (Xizang).

Habitat: Mountain slopes.

Culm uniform, round, hollow, with a diameter of 0.6–0.8 mm. Epidermis thin- to thick-walled. Continuous, 2–3-seriate peripheral sclerenchymatic belt of thin- to thick-walled fibers centripetal of the chlorenchyma. Girders conical to square. Chlorenchyma groups irregularly formed. Intercellulars in the parenchymatic zone very small and triangular. Vascular bundles arranged in two offset circular rows. Large bundles within the thin-walled parenchymatic tissue, small bundles centripetal of the sclerenchymatic belt centripetal of girders. Vascular bundles with large vessels, diameter 10–15 µm. Phloem oval, consisting of uniform cells. A few intercellulars centripetal of the protoxylem. Vascular bundles surrounded by a 1–2-seriate sheath of polygonal, thin- to thick-walled cells.

Analyzed: 2 individuals

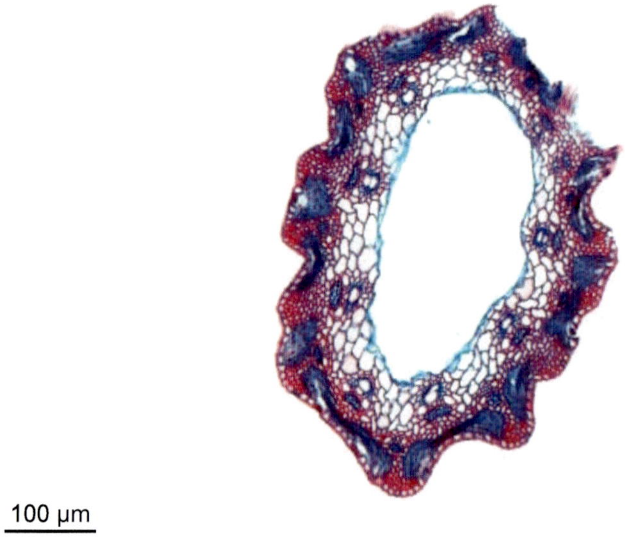

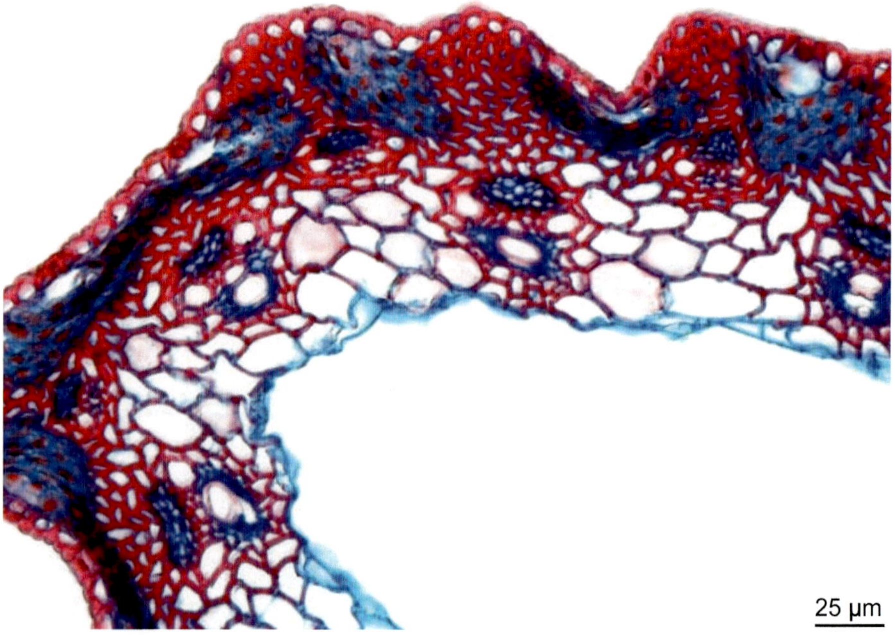

Elymus cognatus (Hack.) Cope
Poaceae

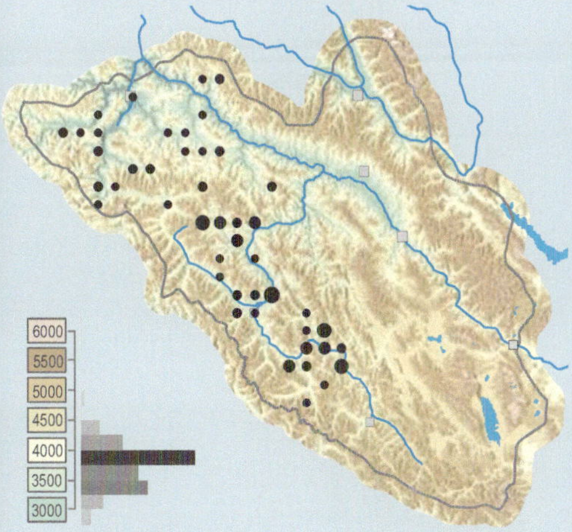

Syn.: *Agropyron cognatum* Hack., *Pseudoroegneria cognata* (Hack.) Á.Löve

Description: Perennial herbs, 25–60 cm tall, loosely cespitose, leaves flat or rolled, 6–15 cm long, 1–3 mm wide, shortly hairy above, spike slender, 6–12 cm long, erect.

Distribution: Russia, Iran, Kazakhstan, Turkmenistan, Uzbekistan, Kyrgyzstan, Afghanistan, Pakistan, NW India, China (Xinjiang).

Habitat: Stony slopes.

Culm uniform, round, hollow, with a diameter of 0.9–1.0 mm. Epidermis thin- to thick-walled. Continuous, 2–3-seriate peripheral sclerenchymatic belt of thin- to thick-walled fibers centripetal of the chlorenchyma. Girders conical to square to rectangular. Chlorenchyma groups tangentially oval. Vascular bundles arranged in two offset circular rows. Large bundles within the thin-walled parenchymatic tissue, small bundles centripetal of the sclerenchymatic belt centripetal of girders. Vascular bundles with large vessels, diameter 30–35 µm. Phloem oval, consisting of sieve tubes and companion cells. A few intercellulars centripetal of the protoxylem. Vascular bundles surrounded by a uniseriate sheath of polygonal, thin- to thick-walled cells.

Analyzed: 1 individual

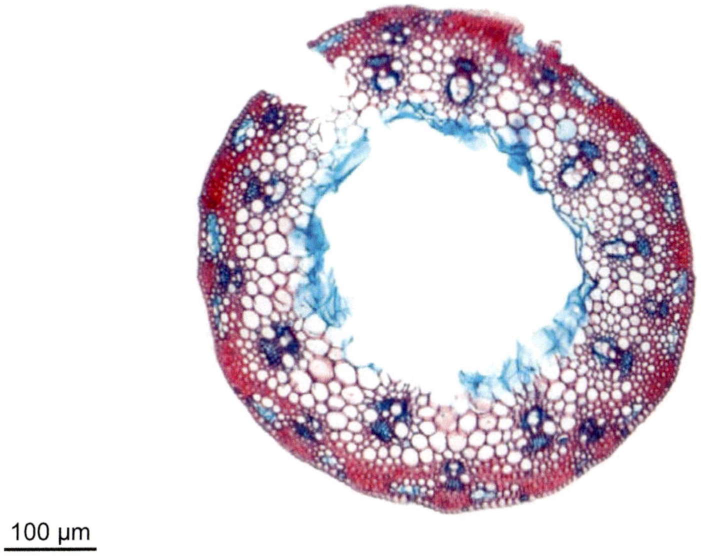

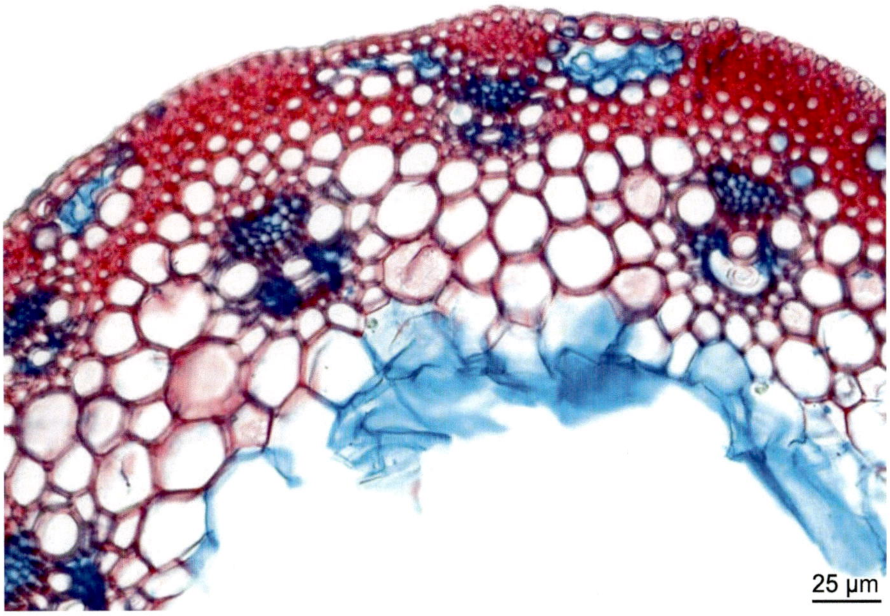

Elymus dahuricus Turcz. ex Griseb.

Poaceae

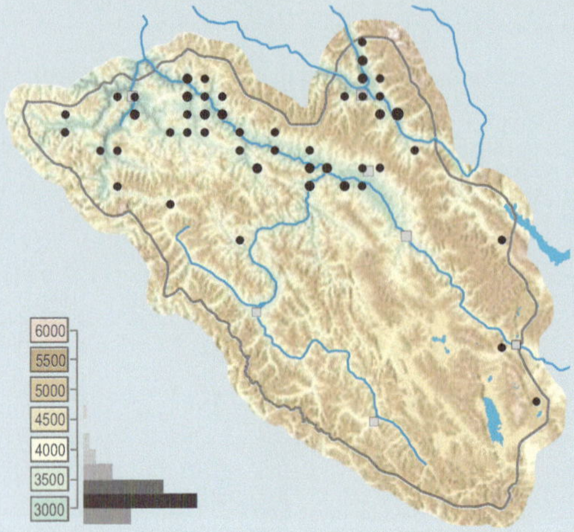

Description: Perennial herbs, 30–130 cm tall, cespitose, leaves usually flat, 10–30 cm long, 3–9 mm wide, spike erect or sometimes nodding, 7–20 cm long, awn straight or slightly curved, 8–15 mm long.

Distribution: Russia (Siberia, Far East), Kazakhstan, Turkmenistan, Uzbekistan, Kyrgyzstan, Afghanistan, N Pakistan, NW India, Nepal, China (Hebei, Heilongjiang, Henan, Nei Mongol, Ningxia, Qinghai, Shaanxi, Shandong, Shanxi, Sichuan, Xinjiang, Xizang, Yunnan), Bhutan, Mongolia, Korea, Japan.

Habitat: Disturbed wet places, field margins.

Culm uniform, round, wavy, hollow, with a diameter of 1.1–1.3 mm. Epidermis thin- to thick-walled. Continuous, 2–3-seriate peripheral sclerenchymatic belt of thin- to thick-walled fibers centripetal of the chlorenchyma. Girders conical to square. Chlorenchyma groups tangentially enlarged. Intercellulars in the parenchymatic zone very small and triangular. Vascular bundles arranged in one circular row. Bundles exist within the thin-walled, parenchymatic tissue, generally centripetal of girders. Vascular bundles with large vessels, diameter 20–30 µm. Phloem oval, consisting of sieve tubes and companion cells. Intercellulars centripetal of the protoxylem absent. Vascular bundles surrounded by a 1–3-seriate sheath of polygonal, thin- to thick-walled cells.

Analyzed: 2 individuals

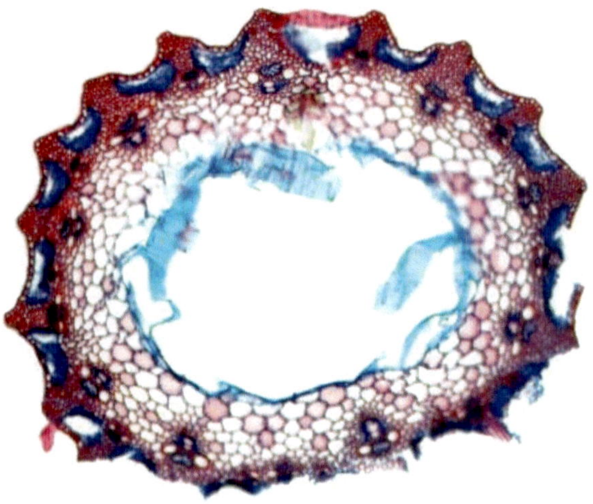

100 µm

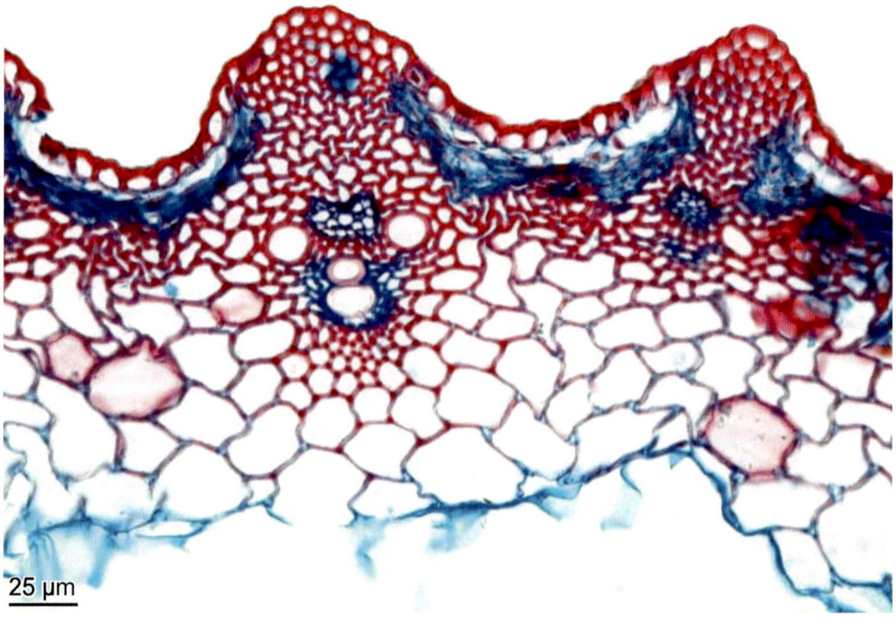

Elymus dentatus (Hook.f.) Tzvelev
Poaceae

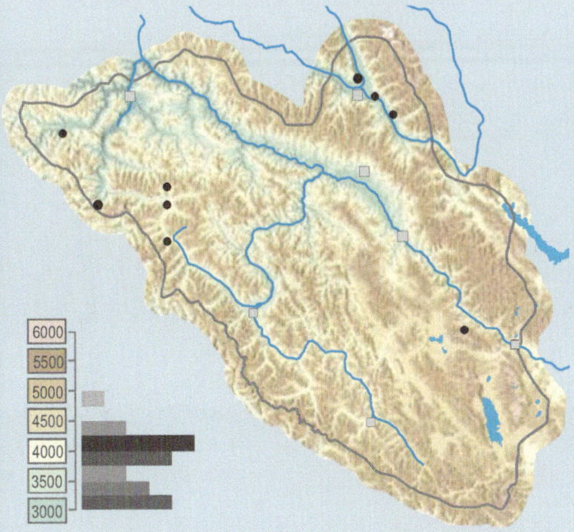

Syn.: *Agropyron dentatum* Hook.f.

Description: Perennial herbs, 25–80 cm tall, cespitose, with creeping rhizome, leaves flat, 10–30 cm long, 4–10 mm wide, spike 6–15 cm long, awns up to 4 mm long.

Distribution: Pakistan, India (Jammu and Kashmir).

Habitat: Steppes.

Culm uniform, round, hollow, with a diameter of 1.4–2.0 mm. Epidermis thin- to thick-walled. Continuous, 2–3-seriate peripheral sclerenchymatic belt of thin- to thick-walled fibers centripetal of the chlorenchyma. Girders square to rectangular. Chlorenchyma groups tangentially oval. Intercellulars in the parenchymatic zone very small and triangular. Vascular bundles arranged in two offset circular rows. Large bundles within the thin-walled parenchymatic tissue, small bundles centripetal of the sclerenchymatic belt. Vascular bundles with large vessels, diameter 12–18 µm. Phloem oval, consisting of sieve tubes and companion cells. Distinct intercellulars centripetal of the protoxylem. Vascular bundles surrounded by a 1–2-seriate sheath of polygonal, thin- to thick-walled cells.

Analyzed: 2 individuals

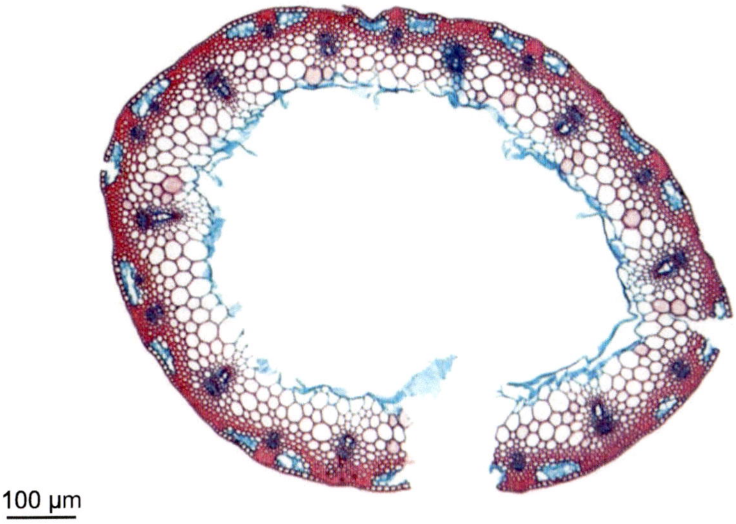

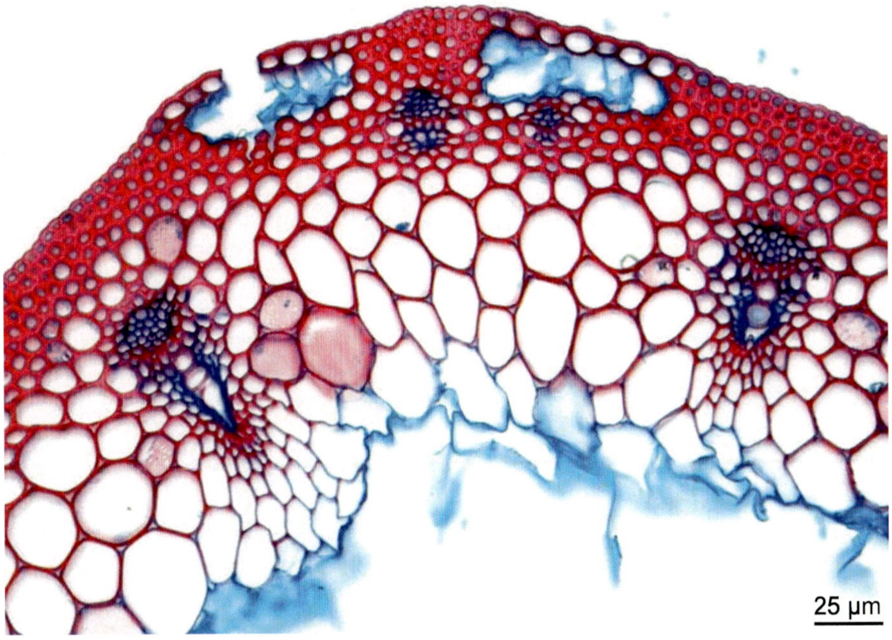

Elymus fedtschenkoi Tzvelev

Poaceae

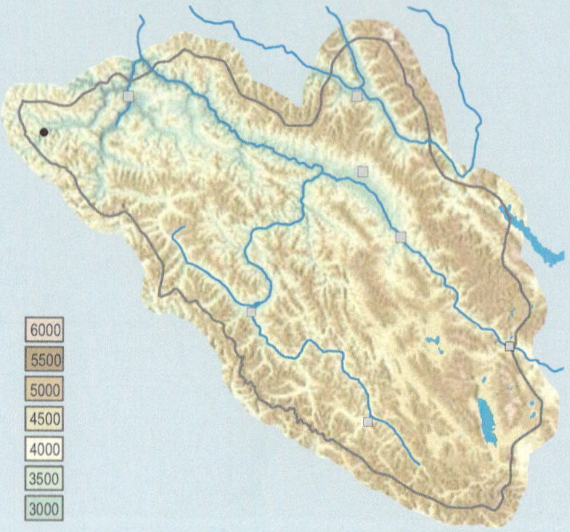

Syn.: *Agropyron curvatum* Nevski

Description: Perennial herbs, 50–100 cm tall, cespitose, with creeping rhizome, leaves flat, 15–30 cm long, 5–12 mm wide, spike 7–15 cm long, awns 25–35 mm long, reflexed.

Distribution: Russia, Kazakhstan, Turkmenistan, Uzbekistan, Kyrgyzstan, Afghanistan, Pakistan, India, China (Xinjiang).

Habitat: Alpine meadows.

Culm uniform, round, hollow, with a diameter of 2–3 mm. Epidermis thin- to thick-walled. Continuous, 2–3-seriate peripheral sclerenchymatic belt of thin- to thick-walled fibers. Girders absent. Small chlorenchyma groups tangentially oval. Intercellulars in the parenchymatic zone small and triangular. Vascular bundles arranged in 2–3 offset circular rows. Large bundles within the thin-walled parenchymatic tissue, small bundles within the sclerenchymatic belt. Vascular bundles with large vessels, diameter 25–35 μm. Phloem oval, consisting of sieve tubes and companion cells. Distinct intercellulars centripetal of the protoxylem. Vascular bundles surrounded by an eccentric, 1–4-seriate sheath of polygonal, thin- to thick-walled cells.

Analyzed: 2 individuals

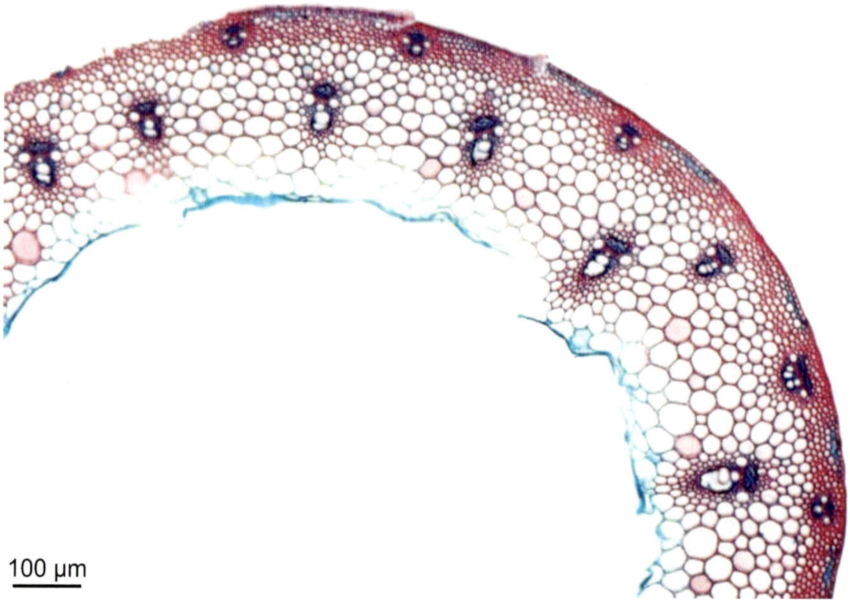

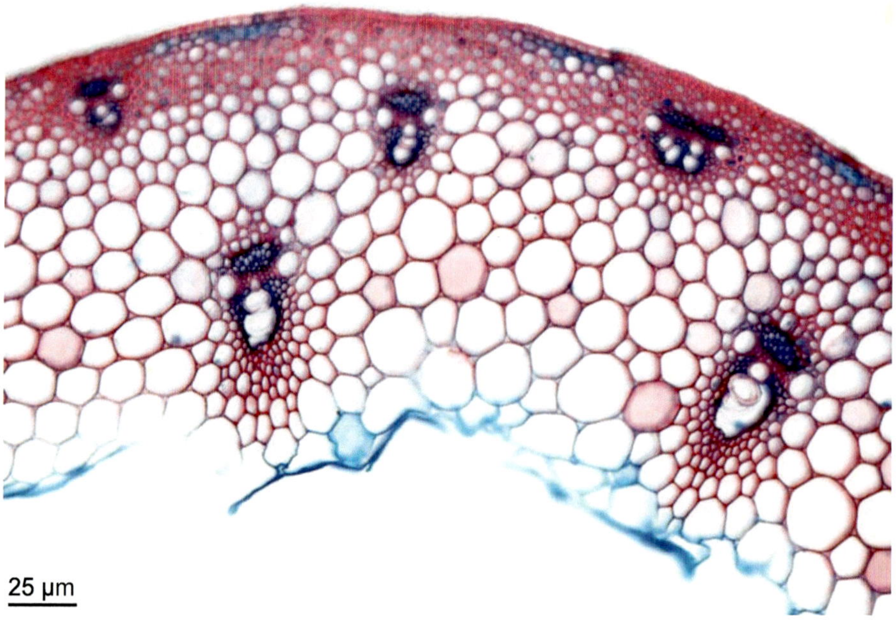

Elymus jacquemontii (Hook.f.) Tzvelev

Poaceae

Syn.: *Agropyron jacquemontii* Hook.f., *Roegneria jacquemontii* (Hook.f.) Nevski

Description: Perennial herbs, 15–20 cm tall, cespitose, leaves glaucous, involute, spike 4–11 × 0.8–2 cm, awns strongly curved, 2–5 cm long.

Distribution: Pakistan, NW India, China (Xinjiang, Xizang).

Habitat: Disturbed and stabilized slopes, mesic turf, pastures, among stones, weedy in barley and oat fields.

Culm uniform, round, hollow, with a diameter of 0.9–1.1 mm. Epidermis thin- to thick-walled. Continuous, 2–3-seriate peripheral sclerenchymatic belt of thin- to thick-walled fibers centripetal of the sclerenchyma belt. Girders conical and square. Chlorenchyma forms an almost continuous belt, interrupted by few girders. Intercellulars in the parenchymatic zone small and triangular. Vascular bundles arranged in 2 offset circular rows. Large bundles within the thin-walled parenchymatic tissue, small bundles within the sclerenchymatic belt. Vascular bundles with large vessels, diameter 2–25 µm. Phloem oval, consisting of uniform cells. Few intercellulars centripetal of the protoxylem. Vascular bundles surrounded by a 1–2-seriate sheath of polygonal, thin- to thick-walled cells.

Analyzed: 2 individuals

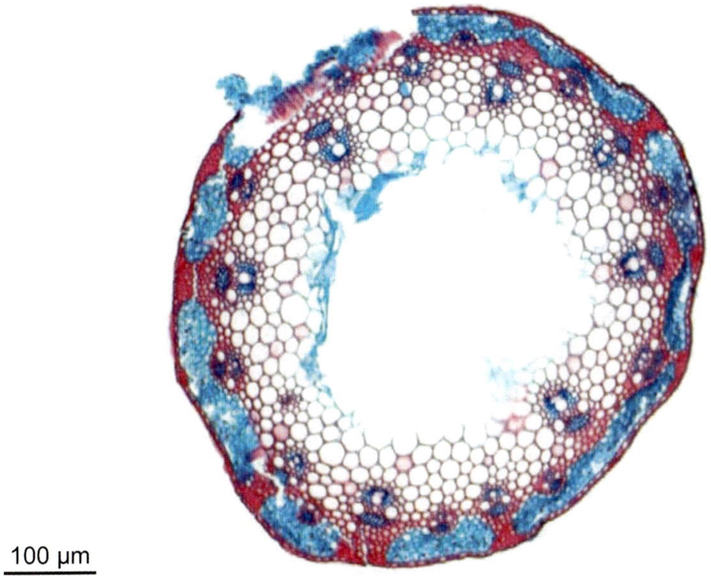

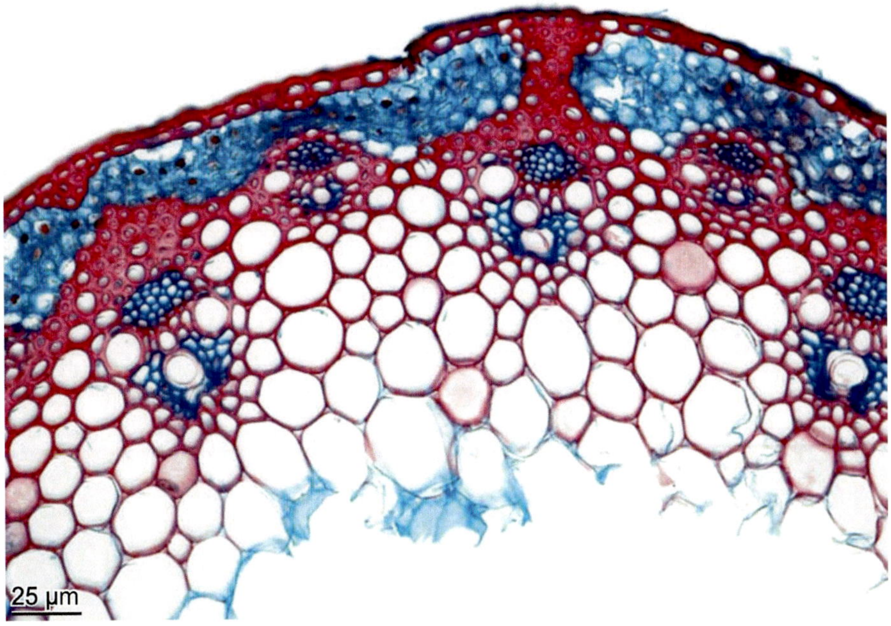

Elymus nutans Griseb.

Poaceae

Syn.: *Clinelymus nutans* (Griseb.) Nevski

Description: Perennial herbs, 25–90 cm tall, cespitose, leaves flat, spike flexuous and often nodding, awns 1.5–3 cm long.

Distribution: Iran, Russia, India, Nepal, China (Gansu, Hebei, Henan, Nei Mongol, Ningxia, Qinghai, Shaanxi, Sichuan, Xinjiang, Xizang, Yunnan), Bhutan, Mongolia, Japan.

Habitat: Dry to mesic gravel of larger rivers.

Culm uniform, round, hollow, with a diameter of 3–4 mm. Epidermis thin- to thick-walled. Continuous, 4–6-seriate peripheral sclerenchymatic belt of thick-walled fibers. Girders tangentially very enlarged. Chlorenchyma groups oval. Intercellulars in the parenchymatic zone small and triangular. Vascular bundles arranged in 2–3 offset circular rows. Large bundles within the thin-walled parenchymatic tissue, small bundles within the sclerenchymatic belt. Vascular bundles with large vessels, diameter 15–20 µm. Phloem oval, consisting of sieve tubes and companion cells. Distinct intercellulars centripetal of the protoxylem. Vascular bundles surrounded by an eccentric, 1–3-seriate sheath of polygonal, thin- to thick-walled cells.

Analyzed: 2 individuals

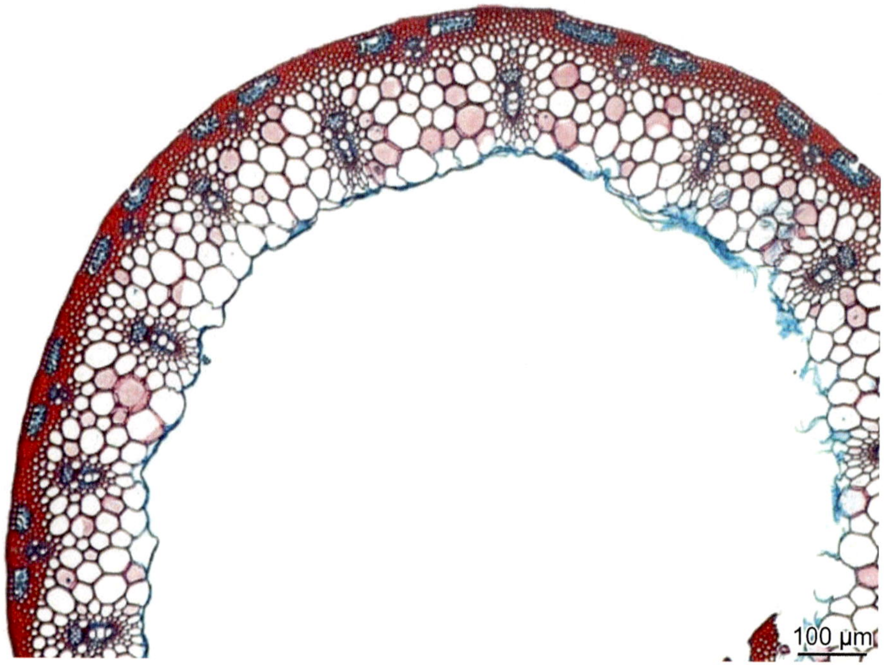

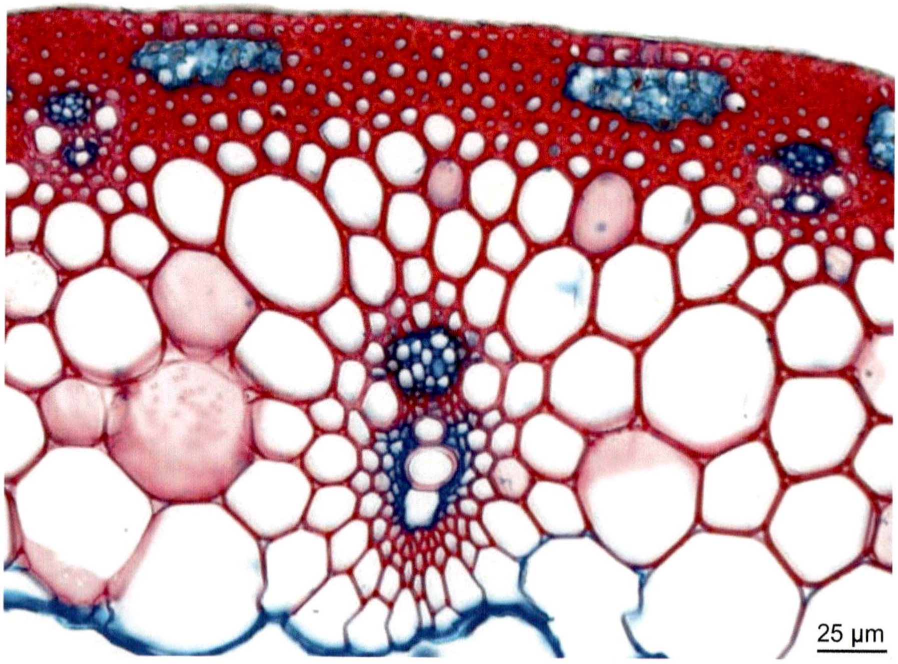

Elymus repens (L.) Gould

Poaceae

Syn.: *Triticum repens* L., *Elytrigia repens* (L.) Nevski, *Agropyron repens* (L.) Beauv.

Description: Perennial herbs, 30–100 cm tall, cespitose, rhizomatous, leaves flat, 6–30 cm long, 3–10 mm wide, spike lax or dense, 5–15 cm long.

Distribution: N Africa, Europe, temperate Asia; naturalized iNorth America and Australasia.

Habitat: Meadows, pastures, weedy in fields.

Culm uniform, round, wavy, hollow, with a diameter of 1.5–2.0 mm. Epidermis thin- to thick-walled. Continuous, 4–6-seriate peripheral sclerenchymatic belt of thin- to thick-walled fibers. Girders absent. Oval to irregularly formed chlorenchyma groups. Intercellulars in the parenchymatic zone small and triangular. Vascular bundles arranged in 2 slightly offset circular rows. Large bundles within the thin-walled parenchymatic tissue, small bundles within the sclerenchymatic belt. Vascular bundles with large vessels, diameter 30–40 µm. Phloem oval, consisting of sieve tubes and companion cells. Distinct intercellulars centripetal of the protoxylem. Vascular bundles surrounded by an eccentric, 1–3-seriate sheath of polygonal, thin- to thick-walled cells.

Analyzed: 2 individuals

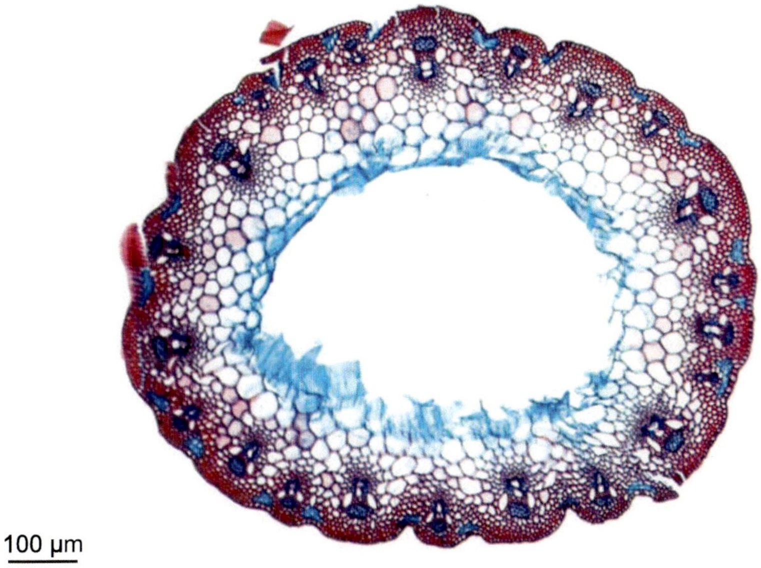

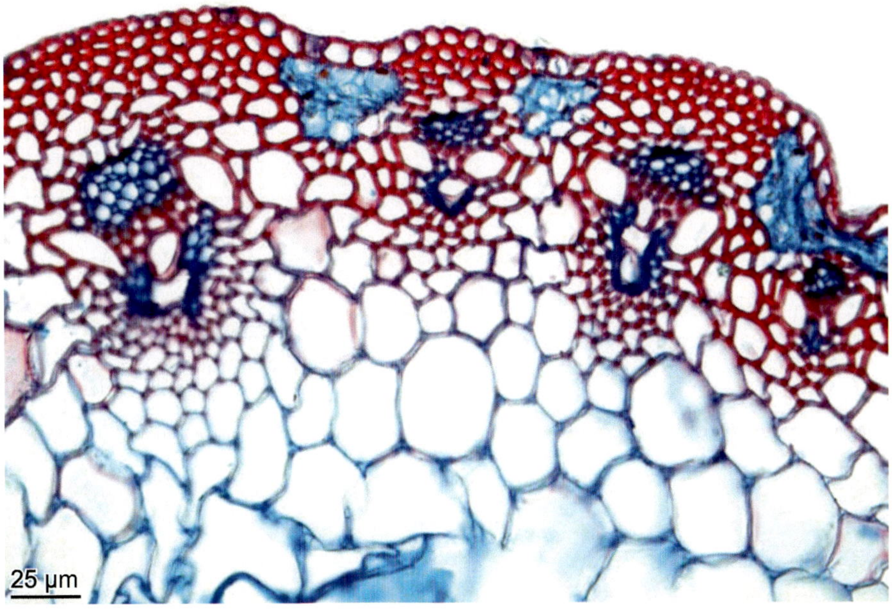

Elymus schrenkianus (Fisch. & C.A.Mey.) Tzvelev

Poaceae

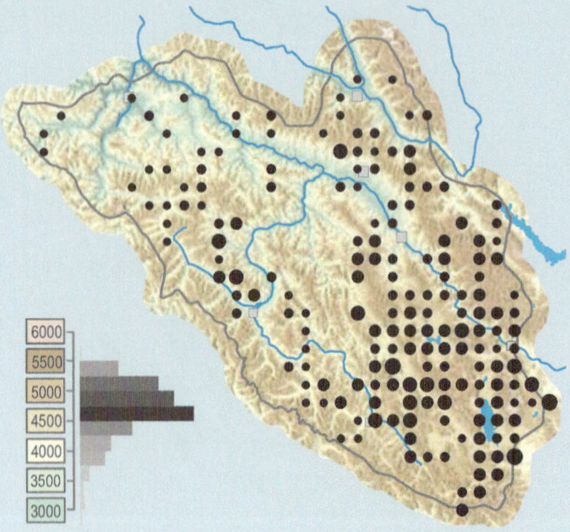

Syn.: *Triticum schrenkianum* Fisch. & C.A.Mey., *Roegneria schrenkiana* (Fisch. & C.A.Mey.) Nevski, *Agropyron schrenkianum* (Fisch. & C.A.Mey.) P.Candargy

Description: Perennial herbs, 20–80 cm tall, cespitose, leaves flat or loosely convolute, spike nodding, awns 1.5–2 cm long.

Distribution: Russia, Kazakhstan, Turkmenistan, Uzbekistan, Kyrgyzstan, NE Pakistan, N India, Nepal, China (Qinghai, Xinjiang, Xizang), Bhutan, Mongolia.

Habitat: Dry to mesic slopes, gravel deposits, unstable screes, stony habitats.

Culm uniform, round, hollow, with a diameter of 1.0–1.3 mm. Epidermis thin- to thick-walled. Continuous, 4–5-seriate peripheral sclerenchymatic belt of thin- to thick-walled fibers. Girders absent. Chlorenchyma groups reduced to tangential uniseriate rows. Intercellulars in the parenchymatic zone small and triangular. Vascular bundles arranged in 2 slightly offset circular rows. Large bundles within the thin-walled parenchymatic tissue, small bundles within the sclerenchymatic belt. Vascular bundles with large vessels, diameter 20–25 μm. Phloem oval, consisting of uniform cells. Distinct intercellulars centripetal of the protoxylem. Vascular bundles surrounded by an eccentric, 1–3-seriate sheath of polygonal, fairly thin-walled cells.

Analyzed: 1 individual

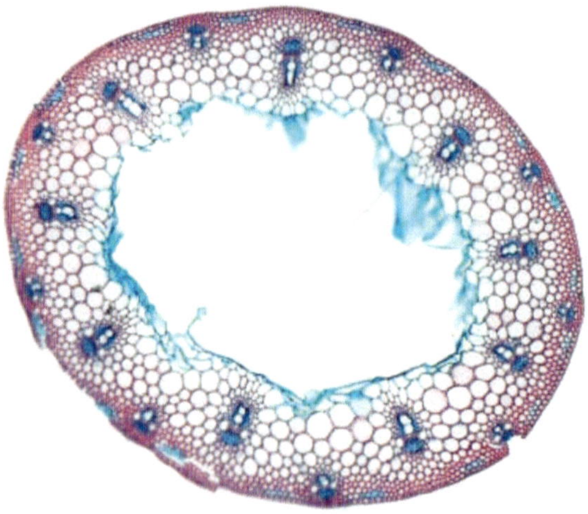

100 μm

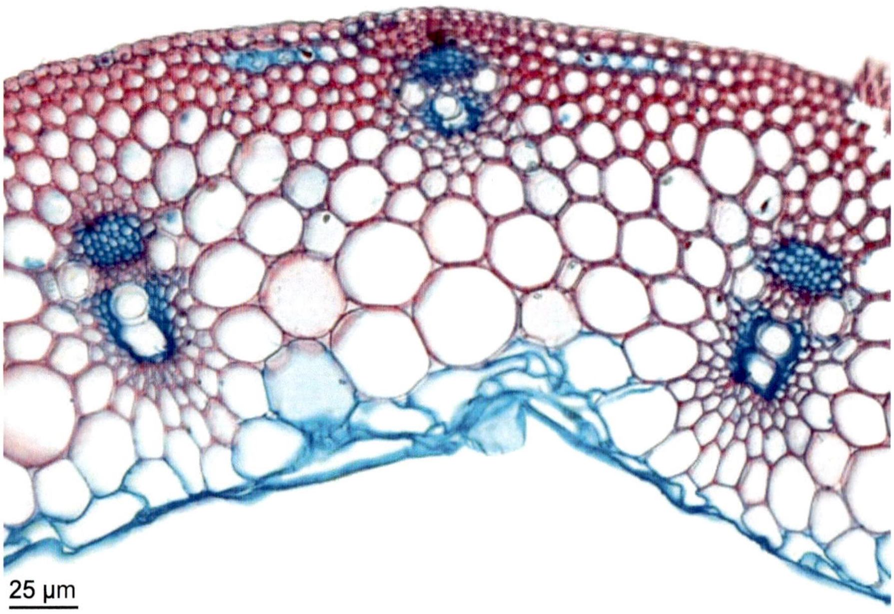

25 μm

Elymus schugnanicus (Nevski) Tzvelev

Poaceae

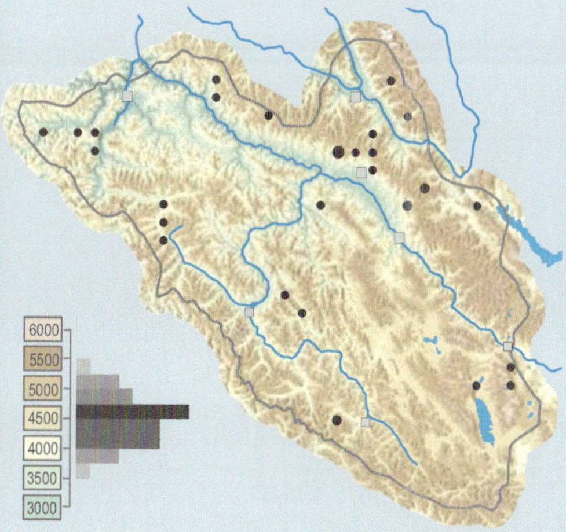

Syn.: *Roegneria schugnanica* (Nevski) Nevski

Description: Perennial herbs, 15–50 cm tall, cespitose, leaves flat or convolute, 4–15 cm long, 2–3 mm wide, densely hairy on both sides, spike lax, 7–10 cm long, nodding, awns strongly curved, ca. 3 cm long.

Distribution: NE Iran, Tajikistan, Afghanistan, N Pakistan, NW India.

Habitat: Stabilized slopes, gravel deposits, stony habitats.

Culm uniform, round, hollow, with a diameter of 1.0–1.2 mm. Epidermis thin- to thick-walled. Continuous, 2–3-seriate peripheral sclerenchymatic belt of thin- to thick-walled fibers centripetal of the sclerenchyma belt. Girders conical. Chlorenchyma forms an almost continuous belt, interrupted by few girders. Intercellulars in the parenchymatic zone small and triangular. Vascular bundles arranged in 2 offset circular rows. Large bundles within the thin-walled parenchymatic tissue, small bundles within the sclerenchymatic belt. Vascular bundles with large vessels, diameter 2–25 µm. Phloem oval, consisting of uniform cells. Few intercellulars centripetal of the protoxylem. Vascular bundles surrounded by a 1–2-seriate sheath of polygonal, thin- to thick-walled cells.

Analyzed: 1 individual

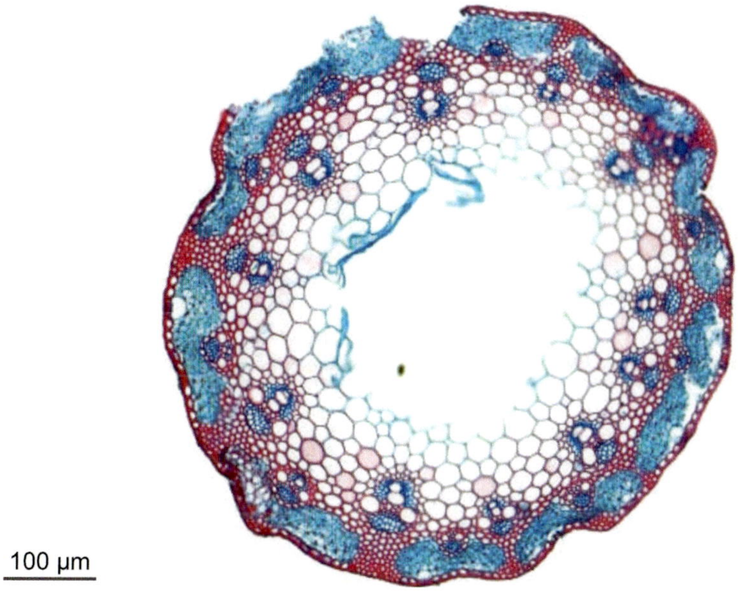

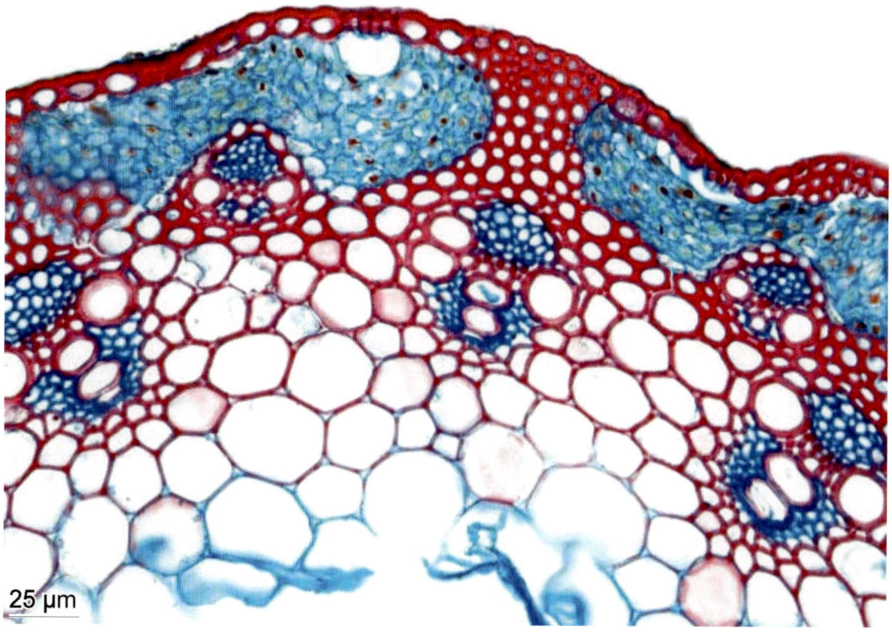

Elymus stewartii (Melderis) Cope
Poaceae

Syn.: *Agropyron stewartii* Melderis

Description: Perennial herbs, 35–70 cm tall, cespitose, leaves flat, 5–15 cm long, 15–25 mm wide, spike lax, 7–11 cm long, awns 1–3 mm long.

Distribution: N Pakistan, India (Jammu and Kashmir).

Habitat: Steppes.

Culm uniform, round, hollow, with a diameter of 3–4 mm. Epidermis thin- to thick-walled. Continuous, 4–8-seriate peripheral sclerenchymatic belt of thick-walled fibers centripetal of the chlorenchyma. Girders and chlorenchyma groups tangentially enlarged. Intercellulars in the parenchymatic zone small and triangular. Vascular bundles arranged in 2 offset circular rows. Large bundles within the thin-walled parenchymatic tissue, small bundles within the sclerenchymatic belt. Vascular bundles with large vessels, diameter 30–40 µm. Phloem oval, consisting of sieve tubes and companion cells. Intercellulars centripetal of the protoxylem absent. Vascular bundles surrounded by a 1–2-seriate sheath of polygonal, thin- to thick-walled cells.

Analyzed: 2 individuals

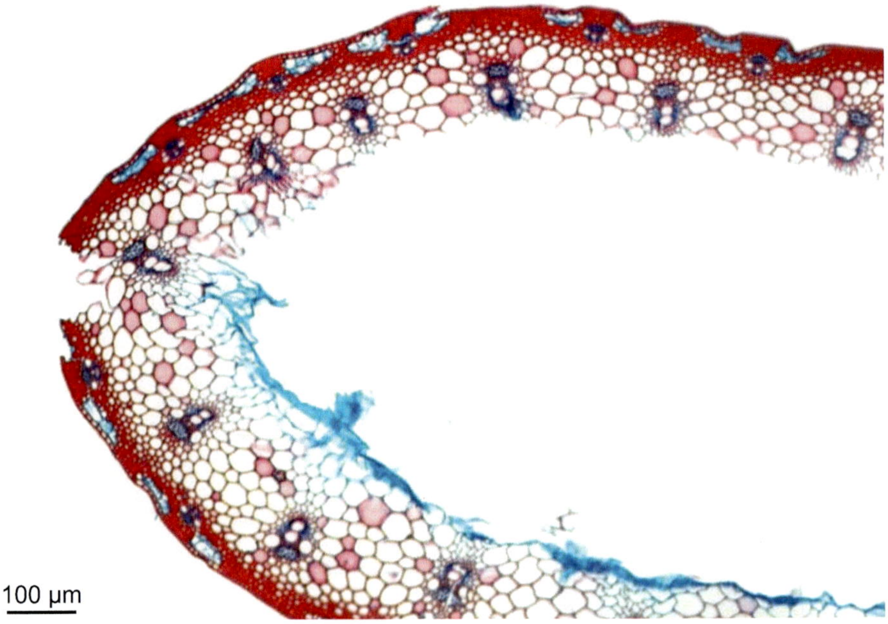

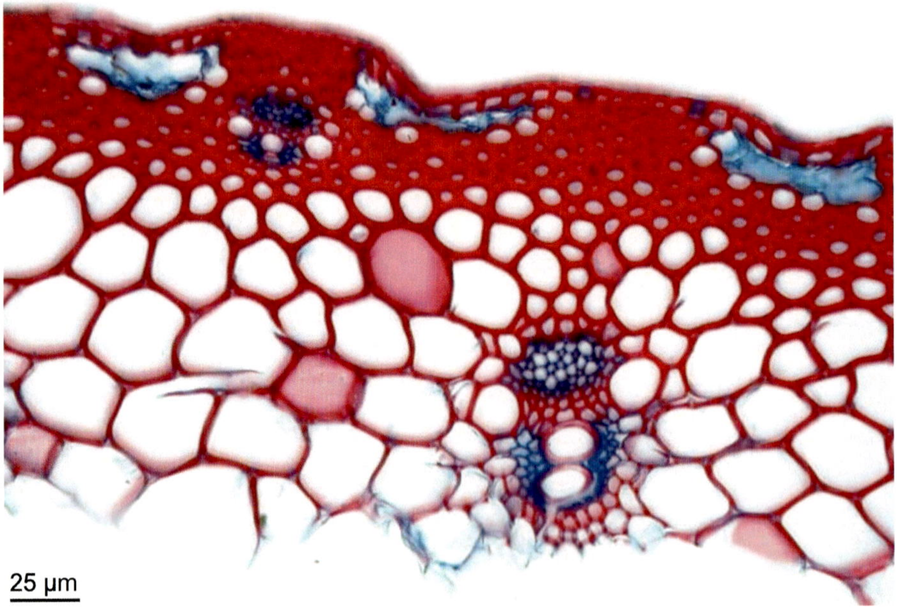

Elymus thomsonii (Hook.f.) Melderis

Poaceae

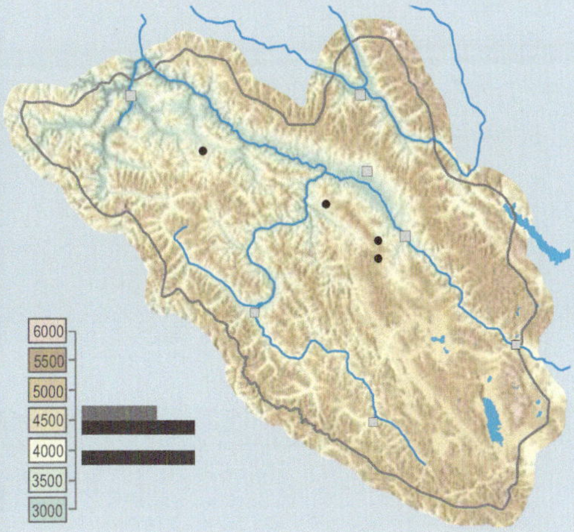

Syn.: *Agropyron thomsonii* Hook.f.

Description: Perennial herbs, 30–90 cm tall, cespitose, leaves flat, spike straight, 5–11 cm long, lemmas silky villous, awns 2–5 mm long.

Distribution: India.

Habitat: Gravel deposits, unstable screes.

Culm uniform, round, hollow, with a diameter of 1.4–1.6 mm. Epidermis thin- to thick-walled. Continuous, 1–2-seriate peripheral sclerenchymatic belt of thin- to thick-walled fibers centripetal of girders and chlorenchyma. Girders square to tangentially rectangular. Chlorenchyma groups round to oval. Vascular bundles arranged in 2 offset circular rows. Large bundles within the thin-walled parenchymatic tissue, small bundles within the sclerenchymatic belt. Vascular bundles with large vessels, diameter 30–40 µm. Phloem oval, consisting of sieve tubes and companion cells. Few intercellulars centripetal of the protoxylem. Vascular bundles surrounded by a uniseriate sheath of polygonal, fairly thin-walled cells.

Analyzed: 2 individuals

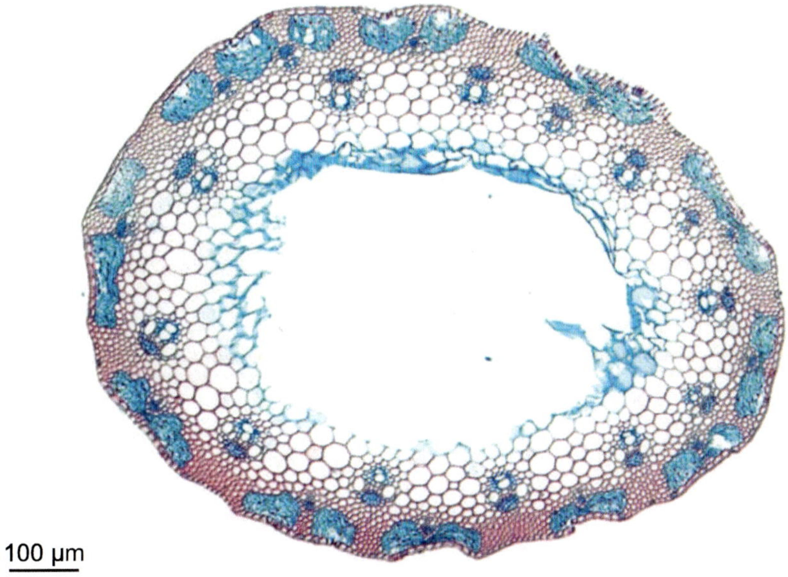

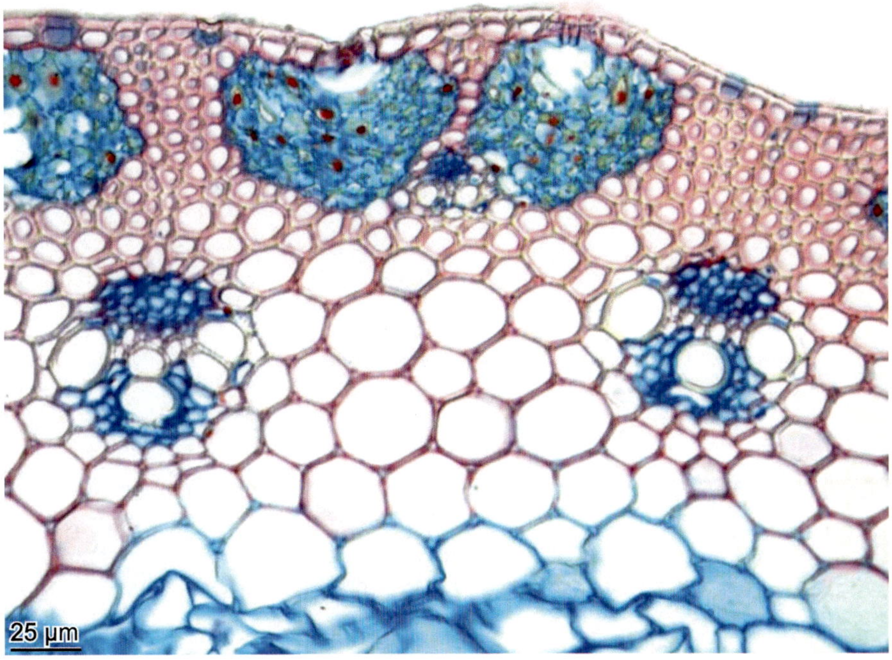

Eremopoa altaica subsp. *songorica*

Poaceae (Schrenk ex Fisch. & C.A.Mey.) Tzvelev

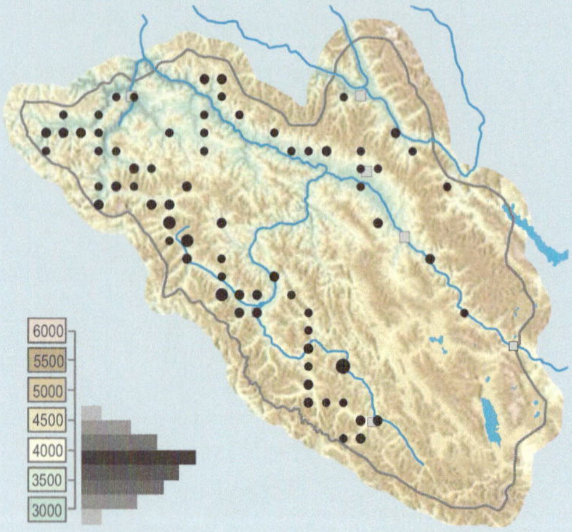

Syn.: *E. songorica* (Schrenk ex Fisch. & C.A.Mey.) Roshev.

Description: Annuals, 5–40 cm tall, cespitose, leaves linear, 2–10 cm long, panicle ovate, dense to very loose.

Distribution: Afghanistan, Iran, Turkey, Armenia, Azerbaijan, Georgia, Russia (Caucasus, Siberia), Kazakhstan, Turkmenistan, Kyrgyzstan, Tajikistan, Pakistan, India (Jammu and Kashmir).

Habitat: Open sandy and moist places.

Culm uniform, round, hollow, with a diameter of 0.9–1.1 mm. Epidermis thick-walled. Continuous, 4–8-seriate peripheral sclerenchymatic belt of thick-walled fibers. Girders conical to tangentially enlarged. Chlorenchyma groups oval. Intercellulars in the parenchymatic tissue absent. Vascular bundles arranged in 2 offset circular rows. Large bundles within the thin- to thick-walled parenchymatic tissue, small bundles within the sclerenchymatic belt. Vascular bundles with large vessels, diameter 5–6 μm. Phloem oval, consisting of uniform cells. Large intercellulars centripetal of the protoxylem. Vascular bundles surrounded by a 1–2-seriate sheath of polygonal, thin- to thick-walled cells.

Analyzed: 2 individuals

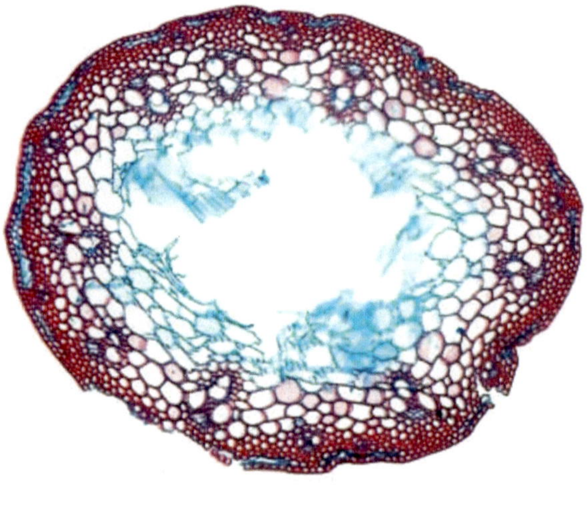

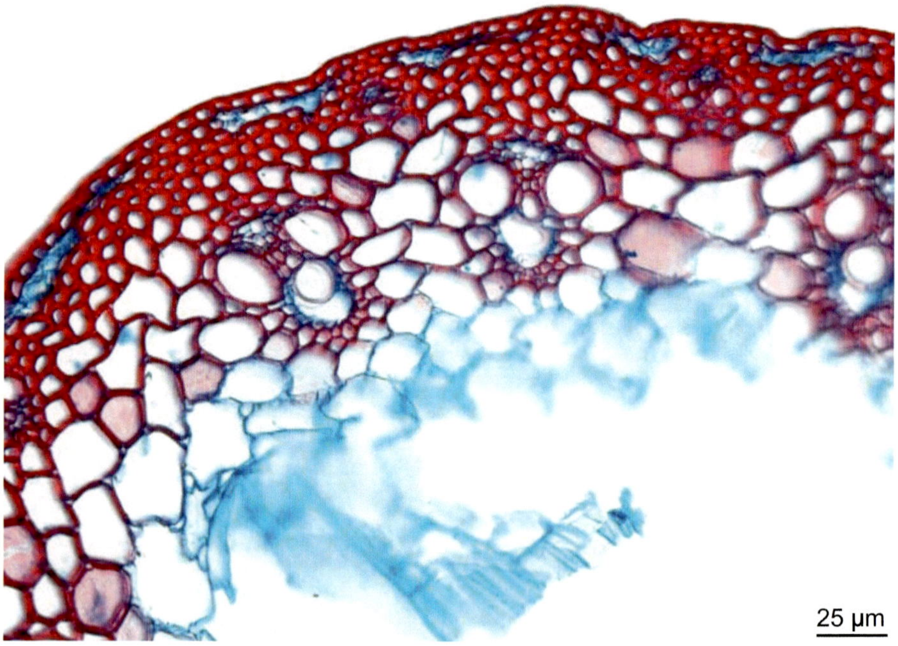

Festuca alaica Drobov

Poaceae

Description: Perennial herbs, 10–30 cm tall, densely cespitose, leaves folded-setaceous, panicle oblong, contracted, 2.5–6 cm long.

Distribution: Turkey, Iran, Kyrgyzstan, Tajikistan, Afghanistan, N Pakistan, N India.

Habitat: Stony slopes.

Culm uniform, round, hollow, with a diameter of 0.5–0.6 mm. Epidermis thin- to thick-walled. Continuous, 4–8-seriate peripheral sclerenchymatic belt of thin- to thick-walled fibers. Girders absent. Chlorenchyma groups tangentially enlarged, 1–2-seriate. Intercellulars in the parenchymatic tissue absent. Vascular bundles arranged in 2 offset circular rows. Large bundles within the thin- to thick-walled parenchymatic tissue, small bundles within the sclerenchymatic belt. Vascular bundles with large vessels, diameter 6–10 μm. Phloem oval, consisting of uniform cells. Intercellulars centripetal of the protoxylem absent. Vascular bundles surrounded by a uniseriate, often indistinct sheath of polygonal, thin- to thick-walled cells.

Analyzed: 2 individuals

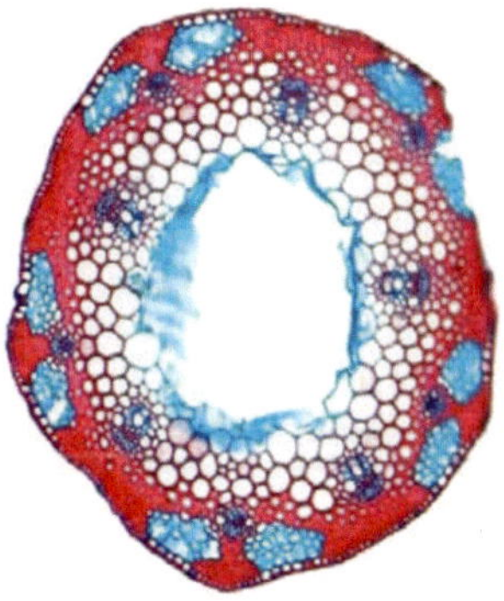

100 μm

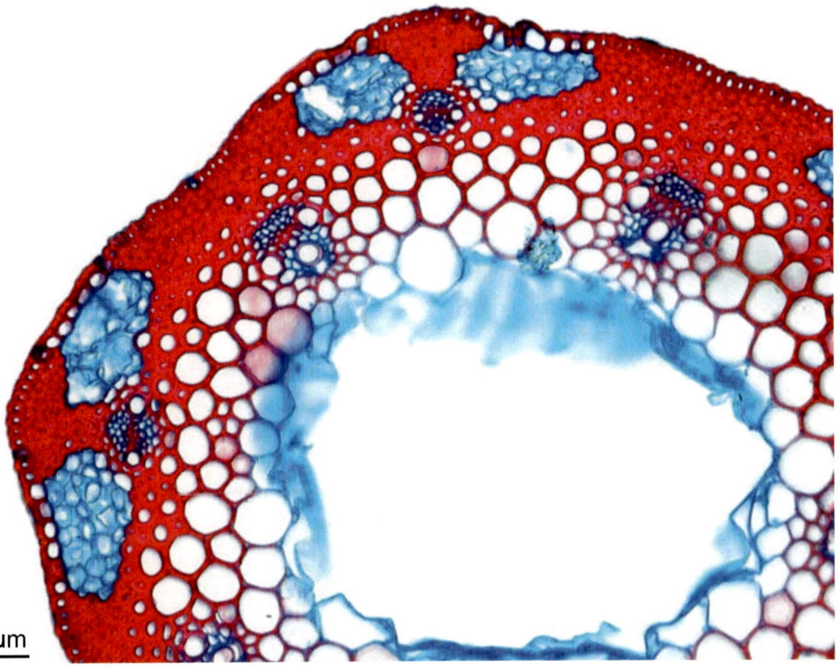

25 μm

Festuca alatavica (St.-Yves) Roshev.

Poaceae

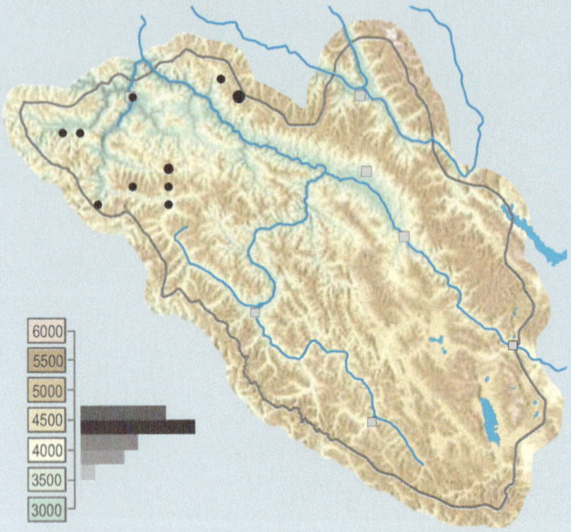

Syn.: *F. tianschanica* Roshev., *F. rubra* subsp. *alatavica* St.-Yves

Description: Perennial herbs, 30–80 cm tall, densely cespitose, leaves flat or involute, up to 20 cm long, 2–3 mm wide, panicle broadly ovate in outline, very loose, 5–15 cm long, branches 3–6 cm long, spreading.

Distribution: E Kazakhstan, Kyrgyzstan, Tajikistan, N Pakistan, N India, China (Xinjiang).

Habitat: Stony slopes, grasslands.

Culm uniform, round, hollow, with a diameter of 1.3–1.5 mm. Epidermis thin- to thick-walled. Continuous, 4–8-seriate peripheral sclerenchymatic belt of thin- to thick-walled fibers. Girders absent. Chlorenchyma groups tangentially enlarged, 1–2-seriate. Intercellulars in the parenchymatic tissue absent. Vascular bundles arranged in 2 offset circular rows. Large bundles within the thin- to thick-walled parenchymatic tissue, small bundles within the sclerenchymatic belt. Vascular bundles with large vessels, diameter 6–10 µm. Phloem oval, consisting of uniform cells. Intercellulars centripetal of the protoxylemare absent. Vascular bundles surrounded by a uniseriate, often indistinct sheath of polygonal, thin- to thick-walled cells.

Analyzed: 2 individuals

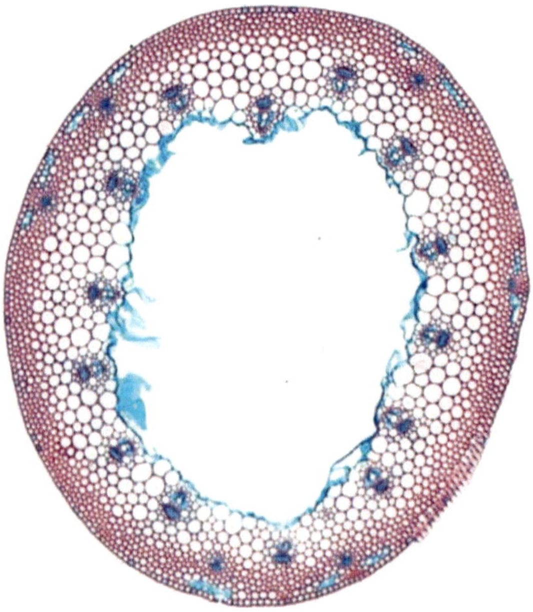

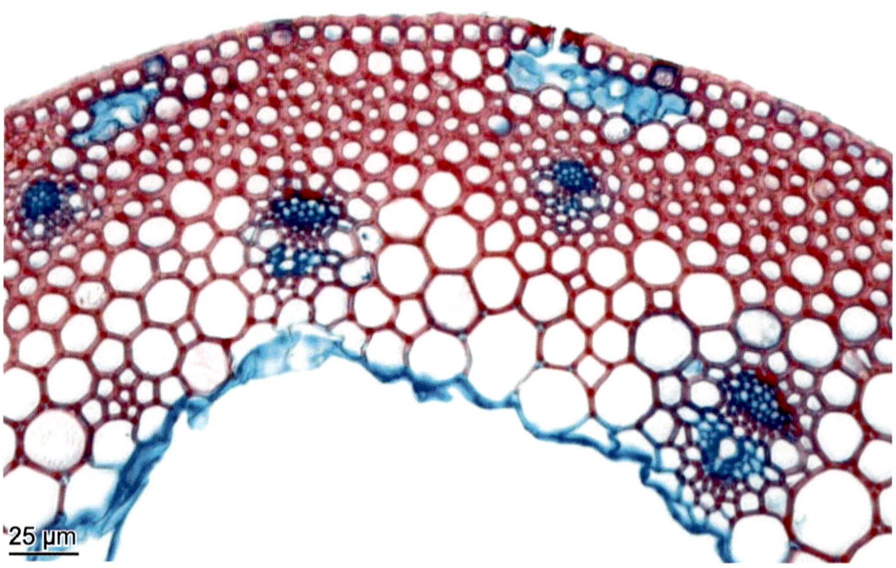

Festuca coelestis (St.-Yves) Krecz. & Bobrov
Poaceae

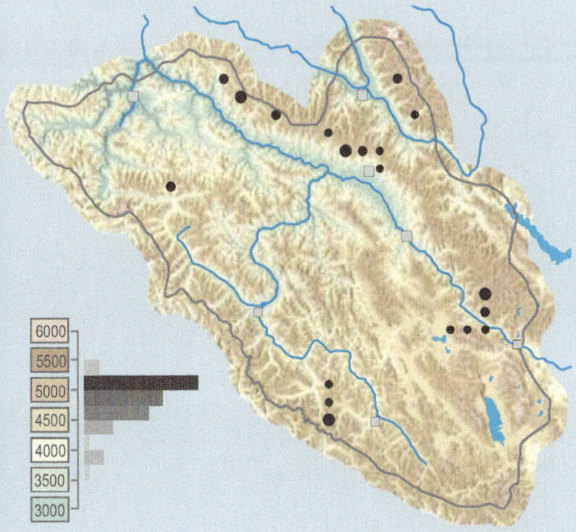

Description: Perennial herbs, 3–30 cm tall, densely cespitose, leaves folded-setaceous, panicle narrowly oblong, open, 1–5 cm long.

Distribution: E Kazakhstan, Kyrgyzstan, Tajikistan, N Pakistan, NW India, China (Gansu, Hubei, Nei Mongol, Qinghai, Sichuan, Xinjiang, Xizang, Yunnan).

Habitat: Alpine meadows.

Culm uniform, round, hollow, with a diameter of 0.6–0.8 mm. Epidermis thin- to thick-walled, with hairs. Continuous, 4–8-seriate peripheral sclerenchymatic belt of thin- to thick-walled fibers. Girders absent. Chlorenchyma groups oval. Intercellulars in the parenchymatic tissue absent. Vascular bundles arranged in 2 offset circular rows. Large bundles within the thin- to thick-walled parenchymatic tissue, small bundles within the sclerenchymatic belt. Vascular bundles with large vessels, diameter 5–8 µm. Phloem oval, consisting of uniform cells. Intercellulars centripetal of the protoxylem absent. Vascular bundles surrounded by a uniseriate, often indistinct sheath of polygonal, thin- to thick-walled cells.

Analyzed: 2 individuals

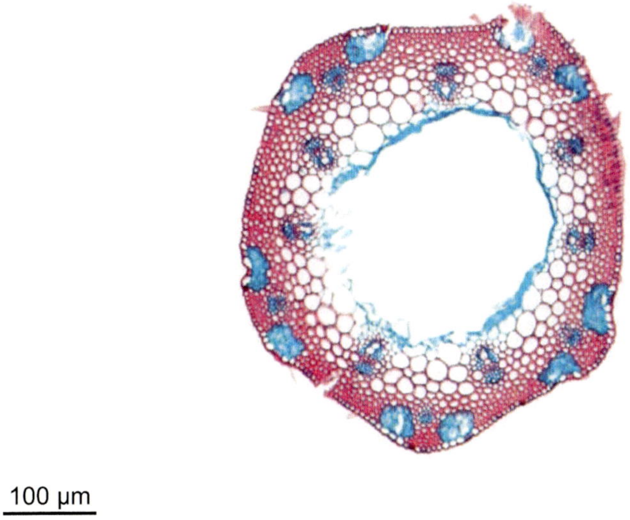

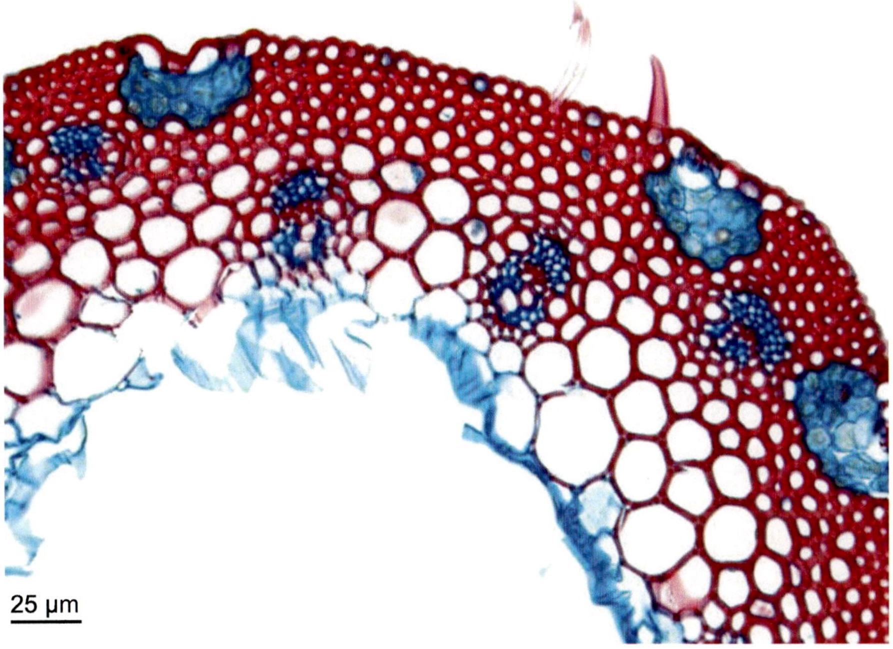

Festuca hartmannii (Markgr.-Dann.) E.B.Alexeev
Poaceae

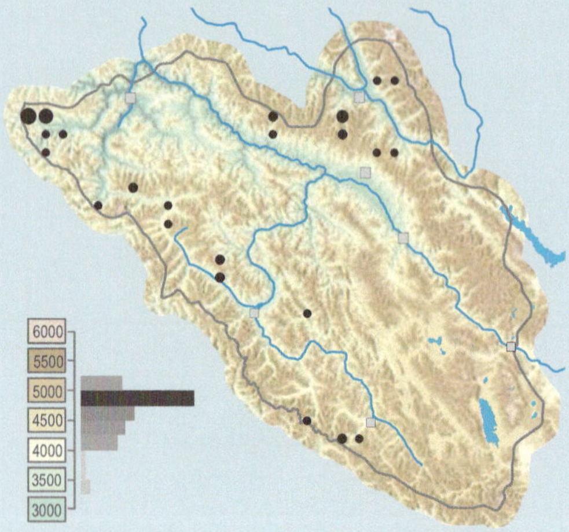

Syn.: *F. ovina* var. *hartmannii* Markgr.-Dannenb.

Description: Perennial herbs, 10–35 cm tall, densely cespitose, leaves folded-setaceous, panicle narrowly oblong, contracted, 2–6 cm long.

Distribution: Pakistan, NW India.

Habitat: Snow beds, spring wetlands.

Culm uniform, round, hollow, with a diameter of 0.5–0.6 mm. Epidermis thin- to thick-walled. Continuous, 4–8-seriate peripheral sclerenchymatic belt of thin- to thick-walled fibers. Girders absent. Chlorenchyma groups tangentially enlarged, 1–2-seriate. Intercellulars in the parenchymatic tissue absent. Vascular bundles arranged in 2 offset circular rows. Large bundles within the thin- to thick-walled parenchymatic tissue, small bundles within the sclerenchymatic belt. Vascular bundles with large vessels, diameter 6–10 μm. Phloem oval, consisting of uniform cells. Intercellulars centripetal of the protoxylem absent. Vascular bundles surrounded by a uniseriate, often indistinct sheath of polygonal, thin- to thick-walled cells.

Analyzed: 2 individuals

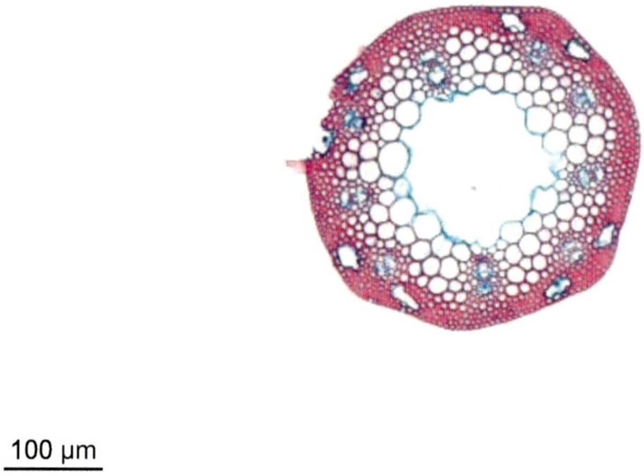

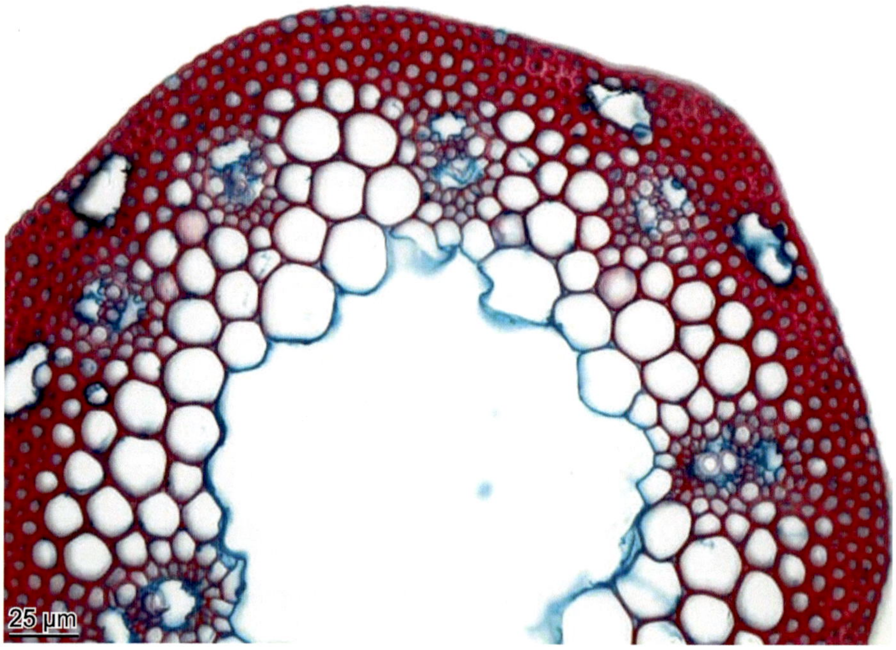

Festuca kashmiriana Stapf

Poaceae

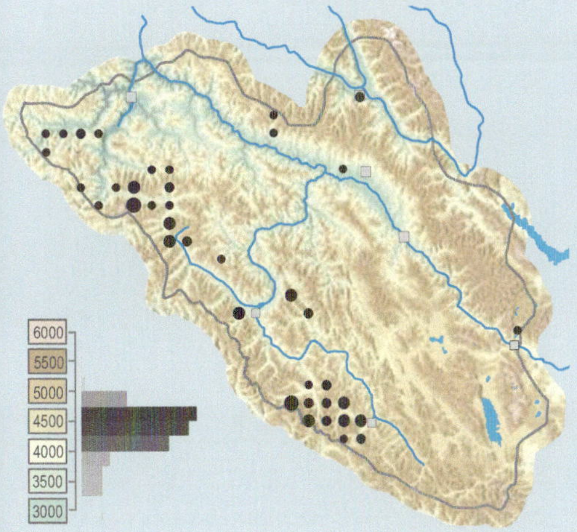

Syn.: *F. rubra* subsp. *kashmiriana* (Stapf) St.-Yves

Description: Perennial herbs, 15–80 cm tall, densely cespitose, leaves filiform, conduplicate, panicle narrowly oblong, loose, 6–10 cm long.

Distribution: N Pakistan, NW India, China (Xizang).

Habitat: Stabilized grassy and stony slopes.

Culm uniform, round, hollow, with a diameter of 0.6–0.8 mm. Epidermis thick-walled. Continuous, 4–6-seriate peripheral sclerenchymatic belt of thick-walled fibers. Girders absent. Chlorenchyma groups oval. Intercellulars in the parenchymatic tissue absent. Vascular bundles arranged in 2 offset circular rows. Large bundles within the thin- to thick-walled parenchymatic tissue, small bundles within the sclerenchymatic belt. Vascular bundles with large vessels, diameter 8–12 µm. Phloem oval, consisting of uniform cells. Few intercellulars centripetal of the protoxylem. Vascular bundles are surrounded by a 1–2-seriate sheath of polygonal, thin- to thick-walled cells.

Analyzed: 2 individuals

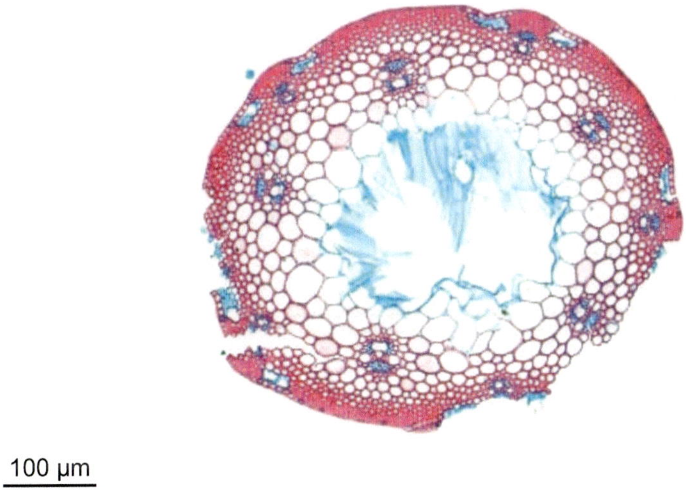

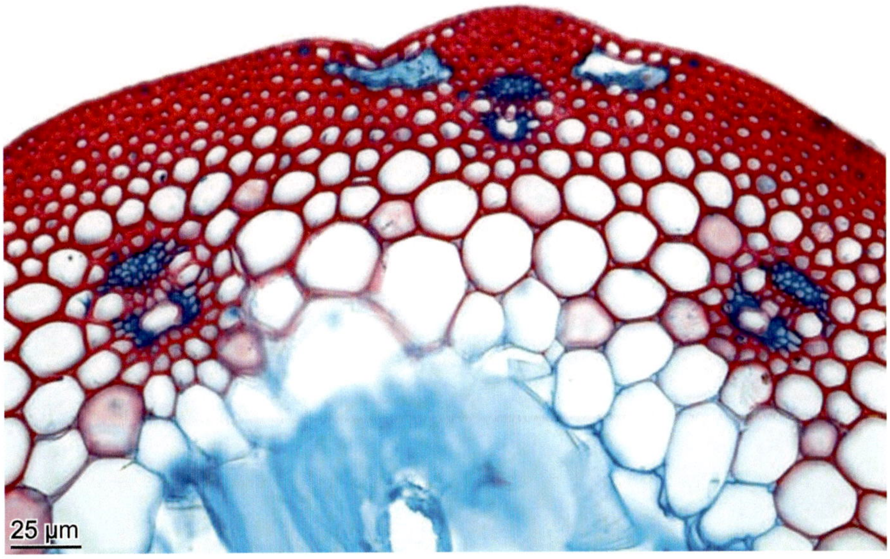

Festuca nitidula Stapf

Poaceae

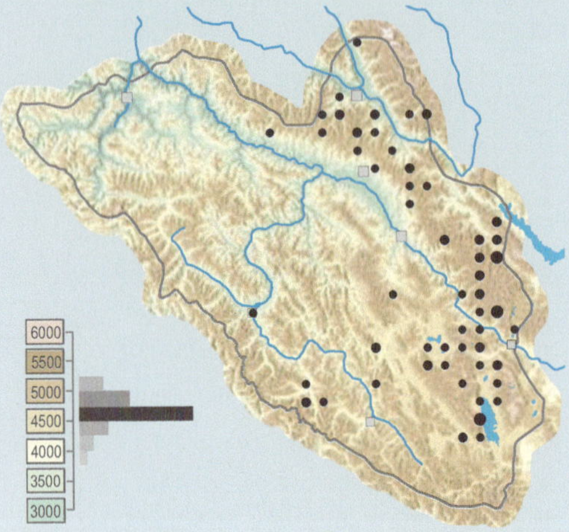

Description: Perennial herbs, 15–50 cm tall, loosely cespitose, leaves setaceous, conduplicate or sometimes flat, panicle loose, open, usually drooping, 4–10 cm long, branches flexuous.

Distribution: Pakistan, NW India, Nepal, China (Gansu, Qinghai, Sichuan, Xizang, Yunnan).

Habitat: Grasslands, sandy and gravel sediments of river banks, wet meadows.

Culm uniform, round, hollow, with a diameter of 0.9–1.1 mm. Epidermis thin- to thick-walled. Continuous, 3–5-seriate peripheral sclerenchymatic belt of thin- to thick-walled fibers. Girders absent. Chlorenchyma groups absent. Intercellulars in the parenchymatic tissue absent. Vascular bundles arranged in 2 offset circular rows. Large bundles within the thin- to thick-walled parenchymatic tissue, small bundles within the sclerenchymatic belt. Vascular bundles with large vessels, diameter 9–12 µm. Phloem oval, consisting of uniform cells. With intercellulars centripetal of the protophloem. Vascular bundles surrounded by a uniseriate sheath of polygonal, thin- to thick-walled cells.

Analyzed: 1 individual

100 µm

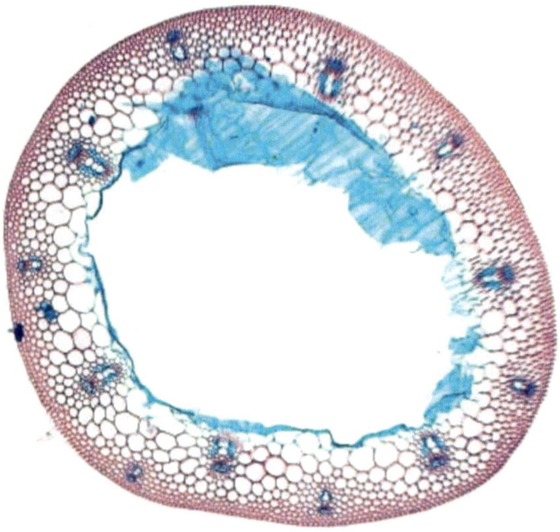

25 µm

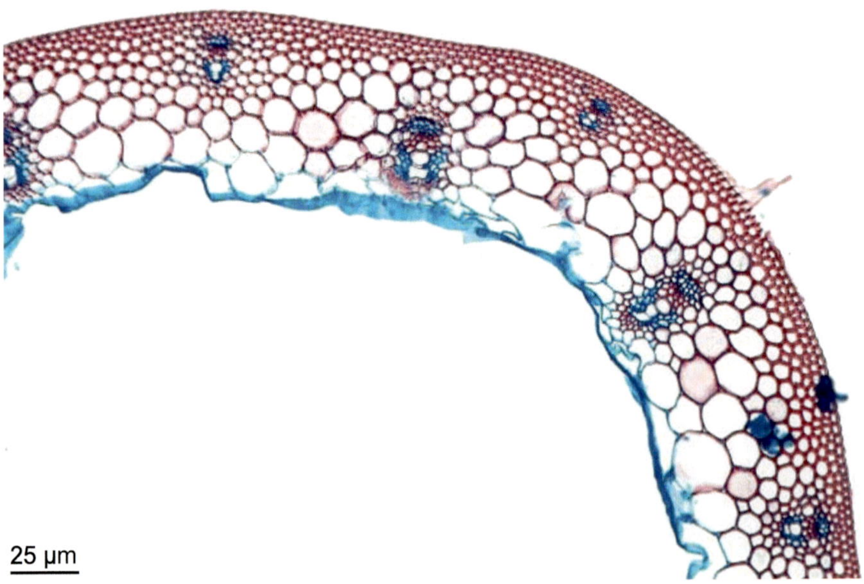

Festuca olgae (Regel) Krivot.

Poaceae

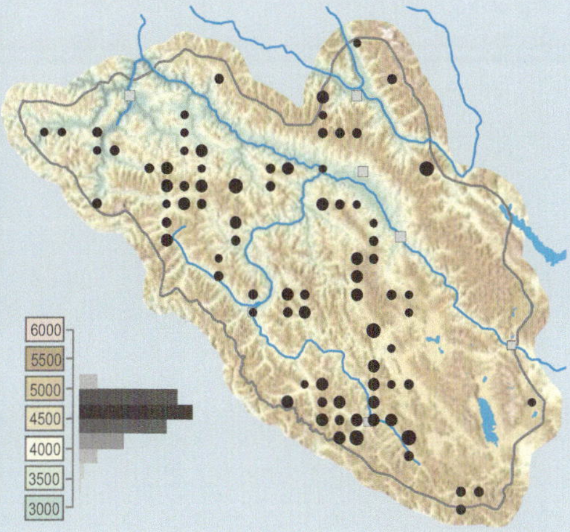

Syn.: *Molinia olgae* Regel, *Colpodium afghanicum* Bor, *F. deasyi* Rendle, *Leucopoa olgae* (Regel) Krecz. & Bobrov

Description: Perennial herbs, 25–80 cm tall, dioecious, densely cespitose, leaves usually flat, panicle lanceolate to ovate, loose, 5–15 cm long.

Distribution: NE Iran, Kyrgyzstan, Tajikistan, Afghanistan, Pakistan, India, China (Xinjiang, Xizang, Yunnan).

Habitat: Stony and gravel slopes.

Culm uniform, round, hollow, with a diameter of 1.4–1.5 mm. Epidermis thin- to thick-walled. Continuous, 6–8-seriate peripheral sclerenchymatic belt of thick-walled fibers. Girders absent. Chlorenchyma groups rare and very small. Intercellulars in the parenchymatic tissue very small. Vascular bundles arranged in 2–3 offset circular rows. Large bundles within the thin- to thick-walled parenchymatic tissue, small bundles within the sclerenchymatic belt. Vascular bundles with large vessels, diameter 20–25 µm. Phloem oval, consisting of sieve tubes and companion cells. Intercellulars centripetal of the protoxylem absent. Vascular bundles surrounded by a 1–2-seriate sheath of polygonal, thin- to thick-walled cells.

Analyzed: 2 individuals

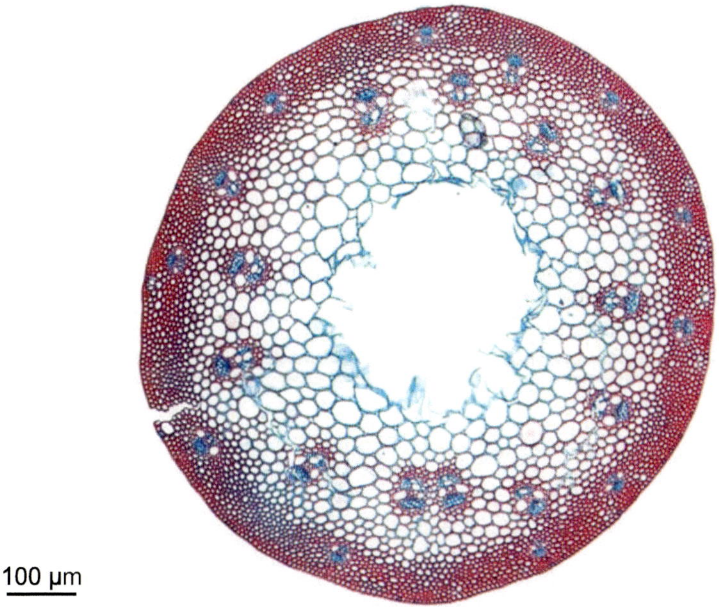

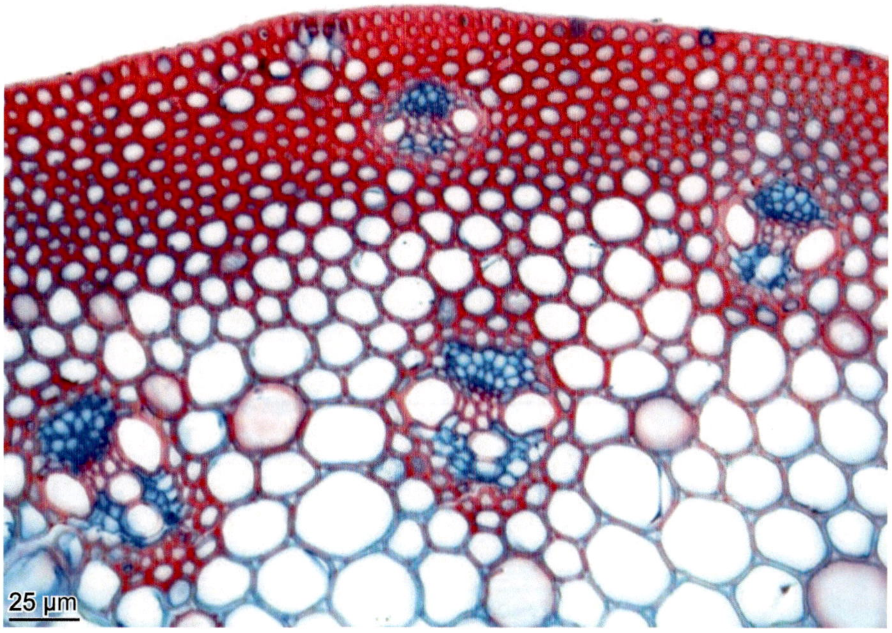

Festuca rubra L.

Poaceae

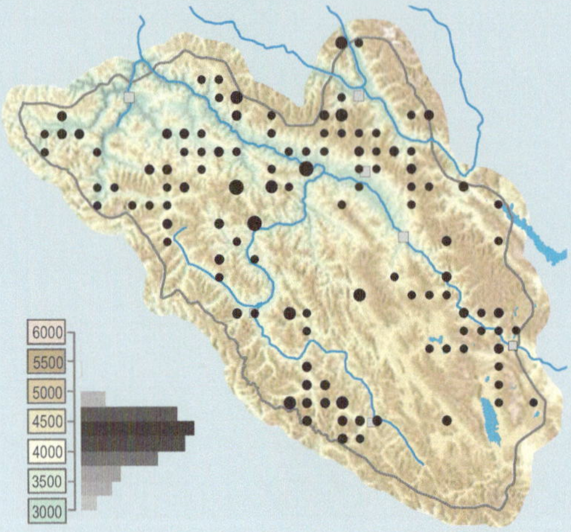

Description: Perennial herbs, 15–60 cm tall, loosely cespitose, shortly rhizomatous, leaves mostly basal, filiform, conduplicate, or flat, 6–30 cm long, 0.4–1 (–4) mm wide, panicle loose, 5–15 cm long, branches 1.5–8 cm long.

Distribution: Widespread in temperate N hemisphere; widely cultivated.

Habitat: Grasslands, alpine meadows.

Culm uniform, round, hollow, with a diameter of 0.6–0.8 mm. Epidermis thin- to thick-walled. Continuous, 3–4-seriate peripheral sclerenchymatic belt of thick-walled fibers. Girders absent. Chlorenchyma groups oval. Intercellulars in the parenchymatic tissue very small. Vascular bundles arranged in one circular row. Large bundles within the thin- to thick-walled parenchymatic tissue, small bundles within the sclerenchymatic belt. Vascular bundles with large vessels, diameter 5–8 µm. Phloem oval, consisting of sieve tubes and companion cells. Intercellulars centripetal of the protoxylem absent. Vascular bundles surrounded by a 1–2-seriate sheath of polygonal, thin- to thick-walled cells.

Analyzed: 1 individual

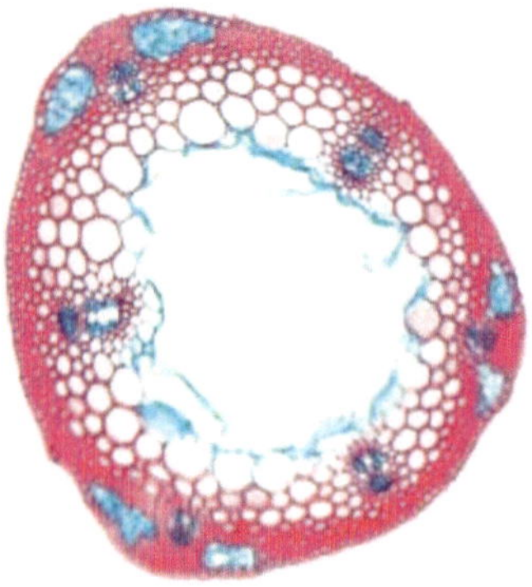

100 µm

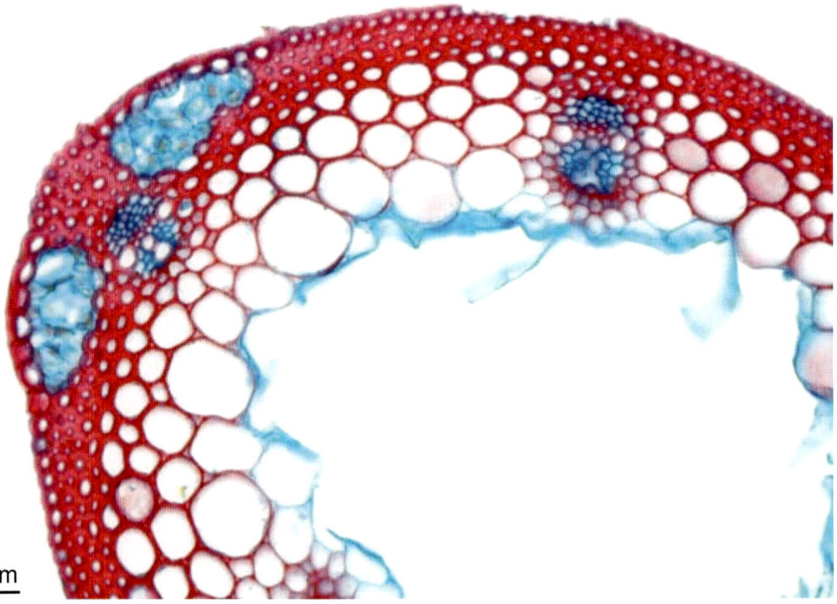

25 µm

Festuca tibetica (Stapf) E.B.Alexeev

Poaceae

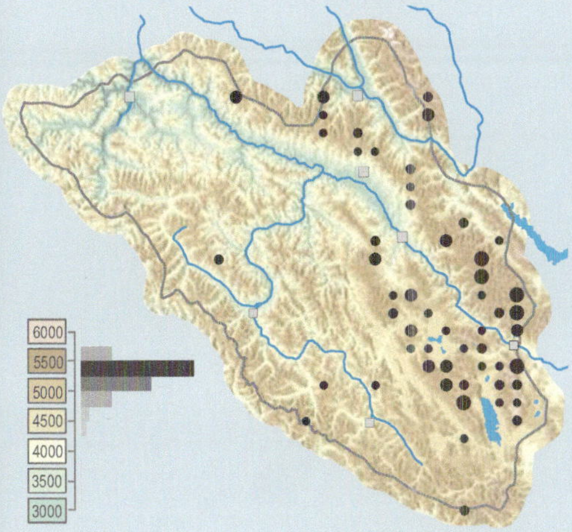

Syn.: *F. valesiaca* var. *tibetica* Stapf

Description: Perennial herbs, 4–15 cm tall, densely cespitose, leaves filiform, conduplicate, panicle oblong, contracted, 1.5–3 cm long, primary branches appressed, awns 1–2 mm long.

Distribution: Pakistan, N India, China (Xizang, Yunnan), Bhutan.

Habitat: Alpine grasslands.

Culm uniform, round, hollow, with a diameter of 0.5–0.7 mm. Epidermis thin- to thick-walled. With a 3–4-seriate continuous peripheral sclerenchymatic belt of thick-walled fibers. Girders absent. Chlorenchyma groups oval to tangentially enlarged. Intercellulars in the parenchymatic tissue very small. Vascular bundles arranged in one circular row. Large bundles within the thin- to thick-walled parenchymatic tissue, small bundles within the sclerenchymatic belt. Vascular bundles with large vessels, diameter 5–7 µm. Phloem oval, consisting of sieve tubes and companion cells. Intercellulars centripetal of the protoxylem absent. Vascular bundles surrounded by a 1–2-seriate sheath of polygonal, thin- to thick-walled cells.

Analyzed: 2 individuals

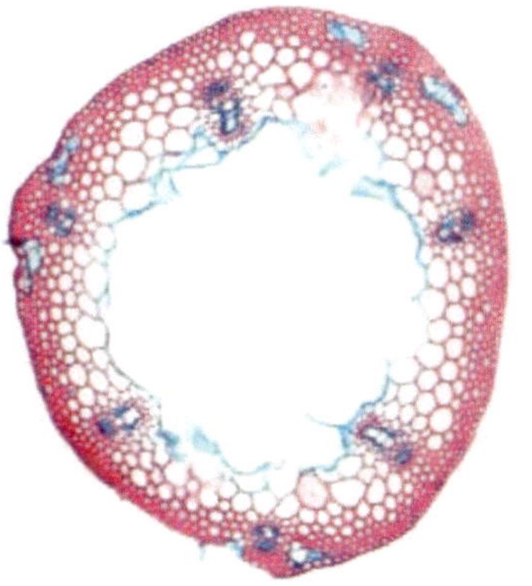

100 µm

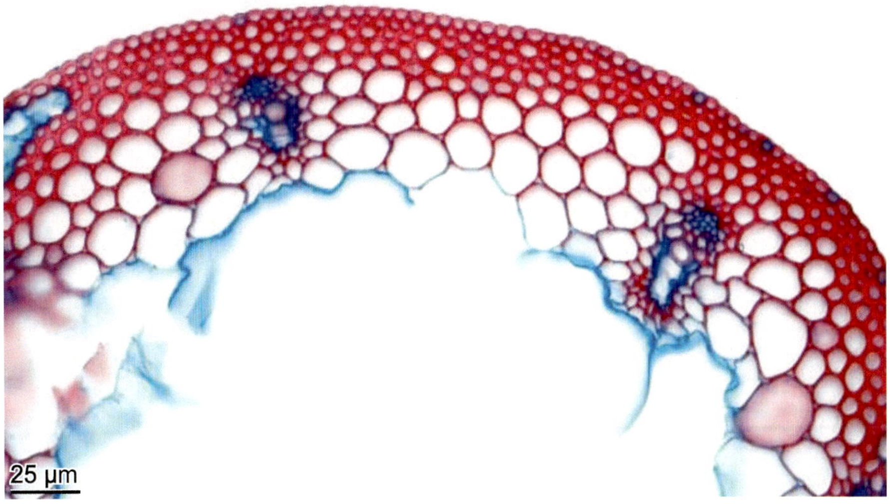

Festuca wallichiana E.B.Alexeev
Poaceae

Description: Perennial herbs, 5–35 cm tall, densely cespitose, leaves filiform, panicle contracted, linear, 1.5–4 cm long.

Distribution: N India, W China.

Habitat: Grassy and stony mountain slopes.

Culm uniform, round, hollow, with a diameter of 0.6–0.7 mm. Epidermis thin- to thick-walled. Continuous, 4–5-seriate peripheral sclerenchymatic belt of thick-walled fibers. Girders absent. Chlorenchyma groups oval. Intercellulars in the parenchymatic tissue absent. Vascular bundles arranged in 2 circular offset rows. Large bundles within the thin- to thick-walled parenchymatic tissue, small bundles within the sclerenchymatic belt. Vascular bundles with large vessels, diameter 5–6 µm. Phloem oval, consisting of sieve tubes and companion cells. Intercellulars centripetal of the protoxylem absent. Vascular bundles surrounded by a 1–2-seriate sheath of polygonal, thin- to thick-walled cells.

Analyzed: 1 individual

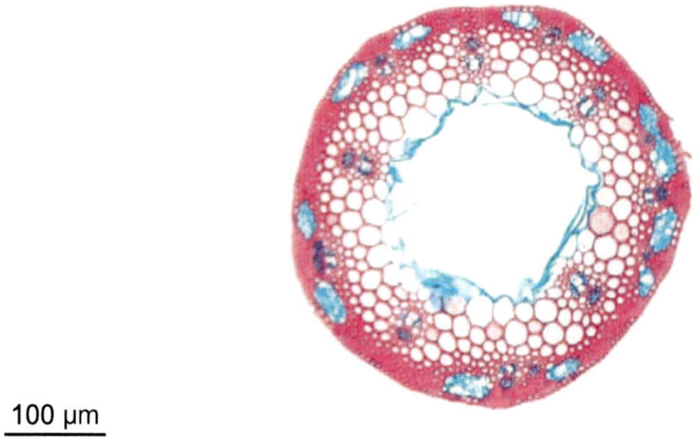

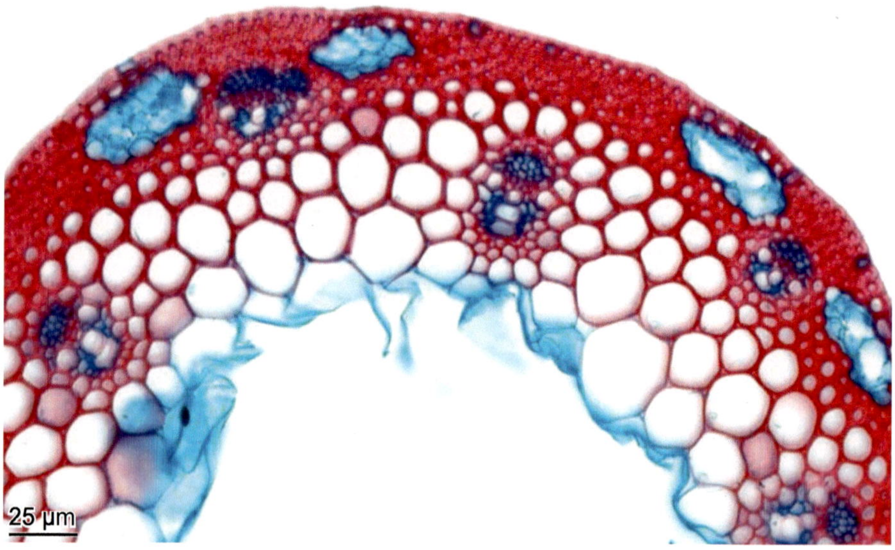

Helictotrichon hookeri (Scribn.) Henrard

Poaceae

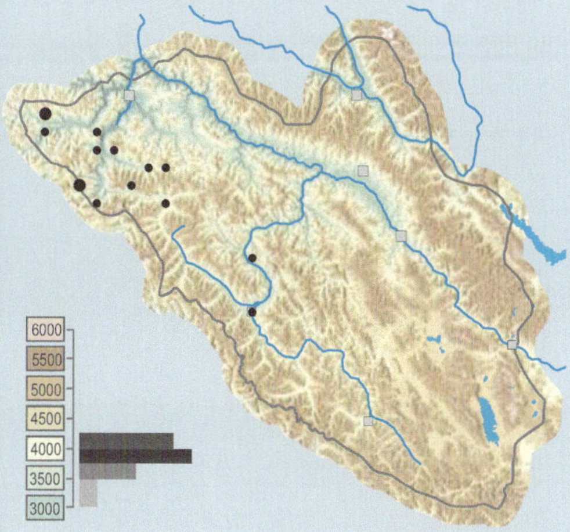

Syn.: *Arrhenatherum hookeri* (Scribn.) Potztal, *Avena hookeri* Scribn., *Avenula hookeri* (Scribn.) Holub

Description: Perennial herbs, 30–40 cm tall, cespitose, leaves flat or conduplicate, 2–4 mm wide, margins cartilaginous, panicle contracted, linear or lanceolate, 5–10 cm long, branches appressed, awns 10–15 mm long, curved.

Distribution: North America, Russia (Siberia, Far East), Azerbaijan, Kazakhstan, Kyrgyzstan, Tajikistan, India (Jammu and Kashmir), China, Mongolia.

Habitat: Steppes, alpine meadows.

Culm uniform, round, hollow, with a diameter of 1.2–1.5 mm. Epidermis thin- to thick-walled. Continuous, 1–3-seriate, peripheral sclerenchymatic belt of thick-walled fibers. Girders absent. Chlorenchyma groups small, round, located at the periphery of the culm. Vascular bundles arranged in two offset circular rows. Large bundles within the thin-walled parenchymatic tissue, small bundles within the sclerenchymatic belt. Vascular bundles with large vessels, diameter 20–25 µm. Intercellulars within vascular bundles absent. Phloem small, oval, consisting of uniform cells. Sclerenchymatic vessel sheaths eccentric, with thin- to thick-walled fibers in centripetal arc-like groups.

Analyzed: 2 individuals

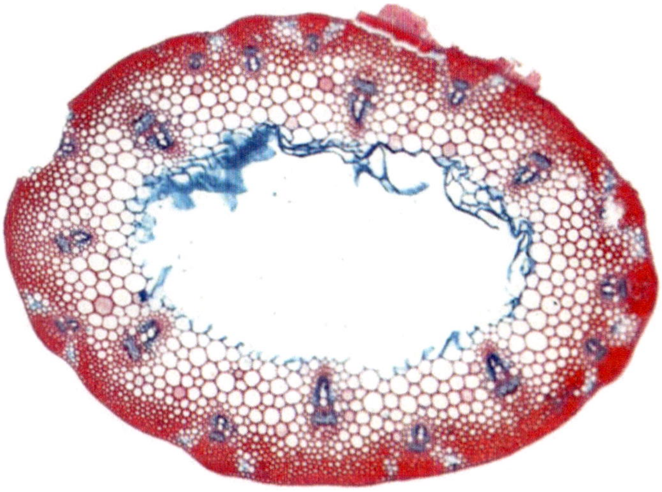

100 µm

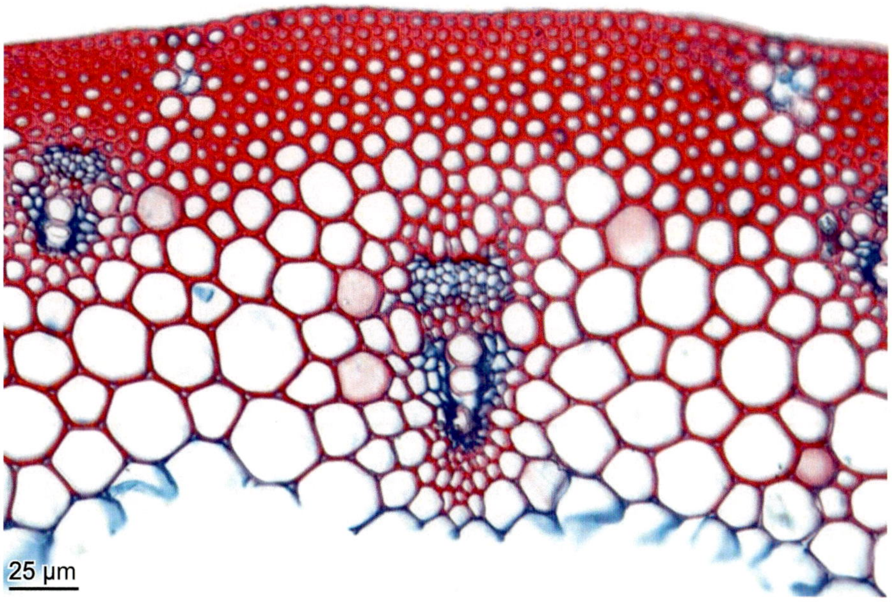

Hierochloe laxa R. Br. ex Hook. f.

Poaceae

Syn.: *Anthoxanthum laxum* (Hook.f.) Veldkamp

Description: Perennial herbs, 20–70 cm tall, aromatic, leaves flat, 2.5–15 cm long, 5–15 mm wide, panicle oblong, 5–15 cm long, open.

Distribution: E Afghanistan, N Pakistan, N India, Nepal.

Habitat: High-mountain ledges, alpine pastures.

Culm uniform, round, hollow, with a diameter of 1.3–1.6 mm. Epidermis thin- to thick-walled. Continuous, 3–6-seriate peripheral sclerenchymatic belt of thin- to thick-walled fibers centripetal of the chlorenchyma. Girders conical and rectangular. Chlorenchyma groups tangentially enlarged. Intercellulars in the parenchymatic tissue very small. Vascular bundles arranged in 2 circular offset rows. Large bundles within the thin- to thick-walled parenchymatic tissue, small bundles within the sclerenchymatic belt. Vascular bundles with large vessels, diameter 25–30 µm. Phloem oval, consisting of sieve tubes and companion cells. With intercellulars centripetal of the protoxylem. Vascular bundles surrounded by a 1–2-seriate sheath of polygonal, thin- to thick-walled cells.

Analyzed: 1 individual

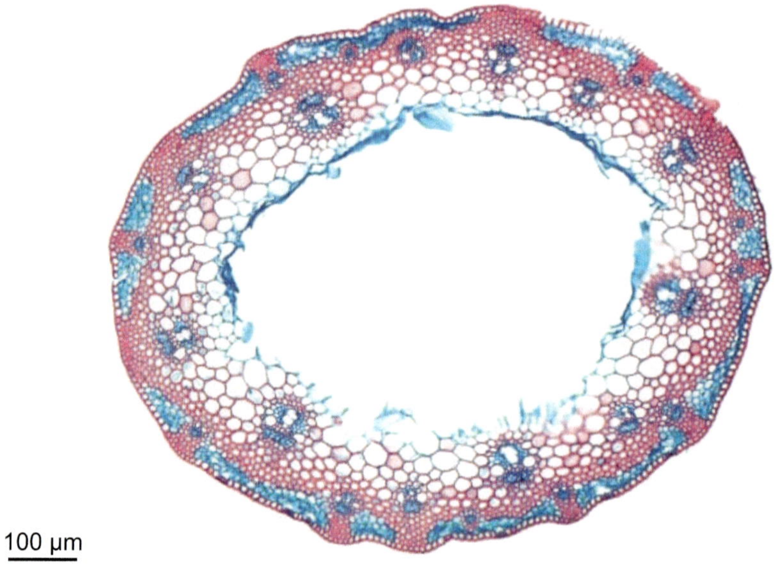

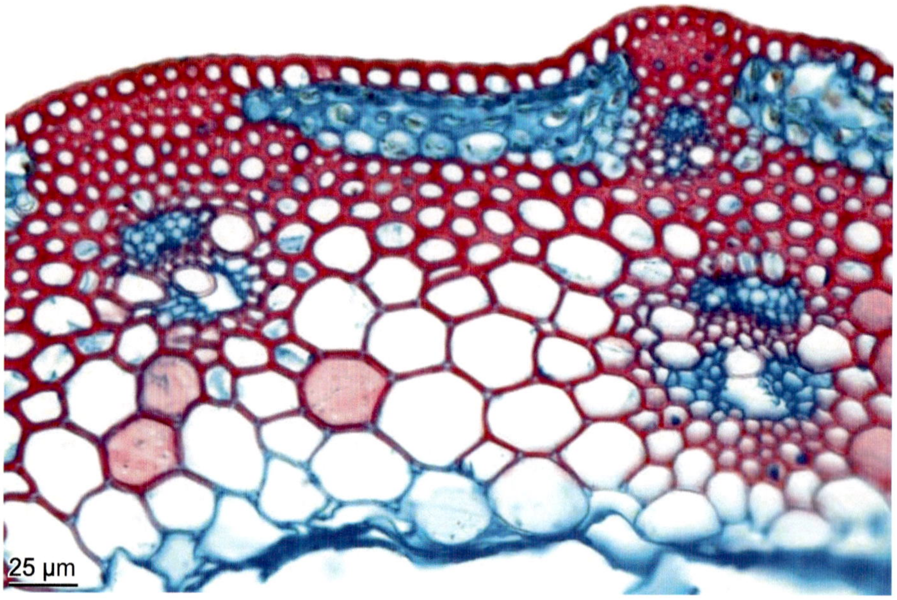

Hordeum brevisubulatum subsp. *turkestanicum*

Poaceae

(Nevski) Tzvelev

Syn.: *H. turkestanicum* Nevski

Description: Perennial herbs, 15–50 cm tall, cespitose, nodes of culms pubescent, leaves flat, spike dense, 2.5–9.5 cm long, usually grayish violet.

Distribution: Russia, Kazakhstan, Turkmenistan, Uzbekistan, Kyrgyzstan, C and NE Afghanistan, N Pakistan, N India, Nepal, China (Xinjiang, Xizang).

Habitat: River beds, loamy and sandy stream sediments, often saline, tolerates high concetrations of salts, saline steppes, common on hummocks covered with a crust of salt.

Culm uniform, round, hollow, with a diameter of 3–4 mm. Epidermis thick-walled. Continuous, 3–6-seriate peripheral sclerenchymatic belt. Chlorenchyma groups absent. Intercellulars in the parenchymatic tissue almost absent. Vascular bundles arranged in 2 circular offset rows. Large bundles within the thin- to thick-walled parenchymatic tissue, small bundles within the sclerenchymatic belt. Vascular bundles with large vessels, diameter 30–35 µm. Phloem oval, consisting of sieve tubes and companion cells. Intercellulars centripetal of the protoxylem absent. Vascular bundles surrounded by a 1–2-seriate sheath of polygonal, thin- to thick-walled cells.

Analyzed: 2 individuals

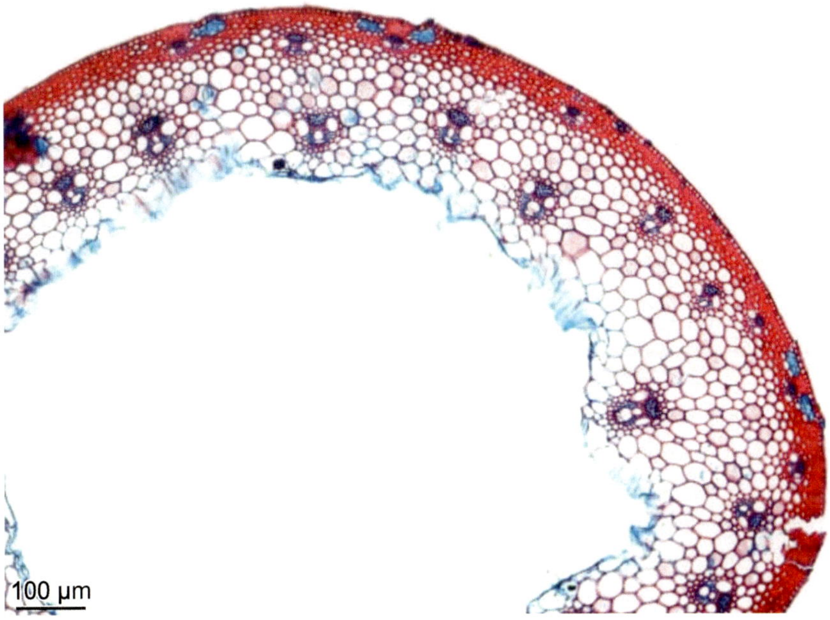

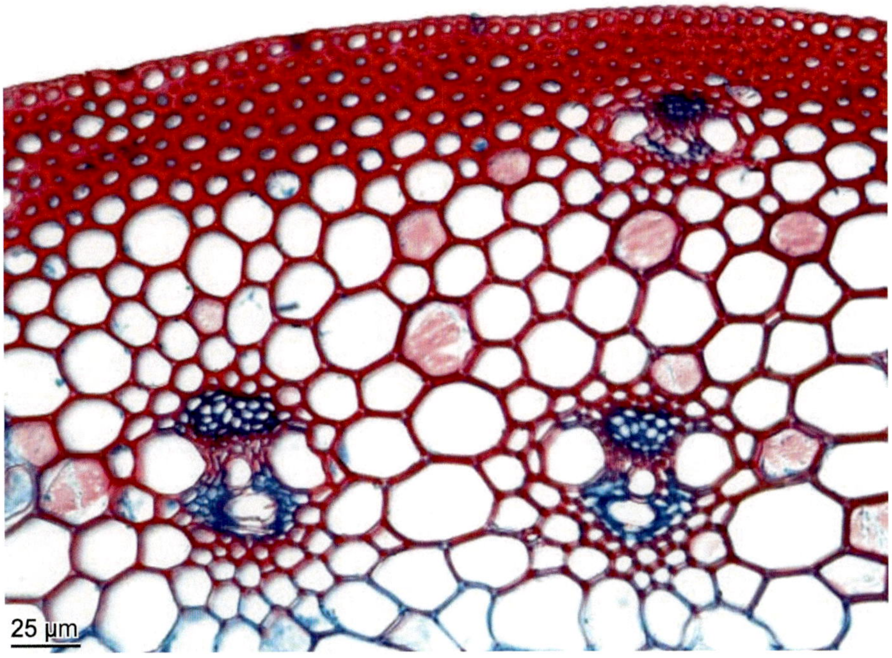

Hordeum murinum L.

Poaceae

Description: Annuals, 5–50 cm tall, culms cespitose or solitary, leaves flat, glabrous or sparsely pubescent, 5– 20 cm long, 2-8 mm wide, spike oblong, 2–7 cm long, strongly compressed, awns 15–50 mm long.

Distribution: Europe, W Asia; naturalized elsewhere (S Africa, Australia, E Asia, North and South America).

Habitat: Ruderal places, weedy in fields, along roads.

Culm uniform, round, hollow, with a diameter of 1.1–1.3 mm. Epidermis thin- to thick-walled. Continuous, 3–4-seriate peripheral sclerenchymatic belt. Chlorenchyma groups few, 1–2-seriate, tangentially enlarged. Intercellulars in the parenchymatic tissue very small. Vascular bundles arranged in 2 circular offset rows. Large bundles within the thin- to thick-walled parenchymatic tissue, small bundles within the sclerenchymatic belt. Vascular bundles with large vessels, diameter 20–35 μm. Phloem oval, consisting of sieve tubes and companion cells. Intercellulars centripetal of the protoxylem absent. Vascular bundles surrounded by a 1–2-seriate sheath of polygonal, thin- to thick-walled cells.

Analyzed: 2 individuals

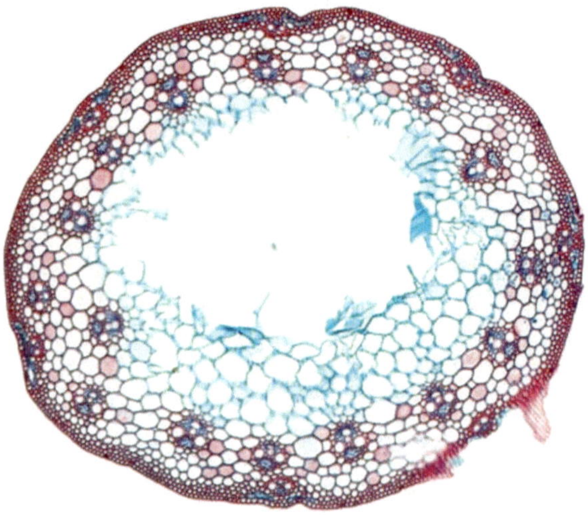

100 μm

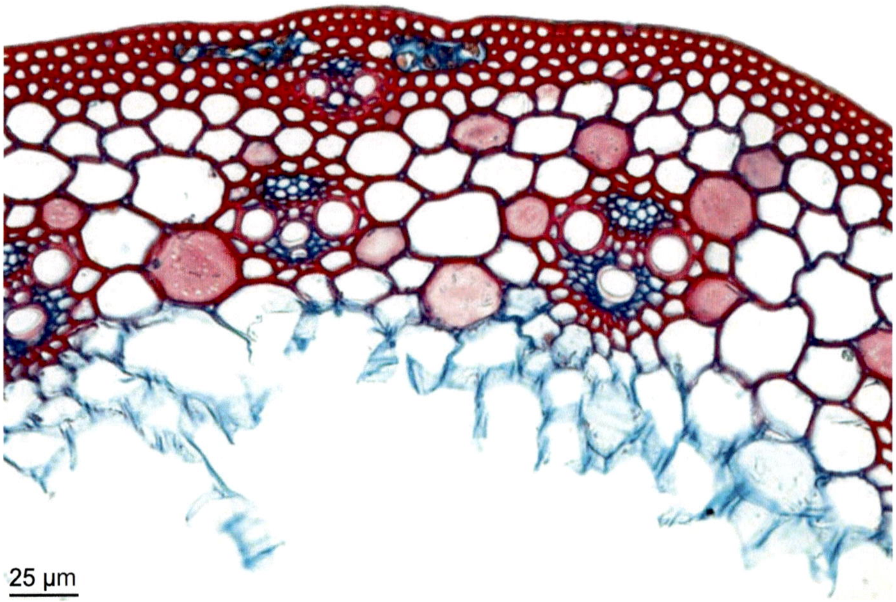

25 μm

Koeleria litvinowii Domin

Poaceae

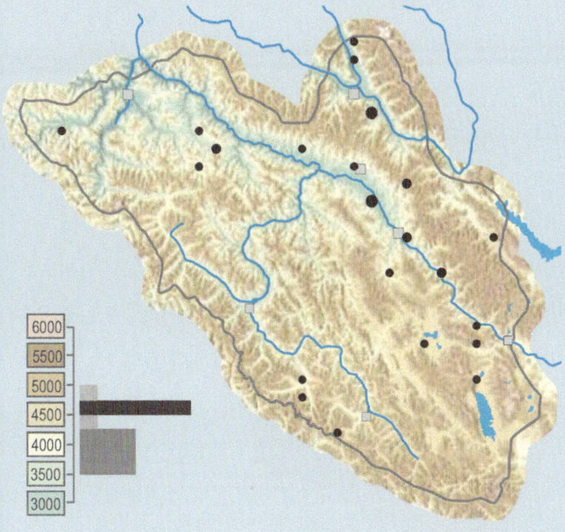

Syn.: *K. argentea* Griseb., *K. litvinowii* subsp. *argentea* (Griseb.) S.M.Phillips & Z.L.Wu, *Trisetum litvinowii* subsp. *argentea* (Griseb.) Tzvelev

Description: Perennial herbs, 25–40 cm tall, loosely cespitose, leaves flat, glabrous or sparsely hairy, 3.5–10 cm long, 2–4 mm wide, panicle oblong, 3–7 cm long, sometimes interrupted in the lower part.

Distribution: Afghanistan, Pakistan, N India, China (Qinghai, Xizang).

Habitat: Moist meadows, gravel river deposits.

Culm uniform, round, hollow, with a diameter of 1.0–1.1 mm. Epidermis thick-walled. Continuous, 3–6-seriate peripheral sclerenchymatic belt of thick-walled fibers. Girders absent. Chlorenchyma groups oval. Intercellulars in the parenchymatic tissue absent. Vascular bundles arranged in 2 circular offset rows. Large bundles within the thin- to thick-walled parenchymatic tissue, small bundles within the sclerenchymatic belt. Vascular bundles with large vessels, diameter 5–6 µm. Phloem oval, consisting of sieve tubes and companion cells. With intercellulars centripetal of the protoxylem. Vascular bundles surrounded by a 1–2-seriate sheath of polygonal, thin- to thick-walled cells.

Analyzed: 2 individuals

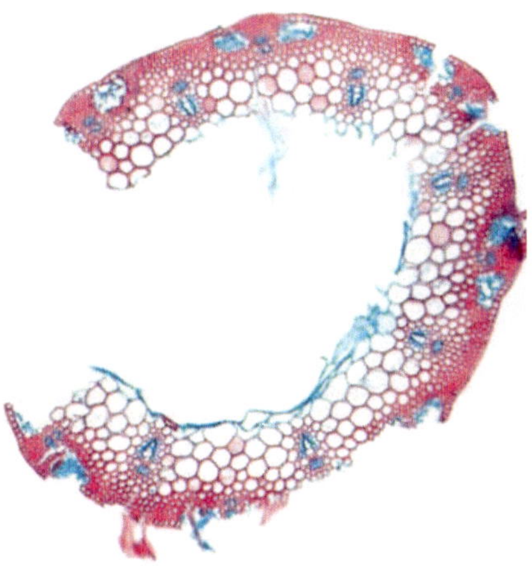

100 µm

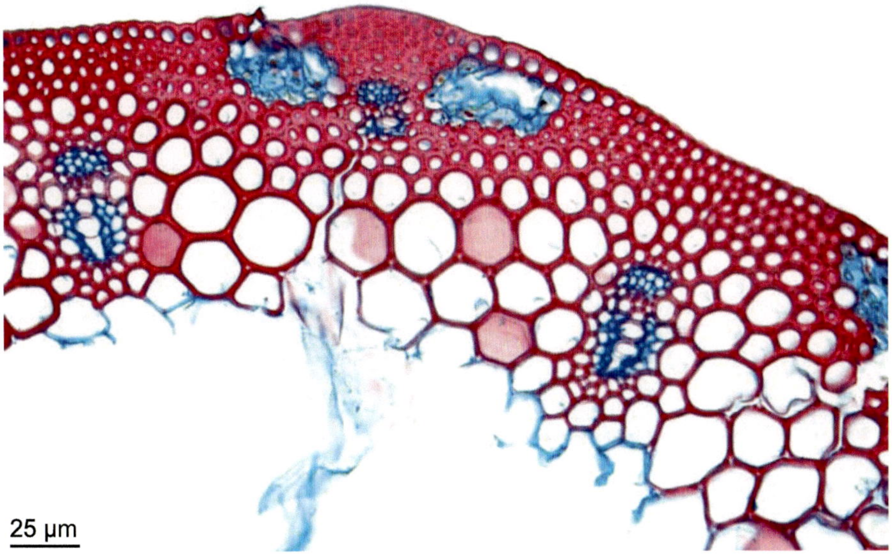

25 µm

Koeleria macrantha (Ledeb.) Schult.

Poaceae

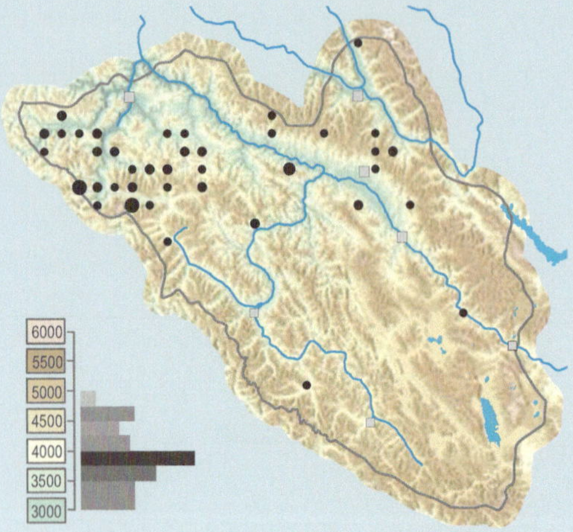

Syn.: *Aira macrantha* Ledeb., *A. cristata* L., *K. cristata* auct. non Pers.

Description: Perennial herbs, 10–60 cm tall, loosely cespitose, leaves mostly basal, linear, flat or convolute, panicle spike-like, linear to lanceolate in outline, 2–10 cm long, usually lobed or interrupted in the lower part, with pubescent branches.

Distribution: North America, Europe, SW Asia, Russia, Kazakhstan, Turkmenistan, Uzbekistan, Kyrgyzstan, Tajikistan, Afghanistan, Pakistan, India, Nepal, China, Mongolia, Japan; introduced to Australia and elsewhere.

Habitat: Dry slopes, steppes, roadsides.

Culm uniform, round, hollow, with a diameter of 1.4–1.6 mm. Epidermis thick-walled. Continuous, 3–6-seriate peripheral sclerenchymatic belt of thick-walled fibers centripetal of the continuous chlorenchyma. Girders small, conical. Almost continuous, 2–3-seriate belt of chlorenchyma groups. Intercellulars in the parenchymatic tissue almost absent. Vascular bundles arranged in 2 circular offset rows. Large bundles within the thin- to thick-walled parenchymatic tissue, small bundles within the sclerenchymatic belt. Vascular bundles with large vessels, diameter 25–30 µm. Phloem oval, consisting of sieve tubes and companion cells. With intercellulars centripetal of the protoxylem. Vascular bundles surrounded by a 1–2-seriate sheath of polygonal, thin- to thick-walled cells.

Analyzed: 1 individual

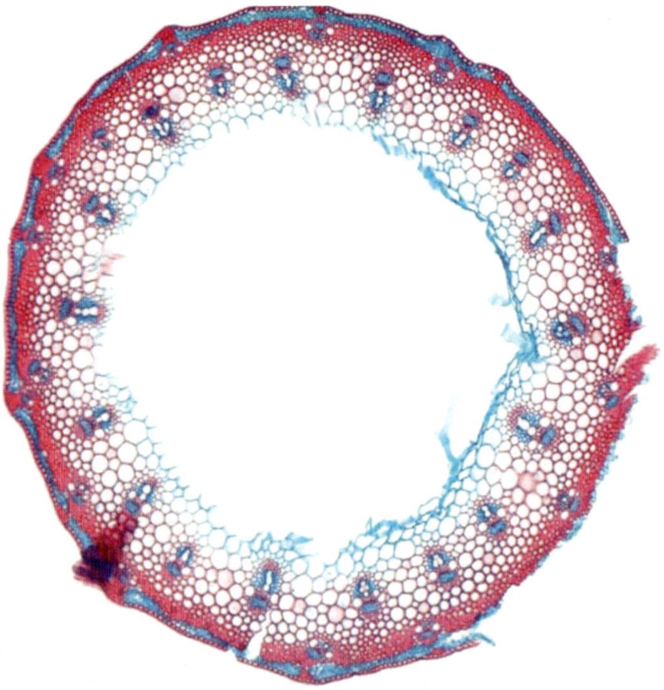

100 µm

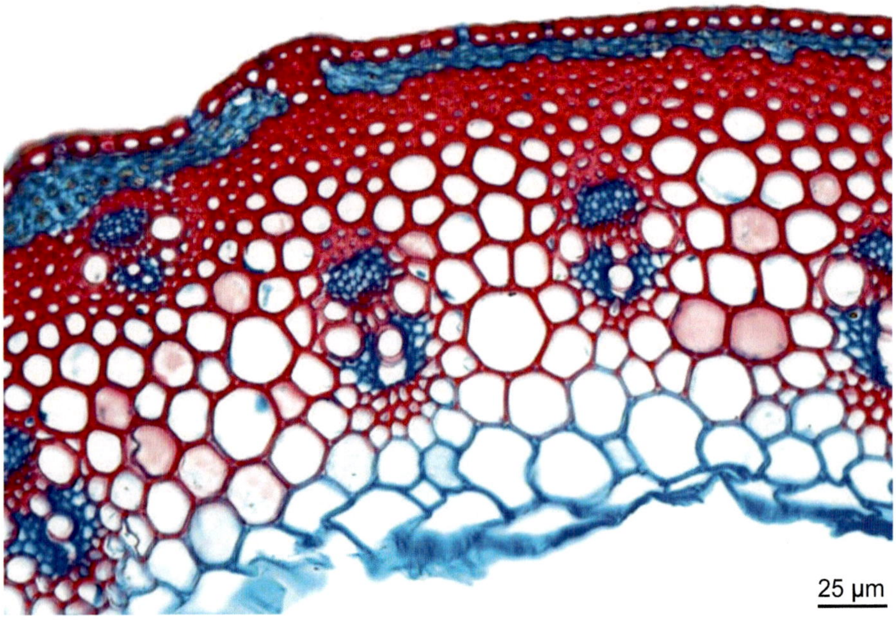

25 µm

Leymus secalinus (Georgi) Tzvelev
Poaceae

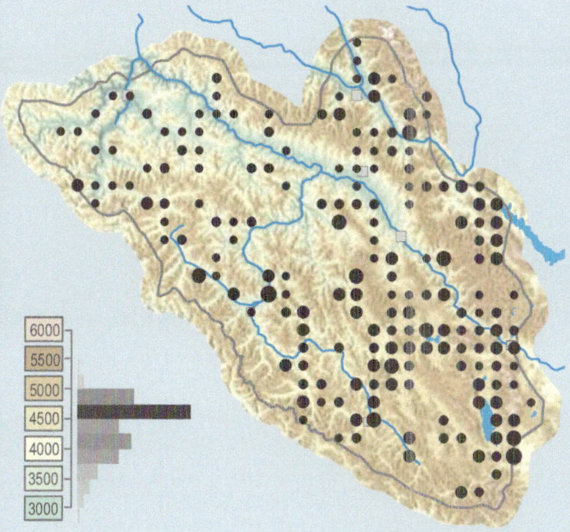

Syn.: *Triticum secalinum* Georgi, *Elymus dasystachys* Trin.

Description: Perennial herbs, 40–100 cm tall, rhizomatous, leaves flat or involute, hairy, spike erect, 8–12 cm long.

Distribution: Russia (Siberia), E Kazakhstan, Turkmenistan, Uzbekistan, Kyrgyzstan, Tajikistan, N Pakistan, NW India, China (Gansu, Hebei, Heilongjiang, Henan, Jilin, Liaoning, Nei Monggol, Ningxia, Qinghai, Shaanxi, Shanxi, Sichuan, Xinjiang, Xizang), Mongolia.

Habitat: River sediments, gravel and sandy plains, saline steppes, abandoned fields.

Culm uniform, round, wavy, hollow, with a diameter of 1.2–1.3 mm. Epidermis thin- to thick-walled. Continuous, 4–8-seriate peripheral sclerenchymatic belt. Chlorenchyma groups few, 2–3-seriate, tangentially enlarged. Intercellulars in the parenchymatic tissue absent. Vascular bundles arranged in 2–3 circular offset rows. Large bundles within the thin- to thick-walled parenchymatic tissue, small bundles within the sclerenchymatic belt. Vascular bundles with large vessels, diameter 15–20 μm. Phloem round, consisting of sieve tubes and companion cells. With intercellulars centripetal of the protoxylem. Vascular bundles surrounded by a 2–3-seriate sheath of polygonal, thin- to thick-walled cells.

Analyzed: 2 individuals

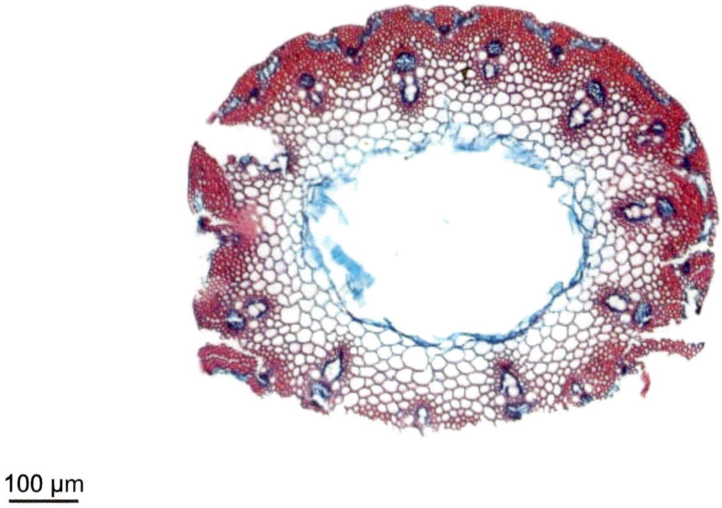

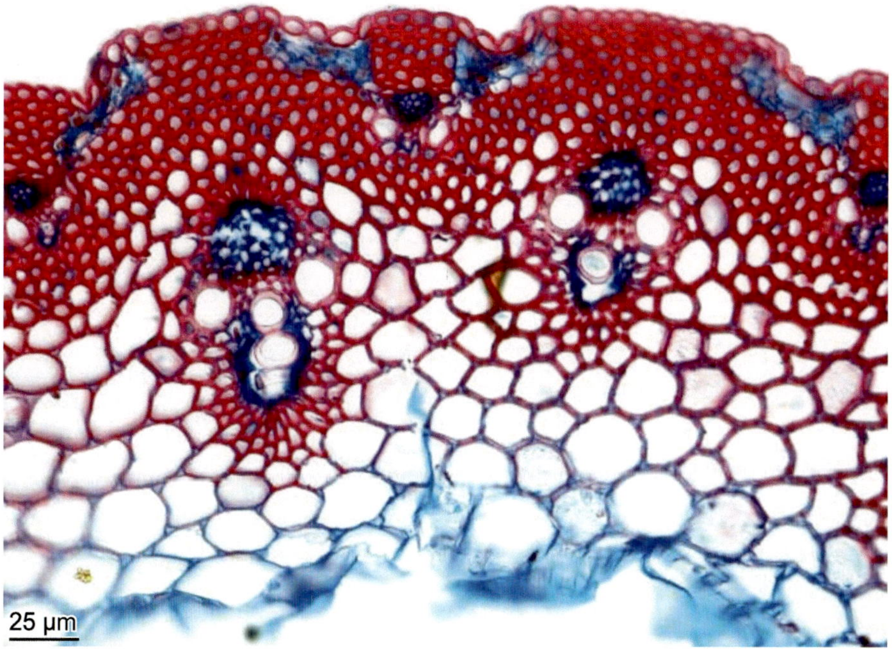

Muhlenbergia huegelii Trin.

Poaceae

Syn.: *M. arisanensis* Hayata, *M. geniculata* Nees ex Steud., *M. longistolon* Ohwi

Description: Perennial herbs, 30–110 cm tall, rhizomatous, culms ascending, leaves flat, 8–15 cm long, 4–8 mm wide, panicle open, lanceolate, interrupted, 10–20 cm long, awns 8–20 mm long, flexuous.

Distribution: Russia, Afghanistan, Pakistan, India, Nepal, China, Bhutan, Korea, Japan, Taiwan, Philippines,

Habitat: Along shaded streams.

Culm uniform, round, hollow, wavy, with a diameter of 1.5–1.8 mm. Epidermis thick-walled. Continuous, 3–4-seriate peripheral sclerenchymatic belt. Chlorenchyma groups absent. Intercellulars in the parenchymatic tissue absent. Vascular bundles arranged in 2–3 circular offset rows. Large bundles within the thin- to thick-walled parenchymatic tissue, small bundles within the sclerenchymatic belt. Vascular bundles with large vessels, diameter 25–35 µm. Phloem oval, consisting of sieve tubes and companion cells. With a few intercellulars centripetal of the protoxylem. Vascular bundles surrounded by a 1–2-seriate sheath of polygonal, thin- to thick-walled cells.

Analyzed: 2 individuals

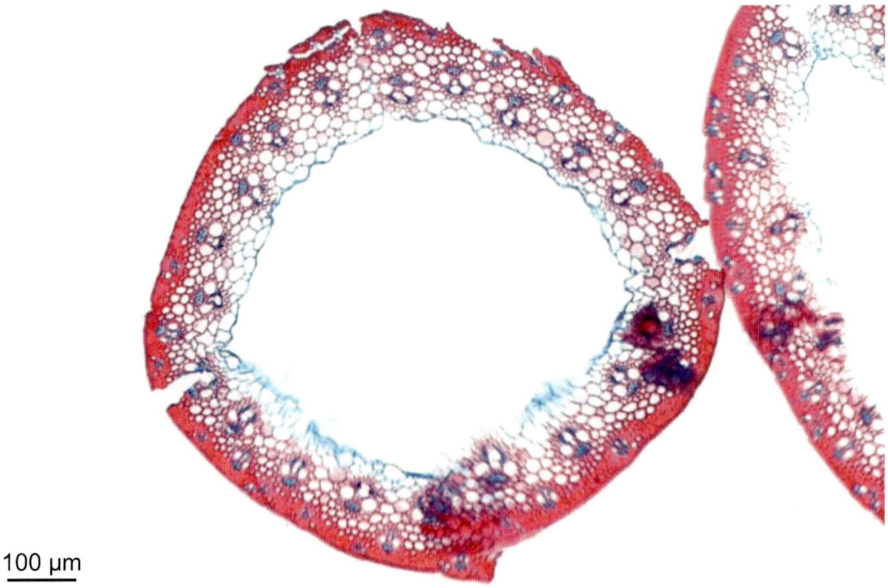

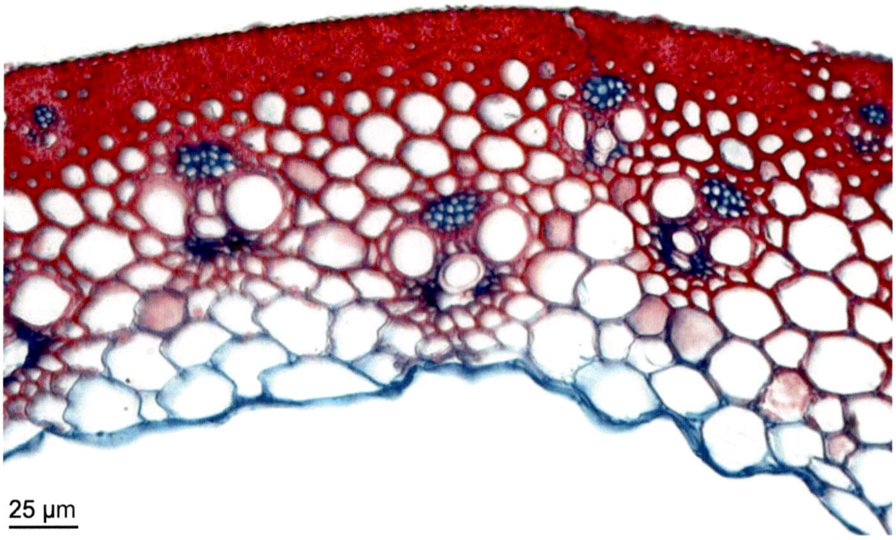

Pennisetum flaccidum Griseb.

Poaceae

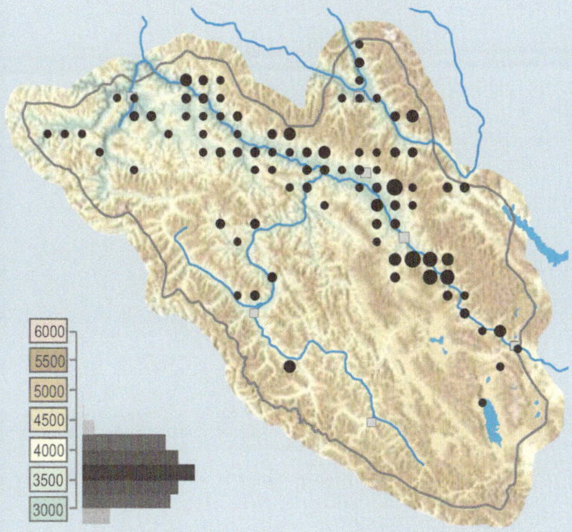

Description: Perennial herbs, 30–120 cm tall, cespitose, rhizomatous, leaves linear, flat or folded, panicle linear, 4–22 cm long, with bristles 15–25 mm long.

Distribution: Iran, Tajikistan, Afghanistan, Pakistan, NW India, Nepal, China (Gansu, Hebei, Heilongjiang, Henan, Hubei, Jilin, Liaoning, Nei Mongol, Ningxia, Qinghai, Shaanxi, Shandong, Shanxi, Sichuan, Xinjiang, Xizang, Yunnan), Bhutan, Mongolia.

Habitat: Along roads, in human-disturbed habitats, often on sand but also on rocks, weedy in fields, tolerates flooding.

Culm uniform, round, wavy, filled, with a diameter of 1–2.5 mm. Epidermis thin- to thick-walled. Continuous, 4–8-seriate peripheral sclerenchymatic belt centripetal of the chlorenchyma. Chlorenchyma bands few, 2–3-seriate, tangentially enlarged. Intercellulars in the parenchymatic zone small and triangular. Vascular bundles arranged in 2–3 circular offset rows. Large bundles within the thin- to thick-walled parenchymatic tissue, small bundles within the sclerenchymatic belt. Vascular bundles with large vessels, diameter 30–35 µm. Phloem oval, consisting of sieve tubes and companion cells. With intercellulars centripetal of the protoxylem. Vascular bundles surrounded by a 1–2-seriate sheath of polygonal, thin- to thick-walled cells.

Analyzed: 1 individual

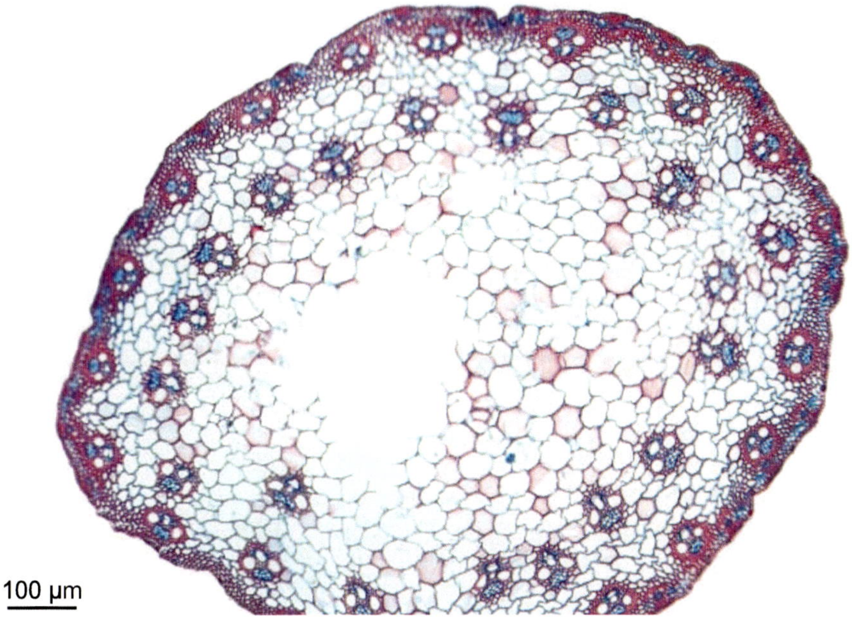

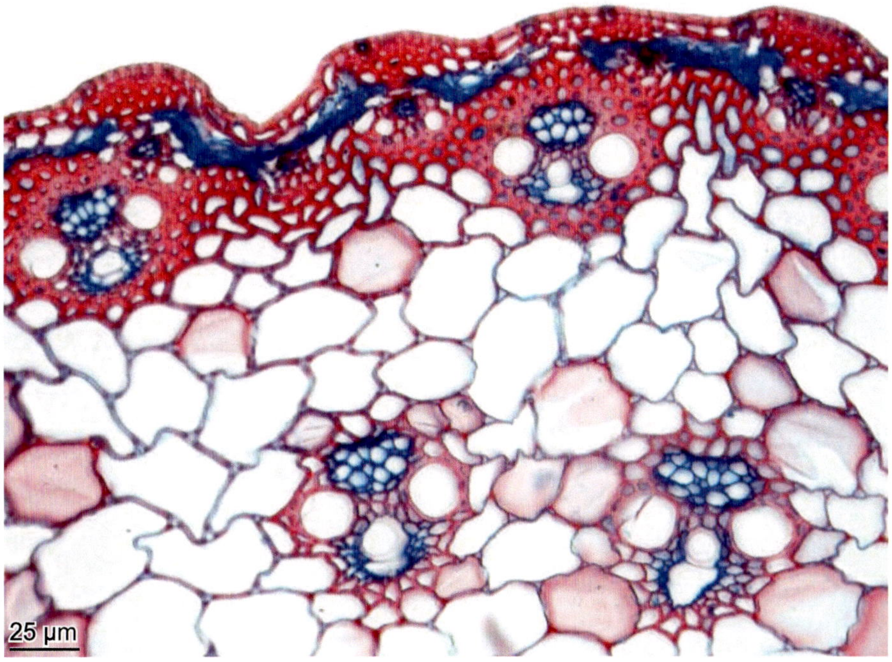

Pennisetum orientale Rich.

Poaceae

Description: Perennial herbs, 20–200 cm tall, shortly rhizomatous, woody and often densely fastigiately branched, leaves flat or convolute, up to 60 cm long and 15 mm wide, panicle linear, 8–30 cm long, often interrupted, with bristles 15–30 mm long.

Distribution: N Africa, W, SW and C Asia, Pakistan, India, Nepal, Bhutan.

Habitat: Along roads, in human-disturbed habitats.

Culm uniform, round, wavy, filled, with a diameter of 1.0–1.1 mm. Epidermis thick-walled. Continuous, 4–6-seriate, wavy peripheral sclerenchymatic belt centripetal of the chlorenchyma. Girders conical. Chlorenchyma groups small. Intercellulars in the parenchymatic zone small and triangular. Vascular bundles arranged in 2 circular offset rows. Large bundles within the thin- to thick-walled parenchymatic tissue, small bundles within the sclerenchymatic belt. Vascular bundles with large vessels, diameter 10–15 µm. Phloem oval, consisting of uniform cells. Few intercellulars centripetal of the protoxylem. Vascular bundles surrounded by a uniseriate sheath of polygonal, thin- to thick-walled cells.

Analyzed: 1 individual

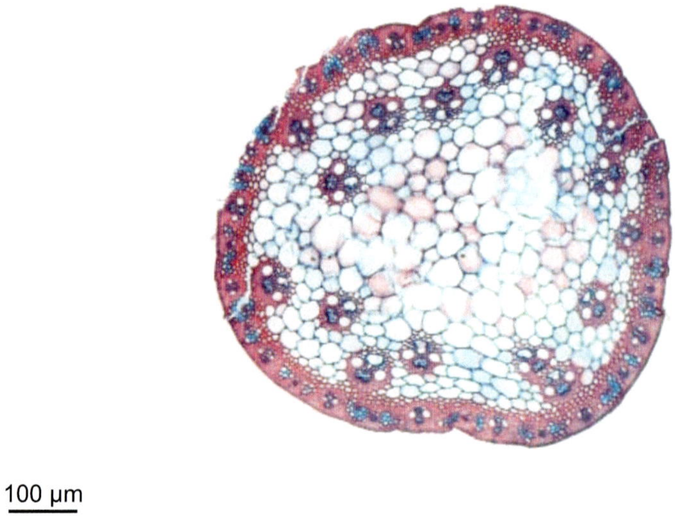

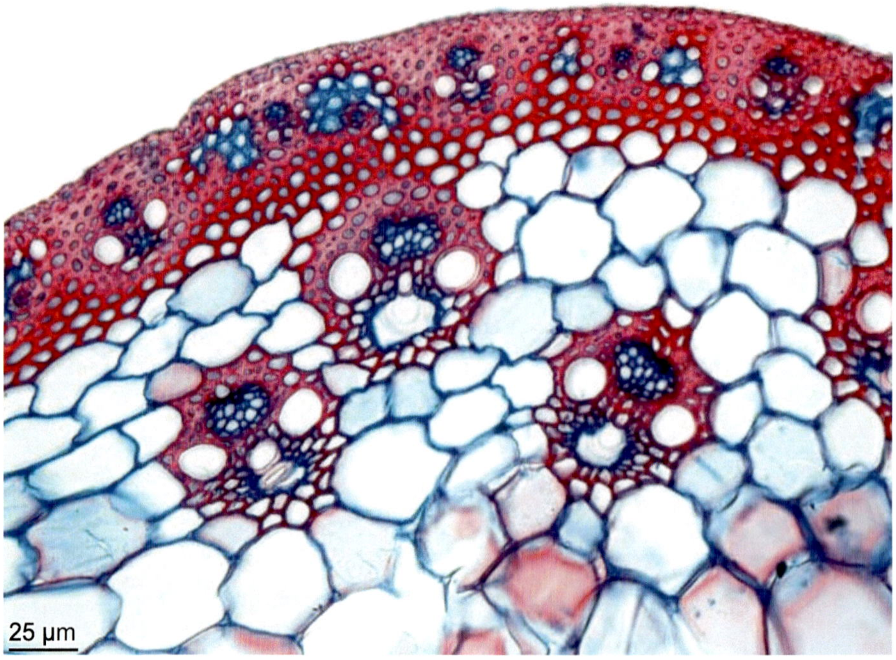

Phalaris canariensis L.

Poaceae

Description: Annuals, 30–60 cm tall, loosely cespitose, leaves flat, 3–10 mm wide, uppermost leaf sheaths inflated, panicle spike-like, dense, ovoid, 1.5–4 cm long.

Distribution: Native to Canary Islands and N Africa; widely introduced and cultivated.

Habitat: Disturbed places, roadsides.

Culm uniform, round, wavy, hollow, with a diameter of 1.0–1.1 mm. Epidermis thick-walled. Continuous, 1–3-seriate, wavy peripheral sclerenchymatic belt centripetal of the chlorenchyma. Girders conical. Chlorenchyma groups oval. Intercellulars in the parenchymatic tissue absent. Vascular bundles arranged in 2 circular offset rows. Large bundles within the thin- to thick-walled parenchymatic tissue, small bundles within the sclerenchymatic belt. Vascular bundles with large vessels, diameter 10–15 µm. Phloem oval, consisting of uniform cells. Few intercellulars centripetal of the protoxylem. Vascular bundles surrounded by a uniseriate sheath of polygonal, thin- to thick-walled cells.

Analyzed: 1 individual

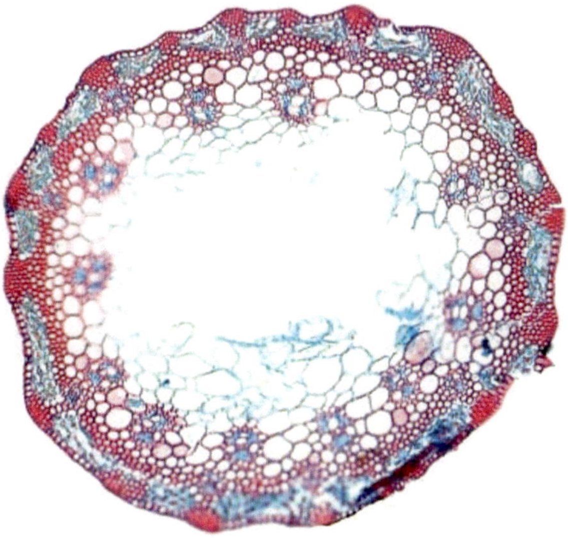

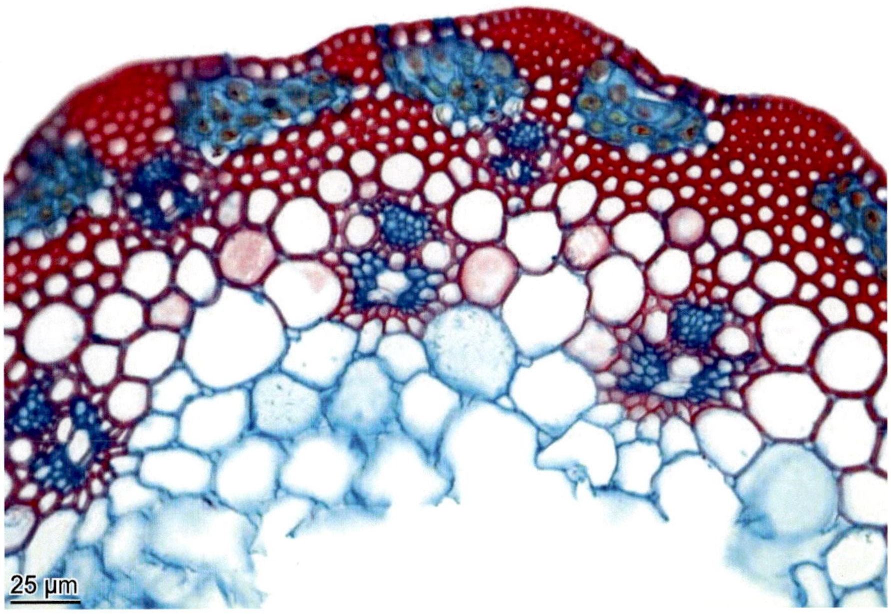

Phleum alpinum L.

Poaceae

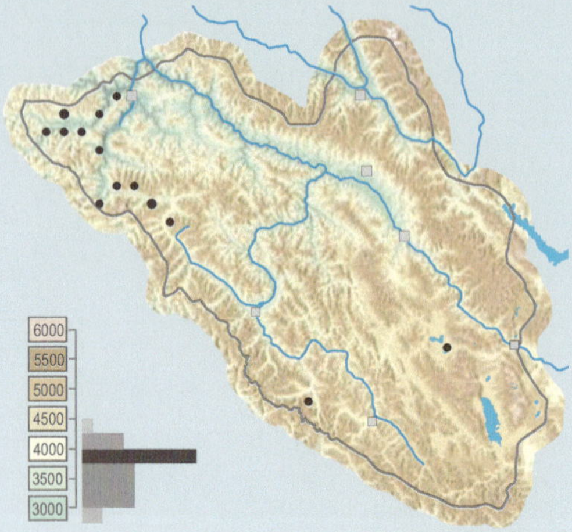

Syn.: *P. commutatum* Gaudich.

Description: Perennial herbs, 5–40 cm tall, shortly rhizomatous, leaves flat, 2–12 cm long, 2–10 mm wide, panicle broadly cylindrical to ovoid, 1–6 cm long, 0.6–1.2 cm wide, usually purplish.

Distribution: Alpine and arctic regions of N hemisphere, also in the Andes.

Habitat: Moist alpine meadows.

Culm uniform, round, hollow, with a diameter of 0.8–0.9 mm. Epidermis thick-walled. Continuous, 2–3-seriate, wavy peripheral sclerenchymatic belt centripetal of the chlorenchyma. With large conical to rectangular girders. Chlorenchyma groups oval. Intercellulars in the parenchymatic tissue absent. Vascular bundles arranged in 2 circular offset rows. Large bundles within the thin- to thick-walled parenchymatic tissue, small bundles within the sclerenchymatic belt. Vascular bundles with large vessels, diameter 5–8 μm. Phloem oval, consisting of uniform cells. Intercellulars centripetal of the protoxylem absent. Vascular bundles surrounded by a 2–3-seriate sheath of polygonal thin- to thick-walled cells.

Analyzed: 1 individual

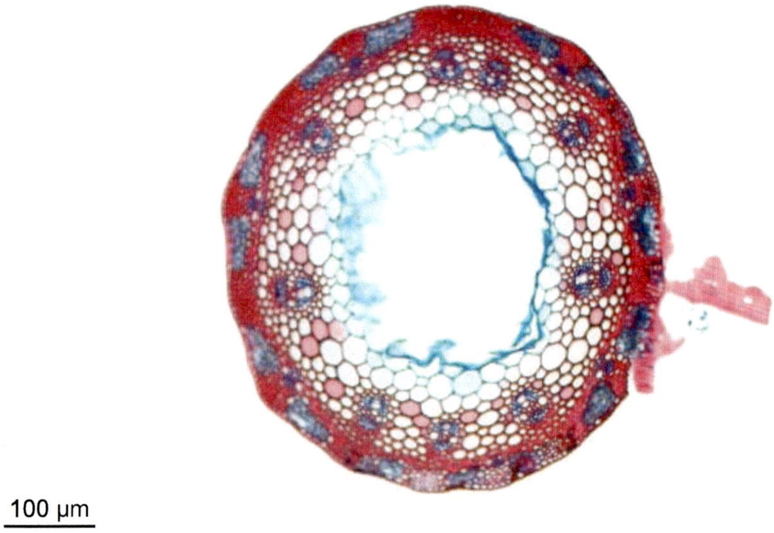

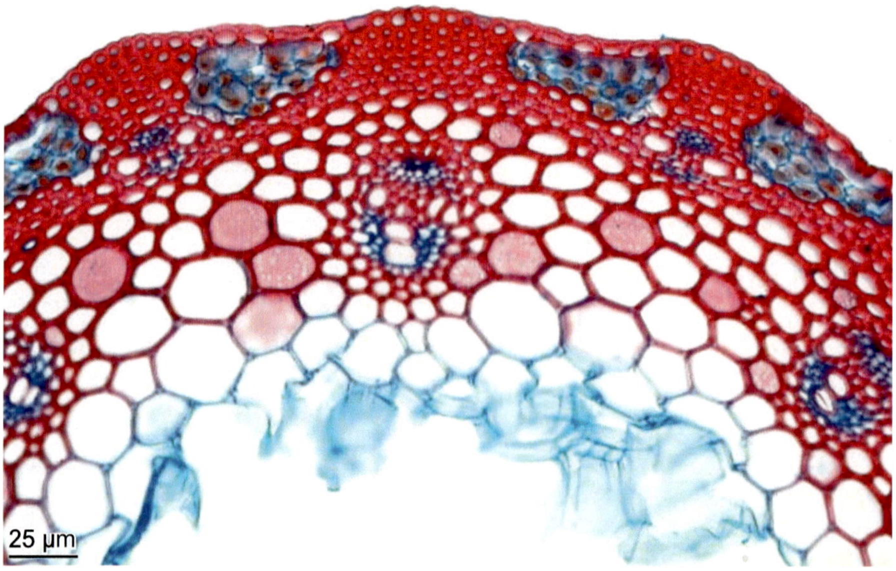

Piptatherum gracile Mez
Poaceae

Syn.: *Oryzopsis gracilis* (Mez) Pilg.

Description: Perennial herbs, 10–50 cm tall, densely cespitose, leaves filiform, usually rolled, panicle contracted, linear, 6–20 cm long, 0.5–1 cm wide, awns 3–6 mm long.

Distribution: Afghanistan, N Pakistan, NW India, Nepal, China (Xizang, Yunnan, Sichuan).

Habitat: Steppes, stony slopes, screes, along roads, dry disturbed sites in settlements.

Culm uniform, round, hollow, with a diameter of 1.2–1.3 mm. Epidermis thin- to thick-walled. Continuous, 3–5-seriate peripheral sclerenchymatic belt. Chlorenchyma groups 1–2-seriate, tangentially enlarged. Intercellulars in the parenchymatic zone small and triangular. Vascular bundles arranged in 2 circular offset rows. Large bundles within the thin- to thick-walled parenchymatic tissue, small bundles within the sclerenchymatic belt. Vascular bundles with large vessels, diameter 20–25 µm. Phloem oval, consisting of sieve tubes and companion cells. Intercellulars centripetal of the protoxylem absent. Vascular bundles surrounded by a 2-seriate sheath of polygonal, thin- to thick-walled cells.

Analyzed: 1 individual

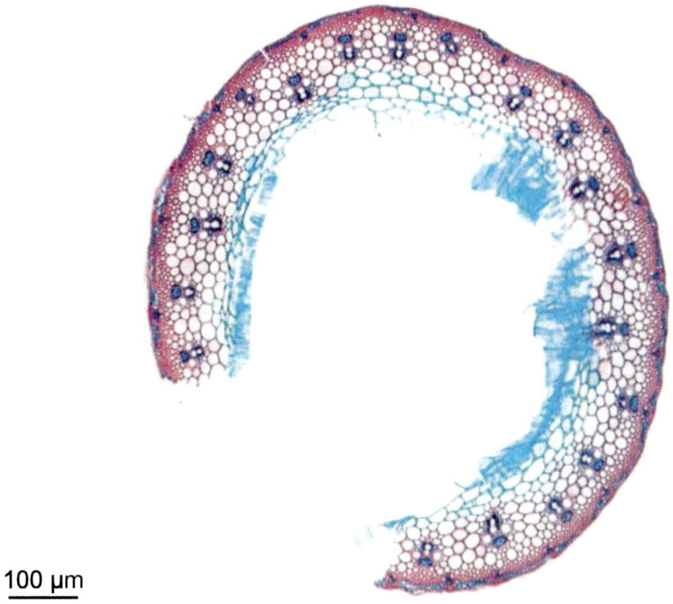

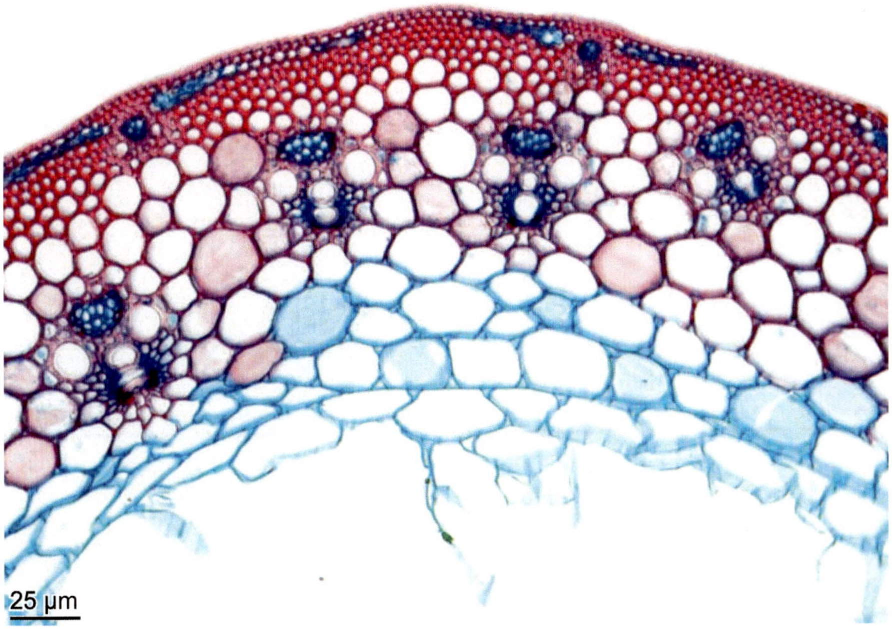

Piptatherum hilariae Pazij
Poaceae

Syn.: *Oryzopsis hilariae* (Pazij) Uniyal, *O. humilis* Bor

Description: Perennial herbs, 10–50 cm tall, densely cespitose, leaves flat or rolled, 5–30 cm long, 1.5–4.5 mm wide, panicle open or contracted, 5–12 cm long, branches spreading at first, erect at maturity.

Distribution: E Afghanistan, Tajikistan, Pakistan, NW India, China (Xizang).

Habitat: River banks.

Culm uniform, round, hollow, with a diameter of 0.7–0.8 mm. Epidermis thin- to thick-walled. Continuous, 2–4-seriate peripheral sclerenchymatic belt centripetal of the chlorenchyma. Girders few, small and square. Continuous, 1–2-seriate belt of chlorenchyma. Intercellulars in the parenchymatic zone small and triangular. Vascular bundles arranged in 2–3 circular offset rows. Large bundles within the thin- to thick-walled parenchymatic tissue, small bundles within the sclerenchymatic belt. Vascular bundles with large vessels, diameter 15–20 µm. Phloem oval, consisting of uniform cells. Intercellulars centripetal of the protoxylem mostly absent. Vascular bundles surrounded by a uniseriate sheath of polygonal, thin- to thick-walled cells.

Analyzed: 2 individuals

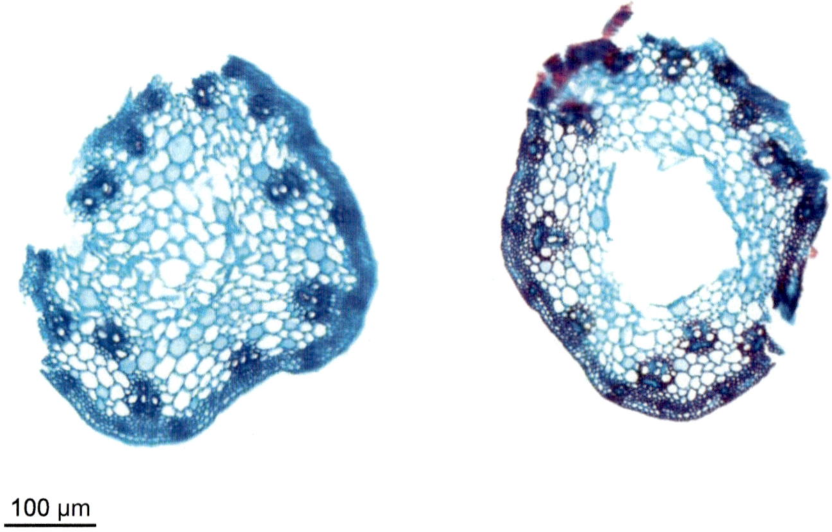

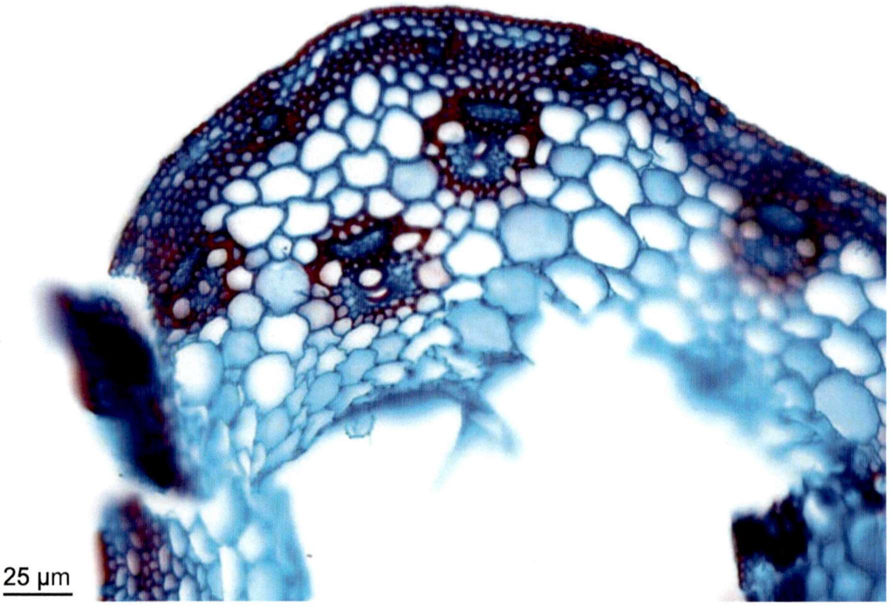

Piptatherum laterale (Munro ex Regel) Roshev.

Poaceae

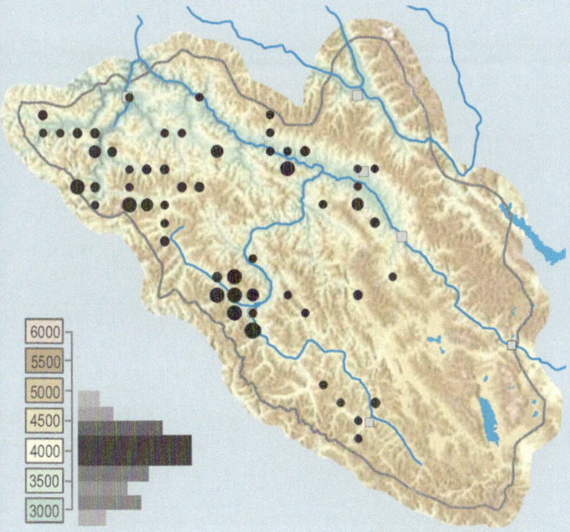

Syn.: *Oryzopsis lateralis* (Regel) Stapf

Description: Perennial herbs, 25–60 cm tall, densely cespitose, leaves setaceous, usually rolled, panicle loosely contracted, linear, 5–20 cm long, 0.5–1 cm wide, awns 2–4 cm tall.

Distribution: E Turkey, Iraq, Iran, Uzbekistan, Kyrgyzstan, Tajikistan, Afghanistan, N Pakistan, NW India, Nepal, China (Sichuan, Xizang).

Habitat: Stony mountain slopes, steppes.

Culm uniform, round, hollow, with a diameter of 1.1–1.2 mm. Epidermis thin- to thick-walled. Continuous, 3–5-seriate peripheral sclerenchymatic belt. Chlorenchyma groups 1–2-seriate, tangentially enlarged. Intercellulars in the parenchymatic zone small and triangular. Vascular bundles arranged in 2–3 circular offset rows. Large bundles within the thin- to thick-walled parenchymatic tissue, small bundles within the sclerenchymatic belt. Vascular bundles with large vessels, diameter 20–25 µm. Phloem oval, consisting of sieve tubes and companion cells. With small intercellulars centripetal of the protoxylem. Vascular bundles surrounded by a 1–2-seriate sheath of polygonal, thin- to thick-walled cells.

Analyzed: 2 individuals

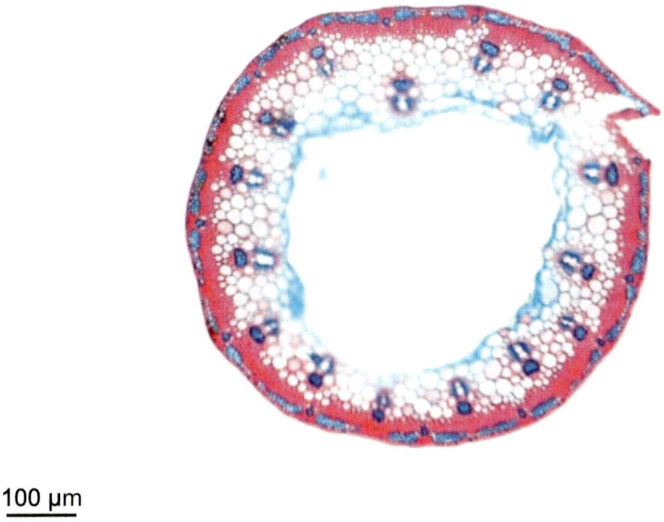

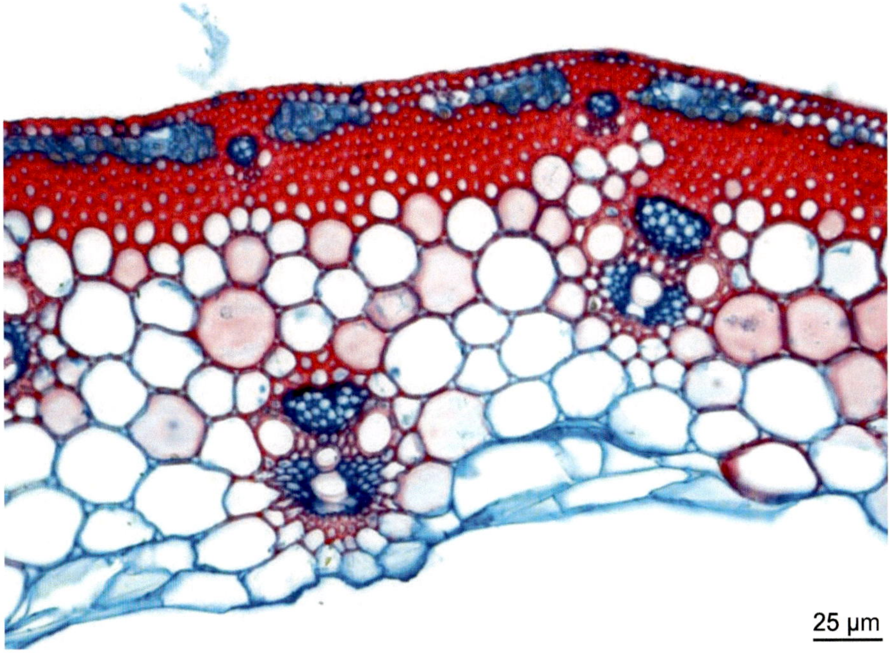

Poa alpina L.

Poaceae

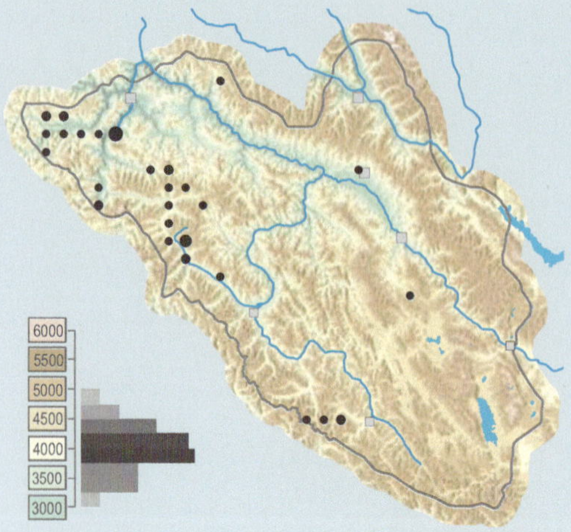

Description: Perennial herbs, 10–30 cm tall, densely cespitose, leaves flat or folded, panicle ovate to pyramidal, open, dense or loose, or nodding, 3–7 cm long, 3–7 cm wide.

Distribution: North America, Europe, Russia (Siberia, Far East), Caucasus, Turkey, Iran, Kazakhstan, Uzbekistan, Kyrgyzstan, Tajikistan, Afghanistan, N Pakistan, N India, Nepal, China (Qinghai, Xinjiang, Xizang), Mongolia, Japan.

Habitat: Alpine meadows.

Culm uniform, round, hollow, with a diameter of 0.6–0.7 mm. Epidermis thick-walled. Continuous, 3–6-seriate peripheral sclerenchymatic belt. With small irregularly formed groups of chlorenchyma. Intercellulars in the parenchymatic tissue absent. Vascular bundles arranged in 2 circular offset rows. Large bundles within the thin- to thick-walled parenchymatic tissue, small bundles within the sclerenchymatic belt. Vascular bundles with large vessels, diameter 10–15 µm. Phloem oval, consisting of uniform cells. Intercellulars centripetal of the protoxylem absent. Vascular bundles are surrounded by a 1–2-seriate sheath of polygonal, thin- to thick-walled cells.

Analyzed: 2 individuals

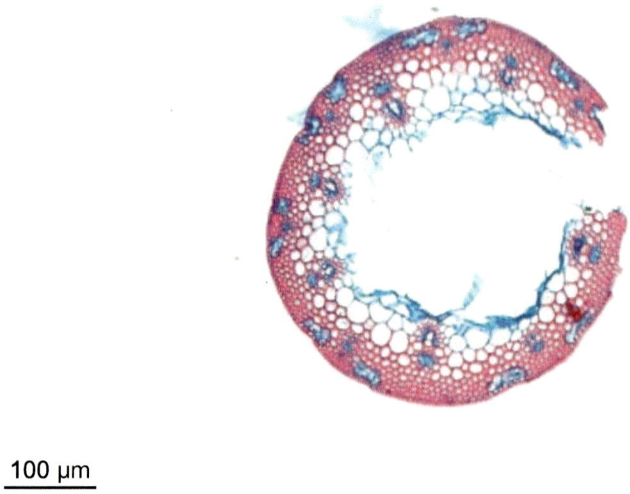

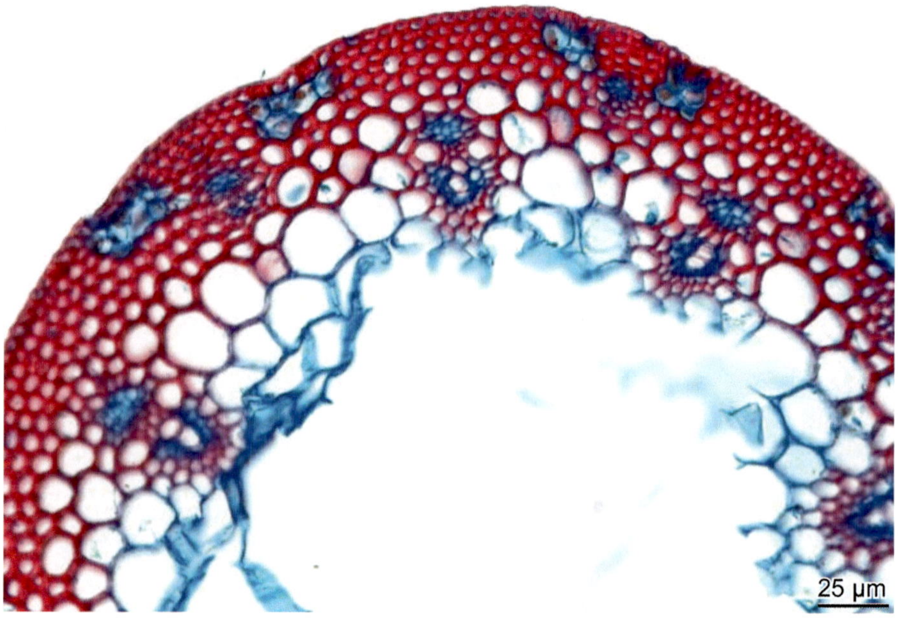

Poa attenuata Trin.

Poaceae

Description: Perennial herbs, 5–50 cm tall, cespitose, leaves mostly basal, linear, flat or convolute, panicle contracted, oblong, 2.5–9 cm long.

Distribution: Russia (Siberia), Kazakhstan, Uzbekistan, Kyrgyzstan, Tajikistan, N Pakistan, N India, Nepal, Bhutan, China (Gansu, Hebei, Nei Mongol, Qinghai, Shaanxi, Sichuan, Xinjiang, Xizang, Yunnan), Mongolia.

Habitat: Stony steppes, dry river beds, screes, mesic turf, alpine slopes.

Culm uniform, round, hollow, with a diameter of 0.9–1.0 mm. Epidermis thick-walled. Continuous, 6–8-seriate peripheral sclerenchymatic belt. Groups of chlorenchyma tangentially enlarged. Girders tangentially rectangular. Intercellulars in the parenchymatic zone small and triangular. Vascular bundles arranged in 2 circular offset rows. Large bundles within the thin- to thick-walled parenchymatic tissue, small bundles within the sclerenchymatic belt. Vascular bundles with large vessels, diameter 12–15 µm. Phloem oval, consisting of sieve tubes and companion cells. With few intercellulars centripetal of the protoxylem. Vascular bundles surrounded by a 1–2-seriate sheath of polygonal, thin- to thick-walled cells.

Analyzed: 2 individuals

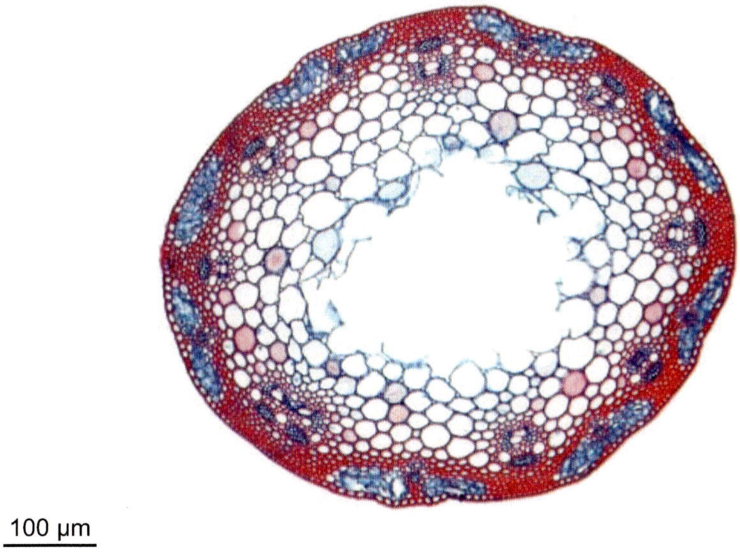

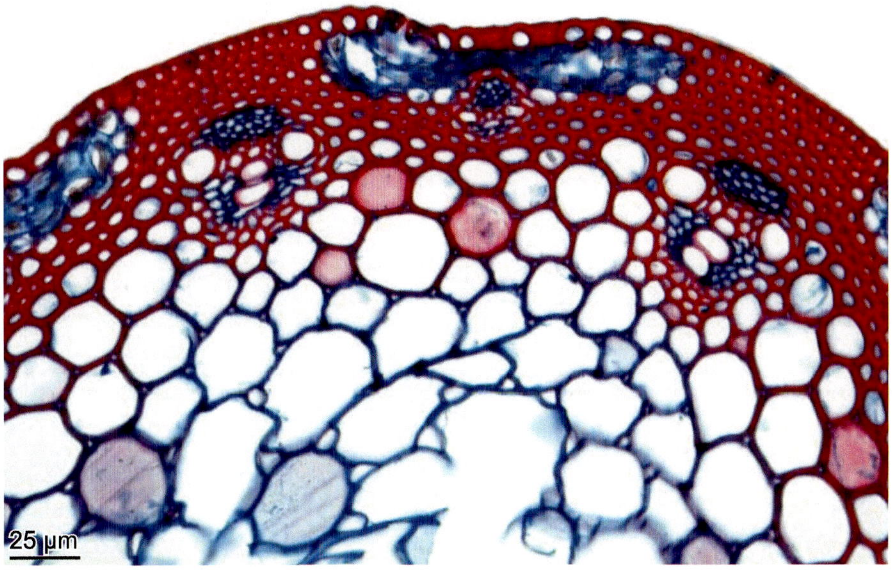

Poa bactriana Roshev.

Poaceae

Description: Perennial herbs, 5–60 cm tall, densely cespitose, culm bases thickened, leaves flat or folded, 2–15 cm long, 1–3 mm wide, panicle contracted, oblong to pyramidal, 2–10 cm long, branches spreading.

Distribution: Iran, Kazakhstan, Turkmenistan, Uzbekistan, Kyrgyzstan, Tajikistan, Afghanistan, Pakistan, N India (Jammu and Kashmir), China (Xinjiang, Xizang).

Habitat: Dry grassy hillsides, screes.

Culm uniform, round, hollow, with a diameter of 0.9–1.0 mm. Epidermis thin- to thick-walled. Continuous, 4–5-seriate peripheral sclerenchymatic belt. Groups of chlorenchyma small, tangentially enlarged. Intercellulars in the parenchymatic zone very small and triangular. Vascular bundles arranged in 2 circular offset rows. Large bundles within the thin- to thick-walled parenchymatic tissue, small bundles within the sclerenchymatic belt. Vascular bundles with large vessels, diameter 25–30 µm. Phloem oval, consisting of uniform cells. With intercellulars centripetal of the protoxylem. Vascular bundles surrounded by a 1–2-seriate sheath of polygonal, thin- to thick-walled cells.

Analyzed: 1 individual

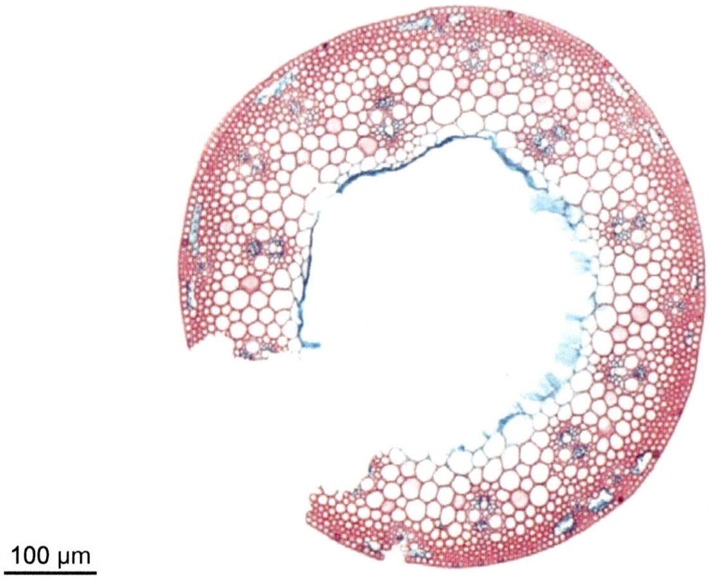

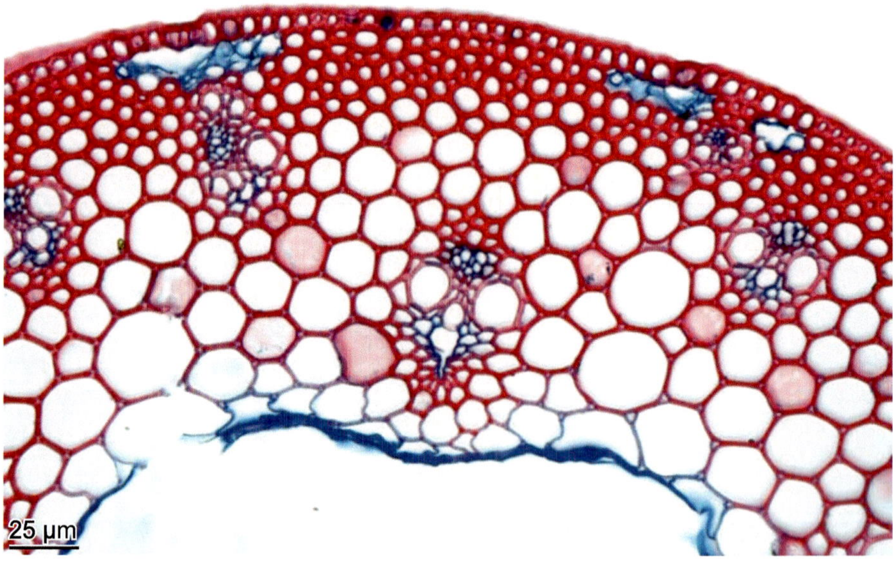

Poa bulbosa L.

Poaceae

Description: Perennial herbs, 5–40 cm tall, cespitose, culm bases thickened, leaves flat or conduplicate, 1–10 cm long, 1–2 mm wide, panicle open, ovate, dense, 2–6 cm long, 1–2.5 cm wide, branches ascending.

Distribution: Africa, Europe, SW Asia, Russia (Siberia), Kazakhstan, Turkmenistan, Uzbekistan, Kyrgyzstan, Tajikistan, Afghanistan, Pakistan, NW India, Nepal, China (Xinjiang, Xizang); introduced elsewhere.

Habitat: Dry grasslands, rocky slopes.

Culm uniform, round, hollow, with a diameter of 0.4–0.6 mm. Epidermis thick-walled. Continuous, 4–5-seriate peripheral sclerenchymatic belt. Chlorenchyma absent. Intercellulars in the parenchymatic zone very small and triangular. Vascular bundles arranged in 1 circular offset row. Large bundles within the thin- to thick-walled parenchymatic tissue, small bundles within the sclerenchymatic belt. Vascular bundles with large vessels, diameter 10–15 µm. Phloem oval, consisting of uniform cells. Intercellulars centripetal of the protoxylem absent. Vascular bundles surrounded by a 1–2-seriate sheath of polygonal, thin- to thick-walled cells.

Analyzed: 2 individuals

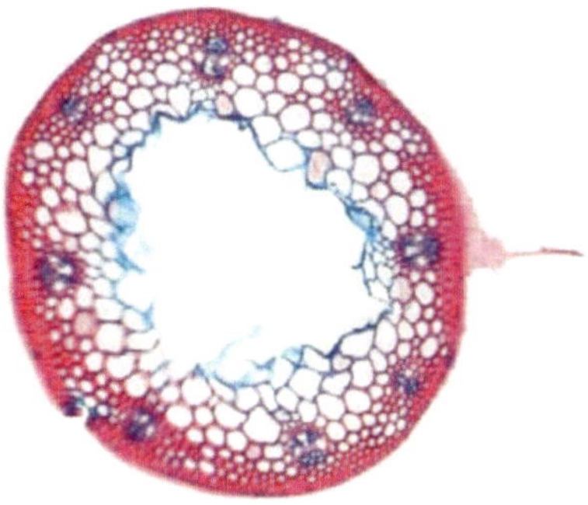

100 µm

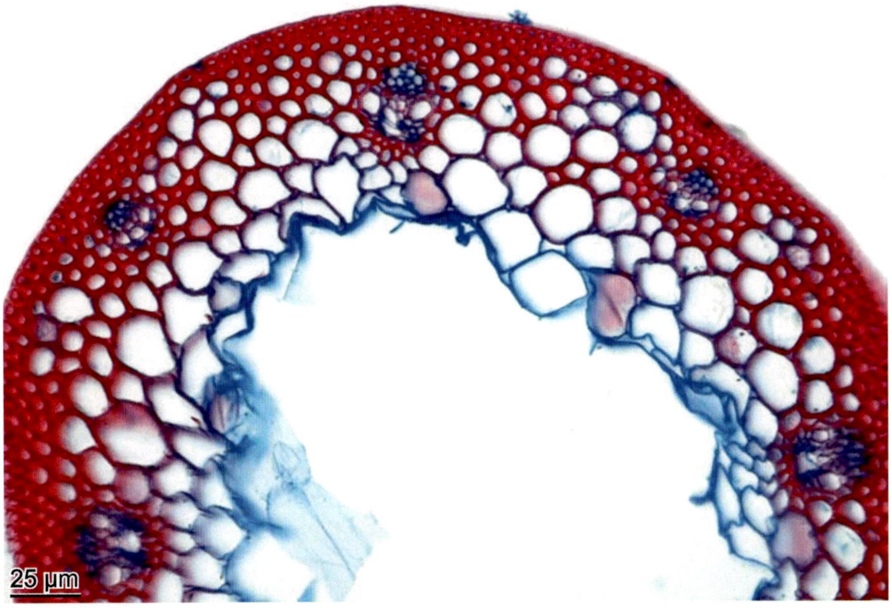

Poa calliopsis Litw. ex Ovcz.

Poaceae

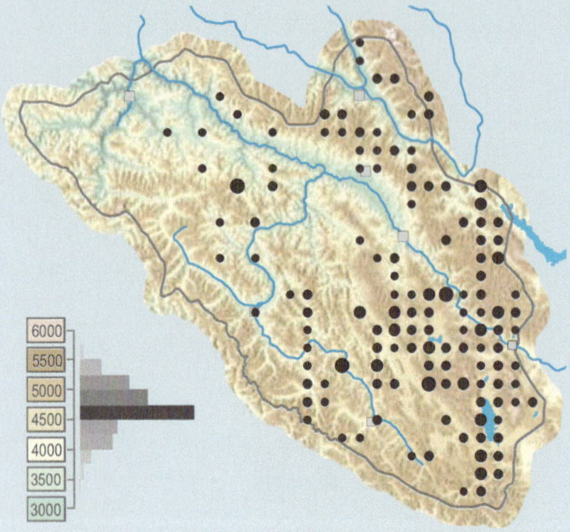

Description: Perennial herbs, 5–15 cm tall, rhizomatous, leaves mostly basal, linear, conduplicate, panicle open, oblong, or pyramidal, 1.5–4.5 cm long, spikelets clustered towards branch tips.

Distribution: Kyrgyzstan, Tajikistan, Pakistan, India, Nepal, China (Gansu, Qinghai, Sichuan, Xinjiang, Xizang, Yunnan), Bhutan.

Habitat: River banks, lake shores, mesic to wet grasslands, along irrigation canals.

Culm uniform, round, hollow, with a diameter of 0.5 mm. Epidermis thick-walled. Continuous, 3–4-seriate peripheral sclerenchymatic belt. Chlorenchyma absent. Intercellulars in the parenchymatic zone very small and triangular. Vascular bundles arranged in 1 circular offset rows. Large bundles within the thin- to thick-walled parenchymatic tissue, small bundles within the sclerenchymatic belt. Vascular bundles with large vessels, diameter 8–10 µm. Phloem oval, consisting of uniform cells. Intercellulars centripetal of the protoxylem absent. Vascular bundles surrounded by a 1–2-seriate sheath of polygonal, thin- to thick-walled cells.

Analyzed: 2 individuals

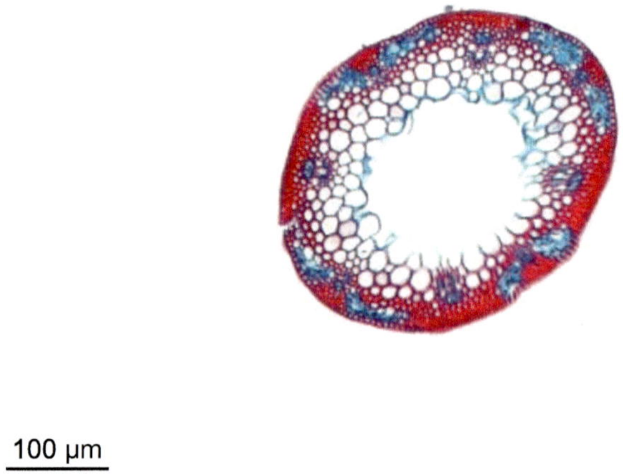

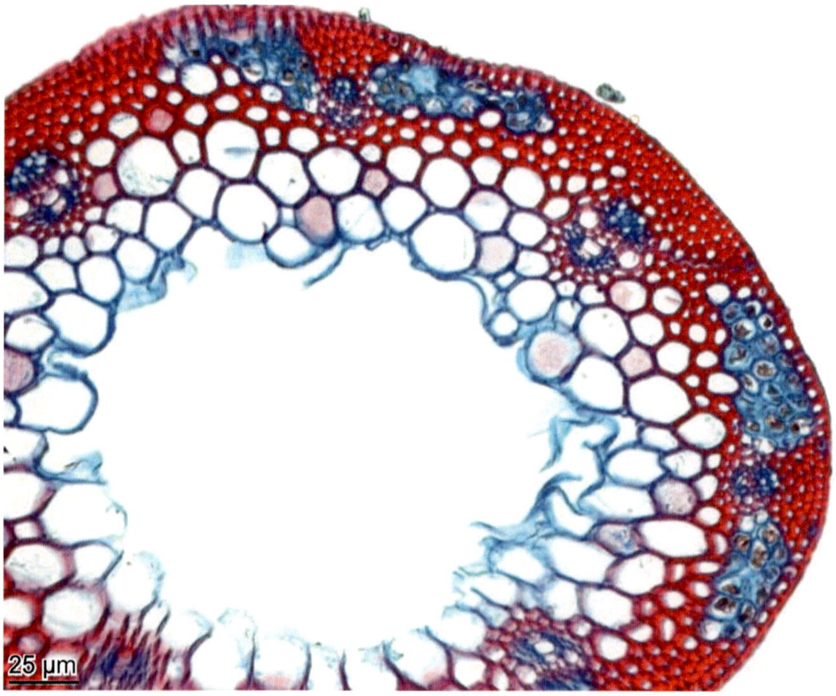

Poa falconeri Hook.f.

Poaceae

Syn.: *P. gilgitica* Dickore

Description: Perennial herbs, 15–80 cm tall, loosely cespitose, culms decumbent, leaves flat, 5–20 cm long, 2–4 mm wide, panicle open, lanceolate, 8–15 cm long, branches appressed.

Distribution: Pakistan, India (Jammu and Kashmir, Himachal Pradesh, Uttar Pradesh), Nepal, China (Xizang).

Habitat: Alpine meadows.

Culm uniform, round, hollow, with a diameter of 0.5–0.7 mm. Epidermis thin- to thick-walled. Continuous, 3–4-seriate peripheral sclerenchymatic belt. Groups of chlorenchyma tangentially enlarged. Small triangular intercellulars in the parenchymatic tissue absent. Vascular bundles arranged in 2 circular offset rows. Large bundles within the thin- to thick-walled parenchymatic tissue, small bundles within the sclerenchymatic belt. Vascular bundles with large vessels, diameter 10–12 µm. Phloem oval, consisting of uniform cells. With a few intercellulars centripetal of the protoxylem. Vascular bundles surrounded by a 1–2-seriate sheath of polygonal, thin- to thick-walled cells.

Analyzed: 1 individual

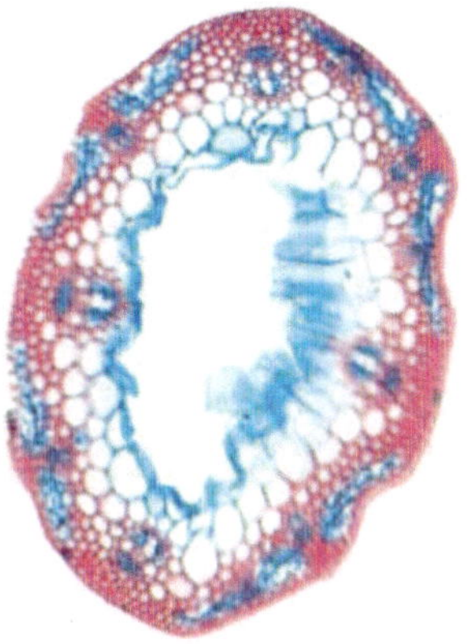

100 µm

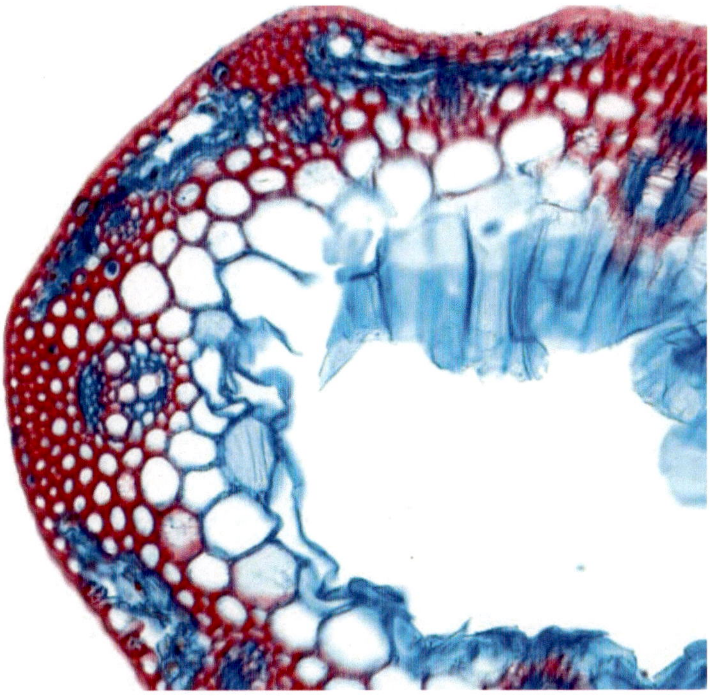

25 µm

Poa pagophila Bor
Poaceae

Syn.: *P. harae* Rajbh., *P. levipes* Keng ex L.Liu; *P. nigropurpurea* C.Ling.

Description: Perennial herbs, 10–40 cm tall, cespitose, leaves flat, or conduplicate, 3–15 cm long. 1.5–2.5 mm wide, panicle open, pyramidal, 6–15 cm long, branches spreading or reflexed.

Distribution: Pakistan, N India, Nepal, China (W Sichuan, Xizang, NW Yunnan), Bhutan.

Habitat: Alpine grasslands.

Culm uniform, round, hollow, with a diameter of 0.7–0.8 mm. Epidermis thin- to thick-walled. Continuous, 2–3-seriate peripheral sclerenchymatic belt centripetal of the chlorenchyma band. With a 1–2-seriate band of chlorenchyma. Intercellulars in the parenchymatic tissue absent. Vascular bundles arranged in 2 circular offset rows. Large bundles within the thin- to thick-walled parenchymatic tissue, small bundles within the sclerenchymatic belt. Vascular bundles with large vessels, diameter 4–5 µm. Phloem oval, consisting of uniform cells. Intercellulars centripetal of the protoxylem absent. Vascular bundles surrounded by a 1–2-seriate sheath of polygonal, thin- to thick-walled cells.

Analyzed: 2 individuals

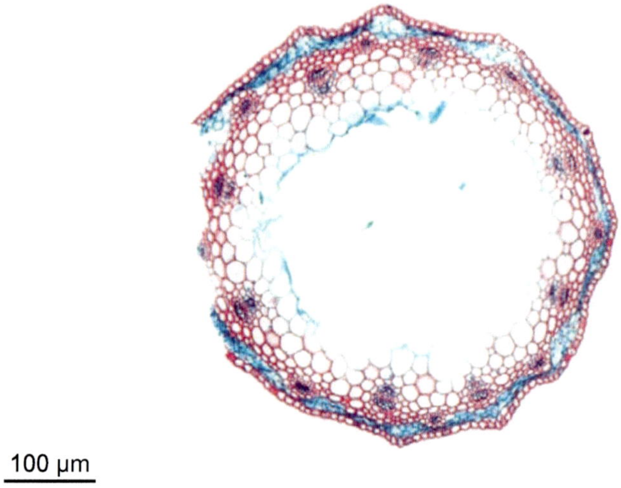

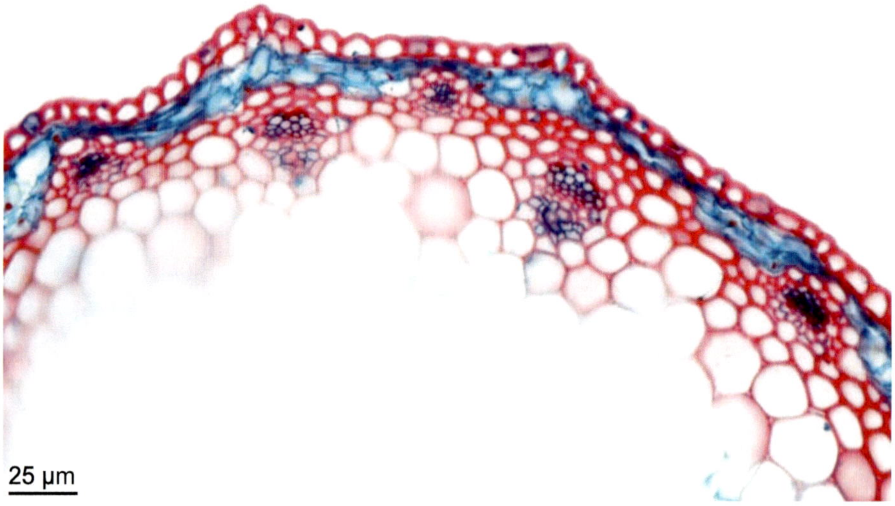

Poa pratensis L.

Poaceae

Description: Perennial herbs, 10–90 cm tall, loosely cespitose, leaves flat or conduplicate, 5–30 cm long, 2–4 mm wide, panicle open or contracted, oblong, 5–15 cm long, branches spreading.

Distribution: Cosmopolitan.

Habitat: Grasslands.

Culm uniform, round, hollow, with a diameter of 0.8–1.0 mm. Epidermis thin- to thick-walled. Continuous, 6–8-seriate peripheral sclerenchymatic belt. Girders tangentially rectangular. Groups of chlorenchyma tangentially enlarged. Intercellulars in the parenchymatic zone small and triangular. Vascular bundles arranged in 2 circular offset rows. Large bundles within the thin- to thick-walled parenchymatic tissue, small bundles within the sclerenchymatic belt. Vascular bundles with large vessels, diameter 15–25 µm. Phloem oval, consisting of uniform cells. With few intercellulars centripetal of the protoxylem. Vascular bundles surrounded by a 1–3-seriate sheath of polygonal, thin- to thick-walled cells.

Analyzed: 4 individuals

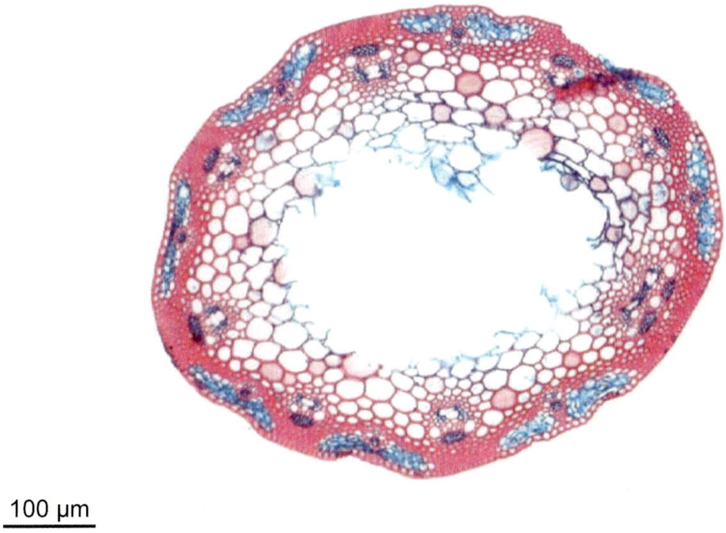

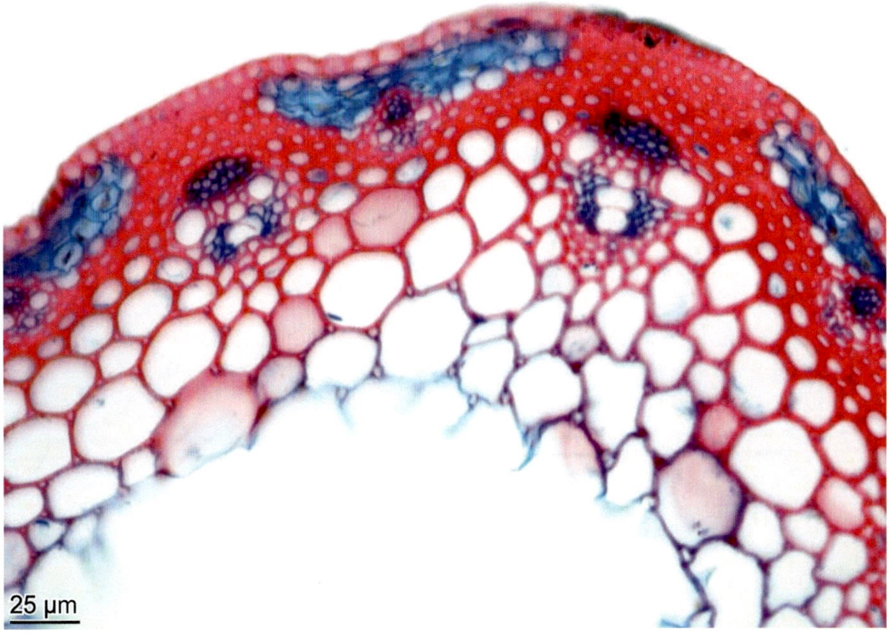

Poa sikkimensis (Stapf) Bor

Poaceae

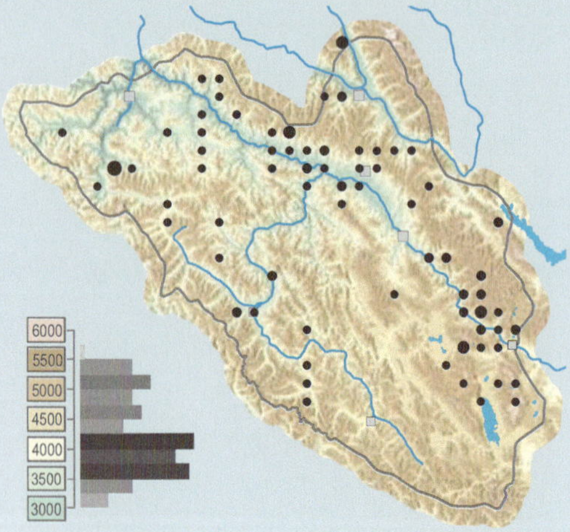

Syn.: *P. annua* var. *sikkimensis* Stapf

Description: Annuals or short-lived perennials, tufted to weakly stoloniferous, culms 5–40 cm tall, leaves flat, panicle loosely contracted to open, oblong to pyramidal, 3–15 cm long.

Distribution: India (Jammu and Kashmir, Sikkim, Assam), Nepal, Bhutan, China (Gansu, Qinghai, Sichuan, Xizang, Yunnan).

Habitat: Wet turf and banks of irrigation canals in settlements, roadsides.

Notes: The Ladakh (Indus valley) populations comprise a probably isolated western outpost of this species' distribution.

Culm uniform, round, hollow, with a diameter of 1.2–1.5 mm. Epidermis thin- to thick-walled. Continuous, 6–10-seriate peripheral sclerenchymatic belt. Girders tangentially rectangular and conical. Chlorenchyma groups round and square. Triangular intercellulars in the parenchymatic tissue absent. Vascular bundles arranged in 2 circular offset rows. Large bundles within the thin- to thick-walled parenchymatic tissue, small bundles within the sclerenchymatic belt. Vascular bundles with large vessels, diameter 12–15 µm. Phloem oval, consisting of uniform cells. With few intercellulars centripetal of the protoxylem. Vascular bundles surrounded by a 1–2-seriate sheath of polygonal, thin- to thick-walled cells.

Analyzed: 2 individuals

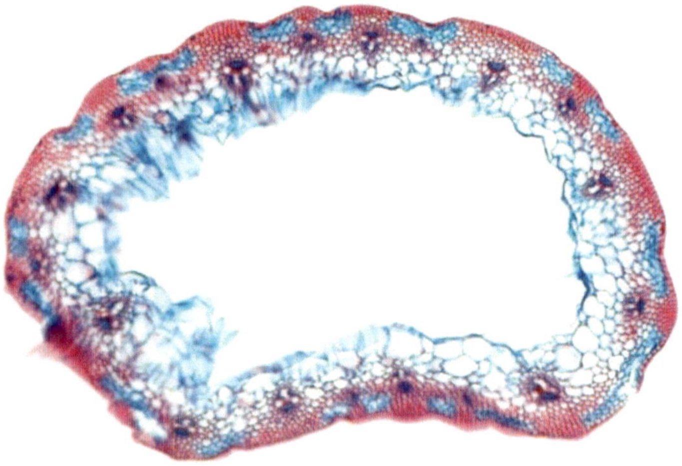

100 µm

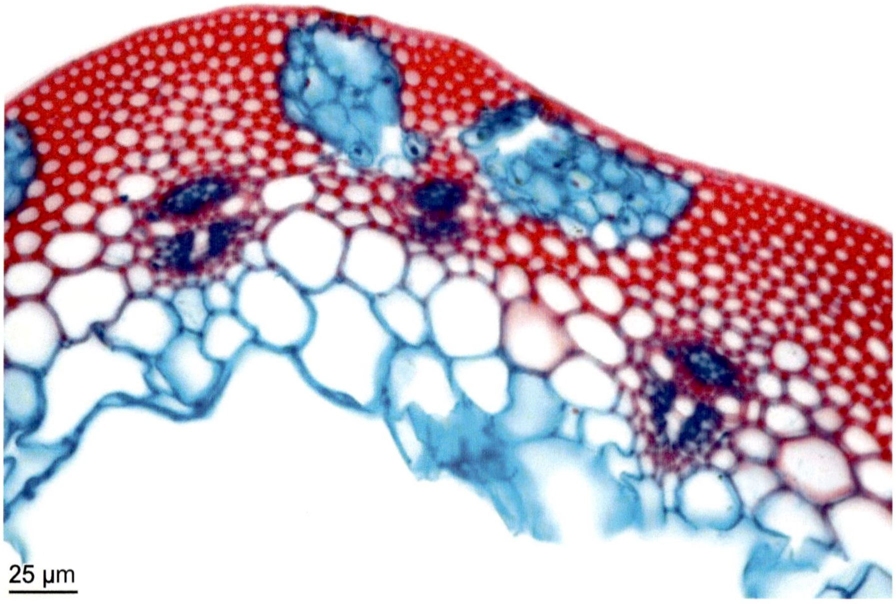

Poa sterilis M.Bieb.

Poaceae

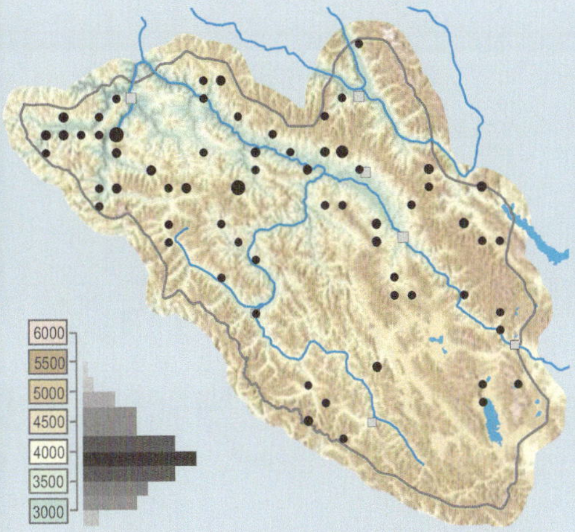

Description: Perennial herbs, 25–40 cm tall, cespitose, leaves usually flat, flaccid, panicle narrow, 7–15 cm long, branches ascending, flexuous.

Distribution: E Europe, Iraq, Syria, Turkey, Iran, Georgia, Afghanistan, W Pakistan, NW India.

Habitat: Among stones, at the foot of stony walls.

Culm uniform, round, hollow, with a diameter of 1.0–1.2 mm. Epidermis thin- to thick-walled. Continuous, 3–4-seriate peripheral sclerenchymatic belt centripetal of the chlorenchyma. Girders radially rectangular and conical. Chlorenchyma groups square to rectangular. Triangular intercellulars in the parenchymatic tissue absent. Vascular bundles arranged in 2 circular offset rows. Large bundles within the thin- to thick-walled parenchymatic tissue, small bundles within the sclerenchymatic belt. Vascular bundles with large vessels, diameter 15–20 µm. Phloem oval, consisting of uniform cells. Intercellulars centripetal of the protoxylem absent. Vascular bundles surrounded by a 1–2-seriate sheath of polygonal, thin- to thick-walled cells.

Analyzed: 2 individuals

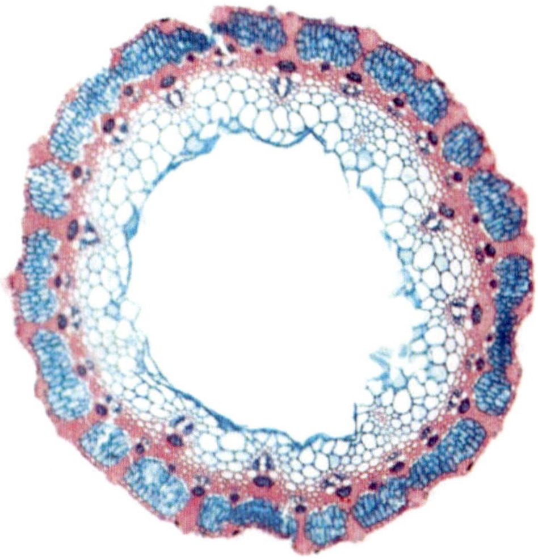

100 µm

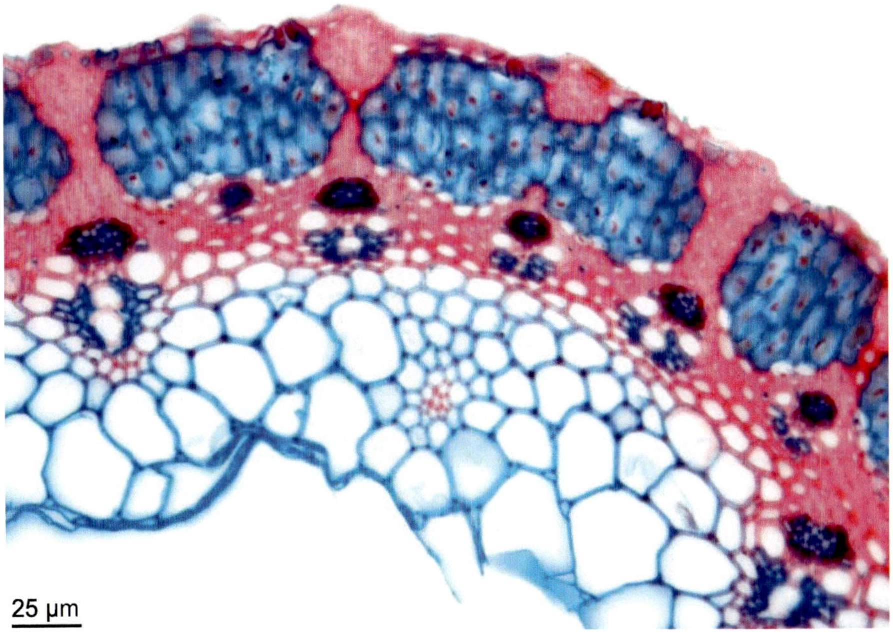

25 µm

Poa supina Schrad.

Poaceae

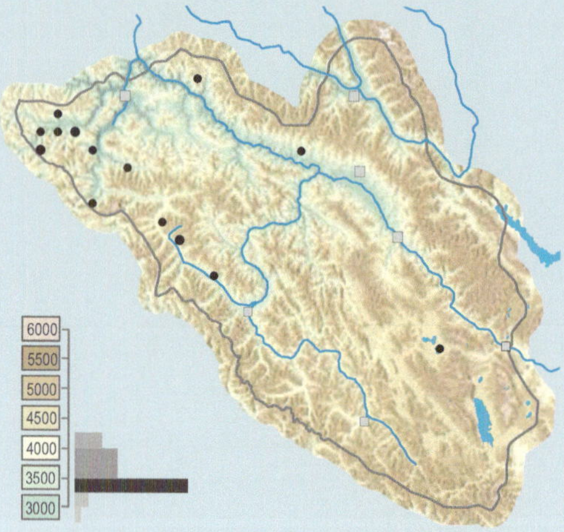

Description: Perennial herbs, 5–25 cm tall, cespitose, culms decumbent at base, leaves flat or conduplicate, 1–7 cm long, 2–3 mm wide, panicle open, pyramidal, 2–3 cm long, branches spreading.

Distribution: Europe, Russia (Siberia, Far East), SW Asia, Tajikistan, Afghanistan, Pakistan, India (Jammu and Kashmir), Nepal, China (Xinjiang, Xizang, Sichuan, Yunnan), Mongolia; introduced to North America.

Habitat: Subalpine meadows, moist places.

Culm uniform, round, hollow, with a diameter of 0.6–0.8 mm. Epidermis thin- to thick-walled. Continuous, 3–4-seriate, wavy peripheral sclerenchymatic belt centripetal of the chlorenchyma. Girders square to tangentially rectangular. Chlorenchyma groups round and tangentially enlarged. Intercellulars in the parenchymatic zone small and triangular. Vascular bundles arranged in 2 circular offset rows. Large bundles within the thin- to thick-walled parenchymatic tissue, small bundles within the sclerenchymatic belt. Vascular bundles with large vessels, diameter 10–15 µm. Phloem oval, consisting of uniform cells. Intercellulars centripetal of the protoxylem absent. Vascular bundles surrounded by a 1–2-seriate sheath of polygonal, thin- to thick-walled cells.

Analyzed: 2 individuals

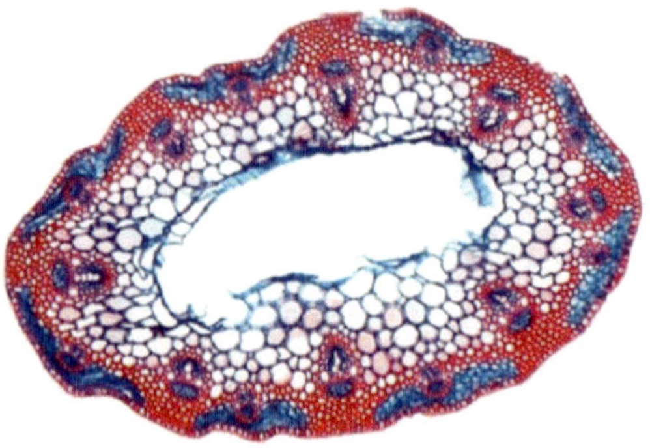

100 µm

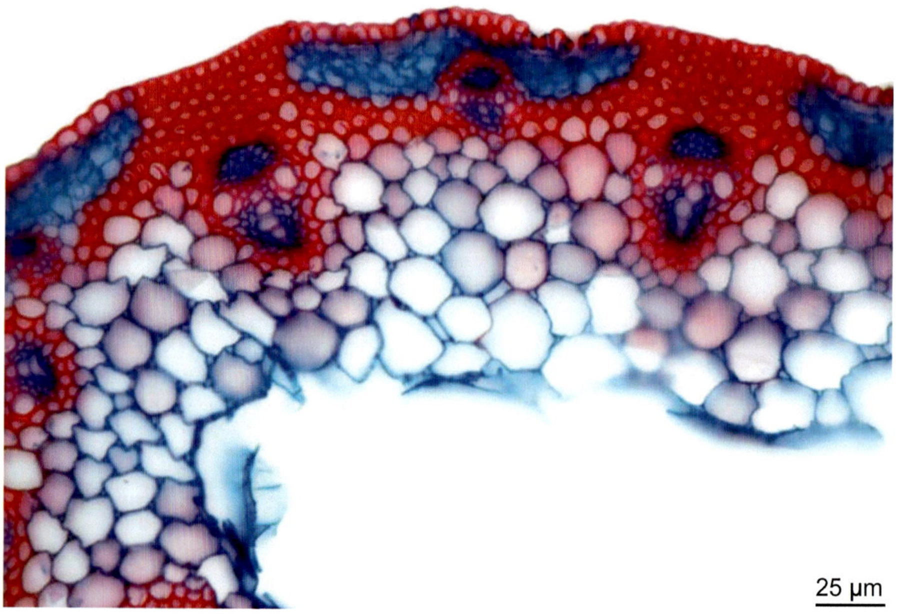

25 µm

Poa suruana H.Hartmann

Poaceae

Description: Perennial herbs, 15–80 cm tall, loosely cespitose, leaves flat, 5–20 cm long, 2–4 mm wide, panicle open, lanceolate, 8–15 cm long, branches appressed.

Distribution: N Pakistan, India (Jammu and Kashmir).

Habitat: Alpine grassy slopes.

Culm uniform, round, hollow, with a diameter of 0.8–1.0 mm. Epidermis thin- to thick-walled. Continuous, 3–4-seriate, wavy peripheral sclerenchymatic belt centripetal of the chlorenchyma. Girders of different size, square and conical. Chlorenchyma groups tangentially enlarged. Triangular intercellulars in the parenchymatic tissue absent. Vascular bundles arranged in 2 circular offset rows. Large bundles within the thin- to thick-walled parenchymatic tissue, small bundles within the sclerenchymatic belt. Vascular bundles with large vessels, diameter 15–20 µm. Phloem oval, consisting of sieve tubes and companion cells. With intercellulars centripetal of the protoxylem. Vascular bundles surrounded by a 1–2-seriate sheath of polygonal, thin- to thick-walled cells.

Analyzed: 2 individuals

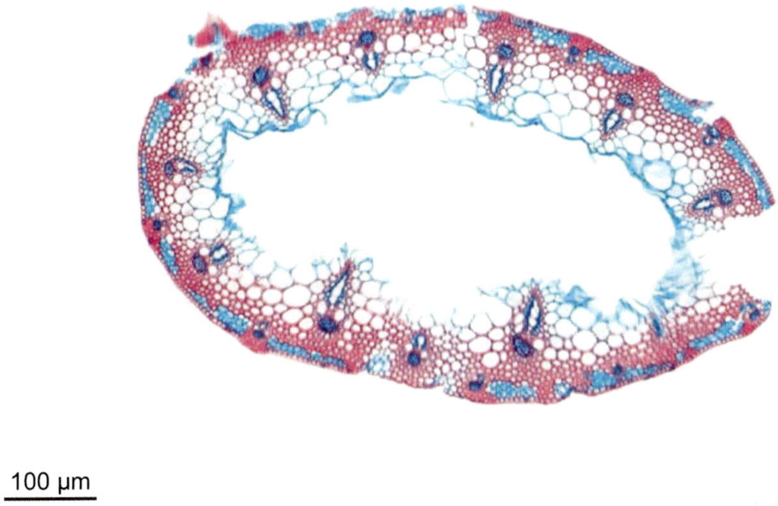

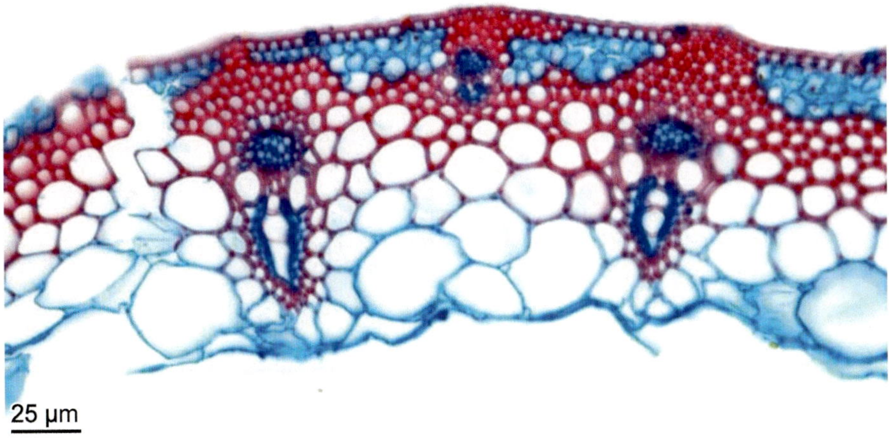

Poa tibetica Munro ex Stapf
Poaceae

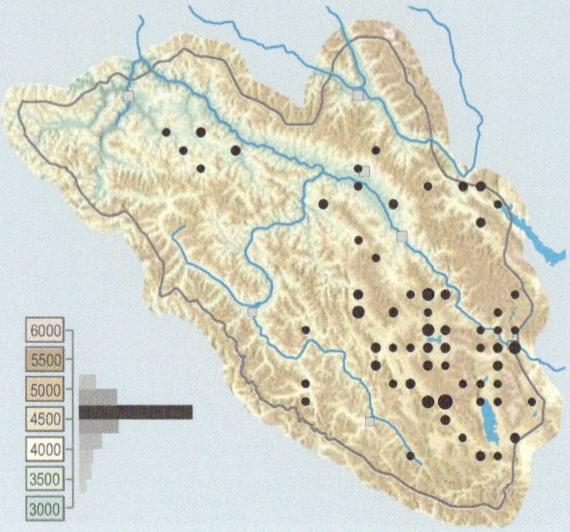

Syn.: *Arctopoa tibetica* (Munro ex Stapf) Prob., *P. pricei* N.D.Simpson

Description: Perennial herbs, 20–60 cm tall, cespitose, rhizomatous, leaves rolled or folded, panicle contracted, often spike-like, linear, 5–12 cm long.

Distribution: Iran, Russia (Siberia), Kazakhstan, Kyrgyzstan, Tajikistan, Pakistan, N India, Nepal, China (Gansu, Nei Mongol, Qinghai, Xinjiang, Xizang), Mongolia.

Habitat: Wet depressions, saline grasslands, river banks, wetlands, along irrigation canals.

Culm uniform, round, hollow, wavy, with a diameter of 1.3–1.5 mm. Epidermis thin- to thick-walled. Continuous, 6–10-seriate, wavy peripheral sclerenchymatic belt centripetal of the chlorenchyma. Girders square. Chlorenchyma groups tangentially enlarged. Triangular intercellulars in the parenchymatic tissue. Vascular bundles arranged in 2–3 circular offset rows. Large bundles within the thin- to thick-walled parenchymatic tissue, small bundles within the sclerenchymatic belt. Vascular bundles with large vessels, diameter 25–30 µm. Phloem oval, consisting of sieve tubes and companion cells. With large intercellulars centripetal of the protoxylem. Vascular bundles surrounded by a 2–3-seriate sheath of polygonal, thin- to thick-walled cells.

Analyzed: 1 individual

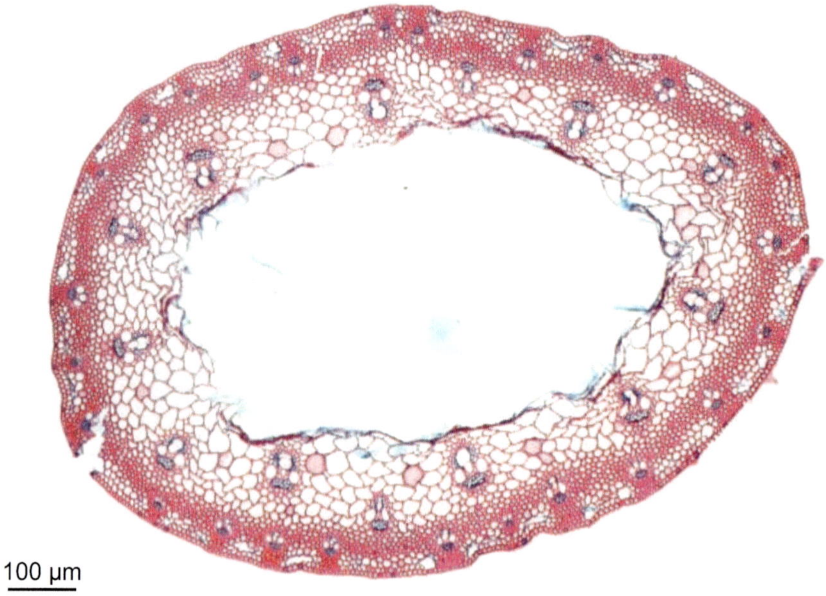

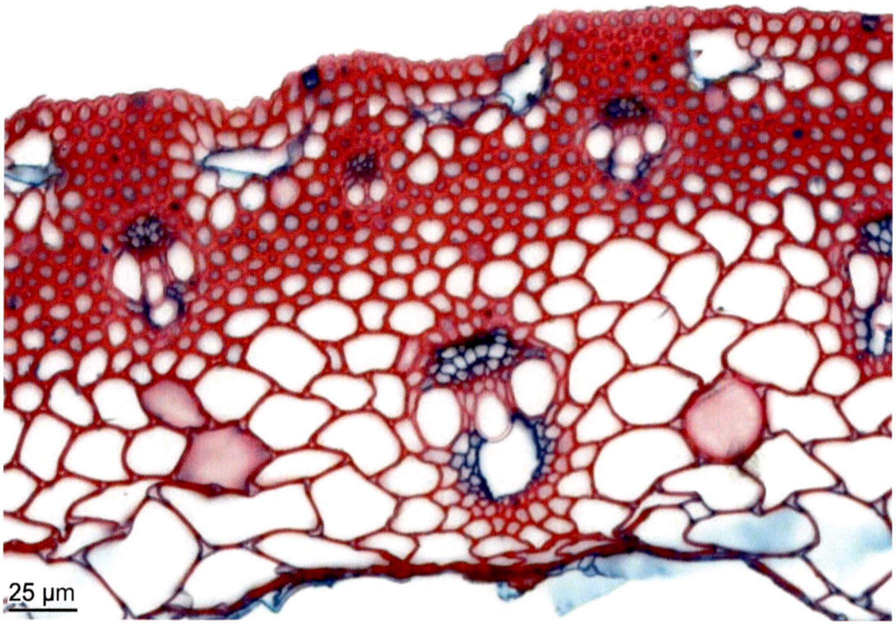

Puccinellia distans (Jacq.) Parl.

Poaceae

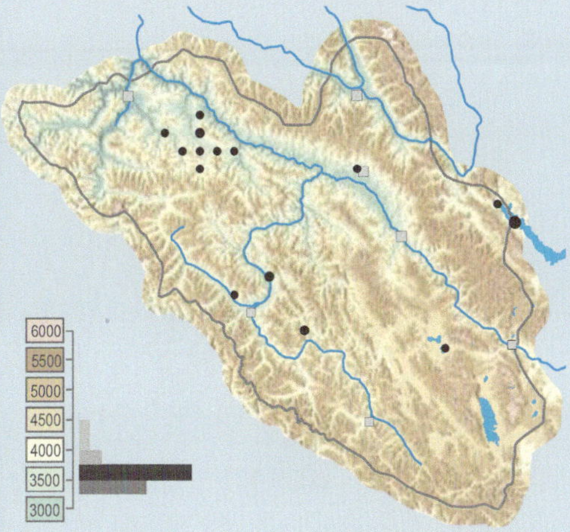

Syn.: *Poa distans* Jacq.

Description: Perennial herbs, 10–60 cm tall, cespitose, culms erect or decumbent, leaves flat, 2–10 cm long, 1.5–4 mm wide, apex hooded, panicle open, ovate, 3–15 cm long, 2–14 cm wide, branches spreading or reflexed.

Distribution: North America, NW Africa, Europe, SW Asia, Russia, Kazakhstan, Turkmenistan, Uzbekistan, Kyrgyzstan, Tajikistan, Pakistan, India, China, Mongolia, Korea Japan.

Habitat: Saline meadows and pastures.

Culm uniform, round, hollow, wavy, with a diameter of 0.7–0.8 mm. Epidermis thick-walled. Continuous, 6–8-seriate peripheral sclerenchymatic belt. Girders square to tangentially rectangular. Chlorenchyma groups oval. Intercellulars in the parenchymatic zone small and triangular. Vascular bundles arranged in 2 circular offset rows. Large bundles within the thin- to thick-walled parenchymatic tissue, small bundles within the sclerenchymatic belt. Vascular bundles with large vessels, diameter 10–15 μm. Phloem oval, consisting of uniform cells. With intercellulars centripetal of the protoxylem. Vascular bundles surrounded by a 1–2-seriate sheath of polygonal, thin- to thick-walled cells.

Analyzed: 1 individual

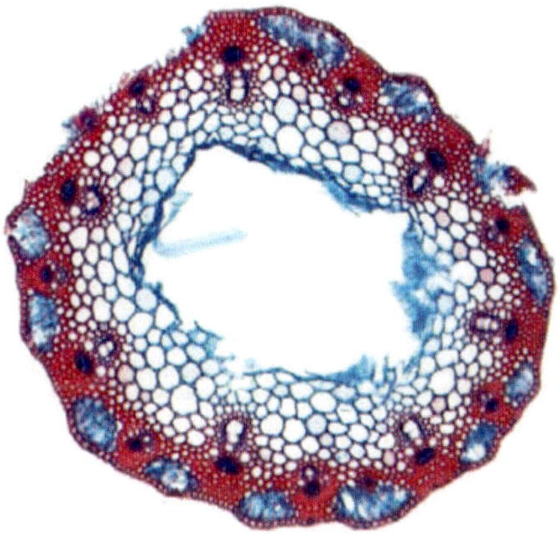

100 μm

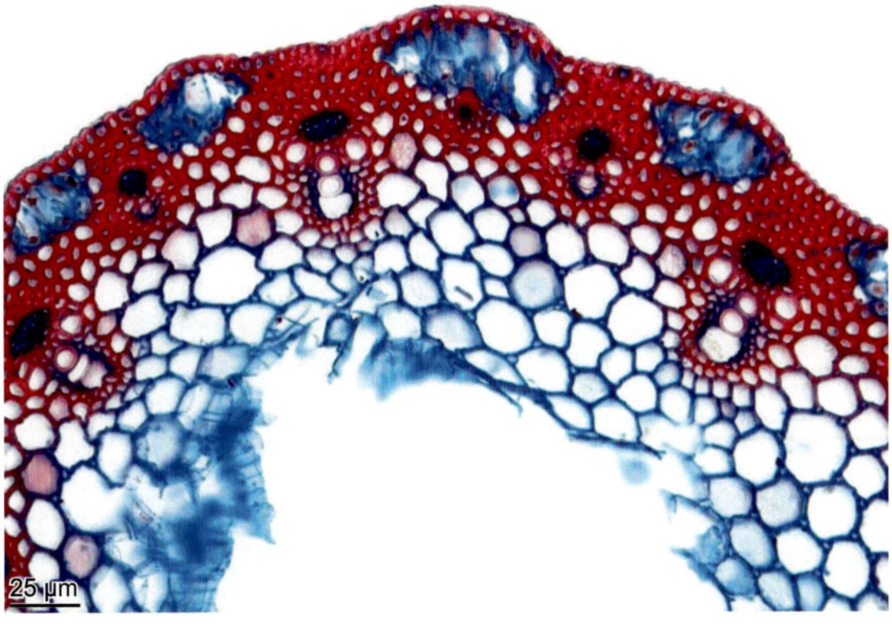

Puccinellia kashmiriana Bor

Poaceae

Description: Perennial herbs, 5–20 cm tall, densely cespitose, culms erect or geniculately ascending, leaves linear, 3–5 cm long, ca. 1 cm wide, upper leaves conduplicate, panicle very narrow, 3–4 cm long, 0.5 cm wide, branches ca. 1.5 cm long, ascending.

Distribution: Afghanistan, Pakistan, India (Jammu and Kashmir), China (Xinjiang, Xizang).

Habitat: Grassy places in open alpine valleys.

Culm uniform, round, hollow, with a diameter of 0.7–0.8 mm. Epidermis thick-walled. Continuous, 3–5-seriate peripheral sclerenchymatic belt. Girders absent. Groups of chlorenchyma absent. Triangular intercellulars in the parenchymatic tissue absent. Vascular bundles arranged in a circular row. Large bundles within the thin- to thick-walled parenchymatic tissue, small bundles within the sclerenchymatic belt. Vascular bundles with large vessels, diameter 15–20 µm. Phloem oval, consisting of uniform cells. Intercellulars centripetal of the protoxylem absent. Vascular bundles surrounded by a uniseriate sheath of polygonal, thin- to thick-walled cells.

Analyzed: 2 individuals

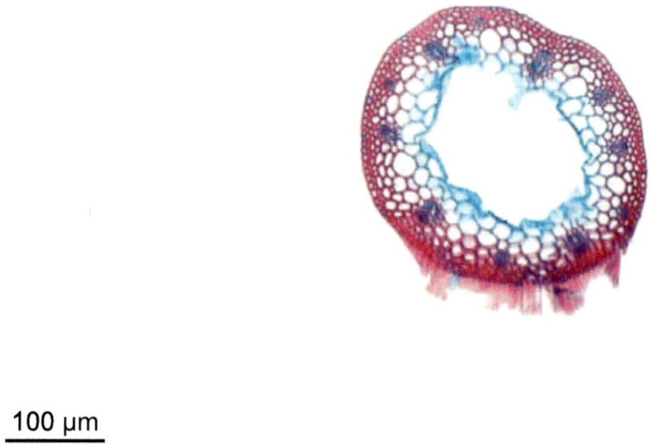

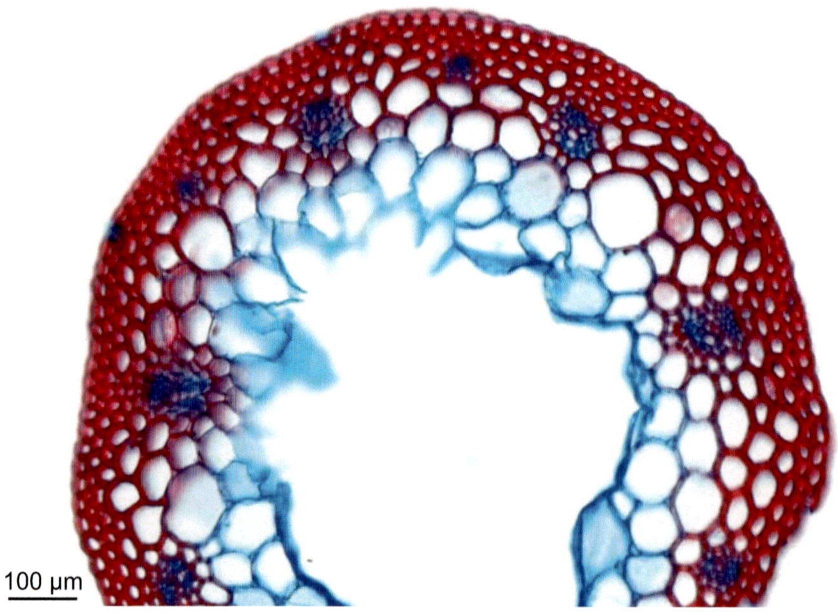

Puccinellia ladakhensis (H.Hartmann) Dickoré

Poaceae

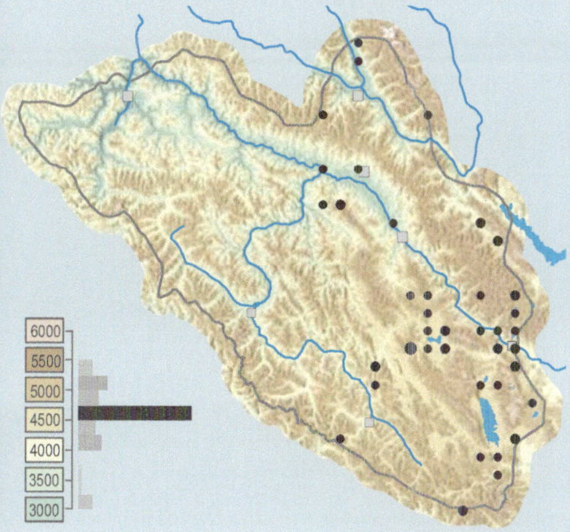

Syn.: *Poa ladakhensis* H.Hartmann

Description: Perennial herbs, 8–20 cm tall, loosely cespitose, leaves conduplicate or rolled, 2–5 cm long, panicle contracted, linear, 5–10 × 1–2 cm.

Distribution: NW India, Nepal, China (Xizang).

Habitat: Gravel and saline loamy grasslands, saline shores of rivers and lakes.

Culm uniform, round, hollow, with a diameter of 0.6–0.7 mm. Epidermis thin- to thick-walled. Continuous, 3–5-seriate peripheral sclerenchymatic belt. Girders absent. With round to oval groups of chlorenchyma. Intercellulars in the parenchymatic zone very small and triangular. Vascular bundles arranged in one circular row. Large bundles within the thin- to thick-walled parenchymatic tissue, small bundles within the sclerenchymatic belt. Vascular bundles with large vessels, diameter 10–12 µm. Phloem oval, consisting of uniform cells. With few intercellulars centripetal of the protoxylem. Vascular bundles surrounded by a 1–2-seriate sheath of polygonal, thin- to thick-walled cells.

Analyzed: 2 individuals

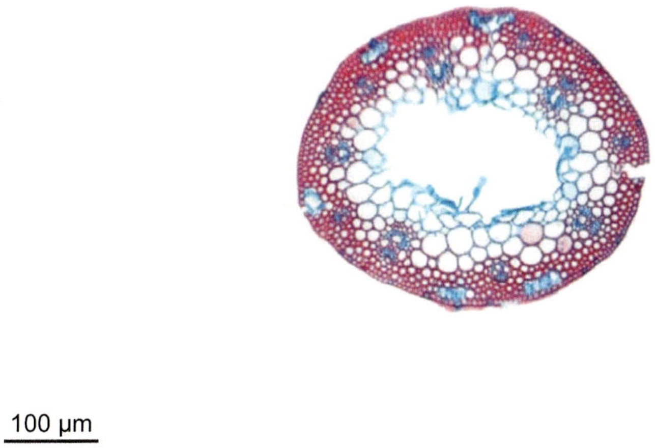

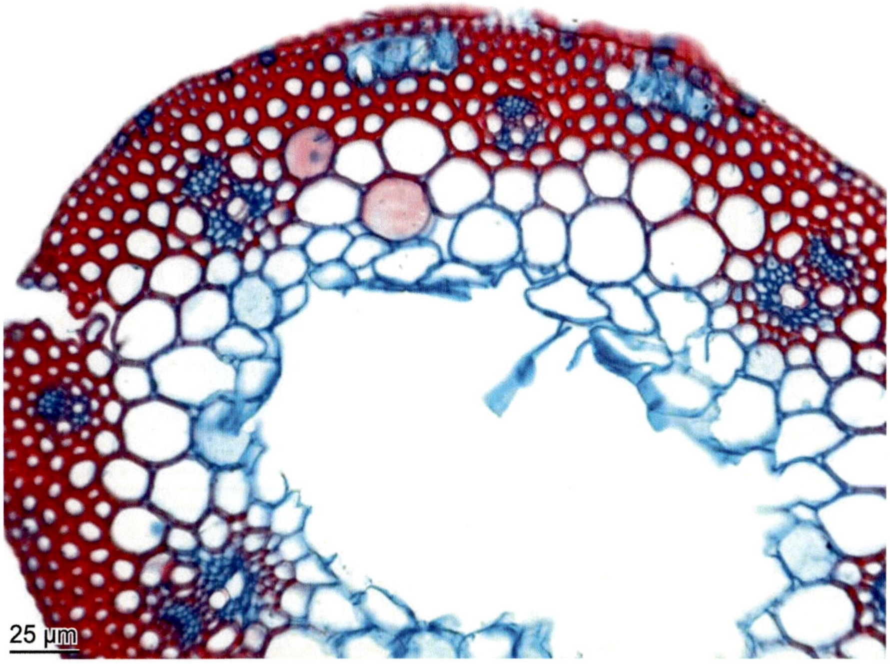

Puccinellia minuta Bor

Poaceae

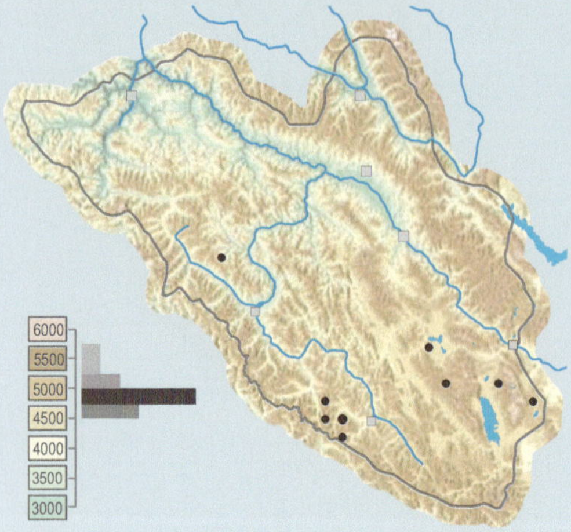

Description: Perennial herbs, 3–8 cm tall, densely cespitose, leaves conduplicate or involute, 1–2 cm long, 0.6–1.2 mm wide, panicle contracted, erect, ca. 2 cm long, 0.5 cm wide, branches ca. 0.5 cm long, ascending.

Distribution: Pakistan, India (Jammu and Kashmir), China (Qinghai, Xizang).

Habitat: Saline meadows, sandy lake shores.

Culm uniform, round, hollow, with a diameter of 0.5–0.7 mm. Epidermis thin- to thick-walled. Continuous, 3–5-seriate peripheral sclerenchymatic belt. Girders square to tangentially rectangular. Groups of chlorenchyma tangentially enlarged. Intercellulars in the parenchymatic zone small and triangular. Vascular bundles arranged in one circular row. Large bundles within the thin- to thick-walled parenchymatic tissue, small bundles within the sclerenchymatic belt. Vascular bundles with large vessels, diameter 8–10 µm. Phloem oval, consisting of uniform cells. Intercellulars centripetal of the protoxylem absent. Vascular bundles surrounded by a uniseriate sheath of polygonal, thin- to thick-walled cells.

Analyzed: 2 individuals

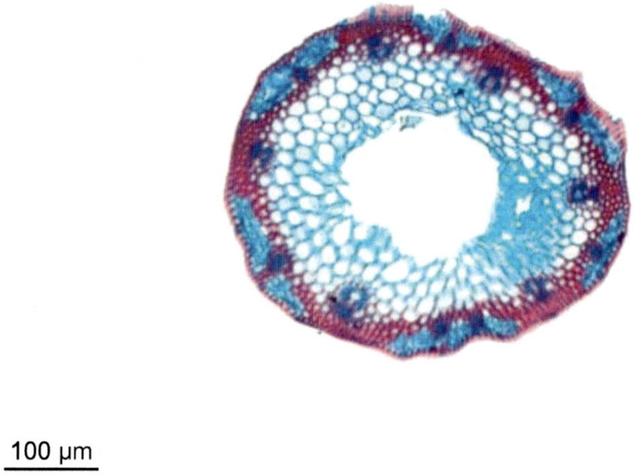

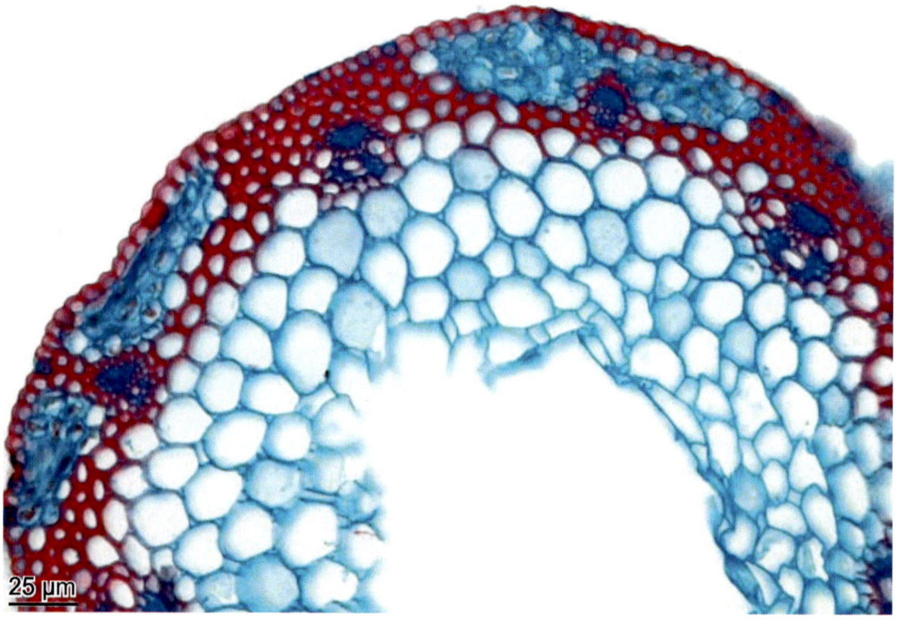

Puccinellia pauciramea (Hack.) Krecz. ex Ovcz. & Czukav.

Poaceae

Syn.: *Atropis pauciramea* (Hack.) Krecz.

Description: Perennial herbs, 15–30 cm tall, cespitose, leaves conduplicate or involute, 4–10 cm long, 1–2 mm wide, panicle widely open, 4–7 cm long.

Distribution: Uzbekistan, Kyrgyzstan, Tajikistan, Afghanistan, India (Jammu and Kashmir), China (Qinghai, Xinjiang, Xizang), Mongolia.

Habitat: Brackish lake banks, gravelly river valleys.

Culm uniform, round, hollow, with a diameter of 0.6–0.7 mm. Epidermis thick-walled. Continuous, 3–5-seriate peripheral sclerenchymatic belt. Girders absent. With a few small groups of chlorenchyma. Triangular intercellulars in the parenchymatic tissue absent. Vascular bundles arranged in one circular row. Large bundles within the thin- to thick-walled parenchymatic tissue, small bundles within the sclerenchymatic belt. Vascular bundles with large vessels, diameter 10–15 µm. Phloem oval, consisting of uniform cells. With a few intercellulars centripetal of the protoxylem. Vascular bundles surrounded by a 1–2-seriate sheath of polygonal, thin- to thick-walled cells.

Analyzed: 2 individuals

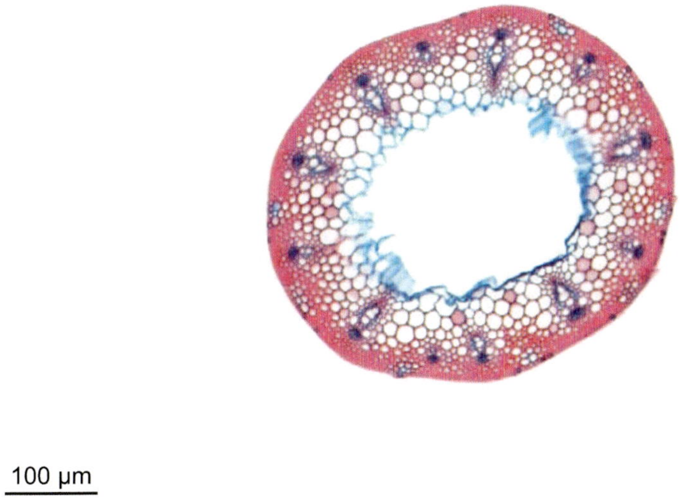

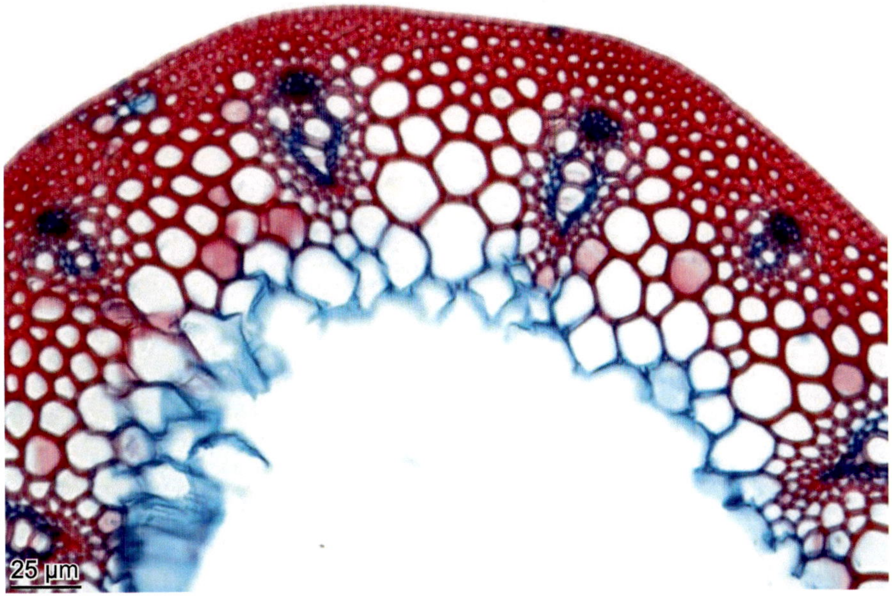

Puccinellia stapfiana R.R.Stewart

Poaceae

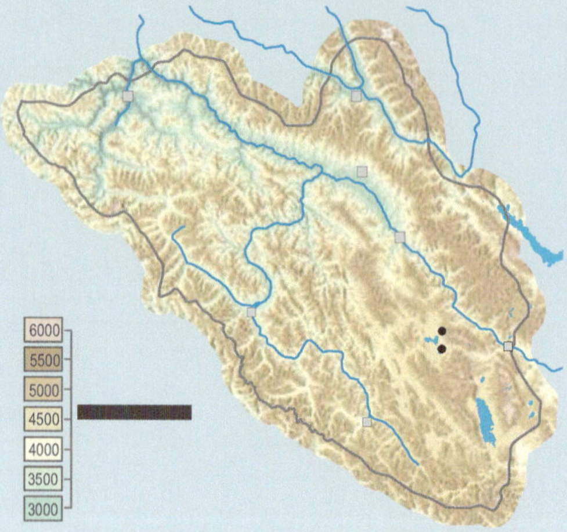

Syn.: *Glyceria poioides* Stapf ex Hook.f.

Description: Perennial herbs, 20–45 cm tall, cespitose, leaves conduplicate, 3–10 cm long, 1–2.5 mm wide, panicle narrow, 5–10 cm long, 1.5 cm wide, branches ascending.

Distribution: N Pakistan, India (Jammu and Kashmir), China (NW Xizang).

Habitat: Saline grasslands.

Culm uniform, round, hollow, with a diameter of 0.7–0.8 mm. Epidermis thick-walled. Continuous, 3–5-seriate peripheral sclerenchymatic belt. Girders square to tangentially rectangular. Chlorenchyma groups oval. Triangular intercellulars in the parenchymatic tissue almost absent. Vascular bundles arranged in 1-2 circular rows. Large bundles within the thin- to thick-walled parenchymatic tissue, small bundles within the sclerenchymatic belt. Vascular bundles with large vessels, diameter 15–20 µm. Phloem oval, consisting of uniform cells. With a few intercellulars centripetal of the protoxylem. Vascular bundles surrounded by a 1–2-seriate sheath of polygonal, thin- to thick-walled cells.

Analyzed: 2 individuals

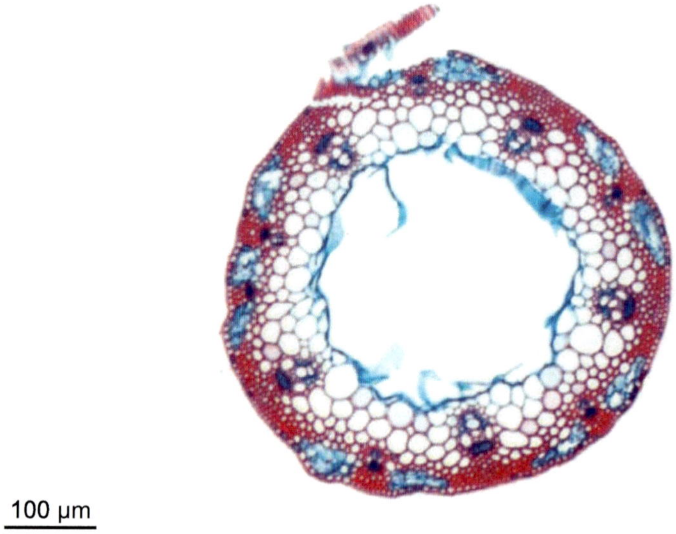

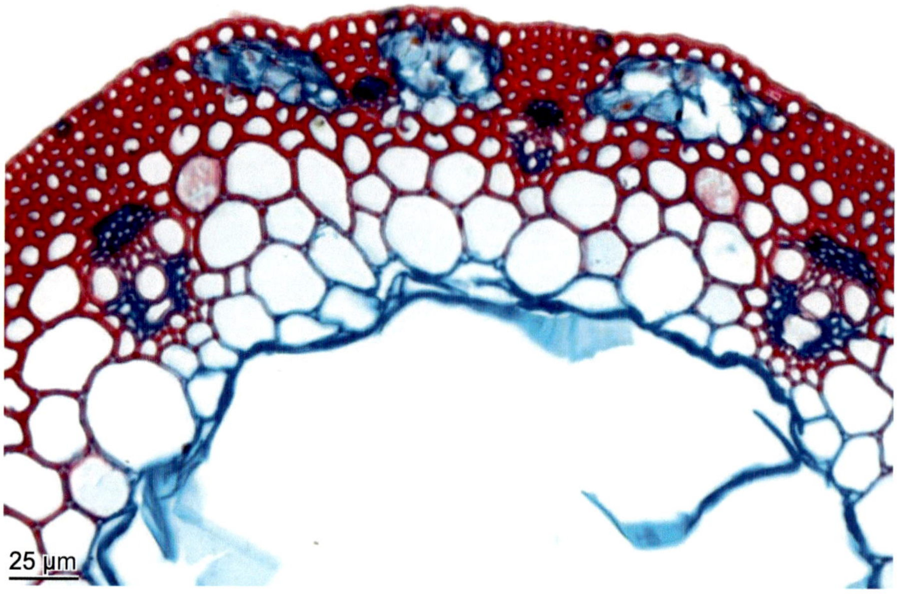

Puccinellia tenuiflora (Griseb.) Scribn. & Merr.

Poaceae

Syn.: *Atropis tenuiflora* Griseb., *Glyceria tenuiflora* (Griseb.) Steud.

Description: Perennial herbs, 30–70 cm tall, cespitose, leaves conduplicate, 2–8 cm long, 1–3 mm wide, panicle loose, 6–15 cm long, branches horizontally spreading.

Distribution: Iran, Russia (Siberia), Kazakhstan, Pakistan, India (Jammu and Kashmir), China (Anhui, Gansu, Hebei, Heilongjiang, Jilin, Liaoning, Nei Mongol, Qinghai, Shanxi, Xinjiang), Mongolia, Japan.

Habitat: Saline grasslands.

Culm uniform, round, hollow, with a diameter of 1.1–1.2 mm. Epidermis thick-walled. Continuous, 2–4-seriate peripheral sclerenchymatic belt. Girders absent. Groups of chlorenchyma absent. Triangular intercellulars in the parenchymatic tissue almost absent. Vascular bundles arranged in 1–2 circular rows. Large bundles within the thin- to thick-walled parenchymatic tissue, small bundles within the sclerenchymatic belt. Vascular bundles with large vessels, diameter 15–20 µm. Phloem oval, consisting of uniform cells. With intercellulars centripetal of the protoxylem. Vascular bundles surrounded by a 1–2-seriate sheath of polygonal, thin- to thick-walled cells.

Analyzed: 2 individuals

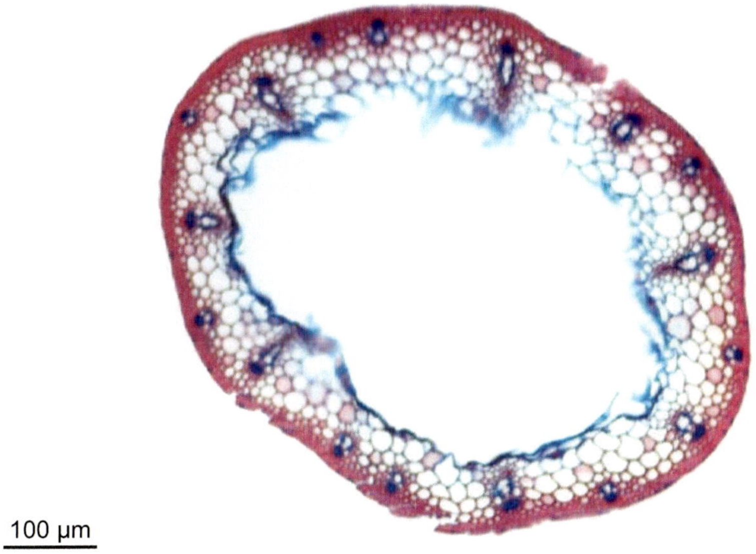

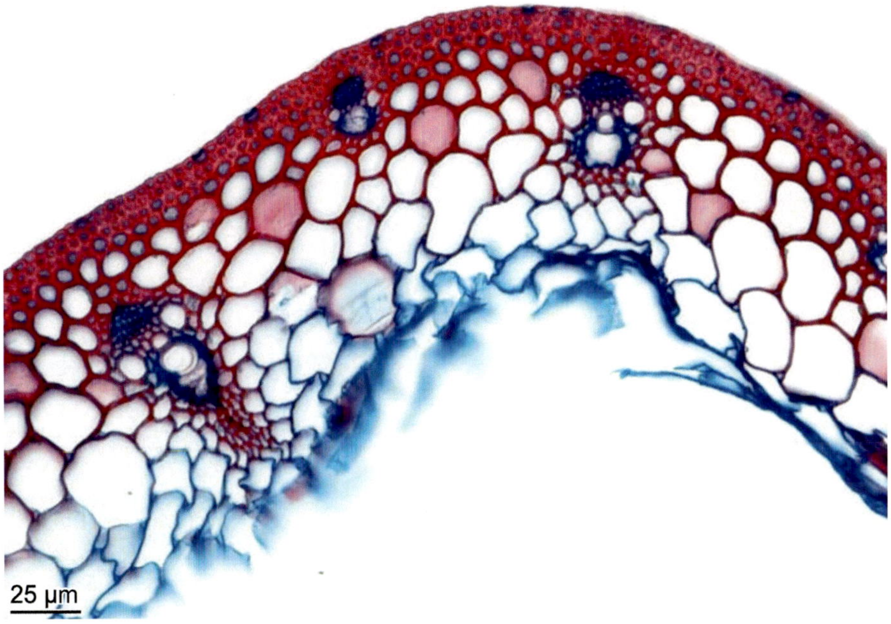

Puccinellia thomsonii (Stapf) R.R.Stewart

Poaceae

Syn.: *Atropis thomsonii* (Stapf) Pamp., *Glyceria thomsonii* Stapf ex Hook.f.

Description: Perennial herbs, 30–50 cm tall, densely cespitose, leaves flat or conduplicate, 5–15 cm long, 1–3 mm wide, panicle open, or contracted, 12–20 cm long, 1–3 cm wide, branches 1–9 cm long, ascending.

Distribution: Pakistan, India (Jammu and Kashmir), China (Xizang).

Habitat: Saline wetlands.

Culm uniform, round, hollow, with a diameter of 0.7–0.8 mm. Epidermis thin- to thick-walled. Continuous, 3–5-seriate peripheral sclerenchymatic belt. Girders square to tangentially rectangular. Groups of chlorenchyma tangentially enlarged. Triangular intercellulars in the parenchymatic tissue almost absent. Vascular bundles arranged in 2 circular rows. Large bundles within the thin- to thick-walled parenchymatic tissue, small bundles within the sclerenchymatic belt. Vascular bundles with large vessels, diameter 40–50 µm. Phloem oval, consisting of uniform cells. Intercellulars centripetal of the protoxylem absent. Vascular bundles surrounded by a uniseriate sheath of polygonal, thin- to thick-walled cells.

Analyzed: 1 individual

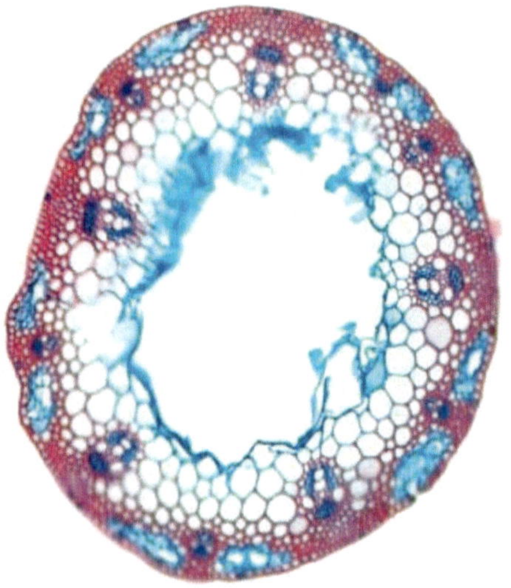

100 µm

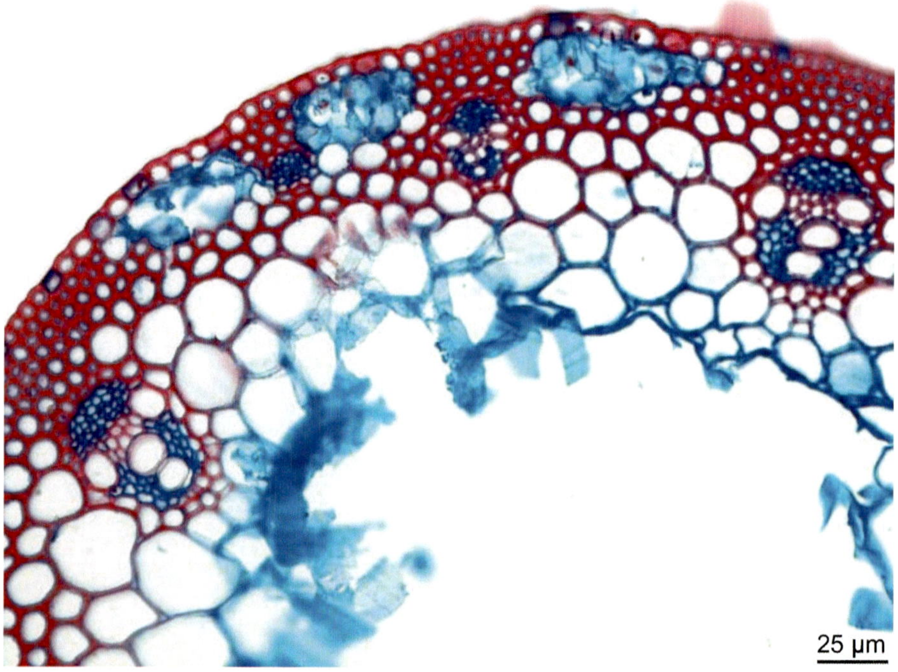

25 µm

Schismus arabicus Nees
Poaceae

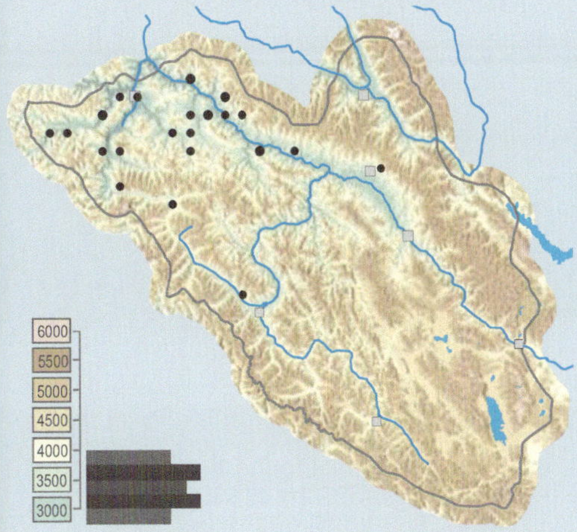

Description: Annuals, 3–15 cm tall, cespitose, leaves linear, flat or convolute, 1.5–10 cm long, 0.5–2 mm wide, glabrous or pilose, panicle contracted, narrow, 1–3 cm long.

Distribution: N Africa, SE Europe, C and SW Asia, Russia (Altay), Afghanistan, Pakistan, NW India, China (Xinjiang, W Xizang), Mongolia; introduced in North America and Australia.

Habitat: Semideserts.

Culm uniform, round, hollow, with a diameter of 0.2–0.5 mm. Epidermis thick-walled. Continuous, 3–5-seriate peripheral sclerenchymatic belt. Girders absent. Groups of chlorenchyma absent. Triangular intercellulars in the parenchymatic tissue absent. Vascular bundles arranged in one circular row. Large bundles within the thin- to thick-walled parenchymatic tissue, small bundles within the sclerenchymatic belt. Vascular bundles with large vessels, diameter 4–6 μm. Phloem oval, consisting of uniform cells. Intercellulars centripetal of the protoxylem absent. Vascular bundles surrounded by a 1–2-seriate sheath of polygonal, thin- to thick-walled cells.

Analyzed: 2 individuals

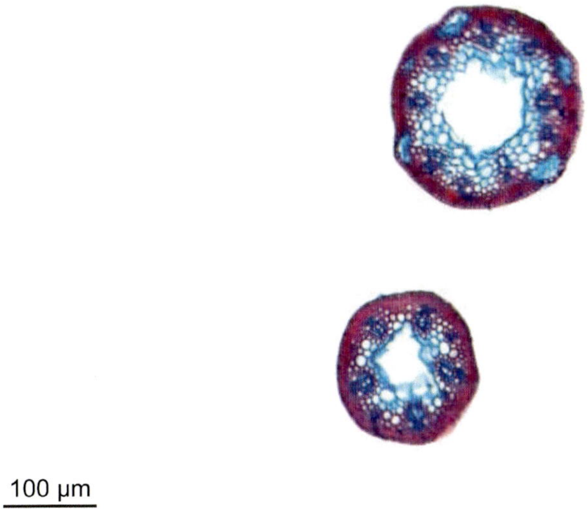

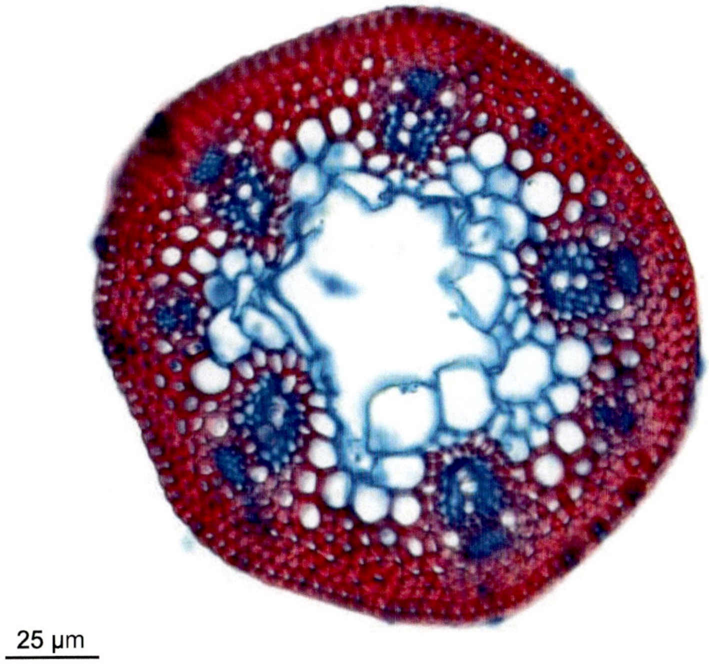

Stipa brandisii Mez

Poaceae

Syn.: *Achnatherum brandisii* (Mez) Z.L.Wu

Description: Perennial herbs, 70–150 cm tall, cespitose, leaves flat, rough above, 20–40 cm long, 4–10 mm wide, panicle open, oblong, 10–30 cm long, branches ascending, or spreading.

Distribution: Afghanistan, Pakistan, NW India, Nepal, China (W Gansu, Qinghai, Sichuan, Xizang, Yunnan), Bhutan.

Habitat: Grassy mountain slopes.

Culm uniform, round, hollow, with a diameter of 0.2–0.5 mm. Epidermis thick-walled. Continuous, 5–7-seriate, thick-walled peripheral sclerenchymatic belt. Girders absent. With small round groups of chlorenchyma. Intercellulars in the parenchymatic zone small and triangular. Vascular bundles arranged in 2 circular rows. Large bundles within the thin- to thick-walled parenchymatic tissue, small bundles within the sclerenchymatic belt. Vascular bundles with large vessels, diameter 10–15 µm. Phloem oval, consisting of sieve tubes and companion cells. With intercellulars centripetal of the protoxylem. Vascular bundles surrounded by an eccentric sheath of polygonal, thin- to thick-walled cells, 1–2-seriate centripetally, and 5–6-seriate in a crescent at the peripheral side.

Analyzed: 2 individuals

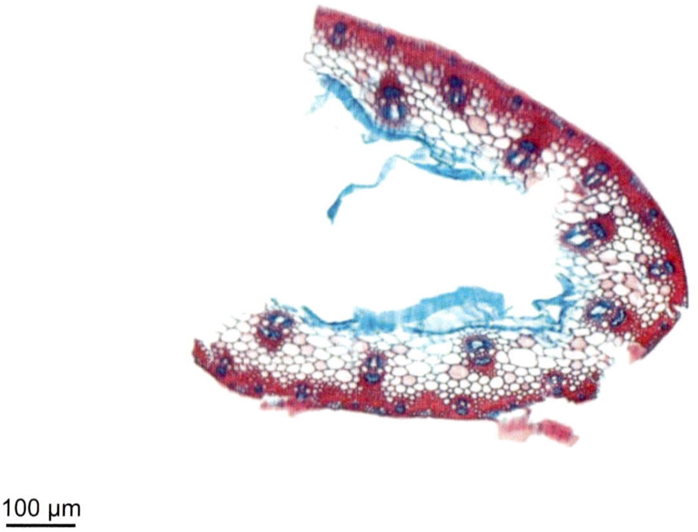

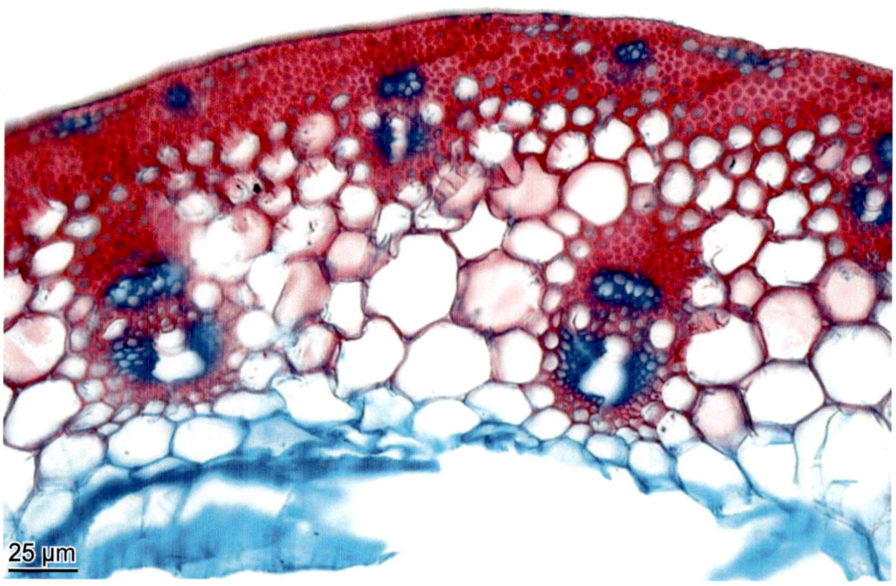

Stipa breviflora Griseb.

Poaceae

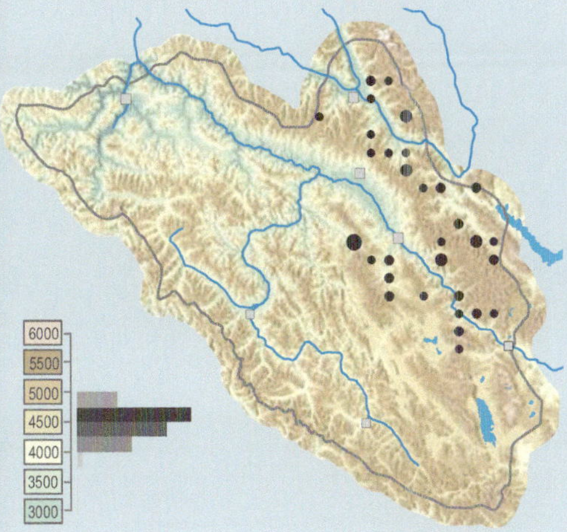

Description: Perennial herbs, 25–40 cm tall, cespitose, leaves filiform, involute, panicle linear, contracted, 10–25 cm long, awns 5.5–8 cm long, plumose with hairs 0.5–0.8 mm long.

Distribution: Kazakhstan, Uzbekistan, Kyrgyzstan, Tajikistan, Pakistan, NW India, Nepal, China (Gansu, Hebei, Nei Mongol, Ningxia, Qinghai, Shaanxi, Shanxi, Sichuan, Xinjiang, Xizang), Mongolia.

Habitat: Gravel and rocky slopes, desert steppes.

Culm uniform, round, hollow, with a diameter of 1.1–1.2 mm. Epidermis thick-walled. Continuous, 4–6-seriate, thick-walled peripheral sclerenchymatic belt. Girders absent. With small round groups of chlorenchyma. Intercellulars in the parenchymatic zone small and triangular. Vascular bundles arranged in 2 circular rows. Large bundles within the thin- to thick-walled parenchymatic tissue, small bundles within the sclerenchymatic belt. Vascular bundles with large vessels, diameter 10–12 µm. Phloem oval, consisting of sieve tubes and companion cells. With intercellulars centripetal of the protoxylem. Vascular bundles surrounded by an eccentric sheath of polygonal cells, 1–2 rows of thin- to thick-walled cells centripetally, and 3–4 rows of thick-walled cells, arranged in a crescent, at the peripheral side.

Analyzed: 2 individuals

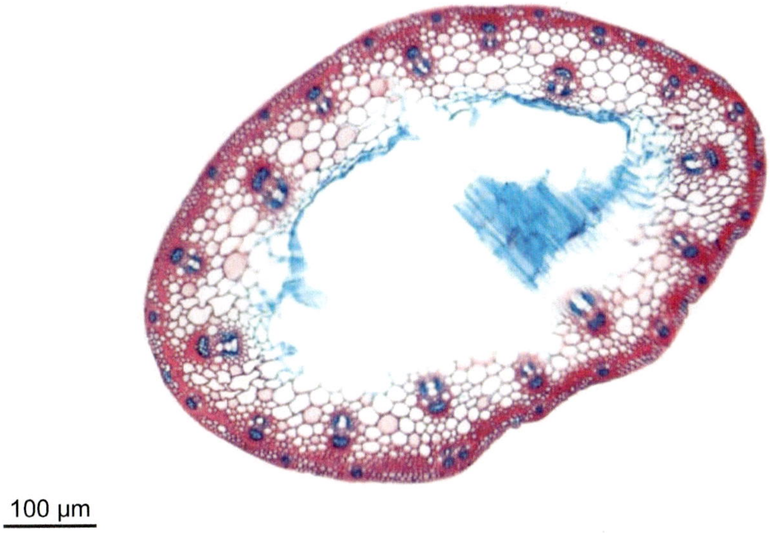

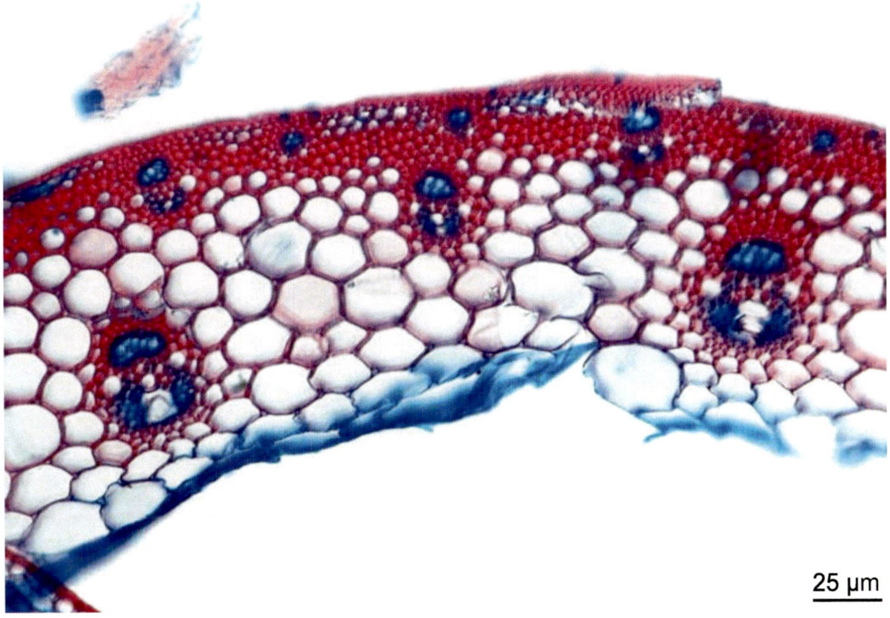

Stipa capillacea Keng

Poaceae

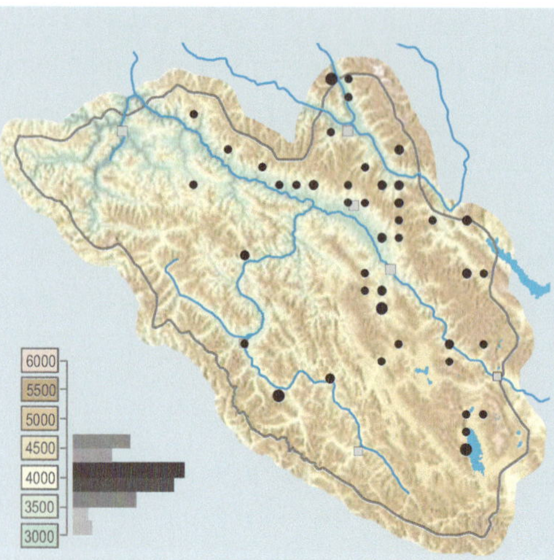

Syn.: *S. koelzii* R.R.Stewart

Description: Perennial herbs, 20–50 cm tall, cespitose, leaves convolute, 6–20 cm long, 1.5–2 mm wide, panicle open, linear, 15–20 cm long, base enclosed by uppermost leaf sheath, awns 5–9 cm long, hairy.

Distribution: Pakistan, N India, Nepal, China (Gansu, Qinghai, Sichuan, Xizang, Yunnan), Bhutan.

Habitat: Grasslands on river banks, moist alpine meadows.

Culm uniform, round, hollow, with a diameter of 1.1–1.2 mm. Epidermis thick-walled. Continuous, 6–8-seriate, thick-walled peripheral sclerenchymatic belt. Girders absent. Chlorenchyma groups small. Intercellulars in the parenchymatic zone very small and triangular. Vascular bundles arranged in 2–3 circular rows. Large bundles within the thin- to thick-walled parenchymatic tissue, small bundles within the sclerenchymatic belt. Vascular bundles with large vessels, diameter 10–15 µm. Phloem oval, consisting of sieve tubes and companion cells. Vascular bundles surrounded by an eccentric sheath of polygonal cells, 1–2 rows of thin- to thick-walled cells centripetally, and 2–3 rows of thick-walled cells at the peripheral side.

Analyzed: 2 individuals

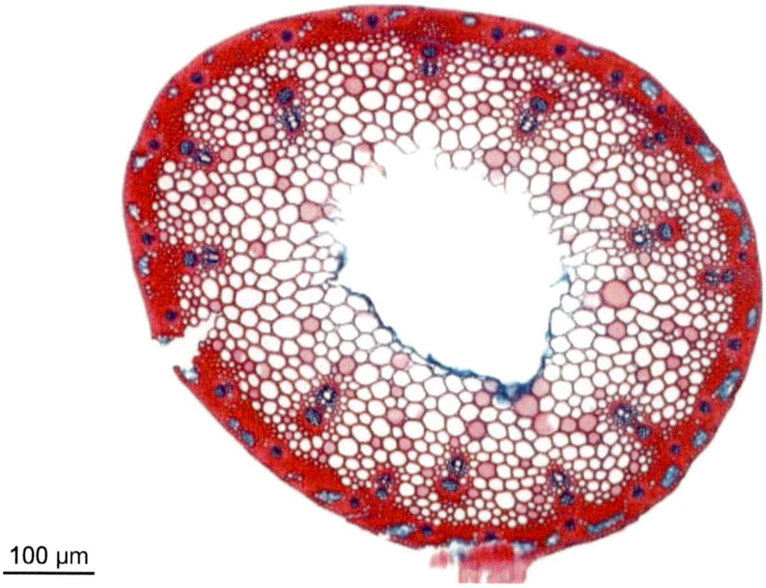

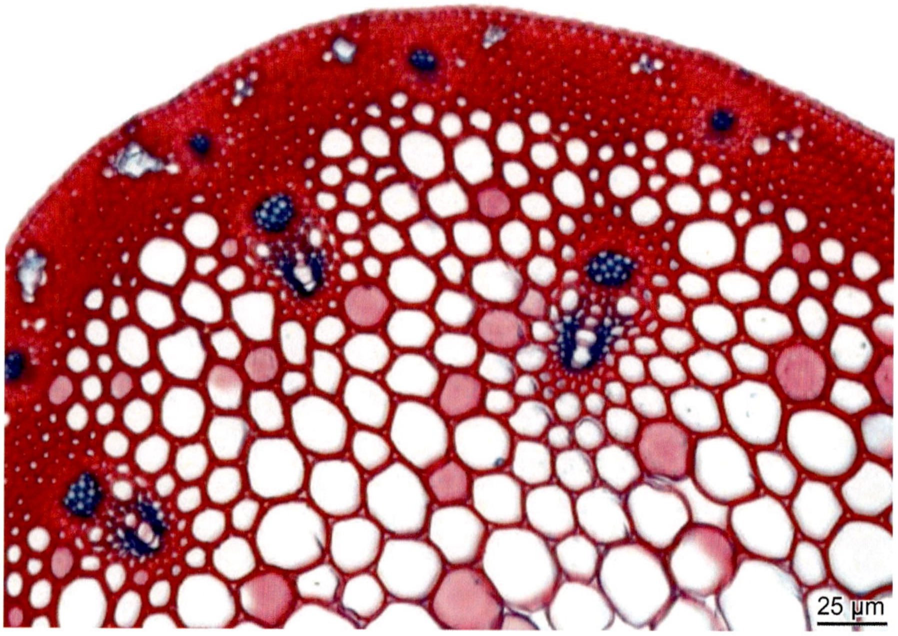

Stipa capillata L.
Poaceae

Syn.: *Aristida avenacea* Houtt.

Description: Perennial herbs, 40–100 cm tall, densely cespitose, leaves convolute, 20–45 cm long, 1–2 mm wide, panicle contracted, linear, 10–25 cm long, base enclosed by uppermost leaf sheath, awn deciduous, 10–20 cm long.

Distribution: Europe, SW Asia, Russia, Kazakhstan, Turkmenistan, Uzbekistan, Kyrgyzstan, Tajikistan, Pakistan, India (Jammu and Kashmir), China (N Gansu, Hebei, Shanxi, N Xinjiang), Mongolia.

Habitat: Rocky slopes.

Culm uniform, round, hollow, with a diameter of 1.0–1.1 mm. Epidermis thick-walled. Continuous, 4–6-seriate, thick-walled peripheral sclerenchymatic belt. Girders absent. With a few small round groups of chlorenchyma. Intercellulars in the parenchymatic zone small and triangular. Vascular bundles arranged in 2 circular rows. Large bundles within the thin- to thick-walled parenchymatic tissue, small bundles within the sclerenchymatic belt. Vascular bundles with large vessels, diameter 20–25 µm. Phloem oval, consisting of sieve tubes and companion cells. With intercellulars centripetal of the protoxylem. Vascular bundles surrounded by an eccentric sheath of polygonal cells, 1–2 rows of thin- to thick-walled cells centripetally, and 2–3 rows of thick-walled cells at the peripheral side.

Analyzed: 2 individuals

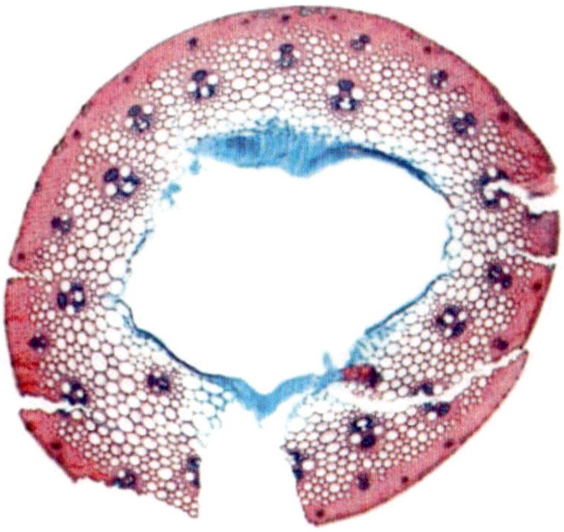

100 µm

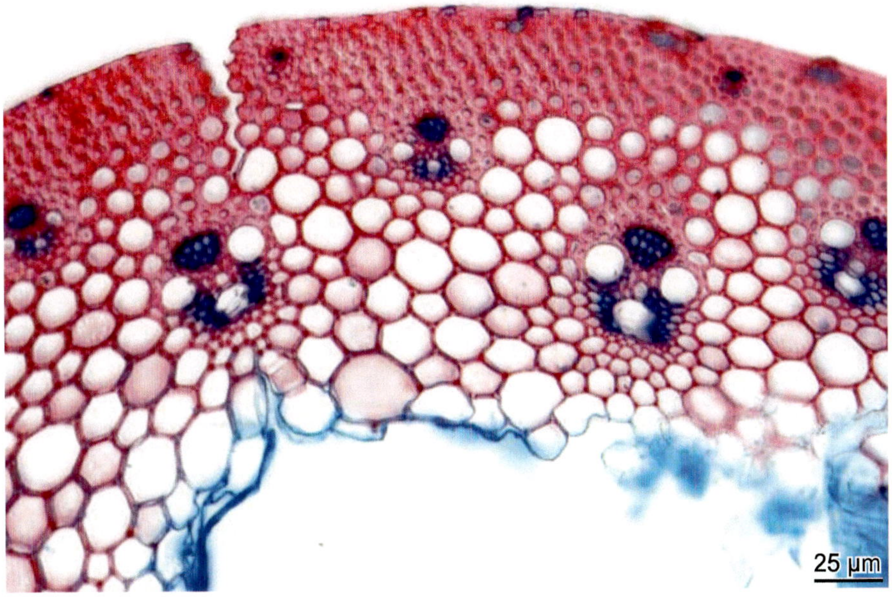

Stipa caucasica Schmalh.

Poaceae

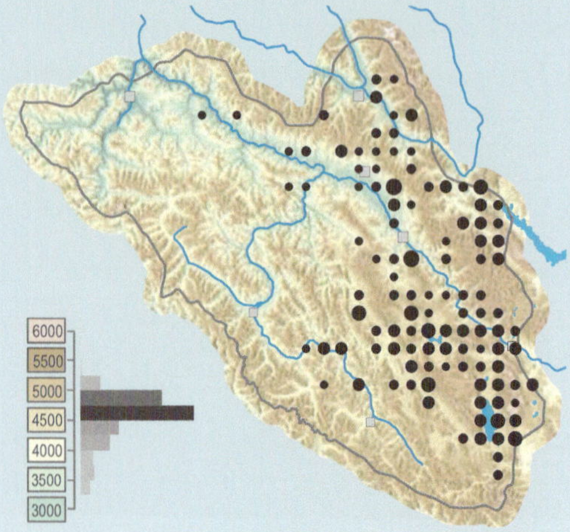

Description: Perennial herbs, 20–70 cm tall, cespitose, leaves linear, involute, panicle narrow, contracted, 5–10 cm long, awns 8–13 cm long, plumose with hairs 4–6 mm long.

Distribution: Caucasus, N Iran, Russia (Siberia), Kazakhstan, Kyrgyzstan, Turkmenistan, Tajikistan, Afghanistan, Pakistan, NW India, China (Gansu, Hebei, Henan, Nei Mongol, Ningxia, Qinghai, Shaanxi, Xinjiang, Xizang), Mongolia.

Habitat: Dry steppes, sandy plains and stony slopes, semideserts.

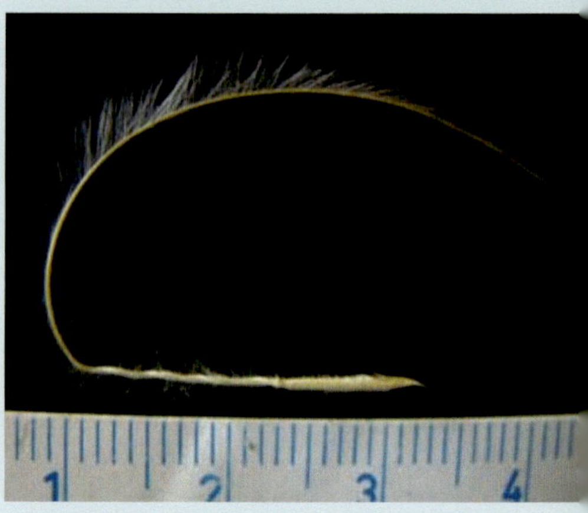

Culm uniform, round, hollow, with a diameter of 0.8–0.9 mm. Epidermis cells thin-walled on the inside, thicker-walled peripherally, with hairs. Continuous, 4–6-seriate, thick-walled peripheral sclerenchymatic belt. Girders square and tangentially rectangular. Chlorenchyma groups tangentially enlarged. With large triangular intercellulars in the parenchymatic tissue. Vascular bundles arranged in 2–3 circular rows. Large bundles within the thin- to thick-walled parenchymatic tissue, small bundles within the sclerenchymatic belt. Vascular bundles with large vessels, diameter 15–20 µm. Phloem oval, consisting of sieve tubes and companion cells. With intercellulars centripetal of the protoxylem. Vascular bundles surrounded by an eccentric sheath of polygonal cells, 1–2 rows of thin- to thick-walled cells centripetally, and 1–2 rows of thick-walled cells at the peripheral side.

Analyzed: 4 individuals

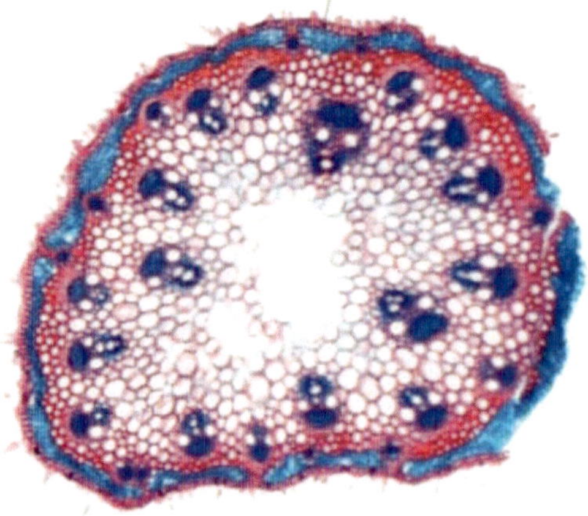

100 µm

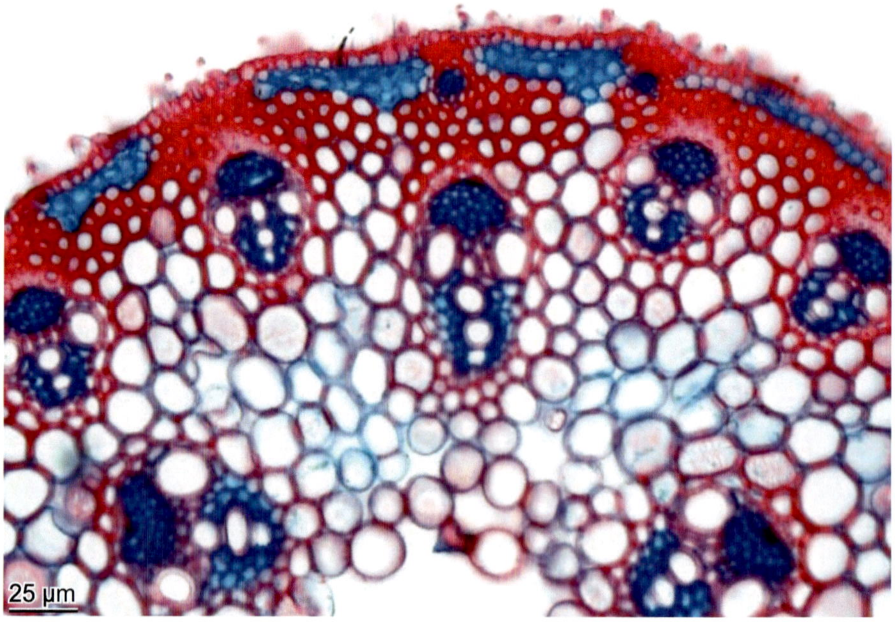

Stipa concinna Hook.f.

Poaceae

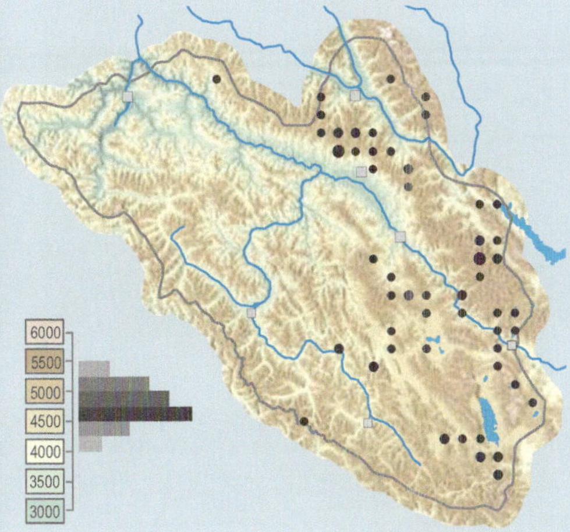

Syn.: *Ptilagrostis concinna* (Hook.) Roshev., *P. schischkinii* (Tzvelev) Czerep.

Description: Perennial herbs, 10–30 cm tall, densely cespitose, leaves setaceous, ca. 0.5 mm wide, panicle contracted, linear, 2.5–5 cm long, awns 10–15 mm long, plumose.

Distribution: Kyrgyzstan, Tajikistan, Pakistan, India (Jammu and Kashmir, Sikkim), China (Gansu, Qinghai, Shaanxi, Sichuan, Xinjiang, Xizang, Yunnan).

Habitat: Alpine meadows, moist grasslands.

Culm uniform, round, hollow, with a diameter of 0.6–0.8 mm. Epidermis thick-walled. Continuous, 4–6-seriate, thick-walled peripheral sclerenchymatic belt. Girders absent. Chlorenchyma groups round to rectangular. Intercellulars in the parenchymatic zone small and triangular. Vascular bundles arranged in 2 circular rows. Large bundles within the thin- to thick-walled parenchymatic tissue, small bundles within the sclerenchymatic belt. Vascular bundles with large vessels, diameter 5–6 µm. Phloem oval, consisting of sieve tubes and companion cells. Intercellulars centripetal of the protoxylem absent. Vascular bundles surrounded by an eccentric sheath of polygonal cells, one row of thin- to thick-walled cells centripetally, and 2–3 rows of thick-walled cells at the peripheral side.

Analyzed: 2 individuals

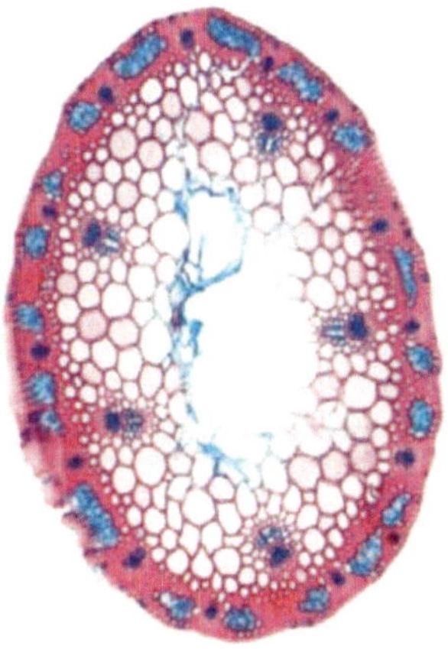

100 µm

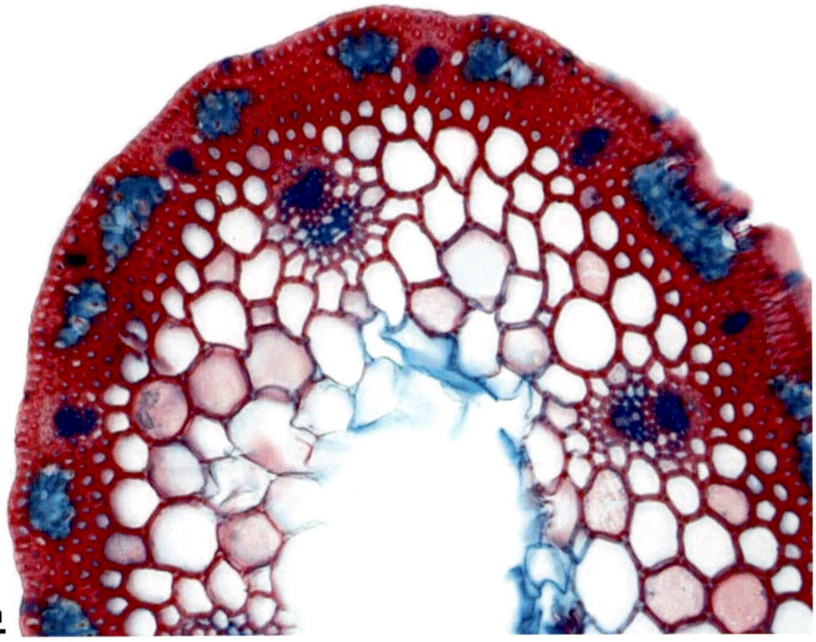

25 µm

Stipa himalaica Roshev.

Poaceae

Description: Perennial herbs, 20–45 cm tall, cespitose, leaves filiform, involute, panicle linear, contracted, 5–10 cm long, awns 10–16 cm long, plumose with hairs 4–5 mm long.

Distribution: N Pakistan, NW India, China (Xizang).

Habitat: Dry steppes.

Culm uniform, round, hollow, with a diameter of 0.9–1.0 mm. Epidermis thick-walled. Continuous, 4–6-seriate, thick-walled peripheral sclerenchymatic belt. Girders absent. Chlorenchyma groups uniseriate, tangentially enlarged. Intercellulars in the parenchymatic zone very small and triangular. Vascular bundles arranged in 2 circular rows. Large bundles within the thin- to thick-walled parenchymatic tissue, small bundles within the sclerenchymatic belt. Vascular bundles with large vessels, diameter 16–20 µm. Phloem oval, consisting of sieve tubes and companion cells. With intercellulars centripetal of the protoxylem. Vascular bundles surrounded by an unlignified, 2–3-seriate sheath of polygonal cells, with thick-walled cells at the peripheral side of the vascular bundles.

Analyzed: 2 individuals

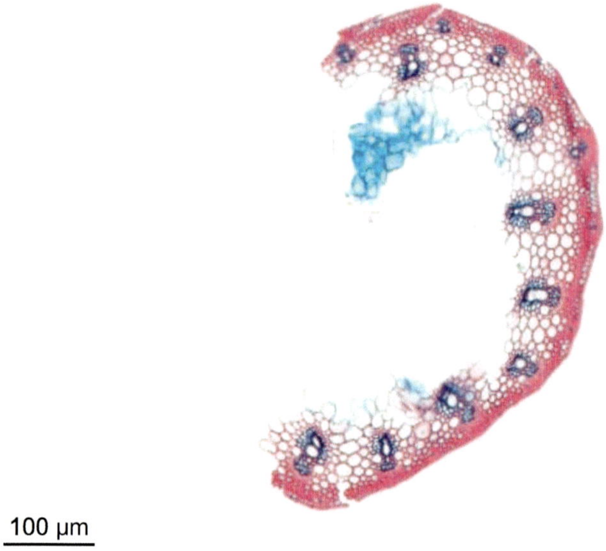

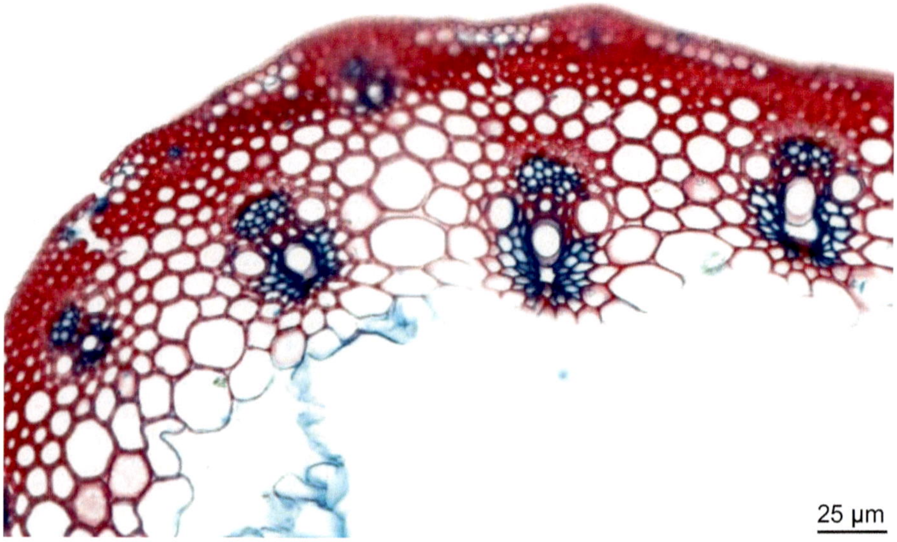

Stipa klimesii M.Nobis
Poaceae

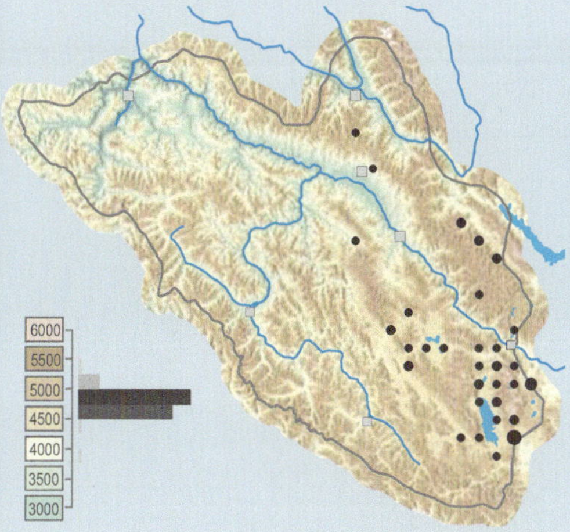

Description: Perennial herbs, 15–45 cm tall, densely cespitose, leaves convolute, green to pale green, 10–20 cm long, 0.20–0.45 mm in diameter, panicle 8–20 cm long, contracted, branches erect, awns 35–45 mm long, bigeniculate.

Distribution: India (Jammu and Kashmir: Ladakh); its occurrence in the neighbouring areas of Tibet is probable.

Habitat: High mountain steppes.

Culm uniform, round, hollow, with a diameter of 1.0–1.1 mm. Epidermis thick-walled. Continuous, 6–8-seriate, thick-walled peripheral sclerenchymatic belt. Girders absent. With small oval groups of chlorenchyma. Triangular intercellulars in the parenchymatic tissue absent. Vascular bundles arranged in 2 circular rows. Large bundles within the thin- to thick-walled parenchymatic tissue, small bundles within the sclerenchymatic belt. Vascular bundles with large vessels, diameter 12–15 µm. Phloem oval, consisting of sieve tubes and companion cells. With intercellulars centripetal of the protoxylem. Vascular bundles surrounded by a 1–2-seriate, thick-walled sheath of polygonal cells

Analyzed: 2 individuals

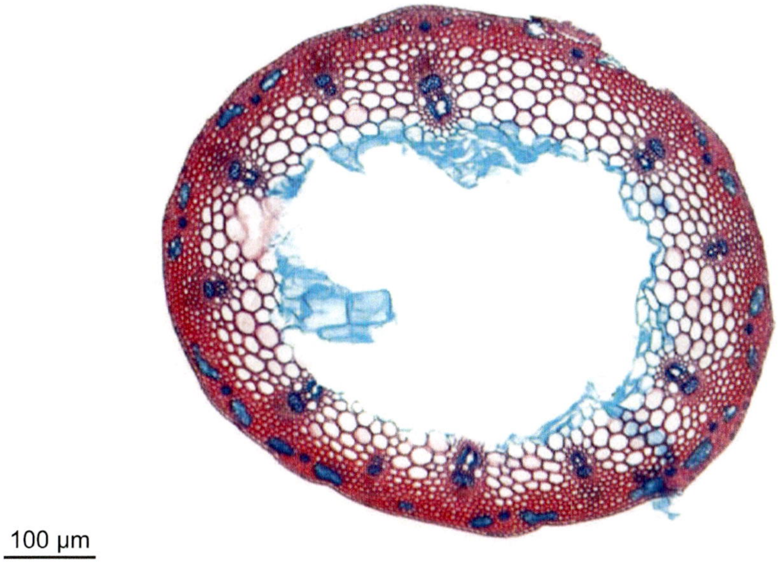

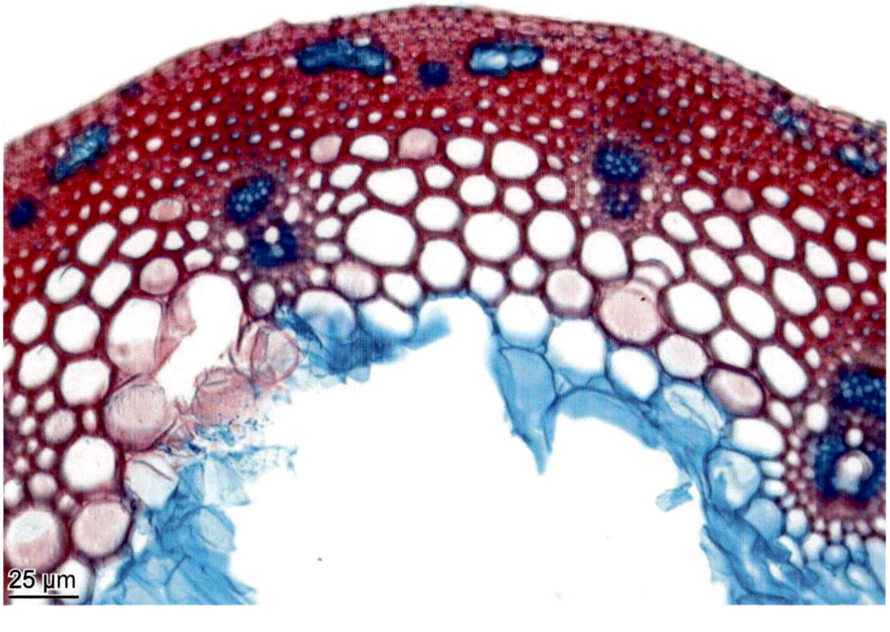

Stipa mongholica Turcz. ex Trin.

Poaceae

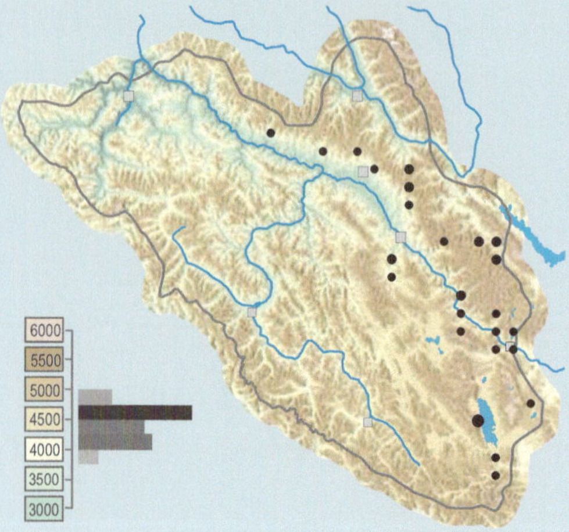

Syn.: *Ptilagrostis mongholica* (Turcz. ex Trin.) Griseb.

Description: Perennial herbs, 15–60 cm tall, densely cespitose, leaves filiform, panicle ovate, loose, 5–15 cm long, with spreading flexuous branches, awns 1.5–3 cm long, plumose with hairs ca. 1.5 mm long.

Distribution: Russia, Pakistan, N India, Nepal, China (Gansu, Hebei, Heilongjiang, Jilin, Liaoning, Nei Mongol, Qinghai, Shaanxi, Shanxi, Sichuan, Xinjiang, Yunnan), Bhutan, Mongolia.

Habitat: Alpine steppes, grasslands on river banks.

Culm uniform, round, hollow, with a diameter of 1.1–1.2 mm. Epidermis thick-walled. Continuous, 3–5-seriate, thick-walled peripheral sclerenchymatic belt. Girders absent. Groups of chlorenchyma absent. Triangular intercellulars in the parenchymatic tissue absent. Vascular bundles arranged in 2–3 circular rows. Large bundles within the thin- to thick-walled parenchymatic tissue, small bundles within the sclerenchymatic belt. Bundles within the sclerenchymatic belt reduced to phloem. Vascular bundles with large vessels, diameter 6–8 µm. Phloem oval, consisting of sieve tubes and companion cells. Vascular bundles surrounded by an eccentric sheath of polygonal cells, 1–2 rows of thin- to thick-walled cells centripetally, and 2–3 rows of thick-walled cells at the peripheral side.

Analyzed: 2 individuals

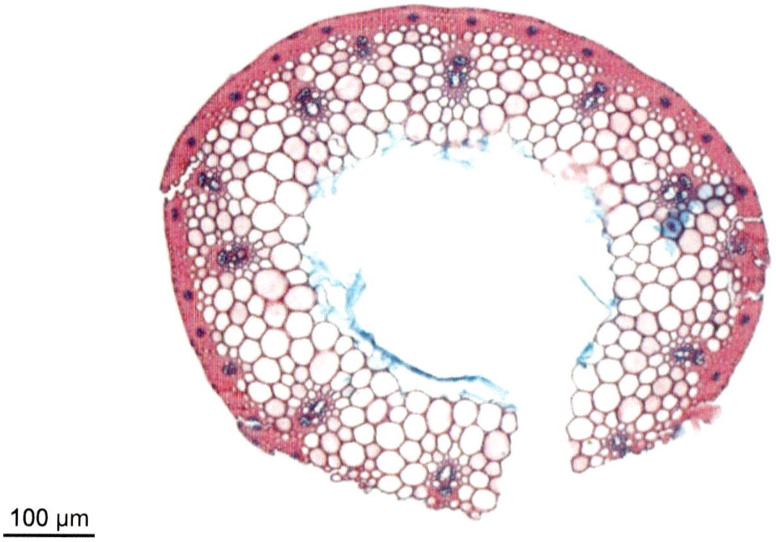

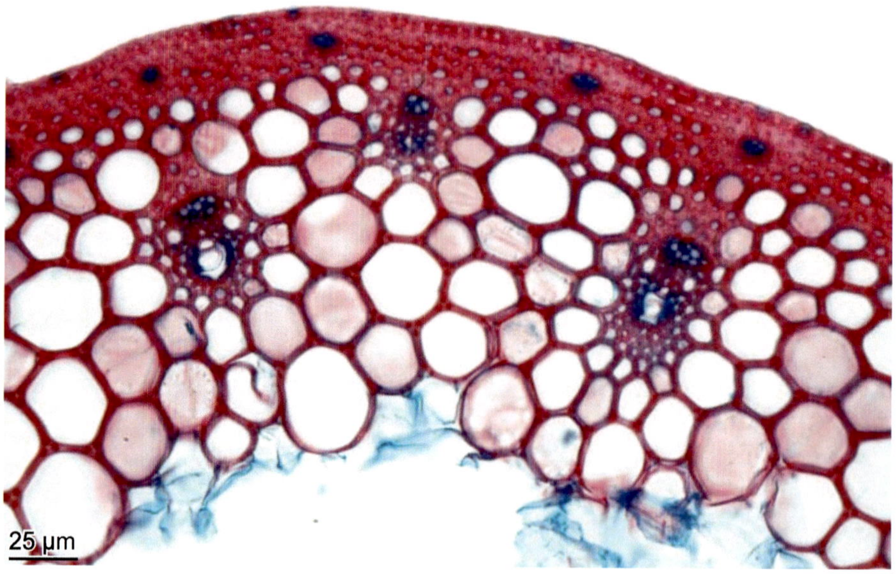

Stipa orientalis Trin. ex Ledeb.

Poaceae

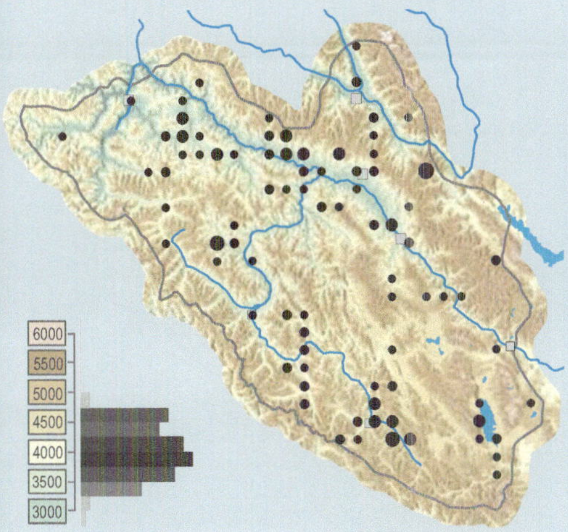

Description: Perennial herbs, 20–35 cm tall, densely cespitose, leaves filiform, involute, 10–20 cm long, 0.5–1 mm wide, panicle contracted, linear, 5–10 cm long, awn bigeniculate, 35–70 mm long, plumose with hairs 3–5 mm long.

Distribution: Russia, Iran, Turkmenistan, Uzbekistan, Kyrgyzstan, Tajikistan, Pakistan, N India, W China, W Mongolia.

Habitat: Desert steppes, rocky slopes.

Culm uniform, round, hollow, with a diameter of 0.7–0.8 mm. Epidermis thick-walled. Continuous, 6–8-seriate, thick-walled peripheral sclerenchymatic belt. Girders absent. Chlorenchyma groups 1–2-seriate, tangentially enlarged. Triangular intercellulars in the parenchymatic tissue absent. Vascular bundles arranged in 2–3 circular rows. Large bundles within the thin- to thick-walled parenchymatic tissue, small bundles within the sclerenchymatic belt. Vascular bundles with large vessels, diameter 6–8 µm. Phloem oval, consisting of sieve tubes and companion cells. Intercellulars centripetal of the protoxylem absent. Vascular bundles surrounded by an unlignified, 2–3-seriate sheath of polygonal cells, with thick-walled cells at the peripheral side of the vascular bundles.

Analyzed: 2 individuals

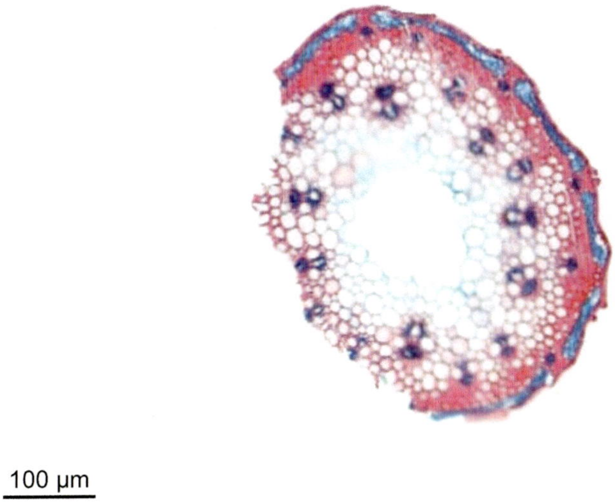

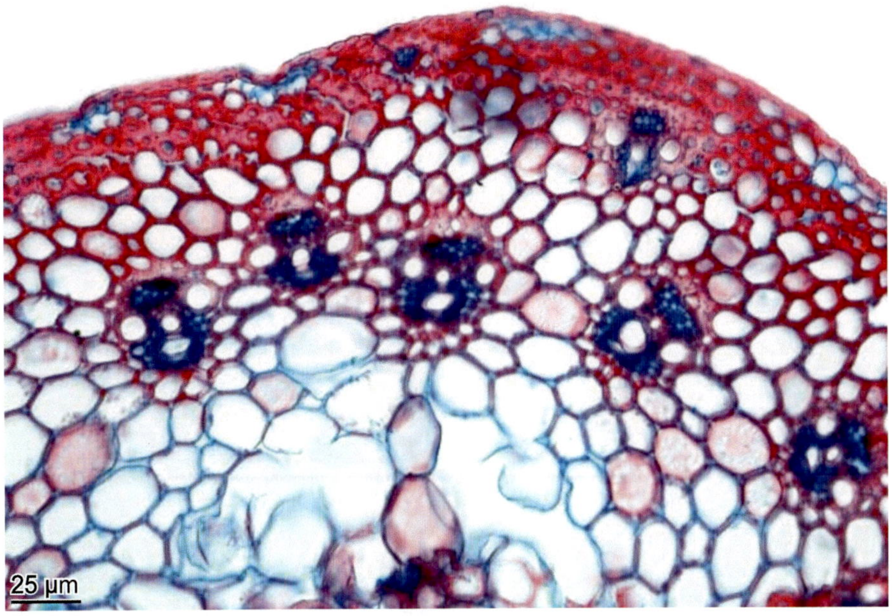

Stipa purpurea Griseb.
Poaceae

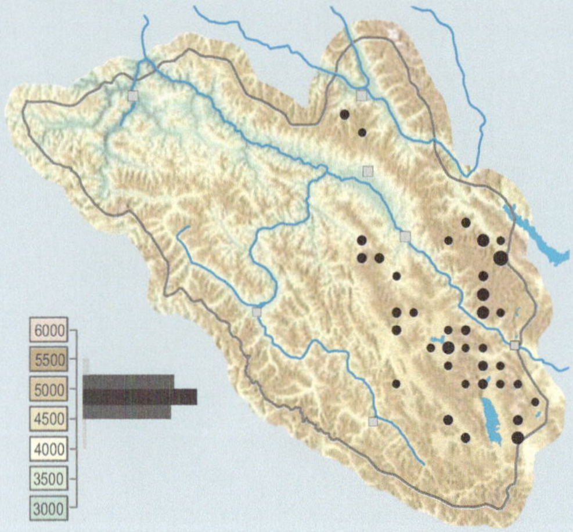

Syn.: *Ptilagrostis purpurea* (Griseb.) Roshev., *Lasiagrostis tremula* Rupr.

Description: Perennial herbs, 15–30 cm tall, cespitose, leaves filiform, involute, panicle lanceolate, open, 6–12 × 1–2 cm, spikelets deep purple, awns 6–8.5 cm long, plumose with hairs 2–3 mm long.

Distribution: Kyrgyzstan, Tajikistan, Pakistan, NW India, China (Gansu, Qinghai, Sichuan, Xinjiang, Xizang).

Habitat: Dry alpine turf and steppes, stony slopes.

Culm uniform, round, hollow, with a diameter of 1.1–1.2 mm. Epidermis thick-walled. Continuous, 6–8-seriate, thick-walled peripheral sclerenchymatic belt. Girders conical to tangentially rectangular. Groups of chlorenchyma round to tangentially enlarged. Triangular intercellulars in the parenchymatic tissue absent. Vascular bundles arranged in 2–3 circular rows. Large bundles within the thin- to thick-walled parenchymatic tissue, small bundles within the sclerenchymatic belt. Vascular bundles with large vessels, diameter 10–12 µm. Phloem oval, consisting of sieve tubes and companion cells. Intercellulars centripetal of the protoxylem absent. Vascular bundles surrounded by an unlignified, 2–3-seriate sheath of polygonal cells, with thick-walled cells at the peripheral side of the vascular bundles.

Analyzed: 2 individuals

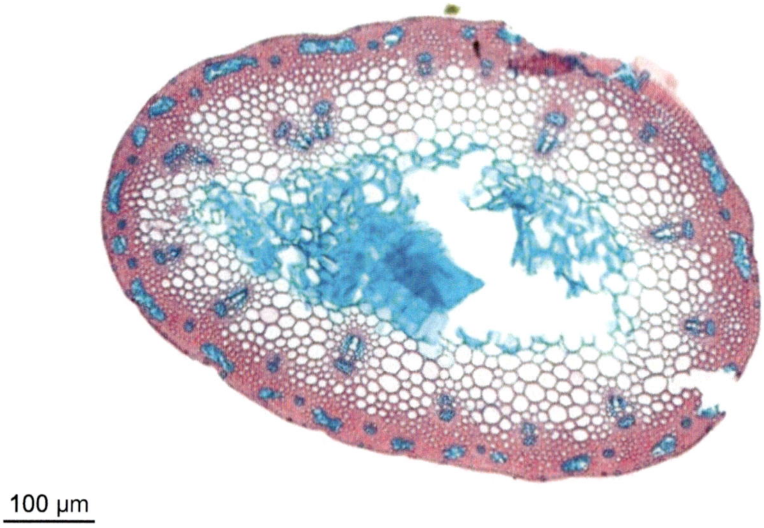

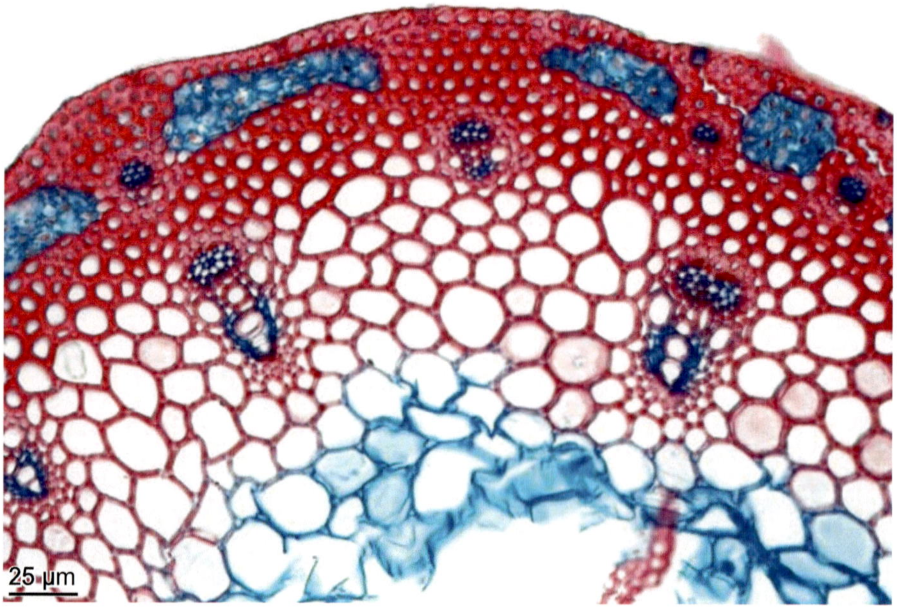

Stipa regeliana Hack.

Poaceae

Syn.: *S. purpurascens* Hitchc., *Achnatherum purpurascens* (Hitchc.) Keng

Description: Perennial herbs, 10–35 cm tall, cespitose, leaves filiform, involute, 5–10 cm long, 0.5–0.8 mm wide, panicle contracted, linear, 5.5–10 cm long, awns 15–20 mm long, pilose.

Distribution: E Kazakhstan, Kyrgyzstan, Tajikistan, Pakistan, N India, China (Gansu, Ningxia, Qinghai, Sichuan, Xinjiang, Xizang, Yunnan).

Habitat: Floodplains, grassy slopes.

Culm uniform, round, hollow, with a diameter of 0.5–0.6 mm. Epidermis thick-walled. Continuous, 6–8-seriate, thick-walled peripheral sclerenchymatic belt. Girders conical to tangentially rectangular. Groups of chlorenchyma round to tangentially enlarged. Very small triangular intercellulars in the parenchymatic tissue. Vascular bundles arranged in 2–3 circular rows. Large bundles within the thin- to thick-walled parenchymatic tissue, small bundles within the sclerenchymatic belt. Bundles within the sclerenchymatic belt reduced to phloem. Vascular bundles with large vessels, diameter 6–8 µm. Phloem oval, consisting of sieve tubes and companion cells. Intercellulars centripetal of the protoxylem absent. Vascular bundles surrounded by an eccentric sheath of polygonal cells, 1–2 rows of thin- to thick walled cells centripetally, and 2–3 rows of thick-walled cells at the peripheral side.

Analyzed: 2 individuals

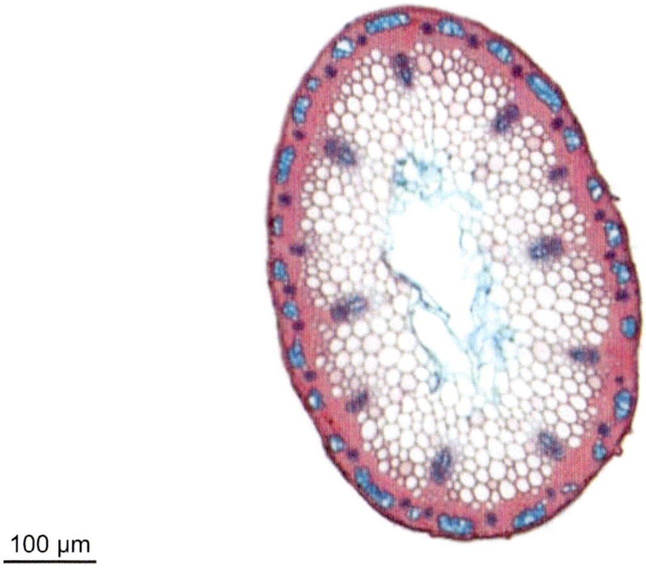

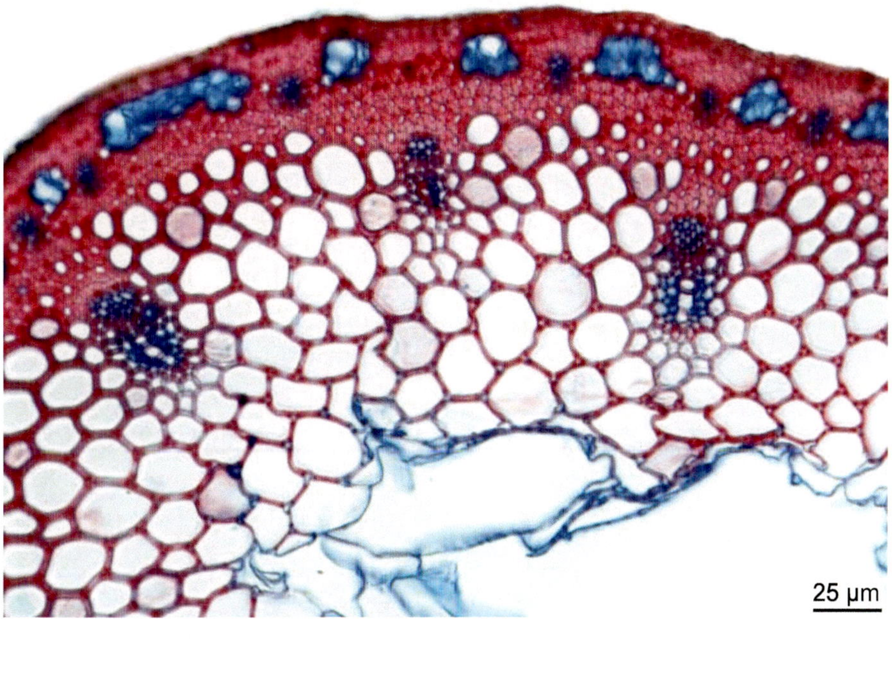

Stipa splendens Trin.

Poaceae

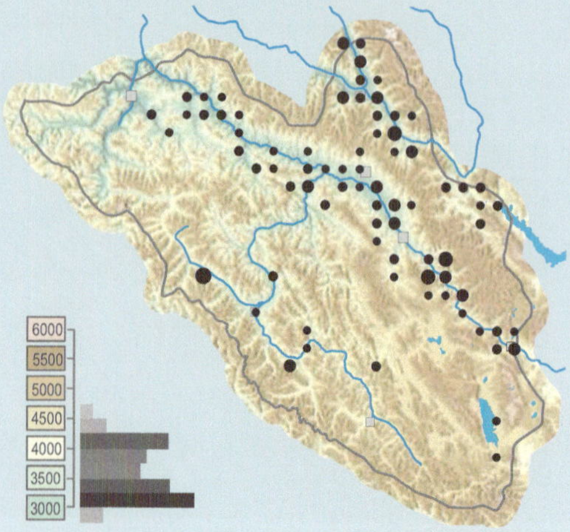

Syn.: *Achnatherum splendens* (Trin.) Nevski, *Lasiagrostis splendens* (Trin.) Kunth, *S. altaica* Trin. ex Ledeb., *S. schlagintweitii* Mez

Description: Perennial herbs, 50–250 cm tall, cespitose, leaves convolute, tough, 30–60 cm long, 5–6 mm wide, panicle pyramidal, 20–60 cm long, awns 6-12 mm long, scabrid.

Distribution: Iran, Russia, Kazakhstan, Turkmenistan, Uzbekistan, Kyrgyzstan, Tajikistan, Afghanistan, Pakistan, NW India, China, Mongolia.

Habitat: Dry mountain slopes, grasslands, tolerates high levels of salinity.

Culm uniform, round, filled, with a diameter of 1.6–2.0 mm. Epidermis thin- to thick-walled. Continuous, 10–20-seriate, thick-walled peripheral sclerenchymatic belt. Girders absent. Groups of chlorenchyma absent. Triangular intercellulars in the parenchymatic tissue absent. Vascular bundles arranged in 3 circular rows. Large bundles within the thin- to thick-walled parenchymatic tissue, small bundles within the sclerenchymatic belt. Vascular bundles with large vessels, diameter 20–25 µm. Phloem oval, consisting of sieve tubes and companion cells. With intercellulars centripetal of the protoxylem. Vascular bundles surrounded by an unlignified, 2–3-seriate sheath of polygonal cells, with thick-walled cells at the peripheral side of the vascular bundles.

Analyzed: 2 individuals

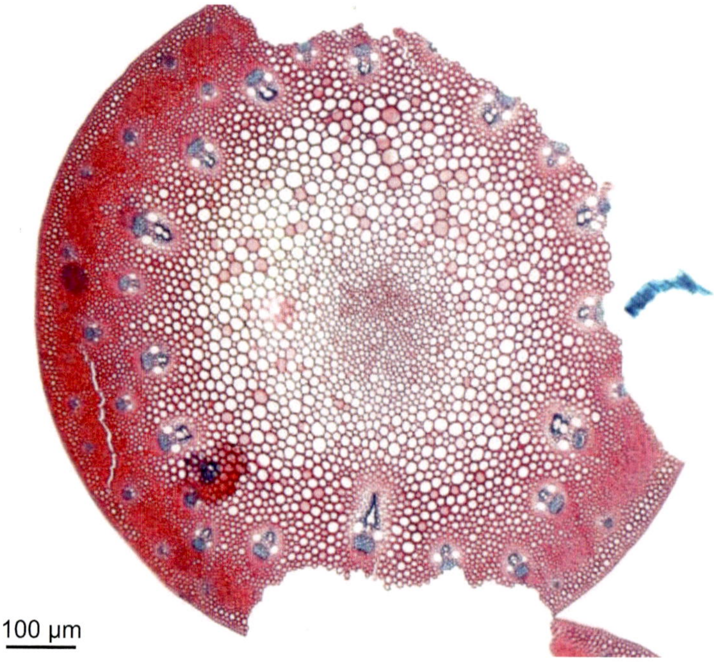

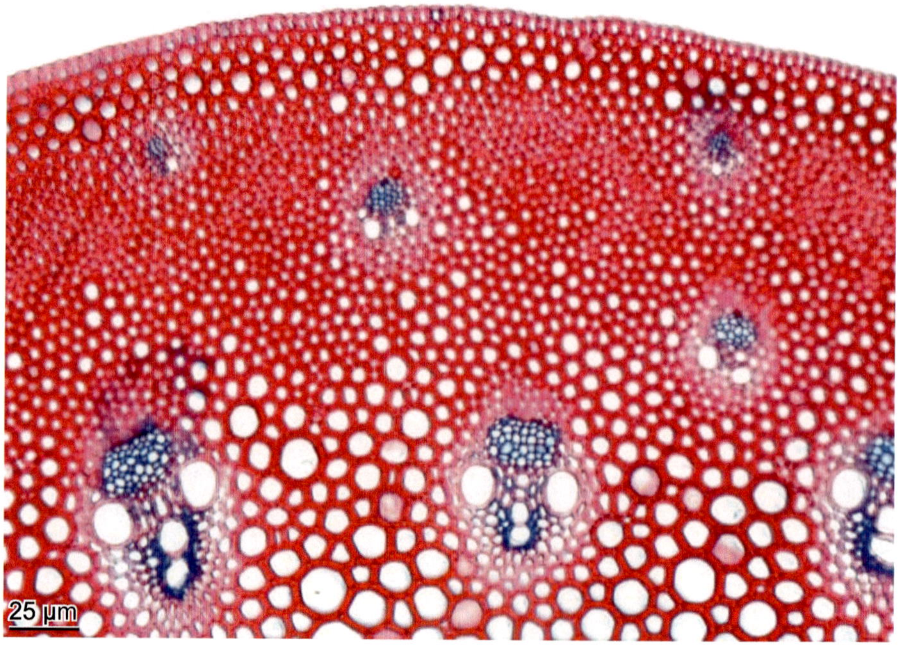

Stipa subsessiliflora (Rupr.) Roshev.

Poaceae

Syn.: *S. basiplumosa* Munro ex Hook.f., *Lasiagrostis subsessiliflora* Rupr., *Ptilagrostis subsessiliflora* (Rupr.) Roshev.

Description: Perennial herbs, 10–35 cm tall, cespitose, leaves setaceous, involute, panicle narrow, contracted, 4–12 cm long, awn plumose, 1–3 cm long.

Distribution: Russia (Altai), E Kazakhstan, Uzbekistan, Kyrgyzstan, Tajikistan, Pakistan, NW India, China (Gansu, Qinghai, Sichuan, Xinjiang, Xizang, Yunnan).

Habitat: Dry and stony steppes, sandy plains, stabilized slopes with scattered vegetation.

Culm uniform, round, hollow, with a diameter of 0.5–0.6 mm. Epidermis thick-walled. Continuous, 4–6-seriate, thick-walled peripheral sclerenchymatic belt. Girders absent. Groups of chlorenchyma absent. Intercellulars in the parenchymatic zone very small and triangular. Vascular bundles arranged in 2 circular rows. Large bundles within the thin- to thick-walled parenchymatic tissue, small bundles within the sclerenchymatic belt. Vascular bundles with large vessels, diameter 6–8 μm. With intercellulars centripetal of the protoxylem. Phloem oval, consisting of sieve tubes and companion cells. Vascular bundles surrounded by a sheath of polygonal cells.

Analyzed: 1 individual

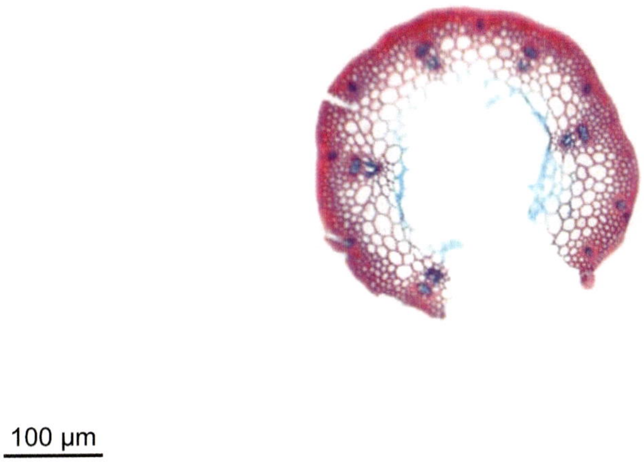

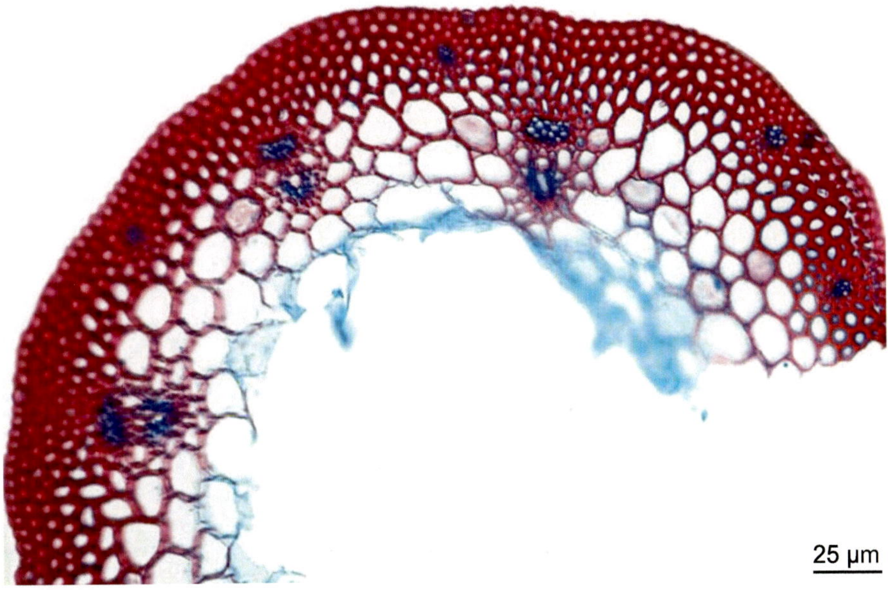

Stipa tianschanica Roshev.

Poaceae

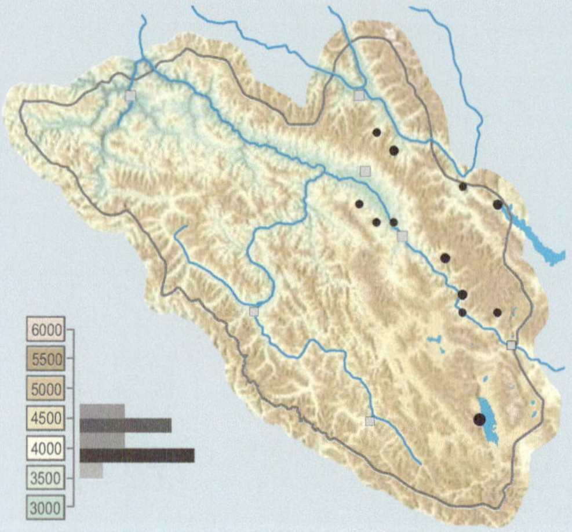

Syn.: *S. gobica* Roshev., *S. sinomongholica* Ohwi, *S. tianschanica* var. *gobica* (Roshev.) P.C.Kuo & Y.H.Sun

Description: Perennial herbs, 10–20 cm tall, cespitose, leaves filiform, involute, up to 10 cm long, 1 mm wide, panicle open, lanceolate, 4–8 cm long, awns 6–8 cm long, plumose.

Distribution: Tajikistan, India (Jammu and Kashmir), China (Gansu, Hebei, Nei Mongol, Ningxia, Qinghai, Shaanxi, Shanxi, Xinjiang), Mongolia.

Habitat: Dry steppes, semideserts, stony slopes.

Culm uniform, round, filled, with a diameter of 0.8–1.0 mm. Epidermis thick-walled. Continuous, 4–6-seriate, thick-walled peripheral sclerenchymatic belt. Girders absent. Subepidermal, 1–2-seriate band of chlorenchyma. Many fairly large, triangular intercellulars in the parenchymatic tissue. Vascular bundles arranged in 2 circular rows. Large bundles within the thin- to thick-walled parenchymatic tissue, small bundles within the sclerenchymatic belt. Vascular bundles with large vessels, diameter 25–30 μm. With intercellulars centripetal of the protoxylem. Phloem oval, consisting of sieve tubes and companion cells. Vascular bundles surrounded by a uniseriate sheath of polygonal cells.

Analyzed: 1 individual

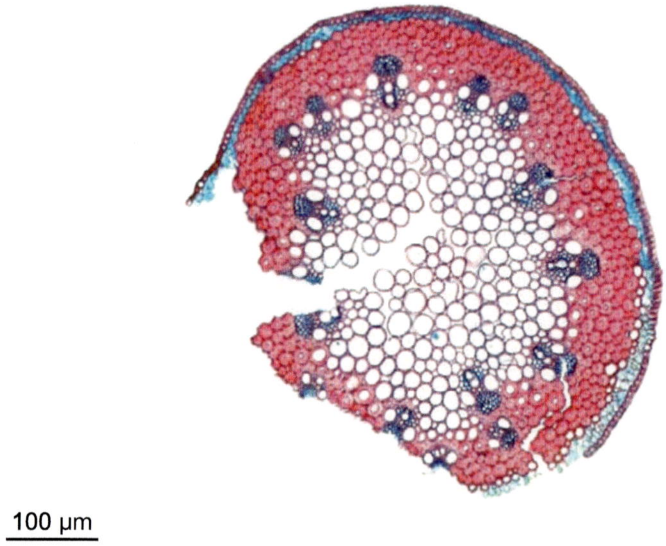

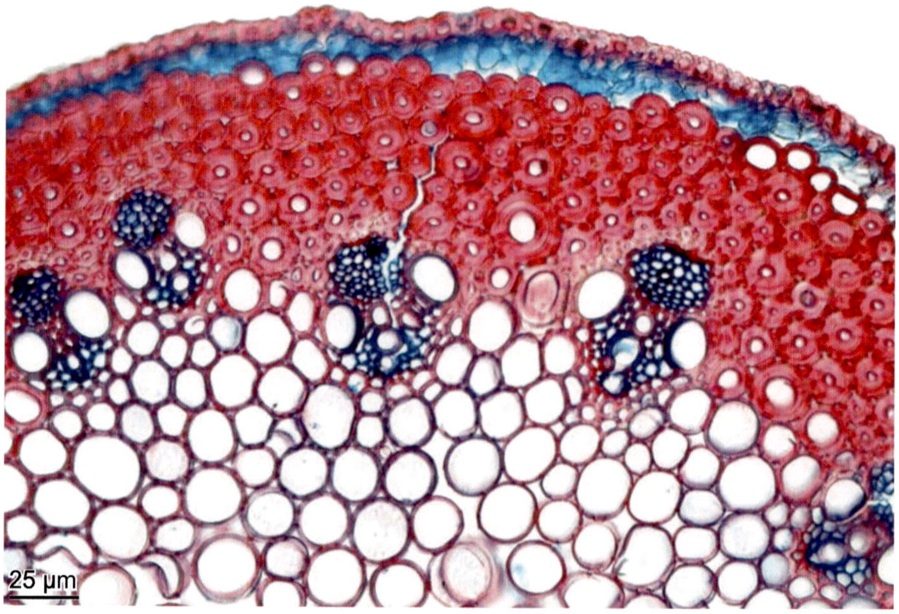

Stipa turkestanica Hack.

Poaceae

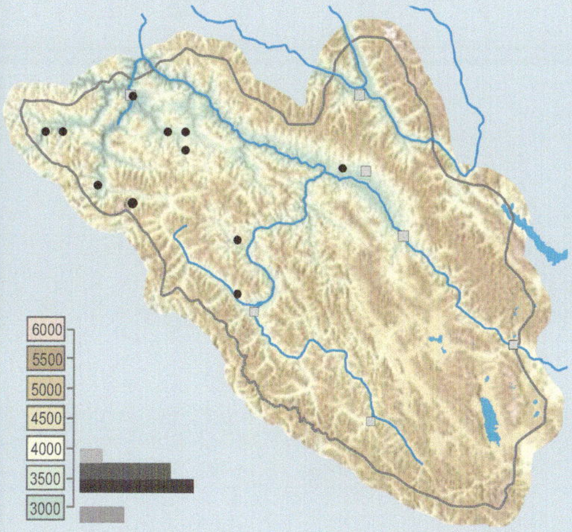

Syn.: *S. trichoides* P.A.Smirn.

Description: Perennial herbs, 30–70 cm tall, cespitose, leaves filiform, involute, 2–4.5 cm long, 1–1.5 mm wide, panicle contracted, linear, 8–14 cm long, base enclosed by uppermost leaf sheath, awns 8–16 (–20) cm long, plumose.

Distribution: Iran, Turkmenistan, Tajikistan, Afghanistan, Pakistan, India (Jammu and Kashmir).

Habitat: Stone mountain steppes, steppe grasslands.

Culm uniform, round, filled, with a diameter of 1.1–1.3 mm. Epidermis thin- to thick-walled. Continuous, 6–8-seriate, thick-walled peripheral sclerenchymatic belt. Girders absent. Groups of chlorenchyma absent. Small triangular intercellulars in the parenchymatic tissue. Vascular bundles arranged in 2–3 circular rows. Large bundles within the thin- to thick-walled parenchymatic tissue, small bundles within the sclerenchymatic belt. Vascular bundles with large vessels, diameter 22–25 µm. With intercellulars centripetal of the protoxylem. Phloem oval, consisting of sieve tubes and companion cells. Vascular bundles surrounded by a uniseriate sheath of polygonal cells.

Analyzed: 2 individuals

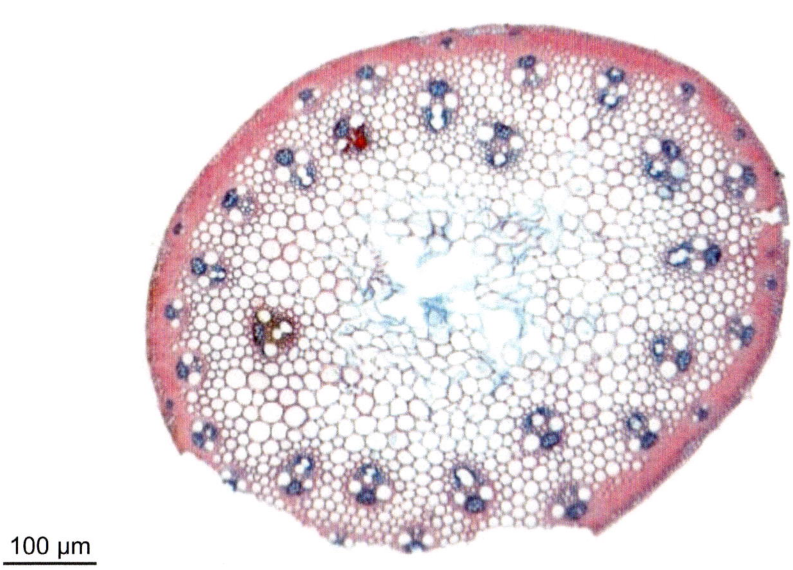

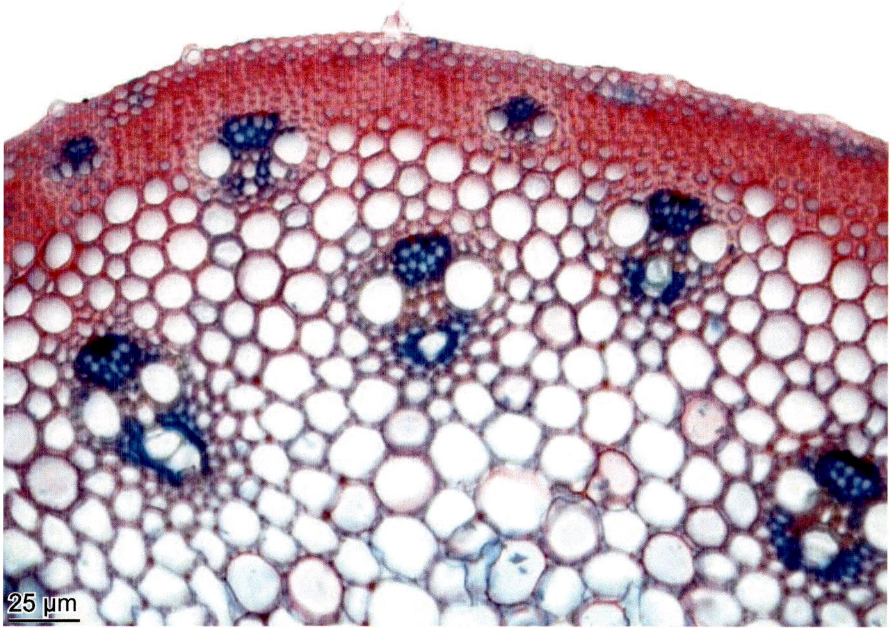

Trikeraia hookeri (Stapf) Bor

Poaceae

Syn.: *Stipa hookeri* Stapf, *Timouria aurita* Hitchc.

Description: Perennial herbs, 60–80 cm tall, with stout rhizomes, leaves involute, 10–40 cm long, 0.4–0.5 cm wide when flattened, panicle contracted or open, 10–30 cm long, 2–10 cm wide, branches 2–12 cm long, awns 1.5 cm long.

Distribution: N Pakistan, India (Jammu and Kashmir), China (Qinghai, Xizang).

Habitat: Sandy steppes.

Culm uniform, round, hollow, with a diameter of 0.8–0.9 mm. Epidermis thick-walled. Continuous, 6–8-seriate, thick-walled peripheral sclerenchymatic belt. Girders square to tangentially rectangular. Chlorenchyma groups oval. Triangular intercellulars in the parenchymatic tissue absent. Vascular bundles arranged in 2 circular rows. Large bundles within the thin- to thick-walled parenchymatic tissue, small bundles within the sclerenchymatic belt. Vascular bundles with large vessels, diameter 8–10 µm. With small intercellulars centripetal of the protoxylem. Phloem oval, consisting of uniform cells. Vascular bundles surrounded by a uniseriate sheath of polygonal cells.

Analyzed: 2 individuals

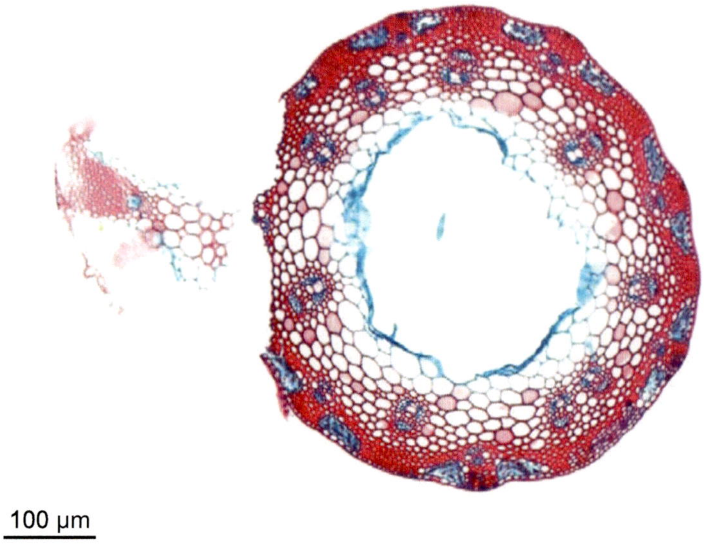

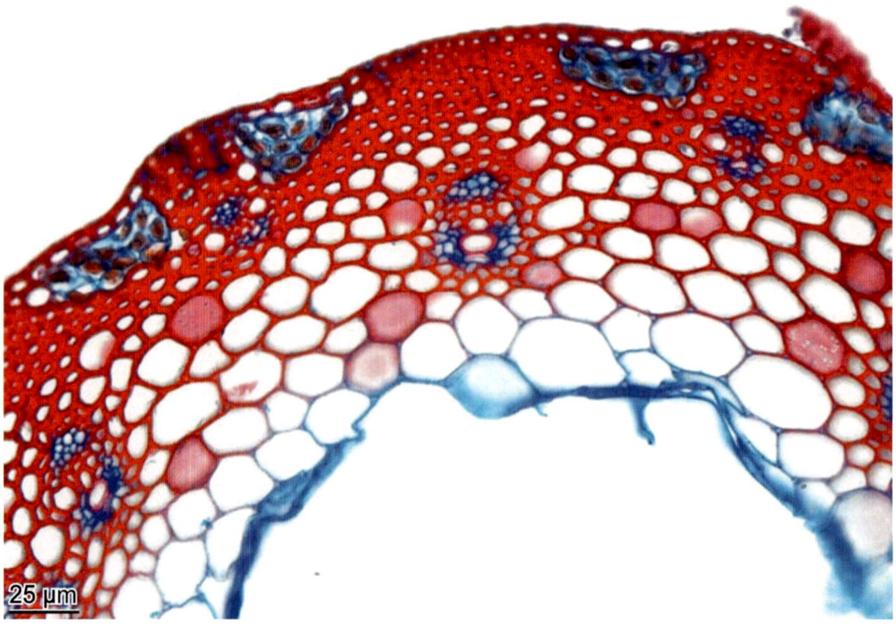

Trikeraia oreophila Cope

Poaceae

Syn.: *T. tianshanica* S. L. Lu & X. F. Lu.

Description: Perennial herbs, 10–50 cm tall, densely cespitose, leaves involute, 3–12 cm long, 1.5–5 mm wide, panicle contracted, narrow, 2.5–8.5 cm long, 1–1.5 cm wide, branches erect, up to 3.5 cm long, awns ca. 1 cm long.

Distribution: India (Jammu and Kashmir, Sikkim), Nepal, China (Xinjiang, Xizang), Bhutan.

Habitat: Dry grasslands, hillsides.

Culm uniform, round, hollow, with a diameter of 0.8–0.9 mm. Epidermis cells internally thick-, externally thin-walled. Continuous, 4–8-seriate, thin- to thick-walled peripheral sclerenchymatic belt centripetal of the chlorenchyma. Girders square to radial rectangular. Groups of chlorenchyma rectangular, tangentially enlarged. Triangular intercellulars in the parenchymatic tissue. Vascular bundles arranged in 2–3 circular rows. Large bundles within the thin- to thick-walled parenchymatic tissue, small bundles within the sclerenchymatic belt. Vascular bundles with large vessels, diameter 12–15 µm. With intercellulars centripetal of the protoxylem. Phloem oval, consisting of sieve tubes and companion cells. Vascular bundles surrounded by a uniseriate sheath of polygonal cells.

Analyzed: 1 individual

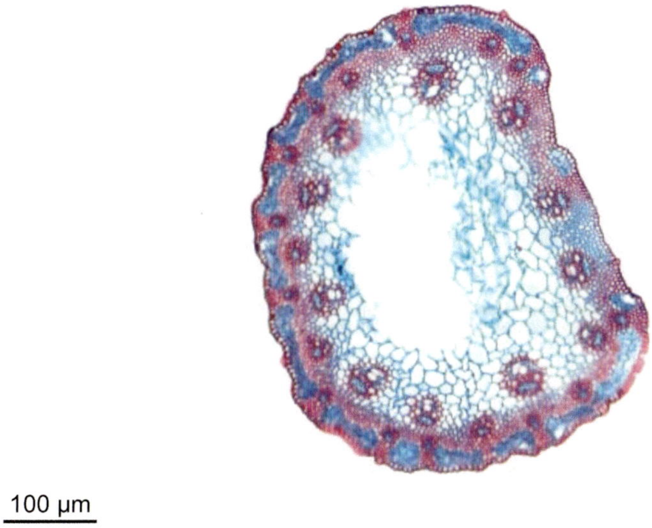

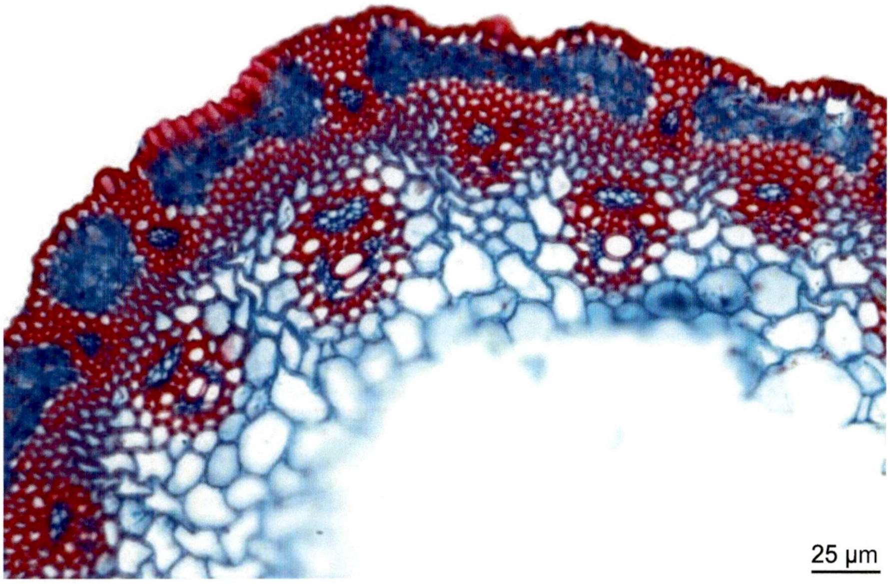

Trisetum clarkei (Hook.f.) R.R.Stewart

Poaceae

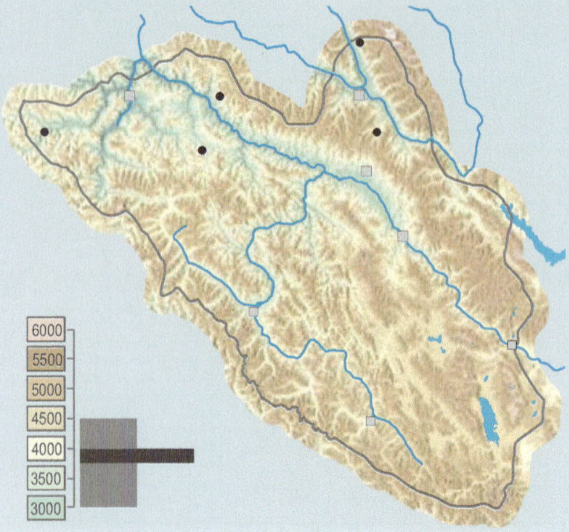

Syn.: *Avena clarkei* Hook.f.

Description: Perennial herbs, 20–70 cm tall, cespitose, culms erect or geniculately ascending, leaf sheaths pubescent, leaves flat, 4–20 cm long, 1.5–2 mm wide, panicle contracted, linear, or lanceolate, 4.5–18 cm long, 1–3 cm wide, branches 1–2 cm long, awns 4–8 mm long.

Distribution: E Afghanistan, Pakistan, NW India, China (Gansu, Hubei, Qinghai, Shaanxi, Sichuan, Xinjiang, Xizang, Yunnan).

Habitat: Moist grassy slopes.

Culm uniform, round, hollow, with a diameter of 0.8–0.9 mm. Epidermis cells thin- to thick-walled. Continuous, 4–8-seriate, thin- to thick-walled peripheral sclerenchymatic belt. Girders absent. Groups of chlorenchyma rectangular, tangentially enlarged. Intercellulars in the parenchymatic zone small and triangular. Vascular bundles arranged in 2 circular rows. Large bundles within the thin- to thick-walled parenchymatic tissue, small bundles within the sclerenchymatic belt. Vascular bundles with large vessels, diameter 8–12 µm. With large intercellulars centripetal of the protoxylem. Phloem oval, consisting of uniform cells. Vascular bundles surrounded by a 1–2-seriate sheath of polygonal cells.

Analyzed: 2 individuals

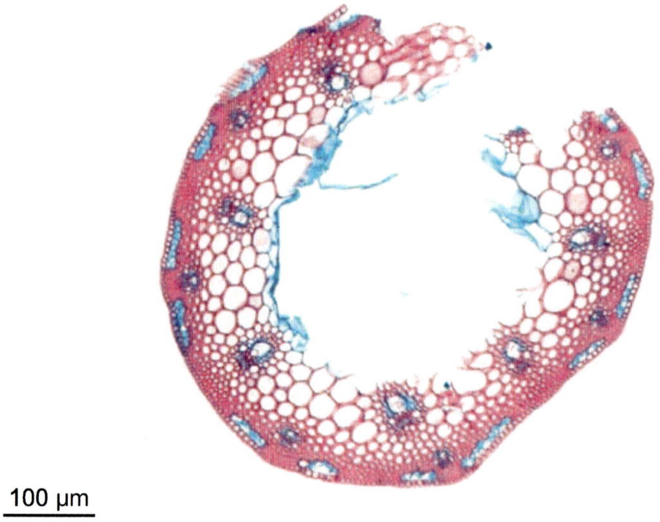

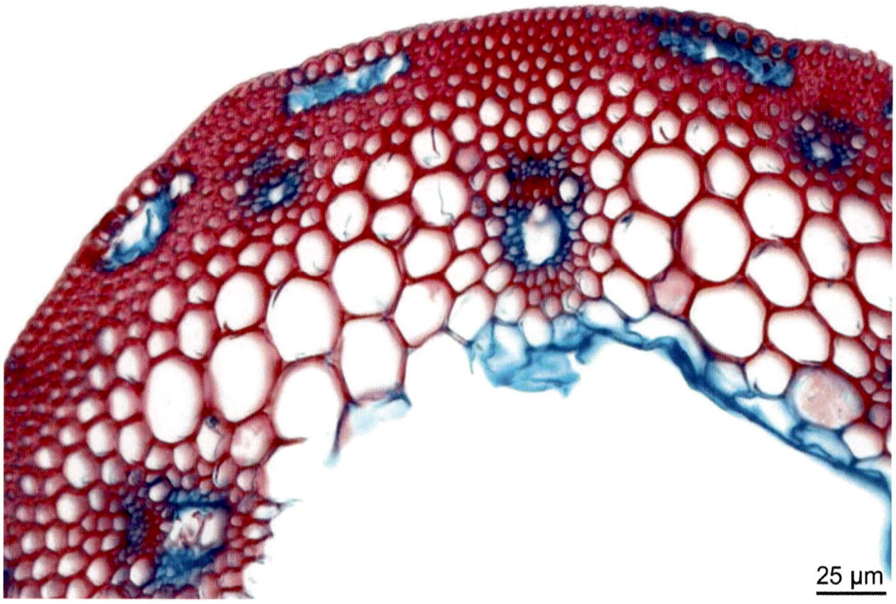

Trisetum spicatum (L.) K.Richt.

Poaceae

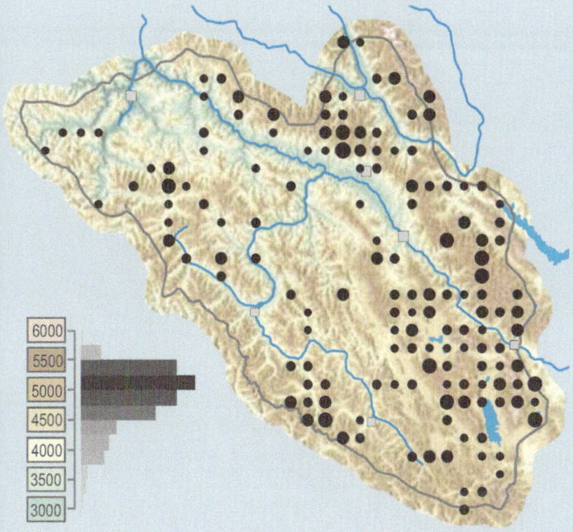

Description: Perennial herbs, 5–40 cm tall, cespitose, tomentose, leaves linear, flat to involute, usually hairy on both surfaces, panicle spike-like, dense, linear to ovate, 1.5–7.5 cm long.

Distribution: North and South America, Europe, temperate Asia, Australia.

Habitat: River banks, along irrigation canals, stabilized stony slopes.

Culm uniform, round, hollow, with a diameter of 0.9–1.0 mm. Epidermis cells internally thick-, externally thin-walled, with hairs. Continuous, 5–8-seriate, thin- to thick-walled peripheral sclerenchymatic belt centripetal of the chlorenchyma. Girders square to conical. Groups of chlorenchyma rectangular, tangentially enlarged. Triangular intercellulars in the parenchymatic tissue absent. Vascular bundles arranged in 2 circular rows. Large bundles within the thin- to thick-walled parenchymatic tissue, small bundles within the sclerenchymatic belt. Vascular bundles with large vessels, diameter 15–20 µm. Intercellulars centripetal of the protoxylem absent. Phloem oval, consisting of uniform cells. Vascular bundles surrounded by a uniseriate sheath of polygonal cells.

Analyzed: 2 individuals

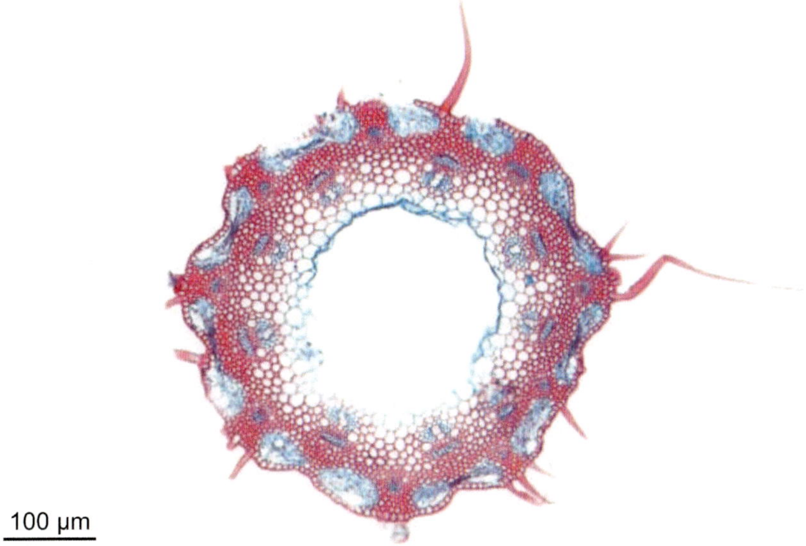

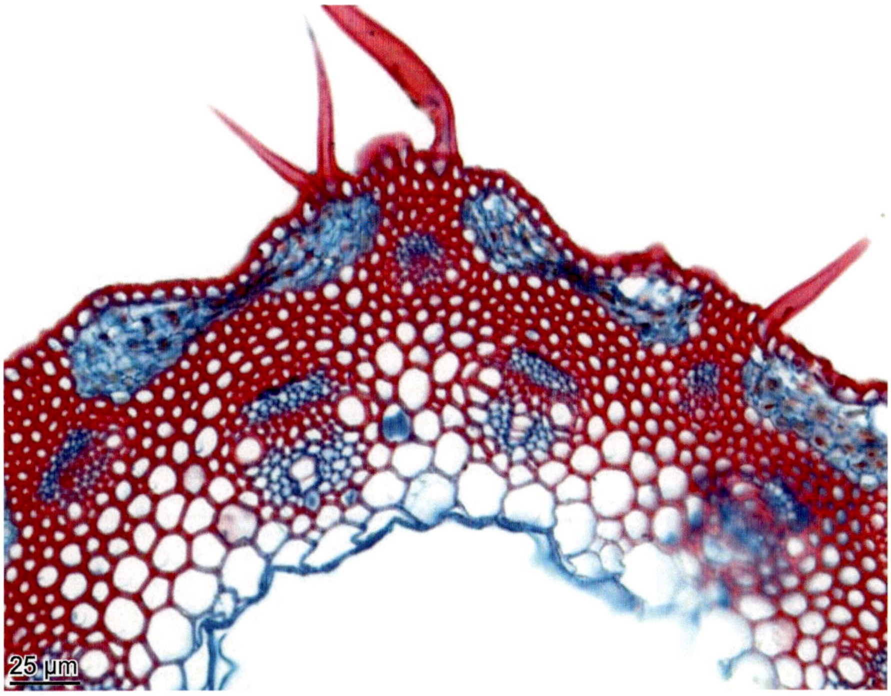

Vulpia myuros (L.) C.C.Gmel.
Poaceae

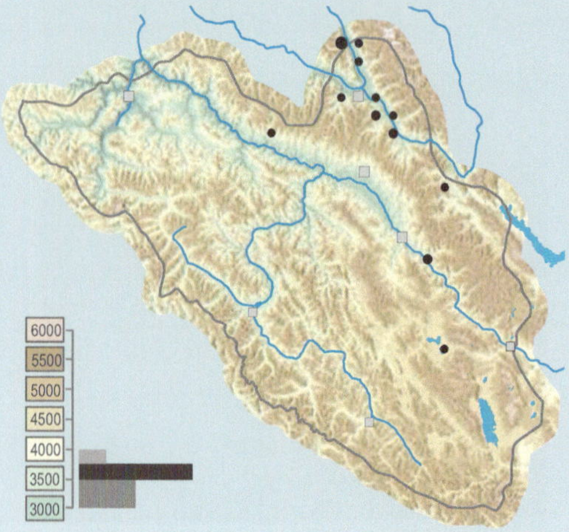

Syn.: *Festuca myuros* L.

Description: Annuals, 5–60 cm tall, culms solitary or cespitose, leaves flat, or involute, pubescent, 2–15 cm long, 0.5–3 mm wide, panicle contracted, linear, nodding, 5–30 cm long, branches appressed, awns 5–15 mm long.

Distribution: N Africa, Europe, SW Asia, S Russia, Turkmenistan, Uzbekistan, Kyrgyzstan, Tajikistan, Afghanistan, Pakistan, India, China, Bhutan; widely introduced in temperate regions.

Habitat: Sandy habitats, roadsides, a weed in fields.

Culm uniform, round, hollow, with a diameter of 0.8–0.9 mm. Epidermis cells thick-walled. Continuous, 2–3-seriate, thin- to thick-walled peripheral sclerenchymatic belt centripetal of the chlorenchyma. Girders absent. Groups of chlorenchyma small, round. Intercellulars in the parenchymatic zone small and triangular. Vascular bundles arranged in 1–2 circular rows. Large bundles within the thin- to thick-walled parenchymatic tissue, small bundles within the sclerenchymatic belt. Vascular bundles with large vessels, diameter 15–25 µm. Large intercellulars centripetal of the protoxylem. Phloem oval, consisting of uniform cells. Vascular bundles surrounded by a 1–2-seriate sheath of polygonal cells.

Analyzed: 2 individuals

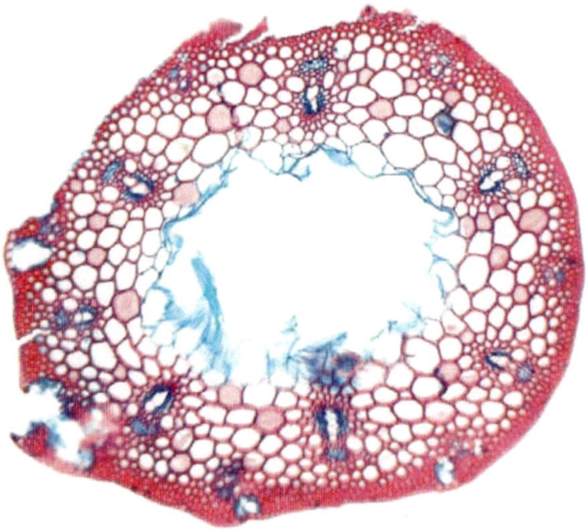

100 µm

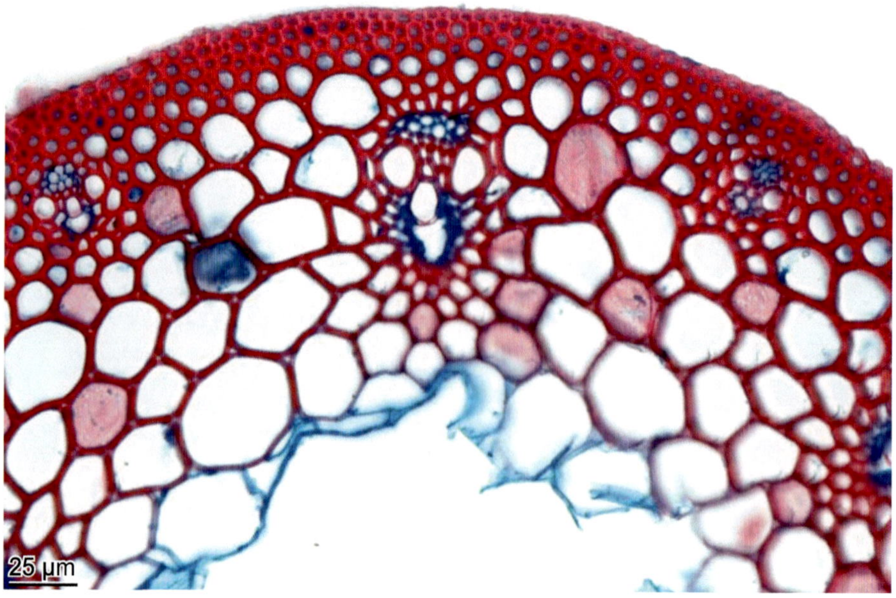

Stuckenia amblyophylla (C.A.Mey.) Holub

Potamogetonaceae

Syn.: *Potamogeton amblyophyllus* C.A.Mey.

Description: Aquatic herbs, submersed, stems branched, 20–60 cm long, leaves filiform, 5–10 cm long, 0.7–2.5 mm wide, flowers minute, in terminal cylindric spikes.

Distribution: SW Asia, Russia, Kazakhstan, Kyrgyzstan, Tajikistan, Afghanistan, India, China.

Habitat: Fresh-water pools, lakes, ponds, marshes.

Stem round to oval, with a diameter of 1–1.5 mm. A large aerenchymatic cortex and an epidermis surround a small central cylinder.

Central cylinder

Central zone consists of large, oval air ducts, surrounded by small cells. Very few, roundish, unlignified vessels inside of the peripheral phloem zone. Sieve tubes large and polygonal. Endodermis consists of round cells with U-shaped cell-wall thickenings.

Cortex

Uniseriate rows of parenchyma cells form a netlike aerenchyma. Epidermis thin-walled.

Analyzed: Submerse annual shoots of 2 individuals.

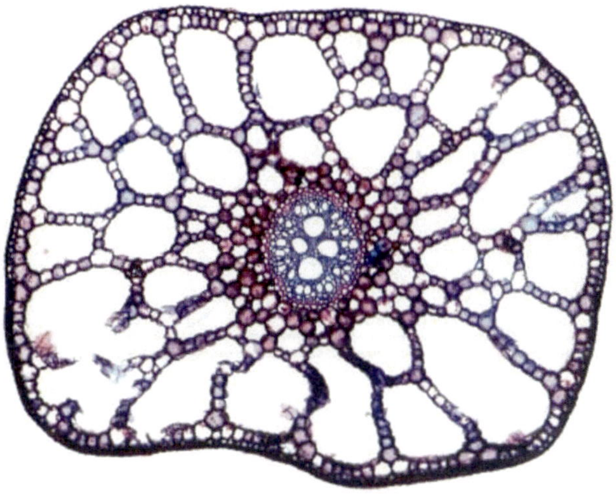

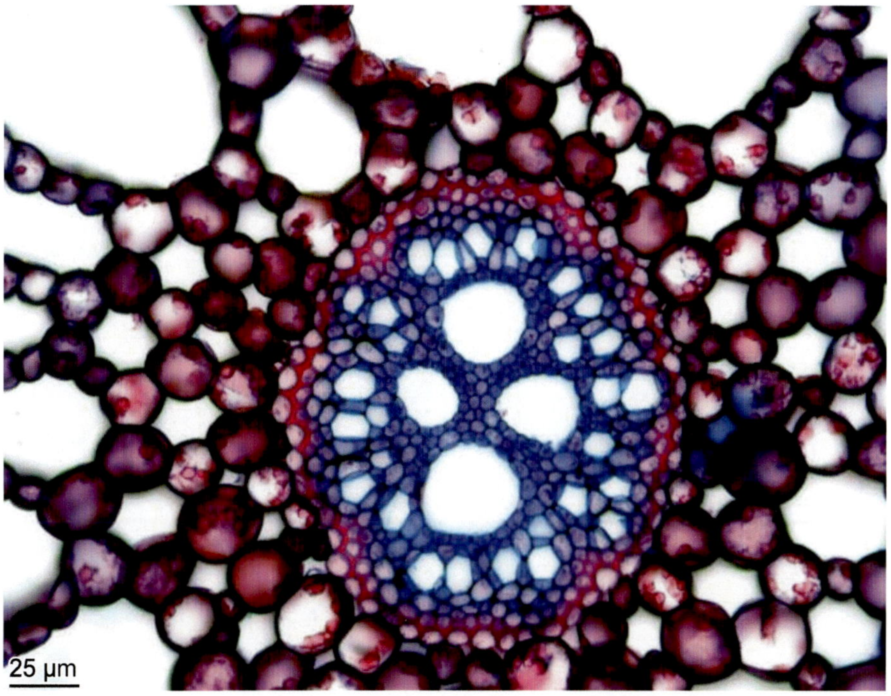

Zannichellia palustris L.

Potamogetonaceae

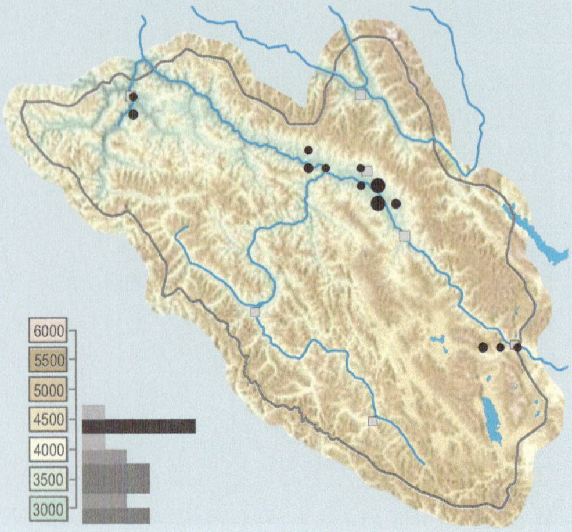

Description: Aquatic submersed herbs, stems 3–50 cm long, leaves opposite, filiform, 2–10 cm × 0.3–0.5 mm, flowers minute, axillary, fruits curved, reniform to semilunate, 2–3 mm long.

Distribution: Cosmopolitan.

Habitat: Fresh or brackish water.

Shoot round, with a diameter of <1 mm. A large aerenchymatic cortex and a thin-walled epidermis surround a small central cylinder.

Central cylinder

An area of poorly differentiated, unlignified cells surrounds a central air duct. Vessels and fibers absent. Sieve tubes represented by unlignified angular cells and small neighboring companion cells. Central cylinder and cortex separated by a poorly developed endodermis.

Cortex

Parenchyma cells and large intercellulars form the cortex. Crystals absent.

Analyzed: 1 individual

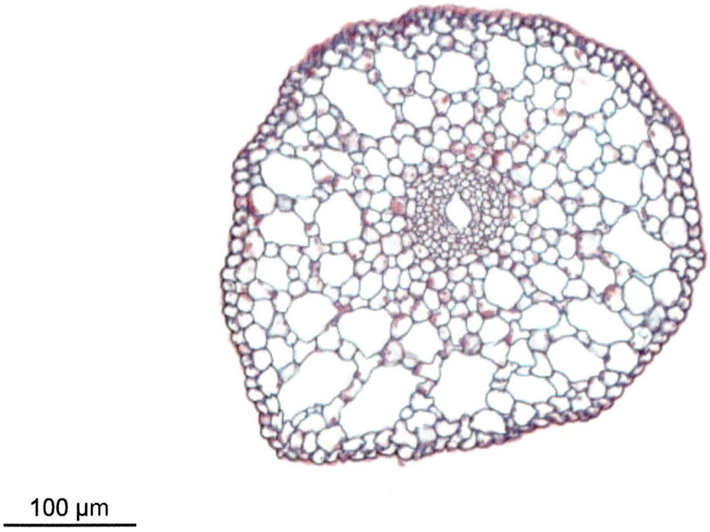

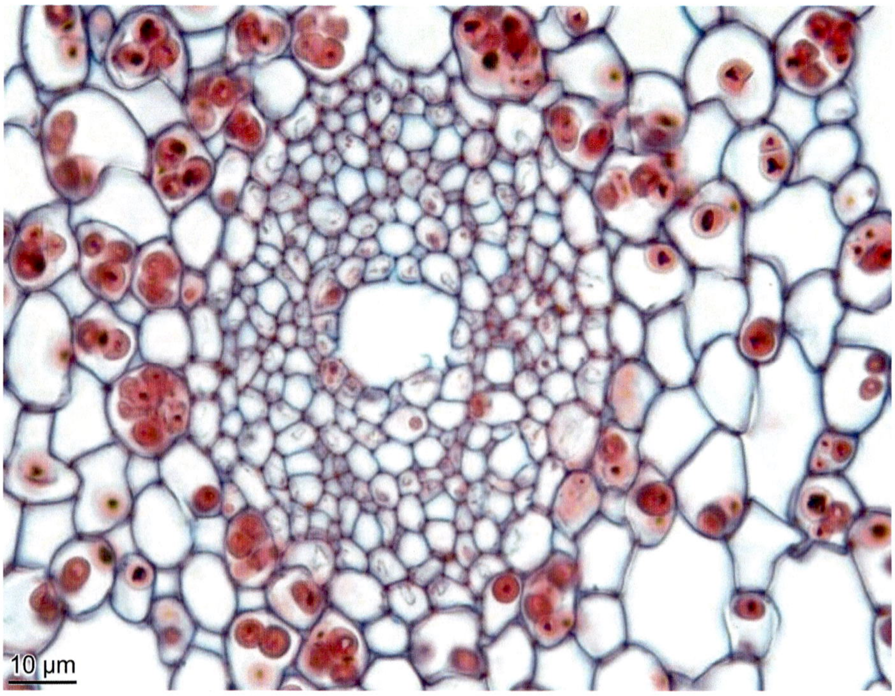

Members of Ladakh expedition in a discussion (from right to left: Fritz Schweingruber, Alan Crivellaro, Jiří Doležal, Pierre Liancourt).

SYNTHESIS OF ANATOMICAL, ECOLOGICAL AND PHYLOGENETICAL DATA

Nature follows two adaptation strategies. First, the floristic composition of a given region is an expression of historical processes, macroclimate, migratory and florogenetic factors (Körner 1999) which can be approached through floristic and phytosociological mapping. Second, each species is anatomically/physiologically adapted to local environmental conditions where natural disturbances and biotic interactions play important roles. Here we combine these two approaches. By comparing the anatomical and age composition of different species in different environments we try to disentangle at least major relationships between environmental and taxonomic factors. Relations between the conductive, stabilization and the storing functions within the plant stem are discussed.

Analyzed were 355 dicotyledonous and 155 monocotyledonous species. Out of 355 dicots, 12 analyzed species (*Astragalus popovii*, *Campanula aristata*, *Cicer arietinum*, *Cynoglossum glochidiatum*, *Epilobium angustifolium*, *Erigeron bellidioides*, *Hippolytia dolichophylla*, *Impatiens thomsonii*, *Potentilla cuneifolia*, *Primula pamirica*, *Rhodiola tibetica* and *Silene nepalensis*) are not presented with a full anatomical description in the book, while 5 species described in the book were excluded from the analysis due to incomplete information (*Bergenia ciliata*, *Juniperus semiglobosa*, *Sagina saginoides*, *Thalictrum alpinum* and *Viola kunawurensis*) .

Anatomy of dicotyledonous species

The 355 studied species belong to 50 families and 184 genera. The most species-rich family is Asteraceae (66 taxa), followed by Fabaceae (30), Brassicaceae (28), Amaranthaceae (21), Polygonaceae (21), Rosaceae (18), Ranunculaceae (16), Caryophyllaceae (15), Lamiaceae (15), Boraginaceae (13), Gentianaceae (12), Apiaceae (9) and Papaveraceae (9). The most species-rich genera are *Artemisia* (14), *Astragalus* (13), *Saussurea* (11), *Oxytropis* (8), *Polygonum* (6), *Potentilla* (6), *Silene* (6), *Corydalis* (5), *Draba* (5) and *Nepeta* (5). Most studied taxa were steppe species (99), followed by alpine grassland (76), scree (69), ruderal (44), wetland (29), desert (19) and subnival (19) species, which corresponds with the distribution of floristic richness reported in previous studies from Ladakh (Dvorský et al. 2011, 2017). Most studied species have their elevation optima between 3500 and 5000 m, with richness decreasing towards both lower elevation deserts and higher elevation subnival zone.

Asteraceae form 50% of studied species with elevation optima at 5500–6000 m, while in the lower elevation belts Asteraceae make up concordantly 14–18%, with exception of deserts where they again prevail with 30%. The fraction of Brassicaceae, Fabaceae and Caryophyllaceae increases with increasing elevation, while that of Amaranthaceae and Lamiaceae decreases with elevation. Similar trends are seen in higher taxonomic units, with Eurosids I and II increasing with elevation, while Asterids and Eusterids I decreasing with elevation. Amaranthaceae make up 37 and 25% in desert and ruderal vegetation, respectively.

The dataset is dominated by hemicryptophytes. The proportion of other life forms such as therophytes, chamaephytes, shrubs, trees or lianas remains generally below 20% in all taxonomic classes and orders. Hemicryptophytic life forms developed the largest diversity under the cold arid environmental conditions. Due to

J. Doležal et al., *Anatomy, Age and Ecology of High Mountain Plants in Ladakh, the Western Himalaya*, https://doi.org/10.1007/978-3-319-78699-5_5

the overall scarcity of water, short, non-clonal herbs with deep tap roots or a pleiocorm with only short below-ground branches (*Pleiocorm* or *Saussurea gnaphalodes* type in Klimešová et al. 2011) prevail (Fig. 2), making up 38% and 15% of the species (mainly from Asteraceae and Fabaceae), followed by annuals and biennials (21%) and clonal plants with short rhizomes (*Srhizome*, 10%) and long rhizomes (*Lrhizome*, 6%) (Fig. 3). The least represented are cushion plants (4%) and woody taxa (6%). Short-lived annual and biennial species are more or less equally represented across all major families, with most representatives in Amaranthaceae (23%), Brassicaceae (12%) and Gentianaceae (12%) (Fig. 3). Cushion plants recruit mostly from Brassicaceae (54%), Caryophyllaceae (15%) and Fabaceae (8%). Long-rhizomatous plants are mainly found in Asteraceae (31%), Rosaceae (23%), Polygonaceae (13%), Ranunculaceae (9%) and Saxifragaceae (9%). Short-rhizomatous plants recruit from Asteraceae (20%), Ranunculaceae (17%), Gentianaceae (9%) and Orobanchaceae (Fig. 4).

Tap-rooted herbs and subshrubs (*Arnebia euchroma* type, see Material and Methods section), dominate in the study regions, prevailing in steppes, *Caragana* shrubs and screes (Fig. 3). Perennial plants with a pleiocorm having long branches (*Saussurea gnaphalodes* type) are abundant in alpine screes (Fig. 3). Annuals and biennials (*Microula tibetica* type) prevail in animal resting places (higher soil N and disturbances), and are also abundant in salt marshes. Compact cushion plants (*Thylacospermum caespitosum* type) occur exclusively in the subnival zone. There is a much higher proportion of Asteraceae taxa among taller plants than among smaller plants, while the opposite is true for Brassicaceae where plants shorter than 5 cm prevail (Fig. 2).

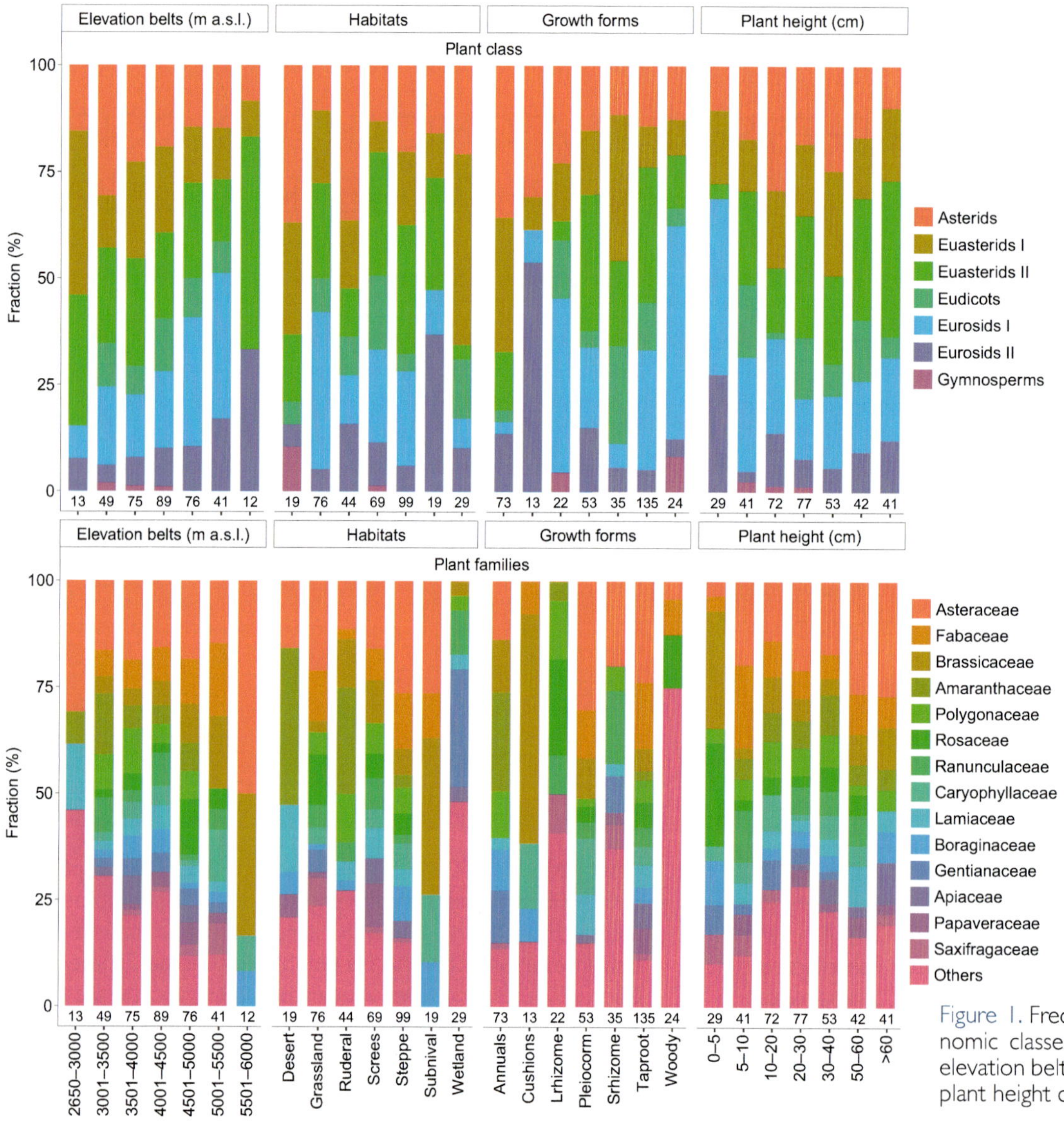

Figure 1. Frequency of main plant taxonomic classes and families in individual elevation belts, habitats, growth form and plant height categories.

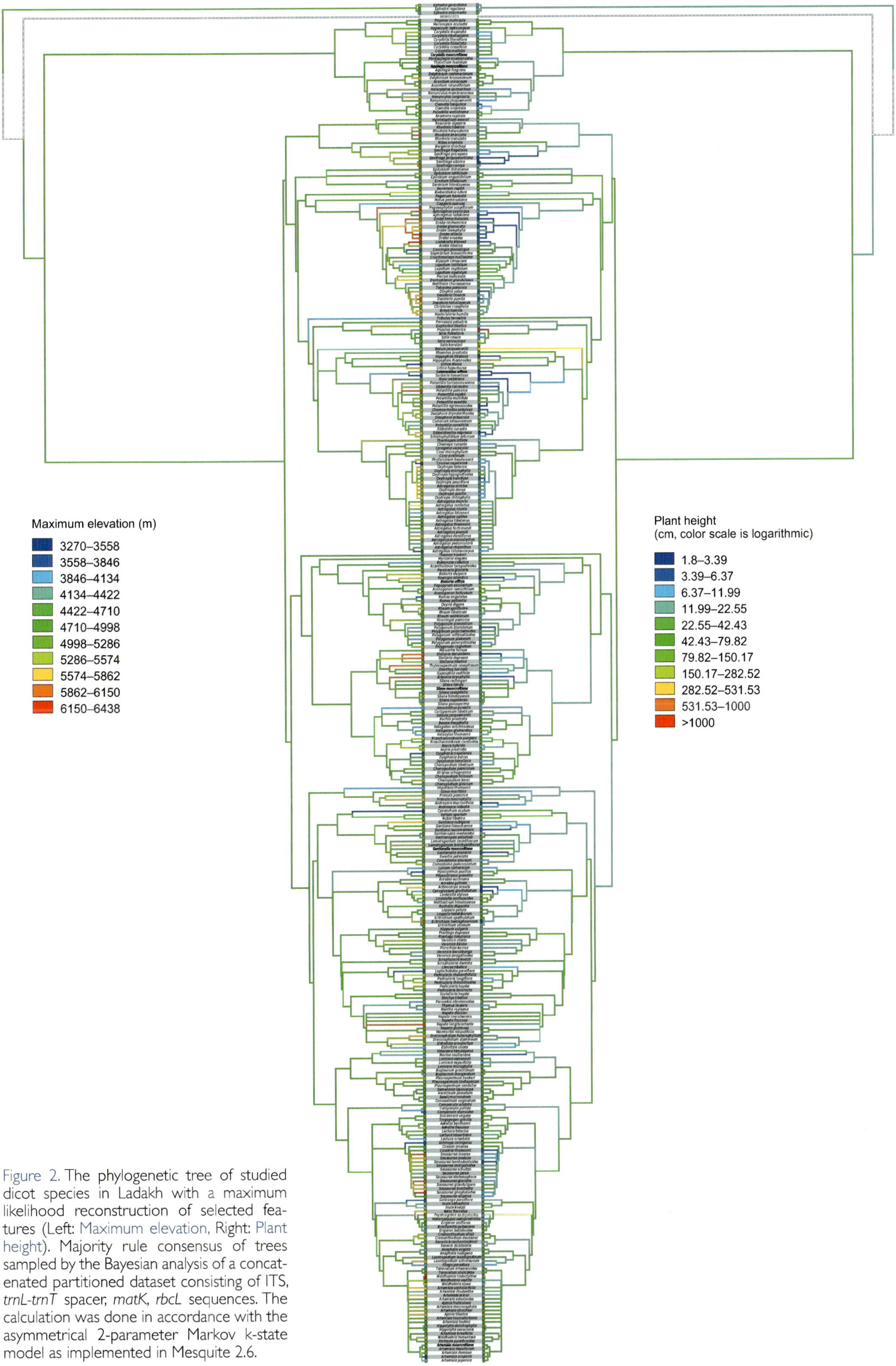

Figure 2. The phylogenetic tree of studied dicot species in Ladakh with a maximum likelihood reconstruction of selected features (Left: Maximum elevation, Right: Plant height). Majority rule consensus of trees sampled by the Bayesian analysis of a concatenated partitioned dataset consisting of ITS, *trnL-trnT* spacer, *matK*, *rbcL* sequences. The calculation was done in accordance with the asymmetrical 2-parameter Markov k-state model as implemented in Mesquite 2.6.

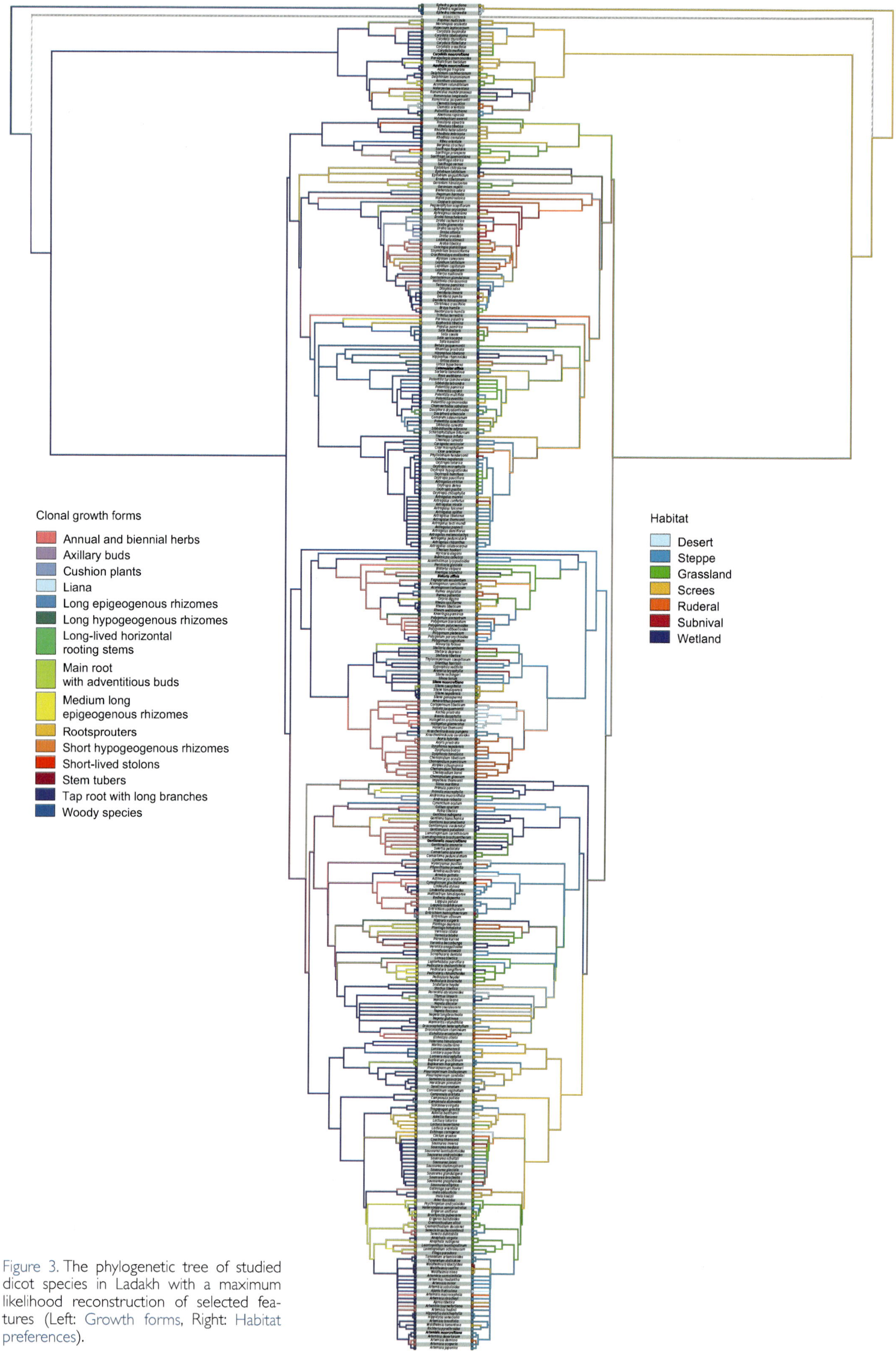

Figure 3. The phylogenetic tree of studied dicot species in Ladakh with a maximum likelihood reconstruction of selected features (Left: Growth forms, Right: Habitat preferences).

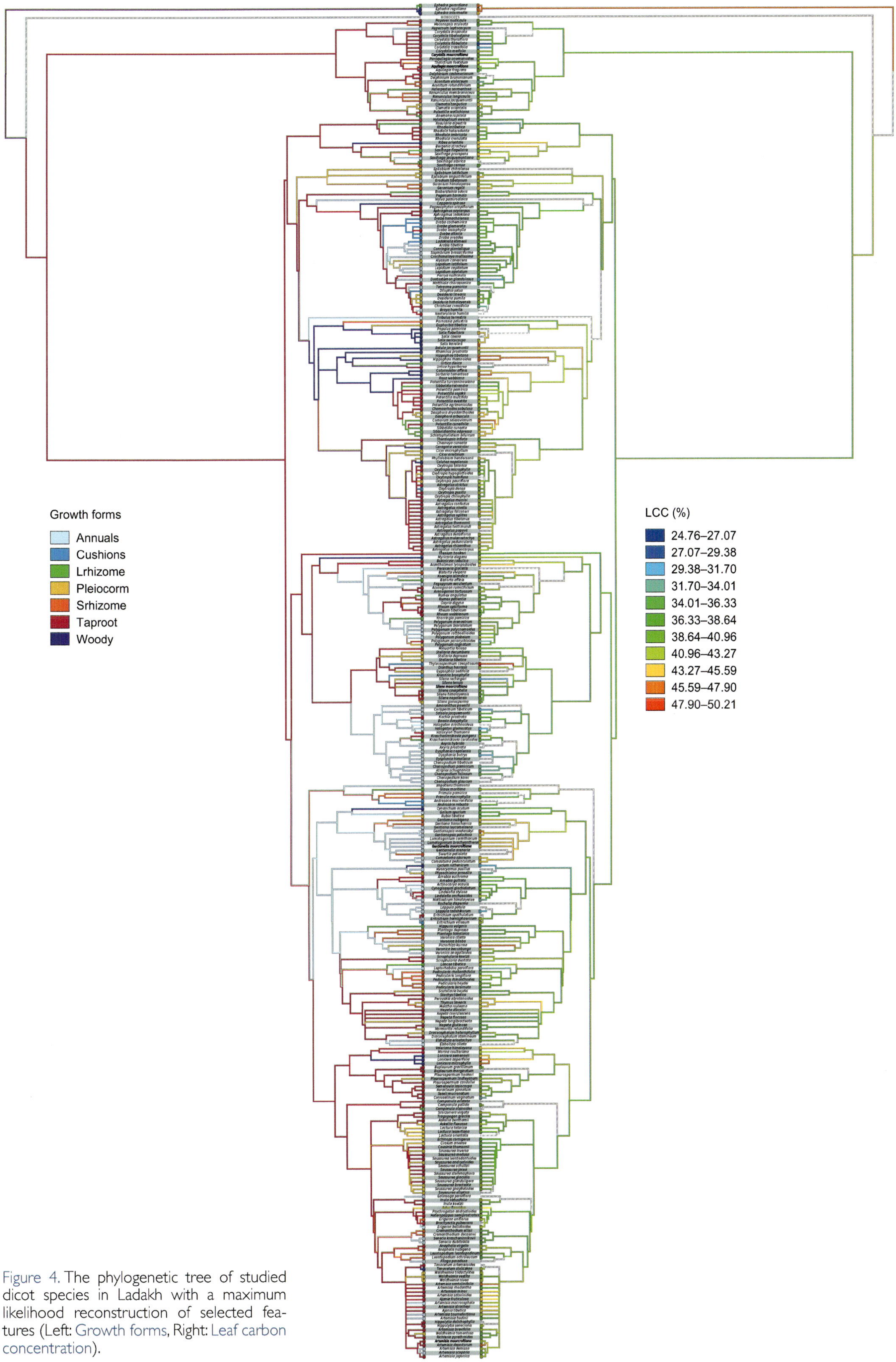

Figure 4. The phylogenetic tree of studied dicot species in Ladakh with a maximum likelihood reconstruction of selected features (Left: Growth forms, Right: Leaf carbon concentration).

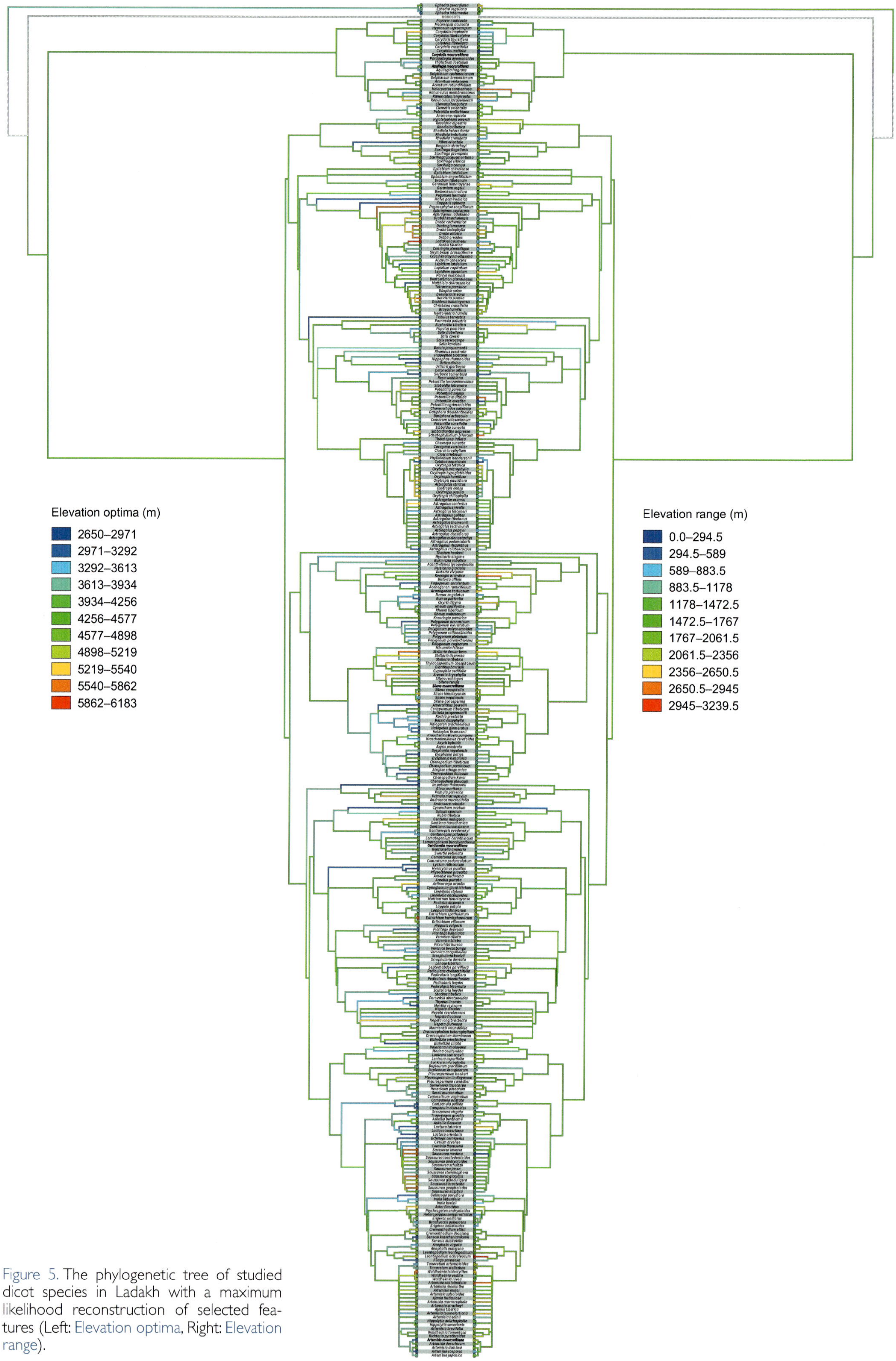

Figure 5. The phylogenetic tree of studied dicot species in Ladakh with a maximum likelihood reconstruction of selected features (Left: Elevation optima, Right: Elevation range).

The principal stem construction is common for all stem-forming dicotyledonous plants—it consists of xylem, phloem and cortex and periderm. The occurrence of axial parenchyma cells, vessels and sieve elements is also common. The presence of periderm with cork cells, fibers, rays and sclereids is not mandatory. In rare cases also vessels might be absent. Due to the small size of the plants (3–60 cm), many features are related to plant size (Fig. 5) but also to elevation (Fig. 6). Characteristic is the large amount of bark in relation to the xylem, very small annual rings (<100 μm), semi-ring porosity, fairly thick-walled vessels with a diameter <20 μm, short vessel elements (80–200 μm), pervasive parenchyma, absence of fibers, and presence of secondary intra-xylary periderm.

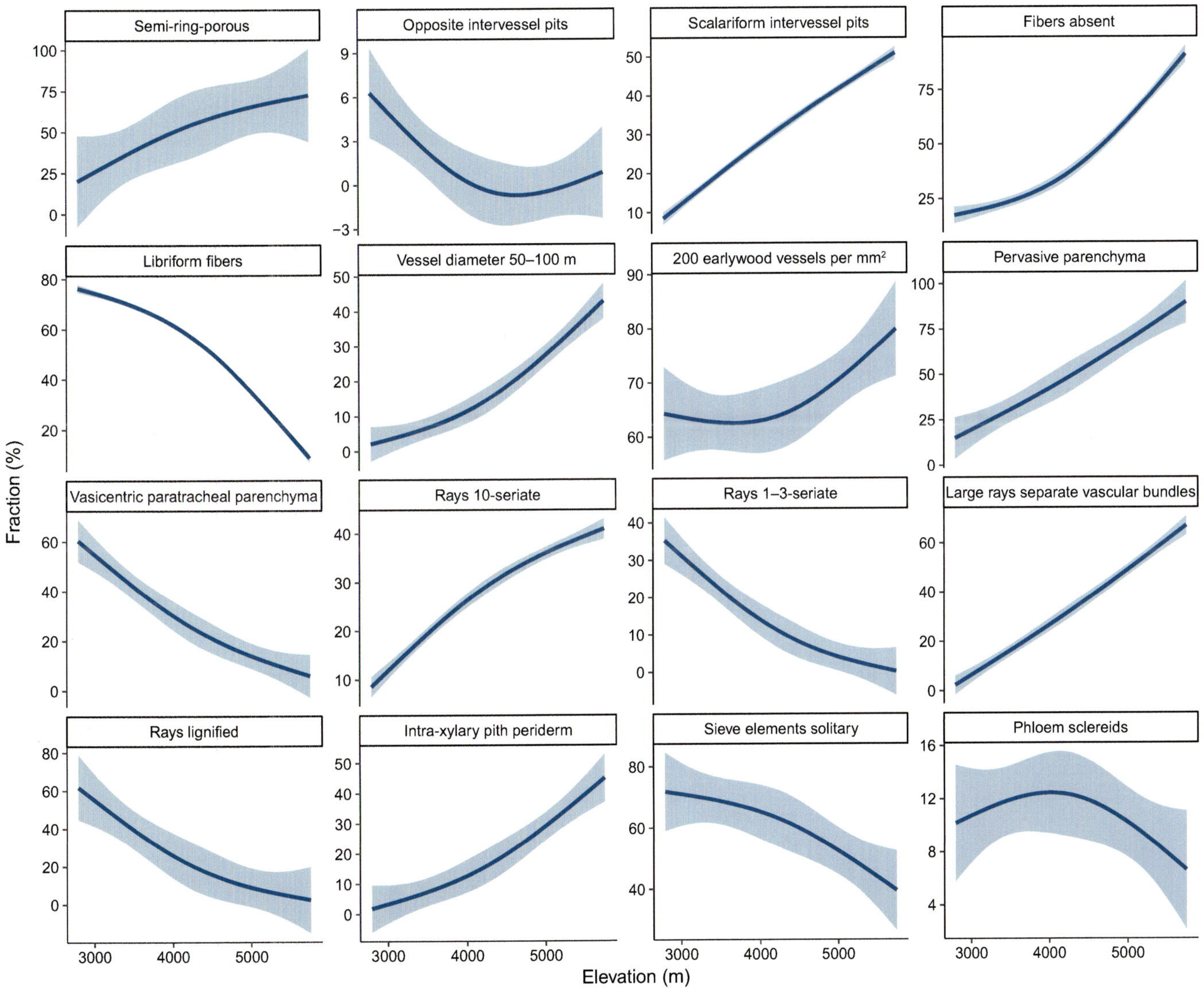

Figure 6. Major elevation-related trends in selected anatomical stem features of dicotyledonous species in Ladakh, W Himalayas, based on 355 taxa distributed from 2650 to 6150 m elevation.

Annual rings are well developed in the majority of plants (Fig. 7), however the degree of distinctness varies, due to taxonomy or environmental conditions. The age can be determined for 60–70% of all dicotyledonous species analyzed. The values vary between families, genera and species. Ages are fairly low in Euasterids I and fairly high in Eurosids I. Many dicotyledonous plants (e.g. Amaranthaceace, Chenopodiacea, *Thylacospermum ceaspitosum*) have no countable rings. Annual plants have one annual ring. Species 2–25 years-old prevail (62%) among the studied species. Short-lived species <2 years old form 70% of species at the lowest elevation, mainly in ruderal vegetation, and their share decreases with elevation, having about 15–30% up to 5500 m. The least represented are short-lived plants in screes and subnival zone, while alpine grassland may have up to 30% (high proportion of Gentianaceae). The fraction of longer-lived species increases with elevation. Most abundant in alpine steppes and screes are plants 2–15 years old, while species 15–25 years old are mainly found in upper alpine and subnival zones above 5200 m. Higher longevity is achieved among alpine cushion plants, long-rhizomatous forbs and subshrubs, plants with pleiocorm and woody species (Fig. 8). Among the oldest plant individuals belong *Artemisia stracheyi* and *Potentilla pamirica* (up to 70 years) and *Arenaria bryophylla* (53 years) in the alpine zone above 5200 m. The oldest plant individual at the highest site with angiosperms at 6150 m was a 22 year-old endemic *Ladakiella klimesii*. Variation in plant age does not seem to be related to plant height.

The majority of annual ring widths varies between 0.05 and 0.4 mm (Fig. 7). Short vegetation periods and low summer temperatures limit growth of all plants. Cell division and cell differentiation occurs only during 4–6 weeks. During this time cambial cells primarily form earlywood vessels and latewood parenchyma cells. Often there is not enough time to form intra-annual tangential structures of even distinct latewood zones. Slow growth and high longevity is a characteristic for most small plants growing in high altitudes of Ladakh (cf. Nobis and Schweingruber 2013). The proportion of species with very narrow annual increments <0.15 mm increases with elevation up to 92% in the subnival zone, while plants with and intermediate growth rate of 0.15–0.3 mm prevail in mid-elevation steppes and screes. The share of species with greater increments of 0.3–1 mm steeply decreases with elevation from 54 to almost 0% above 5500 m. Species with increments >1 mm form only 7% of all studied taxa, with the highest share (23%) at the lowest elevation woody and ruderal species such as *Sisymbrium brassiciforme*, *Amaranthus powellii*, and *Lappula tadshikorum* (Fig. 8). Unlike plant longevity, growth rate is clearly negatively related to plant height, being smallest in tiny cushion and short-rhizomatous plants, and the highest in tall herbaceous tap-rooted forbs and subshrubs and woody shrubs.

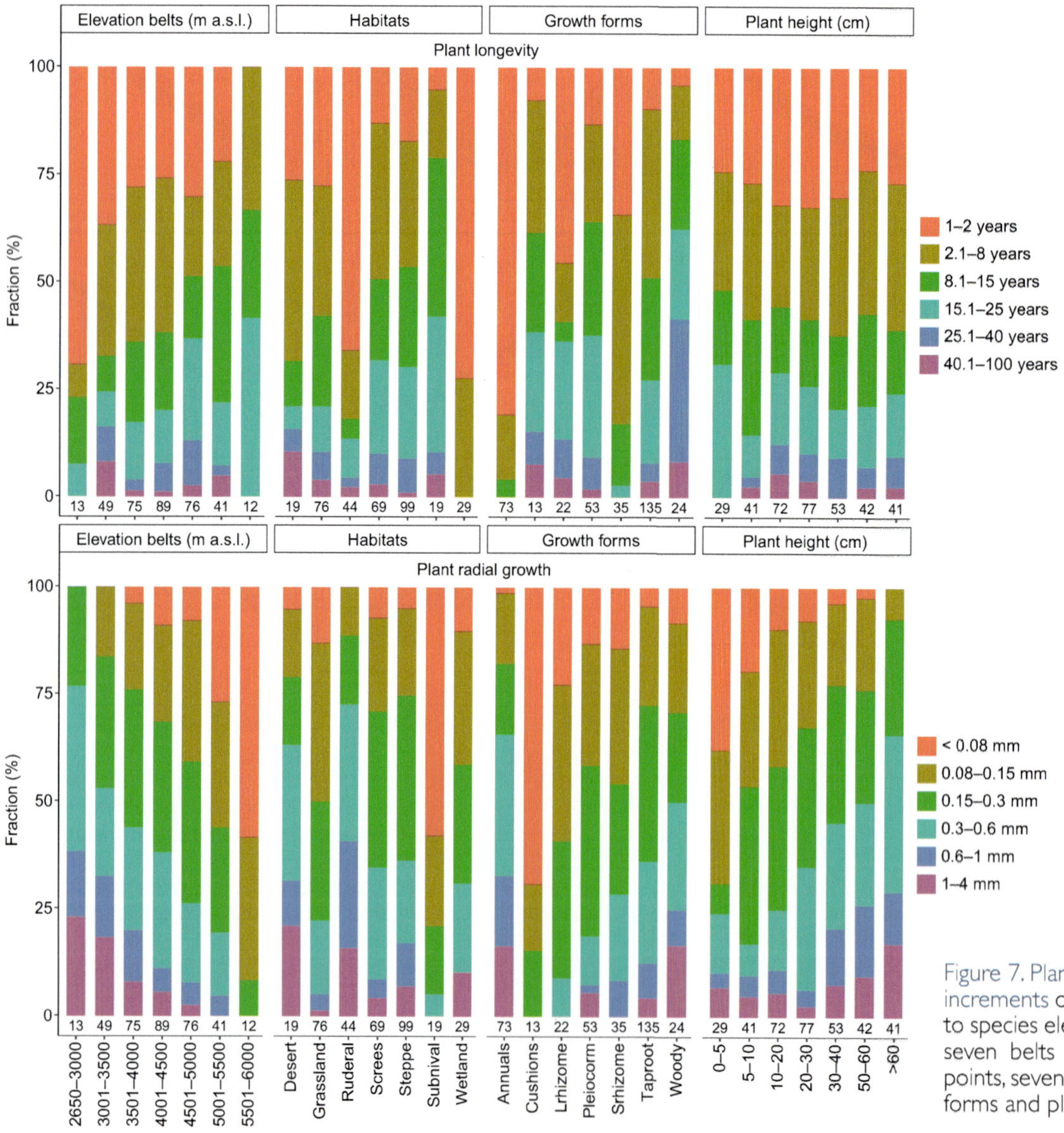

Figure 7. Plant age and annual ring width increments of 355 dicot plants in relation to species elevation optima in one of the seven belts represented by their mid-points, seven major habitat types, growth forms and plant height classes.

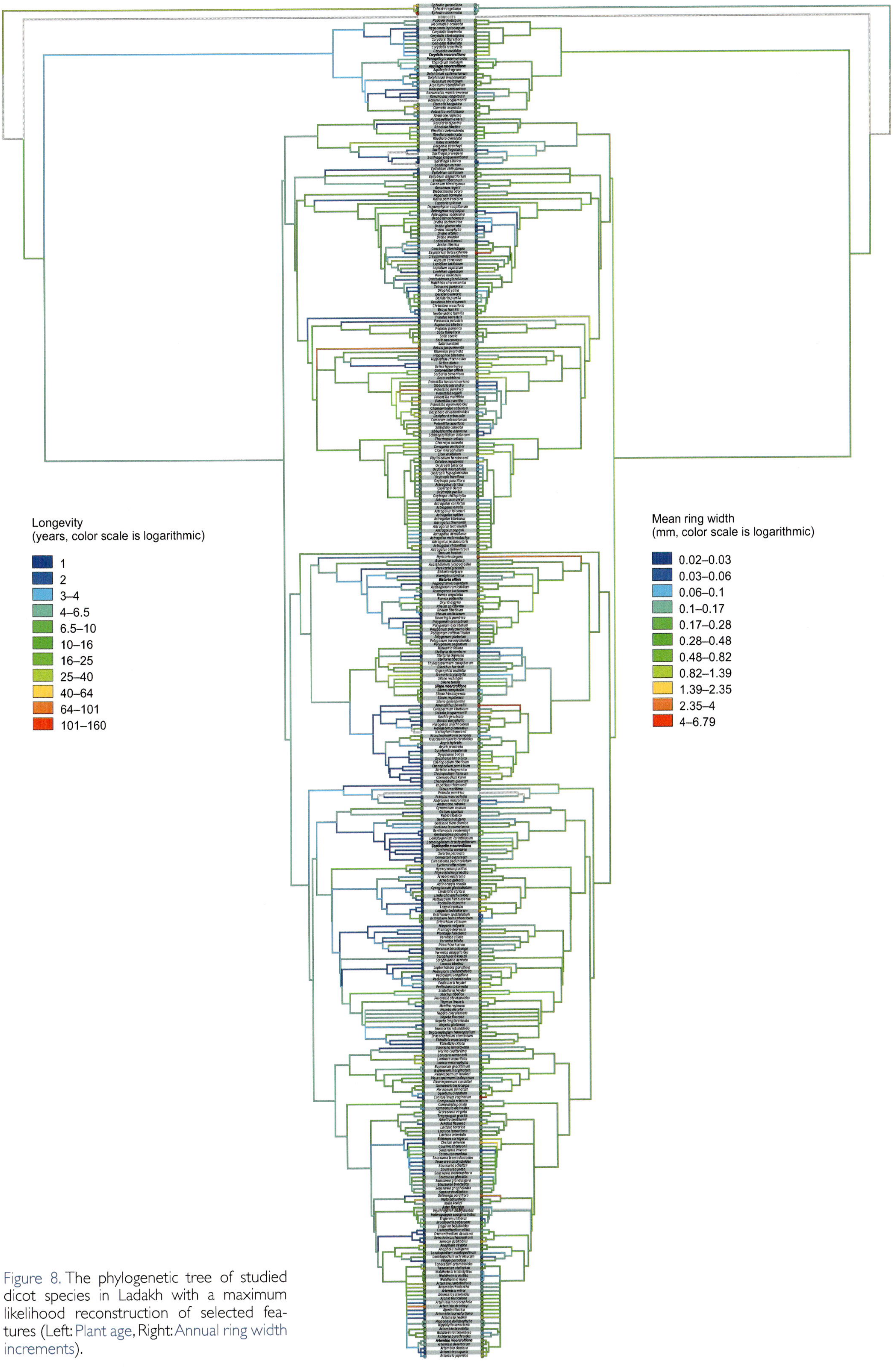

Figure 8. The phylogenetic tree of studied dicot species in Ladakh with a maximum likelihood reconstruction of selected features (Left: Plant age, Right: Annual ring width increments).

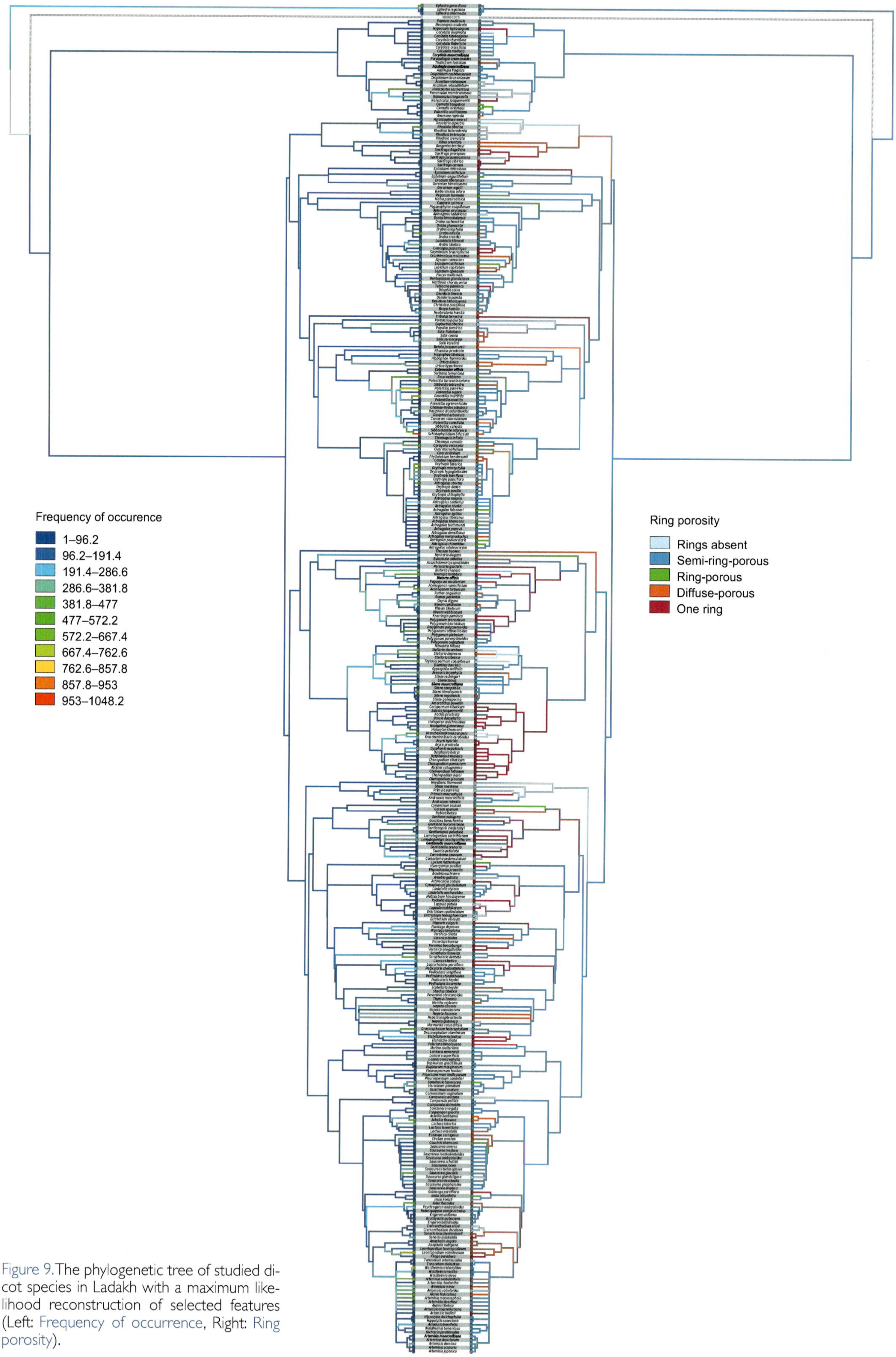

Figure 9. The phylogenetic tree of studied dicot species in Ladakh with a maximum likelihood reconstruction of selected features (Left: Frequency of occurrence, Right: Ring porosity).

The harsh environment homogenizes the structure of the hydrological conductive system. Semi-ring porosity (52%) and diffuse porosity (14%) dominate (Fig. 9). Ring porosity is rare (5%), restricted to a few genera (e.g. *Capparis spinosa*, *Myricaria elegans*, *Clematis* spp., *Lepidium latifolium*), it decreases with elevation and does not occur above 5000 m a.s.l. (Fig. 10). Semi-ring porosity increases with elevation from 8 to 83%, with the highest share among subnival cushion plants (69%), alpine scree (68%) and steppe (63%) plants, and the lowest fraction among ruderal (22%) and wetland plants (34%). Semi-ring porosity is characteristic of the genera *Saussurea*, *Waldheimia* and *Potentilla* (Fig. 9). Diffuse porosity decreases with elevation from 23 to 3%, and increases with plant height, being the lowest among annuals and tap-rooted perennial herbs and the highest in woody species.

Vessel number varies between 50 and 500 per mm^2 (Fig. 10). Species with 200–500 vessels per mm^2 prevail (65%) in all elevation belts and height classes, followed by plants with 100–200 vessels per mm^2 (33%). The smallest number of vessels is found in tall woody plants with optima at the lowest elevation, while the highest vessel numbers are found in cushion plants in subnival zone. The length of vessel elements ranges from <50 to 800 μm (Fig. 11). Plants with small 50–150 μm vessels prevail (46%), followed by 150–350 μm sizes (41%). Large vessels are found in taller woody species (*Salix* spp. *Populus* spp., *Betula jacquemontii*), prostrate shrubs (*Rhamnus prostrata*, *Ephedra* spp.), but also small herbs (*Minuartia foliosa*, *Silene tenuis*, *Erigeron bellidioides*, *Leontopodium ochroleucum*) (Fig. 11). The fraction of species with small vessels <150 μm increases significantly with elevation, while those with vessels longer than 150 μm decreases with increasing elevation (Fig. 12). Vessel diameters vary between <20 to maximal 100 μm. The fraction of species with small diameter vessels <20 μm increases significantly with elevation, while the number of those with bigger vessels >50 μm decreases with increasing elevation. Variation in vessel diameter is significantly related to plant height, with small cushion and short rhizomatous plants having smallest diameters and larger tap-rooted herbs and shrubs having largest diameters.

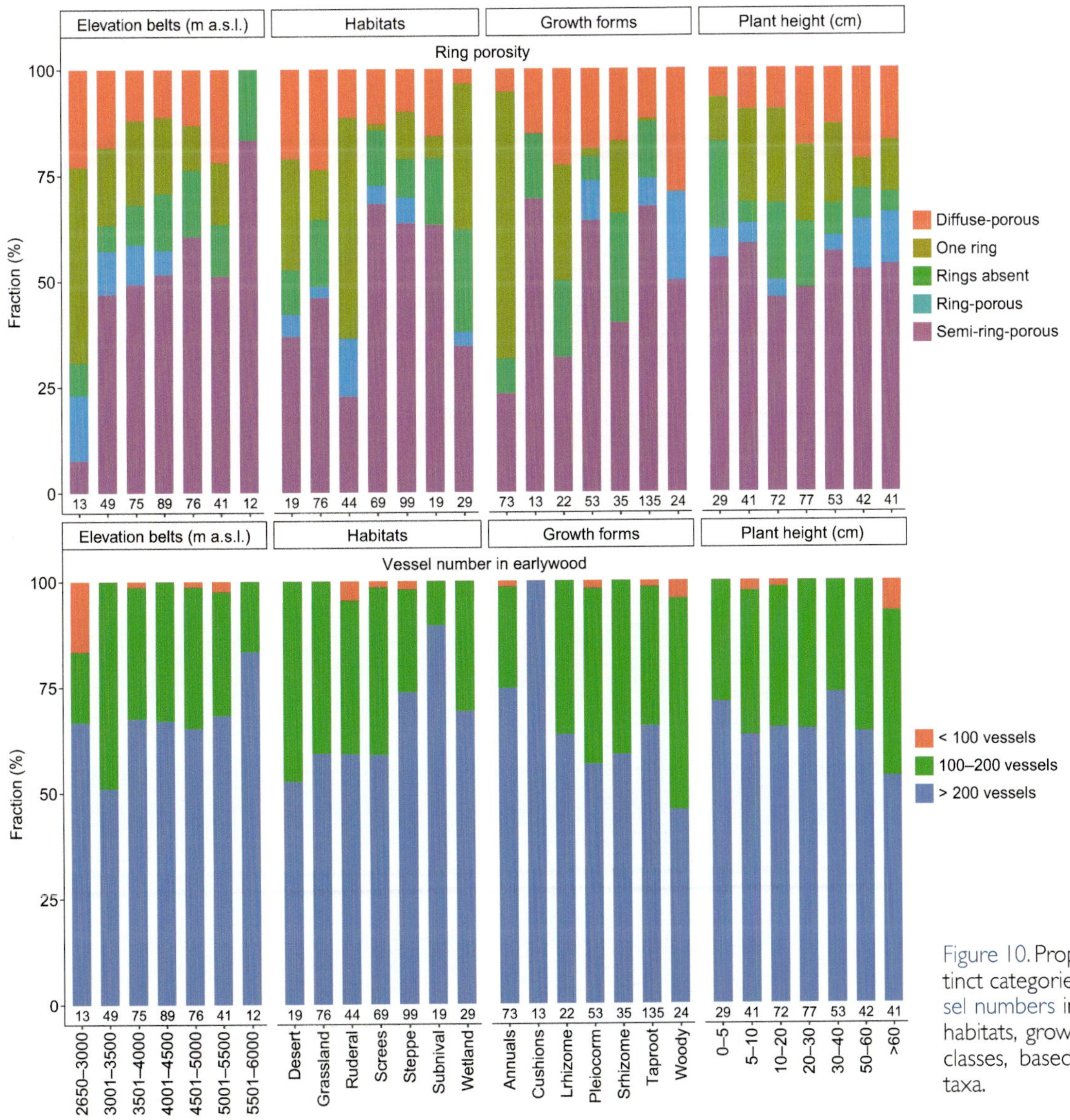

Figure 10. Proportion of species with distinct categories of ring porosity and vessel numbers in individual elevation belts, habitats, growth form types and height classes, based on 355 dicotyledonous taxa.

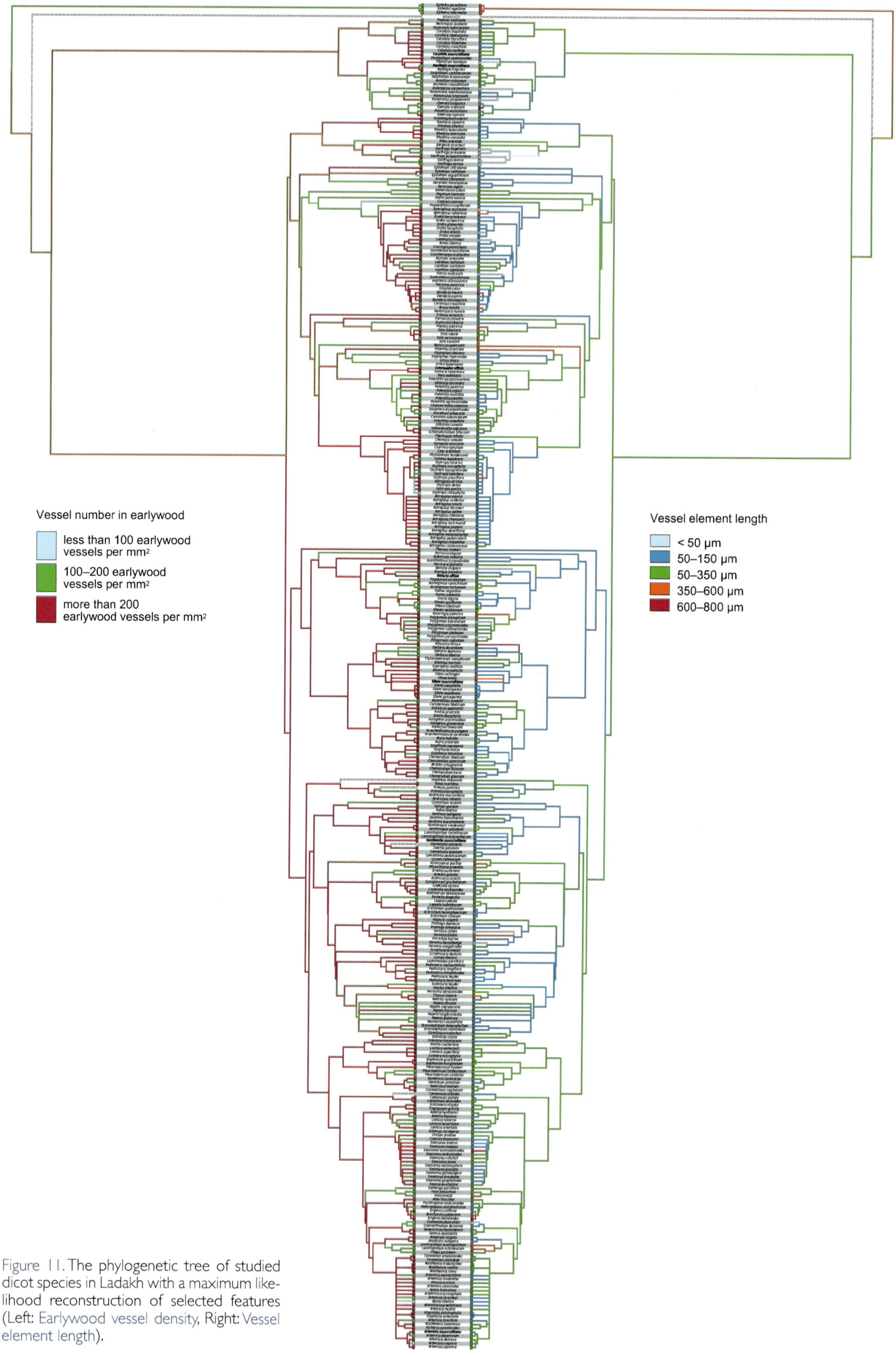

Figure 11. The phylogenetic tree of studied dicot species in Ladakh with a maximum likelihood reconstruction of selected features (Left: Earlywood vessel density, Right: Vessel element length).

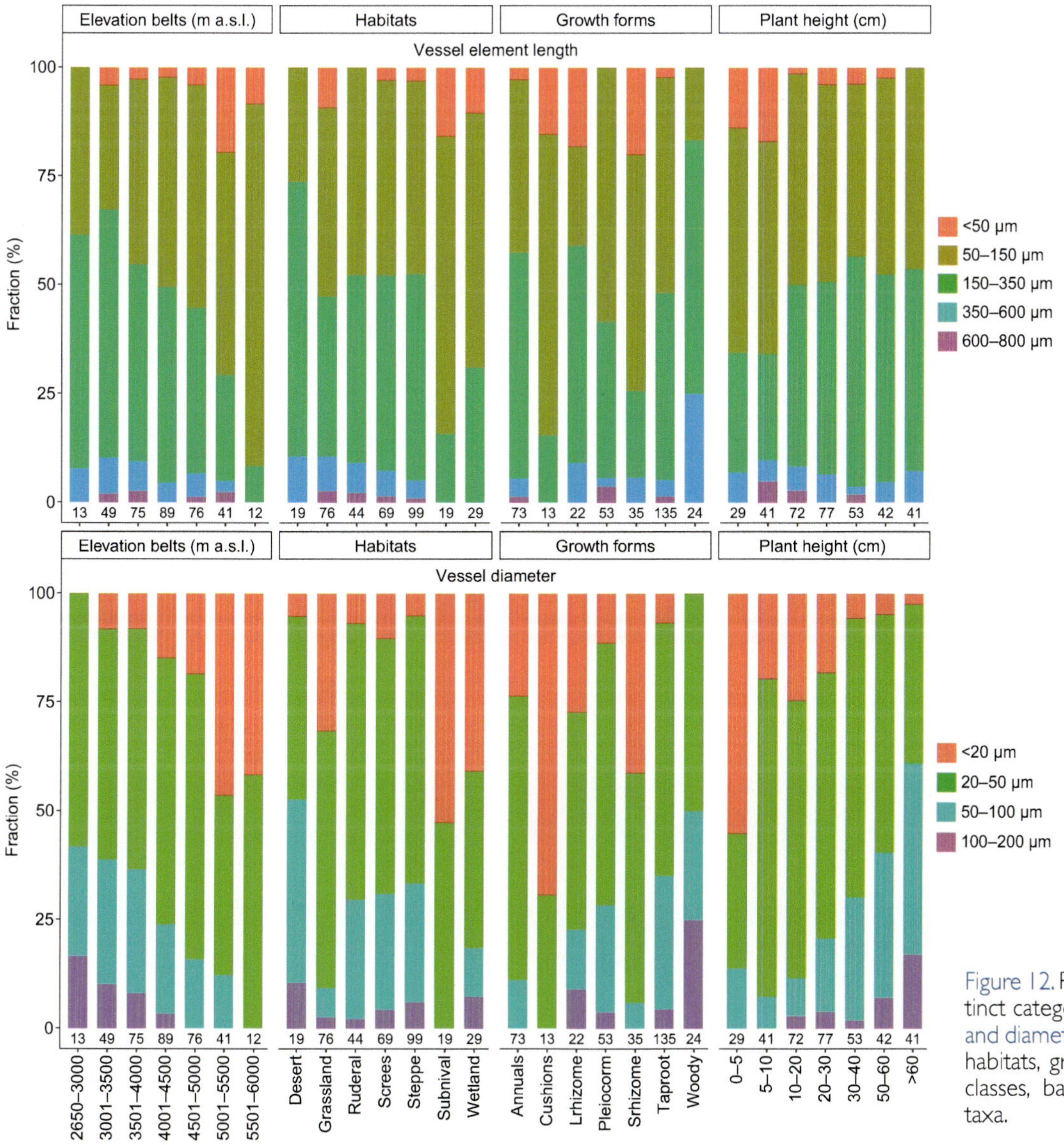

Figure 12. Proportion of species with distinct categories of vessel element length and diameter in individual elevation belts, habitats, growth form types and height classes, based on 355 dicotyledonous taxa.

Vessels are primarily solitary (39%) or in small clusters (34%), and less frequently in short radial multiples (16%) (Fig. 13). The proportion of species with solitary vessels increases with elevation from 33 to 92% between 3800 to 6150 m, and decreases with increasing plant height from 76 to 16%. Most represented are solitary vessels in subnival (63%), scree (43%) and alpine grassland (49%) plants, while vessels in short radial multiples are most represented among ruderal (32%) and saline wetland (30%) plants. Solitary vessels are an exclusive feature in Rosaceae (*Potentilla*, *Sibbaldia*, *Dasiphora*, *Comarum*) and Caryophyllaceae (*Stellaria*, *Silene*, *Arenaria*), vessels arranged in small clusters are found in genera *Astragalus*, *Nepeta*, *Corydalis*, *Artemisia*; vessels in long radial multiples occur in *Polygonum* spp. (Fig. 14).

Simple perforation plates are characteristic for 99% of all species. Scalariform perforation plates are almost absent. Vessels of a few larger plants have helical thickenings (2%), mostly in shrubs such as *Lonicera* spp., *Caragana versicolor*, *Sorbaria tomentosa* and *Colutea nepalensis*. A large amount of species have scalariform intervessel pits (48%), with clear elevation- and height-related variation (Fig. 13). The proportion of species with scalarifom pits increases from 9 to 56% with elevation and decreases from 46% in small cushion and scree plants to 8% in tall woody species in which instead round and alternate pits prevail (Fig. 13). Vessels with scalariform intervessel pits are normally embedded in pervasive parenchyma. It is not clear how these pits function. There is probably a hydraulic continuity between vessels and parenchyma. This contradicts the fact that vessels of 10% of all species are thick-walled (Fig. 15). Hydraulic tension in small plants must be therefore stronger than expected. The role of the pits is not clear. The absence of tyloses or phenolic plugs indicates that all vessels are functional during their time of existence. The absences of frost-rings in all small plants assume that late frost events are avoided by a late start of growth.

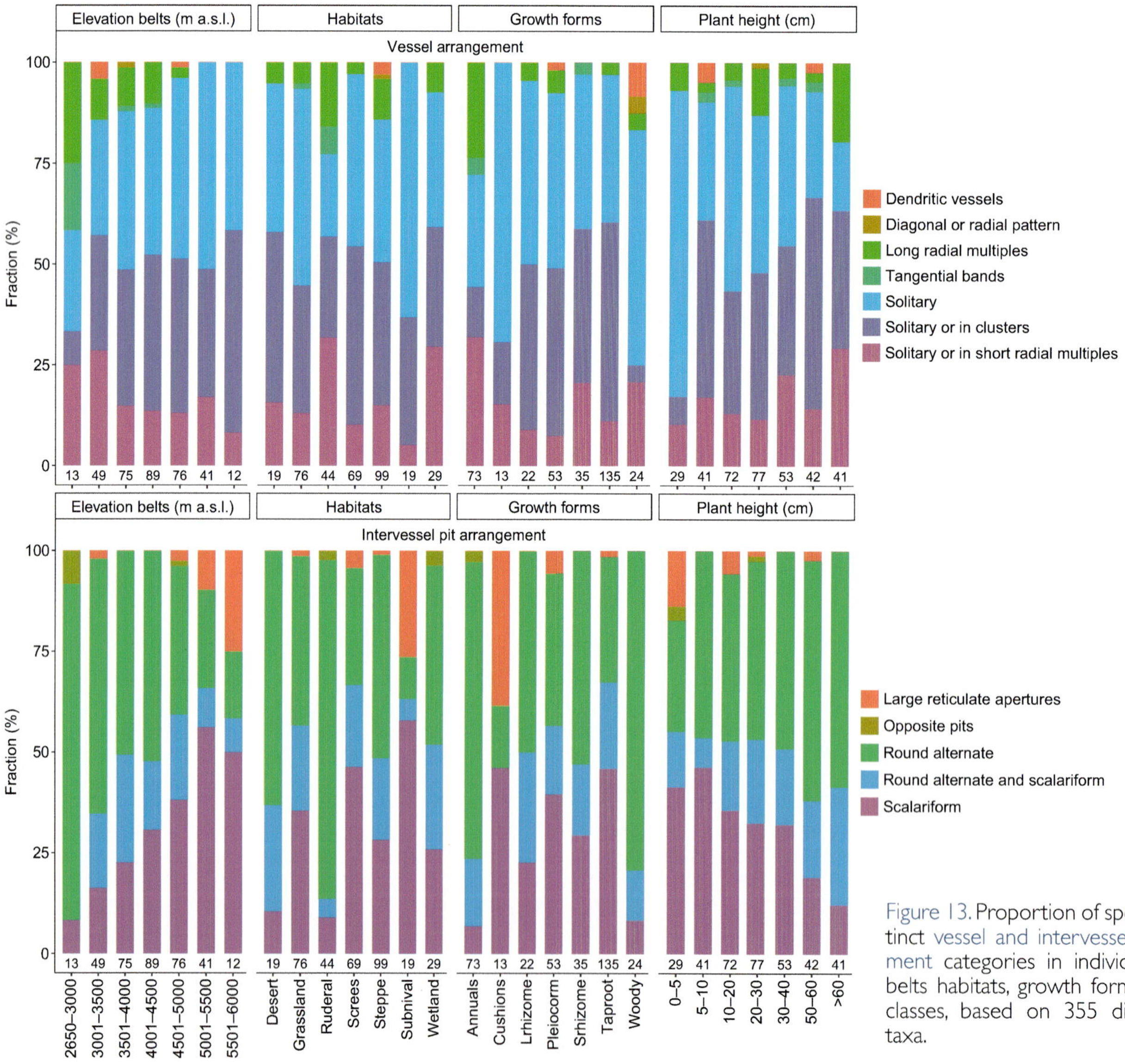

Figure 13. Proportion of species with distinct vessel and intervessel pit arrangement categories in individual elevation belts habitats, growth forms and height classes, based on 355 dicotyledonous taxa.

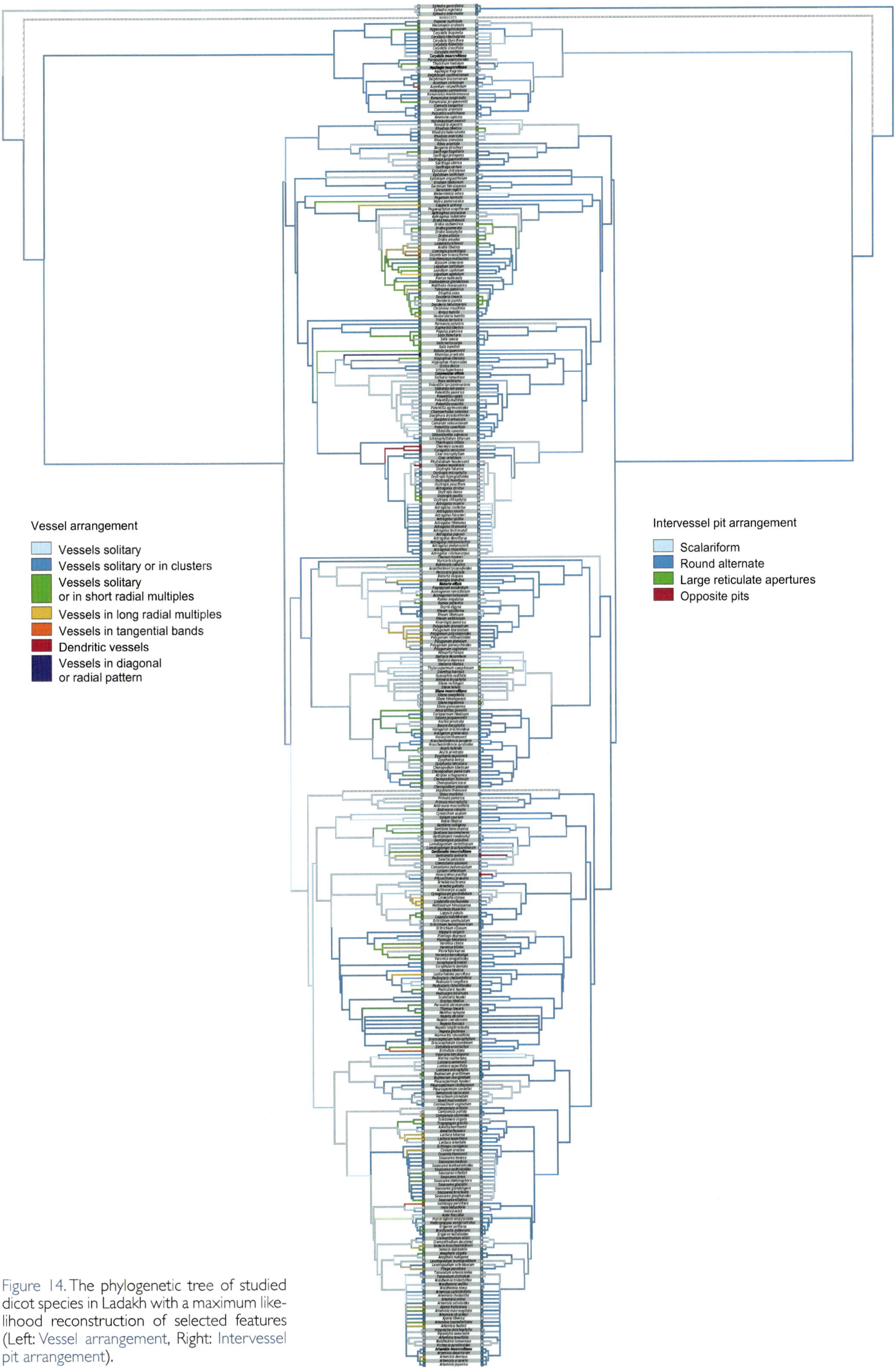

Figure 14. The phylogenetic tree of studied dicot species in Ladakh with a maximum likelihood reconstruction of selected features (Left: Vessel arrangement, Right: Intervessel pit arrangement).

Vessels with thin and lignified walls predominate (82%), in particular at elevations below 5000 m (Fig. 15). In alpine and subnival zone above 5000 m, there is an increasing share of thick-walled vessels (16%), especially among phylogenetically unrelated cushion plants (*Ladakiella klimesii*, *Thylacospermum caespitosum*, *Astragalus confertus*, *Desideria linearis*) (Fig. 16). A high number of plants is without fibers (42%), which can be explained by small size, with the majority of total biomass allocated in below-ground organs and only a small share in aboveground shoots and leaves, resulting in a limited need for stem mechanical support (Fig. 17). There is a clear linear relationship between fiber proportion and elevation and plant size. Sixty-nine percent of plants shorter than 5 cm have no fibers, all alpine cushion plants, but also small plants from screes and wetlands. This value decreases to 19% in plants with a height >60 cm, mainly tall woody species which are heavily lignified (Fig. 17). Ninety-two percent of species with elevation optima above 5500 m have no fibers, while only 19% of species below 3000 m are fiberless. Thin- to thick-walled libriform fibers are found in the majority of plants with fibers, thick-walled fibers in 17% and thin-walled fibers in only 1% of studied species. Fiber-wall thickness is not much related to plant size, though thick-walled fibers are more represented in woody taxa (42%) such as tall shrubs (*Myricaria elegans*, *Rosa webbiana* and *Lonicera microphylla*), prostrate shrubs (*Lonicera asperifolia*, *Dasiphora dryadanthoides*, *Caragana versicolor*, *Rhamnus prostrata*) and taprooted subshrubs (*Artemisia* spp., *Tanacetum stoliczkae*) and tall forbs (e.g. *Lepidium latifolium*, the majority of tall *Astragalus* and *Oxytropis* spp.) (Fig. 18). Since mechanical stress is low in tiny plants, it is difficult to explain why thin-walled fibers are rare while plants with thick-walled fibers are more frequent. The presence of contractile roots is not systematically recorded. Wavy stem centers (radial section) might be a reaction to soil-frost heave.

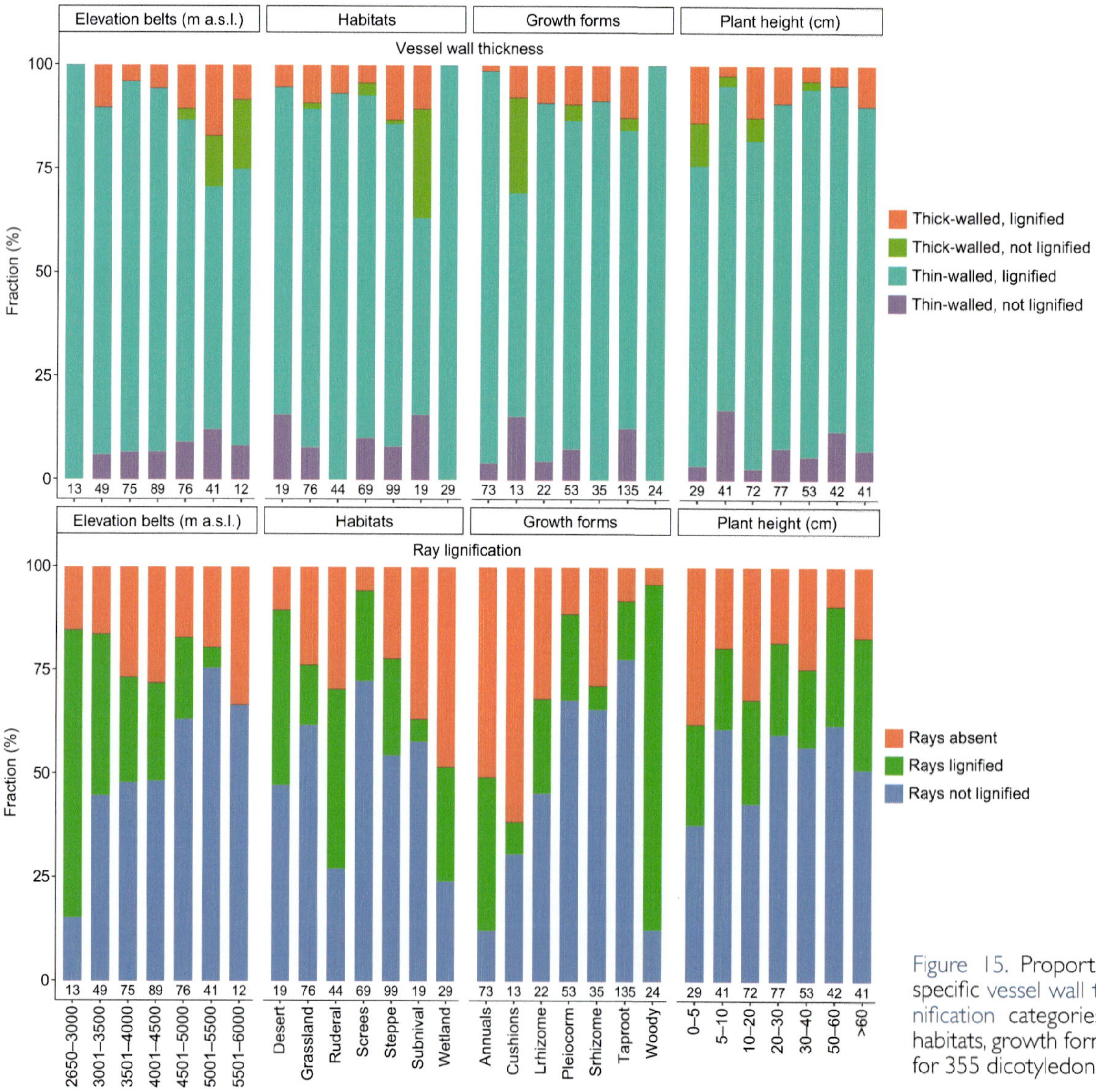

Figure 15. Proportion of species with specific vessel wall thickness and ray lignification categories in elevation belts, habitats, growth forms and height classes for 355 dicotyledonous taxa.

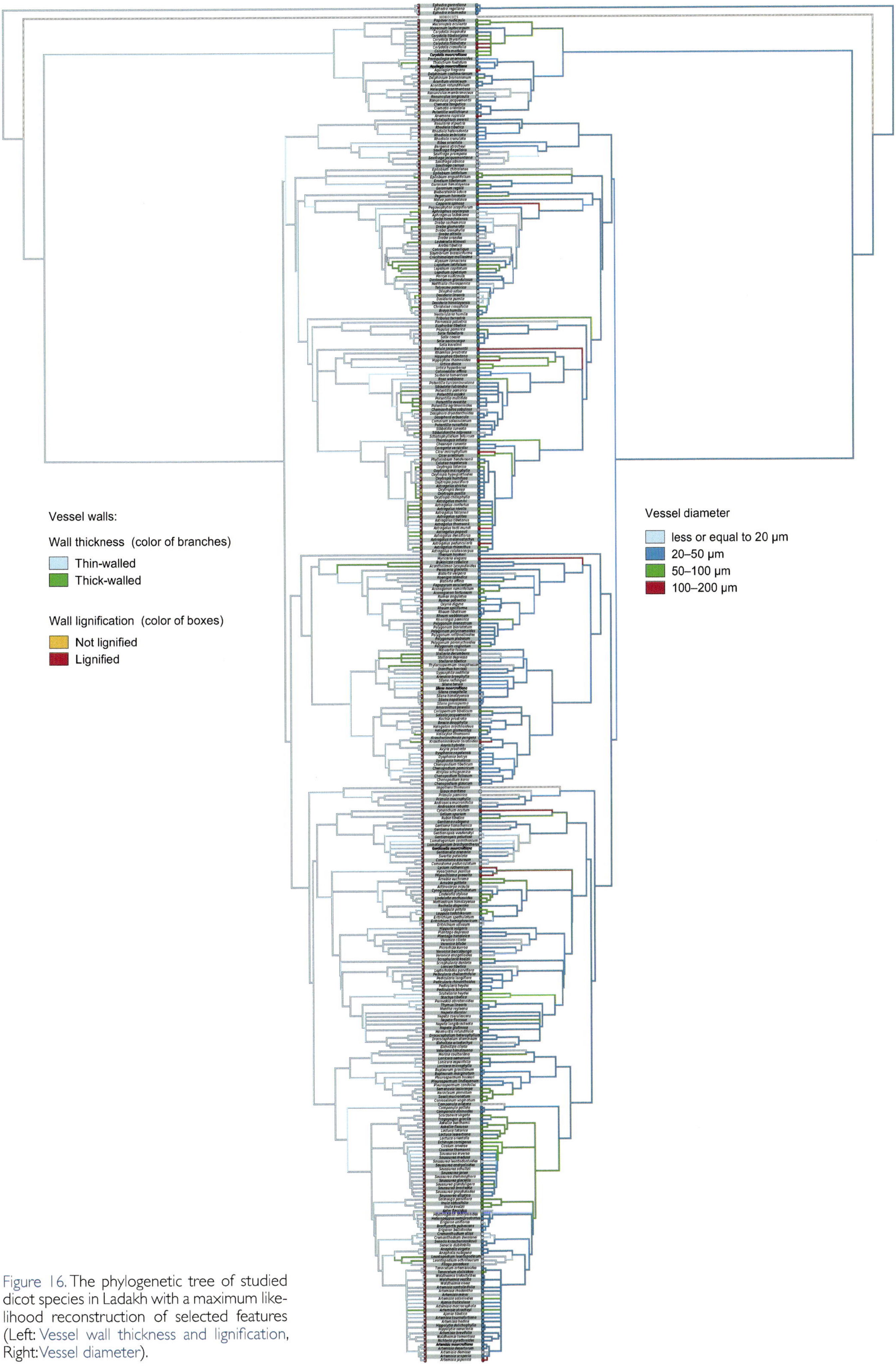

Figure 16. The phylogenetic tree of studied dicot species in Ladakh with a maximum likelihood reconstruction of selected features (Left: Vessel wall thickness and lignification, Right: Vessel diameter).

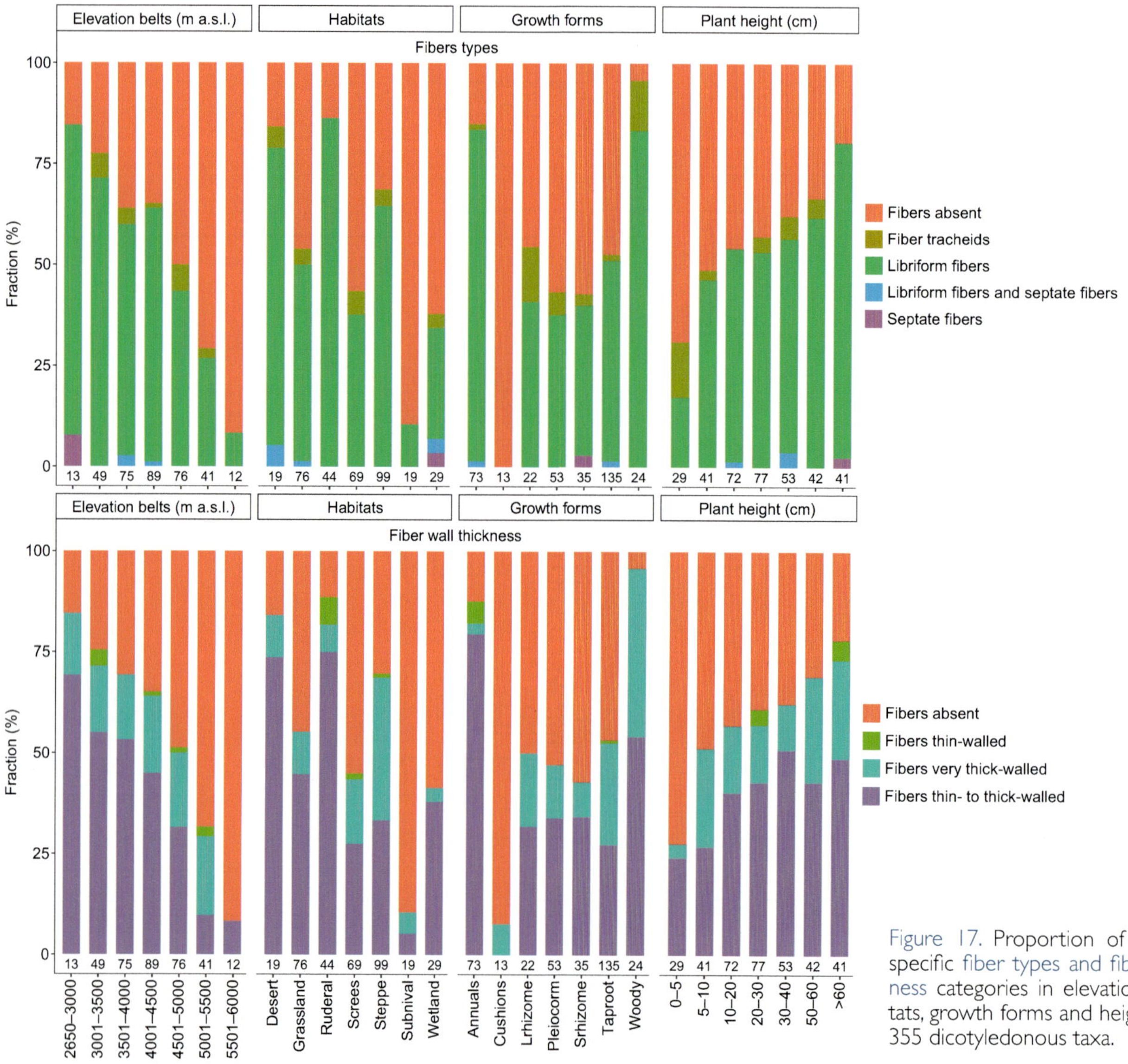

Figure 17. Proportion of species with specific fiber types and fiber wall thickness categories in elevation belts, habitats, growth forms and height classes for 355 dicotyledonous taxa.

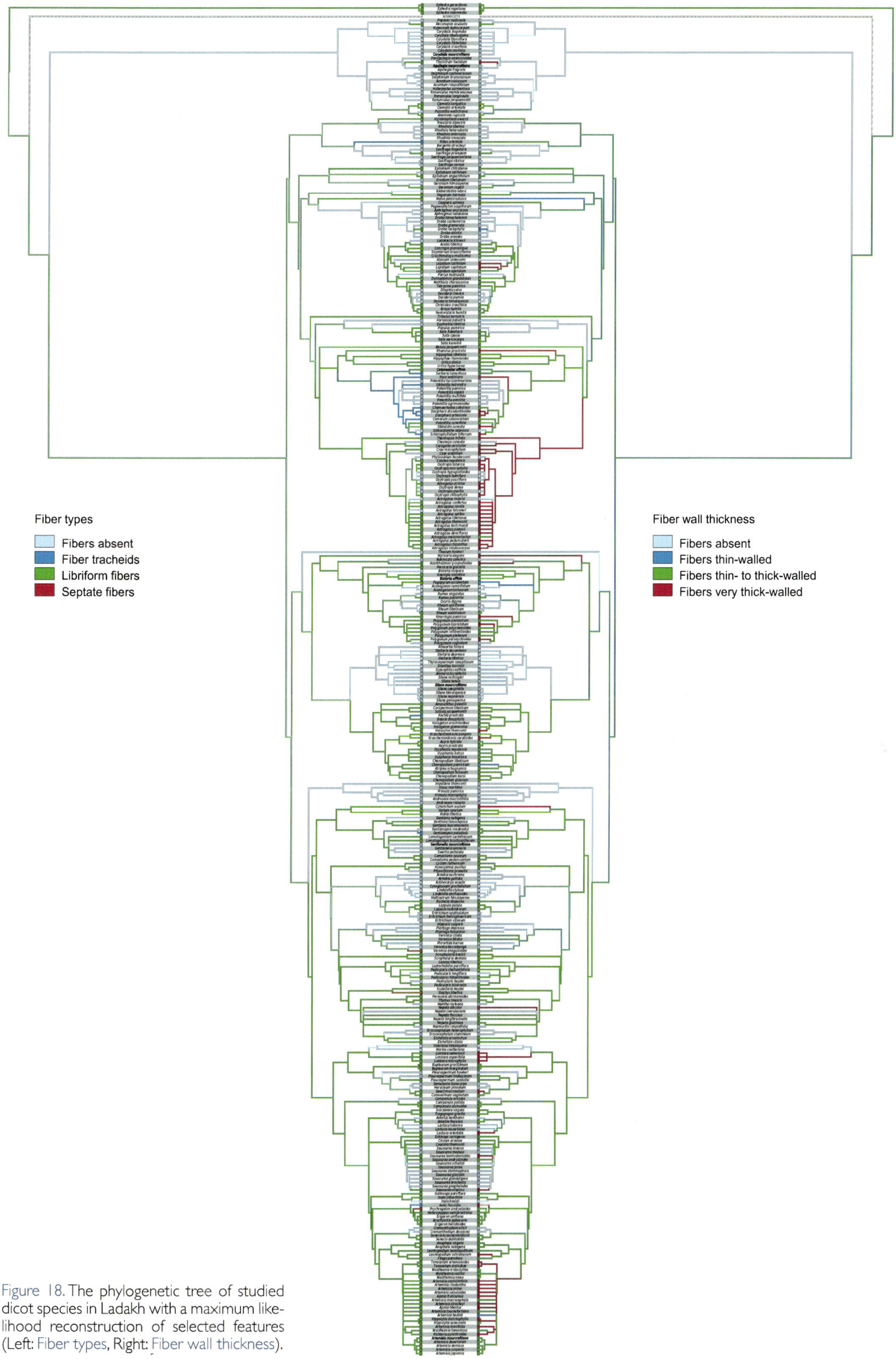

Figure 18. The phylogenetic tree of studied dicot species in Ladakh with a maximum likelihood reconstruction of selected features (Left: Fiber types, Right: Fiber wall thickness).

Related to extreme site conditions and short vegetation periods is the dominance of unlignified tissues (Fig. 15), as the lignification process requires higher temperature (>6 °C) than provided by ambient conditions (Nagelmüller et al. 2017), especially above 5500 m, where mean growing season temperatures are well below 6 °C. This is indicated by the blue aspect of Astrablue/Safranin stained slides. Dominant are plants with very large unlignified cortex zones, pervasive parenchyma, very large, unlignified rays and parenchyma zones between vascular bundles (Fig. 19, 20). Since parenchyma cells store mobile carbohydrates (starch, fructan) we suppose that these systems are optimized to survive intra-seasonal frost events or long-lasting cold periods. If photosynthetic structures were destroyed during the vegetation period or if one vegetation periods failed, the plant would be able to mobilize reserves to form new tissues or to survive below ground for at least two years.

Axial parenchyma cells and rays form the storage system in the xylem (Fig. 21). However, the largest part of the storage system is in the bark, mainly in the cortex. Obvious in the xylem is the high amount of species with much parenchyma, mostly pervasive parenchyma (49%, Fig. 21) or combined pervasive and vasicentric parenchyma (16%), and also large ray-like zones between permanent vascular bundles (34%). Frequent are plants with vasicentric paratracheal parenchyma (28%). Apotracheal parenchyma is almost absent (3%). This is a feature belonging to trees and shrubs. The proportion of species with vasicentric paratracheal parenchyma decreases linearly with elevation from 64 to 8%, while species with pervasive parenchyma increase with elevation from 9 to 92% (Fig. 21). Pervasive parenchyma is an exclusive feature in subnival cushion plants (95% of cases), and very common also among short-rhizomatous alpine grassland (60%) and scree (57%) species. Vasicentric paratracheal parenchyma is widespread in short-lived ruderal (64%) and desert (52%) species (Fig. 22). The fraction of species with large ray-like zones between permanent vascular bundles increases from low values among low-elevation short-lived ruderals to 67% in upper alpine and subnival perennial plants found above 5000 m. The vascular bundles separated by large parenchyma zones are found in many unrelated high altitude taxa including most *Saussurea* spp.,

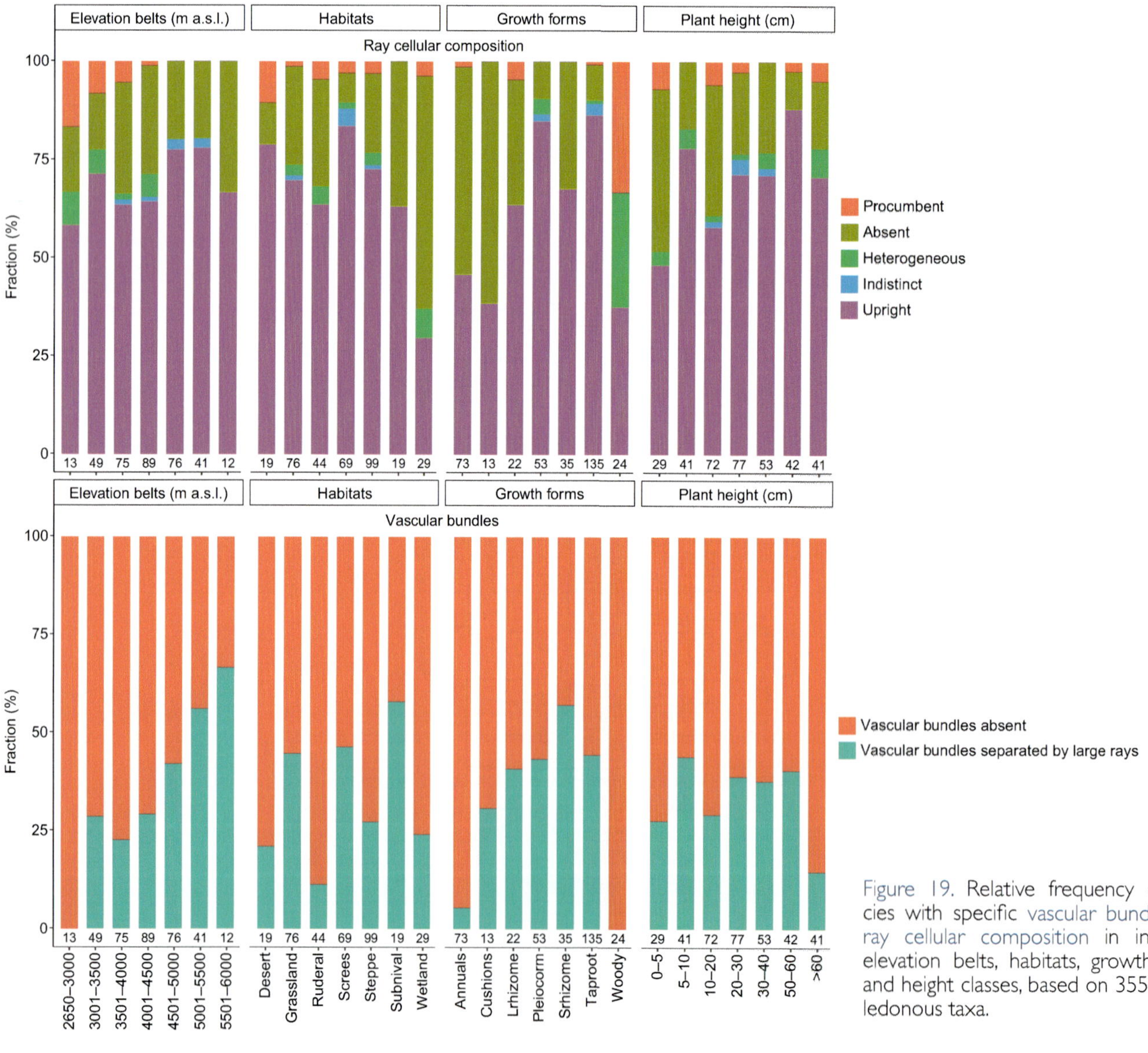

Figure 19. Relative frequency of species with specific vascular bundles and ray cellular composition in individual elevation belts, habitats, growth forms and height classes, based on 355 dicotyledonous taxa.

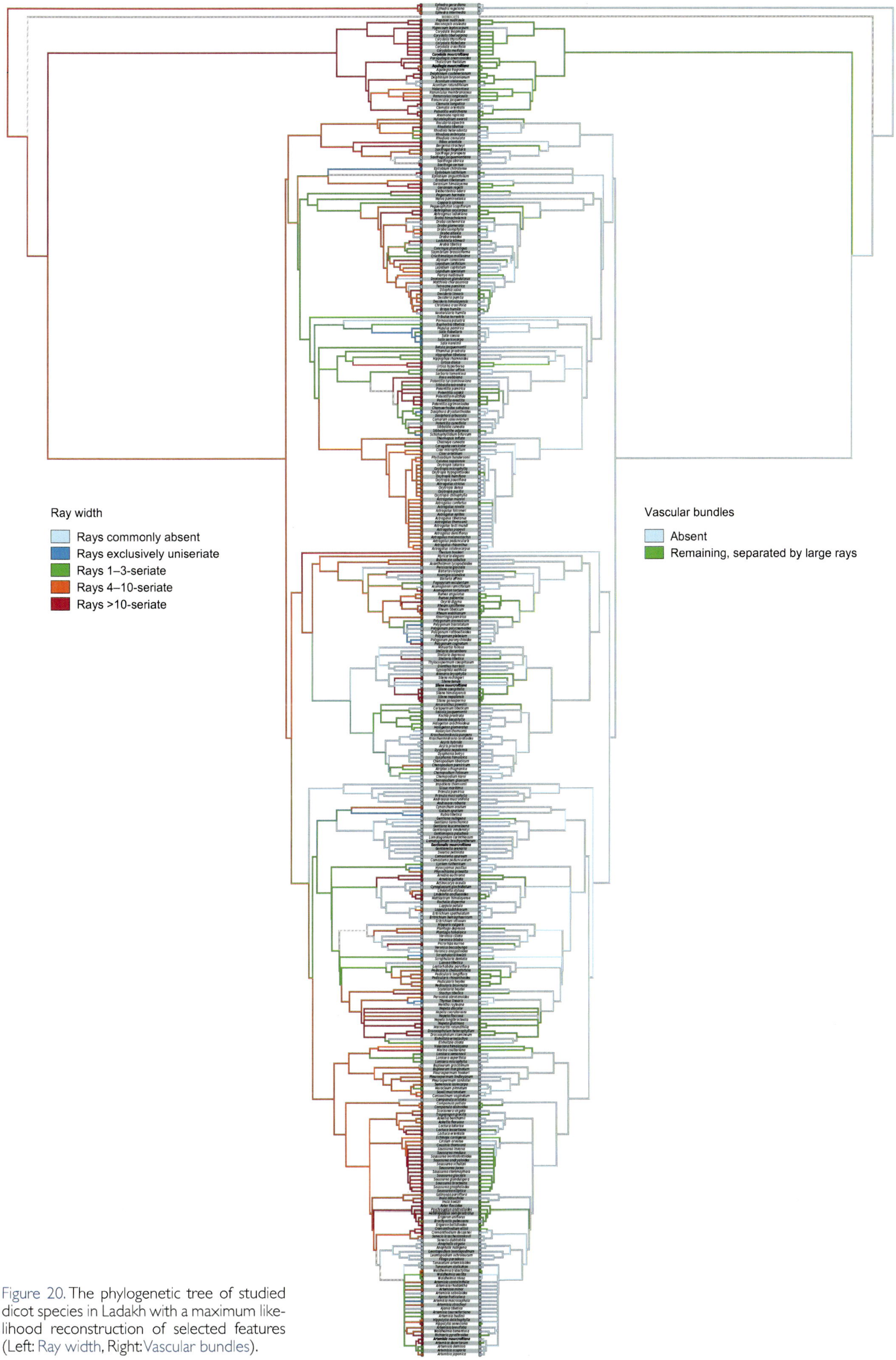

Figure 20. The phylogenetic tree of studied dicot species in Ladakh with a maximum likelihood reconstruction of selected features (Left: Ray width, Right: Vascular bundles).

Rheum spp., *Aphragmus* spp. and also most *Pedicularis* and *Corydalis* species (Fig. 20). Common to all these species is their occurrence in wetter soil and hence an elevated risk of frost damage due to repeated summer nighttime freezing and daytime thawing of soil water. Their xylem, composed of living parenchyma cells, stores a large amount of non-structural carbohydrates, which may function as cryoprotectants similarly as vehicle's antifreeze.

The high values of parenchyma in the xylem in any form just demonstrate that survival of high alpine plants is anatomically based on an extended storage system. The plant's need for storing carbohydrates in cold and arid climate is reflected in the prevalence of multiseriate rays (71%), while exclusively uniseriate rays are rare (4%) (Fig. 21). Most represented are plants with 4–10-seriate rays (29%), >10-seriate (28%), and 1–3-seriate (13%) rays. There is clear a relationship between ray width and elevation, the higher the plant elevation optima the wider the rays. The fraction of species with >10-seriate rays increases from 7 to 42% along the studied altitudinal gradient, while the proportion of 1–3-seriate rays decreases from 38 to 0% (Fig. 21). Wider >10-seriate rays are mostly found in taller alpine grassland and scree forbs from genera *Corydalis*, *Silene*, *Saussurea*, *Lactuca* and *Potentilla*, all having deep desintegrating tap roots (Fig. 20). On the other hand, the fraction of species with narrow uniseriate or 1–3-seriate rays decreases with increasing elevation, being most represented among short-lived ruderal species from Amaranthaceae (e.g. *Halogeton arachnoideus*, *H. glomeratus*), Zygophyllaceae (*Tribulus terrestris*), Nitrariaceae (*Peganum harmala*), Lamiaceae (*Elsholtzia ciliata*), Asteraceae (*Artemisia demissa*, *A. macrocephala*) and Brassicaceae (*Sisymbrium brassiciforme*, *Conringia planisiliqua*) (Fig. 20). Plants without rays form 25%, being most represented among small <15 cm tall taxa from productive wetlands and grasslands (e.g. *Koenigia islandica*, most Gentianaceae, Primulaceae such as *Glaux maritima* or *Primula macrophylla* and water-bound *Veronica* spp.). Most plants have rays with upright cells (70%) and in many cases they are confluent with the axial parenchyma (40%). Since stem-radial stress in tiny plants is minimal there is no need to form prostrate cells. Rays with prostrate (procumbent) ray cells (3%) occur mainly in prostrate shrubs (*Ephedra* spp., *Capparis spinosa*, *Rhamnus prostrata*), large shrubs (*Myricaria elegans*, *Sorbaria tomentosa*) and trees (*Betula jacquemontii*) growing at lower elevations (Fig. 25). Rays with heterogeneous ray cells (4%) occur mainly in shrub genera of *Salix*, *Lonicera* and *Hippophae* and also in climbers (*Clematis orientalis*, *Cynanchum acutum*).

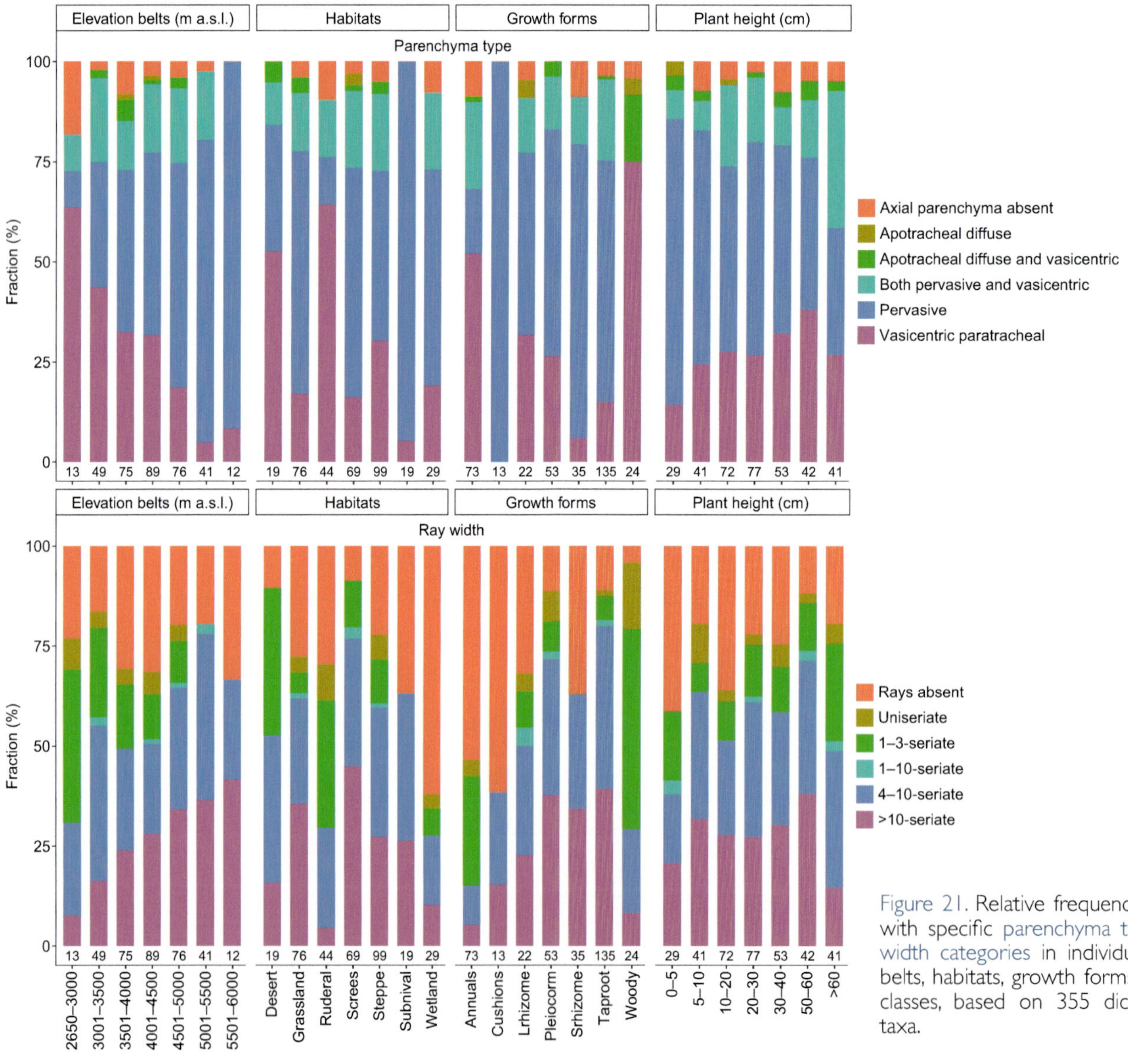

Figure 21. Relative frequency of species with specific parenchyma type and ray width categories in individual elevation belts, habitats, growth forms and height classes, based on 355 dicotyledonous taxa.

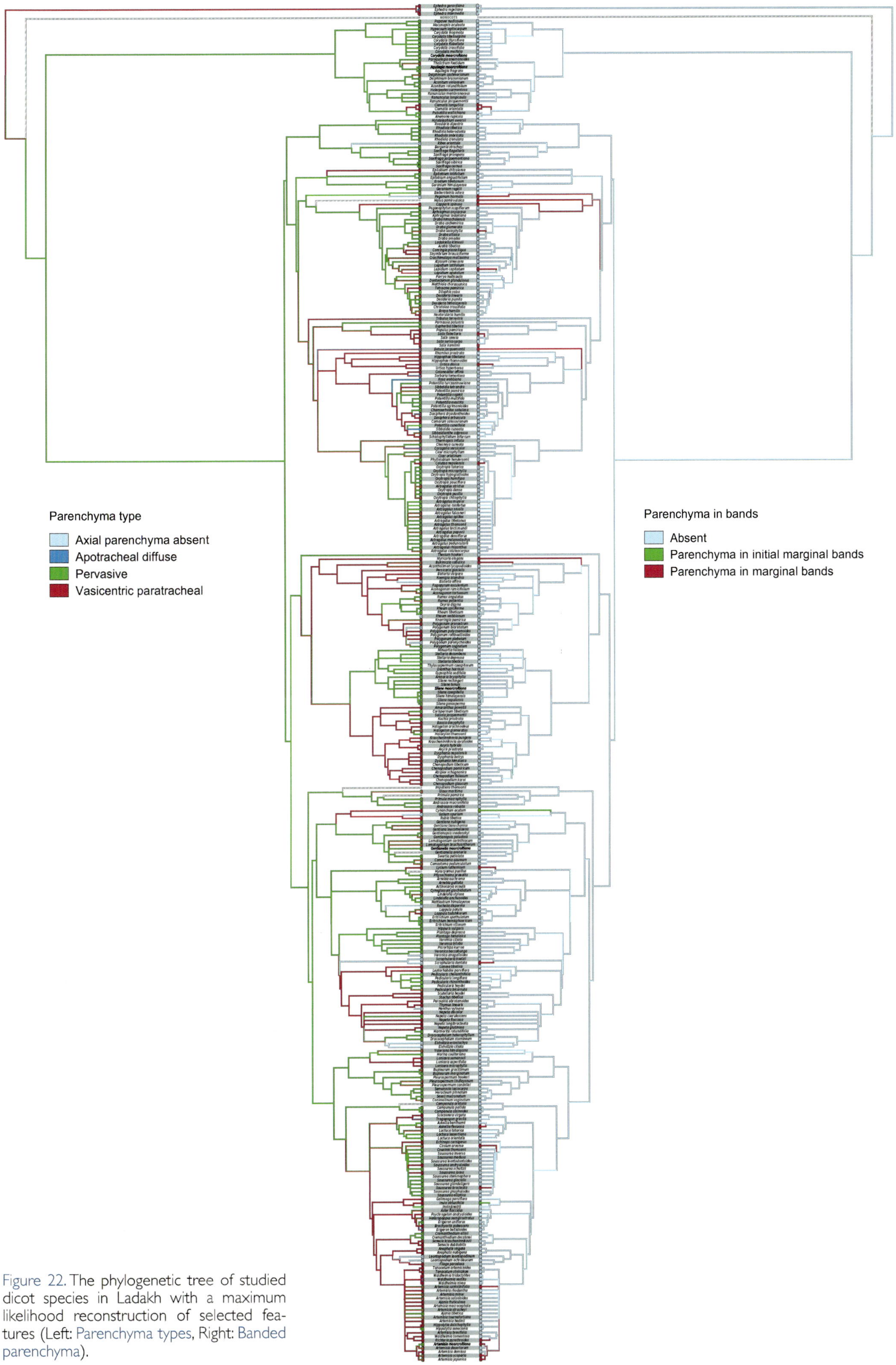

Figure 22. The phylogenetic tree of studied dicot species in Ladakh with a maximum likelihood reconstruction of selected features (Left: Parenchyma types, Right: Banded parenchyma).

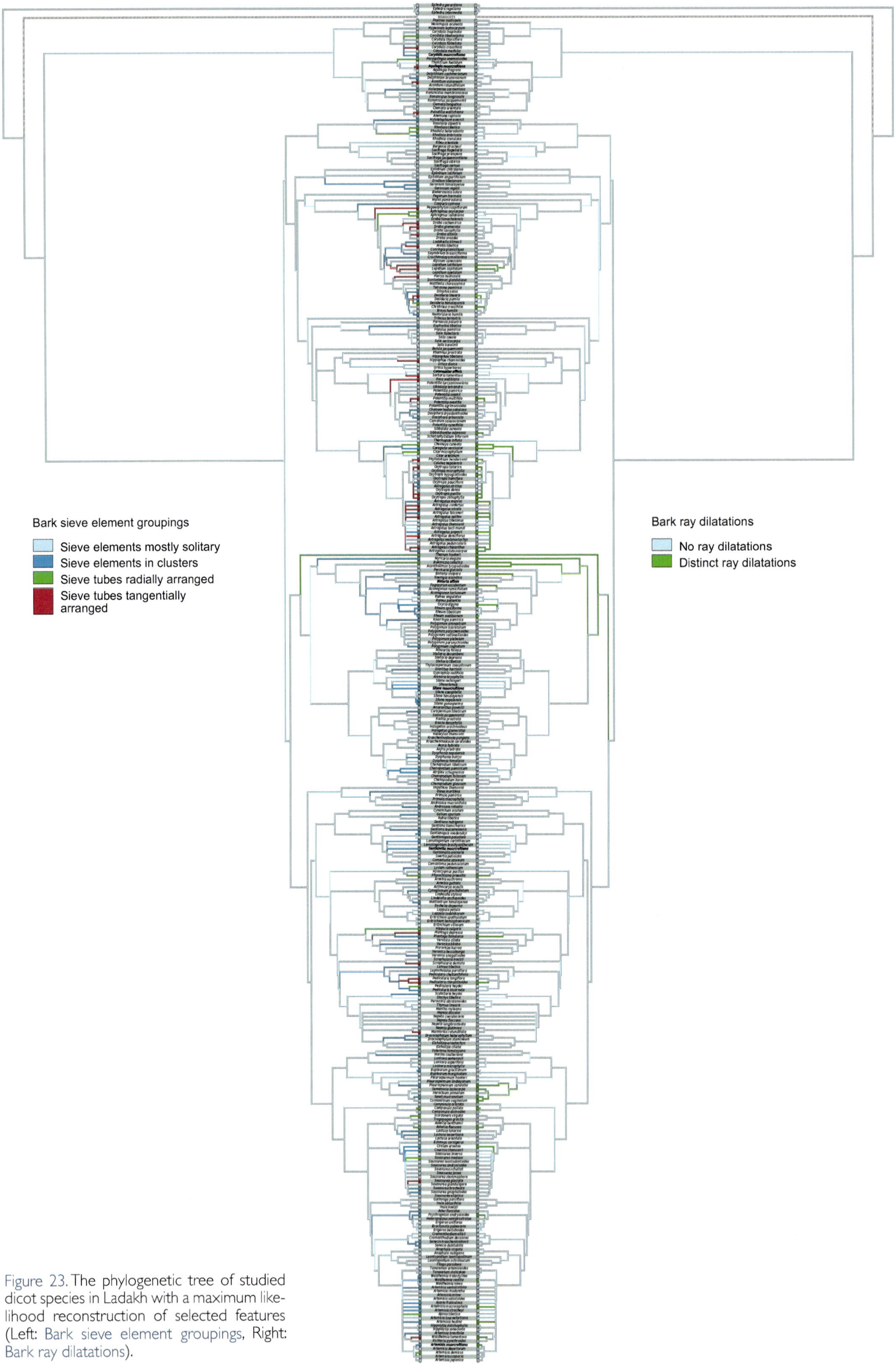

Figure 23. The phylogenetic tree of studied dicot species in Ladakh with a maximum likelihood reconstruction of selected features (Left: Bark sieve element groupings, Right: Bark ray dilatations).

Characteristic to many phylogenetically unrelated, long-lived and tap-rooted species of NW Himalayan cold deserts and steppes is the root segmentation as a result of senescence. These plants evolved special mechanisms to survive. One of those is the formation of secondary intra-xylem periderm (Fig. 24). If the need between the capacity of water transport and the need to support the stem metabolism is out of balance, living parenchyma cells within the xylem change their storing mode to a meristematic mode; they form an inter-xylary periderm towards the pith, in particular plants in cold high elevations (Fig. 24). This layer separates the metabolically active outer part from a dead central part. This process can occur several times. In special cases the inner part of the xylem compartimentalizes the living from the dead xylem without forming a periderm such as in *Delphinium brunonianum* (Fig. 25). Root and stem segregation is combined with periderm formation. All segregated parts form a new phellem around the spitted stem. Segmentation restricts the absolute dating of plant age, just the segment age can be determined which in many species varies between 8 to 15 years.

Tertiary meristematic activities resulting in the formation of secondary intra-xylem periderm towards the pith were observed in 18% of species, intra-xylary cork-band periderm in 4%, intra-xylary phloem in 2% and successive cambia producing simultaneously xylem and phloem in 7% of species. The fraction of species with intra-xylem periderm towards the pith increases from zero values in elevations <3000 m to 50% in elevations >5000 m, the feature is typical in tall steppe and scree species recruiting from genera *Aquilegia*, *Astragalus*, *Cicer*, *Oxytropis*, *Lindelofia*, *Potentilla* and *Rhodiola*. Successive cambia are almost exclusively a feature of Amaranthaceace. Intra-xylary phloem is found in unrelated species such as *Scrophularia* spp., *Menitskia tibetica*, *Mattiastrum himalayense* and *Epilobium latifolium*. Intra-annual cork layers are a feature mostly in Asteraceae of the tribe Anthemideae (*Ajania fruticulosa*, *Artemisia* spp., *Tanacetum artemisioides*) and Cichorieae (*Askellia benthamii*, *Lactuca orientalis*). Sieve elements are mostly solitary (61%), in clusters (24%), radially (6%) or tangentially (9%) arranged. Solitary sieve tubes are common below 4500 m, their occurrence is about 30% above 5500 m, where sieve clusters and tangentially arranged elements predominate. Phloem ray dilatations were found 13% of species, mostly in the genera *Astragalus*, *Oxytropis*, *Lepidium*, *Artemisia* and *Waldheimia* (Fig. 23, 26).

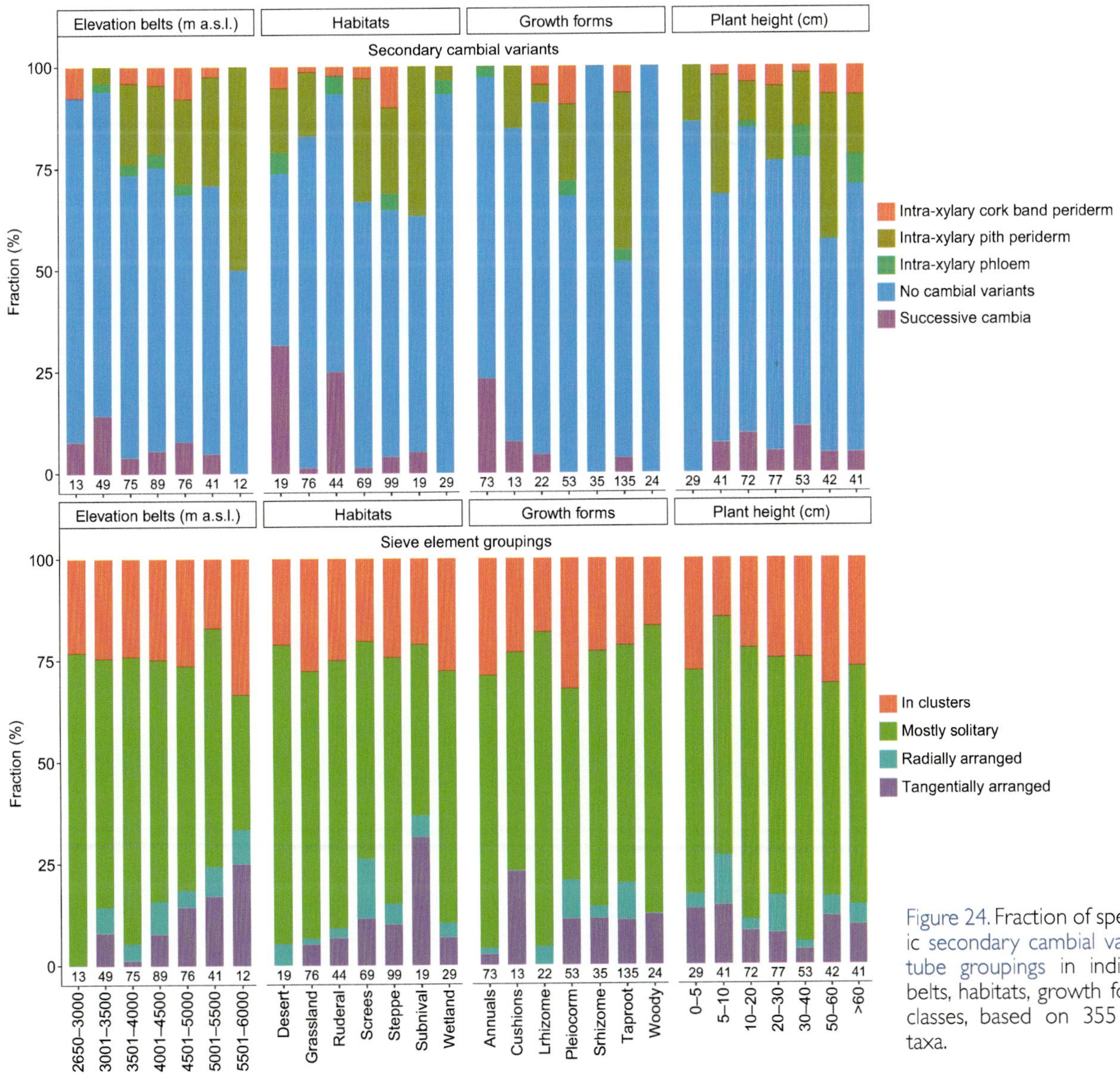

Figure 24. Fraction of species with specific secondary cambial variants and sieve tube groupings in individual elevation belts, habitats, growth forms and height classes, based on 355 dicotyledonous taxa.

Figure 25. The phylogenetic tree of studied dicot species in Ladakh with a maximum likelihood reconstruction of selected features (Left: Ray cellular composition, Right: Secondary cambial variants).

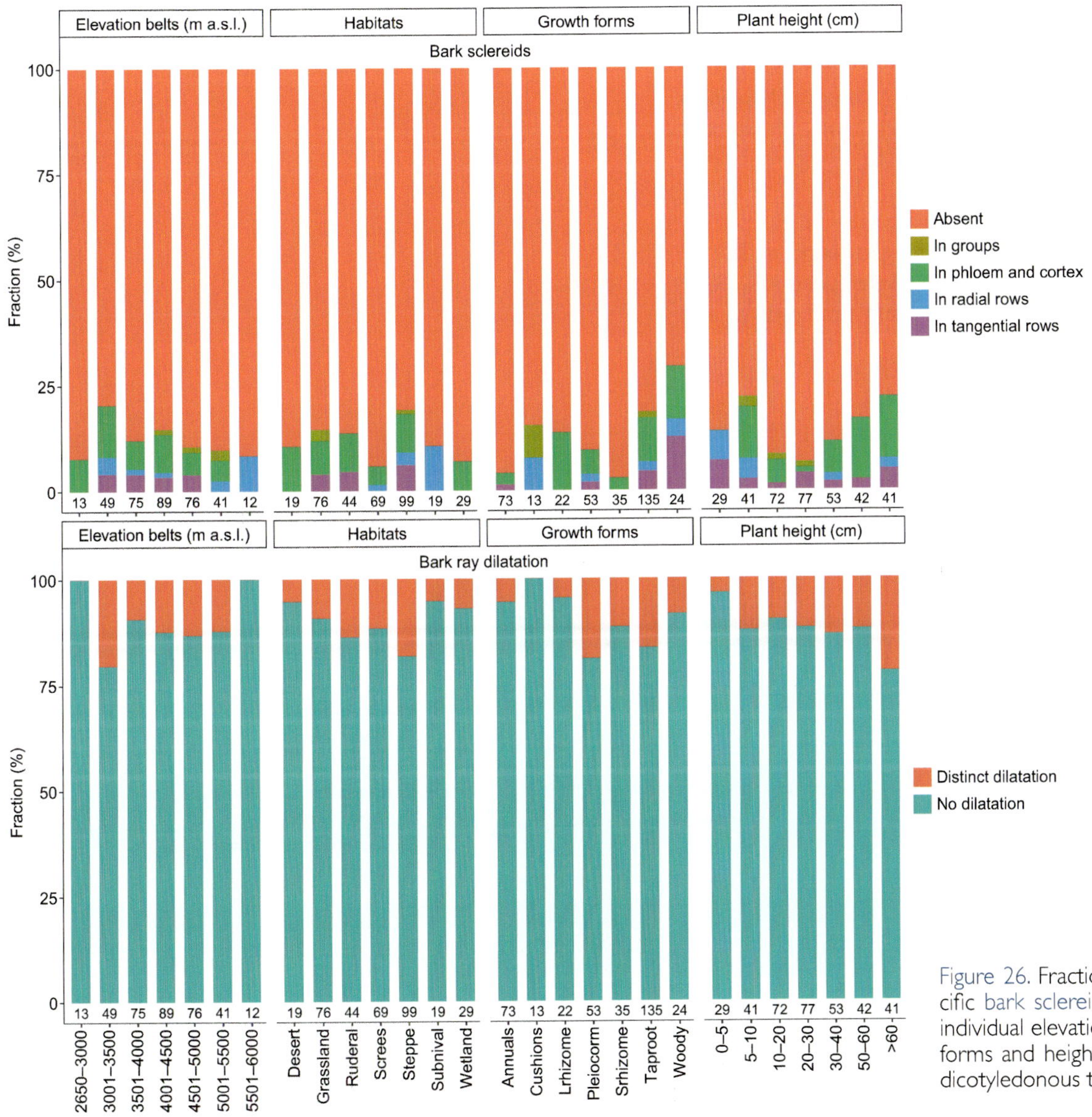

Figure 26. Fraction of species with specific bark sclereids and ray dilatation in individual elevation belts, habitas, growth forms and height classes, based on 355 dicotyledonous taxa.

Crystals, mainly in form of druses, are rare (11%) (Fig. 27). We can only speculate about their significance. They might support the regulation of calcium levels in cells or belong to the defense system (herbivory). Xylem crystals are typically found in species occupying most arid parts of Ladakh (Fig. 27). Acicular crystals are found in desert and dry steppe species such as *Ephedra gerardiana*, *E. intermedia*, *Oxytropis tatarica*, *Peganum harmala*, *Menitskia tibetica* (Fig. 28). Crystal sand is an almost exclusive feature of Amaranthaceae, both short-lived (*Amaranthus powellii*, *Atriplex schugnanica*, *Corispermum tibeticum*, *Dysphania botrys*, *Dysphania nepalensis*) and perennial (*Kochia prostrata*, *Haloxylon thomsonii*), all from most arid lowest elevations (Fig. 28). Prismatic crystals occur in *Halogeton* spp. and *Salsola jacquemontii*. Crystal druses were found in the xylem of Caryophyllaceae (*Silene*, *Arenaria*, *Stellaria*), Polygonaceae (*Rheum*, *Oxyria*, *Bistorta*, *Aconogonon*, *Knorringia*, *Polygonum*), Geraniaceae (*Geranium*, *Erodium*), Morinaceae (*Morina coulteriana*) and Rosaceae (*Potentilla*). Rhaphides were found in *Epilobium* spp. and *Rubia tibetica*.

Phellem secretory and crystal elements were found in 35% of the studied taxa (Fig. 27). Bark sclereids were found in 13% of the species (Fig. 26), mostly in tall forbs and subshrubs of arid and semi-arid biotops from Asteraceae (*Artemisia* spp., *Ajania fruticulosa*, *Anaphalis virgata*, *Richteria pyrethroides*, *Cousinia thomsonii*) Fabaceae (all tall *Astragalus* spp., *Oxytropis chiliophylla*) and Brassicaceae (*Christolea crassifolia*) families. Bark oil ducts were found in most *Artemisia* spp., *Waldheimia* spp., and most Apiaceae (e.g. *Pleurospermum candollei*, *Semenovia lasiocarpa*, *Seseli mucronatum*, *Conioselinum vaginatum*). Latex occurs in *Euphorbia tibetica*, *Inula obtusifolia* and *Polygonum cognatum*. Bark crystals are mostly found in representatives of Polygonaceae, Amaranthaceae, Caryophyllaceae and Rosaceae (Fig. 28).

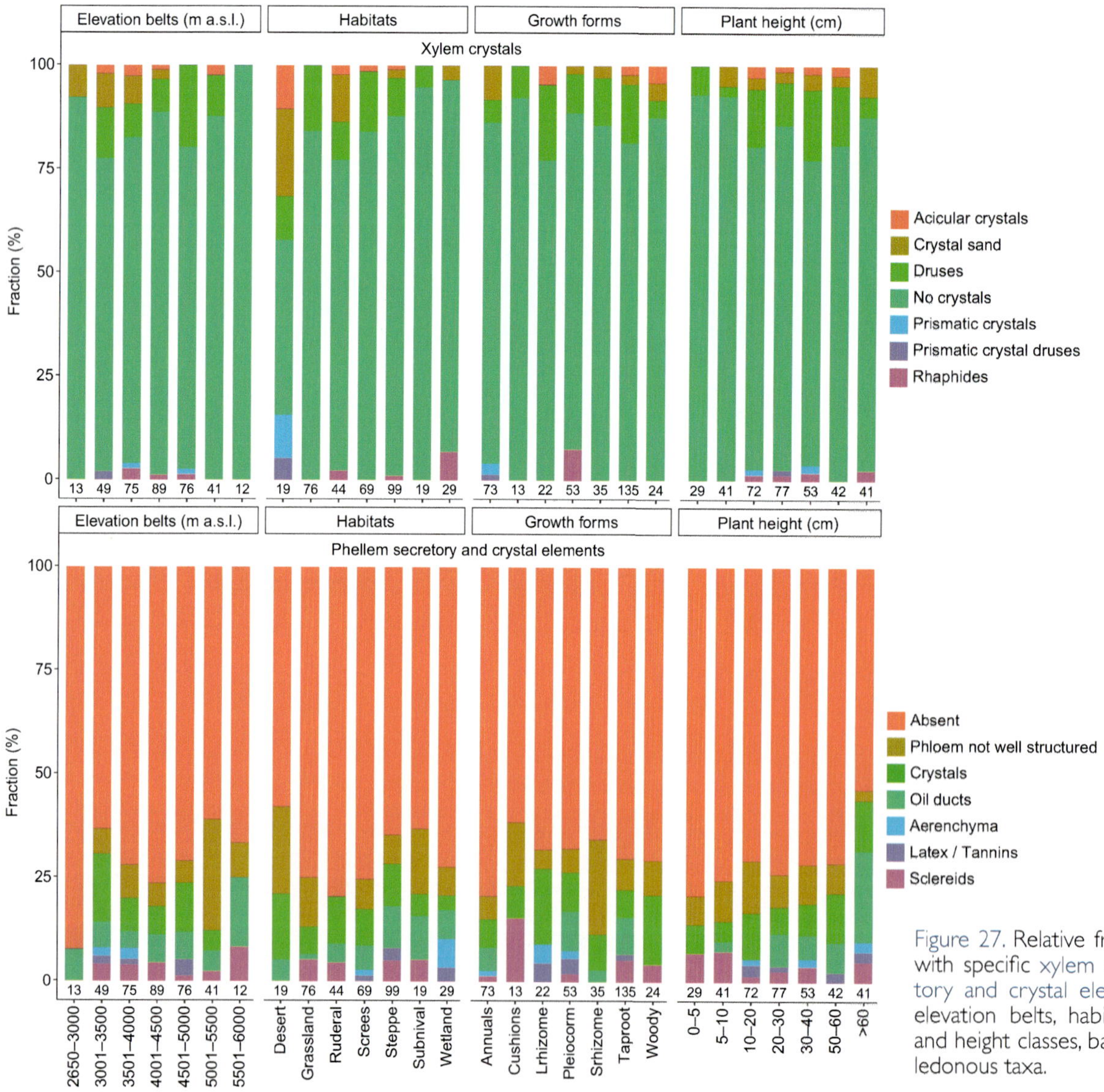

Figure 27. Relative frequency of species with specific xylem and phellem secretory and crystal elements in individual elevation belts, habitats, growth forms and height classes, based on 355 dicotyledonous taxa.

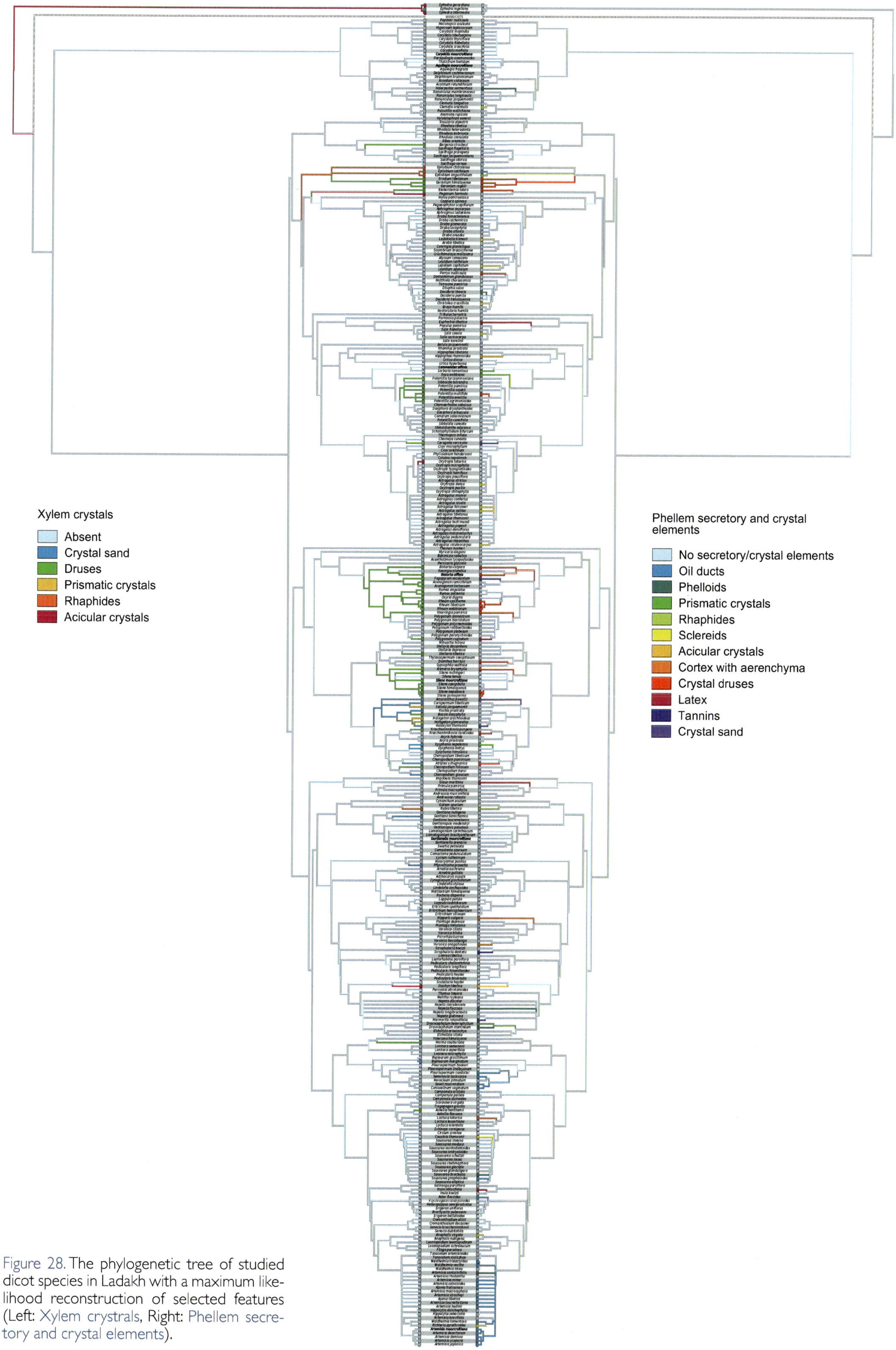

Figure 28. The phylogenetic tree of studied dicot species in Ladakh with a maximum likelihood reconstruction of selected features (Left: Xylem crystrals, Right: Phellem secretory and crystal elements).

Principal component analysis of all anatomical features of Ladakh dicots showed their interrelationships and affinities to plant growth forms and environmental drivers. The ecological attributes (habitat type, elevation) and adult plant height had a significant effect on the divergence of anatomical traits. The main anatomical differences along the first (horizontal) ordination axis are associated with habitat preferences along elevational gradients from cold deserts to alpine and subnival zone, i.e., the first axis separated low-elevation plants (e.g. Amaranthaceae and Salicaceae) from high-elevation taxa belonging mostly to Brassicaceae and Caryophylaceae (Fig. 29). The low-elevation plants have an anatomical structure characterized by large vessels arranged in long radial multiples, paratracheal parenchyma, thick-walled libriform fibers and uniseriate rays. The high-elevation plants have large rays separating vascular bundles, large multiseriate rays, a large number of small vessels, absent fibers, pervasive parenchyma and scalariform pits, high longevity and small annual ring increments. The changes along the second axis seemed to be associated with decreasing adult plant height (Fig. 30) from small wetland species (e.g. Gentianaceae) to large steppe and scree forbs and subshrubs from Asteraceae and Fabaceae families (Fig. 31). Smaller plants have small vessel diameters, radial increments, while tall forbs have vessel clusters, distinct phloem ray dilatations, rays with sheath cells and multiseriate rays.

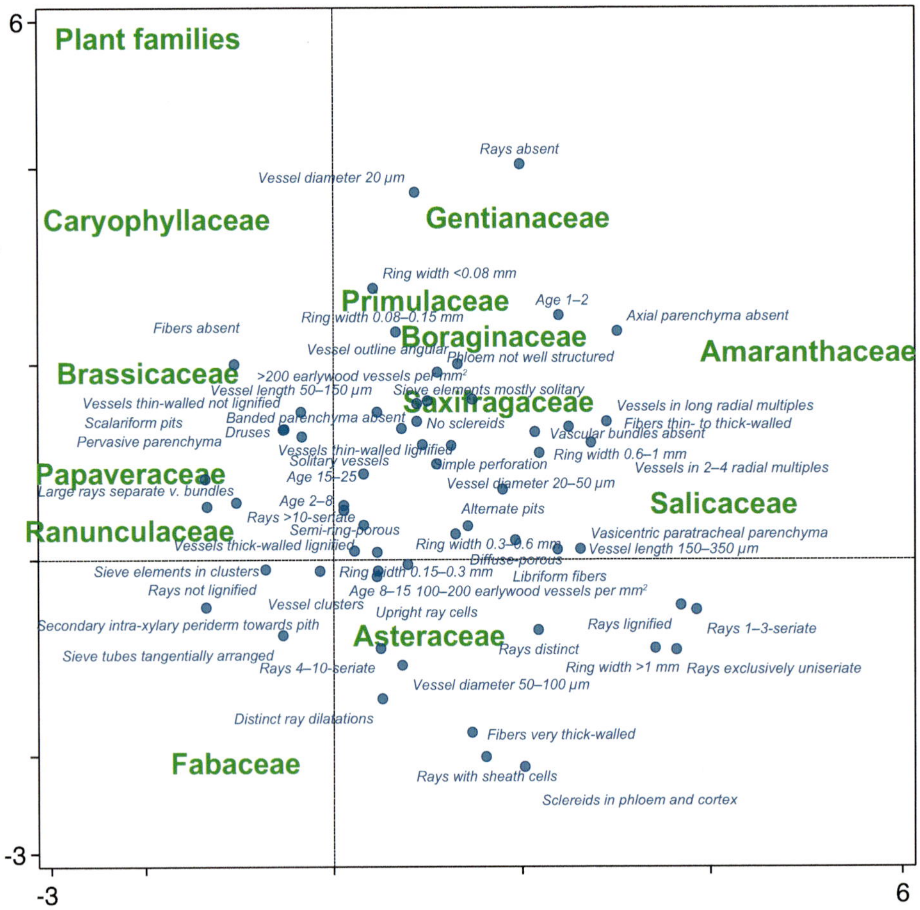

Figure 29. Principal Component Analysis of plant anatomical features in relation to plant families.

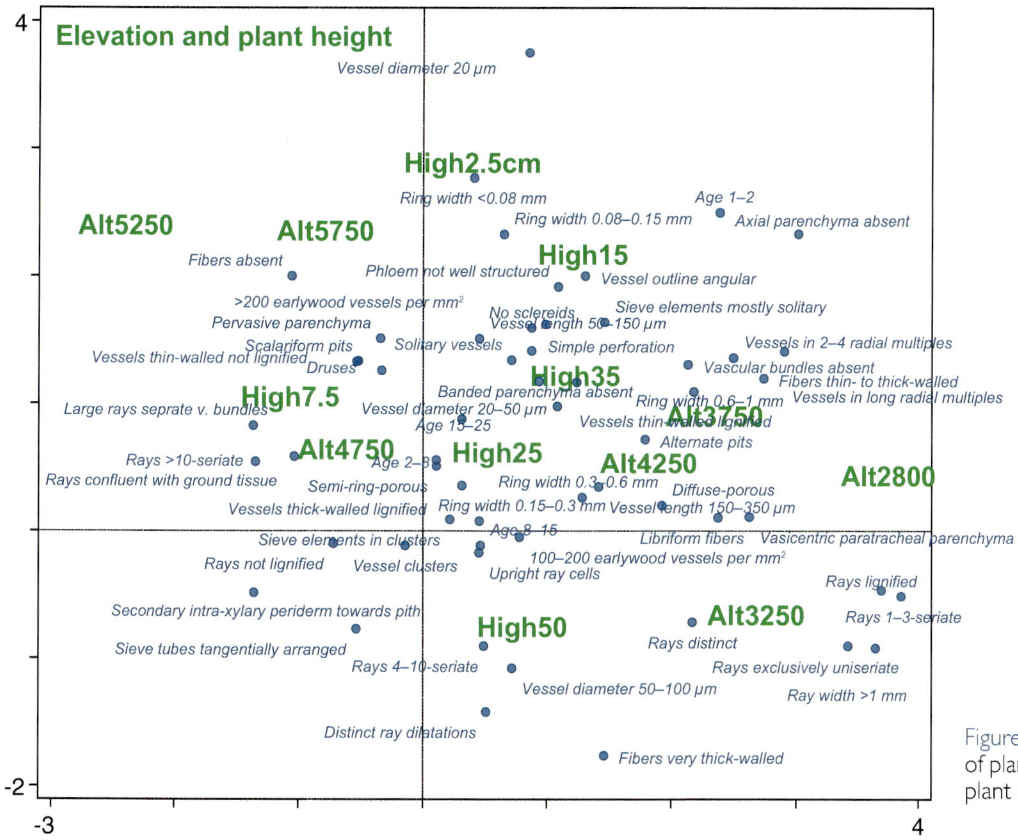

Figure 30. Principal Component Analysis of plant anatomical features in relation to plant height and elevation optima.

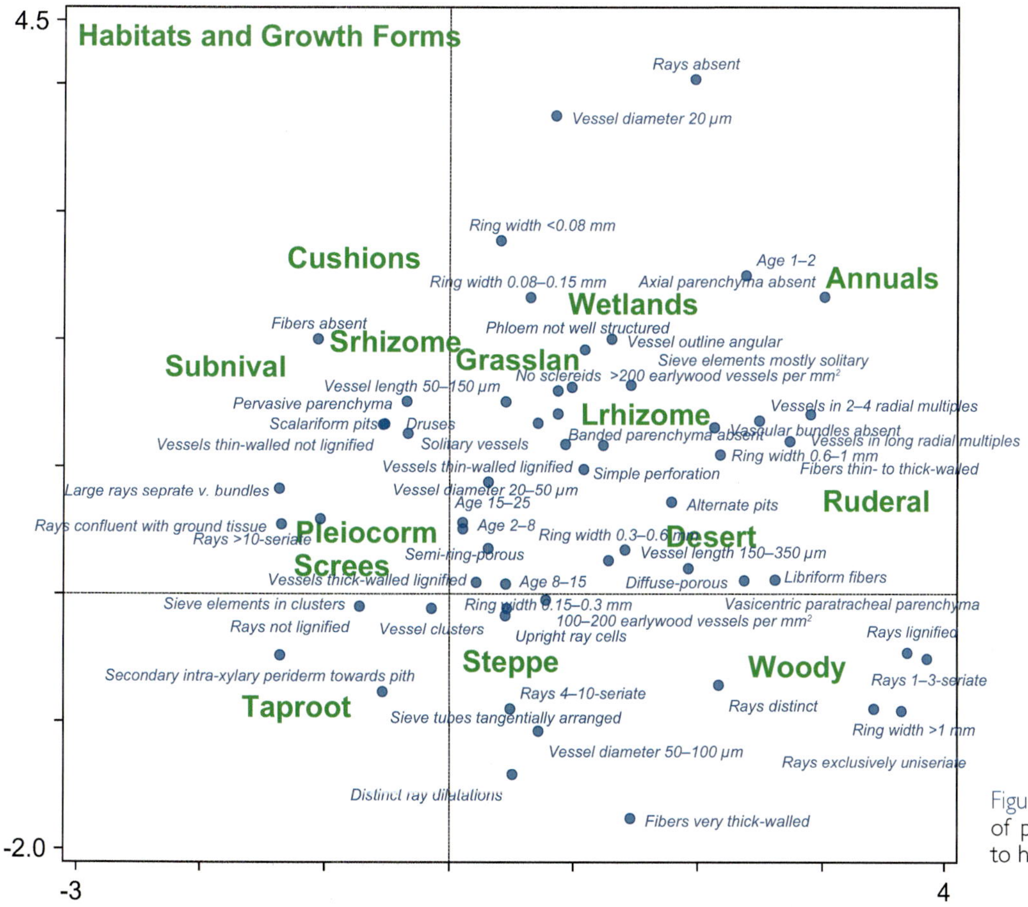

Figure 31. Principal Component Analysis of plant anatomical features in relation to habitat preferences and growth form.

Culm anatomy in monocotyledonous species

In this chapter, we present characteristic anatomical structures of a large number of monocotyledonous species, mostly graminoids, from major habitats in Ladakh, and try to relate microscopic features to plant size, altitude, hydrological conditions and taxonomic origin. The study is based on cross sections of culms. Each section is located at internodes above leaf sheathes, approximately in the middle part of the total length of culms. All major species which occur in the Ladakh region are analyzed, but only two specimens of each species. The material includes species of all vegetation zones, from 3000 to 6150 m a.s.l. of all possible habitats from wet alpine grasslands to arid semideserts and steppes, and from nutrient-rich ruderal sites to salty wetlands (Fig. 32). Included are also rare and introduced species. Since anatomical variation is minimized by the constant internal location of the section within the culm, the whole set of analyzed material is comparable. Double staining gives additional information about the distribution of lignified and unlignified parts.

The 155 monocot species studied belong to 7 families and 43 genera. The most species-rich family is Poaceae (109 taxa, 70%), followed by Cyperaceae (35 species, 23%), Juncaceae (5 taxa, 3%), Liliaceae (3, 2%), Juncagninaceae (1), Potamogetonaceae (1) and Zanichelliaceae (1). The most species-rich genera are *Carex* (20), *Stipa* (16), *Poa* (13), *Elymus* (12), *Kobresia* (11), *Bromus* (10), *Festuca* (10), *Puccinellia* (8), *Juncus* (5) and *Agrostis* (4) (Fig. 32). Poaceae form the majority of studied monocotyledonous species at lower elevations (Fig. 33), and their dominance decreases with increasing elevation from 100% below 3000 m to 30% above 5000 m. The Cyperaceae show the opposite elevational pattern, making up 70% of species above 5000 m. They are most abundant in alpine grasslands (48%) and wetlands (29%), while Poaceae dominate (86–100%) in all dry habitats (deserts, steppes and screes) (Fig. 34). Annuals are exclusively represented among Poaceae (94%) and Juncaceae (6%), short-rhizomatous turf graminoids among Poaceae (78%) and less in Cyperaceae (20%), while plants with long hypogeogenous rhizomes (*Poa tibetica* type) are rather equally distributed among Poaceae (52%) and Cyperaceae (44%). Most Cyperaceae are short species less than 40 cm tall (Fig. 35), while Poaceae dominate among taller graminoids. Alpine grasslands are dominated by *Carex* (31%), *Kobresia* (28%) and *Poa* (16%), ruderal vegetation by *Bromus* (67%), steppes by *Elymus* (21%) and *Stipa* (34%), saline wetlands by *Carex* (30%), *Puccinellia* (26%) and *Juncus* (22%), and screes by *Poa* (21%), *Stipa* (21%) and *Elymus* (28%). The fraction of *Carex* and *Kobresia* species increases significantly towards higher elevation belts, while that of *Stipa* and *Elymus* decreases.

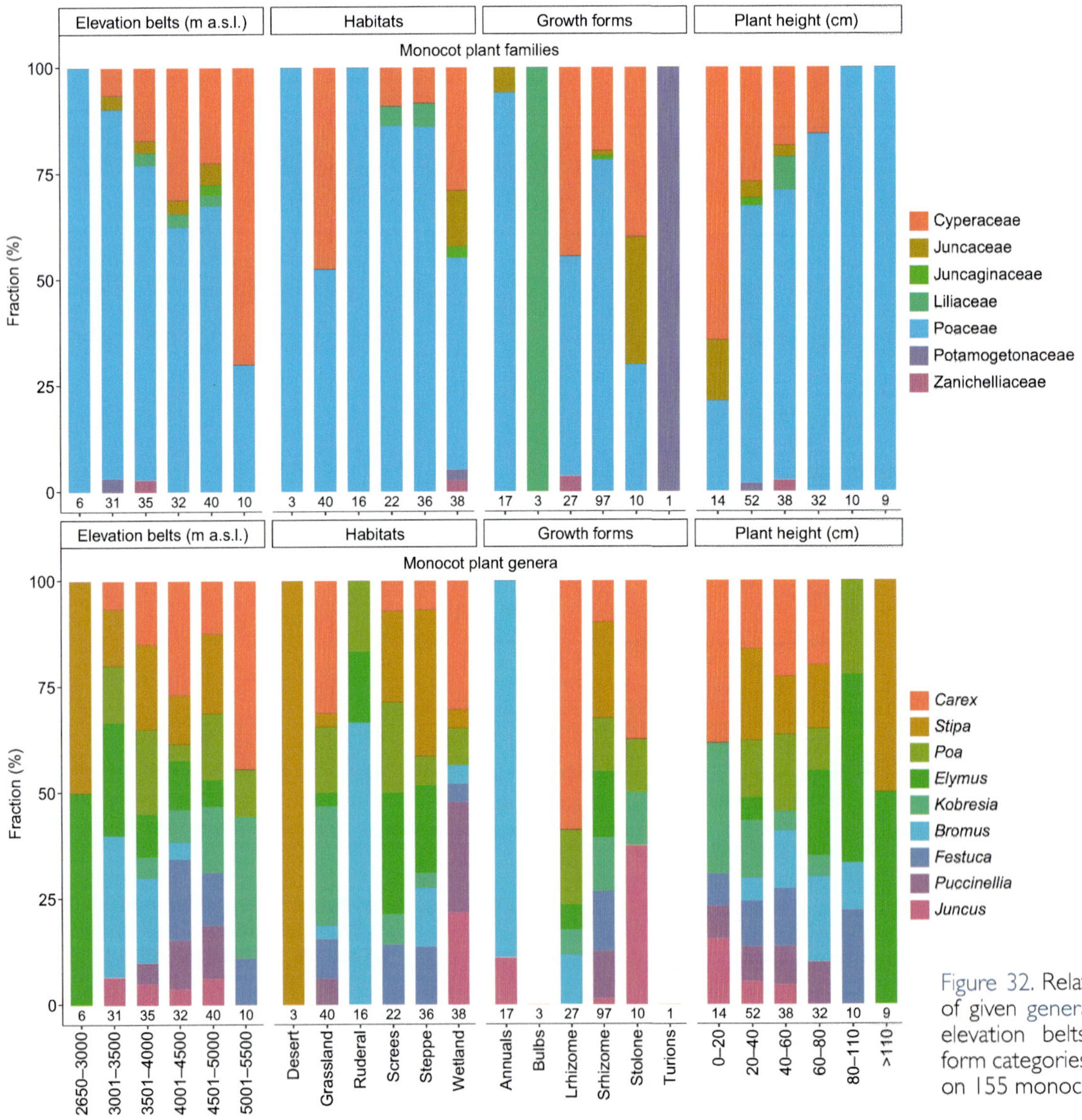

Figure 32. Relative frequency of species of given genera and family in individual elevation belts, habitat types, growth form categories and height classes, based on 155 monocotyledonous species.

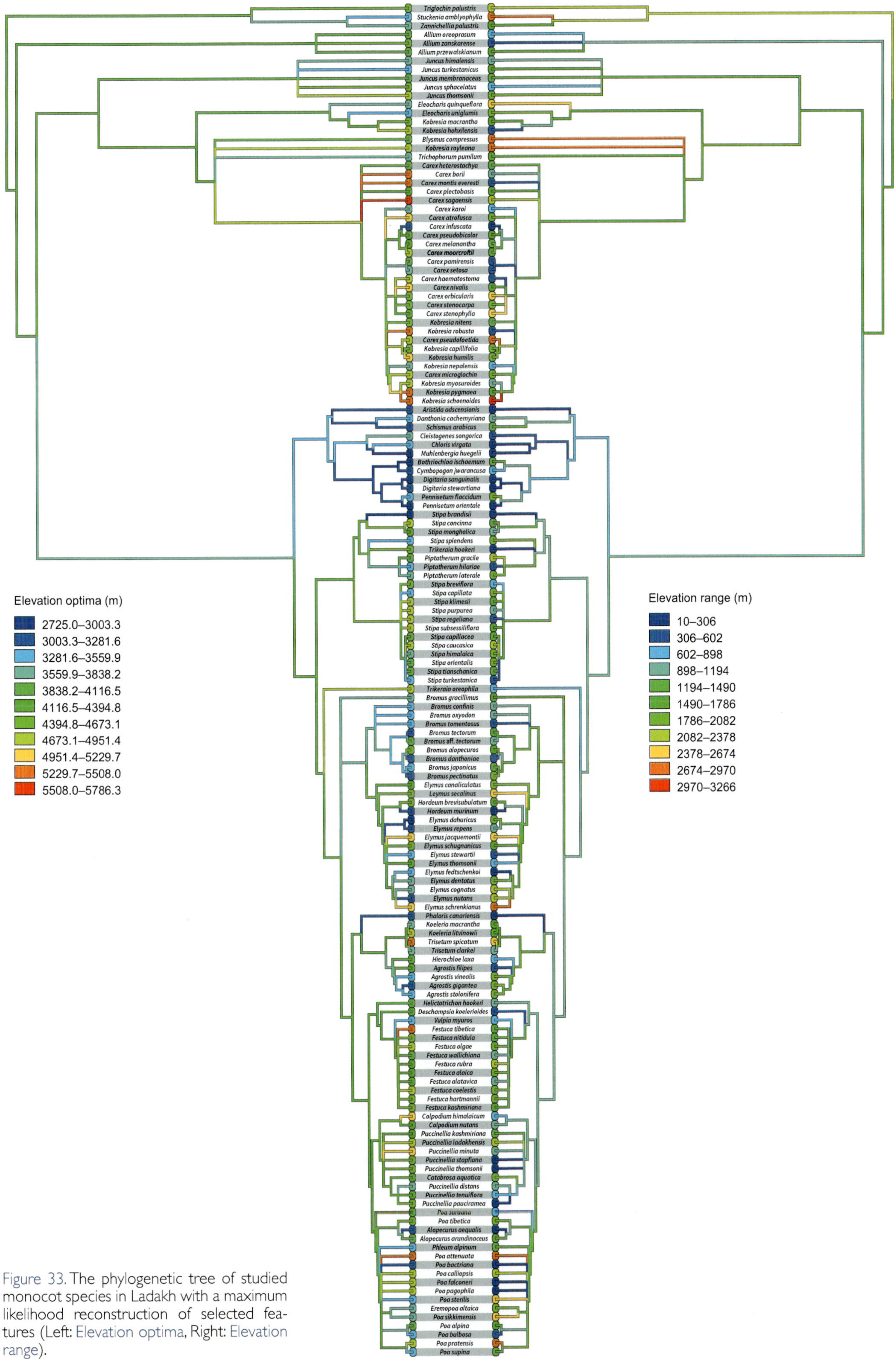

Figure 33. The phylogenetic tree of studied monocot species in Ladakh with a maximum likelihood reconstruction of selected features (Left: Elevation optima, Right: Elevation range).

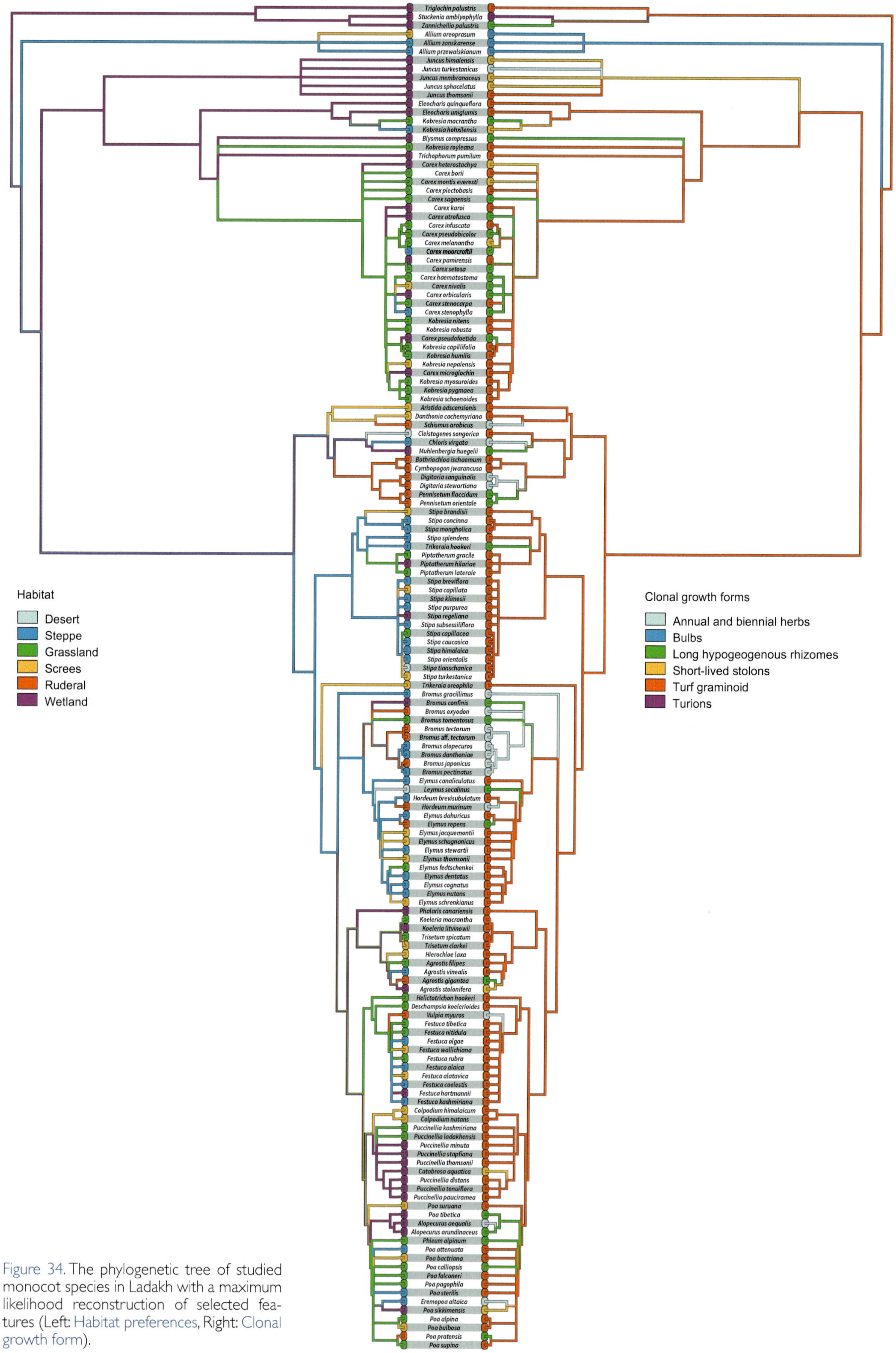

Figure 34. The phylogenetic tree of studied monocot species in Ladakh with a maximum likelihood reconstruction of selected features (Left: Habitat preferences, Right: Clonal growth form).

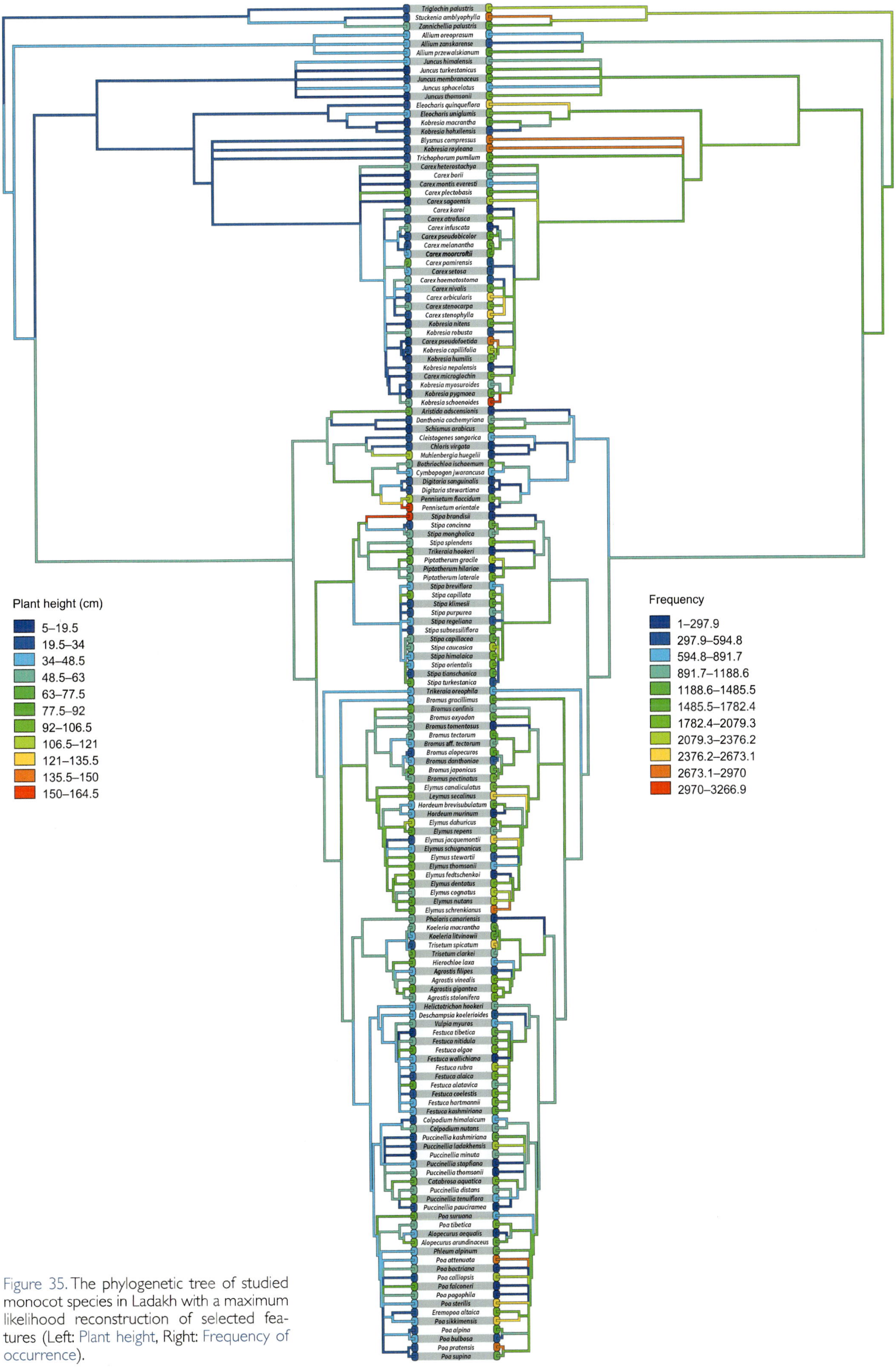

Figure 35. The phylogenetic tree of studied monocot species in Ladakh with a maximum likelihood reconstruction of selected features (Left: Plant height, Right: Frequency of occurrence).

Culm diameter and surface outline

Culms of the majority of studied species have small diameters <0.5 mm (46%) or 1–2 mm (46%), while species with thick culm diameter are rare (7%, e.g. *Allium* spp., *Juncus sphacelatus, Pennisetum flaccidum, Bromus tomentosus*) and their proportion decreases from 16% below 3000 m to zero above 5000 m elevation (Fig. 36). The fraction of species with diameters <0.5 mm increases significantly with increasing elevation from 16 to 60%, being most represented in short <20 cm-tall species with short epigeogenous rhizomes (e.g. *Kobresia pygmaea, Carex microglochin, Trikeraia oreophila*, most *Stipa* and *Puccinellia* spp.) (Fig. 37), while the species with thicker culms of 1–2 mm diameter are taller, often stoloniferous (e.g. *Agrostis stolonifera, Catabrosa aquatica, Poa sikkimensis, Carex heterostachya*) or long-rhizomatous (e.g. *Leymus secalinus, Alopecurus arundinaceus, Blysmus compressus, Elymus repens*) species from productive wetlands. Most species have circular culms with smooth (68%) or wavy (15%) surface. Culms with triangular outline are restricted to Cyperaceae with obtusely (10%, in *Carex* and *Kobresia*) or acutely (4%, in *Carex*) shaped surface (Fig. 37).

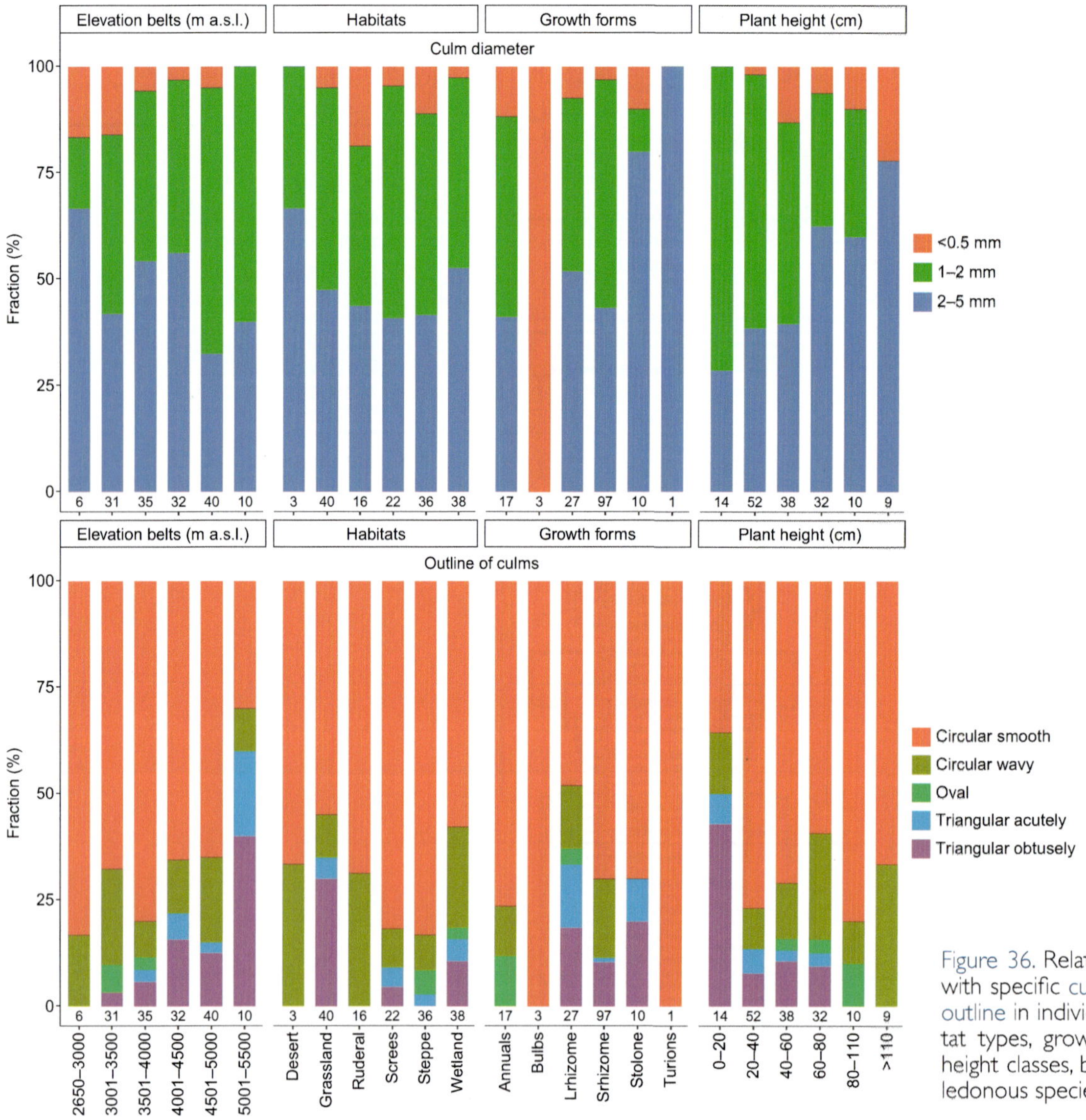

Figure 36. Relative frequency of species with specific culm diameter and surface outline in individual elevation belts, habitat types, growth form categories and height classes, based on 155 monocotyledonous species.

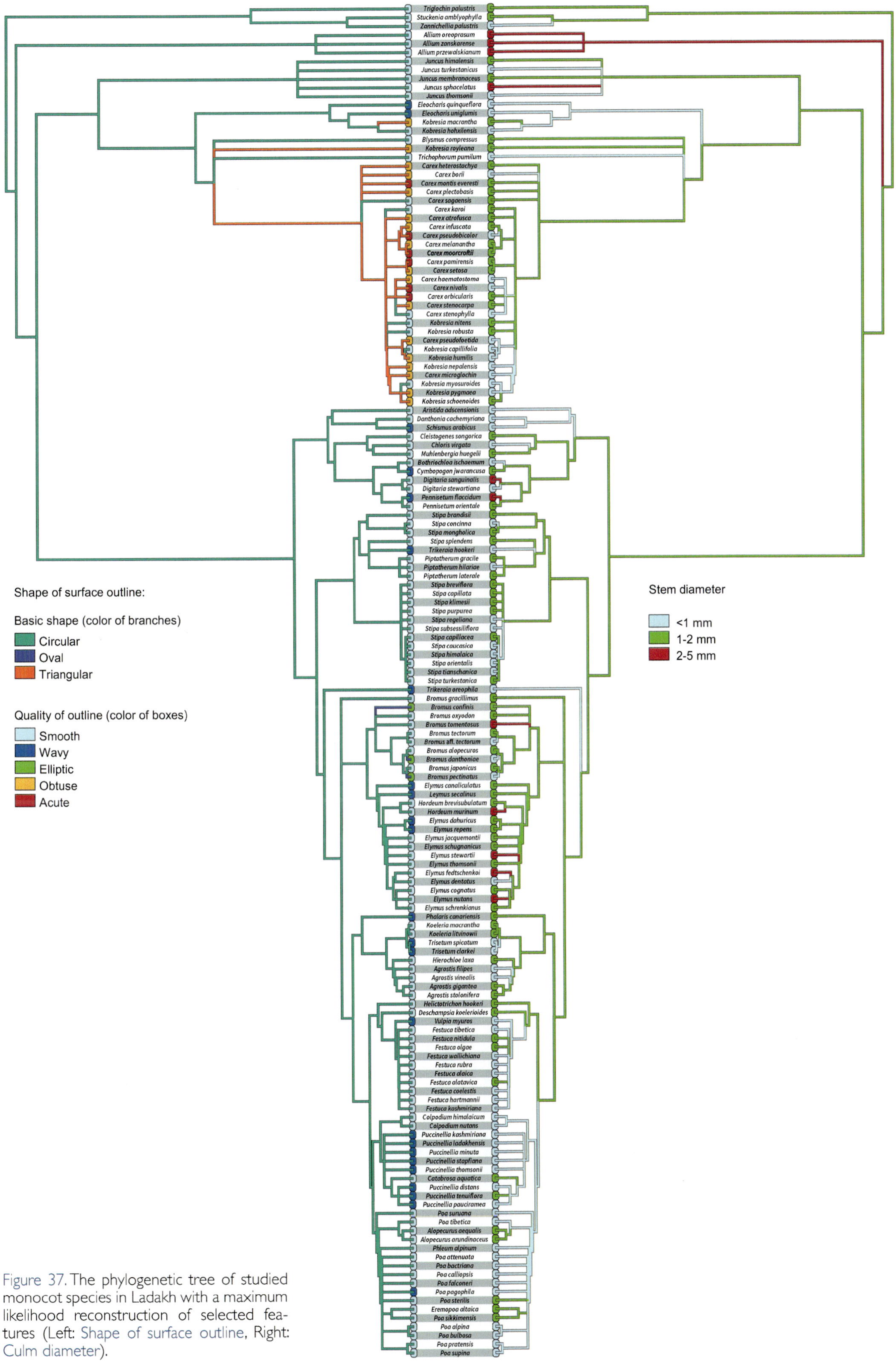

Figure 37. The phylogenetic tree of studied monocot species in Ladakh with a maximum likelihood reconstruction of selected features (Left: Shape of surface outline, Right: Culm diameter).

Species with hollow culms surrounded by many large, thin-walled, unlignified cells prevail (43%), mainly in tall turf graminoids (54%), steppe (64%), scree (54%) and desert (67%) species (majority of *Elymus, Festuca, Piptatherum, Poa* spp.) (Fig. 38). Species with hollow culms surrounded with a few thin-walled, unlignified cells form 21%, mostly among low-elevation saline wetland (29%, *Puccinellia*) and alpine grassland (28%, *Carex* and *Poa*) species at elevation below 4500 m. Species with hollow culms surrounded by a few thin-walled, lignified cells form 19% and include most *Kobresia* and *Bromus* taxa. Full culms containing unlignified cells form 9%, and are found in low-elevation wetland and ruderal species (*Juncus, Eleocharis, Digitaria* and *Pennisetum* spp., *Chloris virgata* and *Danthonia cachemyriana*) (Fig. 39). Full culms containing lignified cells make up 5% in species from lower-elevation steppe and ruderal habitats (*Allium* spp, *Cymbopogon jwarancusa, Bothriochloa ischaemum* and some *Stipa* spp.). Culms with net-like aerenchyma are typical for water-dwelling monocots (*Stuckenia amblyophylla, Triglochin palustris, Zannichellia palustris*). Epidermis cells have mostly thick walls (63%), mainly in lower-elevation ruderal (94%), scree (81%), steppe (80%) and desert (100%) species (Fig. 38). Very thick-walled epidermis is found in *Bromus, Colpodium, Elymus, Piptatherum* and *Leymus* species. Thin epidermis cells (5%, found in water plants) or epidermis cells which are thin on the inside and lignified on the periphery (25%, typical for *Kobresia* and *Carex*) are found in higher elevation grassland (45%) and wetland (53%) taxa. Epidermis with hairs was found in e.g. *Bromus oxyodon* and *Trisetum spicatum*, papillous epidermis in *Carex nivalis, C. moorcroftii* and *C. atrofusca* (Fig. 39).

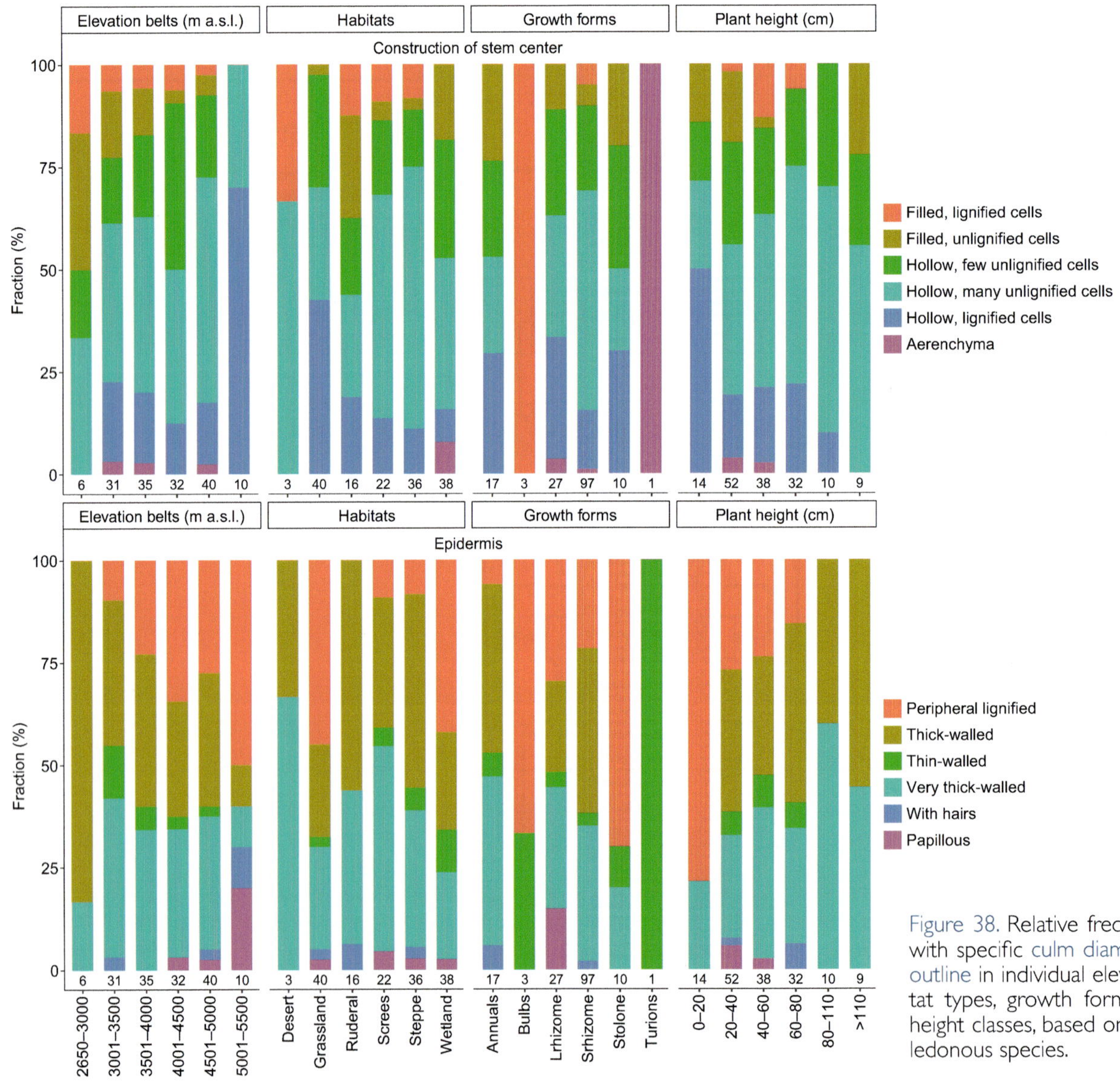

Figure 38. Relative frequency of species with specific culm diameter and surface outline in individual elevation belts, habitat types, growth form categories and height classes, based on 155 monocotyledonous species.

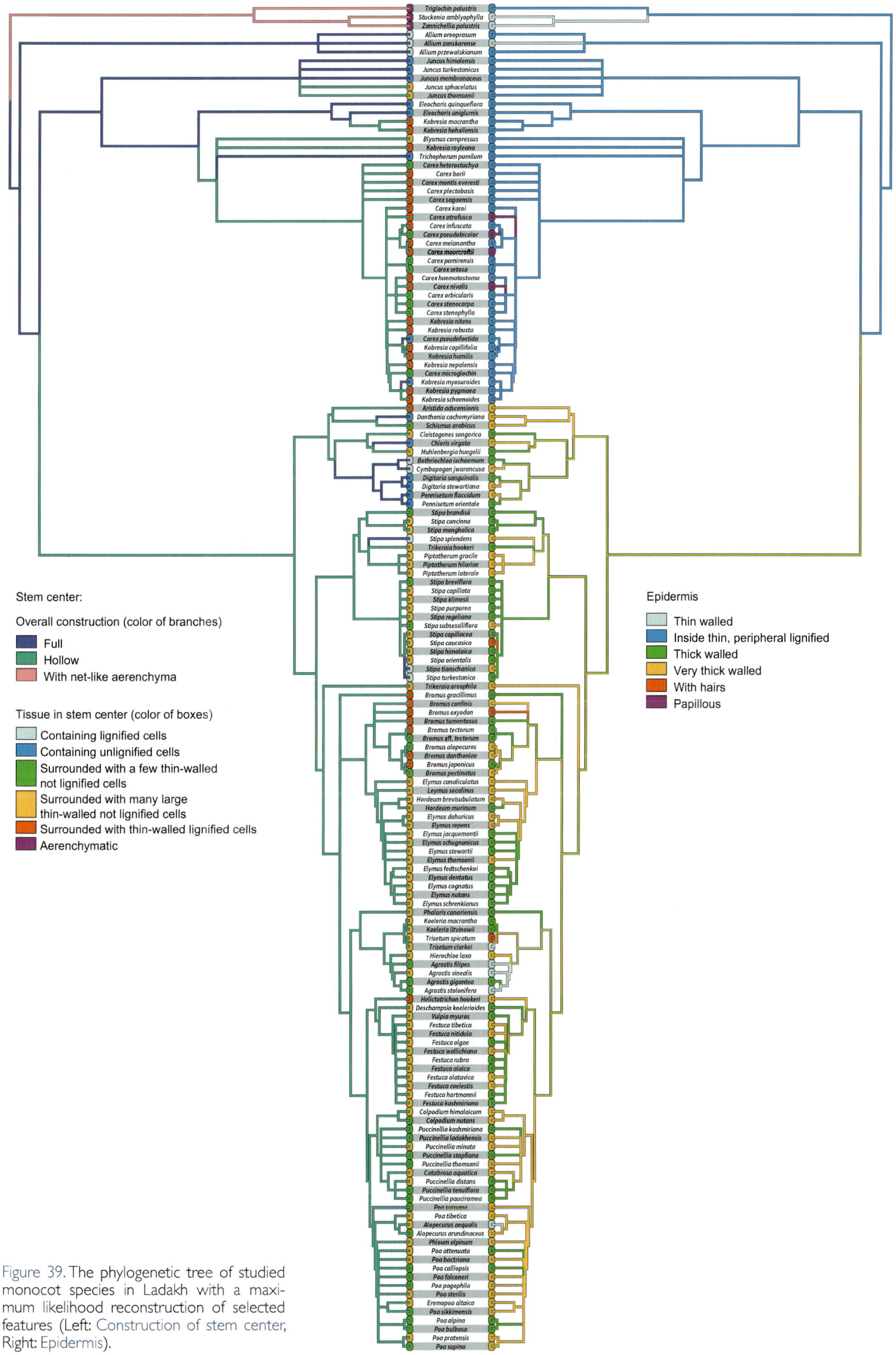

Figure 39. The phylogenetic tree of studied monocot species in Ladakh with a maximum likelihood reconstruction of selected features (Left: Construction of stem center, Right: Epidermis).

Vascular bundles are positioned in 2–3 peripheral rows in most species (82%), or in one peripheral row (16%), rarely in the whole stem (*Juncus himalensis*) (Fig. 40). Vascular bundles in one peripheral row are found in *Allium* spp., short-rhizomatous (*Puccinellia, Eleocharis, Triglochin, Carex microglochin*) and annual (*Bromus oxyodon, Digitaria stewartiana, Schismus arabicus*) ruderal and wetland taxa (Fig. 41). Vascular bundles are missing in water plants (*Stuckenia amblyophylla, Zannichellia palustris*). Vascular bundles are in 96% of species surrounded by a fibrous sheath. Sclerenchymatic sheath composed of one-sided small, 1–2 cells prevail (64%, typical of *Agrostis, Bromus, Elymus, Festuca, Puccinellia*), followed by one-sided large 2–4 centripetal cell sheath (19%, *Carex* and *Kobresia*), and one-sided large 2–4 centrifugal cells (8%, *Stipa* and *Juncus*) (Fig. 42). Sclerenchymatic sheaths are mostly lignified. A sheath of thin-walled, unlignified, large, round cells can be found e.g. in *Blysmus compressus, Allium* spp., *Alopecurus aequalis, Stipa himalaica* (Fig. 43). A sclerenchymatic sheath is entirely absent in *Lloydia serotina*. Rarely represented are species with a sclerenchymatic sheath made of circular large cells (2%, *Carex pseudofoetida, Achnatherum splendens, Elymus fedtschenkoi*) or bilateral large cells (3%, e.g. in *Helictotrichon hookeri, Bromus alopecuros*). Bundles surrounded with a small fibrous sheath dominate in ruderal and dry habitats at lower elevations, and are less frequent (30%) in high-elevation alpine grasslands where species (*Carex, Kobresia*) have sclerenchymatic sheaths composed of large centripetally arranged cells. Sixty-two percent of the studied species have small vessels with a diameter <20 µm, and the fraction of species with small vessels is increasing with elevation from 16% below 3000 m to 90% above 5000 m (Fig. 43). Again, variation in vessel sizes are mostly related to taxonomical identity and plant size. Small vessels are mostly found in small *Kobresia, Festuca, Poa* species from higher-elevation alpine grasslands, while larger vessels with a diameter of 20–50 µm are found in taller grasses (*Bromus, Elymus*) but also tall sedges (*Carex stenocarpa, C. nivalis*) and other monocots (*Allium* spp.) from drier lower-elevation habitats (Fig. 43).

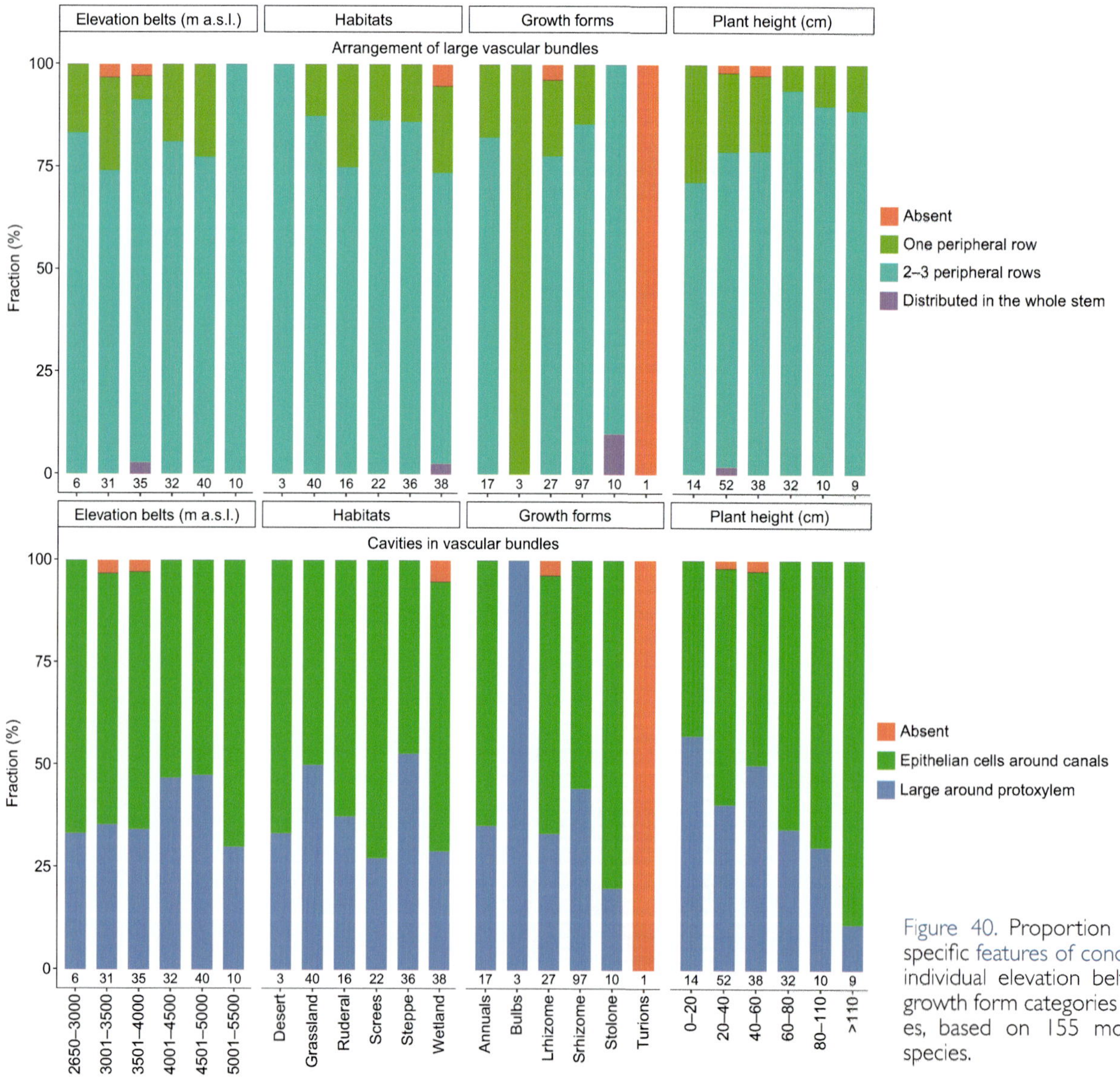

Figure 40. Proportion of species with specific features of conductive system in individual elevation belts, habitat types, growth form categories and height classes, based on 155 monocotyledonous species.

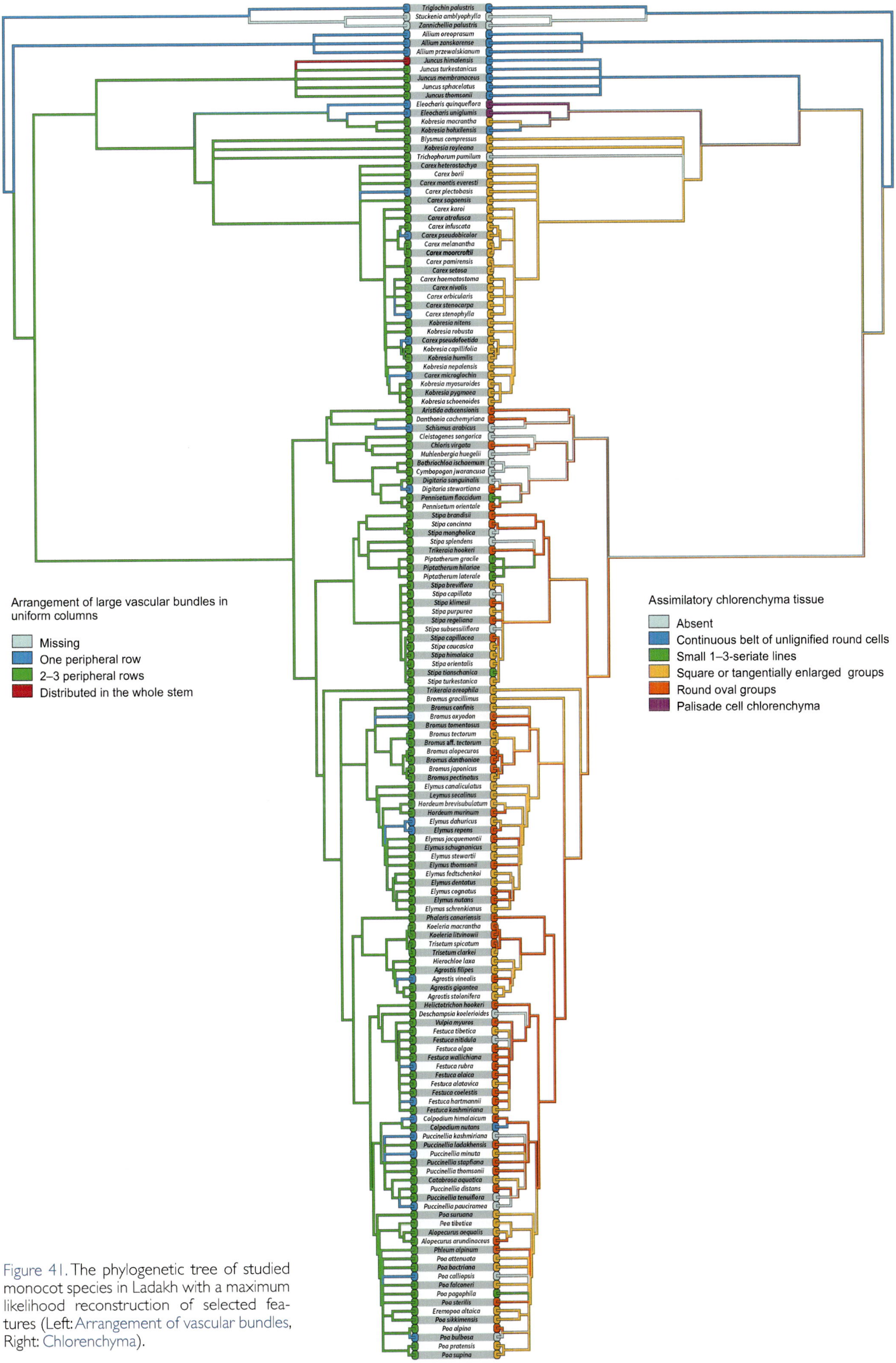

Figure 41. The phylogenetic tree of studied monocot species in Ladakh with a maximum likelihood reconstruction of selected features (Left: Arrangement of vascular bundles, Right: Chlorenchyma).

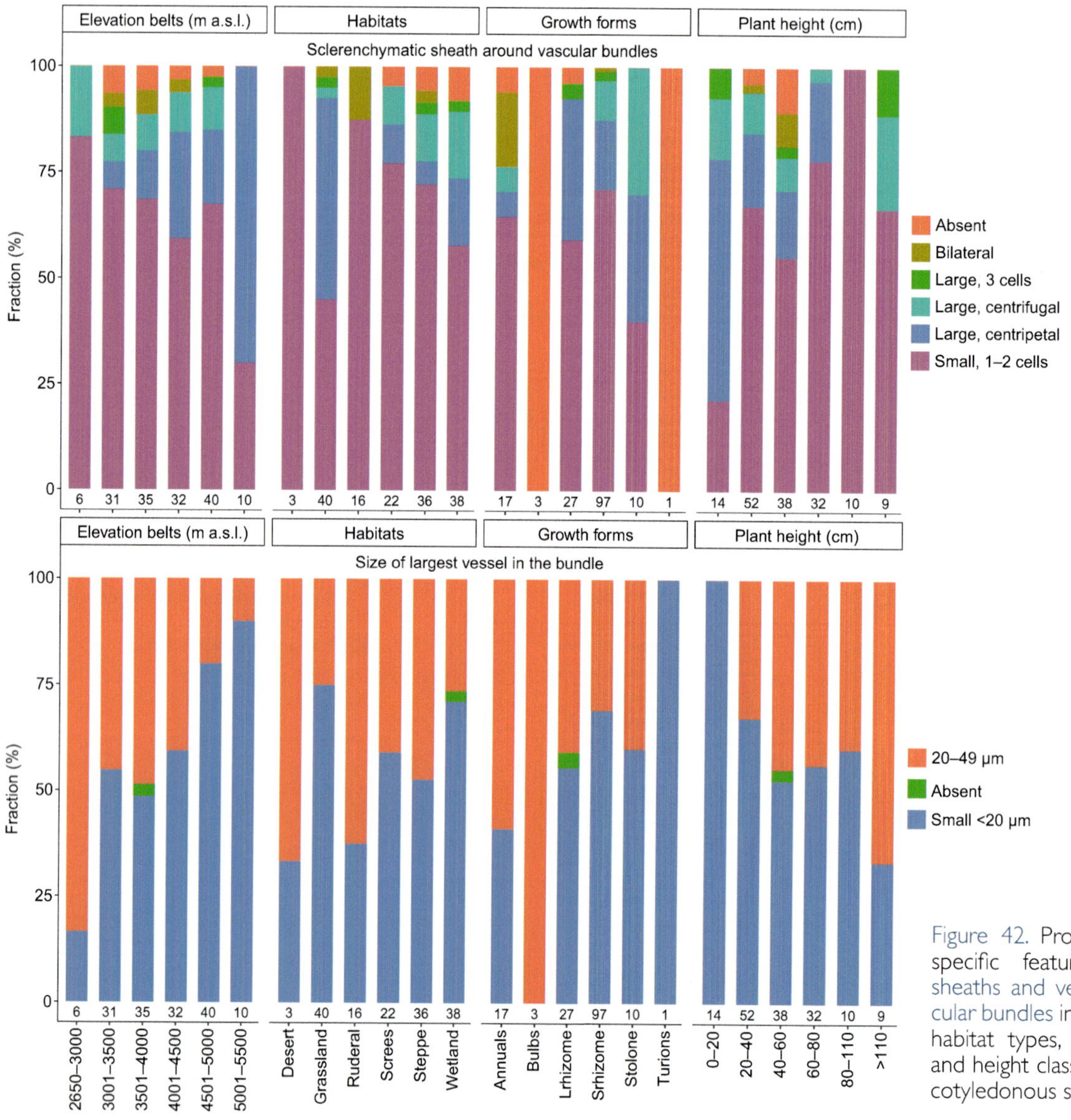

Figure 42. Proportion of species with specific features of sclerenchymatic sheaths and vessel sizes around/in vascular bundles in individual elevation belts, habitat types, growth form categories and height classes, based on 155 monocotyledonous species.

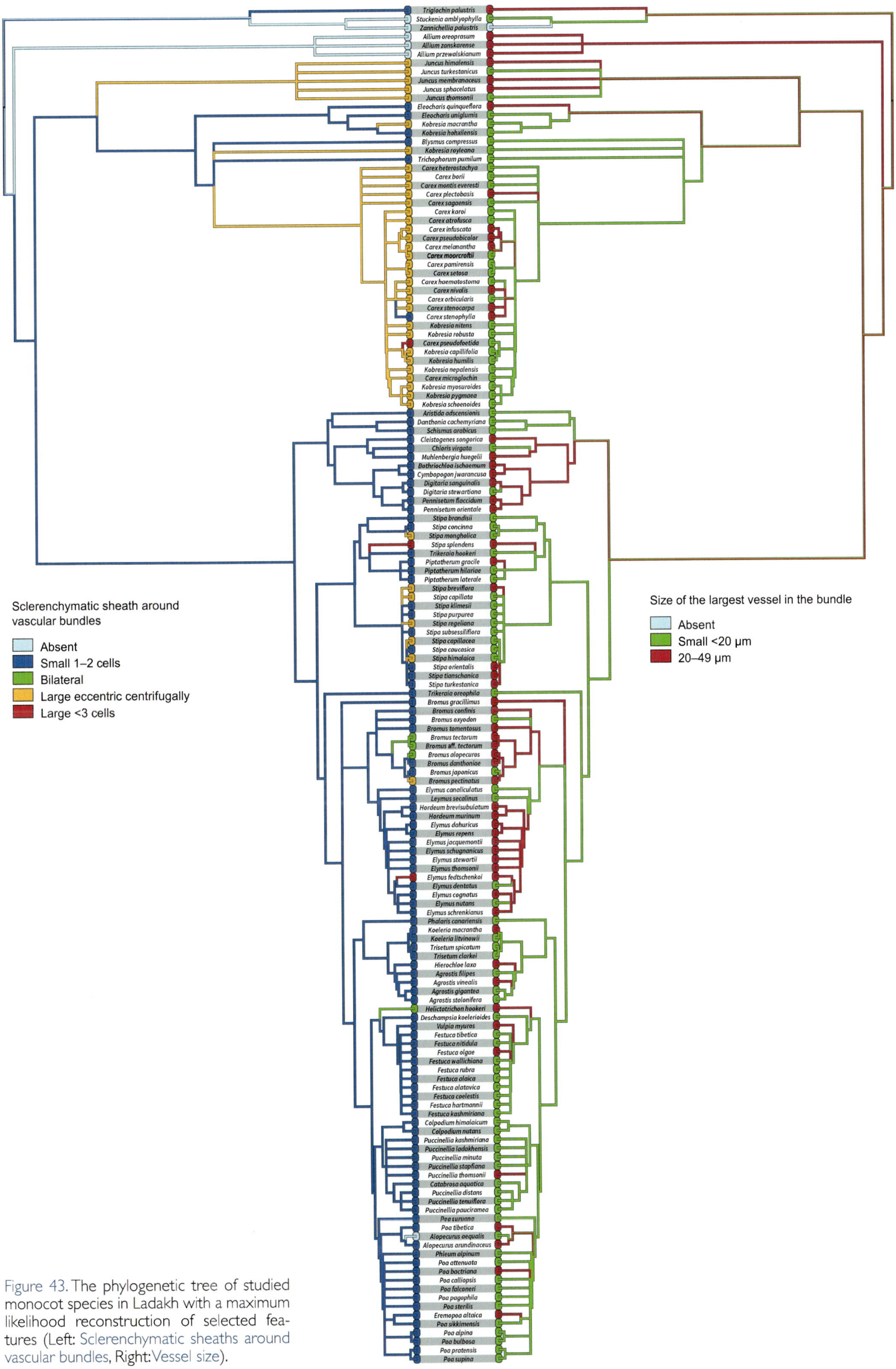

Figure 43. The phylogenetic tree of studied monocot species in Ladakh with a maximum likelihood reconstruction of selected features (Left: Sclerenchymatic sheaths around vascular bundles, Right: Vessel size).

Chlorenchyma is a widespread feature (86%) (Fig. 44). Square or tangentially enlarged chlorenchyma groups are found in almost half of the species (46%), followed by oval or rectangularly arranged groups (27%), and continuous chlorenchymatic belts of unlignified round cells (8%, *Allium*, *Juncus*, *Triglochin*) (Fig. 41). Chlorenchyma is more frequently absent in the lower-elevation ruderal plants (33%). We assume that the amount of culms in relation to leaves is smaller in the alpine graminoids. Therefore the assimilating surface is compensated by a large amount of chlorenchyma in culms of small *Carex* and *Kobresia* species that dominate in high-elevation alpine grasslands above 5000 m more than in lower-elevation grasses of dry habitats. Large tangentially enlarged groups of assimilating chlorenchyma tissues predominate (46%), increasing with elevation from 17% below 3000 m to 90% above 5000 m, and are characteristic for small *Carex* and *Kobresia* species, but also for *Elymus*, *Bromus* and *Agrostis* (Fig. 44). Round oval chlorenchyma groups (27%) occur mainly in short-lived (42%) and perennial short-rhizomatous grasses of dry low-elevation habitats. Palisade cell chlorenchyma (1%) occurs in *Eleocharis* spp., small 1–3-seriate chlorenchymatic lines (4%) in *Piptatherum* spp.

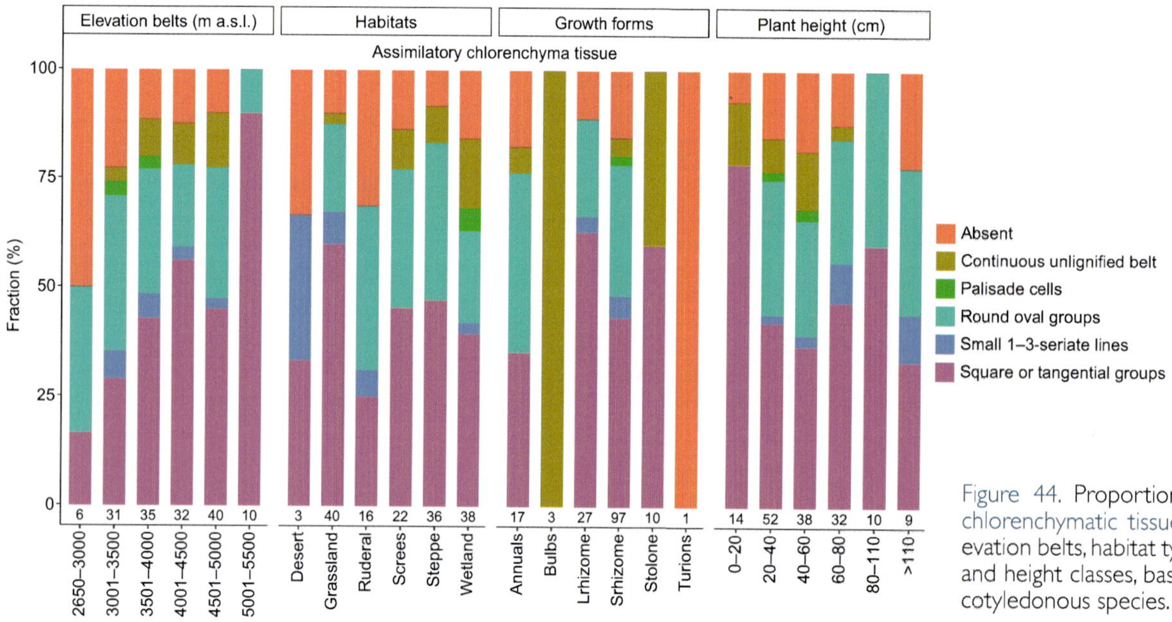

Figure 44. Proportion of species with chlorenchymatic tissues in individual elevation belts, habitat types, growth forms and height classes, based on 155 monocotyledonous species.

Mechanical support is achieved through the formation of a hollow stem, and stabilizing sclerenchymatic elements at the periphery of the culms in form of fairly thick continuous belts or girders composed of thick-walled lignified cells (Fig. 45). Amongst the studied monocotyledons, a ring of supporting or mechanical tissue is often found just beneath the epidermis, or forming part of the outer or inner cortex (sclerenchymatic girders). Thick-walled peripheral >3-seriate (27%) and thick-walled peripheral 1–3-seriate (25%) sclerenchymatic belts prevail, especially in tall turf short-rhizomatous and annual grasses of lower-elevation dry habitats (*Agrostis*, *Bromus*, *Stipa*, *Puccinellia*) (Fig. 46). Wavy peripheral belts of sclerenchyma located inside of the chlorenchyma are found in 13% of the species (*Juncus* and *Elymus* spp., *Leymus secalinus*, *Phalaris canariensis*), while smooth peripheral belts centripetal of chlorenchyma are found in 9% (e.g. *Allium* spp., *Triglochin palustris*, *Poa tibetica*) (Fig. 46). Peripheral sclerenchymatic belts are missing in 26% of the species (all *Carex* and *Kobresia* spp., *Blysmus compressus*, *Colpodium* and *Eleocharis* spp.). Sedges instead contain peripheral sclerenchymatic girders, found in total of all 48% studied taxa. Girders composed of square or radial rectangular cells (28%) are almost exclusively found in *Kobresia*, *Elymus* and *Poa* species, while conic girders (18%) occur in all *Carex* and *Colpodium* species. Tangentially enlarged girders occur in *Catabrosa aquatica*, round girders in *Bromus confinis* and *B. danthoniae*.

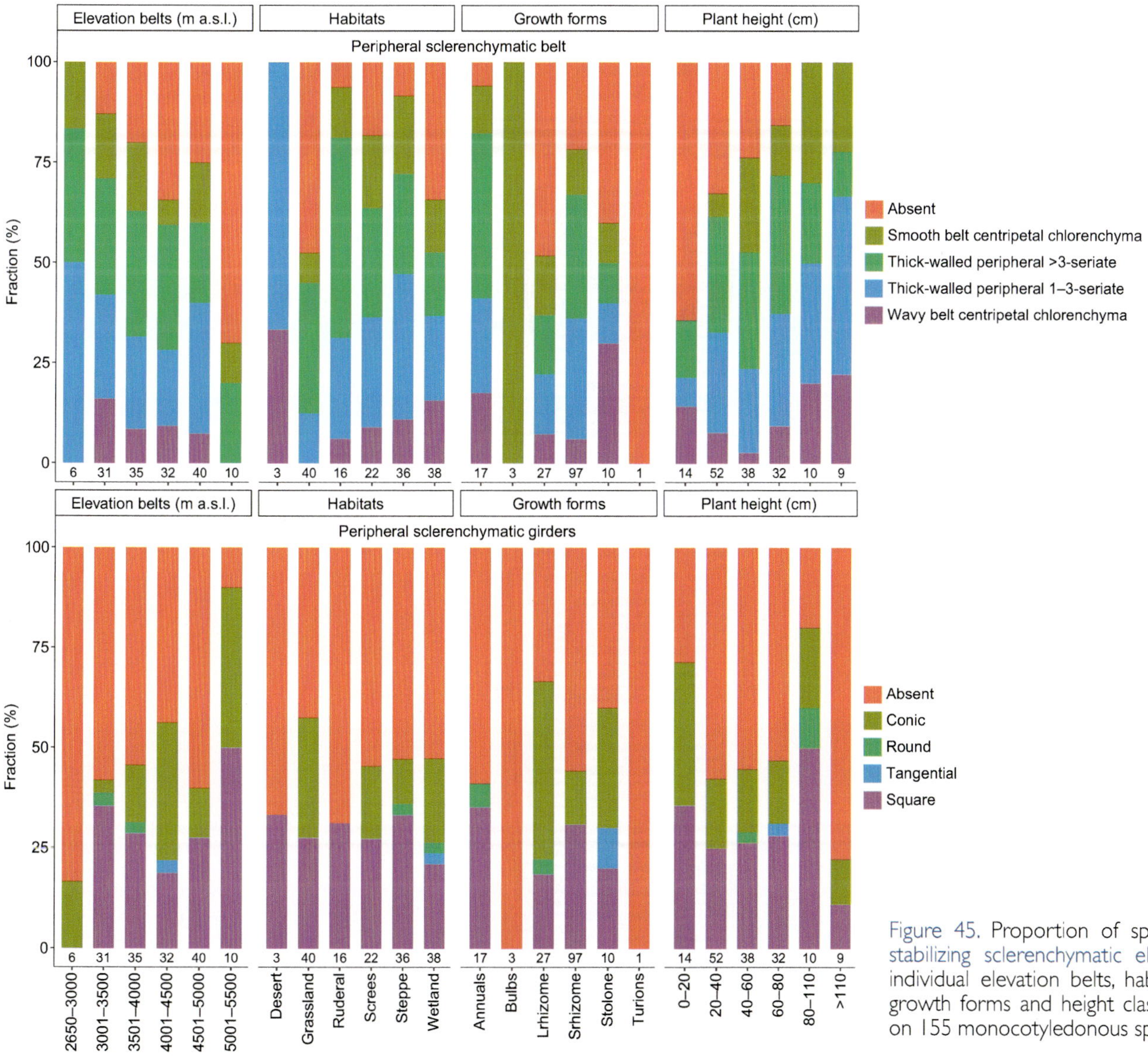

Figure 45. Proportion of species with stabilizing sclerenchymatic elements in individual elevation belts, habitat types, growth forms and height classes, based on 155 monocotyledonous species.

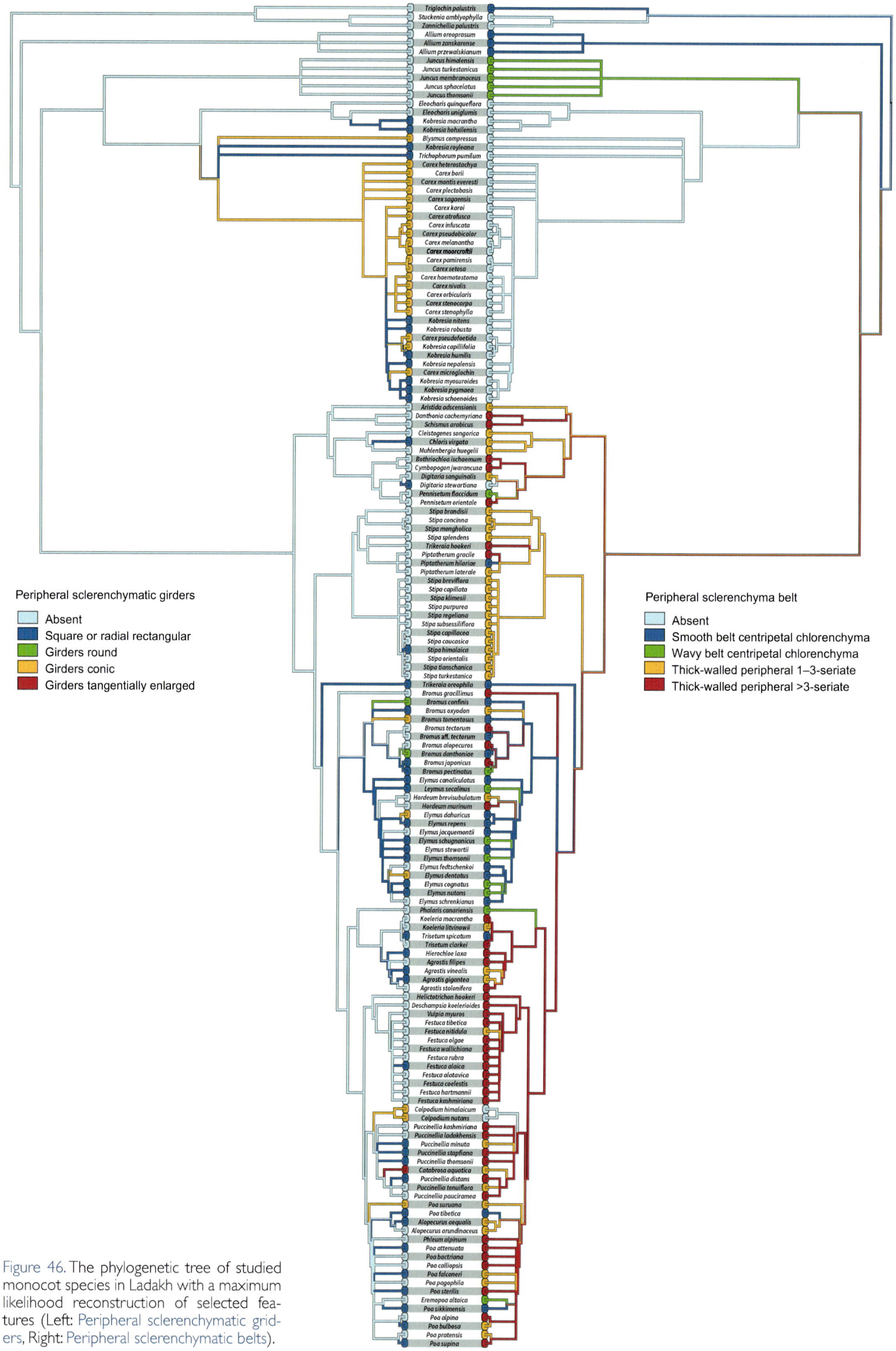

Figure 46. The phylogenetic tree of studied monocot species in Ladakh with a maximum likelihood reconstruction of selected features (Left: Peripheral sclerenchymatic girders, Right: Peripheral sclerenchymatic belts).

Additional mechanical support is achieved through the formation of conspicuous intercellular spaces (cavities) in parenchymatic tissue, present in 66% of studied taxa, or in the protoxylem of vascular bundles (99% of cases) (Fig. 47). Small triangular cavities prevail (39%), typically in tall grasses of dry habitats (*Elymus*, *Bromus*, *Stipa*, *Piptatherum*), round oval to radial intercellulars (14%) are present in *Carex* and *Blysmus*, and large irregular cavities in *Kobresia* species. Intercellulars in the protoxylem of vascular bundles are either smaller canals surrounded by epithelian cells (58%) found in sedges, rushes and tall grasses (*Stipa*, *Bromus*, *Elymus*, *Pennisetum*, *Leymus secalinus*, *Muhlenbergia huegelii*), while large intercellulars around protoxylem in vascular bundles (41%) are more characteristic of smaller graminoids (Fig. 48).

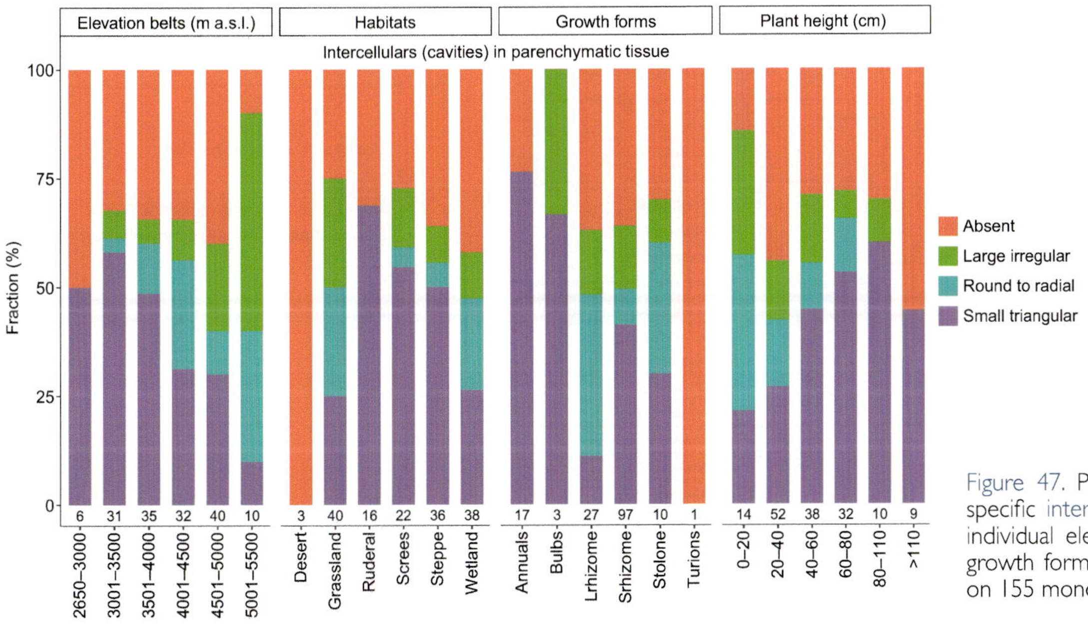

Figure 47. Proportion of species with specific intercellular spaces (cavities) in individual elevation belts, habitat types, growth forms and height classes, based on 155 monocotyledonous species.

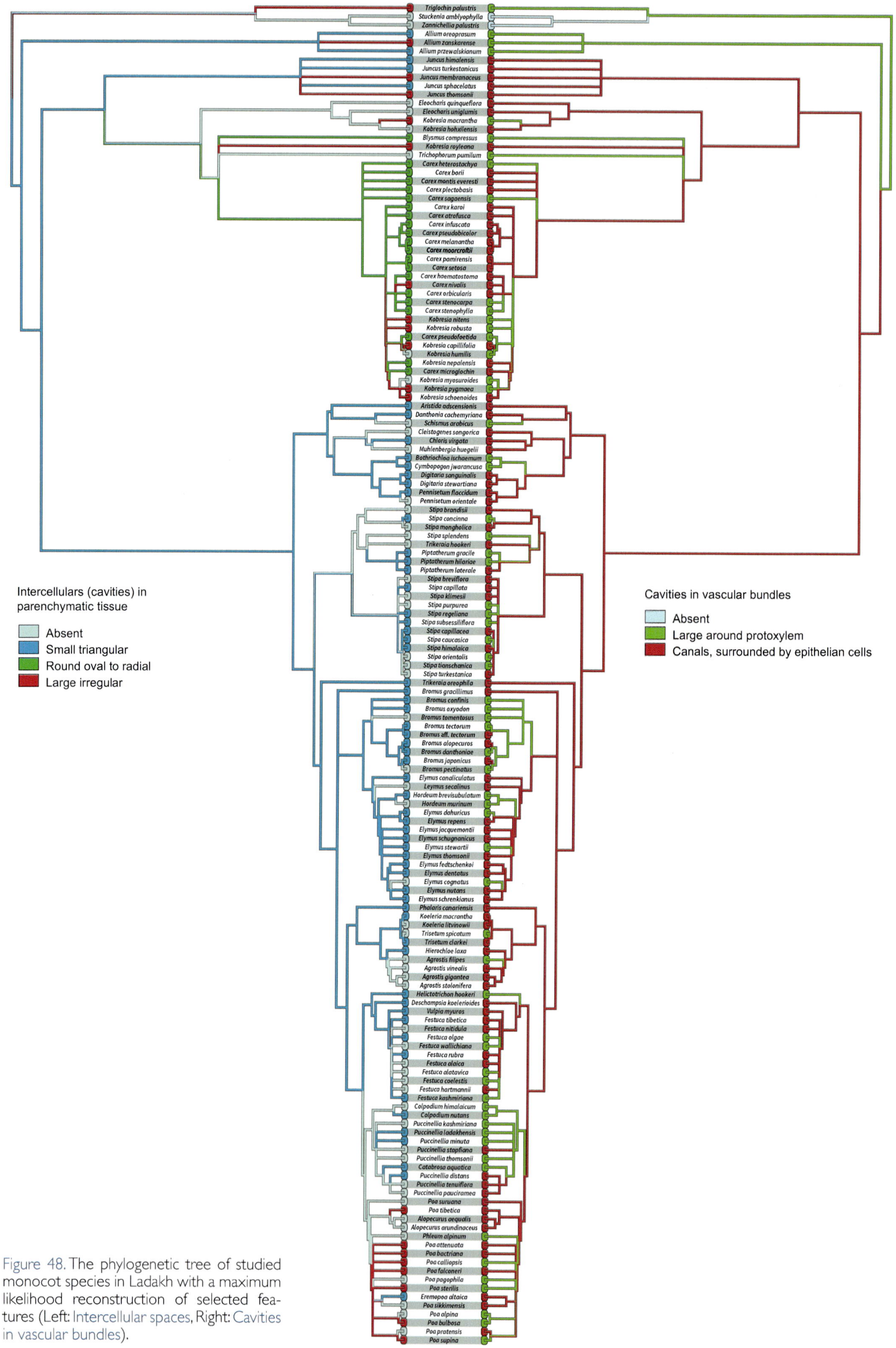

Figure 48. The phylogenetic tree of studied monocot species in Ladakh with a maximum likelihood reconstruction of selected features (Left: Intercellular spaces, Right: Cavities in vascular bundles).

Martin Kopecký and Eliška Lehečková excavating a stout *Lindelofia stylosa* plant.

MATERIAL AND METHODS

Plant age and growth determination

To describe anatomy, age and growth of plants in high elevation communities, we measured more than 2000 individual specimens collected in August–September 2012–2016. Due to dry soils in the study region, perennial plants often have deep tap roots (Klimešová et al. 2011) and distinct annual ring increments. Three to four adult undamaged individuals of each species were excavated and the washed roots placed in plastic bags containing ethanol until further analysis. The samples were analyzed in Birmensdorf, the Swiss Federal Research Institute WSL, in Switzerland and in Třeboň, Institute of Botany, CAS, Czech Republic by a method developed by Gärtner and Schweingruber (2013). Transverse, tangential and radial sections were cut from all studied individuals using a sliding microtome and double stained with Astrablue and Safranin. Microscopic images of optimal ring width sequences were made for counting of the consecutive number of rings, and measuring the width of annual increments (Gärtner and Schweingruber 2013). Plant ages were determined by ring counting in the oldest sections in the transition between the hypocotyl and the primary root (root collar). The possibility of non-transparent rings due to inappreciable growth in extremely cold or snowy summer cannot be excluded, therefore the numbers of counted rings represent minimum plant ages.

Species phylogeny

The phylogenetic relationships between studied species were constructed based on two chloroplast markers (the gene rbcL and the intergenic spacer trnL-trnF) and one nuclear marker (the intron ITS). These regions are commonly used to build species-level phylogeny (Baldwin 1992). Firstly, we obtained all available sequences of these marker regions from NCBI GenBank (about one third of the species presented here). Secondly, for those species not present in GenBank we performed sequencing ourselves on material collected in the field to complete the tree. DNA was isolated from silica-gel dried leaf tissue using Qiagen DNeasy Plant Mini Kit. RbcL, trnL-trnF and ITS regions were amplified using primers rbcLa f and rbcLa rev (Levin et al. 2003), trnL e and trnL f, ITS5 and ITS4 (White et al. 1990), respectively. DNA was sequenced using Sanger sequencing. Sequences were further edited, trimmed and aligned using Sequencher 5.2.3 (Gene Codes Corp) and finally, the phylogeny was constructed implementing the Bayesian approach (Ronquist and Huelsenbeck 2003).

The L-INS-i algorithm implemented in the online version of MAFFT 6 (*http://mafft.cbrc.jp/alignment/server*, Katoh and Toh 2008) was employed to align the sequence datasets. Partial alignments were concatenated, manually adjusted in BioEdit (Hall 1999) and subdued to the automated1 algorithm in trimAll software (Capella-Gutierrez et al. 2009) to exclude highly divergent and gap-rich regions. Prior to the phylogenetic analysis, the best-fit model was selected by Kakusan4 (Tanabe 2011). Baseml software (Adachi and Hagesawa 1996) served as the computational core. Both non-partitioned and partitioned models were evaluated. Based on the Bayesian information criterion (Schwarz 1978),

J. Doležal et al., *Anatomy, Age and Ecology of High Mountain Plants in Ladakh, the Western Himalaya*, https://doi.org/10.1007/978-3-319-78699-5_6

we finally decided to use the GTR model with rate variation across locations simulated by discrete gamma distribution (Γ8), autocorrelated by the AdGamma rates prior and unlinked for particular gene partitions. To reflect the increased probability of transitions over transversions in non-coding loci, the substitution rates prior (revMatPr) was set for the ITS, trnT-L and petB-D partition to the Dirichlet function with values 1 and 3.

The phylogenetic analysis was represented by the Bayesian inference (BI), conducted in MrBayes 3.1.2 (Ronquist and Huelsenbeck 2003). The phylogenetic analysis was provided using two independent runs with four Metropolis-coupled MCMC chains of 1 × 107 generations sampled after every 1000th generation. In each run, one Markov chain was cold and three were incrementally heated by a parameter of 0.3. The first 25% of entries were discarded as burn-in and the rest was used to compute the majority-rule consensus to eliminate trees sampled before reaching apparent stationarity. The resulting tree was edited to be compatible in future ecological studies, since it contained more terminals (unknown multispecies genera were substituted with five terminals). Internal phylogeny of these genera was collapsed to polytomy, branch lengths were averaged and excessive branches were cut out.

Plant growth forms

In order to understand how much is observed anatomical variation among species related to their morphology and clonal propagation strategies, each species was classified into one of 20 clonal growth forms named after typical representatives, and four broader space occupancy strategies, distinguished by Klimešová et al. (2011) based on clonal reproduction traits, i.e. persistence at a spot, lateral spread and clonal propagation. The classification was based on which organ bears buds for iteration of shoots, which organ provides connections between offspring shoots, if this organ is short or long, if a plant is able to form adventitious roots, and if there are special storage organs. There are three exceptions: annuals and biennials (21% among studied species, *Microula tibetica* type in Klimešová et al. 2011), woody plants (7%, *Juniperus semiglobosa* type) and compact alpine cushion plants (4%, *Thylacospermum caespitosum* type, Dvorský et al. 2013) are assessed according to their whole morphology.

Due to overall scarcity of water, non-clonal tap-rooted herbs predominate (38%) among Ladakh's herbaceous plants. They have deep, often segmented roots with caudex and short branches (*Arnebia euchroma* type in Klimešová et al. 2011), their primary root system lacks adventitious roots and buds and serves as a storage organ; the bud bank is situated on perennial stems with short (less than 10 cm) branches serving as vascular link between shoots and primary root. Plants with pleiocorm and long branches (*Saussurea gnaphalodes* type) are also possessing a primary root system lacking adventitious roots and buds; the bud bank is situated on perennial stems with long (more than 10 cm) branches serving as vascular link between shoots and primary root.

Clonality in plants is a strategy of multiplying renewing buds, and clonal plants are generally abundant in wet, cold and nutrient-poor environments. The category of partially clonal plants with main root and adventitious buds (*Parrya nudicaulis* type in Klimešová et al. 2011) is characterized by the main root (include hypocotyle) forming adventitious buds spontaneously or after injury; clonal growth occurs usually only after fragmentation of main root; due to small number of species, this category was merged with *Arnebia euchroma* type into *Tap root* category for presentation of anatomical variation in our dataset. Root sprouters (*Ptilotrichum canescens* type, included into the *Pleiocorm* category used for anatomical comparisons) are plants with horizontal roots with adventitious buds; branches of main root and adventitious roots forming adventitious buds spontaneously or after injury; roots serving as bud bank and vascular connection between offspring shoots; lateral spread usually extensive; persistence differing among species. Plants with long-lived horizontal rooting stems on the soil surface (*Thymus linearis* type, included into the *Pleiocorm* category used in anatomical presentation) are usually woody taxa, clonal growth organ is formed by creeping axis, rooting in the soil; nodes of youngest parts bearing leaves, internodes shorter than in short-lived horizontal stems; stem serving as a storage organ and bud bank; vegetative spreading slow and persistence high.

Clonal plants with short-lived horizontal rooting stems on or above soil surface stolon (*Halerpestes sarmentosa* type, stolons included in *Lrhizome* category) providing connection between offspring plants or formed by creeping axis; nodes on stem are bearing leaves, internodes usually long, stems are serving as a storage organ and bud bank; vegetative spreading usually fast and persistence low. Clonal plants with short (<10 cm long increments) hypogeogenous rhizomes (*Leontopodium ochroleucum* type, included in short *Srhizome* category) possesing perennial organs of stem origin formed below-ground; rhizome usually growing at a species specific depth, periodically becoming orthotropic and forming above-ground shoots; horizontal part of the rhizome bearing bracts, some roots and possessing short internodes; vegetative spreading intermediate; persistence differing considerably among species. Clonal plants with long (>10 cm increments) hypogeogenous rhizomes (*Poa tibetica* type, included in long *Lrhizome* category) having perennial organs of stem origin formed below-ground; rhizome usually growing at a species-specific depth, periodically becoming orthotropic and forming above-ground shoots; horizontal part of the rhizome bearing bracts, some roots possessing long internodes; vegetative spreading fast; persistence differing considerably among species. Clonal turf graminoids with very short epigeogenous rhizomes (*Festuca kashmiriana* type, included in *Srhizome* category) having perennial organ of stem origin formed above-ground; its distal part covered by soil and litter or pulled into the soil by contraction of roots; nodes bearing green leaves; internodes usually short; rhizomes bearing roots and serving as a bud bank and storage organ; vegetative spread low (up to 1 cm per year); persistence usually low; typical of tussock grasses. Clonal forbs with medium (5–10 long increments) epigeogenous rhizomes (*Viola kunawurensis* type, included in *Srhizome* type) having perennial organ of stem origin formed above-ground; its distal part covered by soil and litter or pulled into the soil by contraction of roots; nodes bearing green leaves; internodes usually short; rhizomes bearing roots and serving as a bud bank and storage organ; vegetative spread low (up to a few centimeters per year); persistence usually low. Clonal forbs and subshrubs with long epigeogenous rhizomes (*Biebersteinia odora* type, included in *Lrhizome* category) having perennial organ of stem origin formed above-ground; its distal part covered by soil and litter or pulled into the soil by contraction of roots; nodes bearing green leaves; internodes usually short; rhizomes bearing roots and serving as a bud bank and storage organ; vegetative spread low (up to a few centimeters per year); persistence usually long, resulting in a large preserved rhizome system. Plant with bulbs (*Lloydia serotina* type) having storage and perennation organ consisting of storage leaves and shortened stem base; stem providing bud bank and connection between offspring shoots; lateral spread low. Plants with stem tubers (*Potamogeton filiformis* type, included in *Lrhizome* category) having below-ground, usually short-lived storage and regenerative organ of shoot origin; lateral spread high in the case of offspring tubers formed on hypogeogenous rhizome, persistence usually low. Plant with turions (*Potamogeton amblyphyllus* type) as detachable over-wintering buds (usually in water plants) composed by tightly arranged leaves filled with food reserves; turions, developing axially or apically and needing vernalisation to re-growth. Plants producing fragments of stem origin (*Potamogeton perfoliatus* type, included in *Lrhizome* category) as detached parts of shoot with rooting ability. Plants with plantlets (pseudovivipary, *Bistorta vivipara* type), meristem normally develops into a flower, but forms vegetative buds (plantlets, bulbils, root or stem tubercules), which are sometimes soon detached from the parent plant; alternatively the entire inflorescence falls off and the plantlets root at the soil surface; offspring size similar to seedlings. Plants with axillary buds (*Saxifraga cernua* type) as small vegetative diaspores produced in axils of leaves on stems above-ground formed by axillary buds subtended by small storage organ of stem, root (tubercule) of leaf (bulbil) origin; buds soon shed from mother plant, beginning to grow immediately; resembling seedlings in size. The four space occupancy strategies were based on the rate of lateral spread (spreading – more than 10 cm per year; non-spreading – less than 10 cm per year) and persistence of connections between ramets (splitters – plants producing adventitious roots with main root decaying; integrators – plants not producing adventitious roots and/or with perennial main root, Klimešová et al. 2011).

Habitat preferences

Information on high-altitude species' ecological optima along environmental gradients, in major habitats and elevation belts were derived from floristic mapping (4,150 vegetation plots, each 100 m × 100 m) and vegetation composition of 369 plots (each 100 m^2) sampled in a stratified design over all Ladakh are to cover major vegetation types over the study area (Klimeš 2003; Dvorský et al. 2011, 2015, Doležal et al. 2016). On each sampled plot, five environmental variables were estimated: (1) soil/substrate stability: 1 – unstable (screes, dunes, solifluction soils), 2 – partly stable (grasslands, steppes), 3 – stable (rocky crevices, *Kobresia pygmaea* mats); (2) light availability: 1 – shaded (gorges, shaded rocky crevices, walls of stream banks), 2 – partially shaded (dense vegetation cover), 3 – full light (sparse vegetation cover); (3) soil moisture: 1 – dry (substrate usually without visible traces of water), 2 – mesic, 3 – wet (water level regularly but transiently above soil surface), 4 – permanent surface water; (4) soil fertility: 1 – low (semi-deserts, steppes), 2 – medium (alpine meadows), 3 – high (stables, animal resting places); (5) soil salinity: 1 – no salt deposits on soil surface, 2 – salt deposits scarce, 3 – salts forming a continuous crust. From these data, indicator values of a species were calculated as arithmetical means of indicator values in individual plots in which that species was recorded, weighed by the logarithm of its cover.

To obtain a robust estimate of the elevational optima of the species, we calculated response curves fitted with HOF models (Huisman et al. 1993). Species response curves were derived from 4,150 vegetation plots (each 100 m × 100 m) sampled over the entire Ladakh area. The dataset contains more than 122,000 records of occurrence of vascular plant species along exceptional elevational gradient from 2800 m to 6150 m (Dvorský et al. 2017).

Acknowledgements

We are grateful for the support of many colleagues from the Institute of Botany and students from the University of South Bohemia in České Budějovice and Charles University in Prague, namely Jan Altman, Ondřej Mudrák, Martin Kopecký, Martin Macek, Zuzana Chlumská, Eliška Lehečková, Milan Kotilínek, Petr Koutecký, Eliška Patáčová, Klára Řeháková, Kateřina Čapková and Ines Hofbauerová, who participated in several field expeditions to Ladakh and helped us to collect plants for age and anatomical analyses. We thank our international collaborators Pierre Liancourt (Institute of Botany in Trebon), Inga Hiisalu and Kristina Sohar (University of Tartu), Alan Crivellaro (University of Padova), Roey Angel (University of Vienna), Christian Rixen (WSL, Davos), and Vera Markgraf (University of Colorado Boulder) for helping us with field campaigns. Our special thanks belong to Lucie Jonátová, Pavel Říha and Viktorie Brožová for phylogenetic analyses, and Bernhard Dickoré for plant identification of complicated taxa. We are especially grateful to Marie Šmilauerová, Petr Koutecký and Jiří Košnar from the University of South Bohemia in České Budějovice for laboratory support. We thank our lab technicians Eva Návratová, Veronika Langová, Iveta Kadlecová, Kajina Kunertová for helping with plant root sectioning and ring counting. We would like to thank Anzar Khuroo and Akhtar Malik of the University of Srinagar in Kashmir, India for logistic support. The funding was provided by the Institute of Botany of the Czech Academy of Sciences, the Czech Science Foundation (GACR 13-13368S and 17-19376S), the University of South Bohemia in České Budějovice, the Institute of Forest, Snow and Landscape, WSL, Birmensdorf in Switzerland, and the European Union's Seventh Framework Programme for research, technological development, and demonstration under Grant Agreement No. GA-2010-267243 – PLANT FELLOWS.

J. Doležal et al., *Anatomy, Age and Ecology of High Mountain Plants in Ladakh, the Western Himalaya*, https://doi.org/10.1007/978-3-319-78699-5

References

Ackerly DD (2000) Taxon sampling, correlated evolution, and independent contrasts. Evolution 54:1480–1492.

Adachi J, Hagesawa M (1996) MOLPHY version 2.3: programs for molecular phylogenetics based on maximum likelihood. Institute of Statistical Mathematics. Tokyo, Japan.

Al-Shehbaz IA. 2002. New Species of *Alyssum*, *Aphragmus*, *Arabis* and *Sinosophiopsis* (Brassicaceae) from China and India. Novon 12:309–313.

Angel A, Conrad R, Dvorský M, Kopecký M, Kotilínek M, Hiiesalu I, Schweingruber F, Doležal J (2016) The Root-Associated Microbial Community of the World's Highest Growing Vascular Plants. Microb Ecol 72:394–406.

Aschenbach K, Conrad R, Rehakova K, Dolezal J, Janatkova, Angel R (2013) Methanogens at the top of the world: occurrence and potential activity of methanogens in newly deglaciated soils in high-altitude cold deserts in the Western Himalayas. Front Microbio 4:359.

Baldwin BG (1992) Phylogenetic utility of the internal transcribed spacers of nuclear ribosomal DNA in plants: an example from the compositae. Mol Phylogenet Evol 1(1):3–16.

Capella-Gutiérrez S, Silla-Martínez JM, Gabaldón T (2009) trimAl: a tool for automated alignment trimming in large-scale phylogenetic analyses. Bioinformatics 25(15):1972–1973.

Crivellaro A, Schweingruber FH (2016) Stem anatomical features of Dicotyledons. Xylem, phloem, cortex and periderm characteristics of ecological and taxonomical analysis. Verlag Dr. Kessel, Remagen-Oberwinther, Germany.

Čapková K, Hauer T, Řeháková K, Doležal J (2016) Some like it high! Phylogenic diversity of high-elevation cyanobacterial community from biological soil crusts of Western Himalaya. Microb Ecol 71:113–123.

de Bello F, Doležal J, Ricotta C, Klimešová J (2011) Plant clonal traits, coexistence and turnover in East Ladakh, Trans-Himalaya. Preslia 83:315–327.

Dickoré WB (1995) Systematische Revision und chorologische Analyse der Monocotyledoneae des Karakorum (Zentralasien, West-Tibet). Flora Karakorumensis I. Angiospermae, Monocotyledoneae. Stapfia 39. Linz, 298 pp.

Dickoré WB (2011) Global Mountain Biodiversity Assessment - GMBA. The Himalayan Uplands Plant database (HUP Version 1). Occurrence Dataset *https://doi.org/10.15468/k64rgi* accessed via GBIF.org on 2017-11-02.

Doležal J, Dvorský M, Kopecký M, Liancourt P, Hiiesalu I, Macek M, Altman J, Chlumská Z, Řeháková K, Čapková K, Borovec J, Mudrak O, Wild J, Schweingruber F (2016) Vegetation dynamics at the upper elevational limit of vascular plants in Himalaya. Sci Rep 6:24881.

Dvorský M, Doležal J, De Bello F, Klimešová J, Klimeš L (2011) Vegetation types of East Ladakh: Species and growth form composition along main environmental gradients. Appl Veg Sci 14:132–147.

Dvorský M, Doležal J, Kopecký M, Chlumská Z, Janatková K, Altman J, de Bello F, Řeháková K (2013) Testing the stress-gradient hypothesis at the roof of the world: effects of the cushion plant *Thylacospermum caespitosum* on species assemblages. PLoS ONE 8:e53514.

Dvorský M, Altman J, Kopecký M, Chlumská Z, Řeháková K, Janatková K, Doležal J (2015) Vascular plants at extreme elevations in eastern Ladakh, northwest Himalayas. Plant Ecol Divers 8:571–584.

Dvorský M, Chlumská Z, Altman J, Čapková K, Řeháková K, Macek M, Kopecký M, Liancourt P, Doležal J (2016) Gardening in the zone of death: an experimental assessment of the absolute elevation limit of vascular plants. Sci Rep 6:24440.

Dvorský M, Macek M, Kopecký M, Wild J, Doležal J (2017) Niche asymmetry of vascular plants increases with elevation. J Biogeogr 44:1418–1425.

Gärtner H, Schweingruber FH (2013) Microscopic preparation techniques for plant stem analysis. Verlag Dr. Kessel, Remagen-Oberwinther, Germany.

Good R (1964) The Geography of the Flowering Plants. John Wiley & Sons, New York.

Hall TA (1999) BioEdit: a user-friendly biological sequence alignment editor and analysis program for Windows 95/98/NT. Nucl Acids Symp Ser. 41:95–98.

Hartmann H (1983) Pflanzengesellschaften entlang der Kashmirroute in Ladakh. Jahrb Vereins Schutze Bergwelt 48:131–173.

Huisman J, Olff H, Fresco LFM (1993) A hierarchical set of models for species response analysis. J Veg Sci 4:37–46.

Chlumská Z, Janeček Š, Doležal J (2014) How to Preserve Plant Samples for Carbohydrate Analysis? Test of Suitable Methods Applicable in Remote Areas. Folia Geobot 49:1–15.

J. Doležal et al., *Anatomy, Age and Ecology of High Mountain Plants in Ladakh, the Western Himalaya*, https://doi.org/10.1007/978-3-319-78699-5

Janatková K, Řeháková K, Doležal J, Šimek M, Chlumská Z, Dvorský M, Kopecký M (2013) Community structure of soil phototrophs along environmental gradients in arid Himalaya. Environ Microbiol 15:2505–2516.

Kachroo P (1993) Plant diversity in Northwest Himalaya: A preliminary survey. In: Dhar U (ed) Himalayan biodiversity. Conservation strategies. Nainital. pp. 111–132.

Kachroo P, Sapru BL, Dhar U (1977) Flora of Ladakh: an ecological and taxonomical appraisal. Dehra Dun, Bishen Singh Mahendra Pal Singh, 172 pp.

Karger DN, Conrad O, Böhner J, Kawohl T, Kreft H, Soria-Auza RW, Zimmermann NE, Linder HP, Kessler M (2016) CHELSA climatologies at high resolution for the earth's land surface areas (Version 1.1). World Data Cent Clim.

Katoh K, Toh H (2008) Recent developments in the MAFFT multiple sequence alignment program. Brief Bioinform 9:286–298.

Klimeš L (2003) Life-forms and clonality of vascular plants along an altitudinal gradient in E Ladakh (NW Himalayas). Basic App Ecol 4:317–328.

Klimeš L (2008) Clonal splitters and integrators in harsh environments of the Trans-Himalaya. Evol Ecol 22:351–367.

Klimeš L, Dickoré WB (2006) Flora of Ladakh (Jammu & Kashmir, India). A preliminary checklist published at *http://www.butbn.cas.cz/klimes*. Accessed 5 March 2015.

Klimeš L, Doležal J (2010) An experimental assessment of the upper elevational limit of flowering plants in the Western Himalayas. Ecography 33:590–596.

Klimešová J, Doležal J, Dvorský M, de Bello F, Klimeš L (2011) Clonal growth forms in eastern Ladakh, Western Himalayas: classification and habitat preferences. Folia Geobot 46:191–217.

Körner CH (1999) Alpine plant life: functional plant ecology of high mountain ecosystems. Springer-Verlag. Berlin.

Kotilínek M, Hiiesalu I, Košnar J, Šmilauerová M, Šmilauer P, Altman J, Dvorský M, Kopecký M, Doležal J (2017) Fungal root symbionts of high-altitude vascular plants in the Himalaya. Sci Rep 7:6562.

Levin RA, Wagner WL, Hoch PC, Nepokroeff M, Pires JC, Zimmer EA, Sytsma KJ (2003) Family-level relationships of Onagraceae based on chloroplast rbcL and ndhF data. Am J Bot 90:107–115.

Liancourt P, Le Bagousse-Pinguet Y, Rixen C, Doležal J (2017) SGH: Stress or Strain Gradient Hypothesis? Insights from an elevation gradient on the roof of the world. Ann Bot 120:29–38.

Macek P, Klimeš L, Adamec L, Doležal J, Chlumská Z, de Bello F, Dvorský M, Řeháková K (2012) Plant nutrient content does not simply increase with elevation under the extreme environmental conditions of Ladakh, NW Himalaya. Arct, Antarct, Alp Res 44:62–66.

Metcalfe CR (1960) Anatomy of the Monocotyledons. I Gramineae. Oxford Clarodendron Press, Oxford.

Miehe G, Miehe S, Bach K, Jianquan L (2011) Ecological stability during the LGM and the mid-Holocene in the alpine steppes of Tibet? Quat Res 76:243–252.

Moorcroft W, Trebeck G (1841) Travels in the Himalayan provinces of Hindustan and the Panjab; in Ladakh and Kashmir; in Peshawar, Kabul, Kunduz, and Bokhara. John Murray, London.

Murti SK (2001) Flora of Cold Deserts of Western Himalaya. Volume 1: Monocotyledons. Dehra Dun.

Nagelmüller S, Hiltbrunner E, Körner C (2017) Low temperature limits for root growth in alpine species are set by cell differentiation. AoB PLANTS (in press).

Nasir E, Ali SI (eds) (1970–) Flora of (West-) Pakistan. Vols. 1–204, Karachi.

Niklas KJ (1985) The evolution of tracheid diameter in early vascular plants and its implications on the hydraulic conductance of the primary xylem strand. Evolution 39:1110–1122.

Nobis MP, Schweingruber FH (2013) Adult age of vascular plant species along an elevational land-use and climate gradient. Ecography 36:1076–1085.

Nüsser M, Dickoré WB (2002) A tangle in the triangle: vegetation map of the Eastern Hindukush (Chitral, Northern Pakistan). Erdkunde 56:37–59.

Pampanini R (1930) La Flora del Caracorum. p. 271, Bologna.

Ronquist F, Huelsenbeck JP (2003) MRBAYES 3: Bayesian phylogenetic inference under mixed models. Bioinformatics 19:1572–1574.

Řeháková K, Chlumská Z, Doležal J (2011) Soil cyanobacterial and microalgal diversity in dry mountains of Ladakh, NW Himalaya, as related to site, altitude, and vegetation. Microb Ecol 62:337–346.

Řeháková K, Chroňáková A, Krištůfek V, Kuchtová B, Janatková K, Scharfen J, Čapek P, Doležal J (2015) Bacterial community of cushion plant *Thylacospermum ceaspitosum* on elevational gradient in the Himalayan cold desert. Front Microbiol 6:304.

Řeháková K, Čapková K, Dvorský M, Kopecký M, Altman J, Šmilauer P, Doležal J (2017) Interactions between soil phototrophs and vascular plants in Himalayan cold desert. Soil Biol Biochem 115:568–578.

Schmidt S, Nüsser M (2017) Changes of High Altitude Glaciers in the Trans-Himalaya of Ladakh over the Past Five Decades (1969–2016). Geosciences 7:27.

Schwarz G (1978) Estimating the Dimension of a Model. Ann Statist 6:461–464.

Schweingruber FH, Berger H (2017) Anatomy of Grass Culms. Verlag Dr. Kessel, Remagen-Oberwinther, Germany.

Schweingruber, FH, Börner A (2018) The Plant Stem – A Microscopic Aspect. Springer Verlag, Berlin, Heidelberg.

Schweingruber FH, Büntgen U (2013) What is 'wood' – An anatomical re-definition. Dendrochronologia 31:187–191.

Schweingruber FH, Crivellaro A (2016) Comments in the IAWA Hardwood List now available on the web. IAWA J 37:369-371.

Schweingruber FH, Poschlod P (2005) Growth rings in herbs and shrubs: life span, age determination and stem anatomy. Forest Snow and Landscape Res 79:197–415.

Schweingruber FH, Börner A, Schulze ED (2011) Atlas of Stem Anatomy in Herbs, Shrubs and Trees. Vol. 1. Springer-Verlag, Berlin Heidelberg.

Schweingruber FH, Říha P, Doležal J (2014) Variation in Stem Anatomical Characteristics of Campanuloideae Species in Relation to Evolutionary History and Ecological Preferences. PLoS ONE 9(2):e88199.

Singh NP, Singh DK, Uniyal BP (eds) (2002) Flora of Jammu & Kashmir. Vol. 1., Kolkata.

Sohar K, Altman J, Lehečková E, Doležal J (2017) Growth–climate relationships of Himalayan conifers along elevational and latitudinal gradients. Int J Climatol 37:2593–2605.

Stewart RR (1916–1917) The flora of Ladakh. Bulletin of the Torrey Botanical Club 43:571–588 and 625–650.

Sultan SE (2000) Phenotypic plasticity for plant development, function, and life-history. Trends Plant Sci 5:537–542.

Takhtajan A (1986) Floristic Regions of the World. University of California Press, Berkeley. 544 pp.

Tanabe AS (2011) Kakusan4 and Aminosan: two programs for comparing nonpartitioned, proportional and separate models for combined molecular phylogenetic analyses of multilocus sequence data. Mol Ecol Resour 11(5):914–21. *doi: 10.1111/j.1755-0998.2011.03021*

Wheeler EA, Baas P, Gasson PE (1989) IAWA list of microscopic features for hardwood identification: with an appendix on non-anatomical information. IAWA Bull 10:219–332.

White TJ, Bruns T, Lee S, Taylor J (1990) Amplification and direct sequencing of fungal ribosomal RNA genes for phylogenetics. pp. 315–322. In: PCR Protocols: A Guide to Methods and Applications. Academic Press, New York.

Xu J, Grumbine ER, Shrestha A, Eriksson M, Yang X, Wang Y, Wilkes A (2009) The melting Himalayas: cascading effects of climate change on water, biodiversity, and livelihoods. Conserv Biol 23:520–530.

Species list

J. Doležal et al., *Anatomy, Age and Ecology of High Mountain Plants in Ladakh, the Western Himalaya*, https://doi.org/10.1007/978-3-319-78699-5

GPSR Compliance

The European Union's (EU) General Product Safety Regulation (GPSR) is a set of rules that requires consumer products to be safe and our obligations to ensure this.

If you have any concerns about our products, you can contact us on ProductSafety@springernature.com

In case Publisher is established outside the EU, the EU authorized representative is:

Springer Nature Customer Service Center GmbH
Europaplatz 3
69115 Heidelberg, Germany

Batch number: 10756586

Printed by Printforce, the Netherlands